Earth Science

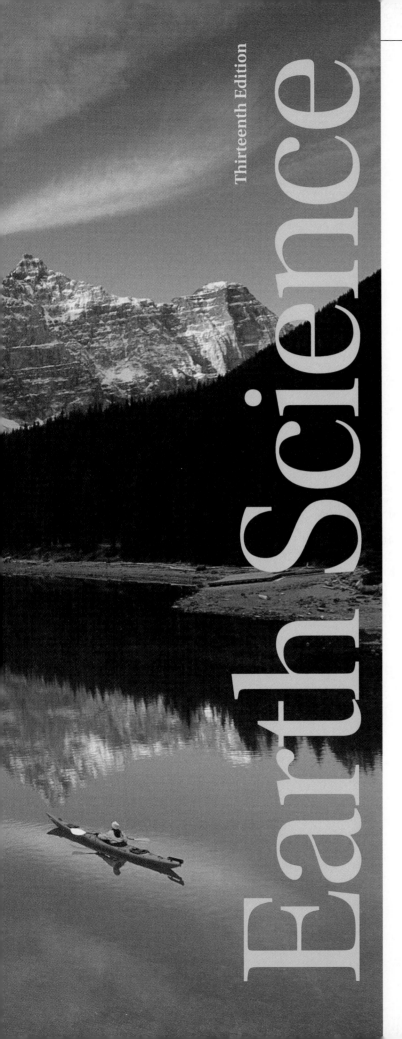

Thirteenth Edition

Earth Science

Edward J. Tarbuck

Frederick K. Lutgens

Illustrated by

Dennis Tasa

Prentice Hall

Boston   Columbus   Indianapolis
New York   San Francisco   Upper Saddle River
Amsterdam   Cape Town   Dubai   London
Madrid   Milan   Munich   Paris   Montréal   Toronto
Delhi   Mexico City   São Paulo   Sydney
Hong Kong   Seoul   Singapore   Taipei   Tokyo

Acquisitions Editor: *Andrew Dunaway*
Marketing Manager: *Maureen McLaughlin*
Project Editor: *Crissy Dudonis*
Assistant Editor: *Sean Hale*
Editorial Assistant: *Michelle White*
Marketing Assistant: *Nicola Houston*
Managing Editor, Geosciences and Chemistry: *Gina M. Cheselka*
Project Manager, Science: *Maureen Pancza*
Art Director: *Derek Bacchus*
Interior and Cover Design: *Gary Hespenheide*
Art Project Manager: *Connie Long*
Senior Operations Supervisor: *Nick Sklitsis*
Operations Specialist: *Maura Zaldivar*
Associate Media Producer: *Tim Hainley*
Media Supervisor: *Liz Winer*
Associate Media Project Managers: *David Chavez, Katherine Foley*
Photo Research Manager: *Elaine Soares*
Photo Researcher: *Kristin Piljay*
Composition/Full Service: *PineTree Laserwords*
Production Editor, Full Service: *Patty Donovan*
Cover Photograph: *Kayaker takes an early morning tour on Lake Moraine, Alberta, Canada;*
© *Charlie Munsey/Corbis*

Library of Congress Cataloging-in-Publication Data
CIP data available upon request.

Printed in the United States
10 9 8 7 6 5 4 3 2 1

ISBN 13: 978-0-321-68850-7
ISBN 10:   0-321-68850-3

Prentice Hall
is an imprint of

www.pearsonhighered.com

To Our Grandchildren

Shannon, Amy, Andy, Ali, and Michael

Allison and Lauren

Each is a bright promise for the future

Brief Contents

GEODe: Earth Science v.3

GEODe: Earth Science is now found within www.masteringgeology.com. This dynamic learning aid reinforces key concepts by using tutorials, animations, and interactive exercises. GEODe: Earth Science in MasteringGeology is easily assignable and student performance is easily assessed.

Contents

UNIT TWO

Sculpturing Earth's Surface   82

4 Weathering, Soil, and Mass Wasting   83

5 Running Water and Groundwater   117

6 Glaciers, Deserts, and Wind   157

UNIT THREE

Forces Within   192

7 Plate Tectonics: Scientific Revolution Unfolds   193

17 Moisture, Clouds, and Precipitation   489

18 Air Pressure and Wind   525

Preface

Earth Science, 13th Edition, is a college-level text designed for an introductory course in Earth science. It consists of seven units that emphasize broad and up-to-date coverage of basic topics and principles in geology, oceanography, meteorology, and astronomy. The book is intended to be a meaningful, nontechnical survey for undergraduate students with little background in science. Usually these students are taking an Earth science class to meet a portion of their college or university's general requirements.

In addition to being informative and up-to-date, *Earth Science, 13th Edition*, strives to meet the need of beginning students for a readable and user-friendly text and a highly usable "tool" for learning basic Earth science principles and concepts.

New to This Edition

- **An enhanced active learning approach.** Each chapter begins with *Focus on Concepts*—a series of questions that alert students to important ideas in the chapter. Within the chapter, each major section concludes with *Concept Checks* that allow students to monitor their understanding and comprehension of significant facts and ideas. Each chapter concludes with a new section called *Give It Some Thought*. These questions and problems challenge learners by involving them in activities that require higher-order thinking skills that include the synthesis, analysis, and application of material in the chapter.
- *MasteringGeology*™. Used by over one million science students each year, the Mastering platform is the most effective and widely used online homework, tutorial, and assessment system for the sciences. *Earth Science, 13th Edition*, is supported by *MasteringGeology* assignable activities that include geoscience animations, *Encounter Earth* Google Earth™ multimedia activities, and GEODe activities, as well as a robust student self Study Area with many digital resources including a Pearson eText version of *Earth Science*. www.masteringgeology.com.
- **A stronger art program.** Dozens of figures are new or redrawn. The result is art that is clearer and easier to understand. Numerous diagrams and maps are paired with photographs for greater effectiveness. Many new and revised art pieces also have additional labels that "narrate" the process being illustrated and/or "guide" students as they examine the figure.
- **More than 150 new, high-quality photos and satellite images.** New "Geologist's Sketch" illustrations accompany many important photographs and satellite images. Each sketch, which resembles what a geologist or other Earth scientist might put in a field notebook, helps the student identify important and sometimes subtle aspects of an image.

This author–artist collaboration helps make an already strong visual component even stronger and more effective.
- **New "Professional Profile" boxes.** Eight chapters include essays that present profiles of working Earth scientists. These special boxes are intended to give students a sense of what Earth scientists do and a perspective on a variety of careers in Earth science.
- **Revised and updated content.** A basic function of a college science textbook is to provide clear, understandable presentations that are accurate, engaging, and up-to-date. Our foremost goal is to keep *Earth Science* current, relevant, and highly readable for the beginning student. With this goal in mind, every part of the book was examined carefully. Significant changes were made throughout, including the chapters on minerals, running water and groundwater, plate tectonics, volcanoes, and earthquakes. Discussions on the ocean floor, global climate change, hurricanes, the solar system, stellar evolution, and the big bang theory also received considerable attention.

Distinguishing Features

Readability

The language of this book is straightforward and *written to be understood*. Clear, readable discussions with a minimum of technical language are the rule. Frequent headings and subheadings help students follow discussions and identify the important ideas presented in each chapter. In this edition, improved readability was achieved by examining chapter organization and flow, and writing in a more personal style. Large portions of the text were substantially rewritten in an effort to make the material more understandable.

Focus on Basic Principles and Instructor Flexibility

Although many topical issues are treated in *Earth Science, 13th Edition*, it should be emphasized that the main focus of this new edition remains the same as its predecessors—to promote student understanding of basic Earth science principles. Whereas student use of the text is a primary concern, the book's adaptability to the needs and desires of the instructor is equally important. Realizing the broad diversity of Earth science courses in both content and approach, we have continued to use a relatively nonintegrated format to allow maximum flexibility for the instructor. Each of the major units stands alone; hence, they can be taught in any order. A unit can be omitted entirely without appreciable loss of continuity, and portions of some chapters may be interchanged or excluded at the instructor's discretion.

A Strong Visual Component

Earth science is highly visual, so art and photographs play a critical role in an introductory textbook. As in previous editions, Dennis Tasa, a gifted artist and respected Earth science illustrator, has worked closely with the authors to plan and produce the diagrams, maps, graphs, and sketches that are so basic to student understanding. The result is art that is clearer and easier to understand.

Our aim is to get the *maximum effectiveness* from the visual component of the book. Michael Collier aided us greatly in this quest. Many of his extraordinary aerial photographs were used in the new edition. Michael is an award-winning geologist-photographer. Among his many awards is the American Geological Institute Award for Outstanding Contribution to the Public Understanding of the Geosciences. We are fortunate to have had Michael's assistance in *Earth Science, 13th Edition.*

Three Important Themes

Chapter 1, "Introduction to Earth Science," presents students with three important themes that recur throughout the book: *Earth as a System, People and the Environment,* and *Understanding Earth.*

Earth as a System

An important occurrence in modern science has been the realization that Earth is a giant multidimensional system. Our planet consists of many separate but interacting parts. A change in any one part can produce changes in any or all of the other parts—often in ways that are neither obvious nor immediately apparent. Although it is not possible to study the entire system at once, it is possible to develop an awareness and appreciation for the concept and for many of the system's important interrelationships. Therefore, starting with the revised discussion of "Earth System Science" in Chapter 1, the theme of "Earth as a System" keeps recurring throughout all major units of the book. It is a thread that "weaves" throughout the chapters and helps tie them together. Several new and revised special interest boxes relate to "Earth as a System." In addition, each chapter concludes with an *Examining the Earth System* section. The questions and problems found here are intended to develop an awareness and appreciation for some of the Earth system's many interrelationships.

People and the Environment

Because knowledge about our planet and how it works is necessary to our survival and well-being, the treatment of environmental issues has always been an important part of *Earth Science.* Such discussions serve to illustrate the relevance and application of Earth science knowledge. With each new edition this focus has been given greater emphasis. This is certainly the case with this edition. The text integrates a great deal of information about the relationship between people and the natural environment and explores the application of the Earth sciences to understanding and solving problems that arise from these interactions. In addition to many basic text discussions, many of the text's special interest boxes involve the "People and the Environment" theme.

Understanding Earth

As members of a modern society, we are constantly reminded of the benefits derived from science. But what exactly is the nature of scientific inquiry? Developing an understanding of how science is done and how scientists work is a third important theme that appears throughout this book, beginning with the section "The Nature of Scientific Inquiry" in Chapter 1. Students will examine some of the difficulties encountered by scientists as they attempt to acquire reliable data about our planet and some of the ingenious methods that have been developed to overcome these difficulties. Students will also explore many examples of how hypotheses are formulated and tested, as well as learn about the evolution and development of some major scientific theories. Many basic text discussions as well as a number of the special interest boxes on "Understanding Earth" provide the reader with a sense of the observational techniques and reasoning processes involved in developing scientific knowledge. The emphasis is not just on what scientists know, but how they figured it out.

For the Instructor

Pearson Prentice Hall continues to improve the instructor resources in this edition with the goal of saving you time in preparing for your classes.

MasteringGeology™ with Pearson eText

Used by over one million science students, the Mastering platform is the most effective and widely used online tutorial, homework, and assessment system for the sciences. Now available with *Earth Science, 13th Edition,* **MasteringGeology™** offers:

- Assignable activities that include geoscience animations, *Encounter Earth* Google Earth multimedia, and *GEODe: Earth Science* activities.
- Additional Concept Check and Give It Some Thought questions, Test Bank questions, and Reading Quizzes
- A student Study Area with geoscience animations, *GEODe: Earth Science* activities, *In the News* RSS feeds, Self Study Quizzes, Web Links, Glossary, and Flashcards
- Pearson eText for *Earth Science, 13th Edition,* which gives students access to the text whenever and wherever they can access the Internet, and includes powerful interactive and customization functions www.masteringgeology.com

Instructor's Resource Center (IRC) on DVD

The IRC on DVD puts all of your lecture resources in one easy-to-reach place:

- Three PowerPoint® presentations for each chapter
- The Geoscience Animation Library
- All of the line art, tables, and photos from the text in .jpg files
- "Images of Earth" photo gallery
- Instructor's Manual in Microsoft Word
- Test Bank in Microsoft Word
- TestGen test generation and management software

PowerPoints®

Found on the IRC are three PowerPoint files for each chapter. Cut down on your preparation time, no matter what your lecture needs.

1. Art–All of your line art, tables, and photos from the text have been preloaded into PowerPoint slides for easy integration into your presentation.
2. Lecture Outline–This set averages 35 slides per chapter and includes customizable lecture outlines with supporting art.
3. Classroom Response System (CRS) Questions–Authored for use in conjunction with any classroom response system. These systems allow you to electronically poll your class for responses to questions, pop quizzes, attendance, and more.

Animations and *Images of Earth*

The Pearson Prentice Hall Geoscience Animation Library includes more than 100 animations illustrating many difficult-to-visualize topics of Earth science. Created through a unique collaboration among five of Pearson Prentice Hall's leading geoscience authors, these animations represent a significant step forward in lecture presentation aids. They are provided both as Flash files and, for your convenience, preloaded into PowerPoint slides.

Images of Earth allows you to supplement your personal and text-specific slides with an amazing collection of more than 300 geologic photos contributed by Marli Miller (University of Oregon) and other professionals in the field. The photos are available on the IRC on DVD.

Instructor's Manual with Test Bank

Authored by Stanley Hatfield (Southwestern Illinois College), the *Instructor's Manual* contains: learning objectives, chapter outlines, answers to end-of-chapter questions, and suggested short demonstrations to spice up your lecture. Authored by Jennifer Cole (Northeastern University), the *Test Bank* incorporates art and averages 75 multiple-choice, true/false, short answer, and critical thinking questions per chapter.

TestGen

Use this electronic version of the Test Bank to customize and manage your tests. Create multiple versions, add or edit questions, add illustrations—your customization needs are easily addressed by this powerful software.

Course Management

Pearson Prentice Hall offers instructor and student media for the 13th edition of *Earth Science* in formats compatible with Blackboard and other course management platforms. Contact your local Pearson representative for more information.

For the Student

The student resources to accompany *Earth Science, 13th Edition*, have been further refined with the goal of focusing the students' efforts and improving their understanding of Earth science concepts.

MasteringGeology™ with Pearson eText

Used by over one million science students, the Mastering platform is the most effective and widely used online tutorial, homework, and assessment system for the sciences. Now available with *Earth Science, 13th Edition*, **MasteringGeology**™ offers students a self Study Area containing:

- Geoscience Animation Library: More than 100 animations illustrating many difficult to understand Earth science concepts.
- *GEODe: Earth Science:* An interactive visual walkthrough of each chapter's content
- *In The News* RSS Feeds: Current Earth science events and news articles are pulled into the site with assessment
- Pearson eText
- Optional Self-Study Quizzes
- Web Links
- Glossary
- Flashcards

Study Guide

Written by experienced educator Stanley Hatfield (Southwestern Illinois College), the *Study Guide* helps students identify the important points from the text, and then provides them with review exercises, study questions, self-check exercises, and vocabulary review.

For the Laboratory

Applications and Investigations in Earth Science, Seventh Edition. Written by Ed Tarbuck, Fred Lutgens, and Ken Pinzke, this full-color laboratory manual contains 23 exercises that provide students with hands-on experience in geology, oceanography, meteorology, astronomy, and Earth science skills. The lab manual is available at a discount when purchased with the text; please contact your local Pearson representative for more details.

Acknowledgments

Writing a college textbook requires the talents and cooperation of many people. We value the excellent work of Mark Watry and Teresa Tarbuck of Spring Hill College whose talents helped us improve Chapters 2, 22, and 24. They helped make these chapters more readable, engaging, and up-to-date.

Working with Dennis Tasa, who is responsible for all of the text's outstanding illustrations and much of the developmental work on the animations and tutorials in *MasteringGeology*, is always special for us. We not only value his outstanding artistic talents and imagination but his friendship as well.

Sincere thanks to Michael Collier whose contributions as an aerial photographer and geologist added greatly to this project. Collaborating with Michael was a special pleasure.

Great thanks also go to those colleagues who prepared in-depth reviews. Their critical comments and thoughtful input helped guide our work and clearly strengthened the text. Special thanks to:

Erin Argyilan, Indiana University Northwest
Jake Armour, University of North Carolina, Charlotte
Natalie Bursztyn, Bakersfield College
Marianne Caldwell, Hillsborough Community College
Daniel Deocampo, California State University, Sacramento
Holly Dodson, Sierra College
Christopher Hooker, Waubonsee Community College
Zoran Kilibarda, Indiana University Northwest
Michael Lewis, University of North Carolina, Greensboro
TinaGayle Osborn, Palm Beach State College
Thomas Sills, Wright City College
David Voorhees, Waubonsee Community College
Jim Wysong, Hillsborough Community College

As always, we want to acknowledge the team of professionals at Pearson Prentice Hall. We sincerely appreciate the company's continuing strong support for excellence and innovation. All are committed to producing the best textbooks possible. Special thanks to our geology editor, Andy Dunaway, and to our conscientious project manager, Crissy Dudonis, for a job well done. The production team led by Patty Donovan at Laserworks Maine did an outstanding job. Kristin Piljay's photo research assistance was also a great help. All are true professionals with whom we are very fortunate to be associated.

Ed Tarbuck

Fred Lutgens

NEW!
ACTIVE LEARNING APPROACH

A new **active learning approach** in the Thirteenth Edition offers students a structured learning path and provides a reliable, consistent framework for mastering the chapter concepts.

FOCUS ON CONCEPTS

To assist you in learning the important concepts in this chapter, focus on the following questions:

● What are minerals, and how are they different from rocks?
● What are the smallest particles of matter?
● How do atoms bond?
● How do isotopes of the same element vary, and why are some isotopes radioactive?
● What are some of the physical and chemical properties of minerals? How can these properties be used to distinguish one mineral from another?
● What are the eight elements that make up most of Earth's continental crust?
● What is the most abundant mineral group?
● What do all silicate minerals have in common?
● What are *renewable* and *nonrenewable resources*?
● When is the term *ore* used with reference to a mineral?

NEW! Focus on Concepts

Each chapter begins with **Focus on Concepts,** which consists of a series of questions that alert students to key concepts in the chapter.

CONCEPT CHECK 2.1

❶ List five characteristics that classify an Earth material as a mineral.
❷ Based on the definition of a mineral, which of the following materials are not classified as minerals, and why: gold, water, synthetic diamonds, ice, and wood.
❸ Define the term *rock*. How do rocks differ from minerals?

NEW! Concept Checks

Within each chapter, every major section concludes with **Concept Checks** that allow students to monitor their understanding and comprehension of significant facts and ideas.

GIVE IT SOME THOUGHT

1. Using the geologic definition of *mineral* as your guide, determine which of the items on the list are minerals and which are not. If an item is not a mineral, explain why not.
    a. gold nugget
    b. seawater
    c. quartz
    d. cubic zirconia
    e. obsidian
    f. ruby
    g. glacial ice
    h. amber
    Refer to the Periodic Table of the Elements (Figure 2.5) to help you answer questions 2, 3, and 4.
2. If the number of protons in a neutral atom is 92 and its mass number is 238:
    a. What is the name of that element?
    b. How many electrons does it have?
    c. How many neutrons does it have?
3. Which element is more likely to form chemical bonds: xenon (Xe) or sodium (Na)? Explain why.
4. The information below refers to three isotopes of the element potassium. Using this information, determine the appropriate number of protons and neutrons for each isotope. Label each isotope in the manner used in the chapter.

    Atomic Number = 19   Atomic Number = 19   Atomic Number = 19
       Atomic Mass = 39   Atomic Mass = 40   Atomic Mass = 41

5. Referring to the accompanying photos of five minerals, determine which of these specimens exhibit a metallic luster and which have a nonmetallic luster.

A.   B.   C.   D.   E.

6. Examine the accompanying photo of a mineral that has several smooth, flat surfaces that resulted when the specimen was broken.
    a. How many flat surfaces are present on this specimen?
    b. How many *different directions* of cleavage does this specimen have?
    c. Do the cleavage directions meet at 90-degree angles?

Cleaved sample

NEW! Give It Some Thought (GIST)

These questions and problems are found at the end of each chapter. They challenge learners by involving them in activities that require higher-order thinking skills such as the synthesis, analysis, and application of material in the chapter.

NEW! Professional Earth Scientist's Perspective

Throughout the new edition, the perspective and tools of the practicing Earth scientist are emphasized as an integral component of the concepts discussed.

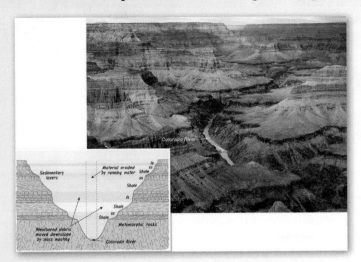

NEW! Geologist Sketches

"Geologist's Sketches" are incorporated into the text's visual program, where particular photographs are shown alongside sketched versions of the same image. This visual feature encourages students to see the world through the eyes of a professional geologist.

NEW! Professional Profile Boxes

Professional Profile boxes are essays that present profiles of working Earth scientists, giving students detailed perspectives on a variety of careers in the field.

Thematic Approach

Three themes that recur throughout the text as boxed essays —"Earth as a System," "People and the Environment," and "Understanding Earth"—help to organize and connect otherwise-dissimilar concepts.

Earth as a System

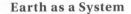

People and the Environment

Understanding Earth

Mastering**GEOLOGY**™ | **www.masteringgeology.com**

The Mastering platform is the most effective and widely used online tutorial, homework, and assessment system for the sciences. The Mastering system empowers students to take charge of their learning through activities aimed at different learning styles, and engages them in learning science through practice and step-by-step guidance—at their convenience, 24/7.

Assignable Content:

- *Encounter Earth* Activities
- Geoscience Animation Activities
- *GEODe* Activities
- Reading Quiz Questions
- Test Bank Questions
- Concept Check and Give It Some Thought Questions

For Student Self Study:

- Geoscience Animations
- *GEODe* Activities
- *In the News* RSS Feeds
- Optional Pearson eText
- Self-Study Quizzes
- Web Links

- Glossary
- Flashcards

Encounter Earth: Interactive Geoscience Explorations

***Encounter Earth* activities** enable students to use the dynamic features of Google Earth™ to visualize and explore geoscience concepts and answer multiple-choice and short answer questions related to core Earth science concepts. Questions include hints and specific wrong-answer feedback to help coach students towards mastery of the concepts. All explorations include corresponding Google Earth KMZ media files.

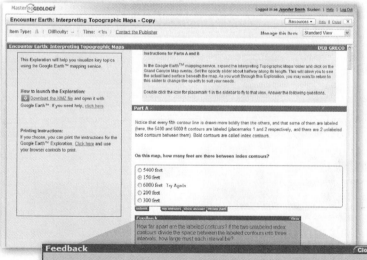

Feedback   Close

How far apart are the labeled contours? If the two unlabeled index contours divide the space between the labeled contours into three intervals, how large must each interval be?

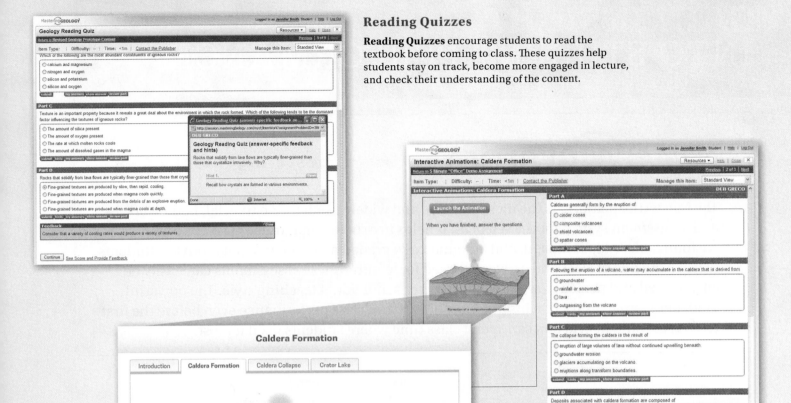

Reading Quizzes

Reading Quizzes encourage students to read the textbook before coming to class. These quizzes help students stay on track, become more engaged in lecture, and check their understanding of the content.

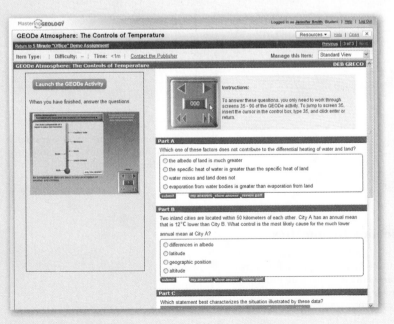

Geoscience Animations

Geoscience Animation Library, the world's largest library of geoscience visualizations, includes over 100 animations that illuminate the most difficult-to-visualize topics from the physical geosciences. Animations include audio narration, a text transcript, and assignable assessments with and specific wrong-answer feedback to help guide students towards mastering the concepts.

GEODe

GEODe Activities provide interactive visual walkthroughs of each chapter's core concepts through animations, videos, illustrations, photographs, and narration. The activities include assessment questions and specific wrong-answer feedback to test those concepts.

GAIN INSIGHTS
INTO STUDENT PERFORMANCE

 | www.masteringgeology.com

The Mastering platform is the most effective and widely used online tutorial, homework, and assessment system for the sciences. It helps instructors maximize class time with customizable, easy-to-assign, and automatically graded assessments that motivate students to learn outside of class and arrive prepared for lecture. These assessments can easily be customized and personalized for an instructor's individual teaching style. The powerful gradebook provides unique insight into student and class performance even before the first test. As a result, instructors can spend class time where students need it most.

The Mastering platform:

- was developed **by scientists** for science students and instructors
- features over **one million** active registrations
- offers data-supported efficacy
- has a proven history with over **9 years** of student use
- includes active users in all **50 states** and in **30 countries**
- offers **99.8%** server reliability

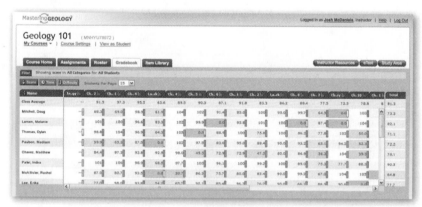

Gradebook

Every assignment is **automatically graded. Shades of red** highlight vulnerable students and challenging assignments.

Gradebook Diagnostics

This screen gives instructors weekly diagnostics. With a single click, charts summarize the most difficult problems, vulnerable students, grade distribution, and even score improvement over the course.

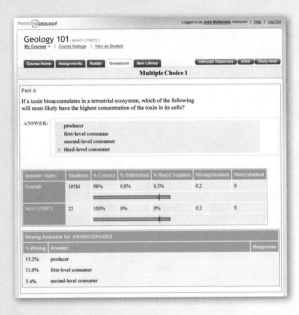

Student Performance Data

At a glance, instructors can identify students who are having difficulty by using the color-coded gradebook. Instructors can also identify the most difficult problem (and step within that problem) in each assignment, or critique the detailed work of anyone who needs more help. They can even compare results on any problem and any step with a previous class, or with the national average.

Continuously Improving Content

MasteringGeology™ offers a dynamic pool of assignable content that improves with student usage. Detailed analysis of student performance statistics—including time spent, answers submitted, solutions requested, and hints used—ensures the highest quality content.

1. **We conduct a thorough analysis** of each question by reviewing student performance data that has been generated by real students.

2. **We make enhancements** to improve the clarity and accuracy of content, answer choices, and instructions for each problem.

3. **We repeat the process.**

This ongoing process helps students learn as students help the system improve.

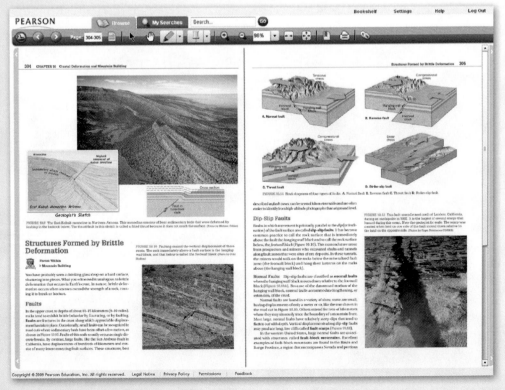

Pearson eText

Pearson eText gives students access to the text whenever and wherever they can access the Internet. The eText pages look exactly like the printed text, and include powerful interactive and customization functions. Users can create notes, highlight text in different colors, create bookmarks, zoom, click hyperlinked words and phrases to view definitions, and view as a single-page or as two-pages. Pearson eText also links students to associated media files, enabling them to view an animation as they read the text and offers a full-text search and the ability to save and export notes.

Introduction to Earth Science

Iceland's Eyjafjallajökull volcano erupting on April 17, 2010. The plume of volcanic ash spread over much of Europe and severely disrupted air traffic.
(Photo by Joanna Vestey/CORBIS)

The spectacular eruption of a volcano, the magnificent scenery of a rocky coast, and the destruction created by a hurricane are all subjects for the Earth scientist. The study of Earth science deals with many fascinating and practical questions about our environment. What forces produce mountains? Why is our daily weather so variable? Is climate really changing? How old is Earth, and how is it related to the other planets in the solar system? What causes ocean tides? What was the Ice Age like? Will there be another? Can a successful well be located at this site?

The subject of this text is *Earth science*. To understand Earth is not an easy task, because our planet is not a static and unchanging mass. Rather, it is a dynamic body with many interacting parts and a long and complex history.

FOCUS ON CONCEPTS

To assist you in learning **the important concepts in this chapter, focus on the following questions:**

- What are the sciences that collectively make up Earth science?
- What are some examples of interactions between people and the natural environment?
- How is a scientific hypothesis different from a scientific theory?
- How old is Earth?
- How did Earth and other planets in our solar system originate?
- What are Earth's four major "spheres"?
- What are the principal divisions of the solid Earth? What criteria were used to establish these divisions?
- What is the theory of plate tectonics?
- What are the major features of the continents and ocean basins?
- Why should Earth be thought of as a system?

What Is Earth Science?

Earth science is the name for all the sciences that collectively seek to understand Earth and its neighbors in space. It includes geology, oceanography, meteorology, and astronomy. In this book, Units 1–4 focus on the science of **geology**, a word that literally means "study of Earth." Geology is traditionally divided into two broad areas: physical and historical.

Physical geology examines the materials composing Earth and seeks to understand the many processes that operate beneath and upon its surface. Earth is a dynamic, ever-changing planet. Internal forces create earthquakes, build mountains, and produce volcanic structures. At the surface, external processes break rock apart and sculpt a broad array of landforms. The erosional effects of water, wind, and ice result in a great diversity of landscapes. Because rocks and minerals form in response to Earth's internal and external processes, their interpretation is basic to an understanding of our planet.

In contrast to physical geology, the aim of *historical geology* is to understand the origin of Earth and the development of the planet through its 4.6-billion-year history. It strives to establish an orderly chronological arrangement of the multitude of physical and biological changes that have occurred in the geologic past (Figure 1.1A). The study of physical geology logically precedes the study of Earth history because we must first understand how Earth works before we attempt to unravel its past.

Unit 5, *The Global Ocean*, is devoted to **oceanography**. Oceanography is actually not a separate and distinct science. Rather, it involves the application of all sciences in a comprehensive and interrelated study of the oceans in all their aspects and relationships. Oceanography integrates chemistry, physics, geology, and biology. It includes the study of the composition and movements of seawater, as well as coastal processes, seafloor topography, and marine life (Figure 1.1B).

Unit 6, *Earth's Dynamic Atmosphere*, examines the mixture of gases that is held to the planet by gravity and thins rapidly with altitude. Acted on by the combined effects of Earth's motions and energy from the Sun, the formless and invisible atmosphere reacts by producing an infinite variety of weather, which in turn creates the basic pattern of global climates. **Meteorology** is the study of the atmosphere and the processes that produce weather and climate. Like oceanography, meteorology involves the application of other sciences in an integrated study of the thin layer of air that surrounds Earth.

Unit 7, *Earth's Place in the Universe*, demonstrates that an understanding of Earth requires that we relate our planet to the larger universe. Because Earth is related to all of the other objects in space, the science of **astronomy**—the study of the universe—is very useful in probing the origins of our own

FIGURE 1.1 A. This hiker is resting atop the Kaibab Formation, the uppermost layer in the Grand Canyon. Hundreds of millions of years of Earth history is contained in the strata that lay beneath him. This is a view of Cape Royal on the Grand Canyon's North Rim. (Photo by Michael Collier) **B.** The *Chikyu* (meaning "Earth" in Japanese), the world's most advanced scientific drilling vessel. In can drill as deep as 7,000 meters (nearly 23,000 feet) below the seabed in water as deep as 2,500 meters (8,200 feet). It is part of the Integrated Ocean Drilling Program (IODP). (AP Photo/Itsuo Inouye)

environment. Because we are so closely acquainted with the planet on which we live, it is easy to forget that Earth is just a tiny object in a vast universe. Indeed, Earth is subject to the same physical laws that govern the many other objects populating the great expanses of space. Thus, to understand explanations of our planet's origin, it is useful to learn something about the other members of our solar system. Moreover, it is helpful to view the solar system as a part of the great assemblage of stars that comprise our galaxy, which in turn is but one of many galaxies.

Understanding Earth science is challenging because our planet is a dynamic body with many interacting parts and a complex history. Throughout its long existence, Earth has been changing. In fact, it is changing as you read this page and will continue to do so into the foreseeable future. Sometimes the changes are rapid and violent, as when severe storms, landslides, or volcanic eruptions occur. Just as often, change takes place so gradually that it goes unnoticed during a lifetime. Scales of size and space also vary greatly among the phenomena studied in Earth science.

Earth science is often perceived as science that is performed in the out of doors, and rightly so. A great deal of what Earth scientists study is based on observations and experiments conducted in the field. But Earth science is also conducted in the laboratory, where, for example, the study of various Earth materials provides insights into many basic processes, and the creation of complex computer models allows for the simulation of our planet's complicated climate system. Frequently, Earth scientists require an understanding and application of knowledge and principles from physics, chemistry, and biology. Geology, oceanography, meteorology, and astronomy are sciences that seek to expand our knowledge of the natural world and our place in it.

CONCEPT CHECK 1.1

❶ List the sciences that make up Earth science.
❷ Name the two broad subdivisions of geology and distinguish between them.

Earth Science, People, and the Environment

The primary focus of this book is to develop an understanding of basic Earth science principles, but along the way we explore numerous important relationships between people and the natural environment. Many of the problems and issues addressed by the Earth sciences are of practical value to people.

Natural hazards are a part of living on Earth. Every day they adversely affect literally millions of people worldwide and are responsible for staggering damages. The chapter opening photo and Figure 1.2 are two examples. Among the hazardous Earth processes studied by Earth scientists are volcanoes, floods, tsunami, earthquakes, landslides, and hurricanes. Of course, these hazards are *natural* processes. They become hazards only when people try to live where these processes occur.

According to the United Nations, in 2008, for the first time, more people lived in cities than in rural areas. This global trend toward urbanization concentrates millions of people into megacities, many of which are vulnerable to natural hazards (Figure 1.3). Coastal sites are becoming more vulnerable because development often destroys natural defenses such as wetlands and sand dunes. In addition, there is a growing threat associated with human influences on the Earth system such as sea level rise that is linked to global climate change.[1] Other megacities are exposed to seismic (earthquake) and volcanic hazards where inappropriate land use and poor construction practices, coupled with rapid population growth, are increasing vulnerability.

Resources represent another important focus that is of great practical value to people. They include water and soil, a great variety of metallic and nonmetallic minerals, and energy (Figure 1.4). Together they form the very foundation of modern civilization. Earth science deals not only with the formation and occurrence of these vital resources but also with maintaining supplies and with the environmental impact of their extraction and use.

Complicating all environmental issues is rapid world population growth *and* everyone's aspiration to a better standard of living. This means a ballooning demand for resources and a growing pressure for people to dwell in environments having significant natural hazards.

Not only do Earth processes have an impact on people but we humans can dramatically influence Earth processes as well. For example, river flooding is natural, but the magnitude and frequency of flooding can be changed significantly by human activities such as clearing forests, building cities, and constructing dams. Unfortunately, natural systems do not always adjust to

---

[1]The idea of the Earth system is explored later in the chapter. Global climate change and its effects are a focus of Chapter 20.

FIGURE 1.2 Crystal Beach, Texas, on September 16, 2008, three days after Hurricane Ike came ashore. At landfall the storm had sustained winds of 165 kilometers (105 miles) per hour. The extraordinary storm surge caused much of the damage pictured here. (Photo by Earl Nottingham/ Associated Press)

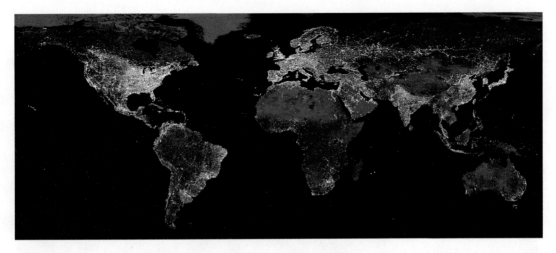

FIGURE 1.3 This composite nighttime image of Earth's city lights shows the geographic distribution of settlements and helps us appreciate the intensity of human presence in many parts of the planet. (NASA)

artificial changes in ways that we can anticipate. Thus, an alteration to the environment that was intended to benefit society often has the opposite effect.

At appropriate places throughout this book, you will have the opportunity to examine different aspects of our relationship with the physical environment. It will be rare to find a chapter that does not address some aspect of natural hazards, environmental issues, or resources. Significant parts of some chapters provide the basic knowledge and principles needed to understand environmental problems. Moreover, a number of the book's special-interest boxes focus on Earth science, people, and the environment by providing case studies or highlighting a topical issue.

CONCEPT CHECK 1.2

① **List at least four phenomena that can be regarded as natural hazards.**

FIGURE 1.4 Wind energy is considered a *renewable* resource. The use of wind turbines for generating electricity is growing rapidly. These wind turbines are operating near Palm Springs, California. (Photo by John Mead/Science Photo Library/Photo Researchers, Inc.)

Students Sometimes Ask...

What is the current world population and how fast is it growing?

It took until about the year 1800 for the world population to reach 1 billion people. In 1970, the number was about 4 billion. According to the U.S. Census Bureau, the world population in mid-2010 was approaching 6.9 billion people. The planet is currently adding people at a rate exceeding 75 million per year.

The Nature of Scientific Inquiry

As members of a modern society, we are constantly reminded of the benefits derived from science. But what exactly is the nature of scientific inquiry? Developing an understanding of how science is done and how scientists work is another important theme that appears throughout this book. You will explore the difficulties in gathering data and some of the ingenious methods that have been developed to overcome these difficulties. You will also see many examples of how hypotheses are formulated and tested, as well as learn about the evolution and development of some major scientific theories.

All science is based on the assumption that the natural world behaves in a consistent and predictable manner that is comprehensible through careful, systematic study. The overall goal of science is to discover the underlying patterns in nature and then to use this knowledge to make predictions about what should or should not be expected, given certain facts or circumstances. For example, by knowing how oil deposits form, geologists are able to predict the most favorable sites for exploration and, perhaps as important, how to avoid regions having little or no potential.

The development of new scientific knowledge involves some basic logical processes that are universally accepted. To determine what is occurring in the natural world, scientists collect scientific "*facts*" through observation and measurement. The

Box 1.1

UNDERSTANDING EARTH

Studying Earth from Space

Scientific facts are gathered in many ways, including laboratory studies and field observations and measurements. Satellite images are another valuable source of data. Such images provide perspectives that are difficult to gain from more traditional sources. Moreover, the high-tech instruments aboard many satellites enable scientists to gather information from remote regions where data are otherwise scarce.

The image in **Figure 1.A** was created using satellite radar data from the Antarctic Mapping Mission. It shows the movement of Antarctica's Lambert Glacier. The smaller glaciers that join Lambert Glacier exhibit low velocities, shown in green, of 100–300 meters (330–980 feet) per year. Most of Lambert Glacier itself moves at rates between 400–800 meters (1,310–2,620 feet) per year. Near its terminus, where the ice spreads out and thins, velocities increase to 1,000–1,200 meters (3,280–3,940 feet) per year. Due to the remoteness and extreme weather conditions associated with this region, only a handful of traditional in-situ

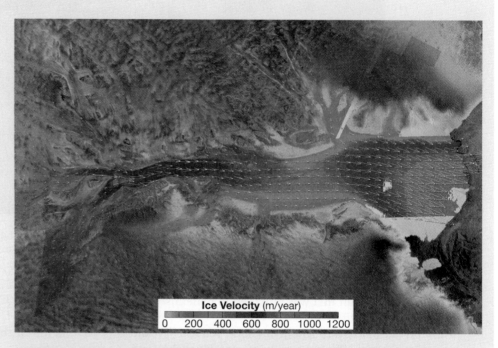

FIGURE 1.A This satellite image provides detailed information about the movement of Antarctica's Lambert Glacier. Such information is basic to understanding changes in the behavior of the glacier over time. The ice velocities are determined from pairs of images obtained 24 days apart, using a technique called radar interferometry. (NASA)

velocity measurements had previously been reported. Now that accurate satellite measurements are available, scientists have a quantitative baseline for future comparisons.

The image in **Figure 1.B** is from NASA's *Tropical Rainfall Measuring Mission (TRMM).*

TRMM's research satellite was designed to expand our understanding of Earth's hydrologic (water) cycle and its role in our climate system. Instruments aboard the *TRMM* satellite have greatly expanded our ability to collect precipitation data. In addition to data for land areas, this satellite provides precise measurements of rainfall over the oceans where conventional land-based instruments cannot see. This is especially important because much of Earth's rain falls in ocean-covered tropical areas, and a great deal of the globe's weather-producing energy comes from heat exchanges involved in the rainfall process. Until the *TRMM*, information on the intensity and amount of rainfall over the tropics was scanty. Such data are crucial to understanding and predicting global climate change.

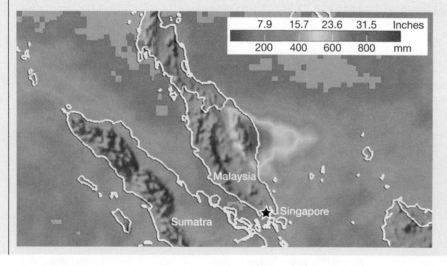

FIGURE 1.B This map of rainfall for December 7–13, 2004, in Malaysia was constructed using *TRMM* data. Over 800 millimeters (32 inches) of rain fell along the east coast of the peninsula (darkest red area). The extraordinary rains caused extensive flooding and triggered many mudflows. (NASA/*TRMM* image)

types of facts or data that are collected generally seek to answer a well-defined question about the natural world. How did this mountain range form? How does rainfall vary in this area? Because some error is inevitable, the accuracy of a particular measurement or observation is always open to question. Nevertheless, these data are essential to science and serve as the springboard for the development of scientific theories (Box 1.1).

Hypothesis

Once facts have been gathered and principles have been formulated to describe a natural phenomenon, investigators try to explain how or why things happen in the manner observed. They often do this by constructing a tentative (or untested) explanation, which is called a scientific **hypothesis**. It is best if an investigator can formulate more than one hypothesis to explain a given set of observations. If an individual scientist is unable to devise multiple hypotheses, others in the scientific community will almost always develop alternative explanations. A spirited debate frequently ensues. As a result, extensive research is conducted by proponents of opposing hypotheses, and the results are made available to the wider scientific community in scientific journals.

Before a hypothesis can become an accepted part of scientific knowledge, it must pass objective testing and analysis. If a hypothesis cannot be tested, it is not scientifically useful, no matter how interesting it might seem. The verification process requires that *predictions* be made based on the hypothesis being considered and that the predictions be tested by comparing them against objective observations of nature. Put another way, hypotheses must fit observations other than those used to formulate them in the first place. Those hypotheses that fail rigorous testing are ultimately discarded. The history of science is littered with discarded hypotheses. One of the best known is the Earth-centered model of the universe—a proposal that was supported by the apparent daily motion of the Sun, Moon, and stars around Earth. As the mathematician Jacob Bronowski so ably stated, "Science is a great many things, but in the end they all return to this: Science is the acceptance of what works and the rejection of what does not."

Theory

When a hypothesis has survived extensive scrutiny and when competing ones have been eliminated, a hypothesis may be elevated to the status of a scientific **theory**. In everyday language we may say, "That's only a theory." But a scientific theory is a well-tested and widely accepted view that the scientific community agrees best explains certain observable facts.

Some theories that are extensively documented and extremely well supported are comprehensive in scope. For example, the theory of plate tectonics provides the framework for understanding the origin of mountains, earthquakes, and volcanic activity. In addition, plate tectonics explains the evolution of the continents and the ocean basins through time—ideas that are explored in some detail in later chapters.

Scientific Methods

The process just described, in which researchers gather facts through observations and formulate scientific hypotheses and theories, is called the *scientific method*. Contrary to popular belief, the scientific method is not a standard recipe that scientists apply in a routine manner to unravel the secrets of our natural world. Rather, it is an endeavor that involves creativity and insight. Rutherford and Ahlgren put it this way: "Inventing hypotheses or theories to imagine how the world works and then figuring out how they can be put to the test of reality is as creative as writing poetry, composing music, or designing skyscrapers."[2]

There is no fixed path that scientists always follow that leads unerringly to scientific knowledge. Nevertheless, many scientific

investigations involve the following steps: (1) a question is raised about the natural world; (2) scientific data are collected that relate to the question (**Figure 1.5**); (3) questions are posed that relate to the data and one or more working hypotheses are developed that

FIGURE 1.5 Gathering data and making careful observations is a basic part of scientific inquiry. **A.** This Automated Surface Observing System (ASOS) installation is one of nearly 900 in use for data gathering as part of the U.S. primary surface observing network. (Photo by Bobbe Christopherson) **B.** These scientists are working with a sediment core recovered from the ocean floor. Such cores often contain useful data about the geologic past and Earth's climate history. (Photo by Science Source/Photo Researchers, Inc.)

A.

B.

[2]F. James Rutherford and Andrew Ahlgren, *Science for All Americans* (New York: Oxford University Press, 1990), p. 7.

may answer these questions; (4) observations and experiments are developed to test the hypotheses; (5) the hypotheses are accepted, modified, or rejected based on extensive testing; (6) data and results are shared with the scientific community for critique and further testing.

Other scientific discoveries may result from purely theoretical ideas, which stand up to extensive examination. Some researchers use high-speed computers to simulate what is happening in the "real" world. These models are useful when dealing with natural processes that occur on very long time scales or take place in extreme or inaccessible locations. Still other scientific advancements are made when a totally unexpected happening occurs during an experiment. These serendipitous discoveries are more than pure luck, for as Louis Pasteur said, "In the field of observation, chance favors only the prepared mind."

Scientific knowledge is acquired through several avenues, so it might be best to describe the nature of scientific inquiry as the methods of science rather than the scientific method. In addition, it should always be remembered that even the most compelling scientific theories are still simplified explanations of the natural world.

In this book, you will discover the results of centuries of scientific work. You will see the end product of millions of observations, thousands of hypotheses, and hundreds of theories. We have distilled all of this to give you a "briefing" on Earth science.

But realize that our knowledge of Earth is changing daily, as thousands of scientists worldwide make satellite observations, analyze drill cores from the seafloor, measure earthquakes, develop computer models to predict climate, examine the genetic codes of organisms, and discover new facts about our planet's long history. This new knowledge often updates hypotheses and theories. Expect to see many new discoveries and changes in scientific thinking in your lifetime.

CONCEPT CHECK 1.3

1 How is a scientific hypothesis different from a scientific theory?
2 List the basic steps followed in many scientific investigations.

Scales of Space and Time in Earth Science

When we study Earth, we must contend with a broad array of space and time scales (Figure 1.6). Some phenomena are relatively easy for us to imagine, such as the size and duration of an afternoon thunderstorm or the dimensions of a sand dune. Other phenomena are so vast or so small that they are difficult to imagine. The number of stars and distances in our galaxy (and beyond!) or the internal arrangement of atoms in a mineral crystal are examples of such phenomena.

Some of the events we study occur in fractions of a second. Lightning is an example. Other processes extend over spans of tens or hundreds of millions of years. The lofty Himalaya Mountains began forming nearly 50 million years ago, and they continue to develop today.

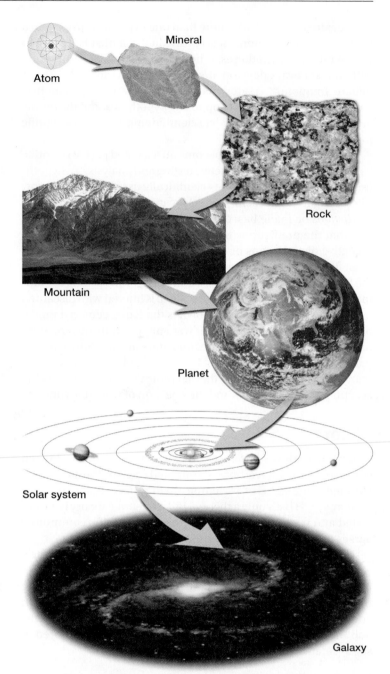

FIGURE 1.6 Earth science involves investigations of phenomena that range in size from atoms to galaxies and beyond.

The concept of geologic time is new to many nonscientists. People are accustomed to dealing with increments of time that are measured in hours, days, weeks, and years. Our history books often examine events over spans of centuries, but even a century is difficult to appreciate fully. For most of us, someone or something that is 90 years old is *very old*, and a 1,000-year-old artifact is *ancient*.

By contrast, those who study Earth science must routinely deal with vast time periods—millions or billions (thousands of millions) of years. When viewed in the context of Earth's 4.6-billion-year history, an event that occurred 100 million years ago may be characterized as "recent" by a geologist, and a rock sample that has been dated at 10 million years may be called "young."

Students Sometimes Ask...

In class you compared a hypothesis to a theory. How is each one different from a scientific law?

A *scientific law* is a basic principle that describes a particular behavior of nature that is generally narrow in scope and can be stated briefly—often as a simple mathematical equation. Because scientific laws have been shown time and time again to be consistent with observations and measurements, they are rarely discarded. Laws may, however, require modifications to fit new findings. For example, Newton's laws of motion are still useful for everyday applications (NASA uses them to calculate satellite trajectories), but they do not work at velocities approaching the speed of light. For these circumstances, they have been supplanted by Einstein's theory of relativity.

An appreciation for the magnitude of geologic time is important in the study of our planet because many processes are so gradual that vast spans of time are needed before significant changes occur.

How long is 4.6 billion years? If you were to begin counting at the rate of one number per second and continued 24 hours a day, seven days a week and never stopped, it would take about two lifetimes (150 years) to reach 4.6 billion!

Over the past 200 years or so, Earth scientists have developed the **geologic time scale** of Earth history. It divides the 4.6-billion-year history of Earth into many different units and provides a meaningful time frame within which the events of the geologic past are arranged (**Figure 1.7**). The principles used to develop the geologic time scale are examined at some length in Chapter 11.

CONCEPT CHECK 1.4

1. List two examples of size/space scales in Earth science that are at opposite ends of the spectrum.
2. How old is Earth?

Students Sometimes Ask...

I've heard scientists use the term "light-year" when discussing astronomy. What is a "light-year"?

At first you might think that a light-year is some sort of time measurement. But, actually, the light-year is a unit for measuring distances to the stars. Such distances are so large that familiar units such as kilometers or miles are too cumbersome to use. A light-year is the distance light travels in one Earth year—about 9.5 trillion kilometers (5.8 trillion miles).

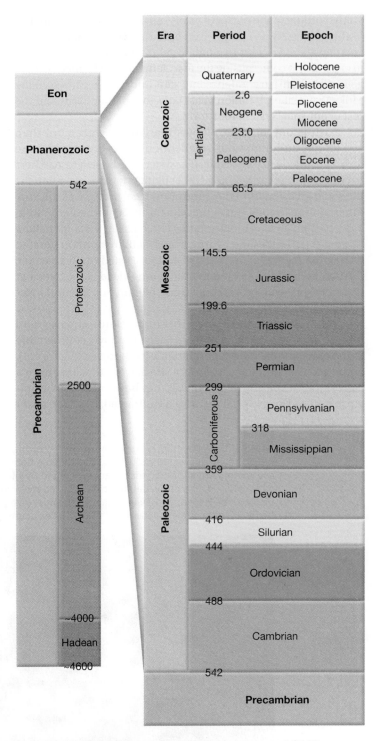

FIGURE 1.7 The geologic time scale divides the vast 4.6-billion-year history of Earth into eons, eras, periods, and epochs. We presently live in the Holocene epoch of the Quaternary period. This period is part of the Cenozoic era, which is the latest era of the Phanerozoic eon. There is much more about geologic time in Chapter 11.

Early Evolution of Earth

This section describes the most widely accepted views on the origin of our solar system. The theory summarized here represents the most consistent set of ideas available to explain what we know about our solar system today.

According to the *big bang theory*, all of the energy and matter of the universe was compressed into an incomprehensibly hot and dense state. About 13.7 billion years ago, our universe began to expand and cool, causing the first elements that formed (hydrogen and helium) to condense into stars and galaxies. It was in the Milky Way Galaxy 9 billion years later that planet Earth and the rest of our solar system took form.

Earth is one of eight planets that, along with more than 160 moons and numerous smaller bodies, revolve around the Sun. The orderly nature of our solar system leads most researchers to conclude that Earth and the other planets formed at essentially the same time and from the same primordial material as the Sun. The **nebular theory** states that the bodies of our solar system evolved from an enormous rotating cloud called the *solar nebula* (Figure 1.8). Besides the hydrogen and helium atoms generated during the Big Bang, the solar nebula consisted of microscopic dust grains and the ejected matter of long-dead stars. (Nuclear fusion in stars converts hydrogen and helium into the other elements found in the universe.)

Nearly 5 billion years ago this huge cloud of gases and minute grains of heavier elements began to slowly contract due to the gravitational interactions among its particles (Figure 1.8). Some external influence, such as a shock wave traveling from a catastrophic explosion (*supernova*), may have triggered the collapse. As this slowly spiraling nebula contracted, it rotated faster and faster for the same reason ice skaters do when they draw their arms toward their bodies. Eventually the inward pull of gravity came into balance with the outward force caused by the rotational motion of the nebula (Figure 1.8). By this time the once vast cloud had assumed a flat disk shape with a large concentration of material at its center called the *protosun* (pre-Sun). (Astronomers are fairly confident that the nebular cloud formed a disk because similar structures have been detected around other stars.)

During the collapse, gravitational energy was converted to thermal energy (heat), causing the temperature of the inner portion of the nebula to dramatically rise. At these high temperatures, the dust grains broke up into molecules and extremely energetic atomic particles. However, at distances beyond the orbit of Mars, the temperatures probably remained quite low. At −200°C, the tiny particles in the outer portion of the nebula were likely covered with a thick layer of ices made of frozen water, carbon dioxide, ammonia, and methane. (Some of this material still resides in the outermost reaches of the solar system in a region called the *Oort cloud*.) The disk-shaped cloud also contained appreciable amounts of the lighter gases hydrogen and helium.

The formation of the Sun marked the end of the period of contraction and thus the end of gravitational heating. Temperatures in the region where the inner planets now reside began

FIGURE 1.8 Formation of the solar system according to the nebular theory. **A.** The birth of our solar system, which began as a cloud of dust and gas called a nebula, started to gravitationally collapse. **B.** The nebula contracted into a rotating disk that was heated by the conversion of gravitational energy into thermal energy. **C.** Cooling of the nebular cloud caused rocky and metallic material to condense into tiny solid particles. **D.** Repeated collisions caused the dust-size particles to gradually coalesce into asteroid-size bodies. Within a few million years these bodies acreted into the planets.

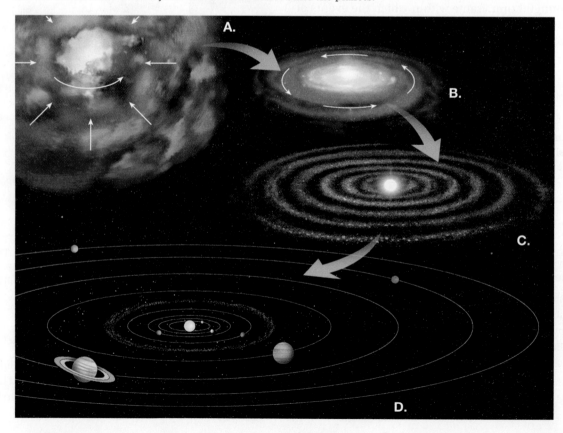

to decline. The decrease in temperature caused those substances with high melting points to condense into tiny particles that began to coalesce (join together). Materials such as iron and nickel and the elements of which the rock-forming minerals are composed—silicon, calcium, sodium, and so forth—formed metallic and rocky clumps that orbited the Sun (Figure 1.8). Repeated collisions caused these masses to coalesce into larger asteroid-size bodies, called *planetesimals*, which in a few tens of millions of years accreted into the four inner planets we call Mercury, Venus, Earth, and Mars. Not all of these clumps of matter were incorporated into the planetesimals. Those rocky and metallic pieces that remained in orbit are called asteroids and become *meteorites* if they impact Earth's surface.

As more and more material was swept up by these growing planetary bodies, the high-velocity impact of nebular debris caused their temperatures to rise. Because of their relatively high temperatures and weak gravitational fields, the inner planets were unable to accumulate much of the lighter components of the nebular cloud. The lightest of these, hydrogen and helium, were eventually whisked from the inner solar system by the solar winds.

At the same time that the inner planets were forming, the larger, outer planets (Jupiter, Saturn, Uranus, and Neptune), along with their extensive satellite systems, were also developing. Because of low temperatures far from the Sun, the material from which these planets formed contained a high percentage of ices—water, carbon dioxide, ammonia, and methane—as well as rocky and metallic debris. The accumulation of ices accounts in part for the large size and low density of the outer planets. The two most massive planets, Jupiter and Saturn, had a surface gravity sufficient to attract and hold large quantities of even the lightest elements—hydrogen and helium.

CONCEPT CHECK 1.5

1 Name and briefly outline the theory that describes the formation of our solar system.
2 List the inner planets and the outer planets. Describe basic differences in size and composition.

Earth's Spheres

The images in **Figure 1.9** are considered to be classics because they let humanity see Earth differently than ever before. Figure 1.9A, known as "Earthrise," was taken when the *Apollo 8* astronauts orbited the Moon for the first time in December 1968. As the spacecraft rounded the Moon, Earth appeared to rise above the lunar surface. Figure 1.9B, referred to as "The Blue Marble," is perhaps the most widely reproduced image of Earth and was taken in December 1972 by the crew of *Apollo 17* during the last lunar mission. These early views profoundly altered our conceptualizations of Earth and remain powerful images decades after they were first viewed. Seen from space, Earth is breathtaking in its beauty and startling in its solitude. The photos remind us

FIGURE 1.9 A. View, called "Earthrise," that greeted the *Apollo 8* astronauts as their spacecraft emerged from behind the Moon. (NASA Headquarters) **B.** Africa and Arabia are prominent in this classic image called "The Blue Marble" taken from *Apollo 17*. The tan cloud-free zones over the land coincide with major desert regions. The band of clouds across central Africa is associated with a much wetter climate that in places sustains tropical rain forests. The dark blue of the oceans and the swirling cloud patterns remind us of the importance of the oceans and the atmosphere. Antarctica, a continent covered by glacial ice, is visible at the south pole. (NASA)

that our home is, after all, a planet—small, self-contained, and in some ways even fragile. Bill Anders, the *Apollo 8* astronaut who took the "Earthrise" photo, expressed it this way: "We came all this way to explore the Moon, and the most important thing is that we discovered the Earth."

As we look closely at our planet from space, it becomes apparent that Earth is much more than rock and soil. In fact, the most conspicuous features in Figure 1.9A are not continents but swirling clouds suspended above the surface of the vast global ocean. These features emphasize the importance of water on our planet.

The closer view of Earth from space shown in Figure 1.9B helps us appreciate why the **physical environment** is traditionally divided into three major spheres: the water portion of our planet, the hydrosphere; Earth's gaseous envelope, the atmosphere; and, of course, the solid Earth, or geosphere.

FIGURE 1.10 The shoreline is one obvious meeting place for rock, water, and air. In this scene along the coast of Newfoundland, ocean waves that were created by the force of moving air break against the rocky shore. The force of the water can be powerful, and the erosional work that is accomplished can be great. (Photo by Radius Images/photolibrary.com)

prominent feature of the hydrosphere, blanketing nearly 71 percent of Earth's surface to an average depth of about 3,800 meters (12,500 feet). It accounts for about 97 percent of Earth's water (Figure 1.11). However, the hydrosphere also includes the fresh water found in streams, lakes, and glaciers, as well as that found underground.

Although these latter sources constitute just a tiny fraction of the total, they are much more important than their meager percentages indicate. In addition to providing the fresh water that is so vital to life on land, streams, glaciers, and groundwater are responsible for sculpturing and creating many of our planet's varied landforms.

Atmosphere

Earth is surrounded by a life-giving gaseous envelope called the **atmosphere**. When we watch a high-flying jet plane cross the sky, it seems that the atmosphere extends upward for a great

It should be emphasized that our environment is highly integrated and is not dominated by rock, water, or air alone. It is instead characterized by continuous interactions as air comes in contact with rock, rock with water, and water with air. Moreover, the biosphere, the totality of life-forms on our planet, extends into each of the three physical realms and is an equally integral part of the planet. Thus, Earth can be thought of as consisting of four major spheres: the hydrosphere, atmosphere, geosphere, and biosphere.

The interactions among the spheres of Earth's environment are incalculable. Figure 1.10 provides us with one easy-to-visualize example. The shoreline is an obvious meeting place for rock, water, and air. In this scene, ocean waves that were created by the drag of air moving across the water are breaking against the rocky shore. The force of the water can be powerful, and the erosional work that is accomplished can be great.

Hydrosphere

Earth is sometimes called the *blue planet* or, as we saw in Figure 1.9B—"The Blue Marble." Water more than anything else makes Earth unique. The **hydrosphere** is a dynamic mass of water that is continually on the move, evaporating from the oceans to the atmosphere, precipitating to the land, and running back to the ocean again. The global ocean is certainly the most

FIGURE 1.11 Distribution of Earth's water. The oceans clearly dominate. When we consider only the nonocean component, ice sheets and glaciers represent nearly 85 percent of Earth's freshwater. Groundwater accounts for just over 14 percent. When only liquid freshwater is considered, the significance of groundwater is obvious. (Glacier photo by Bernhard Edmaier/Photo Researchers, Inc.; stream photo by E. J. Tarbuck; and groundwater photo by Michael Collier)

Oceans 97.2%

2.8%

Freshwater lakes 0.009%
Saline lakes and inland seas 0.008%
Soil moisture 0.005%
Atmosphere 0.001%
Stream channels 0.0001%

Hydrosphere

Glaciers 2.15%

Groundwater 0.62%

Nonocean Component (% of total hydrosphere)

Stream channel

Glaciers

Groundwater (spring)

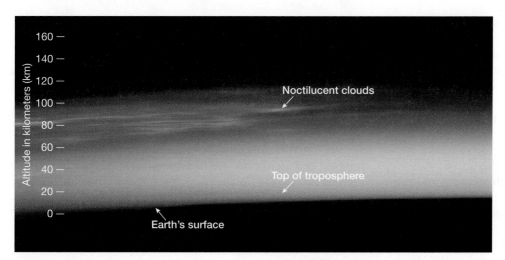

FIGURE 1.12 This unique image of Earth's atmosphere merging with the emptiness of space resembles an abstract painting. It was taken over western China in June 2007 by a Space Shuttle crew member. The thin silvery streaks (called noctilucent clouds) high in the blue area are at a height of about 80 kilometers (50 miles). The atmosphere at this altitude is *very* thin. Air pressure at this height is less than a thousandth of that at sea level. The thin reddish zone in the lower portion of the image is the densest part of the atmosphere. It is here, in a layer called the *troposphere*, that practically all weather and cloud formation occur. Ninety percent of Earth's atmosphere occurs within just 16 kilometers (10 miles) of the surface. (NASA)

Most life on land is also concentrated near the surface, with tree roots and burrowing animals reaching a few meters underground and flying insects and birds reaching a kilometer or so above Earth. A surprising variety of life-forms are also adapted to extreme environments. For example, on the ocean floor, where pressures are extreme and no light penetrates, there are places where vents spew hot, mineral-rich fluids that support communities of exotic life-forms. On land, some bacteria thrive in rocks as deep as 4 kilometers (2.5 miles) and in boiling hot springs. Moreover, air currents can carry micro-organisms many kilometers into the atmosphere. But even when we consider these extremes, life still must be thought of as being confined to a narrow band very near Earth's surface.

Plants and animals depend on the physical environment for the basics of life. However, organisms do more than just respond to their physical environment. Through countless interactions, life-forms help maintain and alter their physical environment. Without life, the makeup and nature of the geosphere, hydrosphere, and atmosphere would be very different.

Geosphere

Beneath the atmosphere and the ocean is the solid Earth or **geosphere**. The geosphere extends from the surface to the center

distance. However, when compared to the thickness (radius) of the solid Earth (about 6,400 kilo-meters, or 4,000 miles), the atmosphere is a very shallow layer. One half lies below an altitude of 5.6 kilometers (3.5 miles), and 90 percent occurs within just 16 kilometers (10 miles) of Earth's surface (**Figure 1.12**). Despite its modest dimensions, this thin blanket of air is nevertheless an integral part of the planet. It not only provides the air that we breathe but also acts to protect us from the Sun's dangerous ultraviolet radiation. The energy exchanges that continually occur between the atmosphere and Earth's surface and between the atmosphere and space produce the effects we call *weather* and *climate*.

If, like the Moon, Earth had no atmosphere, our planet would not only be lifeless but also many of the processes and interactions that make the surface such a dynamic place could not operate. Without weathering and erosion, the face of our planet might more closely resemble the lunar surface, which has not changed appreciably in nearly 3 billion years.

FIGURE 1.13 The hydrosphere contains a significant portion of Earth's biosphere. Modern coral reefs are unique and complex examples and are home to about 25 percent of all marine species. Because of this diversity, they are sometimes referred to as the ocean equivalent of rain forests. (Photo by Darryl Leniuk/age footstock)

Biosphere

The **biosphere** includes all life on Earth. Ocean life is concentrated in the sunlit surface waters of the sea (**Figure 1.13**).

of the planet, a depth of 6,400 kilometers, making it by far the largest of Earth's four spheres. Much of our study of the solid Earth focuses on the more accessible surface features. Fortunately, many of these features represent the outward expressions of the dynamic behavior of Earth's interior. By examining the most prominent surface features and their global extent, we can obtain clues to the dynamic processes that have shaped our planet. A first look at the structure of Earth's interior and at the major surface features of the geosphere comes in the next section of this chapter.

Soil, the thin veneer of material at Earth's surface that supports the growth of plants, may be thought of as part of all four spheres. The solid portion is a mixture of weathered rock debris (geosphere) and organic matter from decayed plant and animal life (biosphere). The decomposed and disintegrated rock debris is the product of weathering processes that require air (atmosphere) and water (hydrosphere). Air and water also occupy the open spaces between the solid particles.

CONCEPT CHECK 1.6

1 Compare the height of the atmosphere to the thickness of the geosphere.
2 How much of the Earth's surface do oceans cover?
3 How much of the planet's total water supply do oceans represent?
4 List and briefly define the four "spheres" that constitute our environment.

A Closer Look at the Geosphere

In this section we make a preliminary examination of the solid Earth. You will become more familiar with the internal and external "anatomy" of our planet and begin to understand that the geosphere is truly dynamic. The diagrams should help a great deal as you begin to develop a mental image of the geosphere's internal structure and major surface features, so study the figures carefully. We begin with a look at Earth's interior—its structure and mobility. Then we conduct a brief survey of the surface of the solid Earth. Although portions of the surface, such as mountains and river valleys, are familiar to most of us, those areas that are out of sight on the floor of the ocean are not so familiar.

Earth's Internal Structure

Early in Earth's history the sorting of material by *compositional* (density) differences resulted in the formation of three layers—the crust, mantle, and core (**Figure 1.14**). In addition to these compositionally distinct layers, Earth is also divided into layers based on *physical properties*. The physical properties that define these zones include whether the layer is solid or liquid and how

weak or strong it is. Knowledge of both types of layers is essential to an understanding of our planet.

Earth's Crust   The **crust**, Earth's relatively thin, rocky outer skin, is of two different types—continental crust and oceanic crust. Both share the word "crust," but the similarity ends there. The oceanic crust is roughly 7 kilometers (5 miles) thick and composed of the dark igneous rock *basalt*. By contrast, the continental crust averages about 35 kilometers (22 miles) thick but may exceed 70 kilometers (40 miles) in some mountainous regions such as the Rockies and Himalayas. Unlike the oceanic crust, which has a relatively homogeneous chemical composition, the continental crust consists of many rock types. Although the upper crust has an average composition of a *granitic rock* called *granodiorite*, it varies considerably from place to place.

Continental rocks have an average density of about 2.7 g/cm^3, and some have been discovered that are 4 billion years old. The rocks of the oceanic crust are younger (180 million years or less) and denser (about 3.0 g/cm^3) than continental rocks.[3]

Earth's Mantle   More than 82 percent of Earth's volume is contained in the **mantle**, a solid, rocky shell that extends to a depth of nearly 2,900 kilometers (1,800 miles). The boundary between the crust and mantle represents a marked change in chemical composition. The dominant rock type in the uppermost mantle is *peridotite*, which is richer in the metals magnesium and iron than the minerals found in either the continental or oceanic crust.

The upper mantle extends from the crust–mantle boundary to a depth of about 660 kilometers (410 miles). The upper mantle can be divided into two different parts. The top portion of the upper mantle is part of the stiff *lithosphere*, and beneath that is the weaker *asthenosphere*.

The **lithosphere** (sphere of rock) consists of the entire crust and uppermost mantle and forms Earth's relatively cool, rigid outer shell. Averaging about 100 kilometers in thickness, the lithosphere is more than 250 kilometers thick below the oldest portions of the continents (Figure 1.14). Beneath this stiff layer to a depth of about 350 kilometers lies a soft, comparatively weak layer known as the **asthenosphere** ("weak sphere"). The top portion of the asthenosphere has a temperature/pressure regime that results in a small amount of melting. Within this very weak zone the lithosphere is mechanically detached from the layer below. The result is that the lithosphere is able to move independently of the asthenosphere, a fact we consider in more detail in Chapter 7.

It is important to emphasize that the strength of various Earth materials is a function of both their composition and of the temperature and pressure of their environment. You should not get the idea that the entire lithosphere behaves like a brittle solid similar to rocks found on the surface. Rather, the rocks of the lithosphere get progressively hotter and weaker (more easily deformed) with increasing depth. At the depth of the uppermost

---

[3]"Liquid water has a density of 1 g/cm^3; therefore, the density of basalt is three times that of water.

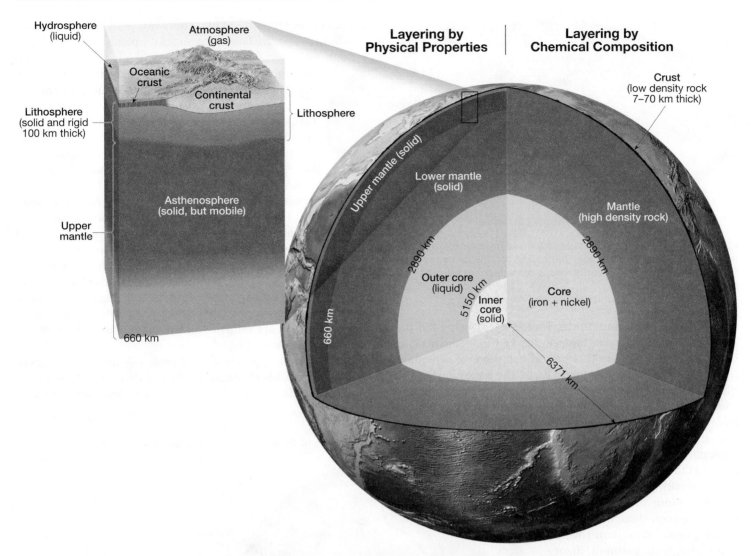

FIGURE 1.14 The right side of the globe shows that Earth's interior is divided into three different layers based on compositional differences—the crust, mantle, and core. The left side of the globe shows the five main layers of Earth's interior based on physical properties and mechanical strength—the lithosphere, asthenosphere, lower mantle, outer core, and inner core. The block diagram on the left shows an enlarged view of the upper portion of Earth's interior.

asthenosphere, the rocks are close enough to their melting temperature (some melting may actually occur) that they are very easily deformed. Thus, the uppermost asthenosphere is weak because it is near its melting point, just as hot wax is weaker than cold wax.

From a depth of 660 kilometers to the top of the core, at a depth of 2,900 kilometers (1,800 miles), is the **lower mantle**. Because of an increase in pressure (caused by the weight of the rock above) the mantle gradually strengthens with depth. Despite their strength, however, the rocks within the lower mantle are very hot and capable of very gradual flow.

Earth's Core The composition of the **core** is thought to be an iron-nickel alloy with minor amounts of oxygen, silicon, and sulfur—elements that readily form compounds with iron. At the extreme pressure found in the core, this iron-rich material has an average density of nearly 11 g/cm³ and approaches 14 times the density of water at Earth's center.

The core is divided into two regions that exhibit very different mechanical strengths. The **outer core** is a *liquid layer* 2,260 kilometers (about 1,400 miles) thick. It is the movement of metallic iron within this zone that generates Earth's magnetic field. The **inner core** is a sphere having a radius of 1,216 kilometers (754 miles). Despite its higher temperature, the iron in the inner core is *solid* due to the immense pressures that exist in the center of the planet.

The Mobile Geosphere

Earth is a dynamic planet! If we could go back in time a few hundred million years, we would find the face of our planet dramatically different from what we see today. There would be no Mount St. Helens, Rocky Mountains, or Gulf of Mexico. Moreover, we would find continents having different sizes and shapes and located in different positions than today's landmasses (**Figure 1.15**).

FIGURE 1.15 Earth as it looked about 200 million years ago, in the late Triassic period. At this time, the modern continents that we are familiar with were joined to form a supercontinent that we call Pangaea ("all land").

Continental Drift and Plate Tectonics   During the past several decades a great deal has been learned about the workings of our dynamic planet. This period has seen an unequaled revolution in our understanding of Earth. The revolution began in the early part of the twentieth century with the radical proposal of *continental drift*—the idea that the continents moved about the face of the planet. This proposal contradicted the established view that the continents and ocean basins are permanent and stationary features on the face of Earth. For that reason, the notion of drifting continents was received with great skepticism and even ridicule. More than 50 years passed before enough data were gathered to transform this controversial hypothesis into a sound theory that wove together the basic processes known to operate on Earth. The theory that finally emerged, called **plate tectonics**, provided geologists with the first comprehensive model of Earth's internal workings.

According to the theory of plate tectonics, Earth's rigid outer shell (*lithosphere*) is broken into numerous slabs called **lithospheric plates**, which are in continual motion. More than a dozen plates exist (**Figure 1.16**). The largest is the Pacific plate, covering much of the Pacific Ocean basin. Notice that several of the large lithospheric plates include an entire continent plus a large area of the seafloor. Note also that none of the plates are defined entirely by the margins of a continent.

Plate Motion   Driven by the unequal distribution of heat within our planet, lithospheric plates move relative to each other at a very slow but continuous rate that averages about 5 centimeters (2 inches) per year—about as fast as your fingernails grow. Because plates move as coherent units relative to all other plates, they

interact along their margins. Where two plates move together, called a *convergent boundary*, one of the plates plunges beneath the other and descends into the mantle (**Figure 1.17**). It is only those lithospheric plates that are capped with relatively dense oceanic crust that sink into the mantle.

Any portion of a plate that is capped by continental crust is too buoyant to be carried into the mantle. As a result, when two plates carrying continental crust converge, a collision of the two continental margins occurs. The result is the formation of a major mountain belt, as exemplified by the Himalayas.

Divergent boundaries are located where plates pull apart (Figure 1.17). Here the fractures created as the plates separate are filled with molten rock that wells up from the mantle. This hot material slowly cools to form solid rock, producing new slivers of seafloor. This process occurs along oceanic ridges where, over spans of millions of years, hundreds of thousands of square kilometers of new seafloor have been generated (Figure 1.17). Thus, while new seafloor is constantly being added at the oceanic ridges, equal amounts are returned to the mantle along boundaries where two plates converge.

At other sites, plates do not push together or pull apart. Instead, they slide past one another, so that seafloor is neither created nor destroyed. These zones are called *transform fault boundaries*.

FIGURE 1.16 Illustration showing some of Earth's lithospheric plates.

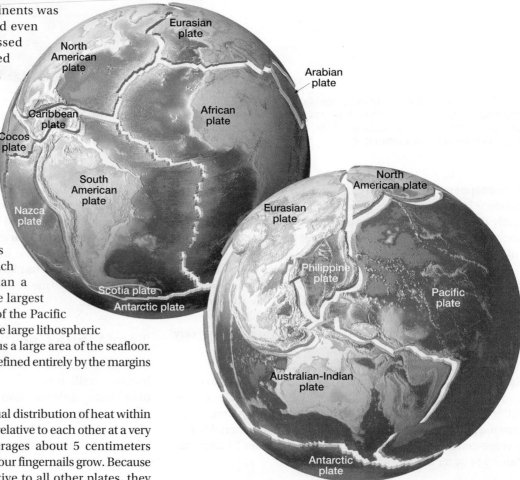

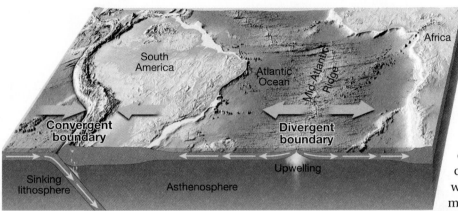

FIGURE 1.17 *Convergent boundaries* occur where two plates move together, as along the western margin of South America. *Divergent boundaries* are located where adjacent plates move away from one another. The Mid-Atlantic ridge is such a boundary.

CONCEPT CHECK 1.7

① List and briefly describe Earth's compositional layers.

② Contrast the lithosphere and the asthenosphere.

③ What are lithospheric plates? List the three types of boundaries that separate plates.

The Face of Earth

The two principal divisions of Earth's surface are the continents and the ocean basins (**Figure 1.18**). A significant difference between these two areas is their relative levels. The continents are remarkably flat features that have the appearance of plateaus protruding above sea level. With an average elevation of about 0.8 kilometer (0.5 mile), continents lie relatively close to sea level, except for limited areas of mountainous terrain. By contrast, the average depth of the ocean floor is about 3.8 kilometers (2.4 miles) below sea level, or about 4.5 kilometers (2.8 miles) lower than the average elevation of the continents.

The elevation difference between the continents and ocean basins is primarily the result of differences in their respective densities and thicknesses. Recall that the continents average about 35 kilometers in thickness and are composed of granitic rocks having a density of about 2.7 g/cm^3. The basaltic rocks that comprise the oceanic crust average only 7 kilometers thick and have an average density of about 3.0 g/cm^3. Thus, the thicker and less dense continental crust is more buoyant than the oceanic crust. As a result, continental crust floats on top of the deformable rocks of the mantle at a higher level than oceanic crust for the same reason that a large, empty (less dense) cargo ship rides higher than a small, loaded (more dense) one.

Major Features of the Continents

The largest features of the continents can be grouped into two distinct categories: extensive, flat, stable areas that have been eroded nearly to sea level, and uplifted regions of deformed rocks that make up present-day mountain belts. Notice in **Figure 1.19**

that young mountain belts tend to be long, narrow features at the margins of continents, and that the flat, stable areas are typically located in the interior of continents.

Mountain Belts The most prominent topographic features of the continents are linear mountain belts. Although the distribution of mountains appears to be random, this is not the case. When the youngest mountains are considered (those less than 100 million years old), we find that they are located principally in two major zones. The circum-Pacific belt (the region surrounding the Pacific Ocean) includes the mountains of the western Americas and continues into the western Pacific in the form of volcanic islands such as the Aleutians, Japan, and the Philippines (Figure 1.18).

The other major mountainous belt extends eastward from the Alps through Iran and the Himalayas and then dips southward into Indonesia. Careful examination of mountainous terrains reveals that most are places where thick sequences of rocks have been squeezed and highly deformed, as if placed in a gigantic vise. Older mountains are also found on the continents. Examples include the Appalachians in the eastern United States and the Urals in Russia. Their once lofty peaks are now worn low, the result of millions of years of erosion.

The Stable Interior Unlike the young mountain belts, which have formed within the last 100 million years, the interiors of the continents have been relatively stable (undisturbed) for the last 600 million years or even longer. Typically, these regions were involved in mountain-building episodes much earlier in Earth's history.

Within the stable interiors are areas known as **shields**, which are expansive, flat regions composed of deformed crystalline rock. Notice in Figure 1.19 that the Canadian Shield is exposed in much of the northeastern part of North America. Age determinations for various shields have shown that they are truly ancient regions. All contain Precambrian-age rocks that are over 1 billion years old, with some samples approaching 4 billion years in age. These oldest-known rocks exhibit evidence of enormous forces that have folded and faulted them and altered them with great heat and pressure. Thus, we conclude that these rocks were once part of an ancient mountain system that has since been eroded away to produce these expansive, flat regions.

Other flat areas of the stable interior exist in which highly deformed rocks, like those found in the shields, are covered by a relatively thin veneer of sedimentary rocks. These areas are called **stable platforms**. The sedimentary rocks in stable platforms are nearly horizontal except where they have been warped to form large basins or domes. In North America a major portion of the stable platform is located between the Canadian Shield and the Rocky Mountains (Figure 1.19).

Major Features of the Ocean Basins

If all water were drained from the ocean basins, a great variety of features would be seen, including linear chains of volcanoes, deep canyons, extensive plateaus, and large expanses of monotonously

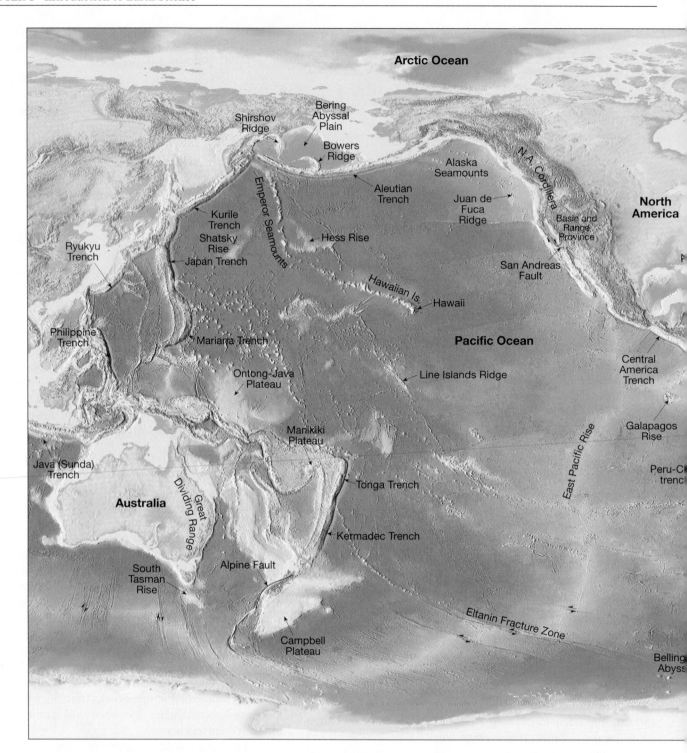

FIGURE 1.18 Major surface features of the geosphere.

flat plains. In fact, the scenery would be nearly as diverse as that on the continents (Figure 1.18).

During the past 70 years, oceanographers using modern depth-sounding equipment have gradually mapped significant portions of the ocean floor. From these studies they have defined three major regions: *continental margins*, *deep-ocean basins*, and *oceanic (mid-ocean) ridges*.

Continental Margins The **continental margin** is that portion of the seafloor adjacent to major landmasses. It may

include the *continental shelf*, the *continental slope*, and the *continental rise*.

Although land and sea meet at the shoreline, this is not the boundary between the continents and the ocean basins. Rather, along most coasts a gently sloping platform of material, called the **continental shelf**, extends seaward from the shore. Because it is underlain by continental crust, it is considered a flooded extension of the continents. A glance at Figure 1.18 shows that the width of the continental shelf is variable. For example, it is broad along the East and Gulf coasts of

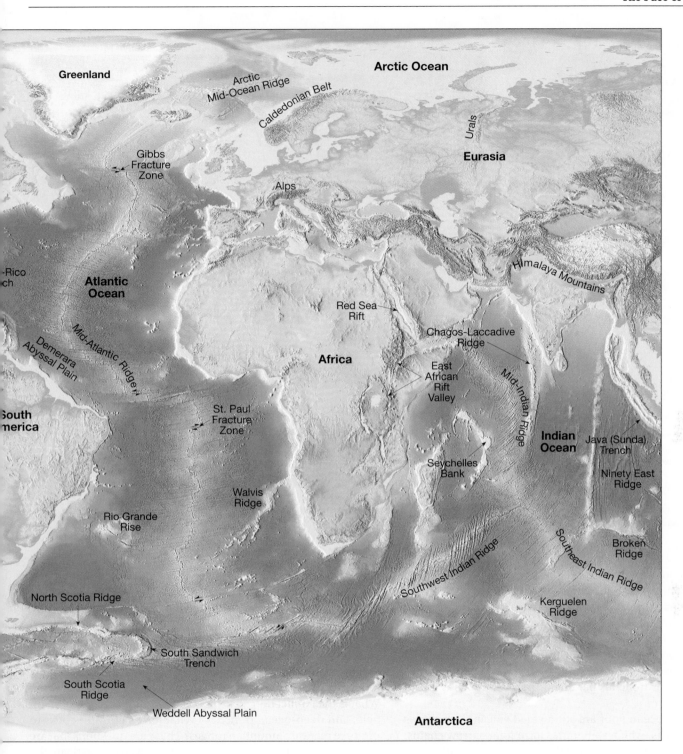

Greenland

Arctic
Mid-Ocean Ridge

Arctic Ocean

Caldedonian Belt

Urals

Eurasia

Gibbs
Fracture
Zone

Alps

-Rico
ch

**Atlantic
Ocean**

Himalaya Mountains

Red Sea
Rift

Chagos-Laccadive
Ridge

Demerara
Abyssal Plain

Mid-Atlantic Ridge

Africa

East
African
Rift
Valley

Mid-Indian Ridge

**Indian
Ocean**

Java (Sunda)
Trench

South
merica

St. Paul
Fracture
Zone

Seychelles
Bank

Ninety East
Ridge

Walvis
Ridge

Rio Grande
Rise

Southwest Indian Ridge

Southeast Indian Ridge

Broken
Ridge

North Scotia Ridge

Kerguelen
Ridge

South Sandwich
Trench

South Scotia
Ridge

Weddell Abyssal Plain

Antarctica

FIGURE 1.18 Continued

the United States but relatively narrow along the Pacific margin of the continent.

The boundary between the continents and the deep-ocean basins lies along the **continental slope**, which is a relatively steep dropoff that extends from the outer edge of the continental shelf to the floor of the deep ocean (Figure 1.18). Using this as the dividing line, we find that about 60 percent of Earth's surface is represented by ocean basins and the remaining 40 percent by continents.

In regions where trenches do not exist, the steep continental slope merges into a more gradual incline known as the **continental rise**. The continental rise consists of a thick accumulation of sediments that moved downslope from the continental shelf to the deep-ocean floor.

Deep-Ocean Basins   Between the continental margins and oceanic ridges lie the **deep-ocean basins**. Parts of these regions consist of incredibly flat features called **abyssal plains**. The ocean

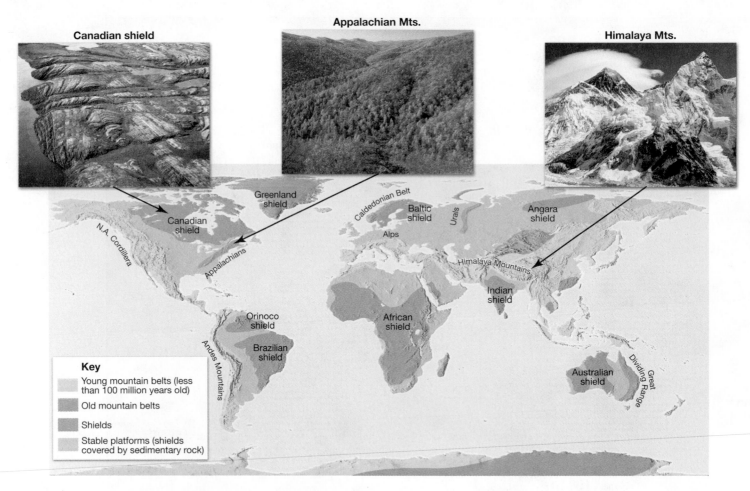

FIGURE 1.19 This map shows the distribution of Earth's mountain belts, stable platforms, and shields. (Photo by Robert Hildebrand [left]; CORBIS [middle]; Image Source Pink/Alamy [right])

floor also contains extremely deep depressions that are occasionally more than 11,000 meters (36,000 feet) deep. Although these **deep-ocean trenches** are relatively narrow and represent only a small fraction of the ocean floor, they are nevertheless very significant features. Some trenches are located adjacent to young mountains that flank the continents. For example, in Figure 1.18 the Peru–Chile trench off the west coast of South America parallels the Andes Mountains. Other trenches parallel linear island chains called *volcanic island arcs.*

Dotting the ocean floor are submerged volcanic structures called **seamounts**, which sometimes form long narrow chains. Volcanic activity has also produced several large *lava plateaus*, such as the Ontong Java Plateau located northeast of New Guinea. In addition, some submerged plateaus are composed of continental-type crust. Examples include the Campbell Plateau southeast of New Zealand and the Seychelles Bank northeast of Madagascar.

Oceanic Ridges   The most prominent feature on the ocean floor is the **oceanic** or **mid-ocean ridge**. As shown in Figure 1.18, the Mid-Atlantic Ridge and the East Pacific Rise are parts of this system. This broad elevated feature forms a continuous belt that winds for more than 70,000 kilometers (43,000 miles) around the globe in a manner similar to the seam of a baseball. Rather than

consisting of highly deformed rock, such as most of the mountains on the continents, the oceanic ridge system consists of layer upon layer of igneous rock that has been fractured and uplifted.

Understanding the topographic features that comprise the face of Earth is critical to our understanding of the mechanisms that have shaped our planet. What is the significance of the enormous ridge system that extends through all the world's oceans? What is the connection, if any, between young, active mountain belts and deep-ocean trenches? What forces crumple rocks to produce majestic mountain ranges? These are questions that are addressed in some of the coming chapters as we investigate the dynamic processes that shaped our planet in the geologic past and will continue to shape it in the future.

CONCEPT CHECK 1.8

❶ Describe the general distribution of Earth's youngest mountains.

❷ What is the difference between shields and stable platforms?

❸ What are the three major regions of the ocean floor and some features associated with each region?

Earth as a System

Anyone who studies Earth soon learns that our planet is a dynamic body with many separate but interacting parts or *spheres*. The hydrosphere, atmosphere, biosphere, and geosphere and all of their components can be studied separately. However, the parts are not isolated. Each is related in some way to the others to produce a complex and continuously interacting whole that we call the *Earth system*.

Earth System Science

A simple example of the interactions among different parts of the Earth system occurs every winter as moisture evaporates from the Pacific Ocean and subsequently falls as rain in the hills of southern California, triggering destructive landslides. A case study in Chapter 4 (p. 106) explores such an event. The processes that move water from the hydrosphere to the atmosphere and then to the solid Earth have a profound impact on the plants and animals (including humans) that inhabit the affected regions. **Figure 1.20** provides another example.

Scientists have recognized that in order to more fully understand our planet they must learn how its individual components (land, water, air, and life-forms) are interconnected. This endeavor, called **Earth system science**, aims to study Earth as a *system* composed of numerous interacting parts, or *subsystems*. Rather than looking through the limited lens of only one of the traditional sciences—geology, atmospheric science, chemistry, biology, and so on—Earth system science attempts to integrate the knowledge of several academic fields. Using this interdisciplinary approach, we hope to achieve the level of understanding necessary to comprehend and solve many of our global environmental problems.

Students Sometimes Ask ...

How do we know about the internal structure of Earth?

You might suspect that the internal structure of Earth has been sampled directly. However, humans have never penetrated beneath the crust! The internal structure of Earth is determined by using indirect observations. Every time there is an earthquake, waves of energy (called *seismic waves*) penetrate Earth's interior. Seismic waves change their speed and are bent and reflected as they move through zones having different properties. An extensive series of monitoring stations around the world detects and records this energy. The data are analyzed and used to work out the structure of Earth's interior.

What Is a System?   Most of us hear and use the term *system* frequently. We may service our car's cooling *system*, make use of the city's transportation *system*, and participate in the political *system*. A news report might inform us of an approaching weather *system*. Furthermore, we know that Earth is just a small part of a larger system known as the *solar system*, which in turn is a *subsystem* of the even larger system called the Milky Way Galaxy.

Loosely defined, a **system** can be any size group of interacting parts that form a complex whole. Most natural systems are driven by sources of energy that move matter and/or energy from one place to another. A simple analogy is a car's cooling system, which contains liquid (usually water and antifreeze) that is driven from the engine to the radiator and back again. The role of this system is to transfer heat generated by combustion in the

FIGURE 1.20 This image provides an example of interactions among different parts of the Earth system. Aerial view of Caraballeda, Venezuela, covered by material from a massive debris flow (popularly called a mud slide in the press). In December 1999, extraordinary rains triggered this debris flow and thousands of others along this mountainous coastal zone. Caraballeda was located at the mouth of a steep canyon. An estimated 19,000 lives were lost. (Photo by Kimberly White/Reuters/CORBIS/Bettmann)

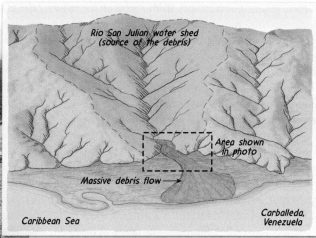

Geologist's Sketch

engine to the radiator, where moving air removes it from the system; hence the term cooling system.

Systems like a car's cooling system are self-contained with regard to matter and are called **closed systems**. Although energy moves freely in and out of a closed system, no matter (liquid in the case of our auto's cooling system) enters or leaves the system. (This assumes you don't get a leak in your radiator.) By contrast, most natural systems are **open systems** and are far more complicated than the foregoing example. In an open system both energy and matter flow into and out of the system. In a weather system such as a hurricane, factors such as the quantity of water vapor available for cloud formation, the amount of heat released by condensing water vapor, and the flow of air into and out of the storm can fluctuate a great deal. At times the storm may strengthen; at other times it may remain stable or weaken.

Feedback Mechanisms   Most natural systems have mechanisms that tend to enhance change, as well as other mechanisms that tend to resist change and thus stabilize the system. For example, when we get too hot, we perspire to cool down. This cooling phenomenon works to stabilize our body temperature and is referred to as a **negative feedback mechanism**. Negative feedback mechanisms work to inhibit change or, in other words, to maintain the status quo. By contrast, mechanisms that enhance or drive change are called **positive feedback mechanisms**.

Most of Earth's systems, particularly the climate system, contain a wide variety of negative and positive feedback mechanisms. For example, substantial scientific evidence indicates that Earth has entered a period of global warming. One consequence of global warming is that some of the world's glaciers and ice caps have begun to melt. Highly reflective snow- and ice-covered surfaces are gradually being replaced by brown soils, green trees, or blue oceans, all of which are darker, so they absorb more sunlight. Therefore, as Earth warms and some snow and ice melt, our planet absorbs more sunlight. The result is a positive feedback that contributes to the warming.

On the other hand, an increase in global temperature also causes greater evaporation of water from Earth's land–sea surface. One result of having more water vapor in the air is an increase in cloud cover. Because cloud tops are white and highly reflective, more sunlight is reflected back to space, which diminishes the amount of sunshine reaching Earth's surface and thus reduces global temperatures. Furthermore, warmer temperatures tend to promote the growth of vegetation. Plants in turn remove carbon dioxide (CO_2) from the air. Since carbon dioxide is one of the atmosphere's *greenhouse gases*, its removal has a negative impact on global warming.[4]

In addition to natural processes, we must also consider the human element. Extensive cutting and clearing of the tropical rain forests and the burning of fossil fuels (oil, natural gas, and coal) result in an increase in atmospheric CO_2. Such activity has been linked to the increase in global temperatures that our planet is experiencing. One of the daunting tasks for Earth system scientists is to predict what the climate will be like in the future by taking into account many variables, including technological changes, population trends, and the overall impact of the numerous competing positive and negative feedback mechanisms.

[4]Greenhouse gases absorb heat energy emitted by Earth and thus help keep the atmosphere warm.

The Earth System

The Earth system has a nearly endless array of subsystems in which matter is recycled over and over again. One example that you will learn about in Chapter 3 traces the movements of carbon among Earth's four spheres. It shows us, for example, that the carbon dioxide in the air and the carbon in living things and in certain sedimentary rocks is all part of a subsystem described by the *carbon cycle*.

Cycles in the Earth System   A more familiar loop or subsystem is the *hydrologic cycle*. It represents the unending circulation of Earth's water among the hydrosphere, atmosphere, biosphere, and geosphere. Water enters the atmosphere by evaporation from Earth's surface and by transpiration from plants. Water vapor condenses in the atmosphere to form clouds, which in turn produce precipitation that falls back to Earth's surface. Some of the rain that falls onto the land sinks in to be taken up by plants or become groundwater, and some flows across the surface toward the ocean.

Viewed over long time spans, the rocks of the geosphere are constantly forming, changing, and reforming. The loop that involves the processes by which one rock changes to another is called the *rock cycle* and is discussed at some length in Chapter 3. The cycles of the Earth system, such as the hydrologic and rock cycles, are not independent of one another. To the contrary, there are many places where they have an interface. An **interface** is a common boundary where different parts of a system come in contact and interact. For example, weathering at the surface gradually disintegrates and decomposes solid rock. The work of gravity and running water may eventually move this material to another place and deposit it. Later, groundwater percolating through the debris may leave behind mineral matter that cements the grains together into solid rock (a rock that is often very different from the rock we started with). This changing of one rock into another, which is part of the rock cycle, could not have occurred without the movement of water through the hydrologic cycle. There are many places where one cycle or loop in the Earth system has an interface with and is a basic part of another.

Energy for the Earth System   The Earth system is powered by energy from two sources. The Sun drives external processes that occur in the atmosphere, hydrosphere, and at Earth's surface. Weather and climate, ocean circulation, and erosional processes such as rivers, glaciers, wind, and waves are driven by energy from the Sun. Earth's interior is the second source of energy. Heat remaining from when our planet formed, and heat that is continuously generated by decay of radioactive elements, powers the internal processes that produce volcanoes, earthquakes, and mountains.

The Parts are Linked   The parts of the Earth system are linked so that a change in one part can produce changes in any or all of the other parts. For example, when a volcano erupts, lava from Earth's interior may flow out at the surface and block a nearby valley. This new obstruction influences the region's drainage system by creating a lake or causing streams to change course. The large quantities of volcanic ash and gases that can be emitted during an eruption might be blown high into the atmosphere and influence the amount of solar energy that can reach Earth's surface. The result could be a drop in air temperatures over the entire hemisphere.

Where the surface is covered by lava flows or a thick layer of volcanic ash, existing soils are buried. This causes the soil-forming

FIGURE 1.21 When Mount St. Helens erupted in May 1980 (inset), the area shown here was buried by a volcanic mudflow. Now, plants are reestablished and new soil is forming. (Photo by Jack Dykinga/Getty Images; inset photo by US Geological Survey)

as a lake, will be created. The potential climate change can also impact sensitive life forms.

The Earth system is characterized by processes that vary on spatial scales from fractions of millimeters to thousands of kilometers. Time scales for Earth's processes range from milliseconds to billions of years. As we learn about Earth, it becomes increasingly clear that despite significant separations in distance or time, many processes are connected, and a change in one component can influence the entire system.

Humans are *part of* the Earth system, a system in which the living and nonliving components are entwined and interconnected. Therefore, our actions produce changes in all of the other parts. When we burn gasoline and coal, build breakwaters along the shoreline, dispose of our wastes, and clear the land, we cause other parts of the system to respond, often in unforeseen ways. Throughout this book you will learn about many of Earth's subsystems: the hydrologic system, the tectonic (mountain-building) system, and the climate system, to name a few. Remember that these components *and we humans* are all part of the complex interacting whole we call the Earth system.

The organization of this text involves traditional groupings of chapters that focus on closely related topics. Nevertheless, the theme of *Earth as a system* keeps recurring through *all* major units of *Earth science*. It is a thread that weaves through the chapters and helps tie them together. At the end of each chapter there is a section titled "Examining the Earth System." The questions and problems found there are intended to help you develop an awareness and appreciation for some of the Earth system's important interrelationships.

processes to begin anew to transform the new surface material into soil (Figure 1.21). The soil that eventually forms will reflect the interactions among many parts of the Earth system—the volcanic parent material, the type and rate of weathering, and the impact of biological activity. Of course, there will also be significant changes in the biosphere. Some organisms and their habitats will be eliminated by the lava and ash, whereas new settings for life, such

CONCEPT CHECK 1.9

1. How is an open system different from a closed system?
2. Contrast positive and negative feedback methanisms.
3. What are the two sources of energy for the Earth system?

GIVE IT SOME THOUGHT

1. After entering a dark room, you turn on a wall switch but the light does not come on. Suggest at least three hypotheses that might explain this observation.

2. Each of the following statements may either be a hypothesis (H), a theory (T), or an observation (O). Use one of these letters to identify each statement. Briefly explain each choice.

   a. A scientist proposes that a recently discovered large ring-shaped structure is the remains of an ancient meteorite crater.

   b. The Redwall Formation in the Grand Canyon is composed primarily of limestone.

   c. The outer part of Earth consists of several large plates that move and interact with each other.

   d. Since 1885, the terminus of Canada's Athabasca Glacier has receded 1.5 kilometers.

   e. The universe originated about 13.7 billion years ago with a period of rapid expansion called the big bang.

3. Consider the possible results of the following scenario and describe one positive and one negative feedback. *Earth is getting warmer, consequently evaporation is increasing.*

4. Making accurate measurements and observations is a basic part of scientific inquiry. The accompanying photo provides one example. Identify at least five additional images in this chapter that illustrate ways in which scientific data are gathered. Suggest advantages that might be associated with each example.

(Photo by Didier Dutheil/Corbis)

5. Refer to Figure 1.20. Which of the four main components of the Earth system (atmosphere, biosphere, geosphere, hydrosphere) were involved in the natural disaster at Caraballeda, Venezuela? Describe how each of the components you list contributed to the debris flow.

6. Look at the concept map linking the four spheres of the Earth system. Between each sphere are arrows representing processes by which these spheres interact and influence each other. For each arrow, describe at least one process.

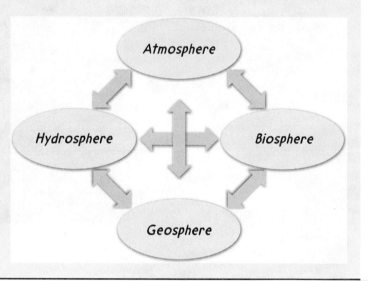

In Review   Chapter 1 Introduction to Earth Science

- *Earth science* is the name for all the sciences that collectively seek to understand Earth and its neighbors in space. It includes *geology, oceanography, meteorology,* and *astronomy.* Geology is traditionally divided into two broad areas: *physical* and *historical.*

- The relationship between people and the natural environment is an important focus of Earth science. This includes natural hazards, resources, and human influences on Earth processes.

- All science is based on the assumption that the natural world behaves in a consistent and predictable manner. The process by which scientists gather facts through observation and careful measurement and formulate scientific *hypotheses* and *theories* is called the *scientific method.* To determine what is occurring in the natural world, scientists often (1) pose questions about the natural world and collect facts that relate to these questions; (2) ask questions and develop hypotheses that may answer these questions, (3) develop observations and experiments to test the hypotheses; (4) accept, modify, or reject hypotheses on the basis of extensive testing; and (5) share results with the broader scientific community. Other discoveries represent purely theoretical ideas that have stood up to extensive examination. Still other scientific advancements have been made when a totally unexpected happening occurred during an experiment.

- One of the challenges for those who study Earth is the great variety of space and time scales. The *geologic time scale* subdivides the 4.6 billion years of Earth history into various units.

- The *nebular theory* describes the formation of the solar system. The planets and Sun began forming about 5 billion years ago from a large cloud of dust and gases. As the cloud contracted, it began to rotate and assume a disk shape. Material that was gravitationally pulled toward the center became the *protosun.* Within the rotating disk, small centers, called *planetesimals,* swept up more and more of the cloud's debris. Because of their high temperatures and weak gravitational fields, the inner planets were unable to accumulate and retain many of the lighter components. Because of the very cold temperatures existing far from the Sun, the large outer planets consist of huge amounts of lighter materials. These gaseous substances account for the comparatively large sizes and low densities of the outer planets.

- Earth's physical environment is traditionally divided into three major parts: the solid Earth or *geosphere*; the water portion of our planet, the *hydrosphere*; and Earth's gaseous envelope, the *atmosphere*. In addition, the *biosphere*, the totality of life on Earth, interacts with each of the three physical realms and is an equally integral part of Earth.

- Earth's internal structure is divided into layers based on differences in chemical composition and on the basis of changes in physical properties. Compositionally, Earth is divided into a thin outer *crust*, a solid rocky *mantle*, and a dense *core*. Other layers, based on physical properties, include the *lithosphere, asthenosphere, lower mantle, outer core,* and *inner core.*

- Two principal divisions of Earth's surface are the *continents* and *ocean basins*. A significant difference is their relative levels. The elevation differences between continents and ocean basins is primarily the result of differences in their respective densities and thicknesses.

- The largest features of the continents can be divided into two categories: *mountain belts* and the *stable interior*. The ocean floor is divided into three major topographic units: *continental margins, deep-ocean basins,* and *oceanic (mid-ocean) ridges.*

- Although each of Earth's four spheres can be studied separately, they are all related in a complex and continuously interacting whole that we call the *Earth system. Earth system science* uses an interdisciplinary approach to integrate the knowledge of several academic fields in the study of our planet and its global environmental problems.

- A *system* is a group of interacting parts that form a complex whole. *Closed systems* are those in which energy moves freely in and out, but matter does not enter or leave the system. In an *open system*, both energy and matter flow into and out of the system.

- The two sources of energy that power the Earth system are (1) *the Sun*, which drives the external processes that occur in the atmosphere, hydrosphere, and at Earth's surface, and (2) *heat from Earth's interior*, which powers the internal processes that produce volcanoes, earthquakes, and mountains.

Key Terms

abyssal plain (p. 19)
asthenosphere (p. 14)
astronomy (p. 2)
atmosphere (p. 12)
biosphere (p. 13)
closed system (p. 22)
continental margin (p. 18)
continental rise (p. 19)
continental shelf (p. 18)
continental slope (p. 19)
core (p. 15)
crust (p. 14)
deep-ocean basin (p. 19)
deep-ocean trench (p. 20)
Earth science (p. 2)

Earth system science (p. 21)
environment (p. 3)
geologic time scale (p. 9)
geology (p. 2)
geosphere (p. 13)
hydrosphere (p. 12)
hypothesis (p. 6)
inner core (p. 15)
interface (p. 22)
lithosphere (p. 14)
lithospheric plate (p. 16)
lower mantle (p. 15)
mantle (p. 14)
meteorology (p. 2)

nebular theory (p. 10)
negative feedback mechanism (p. 22)
oceanic (mid-ocean) ridge (p. 20)
oceanography (p. 2)
open system (p. 22)
outer core (p. 15)
physical environment (p. 11)
plate tectonics (p. 16)
positive feedback mechanism (p. 22)
seamount (p. 20)
shield (p. 17)
stable platform (p. 17)
system (p. 21)
theory (p. 7)

Examining the Earth System

1. Examine the chapter-opening photo, Figure 1.1, and Figure 1.2. Make a list of features in each photo and indicate whether the item belongs to the geosphere, hydrosphere, atmosphere, or biosphere. Are there any features that might belong to more than one of the spheres?

2. Examine Figure 1.20 and describe how all four spheres of the Earth system might have been involved and/or influenced by the event depicted here.

3. Humans are a part of the Earth system. List at least three examples of how you, in particular, influence one or more of Earth's major spheres.

Mastering Geology

Looking for additional review and test prep materials? Visit the Self Study area in **www.masteringgeology.com** to find practice quizzes, study tools, and multimedia that will aid in your understanding of this chapter's content. In **MasteringGeology**™ you will find:

- **GEODe: Earth Science: An interactive visual walkthrough of key concepts**

- **Geoscience Animation Library: More than 100 animations illuminating many difficult-to-understand Earth science concepts**

- **In The News RSS Feeds: Current Earth science events and news articles are pulled into the site with assessment**

- **Pearson eText**

- **Optional Self Study Quizzes**

- **Web Links**

- **Glossary**

- **Flashcards**

Earth Materials

2

Matter and Minerals

Cave of Crystals is a cave connected to Naica Mine in Chihuahua, Mexico. The main chamber contains giant gypsum crystals, some of the largest natural crystals ever found. (Photo by Carsten Peter/Speleoresearch & Films/ National Geographic Stock)

Earth's crust and oceans are the source of a wide variety of useful and essential minerals. Most people are familiar with the common uses of many basic metals, including aluminum in beverage cans, copper in electrical wiring, and gold and silver in jewelry. But some people are not aware that pencil lead contains the greasy-feeling mineral graphite and that bath powders and many cosmetics contain the mineral talc. Moreover, many do not know that drill bits impregnated with diamonds are employed by dentists to drill through tooth enamel, or that the common mineral quartz is the source of silicon for computer chips. In fact, practically every manufactured product contains materials obtained from minerals.

FOCUS ON CONCEPTS

To assist you in learning the important concepts in this chapter, focus on the following questions:

● What are minerals, and how are they different from rocks?
● What are the smallest particles of matter?
● How do atoms bond?
● How do isotopes of the same element vary, and why are some isotopes radioactive?
● What are some of the physical and chemical properties of minerals? How can these properties be used to distinguish one mineral from another?
● What are the eight elements that make up most of Earth's continental crust?
● What is the most abundant mineral group?
● What do all silicate minerals have in common?
● What are *renewable* and *nonrenewable resources*?
● When is the term *ore* used with reference to a mineral?

Minerals: Building Blocks of Rocks

Earth Materials
▶ Minerals

We begin our discussion of Earth materials with an overview of **mineralogy** (*mineral* = mineral, *ology* = the study of) because minerals are the building blocks of rocks. In addition, minerals have been employed by humans for both useful and decorative purposes for thousands of years (Figure 2.1). The first minerals mined were flint and chert, which people fashioned into weapons and cutting tools. As early as 3700 B.C., Egyptians began mining gold, silver, and copper; and by 2200 B.C. humans discovered how to combine copper with tin to make bronze, a strong, hard alloy. Later, humans developed a process to extract iron from minerals such as hematite—a discovery that marked the decline of the Bronze Age. By about 800 B.C., iron-working technology had advanced to the point that weapons and many everyday objects were made of iron rather than copper, bronze, or wood. During the Middle Ages, mining of a variety of minerals

was common throughout Europe and the impetus for the formal study of minerals was in place.

The term *mineral* is used in several different ways. For example, those concerned with health and fitness extol the benefits of vitamins and minerals. The mining industry typically uses the word when referring to anything taken out of the

FIGURE 2.1 Collection of well-developed quartz crystals found near Hot Springs, Arkansas. (Photo by Jeff Scovil)

ground, such as coal, iron ore, or sand and gravel. The guessing game known as "Twenty Questions" usually begins with the question, *Is it animal, vegetable, or mineral?* What criteria do geologists use to determine whether something is a mineral?

Geologists define **mineral** as *any naturally occurring inorganic solid that possesses an orderly crystalline structure and can be represented by a chemical formula.* Thus, Earth materials that are classified as minerals exhibit the following characteristics:

1. **Naturally occurring.** Minerals form by natural, geologic processes. Synthetic materials, meaning those produced in a laboratory or by human intervention, are not considered minerals.

2. **Solid substance.** Only crystalline substances that are solid at temperatures encountered at Earth's surface are considered minerals. Ice (frozen water) fits this criterion and is considered a mineral, whereas liquid water and water vapor do not. The exception is mercury, which is found in its liquid form in nature.

3. **Orderly crystalline structure.** Minerals are crystalline substances, which means their atoms are arranged in an orderly, repetitive manner (**Figure 2.2**). This orderly packing of atoms is reflected in the regularly shaped objects called crystals. Some naturally occurring solids, such as volcanic glass (obsidian), lack a repetitive atomic structure and are not considered minerals.

4. **Generally inorganic.** Inorganic crystalline solids, such as ordinary table salt (halite), that are found naturally in the ground are considered minerals. Organic compounds, on the other hand, are generally not. Sugar, a crystalline solid like salt but which comes from sugarcane or sugar beets, is a common example of such an organic compound. Many marine animals secrete inorganic compounds, such as calcium carbonate (calcite), in the form of shells and coral reefs. If these materials are buried and become part of the rock record, they are considered minerals by geologists.

5. **Can be represented by a chemical formula.** Most minerals are chemical compounds having compositions that can be expressed by a chemical formula. For example, the common mineral quartz has the formula SiO_2, which indicates that quartz consists of silicon (Si) and oxygen (O) atoms in a ratio of one-to-two. This proportion of silicon to oxygen is true for any sample of pure quartz, regardless of its origin. However, the compositions of some minerals vary *within specific, well-defined limits.* This occurs because certain elements can substitute for others of similar size without changing the mineral's internal structure. An example is the mineral olivine, in which either the element magnesium (Mg)

A. Sodium and chlorine ions.

B. Basic building block of the mineral halite.

C. Collection of basic building blocks (crystal).

D. Intergrown crystals of the mineral halite.

FIGURE 2.2 This diagram illustrates the orderly arrangement of sodium and chloride ions in the mineral halite. The arrangement of atoms into basic building blocks having a cubic shape results in regularly shaped cubic crystals. (Photo by Dennis Tasa)

or iron (Fe) may occupy the same site in the crystal structure. Therefore, olivine's formula, $(Mg, Fe)_2SiO_4$, expresses variability in the relative amounts of magnesium and iron. However, the ratio of magnesium plus iron (Mg + Fe) to silicon (Si) and oxygen (O) remains fixed at 2:1:4.

In contrast to minerals, rocks are more loosely defined. Simply a **rock** is any solid that consists of an aggregate of minerals, pieces of preexisting rocks, or a mass of mineral-like matter such as natural glass. Some rocks are composed almost entirely of one mineral. A common example is the sedimentary rock *limestone*, which consists of impure masses of the mineral calcite. However, most rocks, like the common rock granite shown in **Figure 2.3**, occur as aggregates of several different minerals. The term *aggregate* implies that the minerals are joined in such a way that their individual properties are retained. Note that the mineral constituents of granite can be easily identified (Figure 2.3).

Some rocks are composed of nonmineral matter. These include the volcanic rocks *obsidian* and *pumice*, which are noncrystalline glassy substances, and *coal*, which consists of solid organic debris.

Although this chapter deals primarily with the nature of minerals, keep in mind that most rocks are simply aggregates of minerals. Because the properties of rocks are determined largely by the chemical composition and crystalline structure of the minerals contained within them, we will first consider these Earth materials. Then, in Chapter 3, we take a closer look at Earth's major rock groups.

Granite
(Rock)

Quartz
(Mineral)

Hornblende
(Mineral)

Feldspar
(Mineral)

FIGURE 2.3 Most rocks are aggregates of two or more minerals. Shown here is a hand sample of the igneous rock granite and three of its major constituent minerals. (Photos by E. J. Tarbuck)

Properties of Protons, Neutrons, and Electrons

Protons and **neutrons** are very dense particles with almost identical masses. By contrast, **electrons** have a negligible mass, about 1/2000th that of a proton. For comparison, if a proton or a neutron had the mass of a baseball, an electron would have the mass of a single grain of rice.

Both protons and electrons share a fundamental property called *electrical charge*. Protons have an electrical charge of $+1$, and electrons have a charge of -1. Neutrons, as the name suggests, have no charge. The charge of protons and electrons are equal in magnitude but opposite in polarity, so when these two particles are paired, the charges cancel each other. Since matter typically contains equal numbers of positively charged protons and negatively charged electrons, most substances are electrically neutral.

In illustrations, electrons are sometimes shown orbiting the nucleus in a manner that resembles the planets of our solar system orbiting the Sun (Figure 2.4A). However, electrons do not actually behave this way. A more realistic depiction shows electrons as a cloud of negative charges surrounding a nucleus

CONCEPT CHECK 2.1

1. List five characteristics that classify an Earth material as a mineral.
2. Based on the definition of a mineral, which of the following materials are not classified as minerals, and why: gold, water, synthetic diamonds, ice, and wood.
3. Define the term *rock*. How do rocks differ from minerals?

Atoms: Building Blocks of Minerals

When minerals are carefully examined, even under optical microscopes, the innumerable tiny particles of their internal structures are not discernable. Nevertheless, all matter, including minerals, is composed of minute building blocks called **atoms**—the smallest particles that cannot be chemically split. Atoms in turn contain even smaller particles—*protons* and *neutrons* located in a central **nucleus** that is surrounded by *electrons* (Figure 2.4).

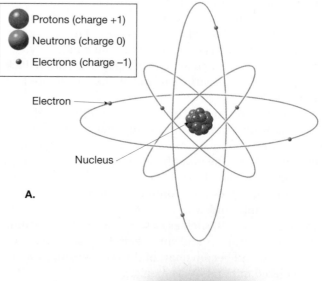

Protons (charge +1)
Neutrons (charge 0)
Electrons (charge –1)

Electron

Nucleus

A.

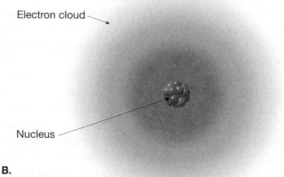

Electron cloud

Nucleus

B.

FIGURE 2.4 Two models of the atom. **A.** A very simplified view of the atom. The central nucleus consists of protons and neutrons encircled by high-speed electrons. **B.** This model of the atom shows electron clouds (shells) surrounding a central nucleus. The nucleus contains virtually all of the mass of the atom. The remainder of the atom is the space in which the light, negatively charged electrons reside. (The relative sizes of the nuclei shown are greatly exaggerated.)

Students Sometimes Ask...

Are the minerals you talked about in class the same as those found in dietary supplements?

Not ordinarily. From a geologic perspective, a mineral must be a *naturally occurring* crystalline solid. Minerals found in dietary supplements are human-made inorganic compounds that contain *elements* needed to sustain life. These dietary minerals typically contain elements that are metals—calcium, potassium, phosphorus, magnesium, and iron. It should also be noted that vitamins are *organic compounds* not *inorganic compounds*, like minerals.

(Figure 2.4B). Studies of the arrangements of electrons show that they move about the nucleus in regions called *principal shells*, each with an associated energy level. In addition, each shell can hold a specific number of electrons, with the outermost shell containing **valence electrons** that interact with other atoms to form chemical bonds.

Most of the atoms in the universe (except hydrogen and helium) were created inside massive stars by nuclear fusion and released into interstellar space during hot, fiery supernova explosions. As this ejected material cooled, the newly formed nuclei attracted electrons to complete their atomic structure. At the temperatures found at Earth's surface, all free atoms (not bonded to other atoms) have a full complement of electrons—one for each proton in the nucleus.

Elements: Defined by Their Number of Protons

The simplest atoms have only one proton in their nuclei, whereas others have more than 100. The number of protons in the nucleus of an atom, called the **atomic number**, determines its chemical nature. All atoms with the same number of protons have the same chemical and physical properties. Together, a group of the same kind of atoms is called an **element**. There are about 90 naturally occurring elements and 23 that have been synthesized. You are probably familiar with the names of many elements including carbon, nitrogen, and oxygen. All carbon atoms have six protons, all nitrogen atoms have seven protons, and all oxygen atoms have eight protons.

Elements are organized so that those with similar properties line up in columns. This arrangement, called the **periodic table**, is shown in **Figure 2.5**. Each element has been assigned a one- or two-letter symbol. The atomic numbers and masses are also included for each element.

FIGURE 2.5 Periodic table of the elements.

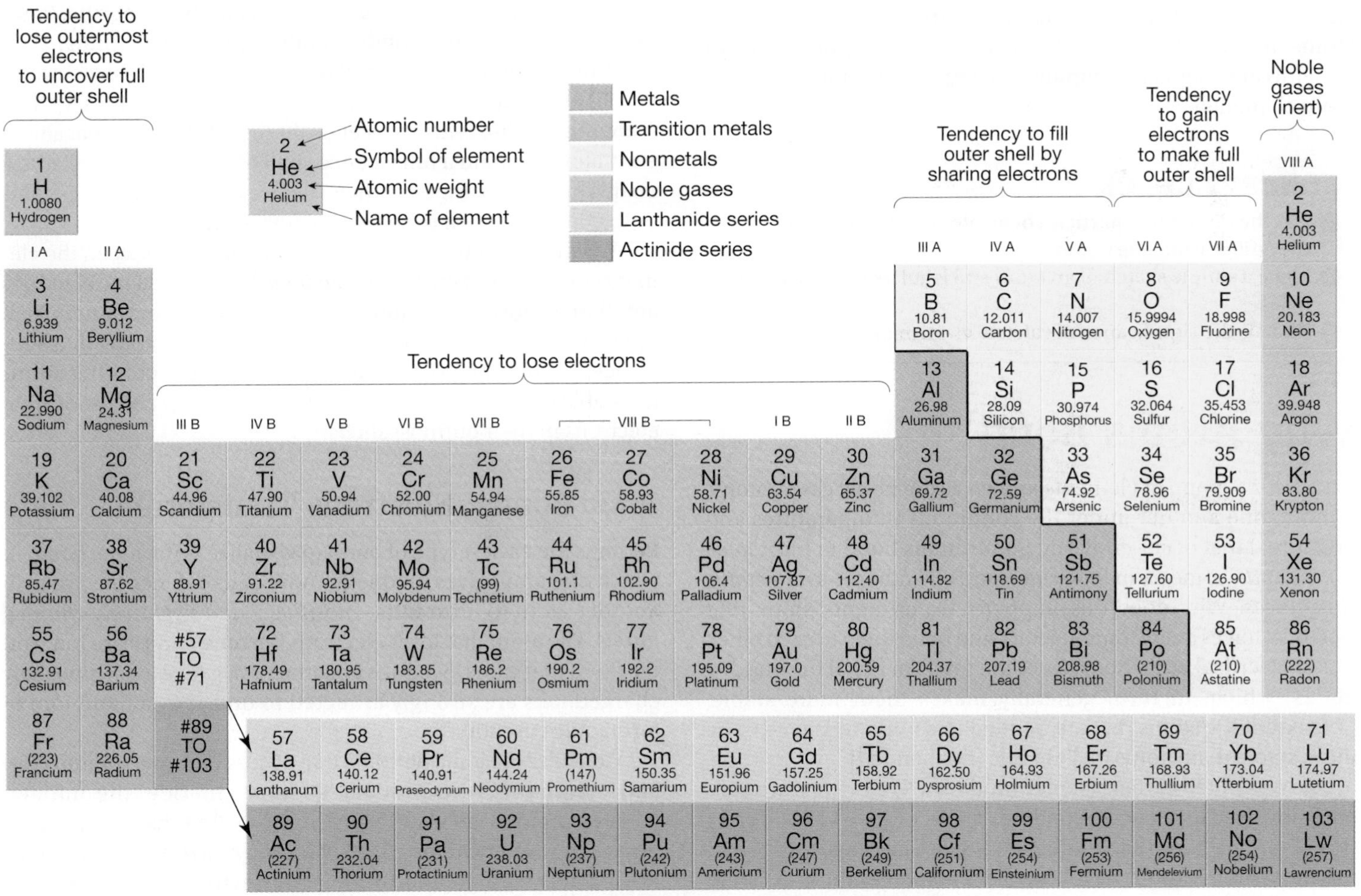

FIGURE 2.6 Gold mixed with quartz. Gold, silver, copper, and diamonds are naturally occurring minerals composed entirely of atoms of a single element.

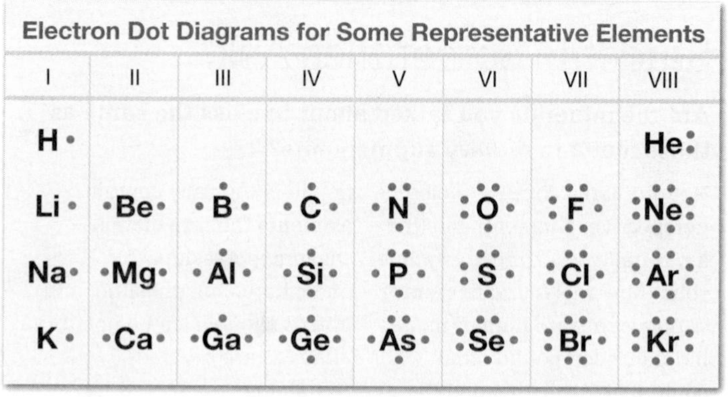

Atoms of the naturally occurring elements are the basic building blocks of Earth's minerals. A few minerals, such as native copper, diamonds, and gold, are made entirely of atoms of only one element (**Figure 2.6**). However, most elements tend to join with atoms of other elements to form **chemical compounds**. Most minerals are chemical compounds composed of atoms of two or more elements.

CONCEPT CHECK 2.2

❶ List the three main particles of an atom and explain how they differ from one another.

❷ Make a simple sketch of an atom and label its three main particles.

❸ What is the significance of valence electrons?

Why Atoms Bond

Except for a group of elements known as the noble gases, atoms bond to one another under the conditions (temperatures and pressures) that occur on Earth. Some atoms bond to form *ionic compounds*, some form *molecules*, and still others form *metallic substances*. Why does this happen? Experiments show that electrical forces hold atoms together and bond them to each other. These electrical attractions lower the total energy of the bonded atoms, which, in turn, generally makes them more stable. Consequently, atoms that are bonded in compounds tend to be more stable than atoms that are free (not bonded).

As was noted earlier, valence (outer shell) electrons are generally involved in chemical bonding. **Figure 2.7** shows a shorthand way of representing the number of valence electrons. Notice that the elements in Group I have one valence electron, those in Group

Electron Dot Diagrams for Some Representative Elements

I	II	III	IV	V	VI	VII	VIII
H·							·:He:
Li·	·Be·	·B·	·C·	·N·	:O·	:F·	:Ne:
Na·	·Mg·	·Al·	·Si·	·P·	:S·	:Cl·	:Ar:
K·	·Ca·	·Ga·	·Ge·	·As·	:Se·	:Br·	:Kr:

FIGURE 2.7 Dot diagrams for some representative elements. Each dot represents a valence electron found in the outermost principal shell.

II have two valence electrons, and so on, up to eight valence electrons in Group VIII.

Octet Rule

The noble gases (except helium) have very stable electron arrangements with eight valence electrons and, therefore, tend to lack chemical reactivity. Many other atoms gain, lose, or share electrons during chemical reactions to end up with electron arrangements of the noble gases. This observation led to a chemical guideline known as the **octet rule**: *Atoms tend to gain, lose, or share electrons until they are surrounded by eight valence electrons.* Although there are exceptions to the octet rule, it is a useful rule of thumb for understanding chemical bonding.

When an atom's outer shell does not contain eight electrons, it is likely to chemically bond to other atoms to fill its shell. A **chemical bond** is the transfer or sharing of electrons that allows each atom to attain a full valence shell of electrons. Some atoms do this by transferring all of their valence electrons to other atoms so that an inner shell becomes the full valence shell.

When the valence electrons are transferred between the elements to form ions, the bond is an *ionic bond*. When the electrons are shared between the atoms, the bond is a *covalent bond*. When the valence electrons are shared among all the atoms in a substance, the bonding is *metallic*. In any case, the bonding atoms get stable electron configurations, which usually consist of eight electrons in their outmost shells.

Ionic Bonds: Electrons Transferred

Perhaps the easiest type of bond to visualize is the *ionic bond*, in which one atom gives up one or more of its valence electrons to another atom to form **ions**—*positively and negatively charged atoms*. The atom that loses electrons becomes a positive ion, and the atom that gains electrons becomes a negative ion. Oppositely charged ions are strongly attracted to one another and join to form ionic compounds.

Consider the ionic bonding that occurs between sodium (Na) and chlorine (Cl) to produce sodium chloride, the mineral halite—common table salt. Notice in **Figure 2.8A** that sodium gives up its single valence electron to chlorine. As a result, sodium now has a stable configuration with eight electrons in its outermost

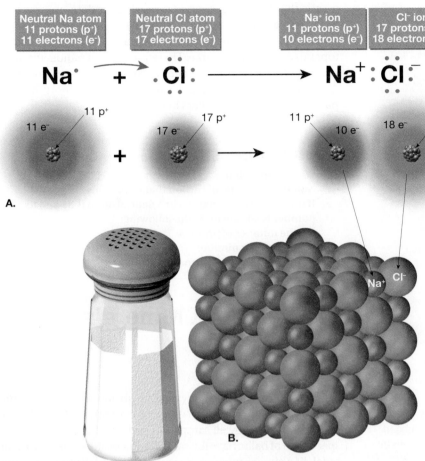

| Neutral Na atom
11 protons (p⁺)
11 electrons (e⁻) | Neutral Cl atom
17 protons (p⁺)
17 electrons (e⁻) | | Na⁺ ion
11 protons (p⁺)
10 electrons (e⁻) | Cl⁻ ion
17 protons (p⁺)
18 electrons (e⁻) |

FIGURE 2.8 Chemical bonding of sodium chloride (table salt). **A.** Through the transfer of one electron in the outer shell of a sodium atom to a chlorine atom, sodium becomes a positive ion and chlorine a negative ion. **B.** Diagram illustrating the arrangement (packing) of sodium and chlorine ions in table salt.

The properties of a chemical compound are dramatically different from the properties of the various elements comprising it. For example, sodium is a soft silvery metal that is extremely reactive and poisonous. If you were to consume even a small amount of elemental sodium, you would need immediate medical attention. Chlorine, a green poisonous gas, is so toxic that it was used as a chemical weapon during World War I. Together, however, these elements produce sodium chloride, a harmless flavor enhancer that we call table salt. Thus, when elements combine to form compounds their properties change significantly.

Covalent Bonds: Electrons Shared

Sometimes the forces that hold atoms together cannot be understood on the basis of the attraction of oppositely charged ions. One example is the hydrogen molecule (H_2), in which the two hydrogen atoms are held together tightly and no ions are present. The strong attractive force that holds two hydrogen atoms together results from a **covalent bond**, *a chemical bond formed by the sharing of a pair of electrons between atoms.*

Imagine two hydrogen atoms (each with one proton and one electron) approaching one another so that their electron clouds overlap (**Figure 2.9**). Once they meet, the electron configuration will change so that both electrons will primarily occupy the space between the atoms. In other words, the two electrons are shared by both hydrogen atoms and attracted simultaneously by the positive charge of the proton in the nucleus of each atom. The attraction between the electrons and both nuclei holds these atoms together. Although ions do not exist in hydrogen molecules, the force that holds these atoms together arises from the attraction of oppositely charged particles—protons in the nuclei and electrons shared by the atoms.

Metallic Bonds: Electrons Free to Move

In **metallic bonds**, the valence electrons are free to move from one atom to another so that all atoms share the available valence electrons. This type of bonding is found in metals such as copper, gold, aluminum, and silver, and in alloys such as brass and bronze. Metallic bonding accounts for the high electrical conductivity of metals, the ease with which metals are shaped, and numerous other special properties.

shell. By acquiring the electron that sodium loses, chlorine (which has seven valence electrons) gains the eighth electron needed to complete its outermost shell. Thus, through the transfer of a single electron, both the sodium and chlorine atoms have acquired a stable electron configuration.

After electron transfer takes place, the atoms are no longer electrically neutral. By giving up one electron, a neutral sodium atom becomes positively charged (with 11 protons and 10 electrons). Similarly, by acquiring one electron, a neutral chlorine atom becomes negatively charged (with 17 protons and 18 electrons). We know that ions with like charges repel, and those with unlike charges attract. Thus, an **ionic bond** is the attraction of oppositely charged ions to one another, producing an electrically neutral compound.

Figure 2.8B illustrates the arrangement of sodium and chlorine ions in ordinary table salt. Notice that salt consists of alternating sodium and chlorine ions, positioned in such a manner that each positive ion is attracted to and surrounded on all sides by negative ions, and vice versa. This arrangement maximizes the attraction between ions with opposite charges while minimizing the repulsion between ions with identical charges. Thus, ionic compounds consist of an orderly arrangement of oppositely charged ions assembled in a definite ratio that provides overall electrical neutrality.

CONCEPT CHECK 2.3

1. What is the difference between an atom and an ion?
2. What occurs in an atom to produce a positive ion? A negative ion?
3. Briefly distinguish between ionic and covalent bonding and the role that electrons play in both.

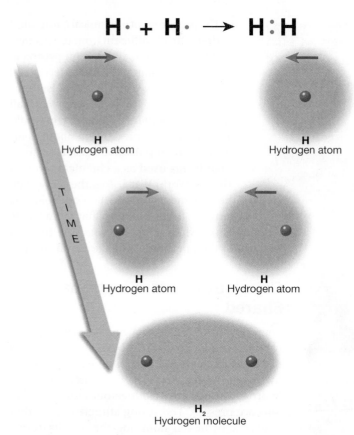

$$H \cdot + H \cdot \longrightarrow H : H$$

T I M E

H
Hydrogen atom

H
Hydrogen atom

H
Hydrogen atom

H
Hydrogen atom

H₂
Hydrogen molecule

FIGURE 2.9 Formation of a covalent bond between two hydrogen atoms (H) to form a hydrogen molecule (H_2). When hydrogen atoms bond, the electrons are shared by both hydrogen atoms and attracted simultaneously by the positive charge of the proton in the nucleus of each atom. The attraction between electrons and both nuclei holds (bonds) these atoms together.

Isotopes and Radioactive Decay

The **mass number** of an atom is simply the total number of its protons and neutrons. All atoms of a particular element have the same number of protons, but they may have varying numbers of neutrons. Atoms with the same number of protons but different numbers of neutrons are **isotopes** of that element. Isotopes of the same element are labeled by placing the mass number after the element's name or symbol. For example, carbon has three well-known isotopes. One has a mass number of 12 (carbon-12), another has a mass number of 13 (carbon-13), and the third, carbon-14, has a mass number of 14. Carbon-12 must also have six neutrons to give it a mass number of 12. Carbon-14, on the other hand, has six protons plus eight neutrons to give it a mass number of 14.

In chemical behavior, all isotopes of the same element are nearly identical. To distinguish among them is like trying to differentiate identical twins, with one weighing slightly more than the other. Because isotopes of the same element exhibit the same chemical behavior, they often become parts of the same mineral. For example, when the mineral calcite ($CaCO_3$) forms, some of its carbon atoms are carbon-12, and some are carbon-14.

The nuclei of most atoms are stable. However, many elements do have isotopes in which the nuclei are unstable—carbon-14 is one example of an unstable isotope. In this context, *unstable*

means that the nuclei change through a random process called **radioactive decay**. During radioactive decay, unstable isotopes radiate energy and emit particles. The rates at which unstable isotopes decay are measurable. Therefore, certain radioactive atoms are used to determine the ages of fossils, rocks, and minerals. A discussion of radioactive decay and its applications in dating past geologic events appears in Chapter 11.

CONCEPT CHECK 2.4

❶ What is an isotope?
❷ Name one isotope of carbon that is unstable.
❸ If the number of electrons in a neutral atom is 35 and its mass number is 80, calculate the following:
   a. the number of protons
   b. the atomic number
   c. the number of neutrons

Properties of Minerals

GEODe
EARTH
SCIENCE

Earth Materials
▶ Minerals

Minerals have definite crystalline structures and chemical compositions that give them unique sets of physical and chemical properties shared by all samples of that mineral. For example, all specimens of halite have the same hardness, the same density, and break in a similar manner. Because a mineral's internal structure and chemical composition are difficult to determine without the aid of sophisticated tests and equipment, the more easily recognized physical properties are frequently used in identification.

Optical Properties

Of the many optical properties of minerals—their luster, their ability to transmit light, their color, and their streak—are most frequently used for mineral identification.

Luster   The appearance or quality of light reflected from the surface of a mineral is known as **luster**. Minerals that have the appearance of metals, regardless of color, are said to have a *metallic luster* (**Figure 2.10**). Some metallic minerals, such as

FIGURE 2.10 The freshly broken sample of galena (right) displays a metallic luster, while the sample on the left is tarnished and has a submetallic luster. (Photo courtesy of E. J. Tarbuck)

FIGURE 2.11 Quartz. Some minerals, such as quartz, occur in a variety of colors. These samples include crystal quartz (colorless), amethyst (purple quartz), citrine (yellow quartz), and smoky quartz (gray to black). (Photo courtesy of E. J. Tarbuck)

native copper and galena, develop a dull coating or tarnish when exposed to the atmosphere. Because they are not as shiny as samples with freshly broken surfaces, these samples are often said to exhibit a *submetallic luster.*

Most minerals have a *nonmetallic luster* and are described using various adjectives such as *vitreous* (*glassy*), *dull* or *earthy* (a dull appearance like soil), *pearly* (such as a pearl or the inside of a clamshell), *silky* (like satin cloth), or *greasy* (as though coated in oil).

The Ability to Transmit Light   Another optical property used in the identification of minerals is the ability to transmit light. When no light is transmitted, the mineral is described as *opaque*; when light but not an image is transmitted through a mineral, it is said to be *translucent*. When both light and an image are visible through the sample, the mineral is described as *transparent*.

Color   Although **color** is generally the most conspicuous characteristic of any mineral, it is considered a diagnostic property of only a few minerals. Slight impurities in the common mineral quartz, for example, give it a variety of tints including pink, purple, yellow, white, gray, and even black (**Figure 2.11**). Other minerals, such as tourmaline, also exhibit a variety of hues, with multiple colors sometimes occurring in the same sample. Thus, the use of color as a means of identification is often ambiguous or even misleading.

Streak   The color of the mineral in powdered form, called **streak**, is often useful in identification. A mineral's streak is obtained by rubbing it across a *streak plate* (a piece of unglazed porcelain) and observing the color of the mark it leaves (**Figure 2.12**). Although the color of a mineral may vary from sample to sample, its streak is usually consistent in color.

Streak can also help distinguish between minerals with metallic luster and those with nonmetallic luster. Metallic minerals generally have a dense, dark streak, whereas minerals with nonmetallic luster typically have a light-colored streak.

It should be noted that not all minerals produce a streak when rubbed across a streak plate. For example, the mineral quartz is harder than a porcelain streak plate. Therefore, no streak is observed using this method.

Crystal Shape or Habit

Mineralogists use the term **crystal shape** or **habit** to refer to the common or characteristic shape of a crystal or aggregate of crystals. A few minerals exhibit somewhat regular polygons that are helpful in their identification. For example, magnetite crystals sometimes occur as octahedrons, garnets often form dodecahedrons, and halite and fluorite crystals tend to grow as cubes or near cubes. While most minerals have only one common habit, a few have two or more characteristic crystal shapes such as the pyrite sample shown in **Figure 2.13**.

By contrast, some minerals rarely develop perfect geometric forms. Many of these, however, develop other characteristic shapes useful for identification. Some minerals tend to grow equally in all three dimensions, whereas others tend to be elongated in one

FIGURE 2.12 Although the color of a mineral is not always helpful in identification, the streak, which is the color of the powdered mineral, can be very useful. (Photo by Dennis Tasa)

FIGURE 2.13 Although most minerals exhibit only one common crystal shape, some, such as pyrite, have two or more characteristic habits. (Photos by Dennis Tasa)

direction, or flattened if growth in one dimension is suppressed. Commonly used terms to describe these and other crystal habits include *equant* (equidimensional), *bladed*, *fibrous*, *tabular*, *prismatic*, *platy*, *blocky*, and *botryoidal*. Some of these habits are pictured in Figure 2.14.

Mineral Strength

How easily minerals break or deform under stress is determined by the type and strength of the chemical bonds that hold the crystals together. Mineralogists use terms including *tenacity*, *hardness*, *cleavage*, and *fracture* to describe mineral strength and how minerals break when stress is applied.

FIGURE 2.14 Some common crystal habits. **A.** *Bladed*. Elongated crystals that are flattened in one direction. **B.** *Prismatic*. Elongated crystals with faces that are parallel to a common direction. **C.** *Banded*. Minerals that have stripes or bands of different color or texture. **D.** *Botryoidal*. Groups of intergrown crystals resembling a bunch of grapes. (Photos by Dennis Tasa)

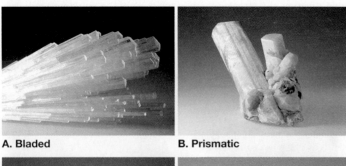

A. Bladed   B. Prismatic

C. Banded   D. Botryoidal

Tenacity   The term **tenacity** describes a mineral's toughness, or its resistance to breaking or deforming. Minerals that are ionically bonded, such as fluorite and halite, tend to be *brittle* and shatter into small pieces when struck. By contrast, minerals with metallic bonds, such as native copper, are *malleable*, or easily hammered into different shapes. Minerals, including gypsum and talc, that can be cut into thin shavings are described as *sectile*. Still others, notably the micas, are *elastic* and will bend and snap back to their original shape after the stress is released.

Hardness   One of the most useful diagnostic properties is **hardness**, a measure of the resistance of a mineral to abrasion or scratching. This property is determined by rubbing a mineral of unknown hardness against one of known hardness, or vice versa. A numerical value of hardness can be obtained by using the **Mohs scale** of hardness, which consists of 10 minerals arranged in order from 1 (softest) to 10 (hardest), as shown in Figure 2.15A. It should be noted that the Mohs scale is a relative ranking, and it does not imply that mineral number 2, gypsum, is twice as hard

FIGURE 2.15 Hardness scales. **A.** Mohs scale of hardness, with the hardness of some common objects. **B.** Relationship between Mohs relative hardness scale and an absolute hardness scale.

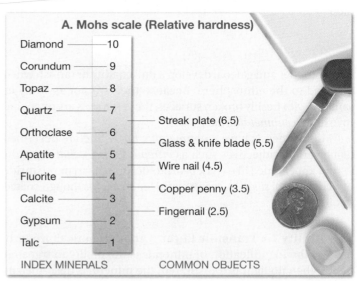
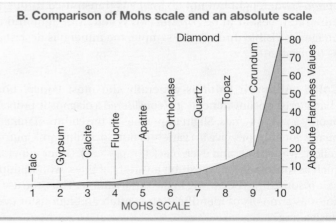

Students Sometimes Ask...

Are there any artificial materials harder than diamonds?

Yes, but you won't be seeing them anytime soon. A hard form of carbon nitride (C_3N_4), described in 1989 and synthesized in a laboratory shortly thereafter, may be harder than diamond but hasn't been produced in large enough amounts for a proper test. In 1999, researchers discovered that a form of carbon made from fused spheres of 20 and 28 carbon atoms—relatives of the famous "buckyballs"—also could be as hard as a diamond. These materials are expensive to produce, so diamonds continue to be used as abrasives and in certain kinds of cutting tools. Synthetic diamonds, produced since 1955, are now widely used in these industrial applications.

FIGURE 2.16 The thin sheets shown here were produced by splitting a mica (muscovite) crystal parallel to its perfect cleavage. (Photo by Chip Clark)

as mineral 1, talc. In fact, gypsum is only slightly harder than talc, as Figure 2.15B indicates.

In the laboratory, other common objects can be used to determine the hardness of a mineral. These include a human fingernail, which has a hardness of about 2.5, a copper penny (3.5), and a piece of glass (5.5). The mineral gypsum, which has a hardness of 2, can be easily scratched with a fingernail. On the other hand, the mineral calcite, which has a hardness of 3, will scratch a fingernail but will not scratch glass. Quartz, one of the hardest common minerals, will easily scratch glass. Diamonds, hardest of all, scratch anything, including other diamonds.

Cleavage   In the crystal structure of many minerals, some atomic bonds are weaker than others. It is along these weak bonds that minerals tend to break when they are stressed. **Cleavage** (*Kleiben* = carve) is the tendency of a mineral to break (cleave) along planes of weak bonding. Not all minerals have cleavage, but those that do can be identified by the relatively smooth, flat surfaces that are produced when the mineral is broken.

The simplest type of cleavage is exhibited by the micas (**Figure 2.16**). Because these minerals have very weak bonds in one direction, they cleave to form thin, flat sheets. Some minerals have excellent cleavage in one, two, three, or more directions, whereas others exhibit fair or poor cleavage, and still others have no cleavage at all. When minerals break evenly in more than one direction, cleavage is described by *the number of cleavage directions and the angle(s) at which they meet* (**Figure 2.17**).

Each cleavage surface that has a different orientation is counted as a different direction of cleavage. For example, some minerals cleave to form six-sided cubes. Because cubes are defined by three different sets of parallel planes that intersect at 90-degree angles, cleavage is described as *three directions of cleavage that meet at 90 degrees.*

Do not confuse cleavage with crystal shape. When a mineral exhibits cleavage, it will break into pieces that all have the same geometry. By contrast, the smooth-sided quartz crystals shown in Figure 2.1 (p. 00) illustrate crystal shape rather than cleavage.

If broken, they fracture into shapes that do not resemble one another or the original crystals.

Fracture   Minerals having chemical bonds that are equally, or nearly equally, strong in all directions exhibit a property called **fracture**. When minerals fracture, most produce uneven surfaces and are described as exhibiting *irregular fracture*. However, some minerals, such as quartz, break into smooth, curved surfaces resembling broken glass. Such breaks are called *conchoidal fractures* (**Figure 2.18**). Still other minerals exhibit fractures that produce splinters or fibers that are referred to as *splintery* and *fibrous fracture*, respectively.

Density and Specific Gravity

Density, an important property of matter, is defined as mass per unit of volume and is often expressed in grams per cubic centimeter (g/cm^3). Mineralogists often use a related measure called **specific gravity** to describe the density of minerals. Specific gravity is a number representing the ratio of a mineral's weight to the weight of an equal volume of water. The specific gravity of water equals 1.

Most common rock-forming minerals have a specific gravity of between 2 and 3. For example, quartz has a specific gravity of 2.65. By contrast, some metallic minerals such as pyrite, native copper, and magnetite are more than twice as dense and thus have more than twice the specific gravity as quartz. Galena, an ore of lead, has a specific gravity of roughly 7.5, whereas the specific gravity of 24-karat gold is approximately 20.

With a little practice, you can estimate the specific gravity of a mineral by hefting it in your hand. Ask yourself, does this mineral

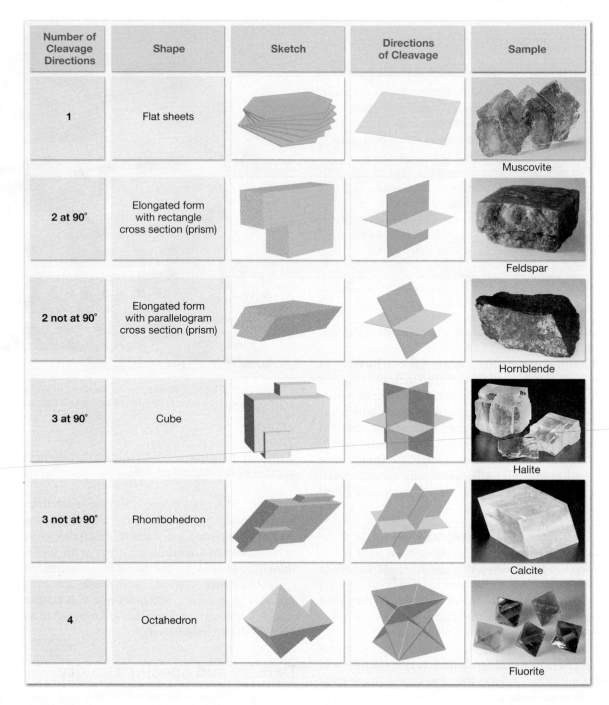

Number of Cleavage Directions	Shape	Sketch	Directions of Cleavage	Sample
1	Flat sheets			Muscovite
2 at 90°	Elongated form with rectangle cross section (prism)			Feldspar
2 not at 90°	Elongated form with parallelogram cross section (prism)			Hornblende
3 at 90°	Cube			Halite
3 not at 90°	Rhombohedron			Calcite
4	Octahedron			Fluorite

FIGURE 2.17 Common cleavage directions exhibited by minerals. (Photos by E. J. Tarbuck and Dennis Tasa)

feel about as "heavy" as similar-sized rocks you have handled? If the answer is "yes," the specific gravity of the sample will likely be between 2.5 and 3.

Other Properties of Minerals

In addition to the properties discussed thus far, some minerals can be recognized by other distinctive properties. For example, halite is ordinary salt, so it can be quickly identified through taste. Talc and graphite both have distinctive feels; talc feels soapy, and graphite feels greasy. Furthermore, the streaks of many sulfur-bearing minerals emit odors like rotten eggs. A few minerals, such as magnetite, have a high iron content and can be picked up with a magnet, while some varieties (lodestone) are natural magnets and will pick up small iron-based objects such as pins and paper clips (see Figure 2.25A, p. 43).

Moreover, some minerals exhibit special optical properties. For example, when a transparent piece of calcite is placed over printed text, the letters appear twice. This optical property is known as *double refraction* (**Figure 2.19**).

One very simple chemical test involves placing a drop of dilute hydrochloric acid from a dropper bottle onto a freshly broken

FIGURE 2.18 Conchoidal fracture. The smooth, curved surfaces result when minerals break in a glasslike manner. (Photo courtesy of E. J. Tarbuck)

mineral surface. Using this technique, certain minerals, called carbonates, will effervesce (fizz) as carbon dioxide gas is released (Figure 2.20). This test is especially useful in identifying the common carbonate mineral calcite.

CONCEPT CHECK 2.5

1. Define *luster*.
2. Why is color not always a useful property in mineral identification? Give an example of a mineral that supports your answer.
3. What is meant when we refer to a mineral's tenacity? List three terms that describe tenacity.
4. What differentiates cleavage from fracture?
5. What simple chemical test is useful in the identification of the mineral calcite?

FIGURE 2.20 Calcite reacting with a weak acid. (Photo by Chip Clark)

Mineral Groups

 Earth Materials
▶ Minerals

Over 4,000 minerals have been named, and several new ones are identified each year. Fortunately, for students who are beginning to study minerals, no more than a few dozen are abundant! Collectively, these few make up most of the rocks of Earth's crust and, as such, are often referred to as the **rock-forming minerals**.

Although less abundant, many other minerals are used extensively in the manufacture of products and are called *economic minerals*. However, rock-forming minerals and economic minerals are not mutually exclusive groups. When found in large deposits, some rock-forming minerals are economically significant. One example is the mineral calcite, which is the primary component of the sedimentary rock limestone and has many uses including being used in the production of cement.

It is worth noting that *only eight elements* make up the vast majority of the rock-forming minerals and represent more than 98 percent (by weight) of the continental crust (Figure 2.21). These elements, in order of abundance from most to least, are oxygen (O), silicon (Si), aluminum (Al), iron (Fe), calcium (Ca), sodium (Na), potassium (K), and magnesium (Mg). As shown in Figure 2.21, silicon and oxygen are by far the most common

FIGURE 2.19 Double refraction illustrated by the mineral calcite. (Photo by Chip Clark)

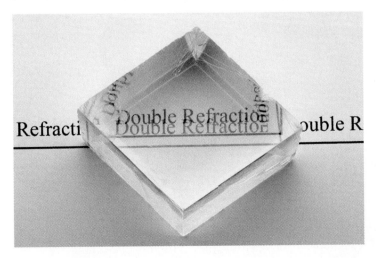

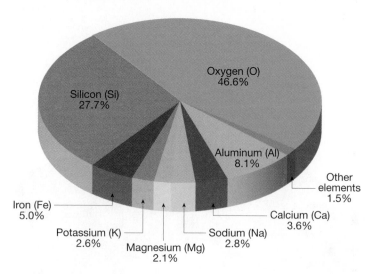

FIGURE 2.21 Relative abundance of the eight most common elements in the continental crust.

elements in Earth's crust. Furthermore, these two elements readily combine to form the basic "building block" for the most common mineral group, the **silicates**. More than 800 silicate minerals are known, and they account for more than 90 percent of Earth's crust.

Because other mineral groups are far less abundant in Earth's crust than the silicates, they are often grouped together under the heading **nonsilicates**. Although not as common as silicates, some nonsilicate minerals are very important economically. They provide us with iron and aluminum to build our automobiles, gypsum for plaster and drywall for home construction, and copper wire that carries electricity and connects us to the Internet. Some common nonsilicate mineral groups include the carbonates, sulfates, and halides. In addition to their economic importance, these mineral groups include members that are major constituents in sediments and sedimentary rocks.

We first discuss the most common mineral group, the silicates, and then consider some of the prominent nonsilicate mineral groups.

Silicate Minerals

Each of the silicate minerals contains oxygen and silicon atoms. Except for a few silicate minerals such as quartz, most silicate minerals also contain one or more additional elements in their crystalline structure. These elements give rise to the great variety of silicate minerals and their varied properties.

All silicates have the same fundamental building block, the **silicon–oxygen tetrahedron** (*tetra* = four, *hedra* = a base). This structure consists of four oxygen atoms surrounding a much smaller silicon atom, as shown in **Figure 2.22**. In some minerals, the tetrahedra are joined into chains, sheets, or three-dimensional networks by sharing oxygen atoms (**Figure 2.23**). These larger silicate structures are then connected to one another by other elements. The primary elements that join silicate structures are

iron (Fe), magnesium (Mg), potassium (K), sodium (Na), and calcium (Ca).

Major groups of silicate minerals and common examples are given in Figure 2.23. The **feldspars** are by far the most plentiful group, comprising over 50 percent of Earth's crust. **Quartz**, the second most abundant mineral in the continental crust, is the only common mineral made completely of silicon and oxygen.

Notice in Figure 2.23 that each mineral *group* has a particular silicate *structure*. A relationship exists between this internal structure of a mineral and the *cleavage* it exhibits. Because the silicon–oxygen bonds are strong, silicate minerals tend to cleave between the silicon–oxygen structures rather than across them. For example, the micas have a sheet structure and thus tend to cleave into flat plates (see muscovite in Figure 2.16). Quartz, which has equally strong silicon–oxygen bonds in all directions, has no cleavage but fractures instead.

How do silicate minerals form? Most crystallize from molten rock as it cools. This cooling can occur at or near Earth's surface (low temperature and pressure) or at great depths (high temperature and pressure). The *environment* during crystallization and the *chemical composition of the molten rock* mainly determine which minerals are produced. For example, the silicate mineral olivine crystallizes at high temperatures (about 1200°C [2200°F]), whereas quartz crystallizes at much lower temperatures (about 700°C [1300°F]).

In addition, some silicate minerals form at Earth's surface from the weathered products of other silicate minerals. Clay minerals are an example. Still other silicate minerals are formed under the extreme pressures associated with mountain building. Each silicate mineral, therefore, has a structure and a chemical composition that *indicate the conditions under which it formed*. Thus, by carefully examining the mineral makeup of rocks, geologists can often determine the circumstances under which the rocks formed.

FIGURE 2.22 Two representations of the silicon–oxygen tetrahedron. **A.** The four large spheres represent oxygen ions, and the blue sphere represents a silicon ion. The spheres are drawn in proportion to the radii of the ions. **B.** An expanded view of the tetrahedron that has an oxygen ion at each of the four corners.

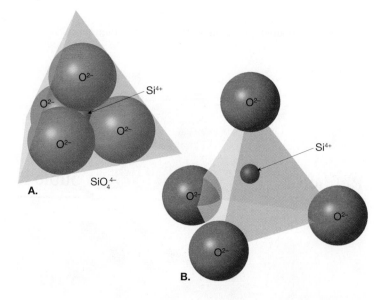

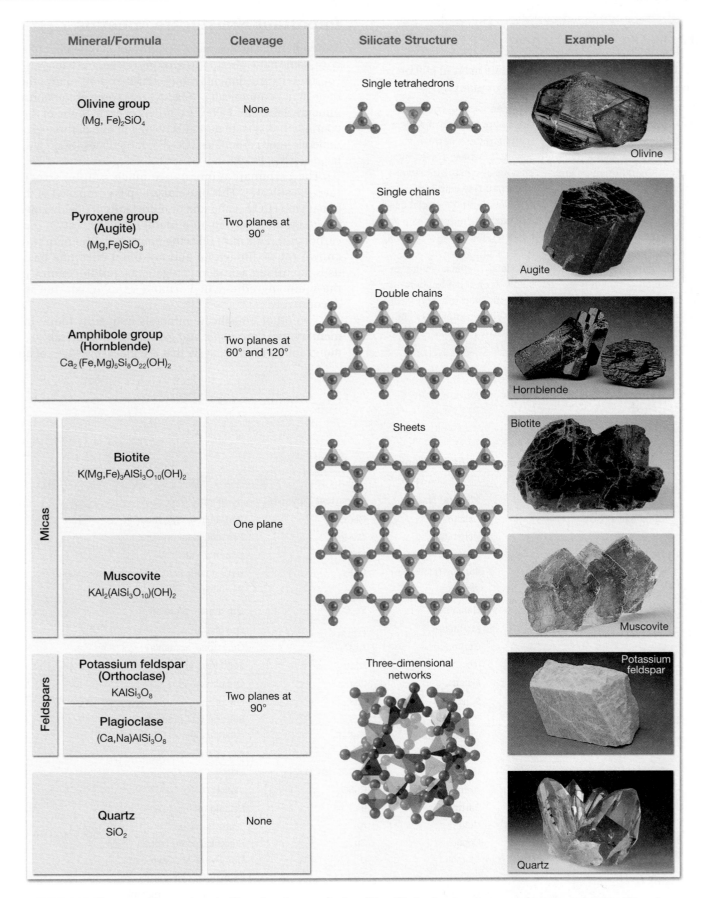

Mineral/Formula	Cleavage	Silicate Structure	Example
Olivine group $(Mg, Fe)_2SiO_4$	None	Single tetrahedrons	Olivine
Pyroxene group (Augite) $(Mg,Fe)SiO_3$	Two planes at 90°	Single chains	Augite
Amphibole group (Hornblende) $Ca_2(Fe,Mg)_5Si_8O_{22}(OH)_2$	Two planes at 60° and 120°	Double chains	Hornblende
Micas — **Biotite** $K(Mg,Fe)_3AlSi_3O_{10}(OH)_2$ / **Muscovite** $KAl_2(AlSi_3O_{10})(OH)_2$	One plane	Sheets	Biotite / Muscovite
Feldspars — **Potassium feldspar (Orthoclase)** $KAlSi_3O_8$ / **Plagioclase** $(Ca,Na)AlSi_3O_8$	Two planes at 90°	Three-dimensional networks	Potassium feldspar
Quartz SiO_2	None		Quartz

FIGURE 2.23 Common silicate minerals. Note that the complexity of the silicate structure increases from top to bottom. (Photos by Dennis Tasa and E. J. Tarbuck)

Students Sometimes Ask...

Are these silicates the same materials used in silicon computer chips and silicone breast implants?

Not really, but all three contain the element silicon (Si). Furthermore, the source of silicon for numerous products, including computer chips and breast implants, comes from silicate minerals. Pure silicon (without the oxygen that silicates have) is used to make computer chips, giving rise to the term "Silicon Valley" for the high-tech region of San Francisco, California's south bay area, where many of these devices are designed.

Manufacturers of computer chips engrave silicon wafers with incredibly narrow conductive lines, squeezing millions of circuits into every fingernail-size chip.

Silicone—the material used in breast implants—is a silicon–oxygen polymer gel that feels rubbery and is water repellent, chemically inert, and stable at extreme temperatures. Concern about the long-term safety of these implants limited their use after 1992.

Important Nonsilicate Minerals

Although nonsilicates make up only about 8 percent of Earth's crust, some minerals, such as gypsum, calcite, and halite, are major constituents in sedimentary rocks. Furthermore, many others are important economically. **Table 2.1** lists some of the nonsilicate mineral classes and a few examples of each. Some of the most common nonsilicate minerals belong to one of three classes of minerals—the carbonates (CO_3^{2-}), the sulfates (SO_4^{2-}), and the halides (Cl^{1-}, F^{1-}, B^{1-}).

The carbonate minerals are much simpler structurally than the silicates. This mineral group is composed of the carbonate ion (CO_3^{2-}) and one or more kinds of positive ions. The most common carbonate mineral is *calcite*, $CaCO_3$ (calcium carbonate). This mineral is the major constituent in two well-known rocks: limestone and marble. Limestone has many uses, including as road aggregate, as building stone, and as the main ingredient in Portland cement. Marble is used decoratively.

Two other nonsilicate minerals frequently found in sedimentary rocks are *halite* and *gypsum*. Both minerals are commonly found in thick layers that are the last vestiges of ancient

TABLE 2.1 Common Nonsilicate Mineral Groups

Mineral Groups [key ion(s) or element(s)]	Mineral Name	Chemical Formula	Economic Use
Carbonates (CO_3^{2-})	Calcite	$CaCO_3$	Portland cement, lime
	Dolomite	$CaMg(CO_3)_2$	Portland cement, lime
Halides (Cl^-, F^-, Br^-)	Halite	$NaCl$	Common salt
	Fluorite (Fluorspar)	CaF_2	Hydrofluoric acid production, steelmaking
	Sylvite	KCl	Fertilizer
Oxides (O^{2-})	Hematite	Fe_2O_3	Ore of iron, pigment
	Magnetite	Fe_3O_4	Ore of iron
	Corundum	Al_2O_3	Gemstone, abrasive
	Ice	H_2O	Solid form of water
Sulfides (S^{2-})	Galena	PbS	Ore of lead
	Sphalerite	ZnS	Ore of zinc
	Pyrite	FeS_2	Sulfuric acid production
	Chalcopyrite	$CuFeS_2$	Ore of copper
	Cinnabar	HgS	Ore of mercury
Sulfates (SO_4^{2-})	Gypsum	$CaSO_4 \cdot 2H_2O$	Plaster
	Anhydrite	$CaSO_4$	Plaster
	Barite	$BaSO_4$	Drilling mud
Native elements (single elements)	Gold	Au	Trade, jewelry
	Copper	Cu	Electrical conductor
	Diamond	C	Gemstone, abrasive
	Sulfur	S	Sulfa drugs, chemicals
	Graphite	C	Pencil lead, dry lubricant
	Silver	Ag	Jewelry, photography
	Platinum	Pt	Catalyst

FIGURE 2.24 Thick bed of halite (salt) at an underground mine in Grand Saline, Texas. Note person for scale. (Photo by Tom Bochsler)

seas that have long since evaporated (**Figure 2.24**). Like limestone, both are important nonmetallic resources. Halite is the mineral name for common table salt (NaCl). Gypsum ($CaSO_4 \cdot 2H_2O$), which is calcium sulfate with water bound into the structure, is the mineral of which plaster and other similar building materials are composed.

Most nonsilicate mineral classes contain members that are prized for their economic value. This includes the oxides, whose members hematite and magnetite are important ores of iron (**Figure 2.25**). Also significant are the sulfides, which are basically compounds of sulfur (S) and one or more metals. Examples of important sulfide minerals include galena (lead), sphalerite (zinc), and chalcopyrite (copper). In addition, native elements, including gold, silver, and carbon (diamonds), plus a host of other nonsilicate minerals—fluorite (flux in making steel), corundum (gemstone, abrasive), and uraninite (a uranium source)—are important economically (see Box 2.1).

FIGURE 2.25 Magnetite (**A**) and hematite (**B**) are both oxides and are both important ores of iron. (Photos by E. J. Tarbuck)

A. Magnetite

B. Hematite

Students Sometimes Ask...

According to the textbook, thick beds of halite and gypsum formed when ancient seas evaporated. Has this happened in the recent past?

Yes. During the past 6 million years, the Mediterranean Sea may have dried up and then refilled several times. When 65 percent of seawater evaporates, the mineral gypsum begins to precipitate, meaning it comes out of solution and settles to the bottom. When 90 percent of the water is gone, halite crystals form, followed by salts of potassium and magnesium. Deep-sea drilling in the Mediterranean has encountered thick deposits of gypsum and salt (mostly halite) sitting one atop the other to a maximum thickness of 2 kilometers (1.2 miles). These deposits are inferred to have resulted from tectonic events that periodically closed and reopened the connection between the Atlantic Ocean and the Mediterranean Sea (the modern-day Straits of Gibraltar) over the past several million years. During periods when the Mediterranean was cut off from the Atlantic, the warm and dry climate in this region caused the Mediterranean to nearly "dry up." Then, when the connection to the Atlantic was opened, the Mediterranean basin would refill with seawater of normal salinity. This cycle was repeated over and over again, producing the layers of gypsum and salt found on the Mediterranean seafloor.

Box 2.1

UNDERSTANDING EARTH

Gemstones

Precious stones have been prized since antiquity. But misinformation abounds regarding gems and their mineral makeup. This stems partly from the ancient practice of grouping precious stones by color rather than mineral makeup. For example, *rubies* and red *spinels* are very similar in color, but they are completely different minerals. Classifying by color led to the more common spinels being passed off to royalty as rubies. Even today, with modern identification techniques, common *yellow quartz* is sometimes sold as the more valuable gemstone *topaz*.

Naming Gemstones

Most precious stones are given names that differ from their parent mineral. For example, *sapphire* is one of two gems that are varieties of the same mineral, *corundum*. Trace elements can produce vivid sapphires of nearly every color (**Figure 2.A**). Tiny amounts of titanium and iron in corundum produce the most prized blue sapphires. When the mineral corundum contains a sufficient quantity of chromium, it exhibits a brilliant red color, and the gem is called *ruby*. Furthermore, if a specimen is not suitable as a gem, it simply goes by the mineral name *corundum*. Because of its hardness, corundum that is not of gem quality is often crushed and sold as an abrasive.

To summarize, when corundum exhibits a red hue, it is called *ruby*, but if it exhibits any other color, the gem is called *sapphire*. Whereas corundum is the base mineral for two gems, quartz is the parent of more than a dozen gems. **Table 2.A** lists some well-known gemstones and their parent minerals.

FIGURE 2.A Australian sapphires depicting variations in cuts and colors. (Photo by Fred Ward, Black Star)

What Constitutes a Gemstone?

When found in their natural state, most gemstones are dull and would be passed over by most people as "just another rock." Gems must be cut and polished by experienced professionals before their true beauty is displayed (Figure 2.A). (One of the methods used to shape a gemstone is *cleaving*, the act of splitting the mineral along one of its planes of weakness, or cleavage.) Only those mineral specimens that are of such quality that they can command a price in excess of the cost of processing are considered gemstones.

Gemstones can be divided into two categories: precious and semiprecious. A *precious* gem has beauty, durability, and rarity, whereas a *semiprecious* gem generally has only one or two of these qualities. The gems traditionally held in highest esteem are diamonds, rubies, sapphires, emeralds, and some varieties of opal (Table 2.A). All other gemstones are classified as semiprecious. However, large high-quality specimens of semiprecious stones often command a very high price.

Today, translucent stones with evenly tinted colors are preferred. The most favored hues are red, blue, green, purple, rose, and yellow. The most prized stones are pigeon-blood rubies, blue sapphires, grass-green emeralds, and canary-yellow diamonds. Colorless gems are generally less than desirable except for diamonds that display "flashes of color" known as *brilliance*.

The durability of a gem depends on its hardness; that is, its resistance to abrasion by objects normally encountered in everyday living. For good durability, gems should be as hard or harder than quartz as defined by the Mohs scale of hardness. One notable exception is opal, which is comparatively soft (hardness 5–6.5) and brittle. Opal's esteem comes from its "fire," which is a display of a variety of brilliant colors, including greens, blues, and reds.

It seems to be human nature to treasure that which is rare. In the case of gemstones, large, high-quality specimens are much rarer than smaller stones. Thus, large rubies, diamonds, and emeralds, which are rare in addition to being beautiful and durable, command the very highest prices.

TABLE 2.A Important Gemstones

Gem	Mineral Name	Prized Hues	Gem	Mineral Name	Prized Hues
Precious			**Semiprecious**		
Diamond	Diamond	Colorless, yellows	Garnet	Garnet	Reds, greens
Emerald	Beryl	Greens	Jade	Jadeite or nephrite	Greens
Opal	Opal	Brilliant hues	Moonstone	Feldspar	Transparent blues
Ruby	Corundum	Reds	Peridot	Olivine	Olive greens
Sapphire	Corundum	Blues	Smoky quartz	Quartz	Browns
Semiprecious			Spinel	Spinel	Reds
Alexandrite	Chrysoberyl	Variable	Topaz	Topaz	Purples, reds
Amethyst	Quartz	Purples	Tourmaline	Tourmaline	Reds, blue-greens
Cat's-eye	Chrysoberyl	Yellows	Turquoise	Turquoise	Blues
Chalcedony	Quartz (agate)	Banded	Zircon	Zircon	Reds
Citrine	Quartz	Yellows			

CONCEPT CHECK 2.6

1. List the eight most common elements in Earth's crust in order of abundance (most to least).
2. Explain the difference between the terms *silicon* and *silicate*.
3. Draw a sketch of the silicon–oxygen tetrahedron.
4. What is the most abundant mineral in Earth's crust?
5. List six common nonsilicate mineral groups. What key ion(s) or element(s) define each group?
6. What is the most common carbonate mineral?
7. List eight common nonsilicate minerals and their economic uses.

Natural Resources

Earth's crust and oceans are the source of a wide variety of useful and essential materials that have played a crucial role in the development of civilization. From the first use of clay to make pottery nearly 10,000 years ago, the use of Earth materials has expanded resulting in more complex societies. Today, practically every manufactured product contains materials obtained from minerals. Table 2.1 lists some of the most economically important mineral groups.

Renewable versus Nonrenewable Resources

Resources are commonly divided into two broad categories. Some are classified as **renewable**, which means that they can be replenished over relatively short time spans. Common examples are plants and animals for food, natural fibers for clothing, and forest products for lumber and paper. Energy from flowing water, wind, and the Sun are also considered renewable (**Figure 2.26**).

By contrast, many other basic resources are classified as **nonrenewable**. Important metals such as iron, aluminum, and copper fall into this category, as do our most important fuels: oil, natural gas, and coal. Although these and other resources continue to form, the processes that create them are so slow that significant deposits take millions of years to accumulate. In essence, Earth contains fixed quantities of these substances. When the present supplies are mined or pumped from the ground, there will be no more. Although some nonrenewable resources, such as aluminum, can be used over and over again, others, such as oil, cannot be recycled.

FIGURE 2.26 Hydroelectric power is one example of a renewable resource. Lake Powell is the reservoir that was created when Glen Canyon Dam was built across the Colorado River. As water in the reservoir is released, it drives turbines and produces electricity. (Photo by Michael Collier)

Mineral Resources

Mineral resources are those occurrences of useful minerals that are formed in such quantities that eventual extraction is reasonably certain. Resources include deposits from which minerals can be presently extracted profitably, as well as known deposits that are not yet economically or technologically recoverable.

An **ore** or **ore deposit** is a naturally occurring concentration of one or more metallic minerals that can be extracted economically (see Figure 2.25). In common usage, the term ore is also applied to some nonmetallic minerals such as fluorite and sulfur. However, materials used for such purposes as building stone, road aggregate, abrasives, ceramics, and fertilizers are not usually called ores; rather, they are classified as industrial rocks and minerals.

Recall that more than 98 percent of Earth's crust is composed of only eight elements, and except for oxygen and silicon, all other elements make up a relatively small fraction of common crustal rocks (see Figure 2.21). Indeed, the natural concentrations of many elements are exceedingly small. A deposit containing the average percentage of a valuable element such as gold has no economic value, because the cost of extracting it greatly exceeds the value of the gold that could be recovered.

To have economic value, an element must be concentrated above the level of its average crustal abundance. For example, copper makes up about 0.0135 percent of the crust. For a deposit to be considered as copper ore, it must contain a concentration that is about 100 times this amount. Aluminum, on the other hand, represents 8.13 percent of the crust and can be extracted profitably when it is found in concentrations only about four times its average crustal percentage.

It is important to realize that a deposit may become profitable to extract or lose its profitability because of economic changes. If demand for a metal increases and prices rise sufficiently, the status of a previously unprofitable deposit changes, and it becomes an ore. The status of unprofitable deposits may also change if a technological advance allows the ore to be extracted at a lower cost than before.

Conversely, changing economic factors can turn a once profitable ore deposit into an unprofitable deposit that can no longer be called an ore. This situation was illustrated at the copper mining operation located at Bingham Canyon, Utah, one of the largest open-pit mines on Earth (**Figure 2.27**). Mining was halted there

FIGURE 2.27 Aerial view of Bingham Canyon copper mine near Salt Lake City, Utah. Although the amount of copper in the rock is less than 1 percent, the huge volume of material removed and processed each day (about 200,000 tons) yields enough metal to be profitable. (Photo by Michael Collier)

in 1985 because outmoded equipment had driven the cost of extracting the copper beyond the current selling price. The owners responded by replacing an antiquated 1,000-car railroad with conveyor belts and pipelines for transporting the ore and waste. These devices achieved a cost reduction of nearly 30 percent and returned this mining operation to profitability.

Over the years, geologists have been keenly interested in learning how natural processes produce localized concentrations of essential minerals. One well-established fact is that occurrences of valuable mineral resources are closely related to the rock cycle. That is, the mechanisms that generate igneous, sedimentary, and metamorphic rocks, including the processes of weathering and erosion, play a major role in producing concentrated accumulations of useful elements.

Moreover, with the development of the theory of plate tectonics, geologists have added another tool for understanding the processes by which one rock is transformed into another. As these rock-forming processes are examined in the following chapters, we consider their role in producing some of our important mineral resources.

CONCEPT CHECK 2.7

1 List three examples of renewable and three examples of non-renewable resources.
2 Compare and contrast a *mineral resource* and an *ore deposit*.
3 What might cause a mineral deposit that previously could not be mined profitably to become reclassified as an ore?

GIVE IT SOME THOUGHT

1. Using the geologic definition of *mineral* as your guide, determine which of the items on the list are minerals and which are not. If an item is not a mineral, explain why not.
   a. gold nugget
   b. seawater
   c. quartz
   d. cubic zirconia
   e. obsidian
   f. ruby
   g. glacial ice
   h. amber
   Refer to the Periodic Table of the Elements (Figure 2.5) to help you answer questions 2, 3, and 4.
2. If the number of protons in a neutral atom is 92 and its mass number is 238:
   a. What is the name of that element?
   b. How many electrons does it have?
   c. How many neutrons does it have?
3. Which element is more likely to form chemical bonds: xenon (Xe) or sodium (Na)? Explain why.
4. The information below refers to three isotopes of the element potassium. Using this information, determine the appropriate number of protons and neutrons for each isotope. Label each isotope in the manner used in the chapter.

   | Atomic Number = 19 | Atomic Number = 19 | Atomic Number = 19 |
   | Mass Number = 39 | Mass Number = 40 | Mass Number = 41 |

5. Referring to the accompanying photos of five minerals, determine which of these specimens exhibit a metallic luster and which have a nonmetallic luster.

A.   B.   C.   D.   E.

6. Examine the accompanying photo of a mineral that has several smooth, flat surfaces that resulted when the specimen was broken.
   a. How many flat surfaces are present on this specimen?
   b. How many *different directions* of cleavage does this specimen have?
   c. Do the cleavage directions meet at 90-degree angles?

Cleaved sample

7. Gold has a specific gravity of almost 20. A 5-gallon bucket of water weighs 40 pounds. How much would a 5-gallon bucket of gold weigh?

8. Do an Internet search to determine what mineral(s) are extracted from the ground during the manufacturing of the following products:
   a. stainless steel utensils
   b. cat litter
   c. Tums brand antacid tablets
   d. lithium batteries
   e. aluminum beverage cans

9. Most states have designated a state mineral, rock, or gemstone to promote interest in the state's natural resources. Describe your state mineral, rock, or gemstone and explain why it was selected. If your state does not have a state mineral, rock, or gemstone, select one from a state adjacent to yours.

In Review   Chapter 2 Matter and Minerals

- A *mineral* is any naturally occurring inorganic solid that possesses an orderly crystalline structure and can be represented by a chemical formula. Most *rocks* are aggregates composed of two or more minerals.

- All matter, including minerals, is composed of minute particles called *atoms*—the building blocks of minerals. Each atom has a *nucleus*, which contains *protons* (particles with positive electrical charges) and *neutrons* (particles with neutral electrical charges). Surrounding the nucleus of an atom in regions called *principal shells* are *electrons*, which have negative electrical charges. The number of protons in an atom's nucleus determines its *atomic number* and the name of the element. An *element* is a large collection of electrically neutral atoms, all having the same atomic number.

- Atoms combine to form ionic compounds, molecules, or metallic substances. Atoms bond by either gaining, losing, or sharing electrons with other atoms. In *ionic bonding*, one or more electrons are transferred from one atom to another, giving the atoms a net positive or negative charge. The resulting electrically charged atoms are called *ions*. Ionic compounds consist of oppositely charged ions assembled in a regular, crystalline structure that allows for the maximum attraction of ions, given their sizes. Another type of bond, the *covalent bond*, is produced when atoms share electrons.

- *Isotopes* are variants of the same element that have a different *mass number* (the total number of neutrons plus protons found in an atom's nucleus). Some isotopes are unstable and disintegrate naturally through a process called *radioactivity*.

- The properties of minerals include *crystal shape* (*habit*), *luster*, *color*, *streak*, *tenacity*, *hardness*, *cleavage*, *fracture*, and *density* or *specific gravity*. In addition, a number of special physical and chemical properties (*taste*, *smell*, *elasticity*, *feel*, *magnetism*, *double refraction*, and *chemical reaction to hydrochloric acid*) are useful in identifying certain minerals. Each mineral has a unique set of properties that can be used for identification.

- Of the nearly 4,000 minerals, no more than a few dozen make up most of the rocks of Earth's crust and, as such, are classified as rock-forming minerals. Eight elements (oxygen, silicon, aluminum, iron, calcium, sodium, potassium, and magnesium) make up the bulk of these minerals and represent over 98 percent (by weight) of Earth's continental crust.

- The most common mineral group is the *silicates*. All silicate minerals have the negatively charged *silicon–oxygen tetrahedron* as their fundamental building block. In some silicate minerals the tetrahedra are joined in chains (the pyroxene and amphibole groups); in others, the tetrahedra are arranged into sheets (the micas—biotite and muscovite), or three-dimensional networks (the feldspars and quartz). The tetrahedra and various silicate structures are often bonded together by the positive ions of iron, magnesium, potassium, sodium, aluminum, and calcium. Each silicate mineral has a structure and a chemical composition that indicates the conditions under which it formed.

- The *nonsilicate* mineral groups, which contain several economically important minerals, include the *oxides* (e.g., the mineral hematite, mined for iron), *sulfides* (e.g., the mineral sphalerite, mined for zinc, and the mineral galena, mined for lead), *sulfates*, *halides*, and *native elements* (e.g., gold and silver). The more common nonsilicate rock-forming minerals include the *carbonate minerals*, calcite and dolomite. Two other nonsilicate minerals frequently found in sedimentary rocks are halite and gypsum.

- Mineral resources are those occurrences of useful minerals ultimately available commercially. Resources include already identified deposits from which minerals can be extracted profitably, called *reserves*, as well as known deposits that are not yet economically or technologically recoverable. Deposits inferred to exist, but not yet discovered, are also considered mineral resources. The term *ore* is used to denote those useful metallic minerals that can be mined for a profit, as well as some nonmetallic minerals, such as fluorite and sulfur, that contain useful substances.

Key Terms

atom (p. 30)
atomic number (p. 31)
chemical bond (p. 32)
chemical compound (p. 32)
cleavage (p. 37)
color (p. 35)
covalent bond (p. 33)
crystal shape (p. 35)
density (p. 37)
electron (p. 30)
element (p. 31)
feldspars (p. 40)
fracture (p. 37)
habit (p. 35)
hardness (p. 36)

ion (p. 32)
ionic bond (p. 33)
isotope (p. 34)
luster (p. 34)
mass number (p. 34)
metallic bond (p. 33)
mineral (p. 29)
mineral resource (p. 46)
mineralogy (p. 28)
Mohs scale (p. 36)
neutron (p. 30)
nonrenewable resources (p. 45)
nonsilicates (p. 40)
nucleus (p. 30)
octet rule (p. 32)

ore (p. 46)
ore deposit (p. 46)
periodic table (p. 31)
proton (p. 30)
quartz (p. 40)
radioactive decay (p. 34)
renewable resources (p. 45)
rock (p. 29)
rock-forming minerals (p. 39)
silicate (p. 40)
silicon–oxygen tetrahedron (p. 40)
specific gravity (p. 37)
streak (p. 35)
tenacity (p. 36)
valence electron (p. 31)

Examining the Earth System

1. Perhaps one of the most significant interrelationships between humans and the Earth system involves the extraction, refinement, and distribution of the planet's mineral wealth. To help you understand these associations, begin by thoroughly researching a mineral commodity that is mined in your local region or state. (You might find useful the information at these United States Geological Survey [USGS] Websites: **http://minerals.er.usgs.gov/minerals/pubs/state/** and **http://minerals.er.usgs.gov:80/minerals/pubs/mcs/**) What products are made from this mineral? Do you use any of these products? Describe the mining and refining of the mineral and the local impact these processes have on each of Earth's spheres (atmosphere, hydrosphere, solid Earth, and biosphere). Are any of the effects negative? If so, what, if anything, is being done to end or minimize the damage?

2. Referring to the mineral you described above, in your opinion does the environmental impact of extracting this mineral outweigh the benefits derived from its products?

Mastering Geology

Looking for additional review and test prep materials? Visit the Self Study area in **www.masteringgeology.com** to find practice quizzes, study tools, and multimedia that will aid in your understanding of this chapter's content. In **MasteringGeology™** you will find:

- **GEODe: Earth Science: An interactive visual walkthrough of key concepts**

- **Geoscience Animation Library: More than 100 animations illuminating many difficult-to-understand Earth science concepts**
- **In The News RSS Feeds: Current Earth science events and news articles are pulled into the site with assessment**
- **Pearson eText**
- **Optional Self Study Quizzes**
- **Web Links**
- **Glossary**
- **Flashcards**

Rocks: Materials of the Solid Earth

Hiker amid the sculpted sedimentary rocks of Fisher Towers near Moab, Utah.

Why study rocks? You have already learned that some rocks and minerals have great economic value. In addition, all Earth processes depend in some way on the properties of these basic Earth materials. Events such as volcanic eruptions, mountain building, weathering, erosion, and even earthquakes involve rocks and minerals. Consequently, a basic knowledge of Earth materials is essential to understanding most geologic phenomena.

Every rock contains clues about the environment in which it formed. For example, some rocks are composed entirely of small shell fragments. This tells Earth scientists that the rock likely originated in a shallow marine environment. Other rocks contain clues that indicate they formed from a volcanic eruption or deep in the Earth during mountain building (Figure 3.1). Thus, rocks contain a wealth of information about events that have occurred over Earth's long history.

FOCUS ON CONCEPTS

To assist you in learning **the important concepts in this chapter, focus on the following questions:**

- What is the rock cycle and why is it important?
- What are the three groups of rocks and the geologic processes involved in the formation of each?
- How does magma differ from lava?
- What two criteria are used to classify igneous rocks?
- How does the rate of cooling influence the crystal size of minerals in igneous rocks?
- What are the names and environments of formation for some common detrital and chemical sedimentary rocks?
- What processes change sediment into sedimentary rock?
- What are two characteristics of sedimentary rocks?
- What are metamorphic rocks, and how do they form?
- What are the agents of metamorphism?
- What are the names, textures, and environments of formation for some common metamorphic rocks?

Earth as a System: The Rock Cycle

Earth Materials
▶ The Rock Cycle

Earth as a system is illustrated most vividly when we examine the rock cycle. The **rock cycle** allows us to see many of the interactions among the many components and processes of the Earth system (Figure 3.2). It helps us understand the origin of igneous, sedimentary, and metamorphic rocks and how they are connected. In addition, the rock cycle demonstrates that any rock type, under the right circumstances, can be transformed into any other type.

The Basic Cycle

We begin our discussion of the rock cycle with molten rock, called *magma,* which forms by melting that occurs primarily within Earth's crust and upper mantle. Once formed, a magma body often rises toward the surface because it is less dense than the surrounding rock. Occasionally magma reaches Earth's surface

where it erupts as *lava.* Eventually, molten rock cools and solidifies, a process called *crystallization* or *solidification.* Molten rock may solidify either beneath the surface or, following a volcanic eruption, at the surface. In either situation, the resulting rocks are called *igneous rocks.*

If igneous rocks are exposed at the surface, they undergo *weathering,* in which the daily influences of the atmosphere slowly disintegrate and decompose rocks. The loose materials that result are often moved downslope by gravity, and then picked up and transported by one or more erosional agents—running water, glaciers, wind, or waves. Eventually, these particles and dissolved substances, called *sediment,* are deposited. Although most sediment ultimately comes to rest in the ocean, other sites of deposition include river floodplains, desert basins, swamps, and sand dunes.

Next, the sediments undergo *lithification,* a term meaning "conversion into rock." Sediment is usually lithified into *sedimentary rock* when compacted by the weight of overlying materials or when cemented as percolating groundwater fills the pores with mineral matter.

If the resulting sedimentary rock becomes deeply buried or is involved in the dynamics of mountain building, it will be subjected to great pressures and intense heat. The sedimentary rock

FIGURE 3.1 Rocks contain information about the processes that produce them. This large exposure of igneous rocks, located in the Sierra Nevada of California, was once a molten mass found deep within Earth. (Photo by Brian Bailey/Getty Images)

may react to the changing environment by turning into the third rock type, *metamorphic rock*. If metamorphic rock is subjected to still higher temperatures, it may melt, creating magma, and the cycle begins again.

Although rocks may appear to be stable, unchanging masses, the rock cycle shows they are not. The changes, however, take time—sometimes millions or even billions of years. In addition, the rock cycle operates continuously around the globe, but in different stages depending on the location. Today, new magma is forming under the island of Hawaii, whereas the rocks that comprise the Colorado Rockies are slowly being worn down by weathering and erosion. Some of this weathered debris will eventually be carried to the Gulf of Mexico, where it will add to the already substantial mass of sediment that has accumulated there.

Alternative Paths

Rocks do not necessarily go through the cycle in the order that was just described. Other paths are also possible. For example, igneous rocks, rather than being exposed to weathering and erosion at Earth's surface, may remain deeply buried. Eventually, these masses may be subjected to the strong compressional forces and high temperature associated with mountain building. When this occurs, they are transformed directly into metamorphic rocks.

Metamorphic and sedimentary rocks, as well as sediment, do not always remain buried. Rather, overlying layers may be eroded away, exposing the once buried rock. When this happens, the material is attacked by weathering processes and turned into new raw materials for sedimentary rocks.

In a similar manner, igneous rocks that formed at depth can be uplifted, weathered, and turned into sedimentary rocks. Alternatively, igneous rocks may remain at depth where the high temperatures and forces associated with mountain building may metamorphose or even melt them. Over time, rocks may be transformed into any other rock type, or even into a different form of its original type. There are many paths that rocks may take through the rock cycle.

What drives the rock cycle? Earth's internal heat is responsible for the processes that form igneous and metamorphic rocks. Weathering and the transport of weathered material are external processes, powered by energy from the Sun. External processes produce sedimentary rocks.

CONCEPT CHECK 3.1

❶ Sketch and label a basic rock cycle. Make sure your sketch includes alternative paths.

❷ Use the rock cycle to explain the statement "One rock is the raw material for another."

ROCK CYCLE

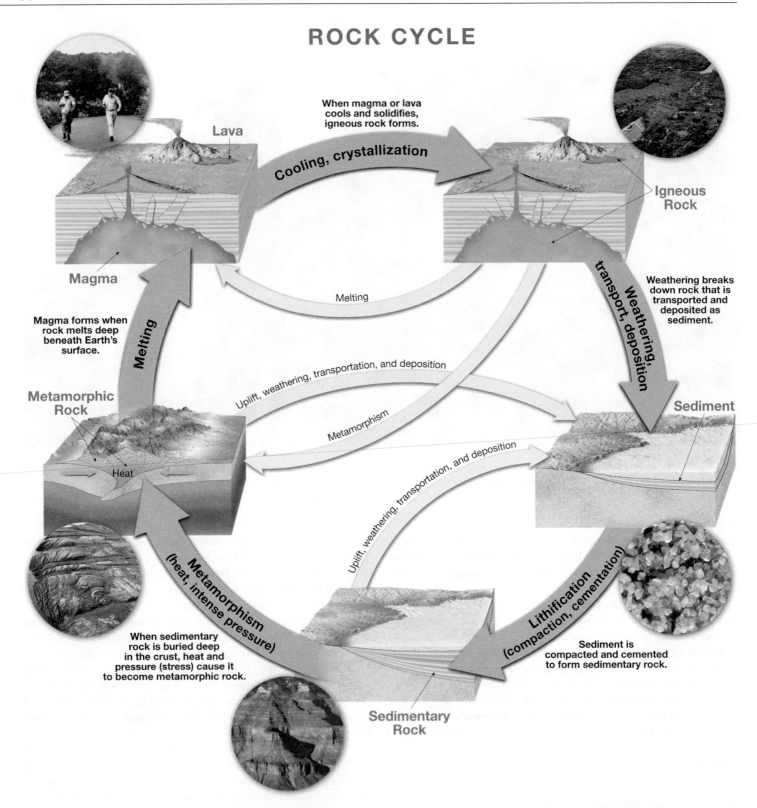

When magma or lava cools and solidifies, igneous rock forms.

Cooling, crystallization

Lava

Magma

Igneous Rock

Magma forms when rock melts deep beneath Earth's surface.

Melting

Melting

Weathering breaks down rock that is transported and deposited as sediment.

Weathering, transport, deposition

Metamorphic Rock

Heat

Uplift, weathering, transportation, and deposition

Metamorphism

Sediment

Uplift, weathering, transportation, and deposition

Metamorphism (heat, intense pressure)

When sedimentary rock is buried deep in the crust, heat and pressure (stress) cause it to become metamorphic rock.

Lithification (compaction, cementation)

Sediment is compacted and cemented to form sedimentary rock.

Sedimentary Rock

FIGURE 3.2 The rock cycle. Viewed over long spans, rocks are constantly forming, changing, and reforming. The rock cycle helps us understand the origin of the three basic rock groups. Arrows represent processes that link each group to the others.

Igneous Rocks: "Formed by Fire"

Earth Materials
▶ **Igneous Rocks**

In the discussion of the rock cycle, we pointed out that **igneous rocks** form as *magma* cools and crystallizes. But what is magma and what is its source? **Magma** is molten rock generated by partial melting of rocks in Earth's mantle and in the lower crust in smaller amounts. This molten material consists mainly of the elements found in silicate minerals. Silicon and oxygen are the main constituents in magma, with lesser amounts of aluminum, iron, calcium, sodium, potassium, magnesium, and others. Magma also contains some gases, particularly water vapor, which are confined within the magma body by the weight (pressure) of the overlying rocks.

Once formed, a magma body buoyantly rises toward the surface because it is less dense than the surrounding rocks. Occasionally molten rock reaches the surface, where it is called **lava** (Figure 3.3). Sometimes, lava is emitted as fountains produced when escaping gases propel molten rock skyward. On other occasions, magma is explosively ejected from vents, producing a spectacular eruption such as the 1980 eruption of Mount St. Helens. However, most eruptions are not violent; rather, volcanoes more often emit quiet outpourings of lava.

Igneous rocks that form when molten rock solidifies *at the surface* are classified as **extrusive** or **volcanic** (after the Roman fire god Vulcan). Extrusive igneous rocks are abundant in western portions of the Americas, including the volcanic cones of the Cascade Range and the extensive lava flows of the Columbia Plateau. In addition, many oceanic islands, including the Hawaiian Islands, are composed almost entirely of volcanic igneous rocks.

Most magma, however, loses its mobility before reaching the surface and eventually crystallizes deep below the surface. Igneous rocks that *form at depth* are termed **intrusive** or **plutonic** (after Pluto, the god of the lower world in classical mythology). Intrusive igneous rocks remain at depth unless portions of the crust are uplifted and the overlying rocks stripped away by erosion. Exposures of intrusive igneous rocks occur in many places, including Mount Washington, New Hampshire; Stone Mountain, Georgia; Mount Rushmore in the Black Hills of South Dakota; and Yosemite National Park, California (Figure 3.4).

From Magma to Crystalline Rock

Magma is a very hot, thick fluid that contains solids (mineral crystals) and gases. The liquid portion of a magma body is composed of atoms that move about freely. As magma cools, random movements slow, and atoms begin to arrange themselves into orderly patterns—a process called *crystallization*. As cooling continues, numerous small crystals develop and

FIGURE 3.3 Fluid basaltic lava emitted from Hawaii's Kilauea Volcano. (Photo by Joe Carini/Photolibrary)

atoms are systematically added to these centers of crystal growth. When the crystals grow large enough for their edges to meet, their growth ceases for lack of space. Eventually, all of the liquid is transformed into a solid mass of interlocking crystals.

The rate of cooling strongly influences crystal size. If a magma cools very slowly, it allows atoms to migrate over great distances. Consequently, *slow cooling results in the formation of large crystals*. On the other hand, if cooling occurs rapidly, the atoms lose their motion and quickly combine. This results in a large number of tiny crystals that all compete for the available atoms. Therefore, *rapid cooling results in the formation of a solid mass of small intergrown crystals*.

If a geologist encounters igneous rock containing crystals large enough to be seen with the unaided eye, it means it formed from molten rock which cooled slowly at depth. But if the crystals can be seen clearly only with a microscope, the geologist knows that the magma cooled quickly, at or near Earth's surface.

FIGURE 3.4 Mount Rushmore National Memorial, located in the Black Hills of South Dakota, is carved from the intrusive igneous rock granite. This massive igneous body cooled very slowly at depth and has since been uplifted and the overlying rocks stripped away by erosion. (Photo by Barbara A. Harvey/Shutterstock)

Students Sometimes Ask...

Are lava and magma the same thing?

No, but their *composition* might be similar. Both are terms that describe molten or liquid rock: Magma exists beneath Earth's surface, and lava is molten rock that has reached the surface.

That's the reason they can be similar in composition: Lava is produced from magma, but it generally has lost materials that escape as a gas, such as water vapor.

If the molten material is quenched almost instantly, there is insufficient time for the atoms to arrange themselves into a crystalline network. Solids produced in this manner consist of randomly distributed atoms. Such rocks are called *glass* and are quite similar to ordinary manufactured glass. "Instant" quenching sometimes occurs during violent volcanic eruptions that produce tiny shards of glass called *volcanic ash.*

In addition to the rate of cooling, the composition of a magma and the amount of dissolved gases influence crystallization. Because magmas differ in each of these aspects, the physical appearance and mineral composition of igneous rocks vary widely. Nevertheless, it is possible to classify igneous rocks based on their *texture* and *mineral composition.*

What Can Igneous Textures Tell Us?

Texture describes the overall appearance of an igneous rock, based on the *size* and *arrangement* of its interlocking crystals. Texture is an important property because it allows geologists to make inferences about a rock's origin based on careful observations of crystal size and other characteristics. You learned that rapid cooling produces small crystals, whereas very slow cooling produces much larger crystals. As you might expect, the rate of cooling is slow in magma chambers lying deep within the crust, whereas a thin layer of lava extruded upon Earth's surface may chill to form solid rock in a matter of hours. Small molten blobs ejected from a volcano during a violent eruption can solidify in mid-air.

Igneous rocks that form rapidly at the surface or as small masses within the upper crust have a **fine-grained texture**, with the individual crystals too small to be seen with the unaided eye (Figure 3.5B). Common in many fine-grained igneous rocks are voids, called *vesicles,* left by gas bubbles that formed as the lava solidified (Figure 3.6). Rocks containing these voids are said to display a **vesicular texture**.

When large masses of magma solidify far below the surface, they form igneous rocks that exhibit a **coarse-grained texture**. These coarse-grained rocks have the appearance of a mass of intergrown crystals, roughly equal in size and large enough that

FIGURE 3.5 Igneous rock textures. **A.** During a volcanic eruption in which silica-rich lava is ejected into the atmosphere, a frothy glass called pumice may form. **B.** Igneous rocks that form at or near Earth's surface cool quickly and often exhibit a fine-grained texture. **C.** A porphyritic texture results when magma that already contains some large crystals migrates to a new location where the rate of cooling increases. The resulting rock consists of large crystals embedded within a matrix of smaller crystals. **D.** Coarse-grained igneous rocks form when magma slowly crystallizes at depth. (Photos courtesy of E. J. Tarbuck)

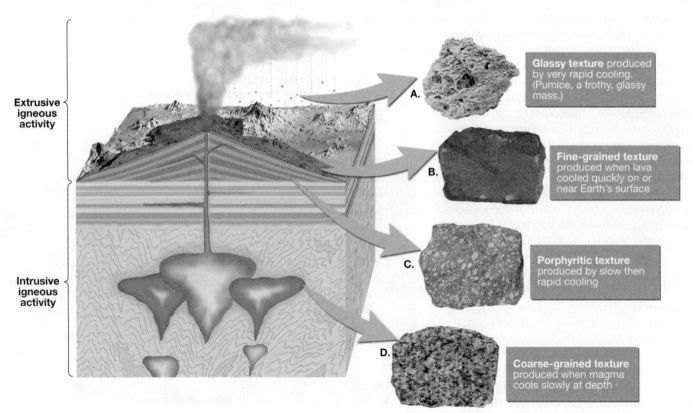

Extrusive igneous activity

Intrusive igneous activity

A. Glassy texture produced by very rapid cooling. (Pumice, a frothy, glassy mass.)

B. Fine-grained texture produced when lava cooled quickly on or near Earth's surface

C. Porphyritic texture produced by slow then rapid cooling

D. Coarse-grained texture produced when magma cools slowly at depth

FIGURE 3.6 Scoria is a volcanic rock that is vesicular. Vesicles form as gas bubbles escape near the top of a lava flow. (Photo courtesy of E. J. Tarbuck)

individual minerals can be identified with the unaided eye. Granite is a classic example (Figure 3.5D).

A large mass of magma located at depth may require tens of thousands, even millions, of years to solidify. Because materials crystallize under different environmental conditions (temperature, pressure), it is possible for crystals of one mineral to become quite large before others even start to form. Should molten rock containing some large crystals move to a different environment—for example, by erupting at the surface—the remaining molten portion of the lava would cool more quickly. The resulting rock, which will have large crystals embedded in a matrix of smaller crystals, is said to have a **porphyritic texture** (Figure 3.5C).

During some volcanic eruptions, molten rock is ejected into the atmosphere, where it is quenched quickly. Rapid cooling may generate rock having a **glassy texture** (Figure 3.5A). Glass results when unordered atoms are "frozen in place" before they are able to unite into an orderly crystalline structure. In addition, magmas containing large amounts of silica (SiO_2) are more likely to form rocks that exhibit a glassy texture than those with a low silica content.

Obsidian, a common type of natural glass, is similar in appearance to a dark chunk of manufactured glass (Figure 3.7). Another volcanic rock that usually exhibits a glassy texture is *pumice*. Often found with obsidian, pumice forms when large amounts of gas escape from molten rock to generate a gray, frothy mass (Figure 3.8). In some samples, the vesicles are quite noticeable, whereas in others, the pumice resembles fine shards of intertwined glass. Because of the large volume of air-filled voids, many samples of pumice will float in water (Figure 3.8).

Igneous Compositions

Igneous rocks are composed mainly of silicate minerals. Chemical analysis shows that silicon and oxygen—usually expressed as the silica (SiO_2) content of a magma—are by far the most abundant

Students Sometimes Ask...

You mentioned that Native Americans used obsidian for making arrowheads and cutting tools. Is this the only material they used?

No. Native Americans used whatever materials were locally available to make tools, including any hard, compact rock material that could be shaped. This includes materials such as the metamorphic rocks slate and quartzite and sedimentary deposits made of silica called jasper, chert, opal, flint, and even jade. Some of these deposits have a limited geographic distribution and can now help anthropologists reconstruct trade routes between different groups of Indians.

constituents of igneous rocks. These two elements, plus ions of aluminum (Al), calcium (Ca), sodium (Na), potassium (K), magnesium (Mg), and iron (Fe), make up roughly 98 percent by weight of most magmas. In addition, magma contains small amounts of many other elements, including titanium and manganese, and trace amounts of much rarer elements such as gold, silver, and uranium.

As magma cools and solidifies, these elements combine to form two major groups of silicate minerals. The *dark silicates* are rich in iron and/or magnesium and are relatively low in silica (SiO_2). *Olivine, pyroxene, amphibole,* and *biotite mica* are the common dark silicate minerals of Earth's crust. By contrast, the *light silicates* contain greater amounts of potassium, sodium, and calcium and are richer in silica than dark silicates. Light silicates include *quartz, muscovite mica,* and the most abundant mineral group, the *feldspars*. Feldspars make up at least 40 percent of most igneous rocks. Thus, in addition to feldspar, igneous rocks contain some combination of the other light and/or dark silicates listed earlier.

FIGURE 3.7 Obsidian, a natural glass, was used by Native Americans for making arrowheads and cutting tools. (Photo by E. J. Tarbuck; inset photo by Jeffrey Scovil)

FIGURE 3.8 Pumice, a glassy rock, is very lightweight because it contains numerous vesicles. (Photo by E. J. Tarbuck; inset photo by Chip Clark)

Classifying Igneous Rocks

Igneous rocks are classified by their texture and mineral composition. The texture of an igneous rock is mainly a result of its cooling history, whereas its mineral composition is largely a consequence of the chemical makeup of the parent magma and the environment of crystallization.

Despite their great compositional diversity, igneous rocks can be divided into broad groups according to their proportions of light and dark minerals. A general classification scheme based on texture and mineral composition is provided in **Figure 3.9**.

Granitic (Felsic) Rocks　Near one end of the continuum are rocks composed almost entirely of light-colored silicates—quartz and potassium feldspar. Igneous rocks in which these are the dominant minerals have a **granitic composition**. Geologists also refer to granitic rocks as being **felsic**, a term derived from *fel*dspar and *si*lica (quartz). In addition to quartz and feldspar, most granitic rocks contain about 10 percent dark silicate minerals, usually biotite mica and amphibole. Granitic rocks are rich in silica (about 70 percent) and are major constituents of the continental crust.

Granite is a coarse-grained igneous rock that forms where large masses of magma slowly solidify at depth. During episodes of mountain building, granite and related crystalline rocks may be uplifted, where the processes of weathering and erosion strip away the overlying crust. Areas where large quantities of granite are exposed at the surface include Pikes Peak in the Rockies, Mount Rushmore in the Black Hills, Stone Mountain in Georgia, and Yosemite National Park in the Sierra Nevada (**Figure 3.10**).

Granite is perhaps the best-known igneous rock (**Figure 3.11**) in part because of its natural beauty, which is enhanced when polished, and partly because of its abundance. Slabs of polished granite are commonly used for tombstones, monuments, and countertops.

Rhyolite is the extrusive equivalent of granite and, likewise, is composed essentially of light-colored silicates (Figure 3.11). This fact accounts for its color, which is usually buff to pink or light gray. Rhyolite is fine-grained and frequently contains glass fragments and voids, indicating rapid cooling in a surface environment. In contrast to granite, which is widely distributed as large intrusive masses, rhyolite deposits are less common and generally less voluminous. Yellowstone Park is one well-known exception where extensive lava flows and thick ash deposits of rhyolitic composition are found.

FIGURE 3.9 Classification of the major groups of igneous rocks based on their mineral composition and texture. Coarse-grained rocks are plutonic, solidifying deep underground. Fine-grained rocks are volcanic, or solidify as shallow, thin plutons. Ultramafic rocks are dark, dense rocks, composed almost entirely of minerals containing iron and magnesium. Although relatively rare on Earth's surface, these rocks are believed to be major constituents of the upper mantle.

Chemical Composition		Granitic (Felsic)	Andesitic (Intermediate)	Basaltic (Mafic)	Ultramafic
Dominant Minerals		Quartz Potassium feldspar Sodium-rich plagioclase feldspar	Amphibole Sodium- and calcium-rich plagioclase feldspar	Pyroxene Calcium-rich plagioclase feldspar	Olivine Pyroxene
TEXTURE	Coarse-grained	 Granite	Diorite	Gabbro	Peridotite
	Fine-grained	 Rhyolite	Andesite	Basalt	Komatiite (rare)
	Porphyritic	 "Porphyritic" precedes any of the above names whenever there are appreciable phenocrysts			Uncommon
	Glassy	 Obsidian (compact glass) Pumice (frothy glass)			
Rock Color (based on % of dark minerals)		0% to 25%	25% to 45%	45% to 85%	85% to 100%

FIGURE 3.10 Half Dome and the surrounding rock in California's Yosemite National Park is granite. The rock formed from magma that crystallized deep beneath the surface. (Photo by Enrique R. Aguirre/Photolibrary; inset photo by E. J. Tarbuck)

FIGURE 3.11 Common igneous rocks. (Photos by E. J. Tarbuck)

Texture	Composition		
	Granitic (Felsic)	**Andesitic** (Intermediate)	**Basaltic** (Mafic)
Course-grained (Intrusive)	Granite	Diorite	Gabbro
Fine-grained (Extrusive)	Rhyolite	Andesite	Basalt

Basaltic (Mafic) Rocks   Rocks that contain substantial dark silicate minerals and calcium-rich plagioclase feldspar (but no quartz) are said to have a **basaltic composition** (Figure 3.11). Basaltic rocks contain a high percentage of dark silicate minerals, so geologists also refer to them as **mafic** (from *ma*gnesium and *fe*rrum, the Latin name for iron). Because of their iron content, basaltic rocks are typically darker and denser than granitic rocks.

Basalt, the most common extrusive igneous rock, is a very dark green to black, fine-grained volcanic rock composed primarily of pyroxene, olivine, and plagioclase feldspar. Many volcanic islands, such as the Hawaiian Islands and Iceland, are composed mainly

FIGURE 3.12   Basaltic lava flowing from Kilauea volcano, Hawaii. (Photo by Roger Ressmeyer/CORBIS)

of basalt (Figure 3.12). Furthermore, the upper layers of the oceanic crust consist of basalt. In the United States, large portions of central Oregon and Washington were the sites of extensive basaltic outpourings.

The coarse-grained, intrusive equivalent of basalt is *gabbro* (Figure 3.11). Although gabbro is not commonly exposed at the surface, it makes up a significant percentage of the oceanic crust.

Andesitic (Intermediate) Rocks   As you can see in Figure 3.11, rocks with a composition between granitic and basaltic rocks are said to have an **andesitic composition**, after the common volcanic rock *andesite,* or **intermediate composition**. Andesitic rocks contain a mixture of both light- and dark-colored minerals, mainly amphibole and plagioclase feldspar. This important category of igneous rocks is associated with volcanic activity typically confined to continental margins. When magma of intermediate composition crystallizes at depth, it forms the coarse-grained rock called *diorite* (Figure 3.11).

Ultramafic Rocks   Another important igneous rock, *peridotite,* contains mostly the dark-colored minerals olivine and pyroxene and thus, falls on the opposite side of the compositional spectrum from granitic rocks (Figure 3.11). Because peridotite is composed almost entirely of dark silicate minerals, its chemical composition is referred to as **ultramafic**. Although ultramafic rocks are rare at Earth's surface, peridotite is believed to be the main constituent of the upper mantle.

How Different Igneous Rocks Form

Because a large variety of igneous rocks exist, it is logical to assume that an equally large variety of magmas also exist. However, geologists have observed that a single volcano may extrude lavas exhibiting quite different compositions. Data of this type led them to examine the possibility that magma might change (evolve) and thus become the parent to a variety of igneous rocks. To explore this idea, a pioneering investigation into the crystallization of magma was carried out by N. L. Bowen in the first quarter of the 20th century.

Bowen's Reaction Series   In a laboratory setting, Bowen demonstrated that magma, with its diverse

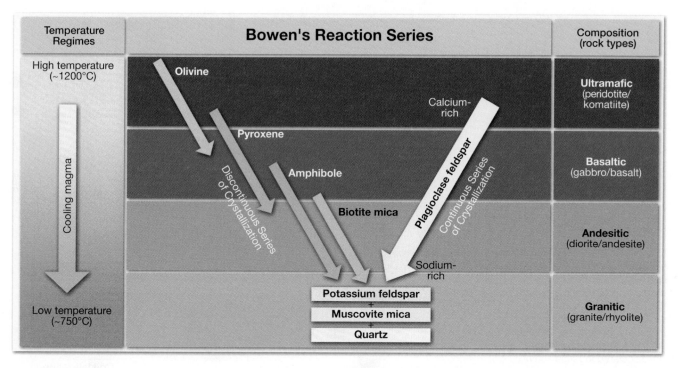

FIGURE 3.13 Bowen's reaction series shows the sequence in which minerals crystallize from a magma. Compare this figure to the mineral composition of the rock groups in Figure 3.9. Note that each rock group consists of minerals that crystallize in the same temperature range.

chemistry, crystallizes over a temperature range of at least 200°C, unlike simple compounds (such as water), which solidify at specific temperatures. As magma cools, certain minerals crystallize first at relatively high temperatures (top of Figure 3.13). At successively lower temperatures, other minerals begin to crystallize. This arrangement of minerals, shown in Figure 3.13, became known as **Bowen's reaction series**.

Bowen discovered that the first mineral to crystallize from a body of magma is *olivine.* Further cooling results in the formation of *pyroxene,* as well as *plagioclase feldspar.* At intermediate temperatures the minerals *amphibole* and *biotite* begin to crystallize.

During the last stage of crystallization, after most of the magma has solidified, the minerals *muscovite* and *potassium feldspar* may form (Figure 3.13). Finally, *quartz* crystallizes from any remaining liquid. Olivine and quartz are seldom found in the same igneous rock, because quartz crystallizes at much lower temperatures than olivine.

Analysis of igneous rocks provides evidence that this crystallization model approximates what can happen in nature. In particular, we find that minerals that form in the same general temperature range on Bowen's reaction series are found together in the same igneous rocks. For example, notice in Figure 3.13 that the minerals quartz, potassium feldspar, and muscovite, located in the same region of Bowen's diagram, are typically found together as major constituents of the igneous rock *granite.*

Magmatic Differentiation Bowen demonstrated that different minerals crystallize from magma systematically. But how do Bowen's findings account for the great diversity of igneous rocks? During the crystallization process, the composition of magma continually changes. This occurs because as crystals form, they selectively remove certain elements from the magma, which

leaves the remaining liquid portion depleted in these elements. Occasionally, separation of the solid and liquid components of magma occurs during crystallization, which creates different mineral assemblages. One such scenario, called **crystal settling**, occurs when the earlier formed minerals are more dense (heavier) than the liquid portion and sink toward the bottom of the magma chamber, as shown in Figure 3.14. When the remaining molten material solidifies—either in place or in another location if it migrates into fractures in the surrounding rocks—it will form a rock with a chemical composition much different from the parent magma (Figure 3.14). The formation of one or more secondary magmas from a single parent magma is called **magmatic differentiation**.

At any stage in the evolution of a magma, the solid and liquid components can separate into two chemically distinct units. Furthermore, magmatic differentiation within the secondary magma can generate other chemically distinct masses of molten rock. Consequently, magmatic differentiation and separation of the solid and liquid components at various stages of crystallization can produce several chemically diverse magmas and, ultimately, a variety of igneous rocks.

Assimilation and Magma Mixing

Strong evidence suggests that the chemical composition of a magma can change by processes other than magmatic differentiation. For example, as magma migrates upward through the crust, it may incorporate some of the surrounding host rock, a process called *assimilation.* Another means by which the composition of a magma body can be altered is called *magma mixing.* This process occurs whenever one magma body intrudes another that has a different composition. Once combined, convective flow may stir the two magmas to generate a fluid with a different composition.

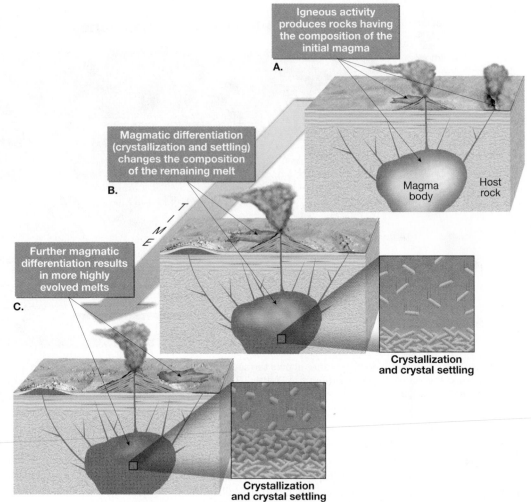

FIGURE 3.14 Illustration of how a magma evolves as the earlier formed minerals (those richer in iron, magnesium, and calcium) crystallize and settle to the bottom of the magma chamber, leaving the remaining liquid richer in sodium, potassium, and silica (SiO_2). **A.** Emplacement of a magma body and associated igneous activity generates rocks having a composition similar to that of the initial magma. **B.** After a period of time, crystallization and settling change the composition of the liquid, while generating rocks having a composition quite different from the original magma. **C.** Further magmatic differentiation results in another more highly evolved mass of molten material with its associated rock types.

In summary, N. L. Bowen successfully demonstrated that through magmatic differentiation, a single parent magma can generate several mineralogically different igneous rocks. This process, in concert with magma mixing and contamination by crustal rocks, accounts in part for the great diversity of magmas and igneous rocks.

CONCEPT CHECK 3.2

1. What is magma? How does magma differ from lava?
2. In what basic settings do intrusive and extrusive igneous rocks originate?
3. How does the rate of cooling influence crystal size? What other factors influence the texture of igneous rocks?
4. What does a porphyritic texture indicate about the history of an igneous rock?
5. List and distinguish among the four basic compositional groups of igneous rocks.
6. How are granite and rhyolite different? In what way are they similar?
7. What is magmatic differentiation? How might this process lead to the formation of several different igneous rocks from a single magma?

Sedimentary Rocks: Compacted and Cemented Sediment

 Earth Materials
▶ **Sedimentary Rocks**

Recall the rock cycle, which shows the origin of **sedimentary rocks**. Weathering begins the process. Next, gravity and erosional agents (running water, wind, waves, and glacial ice) remove the products of weathering and carry them to a new location where they are deposited. Usually, the particles are broken down further during this transport phase. Following deposition, this **sediment** may become lithified, or "turned to rock." Commonly, *compaction* and *cementation* transform the sediment into solid sedimentary rock.

The word *sedimentary* indicates the nature of these rocks, for it is derived from the Latin *sedimentum,* which means "settling," a reference to a solid material settling out of a fluid. Most sediment is deposited in this fashion. Weathered debris is constantly being swept from bedrock and carried away by water, ice, or wind. Eventually, the material is deposited in lakes, river valleys, seas, and countless other places. The particles in a desert sand dune, the mud on the floor of a swamp, the gravels in a streambed, and even household dust are examples of sediment produced by this never-ending process.

The weathering of bedrock and the transport and deposition of the weathering products are continuous. Therefore, sediment is found almost everywhere. As piles of sediment accumulate, the materials near the bottom are compacted by the weight of the overlying layers. Over long periods, these sediments are cemented together by mineral matter deposited from water in the spaces between particles. This forms solid sedimentary rock.

Geologists estimate that sedimentary rocks account for only about 5 percent (by volume) of Earth's outer 16 kilometers (10 miles). However, the importance of this group of rocks is far greater than this percentage implies. If you sampled the rocks exposed at Earth's surface, you would find that the great majority are sedimentary (Figure 3.15). Indeed, about 75 percent of all rock outcrops on the continents are sedimentary. Therefore, we can think of sedimentary rocks as comprising a relatively thin and somewhat discontinuous layer in the uppermost portion of the crust. This makes sense because sediment accumulates at the surface.

It is from sedimentary rocks that geologists reconstruct many details of Earth's history. Because sediments are deposited in a variety of different settings at the surface, the rock layers that they eventually form hold many clues to past surface environments. They may also exhibit characteristics that allow geologists to decipher information about the method and distance of sediment

Students Sometimes Ask...

Why are the sedimentary rocks in Figure 3.15 so colorful?

In the western and southwestern United States, steep cliffs and canyon walls made of sedimentary rocks often exhibit a brilliant display of different colors. Utah's Canyonlands National Park is one good example (Figure 3.15). Another is Arizona's Grand Canyon where we can see layers that may be red, orange, purple, gray, brown, and buff (see Figure 11.2, p. 325). Sedimentary rocks in more humid places are also colorful, but they are usually covered by soil and vegetation.

The most important "pigments" are iron oxides, and only very small amounts are needed to color a rock. Hematite tints rocks red or pink, whereas limonite produces shades of yellow and brown. When sedimentary rocks contain organic matter, it often colors them black or gray.

transport. Furthermore, it is sedimentary rocks that contain fossils, which are vital evidence in the study of the geologic past.

Finally, many sedimentary rocks are important economically. Coal, which is burned to provide a significant portion of U.S. elec-

FIGURE 3.15 Sedimentary rocks exposed in Utah's Canyonlands National Park. Sedimentary rocks occur in layers called *strata*. About 75 percent of all rock exposures on continents are sedimentary rocks. (Photo by Jeff Gnass)

Detrital Sedimentary Rocks		
Clastic Texture (particle size)	Sediment Name	Rock Name
Coarse (over 2 mm)	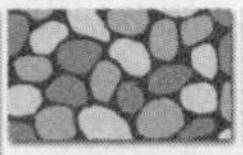 Gravel (Rounded particles)	Conglomerate
 Coarse (over 2 mm)	Gravel (Angular particles)	Breccia
Medium (1/16 to 2 mm)	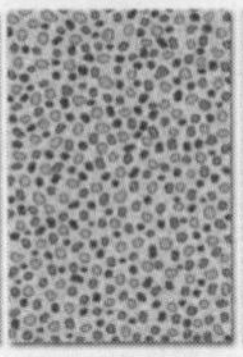 Sand (If abundant feldspar is present the rock is called Arkose)	Sandstone
Fine (1/16 to 1/256 mm)	Mud	Siltstone
Very fine (less than 1/256 mm)	Mud	Shale or Mudstone

Chemical and Organic Sedimentary Rocks			
Composition	Texture	Rock Name	
Calcite, CaCO₃	Nonclastic: Fine to coarse crystalline	Crystalline Limestone	
Calcite, CaCO₃	Nonclastic: Fine to coarse crystalline	Travertine	
Calcite, CaCO₃	Clastic: Visible shells and shell fragments loosely cemented	Coquina	Biochemical Limestone
Calcite, CaCO₃	Clastic: Various size shells and shell fragments cemented with calcite cement	Fossiliferous Limestone	Biochemical Limestone
Calcite, CaCO₃	Clastic: Microscopic shells and clay	Chalk	Biochemical Limestone
Quartz, SiO₂	Nonclastic: Very fine crystalline	Chert (light colored) Flint (dark colored)	
Gypsum CaSO₄•2H₂O	Nonclastic: Fine to coarse crystalline	Rock Gypsum	
Halite, NaCl	Nonclastic: Fine to coarse crystalline	Rock Salt	
Altered plant fragments	Nonclastic: Fine-grained organic matter	Bituminous Coal	

FIGURE 3.16 Identification of sedimentary rocks. Sedimentary rocks are divided into two major groups, detrital and chemical, based on their source of sediment. The main criterion for naming detrital rocks is particle size, whereas the primary basis for distinguishing among chemical rocks is their mineral composition.

trical energy, is classified as a sedimentary rock. Other major energy resources (petroleum and natural gas) occur in pores within sedimentary rocks. Other sedimentary rocks are major sources of iron, aluminum, manganese, and fertilizer, plus numerous materials essential to the construction industry.

Classifying Sedimentary Rocks

Materials accumulating as sediment have two principal sources. First, sediments may originate as solid particles from weathered rocks, such as the igneous rocks earlier described. These particles are called *detritus,* and the sedimentary rocks that they form are called **detrital sedimentary rocks** (Figure 3.16).

The second major source of sediment is soluble material produced largely by chemical weathering. When these dissolved substances are precipitated back as solids, they are called *chemical sediment,* and they form **chemical sedimentary rocks**. We now look at detrital and chemical sedimentary rocks (Figure 3.16).

Detrital Sedimentary Rocks   Though a wide variety of minerals and rock fragments may be found in detrital rocks, clay minerals and quartz dominate. As you learned earlier, clay minerals are the most abundant product of the chemical weathering of silicate minerals, especially the feldspars. Quartz, on the other hand, is abundant because it is extremely durable and very resistant to chemical weathering. Thus, when igneous rocks such as granite are weathered, individual quartz grains are set free.

Geologists use particle size to distinguish among detrital sedimentary rocks. Figure 3.16 presents the four size categories for particles making up detrital rocks. When gravel-size particles predominate, the rock is called *conglomerate* if the sediment is rounded (Figure 3.17A) and *breccia* if the pieces are angular (Figure 3.17B).

Angular fragments indicate that the particles were not transported very far from their source prior to deposition and so have not had corners and rough edges abraded. *Sandstone* is the name given rocks when sand-size grains prevail (Figure 3.18). *Shale,* the most common sedimentary rock, is made of very fine-grained sediment (Figure 3.19). *Siltstone,* another rather fine-grained rock, is sometimes difficult to differentiate from rocks such as shale, which are composed of even smaller clay-size sediment.

Particle size is not only a convenient method of dividing detrital rocks; the sizes of the component grains also provide useful information about the environment in which the sediment was deposited. Currents of water or air sort the particles by size. The stronger the current, the larger the particle size carried. Gravels, for example, are moved by swiftly flowing rivers, rockslides, and glaciers. Less energy is required to transport sand; thus, it is common in windblown dunes, river deposits, and beaches. Because silts and clays settle very slowly, accumulations of these materials are generally associated with the quiet waters of a lake, lagoon, swamp, or marine environment.

Although detrital sedimentary rocks are classified by particle size, in certain cases the mineral composition is also part of naming a rock. For example, most sandstones are predominantly quartz-rich, and they are often referred to as quartz sandstone. In addition, rocks consisting of detrital sediments are rarely composed of grains of just one size. Consequently, a rock containing quantities of both sand and silt can be correctly classified as sandy siltstone or silty sandstone, depending on which particle size dominates.

A.

B.

FIGURE 3.17 Detrital rocks made up of gravel-sized particles.
A. *Conglomerate* (rounded particles) **B.** *Breccia* (angular particles).
(Photos by E. J. Tarbuck)

Chemical Sedimentary Rocks   In contrast to detrital rocks, which form from the solid products of weathering, chemical sediments are derived from material that is carried in solution to lakes and seas. This material does not remain dissolved in the water indefinitely. When conditions are right, it precipitates to form chemical sediments. This precipitation may occur directly as the result of physical processes, or indirectly through life processes of water-dwelling organisms. Sediment formed in this second way has a *biochemical* origin.

An example of a deposit resulting from physical processes is the salt left behind as a body of saltwater evaporates. In contrast, many water-dwelling animals and plants extract dissolved mineral matter to form shells and other hard parts. After the organisms die, their skeletons may accumulate on the floor of a lake or ocean.

Limestone is the most abundant chemical sedimentary rock. It is composed chiefly of the mineral calcite ($CaCO_3$). Ninety per-

FIGURE 3.18 Quartz sandstone. After shale, sandstone is the most abundant sedimentary rock. (Photos by E. J. Tarbuck)

Close up

FIGURE 3.19 Shale is a fine-grained detrital rock that is, by far, the most abundant of all sedimentary rocks. Dark shales containing plant remains are relatively common. (Photo courtesy of E. J. Tarbuck)

cent of limestone is biochemical sediment, while the remaining amount chemically precipitates from seawater.

One easily identified biochemical limestone is *coquina,* a coarse rock composed of loosely cemented shells and shell fragments (**Figure 3.20**). Another less obvious but familiar example is

FIGURE 3.20 Biochemical sedimentary rocks. **A.** This rock, called *coquina,* consists of shell fragments; therefore, it has a biochemical origin. (Photo by E. J. Tarbuck) **B.** Seashells cover the beach at Sanibel Island, Florida. (Photo by Donald R. Frazier/Photolibrary, Inc./Alamy)

$\longleftarrow$ 5 cm $\longrightarrow$

A.

A.

jasper (red), and agate (banded). These chemical sedimentary rocks may have either an inorganic or biochemical origin, but the mode of origin is usually difficult to determine.

Very often, evaporation causes minerals to precipitate from water. Such minerals include halite, the chief component of *rock salt,* and gypsum, the main ingredient of *rock gypsum.* Both materials have significant commercial importance. Halite is familiar to everyone as the common salt used in cooking and seasoning foods. Of course, it has many other uses and has been considered important enough that people have sought, traded, and fought over it for much of human history.

Gypsum is the basic ingredient of plaster of Paris. This material is used most extensively in the construction industry for "drywall" and plaster.

In the geologic past, many areas that are now dry land were covered by shallow arms of the sea that had only narrow

FIGURE 3.22 *Chert* is a name used for a number of dense, hard rocks made of microcrystalline quartz. Three examples are shown here. **A.** *Agate* is the banded variety. (Photo by Jeffrey A. Scovil) **B.** The dark color of *flint* results from organic matter. (Photo by E. J. Tarbuck) **C.** The red variety, called *jasper,* gets its color from iron oxide. (Photo by E. J. Tarbuck) **D.** Native Americans frequently made arrowheads and sharp tools from chert. (Photo by LA VENTA/CORBIS/SYGMA)

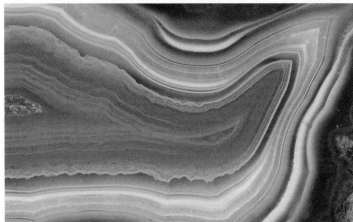

A. Agate

FIGURE 3.21 Chalk is a soft, porous limestone consisting mainly of hard parts of microscopic organisms. **A.** White Chalk Cliffs, East Sussex, England. Thick deposits of chalk are formed in scattered sites in much of western Europe, and were deposited in a marine environment during the Cretacous period. (Photo by Prisma/Superstock) **B.** The majority of fossil debris comprising chalk deposits consists of microscopic plates, called coccoliths, that are mainly dispersed following the death of single-celled algae. (Steve Gschmeissner/Photo Researchers, Inc.)

chalk, a soft, porous rock made up almost entirely of the hard parts of microscopic organisms that are no larger than the head of a pin. Among the most famous chalk deposits are the White Chalk Cliffs exposed along the southeast coast of England (**Figure 3.21**).

Inorganic limestone forms when chemical changes or high water temperatures increase the concentration of calcium carbonate to the point that it precipitates. *Travertine,* the type of limestone that decorates caverns, is one example. Groundwater is the source of travertine that is deposited in caves. As water drops reach the air in a cavern, some of the carbon dioxide dissolved in the water escapes, causing calcium carbonate to precipitate.

Dissolved silica (SiO_2) precipitates to form varieties of microcrystalline quartz (**Figure 3.22**). Sedimentary rocks composed of microcrystalline quartz include chert (light color), flint (dark),

B. Flint   C. Jasper   D. Arrowhead

connections to the open ocean. Under these conditions, water continually moved into the bay to replace water lost by evaporation. Eventually, the waters of the bay became saturated and salt deposition began. Today, these arms of the sea are gone, and the remaining deposits are called **evaporite deposits**.

On a smaller scale, evaporite deposits can be seen in such places as Death Valley, California. Here, following rains or periods of snowmelt in the mountains, streams flow from surrounding mountains into an enclosed basin. As the water evaporates, *salt flats* form from dissolved materials left behind as a white crust on the ground (Figure 3.23).

Coal is quite different from other chemical sedimentary rocks. Unlike other rocks in this category, which are calcite or silica-rich, coal is made mostly of organic matter. Close examination of a piece of coal under a microscope or magnifying glass often reveals plant structures such as leaves, bark, and wood that have been chemically altered but are still identifiable. This supports the conclusion that coal is the end product of the burial of large amounts of plant material over extended periods (Figure 3.24).

The initial stage in coal formation is the accumulation of large quantities of plant remains. However, special conditions are required for such accumulations, because dead plants normally decompose when exposed to the atmosphere. An ideal environment that allows for the buildup of plant material is a swamp. Because stagnant swamp water is oxygen-deficient, complete decay (oxidation) of the plant material is not possible. At various times during Earth history, such environments have been common. Coal undergoes successive stages of formation. With each successive stage, higher temperatures and pressures drive off impurities and volatiles, as shown in Figure 3.24.

Lignite and bituminous coals are sedimentary rocks, but anthracite is a metamorphic rock. Anthracite forms when

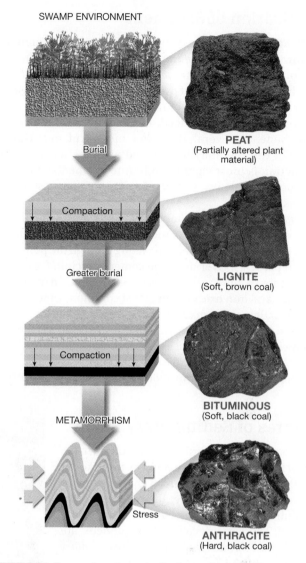

SWAMP ENVIRONMENT

Burial

PEAT
(Partially altered plant material)

Compaction

Greater burial

LIGNITE
(Soft, brown coal)

Compaction

METAMORPHISM

BITUMINOUS
(Soft, black coal)

Stress

ANTHRACITE
(Hard, black coal)

FIGURE 3.24 Successive stages in the formation of coal.

FIGURE 3.23 Bonneville Salt Flats, Utah. (Photo by Nash Photos/Photolibrary)

sedimentary layers are subjected to the folding and deformation associated with mountain building.

In summary, we divide sedimentary rocks into two major groups: detrital and chemical. The main criterion for classifying detrital rocks is particle size, whereas chemical rocks are distinguished by their mineral composition. The categories presented here are more rigid than is the actual state of nature. Many detrital sedimentary rocks are a mixture of more than one particle size. Furthermore, many sedimentary rocks classified as chemical also contain at least small quantities of detrital sediment, and practically all detrital rocks are cemented with material that was originally dissolved in water.

Lithification of Sediment

Lithification refers to the processes by which sediments are transformed into solid sedimentary rocks. One of the most common processes is *compaction.* As sediments accumulate through time, the weight of overlying material compresses the deeper sediments. As the grains are pressed closer and closer, pore space is greatly reduced. For example, when clays are buried beneath several thousand meters of material, the volume of the clay may be reduced as much as 40 percent. Compaction is most significant in fine-grained sedimentary rocks such as shale because sand and other coarse sediments compress little.

Cementation is another important means by which sediments are converted to sedimentary rock. The cementing materials are carried in a solution by water percolating through the pore spaces between particles. Over time, the cement precipitates onto the sediment grains, fills the open spaces, and joins the particles. Calcite, silica, and iron oxide are the most common cements. Identification of the cementing material is simple. Calcite cement will effervesce (fizz) with dilute hydrochloric acid. Silica is the hardest cement and thus produces the hardest sedimentary rocks. When a sedimentary rock has an orange or red color, this usually means iron oxide is present.

Features of Sedimentary Rocks

Sedimentary rocks are particularly important in the study of Earth history. These rocks form at Earth's surface, and as layer upon layer of sediment accumulates, each records the nature of the environment at the time the sediment was deposited. These layers, called **strata**, or **beds**, are the *single most characteristic feature of sedimentary rocks* (see Figure 3.15).

The thickness of beds ranges from microscopically thin to tens of meters thick. Separating the strata are *bedding planes,* flat surfaces along which rocks tend to separate or break. Generally, each bedding plane marks the end of one episode of sedimentation and the beginning of another.

Sedimentary rocks provide geologists with evidence for deciphering past environments. A conglomerate, for example, indicates a high-energy environment, such as a rushing stream, where only the coarse materials can settle out. By contrast, black shale and coal are associated with a low-energy, organic-rich environment, such as a swamp or lagoon. Other features found in some sedimentary rocks also give clues to past environments (**Figure 3.25**).

Fossils, the traces or remains of prehistoric life, are perhaps the most important inclusions found in some sedimentary rock. Knowing the nature of the life-forms that existed at a particular time may help answer many questions about the environment. Was it land or ocean, lake or swamp? Was the climate hot or cold, rainy or dry? Was the ocean water shallow or deep, turbid or clear? Furthermore, fossils are important time indicators and play a key role in matching up rocks from different places that are the same age. Fossils are important tools used in interpreting the geologic past and will be examined in some detail in Chapter 11.

CONCEPT CHECK 3.3

1 Why are sedimentary rocks important?

2 What minerals are most abundant in detrital sedimentary rocks? In which rocks do these sediments predominate?

3 Distinguish between conglomerate and breccia.

4 What are the two categories of chemical sedimentary rock? Give an example of a rock that belongs to each category.

5 How do evaporates form? Give an example.

6 Compaction is an important lithification process with which sediment size?

7 What is the single most characteristic feature of sedimentary rocks?

FIGURE 3.25 Sedimentary environments. **A.** Ripple marks preserved in sedimentary rocks may indicate a beach or stream channel environment. (Photo by Ken Hamblin) **B.** Mud cracks form when wet mud or clay dries and shrinks, perhaps signifying a tidal flat or desert basin. (Photo by Gary Yeowell/Getty Images Inc.—Stone Allstock)

A

B

Box 3.1

EARTH AS A SYSTEM

The Carbon Cycle and Sedimentary Rocks

To illustrate the movement of material and energy in the Earth system, let us take a brief look at the *carbon cycle* (**Figure 3.A**). In nature, most carbon atoms are bonded chemically to other elements to form compounds such as carbon dioxide, calcium carbonate, and the hydrocarbons found in coal and petroleum. Carbon is also the basic building block of life as it readily combines with hydrogen and oxygen to form the fundamental organic compounds that compose living things.

In the atmosphere, carbon is found mainly as carbon dioxide (CO_2). Atmospheric carbon dioxide is significant because it is a greenhouse gas, which means it is an efficient absorber of energy emitted by Earth and thus influences the heating of the atmosphere. Because many of the processes that operate on Earth involve carbon dioxide, this gas is constantly moving into and out of the atmosphere. For example, through the process of photosynthesis, plants absorb carbon dioxide from the atmosphere to produce the essential organic compounds needed for growth. Animals that consume these plants (or consume other animals that eat plants) use these organic compounds as a source of energy and, through the process of respiration, return carbon dioxide to the atmosphere. Furthermore, when plants die and decay or are burned, this biomass is oxidized, and carbon dioxide is returned to the atmosphere.

Not all dead plant material decays. As a result, over long spans of geologic time, considerable biomass is buried with sediment. Under the right conditions, some of these carbon-rich deposits are converted to fossil fuels—coal, petroleum, or natural gas. Eventually, some of the fuels are recovered (mined or pumped from a well) and burned to run factories and fuel our transportation system. One result of fossil-fuel combustion is the release of huge quantities of CO_2 into the atmosphere. Certainly one of the most active parts of the carbon cycle is the movement of CO_2 from the atmosphere to the biosphere and back again.

Carbon also moves from the geosphere and hydrosphere to the atmosphere and back again. For example, volcanic activity early in Earth's history is thought to be the source of much of the carbon dioxide found in the atmosphere. One way that carbon dioxide makes its way back to the hydrosphere and then to the solid Earth is by first combining with water to form carbonic acid (H_2CO_3), which then attacks the rocks that compose the geosphere. One product of this chemical weathering of solid rock is the soluble bicarbonate ion ($2\ HCO_3^-$), which is carried by groundwater and streams to the ocean. Here water-dwelling organisms extract this dissolved material to produce hard parts of calcium carbonate ($CaCO_3$). When the organisms die, these skeletal remains settle to the ocean floor as biochemical sediment and become sedimentary rock. In fact, the geosphere is by far Earth's largest depository of carbon, where it is a constituent of a variety of rocks, the most abundant being limestone. Eventually, the limestone may be exposed at Earth's surface, where chemical weathering will cause the carbon stored in the rock to be released to the atmosphere as CO_2.

In summary, carbon moves among all four of Earth's major spheres. It is essential to every living thing in the biosphere. In the atmosphere carbon dioxide is an important greenhouse gas. In the hydrosphere, carbon dioxide is dissolved in lakes, rivers, and the ocean. In the geosphere, carbon is contained in carbonate sediments and sedimentary rocks and is stored as organic matter dispersed through sedimentary rocks and as deposits of coal and petroleum.

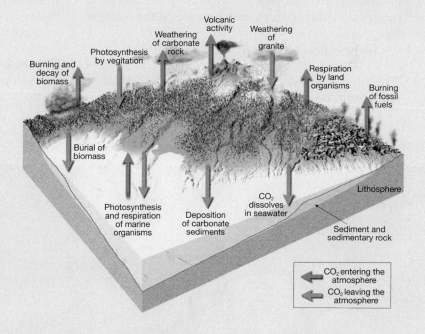

FIGURE 3.A Simplified diagram of the carbon cycle, with emphasis on the flow of carbon between the atmosphere and the hydrosphere, geosphere, and biosphere. The colored arrows show whether the flow of carbon is into or out of the atmosphere.

Metamorphic Rocks: New Rock from Old

Earth Materials
▶ **Metamorphic Rocks**

Recall from the discussion of the rock cycle that metamorphism is the transformation of one rock type into another. **Metamorphic rocks** are produced from preexisting igneous, sedimentary, or even other metamorphic rocks. Thus, every metamorphic rock has a *parent rock*—the rock from which it was formed.

Metamorphism, which means "to change form," is a process that leads to changes in the mineralogy, texture (e.g., grain size), and sometimes the chemical composition of rocks. Metamorphism takes place when preexisting rock is subjected to a physical or chemical environment that is significantly different from that in which it initially formed. In response to these new conditions the rock gradually changes until a state of equilibrium with the new environment is reached. Most metamorphic changes occur at the elevated temperatures and pressures that exist in the zone beginning a few kilometers below Earth's surface and extending into the upper mantle.

Metamorphism often progresses incrementally, from slight changes (*low-grade metamorphism*) to substantial changes (*high-grade metamorphism*). For example, under low-grade metamorphism, the common sedimentary rock *shale* becomes the more compact metamorphic rock called *slate.* Hand samples of these rocks are sometimes difficult to distinguish, illustrating that the transition from sedimentary to metamorphic rock is often gradual and the changes can be subtle.

In more extreme environments, metamorphism causes a transformation so complete that the identity of the parent rock cannot be determined. In high-grade metamorphism, such features as bedding planes, fossils, and vesicles that may have existed in the parent rock are obliterated. Furthermore, when rocks deep in the crust (where temperatures are high) are subjected to directed pressure, the entire mass may deform, producing large-scale structures such as folds (Figure 3.26).

In the most extreme metamorphic environments, the temperatures approach those at which rocks melt. However, *during metamorphism the rock must remain essentially solid,* for if complete melting occurs, we have entered the realm of igneous activity.

Most metamorphism occurs in one of two settings:

1. When rock is intruded by magma, **contact** or **thermal metamorphism** may take place. In such a situation, change is caused by the rise in temperature within the host rock surrounding the mass of molten material.

2. During mountain building, great quantities of rock are subjected to directed pressures and high temperatures associated with large-scale deformation called **regional metamorphism**.

Extensive areas of metamorphic rocks are exposed on every continent. Metamorphic rocks are an important component of many mountain belts, where they make up a large portion of a mountain's crystalline core. Even the stable continental interiors, which are generally covered by sedimentary rocks, are underlain by metamorphic basement rocks. In all of these settings, the metamorphic rocks are usually highly deformed and intruded by igneous masses. Consequently, significant parts of Earth's continental crust are composed of metamorphic and associated igneous rocks.

What Drives Metamorphism?

The agents of metamorphism include *heat, pressure* (stress), and *chemically active fluids.* During metamorphism, rocks are usually subjected to all three metamorphic agents simultaneously. However, the degree of metamorphism and the contribution of each agent vary greatly from one environment to another.

Heat as a Metamorphic Agent Thermal energy (*heat*) is the most important factor driving metamorphism. It triggers chemical reactions that result in the recrystallization of existing

FIGURE 3.26 Folded and metamorphosed rocks in Anza Borrego Desert State Park, California. (Photo by A. P. Trujillo/APT Photos)

minerals and the formation of new minerals. Thermal energy for metamorphism comes mainly from two sources. Rocks experience a rise in temperature when they are intruded by magma rising from below. This is called *contact* or *thermal metamorphism*. In this situation, the adjacent host rock is "baked" by the emplaced magma.

By contrast, rocks that formed at Earth's surface will experience a gradual increase in temperature as they are taken to greater depths. In the upper crust, this increase in temperature averages about 25°C per kilometer. When buried to a depth of about 8 kilometers (5 miles), where temperatures are between 150°C and 200°C, clay minerals tend to become unstable and begin to recrystallize into other minerals, such as chlorite and muscovite, that are stable in this environment. (Chlorite is a micalike mineral formed by the metamorphism of iron- and magnesian-rich silicates.) However, many silicate minerals, particularly those found in crystalline igneous rocks—quartz and feldspar, for example—remain stable at these temperatures. Thus, metamorphic changes in these minerals require much higher temperatures in order to recrystallize.

Confining Pressure and Differential Stress as Metamorphic Agents   Pressure, like temperature, also increases with depth as the thickness of the overlying rock increases. Buried rocks are subjected to *confining pressure*—similar to water pressure in that

Students Sometimes Ask...

How hot is it deep in Earth's crust?

The increase in temperature with depth, based on the geothermal gradient, can be expressed as *the deeper one goes, the hotter it gets*. This relationship has been observed by miners in deep mines and in deeply drilled wells. In the deepest mine in the world (the Western Deep Levels gold mine in South Africa, which is 4 kilometers, or 2.5 miles, deep), the temperature of the surrounding rock is so hot that it can scorch human skin!

The temperature is even higher at the bottom of the deepest well in the world, which was completed in the Kola Peninsula of Russia in 1992 and goes down a record 12.3 kilometers (7.7 miles). At this depth it is 245°C (473°F), much greater than the boiling point of water. What keeps water from boiling is the high confining pressure at depth.

the forces are equally applied in all directions (**Figure 3.27A**). The deeper you go in the ocean, the greater the confining pressure. The same is true for buried rock. Confining pressure causes the spaces between mineral grains to close, producing a more compact rock having greater density. Further, at great

FIGURE 3.27 Pressure (stress) as a metamorphic agent. **A.** In a depositional environment, as confining pressure increases, rocks deform by decreasing in volume. **B.** During mountain building, rocks subjected to differential stress are shortened in the direction that pressure is applied, and lengthened in the direction perpendicular to that force.

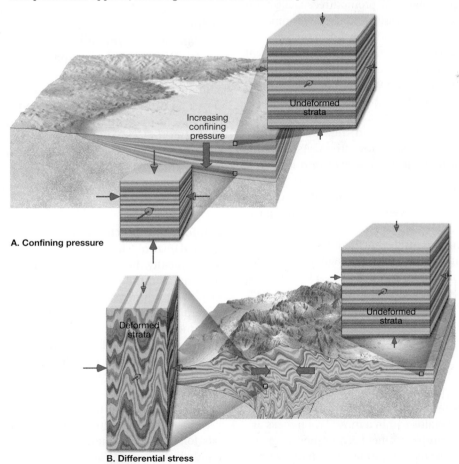

A. Confining pressure

B. Differential stress

depths, confining pressure may cause minerals to recrystallize into new minerals that display more compact crystalline forms.

During episodes of mountain building, large rock bodies become highly crumpled and metamorphosed (Figure 3.27B). The forces that generate mountains are unequal in different directions and are called *differential stress*. Unlike confining pressure, which "squeezes" rock equally in all directions, differential stresses are greater in one direction than in others. As shown in Figure 3.27B, rocks subjected to differential stress are shortened in the direction of greatest stress, and elongated, or lengthened, in the direction perpendicular to that stress. The deformation caused by differential stresses plays a major role in developing metamorphic textures.

In surface environments where temperatures are relatively low, rocks are *brittle* and tend to fracture when subjected to differential stress. Continued deformation grinds and pulverizes the mineral grains into small fragments. By contrast, in high-temperature environments, rocks are *ductile*. When rocks exhibit ductile behavior, their mineral grains tend to flatten and elongate when subjected to differential stress. This accounts for their ability to deform by flowing (rather than fracturing) to generate intricate folds (**Figure 3.28**).

Chemically Active Fluids   Fluids composed mainly of water and other volatiles (materials that readily change to gases at surface conditions), including carbon dioxide, are believed to play an important role in some types of metamorphism. Fluids that surround mineral grains act as catalysts to promote recrystallization by enhancing ion migration. In progressively hotter environments, these ion-rich fluids become correspondingly more reactive.

When two mineral grains are squeezed together, the parts of their crystalline structures that touch are the most highly stressed. Atoms at these sites are readily dissolved by the hot fluids and move to the voids between individual grains. Thus, hot fluids aid in the recrystallization of mineral grains by dissolving material from regions of high stress and then precipitating (depositing) this material in areas of low stress. As a result, *minerals tend to recrystallize and grow longer in a direction perpendicular to compressional stresses.*

When hot fluids circulate freely through rocks, ionic exchange may occur between adjacent rock layers, or ions may migrate great distances before they are finally deposited. The latter situation is particularly common when we consider hot fluids that escape during the crystallization of an intrusive mass of magma. If the rocks surrounding the magma differ markedly in composition from the invading fluids, there may be a substantial exchange of ions between the fluids and host rocks. When this occurs, the overall composition of the surrounding rock changes.

Metamorphic Textures

The degree of metamorphism is reflected in the rock's *texture* and *mineralogy*. (Recall that the term *texture* is used to describe the size, shape, and arrangement of grains within a rock.) When rocks are subjected to low-grade metamorphism, they become more compact and thus denser. A common example is the metamorphic rock slate,

FIGURE 3.28 Deformed metamorphic rocks exposed in a road cut in the Eastern Highland of Connecticut. Imagine the tremendous force required to fold rock in this manner. (Photo by Phil Dombrowski)

which forms when shale is subjected to temperatures and pressures only slightly greater than those associated with the compaction that lithifies sediment. In this case, differential stress causes the microscopic clay minerals in shale to align into the more compact arrangement found in slate.

Under more extreme conditions, stress causes certain minerals to recrystallize. In general, recrystallization encourages the growth of larger crystals. Consequently, many metamorphic rocks consist of visible crystals, much like coarse-grained igneous rocks.

Foliation   The term **foliation** refers to any planer (nearly flat) arrangement of mineral grains or structural features within a rock (**Figure 3.29**). Although foliation may occur in some sedimentary and even a few types of igneous rocks, it is a fundamental characteristic of regionally metamorphosed rocks—that is, rock units that have been strongly deformed, mainly by folding. In metamorphic environments, foliation is ultimately driven by compressional stresses that shorten rock units, causing mineral grains in preexisting rocks to develop parallel, or nearly parallel, alignments. Examples of foliation include the parallel alignment of platy minerals such as the micas; the parallel alignment of flattened pebbles; compositional banding in which dark and light minerals separate generating a layered appearance; and rock cleavage in which rocks can be easily split into tabular slabs.

Nonfoliated Textures   Not all metamorphic rocks exhibit a foliated texture. Those that do not are referred to as **nonfoliated**, and typically develop in environments where deformation is minimal and the parent rocks are composed of minerals that exhibit equidimensional crystals, such as quartz or calcite. For example, when a fine-grained limestone (made of calcite) is metamorphosed by the intrusion of a hot magma body, the small calcite grains recrystallize to form larger interlocking crystals. The resulting rock, *marble*, exhibits large, equidimensional grains that are randomly oriented, similar to those in a coarse-grained igneous rock.

To review, metamorphic processes cause many changes in existing rocks, including increased density, growth of larger crystals, foliation (reorientation of the mineral grains into a layered or banded appearance), and the transformation of low-temperature minerals into high-temperature minerals. Furthermore, the introduction of ions generates new minerals, some of which are economically important.

Common Metamorphic Rocks

A chart depicting the common rocks produced by metamorphic processes is found in **Figure 3.30**, and a description of each follows.

Foliated Rocks   *Slate* is a very fine-grained foliated rock composed of minute mica flakes that are too small to be visible (**Figure 3.31**). A noteworthy characteristic of slate is its excellent rock cleavage, or tendency to break into flat slabs. This property has made slate a useful rock for roof and floor tile, as well as billiard tables (**Figure 3.32**). Slate is usually generated by the low-grade metamorphism of shale. Less frequently, it is produced when volcanic ash is metamorphosed. Slate's color is variable. Black slate contains organic material; red slate gets its color from iron oxide; and green slate is usually composed of chlorite, a greenish micalike mineral.

Schists are strongly foliated rocks formed by regional metamorphism (Figure 3.31). They are platy and can be readily split into thin flakes or slabs. Many schists, like slates, originate from shale parent rock. However, schist forms under more extreme metamorphic conditions. The term *schist* describes the *texture* of a rock regardless of composition. For example, schists composed primarily of muscovite and biotite are called *mica schists.*

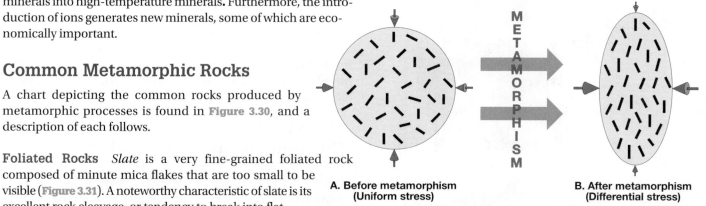

A. Before metamorphism (Uniform stress)

METAMORPHISM

B. After metamorphism (Differential stress)

FIGURE 3.29 Under the pressures of metamorphism, some mineral grains become reoriented and aligned at right angles to the stress. The resulting orientation of mineral grains gives the rock a foliated (layered) texture. If the coarse-grained igneous rock (granite) on the left underwent intense metamorphism, it could end up closely resembling the metamorphic rock on the right (gneiss). (Photos by E. J. Tarbuck)

FIGURE 3.30 Classification of common metamorphic rocks.

Rock Name	Texture			Grain Size	Comments	Original Parent Rock
Slate	Increasing Metamorphism	Foliated		Very fine	Excellent rock cleavage, smooth dull surfaces	Shale, mudstone, or siltstone
Schist				Medium to Coarse	Micaceous minerals dominate, scaly foliation	Shale, mudstone, or siltstone
Gneiss				Medium to Coarse	Compositional banding due to segregation of minerals	Shale, granite, or volcanic rocks
Marble		Nonfoliated		Medium to coarse	Interlocking calcite or dolomite grains	Limestone, dolostone
Quartzite				Medium to coarse	Fused quartz grains, massive, very hard	Quartz sandstone
Anthracite				Fine	Shiny black rock that may exhibit conchoidal fracture	Bituminous coal

Foliated metamorphic rocks

Slate

Schist

Gneiss

Nonfoliated metamorphic rocks

Marble

Quartzite

FIGURE 3.31 Common metamorphic rocks. (Photos by E. J. Tarbuck)

Gneiss (pronounced "nice") is the term applied to banded metamorphic rocks in which elongated and granular (as opposed to platy) minerals predominate (Figure 3.31). The most common minerals in gneisses are quartz and feldspar, with lesser amounts of muscovite, biotite, and hornblende. Gneisses exhibit strong segregation of light and dark silicates, giving them a characteristic

banded texture. While still deep below the surface where temperatures and pressures are great, banded gneisses can be deformed into intricate folds.

Nonfoliated Rocks *Marble* is a coarse, crystalline rock whose parent rock is limestone. Marble is composed of large interlocking calcite crystals, which form from the recrystallization of smaller grains

Students Sometimes Ask...

Is glacial ice considered a metamorphic rock?

Yes! Although metamorphic rock is typically formed in high temperature environments, glacial ice is an exception. Despite its formation in cold climates, glacial ice clearly satisfies the criteria for being classified as a metamorphic rock. The formation of a glacier begins as snow crystals are transformed into a much denser mass of small ice grains called *firn*. As more snow is added to the pile, the pressure on the layers below promotes the recrystallization (metamorphism) of firn producing larger interlocking ice crystals. In addition, glacial movement is an example of solid-state ductile flow, another characteristic of metamorphic rocks. Ductile flow is facilitated by the internal deformation and recrystallization of individual ice crystals. The resulting ductile flow is often made visible because we can see deformed dirty layers within the ice. These structures are similar to the folds exhibited by more "typical" metamorphic rocks.

FIGURE 3.32 Because slate breaks into flat slabs, it has many uses. The larger image shows a quarry near Alta, Norway. (Photo by Fred Bruemmer/Photolibrary). In the inset photo, slate is used to roof a house in Switzerland. (Photo by E. J. Tarbuck)

in the parent rock. Because of its color and relative softness (hardness of only 3 on the Mohs scale), marble is a popular building stone. White marble is particularly prized as a stone from which to carve monuments and statues, such as the Lincoln Memorial in Washington, D.C., and the Taj Mahal in India (**Figure 3.33**). The parent rocks from which various marbles form contain impurities that color the stone. Thus, marble can be pink, gray, green, or even black.

Quartzite is a very hard metamorphic rock most often formed from quartz sandstone. Under moderate- to high-grade metamorphism, the quartz grains in sandstone fuse. Pure quartzite is white, but iron oxide may produce reddish or pinkish stains, and dark minerals may impart a gray color.

CONCEPT CHECK 3.4

1 Metamorphism means to "change form." Describe how a rock may change during metamorphism.

2 Briefly describe what is meant by the statement, "every metamorphic rock has a *parent rock*."

3 List the three agents of metamorphism and describe the role of each.

4 Distinguish between regional and contact metamorphism.

5 Which feature would easily distinguish schist and gneiss from quartzite and marble.

6 In what ways do metamorphic rocks differ from the igneous and sedimentary rocks from which they formed?

Resources from Rocks and Minerals

The outer layer of Earth, which we call the crust, is only as thick when compared to the remainder of the Earth as a peach skin is to a peach, yet it is of supreme importance to us. We depend on it for fossil fuels and as a source of such diverse minerals as the talc for baby powder, salt to flavor food, and gold for world trade. In fact, on occasion, the availability or absence of certain Earth materials has altered the course of history. As the material requirements of modern society grow, the need to locate additional supplies of useful minerals also grows and becomes more challenging.

Metallic Mineral Resources

Some of the most important accumulations of metals, such as gold, silver, copper, mercury, lead, platinum, and nickel, are produced by igneous and metamorphic processes (**Table 3.1**). These mineral resources, like most others, result from processes that concentrate desirable materials to the extent that extraction is economically feasible.

The igneous processes that generate some metal deposits are quite straightforward. For example, as a large magma body cools, the heavy minerals that crystallize early tend to settle to the lower portion of the magma chamber. This type of magmatic differen-

---

FIGURE 3.33 Marble, because of its workability, is a widely used building stone. **A.** The white exterior of the Lincoln Memorial in Washington, D.C., is constructed mainly of marble that was quarried in Marble, Colorado. Inside, pink Tennessee marble was used for the floors, Alabama marble for the ceilings, and Georgia marble for Lincoln's statue. (Photo by Daniel Grill/iStockphoto) **B.** The exterior of India's Taj Mahal is constructed primarily of the metamorphic rock marble. (Photo by Holger Mette/Shutterstock)

A.

B.

TABLE 3.1 Ore Minerals of Important Metals

Metal	Ore Mineral
Aluminum	Bauxite
Chromium	Chromite
Copper	Chalcopyrite
	Bornite
	Chalcocite
Gold	Native gold
Iron	Hematite
	Magnetite
	Limonite
Lead	Galena
Magnesium	Magnesite
	Dolomite
Manganese	Pyrolusite
Mercury	Cinnabar
Molybdenum	Molybdenite
Nickel	Pentlandite
Platinum	Native platinum
Silver	Native silver
	Argentite
Tin	Cassiterite
Titanium	Ilmenite
	Rutile
Tungsten	Wolframite
	Scheelite
Uranium	Uraninite (pitchblende)
Zinc	Sphalerite

FIGURE 3.34 This granite pegmatite is composed mainly of quartz (dark gray) and feldspar (white and salmon color). (Photo by Marli Miller)

tiation is particularly active in large basaltic magmas where chromite (ore of chromium), magnetite, and platinum are occasionally generated. Layers of chromite, interbedded with other heavy minerals, are mined from such deposits in the Bushveld Complex in South Africa, which contains over 70 percent of the world's known reserves of platinum.

Igneous processes are also important in generating other types of mineral deposits. For example, as a granitic magma cools and crystallizes, the residual melt becomes enriched in rare elements and heavy metals, including gold and silver. Furthermore, because water and other volatile substances do not crystallize along with the bulk of the magma body, these fluids make up a high percentage of the melt during the final phase of solidification. Crystallization in a fluid-rich environment enhances the migration of ions and results in the formation of crystals several centimeters, or even a few meters, in length. The resulting rocks, called **pegmatites**, are composed of these unusually large crystals (Figure 3.34).

Most pegmatites are granitic in composition and consist of unusually large crystals of quartz, feldspar, and muscovite. Feldspar is used in the production of ceramics, and muscovite is used for electrical insulation and glitter. In addition to the common silicates, some pegmatites include semiprecious gems such as beryl, topaz, and tourmaline. Moreover, minerals containing the elements lithium, cesium, uranium, and the rare earths are sometimes found. Most pegmatites are located within large igneous masses or as dikes or veins that cut into the host rock surrounding the magma chamber (Figure 3.35).

Among the best-known and most important ore deposits are those generated from **hydrothermal** (*hydra* = water, *therm* = heat) **solutions**. Included in this group are the gold deposits of the Homestake mine in South Dakota; the lead, zinc, and silver ores near Coeur d'Alene, Idaho; the silver deposits of the Comstock Lode in Nevada; and the copper ores of the Keweenaw Peninsula in Michigan.

The majority of hydrothermal deposits are thought to originate from hot, metal-rich fluids that are associated with cooling magma bodies. During solidification, liquids plus various metallic ions accumulate near the top of the magma chamber. Because these hot fluids are very mobile, they can migrate great distances through the surrounding rock before they are eventually

FIGURE 3.35 Illustration of the relationship between a parent igneous body and the associated pegmatite and hydrothermal mineral deposits.

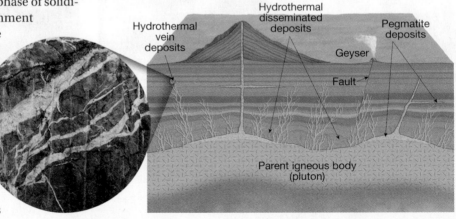

TABLE 3.2 Uses of Nonmetallic Minerals

Mineral	Uses
Apatite	Phosphorous fertilizers
Asbestos (chrysotile)	Incombustible fibers
Calcite	Aggregate; steelmaking; soil conditioning; chemicals; cement; building stone
Clay minerals (kaolinite)	Ceramics; china
Corundum	Gemstones; abrasives
Diamond	Gemstones; abrasives
Fluorite	Steelmaking; aluminum refining; glass; chemicals
Garnet	Abrasives; gemstones
Graphite	Pencil lead; lubricant; refractories
Gypsum	Plaster of Paris; wallboard
Halite	Table salt; chemicals; ice control
Muscovite	Insulator in electrical applications
Quartz	Primary ingredient in glass
Sulfur	Chemicals; fertilizer manufacture
Sylvite	Potassium fertilizers
Talc	Powder used in paints, cosmetics, etc.

deposited. Some of this fluid moves along fractures or bedding planes, where it cools and precipitates the metallic ions to produce **vein deposits** (Figure 3.35). Many of the most productive deposits of gold, silver, and mercury occur as hydrothermal vein deposits.

Another important type of accumulation generated by hydrothermal activity is called a **disseminated deposit**. Rather than being concentrated in narrow veins and dikes, these ores are distributed as minute masses throughout the entire rock mass (Figure 3.35). Much of the world's copper is extracted from disseminated deposits, including the huge Bingham Canyon copper mine in Utah. Because these accumulations contain only 0.4–0.8 percent copper, between 125 and 250 metric tons of ore must be mined for every ton of metal recovered. The environmental impact of these large excavations, including the problems of waste disposal, is significant.

Nonmetallic Mineral Resources

Mineral resources not used as fuels or processed for the metals they contain are referred to as *nonmetallic mineral resources.* These materials are extracted and processed either to make use of the nonmetallic elements they contain or for the physical and chemical properties they possess (**Table 3.2**). Nonmetallic mineral resources are commonly divided into two broad groups: *building materials* and *industrial minerals* (**Figure 3.36**). As some substances have many different uses, they are found in both categories. Limestone, perhaps the most versatile and widely used rock of all, is the best example. As a building material, it is used not only as crushed rock and building stone but also in the making of cement. Moreover, as an industrial mineral, it is an ingredient in the manufacture of steel and is used in agriculture to neutralize acidic soils.

Besides aggregate (sand, gravel, and crushed rock) and cut stone, the other important building materials include gypsum

FIGURE 3.36 Rock of Ages granite quarry in Barre, Vermont. (Photo by Raymond Forbes/Photolibrary)

for plaster and wallboard, clay for tile and bricks, and cement, which is made from limestone and shale. Cement and aggregate go into the making of concrete, a material that is essential to practically all construction.

Many and various nonmetallic resources are classified as industrial minerals. People often do not realize the importance of industrial minerals because they see only the products that resulted from their use and not the minerals themselves. That is, many nonmetallics are used up in the process of creating other products. Examples include fluorite and limestone, which are part of the steelmaking process; corundum and garnet, which are used as abrasives to make machinery parts; and sylvite, which is employed in the production of the fertilizers used to grow food crops.

CONCEPT CHECK 3.5

1. List two general types of hydrothermal deposits.
2. Nonmetallic resources are commonly divided into two broad groups. List the two groups and give some examples of materials that belong to each.

GIVE IT SOME THOUGHT

1. Refer to Figure 3.2. How does the rock cycle diagram, in particular the labeled arrows, support the fact that sedimentary rocks are the most abundant rock type on Earth's surface?
2. Would you expect all of the crystals in an intrusive igneous rock to be the same size? Explain why or why not.
3. Apply your understanding of igneous rock textures to describe the cooling history of each of the igneous rocks pictured here.

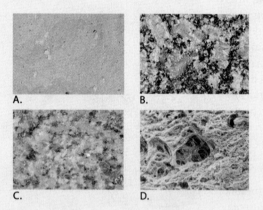

4. Is it possible for two igneous rocks to have the same mineral composition but be different rocks? Use an example to explain your answer.
5. Use your understanding of Bowen's reaction series (Figure 3.13) and the process of magmatic differentiation to explain how partial melting can generate magmas with different compositions.
6. Dust collecting on furniture is an everyday example of a sedimentary process. Provide another example of a sedimentary process that might be observed in or around where you live.
7. Describe two reasons why sedimentary rocks are more likely to contain fossils than igneous rocks.
8. If you hiked to a mountain peak and found limestone at the top, what would that indicate about the likely geologic history of the rock there?

9. The accompanying photos each illustrate either a typical igneous, sedimentary, or metamorphic rock body. Which do you think is a metamorphic rock? Explain why you ruled out the other rock bodies.

A.   B.   C.

10. Examine the accompanying photos, which show the geology of the Grand Canyon. Notice that most of the canyon consists of layers of sedimentary rocks, but if you were to hike down into the inner gorge you would encounter the Vishnu schist, a metamorphic rock.

  a. What process might have been responsible for the formation of the Vishnu schist? How does this process differ from the processes that formed the sedimentary rocks that are atop the Vishnu schist?

  b. What does the Vishnu schist tell you about the history of the Grand Canyon prior to the formation of the canyon itself?

  c. Why is the Vishnu schist visible at Earth's surface?

  d. Is it likely that rocks similar to the Vishnu schist exist elsewhere but are not exposed at Earth's surface? Explain.

A. Grand Canyon

B. Close up of Vishnu Schist (dark color)

In Review   Chapter 3 Rocks: Materials of the Solid Earth

- The three rock groups are igneous, sedimentary, and meta-morphic. *Igneous rock* forms from *magma* that cools and solidifies in a process called *crystallization. Sedimentary rock* forms from the *lithification* of sediment. *Metamorphic rock* forms from rock that has been subjected to great pressure and heat in a process called *metamorphism.*

- Igneous rocks are classified by their *texture* and *mineral com-position.* The rate of cooling of magma greatly influences the size of mineral crystals in igneous rock and thus its texture. The four basic igneous rock textures are (1) *fine-grained,* (2) *coarse-grained,* (3) *porphyritic,* and (4) *glassy.*

- Igneous rocks are divided into broad compositional groups based on the percentage of dark and light silicate minerals they contain. *Felsic rocks* (e.g., granite and rhyolite) are com-posed mostly of the light-colored silicate minerals potassium feldspar and quartz. Rocks of *intermediate* composition (e.g., andesite) contain plagioclase feldspar and amphibole. *Mafic rocks* (e.g., basalt) contain abundant pyroxene and calcium-rich plagioclase feldspar.

- The mineral makeup of an igneous rock is ultimately deter-mined by the chemical composition of the magma from which it crystallized. N. L. Bowen showed that as magma cools, minerals crystallize in an orderly fashion at different temperatures. *Magmatic differentiation* changes the composi-tion of magma and causes more than one rock type to form from a common parent magma.

- *Detrital sediments* originate as solid particles derived from weathering and are transported. *Chemical sediments* are solu-ble materials produced largely by chemical weathering that are precipitated by either inorganic or biological processes. *Detrital sedimentary rocks,* which are classified by particle size, contain a variety of mineral and rock fragments, with clay minerals and quartz the chief constituents. *Chemical sedimen-tary rocks* often contain the products of biological processes or mineral crystals that form as water evaporates and minerals precipitate. *Lithification* refers to the processes by which sedi-ments are transformed into solid sedimentary rocks.

- Common detrital sedimentary rocks include *shale* (the most common sedimentary rock), *sandstone,* and *con-glomerate.* The most abundant chemical sedimentary rock is *limestone,* consisting chiefly of the mineral calcite. *Rock gypsum* and *rock salt* are chemical rocks that form as water evaporates.

- Some features of sedimentary rocks that are often used in the interpretation of Earth history and past environments include *strata* or *beds* (the single most characteristic feature), *bedding planes,* and *fossils.*

- Two types of metamorphism are (1) *regional metamorphism* and (2) *contact or thermal metamorphism.* The agents of metamorphism include *heat, pressure* (stress), and *chemically active fluids.* Heat is the most important because it provides the energy to drive the reactions that result in the *recrystallization* of minerals. Metamorphic processes cause many changes in rocks, including *increased density,* growth of *larger mineral crystals, reorientation of the mineral grains* into a layered or banded appearance known as *foliation,* and the formation of *new minerals.*

- Some common metamorphic rocks with a *foliated texture* include *slate, schist,* and *gneiss.* Metamorphic rocks with a *nonfoliated texture* include *marble* and *quartzite.*

- Some of the most important accumulations of *metallic mineral resources* are produced by igneous and metamorphic processes. *Vein deposits* (deposits in fractures or bedding planes) and *disseminated deposits* (deposits distributed throughout the entire rock mass) are produced from *hydrothermal solutions*—hot, metal-rich fluids associated with cooling magma bodies.

- *Nonmetallic mineral resources* are mined for the nonmetallic elements they contain or for the physical and chemical properties they possess. The two groups of nonmetallic mineral resources are (1) *building materials* (e.g., limestone and gypsum) and (2) *industrial minerals* (e.g., fluorite and corundum).

Key Terms

andesitic (intermediate) composition (p. 60)
basaltic composition (p. 60)
Bowen's reaction series (p. 61)
chemical sedimentary rock (p. 64)
coarse-grained texture (p. 56)
contact (thermal) metamorphism (p. 70)
crystal settling (p. 61)
detrital sedimentary rock (p. 64)
disseminated deposit (p. 77)
evaporite deposit (p. 67)
extrusive (volcanic) (p. 55)
felsic (p. 58)
fine-grained texture (p. 56)

foliation (p. 72)
fossil (p. 68)
glassy texture (p. 57)
granitic composition (p. 58)
hydrothermal solutions (p. 76)
igneous rock (p. 55)
intrusive (plutonic) (p. 55)
lava (p. 55)
lithification (p. 68)
mafic (p. 60)
magma (p. 55)
magmatic differentiation (p. 61)
metamorphic rock (p. 69)
metamorphism (p. 69)

nonfoliated texture (p. 72)
pegmatite (p. 76)
porphyritic texture (p. 57)
regional metamorphism (p. 70)
rock cycle (p. 52)
sediment (p. 62)
sedimentary rock (p. 62)
strata (beds) (p. 68)
texture (p. 56)
ultramafic composition (p. 60)
vein deposit (p. 77)
vesicular texture (p. 56)

Examining the Earth System

1. The sedimentary rocks coquina and shale have each formed in response to interactions among two or more of Earth's spheres. List the spheres associated with the formation of each of the rocks and write a short explanation for each of your choices.

2. Of the two main sources of energy that drive the rock cycle—Earth's internal heat and solar energy—which is primarily responsible for each of the three groups of rocks found on and within Earth? Explain your reasoning.

3. Every year about 20,000 pounds of stone, sand, and gravel are mined for each person in the United States.

   a. Calculate how many pounds of stone, sand, and gravel will be needed for an individual during an 80-year lifespan.

   b. If one cubic yard of rocks weighs roughly 1,700 pounds, calculate (in cubic yards) how large a hole must be dug to supply an individual with 80 years' worth of stone, sand, and gravel.

   c. A typical pickup truck can carry about a half cubic yard of rock. How many pickup truck loads would be necessary during the 80-year span?

Mastering Geology

Looking for additional review and test prep materials? Visit the Self Study area in **www.masteringgeology.com** to find practice quizzes, study tools, and multimedia that will aid in your understanding of this chapter's content. In **MasteringGeology™** you will find:

- **GEODe: Earth Science: An interactive visual walkthrough of key concepts**

- **Geoscience Animation Library: More than 100 animations illuminating many difficult-to-understand Earth science concepts**
- **In The News RSS Feeds: Current Earth science events and news articles are pulled into the site with assessment**
- **Pearson eText**
- **Optional Self Study Quizzes**
- **Web Links**
- **Glossary**
- **Flashcards**

Weathering, Soil, and Mass Wasting

4

Mechanical
and chemical
weathering
contributed to the
formation of the
arches and other rock
formations in Utah's
Arches National Park.
(Photo by Dennis Tasa)

84

E arth's surface is constantly changing. Rock is disintegrated and decomposed, moved to lower elevations by gravity, and carried away by water, wind, or ice. In this manner Earth's physical landscape is sculptured. This chapter focuses on the first two steps of this never-ending process, weathering and mass wasting. What causes solid rock to crumble, and why does the type and rate of weathering vary from place to place? What mechanisms act to move weathered debris downslope? Soil, an important product of the weathering process and a vital resource, is also examined.

FOCUS ON CONCEPTS

To assist you in learning **the important concepts in this chapter, focus on the following questions:**

- What are Earth's external processes, and what roles do they play in the rock cycle?
- What are the two main categories of weathering? In what ways are they different?
- What factors determine the rate at which rock weathers?
- What is soil? What are the factors that control soil formation?
- What factors influence natural rates of soil erosion? What impact have humans had?
- How is weathering related to the formation of ore deposits?
- What is mass wasting and what role does it play in the development of valleys?
- What are the controls and triggers of mass wasting?
- What criteria are used to divide and describe the various types of mass wasting?
- What are the general characteristics of slump, rockslide, debris flow, earthflow, and creep?

Earth's External Processes

Sculpturing Earth's Surface
▶ Weathering and Soil

Weathering, mass wasting, and erosion are called **external processes** because they occur at or near Earth's surface and are powered by energy from the Sun. External processes are a basic part of the rock cycle because they are responsible for transforming solid rock into sediment.

To the casual observer, the face of Earth may appear to be without change, unaffected by time. In fact, 200 years ago most people believed that mountains, lakes, and deserts were permanent features of an Earth that was thought to be no more than a few thousand years old. Today we know that Earth is 4.6 billion years old and that mountains eventually succumb to weathering and erosion, lakes fill with sediment or are drained by streams, and deserts come and go with changes in climate.

Earth is a dynamic body. Some parts of Earth's surface are gradually elevated by mountain building and volcanic activity. These **internal processes** derive their energy from Earth's interior. Meanwhile, opposing external processes are continually breaking rock apart and moving the debris to lower elevations (**Figure 4.1**). The latter processes include:

1. **Weathering**—the physical breakdown (disintegration) and chemical alteration (decomposition) of rocks at or near Earth's surface.
2. **Mass wasting**—the transfer of rock and soil downslope under the influence of gravity.

3. **Erosion**—the physical removal of material by mobile agents such as water, wind, or ice.

We first turn our attention to the process of weathering and the products generated by this activity. However, weathering cannot be easily separated from the other two processes because, as weathering breaks rocks apart, it facilitates the movement of rock debris by mass wasting and erosion. Conversely, the transport of material by mass wasting and erosion further disintegrates and decomposes the rock.

CONCEPT CHECK 4.1

❶ List examples of Earth's external and internal processes.
❷ From where do these processes derive their energy?

Weathering

Sculpturing Earth's Surface
▶ Weathering and Soil

Weathering goes on all around us, but it seems like such a slow and subtle process that it is easy to underestimate its importance. Yet, it is worth remembering that weathering is a basic part of the rock cycle and thus a key process in the Earth system. Weathering is also important to humans—even to those of us who are not studying geology. For example, many of the life-sustaining minerals and elements found in soil, and ultimately in the food we

Colorado River

Geologist's Sketch

Sedimentary layers

Material eroded by running water

ls
ss
Shale
ss
Shale
ls
Shale
ss
Shale

Metamorphic rocks

Weathered debris moved downslope by mass wasting

Colorado River

FIGURE 4.1 Slopes are places where materials are continually moving from higher to lower elevations. Weathering begins the process by attacking the solid rock exposed at the surface. Next, gravity moves the weathered debris downslope. This step, termed mass wasting, may range from a slow and gradual creep to a thundering landslide. Eventually, the material that was once high up the slope reaches the stream at the bottom. The moving water then transports the debris away. The walls of the Grand Canyon extend far from the Colorado River. This results primarily from the transfer of weathered debris downslope to the river and its tributaries by mass-wasting processes. (Photo by Bryan Brazil/Shutterstock)

eat, were freed from solid rock by weathering processes. As the chapter-opening photo and many other images in this book illustrate, weathering also contributes to the formation of some of Earth's most spectacular scenery. Of course, these same processes are also responsible for causing the deterioration of many of the structures we build (Figure 4.2).

All materials are susceptible to weathering. Consider, for example, the fabricated product concrete, which closely resembles the sedimentary rock called conglomerate. A newly poured concrete sidewalk has a smooth, unweathered look. However, not many years later, the same sidewalk will appear chipped, cracked, and rough, with pebbles exposed at the surface. If a tree is nearby, its roots may grow under the concrete, heaving and buckling it. The same natural processes that eventually break apart a concrete sidewalk also act to disintegrate rocks, regardless of their type or strength.

FIGURE 4.2 Temple of Olympian Zeus, Athens, Greece. Even the most "solid" monuments that people erect eventually yield to the day-in/day-out attack of weathering processes. (Photo by CORBIS)

Weathering occurs when rock is mechanically fragmented (disintegrated) and/or chemically altered (decomposed). **Mechanical weathering** is accomplished by physical forces that break rock into smaller and smaller pieces without changing the rock's mineral composition. **Chemical weathering** involves a chemical transformation of rock into one or more new compounds. These two concepts can be illustrated with a piece of paper. The paper can be disintegrated by tearing it into smaller and smaller pieces, whereas decomposition occurs when the paper is set afire and burned.

In the following sections we discuss the various modes of mechanical and chemical weathering. Although we consider these two categories separately, keep in mind that mechanical and chemical weathering processes usually work simultaneously in nature and reinforce each other.

CONCEPT CHECK 4.2

❶ What are the two basic categories of weathering?
❷ How do the products of each category of weathering differ?

Mechanical Weathering

Sculpturing Earth's Surface
▶ Weathering and Soil

When a rock undergoes *mechanical weathering*, it is broken into smaller and smaller pieces, each retaining the characteristics of the original material. The end result is many small pieces from a single large one. **Figure 4.3** shows that breaking a rock into smaller pieces increases the surface area available for chemical attack. Hence, by breaking rocks into smaller pieces, mechanical weathering increases the amount of surface area available for chemical weathering.

In nature, four physical processes are especially important in breaking rocks into smaller fragments: frost wedging, salt crystal growth, expansion resulting from unloading (sheeting), and biological activity. In addition, although the work of erosional agents such as waves, wind, glacial ice, and running water is usually considered separately from mechanical weathering, it is nevertheless important. As these mobile agents move rock debris, they relentlessly disintegrate these materials.

Frost Wedging

If you leave a glass bottle of water in the freezer a bit too long, you will find the bottle fractured. The bottle breaks because water has the unique property of expanding about 9 percent upon freezing. This is also the reason that poorly insulated or exposed water pipes rupture during frigid weather. You might also expect this same process to fracture rocks in nature. This is, in fact, the basis for the traditional explanation of **frost wedging**. After water works its way into the cracks in rock, the freezing water enlarges the cracks and angular fragments are eventually produced (**Figure 4.4**).

For many years, the conventional wisdom was that most frost wedging occurred in the manner just described. Recently, however, research has shown that frost wedging can also occur in a different way.[5] It has long been known that when moist soils freeze, they expand or *frost heave* due to the growth of ice lenses. These masses of ice grow larger because they are supplied with water migrating from unfrozen areas as thin liquid films. As more water accumulates and freezes, the soil is heaved upward. A similar process occurs within the cracks and pore spaces of rocks. Lenses of ice grow larger as they attract liquid water from surrounding pores. The growth of these ice masses gradually weakens the rock, causing it to fracture.

[5]Bernard Hallet, "Why Do Freezing Rocks Break?" *Science* 314 (November 2006): 1092–1099.

FIGURE 4.3 Chemical weathering can occur only to those portions of a rock that are exposed to the elements. Mechanical weathering breaks rock into smaller and smaller pieces, thereby increasing the surface area available for chemical attack.

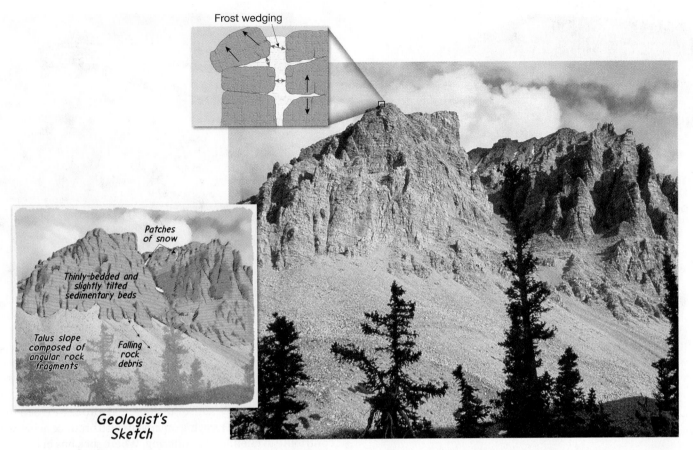

FIGURE 4.4 Traditional depiction of frost wedging. As water freezes, it expands, exerting a force great enough to break rock. When frost wedging occurs in a setting such as this, the broken rock fragments fall to the base of the cliff and create a cone-shaped accumulation known as a talus slope. (Photo by Tom Bean/Corbis)

Salt Crystal Growth

Another expansive force that can split rocks is created by the growth of salt crystals. Rocky shorelines and arid regions are common settings for this process. It begins when sea spray from breaking waves or salty groundwater penetrates crevices and pore spaces in rock. As this water evaporates, salt crystals form. As these crystals gradually grow larger, they weaken the rock by pushing apart the surrounding grains or enlarging tiny cracks.

This same process can also contribute to crumbling roadways where salt is spread to melt snow and ice in winter. The salt dissolves in water and seeps into cracks that quite likely originated from frost action. When the water evaporates, the growth of salt crystals further breaks the pavement.

Sheeting

When large masses of igneous rock, particularly those composed of granite, are exposed by erosion, concentric slabs begin to break loose. The process generating these onion-like layers is called **sheeting** and probably takes place in response to the great reduction in pressure that occurs when the overlying rock is eroded away. Accompanying this *unloading*, the outer layers expand more than the rock below and thus separate from the rock body (**Figure 4.5A, B**). Continued weathering eventually causes the slabs to separate and spall off, creating **exfoliation domes**

(*ex* = off, *folium* = leaf). Excellent examples of exfoliation domes include Stone Mountain in Georgia and Half Dome in Yosemite National Park (Figure 4.5C).

Deep underground mining provides us with another example of how rocks behave once the confining pressure is removed. Large rock slabs sometimes explode off the walls of newly cut mine tunnels because of the abruptly reduced pressure. Evidence of this type, plus the fact that fracturing occurs parallel to the floor of a quarry when large blocks of rock are removed, strongly supports the process of unloading as the cause of sheeting.

Although many fractures are created by expansion, others are produced by contraction as igneous materials cool (see Figure 9.28, p. 278, and still others by tectonic forces during mountain building. Fractures produced by these activities often form a definite pattern and are called joints (see Figure 4.8, p. 91 and Figure 10.15, p. 307). Joints are important rock structures that allow water to penetrate deeply and start the process of weathering long before the rock is exposed.

Biological Activity

Weathering is also accomplished by the activities of organisms, including plants, burrowing animals, and humans. Plant roots in search of minerals and water grow into fractures, and as the roots grow, they wedge the rock apart (**Figure 4.6**). Burrowing animals further break down the rock by moving fresh material to the

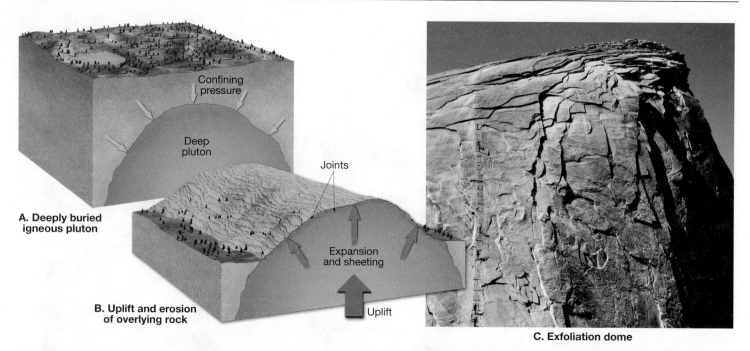

C. Exfoliation dome

FIGURE 4.5 Sheeting is caused by the expansion of crystalline rock as erosion removes the overlying material. When the deeply buried pluton in A is exposed at the surface following uplift and erosion in B, the igneous mass fractures into thin slabs. The photo in C is of the summit of Half Dome in Yosemite National Park, California. It is an exfoliation dome and illustrates the onionlike layers created by sheeting. (Photo by Gary Moon/agefotostock)

FIGURE 4.6 Mt. Sanitas, Boulder, Colorado. Root wedging widens fractures in rock and aids the process of mechanical weathering. (Photo by Kristin Piljay).

surface, where physical and chemical processes can more effectively attack it. Decaying organisms also produce acids, which contribute to chemical weathering. Where rock has been blasted in search of minerals or for road construction, the impact of humans is particularly noticeable.

There are numerous ways that organisms play a role in chemical weathering. For example, plant roots, fungi, and lichens that occupy fractures or that may encrust a rock produce acids that promote decomposition. Moreover, some bacteria are capable of extracting compounds from minerals and using the energy from the compound's chemical bonds to supply their life needs. These primitive "mineral eating" life-forms can live at depths as great as a few kilometers.

CONCEPT CHECK 4.3

1. When a rock is mechanically weathered, how does its surface area change? How does this influence chemical weathering?
2. Explain how water can cause mechanical weathering.
3. How does an exfoliation dome form?
4. How do joints promote weathering?
5. How does biological activity contribute to weathering?

Chemical Weathering

GEODe
Sculpturing Earth's Surface
EARTH
SCIENCE   ▶ **Weathering and Soil**

In the preceding discussion of mechanical weathering you learned that breaking rock into smaller pieces aids chemical weathering by increasing the surface area available for chemical attack. It should also be pointed out that chemical weathering contributes to mechanical weathering. It does so by weakening the outer portions of some rocks, which, in turn, makes them more susceptible to being broken by mechanical weathering processes.

Chemical weathering involves the complex processes that alter the internal structures of minerals by removing and/or adding elements. During this transformation, the original rock decomposes into substances that are stable in the surface environment. Consequently, the products of chemical weathering will remain essentially unchanged as long as they remain in an environment similar to the one in which they formed.

Water and Carbonic Acid

Water is by far the most important agent of chemical weathering. Although pure water is nonreactive, a small amount of dissolved material is generally all that is needed to activate it. Oxygen dissolved in water will *oxidize* some materials. For example, when an iron nail is found in moist soil, it will have a coating of rust (iron oxide), and if the time of exposure has been long, the nail will be so weak that it can be broken as easily as a toothpick. When rocks containing iron-rich minerals oxidize, a yellow to reddish-brown rust will appear on the surface (**Figure 4.7**).

Carbon dioxide (CO_2) dissolved in water (H_2O) forms carbonic acid (H_2CO_3), the same weak acid produced when soft drinks are carbonated. Rain dissolves some carbon dioxide as it falls through the atmosphere, and additional amounts released by decaying organic matter are acquired as the water percolates through the soil. Carbonic acid ionizes to form the very reactive hydrogen ion (H^+) and the bicarbonate ion (HCO_3^-).

Acids such as carbonic acid readily decompose many rocks and produce certain products that are water soluble. For example, the mineral calcite $CaCO_3$, which composes the common building stones marble and limestone, is easily attacked by even a weakly acidic solution.

FIGURE 4.7 Iron reacts with oxygen to form iron oxide, as seen on these rusted barrels. (Photo by Steven Robertson/istockphoto)

How Granite Weathers

To illustrate how rock chemically weathers when attacked by carbonic acid, we now consider the weathering of granite, an abundant continental rock. Recall that granite consists mainly of quartz and potassium feldspar. The weathering of the potassium feldspar component of granite takes place as follows:

$$2 \text{ KAlSi}_3\text{O}_8 + 2(\text{H}^+ + \text{HCO}_3^-) + \text{H}_2\text{O} \rightarrow$$
potassium feldspar   carbonic acid   water

$$\text{Al}_2\text{Si}_2\text{O}_5(\text{OH})_4 + 2\text{ K}^+ + 2\text{ HCO}_3^- + 4\text{ SiO}_2$$
clay mineral

potassium ion   bicarbonite ion   silica

in solution

In this reaction, the hydrogen ions (H^+) attack and replace potassium ions (K^+) in the feldspar structure, thereby disrupting the crystalline network. Once removed, the potassium is available as a nutrient for plants or becomes the soluble salt potassium bicarbonate ($KHCO_3$), which may be incorporated into other minerals or carried to the ocean in dissolved form by streams.

The most abundant products of the chemical breakdown of feldspar are residual clay minerals. Clay minerals are the end product of weathering and are very stable under surface conditions. Consequently, clay minerals make up a high percentage of the inorganic material in soils. Moreover, the most abundant sedimentary rock, shale, contains a high proportion of clay minerals.

In addition to the formation of clay minerals during this reaction, some silica is removed from the feldspar structure and is carried away by groundwater (water beneath Earth's surface). This dissolved silica will eventually precipitate to produce nodules of chert or flint, fill in the pore spaces between sediment grains, or be carried to the ocean, where microscopic animals will remove it to build hard silica shells.

To summarize, the weathering of potassium feldspar generates a residual clay mineral, a soluble salt (potassium bicarbonate), and some silica that enters into solution.

Quartz, the other main component of granite, is *very resistant* to chemical weathering; it remains substantially unaltered when attacked by weakly acidic solutions. As a result, when granite weathers, the feldspar crystals dull and slowly turn to clay, releasing the once interlocked quartz grains, which still retain their fresh, glassy appearance. Although some quartz remains in the soil, much is transported to the sea or to other sites of deposition, where it becomes the main constituent of such features as sandy beaches and sand dunes. In time it may become lithified to form the sedimentary rock *sandstone*.

Weathering of Silicate Minerals

Table 4.1 lists the weathered products of some of the most common silicate minerals. Remember that silicate minerals make up most of Earth's crust and that these minerals are composed essentially of only eight elements. When chemically weathered, silicate minerals yield sodium, calcium, potassium, and magnesium ions, which form soluble products that may be

Box 4.1

UNDERSTANDING EARTH

The Old Man of the Mountain

The Old Man of the Mountain, also known as The Great Stone Face or simply The Profile, was one of New Hampshire's (*the Granite State*) best-known and most enduring symbols (**Figure 4.A** left). Beginning in 1945, it appeared at the center of the official state emblem (see inset between photos). It was a natural rock formation sculpted from Conway Red Granite that, when viewed from the proper location, gave the appearance of an old man. Each year hundreds of thousands of people traveled to view the Old Man, which protruded from high on Cannon Mountain, 360 meters (1,200 feet) above Profile Lake in northern New Hampshire's Franconia Notch State Park.

On Saturday morning, May 3, 2003, the people of New Hampshire learned that the famous landmark had succumbed to nature

FIGURE 4.A The photo on the left shows the Old Man of the Mountain, high above Franconia Notch in New Hampshire's White Mountains, as it appeared prior to May 3, 2003. The inset shows the state emblem of New Hampshire. The photo on the right shows the famous granite outcrop after it collapsed on May 3, 2003. The natural processes that sculpted the Old Man ultimately destroyed it. (Associated Press photos by Jim Cole)

and collapsed (**Figure 4.A** right). The collapse ended decades of efforts to protect the state symbol from the same natural

processes that created it in the first place. Ultimately, frost wedging and other weathering processes prevailed.

TABLE 4.1 Products of Weathering

Mineral	Residual Products	Material in Solution
Quartz	Quartz grains	Silica
Feldspars	Clay minerals	Silica, K^+, Na^+, Ca^{2+}
Amphibole	Clay minerals Iron oxides	Silica, Ca^{2+}, Mg^{2+}
Olivine	Iron oxides	Silica, Mg^{2+}

removed by groundwater. The element iron combines with oxygen, producing relatively insoluble iron oxides, which give soil a reddish-brown or yellowish color. Under most conditions the three remaining elements—aluminum, silicon, and oxygen—join with water to produce residual clay minerals. However, even the highly insoluble clay minerals are very slowly removed by subsurface water.

Spheroidal Weathering

In addition to altering the internal structure of minerals, chemical weathering causes physical changes as well. For instance, when angular rock masses are chemically weathered as water enters along joints, they tend to take on a spherical shape. Gradually the corners and edges of the angular blocks become more rounded. The corners are attacked most readily because of their greater surface area, as compared to the edges and faces. This process, called **spheroidal weathering**, gives the weathered rock a more rounded or spherical shape (**Figure 4.8**).

CONCEPT CHECK 4.4

1 How is carbonic acid formed in nature?
2 What products result when carbonic acid reacts with potassium feldspar?
3 Explain how the rounded boulders in Figure 4.8D were formed.

Rates of Weathering

Sculpturing Earth's Surface
▶ Weathering and Soil

Several factors influence the type and rate of rock weathering. We have already seen how mechanical weathering affects the rate of weathering. By breaking rock into smaller pieces, the amount of surface area exposed to chemical weathering is increased. Other important factors examined here include rock characteristics and climate.

Rock Characteristics

Rock characteristics encompass all of the chemical traits of rocks, including mineral composition and solubility. In addition, any physical features, such as joints (cracks), can be important because they influence the ability of water to penetrate rock.

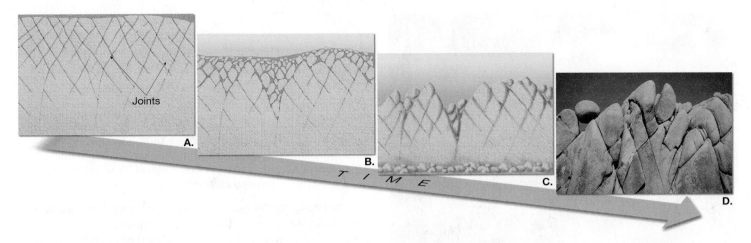

FIGURE 4.8 Joshua Tree National Park, California. Spheroidal weathering of extensively jointed rock. Chemical weathering associated with water moving through the joints enlarges them. Because the rocks are attacked more on the corners and edges, they take on a spherical shape. (Photo by E. J. Tarbuck)

The variations in weathering rates, due to the mineral constituents, can be demonstrated by comparing old headstones made from different rock types. Headstones of granite, which are composed of silicate minerals, are relatively resistant to chemical weathering. We can see this by examining the inscriptions on the headstones shown in Figure 4.9. This is not true of the marble headstone, which shows signs of extensive chemical alteration over a relatively short period. Marble is composed of calcite (calcium carbonate), which readily dissolves even in a weakly acidic solution.

The silicates, the most abundant mineral group, weather in essentially the same sequence as their order of crystallization. By examining Bowen's reaction series (see Figure 3.13, p. 61), you can see that olivine crystallizes first and is therefore the least resistant to chemical weathering, whereas quartz, which crystallizes last, is the most resistant.

FIGURE 4.9 An examination of two headstones in the same cemetary reveals the rate of chemical weathering on diverse rock types. The granite headstone (left) was erected four years before the marble headstone (right). The inscription date of 1872 on the marble monument is nearly illegible. (Photos by E. J. Tarbuck)

Climate

Climatic factors, particularly temperature and moisture, are crucial to the rate of rock weathering. One important example from mechanical weathering is that the frequency of freeze–thaw cycles greatly affects the amount of frost wedging. Temperature and moisture also exert a strong influence on the rates of chemical weathering and on the kind and amount of vegetation present. Regions with lush vegetation generally have a thick mantle of soil rich in decayed organic matter from which chemically active fluids such as carbonic and humic acids are derived.

The optimal environment for chemical weathering is a combination of warm temperatures and abundant moisture. In polar regions chemical weathering is ineffective because frigid temperatures keep the available moisture locked up as ice, whereas in arid regions there is insufficient moisture to foster rapid chemical weathering.

Human activities can influence the composition of the atmosphere, which in turn can impact the rate of chemical weathering. One well-known example is acid rain (Figure 4.10).

Differential Weathering

Masses of rock do not weather uniformly. Take a moment to look at the photo of Shiprock, New Mexico, in Figure 9.25 (p. 276). The durable volcanic neck protrudes high above the surrounding terrain. A glance at the chapter-opening photo shows an additional example of this phenomenon, called **differential weathering**. The results vary in scale from the rough, uneven surface of the marble headstone in Figure 4.9 to the boldly sculpted exposures in Arizona's Monument Valley (Figure 4.11) and the spectacular arches in Arches National Park, Utah, shown in the chapter-opening photo.

Many factors influence the rate of rock weathering. Among the most important are variations in the composition of the rock. More resistant rock protrudes as ridges or pinnacles, or as steeper cliffs on a canyon wall (see Figure 11.3, p. 327). The number and spacing of joints can also be a significant factor (see Figure 4.8 and Figure 10.15, p. 307).

FIGURE 4.10 As a consequence of burning large quantities of coal and petroleum, more than 27 million tons of sulfur and nitrogen oxides are released into the atmosphere each year in the United States. Through a series of complex chemical reactions, some of these pollutants are converted into acids that then fall to Earth's surface as rain or snow. Among its many environmental effects, acid rain accelerates the chemical weathering of stone monuments and structures, including this building facade in Leipzig, Germany. (Photo by Doug Plummer/Photo Researchers, Inc.)

Differential weathering and subsequent erosion are responsible for creating many unusual and sometimes spectacular rock formations and landforms.

CONCEPT CHECK 4.5

❶ **What rock characteristic caused the headstones in Figure 4.9 weather so differently?**

❷ **How does climate influence weathering?**

FIGURE 4.11 Differential weathering is illustrated by these sculpted rock pinnacles in Arizona's Monument Valley. (Photo by jovannig/Shutterstock)

Soil

Soil covers most land surfaces. Along with air and water, it is one of our most indispensable resources. Also, like air and water, soil is taken for granted by many of us. The following quote helps put this vital layer in perspective.

> Science, in recent years, has focused more and more on the Earth as a planet, one that for all we know is unique—where a thin blanket of air, a thinner film of water, and the thinnest veneer of soil combine to support a web of life of wondrous diversity in continuous change.[6]

Soil has accurately been called "the bridge between life and the inanimate world." All life—the entire biosphere—owes its existence to a dozen or so elements that must ultimately come from Earth's crust. Once weathering and other processes create soil, plants carry out the intermediary role of assimilating the necessary elements and making them available to animals, including humans.

An Interface in the Earth System

When Earth is viewed as a system, soil is referred to as an *interface*—a common boundary where different parts of a system interact. This is an appropriate designation because soil forms where the solid Earth, the atmosphere, the hydrosphere, and the biosphere meet. Soil is a material that develops in response to complex environmental interactions among different parts of the Earth system. Over time, soil gradually evolves to a state of equilibrium, or balance, with the environment. Soil is dynamic and sensitive to almost every aspect of its surroundings. Thus, when environmental changes occur—in climate, vegetative cover, or animal (including human) activity—the soil responds. Any such change produces a gradual alteration of soil characteristics until a new balance is reached. Although thinly distributed over the land surface, soil functions as a fundamental interface, providing an excellent example of the integration among many parts of the Earth system.

What Is Soil?

With few exceptions, Earth's land surface is covered by **regolith** (*rhegos* = blanket, *lithos* = stone), the layer of rock and mineral fragments produced by weathering. Some would call this material soil, but soil is more than an accumulation of weathered debris. **Soil** is a combination of mineral and organic matter, water, and air—that portion of the regolith that supports the growth of plants. Although the proportions of the major components in soil vary, the same four components always are present to some extent (**Figure 4.12**). About one half of the total volume of a good-quality surface soil is a mixture of disintegrated and decomposed rock (mineral matter) and *humus*, the decayed remains of animal and plant life (organic matter). The remaining half consists of pore spaces among the solid particles where air and water circulate.

[6]Jack Eddy, "A Fragile Seam of Dark Blue Light," in *Proceedings of the Global Change Research Forum*. U.S. Geological Survey Circular 1086, 1993, p. 15.

FIGURE 4.12 Soil is an essential resource that we often take for granted. Soil is not a living entity, but it contains a great deal of life. Moreover, this complex medium supports nearly all plant life, which in turn supports animal life. The pie chart shows the composition (by volume) of a soil in good condition for plant growth. Although the percentages vary, each soil is composed of mineral and organic matter, water, and air. (Photo by Colin Molyneux/Getty Images)

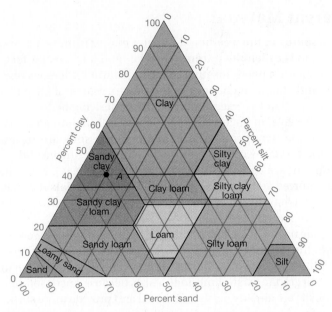

FIGURE 4.13 The texture of any soil can be represented by a point on this soil-texture diagram. Soil texture is one of the most significant factors used to estimate agricultural potential and engineering characteristics. (After U.S. Department of Agriculture)

Although the mineral portion of the soil is usually much greater than the organic portion, humus is an essential component. In addition to being an important source of plant nutrients, humus enhances the soil's ability to retain water. Because plants require air and water to live and grow, the portion of the soil consisting of pore spaces that allow for the circulation of these fluids is as vital as the solid soil constituents.

Soil water is far from "pure" water; instead, it is a complex solution containing many soluble nutrients. Soil water not only provides the necessary moisture for the chemical reactions that sustain life; it also supplies plants with nutrients in a form they can use. The pore spaces not filled with water contain air. This air is the source of necessary oxygen and carbon dioxide for most microorganisms and plants that live in the soil.

Soil Texture and Structure

Most soils are far from uniform and contain particles of different sizes. **Soil texture** refers to the proportions of different particle sizes. Texture is a very basic soil property because it strongly influences the soil's ability to retain and transmit water and air, both of which are essential to plant growth. Sandy soils may drain too rapidly and dry out quickly. At the opposite extreme, the pore spaces of clay-rich soils may be so small that they inhibit drainage, and long-lasting puddles result. Moreover, when the clay and silt content is very high, plant roots may have difficulty penetrating the soil.

Because soils rarely consist of particles of only one size, *textural categories* have been established based on the varying proportions of clay, silt, and sand. The standard system of classes used by the U.S. Department of Agriculture is shown in Figure 4.13. For example, point *A* on this triangular diagram (left center) represents a soil composed of 10 percent silt, 40 percent clay, and 50 percent sand. Such a soil is called a *sandy clay*.

The soils called *loam*, which occupy the central portion of the diagram, are those in which no single particle size predominates over the other two. Loam soils are best suited to support plant life because they generally have better moisture characteristics and nutrient storage ability than do soils composed predominantly of clay or coarse sand.

Soil particles are seldom completely independent of one another. Rather, they usually form clumps called *peds* that give soils a particular structure. Four basic soil structures are recognized: platy, prismatic, blocky, and spheroidal. Soil structure is important because it influences the ease of a soil's cultivation as well as the susceptibility of a soil to erosion. In addition, soil structure affects the porosity and permeability of soil (i.e., the ease with which water can penetrate). This in turn influences the movement of nutrients to plant roots. Prismatic and blocky peds usually allow for moderate water infiltration, whereas platy and spheroidal structures are characterized by slower infiltration rates.

CONCEPT CHECK 4.6

1. Why is soil considered an interface in the Earth system?
2. How is regolith different from soil?
3. Why is texture an important soil property?
4. Using the soil texture diagram (Figure 4.13), name the soil that consists of 60 percent sand, 30 percent silt, and 10 percent clay.

Controls of Soil Formation

Soil is the product of the complex interplay of several factors. The most important of these are parent material, time, climate, plants and animals, and topography. Although all of these factors are interdependent, their roles are examined separately.

Parent Material

The source of the weathered mineral matter from which soils develop is called the **parent material**, and it is a major factor influencing a newly forming soil. Gradually it undergoes physical and chemical changes as the processes of soil formation progress. Parent material may be the underlying bedrock, or it can be a layer of unconsolidated deposits, as in a stream valley. When the parent material is bedrock, the soils are termed *residual soils*. By contrast, those developed on unconsolidated sediment are called *transported soils* (**Figure 4.14**). Note that transported soils form *in place* on parent materials that have been carried from elsewhere and deposited by gravity, water, wind, or ice.

The nature of the parent material influences soils in two ways. First, the type of parent material affects the rate of weathering and thus the rate of soil formation. (Consider the weathering rates of granite versus limestone.) Also, because unconsolidated deposits are already partly weathered and provide more surface area for chemical weathering, soil development on such material usually progresses more rapidly. Second, the chemical makeup of the parent material affects the soil's fertility. This influences the character of the natural vegetation the soil can support.

At one time parent material was thought to be the primary factor causing differences among soils. However, soil scientists came to understand that other factors, especially climate, are more important. In fact, they learned that similar soils often develop from different parent materials and that dissimilar soils can develop from the same parent material. Such discoveries reinforce the importance of the other soil-forming factors.

FIGURE 4.14 The parent material for residual soils is the underlying bedrock, whereas transported soils form on unconsolidated deposits. Also note that as slopes become steeper, soil becomes thinner. (Left and center photos by E. J. Tarbuck; right photo by Grilly Bernard/Getty Images, Inc./Stone Allstock)

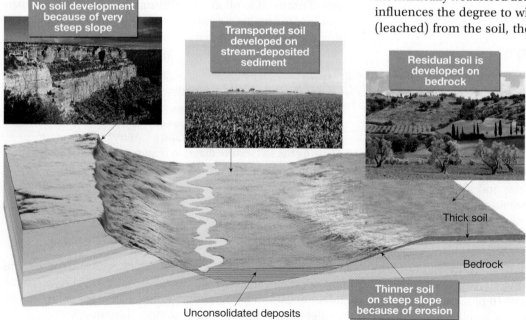

Time

Time is an important component of *every* geological process, and soil formation is no exception. The nature of soil is strongly influenced by the length of time that processes have been operating. If weathering has been going on for a comparatively short time, the parent material strongly influences the characteristics of the soil. As weathering processes continue, the influence of parent material on soil is overshadowed by the other soil-forming factors, especially climate. The amount of time required for various soils to evolve cannot be specified because the soil-forming processes act at varying rates under different circumstances. However, as a rule, the longer a soil has been forming, the thicker it becomes and the less it resembles the parent material.

Climate

Climate is the most influential control of soil formation. Just as temperature and precipitation are the climatic elements that influence people the most, so too are they the elements that exert the strongest impact on soil formation. Variations in temperature and precipitation determine whether chemical or mechanical weathering predominates. They also greatly influence the rate and depth of weathering. For instance, a hot, wet climate may produce a thick layer of chemically weathered soil in the same amount of time that a cold, dry climate produces a thin mantle of mechanically weathered debris. Also, the amount of precipitation influences the degree to which various materials are removed (leached) from the soil, thereby affecting soil fertility. Finally, climatic conditions are important factors controlling the type of plant and animal life present.

Plants and Animals

The biosphere plays a vital role in soil formation. The types and abundance of organisms present have a strong influence on the physical and chemical properties of a soil (**Figure 4.15**). In fact, for well-developed soils in many regions, the significance of natural vegetation is frequently implied in the description used by soil scientists. Such phrases as *prairie soil, forest soil*, and *tundra soil* are common.

FIGURE 4.15 The nature of the vegetation in an area can have a significant influence on soil formation. **A.** The northern coniferous forest. The organic litter received by the soil from the conifers is high in acid resins, which contributes to an accumulation of acid in the soil. As a result, intensive acid leaching is an important soil-forming process. (Photo by Bill Brooks/Alamy) **B.** The relatively meager vegetation in Arizona's Sonoran Desert is very different in character from the northern coniferous forest. Desert soils typically lack much organic matter. (Photo by Russ Bishop/age footstock)

Plants and animals furnish organic matter to the soil. Certain bog soils are composed almost entirely of organic matter, whereas desert soils may contain only a tiny percentage. Although the quantity of organic matter varies substantially among soils, it is the rare soil that completely lacks it.

The primary source of organic matter is plants, although animals and the uncountable microorganisms also contribute. When organic matter decomposes, important nutrients are supplied to plants, as well as to animals and microorganisms living in the soil. Consequently, soil fertility depends in part on the amount of organic matter present. Furthermore, the decay of plant and animal remains causes the formation of various organic acids. These complex acids hasten the weathering process. Organic matter also has a high water-holding ability and thus aids water retention in a soil.

Microorganisms, including fungi, bacteria, and single-celled protozoa, play an active role in the decay of plant and animal remains. The end product is *humus*, a material that no longer resembles the plants and animals from which it is formed. In addition, certain microorganisms aid soil fertility because they have the ability to convert atmospheric nitrogen gas into soil nitrogen compounds.

Earthworms and other burrowing animals act to mix the mineral and organic portions of a soil. Earthworms, for example, feed on organic matter and thoroughly mix soils in which they live, often moving and enriching many tons per acre each year. Burrows and holes also aid the passage of water and air through the soil.

Topography

The lay of the land can vary greatly over short distances. Such variations in topography can lead to the development of a variety of localized soil types. Many of the differences exist because the length and steepness of slopes have a significant impact on the amount of erosion and the water content of soil.

On steep slopes, soils are often poorly developed. In such situations little water can soak in, and as a result, soil moisture may be insufficient for vigorous plant growth. Furthermore, because of accelerated erosion on steep slopes, the soils are thin or nonexistent (see Figure 4.14).

In contrast, waterlogged soils in poorly drained bottomlands have a much different character. Such soils are usually thick and dark. The dark color results from the large quantity of organic matter that accumulates because saturated conditions retard the decay of vegetation. The optimum terrain for soil development is a flat-to-undulating upland surface. Here we find good drainage, minimum erosion, and sufficient infiltration of water into the soil.

Slope orientation, the direction a slope is facing, also is significant. In the mid-latitudes of the Northern Hemisphere, a south-facing slope receives a great deal more sunlight than does a north-facing slope. In fact, a steep north-facing slope may receive no direct sunlight at all. The difference in the amount of solar radiation received causes substantial differences in soil temperature and moisture, which in turn influences the nature of the vegetation and the character of the soil.

Although we have dealt separately with each of the soil-forming factors, remember that *all of them work together* to form soil. No single factor is responsible for a soil being as it is. Rather, it is the combined influence of parent material, time, climate, plants and animals, and topography that determines a soil's character.

CONCEPT CHECK 4.7

1 **List the five basic controls of soil formation.**
2 **Which factor is most influential in soil formation?**
3 **How might the direction a slope is facing influence soil formation?**

The Soil Profile

It is important to realize that soil-forming processes *operate from the surface downward*. Thus, variations in composition, texture, structure, and color gradually evolve at varying depths. These vertical differences, which usually become more pronounced as time passes, divide the soil into zones or layers known as **horizons**. If you were to dig a trench in soil, you would see that its walls are layered. Such a vertical section through all of the soil horizons constitutes the **soil profile** (Figure 4.16).

Figure 4.17 presents an idealized view of a well-developed soil profile in which five horizons are identified. From the surface downward, they are designated as *O, A, E, B,* and *C*, respectively. These five horizons are common to soils in temperate regions. The characteristics and extent of development of horizons vary in different environments. Thus, different localities exhibit soil profiles that can contrast greatly with one another.

The *O* horizon consists largely of organic material. This is in contrast to the layers beneath it, which consist mainly of mineral matter. The upper portion of the *O* horizon is primarily plant litter such as loose leaves and other organic debris that are still recognizable. By contrast, the lower portion of the *O* horizon is

A.

B.

FIGURE 4.16 We all know that a book should not be judged by its cover. Neither should a soil be evaluated only by examining its surface. A soil profile is a vertical cross-section from the surface through all of the soil's horizons and into the parent material. **A.** This profile shows a well-developed soil in southeastern South Dakota. (Photo by E. J. Tarbuck) **B.** The boundaries between horizons in this soil in Puerto Rico are indistinct, giving it a relatively uniform appearance. (Photo courtesy of Soil Science Society of America)

made up of partly decomposed organic matter (humus) in which plant structures can no longer be identified. In addition to plants, the *O* horizon is teeming with microscopic life, including bacteria, fungi, algae, and insects. All of these organisms contribute oxygen, carbon dioxide, and organic acids to the developing soil.

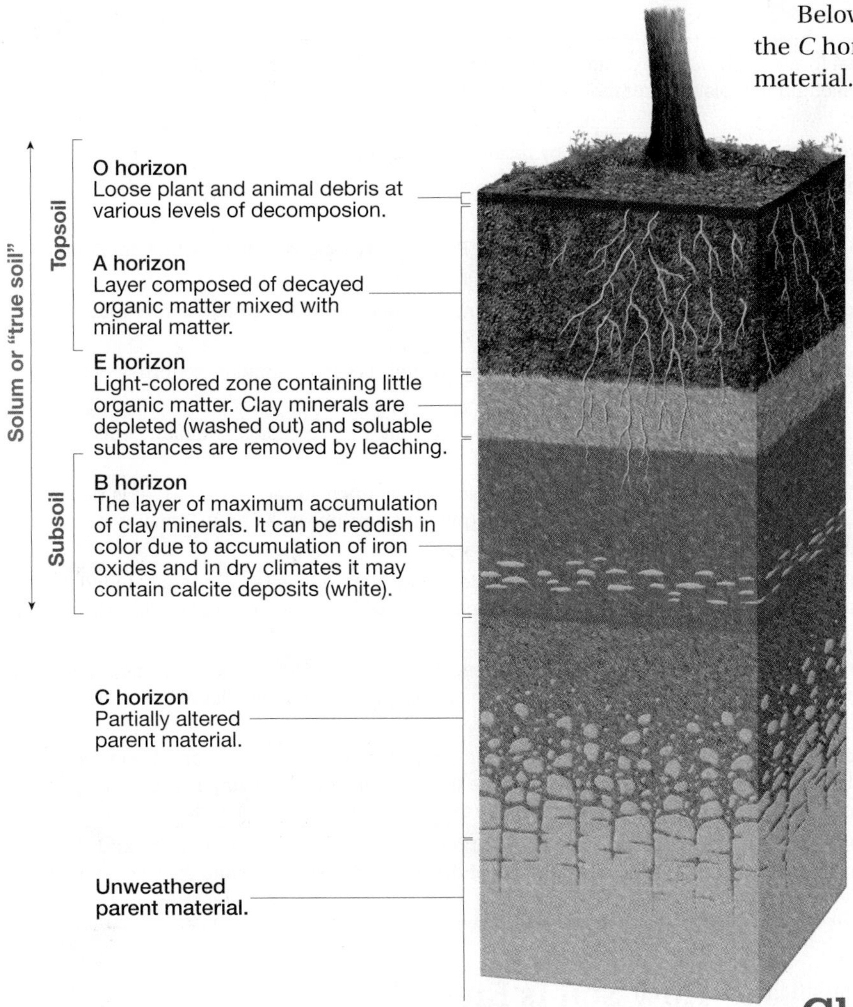

Below the solum and above the unaltered parent material is the *C* horizon, a layer characterized by partially altered parent material. Whereas the *O, A, E,* and *B* horizons bear little resemblance to the parent material, it is easily identifiable in the *C* horizon. Although this material is undergoing changes that will eventually transform it into soil, it has not yet crossed the threshold that separates regolith from soil.

The characteristics and extent of development can vary greatly among soils in different environments. The boundaries between soil horizons can be very distinct or the horizons may blend gradually from one to another. A well-developed soil profile indicates that environmental conditions have been relatively stable over an extended time span and that the soil is *mature*. By contrast, some soils lack horizons altogether. Such soils are called *immature* because soil building has been going on for only a short time. Immature soils are also characteristic of steep slopes, where erosion continually strips away the soil, preventing full development.

FIGURE 4.17 Idealized soil profile from a humid climate in the middle latitudes.

O horizon
Loose plant and animal debris at various levels of decomposition.

A horizon
Layer composed of decayed organic matter mixed with mineral matter.

E horizon
Light-colored zone containing little organic matter. Clay minerals are depleted (washed out) and soluable substances are removed by leaching.

B horizon
The layer of maximum accumulation of clay minerals. It can be reddish in color due to accumulation of iron oxides and in dry climates it may contain calcite deposits (white).

C horizon
Partially altered parent material.

Unweathered parent material.

Solum or "true soil"
Topsoil
Subsoil

CONCEPT CHECK 4.8

❶ Sketch and label the main soil horizons in a well-developed soil profile.

❷ Describe the following features or processes: eluviation, leaching, zone of accumulation, hardpan.

Classifying Soils

There are many variations from place to place and from time to time among the factors that control soil formation. These differences lead to a bewildering variety of soil types. To cope with such variety, it is essential to devise some means of classifying the vast array of data to be studied. By establishing groups consisting of items that have certain important characteristics in common, order and simplicity are introduced. Bringing order to large quantities of information not only aids comprehension and understanding but also facilitates analysis and explanation.

In the United States, soil scientists have devised a system for classifying soils known as the **Soil Taxonomy**. It emphasizes the physical and chemical properties of the soil profile and is organized on the basis of observable soil characteristics. There are six hierarchical categories of classification, ranging from *order*, the broadest category, *to series*, the most specific category. The system recognizes 12 soil orders and more than 19,000 soil series.

The names of the classification units are combinations of syllables, most of which are derived from Latin or Greek. The names are descriptive. For example, soils of the order Aridosol (from the Latin *aridus* = dry, and *solum* = soil) are characteristically dry soils in arid regions. Soils in the order Inceptisols (from the Latin *inceptum* = beginning, and *solum* = soil) are soils with only the beginning or inception of profile development.

Brief descriptions of the 12 basic soil orders are provided in **Table 4.2. Figure 4.18** shows the complex worldwide distribution

Underlying the organic-rich *O* horizon is the *A* horizon. This zone is largely mineral matter, yet biological activity is high and humus is generally present—up to 30 percent in some instances. Together, the *O* and *A* horizons make up what is commonly called *topsoil*. Below the *A* horizon, the *E* horizon is a light-colored layer that contains little organic material. As water percolates downward through this zone, finer particles are carried away. This washing-out of the fine soil components is termed **eluviation** (*elu* = get away from, *via* = a way). Water percolating downward also dissolves soluble inorganic soil components and carries them to deeper zones. This depletion of soluble materials from the upper soil is termed **leaching**.

Immediately below the *E* horizon is the *B* horizon, or *subsoil*. Much of the material removed from the *E* horizon by eluviation is deposited in the *B* horizon, which is often referred to as the *zone of accumulation*. The accumulation of the fine clay particles enhances water retention in the subsoil. However, in extreme cases clay accumulation can form a very compact and impermeable layer called *hardpan*. The *O, A, E,* and *B* horizons together constitute the **solum**, or "true soil." It is in the solum that the soil-forming processes are active and that living roots and other plant and animal life are largely confined.

TABLE 4.2 Basic Soil Orders

Alfisols	Moderately weathered soils that form under boreal forests or broadleaf deciduous forests, rich in iron and aluminum. Clay particles accumulate in a subsurface layer in response to leaching in moist environments. Fertile, productive soils, because they are neither too wet nor too dry.
Andisols	Young soils in which the parent material is volcanic ash and cinders, deposited by recent volcanic activity.
Aridosols	Soils that develop in dry places; insufficient water to remove soluble minerals; may have an accumulation of calcium carbonate, gypsum, or salt in subsoil; low organic content (see Figure 4.15B).
Entisols	Young soils having limited development and exhibiting properties of the parent material. Productivity ranges from very high for some formed on recent river deposits to very low for those forming on shifting sand or rocky slopes.
Gelisols	Young soils with little profile development that occur in regions with permafrost. Low temperatures and frozen conditions for much of the year slow soil-forming processes.
Histosols	Organic soils with little or no climatic implications. Can be found in any climate where organic debris can accumulate to form a bog soil. Dark, partially decomposed organic material commonly referred to as *peat*.
Inceptisols	Weakly developed young soils in which the beginning (inception) of profile development is evident. Most common in humid climates, they exist from the Arctic to the tropics. Native vegetation is most often forest.
Mollisols	Dark, soft soils that have developed under grass vegetation, generally found in prairie areas. Humus-rich surface horizon that is rich in calcium and magnesium. Soil fertility is excellent. Also found in hardwood forests with significant earthworm activity. Climatic range is boreal or alpine to tropical. Dry seasons are normal (see Figure 4.16A).
Oxisols	Soils that occur on old land surfaces unless parent materials were strongly weathered before they were deposited. Generally found in the tropics and subtropical regions. Rich in iron and aluminum oxides, oxisols are heavily leached, hence are poor soils for agricultural activity (see Figure 4.16B).
Spodosols	Soils found only in humid regions on sandy material. Common in northern coniferous forests (see Figure 4.15A) and cool humid forests. Beneath the dark upper horizon of weathered organic material lies a light-colored horizon of leached material, the distinctive property of this soil.
Ultisols	Soils that represent the products of long periods of weathering. Water percolating through the soil concentrates clay particles in the lower horizons (argillic horizons). Restricted to humid climates in the temperate regions and the tropics, where the growing season is long. Abundant water and a long frost-free period contribute to extensive leaching, hence poorer soil quality.
Vertisols	Soils containing large amounts of clay, which shrink upon drying and swell with the addition of water. Found in subhumid to arid climates, provided that adequate supplies of water are available to saturate the soil after periods of drought. Soil expansion and contraction exert stresses on human structures.

pattern of the Soil Taxonomy's 12 soil orders. Like many classification systems, the Soil Taxonomy is not suitable for every purpose. It is especially useful for agricultural and related land-use purposes, but it is not a useful system for engineers who are preparing evaluations of potential construction sites.

CONCEPT CHECK 4.9

① **Why are soils classified?**

Soil Erosion

Soils are just a tiny fraction of all Earth materials, yet they are a vital resource. Because soils are necessary for the growth of rooted plants, they are the very foundation of the human life-support system. Just as human ingenuity can increase the agricultural productivity of soils through fertilization and irrigation, soils can be damaged or destroyed by carelessness. Despite their basic role in providing food, fiber, and other basic materials, soils are among our most abused resources.

Perhaps this neglect and indifference has occurred because a substantial amount of soil seems to remain even where soil erosion is serious. Nevertheless, although the loss of fertile topsoil may not be obvious to the untrained eye, it is a growing problem as human activities expand and disturb more and more of Earth's surface.

How Soil Is Eroded

Soil erosion is a natural process; it is part of the constant recycling of Earth materials that we call the *rock cycle*. Once soil forms, erosional forces, especially water and wind, move soil components from one place to another. Every time it rains, raindrops strike the land with surprising force (Figure 4.19). Each drop acts like a tiny bomb, blasting movable soil particles out of their positions in the soil mass. Then, water flowing across the surface carries away the dislodged soil particles. Because the soil is moved by thin sheets of water, this process is termed *sheet erosion*.

After a thin, unconfined sheet has flowed for a relatively short distance, threads of current typically develop and tiny channels called *rills* begin to form. Still deeper cuts in the soil, known as *gullies*, are created as rills enlarge (Figure 4.20). When normal farm cultivation cannot eliminate the channels, we know the rills have grown large enough to be called gullies. Although most dislodged soil particles move only a short distance during each rainfall, substantial quantities eventually leave the fields and make their way downslope to a stream. Once in the stream channel, these soil particles, which can now be called *sediment*, are transported downstream and eventually deposited.

Rates of Erosion

We know that soil erosion is the ultimate fate of practically all soils. In the past, erosion occurred at slower rates than it does today because more of the land surface was covered and protected

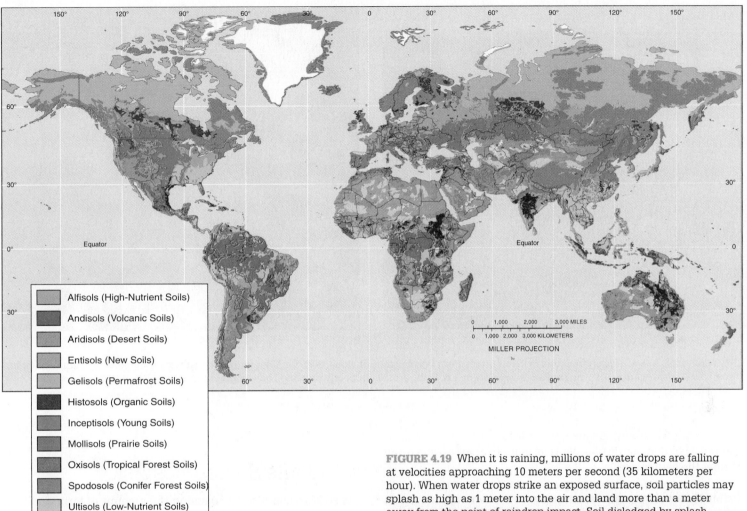

Alfisols (High-Nutrient Soils)

Andisols (Volcanic Soils)

Aridisols (Desert Soils)

Entisols (New Soils)

Gelisols (Permafrost Soils)

Histosols (Organic Soils)

Inceptisols (Young Soils)

Mollisols (Prairie Soils)

Oxisols (Tropical Forest Soils)

Spodosols (Conifer Forest Soils)

Ultisols (Low-Nutrient Soils)

Vertisols (Swelling Clay Soils)

Rock Land

Shifting Sands

Ice/Glacier

FIGURE 4.18 Global soil regions. Worldwide distribution of the Soil Taxonomy's 12 soil orders. (After U.S. Department of Agriculture, Natural Resources Conservation Service, World Soil Resources Staff)

FIGURE 4.19 When it is raining, millions of water drops are falling at velocities approaching 10 meters per second (35 kilometers per hour). When water drops strike an exposed surface, soil particles may splash as high as 1 meter into the air and land more than a meter away from the point of raindrop impact. Soil dislodged by splash erosion is more easily moved by sheet erosion. (Photo courtesy of U.S.D.A/Natural Resources Conservation Service)

by trees, shrubs, grasses, and other plants. However, human activities such as farming, logging, and construction, which remove or disrupt the natural vegetation, have greatly accelerated the rate of soil erosion. Without the stabilizing effect of plants, the soil is more easily swept away by the wind or carried downslope by sheet wash.

Natural rates of soil erosion vary greatly from one place to another and depend on soil characteristics as well as such factors as climate, slope, and type of vegetation. Over a broad area, erosion caused by surface runoff may be estimated by determining the sediment loads of the streams that drain the region. When studies of this kind were made on a global scale, they indicated that prior to the appearance of humans, sediment transport by rivers to the oceans amounted to just over 9 billion metric tons per year (1 metric ton = 1,000 kilograms). By contrast, the amount of material currently transported to the sea by rivers is

A.

B.

FIGURE 4.20 **A.** Soil erosion from this field in northeastern Wisconsin is obvious. Just 1 millimeter of soil lost from a single acre of land amounts to about 5 tons. (Photo by D. P. Burnside/Photo Researchers, Inc.) **B.** Gully erosion in poorly protected soil, southern Colombia. (Photo by Carl Purcell/Photo Researchers, Inc.)

about 24 billion metric tons per year, or more than two and a half times the earlier rate.

It is more difficult to measure the loss of soil due to wind erosion. However, the removal of soil by wind is generally much less significant than erosion by flowing water except during periods of prolonged drought. When dry conditions prevail, strong winds can remove large quantities of soil from unprotected fields (Figure 4.21). Such was the case in the 1930s in the portions of the Great Plains that came to be called the *Dust Bowl*.

In many regions the rate of soil erosion is significantly greater than the rate of soil formation. This means that a renewable resource has become nonrenewable in these places. At present, it is estimated that topsoil is eroding faster than it forms on more than one-third of the world's croplands. The result is lower productivity, poorer crop quality, reduced agricultural income, and an ominous future.

Sedimentation and Chemical Pollution

Another problem related to excessive soil erosion involves the deposition of sediment. Each year in the United States hundreds of millions of tons of eroded soil are deposited in lakes, reservoirs, and streams. The detrimental impact of this process can be significant. For example, as more and more sediment is deposited in a reservoir, the capacity of the reservoir is reduced, limiting its usefulness for flood control, water supply, and/or hydroelectric power generation. In addition, sedimentation in streams and other waterways can restrict navigation and lead to costly dredging operations.

In some cases soil particles are contaminated with pesticides used in farming. When these chemicals are introduced

Students Sometimes Ask...

Is soil erosion actually causing the amount of farmland to diminish?

Yes, indeed. It's been estimated that between 3 and 5 million acres of prime U.S. farmland are lost each year through mismanagement (including soil erosion) and conversion to nonagricultural uses. According to the United Nations, since 1950, more than one-third of the world's farmable land has been lost to soil erosion.

into a lake or reservoir, the quality of the water supply is threatened and aquatic organisms can be endangered. In addition to pesticides, nutrients found naturally in soils as well as those added by agricultural fertilizers make their way into streams and lakes, where they stimulate the growth of plants. Over a period of time, excessive nutrients accelerate the process by which plant growth leads to the depletion of oxygen and an early death of the lake.

The availability of good soils is critical if the world's rapidly growing population is to be fed. On every continent, unnecessary soil loss is occurring because appropriate conservation measures are not being used. Although it is a recognized fact that soil erosion can never be completely eliminated, soil conservation programs can substantially reduce the loss of this basic resource. Windbreaks (rows of trees), terracing, and plowing along the contours of hills are some of the effective measures, as are special tillage practices and crop rotation.

Bauxite

The formation of *bauxite*, the principal ore of aluminum, is one important example of an ore created as a result of enrichment by weathering processes (Figure 4.22). Although aluminum is the third most abundant element in Earth's crust, economically valuable concentrations of this important metal are not common because most aluminum is tied up in silicate minerals from which it is extremely difficult to extract.

Bauxite forms in rainy tropical climates. When aluminum-rich source rocks are subjected to the intense and prolonged chemical weathering of the tropics, most of the common elements, including calcium, sodium, and potassium, are removed by leaching. Because aluminum is extremely insoluble, it becomes concentrated in the soil (as bauxite, a hydrated aluminum oxide). Thus, the formation of bauxite depends on climatic conditions in which chemical weathering and leaching are pronounced, plus, of course, the presence of aluminum-rich source rock. In a similar manner, important deposits of nickel and cobalt develop from igneous rocks rich in silicate minerals such as olivine.

There is significant concern regarding the mining of bauxite and other residual deposits because they tend to occur in environmentally sensitive areas of the tropics. Mining is preceded by the removal of tropical vegetation, thus destroying rain forest ecosystems. Moreover, the thin moisture-retaining layer of organic matter is also disturbed. When the soil dries out in the hot sun, it becomes bricklike and loses its moisture-retaining qualities. Such soil cannot be productively farmed nor can it support significant forest growth. The long-term consequences of bauxite mining are clearly of concern for developing countries in the tropics where this important ore is mined.

FIGURE 4.21 **A.** Windbreaks protecting wheat fields in North Dakota. These flat expanses are susceptible to wind erosion, especially when the fields are bare. The rows of trees slow the wind and deflect it upward, which decreases the loss of fine soil particles. **B.** This windbreak of conifers in Indiana provides year-round protection from wind erosion for this cropland. (Photos by Erwin C. Cole/U.S.D.A./Natural Resources Conservation Service)

CONCEPT CHECK 4.10

① How have human activities affected the rate of soil erosion?
② What are two detrimental effects of soil erosion aside from the loss of topsoil?

Weathering Creates Ore Deposits

Weathering creates many important mineral deposits by concentrating minor amounts of metals that are scattered through unweathered rock into economically valuable concentrations. Such a transformation is often termed **secondary enrichment** and takes place in one of two ways. In one situation, chemical weathering coupled with downward-percolating water removes undesired materials from decomposing rock, leaving the desired elements enriched in the upper zones of the soil. The second way is basically the reverse of the first. That is, the desirable elements that are found in low concentrations near the surface are removed and carried to lower zones, where they are redeposited and become more concentrated.

FIGURE 4.22 Bauxite is the ore of aluminum and forms as a result of weathering processes under tropical conditions. Its color varies from red or brown to nearly white. (Photo by E. J. Tarbuck)

Other Deposits

Many copper and silver deposits result when weathering processes concentrate metals that are dispersed through a low-grade primary ore. Usually such enrichment occurs in deposits containing pyrite (FeS_2), the most common and widespread sulfide mineral. Pyrite is important because when it chemically weathers, sulfuric acid forms, which enables percolating waters to dissolve the ore metals. Once dissolved, the metals gradually migrate downward through the primary ore body until they are precipitated. Deposition takes place because of changes that occur in the chemistry of the solution when it reaches the groundwater zone (the zone beneath the surface where all pore spaces are filled with water). In this manner, the small percentage of dispersed metal can be removed from a large volume of rock and redeposited as a higher-grade ore in a smaller volume of rock.

CONCEPT CHECK 4.11

❶ How can weathering create an ore deposit? What important ore is an example?

Mass Wasting: The Work of Gravity

Landslides are spectacular examples of a basic geologic process called *mass wasting*. Mass wasting refers to the downslope movement of rock, regolith, and soil under the direct influence of gravity. It is distinct from the erosional processes that are examined in subsequent chapters because mass wasting does not require a transporting medium such as water, wind, or glacial ice.

Earth's surface is never perfectly flat but instead consists of slopes. Some are steep and precipitous; others are moderate or gentle. Some are long and gradual; others are short and abrupt. Some slopes are mantled with soil and covered by vegetation; others consist of barren rock and rubble. Their form and variety are great. Although most slopes appear to be stable and unchanging, they are not static features because the force of gravity causes material to move. At one extreme the movement may be gradual and practically imperceptible. At the other extreme it may consist of a roaring debris flow or a thundering rock avalanche. Landslides are a worldwide natural hazard (Figure 4.23). When these natural processes lead to loss of life and property, they become natural disasters.

Most mass wasting, whether spectacular or subtle, is the result of circumstances that are completely independent of human activities. Very few landslides occur where they cannot be anticipated. In places where mass wasting is a recognized threat, steps can often be taken to control downslope movements or limit the damages that such movements can cause. If the potential for mass wasting goes unrecognized or is ignored, the results can be costly and dangerous. We should also note that although most downslope movements occur whether people are present or not, many occurrences are aggravated or even triggered by human actions.

The Role of Mass Wasting

In the evolution of most landforms, mass wasting is the step that follows weathering. By itself, weathering does not produce significant landforms. Rather, landforms develop as the products of weathering are removed from the places where they originate. Once weathering weakens rock and breaks it apart, mass wasting transfers the debris downslope, where a stream, acting as a conveyor belt, usually carries it away (see Figure 4.1, p. 85). Although there may be many intermediate stops along the way, the sediment is eventually transported to its ultimate destination: the sea.

FIGURE 4.23 On August 6, 2006, photographer Herb Dunn witnessed this rockfall along the Merced River in California's Yosemite National Park. It appears to have been an event that had no discernible trigger. In the photo on the left, the impact of falling rock produces an explosion of dust and debris. In the photo on the right, the rockfall triggers a chain reaction. A debris avalanche moves down the slope knocking down trees along the way. (© DunnRight Photograph)

The combined effects of mass wasting and running water produce stream valleys, which are the most common and conspicuous of Earth's landforms. If streams alone were responsible for creating the valleys in which they flow, the valleys would be very narrow features. However, the fact that most stream valleys are much wider than they are deep is a strong indication of the significance of mass-wasting processes in supplying material to streams. The walls of a canyon extend far from the stream because of the transfer of weathered debris downslope to the stream and its tributaries by mass-wasting processes. In this manner, streams and mass wasting combine to modify and sculpt the surface. Of course, glaciers, groundwater, waves, and wind are also important agents in shaping landforms and developing landscapes.

Slopes Change Through Time

It is clear that if mass wasting is to occur, there must be slopes that rock, soil, and regolith can move down. It is Earth's mountain-building and volcanic processes that produce these slopes through sporadic changes in the elevations of landmasses and the ocean floor. If dynamic internal processes did not continually produce regions having higher elevations, the system that moves debris to lower elevations would gradually slow and eventually cease.

Most rapid and spectacular mass-wasting events occur in areas of rugged, geologically young mountains. Newly formed mountains are rapidly eroded by rivers and glaciers into regions characterized by steep and unstable slopes. It is in such settings that massive destructive landslides occur. As mountain building subsides, mass wasting and erosional processes lower the land. Through time, steep and rugged mountain slopes give way to gentler, more subdued terrain. Thus, as a landscape ages, massive and rapid mass-wasting processes give way to smaller, less dramatic downslope movements.

CONCEPT CHECK 4.12

1 What is the controlling force of mass wasting?

2 Sketch or describe how mass wasting contributes to the development of a stream valley.

3 In what kind of setting do most rapid mass-wasting events occur?

Controls and Triggers of Mass Wasting

Sculpturing Earth's Surface
▸ **Mass Wasting: The Work of Gravity**

Gravity is the controlling force of mass wasting, but several factors play an important role in overcoming inertia and creating downslope movements. Long before a landslide occurs, various processes work to weaken slope material, gradually making it more and more susceptible to the pull of gravity. During this span, the slope remains stable but gets closer and closer to being unstable. Eventually, the strength of the slope is weakened to the point that something causes it to cross the threshold from stability to instability. Such an event that initiates downslope movement is called a *trigger*. Remember that the trigger is not the sole cause of the mass-wasting event, but just the last of many causes. Among the common factors that trigger mass-wasting processes are saturation of material with water, oversteepening of slopes, removal of anchoring vegetation, and ground vibrations from earthquakes.

The Role of Water

Mass wasting is sometimes triggered when heavy rains or periods of snowmelt saturate surface materials. This was the case in December 1999 when torrential rains triggered thousands of landslides along the coast of Venezuela (see Figure 1.20, p. 21) and in 1998 when rains from Hurricane Mitch led to the debris flow shown in **Figure 4.24**. A case study of another rain-triggered mass-wasting event that occurred at La Conchita, California, in January 2005, is found in Box 4.2.

When the pores in sediment become filled with water, the cohesion among particles is destroyed, allowing them to slide past one another with relative ease. For example, when sand is slightly moist, it sticks together quite well. However, if enough water is added to fill the openings between the grains, the sand will ooze out in all directions (**Figure 4.25**). Thus, saturation reduces the internal resistance of materials, which are then easily set in motion by the force of gravity. When clay is wetted, it becomes very slick—another example of the "lubricating" effect of water. Water also adds considerable weight to a mass of material. The added weight in itself may be enough to cause the material to slide or flow downslope.

Oversteepened Slopes

Oversteepening of slopes is another trigger of many mass movements. There are many situations in nature where this takes place. A stream undercutting a valley wall and waves pounding against the base of a cliff are two familiar examples. Furthermore, through their activities, people often create oversteepened and unstable slopes that become prime sites for mass wasting (**Figure 4.26**).

Unconsolidated, granular (sand-size or coarser) particles assume a stable slope called the **angle of repose** (*reposen* = to be at rest). This is the steepest angle at which material remains stable (**Figure 4.27**). Depending on the size and shape of the particles, the angle varies from 25 to 40 degrees. The larger, more angular particles maintain the steepest slopes. If the angle is increased, the rock debris will adjust by moving downslope.

Oversteepening is important not only because it triggers movements of unconsolidated granular materials, but it also produces unstable slopes and mass movements in cohesive soils, regolith, and bedrock. The response will not be immediate, as with loose, granular material, but sooner or later one or more mass-wasting processes will eliminate the oversteepening and restore stability to the slope.

Removal of Vegetation

Plants protect against erosion and contribute to the stability of slopes because their root systems bind soil and regolith together. In addition, plants shield the soil surface from the erosional

Debris flow

FIGURE 4.24 Many large debris flows and floods were triggered by the torrential rains that accompanied Hurricane Mitch when it struck Honduras in October 1998. It was that country's worst natural disaster in 200 years. Pictured here is the El Berrinche landslide, one of two that destroyed portions of the city of Tegucigalpa, killing more than 1,000 people and damming the Rio Choluteca. The mass of debris flow material was estimated to be 6 million cubic meters, enough to fill nearly 300,000 average-size dump trucks! (Photos by Michael Collier)

effects of raindrop impact (see Figure 4.19). Where plants are lacking, mass wasting is enhanced, especially if slopes are steep and water is plentiful. When anchoring vegetation is removed by forest fires or by people (for timber, farming, or development), surface materials frequently move downslope.

In July 1994 a severe wildfire swept Storm King Mountain west of Glenwood Springs, Colorado, denuding the slopes of vegetation. Two months later heavy rains resulted in numerous debris flows, one of which blocked Interstate 70 and threatened

FIGURE 4.25 The effect of water on mass wasting can be great. When little or no water is present, friction among the closely packed soil particles on the slope holds them in place. When the soil is saturated, the grains are forced apart and friction is reduced, allowing the soil to move downslope.

to dam the Colorado River. A 5-kilometer (3-mile) length of the highway was inundated with tons of rock, mud, and burned trees. The closure of Interstate 70 imposed costly delays on this major highway. Events such as this are relatively common in the American West where slopes are often steep and summer wildfires may affect millions of acres each year (**Figure 4.28**).

Earthquakes as Triggers

Conditions favoring mass wasting can exist in an area for a long time without movement occurring. An additional factor is sometimes necessary to trigger the movement. Among the more important and dramatic triggers are earthquakes. An earthquake and its aftershocks can dislodge enormous volumes of rock and unconsolidated material. The mass-wasting event shown in **Figure 4.29** was triggered by an earthquake. In many areas that are jolted by earthquakes, it is not ground vibrations directly but landslides and ground subsidence triggered by the vibrations that cause the greatest damage.

Dry sand

Damp sand

Wet sand

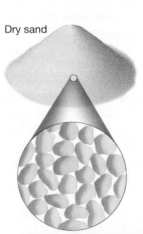
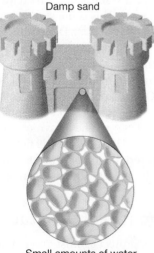
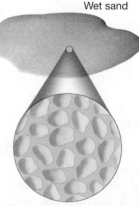

Dry sand grains are bound mainly by friction with one another

Small amounts of water increase the cohesion among sand grains

Saturation reduces friction and causes the sand to flow

CONCEPT CHECK 4.13

❶ How does water affect mass-wasting processes?
❷ Describe the significance of the angle of repose.
❸ How might a forest fire influence mass wasting?
❹ Link earthquakes to landslides.

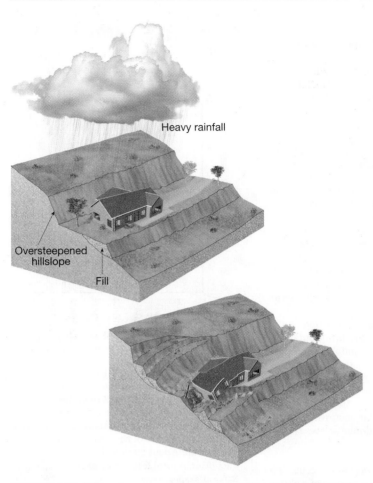

FIGURE 4.26 When slopes are oversteepened they become unstable and are prime sites for mass wasting. Natural processes such as stream and wave erosion can create oversteepened slopes. Changing the slope to accommodate a new house or road can also lead to a destructive mass-wasting event.

FIGURE 4.27 The angle of repose for this granular material is about 30°. (Photo by G. Leavens/Photo Researchers, Inc.)

steep for loose material to remain on the surface (see Figure 4.29). Many falls result when freeze and thaw cycles or the action of plant roots loosen rock to the point that gravity takes over. Rockfall is the primary way in which talus slopes are built and maintained (see Figure 4.4). Sometimes falls may trigger other forms of downslope movement.

Many mass-wasting processes are **slides**, which occur whenever material remains fairly coherent and moves along a well-defined surface. Sometimes the surface is a joint, a fault, or a bedding plane that is roughly parallel to the slope. However, in the movement called *slump*, the descending material moves en masse along a curved surface of rupture.

A third type of movement common to mass wasting is termed **flow**. Flow occurs when material moves downslope as a viscous fluid. Most flows are saturated with water and typically move as lobes or tongues.

Classifying Mass-Wasting Processes

Geologists include several processes under the name of mass wasting. Four are illustrated in Figure 4.30. Generally, each process is defined by the type of material involved, the kind of motion, and the velocity of the movement.

If soil and regolith dominate, terms such as *debris*, *mud*, or *earth* are used. In contrast, when a mass of bedrock breaks loose and moves downslope, the term *rock* may be part of the description. Generally, the kind of motion is described as either a fall, a slide, or a flow.

Type of Motion

When the movement involves the freefall of detached individual pieces of any size, it is termed a **fall**. Falls are common on slopes that are too

FIGURE 4.28 During the summer, wildfires are common occurrences in many parts of the west. Millions of acres are burned each year. The loss of anchoring vegetation sets the stage for accelerated mass wasting. (Photo by Raymond Gehman/National Geographic Stock)

Box 4.2

PEOPLE AND THE ENVIRONMENT

Landslide Hazards at La Conchita, California*

Southern California lies astride a major plate boundary defined by the San Andreas Fault and numerous other related faults that are spread across the region. It is a dynamic environment characterized by rugged mountains and steep-walled canyons. Unfortunately, this scenic landscape presents serious geologic hazards. Just as tectonic forces are steadily pushing the landscape upward, gravity is relentlessly pulling it downward. When gravity prevails, landslides occur.

As you might expect, some of the region's landslides are triggered by earthquakes. Many others, however, are related

*Based in part on material prepared by the U.S. Geological Survey.

to periods of prolonged and intense rainfall. A tragic example of the latter situation occurred on January 10, 2005, when a massive debris flow (popularly called a *mudslide*) swept through La Conchita, California, a small town located about 80 kilometers (50 miles) northwest of Los Angeles (**Figure 4.B**).

Although the rapid torrent of mud took many of the town's inhabitants by surprise, such an event should not have been unexpected. Let's briefly examine the factors that contributed to the deadly debris flow at La Conchita.

The town is situated on a narrow coastal strip about 250 meters (800 feet) wide between the shoreline and a steep 180-meter (600-foot) bluff. The bluff consists of poorly sorted marine sediments and weakly cemented layers of shale, siltstone, and sandstone.

The deadly 2005 debris flow involved little or no newly failed material, but rather consisted of the remobilization of a portion of a large landslide that destroyed several homes in 1995. In fact, historical accounts dating back to 1865 indicate that landslides in the immediate area have been a regular occurrence. Furthermore, geologic evidence shows that

FIGURE 4.C Daily rainfall at the nearby town of Ventura during the weeks leading up to the January 2005 La Conchita event. Each line on the bar graph shows rain for a particular day. The 2005 debris flow occurred at the culmination of the heaviest rainfall of the season. About 80 percent of the season's exceptional total fell in this short span. (After National Weather Service)

landsliding of a variety of types and scales has probably been occurring at La Conchita for thousands of years.

The most significant contributing factor to the tragic 2005 debris flow was prolonged and intense rain. The event occurred at the end of a span that produced near record amounts of rainfall in southern California. Wintertime rainfall at nearby Ventura totaled 49.3 centimeters (19.4 inches) as compared to an average value of just 12.2 centimeters (4.8 inches). As **Figure 4.C** indicates, much of that total fell during the two weeks immediately preceding the debris flow.

This was not the first destructive landslide to strike La Conchita, nor is it likely to be the last. The town's geologic setting and history of rapid mass-wasting events clearly support this notion. When the amount and intensity of rainfall is sufficient, debris flows are to be expected.

FIGURE 4.B Views of the La Conchita debris flow shortly after it occurred in January 2005. The light-colored exposed rock in the upper part of the photo on the left is the main scarp of a slide that occurred 10 years earlier in 1995. The January 2005 event was a remobilization of the 1995 event. The flow was quite thick (viscous) and moved houses in its path rather than flowing around them. (AP Wide World Photos)

Scarp created by 1995 landslide

Area outlined at bottom of image on left

2005 debris flow

FIGURE 4.29 Rockfall blocking the Karakoram Highway in Pakistan's Hunza Valley. (Photo by Robert Holmes/CORBIS)

events called *rock avalanches*, rock and debris can hurtle downslope at speeds exceeding 200 kilometers (125 miles) per hour. Researchers now understand that rock avalanches, such as the one that produced the scene in Figure 4.31, must literally "float on air" as they move downslope. That is, high velocities result when air becomes trapped and compressed beneath the falling mass of debris, allowing it to move as a buoyant, flexible sheet across the surface.

Most mass movements, however, do not move with the speed of a rock avalanche. In fact, a great deal of mass wasting is imperceptibly slow. One process that we will examine later, termed *creep*, results in particle movements that are usually measured in millimeters or centimeters per year. Thus, as you can see, rates of movement can be spectacularly sudden or exceptionally gradual. Although various types of mass wasting are often classified as either rapid or slow, such a distinction is highly subjective because a wide range of rates exists between the two extremes. Even the velocity of a single process at a particular site can vary considerably from one time to another.

Rate of Movement

When mass-wasting events make the news, a large quantity of material has in all likelihood moved rapidly downslope and has had a disastrous effect on people and property. Indeed, during

CONCEPT CHECK 4.14

❶ What terms describe the way material moves during mass wasting?

❷ Why can rock avalanches move at such great speeds?

FIGURE 4.30 The four processes illustrated here are all considered to be relatively rapid forms of mass wasting. Because material in slumps **A.** and rockslides **B.** move along well-defined surfaces, they are said to move by sliding. By contrast, when material moves downslope as a viscous fluid, the movement is described as a flow. Debris flow **C.** and earthflow **D.** advance downslope in this manner.

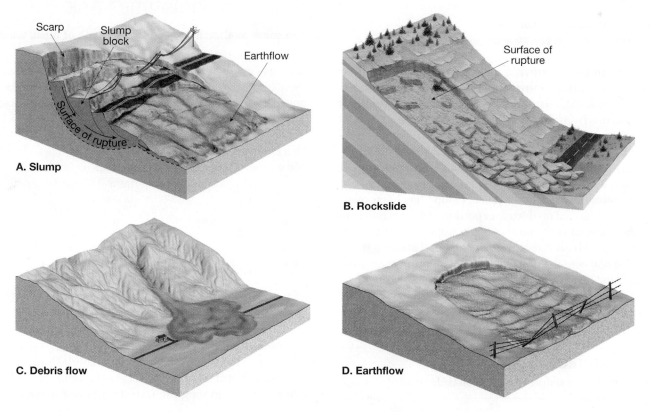

A. Slump

B. Rockslide

C. Debris flow

D. Earthflow

FIGURE 4.31 Aerial view of the Blackhawk landslide, a prehistoric event that occurred on the north slope of California's San Bernardino Mountains. It is one of the largest known landslides in North America. This 8-kilometer-long tongue of rubble is about 3 kilometers wide and 9–30 meters thick. Research has shown that the mass of debris raced to its resting place on a cushion of compressed air. (Photo by Michael Collier)

Slump

Sculpturing Earth's Surface
▶ Mass Wasting: The Work of Gravity

Slump refers to the downward sliding of a mass of rock or unconsolidated material moving as a unit along a *curved* surface (see Figure 4.30A). Usually the slumped material does not travel spectacularly fast nor very far. This is a common form of mass wasting, especially in thick accumulations of cohesive materials such as clay. As the movement occurs, a crescent-shaped scarp (cliff) is created at the head, and the block's upper surface is sometimes tilted backward.

Slump commonly occurs because a slope has been oversteepened. The material on the upper portion of a slope is held in place by the material at the bottom of the slope. As this anchoring material at the base is removed, the material above is made unstable and reacts to the pull of gravity. A common example is a valley wall that becomes oversteepened by a meandering river. Another is a coastal area that has been undercut by wave activity at its base.

CONCEPT CHECK 4.15

❶ Sketch and label a simple cross-section (side view) of a slump.

Students Sometimes Ask...

Are snow avalanches considered a type of mass wasting?

Sure. Sometimes these thundering downslope movements of snow and ice move large quantities of rock, soil, and trees. Of course, snow avalanches are very dangerous, especially to skiers on high mountain slopes and to buildings and roads at the bottoms of slopes in avalanche-prone regions.

About 10,000 snow avalanches occur each year in the mountainous western United States. In an average year they claim between 15 and 25 lives in the United States and Canada. They are a growing problem as more people become involved in winter sports and recreation.

Rockslide

Sculpturing Earth's Surface
▶ Mass Wasting: The Work of Gravity

Rockslides occur when blocks of bedrock break loose and slide down a slope (see Figure 4.30B). If the material is mostly unconsolidated soil and regolith, the term *debris slide* is used instead. Such events are among the fastest and most destructive mass movements. Usually rockslides take place in a geologic setting where the rock strata are inclined or where joints and fractures exist parallel to the slope. When such a rock unit is undercut at the base of the slope, it loses support and the rock eventually gives way.

Sometimes an earthquake is the trigger. On other occasions the rockslide is triggered when rain or melting snow lubricates the underlying surface to the point that friction is no longer sufficient to hold the rock unit in place. As a result, rockslides tend to be more common during the spring, when heavy rains and melting snow are most prevalent. The massive Gros Ventre slide shown in Figure 4.32 is a classic example.

CONCEPT CHECK 4.16

1. Both slump and rockslide move by sliding. How do these processes differ?
2. What factors led to the massive rockslide at Gros Ventre, Wyoming?

Debris Flow

Sculpturing Earth's Surface
▶ Mass Wasting: The Work of Gravity

Debris flow is a relatively rapid type of mass wasting that involves a flow of soil and regolith containing a large amount of water (see Figure 4.30C). Several debris flows are pictured in this book—see Figure 1.20, Figure 4.24, and Figure 4.B. Debris flows are sometimes called **mudflows** when the material is primarily fine-grained. Although they can occur in many different climate settings, they probably occur more frequently in semiarid mountainous regions. Debris flows, called *lahars*, are also common on the steep slopes of some volcanoes. Because of their fluid properties, debris flows frequently follow canyons and stream channels. In populated areas, debris flows can pose a significant hazard to life and property.

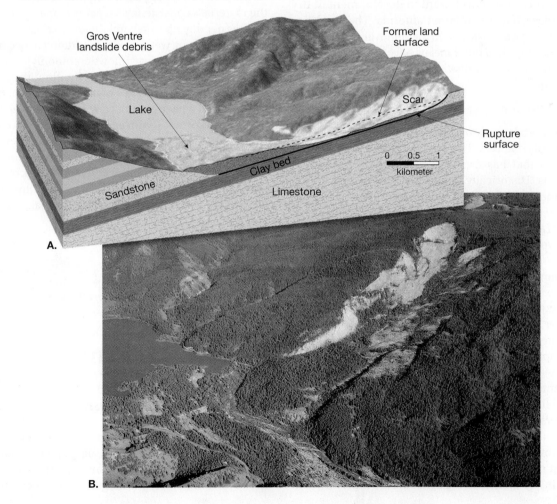

FIGURE 4.32 On June 23, 1925, a massive rockslide took place in the valley of the Gros Ventre River in northwestern Wyoming following heavy spring rains and snowmelt. The volume of debris, estimated at 38 million cubic meters, created a 70-meter-high dam. Later, the lake created by the debris dam overflowed, resulting in a devastating flood downstream. **A.** Cross-sectional view. The slide occurred when the tilted and undercut sandstone bed could no longer maintain its position atop the saturated bed of clay. **B.** Even though the Gros Ventre rockslide occurred in 1925, the scar left on the side of Sheep Mountain is still a prominent feature. (Part A after W. C. Alden, "Landslide and Flood at Gros Ventre, Wyoming," *Transactions* (AIME) 76 (1928): 348. Part B photo by Michael Collier)

Students Sometimes Ask...

I noticed that none of the mass-wasting processes described in this chapter are actually called "landslides." Why?

That's very observant! Although many people, including geologists, frequently use the word *landslide*, the term has no specific definition in geology. Rather, it should be considered as a popular nontechnical term to describe all relatively rapid forms of mass wasting, including those in which sliding does not occur.

Debris Flows in Semiarid Regions

When a cloudburst or rapidly melting mountain snows create a sudden flood in a semiarid region, large quantities of soil and regolith are washed into nearby stream channels because there is usually little vegetation to anchor the surface material. The end product is a flowing tongue of well mixed mud, soil, rock, and water. Its consistency may range from that of wet concrete to a soupy mixture not much thicker than muddy water. The rate of flow therefore depends not only on the slope but also on the water content. When dense, debris flows are capable of carrying or pushing large boulders, trees, and even houses with relative ease.

Debris flows pose a serious hazard to development in dry mountainous areas such as southern California. The construction of homes on canyon hillsides and the removal of anchoring vegetation by brush fires and other means have increased the frequency of these destructive events.

Lahars

Debris flows composed mostly of volcanic materials on the flanks of volcanoes are called **lahars**. The word originated in Indonesia, a volcanic region that has experienced many of these often destructive events. Historically, lahars have been some of the deadliest volcano hazards. They can occur either during an eruption or when a volcano is quiet. They take place when highly unstable layers of ash and debris become saturated with water and flow down steep volcanic slopes, generally following existing stream channels. Heavy rainfalls often trigger these flows. Others are triggered when large volumes of ice and snow are suddenly melted by heat flowing to the surface from within the volcano or by the hot gases and near-molten debris emitted during a violent eruption.

In November 1985 lahars were produced when Nevado del Ruiz, a 5,300-meter (17,400-foot) volcano in the Andes Mountains of Colombia, erupted. The eruption melted much of the snow and ice that capped the uppermost 600 meters (2,000 feet) of the peak, producing torrents of hot, thick mud, ash, and debris. The lahars moved outward from the volcano, following the valleys of three rain-swollen rivers that radiate from the peak. The flow that moved down the valley of the Lagunilla River was the most destructive, devastating the town of Armero, 48 kilometers (30 miles) from the mountain. Most of the more than 25,000 deaths caused by the event occurred in this once-thriving agricultural community.

CONCEPT CHECK 4.17
❶ How is a lahar different from a debris flow that might occur in southern California?

Earthflow

Sculpturing Earth's Surface
▸ Mass Wasting: The Work of Gravity

We have seen that debris flows are frequently confined to channels in semiarid regions. In contrast, **earthflows** most often form on hillsides in humid areas during times of heavy precipitation or snowmelt (see Figure 4.30D). When water saturates the soil and regolith on a hillside, the material may break away, leaving a scar on the slope and forming a tongue- or teardrop-shaped mass that flows downslope (**Figure 4.33**). The materials most commonly involved are rich in clay and silt and contain only small proportions of sand and coarser particles. Earthflows range in size from bodies a few meters long, a few meters wide, and less than 1 meter deep to masses more than 1 kilometer long, several hundred meters wide, and more than 10 meters deep.

Because earthflows are quite viscous, they generally move at slower rates than the more fluid debris flows described in the preceding section. They move slowly and persistently for periods ranging from days to years. Depending on slope steepness and the material's consistency, velocities range from less than 1 millimeter per day up to several meters per day. Movement is typically faster during wet periods. In addition to occurring as isolated hillside phenomena, earthflows commonly take place in association with large slumps. In this situation, they may be seen as tonguelike flows at the base of the slump.

A special type of earthflow, known as *liquefaction*, sometimes occurs in association with earthquakes. Porous clay- to sand-size sediments that are saturated with water are most vulnerable. When shaken suddenly, the grains lose cohesion and the ground flows. Liquefaction can cause buildings to sink or tip on their sides and underground storage tanks and sewer lines to float upward. To say the least, damage can be substantial. You will learn more about this in Chapter 8.

CONCEPT CHECK 4.18
❶ Contrast earthflow with debris flow.

Slow Movements

Sculpturing Earth's Surface
▸ Mass Wasting: The Work of Gravity

Movements such as rockslides, rock avalanches, and lahars are certainly the most spectacular and catastrophic forms of mass wasting. These dangerous events deserve intensive study to enable more effective prediction, timely warnings, and better controls to save lives.

FIGURE 4.33 This small, tongue-shaped earthflow occurred on a newly formed slope along a recently constructed highway. It formed in clay-rich material following a period of heavy rain. Notice the small slump at the head of the earthflow. (Photo by E. J. Tarbuck)

However, because of their large size and spectacular nature, they give us a false impression of their importance as a mass-wasting process. Indeed, sudden movements transport less material than the slow, subtle action of creep. Whereas rapid types of mass wasting are characteristic of mountains and steep hillsides, creep takes place on both steep and gentle slopes and is thus much more widespread.

Creep

Creep is a type of mass wasting that involves the gradual downhill movement of soil and regolith. One factor that contributes to creep is the alternate expansion and contraction of surface material caused by freezing and thawing or wetting and drying. As shown in **Figure 4.34**, freezing or wetting lifts particles at right angles to the slope, and thawing or drying allows the particles to fall back to a slightly lower level. Each cycle therefore moves the material a short distance downhill.

Creep is aided by anything that disturbs the soil. For example, raindrop impact and disturbance by plant roots and burrowing animals may contribute. Creep is also promoted when the ground becomes saturated with water. Following a heavy rain or snowmelt, a waterlogged soil may lose its internal cohesion, allowing gravity to pull the material down-slope. Because creep is imperceptibly slow, the process cannot be observed in action. What can be observed, however, are the effects of creep. Creep causes fences and utility poles to tilt and retaining walls to be displaced.

Solifluction

When soil is saturated with water, the soggy mass may flow downslope at a rate of a few millimeters or a few centimeters per day or per year. Such a process is called **solifluction** (literally, "soil flow"). It is a type of mass wasting that is common wherever water cannot escape from the saturated surface layer by infiltrating to deeper levels. A dense clay hardpan in soil or an impermeable bedrock layer can promote solifluction.

Solifluction is common in regions underlain by **permafrost**. Permafrost refers to the permanently frozen ground that occurs in Earth's harsh tundra and ice-cap climates. It occurs in a zone above the permafrost called the *active layer*, which thaws in summer and refreezes in winter. During summer, water is unable

FIGURE 4.34 The repeated expansion and contraction of the surface material causes a net downslope migration of rock particles—a process called *creep*.

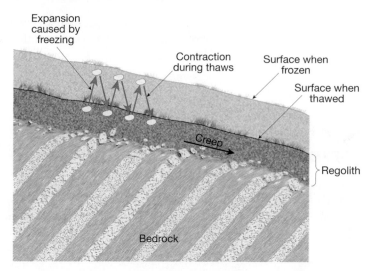

to percolate into the impervious permafrost layer below. As a result, the active layer becomes saturated and slowly flows. The process can occur on slopes as gentle as 2–3 degrees. Where there is a well-developed mat of vegetation, a solifluction sheet may move in a series of well-defined lobes or overriding folds (Figure 4.35).

CONCEPT CHECK 4.19

❶ Describe the basic mechanisms that contribute to creep.
❷ During what season does solifluction occur? Explain why it occurs only during that time of year.

Students Sometimes Ask...

How many deaths and how much damage is caused by mass wasting each year?

The U.S. Geological Survey estimates that between 25 and 50 people are killed by landslides annually in the United States. The worldwide death toll, of course, is much higher. The annual cost of damage by all types of mass wasting in the United States is about $4 billion. However, this figure is probably conservative because slow movements like creep do a substantial amount of damage that goes unreported.

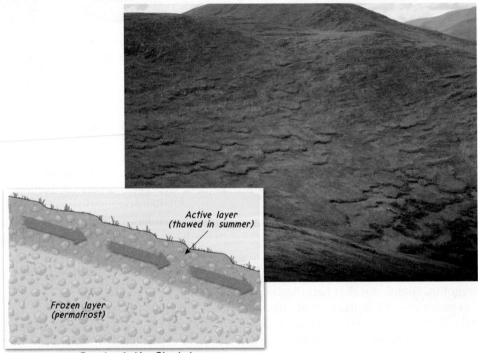

Active layer (thawed in summer)

Frozen layer (permafrost)

Geologist's Sketch

FIGURE 4.35 Solifluction lobes northeast of Fairbanks, Alaska. Solifluction occurs in permafrost regions when the active layer thaws in summer. (Photo by James E. Patterson)

GIVE IT SOME THOUGHT

1. Describe how plants promote mechanical and chemical weathering but inhibit erosion.
2. Granite and basalt are exposed at Earth's surface in a hot, wet region. Will mechanical weathering or chemical weathering predominate? Which rock will weather more rapidly? Why?

3. The accompanying photo shows a footprint on the Moon left by an *Apollo* astronaut in material popularly called *lunar soil*. Does this material satisfy the definition we use for soil on Earth? Explain why or why not. You may want to refer to Figure 4.12.

4. What might cause different soils to develop from the same kind of parent material, or similar soils to form from different parent materials?

5. Refer to Figure 4.18 to identify the main soil order in the region adjacent to South America's Amazon River and the main soil order in the American Southwest. Briefly contrast these soils. Do they have anything in common? Referring to Table 4.2 might be helpful.

6. The concept of external and internal processes was introduced at the beginning of the chapter. In the accompanying photo, external processes are clearly active. Describe the role mass wasting is playing. Identify one feature that is the result of mass wasting.

7. Describe at least one situation in which an internal process might cause or contribute to a mass-wasting process.

8. Mass wasting is influenced by many processes associated with all four spheres of the Earth system. Select three items from the list below. For each, outline a series of events that relate the item to various spheres and to a mass-wasting process. Here is an example that assumes "frost wedging" is an item on the list: *Frost wedging involves rock (geosphere) being broken when water (hydrosphere) freezes. Freeze–thaw cycles (atmosphere) promote frost wedging. When frost wedging loosens a rock on a cliff, the fragment tumbles to the base of the cliff. This event, called a rockfall, is an example of mass wasting.* Now you give it a try. Use your imagination.

  a. Wildfire

  b. Spring thaw/melting snow

  c. Highway road cut

  d. Crashing waves

  e. Cavern formation (see Figures 5.42 and 5.43)

In Review   Chapter 4 Rocks: Weathering, Soil, and Mass Wasting

- External processes include (1) *weathering*—the disintegration and decomposition of rock at or near the surface, (2) *mass wasting*—the transfer of rock material downslope under the influence of gravity, and (3) *erosion*—the incorporation and transportation of material by a mobile agent, usually water, wind, or ice. They are called *external processes* because they occur at or near Earth's surface and are powered by energy from the Sun. By contrast, *internal processes*, such as volcanism and mountain building, derive their energy from Earth's interior.

- *Mechanical weathering* is the physical breaking-up of rock into smaller pieces. *Chemical weathering* alters a rock's chemistry, changing it into different substances. Rocks can be broken into smaller fragments by *frost wedging, salt crystal growth, unloading,* and *biological activity*. Water is by far the most important agent of chemical weathering. Oxygen in water can *oxidize* some materials, while carbon dioxide (CO_2) dissolved in water forms *carbonic acid*. The

chemical weathering of silicate minerals frequently produces (1) soluble products containing sodium, calcium, potassium, and magnesium; (2) insoluble iron oxides; and (3) clay minerals.

- The rate at which rock weathers depends on such factors as (1) *particle size*—small pieces generally weather faster than large pieces; (2) *mineral makeup*—calcite readily dissolves in mildly acidic solutions, and silicate minerals that form first from magma are least resistant to chemical weathering; and (3) *climatic factors*, particularly temperature and moisture. Frequently, rocks exposed at Earth's surface do not weather at the same rate. This *differential weathering* of rocks is influenced by such factors as mineral makeup and degree of jointing.

- *Soil* is a combination of mineral and organic matter, water, and air—that portion of the *regolith* (the layer of rock and mineral fragments produced by weathering) that supports the growth of plants. *Soil texture* refers to the proportions of different particle sizes (clay, silt, and sand) found in soil. The most important *factors that control soil formation are parent material, time, climate, plants and animals, and topography.*

- Soil-forming processes operate from the surface downward and produce zones or layers in the soil called *horizons.* From the surface downward the horizons are designated as O, A, E, B, and C, respectively.

- In the United States, soils are classified using a system known as the *Soil Taxonomy.* It is based on physical and chemical properties of the soil profile and includes six hierarchical categories. The system is especially useful for agricultural and related land-use purposes.

- Soil erosion is a natural process; it is part of the constant recycling of Earth materials that we call the *rock cycle.* Rates of soil erosion vary from one place to another and depend on the soil's characteristics as well as such factors as climate, slope, and type of vegetation. Human activities have greatly accelerated the rate of soil erosion in many areas.

- Weathering creates mineral deposits by concentrating metals into economically valuable deposits. The process, called *secondary enrichment,* is accomplished by either (1) remov-

ing undesirable materials and leaving the desired elements enriched in the upper zones of the soil or (2) removing and carrying the desirable elements to lower soil zones where they are redeposited and thus become more concentrated. *Bauxite,* the principal ore of aluminum, is one important ore created by secondary enrichment.

- In the evolution of most landforms, mass wasting is the step that follows weathering. The combined effect of mass wasting and erosion by running water produce stream valleys. *Gravity is the controlling force of mass wasting.* Other factors that influence or trigger downslope movements are saturation of the material with water, oversteepening of slopes beyond the *angle of repose,* removal of anchoring vegetation, and ground vibrations from earthquakes.

- The various processes included under the name of mass wasting are classified and described on the basis of (1) the type of material involved (debris, mud, earth, or rock), (2) the kind of motion (fall, slide, or flow), and (3) the rate of movement (fast, slow). The various kinds of mass wasting include the more rapid forms called *slump, rockslide, debris flow,* and *earthflow,* as well as the slow movements referred to as *creep* and *solifluction.*

Key Terms

angle of repose (p. 103)
chemical weathering (p. 86)
creep (p. 111)
debris flow (p. 109)
differential weathering (p. 91)
earthflow (p. 110)
eluviation (p. 97)
erosion (p. 84)
exfoliation dome (p. 87)
external processes (p. 84)
fall (p. 105)
flow (p. 105)

frost wedging (p. 86)
horizon (p. 96)
internal processes (p. 84)
lahar (p. 110)
leaching (p. 97)
mass wasting (p. 84)
mechanical weathering (p. 86)
mudflow (p. 109)
parent material (p. 94)
permafrost (p. 111)
regolith (p. 92)
rockslide (p. 108)

secondary enrichment (p. 101)
sheeting (p. 87)
slide (p. 105)
slump (p. 108)
soil (p. 92)
soil profile (p. 96)
Soil Taxonomy (p. 97)
soil texture (p. 93)
solifluction (p. 111)
solum (p. 97)
spheroidal weathering (p. 90)
weathering (p. 84)

Examining the Earth System

1. Which of Earth's spheres are associated with the production of carbonic acid?

2. The level of carbon dioxide (CO_2) in the atmosphere has been increasing for more than a century. Should this increase tend to accelerate or slow down the rate of chemical weathering of Earth's surface rocks? Explain how you arrived at your conclusion.

3. Many gases are emitted into the atmosphere as a result of human activities. List two gases other than carbon dioxide that create acids and therefore accelerate the rate of chemical weathering. Do these acids have other effects on the Earth system? You will find the information about acid rain at this Web site helpful: http://www.epa.gov/acidrain

4. Discuss the interaction of the atmosphere, geosphere, biosphere, and hydrosphere in the formation of soil.

Mastering Geology

Looking for additional review and test prep materials? Visit the Self Study area in **www.masteringgeology.com** to find practice quizzes, study tools, and multimedia that will aid in your understanding of this chapter's content. In **MasteringGeology™** you will find:

- **GEODe: Earth Science: An interactive visual walkthrough of key concepts**

- **Geoscience Animation Library: More than 100 animations illuminating many difficult-to-understand Earth science concepts**
- **In The News RSS Feeds: Current Earth science events and news articles are pulled into the site with assessment**
- **Pearson eText**
- **Optional Self Study Quizzes**
- **Web Links**
- **Glossary**
- **Flashcards**

Running Water and Groundwater

5

Sunset along the Li River in China's Guilin District. This part of southern China exhibits tower karst development in which groundwater has dissolved large volumes of limestone, leaving only these residual towers. For more about this landscape, see the section "Karst Topography" at the end of the chapter.

(Photo by MOODBOARD/ CORBIS)

"**A**ll the rivers run into the sea; yet the sea is not full; unto the place from whence the rivers come, thither they return again." (Ecclesiastes 1:7)

As the perceptive writer of Ecclesiastes indicated, water is continually on the move, from the ocean to the land and back again in an endless cycle. This chapter deals with that part of the hydrologic cycle that returns water to the sea. Some water travels quickly via a rushing stream, and some moves more slowly below the surface. When viewed as part of the Earth system, streams and groundwater represent basic links in the constant cycling of the planet's water. In this chapter, we examine the factors that influence the distribution and movement of water, as well as look at how water sculptures the landscape. To a great extent, the Grand Canyon, Niagara Falls, Old Faithful, and Mammoth Cave all owe their existence to the action of water on its way to the sea.

FOCUS ON CONCEPTS

To assist you in learning **the important concepts in this chapter, focus on the following questions:**

- What is the hydrologic cycle? What is the source of energy that powers the cycle?
- What are the three main zones of a river system?
- What are the factors that determine the flow velocity of a stream?
- The work of a stream includes what three processes?
- What are the two general types of stream valleys and some features associated with each?
- How are deltas, natural levees, and alluvial fans similar? How are they different?
- Why do floods occur? What are some basic flood-control strategies?
- Why is groundwater important?
- What is groundwater, and what factors influence its storage and movement?
- How do springs, geysers, wells, and artesian systems form?
- What are some environmental problems associated with groundwater?
- What features are produced by the geologic work of groundwater? What is karst topography?

Earth as a System: The Hydrologic Cycle

GEODe
Sculpturing Earth's Surface
EARTH SCIENCE ▶ **Running Water**

Water is just about everywhere on Earth—in the oceans, glaciers, rivers, lakes, air, soil, and in living tissue (**Figure 5.1**). All of these reservoirs constitute Earth's hydrosphere. In all, the water content of the hydrosphere comprises about 1.36 billion cubic kilometers (326 million cubic miles). The vast bulk of it, about 97.2 percent, is stored in the global oceans. Ice sheets and glaciers account for another 2.15 percent, leaving only 0.65 percent to be divided among lakes, streams, groundwater, and the atmosphere (see Figure 1.11, p. 12). Although the percentages of Earth's total water found in each of the latter sources is but a small fraction of the total inventory, the absolute quantities are great.

Water can readily change from one state of matter (solid, liquid, or gas) to another at the temperatures and pressures occurring at Earth's surface. Therefore, water is constantly moving among the oceans, the atmosphere, the solid Earth, and the biosphere. This unending circulation of Earth's water supply is called the **hydrologic cycle**. The cycle shows us many critical interrelationships among different parts of the Earth system.

The hydrologic cycle is a gigantic worldwide system powered by energy from the Sun in which the atmosphere provides the vital link between the oceans and continents (**Figure 5.2**). Water evaporates into the atmosphere from the ocean and to a much lesser extent from the continents. Winds transport this moisture-laden air, often great distances, until conditions cause the moisture to condense into clouds and to precipitate and fall. The precipitation that falls into the ocean has completed its cycle and is ready to begin another. The water that falls on the continents, however, must make its way back to the ocean.

What happens to precipitation once it has fallen on land? A portion of the water soaks into the ground (called **infiltration**), slowly moving downward, then laterally, finally seeping into lakes, streams, or directly into the ocean. When the rate of rainfall exceeds Earth's ability to absorb it, the surplus water flows over the surface into lakes and streams, a process called **runoff**. Much of the water that infiltrates or runs off eventually

FIGURE 5.1 Stream in New York's Lake George Wild Forest. Streams are a basic part of the hydrologic cycle. (Photo by Radius Images/Photolibrary)

FIGURE 5.2 The hydrologic cycle. About 320,000 cubic kilometers of water evaporate each year from the oceans, while evaporation from the land (including lakes and streams) contributes 60,000 cubic kilometers of water. Of this total of 380,000 cubic kilometers of water, about 284,000 cubic kilometers fall back to the ocean, and the remaining 96,000 cubic kilometers fall on Earth's land surface. Since 60,000 cubic kilometers of water evaporate from the land, 36,000 cubic kilometers of water remain to erode the land during the journey back to the ocean.

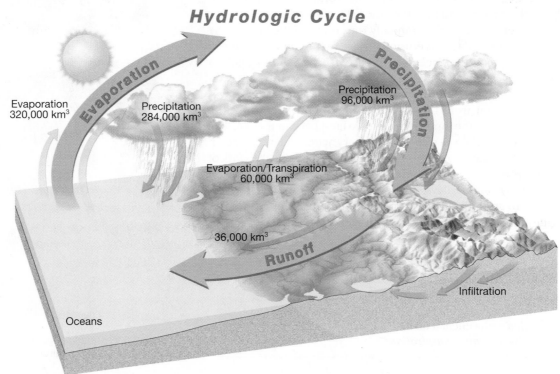

returns to the atmosphere because of evaporation from the soil, lakes, and streams. Also, some of the water that infiltrates the ground surface is absorbed by plants, which then release it into the atmosphere. This process is called **transpiration** (*trans* = across, *spiro* = to breathe). Because we cannot clearly distinguish between the amount of water that is evaporated and the amount that is transpired by plants, the term **evapotranspiration** is often used for the combined effect.

When precipitation falls in very cold areas—at high elevations or high latitudes—the water may not immediately soak in, run off, or evaporate. Instead, it may become part of a snowfield or a glacier. In this way, glaciers store large quantities of water on land. If present-day glaciers were to melt and release all their water, sea level would rise by several dozen meters. This would submerge many heavily populated coastal areas. As you will see in Chapter 6, over the past 2 million years, huge ice sheets have formed and melted on several occasions, each time changing the balance of the hydrologic cycle.

Figure 5.2 also shows Earth's overall *water balance*, or the volume of water that passes through each part of the cycle annually. The amount of water vapor in the air at any one time is just a tiny fraction of Earth's total water supply. But the *absolute* quantities that are cycled through the atmosphere over a one-year period are immense—some 380,000 cubic kilometers—enough to cover Earth's entire surface to a depth of about 1 meter (39 inches). Estimates show that over North America almost six times more water is carried by moving currents of air than is transported by all the continents' rivers.

It is important to know that the hydrologic cycle is *balanced*. Because the total amount of water vapor in the atmosphere remains about the same, the average annual precipitation over Earth must be equal to the quantity of water evaporated. However, for all of the continents taken together, precipitation exceeds evaporation. Conversely, over the oceans, evaporation exceeds precipitation. Because the level of the world ocean is not dropping, the system must be in balance. In Figure 5.2, the 36,000 cubic kilometers of water that annually runs off from the land to the ocean causes enormous erosion. In fact, this immense volume of moving water is the *single most important agent sculpting Earth's land surface*.

To summarize, the hydrologic cycle is the continuous movement of water from the oceans to the atmosphere, from the atmosphere to the land, and from the land back to the sea. The land-back-to-the-sea step is the primary action that wears down Earth's land surface. In this chapter, we first observe the work of water running over the surface, including floods, erosion, and the formation of valleys. Then we look underground at the slow labors of groundwater as it forms springs and caverns and provides drinking water on its long migration to the sea.

CONCEPT CHECK 5.1

1. Describe or sketch the movement of water through the hydrologic cycle. Once precipitation has fallen on land, what paths might it take?
2. What is *evapotranspiration?*

Running Water

Sculpturing Earth's Surface
▸ Running Water

Recall that most of the precipitation that falls on land either enters the soil (infiltration) or remains at the surface, moving downslope as runoff. The amount of water that runs off rather than soaking into the ground depends on several factors: (1) intensity and duration of rainfall, (2) amount of water already in the soil, (3) nature of the surface material, (4) slope of the land, and (5) the extent and type of vegetation. When the surface material is highly impermeable, or when it becomes saturated, runoff is the dominant process. Runoff is also high in urban areas because large areas are covered by impermeable buildings, roads, and parking lots.

Runoff initially flows in broad, thin sheets. This unconfined flow eventually develops threads of current that form tiny channels called rills. Rills meet to form gullies, which join to form streams. At first streams are small, but as one intersects another, larger and larger ones form. Eventually rivers develop that carry water from a broad region.

Drainage Basins

The land area that contributes water to a river system is called a **drainage basin** (**Figure 5.3**). The drainage basin of one stream is separated from the drainage basin of another by an imaginary

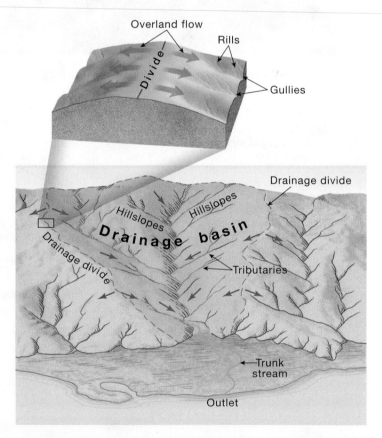

FIGURE 5.3 A *drainage basin* is the area drained by a stream and its tributaries. Boundaries between drainage basins are called *divides*.

Students Sometimes Ask...

Is the amount of water vapor that plants emit into the atmosphere through transpiration a significant amount?

Use this example to judge for yourself. Each year a field of crops may transpire the equivalent of a water layer 60 centimeters (2 feet) deep over the entire field. The same area of trees may pump twice this amount into the atmosphere.

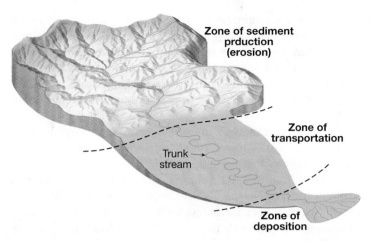

FIGURE 5.5 A river system can be divided into three zones based on the dominant process operating within each zone. These are the zones of sediment production (erosion), sediment transport, and sediment deposition. (After Shumm)

line called a **divide**. Divides range in scale from a ridge separating two small gullies on a hillside to a *continental divide*, which splits whole continents into enormous drainage basins. The Mississippi River has the largest drainage basin in North America (**Figure 5.4**). Extending between the Rocky Mountains in the west and the Appalachian Mountains in the east, the Mississippi River and its tributaries collect water from more than 3.2 million square kilometers (1.2 million square miles) of the continent.

River Systems

River systems involve not only a network of stream channels, but the entire drainage basin. Based on the dominant processes operating within them, river systems can be divided into three zones: *sediment production*—where erosion dominates, *sediment transport*, and *sediment deposition* (**Figure 5.5**). It is important to recognize that sediment is being eroded, transported, and deposited along the entire length of a stream, regardless of which process is dominant within each zone.

The zone of *sediment production*, where most of the water and sediment is derived, is located in the headwater region of the river system. Much of the sediment carried by streams begins as bedrock that is subsequently broken down by weathering, then transported downslope by mass wasting and overland flow. Bank erosion can also contribute significant amounts of sediment. In addition, scouring of the channel bed deepens the channel and adds to the stream's sediment load.

FIGURE 5.4 Drainage basins and divides exist for all streams, regardless of size. The drainage basin of the Yellowstone River is one of many that contribute water to the Missouri River, which in turn is one of many that make up the drainage basin of the Mississippi River. The drainage basin of the Mississippi River, North America's largest, covers about 3 million square kilometers.

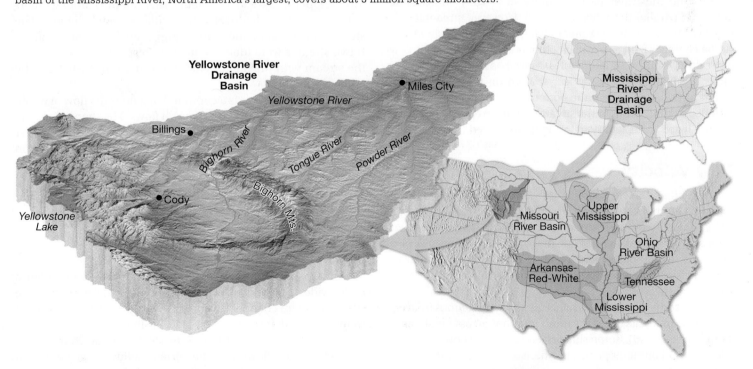

Sediment acquired by a stream is then transported through the channel network along sections referred to as *trunk streams*. When trunk streams are in balance, the amount of sediment eroded from their banks equals the amount deposited elsewhere in the channel. Although trunk streams rework their channels over time, they are not a source of sediment nor do they accumulate or store it.

When a river reaches the ocean, or another large body of water, it slows and the energy to transport sediment is greatly reduced. Most of the sediments either accumulate at the mouth of the river to form a delta, are reconfigured by wave action to form a variety of coastal features, or are moved far offshore by ocean currents. Because coarse sediments tend to be deposited upstream, it is primarily the fine sediments (clay, silt, and fine sand) that eventually reach the ocean. Taken together, erosion, transportation, and deposition are the processes by which rivers move Earth's surface materials and sculpt landscapes.

CONCEPT CHECK 5.2

1. List several factors that influence infiltration.
2. Draw a simple sketch of a drainage basin and divide and label each.
3. What are the three main parts (zones) of a river system?

Streamflow

Sculpturing Earth's Surface
▶ Running Water

Water may flow in one of two ways, either as **laminar flow** or **turbulent flow**. In very slow-moving streams the flow is often laminar and the water particles move in roughly straight-line paths that parallel the stream channel. However, streamflow is usually turbulent, with the water moving in an erratic fashion that can be characterized as a swirling motion. Strong turbulent flow may exhibit whirlpools and eddies, as well as roiling whitewater rapids. Even streams that appear smooth on the surface often exhibit turbulent flow near the bottom and sides of the channel. Turbulence contributes to the stream's ability to erode its channel because it acts to lift sediment from the streambed.

Flow Velocity

Flow velocities can vary significantly from place to place along a stream channel, as well as over time, in response to variations in the amount and intensity of precipitation. If you have ever waded into a stream, you know that velocity increases as you move into deeper parts of the channel. This is the result of frictional resistance, which is greatest near the banks and beds of stream channels.

Scientists determine flow velocities at gaging stations by averaging measurements taken at various locations across the stream's channel (**Figure 5.6**). Some sluggish streams have flow velocities of less than 1 kilometer per hour, whereas stretches of some fast-flowing rivers may exceed 30 kilometers per hour.

The ability of a stream to erode and transport material is directly related to its flow velocity. Even slight variations in flow rate can lead to significant changes in the sediment load transported by a stream. Several factors influence flow velocities and, therefore, control a stream's potential to do "work." These factors include (1) channel slope or gradient; (2) channel size and cross-sectional shape; (3) channel roughness; and (4) the amount of water flowing in the channel.

Gradient and Channel Characteristics

The slope of a stream channel expressed as the vertical drop of a stream over a specified distance is **gradient**. Portions of the lower Mississippi River have very low gradients of 10 centimeters per kilometer or less. By contrast, some mountain stream channels decrease in elevation at a rate of more than 40 meters per kilometer—a gradient 400 times steeper than the lower Mississippi (**Figure 5.7**). Gradient varies not only among different streams but also over a particular stream's length. The steeper the gradient, the more energy available for streamflow. If two streams were identical in every respect except gradient, the stream with the higher gradient would obviously have the greater velocity.

A stream's channel is a conduit that guides the flow of water, but the water encounters friction as it flows. The shape, size, and roughness of the channel affect the amount of friction. Larger channels have more efficient flow because a smaller proportion of water is in contact with the channel. A smooth channel promotes a more uniform flow, whereas an irregular channel filled with boulders creates enough turbulence to slow the stream significantly.

Discharge

Streams vary in size from small headwater creeks less than a meter wide, to large rivers with widths of several kilometers. The size of a stream channel is largely determined by the amount of water supplied from the drainage basin. The measure most often used to compare the size of streams is **discharge**—the volume of water flowing past a certain point in a given unit of time. Discharge, usually measured in cubic meters per second

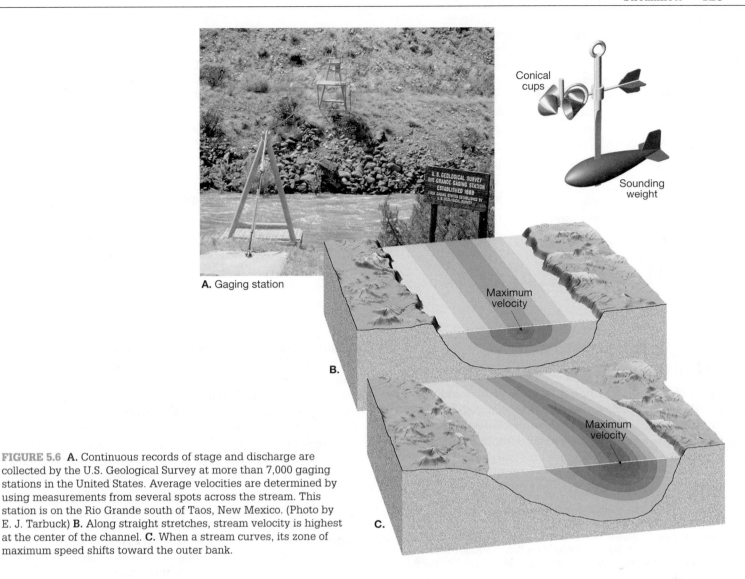

A. Gaging station

FIGURE 5.6 A. Continuous records of stage and discharge are collected by the U.S. Geological Survey at more than 7,000 gaging stations in the United States. Average velocities are determined by using measurements from several spots across the stream. This station is on the Rio Grande south of Taos, New Mexico. (Photo by E. J. Tarbuck) **B.** Along straight stretches, stream velocity is highest at the center of the channel. **C.** When a stream curves, its zone of maximum speed shifts toward the outer bank.

FIGURE 5.7 Rapids are common in mountain streams where the gradient is steep and the channel is rough and irregular. Although most streamflow is turbulent, it is usually not as rough as that experienced by these river runners. (Photo by Fogstock Lic/Photolibrary)

or cubic feet per second, is determined by multiplying a stream's cross-sectional area by its velocity.

Table 5.1 lists the world's largest rivers in terms of discharge. The largest river in North America, the Mississippi, has a discharge that averages 17,300 cubic meters per second. Nevertheless, that amount is dwarfed by South America's mighty Amazon, which discharges 12 times more water than the Mississippi.

The discharge of a river system changes over time because of variations in the amount of precipitation received by the drainage basin. Studies show that when discharge increases, the width, depth, and flow velocity of the channel all increase predictably. When the size of the channel increases, proportionally less water is in contact with the bed and banks of the channel. Thus, friction is reduced, resulting in an increase in the rate of flow.

TABLE 5.1 World's Largest Rivers Ranked by Discharge

Rank	River	Country	Drainage Area		Average Discharge	
			Square kilometers	Square miles	Cubic meters per second	Cubic feet per second
1	Amazon	Brazil	5,778,000	2,231,000	212,400	7,500,000
2	Congo	Rep. of Congo	4,014,500	1,550,000	39,650	1,400,000
3	Yangtze	China	1,942,500	750,000	21,800	770,000
4	Brahmaputra	Bangladesh	935,000	361,000	19,800	700,000
5	Ganges	India	1,059,300	409,000	18,700	660,000
6	Yenisei	Russia	2,590,000	1,000,000	17,400	614,000
7	Mississippi	United States	3,222,000	1,244,000	17,300	611,000
8	Orinoco	Venezuela	880,600	340,000	17,000	600,000
9	Lena	Russia	2,424,000	936,000	15,500	547,000
10	Parana	Argentina	2,305,000	890,000	14,900	526,000

Changes from Upstream to Downstream

One useful way of studying a stream is to examine its **longitudinal profile**. Such a profile is simply a cross-sectional view of a stream from its source area (called the *head* or *headwaters*) to its *mouth*, the point downstream where it empties into another water body—a river, lake, or ocean. By examining **Figure 5.8**, the most obvious feature of a typical longitudinal profile is its concave shape—a result of the decrease in slope that occurs from the headwaters to the mouth. In addition, local irregularities exist in the profiles of most streams—the flatter sections may be associated with lakes or reservoirs, and the steeper sections are sites of rapids or waterfalls.

The change in slope observed on most stream profiles is usually accompanied by an increase in discharge and channel size, and a reduction in sediment particle size (**Figure 5.9**). For example, data from successive gaging stations along most rivers shows that, in humid regions, discharge increases toward the mouth. This should come as no surprise because, as we move downstream, more and more tributaries contribute water to the main channel. In the case of the Amazon, for example, about 1,000 tributaries join the main river along its 6,500-kilometer course across South America.

In order to accommodate the growing volume of water, channel size typically increases downstream as well. Recall that flow velocities are higher in large channels compared to small channels. Furthermore, observations show a general decline in sediment size downstream, making the channel smoother and more efficient.

Although the channel slope decreases toward a stream's mouth, the flow velocity generally increases. This fact contradicts our intuitive assumptions of swift, narrow headwater streams and wide, placid rivers flowing across more subtle topography. Increases in channel size and discharge, and decreases in channel roughness that occur downstream, compensate for the decrease in slope—thereby making the stream more efficient.

FIGURE 5.8 A longitudinal profile is a cross-section along the length of a stream. Note the concave-upward curve of the profile, with a steeper gradient upstream and a gentler gradient downstream. Longitudinal profile of California's Kings River. It originates in the Sierra Nevada and flows westward into the San Joaquin Valley.

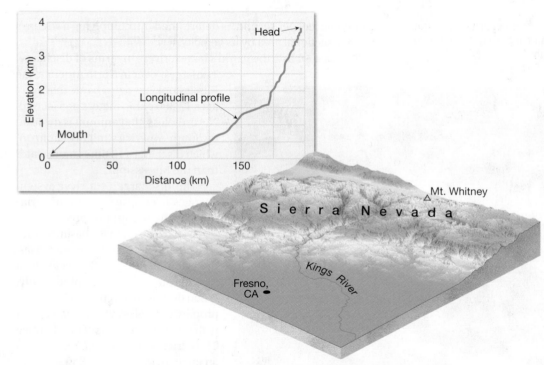

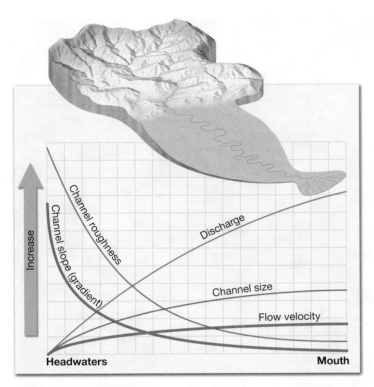

FIGURE 5.9 Graph showing how various properties of a stream change from its headwaters to its mouth. Although the gradient decreases toward the mouth, increases in channel size and discharge, combined with decreased roughness, more than offset the decrease in channel slope. Consequently, flow velocity usually increases toward the mouth.

CONCEPT CHECK 5.3

1 Contrast laminar and turbulent flow.
2 Define stream *discharge*. How is discharge measured?
3 List four factors that influence flow velocity.
4 What typically happens to channel width, channel depth, flow velocity, and discharge between the head and mouth of a stream? Briefly explain why these changes occur.

The Work of Running Water

Streams are Earth's most important erosional agents. Not only do they have the ability to downcut and widen their channels but streams also have the capacity to transport the enormous quantities of sediment that are delivered to the stream by sheet flow, mass wasting, and groundwater. Eventually much of this material is deposited to create a variety of landforms.

Stream Erosion

A stream's ability to accumulate and transport soil and weathered rock is aided by the work of raindrops, which knock sediment particles loose (see Figure 4.19, p. 99). When the ground is saturated, rainwater cannot infiltrate so it flows downslope, transporting some of the material it has dislodged. On barren slopes the sheet flow will often erode small channels, or *rills*, which in time may evolve into larger *gullies* (see Figure 4.20 on p. 100).

Once flow is confined in a channel, the erosional power of a stream is related to its slope and discharge. When the flow of water is sufficiently strong, it can dislodge particles from the channel and lift them into the moving water. In this manner, the force of running water swiftly erodes poorly consolidated materials on the bed and sides of a stream channel. On occasion, the banks of the channel may be undercut, dumping even more loose debris into the water to be carried downstream.

In addition to eroding unconsolidated materials, the hydraulic force of streamflow can also cut a channel into solid bedrock. A stream's ability to erode bedrock is greatly enhanced by the particles it carries. These particles can be any size, from large boulders in very fast-flowing waters to sand and gravel-size particles in somewhat slower flow. Just as the particles of grit on sandpaper can wear away a piece of wood, so too can the sand and gravel carried by a stream abrade a bedrock channel. Moreover, pebbles caught in swirling eddies can act like "drills" and bore circular *potholes* into the channel floor (**Figure 5.10**).

FIGURE 5.10 Potholes in the bed of a stream. The rotational motion of swirling pebbles acts like a drill to create potholes. (Photo by Elmari Joubert/Alamy)

Transportation of Sediment

All streams, regardless of size, transport some rock material. Streams also sort the solid sediment they transport because finer, lighter material is carried more readily than larger particles. Streams transport their load of sediment in three ways: (1) in solution (**dissolved load**), (2) in suspension (**suspended load**), and (3) sliding or rolling along the bottom (**bed load**).

Dissolved Load   Most of the *dissolved load* is brought to a stream by groundwater and is dispersed throughout the flow. Usually the amount of dissolved load is small and therefore is expressed as parts of dissolved material per million parts of water (parts per million, or ppm). Although some rivers may have a dissolved load of 1,000 ppm or more, the average figure for the world's rivers is estimated at 115 to 120 ppm.

The velocity of streamflow has essentially no effect on a stream's ability to carry its dissolved load. Once material is in solution, it goes wherever the stream goes, regardless of velocity. Precipitation occurs only when the chemistry of the water changes.

Suspended Load   Most large rivers carry the largest part of their load in *suspension*. Indeed, the muddy appearance created by suspended sediment is the most obvious portion of a stream's load (**Figure 5.11**). Usually only fine particles consisting of silt and clay can be carried this way, but during a flood, sand and even gravel-size particles are transported as well. Also, during a flood, the total quantity of material carried in suspension increases dramatically, as can be verified by anyone whose home has been a site for the deposition of this material.

The type and amount of material carried in suspension are controlled by two factors: the flow velocity and the settling velocity of each sediment grain. **Settling velocity** is defined as the speed at which a particle falls through a still fluid. The larger the particle, the more rapidly it settles toward the stream bed. In addition to size, the shape and specific gravity of particles also influence settling velocity. Flat grains sink through water more slowly than do spherical grains, and dense particles fall toward the bottom more rapidly than do less dense particles. The slower the settling velocity and the higher the flow velocity, the longer a sediment particle will stay in suspension, and the farther it will be carried downstream.

Bed Load   A portion of a stream's load of solid material consists of sediment that is too large to be carried in suspension. These coarser particles move along the bottom (bed) of the stream and constitute the *bed load*. In terms of the erosional

FIGURE 5.11 The Colorado River in Grand Canyon National Park. The muddy appearance is a result of suspended sediment. (Photo by Michael Collier)

work accomplished by a downcutting stream, the grinding action of the bed load is of great importance.

The particles that make up the bed load move by rolling, sliding, and saltation. Sediment moving by **saltation** (*saltare* = to leap) appears to jump or skip along the stream bed. This occurs as particles are propelled upward by collisions or lifted by the current and then carried downstream a short distance until gravity pulls them back to the bed of the stream. Particles that are too large or heavy to move by saltation either roll or slide along the bottom, depending on their shapes.

The bed load usually does not exceed 10 percent of a stream's total load. Estimates of a stream's bed load should be viewed cautiously, however, because the transport of sediment as bed load or as suspended load changes frequently. The proportions of each depend on the characteristics of streamflow at any time and these may fluctuate over short intervals. With an increase in velocity, parts of the bed load are thrown into suspension. Conversely, when velocity decreases, a portion of the bed load is deposited and some sediment that had been suspended changes to moving as bed load.

Competence and Capacity   A stream's ability to carry solid particles is described using two criteria: *capacity* and *competence*. **Capacity** is the maximum load of solid particles a stream can transport per unit of time. The greater the discharge, the greater the stream's capacity for hauling sediment. Consequently, large rivers with high flow velocities have large capacities.

Competence is a measure of a stream's ability to transport particles based on size rather than quantity. Flow velocity is the key—swift streams have greater competencies than slow streams, regardless of channel size. A stream's competence increases proportionately to the square of its velocity. Thus, if the velocity of a stream doubles, the impact force of the water increases four times; if the velocity triples, the force increases nine times, and so forth. Hence, large boulders that are often visible during low water and seem immovable can, in fact, be transported during exceptional floods because of the stream's increased competence.

By now it should be clear why the greatest erosion and transportation of sediment occur during floods. The increase in discharge results in greater capacity and the increased velocity produces greater competency. Rising velocity makes the water more turbulent, and larger particles are set in motion. In just a few days, or perhaps a few hours, a stream at flood stage can erode and transport more sediment than it does during many months of normal flow.

Deposition of Sediment

Whenever a stream slows down, the situation reverses. As its velocity decreases, its competence is reduced and sediment begins to drop out, largest particles first. Recall that each particle size has a settling velocity. As streamflow drops below the setting velocity of a certain particle size, sediment in that category begins to settle out. Thus, stream transport provides a mechanism by which solid particles of various sizes are separated. This process, called **sorting**, explains why particles of similar size are deposited together.

The material deposited by a stream is called **alluvium**, the general term for any stream-deposited sediment. Many different depositional features are composed of alluvium. Some occur within stream channels, some occur on the valley floor adjacent to the channel, and some exist at the mouth of the stream. We consider the nature of these features later in the chapter.

CONCEPT CHECK 5.4

❶ **List two ways in which streams erode their channels.**
❷ **In what three ways does a stream transport its load?**
❸ **What is the difference between capacity and competency?**
❹ **What is settling velocity? What factors influence settling velocity?**

Stream Channels

A basic characteristic of streamflow that distinguishes it from sheet flow is that it is usually confined to a channel. A stream channel can be thought of as an open conduit that consists of the streambed and banks that act to confine the flow, except during floods.

Although somewhat oversimplified, we can divide stream channels into two types. *Bedrock channels* are those in which the streams are actively cutting into solid rock. In contrast, when the bed and banks are composed mainly of unconsolidated sediment, the channel is called an *alluvial channel*.

Bedrock Channels

In their headwaters, where the gradient is usually steepest, many rivers cut into bedrock. Such streams (often mountain streams) typically transport coarse particles that actively abrade the bedrock channel. Potholes are often visible evidence of the erosional forces at work.

Bedrock channels often alternate between relatively gently sloping segments where alluvium tends to accumulate, and steeper segments where bedrock is exposed. These steeper areas may contain rapids or occasionally a waterfall. The channel pattern exhibited by streams cutting into bedrock is controlled by the underlying geologic structure. Even when flowing over rather uniform bedrock, streams tend to exhibit a winding or irregular pattern rather than flowing in a straight channel. Anyone who has gone on a whitewater rafting trip has observed the steep, winding nature of a stream flowing in a bedrock channel.

Alluvial Channels

Many stream channels are composed of loosely consolidated sediment (alluvium) and therefore can undergo significant changes in shape because the sediments are continually being eroded, transported, and redeposited. The major factors affecting the shapes of these channels is the average size of the sediment being transported, the channel gradient, and the discharge.

Alluvial channel patterns reflect a stream's ability to transport its load at a uniform rate, while expending the least amount of energy. Thus, the size and type of sediment being carried help determine the nature of the stream channel. Two common types of alluvial channels are *meandering channels* and *braided channels*.

Meandering Streams   Streams that transport much of their load in suspension generally move in sweeping bends called **meanders**. These streams flow in relatively deep, smooth channels and transport mainly mud (silt and clay), sand, and occasionally fine gravel. The lower Mississippi River exhibits a channel of this type.

Meandering channels evolve over time as individual bends migrate across the floodplain. Most of the erosion is focused at the outside of the meander, where velocity and turbulence are greatest. In time, the outside bank is undermined, especially during periods of high water. Because the outside of a meander is a

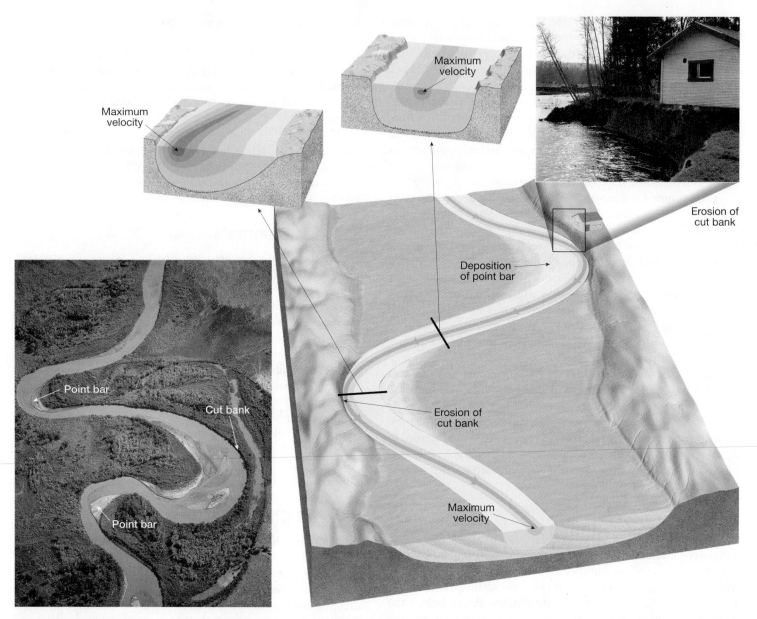

FIGURE 5.12 When a stream meanders, its zone of maximum speed shifts toward the outer bank. A point bar is deposited where the water on the inside of a meander slows. The point bar shown here is on the White River near Vernal, Utah. The black-and-white photo shows erosion of a cut bank along the Newaukum River in Washington State. By eroding the outer bank and depositing material on the inside of the bend, a stream is able to shift its channel. (Point bar photo by Michael Collier; cut bank photo by P. A. Glancy, U.S. Geological Survey)

zone of active erosion, it is often referred to as the **cut bank** (Figure 5.12). Debris acquired by the stream at the cut bank moves downstream where the coarser material is generally deposited as **point bars** on the insides of bends. In this manner, meanders migrate laterally by eroding the outside of the bends and depositing sediment on the inside without appreciably changing their shape.

In addition to migrating laterally, the bends in a channel also migrate down the valley. This occurs because erosion is more effective on the downstream (downslope) side of the meander. Sometimes the downstream migration of a meander is slowed when it reaches a more resistant bank material. This allows the next meander upstream to gradually erode the material between the two meanders, as shown in Figure 5.13. Eventually, the river may erode through the narrow neck of land, forming a new,

shorter channel segment called a **cutoff**. Because of its shape, the abandoned bend is called an **oxbow lake**.

Braided Streams Some streams consist of a complex network of converging and diverging channels that thread their way among numerous islands or gravel bars (Figure 5.14). Because these channels have an interwoven appearance, these streams are said to be **braided**. Braided channels form where a large proportion of the stream's load consists of coarse material (sand and gravel) and the stream has a highly variable discharge. Because the bank material is readily erodable, braided channels are wide and shallow.

One setting in which braided streams form is at the end of a glacier where there is a large seasonal variation in discharge. During the summer, large amounts of ice-eroded sediment are

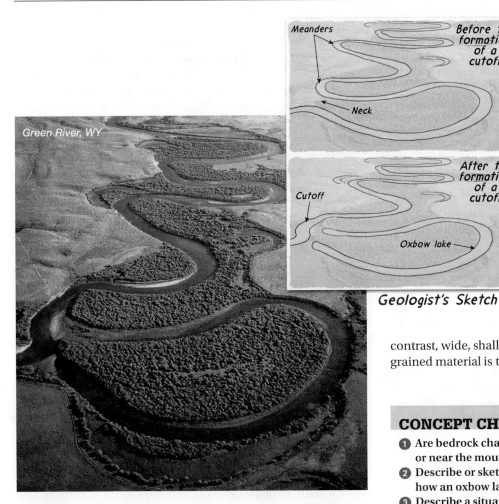

Before the formation of a cutoff

Meanders

Neck

After the formation of a cutoff

Cutoff

Oxbow lake

Geologist's Sketch

FIGURE 5.13 Formation of a cutoff and oxbow lake. Oxbow lakes occupy abandoned meanders. As they fill with sediment, oxbow lakes gradually become swampy meander scars. Aerial view of an oxbow lake created by the meandering Green River near Bronx, Wyoming. (Photo by Michael Collier)

dumped into the meltwater streams flowing away from the glacier. However, when flow is sluggish, the stream is unable to move all of the sediment and therefore deposits the coarsest material as bars that force the flow to split and follow several paths. Usually the laterally shifting channels completely rework most of the surface sediments each year, thereby transforming the entire streambed. In some braided streams, however, the bars have built up to form islands that are anchored by vegetation.

In summary, meandering channels develop where the load consists largely of fine-grained particles that are transported as suspended load in deep, relatively smooth channels. By contrast, wide, shallow braided channels develop where coarse-grained material is transported as bedload.

CONCEPT CHECK 5.5

1. Are bedrock channels more likely to be found near the head or near the mouth of a stream?
2. Describe or sketch the development of a meander, including how an oxbow lake forms.
3. Describe a situation that might cause a stream to become braided.

FIGURE 5.14 The Knik River is a classic braided stream with multiple channels separated by migrating gravel bars. The river is choked with sediment contributed by meltwater from four glaciers in the Chugach Mountains north of Anchorage, Alaska. (Photo by Michael Collier)

Base Level and Stream Erosion

Streams cannot endlessly erode their channels deeper and deeper. There is a lower limit to how deep a stream can erode, and that limit is called **base level**. Although the idea is relatively straightforward, it is nevertheless a key concept in the study of stream activity. Base level is defined as the lowest elevation to which a stream can erode its channel. Essentially this is the level at which the mouth of a stream enters the ocean, a lake, or another stream. Base level accounts for the fact that most stream profiles have low gradients near their mouths, because the streams are approaching the elevation below which they cannot erode their beds.

Two general types of base level are recognized. Sea level is considered the *ultimate base level*, because it is the lowest level to which stream erosion could lower the land. *Temporary*, or *local*, *base levels* include lakes, resistant layers of rock, and main streams that act as base levels for their tributaries. For example, when a stream enters a lake, its velocity quickly approaches zero and its ability to erode ceases. Thus, the lake prevents the stream from eroding below its level at any point upstream from the lake. However, because the outlet of the lake can cut downward and drain the lake, the lake is only a temporary hindrance to the stream's

ability to downcut its channel. In a similar manner, the layer of resistant rock at the lip of the waterfall in **Figure 5.15** acts as a temporary base level. Until the ledge of hard rock is eliminated, it will limit the amount of downcutting upstream.

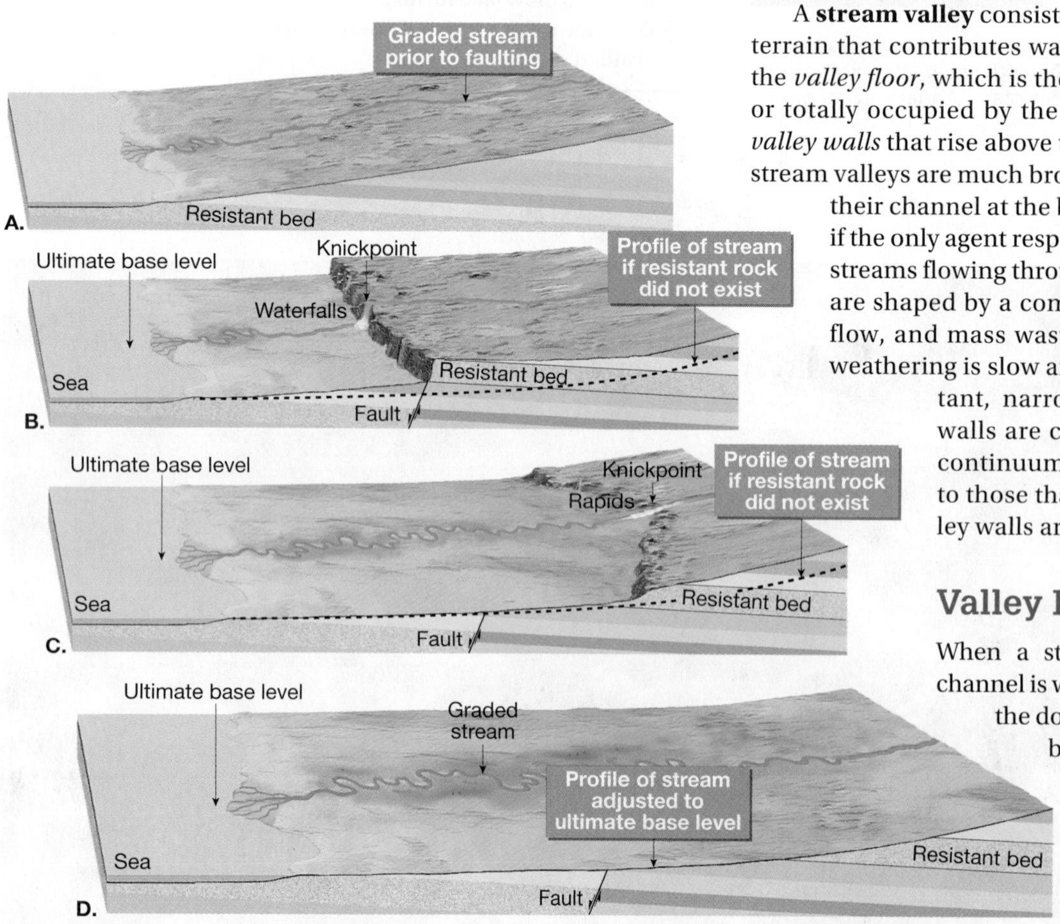

Any change in base level will cause a corresponding readjustment of stream activities. When a dam is built along a stream, the reservoir that forms behind it raises the base level of the stream (**Figure 5.16**). Upstream from the dam the gradient is reduced, lowering the stream's velocity and, hence, its sediment-transporting ability. The stream, now having too little energy to transport its entire load, will deposit sediment. This builds up its channel. Deposition will be the dominant process until the stream's gradient increases sufficiently to transport its load.

CONCEPT CHECK 5.6

1 What is base level and how does it influence stream activity?

2 Distinguish between ultimate and temporary (local) base level.

Shaping Stream Valleys

Sculpturing Earth's Surface
▸ Running Water

Streams, with the aid of weathering and mass wasting, shape the landscape through which they flow. As a result, streams continuously modify the valleys that they occupy.

A **stream valley** consists of a channel and the surrounding terrain that contributes water to the stream. Thus it includes the *valley floor*, which is the lower, flatter area that is partially or totally occupied by the stream channel, and the sloping *valley walls* that rise above the valley floor on both sides. Most stream valleys are much broader at the top than is the width of their channel at the bottom. This would not be the case if the only agent responsible for eroding valleys were the streams flowing through them. The sides of most valleys are shaped by a combination of weathering, overland flow, and mass wasting. In some arid regions, where weathering is slow and where rock is particularly resistant, narrow valleys having nearly vertical walls are common. Stream valleys exist in a continuum from narrow, steep-sided valleys to those that are so flat and wide that the valley walls are not discernable.

Valley Deepening

When a stream's gradient is steep and the channel is well above base level, downcutting is the dominant activity. Abrasion caused by bed load sliding and rolling along the bottom, and the hydraulic power of fast-moving water, slowly lowers the streambed. The result is usually a V-shaped valley with steep sides. A classic example of

FIGURE 5.15 A resistant layer of rock can act as a local (temporary) base level. Because the durable layer is eroded more slowly, it limits the amount of downcutting upstream.

A.
Graded stream prior to faulting
Resistant bed

B.
Ultimate base level
Knickpoint
Waterfalls
Sea
Resistant bed
Fault
Profile of stream if resistant rock did not exist

C.
Ultimate base level
Knickpoint
Rapids
Sea
Resistant bed
Fault
Profile of stream if resistant rock did not exist

D.
Ultimate base level
Graded stream
Sea
Resistant bed
Fault
Profile of stream adjusted to ultimate base level

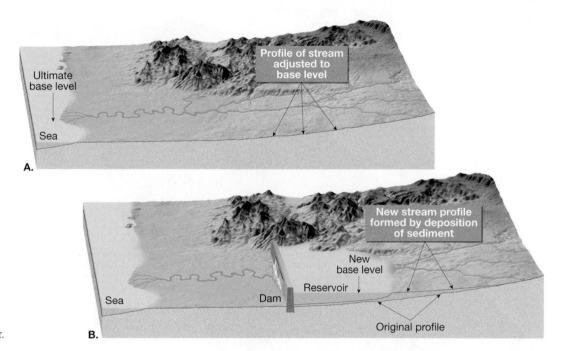

A.

B.

FIGURE 5.16 When a dam is built and a reservoir forms, the stream's base level is raised. This reduces the stream's velocity and leads to deposition and a reduction of the gradient upstream from the reservoir.

a V-shaped valley is located in the section of Yellowstone River shown in **Figure 5.17**.

The most prominent features of a V-shaped valley are *rapids* and *waterfalls*. Both occur where the stream's gradient increases significantly, a situation usually caused by variations in the erodability of the bedrock into which a stream channel is cutting. Resistant beds create rapids by acting as a temporary base level upstream while allowing downcutting to continue downstream. In time erosion usually eliminates the resistant rock. Waterfalls are places where the stream makes an abrupt vertical drop.

Valley Widening

Once a stream has cut its channel closer to base level, downward erosion becomes less dominant. At this point the stream's channel takes on a meandering pattern, and more of the stream's energy is directed from side to side. The result is a widening of the valley as the river cuts away first at one bank and then at the other (**Figure 5.18**). The continuous lateral erosion caused by shifting of the stream's meanders produces an increasingly broader, flat valley floor covered with alluvium. This feature, called a **floodplain**, is appropriately named because when a river overflows its banks during flood stage, it inundates the floodplain.

Over time the floodplain will widen to the point that the stream is only actively eroding the valley walls in a few places. In fact, in large rivers such as the lower Mississippi River valley, the distance from one valley wall to another can exceed 100 miles.

FIGURE 5.17 V-shaped valley of the Yellowstone River. The rapids and waterfalls indicate that the river is vigorously downcutting. (Photo by Lane V. Erickson/Shutterstock)

Students Sometimes Ask...

What's the highest waterfall in the world?

The world's highest uninterrupted waterfall is Angel Falls on Venezuela's Churun River. Named for American aviator Jimmie Angel, who first sighted the falls from the air in 1933, the river plunges 979 meters (3,212 feet, or more than 0.6 mile).

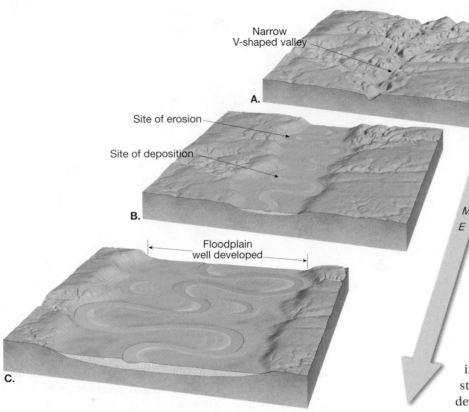

Narrow
V-shaped valley

A.

Site of erosion

Site of deposition

B.

Floodplain
well developed

C.

*T
I
M
E*

FIGURE 5.18 Stream eroding its floodplain.

Changing Base Level and Incised Meanders

We usually expect a stream with a highly meandering course to be on a floodplain in a wide valley. However, certain rivers exhibit meandering channels that flow in steep, narrow valleys. Such meanders are called **incised** (*incisum* = to cut into) **meanders** (**Figure 5.19**). How do such features form?

Originally, the meanders probably developed on the floodplain of a stream that was relatively near base level. Then, a change in base level caused the stream to begin downcutting. One of two events could have occurred. Either base level dropped or the land upon which the river was flowing was uplifted.

An example of the first circumstance happened during the Ice Age when large quantities of water were withdrawn from the ocean and locked up in glaciers on land. The result was that sea level (ultimate base level) dropped, causing rivers flowing into the ocean to begin to downcut. Of course, this activity ceased at the close of the Ice Age when ice sheets melted and sea level rose.

Regional uplift of the land, the second cause for incised meanders, is exemplified by the Colorado Plateau in the southwestern United States. Here, as the plateau was gradually uplifted, numerous meandering rivers adjusted to being higher above base level by downcutting.

CONCEPT CHECK 5.7

❶ Explain why V-shaped valleys often contain rapids and/or waterfalls.
❷ Describe or sketch how an erosional floodplain develops.
❸ Relate incised meanders to changes in base level.

Depositional Landforms

Recall that streams continually pick up sediment in one part of their channel and deposit it downstream. These small-scale channel deposits are most often composed of sand and gravel and are commonly referred to as **bars**. Such features, however, are only temporary, because the material will be picked up again and eventually carried to the ocean. In addition to sand and gravel bars, streams also create other depositional features that have a somewhat longer life span. These include *deltas, natural levees,* and *alluvial fans.*

Deltas

Deltas form where sediment-charged streams enter the relatively still waters of a lake, an inland sea, or the ocean (**Figure 5.20**). As the stream's forward motion slows, sediments are deposited by the dying current. As the delta grows outward, the stream's gradient continually lessens. This circumstance eventually causes the channel to become choked with sediment deposited from the slowing water. As a consequence, the river seeks a shorter, higher-gradient route to base level, as illustrated in Figure 5.20B. This illustration shows the main channel dividing into several smaller ones, called **distributaries**. Most deltas are characterized by these shifting channels that act in an opposite way to that of tributaries.

FIGURE 5.19 Incised meanders of the Colorado River in Canyonlands National Park, Utah. As the Colorado Plateau was gradually uplifted, the meandering river adjusted to being higher above base level by downcutting. (Photo by Michael Collier)

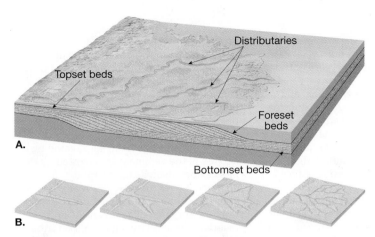

FIGURE 5.20 **A.** Structure of a simple delta that forms in the relatively quiet waters of a lake. **B.** Growth of a simple delta. As a stream extends its channel, the gradient is reduced. Frequently, during flood stage the river is diverted to a higher-gradient route, forming a new distributary. Old, abandoned distributaries are gradually invaded by aquatic vegetation and fill with sediment. (After Ward's Natural Science Establishment, Inc., Rochester, N.Y.)

Rather than carrying water into the main channel, distributaries carry water away from the main channel. After numerous shifts of the channel, a delta may grow into a roughly triangular shape like the Greek letter delta (Δ), for which it is named. Note, however, that many deltas do not exhibit the idealized shape. Differences in the configurations of shorelines and variations in the nature and strength of wave activity result in many shapes. Many

large rivers have deltas extending over thousands of square kilometers. The delta of the Mississippi River is one example (see Box 5.1). It resulted from the accumulation of huge quantities of sediment derived from the vast region drained by the river and its tributaries (see Figure 5.4, p. 121). Today, New Orleans rests where there was ocean less than 5,000 years ago. **Figure 5.21** shows that portion of the Mississippi delta that has been built over the past 6,000 years. As you can see, the delta is actually a series of seven coalescing subdeltas. Each formed when the river left its existing channel in favor of a shorter, more direct path to the Gulf of Mexico. The individual subdeltas interfinger and partially cover one another to produce a very complex structure. The present subdelta, called a *bird-foot* delta because of the configuration of its distributaries, has been built by the Mississippi in the last 500 years.

Natural Levees

Some rivers occupy valleys with broad floodplains and build **natural levees** that parallel their channels on both banks (**Figure 5.22**). Natural levees are built by successive floods over many years. When a stream overflows its banks, its velocity immediately diminishes, leaving coarse sediment deposited in strips bordering the channel. As the water spreads out over the valley, a lesser amount of fine sediment is deposited over the valley floor. This uneven distribution of material produces the very gentle slope of the natural levee.

The natural levees of the lower Mississippi rise 6 meters (20 feet) above the floodplain. The area behind the levee is characteristically poorly drained for the obvious reason that water cannot flow up the levee and into the river. Marshes called

FIGURE 5.21 During the past 6,000 years, the Mississippi River has built a series of seven coalescing subdeltas. The numbers indicate the order in which the subdeltas were deposited. The present birdfoot delta (number 7) represents the activity of the past 500 years. (Image courtesy of JPL/Cal Tech/NASA) Without ongoing human efforts, the present course will shift and follow the path of the Atchafalaya River. The inset on the left shows the point where the Mississippi may someday break through (arrow) and the shorter path it would take to the Gulf of Mexico. (After C. R. Kolb and J. R. Van Lopik)

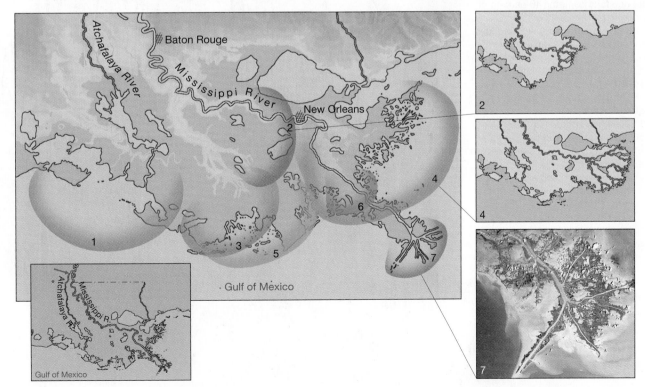

Box 5.1

PEOPLE AND THE ENVIRONMENT

Coastal Wetlands Are Vanishing on the Mississippi Delta

Coastal wetlands form in sheltered environments that include swamps, tidal flats, coastal marshes, and bayous. They are rich in wildlife and provide nesting grounds and important stopovers for waterfowl and migratory birds, as well as spawning areas and valuable habitats for fish.

The delta of the Mississippi River in Louisiana contains about 40 percent of all coastal wetlands in the lower 48 states. Louisiana's wetlands are sheltered from the wave action of hurricanes and winter storms by low-lying offshore barrier islands. Both the wetlands and the barrier islands have formed as a result of the shifting of the Mississippi River during the past 6,000 years.

The dependence of Louisiana's coastal wetlands and offshore islands on the Mississippi River and its distributaries as a direct source of sediment leaves them vulnerable to changes in the river system. Moreover, the reliance on barrier islands for protection from storm waves leaves coastal wetlands vulnerable when these narrow offshore islands are eroded.

Today, the coastal wetlands of Louisiana are disappearing at an alarming rate. Although Louisiana contains 40 percent of the wetlands in the lower 48 states, it accounts for 80 percent of the wetland loss. According to the U.S. Geological Survey, Louisiana lost nearly 5,000 square kilometers (1,900 square miles) of coastal land between 1932 and 2000. The state continues to lose between 65 and 91 square kilometers (25–35 square miles) each year. At this rate another 1,800–4,500 square kilometers (700–1,750 square miles) will vanish under the Gulf of Mexico by the year 2050.* Global climate change could increase the severity of the problem because rising sea level and stronger tropical storms accelerate rates of coastal erosion.** Unfortunately, this was

*See "Louisiana's Vanishing Wetlands: Going, Going . . ." in *Science* 289 (September 15, 2000): 1860–1863. Also see Elizabeth Kolbert, "Watermark—Can Southern Louisiana be Saved?" *The New Yorker*, February 27, 2006, pp. 46–57.

**For more on this possibility, see "Some Possible Consequences of Global Warming" in Chapter 20.

observed firsthand during the extraordinary 2005 hurricane season when hurricanes Katrina and Rita devastated portions of the Gulf Coast.

By nature, the delta, its wetlands, and the adjacent barrier islands are dynamic features. Over the millennia, as sediment accumulated and built the delta in one area, erosion and subsidence caused losses elsewhere. Whenever the river shifted, the zones of delta growth and destruction also shifted. However, with the arrival of people, this relative balance between formation and destruction changed—the rate at which the delta and its wetlands were destroyed accelerated and now greatly exceeds the rate of formation. Why are Louisiana's wetlands shrinking?

Before Europeans settled the delta, the Mississippi River regularly overflowed its banks in seasonal floods. The huge quantities of sediment that were deposited renewed the soil and kept the delta from sinking below sea level. However, with settlement came flood-control efforts and the desire to maintain and improve navigation on the river. Artificial levees were constructed to contain the rising river during flood stage. Over time the levees were extended all the way to the mouth of the Mississippi to keep the channel open for navigation.

The effects have been straightforward. The levees prevent sediment and freshwater from being dispersed into the wetlands. Instead, the river is forced to carry its load to the deep waters at the mouth. Meanwhile, the processes of compaction, subsidence, and wave erosion continue. Because not enough sediment is added to offset these forces, the size of the delta and the extent of its wetlands gradually shrink.

The problem has been aggravated by a decline in the sediment transported by the Mississippi, decreasing by approximately 50 percent over the past 100 years. A substantial portion of the reduction results from the trapping of sediment in large reservoirs created by dams built on tributaries to the Mississippi.

Another factor contributing to wetland decline is the fact that the delta is laced with 13,000 kilometers (8,000 miles) of navigation channels and canals. These artificial openings to the sea allow salty Gulf waters to flow far inland. The invasion of saltwater and tidal action causes massive "brownouts" or marsh die-offs (**Figure 5.A**).

Understanding and modifying the impact of people is a necessary basis for any plan to reduce the loss of wetlands in the Mississippi delta. The U.S. Geological Survey estimates that restoring Louisiana's coasts will require about $14 billion over the next 40 years. What if nothing is done? State and federal officials estimate that costs of inaction could exceed $100 billion.

FIGURE 5.A Dead cypress trees, known as a ghost forest, killed by encroaching salt water in Terrebonne Parish, Louisiana. (Photo by Robert Caputo/Aurora Photos)

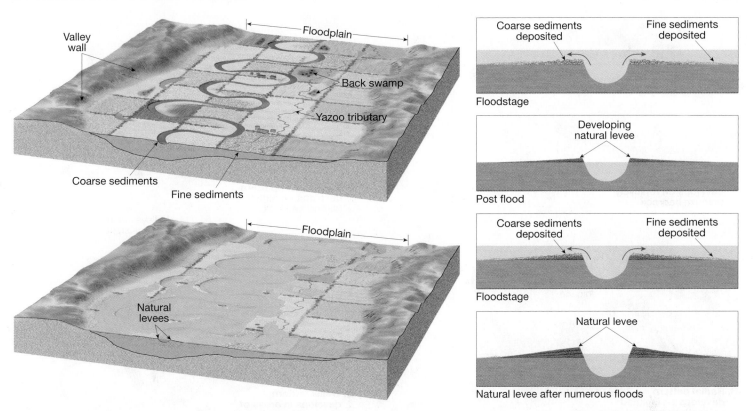

FIGURE 5.22 Natural levees are gently sloping deposits that are created by repeated floods. The diagrams on the right show the sequence of development. Because the ground next to the stream channel is higher than the adjacent floodplain, back swamps and yazoo tributaries may develop.

backswamps result. A tributary stream that cannot enter a river because levees block the way often has to flow parallel to the river until it can breach the levee. Such streams are called **yazoo tributaries** after the Yazoo River, which parallels the Mississippi for over 300 kilometers (about 190 miles).

Alluvial Fans

Alluvial fans typically develop where a high-gradient stream leaves a narrow valley in mountainous terrain and comes out suddenly onto a broad, flat plain or valley floor (see Figure 6.31, p. 183). Alluvial fans form in response to the abrupt drop in gradient combined with the change from a narrow channel of a mountain stream to less confined channels at the base of the mountains. The sudden drop in velocity causes the stream to dump its load of sediment quickly in a distinctive cone- or fan-shaped accumulation. As illustrated by Figure 6.30, the surface of the fan slopes outward in a broad arc from an apex at the mouth of the steep valley. Usually, coarse material is dropped near the apex of the fan, while fine material is carried toward the base of the deposit.

CONCEPT CHECK 5.8

1. What happens when a stream enters the relatively still waters of a lake, an inland sea, or the ocean?
2. Briefly describe the formation of a natural levee. How is this feature related to back swamps and yazoo tributaries?
3. How does an alluvial fan differ from a delta?

Drainage Patterns

 Sculpturing Earth's Surface
▶ **Running Water**

Drainage systems are networks of streams that together form distinctive patterns. The nature of a drainage pattern can vary greatly from one type of terrain to another, primarily in response to the kinds of rock on which the streams developed or the structural pattern of faults and folds.

The most commonly encountered drainage pattern is the **dendritic pattern** (**Figure 5.23A**). This pattern of irregularly branching tributary streams resembles the branching pattern of a deciduous tree. In fact, the word *dendritic* means "treelike." The dendritic pattern forms where the underlying material is relatively uniform. Because the surface material is essentially uniform in its resistance to erosion, it does not control the pattern of streamflow. Rather, the pattern is determined chiefly by the direction of slope of the land.

When streams diverge from a central area like spokes from the hub of a wheel, the pattern is said to be **radial** (Figure 5.23B). This pattern typically develops on isolated volcanic cones and domal uplifts.

Figure 5.23C illustrates a **rectangular** pattern, in which many right-angle bends can be seen. This pattern develops when the bedrock is crisscrossed by a series of joints and/or faults. Because these structures are eroded more easily than unbroken rock, their geometric pattern guides the directions of valleys.

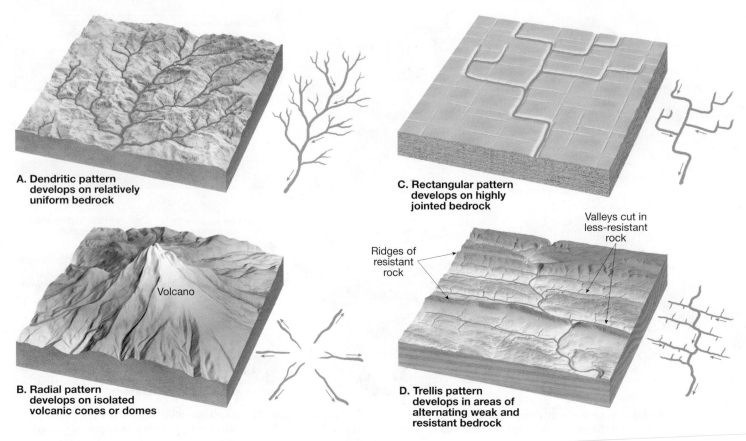

A. Dendritic pattern
develops on relatively
uniform bedrock

B. Radial pattern
develops on isolated
volcanic cones or domes

Volcano

C. Rectangular pattern
develops on highly
jointed bedrock

Valleys cut in
less-resistant
rock

Ridges of
resistant
rock

D. Trellis pattern
develops in areas of
alternating weak and
resistant bedrock

FIGURE 5.23 Drainage patterns. **A.** Dendritic. **B.** Radial. **C.** Rectangular. **D.** Trellis.

Figure 5.23D illustrates a **trellis** drainage pattern, a rectangular pattern in which tributary streams are nearly parallel to one another and have the appearance of a garden trellis. This pattern forms in areas underlain by alternating bands of resistant and less-resistant rock.

CONCEPT CHECK 5.9

❶ Make a simple sketch of the four basic drainage patterns.

Floods and Flood Control

When the discharge of a stream becomes so great that it exceeds the capacity of its channel, it overflows its banks as a **flood**. Floods are among the most deadly and most destructive of all geologic hazards. They are, nevertheless, simply part of the *natural* behavior of streams.

Causes of Floods

Rivers flood because of the weather. Rapid melting of snow in the spring and/or major storms that bring heavy rains over a large region cause most floods. The extensive 1997 flood along the Red River of the north is an example of an event triggered by rapid snowmelt. Exceptional rains caused the devastating floods in the upper Mississippi River Valley during the summer of 1993 (**Figure 5.24**).

Unlike the extensive *regional floods* just mentioned, *flash floods* are more limited in extent. Flash floods occur with little warning and can be deadly because they produce a rapid rise in water levels and can have a devastating flow velocity. Several factors influence flash flooding. Among them are rainfall intensity and duration, topography, and surface conditions. Mountainous areas are susceptible because steep slopes can quickly funnel runoff into narrow canyons (**Figure 5.25**). Urban areas are susceptible to flash floods because a high percentage of the surface area is composed of impervious surfaces such as roofs, streets, and parking lots where runoff is very rapid. In fact, a study indicated that the area of impervious surfaces in the United States (excluding Alaska and Hawaii) amounts to more than 112,600 square kilometers (nearly 44,000 square miles), which is slightly less than the area of the state of Ohio.[7]

Human interference with the stream system can worsen or even cause floods. A prime example is the failure of a dam or an artificial levee. These structures are built for flood protection. They are designed to contain floods of a certain magnitude. If a larger flood occurs, the dam or levee is overtopped. If the dam or levee fails or is washed out, the water behind it is released to become a flash flood. The bursting of a dam in 1889 on the Little Conemaugh River caused the devastating Johnstown, Pennsylvania, flood that took some 3,000 lives. A second dam failure occurred there again in 1977 and caused 77 fatalities.

[7]C. D. Elvidge, et al. "U.S. Constructed Area Approaches the Size of Ohio," in *EOS, Transactions, American Geophysical Union* 85, No. 24 (June 15, 2004): 233.

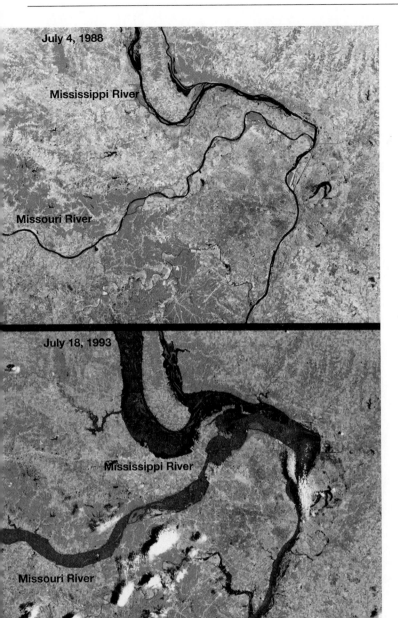

July 4, 1988

Mississippi River

Missouri River

July 18, 1993

Mississippi River

Missouri River

FIGURE 5.24 Satellite views of the Missouri River flowing into the Mississippi River. St. Louis is just south of their confluence. The upper image shows the rivers during a drought that occurred in summer 1988. The lower image depicts the peak of the record-breaking 1993 flood. Exceptional rains produced the wettest spring and early summer of the 20th century in the upper Mississippi River basin. In all, nearly 14 million acres were inundated, displacing at least 50,000 people. (Photos courtesy of Spaceimaging.com)

Flood Control

Several strategies have been devised to eliminate or lessen the catastrophic effects of floods. Engineering efforts include the construction of artificial levees, the building of flood-control dams, and river channelization.

Artificial Levees   *Artificial levees* are earthen mounds built on the banks of a river to increase the volume of water the channel can hold. These most common of stream-containment structures have been used since ancient times and continue to be used today.

Artificial levees are usually easy to distinguish from natural levees because their slopes are much steeper. In some locations, especially urban areas, concrete floodwalls are sometimes constructed that serve the same purpose as artificial levees.

Such structures do not always provide the flood protection that was intended. Many artificial levees were not built to withstand periods of extreme flooding. For example, levee failures were numerous in the Midwest during the summer of 1993, when the upper Mississippi and many of its tributaries experienced record floods (**Figure 5.26**). During that same event, floodwalls at St. Louis, Missouri, created a bottleneck for the river that led to increased flooding upstream of the city.

Flood-Control Dams   *Flood-control dams* are built to store floodwater and then let it out slowly. This lowers the flood crest by spreading it out over a longer time span. Since the 1920s, thousands of dams have been built on nearly every major river in the United States. Many dams have significant nonflood-related functions, such as providing water for irrigated agriculture and for hydroelectric power generation. Many reservoirs are also major regional recreational facilities.

Although dams may reduce flooding and provide other benefits, building these structures also has significant costs and consequences. For example, reservoirs created by dams may cover fertile farmland, useful forests, historic sites, and scenic valleys. Of course, dams trap sediment. Therefore, deltas and floodplains downstream erode because they are no longer replenished with silt during floods (see Box 5.1). Large dams can also cause significant ecological damage to river environments that took thousands of years to establish.

FIGURE 5.25 The disastrous nature of flash floods is illustrated by the Big Thompson River flood of July 31, 1976, in Colorado. During a 4 hour span more than 30 centimeters (12 inches) of rain fell on portions of the river's small drainage basin. This amounted to nearly three-quarters of the average yearly total. The flash flood in the narrow canyon lasted only a few hours but cost 130 people their lives. (Photo by U.S. Geological Survey)

FIGURE 5.26 Water rushes through a break in an artificial levee in Monroe County, Illinois. During the record-breaking 1993 Midwest floods, many artificial levees could not withstand the force of the floodwaters. Sections of many weakened structures were overtopped or simply collapsed. (Photo by James A. Finley/AP/Wide World Photos)

Students Sometimes Ask...

Sometimes when there is a major flood, it is described as a 100-year flood. What does that mean?

The phrase "100-year flood" is misleading because it leads people to believe that such an event happens only once every 100 years. The truth is that an uncommonly big flood can happen any year. The phrase "100-year flood" is really a statistical designation, indicating that there is a 1-in-100 chance that a flood this size will happen during any year. Perhaps a better term would be the "1-in-100-chance flood."

Many flood designations are reevaluated and changed over time as more data are collected or when a river basin is altered in a way that affects the flow of water. Dams and urban development are examples of some human influences in a basin that affect floods.

Building a dam is not a permanent solution to flooding. Sedimentation behind a dam means that the volume of its reservoir will gradually diminish, reducing the effectiveness of this flood-control measure.

Channelization *Channelization* involves altering a stream channel in order to speed the flow of water to prevent it from reaching flood height. This may simply involve clearing a channel of obstructions or dredging a channel to make it wider and deeper.

Another alteration involves straightening a channel by creating *artificial cutoffs*. The idea is that by shortening the stream, the gradient and hence the velocity are both increased. By increasing velocity, the larger discharge associated with flooding can be dispersed more rapidly.

Since the early 1930s, the U.S. Army Corps of Engineers has created many artificial cutoffs on the Mississippi for the purpose of increasing the efficiency of the channel and reducing the threat of flooding. In all, the river has been shortened more than 240 kilometers (150 miles). The program has been somewhat successful in reducing the height of the river in flood stage. However, because the river's tendency toward meandering still exists, preventing the river from returning to its previous course has been difficult.

A Nonstructural Approach All of the flood-control measures described so far have involved structural solutions aimed at "controlling" a river. These solutions are expensive and often give people residing on the floodplain a false sense of security.

Today, many scientists and engineers advocate a nonstructural approach to flood control. They suggest that an alternative to artificial levees, dams, and channelization is sound floodplain management. By identifying high-risk areas, appropriate zoning regulations can be implemented to minimize development and promote more appropriate land use.

CONCEPT CHECK 5.10

① Contrast regional floods and flash floods. Which is more deadly?

② List and briefly describe three basic flood-control strategies. What are some drawbacks of each?

Groundwater: Water Beneath the Surface

Sculpturing Earth's Surface
▶ **Groundwater**

Groundwater is one of our most important and widely available resources. Yet people's perceptions of groundwater are often unclear and incorrect. The reason is that groundwater is hidden from view except in caves and mines, and the impressions people gain from these subsurface openings are often misleading. Observations on the land surface give an impression that Earth is solid. This view is not changed very much when we enter a cave and see water flowing in a channel that appears to have been cut into solid rock. Because of such observations, many people believe that groundwater occurs only in underground "rivers." But actual rivers underground are rare.

In reality, most of the subsurface environment is not solid at all. It includes countless tiny *pore spaces* between grains of soil and sediment, plus narrow joints and fractures in bedrock. Together, these spaces add up to an immense volume. It is in these tiny openings that groundwater collects and moves.

The Importance of Groundwater

Considering the entire hydrosphere, or all of Earth's water, only about six-tenths of 1 percent occurs underground (See Figure 1.11 p. 12). Nevertheless, this small percentage, stored in the rocks and

TABLE 5.2 Freshwater of the Hydrosphere

Parts of the Hydrosphere	Volume of Freshwater (km³)	Share of Total Volume of Freshwater (percent)
Ice sheets and glaciers	24,000,000	84.945
Groundwater	4,000,000	14.158
Lakes and reservoirs	155,000	0.549
Soil moisture	83,000	0.294
Water vapor in the atmosphere	14,000	0.049
River water	1,200	0.004
Total	28,253,200	100.00

Source: U.S. Geological Survey Water Supply Paper 2220, 1987.

sediments beneath Earth's surface, is a vast quantity. When the oceans are excluded and only sources of freshwater are considered, the significance of groundwater becomes more apparent.

Table 5.2 contains estimates of the distribution of freshwater in the hydrosphere. Clearly, the largest volume occurs as glacial ice. Second in rank is groundwater, with slightly more than 14 percent of the total. However, when ice is excluded and just liquid water is considered, more than 94 percent is groundwater. Without question, *groundwater represents the largest reservoir of freshwater that is readily available to humans.* Its value in terms of economics and human well-being is incalculable.

Worldwide, wells and springs provide water for cities, crops, livestock, and industry. In the United States, groundwater supplies nearly one-quarter of our freshwater needs—about 80 billion gallons per day. In some areas, however, overuse of this basic resource has caused serious problems, including streamflow depletion, land subsidence, and increased pumping costs. In addition, groundwater contamination due to human activities is a real and growing threat in many places.

Groundwater's Geological Roles

Geologically, groundwater is important as an erosional agent. The dissolving action of groundwater slowly removes soluble rock, allowing surface depressions known as sinkholes to form as well as creating subterranean caverns (**Figure 5.27**). Groundwater is also an equalizer of streamflow. Much of the water that flows in rivers is not direct runoff from rain and snowmelt. Rather, a large percentage of precipitation soaks in and then moves slowly underground to stream channels. Groundwater is thus a form of storage that sustains streams during periods when rain does not fall. When we observe water flowing in a river during a dry period, it is water from rain that fell at some earlier time and was stored underground.

CONCEPT CHECK 5.11

1. What percentage of freshwater is groundwater? How does this change if glacial ice is excluded?
2. What are two geological roles for groundwater?

A.

B.

FIGURE 5.27 A. A view of the interior of New Mexico's Carlsbad Caverns. The dissolving action of groundwater created the caverns. Later, groundwater deposited the limestone decorations. (Photo by Clint Farlinger/Alamy) **B.** Groundwater was responsible for creating these sinkholes west of Timaru on New Zealand's South Island. (Photo by David Wall/Alamy)

Students Sometimes Ask...

About how much of a river's flow is contributed by groundwater?

In one study of 54 streams in all parts of the United States, the analysis indicated that 52 percent of the streamflow was contributed by groundwater. The groundwater contribution ranged from a low of 14 percent to a maximum of 90 percent. Groundwater is also a major source of water for lakes and wetlands.

Distribution and Movement of Groundwater

Sculpturing Earth's Surface
▶ Groundwater

When rain falls, some of the water runs off, some returns to the atmosphere by evaporation and transpiration, and the remainder soaks into the ground. This last path is the primary source of practically all subsurface water. The amount of water that takes each of these paths, however, varies considerably both in time and space. Influential factors include the steepness of the slopes, the nature of the surface materials, the intensity of the rainfall, and the type and amount of vegetation. Heavy rains falling on steep slopes underlain by impervious materials will result in a high percentage of the water running off. Conversely, if rain falls steadily and gently on more gradual slopes composed of materials that are easily penetrated by water, a much larger percentage of the water soaks into the ground.

Distribution

Some of the water that soaks in does not travel far, because it is held by molecular attraction as a surface film on soil particles. This near-surface zone is called the *zone of soil moisture*. It is crisscrossed by roots, voids left by decayed roots, and animal and worm burrows that enhance the infiltration of rainwater into the soil. Soil water is used by plants in life functions and transpiration. Some water also evaporates directly back into the atmosphere.

Water that is not held as soil moisture percolates downward until it reaches a zone where all of the open spaces in sediment and rock are completely filled with water. This is the **zone of saturation**.

FIGURE 5.28 This diagram illustrates the relative positions of many features associated with subsurface water.

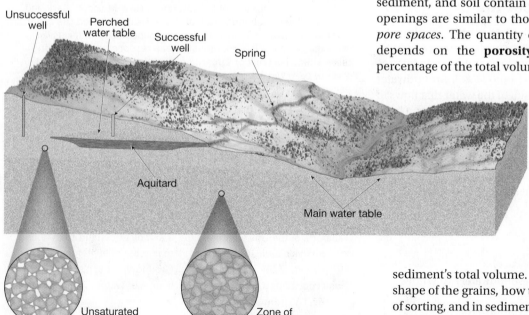

Water within it is called **groundwater**. The upper limit of this zone is known as the **water table**. The area above the water table where the soil, sediment, and rock are not saturated is called the **unsaturated zone** (Figure 5.28). Although a considerable amount of water can be present in the unsaturated zone, this water cannot be pumped by wells, because it clings too tightly to rock and soil particles. By contrast, below the water table, the water pressure is great enough to allow water to enter wells, thus permitting groundwater to be withdrawn for use. We examine wells more closely later in the chapter.

The water table is rarely level as we might expect a table to be. Instead, its shape is usually a subdued replica of the surface, reaching its highest elevations beneath hills and decreasing in height toward valleys. When you see a wetland (swamp), it indicates that the water table is right at the surface. Lakes and streams generally occupy areas low enough that the water table is above the land surface.

Several factors contribute to the irregular surface of the water table. One important influence is the fact that groundwater moves very slowly. Because of this, water tends to "pile up" beneath high areas between stream valleys. If rainfall were to cease completely, these water "hills" would slowly subside and gradually approach the level of the valleys. However, new supplies of rainwater are usually added often enough to prevent this. Nevertheless, in times of extended drought, the water table may drop enough to dry up shallow wells. Other causes for the uneven water table are variations in rainfall and permeability from place to place.

Factors Influencing the Storage and Movement of Groundwater

The nature of subsurface materials strongly influences the rate of groundwater movement and the amount of groundwater that can be stored. Two factors are especially important: porosity and permeability.

Porosity Water soaks into the ground because bedrock, sediment, and soil contain countless voids or openings. These openings are similar to those of a sponge and are often called *pore spaces*. The quantity of groundwater that can be stored depends on the **porosity** of the material, which is the percentage of the total volume of rock or sediment that consists of pore spaces. Voids most often are spaces between sedimentary particles, but also common are joints, faults, cavities formed by the dissolving of soluble rock such as limestone, and vesicles (voids left by gases escaping from lava).

Variations in porosity can be great. Sediment is commonly quite porous, and open spaces may occupy 10–50 percent of the sediment's total volume. Pore space depends on the size and shape of the grains, how they are packed together, the degree of sorting, and in sedimentary rocks, the amount of cementing material. Where sediments are poorly sorted, the porosity is reduced because finer particles tend to fill the openings among the larger grains. Most igneous and metamorphic rocks, as well as some sedimentary rocks, are composed of tightly interlocking crystals so the voids between grains may be negligible. In these rocks, fractures must provide the voids.

Permeability Porosity alone cannot measure a material's capacity to yield groundwater. Rock or sediment may be very porous yet still not allow water to move through it. The pores must be *connected* to allow water flow, and they must be *large enough* to allow flow. Thus, the **permeability** of a material, its ability to *transmit* a fluid, is also very important.

Groundwater moves by twisting and turning through small, interconnected openings. The smaller the pore spaces, the slower the groundwater moves. If the spaces between particles are too small, water cannot move at all. For example, clay's ability to store water can be great, owing to its high porosity, but its pore spaces are so small that water is unable to move through it. Thus, we say that clay is *impermeable*.

Aquitards and Aquifers Impermeable layers that hinder or prevent water movement are termed **aquitards** (*aqua* = water, *tard* = slow). Clay is a good example. In contrast, larger particles, such as sand or gravel, have larger pore spaces. Therefore, the water moves with relative ease. Permeable rock strata or sediments that transmit groundwater freely are called **aquifers**

Students Sometimes Ask...

How fast does groundwater move?

The rate of groundwater movement is highly variable. One method of measuring this movement involves introducing dye into a well. The time is measured until the coloring agent appears in another well at a known distance from the first. A typical rate is about 15 meters per year (about 4 centimeters per day).

("water carriers"). Sands and gravels are common examples. Aquifers are important because they are the water-bearing layers sought after by well drillers.

Groundwater Movement

The movement of most groundwater is exceedingly slow, from pore to pore. A typical rate is a few centimeters per day. The energy that makes the water move is provided by the force of gravity. In response to gravity, water moves from areas where the water table is high to zones where the water table is lower (see Box 5.2). This means that water usually gravitates toward a

---

Box 5.2

UNDERSTANDING EARTH

Measuring Groundwater Movement

The foundations of our modern understanding of groundwater movement began in the mid-19th century with the work of French scientist-engineer Henri Darcy. Among the experiments carried out by Darcy was one that showed that the velocity of groundwater flow is proportional to the slope of the water table—the steeper the slope, the faster the water moves (because the steeper the slope, the greater the pressure difference between two points). The water-table slope is known as the hydraulic gradient and can be expressed as follows:

$$\text{hydraulic gradient} = \frac{h_1 - h_2}{d}.$$

Where h_1 is the elevation of one point on the water table, h_2 is the elevation of a second point, and d is the horizontal distance between the two points (**Figure 5.B**).

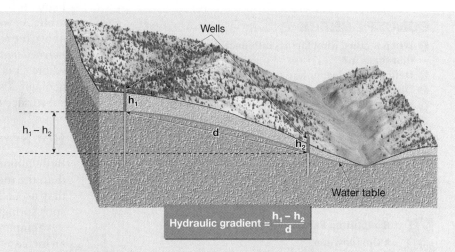

FIGURE 5.B The hydraulic gradient is determined by measuring the difference in elevation between two points on the water table ($h_1 - h_2$) divided by the distance between them, d. Wells are used to determine the height of the water table.

Darcy also discovered that the flow velocity varied with the permeability of the sediment—groundwater flows more rapidly through sediments having greater permeability than through materials having lower permeability. This factor is known as *hydraulic conductivity* and is a coefficient that takes into account the permeability of the aquifer and the viscosity of the fluid.

To determine discharge (Q)—that is, the actual volume of water that flows through an aquifer in a specified time—the following equation is used:

$$Q = \frac{K\,A(h_1 - h_2)}{d}.$$

Where $\frac{h_1 - h_2}{d}$ is the hydraulic gradient, K is the coefficient that represents hydraulic conductivity, and A is the cross-sectional area of the aquifer. This expression has come to be called *Darcy's law*.

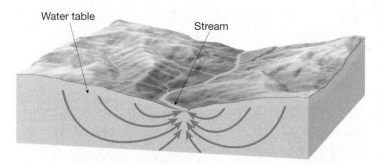

FIGURE 5.29 Arrows indicate groundwater movement through uniformly permeable material. The looping curves may be thought of as a compromise between the downward pull of gravity and the tendency of water to move toward areas of reduced pressure.

stream channel, lake, or spring. Although some water takes the most direct path down the slope of the water table, much of the water follows long, curving paths toward the zone of discharge.

Figure 5.29 shows how water percolates into a stream from all possible directions. Some paths clearly turn upward, apparently against the force of gravity, and enter through the bottom of the channel. This is easily explained: The deeper you go into the zone of saturation, the greater the water pressure. Thus, the looping curves followed by water in the saturated zone may be thought of as a compromise between the downward pull of gravity and the tendency of water to move toward areas of reduced pressure.

CONCEPT CHECK 5.12

1. When it rains, what factors influence the amount of water that soaks in?
2. Define groundwater and relate it to the water table.
3. Distinguish between porosity and permeability.
4. What is the difference between an aquifer and an aquitard?
5. What factors cause water to follow the paths shown in Figure 5.29?

Springs

Sculpturing Earth's Surface
▶ Groundwater

Springs have aroused the curiosity and wonder of people for thousands of years. The fact that springs were (and to some people still are) rather mysterious phenomena is not difficult to understand, for here water is flowing freely from the ground in all kinds of weather in seemingly inexhaustible supply but with no obvious source. Today, we know that the source of springs is water from the zone of saturation and that the ultimate source of this water is precipitation.

Whenever the water table intersects the ground surface, a natural flow of groundwater results, which we call a **spring**. Springs such as the one in Figure 5.30 form when an aquitard blocks the downward movement of groundwater and forces it to move laterally. When the permeable bed (aquifer) outcrops in a valley, a spring or series of springs results.

FIGURE 5.30 Thousand Springs along the Snake River in Hagerman Valley, Idaho. (Photo by David R. Frazier/Alamy)

Another situation that can produce a spring is illustrated in Figure 5.28. Here an aquitard is situated above the main water table. As water percolates downward, a portion accumulates above the aquitard to create a localized zone of saturation and a *perched water table*. Springs, however, are not confined to places where a perched water table creates a flow at the surface. Many geological situations lead to the formation of springs because subsurface conditions vary greatly from place to place.

Hot Springs

By definition, the water in **hot springs** is 6–9°C (10–15°F) warmer than the mean annual air temperature for the localities where they occur. In the United States alone, there are well over 1,000 such springs.

Temperatures in deep mines and oil wells usually rise with an increase in depth averaging about 2°C per 100 meters (1°F per 100 feet). Therefore, when groundwater circulates at great depths, it becomes heated, and if it rises to the surface, the water may emerge as a hot spring. The water of some hot springs in the United States, particularly in the east, is heated in this manner. However, the great majority (over 95 percent) of the hot springs (and geysers) in the United States are found in the west. The reason for such a distribution is that the source of heat for most hot springs is cooling igneous rock, and it is in the west that igneous activity has been most recent.

Geysers

Geysers are intermittent hot springs or fountains in which columns of water are ejected with great force at various intervals, often rising 30–60 meters (100–200 feet). After the jet of water

a temperature of nearly 230°C (450°F) before it will boil. The heating causes the water to expand, with the result that some is forced out at the surface. This loss of water reduces the pressure on the remaining water in the chamber, which lowers the boiling point. A portion of the water deep within the chamber quickly turns to steam and causes the geyser to erupt. Following the eruption, cool groundwater again seeps into the chamber, and the cycle begins anew.

FIGURE 5.31 Yellowstone's Old Faithful is one of the world's most famous geysers. Each eruption lasts roughly 1-1/2 minutes and emits up to 32,000 liters (8,400 gallons) of hot water and steam skyward as high as 55 meters (more than 180 feet). Time spans between eruptions vary from about 65 to more than 90 minutes, and have generally increased over the years thanks to changes in the geyser's plumbing. (Photo by Art Director & Trip/Alamy)

ceases, a column of steam rushes out, usually with a thundering roar. Figure 5.31 shows a wintertime eruption of Old Faithful in Yellowstone National Park, perhaps the most famous geyser in the world, which erupts about once each hour. Geysers are also found in other parts of the world, including New Zealand and Iceland. In fact, the Icelandic word *geysa*—to gush—gives us the name *geyser*.

Geysers occur where extensive underground chambers exist within hot igneous rocks. How they operate is shown in Figure 5.32. As relatively cool groundwater enters the chambers, it is heated by the surrounding rock. At the bottom of the chamber, the water is under great pressure because of the weight of the overlying water. This great pressure prevents the water from boiling at the normal surface temperature of 100°C (212°F). For example, at the bottom of a 300-meter (1,000-foot) water-filled chamber, water must attain

CONCEPT CHECK 5.13

1 Describe the circumstances that created the springs shown in Figure 5.28 and Figure 5.30.
2 What is the source of heat for most hot springs and geysers?
3 Describe what causes a geyser to erupt.

Wells

Sculpturing Earth's Surface
▶ Groundwater

The most common method for removing groundwater is the **well**, a hole bored into the zone of saturation (see Figure 5.28). Wells serve as small reservoirs into which groundwater moves and from which it can be pumped to the surface. The use of wells dates back many centuries and continues to be an important method of obtaining water today. By far the single greatest use of this water in the United States is irrigation for agriculture. More than 68 percent of the groundwater used each year is for this purpose (Figure 5.33).

The level of the water table may fluctuate considerably during the course of a year, dropping during dry seasons and rising following periods of rain. Therefore, to ensure a continuous supply of water, a well must penetrate far below the water table. Whenever a substantial amount of water is withdrawn from a well, the water table around the well is lowered. This effect, termed **drawdown**, decreases with increasing distance from the well. The result is a depression in the water table, roughly conical in shape,

Students Sometimes Ask...

I have heard people say that supplies of groundwater can be located using a forked stick. Can this actually be done?

What you describe is a practice called "water dowsing." In the classic method, a person holding a forked stick walks back and forth over an area. When water is detected, the bottom of the "Y" is supposed to be attracted downward.

Geologists and engineers are dubious, to say the least. Case histories and demonstrations may seem convincing, but when dowsing is exposed to scientific scrutiny, it fails. Most "successful" examples of water dowsing occur in places where water would be hard to miss. In a region of adequate rainfall and favorable geology, it is difficult to drill and *not* find water!

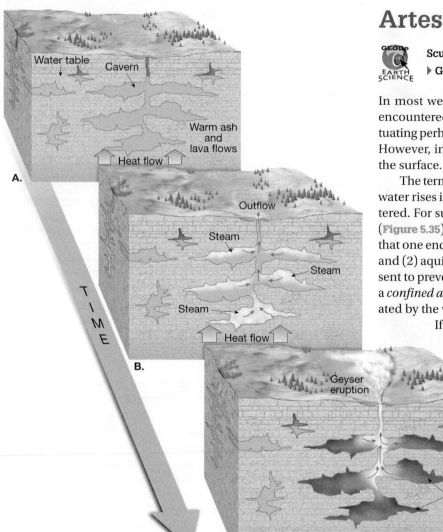

Artesian Wells

GEODe
EARTH SCIENCE
Sculpting Earth's Surface
▶ Groundwater

In most wells, water does not rise on its own. If water is first encountered at 30 meters of depth, it remains at that level, fluctuating perhaps a meter or two with seasonal wet and dry periods. However, in some wells, water rises, sometimes overflowing at the surface.

The term **artesian** is applied to any situation in which groundwater rises in a well above the level where it was initially encountered. For such a situation to occur, two conditions usually exist (**Figure 5.35**): (1) Water is confined to an aquifer that is inclined so that one end is exposed at the surface, where it can receive water; and (2) aquitards both above and below the aquifer must be present to prevent the water from escaping. Such an aquifer is called a *confined aquifer*. When such a layer is tapped, the pressure created by the weight of the water above will force the water to rise. If there were no friction, the water in the well would rise to the level of the water at the top of the aquifer. However, friction reduces the height of this pressure surface. The greater the distance from the recharge area (area where water enters the inclined aquifer), the greater the friction and the less the rise of water.

FIGURE 5.33 Each day in the United States we use an average of 349 billion gallons of freshwater. Groundwater is the source of nearly one-quarter of the total. More groundwater is used for irrigation than for all other uses combined. (Data from U.S. Geological Survey)

FIGURE 5.32 Idealized diagrams illustrating the stages in the eruption cycle of a geyser. **A.** Groundwater enters underground caverns and fractures in hot igneous rock, where it is heated to near its boiling point. **B.** Heating causes the water to expand, with some being forced out at the surface. The loss of water reduces the pressure on the remaining water, thus reducing its boiling temperature. Some of the water flashes to steam. **C.** The rapidly expanding steam forces the hot water out of the chambers to produce a geyser. The empty chambers fill again, and the cycle starts anew.

known as a **cone of depression** (**Figure 5.34**). For most small domestic wells, the cone of depression is negligible. However, when wells are used for irrigation or for industrial purposes, the withdrawal of water can be great enough to create a very wide and steep cone of depression that may substantially lower the water table in an area and cause nearby shallow wells to become dry. Figure 5.34 illustrates this situation.

CONCEPT CHECK 5.14

❶ How does a heavily pumping well affect the water table?
❷ In Figure 5.28, two wells are at the same level. Why was one successful and the other not?

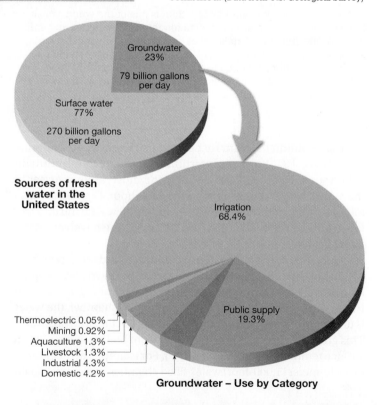

Sources of fresh water in the United States

Groundwater – Use by Category

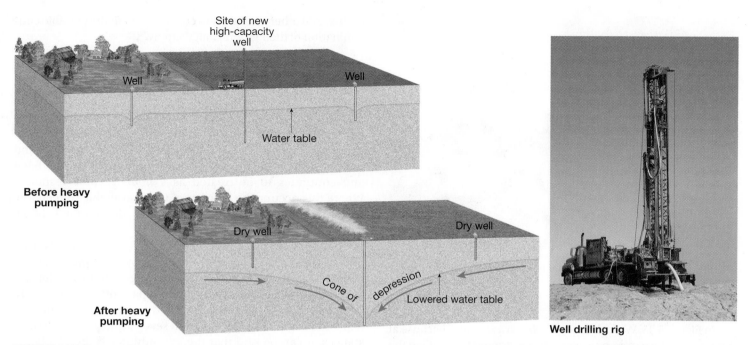

FIGURE 5.34 A cone of depression in the water table often forms around a pumping well. If heavy pumping lowers the water table, some wells may be left dry. (ASP/YPP/agefotostock)

FIGURE 5.35 Artesian systems occur when an inclined aquifer is surrounded by impermeable beds. Such an aquifer is called a *confined aquifer*. The photo shows a flowing artesian well. (Photo by James E. Patterson)

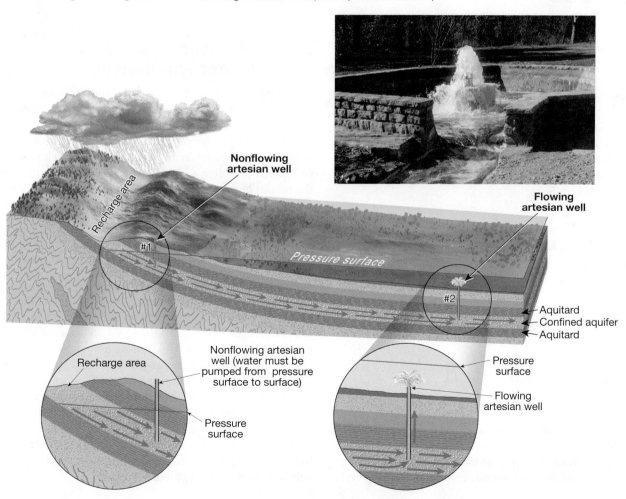

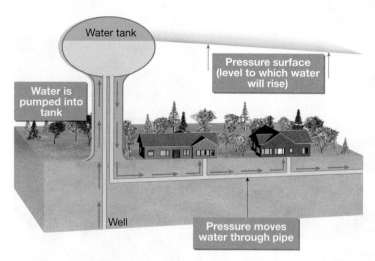

FIGURE 5.36 City water systems can be considered to be artificial artesian systems.

In Figure 5.35, Well 1 is a *nonflowing artesian well* because at this location the pressure surface is below ground level. When the pressure surface is above the ground and a well is drilled into the aquifer, a *flowing artesian well* is created (Well 2, Figure 5.35). Not all artesian systems are wells. *Artesian springs* also exist. In such situations groundwater may reach the surface by rising along a natural fracture such as a fault rather than through an artificially produced hole. In deserts, artesian springs are sometimes responsible for creating an oasis.

Artesian systems act as conduits, transmitting water from remote areas of recharge great distances to the points of discharge. In this manner, water that fell in central Wisconsin years ago is now taken from the ground and used by communities many kilometers to the south in Illinois. In South Dakota, such a system brings water eastward across the state from the Black Hills in the west.

On a different scale, city water systems may be considered examples of artificial artesian systems (**Figure 5.36**). The water tower, into which water is pumped, may be considered the area of recharge, the pipes the confined aquifer, and the faucets in homes the flowing artesian wells.

CONCEPT CHECK 5.15

❶ **Sketch a simple cross-section of an artesian system with a flowing well. Label the aquitards, aquifer, and pressure surface.**

Environmental Problems Associated with Groundwater

As with many of our valuable natural resources, groundwater is being exploited at an increasing rate. In some areas, overuse threatens the groundwater supply. In other places, groundwater withdrawal has caused the ground and everything resting upon it

to sink. Still other localities are concerned with the possible contamination of their groundwater supply.

Treating Groundwater as a Nonrenewable Resource

For many, groundwater appears to be an endlessly renewable resource, for it is continually replenished by rainfall and melting snow. But in some regions, groundwater has been and continues to be treated as a *nonrenewable* resource. Where this occurs, the amount of water available to recharge the aquifer is significantly less than the amount being withdrawn.

The High Plains, a relatively dry region that extends from South Dakota to western Texas, provides one example (**Figure 5.37**). In this region an extensive agricultural economy is largely dependent on irrigation. In some parts of the region where intense irrigation has been practiced for an extended period, depletion of groundwater has been severe. Under these circumstances, it can be said that the groundwater is literally being "mined." Even if pumping were to cease immediately, it would take thousands of years for the groundwater to be fully replenished. Groundwater depletion has been a concern in the High Plains and other areas of the west for many years, but the problem is not confined to this part of the country. Increasing demands on groundwater resources have overstressed aquifers in many areas, not just in arid and semiarid regions.

Land Subsidence Caused by Groundwater Withdrawal

As you will see later in this chapter, surface subsidence can result from natural processes related to groundwater. However, the ground may also sink when water is pumped from wells faster than natural recharge processes can replace it. This effect is particularly pronounced in areas underlain by thick layers of loose sediments. As water is withdrawn, the water pressure drops and the weight of the overburden is transferred to the sediment. The greater pressure packs the sediment grains more tightly together and the ground subsides.

Many areas can be used to illustrate such land subsidence. A classic example in the United States occurred in the San Joaquin Valley of California (**Figure 5.38**). Other well-known cases of land subsidence resulting from groundwater pumping in the United States include Las Vegas, Nevada; New Orleans and Baton Rouge, Louisiana; portions of southern Arizona; and the Houston–Galveston area of Texas. In the low-lying coastal area between Houston and Galveston, land subsidence ranges from 1.5–3 meters (5–10 feet). The result is that about 78 square kilometers (30 square miles) are permanently flooded.

Outside the United States, one of the most spectacular examples of subsidence occurred in Mexico City, a portion of which is built on a former lake bed. In the first half of the 20th century, thousands of wells were sunk into the water-saturated sediments beneath the city. As water was withdrawn, portions of the city subsided by 6 meters (20 feet) or more.

FIGURE 5.37 A. The High Plains Aquifer underlies about 111 million acres in parts of eight western states. It is one of the largest and most agriculturally significant aquifers in the United States. **B.** In some agricultural regions, water is pumped from the ground faster than it is replenished. In such instances, groundwater is being treated as a nonrenewable resource. This aerial view shows circular crop fields irrigated by center pivot irrigation systems in semiarid eastern Colorado. (Photo by James L. Amos/CORBIS/Bettmann) **C.** Groundwater provides more than 54 billion gallons per day in support of the agricultural economy of the United States. (Photo by Michael Collier)

FIGURE 5.38 The shaded area on the map shows California's San Joaquin Valley. The marks on the utility pole in the photo indicate the level of the surrounding land in preceding years. Between 1925 and 1975 this part of the San Joaquin Valley subsided almost 9 meters because of the withdrawal of ground-water and the resulting compaction of sediments. (Photo courtesy of U.S. Geological Survey)

Groundwater Contamination

The pollution of groundwater is a serious matter, particularly in areas where aquifers provide a large part of the water supply. One common source of groundwater pollution is sewage. Its sources include an ever increasing number of septic tanks, as well as farm wastes and inadequate or broken sewer systems.

If sewage water that is contaminated with bacteria enters the groundwater system, it may become purified through natural processes. The harmful bacteria can be mechanically filtered by the sediment through which the water percolates, destroyed by chemical oxidation, and/or assimilated by other organisms. For purification to occur, however, the aquifer must be of the correct composition. For example, extremely permeable aquifers (such as highly fractured crystalline rock, coarse gravel, or cavernous limestone) have such large openings that contaminated groundwater may travel long distances without being cleansed. In this case, the water flows too rapidly and is not in contact with the surrounding

Students Sometimes Ask...

Does land subsidence caused by groundwater withdrawal affect a very large area?

According to an estimate by the U.S. Geological Survey, it is substantial. In the contiguous 48 states, the area amounts to approximately 26,000 square kilometers (more than 10,000 square miles)—an area about the same size as the state of Massachusetts!

PROFESSIONAL PROFILE

Michael Collier— Feet in the Fire: Communicating Geologist

Once, after an eight-hour flight to Chile and another two hours to the Falkland Islands, I found myself aboard a boat headed toward Antarctica. It was early evening and even though the December sky was dark, familiar Northern Hemisphere landmarks were nowhere to be seen. I was on the top deck, leaning backward over the railing, when a fierce gust of loneliness broadsided me: home was seven thousand miles away. Leaning precariously over that rail, I took comfort in the sight of Orion, 'upright' because I was upside down. I've always needed to know where I fit in, either upon the Earth or in the sky.

A few years earlier, on a winter river trip through Utah's Cataract Canyon, we were boating alongside ice on the Colorado when I realized for the first time that I was exactly in the world where I wanted to be. I was a budding photographer, capturing landscape images throughout the American West. I was learning to love the color and cadence of stories that inevitably shine through the photographs—rivers rolling, waves crashing, mountains gleaming. These stories had grand themes of creation, metamorphosis, and destruction. I had unexpectedly warmed to a geologic way of looking at the world.

I first studied "gee-whiz geology" at a college near Grand Canyon, where all we undergraduates wanted to do was hike our brains out. Then, as a graduate student in California, I was thrown into an unexpectedly mathematical appreciation of structural folds and faults. Feet in the fire, head in the freezer: it was a balanced education that nicely prepared me for life as a communicating geologist.

After finishing school, I rowed boats in Grand Canyon for a living. We would spend almost two weeks drifting from Lees Ferry to the takeout at Diamond Creek. My masters thesis had examined a curious fold in the Muav Limestone created when lava had temporarily flooded the Canyon a few thousand years earlier. My passengers, leaning over the raft's rubber side and trailing their fingers absentmindedly in the water, were a captive audience to my geologic stories about time and rock and those Muav folds. They didn't seem to mind.

Along the way, I learned to fly and kayak. The two motions resonated: three-dimensional activities in fluid media. Like a paddling friend says, sometimes you're supposed to be upside down in a rapid. And now, after 5500 hours aloft, I've found the same to be true in the air. It doesn't seem unnatural when my plane is sideways howling around a curve at the bottom of some narrow canyon. You're supposed to be sideways. Otherwise you'll smack into the wall that's filling the windshield.

My plane is a fifty five-year-old Cessna 180 that's happiest when operating off dirt strips in the middle of nowhere. My wife swears that the plane's tail wags when I walk up to it. Together my plane and I have seen the continent from Tegucigalpa to Tanana, from Portland to Portland, from Baja to Boston. I use the plane to study and photograph landscapes. From the sky, I've grown intimate with the ground. From above, I see stories that the Earth has to tell; I've learned where I fit in. Once the pilot of a large business plane was shocked when I told him that upon take-off, I might know where I'm going to land only two times out of three. I go where the stories lead, where the light draws my camera.

I started out publishing small books about the geology of national parks—Grand Canyon, of course, Capitol Reef, Denali, and Death Valley. Regardless of how much scientific research I had found in libraries, I always insisted on developing a personal sense of each new landscape—hiking, scrambling, driving, and flying before beginning to write. Without a feel for the land, my words and photographs were doomed to be vacuous.

I now write books about geologic processes for lay audiences— how mountains are built, how glaciers move, how faults whipsaw the Earth, how climate behaves, how rivers flow. Each requires the assimilation of new information about unfamiliar topics. Each book introduces me to accomplished scientists, once again humbling me in the presence of experts in their chosen work. Each new project is a great opportunity to learn and grow. Each makes *my* tail wag.

> **"Feet in the fire, head in the freezer: it was a balanced education that nicely prepared me for life as a communicating geologist."**

But here's an odd twist . . . after a while I became dissatisfied with the mercurial life of an itinerant geologist, boatman, pilot, writer, and freelance photographer. I wanted to be able to walk up and shake a person's hand, look them in the eye and ask how could I help. So, in the 1980s I detoured into medical school and, after residency, became

> **"It doesn't seem unnatural when my plane is sideways howling around a curve at the bottom of some narrow canyon."**

a family practice physician. Thirty years later, I still fly and write and photograph, but now I also spend half my life as a doctor. We only go around once, so we might as well pack it in while we can. Doctoring has been profoundly satisfying because it balances a life spent outdoors, studying beautiful landscapes and listening to stories that are as large as the Earth itself.

—Michael Collier

Geologist-pilot Michael Collier and his Cessna 180. He is also an award-winning photographer and writer. Michael classifies himself as a "communicating geologist." You will see dozens of his outstanding images in the pages of this book. (Photo by Patty DiRienzo)

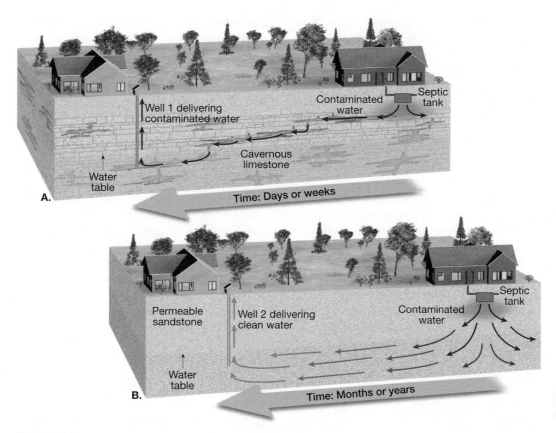

FIGURE 5.39 **A.** Although the contaminated water has traveled more than 100 meters before reaching Well 1, the water moves too rapidly through the cavernous limestone to be purified. **B.** As the discharge from the septic tank percolates through the permeable sandstone, it is purified in a relatively short distance.

material long enough for purification to occur. This is the problem at Well 1 in Figure 5.39A.

Conversely, when the aquifer is composed of sand or permeable sandstone, the water can sometimes be purified after traveling only a few dozen meters through it. The openings between sand grains are large enough to permit water movement, yet the movement of the water is slow enough to allow ample time for its purification (Well 2, Figure 5.39B).

Other sources and types of contamination also threaten groundwater supplies (Figure 5.40). These include widely used substances such as highway salt, fertilizers that are spread across the land surface, and pesticides. In addition, a wide array of chemicals and industrial materials may leak from pipelines, storage tanks, landfills, and holding ponds. Some of these pollutants are classified as *hazardous*, meaning that they are either flammable, corrosive, explosive, or toxic. As rainwater oozes through the

FIGURE 5.40 Sometimes agricultural chemicals **A.** and materials leached from landfills **B.** find their way into the groundwater. These are two potential sources of groundwater contamination. (Photo A by Michael Collier; Photo B by F. Rossotto/The Stock Market)

A.

B.

refuse, it may dissolve a variety of potential contaminants. If the leached material reaches the water table, it will mix with the groundwater and contaminate the supply.

Because groundwater movement is usually slow, polluted water might go undetected for a long time. In fact, contamination is sometimes discovered only after drinking water has been affected and people become ill. By this time, the volume of polluted water might be very large, and even if the source of contamination is removed immediately, the problem is not solved. Although the sources of groundwater contamination are numerous, there are relatively few solutions.

Once the source of the problem has been identified and eliminated, the most common practice is simply to abandon the water supply and allow the pollutants to be flushed away gradually. This is the least costly and easiest solution, but the aquifer must remain unused for many years. To accelerate this process, polluted water is sometimes pumped out and treated. Following removal of the tainted water, the aquifer is allowed to recharge naturally or, in some cases, the treated water or other freshwater is pumped back in. This process is costly, time-consuming, and it may be risky because there is no way to be certain that all of the contamination has been removed. Clearly, the most effective solution to groundwater contamination is prevention.

CONCEPT CHECK 5.16

1. **Describe the problem associated with pumping groundwater for irrigation in parts of the High Plains.**
2. **What happened in the San Joaquin Valley as the result of excessive groundwater pumping?**
3. **Which aquifer would be most effective in purifying polluted groundwater: coarse gravel, sand, or cavernous limestone?**

The Geologic Work of Groundwater

Groundwater dissolves rock. This fact is key to understanding how caverns and sinkholes form. Because soluble rocks, especially limestone, underlie millions of square kilometers of Earth's surface, it is here that groundwater carries on its important role as an erosional agent. Limestone is nearly insoluble in pure water but is quite easily dissolved by water containing small quantities of carbonic acid. Most natural water contains this weak acid because rainwater readily dissolves carbon dioxide from the air and from decaying plants. Therefore, when groundwater comes in contact with limestone, the carbonic acid reacts with calcite in the rocks to form calcium bicarbonate, a soluble material that is then carried away in solution.

Caverns

The most spectacular results of groundwater's erosional handiwork are limestone **caverns**. In the United States alone about 17,000 caves have been discovered. Although most are relatively small, some have spectacular dimensions. Carlsbad Caverns in southeastern New Mexico and Mammoth Cave in Kentucky are famous examples. One chamber in Carlsbad Caverns has an area equivalent to 14 football fields and enough height to accommodate the U.S. Capitol Building. At Mammoth Cave, the total length of interconnected caverns extends for more than 540 kilometers (340 miles).

Most caverns are created at or below the water table in the zone of saturation. Here acidic groundwater follows lines of weakness in the rock, such as joints and bedding planes. As time passes, the dissolving process slowly creates cavities and gradually enlarges them into caverns. Material that is dissolved by the groundwater is eventually discharged into streams and carried to the ocean.

Certainly the features that arouse the greatest curiosity for most cavern visitors are the stone formations that give some caverns a wonderland appearance. These are not erosional features, like the caverns in which they reside, but depositional features. They are created by the seemingly endless dripping of water over great spans of time. The calcium carbonate that is left behind produces the limestone we call *travertine*. These cave deposits, however, are also commonly called *dripstone*, an obvious reference to their mode of origin.

Although the formation of caverns takes place in the zone of saturation, the deposition of dripstone is not possible until the caverns are above the water table in the unsaturated zone. This commonly occurs as nearby streams cut their valleys deeper, lowering the water table as the elevation of the rivers drops. As soon as the chamber is filled with air, the conditions are right for the decoration phase of cavern building to begin.

Of the various dripstone features found in caverns, perhaps the most familiar are **stalactites**. These icicle-like pendants hang from the ceiling of the cavern and form where water seeps through cracks above. When water reaches air in the cave, some of the dissolved carbon dioxide escapes from the drop and calcite begins to precipitate. Deposition occurs as a ring around the edge of the water drop. As drop after drop follows, each leaves an infinitesimal trace of calcite behind, and a hollow limestone tube is created. Water then moves through the tube, remains suspended momentarily at the end, contributes a tiny ring of calcite, and falls to the cavern floor. The stalactite just described is appropriately called a *soda straw* (Figure 5.41A). Often the hollow tube of the soda straw becomes plugged or its supply of water increases. In either case, the water is forced to flow and deposit along the outside of the tube. As deposition continues, the stalactite takes on the more common conical shape.

Formations that develop on the floor of a cavern and reach upward toward the ceiling are called **stalagmites** (Figure 5.41B). The water supplying the calcite for stalagmite growth falls from the ceiling and splatters over the surface. As a result, stalagmites do not have a central tube and are usually more massive in appearance and more rounded on their upper ends than stalactites. Given enough time, a downward-growing stalactite and an upward-growing stalagmite may join to form a *column*.

Karst Topography

Many areas of the world have landscapes that to a large extent have been shaped by the dissolving power of groundwater. Such areas are said to exhibit **karst topography**, named for the *Krs* region in Slovenia where such topography is strikingly developed. In the

A.   **B.**

FIGURE 5.41 **A.** Live soda-straw stalactite in Chinn Springs Cave, Independence County, Arkansas. (Photo by Dante Fenolio/Photo Researchers, Inc.) **B.** Stalactites, stalagmites, and columns in New Mexico's Carlsbad Caverns National Park. (Photo by Tbkmedia.de/Alamy)

United States, karst landscapes occur in many areas that are underlain by limestone, including portions of Kentucky, Tennessee, Alabama, southern Indiana, and central and northern Florida (Figure 5.42). Generally, arid and semiarid areas do not develop karst topography because there is insufficient groundwater. When solution features exist in such regions, they are likely to be remnants of a time when rainier conditions prevailed.

Karst areas typically have irregular terrain punctuated with many depressions called **sinkholes** or, simply, **sinks** (see Figure 5.27B, p. 139). In the limestone areas of Florida, Kentucky, and southern Indiana, there are literally tens of thousands of these depressions varying in depth from just a meter or two to a maximum of more than 50 meters.

Sinkholes commonly form in one of two ways. Some develop gradually over many years without any physical disturbance to the rock. In these situations, the limestone immediately below the soil is dissolved by downward-seeping rainwater that is freshly charged with carbon dioxide. These depressions are usually not deep and are characterized by relatively gentle slopes. By contrast, sinkholes can also form suddenly and without warning when the roof of a cavern collapses under its own weight. Typically, the depressions created in this manner are steep-sided and deep. When they form in populous areas, they may represent a serious geologic hazard. Such a situation is clearly the case in Figure 5.43.

In addition to a surface pockmarked by sinkholes, karst regions characteristically show a striking lack of surface drainage (streams). Following a rainfall, runoff is quickly funneled below ground through sinks. It then flows through caverns until it finally reaches the water table. Where streams do exist at the surface, their paths are usually short. The names of such streams often give a clue to their fate. In the Mammoth Cave area of Kentucky, for example, there is Sinking Creek, Little Sinking Creek, and Sinking Branch. Some sinkholes become plugged with clay and debris, creating small lakes or ponds.

Some regions of karst development exhibit landscapes that look very different from the sinkhole-studded terrain depicted in Figure 5.42. One striking example is an extensive region in southern China that is described as exhibiting *tower karst*. As the chapter-opening photo shows, the term *tower* is appropriate because the landscape consists of a maze of isolated steep-sided hills that rise abruptly from the ground. Each is riddled with interconnected caves and passageways. This type of karst topography forms in wet tropical and subtropical regions having thick beds of highly jointed limestone. Here groundwater has dissolved large volumes of limestone, leaving only these residual towers. Karst development is

Students Sometimes Ask...

Is limestone the only rock type that develops karst features?

No. For example, karst development can occur in other carbonate rocks such as marble and dolostone. In addition, evaporates such as gypsum and salt (halite) are highly soluble and are readily dissolved to form karst features, including sinkholes, caves, and disappearing streams. This latter situation is termed *evaporite karst.*

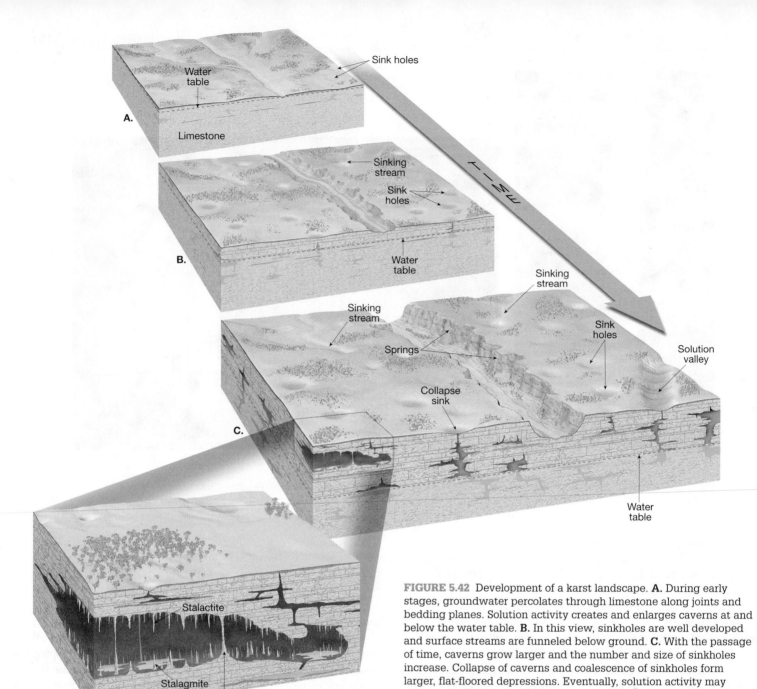

FIGURE 5.42 Development of a karst landscape. **A.** During early stages, groundwater percolates through limestone along joints and bedding planes. Solution activity creates and enlarges caverns at and below the water table. **B.** In this view, sinkholes are well developed and surface streams are funneled below ground. **C.** With the passage of time, caverns grow larger and the number and size of sinkholes increase. Collapse of caverns and coalescence of sinkholes form larger, flat-floored depressions. Eventually, solution activity may remove most of the limestone from the area, leaving only isolated remnants as in the chapter-opening photo.

more rapid in tropical climates due to the abundant rainfall and the greater availability of carbon dioxide from the decay of lush tropical vegetation. The extra carbon dioxide in the soil means there is more carbonic acid for dissolving limestone. Other tropical areas of advanced karst development include portions of Puerto Rico, western Cuba, and northern Vietnam.

CONCEPT CHECK 5.17

❶ How does groundwater create caverns?
❶ How do stalactites and stalagmites form?
❶ Describe two ways in which sinkholes form.

FIGURE 5.43 This small sinkhole formed suddenly in 1991 when the roof of a cavern collapsed, destroying this home in Frostproof, Florida. (Photo by *St. Petersburg Times*/Liaison Agency, Inc.)

GIVE IT SOME THOUGHT

1. A river system consists of three zones based on the dominant process operating in that part of the river system. On the accompanying illustration, match each process with one of the three zones:
   a. sediment production (erosion)   c. sediment transport
   b. sediment deposition

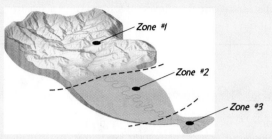

2. Match the description that best fits each of the accompanying photos.
   a. bedrock channel   c. alluvial channel (non meandering)
   b. alluvial channel (meandering)   d. alluvial channel (braided)

River #1   River #2   River #3   River #4

 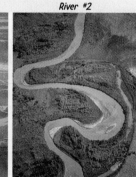  

3. Identify the main river in your area. For what streams does it act as base level? What is base level for the Mississippi River? Missouri River?

4. The accompanying diagram shows the time relationship between a rainstorm and flooding. Based on this graph, when does the flooding occur in relation to the rainstorm?

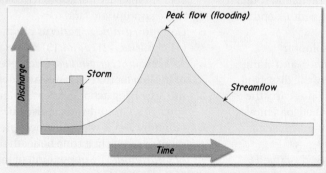

5. The accompanying graphs show lag times between rainfall and peak flow (flooding) for an urban area and a rural area. Which graph (A or B) represents a rural area? Explain your choice.

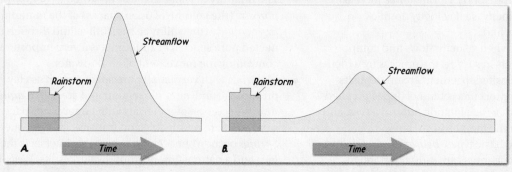

6. Which one of the three basic rock types (igneous, sedimentary, or metamorphic) has the likelihood of being a good aquifer? Why?

7. During a trip to the grocery store, your friend wants to buy some bottled water. Some brands promote the fact that their product is artesian. Other brands boast that their water comes from a spring. Your friend asks, "Is artesian water or spring water necessarily better than water from other sources?" How would you answer?

8. What is the likely difference between an intermittent stream (one that flows off and on) and a stream that flows all the time, even during extended dry periods?

In Review   Chapter 5 Running Water and Groundwater

- The *hydrologic cycle* describes the continuous interchange of water among the oceans, atmosphere, and continents. Powered by energy from the Sun, it is a global system in which the atmosphere provides the link between the oceans and continents. The processes involved in the hydrologic cycle include *precipitation, evaporation, infiltration* (the movement of water into rocks or soil through cracks and pore spaces), *runoff* (water that flows over the land rather than infiltrating into the ground), and *transpiration* (the release of water vapor to the atmosphere by plants). *Running water is the single most important agent sculpturing Earth's land surface.*

- The land area that contributes water to a stream is its *drainage basin*. Drainage basins are separated by imaginary lines called *divides*.

- River systems consist of three main parts: the zones of erosion, transportation, and deposition.

- The factors that determine a stream's *velocity* are *gradient* (slope of the stream channel); *shape, size*, and *roughness* of the channel; and the stream's *discharge* (amount of water passing a given point per unit of time, frequently measured in cubic feet per second). Most often, the gradient and roughness of a stream decrease downstream, while width, depth, discharge, and velocity increase.

- Streams transport their load of sediment in solution (*dissolved load*), in suspension (*suspended load*), and along the bottom of the channel (*bed load*). Much of the dissolved load is contributed by groundwater. Most streams carry the greatest part of their load in suspension. The bed load moves only intermittently and is usually the smallest portion of a stream's load.

- A stream's ability to transport solid particles is described using two criteria: *capacity* (the maximum load of solid particles a stream can carry) and *competence* (the maximum particle size a stream can transport). Competence increases to the square of stream velocity, so if velocity doubles, water's force increases fourfold.

- Streams deposit sediment when velocity slows and competence is reduced. This results in *sorting*, the process by which like-sized particles are deposited together. Stream deposits are called *alluvium* and may occur as channel deposits called *bars*; as floodplain deposits, which include *natural levees*; and as *deltas* or *alluvial fans* at the mouths of streams.

- Stream channels are of two basic types: *bedrock channels* and *alluvial channels*. Bedrock channels are most common in headwater regions where gradients are steep. Rapids and waterfalls are common features. Two types of alluvial channels are *meandering channels* and *braided channels*.

- The two general types of *base level* (the lowest point to which a stream may erode its channel) are (1) *ultimate base level* and (2) *temporary*, or *local, base level*. Any change in base level will cause a stream to adjust and establish a new balance. Lowering base level will cause a stream to downcut, whereas raising base level results in deposition of material in the channel.

- When a stream has cut its channel closer to base level, its energy is directed from side to side, and erosion produces a flat valley floor, or *floodplain*. Streams that flow upon floodplains often move in sweeping bends called *meanders*. Widespread meandering may result in shorter channel segments, called *cutoffs*, and/or abandoned bends, called *oxbow lakes*.

- *Floods* are triggered by heavy rains and/or snowmelt. Sometimes human interference can worsen or even cause floods. Flood-control measures include the building of *artificial levees* and dams, as well as *channelization*, which could involve creating *artificial cutoffs*. Many scientists and engineers advocate a nonstructural approach to flood control that involves more appropriate land use.

- Common *drainage patterns* produced by streams include (1) *dendritic*, (2) *radial*, (3) *rectangular*, and (4) *trellis*.

- As a resource, *groundwater* represents the largest reservoir of freshwater that is readily available to humans. Geologically, the dissolving action of groundwater produces *caves* and *sinkholes*. Groundwater is also an equalizer of streamflow.

- Groundwater is water that occupies the pore spaces in sediment and rock in a zone beneath the surface called the *zone of saturation*. The upper limit of this zone is the *water table*. The *unsaturated zone* is above the water table where the soil, sediment, and rock are not saturated.

- The quantity of water that can be stored depends on the *porosity* (the volume of open spaces) of the material. The *permeability* (the ability to transmit a fluid through interconnected pore spaces) of a material is a very important factor controlling the movement of groundwater.

- Materials with very small pore spaces (such as clay) hinder or prevent groundwater movement and are called *aquitards*. *Aquifers* consist of materials with larger pore spaces (such as sand) that are permeable and transmit groundwater freely.

- *Springs* occur whenever the water table intersects the land surface and a natural flow of groundwater results. *Wells*, openings drilled into the zone of saturation, withdraw groundwater and create roughly conical depressions in the water table known as *cones of depression*. *Artesian wells* occur when water rises above the level at which it was initially encountered.

- When groundwater circulates at great depths, it becomes heated. If it rises, the water may emerge as a *hot spring*. *Geysers* occur when groundwater is heated in underground chambers, expands, and some water quickly changes to steam, causing the geyser to erupt. The source of heat for most hot springs and geysers is hot igneous rock.

- Some of the current environmental problems involving groundwater include (1) *overuse* by intense irrigation, (2) *land subsidence* caused by groundwater withdrawal, and (3) *contamination* by pollutants.

- Most *caverns* form in limestone at or below the water table when acidic groundwater dissolves rock along lines of weakness, such as joints and bedding planes. *Karst topography* exhibits an irregular terrain punctuated with many depressions, called *sinkholes*.

Key Terms

alluvial fan (p. 135)
alluvium (p. 127)
aquifer (p. 141)
aquitard (p. 141)
artesian well (p. 144)
backswamp (p. 135)
bar (p. 132)
base level (p. 130)
bed load (p. 126)
braided stream (p. 128)
capacity (p. 127)
cavern (p. 150)
competence (p. 127)
cone of depression (p. 144)
cut bank (p. 128)
cutoff (p. 128)
dendritic pattern (p. 135)
discharge (p. 122)
dissolved load (p. 126)
distributary (p. 132)
divide (p. 121)

drainage basin (p. 120)
drawdown (p. 143)
evapotranspiration (p. 120)
flood (p. 136)
floodplain (p. 131)
geyser (p. 142)
gradient (p. 122)
groundwater (p. 140)
hot spring (p. 142)
hydrologic cycle (p. 118)
incised meander (p. 132)
infiltration (p. 118)
karst topography (p. 150)
laminar flow (p. 122)
longitudinal profile (p. 124)
meander (p. 127)
natural levee (p. 133)
oxbow lake (p. 128)
permeability (p. 141)
point bar (p. 128)
porosity (p. 140)

radial pattern (p. 135)
rectangular pattern (p. 135)
runoff (p. 118)
saltation (p. 126)
settling velocity (p. 126)
sinkhole (sink) (p. 151)
sorting (p. 127)
spring (p. 142)
stalactite (p. 150)
stalagmite (p. 150)
stream valley (p. 130)
suspended load (p. 126)
transpiration (p. 120)
trellis pattern (p. 136)
turbulent flow (p. 122)
unsaturated zone (p. 140)
water table (p. 140)
well (p. 143)
yazoo tributary (p. 135)
zone of saturation (p. 140)

Examining the Earth System

1. List the process(es) involved in moving water through the hydrologic cycle from the (a) hydrosphere to the atmosphere, (b) atmosphere to the geosphere, (c) biosphere to the atmosphere, and (d) hydrosphere (land) to the hydrosphere (ocean).

2. Over the oceans, evaporation exceeds precipitation, yet sea level does not drop. Why?

3. List at least three specific examples of interactions between humans and the hydrologic cycle (e.g., the construction of a dam). Briefly describe the consequence(s) of each of these interactions.

Mastering Geology

Looking for additional review and test prep materials? Visit the Self Study area in **www.masteringgeology.com** to find practice quizzes, study tools, and multimedia that will aid in your understanding of this chapter's content. In **MasteringGeology™** you will find:

- **GEODe: Earth Science: An interactive visual walkthrough of key concepts**

- **Geoscience Animation Library: More than 100 animations illuminating many difficult-to-understand Earth science concepts**
- **In The News RSS Feeds: Current Earth science events and news articles are pulled into the site with assessment**
- **Pearson eText**
- **Optional Self Study Quizzes**
- **Web Links**
- **Glossary**
- **Flashcards**

Glaciers, Deserts, and Wind

Landslide debris atop Buckskin Glacier in Alaska's Denali National Park.
(Photo by Michael Collier)

Like the running water and groundwater that were the focus of Chapter 5, glaciers and wind are significant erosional processes. They are responsible for creating many different landforms and are part of an important link in the rock cycle in which the products of weathering are transported and deposited as sediment.

Climate has a strong influence on the nature and intensity of Earth's external processes. This fact is dramatically illustrated in this chapter. The existence and extent of glaciers is largely controlled by Earth's changing climate. Another excellent example of the strong link between climate and geology is seen when we examine the development of arid landscapes.

FOCUS ON CONCEPTS

To assist you in learning **the important concepts in this chapter, focus on the following questions:**

- What is a glacier? What are the different types of glaciers?
- How do glaciers move? What determines whether the front of a glacier advances, retreats, or remains stationary?
- What are the various processes of glacial erosion? What are some erosional features?
- What features are created by glacial deposition? What materials make up these depositional features?
- What is some evidence of the Ice Age? What are some indirect effects of Ice Age glaciers?
- What are some proposals that attempt to explain the causes of glacial ages?
- What are the roles of weathering, water, and wind in arid and semiarid climates?
- How do landscapes in the dry Basin and Range region of the United States evolve?
- How does wind erode?
- What are some depositional features produced by wind?

Glaciers: A Part of Two Basic Cycles in the Earth System

Sculpturing Earth's Surface
▶ Glaciers

Today, glaciers cover nearly 10 percent of Earth's land surface; however, in the recent geologic past, ice sheets were three times more extensive, covering vast areas with ice thousands of meters thick. Many regions still bear the mark of these glaciers. The basic character of such diverse places as the Alps, Cape Cod, and Yosemite Valley was fashioned by now vanished masses of glacial ice. Moreover, Long Island, the Great Lakes, and the fiords of Norway and Alaska all owe their existence to glaciers. Glaciers, of course, are not just a phenomenon of the geologic past. As you will see, they are still modifying the physical landscapes of many regions today.

Glaciers are a part of two of Earth's basic cycles: the hydrologic cycle and the rock cycle. In the hydrologic cycle, when precipitation falls on land at high elevations or high latitudes, the water may not immediately make its way back toward the sea. Instead, it may become part of a glacier. Although the ice will eventually melt, allowing the water to continue its path to the sea, water can be stored as glacial ice for tens, hundreds, or even thousands of years. During the time that the water is part of a glacier, the moving mass of ice can do enormous amounts of work—scouring the land surface and acquiring, transporting, and depositing great quantities of sediment. This activity is a basic part of the rock cycle (**Figure 6.1**).

A **glacier** is a thick ice mass that forms over hundreds or thousands of years. It originates on land from the accumulation, compaction, and recrystallization of snow. A glacier appears to be motionless, but it is not—glaciers move very slowly. Like running water, groundwater, wind, and waves, glaciers are dynamic erosional agents that accumulate, transport, and deposit sediment. Although glaciers are found in many parts of the world, most are located in remote areas, either near Earth's poles or in high mountains.

Valley (Alpine) Glaciers

Literally thousands of relatively small glaciers exist in lofty mountain areas, where they usually follow valleys originally occupied by streams. Unlike the rivers that previously flowed in these valleys, the glaciers advance slowly, perhaps only a few centimeters each day. Because of their setting, these moving ice masses are termed **valley glaciers** or **alpine glaciers** (Figure 6.1). Each glacier is a stream of ice, bounded by precipitous rock walls, that flows downvalley from a snow accumulation center near its head. Like rivers, valley glaciers can be long or short, wide or narrow, single or with branching tributaries. Generally, the widths of alpine glaciers are small compared to their lengths; some extend for just a fraction of a kilometer, whereas others go on for many dozens of kilometers. The west branch of the Hubbard Glacier, for example, runs through 112 kilometers (nearly 70 miles) of mountainous terrain in Alaska and the Yukon Territory.

FIGURE 6.1 Kennicott Glacier is a 43-kilometer- (27-mile-) long valley glacier that is sculpting the mountains in Alaska's Wrangell–St. Elias National Park. The dark stripes of sediment are called *medial moraines* (see p. 168). Because glaciers transport and deposit sediment, they are a basic part of the rock cycle. (Photo by Michael Collier)

Ice Sheets

In contrast to valley glaciers, **ice sheets** exist on a *much* larger scale. In fact, they are often referred to as *continental ice sheets*. These enormous masses flow out in all directions from one or more centers and completely obscure all but the highest areas of underlying terrain. Although many ice sheets have existed in the past, just two achieve this status at present (**Figure 6.2**). Nevertheless, their combined areas represent almost 10 percent of Earth's land area. In the Northern Hemisphere, Greenland is covered by an imposing ice sheet that occupies 1.7 million square kilometers (0.7 million square miles), or about 80 percent of this large island. Averaging nearly 2.3 kilometers (1.6 miles) thick, in places the ice extends 3,000 meters (10,000 feet) above the island's bedrock floor.

In the Southern Hemisphere, the huge Antarctic Ice Sheet attains a maximum thickness of nearly 4.3 kilometers (2.7 miles) and covers nearly the entire continent, an area of more than 13.9 million square kilometers (5.4 million square miles). The Antarctic Ice Sheet represents 80 percent of the world's ice and nearly two-thirds of Earth's freshwater. If this ice melted, sea level would rise an estimated 60–70 meters (200–230 feet) and the ocean would inundate many densely populated coastal areas (**Figure 6.3**).

Students Sometimes Ask...

It's hard to imagine how large the Greenland ice sheet really is. Could you provide a comparison that we might be able to relate to?

Try this. The ice sheet is long enough to extend from Key West, Florida, to 100 miles beyond Portland, Maine. Its width could reach from Washington, D.C., to Indianapolis, Indiana. Put another way, the ice sheet is 80 percent as big as the entire United States east of the Mississippi River. The area of Antarctica's ice sheet is more than eight times as great!

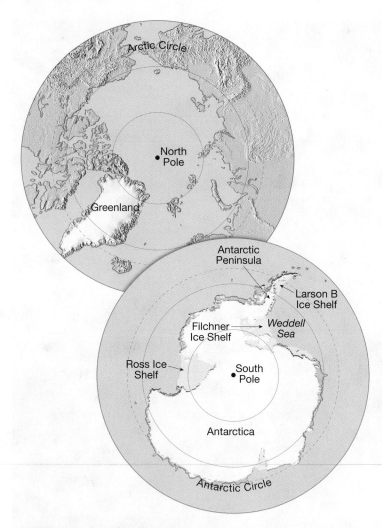

FIGURE 6.2 The only present-day continental ice sheets are those covering Greenland and Antarctica. Their combined areas represent almost 10 percent of Earth's land area. Greenland's ice sheet occupies 1.7 million square kilometers, or about 80 percent of the island. The area of the Antarctic Ice Sheet is almost 14 million square kilometers. Along portions of the Antarctic coast, glacial ice flows into bays, creating features called *ice shelves.* These large masses of floating ice remain attached to the land on one or more sides. Ice shelves occupy an additional 1.4 million square kilometers adjacent to the Antarctic Ice Sheet.

These enormous masses flow out in all directions from one or more snow-accumulation centers and completely obscure all but the highest areas of underlying terrain. Even sharp variations in the topography beneath the glacier usually appear as relatively subdued undulations on the surface of the ice. Such topographic differences, however, do affect the behavior of the ice sheets, especially near their margins, by guiding flow in certain directions and creating zones of faster and slower movement.

Along portions of the Antarctic coast, glacial ice flows into the adjacent ocean, creating features called **ice shelves**. They are large, relatively flat masses of floating ice that extend seaward from the coast but remain attached to the land along one or more sides. The shelves are thickest on their landward sides, and they become thinner seaward. They are sustained by ice from the adjacent ice sheet as well as being nourished by snowfall and the freezing of seawater to their bases. Antarctica's ice shelves extend over

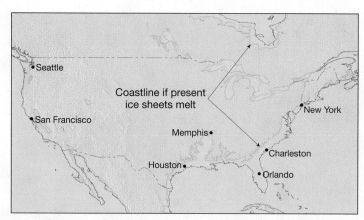

FIGURE 6.3 This map of a portion of North America shows the present-day coastline compared to the coastline that would exist if present ice sheets on Greenland and Antarctica melted.

approximately 1.4 million square kilometers (0.6 million square miles). The Ross and Filchner ice shelves are the largest, with the Ross Ice Shelf alone covering an area approximately the size of Texas (see Figure 6.2).

Other Types of Glaciers

In addition to valley glaciers and ice sheets, other types of glaciers are also identified. Covering some uplands and plateaus are masses of glacial ice called **ice caps**. Like ice sheets, ice caps completely bury the underlying landscape but are much smaller than the continental-scale features. Ice caps occur in many places, including Iceland and several of the large islands in the Arctic Ocean (**Figure 6.4A**). Another type, known as **piedmont glaciers**, occupy broad lowlands at the bases of steep mountains and form when one or more valley glaciers emerge from the confining walls of mountain valleys. Here the advancing ice spreads out to form a broad sheet. The size of individual piedmont glaciers varies greatly. Among the largest is the broad Malaspina Glacier along the coast of southern Alaska. It covers more than 5,000 square kilometers (2,000 square miles) of the flat coastal plain at the foot of the lofty St. Elias range (Figure 6.4B).

CONCEPT CHECK 6.1

❶ Where are glaciers found today and what percentage of Earth's land surface do they cover?

❷ Describe how glaciers fit into the hydrologic cycle. What role do they play in the rock cycle?

❸ List and briefly distinguish among four types of glaciers.

Students Sometimes Ask...

Is the ice at the North Pole part of a glacier?

The Arctic Ocean surrounds the North Pole, so the ice in this region is sea ice (frozen seawater), *not* glacial ice. Remember, glaciers form on land from the accumulation of snow.

is complex and is of two basic types. The first of these, *plastic flow*, involves movement *within* the ice. Ice behaves as a brittle solid until the pressure upon it is equivalent to the weight of about 50 meters (165 feet) of ice. Once that load is surpassed, ice behaves as a plastic material and flow begins. A second and often equally important mechanism of glacial movement consists of the entire ice mass slipping along the ground. The lowest portions of most glaciers are thought to move by this sliding process.

The upper 50 meters or so of a glacier is not under sufficient pressure to exhibit plastic flow. Rather, the ice in this uppermost zone is brittle and is appropriately referred to as the *zone of fracture*. The ice in this zone is carried along "piggyback" style by the ice below. When the glacier moves over irregular terrain, the zone of fracture is subjected to tension, resulting in cracks called **crevasses** (Figure 6.5). These gaping cracks can make travel across glaciers dangerous and can extend to depths of 50 meters (165 feet). Beyond this depth, plastic flow seals them off.

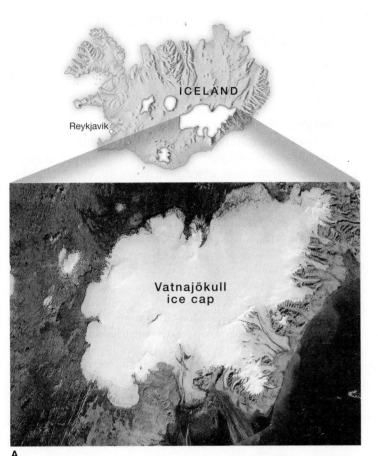

A.

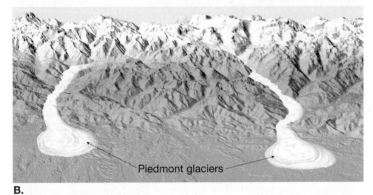

Piedmont glaciers

B.

FIGURE 6.4 A. Iceland's Vatnajökull ice cap. In 1996 the Grimsvötn volcano erupted beneath the ice cap, an event that triggered melting and floods. (NASA) **B.** Piedmont glaciers occur where valley glaciers exit a mountain range onto broad lowlands, are no longer laterally confined, and spread to become wide lobes.

FIGURE 6.5 Crevasses form in the brittle ice of the zone of fracture. They can extend to depths of 50 meters and can obviously make travel across glaciers dangerous. (Photo by Bo Tornvig/agefotostock)

How Glaciers Move

Sculpting Earth's Surface
▶ **Glaciers**

The movement of glacial ice is generally referred to as *flow*. The fact that glacial movement is described in this way seems paradoxical—how can a solid flow? The way in which ice flows

Observing and Measuring Movement

Unlike streamflow, glacial movement is not obvious. If we could watch a valley glacier move, we would see that like the water in a river, all of the ice does not move downstream at the same rate. Flow is greatest in the center of the glacier because of the drag created by the walls and floor of the valley.

Early in the 19th century, the first experiments involving the movement of glaciers were designed and carried out in the Alps. Markers were placed in a straight line across an alpine glacier. The position of the line was marked on the valley walls so that if the ice moved, the change in position could be detected. Periodically the positions of the markers were noted, revealing the movement just described. Although most glaciers move too slowly for direct visual detection, the experiments succeeded in demonstrating that movement nevertheless occurs. The experiment illustrated in Figure 6.6 was carried out at Switzerland's Rhone Glacier later in the 19th century. It not only traced the movement of markers within the ice but also mapped the position of the glacier's terminus.

For many years, time lapse photography has allowed us to observe glacial movement. Images are taken from the same vantage point on a regular basis (e.g., once per day) over an extended span and then played back like a movie. More recently, satellites let us track the movement of glaciers and observe glacial behavior (see Figure 1.A, p. 6). This is especially useful because the remoteness and extreme weather associated with many glacial areas limits on-site study.

How rapidly does glacial ice move? Average rates vary considerably from one glacier to another. Some move so slowly that trees and other vegetation may become well established in the debris that accumulates on the glacier's surface. Others advance up to several meters per day. The movement of some glaciers is characterized by periods of extremely rapid advance followed by periods during which movement is practically nonexistent.

FIGURE 6.6 Ice movement and changes in the terminus at Rhone Glacier, Switzerland. In this classic study of a valley glacier, the movement of stakes clearly showed that ice along the sides of the glacier moves slowest. Also notice that even though the ice front was retreating, the ice within the glacier was advancing.

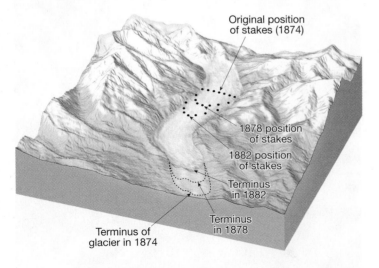

Original position of stakes (1874)

1878 position of stakes

1882 position of stakes

Terminus in 1882

Terminus in 1878

Terminus of glacier in 1874

Budget of a Glacier: Accumulation Versus Wastage

Snow is the raw material from which glacial ice originates. Therefore, glaciers form in areas where more snow falls in winter than can melt during the summer. Glaciers are constantly gaining and losing ice.

Glacial Zones Snow accumulation and ice formation occur in the **zone of accumulation** (Figure 6.7). The addition of snow thickens the glacier and promotes movement. Beyond this area of ice formation is the **zone of wastage** where there is a net loss to the glacier when the snow from the previous winter melts, as does some of the glacial ice (Figure 6.7).

In addition to melting, glaciers also waste as large pieces of ice break off the front of a glacier in a process called *calving*. Where glaciers reach the sea, calving creates *icebergs* (Figure 6.8). Because icebergs are just slightly less dense than seawater, they float very low in the water, with more than 80 percent of their mass submerged. The margins of the Greenland Ice Sheet produce thousands of icebergs each year. Many drift southward and find their way into the North Atlantic, where they are a hazard to navigation.

Glacial Budget Whether the margin of a glacier is advancing, retreating, or remaining stationary depends on the *budget* of the glacier. The glacial budget is the balance or lack of balance between accumulation at the upper end of a glacier and loss at the lower end. If ice accumulation exceeds wastage, the glacial front advances until the two factors balance. At this point, the terminus of the glacier becomes stationary.

If a warming trend increases wastage and/or if a drop in snowfall decreases accumulation, the ice front will retreat. As the terminus of the glacier retreats, the extent of the zone of wastage diminishes. Therefore, in time a new balance will be reached between accumulation and wastage, and the ice front will again become stationary.

Whether the margin of a glacier is advancing, retreating, or stationary, the ice within the glacier continues to flow forward. In the case of a receding glacier, the ice still flows forward, but not rapidly enough to offset wastage. This point is illustrated in Figure 6.6. As the line of stakes within the Rhone Glacier continued to move downvalley, the terminus of the glacier slowly retreated upvalley.

Glaciers in Retreat Because glaciers are sensitive to changes in temperature and precipitation, they provide clues about changes in climate. With few exceptions, valley glaciers around the world have been retreating at unprecedented rates over the last century (Figure 6.9). Many valley glaciers have disappeared altogether. For example, 150 years ago, there were 147 glaciers in Montana's glacier National Park. Today, only 37 remain and these may vanish by 2030.

CONCEPT CHECK 6.2

❶ Describe the two components of glacial movement. How fast do glaciers move?

❷ Under what circumstances will the front of a glacier advance? Retreat? Remain stationary?

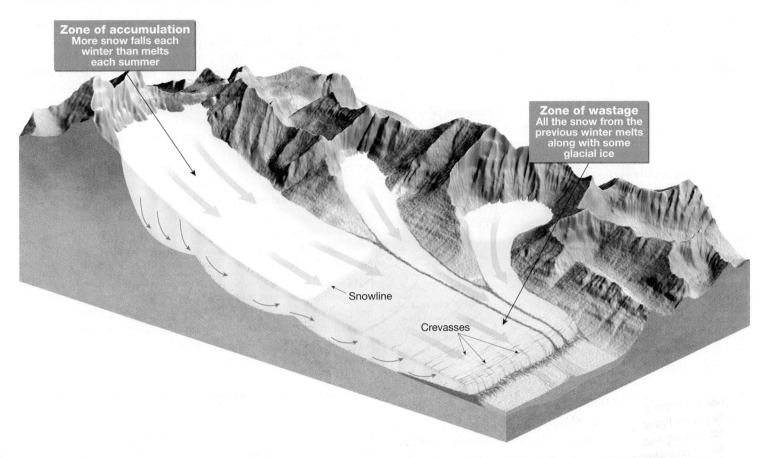

Zone of accumulation
More snow falls each winter than melts each summer

Zone of wastage
All the snow from the previous winter melts along with some glacial ice

Snowline

Crevasses

FIGURE 6.7 The snowline separates the zone of accumulation and the zone of wastage. Above the snowline, more snow falls each winter than melts each summer. Below the snowline, the snow from the previous winter completely melts, as does some of the underlying ice. Whether the margin of a glacier advances, retreats, or remains stationary depends on the balance or lack of balance between accumulation and wastage. When a glacier moves across irregular terrain, *crevasses* form in the brittle portion.

FIGURE 6.9 Two images taken 63 years apart from the same vantage point in Alaska's Glacier Bay National Park. Muir Glacier, which is prominent in the 1941 photo, has retreated out of the field of view in the 2004 image. Also Riggs Glacier (upper right) has thinned and retreated significantly. (Photos courtesy of National Snow and Ice Data Center)

FIGURE 6.8 Icebergs are created when large pieces break off the front of a glacier after it reaches a water body, a process known as *calving.* The photo shows an iceberg off the coast of Newfoundland. As the drawing illustrates, only about 20 percent (or less) of an iceberg protrudes above the waterline. (Photo by Radius Images/ Photolibrary)

Geologist's Sketch

1941

2004

Glacial Erosion

Sculpturing Earth's Surface
▶ Glaciers

Glaciers erode tremendous volumes of rock. For anyone who has observed the terminus of an alpine glacier, the evidence of its erosive force is clear. You can witness firsthand the release of rock fragments of various sizes from the ice as it melts (**Figure 6.10**). All signs lead to the conclusion that the ice has scraped, scoured, and torn rock debris from the floor and walls of the valley and carried it downvalley. It should also be pointed out that in mountainous regions mass-wasting processes also make substantial contributions to the sediment load of a glacier. The chapter-opening photo (p. 156) provides on excellent example.

Once rock debris is acquired by the glacier, it cannot settle out as does the load carried by a stream or by the wind. Consequently, glaciers can carry huge blocks that no other erosional agent could possibly budge. Although today's glaciers are of limited importance as erosional agents, many landscapes that were modified by the widespread glaciers of the recent Ice Age still reflect to a high degree the work of ice.

How Glaciers Erode

Glaciers erode land primarily in two ways: plucking and abrasion. First, as a glacier flows over a fractured bedrock surface, it loosens and lifts blocks of rock and incorporates them into the ice. This process, known as **plucking**, occurs when meltwater penetrates the cracks and joints along the rock floor of the glacier and freezes. When water freezes it expands, exerting tremendous leverage that pries the rock loose. In this manner, sediment of all sizes becomes part of the glacier's load.

The second major erosional process is **abrasion**. As the ice and its load of rock fragments slide over bedrock, they function like sandpaper to smooth and polish the surface below. The pulverized rock produced by the glacial gristmill is appropriately called **rock flour**. So much rock flour may be produced that meltwater streams leaving a glacier often have the grayish appearance of skim milk—visible evidence of the grinding power of the ice.

When the ice at the bottom of a glacier contains large rock fragments, long scratches and grooves called **glacial striations** may be gouged into the bedrock (**Figure 6.11A**). These linear scratches on the bedrock surface provide clues to the direction of glacial movement. By mapping the striations over large areas, glacial flow patterns can often be reconstructed.

Not all abrasive action produces striations. The rock surface over which the glacier moves may also become highly polished by the ice and its load of finer particles. The broad expanses of smoothly polished granite in California's Yosemite National Park provide an excellent example (Figure 6.11B).

As is the case with other agents of erosion, the rate of glacial erosion is highly variable. This differential erosion by ice is largely controlled by four factors: (1) rate of glacial movement; (2) thickness of the ice; (3) shape, abundance, and hardness of the rock fragments contained in the ice at the base of the glacier; and (4) the erodibility of the surface beneath the glacier. Variations in any or all of these factors from time to time and/or from place to place mean that the features, effects, and degree of landscape modification in glaciated regions can vary greatly.

Landforms Created by Glacial Erosion

Although the erosional accomplishments of ice sheets can be tremendous, landforms carved by these huge ice masses usually do not inspire the same awe as do the erosional features created by valley glaciers. In regions where the erosional effects of ice sheets are significant, glacially scoured surfaces and subdued terrain are the rule. By contrast, in mountainous areas, erosion by valley glaciers produces many truly spectacular features. Much of the rugged mountain scenery so celebrated for its majestic beauty is the product of erosion by valley glaciers.

Take a moment to study **Figure 6.12**, which shows a mountain setting before, during, and after glaciation. You will refer to this often in the following discussion.

Glaciated Valleys   Prior to glaciation, alpine valleys are characteristically V-shaped because streams are well above base level and are therefore downcutting (Figure 6.12A). However, in mountainous regions that have been glaciated, the valleys are no longer narrow. As a glacier moves down a valley once occupied by a stream, the ice modifies it in three ways: The glacier widens, deepens, and straightens the valley, so that what was once a narrow V-shaped valley is transformed into a U-shaped **glacial trough** (Figures 6.12C).

FIGURE 6.10 Glaciers are capable of great erosion. As the terminus of this Alaskan glacier wastes away, it deposits large quantities of unsorted sediment called *till*. (Photos by Michael Collier)

FIGURE 6.11 **A.** Cove Glacier, Prince William Sound, northeast of Whittier, Alaska. Glacial abrasion created the scratches and grooves in this bedrock. **B.** Glacially polished granite in California's Yosemite National Park. (Photos by Michael Collier)

The amount of glacial erosion that takes place in different valleys in a mountainous area varies. Prior to glaciation, the mouths of tributary streams join the main (*trunk*) valley at the elevation of the stream in that valley. During glaciation, the amount of ice flowing through the main valley can be much greater than the amount advancing down each tributary. Consequently, the valley containing the trunk glacier is eroded deeper than the smaller valleys that feed it. Thus, after the ice has receded, the valleys of tributary glaciers are left standing above the main glacial trough and are termed **hanging valleys** (Figure 6.12C). Rivers flowing through hanging valleys can produce spectacular waterfalls, such as those in Yosemite National Park, California.

Cirques   At the head of a glacial valley is a characteristic and often imposing feature associated with an alpine glacier—a **cirque**. As Figure 6.12 illustrates, these bowl-shaped depressions have precipitous walls on three sides but are open on the downvalley side. The cirque is the focal point of the glacier's growth because it is the area of snow accumulation and ice formation. Cirques begin as irregularities in the mountainside that are subsequently enlarged by frost wedging and plucking

along the sides and bottom of the glacier. The glacier in turn acts as a conveyor belt that carries away the debris. After the glacier has melted away, the cirque basin is sometimes occupied by a small lake called a *tarn* (Figure 6.12C).

Arêtes and Horns   The Alps, Northern Rockies, and many other mountain landscapes sculpted by valley glaciers reveal more than glacial troughs and cirques. In addition, sinuous, sharp-edged ridges called **arêtes** and sharp, pyramid-like peaks termed **horns** project above the surroundings (Figure 6.12C). Both features can originate from the same basic process: the enlargement of cirques produced by plucking and frost action. Several cirques around a single high mountain create the spires of rock called *horns*. As the cirques enlarge and converge, an isolated horn is produced. A famous example is the Matterhorn in the Swiss Alps (Figure 6.13).

Arêtes can form in a similar manner except that the cirques are not clustered around a point but rather exist on opposite sides of a divide. As the cirques grow, the divide separating them is reduced to a very narrow, knifelike partition. An arête can also be created in another way. When two glaciers occupy parallel

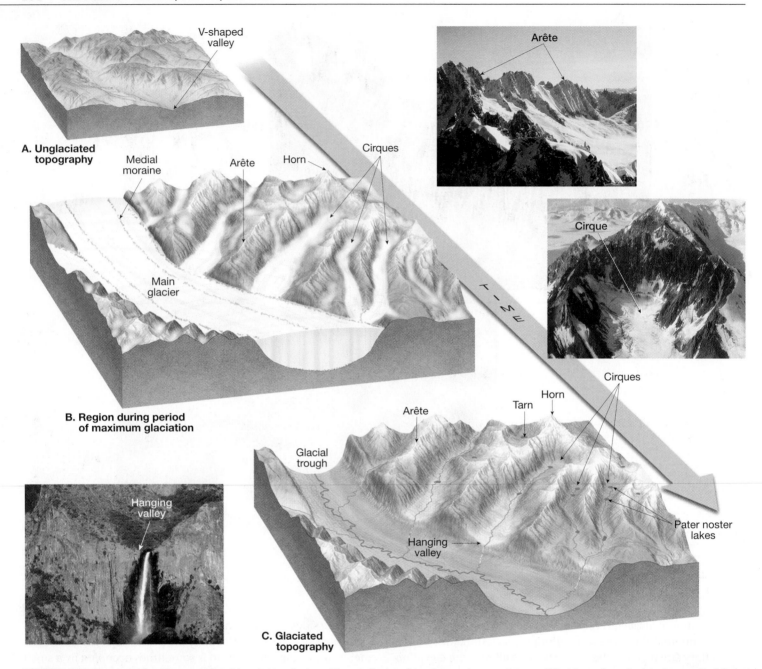

FIGURE 6.12 Erosional landforms created by alpine glaciers. The unglaciated landscape in part **A** is modified by valley glaciers in part **B**. After the ice recedes, in part **C**, the terrain looks very different than before glaciation. (Arête photo from James E. Patterson Collection, cirque photo by Marli Miller, Hanging Valley photo by John Warden/Superstock)

valleys, an arête can form when the land separating the moving tongues of ice is progressively narrowed as the glaciers scour and widen their valleys.

Fiords Fiords are deep, often spectacular, steep-sided inlets of the sea that exist in many high-latitude areas of the world where mountains are adjacent to the ocean (**Figure 6.14**). Norway, British Columbia, Greenland, New Zealand, Chile, and Alaska all have coastlines characterized by fiords. They are glacial troughs that became submerged as the ice left the valley and sea level rose following the Ice Age.

The depths of some fiords can exceed 1,000 meters (3,300 feet). However, the great depths of these flooded troughs are only partly explained by the post–Ice Age rise in sea level. Unlike the situation

governing the downward erosional work of rivers, sea level does not act as a base level for glaciers. As a consequence, glaciers are capable of eroding their beds far below the surface of the sea. For example, a valley glacier 300 meters (1,000 feet) thick can carve its valley floor more than 250 meters (800 feet) below sea level before downward erosion ceases and the ice begins to float.

CONCEPT CHECK 6.3

1 How do glaciers acquire their load of sediment?

2 How does a glaciated mountain valley differ in appearance from a mountain valley that was not glaciated?

3 Describe the features created by glacial erosion that you might see in an area where valley glaciers recently existed.

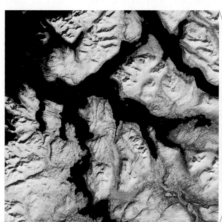

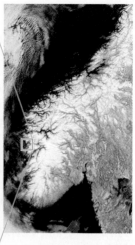

FIGURE 6.13 Horns are sharp, pyramid-like peaks that are fashioned by alpine glaciers. This example is the famous Matterhorn in the Swiss Alps. (Photo by Andy Selinger/agefotostock)

Glacial Deposits

Sculpturing Earth's Surface
▶ Glaciers

Glaciers pick up and transport a huge load of debris as they slowly advance across the land. Ultimately these materials are deposited when the ice melts. In regions where glacial sediment is deposited, it can play a truly significant role in forming the physical landscape. For example, in many areas once covered by the ice sheets of the recent Ice Age, the bedrock is rarely exposed, because glacial deposits that are dozens or even hundreds of meters thick completely mantle the terrain. The general effect of these deposits is to reduce the local relief and thus level the topography. Indeed, rural country scenes that are familiar to many of us—rocky pastures in New England, wheat fields in the Dakotas, rolling farmland in the Midwest—result directly from glacial deposition.

FIGURE 6.14 The coast of Norway is known for its many fiords. Frequently these ice-sculpted inlets of the sea are hundreds of meters deep. (Satellite images courtesy of NASA; photo by Wolfgang Meier Photography)

Types of Glacial Drift

Long before the theory of an extensive Ice Age was proposed, much of the soil and rock debris covering portions of Europe was recognized as coming from elsewhere. At the time, these foreign materials were believed to have been "drifted" into their present positions by floating ice during an ancient flood. As a consequence, the term *drift* was applied to this sediment. Although rooted in a concept that was not correct, this term was so well established by the time the true glacial origin of the debris became widely recognized that it remained in the glacial vocabulary. Today, **glacial drift** is an all-embracing term for sediments of glacial origin, no matter how, where, or in what form they were deposited.

Glacial drift is divided into two distinct types: (1) materials deposited directly by the glacier, which are known as *till*, and (2) sediments laid down by glacial meltwater, called *stratified drift*. **Till** is deposited as glacial ice melts and drops its load of rock debris. Unlike moving water and wind, ice cannot sort the sediment it carries; therefore, deposits of till are characteristically unsorted mixtures of many particle sizes (**Figure 6.15**). **Stratified drift** is sorted according to the size and weight of the fragments. As ice is not capable of such sorting activity, these sediments are not deposited directly by the glacier. Rather, they reflect the sorting action of glacial meltwater.

Some deposits of stratified drift are made by streams issuing directly from the glacier. Other stratified deposits involve sediment that was originally laid down as till and later picked up, transported, and redeposited by meltwater beyond the margin of the ice. Accumulations of stratified drift often consist largely of sand and gravel because the meltwater is not capable of moving larger material and because the finer rock flour remains suspended and is commonly carried far from the glacier. An indication that stratified drift consists primarily of sand and gravel can be seen in many areas where these deposits are actively mined as aggregate for road work and other construction projects.

Boulders found in the till or lying free on the surface are called **glacial erratics** if they are different from the bedrock below. Of course, this means that they must have been derived from a source outside the area where they are found. Although the locality of origin for most erratics is unknown, the origin of some can be determined. Therefore, by studying glacial erratics as well as the mineral composition of the till, geologists can sometimes trace the path of a lobe of ice. In portions of New England as well as other areas, erratics can be seen dotting pastures and farm fields. In some places, these rocks were cleared from fields and piled to make fences and walls.

Moraines, Outwash Plains, and Kettles

Perhaps the most widespread features created by glacial deposition are *moraines*, which are simply layers or ridges of till. Several types of moraines are identified; some are common only to mountain valleys, and others are associated with areas affected by either ice sheets or valley glaciers. Lateral and medial moraines fall in the first category, whereas end moraines and ground moraines are in the second.

Lateral and Medial Moraines   The sides of a valley glacier accumulate large quantities of debris from the valley walls. When the glacier wastes away, these materials are left as ridges, called **lateral moraines**, along the sides of the valley. **Medial moraines** are formed when two valley glaciers coalesce to form a single ice stream. The till that was once carried along the edges of each glacier joins to form a single dark stripe of debris within the newly enlarged glacier. Creation of these dark stripes within the ice stream is obvious proof that glacial ice moves, because the medial moraine could not form if the ice did not flow downvalley. It is common to see several medial moraines within a large alpine glacier, because a streak will form whenever a tributary glacier joins the main valley. Kennicott Glacier in Figure 6.1 (p. 159) provides an excellent example.

Close up of cobble

FIGURE 6.15 Glacial till is an unsorted mixture of many different sediment sizes. A close examination often reveals cobbles that have been scratched as they were dragged along by the glacier. (Photos by E. J. Tarbuck)

End Moraines and Ground Moraines   An **end moraine** is a ridge of till that forms at the terminus of a glacier and is characteristic of ice sheets and valley glaciers alike. These relatively common landforms are deposited when a state of equilibrium is attained between wastage and ice accumulation. That is, the end moraine forms when the ice is melting near the end of the glacier at a rate equal to the forward advance of the glacier from its region of nourishment. Although the terminus of the glacier is stationary, the ice continues to flow forward, delivering a continuous supply of sediment in the same manner a conveyor belt delivers goods to the end of a production line. As the ice melts, the till is dropped and the end moraine grows. The longer the ice front remains stable, the larger the ridge of till will become.

Eventually the time comes when wastage exceeds nourishment. At this point, the front of the glacier begins to recede in the direction from which it originally advanced. However, as the ice front retreats, the conveyor-belt action of the glacier continues to provide fresh supplies of sediment to the terminus. In this manner a large quantity of till is deposited as the ice melts away, creating a rock-strewn, undulating plain. This gently rolling layer of till deposited as the ice front recedes is termed **ground moraine**. Ground moraine has a leveling effect, filling in low spots and clogging old stream channels, often leading to a derangement of the existing drainage system. In areas where this layer of till is still relatively fresh, such as the northern Great Lakes region, poorly drained swampy lands are quite common.

Periodically, a glacier will retreat to a point where wastage and nourishment once again balance. When this happens, the ice front stabilizes and a new end moraine forms.

The pattern of end moraine formation and ground moraine deposition may be repeated many times before the glacier has completely vanished. Such a pattern is illustrated in Figure 6.16. The very first end moraine to form marks the farthest advance of the glacier and is called the *terminal end moraine*. Those end moraines that form as the ice front occasionally stabilizes during retreat are termed *recessional end moraines*. Terminal and recessional moraines are essentially alike; the only difference between them is their relative positions.

End moraines deposited by the most recent major stage of Ice Age glaciation are prominent features in many parts of the Midwest and Northeast. In Wisconsin, the wooded, hilly terrain of the Kettle Moraine near Milwaukee is a particularly picturesque example. A well-known example in the Northeast is Long Island. This linear strip of glacial sediment that extends northeastward from New York City is part of an end moraine complex that stretches from eastern Pennsylvania to Cape Cod, Massachusetts (Figure 6.17).

Figure 6.18 represents a hypothetical area during and following glaciation. It shows the end moraines that were just described as well as the depositional features that are discussed in the sections that follow. This figure depicts landscape features similar to what might be encountered if you were traveling in the upper Midwest or New England. As you read about other glacial deposits, you will be referred to this figure again.

Outwash Plains and Valley Trains   At the same time that an end moraine is forming, meltwater emerges from the ice in rapidly moving streams. Often they are choked with suspended material and carry a substantial bed load. As the water leaves the glacier, it rapidly loses velocity and much of its bed load is dropped. In this way a broad, ramplike accumulation of stratified drift is built adjacent to the downstream edge of most end moraines. When the feature is formed in association with an ice sheet, it is termed an **outwash plain**, and when it is confined to a mountain valley, it is usually referred to as a **valley train** (Figure 6.18).

FIGURE 6.16 End moraines of the Great Lakes region. Those deposited during the most recent (Wisconsinan) stage are most prominent.

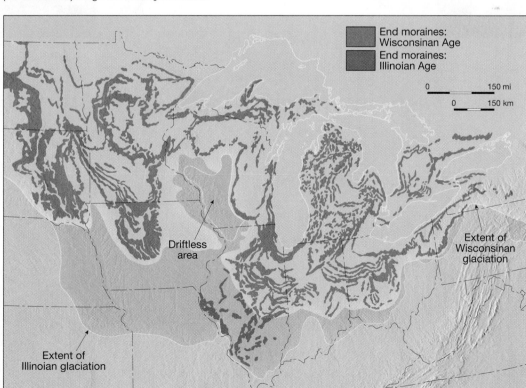

End moraines: Wisconsinan Age
End moraines: Illinoian Age

0   150 mi
0   150 km

Driftless area

Extent of Wisconsinan glaciation

Extent of Illinoian glaciation

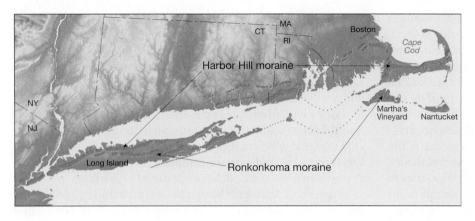

FIGURE 6.17 End moraines make up substantial parts of Long Island, Cape Cod, Martha's Vineyard, and Nantucket. Although portions are submerged, the Ronkonkoma moraine (a terminal end moraine) extends through central Long Island, Martha's Vineyard, and Nantucket. It was deposited about 20,000 years ago. The recessional Harbor Hill moraine, which formed about 14,000 years ago, extends along the north shore of Long Island, through southern Rhode Island, and through Cape Cod.

Kettles Often end moraines, outwash plains, and valley trains are pockmarked with basins or depressions known as **kettles** (Figure 6.18). Kettles form when blocks of stagnant ice become buried in drift and eventually melt, leaving pits in the glacial sediment. Most kettles do not exceed 2 kilometers in diameter, and the typical depth of most kettles is less than 10 meters (33 feet). Water often fills the depression and forms a pond or lake. One well-known example is Walden Pond near Concord, Massachusetts. It is here that Henry David Thoreau lived alone for 2 years in the 1840s and about which he wrote *Walden*, his classic of American literature.

Drumlins, Eskers, and Kames

Moraines are not the only landforms deposited by glaciers. Some landscapes are characterized by numerous elongate parallel hills made of till. Other areas exhibit conical hills and relatively narrow winding ridges composed mainly of stratified drift.

FIGURE 6.18 This hypothetical area illustrates many common depositional landforms. The outermost end moraine marks the limit of glacial advance and is called the *terminal end moraine.* End moraines that form as the ice front occasionally becomes stationary during retreat are called *recessional end moraines.* (Drumlin photo courtesy of Ward's Natural Science Establishment; kame, esker, and kettle photos by Richard P. Jacobs/JLM Visuals)

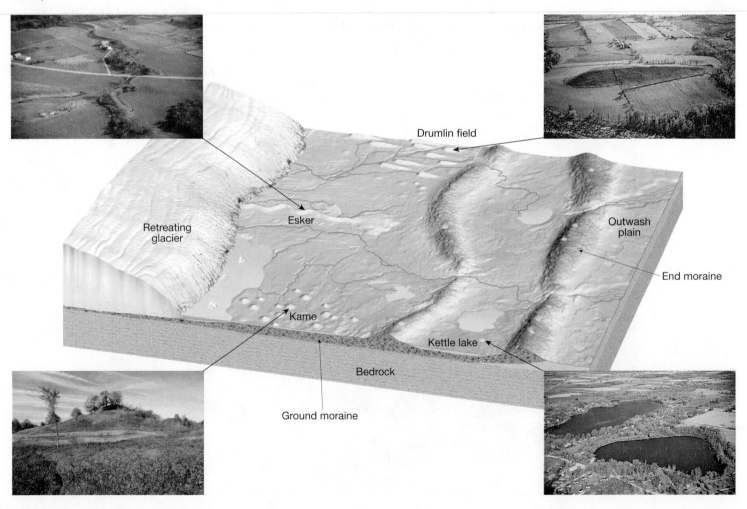

Drumlins **Drumlins** are streamlined asymmetrical hills composed of till (Figure 6.18). They range in height from 15 to 60 meters (50–200 feet) and average 0.4–0.8 kilometer (0.25–0.50 mile) in length. The steep side of the hill faces the direction from which the ice advanced, whereas the gentler slope points in the direction the ice moved. Drumlins are not found singly but rather occur in clusters, called *drumlin fields*. One such cluster, east of Rochester, New York, is estimated to contain about 10,000 drumlins. Their streamlined shape indicates that they were molded in the zone of flow within an active glacier. It is thought that drumlins originate when glaciers advance over previously deposited drift and reshape the material.

Eskers and Kames   In some areas that were once occupied by glaciers, sinuous ridges composed largely of sand and gravel might be found. These ridges, called **eskers**, are deposits made by streams flowing in tunnels beneath the ice, near the terminus of a glacier (Figure 6.18). They may be several meters high and extend for many kilometers. In some areas they are mined for sand and gravel, and for this reason, eskers are disappearing in some localities.

Kames are steep-sided hills that, like eskers, are composed of sand and gravel (Figure 6.18). Kames originate when glacial meltwater washes sediment into openings and depressions in the stagnant wasting terminus of a glacier. When the ice eventually melts away, the stratified drift is left behind as mounds or hills.

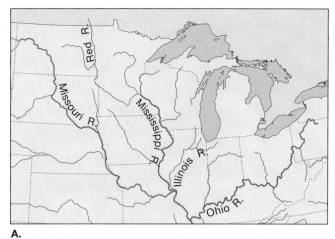

A.

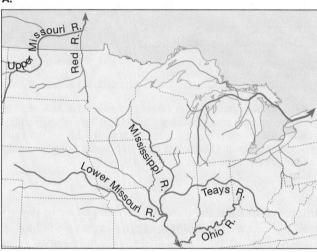

B.

FIGURE 6.19 **A.** This map shows the Great Lakes and the familiar present-day pattern of rivers in the central United States. Pleistocene ice sheets played a major role in creating this pattern. **B.** Reconstruction of drainage systems in the central United States prior to the Ice Age. The pattern was very different from today, and there were no Great Lakes.

CONCEPT CHECK 6.4

1. What is the difference between till and stratified drift?
2. Distinguish between a terminal end moraine and a recessional end moraine.
3. Describe the formation of a medial moraine.
4. List four depositional features other than moraines.

Other Effects of Ice Age Glaciers

In addition to the massive erosional and depositional work carried on by Ice Age glaciers, the ice sheets had other, sometimes profound, effects on the landscape. For example, as the ice advanced and retreated, animals and plants were forced to migrate. This led to stresses that some organisms could not tolerate. Furthermore, many present-day stream courses bear little resemblance to their preglacial routes. The Missouri River once flowed northward toward Hudson Bay in Canada. The Mississippi River followed a path through central Illinois, and the head of the Ohio River reached only as far as Indiana (Figure 6.19). A comparison of the two parts of Figure 6.19 shows that the Great Lakes were created by glacial erosion during the Ice Age. Prior to the Pleistocene, the basins occupied by these huge lakes were lowlands with rivers that ran eastward to the Gulf of St. Lawrence.

In areas that were centers of ice accumulation, such as Scandinavia and northern Canada, the land has been slowly rising for the past several thousand years. The land had downwarped under the tremendous weight of 3-kilometer-thick (almost 2-mile-thick)

masses of ice. Following the removal of this immense load, the crust has been adjusting by gradually rebounding upward ever since.[8]

Ice sheets and alpine glaciers can act as dams to create lakes by trapping glacial meltwater and blocking the flow of rivers. Some of these lakes are relatively small and short-lived. Others can be large and exist for hundreds or thousands of years.

Figure 6.20 is a map of Lake Agassiz—the largest lake to form during the Ice Age in North America. With the retreat of the ice sheet came enormous volumes of meltwater. The Great Plains generally slope upward to the west. As the terminus of the ice sheet receded northeastward, meltwater was trapped between the ice on one side and the sloping land on the other, causing Lake Agassiz to deepen and spread across the landscape. It came into existence about 12,000 years ago and lasted for about 4,500 years. Such water bodies are termed *proglacial lakes*, referring to their

---

[8]For a more complete discussion of this concept, termed *isostatic adjustment*, see the section "Principle of Isostasy" in Chapter 10, p. 317.

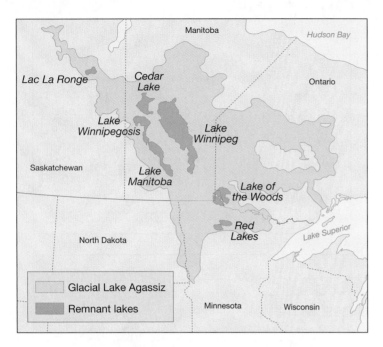

FIGURE 6.20 Map showing the extent of glacial Lake Agassiz. This proglacial lake came into existence about 12,000 years ago and lasted for about 4,500 years. It was an immense feature—bigger than all of the present-day Great Lakes combined. The modern-day remnants of the water body are still major landscape features.

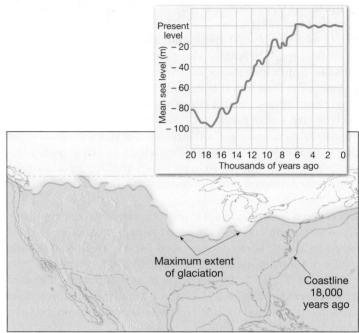

FIGURE 6.21 Changing sea level during the past 20,000 years. About 18,000 years ago, when the most recent ice advance was at a maximum, sea level was nearly 100 meters (330 feet) lower than at present. Thus, land that is presently covered by the ocean was exposed and the shoreline looked very different than today. As the ice sheets melted, sea level rose and the shoreline shifted

position just beyond the outer limits of a glacier or ice sheet. Research shows that the shifting of glaciers and the failure of ice dams can cause the rapid release of huge volumes of water. Such events occurred during the history of Lake Agassiz. One of the most dramatic examples of such glacial outbursts occurred in the Pacific Northwest and is described in Box 6.1.

A far-reaching effect of the Ice Age was the worldwide change in sea level that accompanied each advance and retreat of the ice sheets. The snow that nourishes glaciers ultimately comes from moisture evaporated from the oceans. Therefore, when the ice sheets increased in size, sea level fell and the shoreline moved seaward (**Figure 6.21**). Estimates suggest that sea level was as much as 100 meters (330 feet) lower than it is today. Consequently, the Atlantic coast of the United States was located more than 100 kilometers (60 miles) to the east of New York City. Moreover, France and Britain were joined where the English Channel is today. Alaska and Siberia were connected across the Bering Strait, and Southeast Asia was tied by dry land to the islands of Indonesia.

The formation and growth of ice sheets was an obvious response to significant changes in climate. But the existence of the glaciers themselves triggered climatic changes in the regions beyond their margins. In arid and semiarid areas on all continents, temperatures were lower, which meant evaporation rates were also lower. At the same time, precipitation was moderate. This cooler, wetter climate resulted in the formation of many lakes called **pluvial lakes** (from the Latin term *pluvia* meaning "rain"). In North America, pluvial lakes were concentrated in the vast Basin and Range region of Nevada and Utah (**Figure 6.22**). Although most are now gone, a few remnants remain, the largest being Utah's Great Salt Lake.

Students Sometimes Ask...

How large are the Great Lakes?

The Great Lakes constitute the largest body of fresh water on Earth. Formed between about 10,000 and 12,000 years ago, the lakes currently contain about 20 percent of Earth's surface freshwater.

CONCEPT CHECK 6.5

❶ Describe at least four effects of Ice Age glaciers aside from the formation of major erosional and depositional features.

Glaciers of the Ice Age

In the preceding pages, we mentioned the Ice Age, a time when ice sheets and alpine glaciers were far more extensive than they are today. There was a time when the most popular explanation for what we now know to be glacial deposits was that the material had been drifted in by means of icebergs or perhaps simply swept across the landscape by a catastrophic flood. However, during the 19th century, field investigations by many scientists provided convincing evidence that an extensive Ice Age was responsible for these deposits and for many other features.

By the beginning of the 20th century, geologists had largely determined the extent of Ice Age glaciation. Furthermore, they discovered that many glaciated regions had not one layer of drift

Box 6.1

EARTH AS A SYSTEM

Glacial Lake Missoula, Megafloods, and the Channeled Scablands

Lake Missoula was a prehistoric proglacial lake in western Montana that existed as the Pleistocene ice age was drawing to a close between about 15,000 and 13,000 years ago. It was a time when the climate was gradually warming and the ice sheet that covered much of western Canada and portions of the Pacific Northwest was melting and retreating.

The lake was the result of an ice dam that formed when a southward protruding mass of glacial ice called the Purcell Lobe blocked the Clark Fork River (**Figure 6.A**). As the water rose behind the 600-meter-(2,000-foot-) high dam, it flooded the valleys of western Montana. At its greatest extent, Lake Missoula extended eastward for more than 300 kilometers (200 miles). Its volume exceeded 2,500 cubic kilometers (600 cubic miles)—greater than present-day Lake Ontario.

Eventually, the lake became so deep that the ice dam began to float. That is, the rising waters behind the dam lifted the buoyant ice so that it no longer functioned as a dam. The result was a catastrophic flood as Lake Missoula's waters suddenly poured out under the failed dam. The outburst rushed across the lava plains of eastern Washington and down the Columbia River to the Pacific Ocean. Based on the sizes of boulders

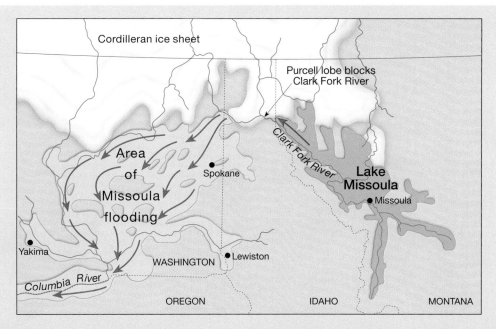

FIGURE 6.A Lake Missoula was a proglacial lake created when the Purcell Lobe of the Cordilleran ice sheet formed an ice dam on the Clark Fork River. Periodically, the ice dam would fail, sending a huge torrent of water flooding across the landscape of eastern Washington.

moved during the event, the velocity of the torrent approached or exceeded 70 kilometers (nearly 45 miles) per hour. The entire lake was emptied in a matter of a few days. Due to the temporary impounding of water behind narrow gaps along the flood's path, flooding probably continued through the devastated region for a few weeks.

The erosional and depositional results of such a megaflood were dramatic. The towering mass of rushing water stripped away thick layers of sediment and soil and cut deep canyons (*coulees*) into the underlying basalt. Today the region is called the *Channeled Scablands*—a landscape consisting of a bizarre assemblage of landforms. Perhaps the most striking features are the table-like lava mesas called *scabs* left between braided, interlocking channels (**Figure 6.B**).

It was not a single flood from glacial Lake Missoula that created this extraordinary landscape. What makes Lake Missoula remarkable is that it alternately filled and emptied in a cycle that was repeated more than 40 times over a span of 1,500 years. After each flood, the glacial lobe would again block the valley and create a new ice dam, and the cycle of lake growth, dam failure, and megaflood would follow at intervals of 20 to 60 years.

FIGURE 6.B The landscape of the Channeled Scablands was sculpted by the megafloods associated with glacial Lake Missoula. The table-like mesas carved from basalt are called *scabs*. (Photo by John S. Shelton)

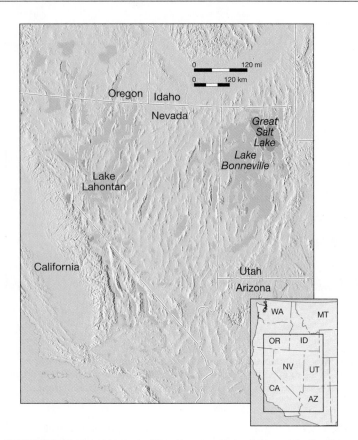

FIGURE 6.23 Maximum extent of ice sheets in the Northern Hemisphere during the Ice Age.

FIGURE 6.22 Pluvial lakes of the western United States. By far the largest pluvial lake in the vast Basin and Range region of Nevada and Utah was Lake Bonneville. With maximum depths exceeding 300 meters and an area of 50,000 square kilometers, Lake Bonneville was nearly the same size as present-day Lake Michigan. The Great Salt Lake is a remnant of this huge pluvial lake. (After R. F. Flint, *Glacial and Quaternary Geology*, New York: John Wiley & Sons.)

but several. Close examination of these older deposits showed well-developed zones of chemical weathering and soil formation as well as the remains of plants that require warm temperatures. The evidence was clear: There had not been just one glacial advance but many, each separated by extended periods when climates were as warm or warmer than the present. The Ice Age had not simply been a time when the ice advanced over the land, lingered for a while, and then receded. Rather, the period was a very complex event characterized by a number of advances and withdrawals of glacial ice.

The glacial record on land is punctuated by many erosional gaps. This makes it difficult to reconstruct the episodes of the Ice Age. But sediment on the ocean floor provides an uninterrupted record of climate cycles for this period. Studies of cores drilled from these seafloor sediments show that glacial/interglacial cycles have occurred about every 100,000 years. Approximately 20 such cycles of cooling and warming were identified for the span we call the Ice Age.

During the glacial age, ice left its imprint on almost 30 percent of Earth's land area, including about 10 million square kilometers of North America, 5 million square kilometers of Europe, and 4 million square kilometers of Siberia (**Figure 6.23**). The amount of glacial ice in the Northern Hemisphere was roughly twice that of the Southern Hemisphere. The primary reason is that the Southern Hemisphere has little land in the middle latitudes, and therefore the southern polar ice could not spread far beyond the margins of Antarctica. By contrast, North America and Eurasia provided great expanses of land for the spread of ice sheets.

Today we know that the Ice Age began between 2 million and 3 million years ago. This means that most of the major glacial episodes occurred during a division of the geologic time scale called the **Pleistocene epoch**. Although the Pleistocene is commonly used as a synonym for the Ice Age, this epoch does not encompass it all. The Antarctic Ice Sheet, for example, formed at least 30 million years ago.

CONCEPT CHECK 6.6

❶ **About what percentage of Earth's land surface was covered at one time by Ice Age glaciers?**

Causes of Glaciation

A great deal is known about glaciers and glaciation. Much has been learned about glacier formation and movement, the extent of glaciers past and present, and the features created by glaciers, both erosional and depositional. However, the causes of glacial ages are not completely understood.

Although widespread glaciation has been rare in Earth's history, the Pleistocene Ice Age is not the only glacial period for which a record exists. Earlier glaciations are indicated by deposits called *tillite*, a sedimentary rock formed when glacial till becomes lithified. Such strata usually contain striated rock fragments, and some lie atop grooved and polished rock surfaces or are associated with sandstones and conglomerates that show features indicating they were deposited as stratified drift. Two Precambrian glacial episodes have been identified in the geologic record, the first approximately 2 billion years ago and the second about 600 million years ago. Furthermore, a well-documented record of an earlier glacial age is found in late Paleozoic rocks that are about 250 million years old and that exist on several landmasses.[9]

Any theory that attempts to explain the causes of glacial ages must successfully answer two basic questions: (1) *What causes the onset of glacial conditions?* For continental ice sheets to have formed, average temperatures must have been somewhat lower than at present and perhaps substantially lower than throughout much of geologic time. Thus, a successful theory would have to account for the cooling that finally leads to glacial conditions. (2) *What caused the alternation of glacial and interglacial stages that have been documented for the Pleistocene epoch?* The first question deals with long-term trends in temperature on a scale of millions of years, but this second question relates to much shorter-term changes.

Although the literature of science contains many hypotheses relating to the possible causes of glacial periods, we discuss only a few major ideas to summarize current thought.

Plate Tectonics

Probably the most attractive proposal for explaining the fact that extensive glaciations have occurred only a few times in the geologic past comes from the theory of plate tectonics.[10] Because glaciers can form only on land, we know that landmasses must exist somewhere in the higher latitudes before an ice age can commence. Many scientists suggest that ice ages have occurred only when Earth's shifting crustal plates have carried the continents from tropical latitudes to more poleward positions.

Glacial features in present-day Africa, Australia, South America, and India indicate that these regions, which are now tropical or subtropical, experienced an ice age near the end of the Paleozoic era, about 250 million years ago. However, there is no evidence that ice sheets existed during this same period in what are today the higher latitudes of North America and Eurasia. For many years this puzzled scientists. Was the climate in these relatively tropical latitudes once like it is today in Greenland and Antarctica? Why did glaciers not form in North America and Eurasia? Until the plate tectonics theory was formulated, there had been no reasonable explanation.

Today, scientists realize that the areas containing these ancient glacial features were joined together as a single supercontinent (Pangaea) located at latitudes far to the south of their present positions. Later, this landmass broke apart, and its pieces, each moving on a different plate, migrated toward their present locations (Figure 6.24). It is now understood that during the geologic past, plate movements accounted for many dramatic climatic changes as landmasses shifted in relation to one another and moved to different latitudinal positions. Changes in oceanic

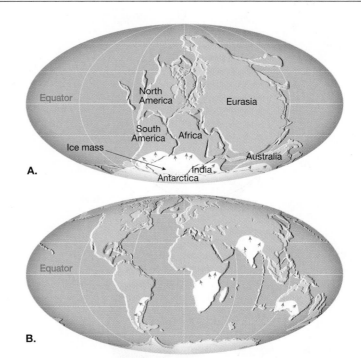

FIGURE 6.24 **A.** The supercontinent Pangaea showing the area covered by glacial ice 300 million years ago. **B.** The continents as they are today. The white areas indicate regions where evidence of the old ice sheets exists.

circulation also must have occurred, altering the transport of heat and moisture and consequently the climate as well. Because the rate of plate movement is very slow—a few centimeters per year—appreciable changes in the positions of the continents occur only over great spans of geologic time. Thus, climatic changes brought about by shifting plates are extremely gradual and occur on a scale of millions of years.

Variations in Earth's Orbit

Because climatic changes brought about by moving plates are extremely gradual, the plate tectonics theory cannot be used to explain the alternation between glacial and interglacial climates that occurred during the Pleistocene epoch. Therefore, we must look to some other triggering mechanism that may cause climatic change on a scale of thousands rather than millions of years. Today, many scientists strongly suspect that the climatic oscillations that characterized the Pleistocene may be linked to variations in Earth's orbit. This hypothesis was first developed and strongly advocated by Serbian scientist Milutin Milankovitch and is based on the premise that variations in incoming solar radiation are a principal factor in controlling Earth's climate.

Milankovitch formulated a comprehensive mathematical model based on the following elements (Figure 6.25):

1. Variations in the shape (*eccentricity*) of Earth's orbit about the Sun.
2. Changes in *obliquity*—that is, changes in the angle that the axis makes with the plane of Earth's orbit.
3. The wobbling of Earth's axis, called *precession*.

---

[9]The terms *Precambrian* and *Paleozoic* refer to time spans on the geologic time scale of Earth history. For more on the geologic time scale, see Chapter 11 and Figure 11.19.

[10]A complete discussion of plate tectonics is presented in Chapter 7.

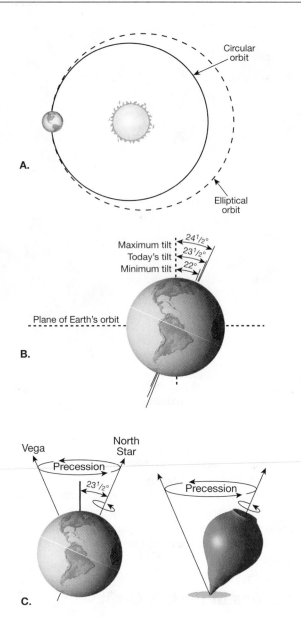

FIGURE 6.25 Orbital variations. **A.** The shape of Earth's orbit changes during a cycle that spans about 100,000 years. It gradually changes from nearly circular to one that is more elliptical and then back again. This diagram greatly exaggerates the amount of change. **B.** Today, the axis of rotation is tilted about 23.5° to the plane of Earth's orbit. During a cycle of 41,000 years, this angle varies from 22° to 24.5°. **C.** Precession. Earth's axis wobbles like that of a spinning top. Consequently, the axis points to different spots in the sky during a cycle of about 26,000 years.

Using these factors, Milankovitch calculated variations in the receipt of solar energy and the corresponding surface temperature of Earth back into time in an attempt to correlate these changes with the climatic fluctuations of the Pleistocene. In explaining climatic changes that result from these three variables, note that they cause little or no variation in the total solar energy reaching the ground. Instead, their impact is felt because they change the degree of contrast between the seasons. Somewhat milder winters in the middle to high latitudes means greater snowfall totals, whereas cooler summers would bring a reduction in snowmelt.

Among the studies that added considerable credibility to this astronomical hypothesis is one in which deep-sea sediments containing certain climatically sensitive microorganisms were analyzed to establish a chronology of temperature changes going back nearly 500,000 years.[11] This time scale of climatic change was then compared to astronomical calculations of eccentricity, obliquity, and precession to determine if a correlation did indeed exist. Although the study was very involved and mathematically complex, the conclusions were straightforward. The researchers found that major variations in climate over the past several hundred thousand years were closely associated with changes in the geometry of Earth's orbit; that is, cycles of climatic change were shown to correspond closely with the periods of obliquity, precession, and orbital eccentricity. More specifically, the researchers stated: "It is concluded that changes in the earth's orbital geometry are the fundamental cause of the succession of Quaternary ice ages."[12]

Let us briefly summarize the ideas that were just described. The theory of plate tectonics provides an explanation for the widely spaced and nonperiodic onset of glacial conditions at various times in the geologic past, whereas the astronomical model proposed by Milankovitch and supported by the work of J. D. Hays and his colleagues furnishes an explanation for the alternating glacial and interglacial episodes of the Pleistocene.

Other Factors

Variations in Earth's orbit correlate closely with the timing of glacial–interglacial cycles. However, the variations in solar energy reaching Earth's surface caused by these orbital changes do not adequately explain the magnitude of the temperature changes that occurred during the most recent ice age. Other factors must also have contributed. One factor involves variations in the chemical composition of the atmosphere. Other influences involve changes in the reflectivity of Earth's surface and in ocean circulation. Let's take a brief look at these factors.

Chemical analyses of air bubbles that become trapped in glacial ice at the time of ice formation indicate that the Ice-Age atmosphere contained less of the gases carbon dioxide and methane than the post–Ice Age atmosphere (**Figure 6.26**). Carbon dioxide and methane are important "greenhouse" gases, which means that they trap radiation emitted by Earth and contribute to the heating of the atmosphere.[13] When the amount of carbon dioxide and methane in the atmosphere increases, global temperatures rise, and when there is a reduction in these gases, as occurred during the Ice Age, temperatures fall. Therefore, reductions in the concentrations of greenhouse gases help explain the magnitude of the temperature drop that occurred during glacial times. Although scientists know that concentrations of carbon dioxide and methane dropped, they do not know what caused the drop. As often occurs in science, observations gathered during one investigation yield information and raise questions that require further analysis and explanation.

[11]J. D. Hays, John Imbrie, and N. J. Shackelton, "Variations in the Earth's Orbit: Pacemaker of the Ice Ages," *Science* 194 (1976): 1121–32.

[12]J. D. Hays et al., p. 1131. The term *Quaternary* refers to the period on the geologic time scale that encompasses the last 2.6 million years.

[13]For more on this idea, see the section, "Heating the Atmosphere: The Greenhouse Effect" in Chapter 16 and the discussion, "Carbon Dioxide, Trace Gases, and Global Warming" in Chapter 20.

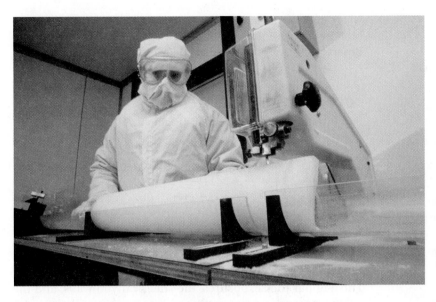

FIGURE 6.26 Preparing an ice core sample from Antarctica for analysis. This scientist is wearing protective clothing and a mask to minimize contamination of the sample. Chemical analysis of ice cores can provide important data about past climates. (Photo by British Antarctic Survey/Photo Researchers, Inc.)

Obviously, whenever Earth enters an ice age, extensive areas of land that were once ice free are covered with ice and snow. In addition, a colder climate causes the area covered by sea ice (frozen surface sea water) to expand as well. Ice and snow reflect a large portion of incoming solar energy back to space. Thus, energy that would have warmed Earth's surface and the air above is lost and global cooling is reinforced.[14]

Yet another factor that influences climate during glacial times relates to ocean currents, which, as you will learn in Chapter 15, are a complex matter. Research has shown that ocean circulation changes during ice ages. For example, studies suggest that the warm current that transports large amounts of heat from the tropics toward higher latitudes in the North Atlantic was significantly weaker during the ice age. This would lead to a colder climate in Europe, amplifying the cooling attributable to orbital variations.

In conclusion, it should be noted that our understanding of the causes of glacial ages is not complete. The ideas that were just discussed do not represent all of the possible explanations. Additional factors may be, and probably are, involved.

CONCEPT CHECK 6.7

① How does the theory of plate tectonics help us understand the cause of ice ages?

② Does the theory of plate tectonics explain alternating glacial–interglacial climates during the Pleistocene? Why or why not?

[14]Recall from Chapter 1 that something that reinforces (adds to) the initial change is called a *positive feedback mechanism*. To review this idea, see the discussion on feedback mechanisms in the section, "Earth as a System" in Chapter 1.

Deserts

 GEODe   Sculpturing Earth's Surface
EARTH SCIENCE   ▶ Deserts and Winds

The dry regions of the world encompass about 42 million square kilometers—a surprising 30 percent of Earth's land surface (Figure 6.27). No other climatic group covers so large a land area.[15] The word *desert* literally means deserted or unoccupied. For many dry regions this is a very appropriate description. Yet where water is available in deserts, plants and animals thrive. Nevertheless, the world's dry regions are among the least familiar land areas on Earth outside of the polar realm.

Desert landscapes frequently appear stark. Their profiles are not softened by a carpet of soil and abundant plant life. Instead, barren rocky outcrops with steep, angular slopes are common. At some places the rocks are tinted orange and red. At others they are gray and brown and streaked with black. For many visitors desert scenery exhibits a striking beauty; to others the terrain seems bleak. No matter which feeling is elicited, it is clear that deserts are very different from the more humid places where most people live.

As you will see, arid regions are not dominated by a single geologic process. Rather, the effects of tectonic forces, running

[15]An examination of dry climates is found in Chapter 20.

FIGURE 6.27 In this view of Earth from space, North Africa's Sahara Desert, the adjacent Arabian Desert, and the Kalahari and Namib deserts in southern Africa are clearly visible as tan-colored, cloud-free zones. These low-latitude deserts are dominated by the dry, subsiding air associated with pressure belts known as the *subtropical highs*. By contrast, the band of clouds that extends across central Africa and the adjacent oceans coincides with the equatorial low-pressure belt, the rainiest region on Earth. (Image courtesy of NASA)

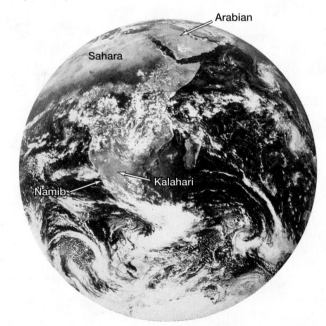

Students Sometimes Ask...

I heard somewhere that deserts are expanding. Is that actually occurring?

Yes. The problem is called *desertification*, and it refers to the alteration of land to desertlike conditions as the result of human activities. It commonly takes place on the margins of deserts and results primarily from inappropriate land use. It is triggered when the modest natural vegetation in marginal areas is removed by plowing or grazing. When drought occurs, as it inevitably does in these regions, and the vegetative cover has been destroyed beyond the minimum to hold the soil against erosion, the destruction becomes irreversible. Desertification is occurring in many places but is particularly serious in the region south of the Sahara Desert known as the Sahel.

water, and wind are all apparent. Because these processes combine in different ways from place to place, the appearance of desert landscapes varies a great deal as well (Figure 6.28).

CONCEPT CHECK 6.8

❶ How extensive are Earth's dry regions?

Geologic Processes in Arid Climates

Sculpturing Earth's Surface
▸ **Deserts and Winds**

The angular rock exposures, the sheer canyon walls, and the pebble- or sand-covered surface of deserts contrast sharply with the rounded hills and curving slopes of more humid places. To

FIGURE 6.28 A scene in Organ Pipe Cactus National Monument near the Arizona–Mexico border. The appearance of desert landscapes varies a great deal from place to place. (Photo by Marek Zak/Alamy)

a visitor from a humid region, a desert landscape may seem to have been shaped by forces different from those operating in wetter areas. However, although the contrasts might be striking, they do not reflect different processes. They merely disclose the differing effects of the same processes that operate under contrasting climatic conditions.

Weathering

In humid regions, relatively well-developed soils support an almost continuous cover of vegetation. Here the slopes and rock edges are rounded. Such a landscape reflects the strong influence of chemical weathering in a humid climate. By contrast, much of the weathered debris in deserts consists of unaltered rock and mineral fragments—the result of mechanical weathering processes. In dry lands rock weathering of any type is greatly reduced because of the lack of moisture and the scarcity of organic acids from decaying plants. Chemical weathering, however, is not completely lacking in deserts. Over long spans of time, clays and thin soils do form, and many iron-bearing silicate minerals oxidize, producing the rust-colored stain found tinting some desert landscapes.

The Role of Water

Permanent streams are normal in humid regions, but almost all desert streams are dry most of the time (Figure 6.29A). Deserts have **ephemeral streams**, which means that they carry water only in response to specific episodes of rainfall. A typical ephemeral stream might flow only a few days or perhaps just a few hours during the year. In some years the channel may carry no water at all.

This fact is obvious even to the casual observer who, while traveling in a dry region, notices the number of bridges with no streams beneath them or the number of dips in the road where dry channels cross. However, when the rare heavy showers do occur, so much rain falls in such a short time that all of it cannot soak in. Because the vegetative cover is sparse, runoff is largely unhindered and consequently rapid, often creating flash floods along valley floors (Figure 6.29B). Such floods, however, are quite unlike floods in humid regions. A flood on a river such as the Mississippi may take many days to reach its crest and then subside, but desert floods arrive suddenly and subside quickly. Because much of the surface material is not anchored by vegetation, the amount of erosional work that occurs during a single short-lived rain event is impressive.

In the dry western United States a number of different names are used for ephemeral streams. Two of the most common are *wash* and *arroyo*. In other parts of the world, a dry desert stream may be called a *wadi* (Arabia and North Africa), a *donga* (South America), or a *nullah* (India).

Humid regions are notable for their integrated drainage systems. But in arid regions streams usually lack an extensive system of tributaries. In fact, a basic characteristic of desert streams is that they are small and die out before reaching the sea. Because the water table is usually far below the surface, few desert streams can draw upon it as streams do in humid regions. Without a steady supply of water, the combination of evaporation and infiltration soon depletes the stream.

The few permanent streams that do cross arid regions, such as the Colorado and Nile rivers, originate *outside* the desert, often in well-watered mountains. In these situations the water supply must be great to compensate for the losses occurring as the stream crosses the desert (Box 6.2). For example, after the Nile leaves the lakes and mountains of central Africa that are its source, it traverses almost 3,000 kilometers (nearly 1,900 miles) of the Sahara *without a single tributary*. By contrast, in humid

FIGURE 6.29 A. Most of the time, desert stream channels are dry. **B.** An ephemeral stream shortly after a heavy shower. Although such floods are short-lived, large amounts of erosion occur. (Photos by E. J. Tarbuck)

A.

B.

Box 6.2

PEOPLE AND THE ENVIRONMENT

The Disappearing Aral Sea—A Large Lake Becomes a Barren Wasteland

The Aral Sea lies on the border between Uzbekistan and Kazakhstan in central Asia (**Figure 6.C**). The setting is the Turkestan desert, a middle-latitude desert in the rainshadow of Afghanistan's high mountains. In this region of interior drainage, two large rivers, the Amu Darya and the Syr Darya, carry water from the mountains of northern Afghanistan across the desert to the Aral Sea. Water leaves the sea by evaporation. Thus, the size of the water body depends on the balance between river inflow and evaporation.

In 1960 the Aral Sea was one of the world's largest inland water bodies, with an area of about 67,000 square kilometers (26,000 square miles). Only the Caspian Sea, Lake Superior, and Lake Victoria were larger. By the year 2008 the area of the Aral Sea was about 10 percent of its 1960 size, and its volume was reduced by nearly 90 percent. The shrinking Aral Sea is depicted in **Figure 6.D**. All that remains are three shallow remnants.

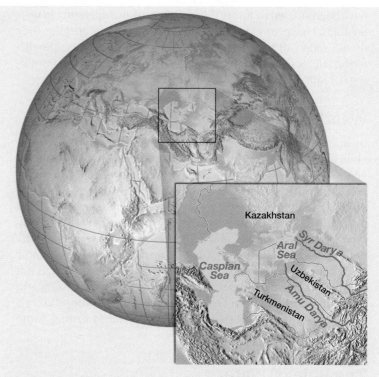

FIGURE 6.C The Aral Sea lies east of the Caspian sea in the Turkestan Desert. Two rivers, the Amu Darya and Syr Darya, bring water from the mountains to the sea.

What caused the Aral Sea to dry up? The answer is that the flow of water from the mountains that supplied the sea was significantly reduced and then all but eliminated. The waters of the Amu Darya and Syr Darya were diverted to supply a major expansion of irrigated agriculture in this dry realm.

The intensive irrigation greatly increased agricultural productivity, but not without significant costs. The deltas of the two major rivers have lost their wetlands, and wildlife has disappeared. The once thriving fishing industry is dead, and the 24 species of fish that once lived in the Aral Sea are no longer

---

regions the discharge of a river usually increases in the downstream direction because tributaries and groundwater contribute additional water along the way.

It should be emphasized that *running water, although infrequent, nevertheless does most of the erosional work in deserts*. This is contrary to a common belief that wind is the most important erosional agent sculpting desert landscapes. Although wind erosion is indeed more significant in dry areas than elsewhere, most desert landforms are nevertheless carved by running water. As you will see shortly, the main role of wind is in the transportation and deposition of sediment, which creates and shapes the ridges and mounds we call *dunes*.

CONCEPT CHECK 6.9

1 How does the rate of rock weathering in dry climates compare to the rate in humid regions?

2 When a permanent stream such as the Nile River crosses a desert, does discharge increase or decrease? How does this compare to a river in a humid area?

3 What is the most important agent of erosion in deserts?

Basin and Range: The Evolution of a Mountainous Desert Landscape

 Sculpturing Earth's Surface
▶ **Deserts and Winds**

Because arid regions typically lack permanent streams, they are characterized as having **interior drainage**. This means that they have a discontinuous pattern of intermittent streams that do not flow out of the desert to the ocean. In the United States, the dry Basin and Range region provides an excellent example. The region includes southern Oregon, all of Nevada, western Utah, southeastern California, southern Arizona, and southern New Mexico. The name Basin and Range is an apt description for this almost 800,000-square-kilometer (more than 300,000-square-mile) region, because it is characterized by more than 200 relatively small mountain ranges that rise 900–1,500 meters (3,000–5,000 feet) above the basins that separate them.

FIGURE 6.D The shrinking Aral Sea.

The shrinking Aral Sea has had a noticeable impact on the region's climate. Without the moderating effect of a large water body, there are greater extremes of temperature, a shorter growing season, and reduced local precipitation. These changes have caused many farms to switch from growing cotton to growing rice, which demands even more diverted water.

Could this crisis be reversed if enough freshwater were to once again flow into the Aral Sea? Prospects appear grim. Experts estimate that restoring the Aral Sea to about twice its present size would require stopping all irrigation from the two major rivers for decades. This could not be done without ruining the economies of the countries that rely on that water.

One effort has improved the situation in the northern portion of the Aral. In November 2005, a large earthen dam was completed that blocked the southward outflow of water. Prior to this structure, the modest amount of water contributed by the Syr Darya was lost as it flowed southward and evaporated. The dam allowed the water from the Syr Darya to recharge and partially restore this portion of the water body. The North Aral Sea is now larger and less salty than before.

The decline of the Aral Sea is a major environmental disaster that sadly is of human making.

there. The shoreline is now tens of kilometers from the towns that were once fishing centers.

The shrinking sea has exposed millions of acres of former seabed to sun and wind. The surface is encrusted with salt and with agrochemicals brought by the rivers. Strong winds routinely pick up and deposit thousands of tons of newly exposed material every year. This process has not only contributed to a significant reduction in air quality for people living in the region but has also appreciably affected crop yields due to the deposition of salt-rich sediments on arable land.

In this region, as in others like it around the world, most erosion occurs without reference to the ocean (ultimate base level), because the interior drainage never reaches the sea. Even where permanent streams flow to the ocean, few tributaries exist, and thus only a narrow strip of land adjacent to the stream has sea level as its ultimate level of land reduction.

The block models in **Figure 6.30** depict how the landscape has evolved in the Basin and Range region. During and following uplift of the mountains, running water begins carving the elevated mass and depositing large quantities of debris in the basin. In this early stage, relief is greatest, and as erosion lowers the mountains and sediment fills the basins, elevation differences diminish.

When the occasional torrents of water produced by sporadic rains move down the mountain canyons, they are heavily loaded with sediment. Emerging from the confines of the canyon, the runoff spreads over the gentler slopes at the base of the mountains and quickly loses velocity. Consequently, most of its load is dumped within a short distance. The result is a cone of debris known as an **alluvial fan** at the mouth of a canyon. Over the years, a fan enlarges, eventually coalescing with fans from adjacent canyons to produce an apron of sediment (*bajada*) along the mountain front.

On the rare occasions of abundant rainfall, or snowmelt in the mountains, streams may flow across the alluvial fans to the center of the basin, converting the basin floor into a shallow **playa lake**. Playa lakes last only a few days or weeks, before evaporation and infiltration remove the water. The dry, flat lake bed that remains is termed a *playa*. Playas occasionally become encrusted with salts (*salt flats*) that are left behind when the water in which they were dissolved evaporates. **Figure 6.31** includes a satellite view (larger image) and an aerial view of a portion of California's Death Valley, a classic Basin and Range landscape. Many of the features just described are prominent, including a bajada, alluvial fans, a playa lake, and extensive salt flats.

With the ongoing erosion of the mountain mass and the accompanying sedimentation, the local relief continues to diminish. Eventually nearly the entire mountain mass is gone. Thus, by the late stages of erosion, the mountain areas are reduced to a few large bedrock knobs (called *inselbergs*) projecting above the sediment-filled basin.

Each of the stages of landscape evolution in an arid climate depicted in Figure 6.30 can be observed in the Basin and Range region. Recently, uplifted mountains in an early stage of erosion

Students Sometimes Ask...

Do the terms "arid" and "drought" mean about the same thing?

The concept of drought differs from that of aridity. Drought is a *temporary* happening that is considered an atmospheric hazard. By contrast, aridity describes regions where low rainfall is a *permanent* feature of the climate. Although natural disasters such as floods and hurricanes generate more attention, droughts can be just as devastating and carry a bigger price tag.

were found in southern Oregon and northern Nevada. Death Valley, California, and southern Nevada fit into the more advanced middle stage, whereas the late stage, with its inselbergs, can be seen in southern Arizona.

CONCEPT CHECK 6.10

1. Describe the features and characteristics associated with each stage in the evolution of a mountainous desert landscape. Where in the United States can each stage be observed?

Wind Erosion

Sculpturing Earth's Surface
▶ Deserts and Winds

Moving air, like moving water, is turbulent and able to pick up loose debris and transport it to other locations. Just as in a stream, the velocity of wind increases with height above the surface. Also like a stream, wind transports fine particles in suspension while heavier ones are carried as bed load (**Figure 6.32**). However, the transport of sediment by wind differs from that of running water in two significant ways. First, wind's lower density compared to water renders it less capable of picking up and transporting coarse materials. Second, because wind is not confined to channels, it can spread sediment over large areas, as well as high into the atmosphere.

Compared to running water and glaciers, wind is a relatively insignificant erosional agent. Recall that even in deserts, most erosion is performed by intermittent running water, not by the wind. Wind erosion is more effective in arid lands than in humid areas because in humid places moisture binds particles together and vegetation anchors the soil. For wind to be an effective erosional force, dryness and scanty vegetation are essential. When such circumstances exist, wind may pick up,

FIGURE 6.30 Stages of landscape evolution in a mountainous desert such as the Basin and Range region of the West. As erosion of the mountains and deposition in the basins continue, relief diminishes. **A.** Early stage. **B.** Middle stage. **C.** Late stage.

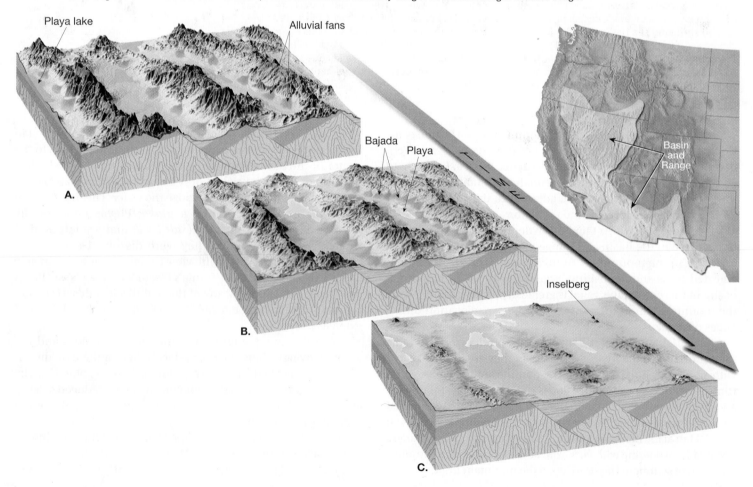

FIGURE 6.31 The satellite image shows a portion of Death Valley, California, a classic Basin and Range landscape. Shortly before this image was taken in February 2005, heavy rains led to the formation of a playa lake—the pool of greenish water on the basin floor. By May 2005, the lake had reverted to a salt-covered playa. (NASA) The small photo is a closer aerial view of one of Death Valley's many alluvial fans. (Photo by Michael Collier)

transport, and deposit great quantities of fine sediment. During the 1930s, parts of the Great Plains experienced vast dust storms. The plowing-under of the natural vegetative cover for farming, followed by severe drought, exposed the land to wind erosion and led to the area being labeled the Dust Bowl (Figure 6.33).

Deflation, Blowouts, and Desert Pavement

One way that wind erodes is by **deflation** (*de* = out, *flat* = blow), the lifting and removal of loose material. Because the competence (ability to transport different-sized particles) of moving air is low, it can suspend only fine sediment, such as clay and silt. Larger grains of sand are rolled or skipped along the surface (a process called *saltation*) and comprise the bed load. Particles larger than sand are usually not transported by wind. Deflation sometimes is difficult to notice because the entire surface is being lowered at the same time, but it can be significant.

The most noticeable results of deflation in some places are shallow depressions called **blowouts** (Figure 6.34). In the Great Plains region, from Texas north to Montana, thousands of blowouts can be seen. They range from small dimples less than 1 meter deep and 3 meters wide to depressions that are over 45 meters deep and several kilometers across.

In portions of many deserts, the surface is characterized by a layer of coarse pebbles and cobbles that are too large to be moved by the wind. This stony veneer, called **desert pavement**, may form as deflation lowers the surface by removing sand and silt from poorly sorted materials. As Figure 6.35A illustrates, the concentration of larger particles at the surface gradually increases as the finer particles are blown away. Eventually, a continuous cover of coarse particles remains.

Studies have shown that the process depicted in Figure 6.34A is not an adequate explanation for all environments in which desert pavement exists. As a result, an alternate explanation was formulated and is illustrated in Figure 6.35B. This hypothesis suggests that pavement develops on a surface that initially consists of coarse pebbles. Over time, protruding cobbles trap fine wind-blown grains

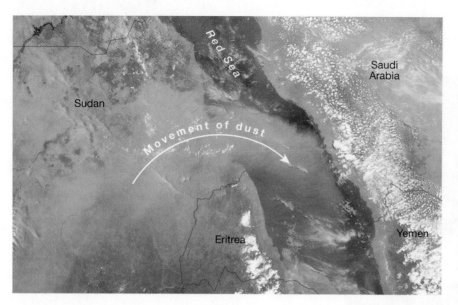

FIGURE 6.32 This satellite image shows thick plumes of dust from the Sahara Desert blowing across the Red Sea on June 30, 2009. Such dust storms are common in arid North Africa. In fact, this region is the largest dust source in the world. Satellites are excellent tools for studying the transport of dust on a global scale. They show us that dust storms can cover huge areas and that dust can be transported great distances. (NASA)

that settle and sift downward through the spaces between the larger surface stones. The process is aided by infiltrating rainwater.

Once desert pavement becomes established, a process that might take hundreds of years, the surface is effectively protected from further deflation if left undisturbed. However, as the layer is only one or two stones thick, the passage of vehicles or animals can dislodge the pavement and expose the fine-grained material

FIGURE 6.33 Dust blackens the sky on May 21, 1937, near Elkhart, Kansas. It was because of storms like this that portions of the Great Plains were called the Dust Bowl in the 1930s. (Photo reproduced from the collection of the Library of Congress)

below. If this happens, the surface is no longer protected from deflation.

Wind Abrasion

Like glaciers and streams, wind erodes in part by *abrasion*. In dry regions as well as along some beaches, windblown sand will cut and polish exposed rock surfaces. Abrasion is often credited for accomplishments beyond its actual capabilities. Such features as balanced rocks that stand high atop narrow pedestals, and intricate detailing on tall pinnacles, are *not* the results of abrasion. Sand seldom travels more than a meter above the surface, so the wind's sandblasting effect is obviously limited in vertical extent. However, in areas prone to such activity, telephone poles have actually been cut through near their bases. For this reason, collars may be fitted on the poles to protect them from being "sawed" down.

CONCEPT CHECK 6.11

1 Why is wind erosion relatively more effective in arid regions than in humid areas?

2 What factor limits the depths of blowouts?

3 Briefly describe two hypotheses used to explain the formation of desert pavement.

Wind Deposits

Sculpturing Earth's Surface
▸ Deserts

Although wind is relatively unimportant in producing *erosional* landforms, significant *depositional* landforms are created by the wind in some regions. Accumulations of windblown sediment are particularly conspicuous in the world's dry lands and along many sandy coasts. Wind deposits are of two distinctive types: (1) extensive blankets of silt, called *loess*, which once were carried in suspension, and (2) mounds and ridges of sand from the wind's bed load, which we call *dunes*.

Loess

In some parts of the world the surface topography is mantled with deposits of windblown silt termed **loess**. Over thousands of years dust storms deposited this material. When loess is breached by streams or road cuts, it tends to maintain vertical cliffs and lacks any visible layers, as you can see in Figure 6.36.

FIGURE 6.34 A. Blowouts are depressions created by deflation. Land that is dry and largely unprotected by anchoring vegetation is particularly susceptible. **B.** In this example, deflation has removed about 4 feet of soil— the distance from the man's outstretched arm to his feet. (Photo courtesy of U.S.D.A./Natural Resources Conservation Service)

FIGURE 6.35 A. This model portrays an area with poorly sorted surface deposits. Coarse particles gradually become concentrated into a tightly packed layer as deflation lowers the surface by removing sand and silt. In this situation, desert pavement is the result of deflation. **B.** This model shows the formation of desert pavement on a surface initially covered with coarse pebbles and cobbles. Windblown dust accumulates at the surface and gradually sifts downward through spaces between coarse particles. Infiltrating rainwater aids the process. This depositional process raises the surface and produces a layer of coarse pebbles and cobbles underlain by a substantial layer of fine sediment.

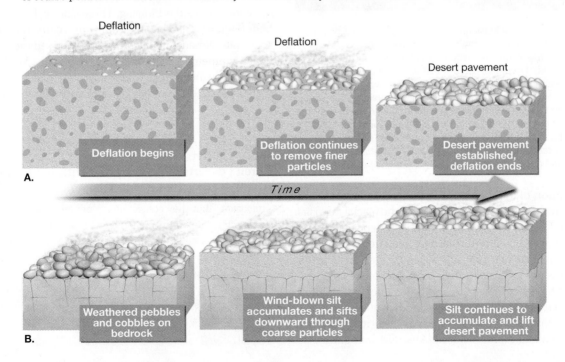

A.

B.

C.

Windblown dust

FIGURE 6.36 **A.** This vertical loess bluff near the Mississippi River in southern Illinois is about 3 meters high. (Photo by James E. Patterson) **B.** In parts of China, loess has sufficient structural strength to permit the excavation of cave-like dwellings. (Photo by Christopher Liu/ChinaStock Photo Library) **C.** This satellite image from March 13, 2003, shows streamers of windblown dust moving southward into the Gulf of Alaska. It illustrates a process similar to the one that created many loess deposits in the American Midwest during the Ice Age. Fine silt is produced by the grinding action of glaciers, then transported beyond the margin of the ice by meltwater streams and deposited. Later, the fine silt is picked up by strong winds and deposited as loess. (Image courtesy of NASA)

The distribution of loess worldwide indicates that there are two primary sources for this sediment: deserts and glacial deposits of stratified drift. The thickest and most extensive deposits of loess on Earth occur in western and northern China. They were blown there from the extensive desert basins of central Asia. Accumulations of 30 meters are not uncommon, and thicknesses of more than 100 meters have been measured. It is this fine, buff-colored sediment that gives the Yellow River (Hwang Ho) its name.

In the United States, deposits of loess are significant in many areas, including South Dakota, Nebraska, Iowa, Missouri, and Illinois, as well as portions of the Columbia Plateau in the Pacific Northwest. Unlike the deposits in China, the loess in the United States, as well as in Europe, is an indirect product of glaciation. Its source is deposits of stratified drift. During the retreat of the ice sheets, many river valleys were choked with sediment deposited by meltwater. Strong westerly winds sweeping across the barren floodplains picked up the finer sediment and dropped it as a blanket on the eastern sides of the valleys.

Sand Dunes

Like running water, wind releases its load of sediment when its velocity falls and the energy available for transport diminishes. Thus, sand begins to accumulate wherever an obstruction across the path of the wind slows its movement. Unlike deposits of loess, which form blanketlike layers over broad areas, winds commonly deposit sand in mounds or ridges called *dunes* (Figure 6.36).

As moving air encounters an object, such as a clump of vegetation or a rock, the wind sweeps around and over it, leaving a shadow of more slowly moving air behind the obstacle as well as a smaller zone of quieter air just in front of the obstacle. Some of the saltating sand grains moving with the wind come to rest in these wind shadows. As the accumulation of sand continues, it forms an increasingly efficient wind barrier to trap even more sand. If there is a sufficient supply of sand and the wind blows steadily long enough, the mound of sand grows into a dune.

Many dunes have an asymmetrical profile, with the leeward (sheltered) slope being steep and the windward slope being more gently inclined. The dunes in Figure 6.37 are a good example. Sand moves up the gentler slope on the windward side by saltation. Just beyond the crest of the dune, where wind velocity is reduced, the sand accumulates. As more sand collects, the slope steepens, and eventually some of it slides or slumps under the pull of gravity (Figure 6.37). In this way, the leeward slope of the dune, called the **slip face**, maintains an angle of about 34 degrees. Continued sand accumulation, coupled with periodic slides down the slip face, results in the slow migration of the dune in the direction of air movement.

As sand is deposited on the slip face, it forms layers inclined in the direction the wind is blowing. These sloping layers are called **cross beds** (Figure 6.38). When the dunes are eventually buried under other layers of sediment and become part of the sedimentary rock record, their asymmetrical shape is destroyed, but the cross beds remain as a testimony to their origin. Nowhere is cross-bedding more prominent than in the sandstone walls of Zion Canyon in Utah (Figure 6.38D).

Students Sometimes Ask...

Aren't deserts mostly covered with sand dunes?

A common misconception about deserts is that they consist of mile after mile of drifting sand dunes. It is true that sand accumulations do exist in some areas and may be striking features. But, perhaps surprisingly, sand accumulations worldwide represent only a small percentage of the total desert area. For example, in the Sahara—the world's largest desert—accumulations of sand cover only *one-tenth* of its area. The sandiest of all deserts is the Arabian, one-third of which consists of sand.

FIGURE 6.37 A. Dunes composed of gypsum sand at White Sands National Monument in southeastern New Mexico. **B.** Sand sliding down the steep slip face of a dune in White Sands National Monument. (Photos by Michael Collier)

Types of Sand Dunes

Dunes are not just random heaps of windblown sediment. Rather, they are accumulations that usually assume patterns that are surprisingly consistent (**Figure 6.39**). A broad assortment of dune forms exist, generally simplified to a few major types for discussion. Of course, gradations exist among different forms as well as irregularly shaped dunes that do not fit easily into any category. Several factors influence the form and size that dunes ultimately assume. These include wind direction and velocity, availability of sand, and the amount of vegetation present. Six basic dune types are shown in Figure 6.39, with arrows indicating wind directions.

Barchan Dunes   Solitary sand dunes shaped like crescents and with their tips pointing downwind are called **barchan dunes**

(Figure 6.39A). These dunes form where supplies of sand are limited and the surface is relatively flat, hard, and lacking vegetation. They migrate slowly with the wind at a rate of up to 15 meters annually. Their size is usually modest, with the largest barchans reaching heights of about 30 meters while the maximum spread between their horns approaches 300 meters. When the wind direction is nearly constant, the crescent form of these dunes is nearly symmetrical. However, when the wind direction is not perfectly fixed, one tip becomes larger than the other.

Transverse Dunes   In regions where the prevailing winds are steady, sand is plentiful, and vegetation is sparse or absent, the dunes form a series of long ridges that are separated by troughs and oriented at right angles to the prevailing wind. Because of

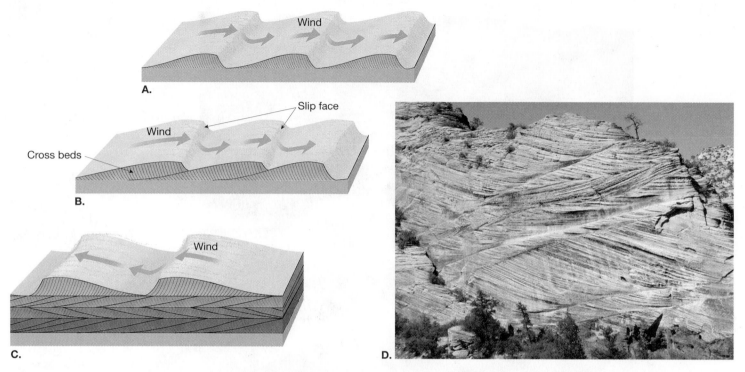

FIGURE 6.38 As parts **A** and **B** illustrate, dunes commonly have an asymmetrical shape. The steeper leeward side is called the *slip face*. Sand grains deposited on the slip face create the cross-bedding of the dunes. **C.** A complex pattern develops in response to changes in wind direction. Also notice that when dunes are buried and become part of the sedimentary record, the cross-bedded structure is preserved. **D.** Cross beds are an obvious characteristic of the Navajo Sandstone in Zion National Park, Utah. (Photo by Dennis Tasa)

FIGURE 6.39 Sand dune types. **A.** Barchan dunes. **B.** Transverse dunes. **C.** Barchanoid dunes. **D.** Longitudinal dunes. **E.** Parabolic dunes. **F.** Star dunes.

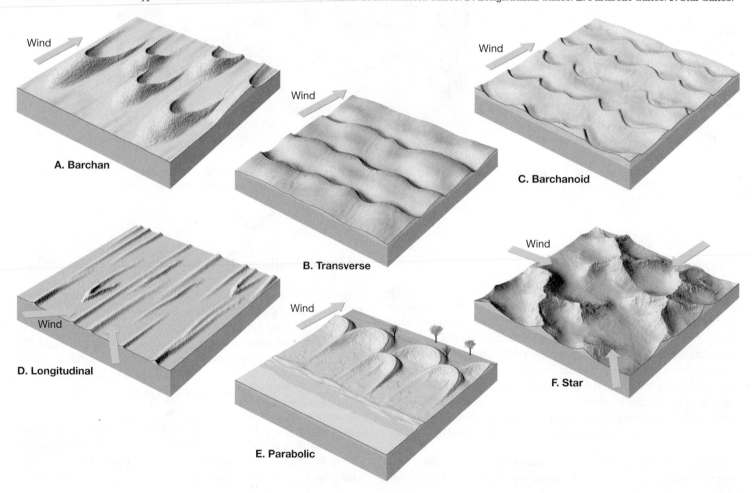

this orientation, they are termed **transverse dunes** (Figure 6.39B). Typically, many coastal dunes are of this type. In addition, transverse dunes are common in many arid regions where the extensive surface of wavy sand is sometimes called a *sand sea*. In some parts of the Sahara and Arabian deserts, transverse dunes reach heights of 200 meters, are 1–3 kilometers across, and can extend for distances of 100 kilometers or more.

Barchanoid Dunes.   There is a relatively common dune form that is intermediate between isolated barchans and extensive waves of transverse dunes. Such dunes, called **barchanoid dunes**, form scalloped rows of sand oriented at right angles to the wind (Figure 6.39C). The rows resemble a series of barchans that have been positioned side by side. Visitors exploring the gypsum dunes at White Sands National Monument, in New Mexico, will recognize this form (see Figure 6.37).

Longitudinal Dunes   **Longitudinal dunes** are long ridges of sand that form more or less parallel to the prevailing wind and where sand supplies are moderate (Figure 6.39D). Apparently the prevailing wind direction varies somewhat but

remains in the same quadrant of the compass. Although the smaller types are only 3 or 4 meters high and several tens of meters long, in some large deserts longitudinal dunes can reach great size. For example, in portions of North Africa, Arabia, and central Australia, these dunes may approach a height of 100 meters and extend for distances of more than 100 kilometers (62 miles).

Parabolic Dunes   Unlike the other dunes that have been described thus far, **parabolic dunes** form where vegetation partially covers the sand. The shape of these dunes resembles the shape of barchans except that their tips point into the wind rather than downwind (Figure 6.39E). Parabolic dunes often form along coasts where there are strong onshore winds and abundant sand. If the sand's sparse vegetative cover is disturbed at some spot, deflation creates a blowout. Sand is then transported out of the depression and deposited as a curved rim that grows higher as deflation enlarges the blowout.

Star Dunes   Confined largely to parts of the Sahara and Arabian deserts, **star dunes** are isolated hills of sand that exhibit a complex form (Figure 6.39F). Their name is derived from the fact that the bases of these dunes resemble multipointed stars. Usually three or four sharp-crested ridges diverge from a central high point that in some cases may approach a height of 90 meters. As their form suggests, star dunes develop where wind directions are variable.

Students Sometimes Ask...

Where are the largest sand dunes located, and how big are they?

The highest dunes in the world are along the southwest coast of Africa in the Namib Desert. In places, these huge dunes reach heights of 300–350 meters (100–1,157 feet). The dunes at Great Sand Dunes National Park in southern Colorado are the highest in North America, rising more than 210 meters (700 feet) above the surrounding terrain.

CONCEPT CHECK 6.12

1. How is loess different from sand? How are some loess deposits related to glaciers?
2. How do sand dunes migrate?
3. List and briefly distinguish among basic dune types.

GIVE IT SOME THOUGHT

1. The accompanying diagram shows the results of a classic experiment used to determine how glacial ice moves in a mountain valley. The experiment was carried out over an 8-year span. Refer to this diagram and answer the following:
   a. What was the average yearly rate of ice advance in the center of the glacier?
   b. About how fast was the center of the glacier advancing *per day*?
   c. What was the average rate at which ice advanced along the sides of the glacier?
   d. Why was the rate at the center different than the rate along the sides?

2. Studies have shown that during the Ice Age the margins of some ice sheets advanced southward from the Hudson Bay region at rates ranging from about 50 to 320 meters per year.
   a. Determine the maximum amount of time required for an ice sheet to move from the southern end of Hudson Bay to the south shore of present-day Lake Erie, a distance of 1,600 kilometers.
   b. Calculate the minimum number of years required for an ice sheet to move this distance.

320 meters

920 meters

3. If the budget of a valley glacier were balanced for an extended span of time, what feature would you expect to find at the terminus of the glacier? Now assume the glacier's budget changes so that wastage exceeds accumulation. How would the terminus of the glacier change? Describe the deposit you would expect to form under these conditions.

4. Assume you and a non-geologist friend are visiting Alaska's Hubbard Glacier shown in Figure 6.1, p. 159. After studying the glacier for quite a long time, your friend asks, "Do these things really move?" How would you convince your companion that this glacier does indeed move using evidence that is clearly visible in this image?

5. Are either or both of the following statements true? Explain your answer.
   a. Wind does its most effective erosional work in dry places.
   b. Wind is the most important agent of erosion in deserts.

6. Refer to Figure 6.37 (p. 187), an image of dunes at New Mexico's White Sands National Monument. What type of dunes are these? From what direction (top, bottom, left, or right) is the prevailing wind? How were you able to determine the wind direction?

7. Compare the sediment deposited by a stream, the wind, and a glacier. Which deposit should have the most uniform grain size? Which one would exhibit the poorest sorting? Explain your choices.

In Review   Chapter 6 Glaciers, Deserts, and Wind

- A *glacier* is a thick mass of ice originating on land from the compaction and recrystallization of snow, and it shows evidence of past or present flow. Today, *valley* or *alpine glaciers* are found in mountain areas where they usually follow valleys that were originally occupied by streams. *Ice sheets* exist on a much larger scale, covering most of Greenland and Antarctica.

- Near the surface of a glacier, in the *zone of fracture*, ice is brittle. However, below about 50 meters, pressure is great, causing ice to *flow* like a *plastic material*. A second important mechanism of glacial movement consists of the whole ice mass *slipping* along the ground.

- Glaciers form in areas where more snow falls in winter than melts during summer. Snow accumulation and ice formation occur in the *zone of accumulation*. Beyond this area is the *zone of wastage*, where there is a net loss to the glacier. The *glacial budget* is the balance, or lack of balance, between accumulation at the upper end of the glacier, and loss at the lower end.

- Glaciers erode land by *plucking* (lifting pieces of bedrock out of place) and *abrasion* (grinding and scraping of a rock surface). Erosional features produced by valley glaciers include *glacial troughs, hanging valleys, cirques, arêtes, horns*, and *fiords*.

- Any sediment of glacial origin is called *drift*. The two distinct types of glacial drift are (1) *till*, which is unsorted sediment deposited directly by the ice, and (2) *stratified drift*, which is relatively well-sorted sediment laid down by glacial meltwater.

- The most widespread features created by glacial deposition are layers or ridges of till, called *moraines*. Associated with valley glaciers are *lateral moraines*, formed along the sides of the valley, and *medial moraines*, formed between two valley glaciers that have joined. *End moraines*, which mark the former position of the front of a glacier, and *ground moraine*, an undulating layer of till deposited as the ice front retreats, are common to both valley glaciers and ice sheets.

- Perhaps the most convincing evidence for the occurrence of several glacial advances during the *Ice Age* is the widespread existence of *multiple layers of drift* and an uninterrupted record of climate cycles preserved in *seafloor sediments*. In addition to massive erosional and depositional work, other effects of Ice Age glaciers included the *migration of organisms, changes in stream courses, adjustment of the crust* by rebounding after the removal of the immense load of ice, and *climate changes* caused by the existence of the glaciers themselves. In the sea, the most far-reaching effect of the Ice Age was the *worldwide change in sea level* that accompanied each advance and retreat of the ice sheets.

- Any theory that attempts to explain the causes of glacial ages must answer two basic questions: (1) What causes the onset of glacial conditions? and (2) What caused the alternating glacial and interglacial stages that have been documented for the Pleistocene epoch? Two of the many hypotheses for the cause of glacial ages involve (1) plate tectonics and (2) variations in Earth's orbit. Other factors that are related to climate change during glacial ages include changes in atmospheric composition, variations in the amount of sunlight reflected by Earth's surface, and changes in ocean circulation.

- Practically all desert streams are dry most of the time and are said to be *ephemeral*. Nevertheless, *running water is responsible for most of the erosional work in a desert*. Although wind erosion is more significant in dry areas than elsewhere, the main role of wind in a desert is in the transportation and deposition of sediment.

- Many of the landscapes of the Basin and Range region of the western and southwestern United States are the result of streams eroding uplifted mountain blocks and depositing the sediment in interior basins. *Alluvial fans, playas,* and *playa lakes* are features often associated with these landscapes.

- For wind erosion to be effective, dryness and scant vegetation are essential. *Deflation*, the lifting and removal of loose material, often produces shallow depressions called *blowouts*.
- *Desert pavement* is a thin layer of coarse pebbles and cobbles that covers some desert surfaces. Once established, it protects the surface from further deflation. Depending on

circumstances, it may develop as a result of deflation or deposition of fine particles.

- *Abrasion*, the sandblasting effect of wind, is often given too much credit for producing desert features. However, abrasion does cut and polish rock near the surface.
- Wind deposits are of two distinct types: (1) extensive *blankets of silt*, called *loess*, carried by wind in *suspension*, and (2) *mounds and ridges of sand*, called *dunes*, which are formed from sediment that is carried as part of the wind's *bed load*.

Key Terms

abrasion (p. 164)
alluvial fan (p. 181)
alpine glacier (p. 158)
arête (p. 165)
barchan dune (p. 187)
barchanoid dune (p. 189)
blowout (p. 183)
cirque (p. 165)
crevasse (p. 161)
cross beds (p. 186)
deflation (p. 183)
desert pavement (p. 183)
drumlin (p. 171)
end moraine (p. 168)
ephemeral stream (p. 179)
esker (p. 171)
fiord (p. 166)
glacial drift (p. 167)

glacial erratic (p. 168)
glacial striations (p. 164)
glacial trough (p. 164)
glacier (p. 158)
ground moraine (p. 169)
hanging valley (p. 165)
horn (p. 165)
ice cap (p. 160)
ice sheet (p. 159)
ice shelf (p. 160)
interior drainage (p. 180)
kame (p. 171)
kettle (p. 170)
lateral moraine (p. 168)
loess (p. 184)
longitudinal dune (p. 189)
medial moraine (p. 168)
outwash plain (p. 169)

parabolic dune (p. 189)
piedmont glacier (p. 160)
playa lake (p. 181)
Pleistocene epoch (p. 174)
plucking (p. 164)
pluvial lake (p. 172)
rock flour (p. 164)
slip face (p. 186)
star dune (p. 189)
stratified drift (p. 168)
till (p. 168)
transverse dune (p. 189)
valley glacier (p. 158)
valley train (p. 169)
zone of accumulation (p. 162)
zone of wastage (p. 162)

Examining the Earth System

1. Assume you are teaching an introductory Earth science class and that you have just assigned Chapter 6 of this text. A student in the class asks why glaciers, deserts, and wind are treated in the same chapter. Formulate a response that connects these topics.

2. Glaciers are solid, but they are a basic part of the hydrologic cycle. Should glaciers be considered a part of the sphere we call the geosphere, or do they belong to the hydrosphere?

3. Some scientists think that ice should be a separate sphere of the Earth system, called the *cryosphere*. Does such an idea have merit?

4. Wind erosion occurs at the interface of the atmosphere, geosphere, and biosphere and is influenced by the hydrosphere and human activity. With this in mind, describe how human activity contributed to the Dust Bowl, the period of intense wind erosion in the Great Plains in the 1930s. You might find it helpful to research Dust Bowl on the Internet using a search engine such as Google (**http://www.google.com/**) or Yahoo (**http://www.yahoo.com/**). You may also find the Wind Erosion Research Unit site to be informative: **http://www.weru.ksu.edu/**.

Mastering Geology

Looking for additional review and test prep materials? Visit the Self Study area in **www.masteringgeology.com** to find practice quizzes, study tools, and multimedia that will aid in your understanding of this chapter's content. In **MasteringGeology™** you will find:

- **Geoscience Animation Library: Over 110 animations illuminating the most difficult-to-understand Earth science concepts**

- **GEODe: Earth Science: An interactive visual walkthrough of each chapter's content**
- **In The News RSS Feeds: Current Earth science events and news articles are pulled into the site with assessment**
- **Pearson eText**
- **Optional Self Study Quizzes**
- **Web Links**
- **Glossary**
- **Flashcards**

7

Plate Tectonics: A Scientific Revolution Unfolds

Hiker below
Mt. Everest's north
face.
(Photo by Jimmy Chin/NGS
Images Collection)

Plate tectonics is the first theory to provide a comprehensive view of the processes that produced Earth's major surface features, including the continents and ocean basins. Within the framework of this theory, geologists have found explanations for the basic causes and distribution of earthquakes, volcanoes, and mountain belts. Furthermore, we are now better able to explain the distribution of plants and animals in the geologic past, as well as the distribution of economically significant mineral deposits.

FOCUS ON CONCEPTS

To assist you in learning **the important concepts in this chapter, focus on the following questions:**

● What evidence was used to support the continental drift hypothesis?
● What was one of the main objections to the continental drift hypothesis?
● What is the theory of plate tectonics?
● In what major way does the plate tectonics theory depart from the continental drift hypothesis?
● What are the three types of plate boundaries?
● Where does new lithosphere form?
● How do mountain systems such as the Himalayas form?
● What type of plate motion occurs along a transform fault boundary?
● What evidence is used to support the plate tectonics theory?
● What are the major driving forces for plate tectonics?
● What models have been proposed to explain the driving mechanism for plate motion?

From Continental Drift to Plate Tectonics

Prior to the 1960s most geologists held the view that the ocean basins and continents had fixed geographic positions and were of great antiquity. Less than a decade later researchers came to realize that Earth's continents are not static; instead, they gradually migrate across the globe. Because of these movements, blocks of continental material collide, deforming the intervening crust, thereby creating Earth's great mountain chains (**Figure 7.1**). Furthermore, landmasses occasionally split apart. As the continental blocks separate, a new ocean basin emerges between them. Meanwhile, other portions of the seafloor plunge into the mantle. In short, a dramatically different model of Earth's tectonic processes emerged.[16]

This profound reversal in scientific thought has been appropriately described as a *scientific revolution*. The revolution began early in the 20th century as a relatively straightforward proposal called the *continental drift hypothesis*. For more than 50 years the idea that continents were capable of movement was categorically rejected by the scientific establishment. Continental drift was particularly distasteful to North American geologists, perhaps because much of the supporting evidence had been gathered from the continents of Africa, South America, and Australia, with which most North American geologists were unfamiliar.

Following World War II, modern instruments replaced rock hammers as the tools of choice for many researchers. Armed with these more advanced tools, geologists and a new breed of researchers, including *geophysicists* and *geochemists*, made several surprising discoveries that began to rekindle interest in the drift hypothesis. By 1968 these developments led to the unfolding of a far more encompassing explanation known as the *theory of plate tectonics*.

In this chapter, we examine the events that led to this dramatic reversal of scientific opinion in an attempt to provide insight into how science works. We also briefly trace the development of the continental drift hypothesis, examine why it was first rejected, and consider the evidence that finally led to the acceptance of its direct descendant—the theory of plate tectonics.

CONCEPT CHECK 7.1

❶ Briefly describe the view held by most geologists regarding the ocean basins and continents prior to the 1960s.
❷ What group of geologists were the least receptive to the continental drift hypothesis? Explain.

Continental Drift: An Idea Before Its Time

The idea that continents, particularly South America and Africa, fit together like pieces of a jigsaw puzzle came about during the 1600s as better world maps became available. However, little

---

[16]Tectonic processes are those that deform Earth's crust to create major structural features such as mountains, continents, and ocean basins.

FIGURE 7.1 Climbers camping on a sheer rock face of a mountain known as K7 in Pakistan's Karakoram, a part of the Himalayas. These mountains formed as India collided with Eurasia. (Photo by Jimmy Chin/National Geographic/Getty)

significance was given to this notion until 1915, when Alfred Wegener (1880–1930), a German meteorologist and geophysicist, wrote *The Origin of Continents and Oceans*. This book, published in several editions, set forth the basic outline of Wegener's hypothesis, called **continental drift**—which dared to challenge the long-held assumption that the continents and ocean basins had fixed geographic positions.

Wegener suggested that a single **supercontinent** consisting of all Earth's landmasses once existed.[17] He named this giant landmass **Pangaea** (pronounced Pan-jee-ah; meaning "all lands") (**Figure 7.2A**). Wegener further hypothesized that about 200 million years ago, during the early part of the Mesozoic era, this supercontinent began to fragment into smaller landmasses. These continental blocks then "drifted" to their present positions over a span of millions of years. The inspiration for continental drift is believed to have come to Wegener when he observed the breakup of sea ice during a Danish-led expedition to Greenland.

Wegener and others who advocated the continental drift hypothesis collected substantial evidence to support their point of view. The fit of South America and Africa and

FIGURE 7.2 Reconstructions of Pangaea as it is thought to have appeared 200 million years ago. **A.** Modern reconstruction. **B.** Wegener's reconstruction redrawn from his book published in 1915.

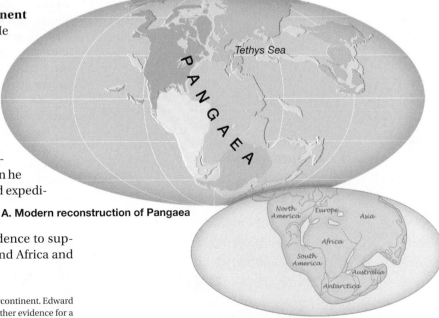

A. Modern reconstruction of Pangaea

B. Wegener's Pangaea

[17]Wegener was not the first person to conceive of a long-vanished supercontinent. Edward Suess (1831–1914), a distinguished 19th-century geologist, pieced together evidence for a giant landmass consisting of the continents of South America, Africa, India, and Australia.

the geographic distribution of fossils and ancient climates all seemed to buttress the idea that these now separate landmasses were once joined. Let us examine some of this evidence.

Evidence: The Continental Jigsaw Puzzle

Like a few others before him, Wegener suspected that the continents might once have been joined when he noticed the remarkable similarity between the coastlines on opposite sides of the Atlantic Ocean. However, Wegener's use of present-day shorelines to fit these continents together was challenged immediately by other Earth scientists. These opponents correctly argued that shorelines are continually modified by wave erosion and depositional processes. Even if continental displacement had taken place, a good fit today would be unlikely. Because Wegener's original jigsaw fit of the continents was crude, it is assumed that he was aware of this problem (Figure 7.2B).

Scientists later determined that a much better approximation of the outer boundary of a continent is the seaward edge of its continental shelf, which lies submerged a few hundred meters below sea level. In the early 1960s, Sir Edward Bullard and two associates constructed a map that pieced together the edges of the continental shelves of South America and Africa at a depth of about 900 meters (Figure 7.3). The remarkable fit that was obtained was more precise than even these researchers had expected. As shown in Figure 7.3 there are a few places where the continents overlap. Some of these overlaps are related to the process of stretching and thinning of the continental margins as they drifted apart. Others can be explained by the work of major river systems. For example, since the breakup of Pangaea the Niger River has built an extensive delta that enlarged the continental shelf of Africa.

FIGURE 7.3 Drawing that shows the best fit of South America and Africa along the continental slope at a depth of 500 fathoms (about 900 meters). The areas where continental blocks overlap appear in orange. (After A. G. Smith, "Continental Drift," in *Understanding the Earth*, edited by I. G. Gass.)

Continental shelf

Modern Equator

SOUTH AMERICA

AFRICA

Modern Equator

Overlap

Evidence: Fossils Match Across the Seas

Although the seed for Wegener's hypothesis came from the remarkable similarities of the continental margins on opposite sides of the Atlantic, it was when he learned that identical fossil organisms had been discovered in rocks from both South America and Africa that his pursuit of continental drift became more focused. Through a review of the literature, Wegener learned that most paleontologists (scientists who study the fossilized remains of ancient organisms) were in agreement that some type of land connection was needed to explain the existence of similar Mesozoic-age life-forms on widely separated landmasses. Just as modern life-forms native to North America are quite different from those of Africa and Australia, one would expect that during the Mesozoic era, organisms on widely separated continents would be distinct.

Mesosaurus   To add credibility to his argument, Wegener documented cases of several fossil organisms that were found on different landmasses despite the unlikely possibility that their living forms could have crossed the vast ocean presently separating them (Figure 7.4). A classic example is *Mesosaurus*, an aquatic fish-catching reptile whose fossil remains are limited to black shales of the Permian period (about 260 million years ago) in eastern South America and southwestern Africa. If *Mesosaurus* had been able to make the long journey across the South Atlantic, its remains would likely be more widely distributed. As this is not the case, Wegener asserted that South America and Africa must have been joined during that period of Earth history.

How did opponents of continental drift explain the existence of identical fossil organisms in places separated by thousands of kilometers of open ocean? Rafting, transoceanic land bridges (isthmian links), and island stepping stones were the most widely invoked explanations for these migrations (Figure 7.5). We know, for example, that during the Ice Age that ended about 8,000 years ago the lowering of sea level allowed mammals (including humans) to cross the narrow Bering Strait that separates Russia and Alaska. Was it possible that land bridges once connected Africa and South America but later subsided below sea level? Modern maps of the seafloor substantiate Wegener's contention that if land bridges of this magnitude once existed, their remnants would still lie below sea level.

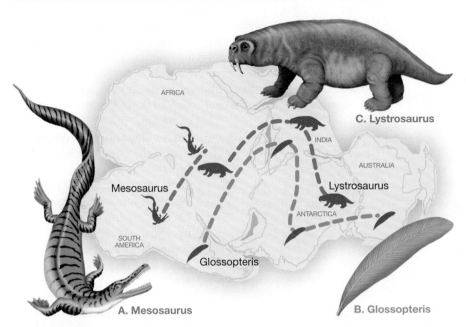

FIGURE 7.4 Fossil evidence supporting continental drift. **A.** Fossils of *Mesosaurus* are found only in nonmarine deposits in eastern South America and western Africa. *Mesosaurus* was a freshwater reptile incapable of swimming the 5,000 kilometers of open ocean that now separate these continents. **B.** Remains of *Glossopteris* and related flora are found in Australia, Africa, South America, Antarctica, and India, landmasses that currently have quite varied climates. However, when *Glossopteris* inhabited these regions during the late Paleozoic era, their climates were all subpolar. **C.** Fossils of *Lystrosaurus*, a land-dwelling reptile, are also found on three of these landmasses.

Glossopteris Wegener also cited the distribution of the fossil "seed fern" *Glossopteris* as evidence for the existence of Pangaea (see Figure 7.4). This plant, identified by its tongue-shaped leaves and seeds that were too large to be carried by the wind, was known to be widely dispersed among Africa, Australia, India, and South America. Later, fossil remains of *Glossopteris* were also discovered in Antarctica.[18] Wegener also learned that these seed ferns and associated flora grew only in a subpolar climate. Therefore, he concluded that when these landmasses were joined, they were located much closer to the South Pole.

Evidence: Rock Types and Geologic Features

Anyone who has worked a jigsaw puzzle knows that its successful completion requires that you fit the pieces together while maintaining the continuity of the picture. The "picture" that must match in the "continental drift puzzle" is one of

[18]In 1912 Captain Robert Scott and two companions froze to death lying beside 35 pounds of rock on their return from a failed attempt to be the first to reach the South Pole. These samples, collected on the moraines of Beardmore Glacier, contained fossil remains of *Glossopteris*.

rock types and geologic features such as mountain belts. If the continents were once together, the rocks found in a particular region on one continent should closely match in age and type those found in adjacent positions on the once adjoining continent. Wegener found evidence of 2.2-billion-year-old igneous rocks in Brazil that closely resembled similarly aged rocks in Africa.

Similar evidence can be found in mountain belts that terminate at one coastline, only to reappear on landmasses across the ocean. For instance, the mountain belt that includes the Appalachians trends northeastward through the eastern United States and disappears off the coast of Newfoundland (**Figure 7.6A**). Mountains of comparable age and structure are found in the British Isles and Scandinavia. When these landmasses are reassembled, as in Figure 7.6B, the mountain chains form a nearly continuous belt.

Wegener described how the similarities in geologic features on both sides of the Atlantic linked these landmasses when he said, "It is just as if we were to refit the torn pieces of a newspaper by matching their edges and then check whether the lines of print run smoothly across. If they do, there is nothing left but to conclude that the pieces were in fact joined in this way."[19]

[19]Alfred Wegener, *The Origin of Continents and Oceans,* translated from the 4th revised German ed. of 1929 by J. Birman (London: Methuen, 1966).

FIGURE 7.5 These sketches by John Holden illustrate various explanations for the occurrence of similar species on landmasses that are presently separated by vast oceans. (Reprinted with permission of John Holden)

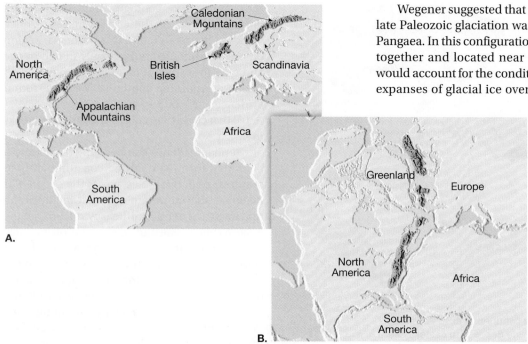

FIGURE 7.6 Matching mountain ranges across the North Atlantic. **A.** The Appalachian Mountains trend along the eastern flank of North America and disappear off the coast of Newfoundland. Mountains of comparable age and structure are found in the British Isles and Scandinavia. **B.** When these landmasses are placed in their pre-drift locations, these ancient mountain chains form a nearly continuous belt.

Wegener suggested that a more plausible explanation for the late Paleozoic glaciation was provided by the supercontinent of Pangaea. In this configuration the southern continents are joined together and located near the South Pole (Figure 7.7C). This would account for the conditions necessary to generate extensive expanses of glacial ice over much of these landmasses. At the same time, this geography would place today's northern continents nearer the equator and account for the tropical swamps that generated the vast coal deposits. Wegener was so convinced that his explanation was correct that he wrote, "This evidence is so compelling that by comparison all other criteria must take a back seat."

How does a glacier develop in hot, arid central Australia? How do land animals migrate across wide expanses of the ocean? As compelling as this evidence may have been, 50 years passed before most of the scientific community accepted the concept of continental drift and the logical conclusions to which it led.

Evidence: Ancient Climates

Because Alfred Wegener was a student of world climates, he suspected that paleoclimatic (*paleo* = ancient, *climatic* = climate) data might also support the idea of mobile continents. His assertion was bolstered when he learned that evidence for a glacial period that dated to the late Paleozoic had been discovered in southern Africa, South America, Australia, and India (Figure 7.7A). This meant that about 300 million years ago, vast ice sheets covered extensive portions of the Southern Hemisphere as well as India (Figure 7.7B). Much of the land area that contains evidence of this period of Paleozoic glaciation presently lies within 30 degrees of the equator in subtropical or tropical climates.

How could extensive ice sheets form near the Equator? One proposal suggested that our planet experienced a period of extreme global cooling. Wegener rejected this explanation because during the same span of geologic time, large tropical swamps existed in several locations in the Northern Hemisphere. The lush vegetation in these swamps was eventually buried and converted to coal. Today, these deposits comprise major coal fields in the eastern United States, northern Europe, and Asia. Many of the fossils found in these coal-bearing rocks were produced by tree ferns that possessed large fronds, a fact consistent with a warm, moist climate. Furthermore, these fern trees lacked growth rings, a characteristic of tropical plants that grow in regions having minimal yearly fluctuations in temperature. By contrast, trees that inhabit the middle latitudes, like those found in most of the United States, develop multiple tree rings—one for each growing season.

CONCEPT CHECK 7.2

1 What was the first line of evidence that led early investigators to suspect the continents were once connected?

2 Explain why the discovery of the fossil remains of *Mesosaurus* in both South America and Africa, but nowhere else, supports the continental drift hypothesis.

3 Early in the 20th century, what was the prevailing view of how land animals migrated across vast expanses of open ocean?

4 How did Wegener account for the existence of glaciers in the southern landmasses at a time when areas in North America, Europe, and Asia supported lush tropical swamps?

Students Sometimes Ask...

Someday will the continents come back together and form a single landmass?

Yes. It is very likely that the continents will come back together, but not anytime soon. Since all of the continents are on the same planetary body, there is only so far a continent can travel before it collides with other continents. Recent research suggests that the continents may form a supercontinent about once every 500 million years or so. Since it has been about 200 million years since Pangaea broke up, we have only about 300 million years before the next supercontinent is completed.

A.

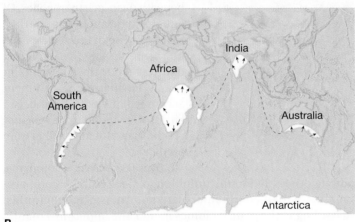

B.

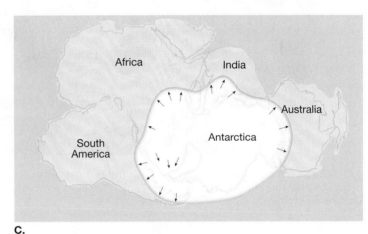

C.

FIGURE 7.7 Paleoclimatic evidence for continental drift. **A.** Glacial striations (scratches) and grooves like these are produced as glaciers drag rock debris across the underlying bedrock. The direction of glacial movement can be deduced from the distinctive patterns of aligned scratches and grooves. (Photo by Gregory S. Springer) **B.** Near the end of the Paleozoic era (about 300 million years ago) ice sheets covered extensive areas of the Southern Hemisphere and India. Arrows show the direction of ice movement that can be inferred from the pattern of glacial striations and grooves found in the bedrock. **C.** The continents restored to their pre-drift positions when they were part of Pangaea. This configuration accounts for the conditions necessary to generate a vast ice sheet and also explains the directions of ice movement that radiated away from an area near the present position of the South Pole.

The Great Debate

Wegener's proposal did not attract much open criticism until 1924, when his book was translated into English, French, Spanish, and Russian. From that point until his death in 1930, the drift hypothesis encountered a great deal of hostile criticism. Respected American geologist R. T. Chamberlain stated, "Wegener's hypothesis in general is of the foot-loose type, in that it takes considerable liberty with our globe, and is less bound by restrictions or tied down by awkward, ugly facts than most of its rival theories. Its appeal seems to lie in the fact that it plays a game in which there are few restrictive rules and no sharply drawn code of conduct."

One of the main objections to Wegener's hypothesis stemmed from his inability to identify a credible mechanism for continental drift. Wegener proposed that gravitational forces of the Moon and Sun that produce Earth's tides were also capable of gradually moving the continents across the globe. However, prominent physicist Harold Jeffreys correctly countered that tidal forces of the magnitude needed to displace the continents would bring Earth's rotation to a halt in a matter of a few years.

Wegener also incorrectly suggested that the larger and sturdier continents broke through thinner oceanic crust, much like ice breakers cut through ice. However, no evidence existed to suggest that the ocean floor was weak enough to permit passage of the continents without the continents being appreciably deformed in the process.

In 1930 Wegener made his fourth and final trip to the Greenland ice sheet. Although the primary focus of this expedition was to study the harsh winter polar climate on the ice-covered island, Wegener continued to test his continental drift hypothesis. As in earlier expeditions, he used astronomical methods in an attempt to verify that Greenland had drifted westward with respect to Europe. While returning from Eismitte (an experimental station located in the center of Greenland), Wegener perished along with a companion. His intriguing idea, however, did not die.

What went wrong? Why was Wegener unable to overturn the established scientific views of his day? Foremost was the fact that, although the central theme of Wegener's drift hypothesis was correct, it contained some incorrect details. For example, continents do not break through the ocean floor, and tidal energy is much

FIGURE 7.8 The movement of Earth's tectonic plates is the cause of our most destructive earthquakes. Pisco, Peru, following a powerful earthquake on August 16, 2007. (Sergio Erday/epa/CORBIS)

too weak to cause continents to be displaced. Moreover, in order for any comprehensive scientific theory to gain wide acceptance, it must stand up to critical testing from all areas of science. Wegener's great contribution to our understanding of Earth notwithstanding, not *all* of the evidence supported the continental drift hypothesis as he had formulated it.

Although many of Wegener's contemporaries opposed his views, even to the point of open ridicule, some considered his ideas plausible. For those geologists who continued the search, the exciting concept of continents adrift held their interest. Others viewed continental drift as a solution to previously unexplainable observations (**Figure 7.8**). Nevertheless, most of the scientific community, particularly in North America, either categorically rejected continental drift or at least treated it with considerable skepticism.

CONCEPT CHECK 7.3

❶ What two aspects of Wegener's continental drift hypothesis were objectionable to most Earth scientists?

Plate Tectonics

Forces Within
▸ Plate Tectonics

Following World War II, oceanographers equipped with new marine tools and ample funding from the U.S. Office of Naval Research embarked on an unprecedented period of oceanographic

exploration. Over the next two decades a much better picture of large expanses of the seafloor slowly and painstakingly began to emerge. From this work came the discovery of a global **oceanic ridge system** that winds through all of the major oceans in a manner similar to the seams on a baseball.

In other parts of the ocean, new discoveries were also being made. Earthquake studies conducted in the western Pacific demonstrated that tectonic activity was occurring at great depths beneath deep-ocean trenches. Of equal importance was the fact that dredging of the seafloor did not bring up any oceanic crust that was older than 180 million years. Furthermore, sediment accumulations in the deep-ocean basins were found to be thin, not the thousands of meters that were predicted.

By 1968, these developments, among others, led to the unfolding of a far more encompassing theory than continental drift, known as **plate tectonics**. According to the plate tectonics model, the uppermost mantle and the overlying crust behave as a strong, rigid layer, known as the **lithosphere**, which is broken into segments commonly referred to as *plates* (**Figure 7.9**). The lithosphere is thinnest in the oceans where it varies from as little as a few kilometers along the axis of the oceanic ridge system to about 100 kilometers (60 miles) in the deep-ocean basins. By contrast, continental lithosphere is generally thicker than 100 kilometers and may extend to a depth of 200–300 kilometers beneath continents.

The lithosphere, in turn, overlies a weak region in the mantle known as the **asthenosphere**. The temperatures and pressures in the upper asthenosphere (100–200 kilometers in depth) are such that the rocks there are very near their melting temperatures and, hence, respond to stress by flowing. As a result, Earth's rigid outer shell is effectively detached from the layers below, which permits it to move independently.

Earth's Major Plates

The lithosphere is composed of about two dozen segments having irregular sizes and shapes called **lithospheric plates** or **tectonic plates** that are in constant motion with respect to one another. As shown in **Figure 7.10**, seven major lithospheric plates are recognized. These plates, which account for 94 percent of Earth's surface area, include the *North American, South American, Pacific, African, Eurasian, Australian-Indian,* and *Antarctic plates*. The largest is the Pacific plate, which encompasses a significant portion of the Pacific Ocean basin. The six other large plates include an entire continent plus a significant amount of ocean floor. Notice in Figure 7.10 that the South American plate encompasses almost all of South America and about one half of the floor of the South Atlantic. This is a major departure from Wegener's continental drift hypothesis, which proposed that the continents moved through the ocean floor, not with it. Note also that none of the plates are defined entirely by the margins of a single continent.

Intermediate-sized plates include the *Caribbean, Nazca, Philippine, Arabian, Cocos, Scotia,* and *Juan de Fuca plates*. These plates, with the exception of the Arabian plate, are composed mostly of oceanic lithosphere. In addition, there are several smaller plates (*microplates*) that have been identified but are not shown in Figure 7.10.

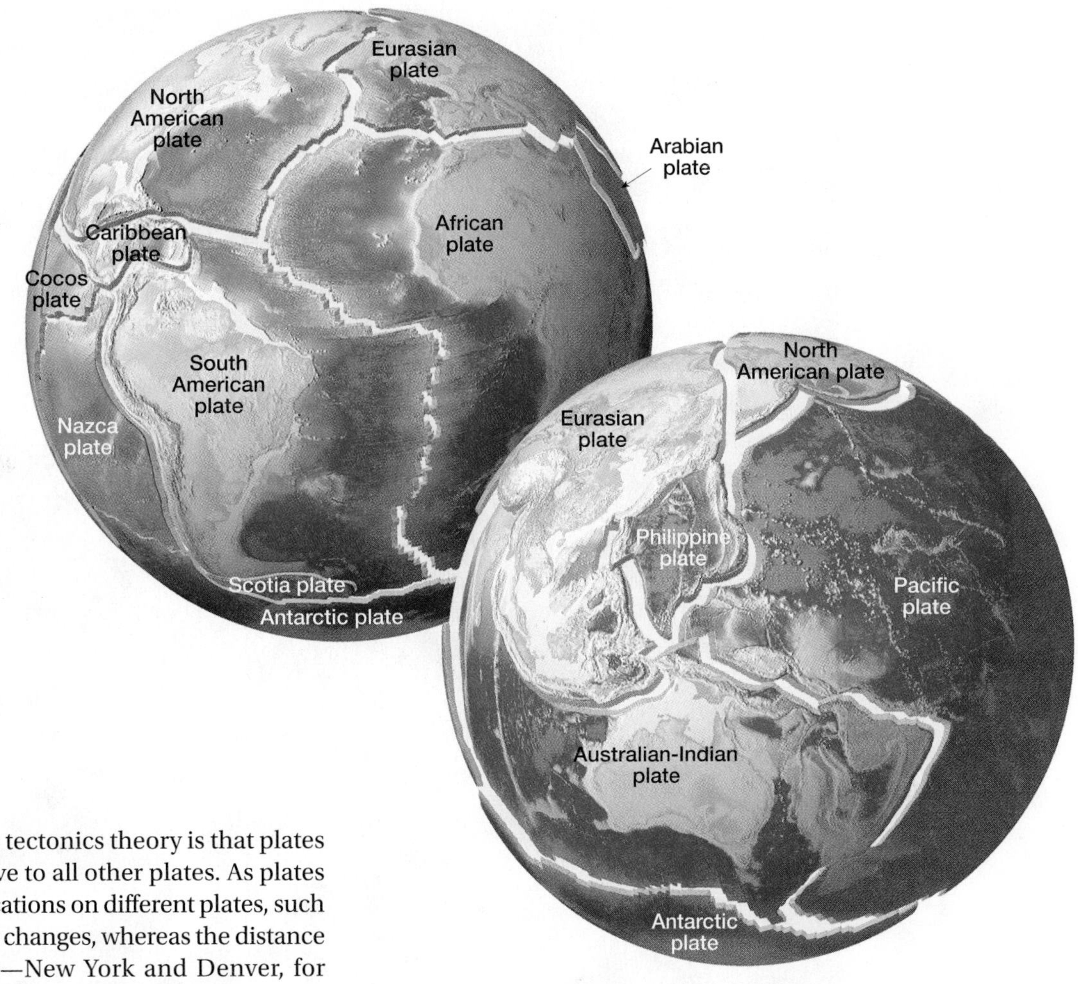

Plate Boundaries

One of the main tenets of the plate tectonics theory is that plates move as semicoherent units relative to all other plates. As plates move, the distance between two locations on different plates, such as New York and London, gradually changes, whereas the distance between sites on the same plate—New York and Denver, for example—remains relatively constant.

Because plates are in constant motion relative to each other, most major interactions among them (and, therefore, most deformation) occur along their *boundaries.* In fact, plate boundaries were first established by plotting the locations of earthquakes and volcanoes. Plates are bounded by three distinct types of boundaries, which are differentiated by the type of movement they exhibit. These boundaries are depicted at the bottom of Figure 7.10 and are briefly described here.

1. **Divergent boundaries** (*constructive margins*)—where two plates move apart, resulting in upwelling of hot material from the mantle to create new seafloor (Figure 7.10A).
2. **Convergent boundaries** (*destructive margins*)—where two plates move together, resulting in oceanic lithosphere descending beneath an overriding plate, eventually to be reabsorbed into the mantle or possibly in the collision of two continental blocks to create a mountain system (Figure 7.10B).
3. **Transform fault boundaries** (*conservative margins*)—where two plates grind past each other without the production or destruction of lithosphere (Figure 7.10C).

Divergent and convergent plate boundaries each account for about 40 percent of all plate boundaries. Transform faults account for the remaining 20 percent. In the following sections we summarize the nature of the three types of plate boundaries.

FIGURE 7.9 Illustration of Earth's major lithospheric plates.

CONCEPT CHECK 7.4

1. What major ocean floor feature was discovered by oceanographers following World War II?
2. Compare and contrast the lithosphere and the asthenosphere.
3. List the seven largest lithospheric plates.
4. List the three types of plate boundaries and describe the relative motion at each of them.

Divergent Boundaries

Forces Within
▶ **Plate Tectonics**

Most **divergent boundaries** are located along the crests of oceanic ridges and can be thought of as *constructive plate margins* because this is where new ocean floor is generated (**Figure 7.11**). Divergent boundaries are also called **spreading centers** because seafloor spreading occurs at these boundaries. Here, two adjacent plates are moving away from each other, producing long, narrow fractures in the ocean crust. As a result, hot rock from the mantle below migrates upward to fill the voids left as the crust is being

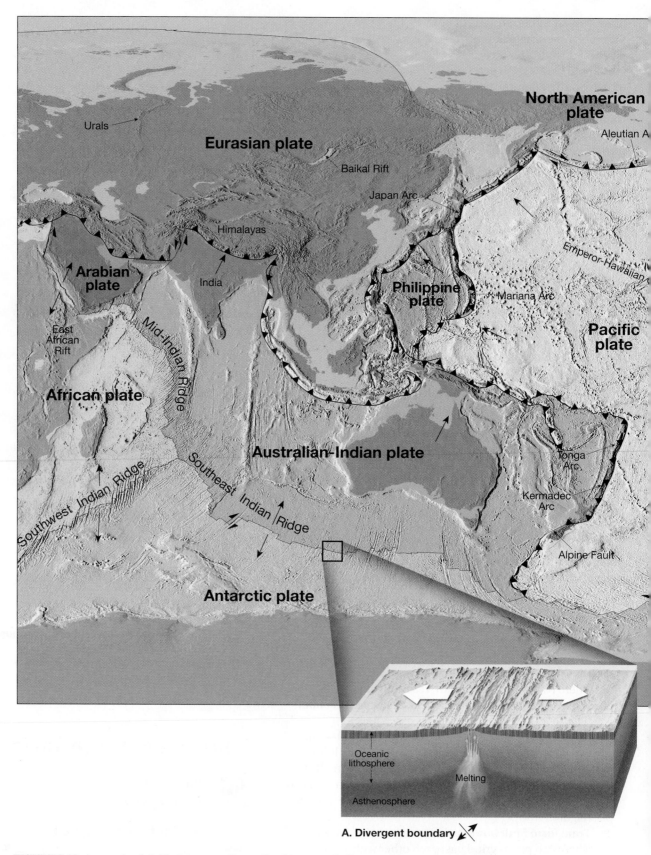

FIGURE 7.10 A mosaic of rigid plates constitutes Earth's outer shell. (After W. B. Hamilton, U.S. Geological Survey)

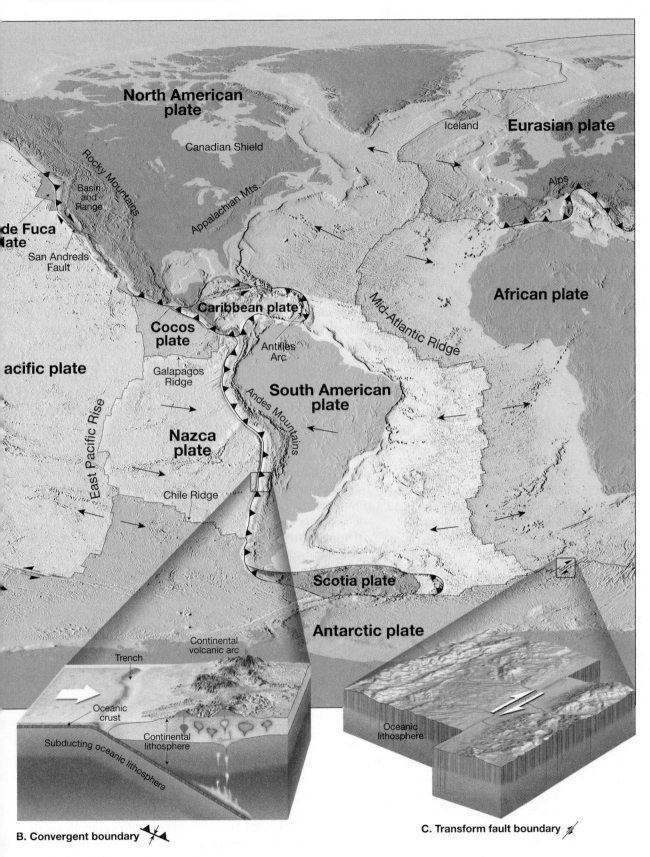

North American
plate

Eurasian plate

Iceland

Canadian Shield

Rocky Mountains

Basin
and
Range

Appalachian Mts.

Alps

de Fuca
late

San Andreas
Fault

African plate

Mid-Atlantic Ridge

Caribbean plate

Cocos
plate

Antilles
Arc

acific plate

Galapagos
Ridge

South American
plate

Andes Mountains

East Pacific Rise

Nazca
plate

Chile Ridge

Scotia plate

Antarctic plate

Continental
volcanic arc

Trench

Oceanic
crust

Continental
lithosphere

Subducting oceanic lithosphere

Oceanic
lithosphere

B. Convergent boundary

C. Transform fault boundary

FIGURE 7.10 Continued

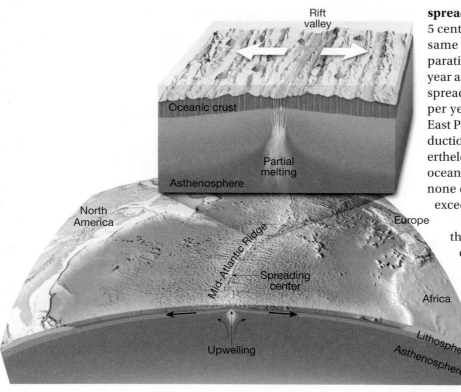

FIGURE 7.11 Most divergent plate boundaries are situated along the crests of oceanic ridges.

spreading. Typical rates of spreading average around 5 centimeters (2 inches) per year. This is roughly the same rate at which human fingernails grow. Comparatively slow spreading rates of 2 centimeters per year are found along the Mid-Atlantic Ridge, whereas spreading rates exceeding 15 centimeters (6 inches) per year have been measured along sections of the East Pacific Rise. Although these rates of seafloor production are slow on a human time scale, they are nevertheless rapid enough to have generated all of Earth's ocean basins within the last 200 million years. In fact, none of the ocean floor that has been dated thus far exceeds 180 million years in age.

The primary reason for the elevated position of the oceanic ridge is that newly created oceanic crust is hot, making it less dense than cooler rocks found away from the ridge axis. As soon as new lithosphere forms, it is slowly yet continually displaced away from the zone of upwelling. Thus, it begins to cool and contract, thereby increasing in density. This thermal contraction accounts for the increase in ocean depths away from the ridge crest. It takes about 80 million years for the temperature of the crust to stabilize and contraction to cease. By this time, rock that was once part of the elevated oceanic ridge system is located in the deep-ocean basin, where it may be buried by substantial accumulations of sediment.

In addition, cooling strengthens the hot material directly below the oceanic crust, thereby adding to the plate's thickness. Stated another way, the thickness of oceanic lithosphere is age-dependent. The older (cooler) it is, the greater its thickness. Oceanic lithosphere that exceeds 80 million years in age is about 100 kilometers thick, about its maximum thickness.

ripped apart. This molten material gradually cools to produce new slivers of seafloor. In a slow, yet unending manner, adjacent plates spread apart and new oceanic lithosphere forms between them.

Oceanic Ridges and Seafloor Spreading

Most divergent plate boundaries are associated with *oceanic ridges:* elevated areas of the seafloor that are characterized by high heat flow and volcanism. The global ridge system is the longest topographic feature on Earth's surface, exceeding 70,000 kilometers (43,000 miles) in length. As shown in Figure 7.10 various segments of the global ridge system have been named, including the Mid-Atlantic Ridge, East Pacific Rise, and Mid-Indian Ridge.

Representing 20 percent of Earth's surface, the oceanic ridge system winds through all major ocean basins like the seam on a baseball. Although the crest of the oceanic ridge is commonly 2–3 kilometers higher than the adjacent ocean basins, the term "ridge" may be misleading because this feature is not narrow but has widths that vary from 1,000 to more than 4,000 kilometers. Furthermore, along the axis of some ridge segments is a deep down-faulted structure called a **rift valley**. This structure is evidence that tensional forces are actively pulling the ocean crust apart at the ridge crest.

The mechanism that operates along the oceanic ridge system to create new seafloor is appropriately called **seafloor**

Continental Rifting

Divergent boundaries can also develop within a continent, in which case the landmass may split into two or more smaller segments separated by an ocean basin. Continental rifting occurs where opposing tectonic forces act to pull the lithosphere apart. The initial stage of rifting tends to include mantle upwelling that is associated with broad upwarping of the overlying lithosphere (**Figure 7.12A**). As a result, the lithosphere is stretched, causing the brittle crustal rocks to break into large slabs. As the tectonic forces continue to pull the crust apart, these crustal fragments sink, generating an elongated depression called a **continental rift** (Figure 7.12B).

A modern example of an active continental rift is the East African Rift (**Figure 7.13**). Whether this rift will eventually result in the breakup of Africa is a topic of continued research. Nevertheless, the East African Rift is an excellent model of the initial stage in the breakup of a continent. Here, tensional forces have stretched and thinned the crust, allowing molten rock to ascend from the mantle. Evidence for recent volcanic activity includes several large volcanic mountains including Mount Kilimanjaro

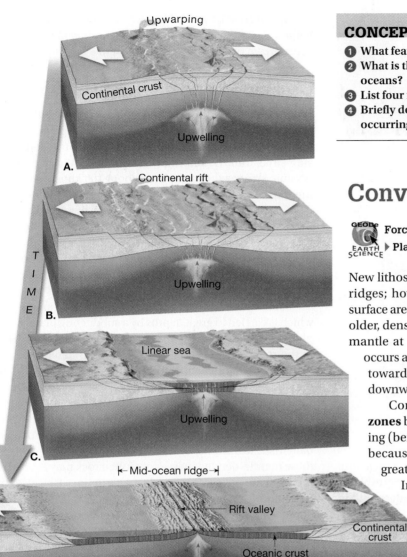

FIGURE 7.12 Continental rifting and the formation of a new ocean basin. **A.** The initial stage of continental rifting tends to include upwelling in the mantle that is associated with broad upwarping of the lithosphere. Tensional forces and buoyant uplifting of the heated lithosphere cause the crust to be broken into large slabs. **B.** As the crust is pulled apart, these large blocks sink, generating a continental rift valley. **C.** Further spreading generates a narrow sea similar to the present-day Red Sea. **D.** Eventually, an expansive deep-ocean basin and oceanic ridge are created.

and Mount Kenya, the tallest peaks in Africa. Research suggests that if rifting continues, the rift valley will lengthen and deepen, eventually extending out to the margin of the landmass (Figure 7.12C). At this point, the rift will become a narrow sea with an outlet to the ocean. The Red Sea, which formed when the Arabian Peninsula split from Africa, is a modern example of such a feature. Consequently, the Red Sea provides us with a view of how the Atlantic Ocean may have looked in its infancy (Figure 7.12D).

CONCEPT CHECK 7.5

1 What feature is generated along *spreading centers*?
2 What is the average rate of seafloor spreading in modern oceans?
3 List four facts that characterize the oceanic ridge system.
4 Briefly describe the process of continental rifting. Where is it occurring today?

Convergent Boundaries

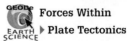 **Forces Within**
▸ **Plate Tectonics**

New lithosphere is constantly being produced at the oceanic ridges; however, our planet is not growing larger—its total surface area remains constant. A balance is maintained because older, denser portions of oceanic lithosphere descend into the mantle at a rate equal to seafloor production. This activity occurs along convergent boundaries, where two plates move toward each other and the leading edge of one is bent downward, as it slides beneath the other.

Convergent boundaries are also called **subduction zones** because they are sites where lithosphere is descending (being subducted) into the mantle. Subduction occurs because the density of the descending tectonic plate is greater than the density of the underlying asthenosphere. In general, oceanic lithosphere is more dense than the asthenosphere, whereas continental lithosphere is less dense and resists subduction. As a consequence, only oceanic lithosphere will subduct to great depths.

Deep-ocean trenches are the surface manifestations produced as oceanic lithosphere descends into the mantle (**Figure 7.14**). These large linear depressions are remarkably long and deep. The Peru–Chile trench along the west coast of South America is more than 4,500 kilometers (3,000 miles) in length and its base is as much as 8 kilometers (5 miles) below sea level. The trenches in the western Pacific, including the Mariana and Tonga trenches, tend to be even deeper than those of the eastern Pacific.

Slabs of oceanic lithosphere descend into the mantle at angles that vary from a few degrees to nearly vertical (90 degrees). The angle at which oceanic lithosphere descends depends largely on its density. For example, when a spreading center is located near a subduction zone, as is the case along the Peru–Chile trench, the subducting lithosphere is young and, therefore, warm and buoyant. Because of this, the angle of descent is small, which results in considerable interaction between the descending slab and the overriding plate. Consequently, the region around the Peru–Chile trench experiences great earthquakes, including the 2010 Chilean earthquake—one of the 10 largest on record.

As oceanic lithosphere ages (gets farther from the spreading center), it gradually cools, which causes it to thicken and increase in density. In parts of the western Pacific, some oceanic lithosphere

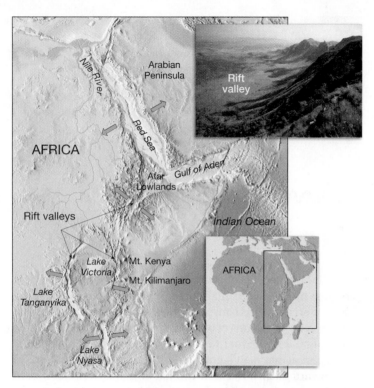

FIGURE 7.13 East African rift valleys and associated features. Photo shows steep cliffs along the East African Rift in Kenya. (Photo by David Keith Jones/Images of Africa Photobank/Alamy)

is 180 million years old. This is the thickest and densest in today's oceans. The very dense slabs in this region typically plunge into the mantle at angles approaching 90 degrees. This largely explains the fact that most trenches in the western Pacific are deeper than trenches in the eastern Pacific.

Although all convergent zones have the same basic characteristics, they are highly variable features. Each is controlled by

the type of crustal material involved and the tectonic setting. Convergent boundaries can form between *two oceanic plates, one oceanic and one continental plate,* or *two continental plates.*

Oceanic–Continental Convergence

Whenever the leading edge of a plate capped with continental crust converges with a slab of oceanic lithosphere, the buoyant continental block remains "floating," while the denser oceanic slab sinks into the mantle (**Figure 7.15A**). When a descending oceanic slab reaches a depth of about 100 kilometers (60 miles), melting is triggered within the wedge of hot asthenosphere that lies above it. But how does the subduction of a cool slab of oceanic lithosphere cause mantle rock to melt? The answer lies in the fact that water contained in the descending plates acts like salt does to melt ice. That is, "wet" rock in a high-pressure environment melts at substantially lower temperatures than does "dry" rock of the same composition.

Sediments and oceanic crust contain a large amount of water, which is carried to great depths by a subducting plate. As the plate plunges downward, heat and pressure drive water from the voids in the rock. At a depth of roughly 100 kilometers, the wedge of mantle rock is sufficiently hot that the introduction of water from the slab below leads to some melting. This process, called **partial melting**, is thought to generate about 10 percent molten material, which is intermixed with unmelted mantle rock. Being less dense than the surrounding mantle, this hot mobile material gradually rises toward the surface. Depending on the environment, these mantle-derived masses of molten rock may ascend through the crust and give rise to a volcanic eruption. However, much of this material never reaches the surface; rather, it solidifies at depth—a process that thickens the crust.

The volcanoes of the towering Andes are the product of molten rock generated by the subduction of the Nazca plate beneath the

FIGURE 7.14 Distribution of the world's oceanic trenches.

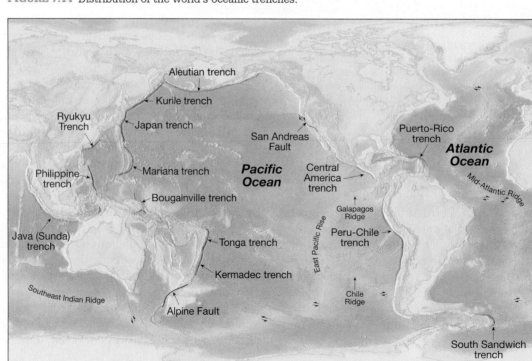

South American continent (Figure 7.15B). Mountain systems, such as the Andes, which are produced in part by volcanic activity associated with the subduction of oceanic lithosphere, are called **continental volcanic arcs**. The Cascade Range in Washington, Oregon, and California is another that consists of several well-known volcanic mountains, including Mount Rainier, Mount Shasta, and Mount St. Helens. This active volcanic arc also extends into Canada, where it includes Mount Garibaldi, Mount Silverthrone, and others.

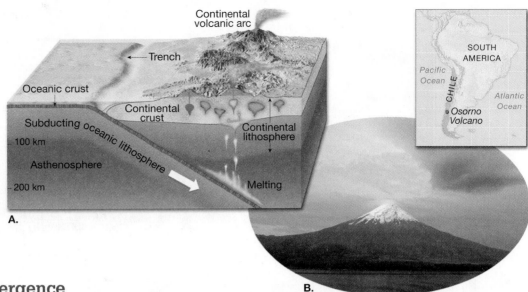

FIGURE 7.15 Oceanic–continental convergent plate boundary. **A.** Illustration of dense, oceanic lithosphere subducting beneath a buoyant continental block. Melting in the asthenosphere generates molten rock that rises toward the surface. This activity produces a chain of structures built on the overriding landmass, called a continental volcanic arc. **B.** Osorno Volcano is one of the most active volcanoes of the southern Chilean Andes, having erupted 11 times between 1575 and 1869. Located on the shore of Lake Llanquihue, Osorno is similar in appearance to Mount Fuji, Japan. (Photo by Michael Collier)

Oceanic–Oceanic Convergence

An *oceanic–oceanic convergent boundary* has many features in common with oceanic–continental plate margins. Where two oceanic slabs converge, one descends beneath the other, initiating volcanic activity by the same mechanism that operates at all subduction zones (**Figure 7.16A**). Water squeezed from the subducting slab of oceanic lithosphere triggers melting in the hot wedge of mantle rock above. In this setting, volcanoes grow up from the ocean floor, rather than upon a continental platform. When subduction is sustained, it will eventually build a chain of volcanic structures large enough to emerge as islands. The volcanic islands tend to be spaced at intervals of about 80 kilometers (50 miles). This newly formed land consisting of an arc-shaped chain of volcanic islands is called a **volcanic island arc**, or simply an **island arc** (Figure 7.16B).

The Aleutian, Mariana, and Tonga islands are examples of relatively young volcanic island arcs. Island arcs are generally located 100–300 kilometers (60–200 miles) from a deep-ocean trench. Located adjacent to the island arcs just mentioned are the Aleutian trench, the Mariana trench, and the Tonga trench.

Most volcanic island arcs are located in the western Pacific. Only two are located in the Atlantic—the Lesser Antilles arc, on the eastern margin of the Caribbean Sea, and the Sandwich Islands located off the tip of South America. The Lesser Antilles are a product of the subduction of the Atlantic seafloor beneath the Caribbean plate. Located within this volcanic arc are the United States and British Virgin Islands as well as the island of Martinique, where Mount Pelée erupted in 1902, destroying the town of St. Pierre and killing an estimated 28,000 people. This chain of islands also includes Montserrat, where there has been recent volcanic activity. (More on these volcanic events is found in Chapter 9.)

Relatively young island arcs are fairly simple structures made of numerous volcanic cones that are underlain by oceanic crust that is

FIGURE 7.16 Oceanic–oceanic convergent plate boundaries. **A.** When oceanic plates converge, one descends beneath the other, initiating volcanic activity in the overriding plate. In this setting a volcanic island arc is built. **B.** These four volcanic structures are part of the Aleutian Islands, a string of both active and dormant volcanoes fed by magma created by the subduction of the Pacific Plate. Steam plumes have recently been observed emanating from Cleveland Volcano (center), evidence of recent activity. (Photo courtesy of NASA)

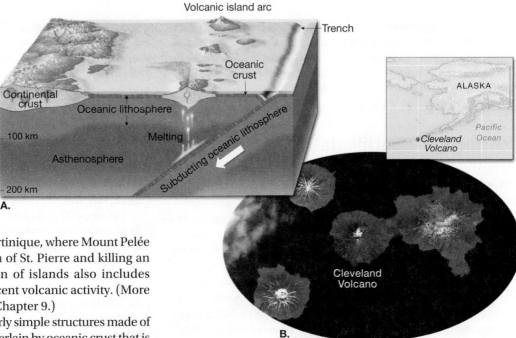

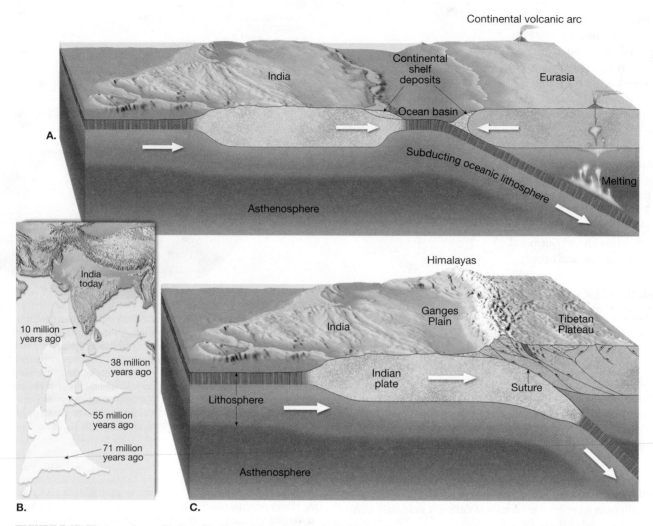

FIGURE 7.17 The ongoing collision of India and Asia began about 50 million years ago—producing the majestic Himalayas. **A.** As India migrated northward the intervening ocean closed as the seafloor was subducted beneath Eurasia. **B.** Position of India in relation to Eurasia at various times. (Modified after Peter Molnar) **C.** Eventually the two landmasses collided, deforming and elevating the sediments that had been deposited along their continental margins. In addition, slices of the crustal rocks were thrust onto the colliding plates.

generally less than 20 kilometers (12 miles) thick. By contrast, older island arcs are more complex and are underlain by highly deformed crust that may reach 35 kilometers in thickness. Examples include the islands that make up the countries of Japan, Indonesia, and the Philippines. These island arcs are built upon material generated by earlier episodes of subduction or on small slivers of continental crust.

Continental–Continental Convergence

The third type of convergent boundary results when one landmass moves toward the margin of another because of subduction of the intervening seafloor (Figure 7.17A). Whereas oceanic lithosphere tends to be dense and sink into the mantle, the buoyancy of continental material inhibits it from being subducted. Consequently, a collision between two converging continental fragments ensues (Figure 7.17C). This event folds and deforms the accumulation of sediments and sedimentary rocks along the continental margins as if they had been placed in a gigantic vise. The result is the formation of a new mountain range composed of deformed sedimentary and metamorphic rocks that often contain slivers of oceanic crust.

Such a collision began about 50 million years ago when the subcontinent of India "rammed" into Asia, producing the Himalayas—the most spectacular mountain range on Earth (Figure 7.17B). During this collision, the continental crust buckled and fractured and was generally shortened and thickened. In addition to the Himalayas, several other major mountain systems, including the Alps, Appalachians, and Urals, formed as continental fragments collided (Figure 7.18). (This topic is considered further in Chapter 10.)

CONCEPT CHECK 7.6

1. Explain why the rate of lithosphere production roughly balances the rate at which it is destroyed?
2. Compare a continental volcanic arc and a volcanic island arc.
3. Describe the process that leads to the formation of deep-ocean trenches.
4. Why does oceanic lithosphere subduct while continental lithosphere does not?
5. Briefly describe how mountain systems such as the Himalayas form.

FIGURE 7.18 The Alps were created by the collision of the African and Eurasian plates. This peak, called Aiguille du Midi, is in the French Alps. (Photo by Radius Images/Photolibrary)

Transform Fault Boundaries

Forces Within
▶ **Plate Tectonics**

Along a transform fault boundary, plates slide horizontally past one another without the production or destruction of lithosphere (*conservative plate margins*). The nature of transform faults was discovered in 1965 by Canadian geologist J. Tuzo Wilson, who proposed that these large faults connected two spreading centers (divergent boundaries) or less commonly two trenches (convergent boundaries). Most transform faults are found on the ocean floor (Figure 7.19A). Here they offset segments of the

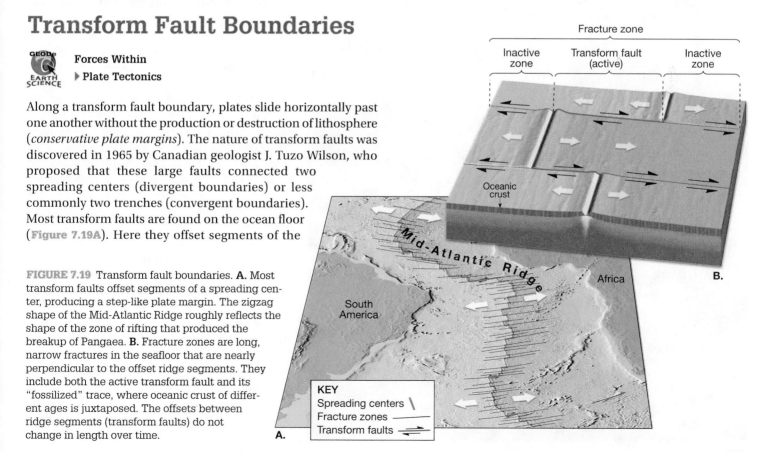

FIGURE 7.19 Transform fault boundaries. **A.** Most transform faults offset segments of a spreading center, producing a step-like plate margin. The zigzag shape of the Mid-Atlantic Ridge roughly reflects the shape of the zone of rifting that produced the breakup of Pangaea. **B.** Fracture zones are long, narrow fractures in the seafloor that are nearly perpendicular to the offset ridge segments. They include both the active transform fault and its "fossilized" trace, where oceanic crust of different ages is juxtaposed. The offsets between ridge segments (transform faults) do not change in length over time.

209

oceanic ridge system, producing a step-like plate margin. Notice that the zigzag shape of the Mid-Atlantic Ridge in Figure 7.10 roughly reflects the shape of the original rifting that caused the breakup of the supercontinent of Pangaea. (Compare the shapes of the continental margins of the landmasses on both sides of the Atlantic with the shape of the Mid-Atlantic Ridge.)

Typically, transform faults are part of prominent linear breaks in the seafloor known as **fracture zones**, which include both the active transform faults as well as their inactive extensions into the plate interior (Figure 7.19B). Active transform faults lie *only between* the two offset ridge segments and are generally defined by weak, shallow earthquakes. Here seafloor produced at one ridge axis moves in the opposite direction of seafloor produced at an opposing ridge segment. Thus, between the ridge segments these adjacent slabs of oceanic crust are grinding past each other along a transform fault. Beyond the ridge crests are inactive zones, where the fractures are preserved as linear topographic depressions. The trend of these fracture zones roughly parallels the direction of plate motion at the time of their formation. Thus, these structures are useful in mapping the direction of plate motion in the geologic past.

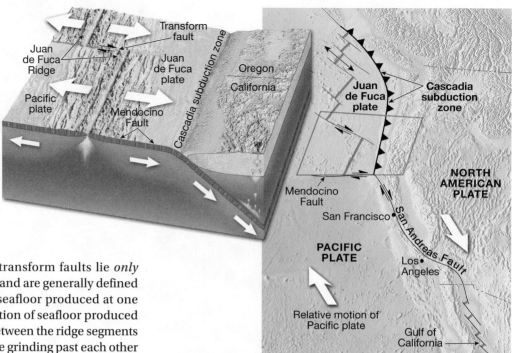

FIGURE 7.20 The Mendocino transform fault permits seafloor generated at the Juan de Fuca ridge to move southeastward past the Pacific plate and beneath the North American plate. Thus, this transform fault connects a spreading center (divergent boundary) to a subduction zone (convergent boundary). Furthermore, the San Andreas Fault, also a transform fault, connects two spreading centers: the Juan de Fuca ridge and a divergent zone located in the Gulf of California.

FIGURE 7.21 Along the San Andreas Fault, the Pacific plate is moving toward the northwest, relative to the North American plate. This aerial view shows the offset in the dry channel of Wallace Creek near Taft, California. (Photo by Michael Collier)

In another role, transform faults provide the means by which the oceanic crust created at ridge crests can be transported to a site of destruction—the deep-ocean trenches. **Figure 7.20** illustrates this situation. Notice that the Juan de Fuca plate moves in a southeasterly direction, eventually being subducted under the west coast of the United States. The southern end of this plate is bounded by a transform fault called the Mendocino Fault. This transform boundary connects the Juan de Fuca ridge to the Cascadia subduction zone. Therefore, it facilitates the movement of the crustal material created at the Juan de Fuca ridge to its destination beneath the North American continent.

Like the Mendocino Fault, most transform fault boundaries are located within the ocean basins; however, a few cut through continental crust. Two examples are the earthquake-prone San Andreas Fault of California and New Zealand's Alpine Fault. Notice in Figure 7.20 that the San Andreas Fault connects a spreading center located in the Gulf of California to the Cascadia subduction zone and the Mendocino Fault located along the northwest coast of the United States. Along the San Andreas Fault, the Pacific plate is moving toward the northwest, past the North American plate (**Figure 7.21**). If this movement continues,

If the continents move, do other features like segments of the oceanic ridge also move?

That's a good observation, and yes, they do! It is interesting to note that very little is really fixed in place on Earth's surface. When we talk about movement of features on Earth, we must consider the question, moving relative to what? Certainly, the oceanic ridge does move relative to the continents (which sometimes causes segments of the oceanic ridges to be subducted beneath the continents). In addition, the oceanic ridge is moving relative to a fixed location beyond Earth. This means that an observer orbiting above Earth would notice, after only a few million years, that all continental and seafloor features—as well as plate boundaries—are indeed moving.

that part of California west of the fault zone, including the Baja Peninsula of Mexico, will become an island off the West Coast of the United States and Canada. It could eventually reach Alaska. However, a more immediate concern is the earthquake activity triggered by movements along this fault system.

CONCEPT CHECK 7.7

❶ Sketch or describe how two plates move in relation to each other along a transform fault boundary.

❷ Differentiate between transform faults and the two other types of plate boundaries.

Testing the Plate Tectonics Model

Some of the evidence supporting continental drift and seafloor spreading has already been presented. With the development of the theory of plate tectonics, researchers began testing this new model of how Earth works. Although new supporting data were obtained, it was often new interpretations of already existing data that swayed the tide of opinion.

Evidence: Ocean Drilling

Some of the most convincing evidence for seafloor spreading came from the Deep Sea Drilling Project, which operated from 1968 until 1983. One of the early goals was to gather samples of the ocean floor in order to establish its age. To accomplish this, the *Glomar Challenger*, a drilling ship capable of working in water thousands of meters deep, was built. Hundreds of holes were drilled through the layers of sediments that blanket the ocean crust, as well

as into the basaltic rocks below. Rather than radiometrically dating the crustal rocks, researchers used the fossil remains of microorganisms found in the sediments resting directly on the crust to date the seafloor at each site.[20]

When the oldest sediment from each drill site was plotted against its distance from the ridge crest, the plot showed that the sediments increased in age with increasing distance from the ridge (**Figure 7.22**). This finding supported the seafloor-spreading hypothesis, which predicted that the youngest oceanic crust would be found at the ridge crest, the site of seafloor production, and the oldest oceanic crust would be located adjacent to the continents.

The data collected by the Deep Sea Drilling Project also reinforced the idea that the ocean basins are geologically young because no seafloor with an age in excess of 180 million years was ever found. By comparison, most continental crust exceeds several hundred million years in age and some has been located that exceeds 4 billion years in age.

The thickness of ocean-floor sediments provided additional verification of seafloor spreading. Drill cores from the *Glomar Challenger* revealed that sediments are almost entirely absent on the ridge crest and that sediment thickness increases with increasing distance from the ridge (Figure 7.22). This pattern of sediment distribution should be expected if the seafloor-spreading hypothesis is correct.

The Ocean Drilling Program, the successor to the Deep Sea Drilling Project, employed a more technologically advanced drilling ship, the *JOIDES Resolution*, to continue the work of the *Glomar Challenger* (**Figure 7.23**). While the Deep Sea Drilling Project validated many of the major tenets of the theory of plate tectonics, the *JOIDES Resolution* was able to probe deeper into the oceanic crust. This allowed for the study of earthquake-generating zones

---

[20]Radiometric dates of the ocean crust itself are unreliable because of the alteration of basalt by seawater.

FIGURE 7.22 Since 1968 drilling ships have gathered core samples of seafloor sediment and crustal rocks at hundreds of sites. Results from these efforts showed that the ocean floor is indeed youngest at the ridge axis. This was the first direct evidence supporting the seafloor spreading hypothesis and the broader theory of plate tectonics.

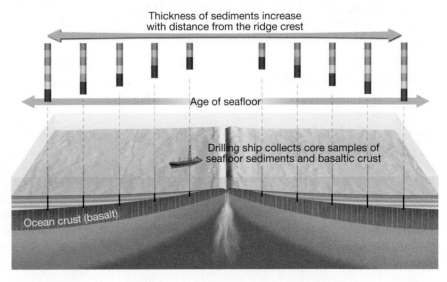

Thickness of sediments increase with distance from the ridge crest

Age of seafloor

Drilling ship collects core samples of seafloor sediments and basaltic crust

Ocean crust (basalt)

A.

B.

FIGURE 7.23 The *JOIDES Resolution*. **A.** *JOIDES Resolution*, one of the drilling ships of the Integrated Ocean Drilling Program. (Photo courtesy of Ocean Drilling Program) **B.** The *JOIDES Resolution* has a tall metal derrick that is used to conduct rotary drilling, while the ship's thrusters hold it in a fixed position at sea. Individual sections of drill pipe are fitted together to make a single string of pipe up to 8,200 meters (27,000 feet) long. The drill bit, located at the end of the pipe string, rotates as it is pressed against the ocean bottom and can drill up to 2,100 meters (6,900 feet) into the seafloor. Like twirling a drinking straw into a layer cake, the drilling operation cuts through sediments and rock and retains a cylinder of material (a core sample) on the inside of the hollow pipe, which can then be raised onboard the ship and analyzed in state-of-the-art laboratory facilities.

at convergent plate margins and for the direct examination of oceanic plateaus and seamounts. Sediment cores from the Ocean Drilling Program have also extended our knowledge of long- and short-term climatic changes.

In October 2003, the *JOIDES Resolution* became part of a new program, the Integrated Ocean Drilling Program (IODP). This new international effort uses multiple vessels for exploration, including the massive 210-meter-long (nearly 770-foot-long) *Chikyu* (meaning "planet Earth" in Japanese), which began operations in 2007. One of the goals of the IODP is to recover a complete section of the ocean crust, from top to bottom.

Evidence: Hot Spots

Mapping volcanic islands and seamounts (submarine volcanoes) in the Pacific Ocean revealed several linear chains of volcanic structures. One of the most studied chains consists of at least 129 volcanoes that extend from the Hawaiian Islands to Midway Island and continue northward toward the Aleutian trench (Figure 7.24). Radiometric dating of this structure, called the *Hawaiian Island–Emperor Seamount chain*, showed that the volcanoes increase in age with increasing distance from the "big island" of Hawaii. The youngest volcanic island in the chain (Hawaii), rose from the ocean floor less than a million years ago, whereas Midway Island is 27 million years old, and Suiko Seamount,

near the Aleutian trench, is about 65 million years old (Figure 7.24).

Most researchers are in agreement that a cylindrically shaped upwelling of hot rock, called a **mantle plume**, is located beneath the island of Hawaii. As the hot, rocky plume ascends through the mantle, the confining pressure drops, which triggers partial melting. (This process, called *decompression melting*, is discussed in Chapter 9.) The surface manifestation of this activity is a **hot spot**, an area of volcanism, high heat flow, and crustal uplifting that is a few hundred kilometers across (Figure 7.25). As the Pacific plate moved over the hot spot, a chain of volcanic structures known as a **hot spot track** was built. As shown in Figure 7.24, the age of each volcano indicates how much time has elapsed since it was situated over the mantle plume.

Taking a closer look at the five largest Hawaiian Islands, we see a similar pattern of ages from the volcanically active island of Hawaii, to the inactive volcanoes that make up the oldest island, Kauai (Figure 7.24). Five million years ago, when Kauai was positioned over the hot spot, it was the *only* Hawaiian Island in existence. Visible evidence of the age of Kauai can be seen by examining its extinct volcanoes, which have been eroded into jagged peaks and vast canyons. By contrast, the relatively young island of Hawaii exhibits many fresh lava flows, and one of its five major volcanoes, Kilauea, remains active today.

Research suggests that at least some mantle plumes originate at great depth, perhaps at the core–mantle boundary. Others, however, may have a much shallower origin. Of the 40 or so hot spots that have been identified worldwide, more than a dozen are located near spreading centers. For example, the mantle plume located beneath Iceland is responsible for the vast accumulation of volcanic rocks found along this exposed section of the Mid-Atlantic Ridge.

Evidence: Paleomagnetism

Anyone who has used a compass to find direction knows that Earth's magnetic field has a north and south magnetic pole. Today, these magnetic poles align closely, but not exactly, with the geographic poles. (The geographic poles are located where Earth's rotational axis intersects the surface.) Earth's magnetic field is similar to that produced by a simple bar magnet. Invisible lines of force pass through the planet and extend from one magnetic pole to the other (Figure 7.26). A compass needle, itself a small magnet free to rotate on an axis, becomes aligned with the magnetic lines of force and points to the magnetic poles.

Unlike the pull of gravity, we cannot feel Earth's magnetic field, yet its presence is revealed because it deflects a compass needle.

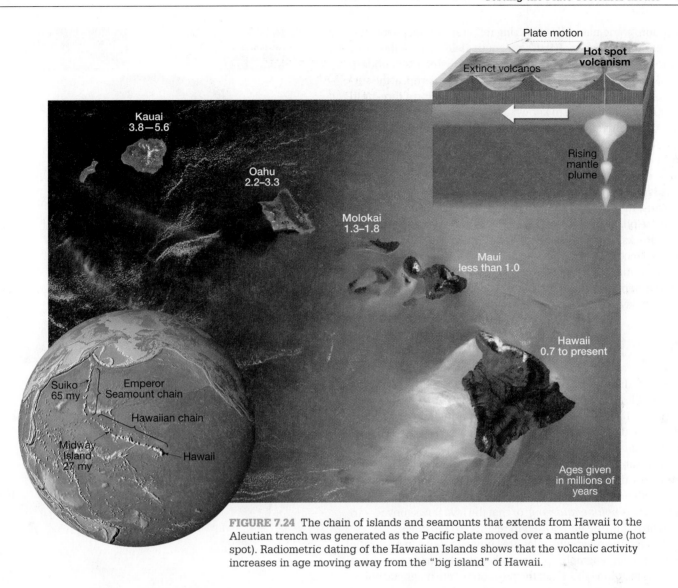

Plate motion

Hot spot volcanism

Extinct volcanos

Rising mantle plume

Kauai
3.8 – 5.6

Oahu
2.2–3.3

Molokai
1.3–1.8

Maui
less than 1.0

Hawaii
0.7 to present

Suiko
65 my

Emperor
Seamount chain

Hawaiian chain

Midway
Island
27 my

Hawaii

Ages given
in millions of
years

FIGURE 7.24 The chain of islands and seamounts that extends from Hawaii to the Aleutian trench was generated as the Pacific plate moved over a mantle plume (hot spot). Radiometric dating of the Hawaiian Islands shows that the volcanic activity increases in age moving away from the "big island" of Hawaii.

FIGURE 7.25 Hot-spot volcanism, Kilauea, Hawaii. (U.S. Geological Survey)

FIGURE 7.26 Earth's magnetic field consists of lines of force much like those a giant bar magnet would produce if placed at the center of Earth.

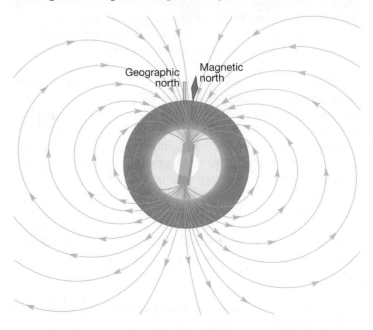

Geographic
north

Magnetic
north

In addition, some naturally occurring minerals are magnetic and hence are influenced by Earth's magnetic field. One of the most common is the iron-rich mineral *magnetite*, which is abundant in lava flows of basaltic composition.[21] Basaltic lavas erupt at the surface at temperatures greater than 1000° C, exceeding a threshold temperature for magnetism known as the **Curie point** (about 585° C). Consequently, the magnetite grains in molten lava are non-magnetic. However, as the lava cools these iron-rich grains become magnetized and align themselves in the direction of the existing magnetic lines of force. Once the minerals solidify, the magnetism they possess will usually remain "frozen" in this position. Thus, they act like a compass needle because they "point" toward the position of the magnetic poles at the time of their formation. Rocks that formed thousands or millions of years ago and contain a "record" of the direction of the magnetic poles at the time of their formation are said to possess **fossil magnetism**, or **paleomagnetism**. During the 1950s paleomagnetic data was collected from lava flows around the globe.

Apparent Polar Wandering

A study of rock magnetism conducted in Europe led to an interesting discovery. The magnetic alignment of iron-rich minerals in lava flows of different ages indicated that the position of the paleomagnetic poles had changed through time. A plot of the location of the magnetic north pole with respect to Europe revealed that during the past 500 million years, the pole had gradually wandered from a location near Hawaii northward to its present location near the North Pole (**Figure 7.27A**). This was strong evidence that either the magnetic poles had migrated, an idea known as *polar wandering*, or that the lava flows moved—in other words, Europe had drifted in relation to the poles.

Although the magnetic poles are known to move in an erratic path around the geographic poles, studies of paleomagnetism from numerous locations show that the positions of the magnetic poles, averaged over thousands of years, correspond closely to the positions of the geographic poles. Therefore, a more acceptable explanation for the apparent polar wandering paths was provided by Wegener's hypothesis. If the magnetic poles remain stationary, their *apparent movement* is produced by continental drift.

Further evidence for continental drift came a few years later when a polar-wandering path was constructed for North America (Figure 7.27A). For the first 300 million years or so, the paths for North America and Europe were found to be similar in direction, but separated by about 5,000 kilometers (3,000 miles). Then, during the middle of the Mesozoic era (180 million years ago) they began to converge on the present North Pole. Because the North Atlantic Ocean is about 5,000 kilometers wide, the explanation for these curves was that North America and Europe were joined until the Mesozoic, when the Atlantic began to open. From this time forward, these continents continuously moved apart. When North America and Europe are moved back to their pre-drift positions, as shown in Figure 7.27B, these apparent wandering paths coincide. This is evidence that North America and Europe were once joined and moved relative to the poles as part of the same continent.

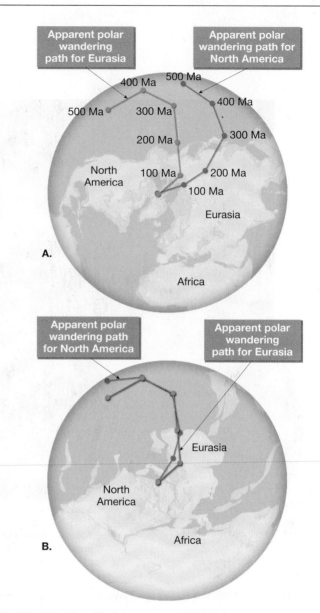

FIGURE 7.27 Simplified apparent polar-wandering paths as established from North American and Eurasian paleomagnetic data. **A.** The more westerly path determined from North American data was caused by the westward movement of North America from Eurasia during the breakup of Pangaea. **B.** The positions of the wandering paths when the landmasses are reassembled.

Magnetic Reversals and Seafloor Spreading

Another discovery came when geophysicists learned that over periods of hundreds of thousands of years, Earth's magnetic field periodically reverses polarity. During a **magnetic reversal** the north magnetic pole becomes the south magnetic pole, and vice versa. Lava solidifying during a period of reverse polarity will be magnetized with the polarity opposite that of volcanic rocks being formed today. When rocks exhibit the same magnetism as the present magnetic field, they are said to possess **normal polarity**, whereas rocks exhibiting the opposite magnetism are said to have **reverse polarity**.

Once the concept of magnetic reversals was confirmed, researchers set out to establish a time scale for these occurrences. The task was to measure the magnetic polarity of hundreds of lava

---

[21]Some sediments and sedimentary rocks contain enough iron-bearing mineral grains to acquire a measurable amount of magnetization.

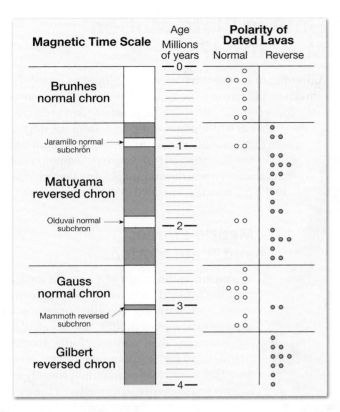

FIGURE 7.28 Time scale of Earth's magnetic field in the recent past. This time scale was developed by establishing the magnetic polarity for lava flows of known age. (Data from Allen Cox and G. B. Dalrymple)

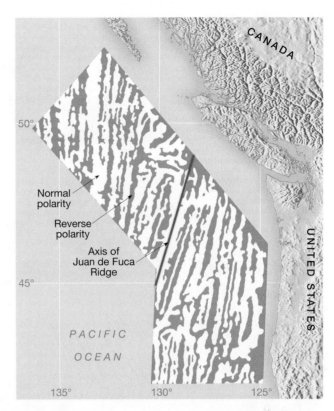

FIGURE 7.29 Pattern of alternating stripes of high- and low-intensity magnetism discovered off the Pacific Coast of North America.

FIGURE 7.30 The ocean floor as a magnetic tape recorder. Magnetic intensities are recorded as a magnetometer is towed across a segment of the oceanic ridge. Notice the symmetrical stripes of low- and high-intensity magnetism that parallel the ridge crest. Vine and Matthews suggested that the stripes of high-intensity magnetism occur where normally magnetized oceanic rocks enhanced the existing magnetic field. Conversely, the low-intensity stripes are regions where the crust is polarized in the reverse direction, which weakens the existing magnetic field.

flows and use radiometric dating techniques to establish the age of each flow. **Figure 7.28** shows the **magnetic time scale** established for the past few million years. The major divisions of the magnetic time scale are called *chrons* and last for roughly 1 million years. As more measurements became available, researchers realized that several, short-lived reversals (less than 200,000 years long) often occurred during a single chron.

Meanwhile, oceanographers had begun to do magnetic surveys of the ocean floor in conjunction with their efforts to construct detailed maps of seafloor topography. These magnetic surveys were accomplished by towing very sensitive instruments, called **magnetometers**, behind research vessels. The goal of these geophysical surveys was to map variations in the strength of Earth's magnetic field that arise from differences in the magnetic properties of the underlying crustal rocks.

The first comprehensive study of this type was carried out off the Pacific coast of North America and had an unexpected outcome. Researchers discovered alternating stripes of high-and low-intensity magnetism, as shown in **Figure 7.29**. This relatively simple pattern of magnetic variation defied explanation until 1963, when Fred Vine and D. H. Matthews demonstrated that the high- and low-intensity stripes supported the concept of seafloor spreading. Vine and Matthews suggested that the stripes of high-intensity magnetism are regions where the paleomagnetism of the ocean crust exhibits normal polarity (**Figure 7.30**). Consequently, these rocks *enhance* (reinforce) Earth's magnetic field. Conversely, the low-intensity stripes are regions where the ocean crust is polarized in the reverse direction and therefore *weaken* the existing magnetic field. But how do parallel stripes of normally and reversely magnetized rock become distributed across the ocean floor?

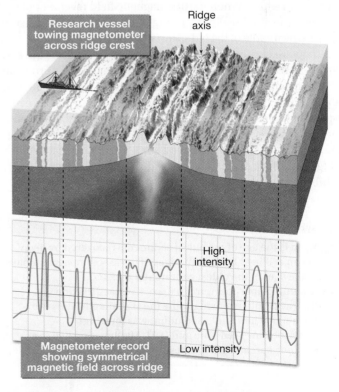

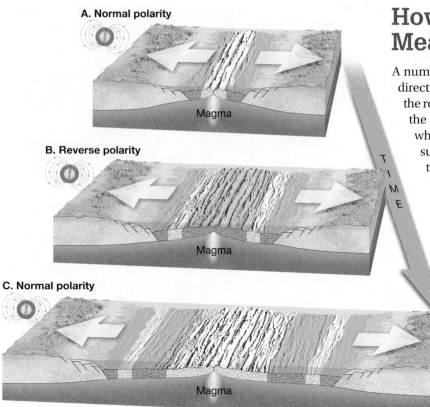

A. Normal polarity

B. Reverse polarity

C. Normal polarity

Magma

TIME

FIGURE 7.31 As new basalt is added to the ocean floor at mid-ocean ridges, it is magnetized according to Earth's existing magnetic field. Hence, it behaves much like a tape recorder as it records each reversal of our planet's magnetic field.

Vine and Matthews reasoned that as magma solidifies along narrow rifts at the crest of an oceanic ridge, it is magnetized with the polarity of the existing magnetic field (**Figure 7.31**). Because of seafloor spreading, this strip of magnetized crust would gradually increase in width. When Earth's magnetic field reverses polarity, any newly formed seafloor (having the opposite polarity) would form in the middle of the old strip. Gradually, the two halves of the old strip are carried in opposite directions away from the ridge crest. Subsequent reversals would build a pattern of normal and reverse magnetic strips, as shown in Figure 7.31. Because new rock is added in equal amounts to both trailing edges of the spreading ocean floor, we should expect the pattern of strips (size and polarity) found on one side of an oceanic ridge to be a mirror image of the other side. A few years later a survey across the Mid-Atlantic Ridge just south of Iceland revealed a pattern of magnetic strips exhibiting a remarkable degree of symmetry to the ridge axis.

CONCEPT CHECK 7.8

1. What is the age of the oldest sediments recovered by deep-ocean drilling? How do the ages of these sediments compare to the ages of the oldest continental rocks?
2. Assuming hot spots remain fixed, in what direction was the Pacific plate moving while the Hawaiian Islands were forming? When Suiko Seamount was forming?
3. Describe how Fred Vine and D. H. Matthews related the seafloor-spreading hypothesis to magnetic reversals.

How is Plate Motion Measured?

A number of methods have been employed to establish the direction and rate of plate motion. Paleomagnetism stored in the rocks of the ocean floor is one method used to determine the speeds at which plates move away from the ridge axes where they were generated. In addition, hot spot tracks, such as the Hawaiian Island–Emperor Seamount chain, trace the speed and direction of plate movement relative to the hot plume embedded in the mantle below.

Mantle Plumes and Plate Motions

By measuring the length of a hot spot track and the time interval between the formation of its oldest and youngest volcanic structures, an average rate of plate motion can be calculated. For example, that portion of the Hawaiian Island–Emperor Seamount chain that extends from Hawaii to Suiko Seamount is roughly 6,000 kilometers in length and formed over the past 65 million years. Thus, the average rate of movement of the Pacific plate, relative to the mantle plume, was about 9 centimeters (4 inches) per year.

Hot spot tracks can also be useful when establishing the direction a plate is moving. Notice in Figure 7.24 that there is a bend in the Hawaiian Island–Emperor Seamount chain. This bend occurred about 50 million years ago when the motion of the Pacific plate changed from one that was nearly due north to a more northwesterly path. Similarly, hot spots found on the floor of the Atlantic have increased our understanding of the migration of landmasses following the breakup of Pangaea.

The existence of mantle plumes and their association with hot spots is well documented. Most mantle plumes are long-lived features that appear to maintain relatively fixed positions within the mantle. However, recent evidence has shown that some hot spots may slowly migrate. Preliminary results suggest that the Hawaiian hot spot may have migrated southward by as much as 20 degrees latitude. If this is the case, models of past plate motion that were based on a "fixed hot spot" frame of reference will need to be reevaluated.

Measuring Plate Motion from Space

Plates are not flat surfaces; instead, they are curved sections of a sphere, which greatly complicates how plate motion is described. In addition, plates usually exhibit some degree of rotational motion, which can cause two locations on the same plate to move at different speeds and in different directions. The latter fact can be illustrated by rotating your dinner plate in a clockwise matter. When doing so you will notice that the items on the left side of the plate move away from you (divergence) as the items on the right side move toward you (convergence). Items in the center will rotate, but their position relative to yours will not change. The complex nature of plate motion makes the task of describing plate

Box 7.1

UNDERSTANDING EARTH

The Breakup of Pangaea

Wegener used evidence from fossils, rock types, and ancient climates to create a jigsaw-puzzle fit of the continents—thereby creating his supercontinent of Pangaea. In a similar manner, but employing modern tools not available to Wegener, geologists have recreated the steps in the breakup of this supercontinent, an event that began nearly 200 million years ago. From this work, the dates when individual crustal fragments separated from one another and their relative motions have become well established (**Figure 7.A**).

An important consequence of the breakup of Pangaea was the creation of a "new" ocean basin: the Atlantic. As you can see in Figure 7.A, part B, splitting of the supercontinent did not occur simultaneously along the margins of the Atlantic. The first split developed between North America and Africa.

Here, the continental crust was highly fractured, providing pathways for huge quantities of fluid lavas to reach the surface. Remnants of these lavas are found along the Eastern Seaboard of the United States—primarily buried beneath younger sedimentary rocks that form the continental shelf. Radiometric dating of these solidified lavas indicates that rifting began in various stages between 180 million and 165 million years ago. This time span can be used as the "birth date" for this section of the North Atlantic.

About 130 million years ago, the South Atlantic began to open near the tip of what is now South Africa. As this zone of rifting migrated northward, it gradually opened the South Atlantic (compare Figure 7.A, parts B and C). Continued breakup in the Southern Hemisphere led to the separation of Africa and Antarctica and sent India on a northward journey. By the early Cenozoic, about 50 million years ago, Australia had separated from Antarctica, and the South Atlantic had emerged as a full-fledged ocean (Figure 7.A, part D).

A modern map (Figure 7.A, part F) shows that India eventually collided with Asia, an event that began about 50 million years ago and created the Himalayas as well as the Tibetan Highlands. About the same time, the separation of Greenland from Eurasia completed the break-up of the northern landmass. During the last 20 million years or so of Earth history, Arabia rifted from Africa to form the Red Sea, and Baja California separated from Mexico to form the Gulf of California (Figure 7.A, part E). Meanwhile, a sliver of land (now known as Central America) was trapped between North America and South America to produce our globe's familiar, modern appearance.

FIGURE 7.A Several views of the breakup of Pangaea over a period of 200 million years.

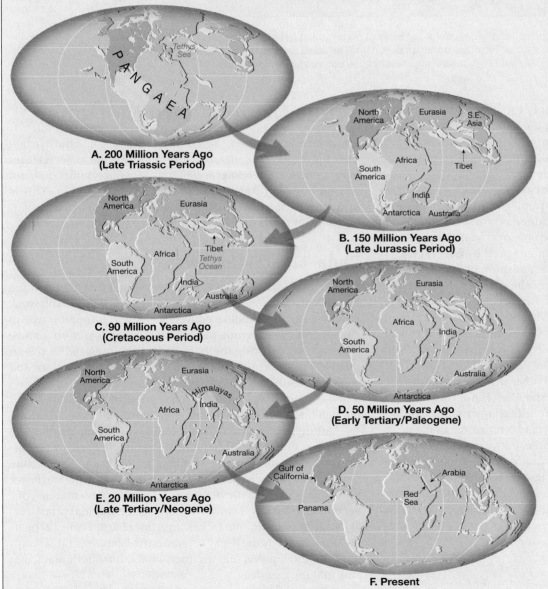

A. 200 Million Years Ago
(Late Triassic Period)

B. 150 Million Years Ago
(Late Jurassic Period)

C. 90 Million Years Ago
(Cretaceous Period)

D. 50 Million Years Ago
(Early Tertiary/Paleogene)

E. 20 Million Years Ago
(Late Tertiary/Neogene)

F. Present

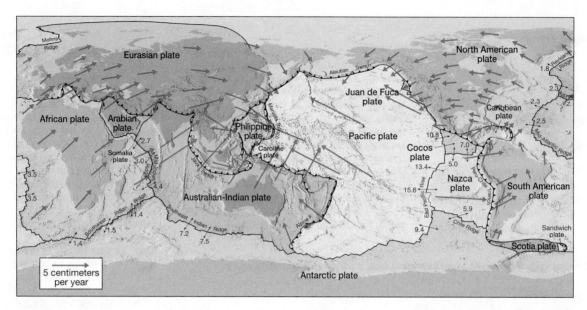

FIGURE 7.32 This map illustrates directions and rates of plate motion in centimeters per year. The red arrows show plate motion at selected locations based on GPS data. The small black arrows and labels show seafloor spreading velocities. (Seafloor data from DeMets and others, GPS data from Jet Propulsion Laboratory)

motions more difficult than simply establishing the relative motion between two plates along the boundary that separates them. Fortunately, using space-age technology, researchers have recently been able to accurately calculate the absolute motion of hundreds of locations across the globe.

You may be familiar with the Global Positioning System (GPS), which is part of the navigation system used in automobiles to locate one's position and to provide directions to some other location. The Global Positioning System employs two dozen satellites that send radio signals that are intercepted by GPS receivers located at Earth's surface. The exact position of the receiver is determined by simultaneously establishing the distance from the receiver to four or more satellites. Researchers use specifically designed equipment that is able to locate the position of a point on Earth to within a few millimeters (about the diameter of a small pea). To establish plate motion, a particular site is surveyed repeatedly over a number of years.

Data obtained from these and other similar techniques are shown in **Figure 7.32**. Calculations show that Hawaii is moving in a northwesterly direction and approaching Japan at 8.3 centimeters per year. A site located in Maryland is retreating from one in England at a speed of 1.7 centimeters per year—a value that is close to the 2.0-centimeters-per-year spreading rate that was established from paleomagnetic evidence. Techniques using GPS devices have also been useful in establishing small-scale crustal movements such as those that occur along faults in regions known to be tectonically active.

CONCEPT CHECK 7.9

❶ Briefly describe how hot spot tracks can be used to determine the rate of plate motion.

❷ Refer to Figure 7.32 and determine the three plates that appear to exhibit the highest rates of motion.

What Drives Plate Motions?

The plate tectonics theory *describes* plate motion and the role that this motion plays in generating and modifying the major features of Earth's crust. Therefore, acceptance of plate tectonics does not rely on knowing precisely what drives plate motion. This is fortunate, because none of the models yet proposed can account for all major facets of plate tectonics.

Plate–Mantle Convection

From geophysical evidence, we have learned that although the mantle consists almost entirely of solid rock, it is hot and weak enough to exhibit fluidlike convective flow. The simplest type of convection is analogous to heating a pot of water on a stove (**Figure 7.33**). Heating the base causes the material to rise in relatively thin sheets or blobs that spread out at the surface and cool. Eventually, the surface layer thickens (increases in density) and sinks back to the bottom where it is reheated until it achieves enough buoyancy to rise again.

Mantle convection is considerably more complex than the model just described. The shape of the mantle does not resemble that of a cooking pot. Rather, it is a spherically shaped zone with a much larger upper boundary (Earth's surface) than lower boundary (core–mantle boundary). Furthermore, mantle convection is driven by a combination of three thermal processes: (1) heating at the bottom by heat loss from Earth's core; (2) heating from within by the decay of radioactive isotopes; and (3) cooling from the top that creates thick, cold lithospheric slabs that sink into the mantle.

When seafloor spreading was first introduced, geologists proposed that the main driving force for plate motion was upwelling that came from deep in the mantle. Upon reaching the base of the lithosphere, this flow was thought to spread laterally and drag the plates along. Thus, plates were viewed as

FIGURE 7.33 Convection is a type of heat transfer that involves the actual movement of a substance. Here the stove warms the water in the bottom of a cooking pot. The heated water expands, becomes less dense (more buoyant), and rises. Simultaneously, the cooler, denser water near the top sinks.

being carried passively by convective flow in the mantle. Based on physical evidence, however, it became clear that upwelling beneath oceanic ridges is quite shallow and not related to deep circulation in the lower mantle. It is the horizontal movement of lithospheric plates away from the ridge that causes mantle upwelling, not the other way around. Thus, modern models have plates being an integral part of mantle convection and perhaps even its most active component.

Although convection in the mantle is still poorly understood, researchers generally agree on the following:

1. Convective flow in the rocky 2,900-kilometer-thick mantle—in which warm, buoyant rock rises and cooler, denser material sinks—is the underlying driving force for plate movement.

2. Mantle convection and plate tectonics are part of the same system. Subducting oceanic plates drive the cold downward-moving portion of convective flow while shallow upwelling of hot rock along the oceanic ridge and buoyant mantle plumes are the upward-flowing arms of the convective mechanism.

3. Convective flow in the mantle is the primary mechanism for transporting heat away from Earth's interior to the surface, where it is eventually radiated into space.

What is not known with any high degree of certainty is the exact structure of this convective flow. First, we look at

some of the forces that contribute to plate motion and then we examine two models that have been proposed to describe plate–mantle convection.

Forces that Drive Plate Motion

There is general agreement that the subduction of cold, dense slabs of oceanic lithosphere is a major driving force of plate motion (Figure 7.34). As these slabs sink into the asthenosphere, they "pull" the trailing plate along. This phenomenon, called **slab pull**, occurs because cold slabs of oceanic lithosphere are more dense than the underlying asthenosphere and hence "sink like a rock."

Another important driving force is **ridge push** (Figure 7.34). This gravity-driven mechanism results from the elevated position of the oceanic ridge, which causes slabs of lithosphere to "slide" down the flanks of the ridge. Ridge push appears to contribute far less to plate motions than slab pull. The primary evidence for this comes from comparing rates of seafloor spreading along ridge segments having different elevations. For example, despite its greater average height above the seafloor, spreading rates along the Mid-Atlantic Ridge are considerably less than spreading rates along the less steep East Pacific Rise (see Figure 7.32). In addition, fast-moving plates are being subducted along a larger percentage of their margins than slow-moving plates. This fact supports the notion that slab pull is a more significant driving force than ridge push. Examples of fast-moving plates that have extensive subduction zones along their margins include the Pacific, Nazca, and Cocos plates.

Although slab pull and ridge push appear to be the dominant forces acting on plates, they are not the only forces that influence plate motion. Beneath plates, convective flow in the mantle exerts a force, perhaps best described as "mantle drag"

FIGURE 7.34 Illustration of some of the forces that act on tectonic plates.

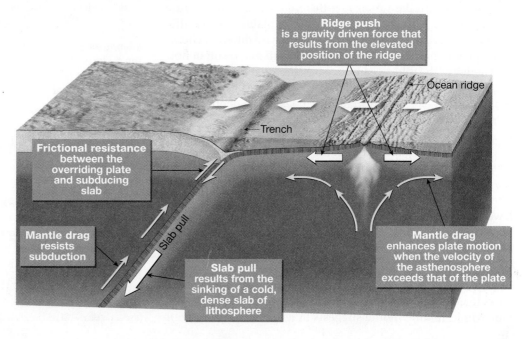

Ridge push is a gravity driven force that results from the elevated position of the ridge

Ocean ridge

Trench

Frictional resistance between the overriding plate and subducing slab

Mantle drag resists subduction

Slab pull

Slab pull results from the sinking of a cold, dense slab of lithosphere

Mantle drag enhances plate motion when the velocity of the asthenosphere exceeds that of the plate

(Figure 7.34). When flow in the asthenosphere is moving at a velocity that exceeds that of the plate, mantle drag enhances plate motion. However, if the asthenosphere is moving more slowly than the plate, or in the opposite direction, this force tends to resist plate motion. Another type of resistance to plate motion occurs along subduction zones. Here friction between the overriding plate and the descending slab generates significant earthquake activity.

Models of Plate–Mantle Convection

Any acceptable model for plate–mantle convection must explain compositional variations known to exist in the mantle. For example, the basaltic lavas that erupt along oceanic ridges, as well as those that are generated by hot-spot volcanism, such as those found in Hawaii, have mantle sources. Yet ocean ridge basalts are very uniform in composition and depleted in certain elements. Hot-spot eruptions, on the other hand, have high concentrations of these elements and tend to have varied compositions. Because basaltic lavas that arise from different tectonic settings have different compositions, they are assumed to be derived from chemically distinct mantle reservoirs.

Layering at 660 Kilometers Some researchers argue that the mantle resembles a "giant layer cake" divided at a depth of 660 kilometers. As shown in **Figure 7.35A**, this layered model has two zones of convection—a thin, dynamic layer in the upper mantle and a thick, sluggish one located below. This model successfully explains why basaltic lavas that erupt along the oceanic ridges have a different chemical makeup than those that erupt in Hawaii as a result of hot-spot activity. The mid-ocean ridge basalts come from the upper convective layer, which is well mixed, whereas the mantle plume that feeds the Hawaiian volcanoes taps a deeper, more primitive magma source that resides in the lower convective layer.

However, data gathered from the study of earthquake waves have shown that at least some subducting oceanic slabs penetrate the 660-kilometer boundary and descend deep into the mantle. The subducting lithosphere should serve to mix the upper and lower layers together, thereby destroying the layered structure proposed in this model.

Whole-Mantle Convection Other researchers favor some type of *whole-mantle convection* in which cold oceanic lithosphere sinks to great depths and stirs the entire mantle (Figure 7.35B). One whole-mantle model suggests that the burial ground for

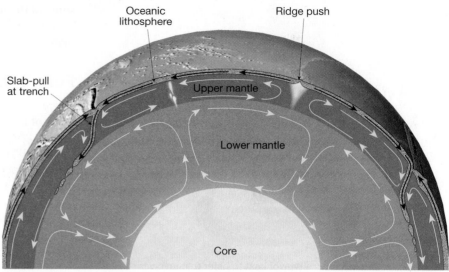

A. Layering at 660 kilometers

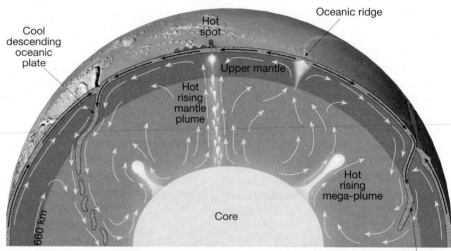

B. Whole mantle convection

FIGURE 7.35 Proposed models for mantle convection. **A.** The "layer cake" model consists of two convection layers—a thin, convective layer above 660 kilometers and a thick one below. **B.** In this whole-mantle convection model, cold oceanic lithosphere descends into the lowermost mantle, while hot mantle plumes transport heat toward the surface.

Students Sometimes Ask...

Will plate tectonics eventually "turn off" and cease to operate on Earth?

Because plate tectonic processes are powered by heat from Earth's interior (which is finite), the forces will slow sometime in the distant future to the point that the plates will cease to move. The work of external processes (wind, water, ice), however, will continue to attack Earth's surface features, most of which will eventually be eroded nearly flat. What a different world it will be—a planet with no earthquakes, no volcanoes, and no mountains. Flatness will prevail!

subducting slabs is the core–mantle boundary. Over time, this material melts and buoyantly rises toward the surface as a mantle plume, thereby transporting hot material toward the surface (Figure 7.35B).

Recent work has predicted that whole-mantle convection would cause the entire mantle to completely mix in a matter of a few hundred million years. This, in turn, would eliminate chemically distinct magma sources—those that are observed in hot-spot volcanism and those associated with volcanic activity along oceanic ridges. Thus, the whole-mantle model also has shortcomings.

Although there is still much to be learned about the mechanisms that cause Earth's tectonic plates to migrate across the globe, one thing is clear. The unequal distribution of heat in Earth's interior generates some type of thermal convection that ultimately drives plate–mantle motion.

CONCEPT CHECK 7.10

❶ **Describe slab pull and ridge push. Which of these forces contributes more to plate motion?**

❷ **What role are mantle plumes thought to play in the convective flow of the mantle?**

❸ **Briefly describe the two models proposed for mantle–plate convection. What is lacking in each of these models?**

Box 7.2

UNDERSTANDING EARTH

Plate Tectonics in the Future

Geologists have extrapolated present-day plate movements into the future. **Figure 7.B** illustrates where Earth's landmasses may be 50 million years from now if present plate movements persist during this time span.

In North America we see that the Baja Peninsula and the portion of southern California that lies west of the San Andreas Fault will have slid past the North American plate. If this northward migration takes place, Los Angeles and San Francisco will pass each other in about 10 million years, and in about 60 million years Los Angeles will begin to descend into the Aleutian Trench.

If Africa continues on a northward path, it will collide with Eurasia, closing the Mediterranean and initiating a major mountain-building episode (Figure 7.B). In other parts of the world, Australia will be astride the equator and, along with New Guinea, will be on a collision course with Asia. Meanwhile, North and South America will begin to separate, while the Atlantic and Indian oceans continue to grow at the expense of the Pacific Ocean.

A few geologists have even speculated on the nature of the globe 250 million years into the future. As shown in **Figure 7.C**, the next supercontinent may form as a result of subduction of the floor of the Atlantic Ocean, resulting in the collision of the Americas with the Eurasian–African landmass. Support for the possible closing of the Atlantic comes from a similar event when the proto-Atlantic closed to form the Appalachian and Caledonian mountains. During the next 250 million years, Australia is also projected to collide with Southeast Asia. If this scenario is accurate, the dispersal of Pangaea will end when the continents reorganize into the next supercontinent.

Such projections, although interesting, must be viewed with considerable skepticism because many assumptions must be correct for these events to unfold as just described.

Nevertheless, changes in the shapes and positions of continents that are equally profound will undoubtedly occur for many hundreds of millions of years to come. Only after much more of Earth's internal heat has been lost will the engine that drives plate motions cease.

FIGURE 7.B **The world as it may look 50 million years from now.** (Modified after Robert S. Dietz, John C. Holden, C. Scotese, and others)

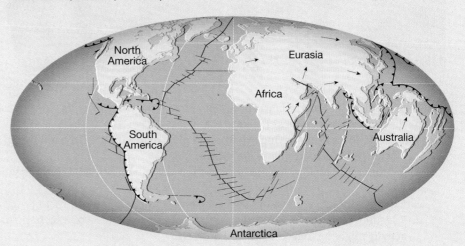

FIGURE 7.C **Reconstruction of Earth as it may appear 250 million years into the future.**

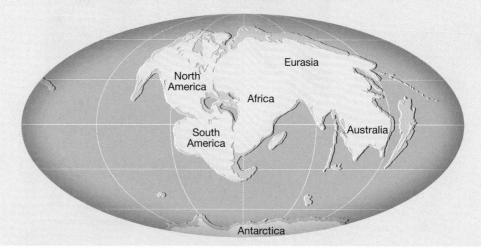

GIVE IT SOME THOUGHT

1. After referring to the section titled "The Nature of Scientific Inquiry," in Chapter 1 complete the following:
   a. What observation led Alfred Wegener to develop his continental drift hypothesis?
   b. What evidence did he gather to support his proposal?
   c. Why was the continental drift hypothesis rejected by the majority of the scientific community?
   d. Do you think Wegner followed the basic steps of scientific inquiry? Support your answer.
2. Referring to the accompanying diagrams that illustrate the three types of convergent plate boundaries, complete the following:
   a. Identify each type of convergent boundary.
   b. Volcanic island arcs develop on what type of crust?
   c. Why are volcanoes largely absent where two continental blocks collide?
   d. Describe two ways that oceanic–oceanic convergent boundaries are different from oceanic–continental convergent boundaries? How are they similar?

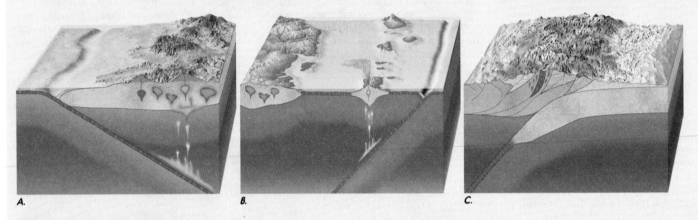

A.   B.   C.

3. Some predict that California will eventually sink into the ocean. Is this idea consistent with the theory of plate tectonics? Explain.
4. Referring to the accompanying hypothetical plate map, complete the following:
   a. Portions of how many plates are shown?
   b. Is continent C moving toward or away from continents A and B? How did you determine your answer?
   c. Explain why active volcanoes are found on continent A and continent B.
   d. Why does continent C lack active volcanoes? Provide at least one scenario in which volcanic activity might be triggered on this continent.

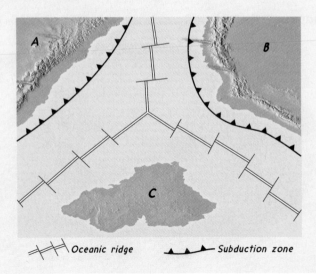

Oceanic ridge   Subduction zone

5. Volcanoes that form over mantle plumes such as the Hawaiian chain are some of the largest on Earth. However, several volcanoes on Mars are gigantic compared to those on Earth. What does this difference tell us about how, or if, plate tectonics operates on Mars? Explain.

6. Imagine you are studying seafloor spreading along two different oceanic ridges. Along the first ridge the magnetic stripes are uniformly narrow. Along the second ridge they are wide near the ridge crest, but they become narrower as you move away from the crest. What can you say about the history of motion in each example?

7. Australian marsupials (kangaroos, koala bears, etc.) have direct fossil links to marsupial opossums found in the Americas. Yet the modern marsupials in Australia are markedly different from their American relatives. How does the breakup of Pangaea help to explain these differences (see Figure 7.A)?

8. During the breakup of Pangaea, which continent was actually growing in size through the accumulation of other landmasses? (see Figure 7.A)

9. Density is a key component in the behavior of Earth materials and is especially important in understanding key aspects of plate tectonics. Describe three ways that density and/or density differences play a role in plate tectonics.

10. Refer to the accompanying map to complete the following:
    a. List the cities (in pairs) that are moving farther apart as a result of plate motion.
    b. List the cities (in pairs) that are moving closer together as a result of plate motion.
    c. List the cities (in pairs) that are presently not moving relative to each other.

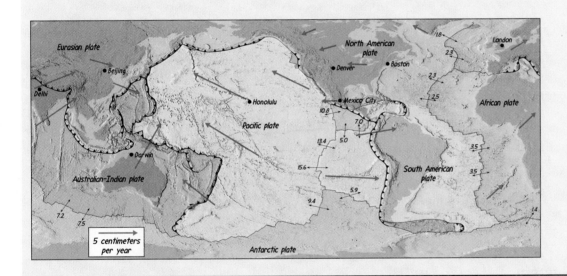

In Review   Chapter 7 Plate Tectonics

- In the early 1900s *Alfred Wegener* set forth his *continental drift* hypothesis. One of its major tenets was that a supercontinent called *Pangaea* began breaking apart about 200 million years ago. The rifted continental fragments then "drifted" to their present positions. To support his hypothesis, Wegener used the *fit of South America and Africa, fossil evidence, rock types and structures,* and *ancient climates.* One of the main objections to the continental drift hypothesis was its inability to provide an acceptable mechanism for the movement of continents.

- By 1968, continental drift was replaced by a far more encompassing theory known as *plate tectonics.* According to plate tectonics, Earth's rigid outer layer (*lithosphere*) overlies

a weaker region called the *asthenosphere.* Furthermore, the lithosphere is broken into several large and numerous smaller segments, called *plates,* that are in motion and continually changing in shape and size. Plates move as relatively coherent units and are deformed mainly along their boundaries.

- *Divergent plate boundaries* occur where plates move apart, resulting in upwelling of material from the mantle to create new seafloor. Most divergent boundaries occur along the axis of the oceanic ridge system and are associated with seafloor spreading. New divergent boundaries may form within a continent (e.g., the East African Rift Valleys), where they may fragment a landmass and develop a new ocean basin.

- *Convergent plate boundaries* occur where plates move together, resulting in the subduction of oceanic lithosphere into the mantle along a deep-ocean trench. Convergence of an oceanic and continental block results in subduction of the oceanic slab and the formation of a *continental volcanic arc* such as the Andes of South America. Oceanic–oceanic convergence results in an arc-shaped chain of volcanic islands called a *volcanic island arc*. When two plates carrying continental crust converge, the buoyant continental blocks collide, resulting in the formation of a mountain belt as exemplified by the Himalayas.
- *Transform fault boundaries* occur where plates grind past each other without the production or destruction of lithosphere. Most transform faults join two segments of a mid-ocean ridge where they provide the means by which oceanic crust created at a ridge crest can be transported to its site of destruction—a deep-ocean trench. Still others, like the San Andreas Fault, cut through continental crust.
- The theory of plate tectonics is supported by (1) the ages of *sediments* from the floors of the deep-ocean basins; (2) the existence of island groups that formed over *hot spots* and that provide a frame of reference for tracing the direction of plate motion; and (3) *paleomagnetism*, the direction and intensity of Earth's magnetism in the geologic past.
- Two basic models for mantle convection are currently being evaluated. Mechanisms that contribute to this convective flow are slab pull and ridge push. *Slab pull* occurs where cold, dense oceanic lithosphere is subducted and pulls the trailing lithosphere along. *Ridge push* results when gravity sets the elevated slabs astride oceanic ridges in motion. Hot, buoyant *mantle plumes* are considered the upward flowing arms of mantle convection. One model suggests that mantle convection occurs in two layers separated at a depth of 660 kilometers (410 miles). Another model proposes whole-mantle convection that stirs the entire 2,900-kilometer-thick (1,800-mile-thick) rocky mantle.

Key Terms

asthenosphere (p. 200)
continental drift (p. 195)
continental rift (p. 204)
continental volcanic arc (p. 207)
convergent boundary (p. 201)
Curie point (p. 214)
deep-ocean trench (p. 205)
divergent boundary (p. 201)
fossil magnetism (p. 214)
fracture zone (p. 210)
hot spot (p. 212)
hot spot track (p. 212)

island arc (p. 207)
lithosphere (p. 200)
lithospheric plate (p. 200)
magnetic reversal (p. 214)
magnetic time scale (p. 215)
magnetometer (p. 215)
mantle plume (p. 212)
normal polarity (p. 214)
oceanic ridge system (p. 200)
paleomagnetism (p. 214)
Pangaea (p. 195)
partial melting (p. 206)

plate tectonics (p. 200)
reverse polarity (p. 214)
ridge push (p. 219)
rift valley (p. 204)
seafloor spreading (p. 204)
slab pull (p. 219)
spreading center (p. 201)
subduction zone (p. 205)
supercontinent (p. 195)
tectonic plate (p. 200)
transform fault boundary (p. 201)
volcanic island arc (p. 207)

Examining the Earth System

1. As an integral subsystem of the Earth system, plate tectonics has played a major role in determining the events that have taken place on Earth since its formation about 4.6 billion years ago. Briefly comment on the general effect that the changing positions of the continents and the redistribution of land and water over Earth's surface have had on the atmosphere, hydrosphere, and biosphere through time. You may want to review plate tectonics by investigating the topic on the Internet using a search engine such as Google (**http://www.google.com/**) or Yahoo (**http://www.yahoo.com/**) and visiting the United States Geological Survey's (USGS)

This Dynamic Earth: The Story of Plate Tectonics Website at **http://pubs.usgs.gov/publications/text/dynamic.html**.

2. Assume that plate tectonics did not cause the breakup of the supercontinent Pangaea. Using Figure 7.2, describe how the climate (atmosphere), vegetation and animal life (biosphere), and geological features (geosphere) of those locations currently occupied by the cities of Seattle, WA; Chicago, IL; New York, NY; and your college campus location would be different from the conditions that exist today.

Mastering Geology

Looking for additional review and test prep materials? Visit the Self Study area in **www.masteringgeology.com** to find practice quizzes, study tools, and multimedia that will aid in your understanding of this chapter's content. In **MasteringGeology™** you will find:

- **GEODe: Earth Science: An interactive visual walkthrough of key concepts**

- **Geoscience Animation Library: More than 100 animations illuminating many difficult-to-understand Earth science concepts**
- **In The News RSS Feeds: Current Earth science events and news articles are pulled into the site with assessment**
- **Pearson eText**
- **Optional Self Study Quizzes**
- **Web Links**
- **Glossary**
- **Flashcards**

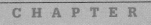

Earthquakes and Earth's Interior

8

Llollero, Chile, in the aftermath of a tsunami that was generated by a major earthquake on February 27, 2010.
(Photo by Eliseo Fernandez.Reuters)

O n February 27, 2010, at 3:34 AM local time, an earthquake measuring 8.8 occurred approximately 11 kilometers (7 miles) off the coast of Chile, resulting in 486 fatalities and widespread property damage. This powerful quake, the fifth largest in the world since 1900, shortened the day's length by 1.26 microseconds and moved Earth's axis by 8 centimeters (3 inches), as estimated by seismologists. In addition, the event laterally displaced cities, including Concepción by over 3 meters (10 feet) and Santiago approximately 24 centimeters (10 inches).

FOCUS ON CONCEPTS

To assist you in learning the important concepts in this chapter, focus on the following questions:

- What is an earthquake?
- What are the types of earthquake waves?
- How is the epicenter of an earthquake determined?
- Where are the principal earthquake zones on Earth?
- How is earthquake strength expressed?
- What are the main factors that affect the amount of destruction caused by seismic shaking?
- What is a tsunami?
- How did Earth acquire its layered structure?
- What are the major zones of Earth's interior?
- How do continental crust and oceanic crust differ?

An Earthquake Disaster in Haiti

On Tuesday, January 12, 2010, an estimated 230,000 people lost their lives when a magnitude 7.0 earthquake struck the small Caribbean nation of Haiti, the poorest country in the Western Hemisphere (**Figure 8.1**).

The quake originated only 25 kilometers (15 miles) from the country's densely populated capital city of Port-au-Prince. It occurred along a San Andreas–like fault at a depth of just 10 kilometers (6 miles). Because of the quake's shallow depth, ground shaking was exceptional for an event of this magnitude. Other factors also contributed to the Port-au-Prince disaster, including the city's geologic setting and the nature of its buildings. The city is not built on solid bedrock but on sediment, which is more susceptible to ground shaking by earthquake waves. More importantly, inadequate or nonexistent building codes meant that buildings collapsed far more readily than they should have. At least 52 aftershocks, measuring 4.5 or greater, jolted the area and added to the trauma survivors experienced for days after the original quake.

The 230,000 death toll rivals the loss of life in the tragic 2004 Indonesian tsunami, an earthquake-generated event that is described later in the chapter. Both of these natural disasters produced death tolls equivalent to the populations of cities the size of Madison, Wisconsin, or Orlando, Florida. The loss of life associated with the 2010 Haitian event is even more extraordinary when it is compared with the 1989 Loma Prieta earthquake in southern California, which had a similar magnitude (M_W 7.1) but claimed just 67 lives. In addition to the staggering death toll, there were more than 300,000 injuries and the destruction of 250,000 residences. Nearly a million people were left homeless.

Relief agencies from around the globe stepped in to distribute food and water and provide for the enormous medical, security, and social needs of the injured and displaced. For a lengthy period following the quake, inadequate infrastructure coupled with devastating damages inhibited the timely delivery of essential services. This human crisis is expected to continue for an extended period due to the likely spread of disease and complications from minimally or untreated injuries. The rebuilding of Haiti will likely take a decade or more and billions of dollars in assistance.

CONCEPT CHECK 8.1

❶ List three factors that contributed to making the Haiti earthquake a serious natural disaster.

What is an Earthquake?

Forces Within
▶ Earthquakes

Earthquakes are natural geologic phenomena caused by the sudden and rapid movement of a large volume of rock. The violent shaking and destruction caused by earthquakes are the result of rupture and slippage along fractures in Earth's crust called **faults**. Larger quakes result from the rupture of larger fault segments. The origin of an earthquake occurs at depths between 5 and

FIGURE 8.1 Haitian women stand amid rubble following the 7.0 magnitude earthquake that devastated Port au Prince, Haiti on January 12, 2010. (Photo by Chip Somodevilla/Getty Images)

700 kilometers, at the **focus**. The point at the surface directly above the focus is called the **epicenter** (Figure 8.2).

During large earthquakes, a massive amount of energy is released as **seismic waves**—a form of elastic energy that causes vibrations in the material that transmits them. Seismic waves are analogous to waves produced when a stone is dropped into a calm pond. Just as the impact of the stone creates a pattern of waves in motion, an earthquake generates waves that radiate outward in all directions from the focus. Even though seismic energy dissipates rapidly with increasing distance, sensitive instruments located around the world detect and record these events.

Thousands of earthquakes occur around the world every day. Fortunately, most are so small that they can only be detected by sensitive instruments. Of these, only about 75 strong quakes are recorded each year and many of these occur in remote regions.

FIGURE 8.2 Earthquake focus and epicenter. The *focus* is the zone within Earth where the initial displacement occurs. The *epicenter* is the surface location directly above the focus.

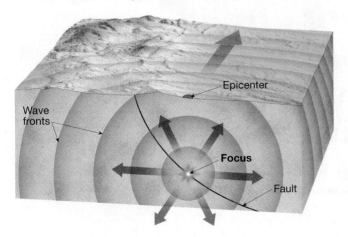

Occasionally, a large earthquake is triggered near a major population center. Such events are among the most destructive natural forces on Earth. The shaking of the ground, coupled with the liquefaction of soils, wreaks havoc on buildings, roadways, and other structures. In addition, when a quake occurs in a populated area, power and gas lines are often ruptured, causing numerous fires. In the famous 1906 San Francisco earthquake, much of the damage was caused by fires that became uncontrollable when broken water mains left firefighters with only trickles of water (Figure 8.3).

Discovering the Causes of Earthquakes

The energy released by atomic explosions or by the movement of magma in Earth's crust can generate earthquake-like waves, but these events are generally quite weak. What mechanism produces a destructive earthquake? As you have learned, Earth is not a static planet. We know that large sections of Earth's crust have been thrust upward, because fossils of marine organisms have been discovered thousands of meters above sea level. Other regions exhibit evidence of extensive subsidence. In addition to these vertical displacements, offsets in fence lines, roads, and other structures indicate that horizontal movements are also common (Figure 8.4).

The actual mechanism of earthquake generation eluded geologists until H. F. Reid of Johns Hopkins University conducted a study following the great 1906 San Francisco earthquake. The earthquake was accompanied by horizontal surface displacements of several meters along the northern portion of the San Andreas Fault. Field studies determined that during this single earthquake, the Pacific plate lurched as much as 4.7 meters (15 feet) northward past the adjacent North American plate.

FIGURE 8.3 San Francisco in flames after the 1906 earthquake. (Reproduced from the collection of the Library of Congress) Inset photo shows fire triggered when a gas line ruptured during the Northridge earthquake in Southern California in 1994. (Photo by AFP/Getty Images)

What Reid concluded from his investigations is illustrated in **Figure 8.5**. Tectonic stresses acting over tens to hundreds of years slowly deform the crustal rocks on both sides of a fault. When deformed by differential stress, rocks bend and store elastic energy, much like a wooden stick does if bent (Figure 8.5B). Eventually, the frictional resistance holding the rocks in place is overcome. Slippage allows the deformed (strained) rock to "snap back" to its original, stress-free shape (Figure 8.5C, D). The "springing back" was termed **elastic rebound** by Reid because the rock behaves elastically, much like a stretched rubber band does when it is released. The vibrations we know as an earthquake are generated by the rock elastically returning to its original shape.

In summary, *earthquakes are produced by the rapid release of elastic energy stored in rock that has been deformed by differential stresses.* Once the strength of the rock is exceeded, it suddenly ruptures, causing the vibrations of an earthquake.

Aftershocks and Foreshocks

Strong earthquakes are followed by numerous smaller tremors, called **aftershocks**, that gradually diminish in frequency and intensity over a period of

FIGURE 8.4 Slippage along a fault produced an offset in this orange grove east of Calexico, California. (Photo by John S. Shelton) Inset photo shows a fence offset 2.5 meters (8.5 feet) during the 1906 San Francisco earthquake. (Photo by G. K. Gilbert, U.S. Geological Survey)

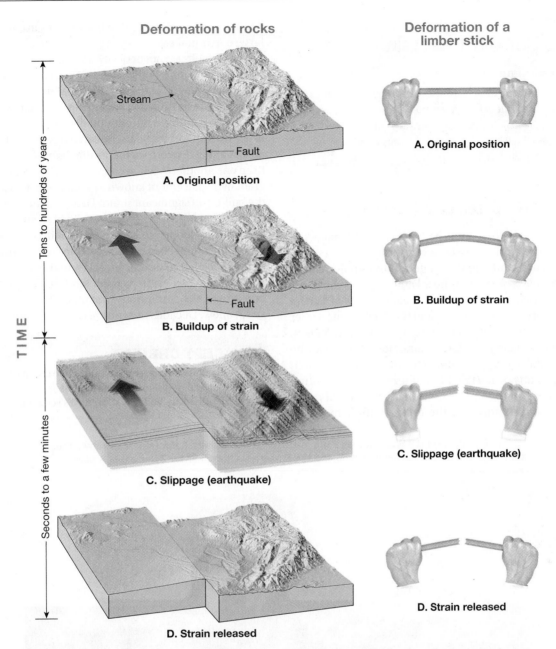

Deformation of rocks

A. Original position
Stream ← Fault

B. Buildup of strain
Fault

C. Slippage (earthquake)

D. Strain released

TIME — Tens to hundreds of years — Seconds to a few minutes

Deformation of a limber stick

A. Original position
B. Buildup of strain
C. Slippage (earthquake)
D. Strain released

FIGURE 8.5 Elastic rebound. As rock is deformed, it bends, storing elastic energy. Once strained beyond its breaking point, the rock cracks, releasing the stored-up energy in the form of earthquake waves.

several months. Within 24 hours of the massive 1964 Alaskan earthquake, 28 aftershocks were recorded, 10 of which had magnitudes that exceeded 6. More than 10,000 aftershocks with magnitudes of 3.5 or above occurred in the following 69 days and thousands of minor tremors were recorded over a span of 18 months. Because aftershocks happen mainly on the section of the fault that has slipped, they provide geologists with data that is useful in establishing the dimensions of the rupture surface.

Although aftershocks are weaker than the main earthquake, they can trigger the destruction of already weakened structures. This occurred in northwestern Armenia (1988) where many people lived in large apartment buildings constructed of brick and concrete slabs. After a moderate earthquake of magnitude 6.9 weakened the buildings, a strong aftershock of magnitude 5.8 completed the demolition.

In contrast to aftershocks, small earthquakes called **foreshocks** often precede a major earthquake by days or in some cases by several years. Monitoring of foreshocks to predict forthcoming earthquakes has been attempted with limited success.

CONCEPT CHECK 8.2

1 What is an *earthquake*? Under what circumstances do earthquakes occur?
2 How are *faults*, *foci*, and *epicenters* related?
3 Who was first to explain the actual mechanism by which earthquakes are generated?
4 Sketch or explain what is meant by *elastic rebound*.

Earthquakes and Faults

Earthquakes take place along faults both new and old that occur in places where differential stresses have ruptured Earth's crust. Some faults are large and capable of generating major earthquakes. One example is the San Andreas Fault, which is the transform fault boundary that separates two great sections of Earth's lithosphere: the North American plate and the Pacific plate. Other faults are small and capable of producing only minor earthquakes.

Most of the displacement that occurs along faults can be satisfactorily explained by the plate tectonics theory, which states that large slabs of Earth's lithosphere are in continual slow motion. These mobile plates interact with neighboring plates, straining and deforming the rocks at their margins.

Faults associated with plate boundaries are the source of most large earthquakes.

Large faults are not perfectly straight or continuous; instead, they consist of numerous branches and smaller fractures that display kinks and offsets. Such a pattern is displayed in **Figure 8.6**, which shows the San Andreas Fault as a system of several faults of various sizes.

The San Andreas is undoubtedly the most studied fault system in the world. Over the years, research has shown that displacement occurs along discrete segments that behave somewhat differently from one another. A few sections of the San Andreas exhibit a slow, gradual displacement known as *fault creep*, which occurs without the buildup of significant strain. These sections produce only minor seismic shaking. Other segments slip at regular intervals, producing small-to-moderate earthquakes. Still other segments remain locked and store energy for a few hundred years before rupturing in great earthquakes. Earthquakes that occur along locked segments of the San Andreas Fault tend to be repetitive. As soon as one is over, the continuous motion of the plates begins building strain anew. Decades or centuries later, the fault fails again.

CONCEPT CHECK 8.3

1. Faults that are experiencing no active creep may be considered safe. Rebut or defend this statement.
2. Why is the San Andreas sometimes referred to as a *fault system*?

FIGURE 8.6 Trace of the San Andreas Fault, Carrizo Plain, Southern California. Map shows the extent of the San Andreas Fault system and the Eastern California Shear Zone. (Photo by Dewitt Jones/CORBIS)

PEOPLE AND THE ENVIRONMENT

Damaging Earthquakes East of the Rockies

When you think earthquake, you probably think of California and Japan. However, six major earthquakes have occurred in the central and eastern United States since colonial times. Three of these had estimated Richter magnitudes of 7.5, 7.3, and 7.8, and they were centered near the Mississippi River Valley in southeastern Missouri. Occurring on December 16, 1811, January 23, 1812, and February 7, 1812, these earthquakes, plus numerous smaller tremors, destroyed the town of New Madrid, Missouri, triggered massive landslides, and caused damage over a six-state area. The course of the Mississippi River was altered, and Tennessee's Reelfoot Lake was enlarged. The distances over which these earthquakes were felt are truly remarkable. Chimneys were reported downed in Cincinnati, Ohio, and Richmond, Virginia, while Boston residents, located 1,770 kilometers (1,100 miles) to the northeast, felt the tremor.

Despite the history of the New Madrid earthquake, Memphis, Tennessee, the largest population center in the area today, does not have adequate earthquake provisions in its building code. Furthermore, because Memphis is located on unconsolidated floodplain deposits, buildings are more susceptible to damage than similar structures built on bedrock. It has been estimated that if an earthquake the size of the 1811–1812 New Madrid event were to strike in the next decade, it would result in casualties in the thousands and damages in tens of billions of dollars.

Damaging earthquakes that occurred in Aurora, Illinois (1909), and Valentine, Texas (1931), remind us that other areas in the central United States are vulnerable.

The greatest historical earthquake in the eastern states occurred August 31, 1886, in Charleston, South Carolina. The event, which spanned 1 minute, caused 60 deaths, numerous injuries, and great economic loss within a radius of 200 kilometers (120 miles) of Charleston. Within 8 minutes, effects were felt as far away as Chicago and St. Louis, where strong vibrations shook the upper floors of buildings, causing people to rush outdoors. In Charleston alone, over 100 buildings were destroyed, and 90 percent of the remaining structures were damaged (Figure 8.A).

Numerous other strong earthquakes have been recorded in the eastern United States. New England and adjacent areas have experienced sizable shocks since colonial times. The first reported earthquake in the Northeast took place in Plymouth, Massachusetts, in 1683, and was followed in 1755 by the destructive Cambridge, Massachusetts, earthquake. Moreover, ever since records have been kept, New York State alone has experienced over 300 earthquakes large enough to be felt.

Earthquakes in the central and eastern United States occur far less frequently than in California. Yet history indicates that the East is vulnerable. Furthermore, these shocks east of the Rockies have generally produced structural damage over a larger area than counterparts of similar magnitude in California. The reason is that the underlying bedrock in the central and eastern United States is older and more rigid. As a result, seismic waves are able to travel greater distances with less weakening than in the western United States. It is estimated that for earthquakes of similar magnitude, the region of maximum ground motion in the east may be up to 10 times larger than in the west. Consequently, the higher rate of earthquake occurrence in the western United States is balanced somewhat by the fact that central and eastern U.S. quakes can damage larger areas.

FIGURE 8.A Damage to Charleston, South Carolina, caused by the August 31, 1886, earthquake. Damage ranged from toppled chimneys and broken plaster to total collapse. (Photo courtesy of U.S. Geological Survey)

Seismology: The Study of Earthquake Waves

Forces Within
▶ **Earthquakes**

The study of earthquake waves, **seismology** dates back to attempts made by the Chinese almost 2,000 years ago to determine the direction from which these waves originated. Modern **seismographs**, instruments that record earthquake waves, are similar to the instruments used by the Chinese. Seismographs have a weight freely suspended from a support that is securely attached to bedrock (Figure 8.7). When vibrations from an earthquake reach the instrument, the inertia of the weight keeps it relatively stationary while Earth and the support move. (*Inertia* is the tendency of objects at rest to stay at rest and objects in motion to remain in motion.)

To detect very weak earthquakes, or a great earthquake that has occurred in another part of the world, most seismographs are designed to amplify ground motion. Other instruments are designed to withstand the violent shaking that occurs very near the focus.

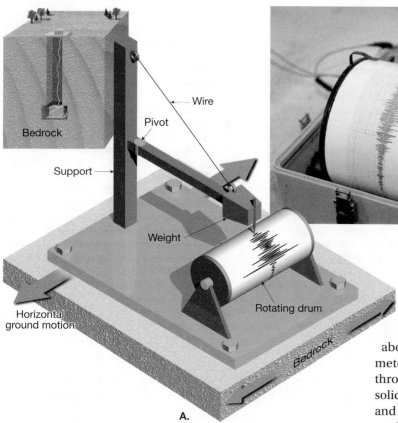

FIGURE 8.7 Principle of the seismograph. **A.** The inertia of the suspended weight tends to keep it motionless while the recording drum, which is anchored to bedrock, vibrates in response to seismic waves. The stationary weight provides a reference point from which to measure the amount of displacement occurring as a seismic wave passes through the ground. **B.** Seismograph recording earthquake tremors. (Photo courtesy of Zephyr/Photo Researchers, Inc.)

The records obtained from seismographs, called **seismograms**, provide useful information about the nature of seismic waves. Seismograms reveal that two main groups of seismic waves are generated by the slippage of a rock mass. One group is called **surface waves** because their motion is restricted to near Earth's surface. Others travel through Earth's interior and are called **body waves**. Body waves are divided into two types: **primary**, or **P**, **waves** and **secondary**, or **S**, waves.

Body waves are identified by their mode of travel through intervening materials. P waves are "push–pull" waves—they momentarily push (squeeze) and pull (stretch) rocks in the direction the wave is traveling (**Figure 8.8A**). This wave motion is similar to that generated by human vocal cords as they move air to create sound. Solids, liquids, and gases resist a change in volume when compressed and will elastically spring back once the force is removed. Therefore, P waves can travel through all of these materials.

On the other hand, S waves "shake" the particles at right angles to their direction of travel. This can be illustrated by fastening one end of a rope and shaking the other end, as shown in Figure 8.8B. Unlike P waves, which temporarily change the *volume* of intervening material by alternately squeezing and stretching it, S waves change the *shape* of the material that transmits them.

Because fluids (gases and liquids) do not resist stresses that cause changes in shape—meaning fluids will not return to their original shape once the stress is removed—they will not transmit S waves.

The motion of surface waves is somewhat more complex. As surface waves travel along the ground, they cause the ground and anything resting upon it to move, much like ocean swells toss a ship (Figure 8.8C). In addition to their up-and-down motion, surface waves have a side-to-side motion similar to an S wave oriented in a horizontal plane (Figure 8.8D). This latter motion is particularly damaging to the foundations of structures.

By examining the "typical" seismic record shown in **Figure 8.9**, you can see a major difference among seismic waves—their speed of travel. P waves are the first to arrive at a seismic station, then S waves, and finally surface waves. The velocity of P waves through the crustal rock granite is about 6 kilometers per second and increases to nearly 13 kilometers per second at the base of the mantle. P waves can pass through Earth's mantle in about 20 minutes. Generally, in any solid material, P waves travel about 1.7 times faster than S waves, and surface waves are roughly 10 percent slower than S waves.

In addition to velocity differences, notice in Figure 8.9 that the height, or *amplitude*, of these wave types also varies. S waves have slightly greater amplitudes than P waves, while surface waves exhibit even greater amplitudes. Surface waves also retain their maximum amplitude longer than P and S waves. As a result, surface waves tend to cause greater ground shaking and hence greater destruction, than either P or S waves.

Seismic waves are useful in determining the location and magnitude of earthquakes. In addition, seismic waves provide an important tool for probing Earth's interior.

CONCEPT CHECK 8.4
1. Describe the principle of a *seismograph*.
2. List the major differences between *P*, *S*, and *surface waves*.
3. Which type of seismic waves tend to cause the greatest destruction to buildings?

Locating the Source of an Earthquake

 Forces Within
▶ **Earthquakes**

When analyzing an earthquake, the first task seismologists undertake is determining its *epicenter*, the point on Earth's surface directly above the focus (see Figure 8.2). The method used for locating an earthquake's epicenter relies on the fact that P waves travel faster than S waves.

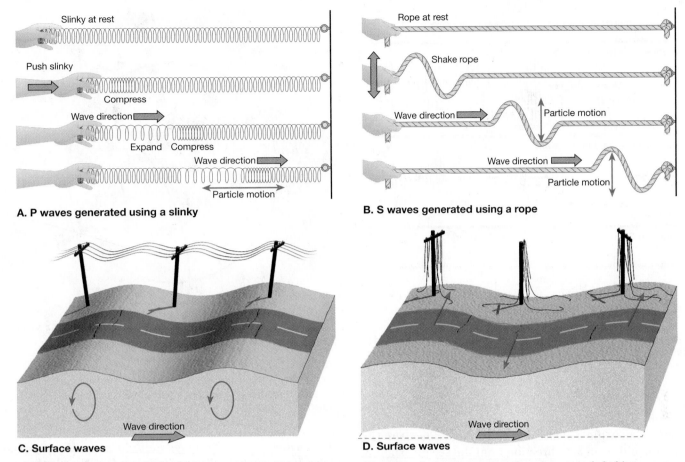

A. P waves generated using a slinky

B. S waves generated using a rope

C. Surface waves

D. Surface waves

FIGURE 8.8 Types of seismic waves and their characteristic motion. (Note that during a strong earthquake, ground shaking consists of a combination of various kinds of seismic waves.) **A.** As illustrated by a slinky, P waves are compressional waves that alternately compress and expand the material through which they pass. **B.** S waves cause material to oscillate at right angles to the direction of wave motion. **C.** One type of surface wave travels along Earth's surface much like rolling ocean waves. The red arrows show the elliptical movement of rock as the wave passes. **D.** Another type of surface wave moves the ground from side to side and can be particularly damaging to the foundations of buildings.

FIGURE 8.9 Typical seismogram. Note the time interval (about 5 minutes) between the arrival of the first P wave and the arrival of the first S wave.

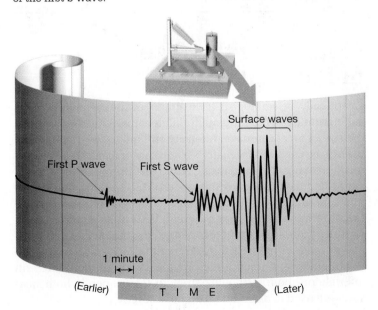

Surface waves

First P wave   First S wave

1 minute

(Earlier)   T I M E   (Later)

The method is analogous to the results of a race between two autos, one faster than the other. The first P wave, like the faster auto, always wins the race, arriving ahead of the first S wave. The greater the length of the race, the greater the difference in their arrival times at the finish line (the seismic station). Therefore, the greater the interval between the arrival of the first P wave and the arrival of the first S wave, the greater the distance to the epicenter. **Figure 8.10** shows three simplified seismograms for the same earthquake. Based on the P–S interval, which city—Nagpur, Darwin, or Paris—is farthest from the epicenter?

The system for locating earthquake epicenters was developed by using seismograms from earthquakes whose epicenters could be easily pinpointed from physical evidence. From these seismograms, travel-time graphs were constructed (**Figure 8.11**). Using the sample seismogram for Nagpur, India, in Figure 8.10A and the travel-time curve in Figure 8.11, we can determine the distance separating the recording station from the earthquake in two steps: (1) Using the seismogram, determine the time interval between the arrival of the first P wave and the arrival of the first S wave, and (2) using the travel-time graph, find the P–S interval on the vertical axis and use that information to determine the distance to the epicenter on the horizontal axis. Following this procedure, we can determine that the earthquake

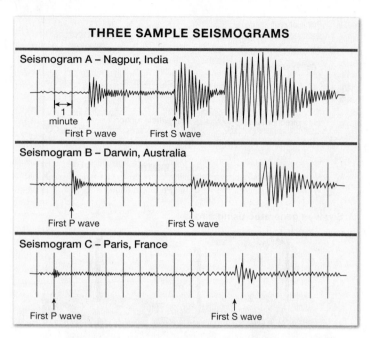

THREE SAMPLE SEISMOGRAMS

Seismogram A – Nagpur, India

1 minute

First P wave   First S wave

Seismogram B – Darwin, Australia

First P wave   First S wave

Seismogram C – Paris, France

First P wave   First S wave

FIGURE 8.10 Simplified seismograms of the same earthquake recorded in three different cities. **A.** Nagpur, India. **B.** Darwin, Australia. **C.** Paris, France.

occurred 3,400 kilometers (2,100 miles) from the recording instrument in Nagpur, India.

Now we know the *distance*, but what about *direction*? The epicenter could be in any direction from the seismic station.

FIGURE 8.11 A travel-time graph is used to determine the distance to an earthquake epicenter. The difference in arrival time between the first P wave and the first S wave in the example is 5 minutes. Thus, the epicenter is roughly 3,400 kilometers (2,100 miles) away.

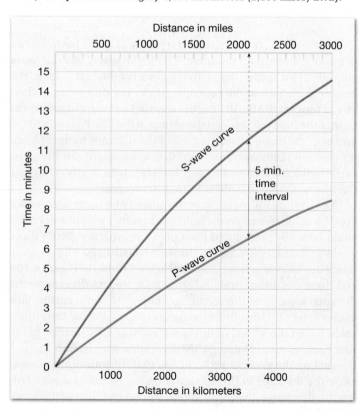

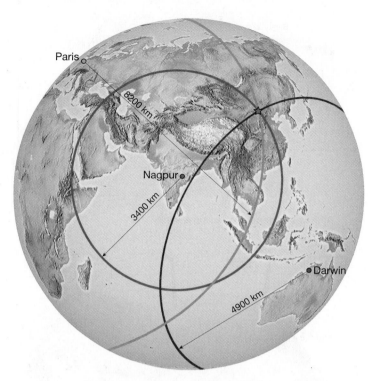

FIGURE 8.12 Determining an earthquake's epicenter using the distances obtained from three or more seismic stations—a method called *triangulation*.

Using a method called *triangulation*, the precise location can be determined when the distance is known from three or more seismic stations (**Figure 8.12**). On a globe, a circle is drawn around each seismic station. The radius of these circles is equal to the distance from the seismic station to the epicenter. The point where the three circles intersect is the epicenter of the quake.

CONCEPT CHECK 8.5

❶ What information does a travel-time graph provide?
❷ Briefly describe the *triangulation* method used to determine the epicenter of an earthquake.

Measuring the Size of Earthquakes

Historically, seismologists have employed a variety of methods to determine two fundamentally different measures that describe the size of an earthquake: intensity and magnitude. The first of these to be used was **intensity**—a measure of the degree of earthquake shaking at a given locale based on observed effects. Later, with the development of seismographs, it became possible to measure ground motion using instruments. This quantitative measurement, called **magnitude**, relies on data gleaned from seismic records and other techniques to estimate the amount of energy released at an earthquake's source.

Intensity and magnitude provide useful, though different, information about earthquake strength. Consequently, both measures are used to describe earthquake severity.

Modified Mercalli Intensity Scale

Numerous intensity scales have been developed over the last 150 years. The one widely used is the **Modified Mercalli Intensity Scale**—named after Giuseppe Mercalli who initially developed it in 1902 (**Table 8.1**). This intensity scale is divided into 12 levels of severity based on observed effects such as people awakening from sleep, furniture moving, plaster cracking and falling, and finally total destruction. As Table 8.1 illustrates, the lower numbers on the Mercalli scale (I–V) refer to what people in various locations felt during the quake, whereas the higher numbers (VI–XII) are based on observable damage to buildings and other structures. **Figure 8.13** shows shaking intensity maps for two San Francisco Bay area quakes: the 1989 Loma Prieta and 1906 San Francisco earthquakes. Although the 1989 Loma Prieta quake caused billions of dollars in damage and claimed more than 60 lives, these shaking maps show that a repeat of the 1906 San Francisco earthquake would certainly be more catastrophic.

Despite their usefulness in providing a tool to compare earthquake severity, intensity scales have significant drawbacks. These scales are based on effects (largely destruction) that depend not only on the severity of ground shaking but also on factors such as building design and the nature of surface materials. For example, the modest 7.0-magnitude 2010 Haiti earthquake mentioned earlier was extremely destructive, mainly because of inferior building practices. Thus, the destruction wrought by an earthquake is frequently not a good measure of the amount of energy that was unleashed.

Magnitude Scales

In order to more accurately compare earthquakes across the globe, a measure was needed that does not rely on parameters that vary considerably from one part of the world to another. As a consequence, several magnitude scales were developed.

Richter Magnitude   In 1935 Charles Richter of the California Institute of Technology developed the first magnitude scale using seismic records. As shown in **Figure 8.14** (top), the **Richter scale**

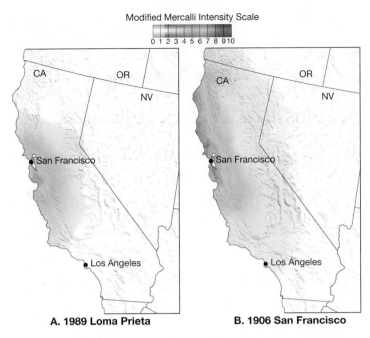

FIGURE 8.13 Comparison of shaking intensity for two San Francisco Bay area earthquakes: the 1989 Loma Prieta and the 1906 San Francisco earthquakes. The colors on the map represent levels of shaking based on the Modified Mercalli Intensity Scale. (Data from U.S. Geological Survey)

is based on the amplitude of the largest seismic wave (P, S, or surface wave) recorded on a seismogram. Because seismic waves weaken as the distance between the focus and the seismograph increases, Richter developed a method that accounts for the decrease in wave amplitude with increasing distance. Theoretically, as long as equivalent instruments are used, monitoring stations at various locations will obtain the same Richter magnitude for each recorded earthquake. In practice, however, different recording stations often obtain slightly different Richter magnitudes for the same earthquake—a consequence of the variations in rock types through which the waves travel.

Earthquakes vary enormously in strength, and great earthquakes produce wave amplitudes that are thousands of times larger than

TABLE 8.1 Modified Mercalli Intensity Scale

I	Not felt except by a very few under especially favorable circumstances.
II	Felt only by a few persons at rest, especially on upper floors of buildings.
III	Felt quite noticeably indoors, especially on upper floors of buildings, but many people do not recognize it as an earthquake.
IV	During the day felt indoors by many, outdoors by few. Sensation like heavy truck striking building.
V	Felt by nearly everyone, many awakened. Disturbances of trees, poles, and other tall objects sometimes noticed.
VI	Felt by all; many frightened and run outdoors. Some heavy furniture moved; few instances of fallen plaster or damaged chimneys. Damage slight.
VII	Everybody runs outdoors. Damage negligible in buildings of good design and construction; slight-to-moderate in well-built ordinary structures; considerable in poorly built or badly designed structures.
VIII	Damage slight in specially designed structures; considerable in ordinary substantial buildings with partial collapse; great in poorly built structures (fall of chimneys, factory stacks, columns, monuments, walls).
IX	Damage considerable in specially designed structures. Buildings shifted off foundations. Ground cracked conspicuously.
X	Some well-built wooden structures destroyed. Most masonry and frame structures destroyed. Ground badly cracked.
XI	Few, if any (masonry) structures remain standing. Bridges destroyed. Broad fissures in ground.
XII	Damage total. Waves seen on ground surfaces. Objects thrown upward into air.

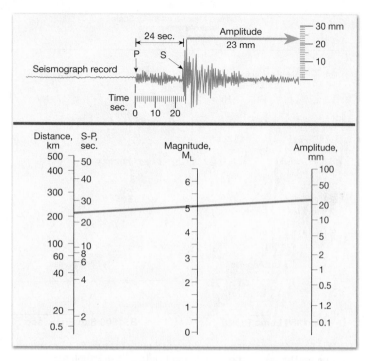

FIGURE 8.14 Illustration showing how the Richter magnitude of an earthquake can be determined graphically using a seismograph record from a Wood–Anderson instrument. First, measure the height (amplitude) of the largest wave on the seismogram (23 mm) and then the distance to the epicenter using the time interval between S and P waves (24 seconds). Next, draw a line between the distance scale (left) and the wave amplitude scale (right). By doing this, you should obtain the Richter magnitude (M_L) of 5. (Data from California Institute of Technology)

those generated by weak tremors (**Figure 8.15**). To accommodate this wide variation, Richter used a *logarithmic scale* to express magnitude, in which a *tenfold* increase in wave amplitude corresponds to an increase of 1 on the magnitude scale. Thus, the degree of ground shaking for a 5-magnitude earthquake is 10 times greater than that produced by an earthquake having a Richter magnitude of 4.

In addition, each unit of Richter magnitude equates to roughly a *32-fold energy increase*. Thus, an earthquake with a magnitude of 6.5 releases 32 times more energy than one with a magnitude of 5.5, and roughly 1,000 times (32×32) more energy than a 4.5-magnitude quake. Furthermore, a major earthquake with a magnitude of 8.5 releases millions of times more energy than the smallest earthquakes felt by humans (Figure 8.15).

Although the Richter scale has no upper limit, the largest magnitude recorded was 8.9. Great shocks such as these release an amount of energy that is roughly equivalent to the detonation of 1 billion tons of explosives. Conversely, earthquakes with a Richter magnitude of less than 2.0 are generally not felt by humans.

Richter's original goal was modest in that he only attempted to rank shallow earthquakes in southern California into groups of large, medium, and small magnitude. Hence, Richter magnitude was designed to classify relatively local earthquakes and is designated by the symbol (M_L)—where M is for *magnitude* and L is for *local*.

The convenience of describing the size of an earthquake by a single number that can be calculated quickly from seismograms makes the Richter scale a powerful tool. Furthermore, unlike intensity scales that can only be applied to populated areas of the globe, Richter magnitudes can be assigned to earthquakes in more

FIGURE 8.15 The size or magnitude of an earthquake (left side) compared to the number of earthquakes of various magnitudes that occur worldwide each year. The largest earthquakes occur less than once a year, whereas strong earthquakes happen more than once a month and weak quakes, those less than magnitude 2, occur hundreds of times per day. (Data from IRIS Consortium, www.iris.edu)

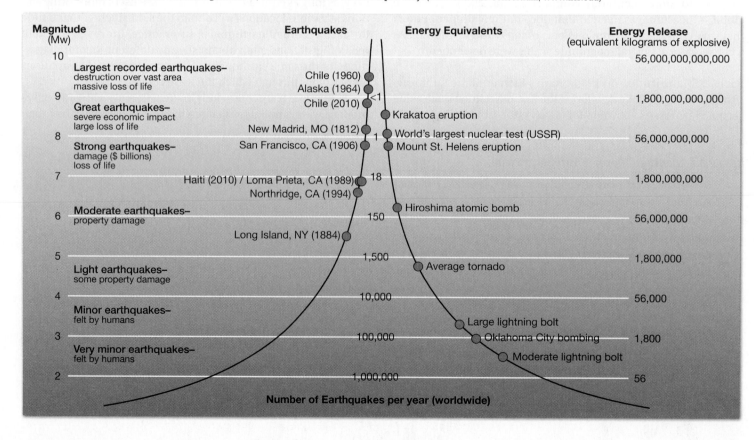

remote regions and even to events that occur in the ocean basins. In time, seismologists modified Richter's work and developed new Richter-like magnitude scales.

Despite its usefulness, the Richter scale is not adequate for describing very large earthquakes. For example, the 1906 San Francisco earthquake and the 1964 Alaskan earthquake had roughly the same Richter magnitudes. However, based on the relative size of the affected areas and the associated tectonic changes, the Alaskan earthquake released considerably more energy than the San Francisco quake. As a result, the Richter scale is said to be *saturated* for major earthquakes because it cannot distinguish among them.

Moment Magnitude   In recent years, seismologists have come to favor a newer measure called **moment magnitude** (M_W), which determines the strain energy released along the entire fault surface. Because moment magnitude estimates the total energy released, it is better for measuring or describing very large earthquakes. In light of this, seismologists have recalculated the magnitudes of older, strong earthquakes using the moment magnitude scale. For example, the 1964 Alaskan earthquake was originally given a Richter magnitude of 8.3, but a recent recalculation using the moment magnitude scale resulted in an upgrade to 9.2. Similarly, the 1906 San Francisco earthquake, which had a Richter magnitude of 8.3, was downgraded to a M_W 7.9. The strongest earthquake on record is the 1960 Chilean subduction zone earthquake, with a moment magnitude of 9.5.

Moment magnitude can be calculated from geologic fieldwork by measuring the average amount of slip on the fault, the area of the fault surface that slipped, and the strength of the faulted rock. The area of the fault plane can be roughly calculated by multiplying the surface-rupture length by the depth of the aftershocks. This method is most effective for determining the magnitude of large earthquakes generated along large faults in which the ruptures reach the surface. Moment magnitude can also be calculated using data from seismograms.

CONCEPT CHECK 8.6

1. What information does the Modified Mercalli Intensity Scale provide about an earthquake? What information is used to establish the lower numbers on the Mercalli scale?
2. How much more energy does an earthquake measuring 7.0 on the Richter scale release compared to an earthquake with a magnitude of 6.0.
3. Why is the *moment magnitude scale* favored over the Richter scale?

Students Sometimes Ask...

Do moderate earthquakes decrease the chances of a major quake in the same region?

No. This is due to the vast increase in release of energy associated with higher-magnitude earthquakes. For instance, an earthquake with a magnitude of 8.5 releases millions of times more energy than the smallest earthquakes felt by humans. Similarly, thousands of moderate tremors would be needed to release the huge amount of energy equal to one "great" earthquake.

FIGURE 8.16 Distribution of nearly 15,000 earthquakes with magnitudes equal to or greater than 5 for a 10-year period. (Data from U.S. Geological Survey)

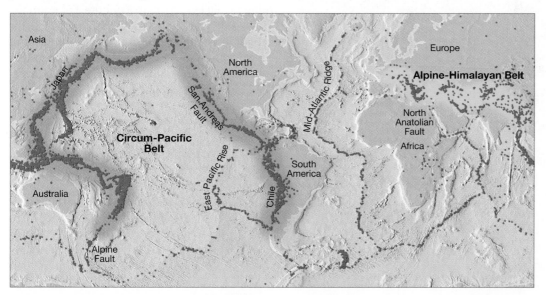

Earthquake Belts and Plate Boundaries

About 95 percent of the energy released by earthquakes originates in the few relatively narrow zones shown in Figure 8.16. The zone of greatest seismic activity, called *circum-Pacific Belt*, encompasses the coastal regions of Chile, Central America, Indonesia, Japan, and Alaska, including the Aleutian Islands. Most earthquakes in the circum-Pacific Belt occur along convergent plate boundaries where one plate slides at a low angle beneath another. The zone of contact between the subducting and overlying plates forms a huge fault called a *megathrust*, along which Earth's largest earthquakes are generated. Because subduction zone earthquakes usually happen beneath the ocean, they may also generate destructive waves called *tsunami*. For example, the 2004 quake off the coast of Sumatra produced a tsunami that claimed an estimated 230,000 lives.

Another major concentration of strong seismic activity, referred to as the *Alpine-Himalayan Belt*, runs through the mountainous regions that flank the Mediterranean Sea and extends past the Himalayan Mountains (Figure 8.16). Tectonic activity in this region is mainly attributed to the collision of the African plate with Eurasia and the collision of the Indian plate with southeast Asia. These plate interactions created many faults that remain active. In addition, numerous faults located away from these plate boundaries have been reactivated as India continues its northward advance into Asia. For example, slippage on a complex fault system in 2008 in the Sichuan Province of China killed at least 70,000 people and left 1.5 million others homeless. The "culprit" is the Indian subcontinent, which shoves the Tibetan Plateau northeastward against the rocks of the Sichuan Basin.

Figure 8.16 shows another continuous earthquake belt that extends for thousands of kilometers through the world's oceans. This zone coincides with the oceanic ridge system, which is an area of frequent but low-intensity seismic activity. As tensional forces pull the plates apart during seafloor spreading, displacement along normal faults generates most of the earthquakes. The remaining seismic activity in this zone is associated with slippage along transform faults located between ridge segments.

Transform faults also run through continental crust where they may generate large earthquakes that tend to occur on a cyclical basis. Examples include California's San Andreas Fault, New Zealand's Alpine Fault, and Turkey's North Anatolian Fault that produced a deadly earthquake in 1999.

CONCEPT CHECK 8.7

1. Where does the greatest amount of seismic activity occur?
2. What type of plate boundary is associated with Earth's largest earthquakes?
3. Name another major concentration of strong earthquake activity.

FIGURE 8.17 Region most affected by the Good Friday earthquake of 1964, the strongest earthquake ever recorded in North America. Note the location of the epicenter (red dot). Inset photo shows the collapse of a street in Anchorage, Alaska. (Photo courtesy of USGS)

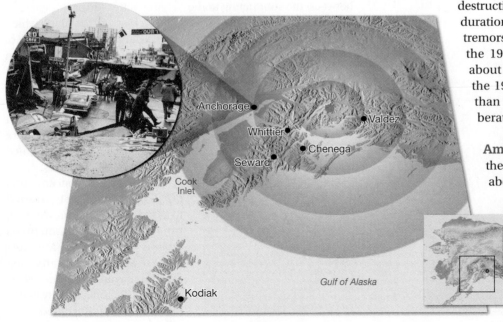

Earthquake Destruction

The most violent earthquake ever recorded in North America—the Good Friday Alaskan earthquake—occurred at 5:36 PM on March 27, 1964. Felt throughout that state, the earthquake had a moment magnitude (M_W) of 9.2 and lasted 3–4 minutes. This event left 131 people dead, thousands homeless, and the economy of the state badly disrupted. Had schools and business districts been open, the toll surely would have been higher. Within 24 hours of the initial shock, 28 aftershocks were recorded, 10 of which exceeded a magnitude of 6.0. The location of the epicenter and the towns that were hardest hit by the quake are shown in Figure 8.17.

Destruction from Seismic Vibrations

The 1964 Alaskan earthquake provided geologists with insights into the role of ground shaking as a destructive force. As the energy released by an earthquake travels along Earth's surface, it causes the ground to vibrate in a complex manner by moving up and down as well as from side to side. The amount of damage to man-made structures attributable to the vibrations depends on several factors, including (1) the intensity and (2) the duration of the shaking, (3) the nature of the material upon which the structure rests, and (4) the nature of building materials and the construction practices of the region.

All of the multistory structures in Anchorage were damaged by the vibrations. The more flexible wood-frame residential buildings fared best. However, many homes were destroyed when the ground failed. A striking example of how construction variations affect earthquake damage is shown in Figure 8.18. You can see that the steel-frame building on the left withstood the vibrations, whereas the poorly designed J.C. Penney building was badly damaged. Engineers have learned that buildings built of blocks and bricks that are not reinforced with steel rods are the most serious safety threat in earthquakes.

Most large structures in Anchorage were damaged, even though they were built according to the earthquake provisions of the Uniform Building Code. Perhaps some of that destruction can be attributed to the unusually long duration of this earthquake. Most quakes involve tremors that last less than a minute. For example, the 1994 Northridge earthquake was felt for about 40 seconds, and the strong vibrations of the 1989 Loma Prieta earthquake lasted less than 15 seconds, but the Alaska quake reverberated for 3–4 minutes.

Amplification of Seismic Waves Although the region near the epicenter will experience about the same intensity of ground shaking, destruction may vary considerably within this area. Such differences are usually attributable to the nature of the ground on which the structures are built. Soft sediments, for example, generally amplify the vibrations more than solid bedrock. Thus, the buildings

FIGURE 8.18 Damage caused to the five-story J.C. Penney Co. building, Anchorage, Alaska. Very little structural damage was incurred by the adjacent building. (Courtesy of NOAA/Seattle)

Students Sometimes Ask...

I've heard that the safest place to be in a house during an earthquake is in a doorframe. Is that really the best place?

No! An enduring earthquake image of California is a collapsed adobe home with the door frame as the only standing part. From this came the belief that a doorway is the safest place to be during an earthquake. In modern homes, doorways are no stronger than any other part of the house and usually have doors that will swing and can injure you.

If you're inside, the best advice is to *duck*, *cover*, and *hold*. When you feel an earthquake, *duck* under a desk or sturdy table. Stay away from windows, bookcases, file cabinets, heavy mirrors, hanging plants, and other heavy objects that could fall. Stay under *cover* until the shaking stops. And, *hold* on to the desk or table: If it moves, move with it.

in Anchorage that were situated on unconsolidated sediments experienced heavy structural damage. By contrast, most of the town of Whittier, although much nearer the epicenter, rested on a firm foundation of solid bedrock and suffered much less damage from seismic shaking. Following the quake, however, Whittier was damaged by a tsunami—a phenomenon that is described later in the chapter.

Liquefaction   In areas where unconsolidated materials are saturated with water, earthquake vibrations can turn stable soil into a mobile fluid, a phenomenon known as **liquefaction** (Figure 8.19). As a result, the ground is not capable of supporting buildings and underground storage tanks, and sewer lines may literally float toward the surface. During the 1989 Loma Prieta earthquake in San Francisco's Marina District, foundations failed and geysers of sand and water shot from the ground, indicating that liquefaction had occurred (Figure 8.20).

FIGURE 8.19 Effects of liquefaction. This building rested on unconsolidated sediment that behaved like quicksand during the 1985 Mexican earthquake. (Photo by James L. Beck)

Much of the damage in the city of Anchorage was attributed to landslides. Homes were destroyed in Turnagain Heights when a layer of clay lost its strength and more than 200 acres of land slid toward the ocean (Figure 8.21). A portion of this spectacular landslide was left in its natural condition as a reminder of this destructive event. The site was appropriately named "Earthquake Park." Downtown Anchorage was also disrupted as sections of the main business district dropped by as much as 3 meters (10 feet).

Fire

More than 100 years ago, San Francisco was the economic center of the western United States, largely because of gold and silver mining. Then, at dawn on April 18, 1906, a violent earthquake struck unexpectedly, triggering an enormous firestorm. Much of the city was reduced to ashes and ruins. It is estimated that 3,000 people died and 225,000 of the city's 400,000 residents were left homeless (see Figure 8.3).

FIGURE 8.20 Liquefaction. **A.** These "mud volcanoes" were produced by the Loma Prieta earthquake of 1989. They formed when geysers of sand and water shot from the ground, an indication that liquefaction occurred. (Photo by Richard Hilton, courtesy of Dennis Fox) **B.** Students experiencing the nature of liquefaction. (Photo by Marli Miller)

FIGURE 8.21 Turnagain Heights slide caused by the 1964 Alaskan earthquake. **A.** Vibrations from the earthquake caused cracks to appear near the edge of the bluff. **B.** Within seconds blocks of land began to slide toward the sea on a weak layer of clay. In less than 5 minutes, as much as 200 meters of the Turnagain Heights bluff area had been destroyed. **C.** Photo of a small portion of the Turnagain Heights slide. (Photo courtesy of U.S. Geological Survey)

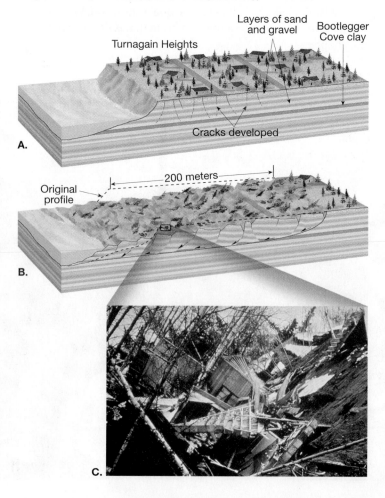

Landslides and Ground Subsidence

The greatest damage to structures is often caused by landslides and ground subsidence triggered by earthquake vibrations. This was the case during the 1964 Alaskan earthquake in Valdez and Seward, where the violent shaking caused deltaic sediments to slump, carrying both waterfronts away. In Valdez, 31 people on a dock died when it slid into the sea. Because of the threat of recurrence, the entire town of Valdez was relocated to more stable ground about 7 kilometers (4 miles) away.

That historic earthquake reminds us of the formidable threat of fire. The central city contained mostly large, older wooden structures and brick buildings. Although many of the unreinforced brick buildings were extensively damaged by vibrations, the greatest destruction was caused by fires, which started when gas and electrical lines were severed. The fires raged out of control for three days and devastated more than 500 blocks of the city. The initial ground shaking, which broke the city's water lines into hundreds of disconnected pieces, made controlling the fires virtually impossible.

The fires were finally contained when buildings were dynamited along a wide boulevard to provide a fire break, similar to the strategy used in fighting forest fires. Only a few deaths were attributed to the San Francisco fires, but other earthquake-initiated fires have been more destructive and claimed many more lives. For example, a 1923 earthquake in Japan triggered an estimated 250 fires, which devastated the city of Yokohama and destroyed more than half the homes in Tokyo. More than 100,000 deaths were attributed to the fires, which were driven by unusually high winds.

What is a Tsunami?

Large undersea earthquakes occasionally set in motion massive waves that scientists call **seismic sea waves**. You may be more familiar with the Japanese term **tsunami**, which is frequently used to describe these destructive phenomena. Because of Japan's location along the circum-Pacific Belt and its expansive coastline, it is especially vulnerable to tsunami destruction.

Most tsunami are caused by the vertical displacement of a slab of seafloor along a fault on the ocean floor, or less often by a large submarine landslide triggered by an earthquake (Figure 8.22). Once generated, a tsunami resembles the ripples formed when a pebble is dropped into a pond. In contrast to ripples, tsunami advance across the ocean at amazing speeds, between 500 and 950 kilometers per hour. Despite this striking characteristic, a tsunami in the open ocean can pass undetected because its height (amplitude) is usually less than 1 meter and the distance between wave crests is great, ranging from 100 to 700 kilometers. However, upon entering shallow coastal waters, these destructive waves "feel bottom" and slow, causing the water to pile up (Figure 8.22). A few exceptional tsunami have reached 30 meters (100 feet) in height. As the crest of a tsunami approaches the shore, it appears as a rapid rise in sea level with a turbulent and chaotic surface (Figure 8.23A).

The first warning of an approaching tsunami is a rapid withdrawal of water from beaches. Some inhabitants of the Pacific basin have learned to heed this warning and move to higher ground. Approximately 5–30 minutes after the retreat of water, a surge capable of extending hundreds of meters inland occurs. In a successive fashion, each surge is followed by a rapid oceanward retreat of the sea.

Tsunami Damage from the 2004 Indonesian Earthquake

A massive undersea earthquake of moment magnitude 9.1 occurred near the island of Sumatra on December 26, 2004, and sent waves of water racing across the Indian Ocean and Bay of Bengal. It was one of the deadliest natural disasters of any kind in modern times, claiming more than 230,000 lives. As water surged several kilometers inland, cars and trucks were flung around like toys in a bathtub, and fishing boats were rammed into homes. In some locations, the backwash of water dragged bodies and huge amounts of debris out to sea.

The destruction was indiscriminate, destroying luxury resorts and poor fishing hamlets on the Indian Ocean coast (see Figure 8.23B). Damages were reported as far away as the Somalia coast of Africa, 4,100 kilometers (2,500 miles) west of the earthquake epicenter.

The killer waves generated by this massive quake achieved heights as great as 10 meters (33 feet) and struck many unprepared areas during a 3-hour span following the earthquake. Although the Pacific basin had a tsunami warning system in place, the Indian Ocean unfortunately did not. The rarity of tsunami in the Indian Ocean also contributed to a lack of preparedness. It should come as no surprise that a tsunami warning system for the Indian Ocean was subsequently established.

FIGURE 8.22 Schematic drawing of a tsunami generated by displacement of the ocean floor. The speed of a wave correlates with ocean depth. Waves moving in deep water advance at speeds in excess of 800 kilometers per hour. Speed gradually slows to 50 kilometers per hour at depths of 20 meters. Decreasing depth slows the movement of the wave. As waves slow in shallow water, they grow in height until they topple and rush onto shore with tremendous force. The size and spacing of these swells are not to scale.

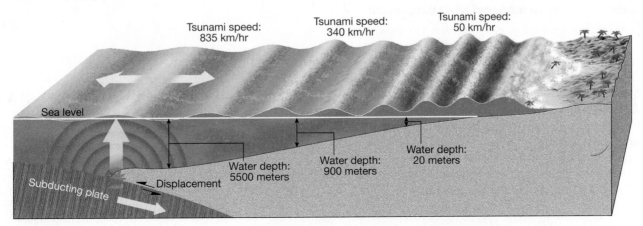

A.

B.

FIGURE 8.23 A massive earthquake (M$_W$ 9.1) off the Indonesian island of Sumatra sent a tsunami racing across the Indian Ocean and Bay of Bengal on December 26, 2004. **A.** Unsuspecting foreign tourists, who at first walked on the sand after the water receded, now rush toward shore as the first of six tsunami roll toward Hat Rai Lay Beach near Krabi in southern Thailand. (AFP/Getty Images Inc.) **B.** Tsunami survivors walk among the debris from this earthquake-triggered event. (Photo by Kimmasa Mayama/Reuters/CORBIS)

Students Sometimes Ask...

What is the largest wave triggered by an earthquake?

The largest wave ever recorded occurred in Lituya Bay, about 200 kilometers (125 miles) west of Juneau, Alaska. On July 9, 1958, an earthquake triggered an enormous rockslide that dumped 90 million tons of rock into the upper part of the bay. The rock-slide created a huge *splash wave* (different than a tsunami, these waves are produced when an object splashes into water) that swept over the ridge facing the rockslide and uprooted or snapped off trees 1,740 feet above the bay. Even larger splash waves have occurred in prehistoric times, including an estimated 3,000-foot wave that is thought to have resulted from a meteorite impact in the Gulf of Mexico about 65 million years ago.

Tsunami Warning System   In 1946, a large tsunami struck the Hawaiian Islands without warning. A wave more than 15 meters (50 feet) high left several coastal villages in shambles. This destruction motivated the U.S. Coast and Geodetic Survey to establish a tsunami warning system for coastal areas of the Pacific. Seismic observatories throughout the region report large earthquakes to the Tsunami Warning Center in Honolulu. Scientists at the Center use deep-sea buoys equipped with pressure sensors to detect energy released by an earthquake. In addition, tidal gauges measure the rise and fall in sea level that accompany tsunami, resulting in warnings issued within the hour. Although tsunami travel very rapidly, there is sufficient time to evacuate all but the areas nearest the epicenter. For example, a tsunami generated off the coast of Chile in 2010 took about 15 hours to reach the Hawaiian Islands (**Figure 8.24**).

CONCEPT CHECK 8.8

1. List four factors that affect the amount of destruction caused by seismic vibrations.
2. In addition to the destruction created directly by seismic vibrations, list three other types of destruction associated with earthquakes.
3. What is a *tsunami*? How is one generated?
4. Cite at least three reasons an earthquake with a moderate magnitude might cause more extensive damage than a quake with a high magnitude.

FIGURE 8.24 Tsunami travel times to Honolulu, Hawaii, from selected locations throughout the Pacific. (Data from NOAA)

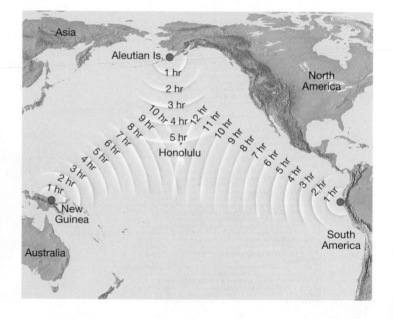

Can Earthquakes be Predicted?

The vibrations that shook Northridge, California, in 1994 caused 60 deaths and more than $40 billion in damage (**Figure 8.25**). This level of destruction was the result of an earthquake of moderate intensity (M_W 6.7). Seismologists warn that other earthquakes of comparable or greater strength can be expected along the San Andreas system, which cuts a 1,300-kilometer (800-mile) path through the state. Can these earthquakes be predicted?

Short-Range Predictions

The goal of short-range earthquake prediction is to provide a warning of the location and magnitude of a large earthquake within a narrow time frame. Substantial efforts to achieve this objective have been attempted in Japan, the United States, China, and Russia—countries where earthquake risks are high. This research has concentrated on monitoring possible *precursors*—events or changes that precede a forthcoming earthquake and thus may provide a warning. In California, for example, seismologists are monitoring changes in ground elevation and variations in strain levels near active faults. Other

researchers are measuring changes in groundwater levels, while still others are trying to predict earthquakes based on an increase in the frequency of foreshocks that precede some, but not all, earthquakes.

One claim of a successful short-range prediction was made by the Chinese government after the February 4, 1975, earthquake in Liaoning Province. According to reports, very few people were killed—even though more than 1 million lived near the epicenter—because the earthquake was predicted and the residents were evacuated. Some Western seismologists have questioned this claim and suggest instead that an intense swarm of foreshocks, which began 24 hours before the main earthquake, may have caused many people to evacuate of their own accord.

One year after the Liaoning earthquake, an estimated 240,000 people died in the Tangshan, China, earthquake, which was *not* predicted (**Table 8.2**). There were no foreshocks. Predictions can also lead to false alarms. In a province near Hong Kong, people reportedly evacuated their dwellings for over a month, but no earthquake followed.

In order for a short-range prediction scheme to warrant general acceptance, it must be both accurate and reliable. Thus, *it must have a small range of uncertainty in regard to location and timing, and it must produce few failures or false alarms.* Can you imagine the debate that would precede an order to evacuate a

FIGURE 8.25 Damage to Interstate 5 caused by the January 17, 1994, Northridge earthquake. (Photo by Tom McHugh/Photo Researchers, Inc.)

TABLE 8.2 Some Notable Earthquakes

Year	Location	Deaths (est.)	Magnitude[†]	Comments
1556	Shensi, China	830,000		Possibly the greatest natural disaster
1755	Lisbon, Portugal	70,000		Tsunami damage extensive
*1811–1812	New Madrid, Missouri	Few	7.9	Three major earthquakes
*1886	Charleston, South Carolina	60		Greatest historical earthquake in the eastern United States
*1906	San Francisco, California	3,000	7.8	Fires caused extensive damage
1908	Messina, Italy	120,000		Deadliest European earthquake; devastating tsunami
1923	Tokyo, Japan	143,000	7.9	Fire caused extensive destruction
1960	Southern Chile	5,700	9.5	The largest-magnitude earthquake ever recorded
*1964	Alaska	131	9.2	Greatest North American earthquake
1970	Peru	70,000	7.9	Great rockslide
*1971	San Fernando, California	65	6.5	Damage exceeded $1 billion
1975	Liaoning Province, China	1,328	7.5	First major earthquake to be predicted
1976	Tangshan, China	255,000	7.5	Not predicted
1985	Mexico City	9,500	8.1	Major damage occurred 400 km from epicenter
1988	Armenia	25,000	6.9	Poor construction practices
*1989	San Francisco Bay area	62	7.1	Damages exceeded $6 billion
1990	Iran	50,000	7.4	Landslides and poor construction practices caused great damage
1993	Latur, India	10,000	6.4	Located in stable continental interior
*1994	Northridge, California	60	6.7	Damages in excess of $15 billion
1995	Kobe, Japan	5,472	6.9	Damages estimated to exceed $100 billion
1999	Izmit, Turkey	17,127	7.4	Nearly 44,000 injured and more than 250,000 displaced
1999	Chi-Chi, Taiwan	2,300	7.6	Severe destruction; 8,700 injuries
2001	Bhuj, India	25,000+	7.9	Millions homeless
2003	Bam, Iran	41,000+	6.6	Ancient city with poor construction
2004	Indian Ocean (Sumatra)	230,000	9.1	Devastating tsunami damage
2005	Pakistan/Kashmir	86,000	7.6	Many landslides; 4 million homeles
2008	Sichuan, China	70,000	7.9	Millions homeless, some towns will not be rebuilt
2010	Port-au-Prince, Haiti	230,000	7.0	More than 300,000 injured and a million homeless
2010	Maule, Chile	486	8.8	One of the 10 largest earthquakes by magnitude

Source: U.S. Geological Survey

*U.S. earthquakes.

[†]Widely differing magnitudes have been estimated for some of these earthquakes. When available, moment magnitudes are used.

large city in the United States, such as Los Angeles or San Francisco? The cost of evacuating millions of people, arranging for living accommodations, and providing for their lost work time and wages would be staggering.

Currently, no reliable method exists for making short-range earthquake predictions. In fact, except for a brief period of optimism during the 1970s, the leading seismologists of the past 100 years have generally concluded that short-range earthquake prediction is not feasible.

Long-Range Forecasts

In contrast to short-range predictions, which aim to predict earthquakes within a time frame of hours or at most days, long-range forecasts give the probability of a certain magnitude earthquake occurring on a time scale of 30–100 years or more. These forecasts give statistical estimates of the expected intensity of ground

motion for a given area over a specified time frame. Although long-range forecasts are not as informative as we might like, these data are useful for providing important guides for building codes so that buildings, dams, and roadways are constructed to withstand expected levels of ground shaking.

For example, in the 1970s, before the 800-mile-long Trans-Alaskan oil pipeline was built, geologists did a hazards study of the Denali Fault system—a major tectonic structure across Alaska. It was determined that during a magnitude-8 earthquake on the Denali Fault, it would experience a 6-meter (20-foot) horizontal displacement. As a result of this investigation, the pipeline was designed to allow it to slide horizontally without breaking (**Figure 8.26**). In 2002, the Denali Fault ruptured, producing a 7.9-magnitude earthquake. Although the total displacement along the fault was about 5 meters (18 feet), there was no oil spill. The Trans-Alaskan pipeline carries nearly 20 percent of the domestic oil supply of the United States—roughly 600,000 barrels per day—

FIGURE 8.26 The Trans-Alaskan oil pipeline was designed and built to withstand several meters of horizontal displacement where it crosses the Denali Fault. During a magnitude-7.9 earthquake in 2002, the pipeline moved as predicted and no oil spill occurred. This illustrates the importance of estimating potential ground motion and designing structures to mitigate the risks. (Photo courtesy of U.S. Geological Survey)

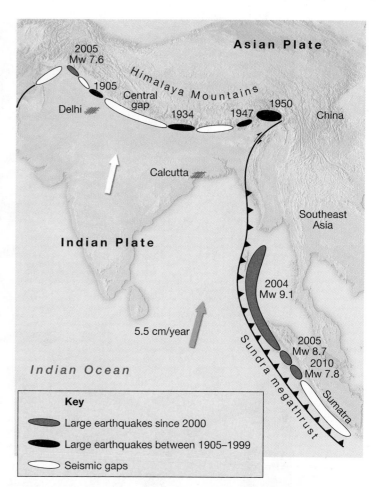

FIGURE 8.27 Seismic gaps. Map of the northern boundary of the Indian plate, where it is moving toward Asia. Shown in red are the rupture zones for the three large earthquakes that have occurred along this plate boundary since 2000. Large earthquakes that occurred between 1905 and 1999 are shown in black and seismic gaps are shown in white. Seismic gaps are "quiet zones" thought to be inactive zones that are storing elastic strain that will eventually produce major earthquakes.

with a degree of scientific reassurance that it will withstand future displacement.

Long-range forecasts are based on evidence that many large faults break repeatedly, producing similar quakes at roughly similar intervals. In other words, as soon as a section of a fault ruptures, the continuing motions of Earth's plates begin to build strain in the rocks again until they fail once more. This led seismologists to study historical records of earthquakes to see if there are any discernible patterns so that the probability of recurrence might be established.

With this concept in mind, seismologists began to plot the distribution of rupture zones associated with great earthquakes around the globe. The maps revealed that individual rupture zones tended to occur adjacent to one another without appreciable overlap, thereby tracing out a plate boundary. Because plates are moving at known velocities, the rate at which strain builds can also be estimated.

When these researchers studied historical records, they discovered that some seismic zones had not produced a large earthquake in more than a century and in some locations for several centuries. These quiet zones, called **seismic gaps**, are believed to be inactive zones that are storing strain for future major quakes.

An area of recent interest to seismologists is the northern edge of the Indian plate, which is colliding with Asia (**Figure 8.27**). Although this area had historically been seismically quiet, four major earthquakes have struck the plate boundary since 2004.

The most destructive was the previously described December 2004 Sumatra earthquake (M_W 9.1). Then in March 2005 and again in April 2010, two strong earthquakes (M_W 8.6) struck Indonesia on the same fault system directly south of the deadly 2004 event. Fortunately, the later events were much less destructive, because no substantial tsunami was generated.

In October 2005, the Pakistan/Kashmir earthquake struck, claiming 86,000 lives. The severity of the destruction caused by this quake was attributed to severe thrusting, coupled with poor construction practices (**Figure 8.28**).

Regrettably, as the map in Figure 8.27 illustrates, several mature seismic gaps (shown in white) are located along this plate margin. One of these lies on the Sunda megathrust, just south of the March 2005/April 2010 ruptures. Did the displacement in 2005 transfer sufficient stress to nudge the neighboring region toward failure?

Other seismic gaps are located within the continent along the margins of the Himalayan Mountains. One of these sites is located

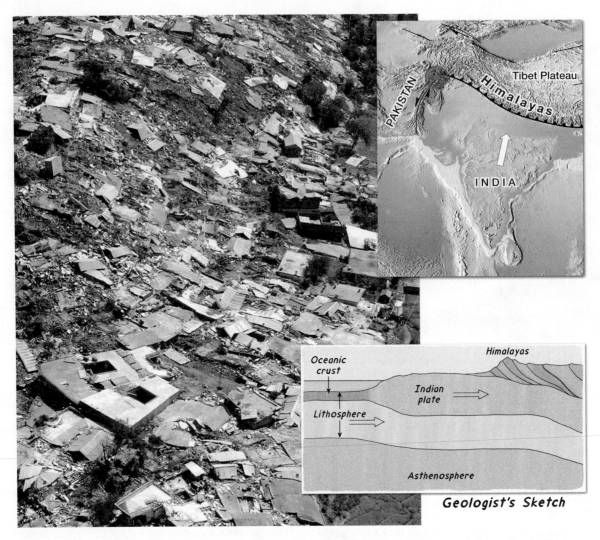

FIGURE 8.28 Destruction caused by the Pakistan/Kashmir earthquake (M_W 7.6) that struck the region in October 2005. The severity of the destruction was attributed to severe thrusting, coupled with poor construction practices. More than 86,000 fatalities occurred. (Photo by AP Wide World Photos)

Students Sometimes Ask...

As compared to continental crust, oceanic crust is quite thin. Has there ever been an attempt to drill through it to obtain a sample of the mantle?

Yes. Project Mohole was initiated in 1958 to retrieve a sample of material from Earth's mantle by drilling a hole through Earth's crust to the *Mohorovičič discontinuity*, or *Moho.* The plan was to drill to the Moho to gain valuable information on Earth's age, makeup, and internal processes. Despite a successful test phase, drilling was halted as control of the project was shifted from one governmental organization to another. Although Project Mohole failed in its intended purpose, it did show that deep-ocean drilling is a viable means of obtaining geological samples.

adjacent to the area of slippage that produced the October 2005 Pakistan/Kashmir quake. Another is a 600-kilometer-long region on the central Himalaya that has apparently not ruptured since 1505.

In summary, *the best prospects for making useful earthquake predictions involve forecasting magnitudes and locations on time scales of years or perhaps even decades. These forecasts are important because they provide information that can be used in the design of structures and to assist in land-use planning in order to reduce injuries and loss of life and property.*

CONCEPT CHECK 8.9

❶ Are accurate, short-range earthquake predictions possible using modern seismic instruments?

❷ What is the value of long-range earthquake forecasts?

Andrea Donnellan: Earthquake Forecaster

Andrea Donnellan talks about earthquakes the same way meteorologists talk about the weather: as dynamic, interconnected systems with their own peculiar set of rules. So it is only appropriate that she refers to her work as earthquake "forecasting."

> **"I look at the quiet part of the earthquake cycle," Donnellan said. "Each earthquake changes where the next earthquake will be."**

Specifically, Donnellan and her colleagues have installed hundreds of high-precision global positioning system (GPS) receivers across southern California, a network known as the Southern California Integrated GPS Network (SCIGN), in order to study the movements of tectonic plates. The receivers, which Donnellan says are more highly sophisticated than commercial GPS devices such as those used in cell phones, can measure millimeter-scale slips for faults and give scientists valuable data for understanding how and why earthquakes take place. Donnellan's work goes far beyond studying just the earthquakes themselves. In fact, most of the work occurs in between all the moving and shaking.

"I look at the quiet part of the earthquake cycle," Donnellan said. "Each earthquake changes where the next earthquake will be."

Like meteorologists modeling a weather system, Donnellan and her colleagues have reproduced earthquake systems using supercomputers in order to understand how they change over time.

"My focus is on modeling earthquake systems," she said. "We want to treat it like weather, where the system is always changing."

But Donnellan's research consists of far more than just sitting in front of a computer all day. In fact, her studies have taken her to places around the globe, from Antarctica to Mongolia to Bolivia, to name a few. Her current research has taken her across much of southern California, where she and her team are studying the San Andreas Fault.

Back in the lab, Donnellan is also involved in an ambitious project called QuakeSim, in which scientists are developing sophisticated, state-of-the-art computer models of earthquake systems.

Donnellan said the results of the project will eventually be accessible by anyone, and that schools are already using some of the software for educational purposes.

> **"My focus is on modeling earthquake systems," she said. "We want to treat it like weather, where the system is always changing."**

In terms of the practical implications for Donnellan's work, her research is already giving scientists a much clearer idea of where earthquakes may strike next, allowing localities to prepare much further in advance. Although, like the weather, earthquakes will probably never be totally predictable, a close study of the ever-shifting activity beneath the ground by Donnellan and her colleagues is making the picture a lot clearer.

—Chris Wilson

Dr. Andrea Donnellan, Deputy Manager of NASA's Jet Propulsion Laboratory Earth and Space Sciences Division, leads a team of scientists working on earthquake forecasting. She was co-winner of the Women in Aerospace 2003 Outstanding Achievement Award and twice a finalist in the NASA astronaut selection process. (NASA Headquarters)

Earth's Interior

GEODe EARTH SCIENCE

Forces Within
▶ Earth's Layered Structure

If you could slice Earth in half, the first thing you would notice is that it has distinct layers. The heaviest materials (metals) would be in the center. Lighter solids (rocks) would be in the middle and liquids and gases would be on top. Within Earth we know these layers as the iron core, the rocky mantle and crust, the liquid ocean, and the gaseous atmosphere. More than 95 percent of the variations in composition and temperature within Earth are due to layering. However, this is not the end of the story. If it were, Earth would be a dead, lifeless cinder floating in space.

There are also variations in composition and temperature with depth that indicate the interior of our planet is very dynamic. The rocks of the mantle and crust are in constant motion, not only moving about through plate tectonics, but also continuously recycling between the surface and the deep interior. Furthermore, it is from Earth's deep interior that the water and air of our oceans and atmosphere are replenished, allowing life to exist at the surface.

Formation of Earth's Layered Structure

As material accumulated to form Earth (and for a short period afterward), the high-velocity impact of nebular debris and the decay of radioactive elements caused the temperature of our planet to steadily increase. During this time of intense heating, Earth became hot enough that iron and nickel began to melt. Melting produced liquid blobs of heavy metal that sank toward the center of the planet. This process occurred rapidly on the scale of geologic time and produced Earth's dense iron-rich core.

The early period of heating resulted in another process of chemical differentiation, whereby melting formed buoyant masses of molten rock that rose toward the surface, and solidified to produce a primitive crust. These rocky materials were rich in oxygen and "oxygen-seeking" elements, particularly silicon and aluminum, along with lesser amounts of calcium, sodium, potassium, iron, and magnesium. In addition, some heavy metals such as gold, lead, and uranium, which have low melting points or were highly soluble in the ascending molten masses, were scavenged from Earth's interior and concentrated in the developing crust. This early period of chemical segregation established the three basic divisions of Earth's interior: (1) the iron-rich *core*, (2) the thin *primitive crust*, and (3) Earth's largest layer, called the *mantle*, which is located between the core and crust.

Probing Earth's Interior: "Seeing" Seismic Waves

Discovering the structure and properties of Earth's deep interior has not been easy. Light does not travel through rock, so we must find other ways to "see" into our planet. The best way to learn about Earth's interior is to dig or drill a hole and examine it directly. Unfortunately, this is only possible at shallow depths. The deepest a drilling rig has ever penetrated is only 12.3 kilometers (8 miles), which is about 1/500 of the way to Earth's center! Even this was an extraordinary accomplishment because temperature and pressure increase rapidly with depth.

Fortunately, many earthquakes are large enough that their seismic waves travel all the way through Earth and can be recorded on the other side (**Figure 8.29**). This means that the seismic waves act like medical x-rays used to take images of a person's insides.

FIGURE 8.29 Slice through Earth's interior showing some of the ray paths that seismic waves from an earthquake would take. Notice that in the mantle, the rays follow curved (refracting) paths rather than straight paths because the seismic velocity of rocks increases with depth, a result of pressure increasing with depth.

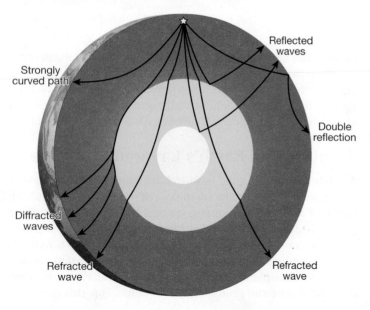

There are about 100–200 earthquakes each year that are large enough (about $M_w > 6$) to be well recorded by seismographs all around the globe. These large earthquakes provide the means to "see" into our planet and have been the source of most of the data that allowed us to figure out the nature of Earth's interior.

Interpreting the waves recorded on seismograms in order to identify Earth structure is challenging. Seismic waves do not travel along straight paths; instead, seismic waves are *reflected*, *refracted*, and *diffracted* as they pass through our planet. They reflect off boundaries between different layers, they refract (or bend) when passing from one layer to another layer, and they diffract around any obstacles they encounter (Figure 8.29). These different wave behaviors have been used to identify the boundaries that exist within Earth.

One of the most noticeable behaviors of seismic waves is that they follow strongly curved paths (Figure 8.29). This occurs because the velocity of seismic waves generally increases with depth. In addition, seismic waves travel faster when rock is stiffer or less compressible. These properties of stiffness and compressibility are then used to interpret the composition and temperature of the rock. For instance, when rock is hotter, it becomes less stiff (imagine taking a frozen chocolate bar and then heating it up!), and waves travel more slowly. Waves also travel at different speeds through rocks of different compositions. Thus, the speed that seismic waves travel can help determine both the kind of rock that is inside Earth and how hot it is.

Earth's Internal Structure

Earth's three compositionally distinct layers—the crust, mantle, and core—can be further subdivided into zones based on physical properties. The physical properties used to define such regions include whether the layer is solid or liquid and how weak or strong it is. Knowledge of both types of layers is essential to our understanding of basic geologic processes, such as volcanism, earthquakes, and mountain building (**Figure 8.30**).

Earth's Crust   The **crust**, Earth's relatively thin, rocky outer skin, is comprised of two types: continental crust and oceanic crust. Both share the word *crust*, but the similarity ends there. The oceanic crust is roughly 7 kilometers (4 miles) thick and composed of the dark igneous rock *basalt*. By contrast, the continental crust averages 35–40 kilometers (22–25 miles) thick but may exceed 70 kilometers (40 miles) in some mountainous regions such as the Rockies and Himalayas. Unlike the oceanic crust, which has a relatively homogeneous chemical composition, the continental crust consists of many rock types. Although the upper crust has an average composition of a *granitic rock* called *granodiorite*, it varies considerably from place to place.

Continental rocks have an average density of about 2.7 grams per cubic centimeter, and some are 4 billion years old. The rocks of the oceanic crust are younger (180 million years or less) and denser (about 3.0 grams per cubic centimeter) than continental rocks.[22]

---

[22]Liquid water has a density of 1 gram per cubic centimeter; therefore, the density of basalt is three times that of water.

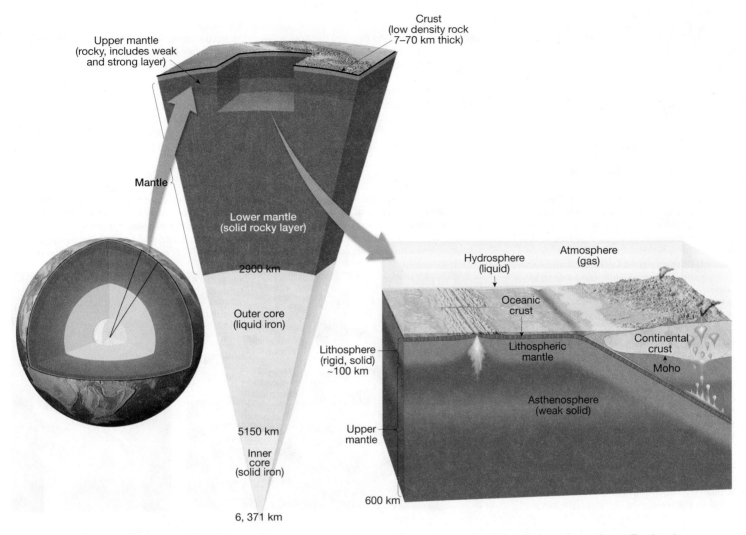

FIGURE 8.30 Views of Earth's layered structure. The study of seismic waves and other geophysical techniques have shown Earth to be a dynamic planet with many interacting parts. The properties of Earth's layers include the physical state of the material (solid, liquid, or gas) as well as how stiff the material is (e.g., the distinction between the lithosphere and asthenosphere). These studies have shown that Earth's layers are mainly determined by density, with the heaviest materials (iron) at the center and the lightest ones on the outside (gases and liquids).

Earth's Mantle   More than 82 percent of Earth's volume is contained in the **mantle**, a solid, rocky shell that extends to a depth of about 2,900 kilometers (1,800 miles). The boundary between the crust and mantle represents a marked change in chemical composition. The dominant rock type in the uppermost mantle is *peridotite*, which is richer in the metals magnesium and iron than the minerals found in either the continental or oceanic crust.

The upper mantle extends from the crust–mantle boundary down to a depth of about 660 kilometers (410 miles). The upper mantle can be divided into two different parts. The top portion of the upper mantle is part of the stiff *lithosphere*, and beneath that is the weaker *asthenosphere*. The **lithosphere** (sphere of rock) consists of the entire crust and uppermost mantle and forms Earth's relatively cool, rigid outer shell. Averaging about 100 kilometers (62 miles) in thickness, the lithosphere is more than 250 kilometers (155 miles) thick below the oldest portions of the continents (Figure 8.30). Beneath this stiff layer to a depth of about 350 kilometers (217 miles) lies a soft, comparatively weak layer known as the **asthenosphere** (weak sphere). The top portion

of the asthenosphere has a temperature/pressure regime that results in a small amount of melting. Within this very weak zone, the lithosphere is mechanically detached from the layer below. The result is that the lithosphere is able to move independently of the asthenosphere, a fact we consider in the next chapter.

It is important to emphasize that the strength of various Earth materials is a function of both their composition and the temperature and pressure of their environment. The entire lithosphere does *not* behave like a brittle solid similar to rocks found on the surface. Rather, the rocks of the lithosphere get progressively hotter and weaker (more easily deformed) with increasing depth. At the depth of the uppermost asthenosphere, the rocks are close enough to their melting temperature that they are very easily deformed, and some melting may actually occur. Thus, the uppermost asthenosphere is weak because it is near its melting point, just as hot wax is weaker than cold wax.

From 660 kilometers (410 miles) deep to the top of the core, at a depth of 2,900 kilometers (1,800 miles), is the lower mantle. Because of an increase in pressure (caused by the weight of the rock above), the mantle gradually strengthens with depth. Despite

UNDERSTANDING EARTH

Recreating the Deep Earth

Seismology alone cannot determine what Earth is made of. Additional information must be obtained by some other means so that the seismic velocities can be interpreted in terms of rock type. This is done using *mineral physics* experiments performed in laboratories. By squeezing and heating minerals and rocks, physical properties like stiffness, compressibility, and density (and therefore seismic velocities) can be directly measured. This means that the conditions of the mantle and core can be simulated, and the results compared to seismic modeling.

Most mineral physics experiments are done using giant presses involving very hard carbonized steel. The highest pressures, however, are obtained using diamond-anvil presses like the one shown in **Figure 8.B**. These take advantage of two important characteristics of diamonds: their hardness and transparency. The tips of two diamonds are cut off, and a small sample of mineral or rock is placed in between. Pressures as great as those in the interior of Jupiter have been

obtained by squeezing the two diamonds together. High temperatures are achieved by firing a laser beam through the diamond and into the mineral sample.

Besides measuring seismic velocities at the conditions of different depths within Earth, there are other important mineral physics experiments. One experiment determines the temperature at which minerals will begin to

melt under various pressures. Another experiment determines (at different temperatures) the pressures at which one mineral phase will become unstable and convert into a new high-pressure phase. Yet another involves making these same tests for slightly different mineral compositions. All of these experiments are needed because there are changes in composition and temperature within Earth.

FIGURE 8.B High-pressure experiments inside a diamond anvil cell (left photo) can recreate the conditions at the center of a planet. The whole apparatus is small and can fit on top of a table. High pressures are generated by cutting the tips off high-quality diamonds (right photo), putting a small sample of rock between, squeezing the diamonds together, and heating the sample with a laser. (Left photo courtesy of Lawrence Livermore National Laboratory; right photo by Douglass L. Peck Photography)

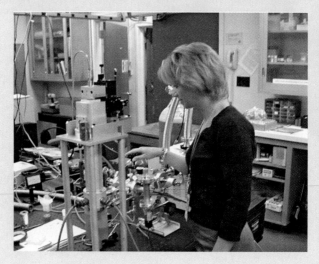

their strength however, the rocks within the lower mantle are very hot and capable of very gradual flow.

Earth's Core   The composition of the **core** is thought to be an iron–nickel alloy with minor amounts of oxygen, silicon, and sulfur—elements that readily form compounds with iron. At the extreme pressure found in the core, this iron-rich material has an average density of nearly 11 grams per cubic centimeter and approaches 14 times the density of water at Earth's center.

The core is divided into two regions that exhibit very different mechanical strengths. The **outer core** is a *liquid layer* 2,270 kilometers (1,410 miles) thick. It is the movement of metallic iron within this zone that generates Earth's magnetic field. The **inner core** is a sphere with a radius of 1,216 kilometers (754 miles).

Despite its higher temperature, the iron in the inner core is *solid* due to the immense pressures that exist in the center of the planet.

CONCEPT CHECK 8.10

1. How did Earth acquire its layered structure?
2. Briefly describe how seismic waves are used to probe Earth's interior.
3. How do continental crust and oceanic crust differ?
4. Contrast the physical makeup of the *asthenosphere* and the *lithosphere*.
5. How are Earth's inner and outer cores different? How are they similar?

GIVE IT SOME THOUGHT

1. Briefly describe the concept of *elastic rebound*. Develop an analogy other than a rubber band to illustrate this concept.
2. The accompanying map shows the locations of the 15 largest earthquakes in the world since 1900. Refer to the map of Earth's plates (Figure 7.10, pp. 202–203) and determine which type of plate boundary is most often associated with these events.

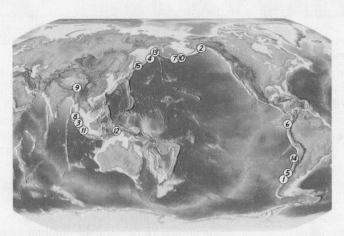

3. Use the accompanying seismogram to answer the following questions:
   a. Which of the three types of seismic waves reached the seismograph first?
   b. What is the time interval between the arrival of the first P wave and the arrival of the first S wave?
   c. Using your answer from question b, and Figure 8.11, determine the distance from the seismic station to the earthquake.
   d. Which of the three types of seismic waves had the highest amplitude when they reached the seismic station?

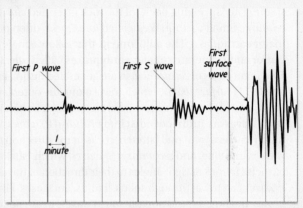

4. You go for a jog on a beach and choose to run near the water where the sand is well packed and solid under your feet. With each step, you notice that your footprint quickly fills with water but not water coming in from the ocean. What is this water's source? For what earthquake-related hazard is this phenomenon a good analogy?

5. Explain, in your own words, why a tsunami often causes a rapid withdrawal of water from beaches before the first surge.

6. Why is it possible to issue a tsunami warning but not provide a warning for an earthquake? Describe a scenario where a tsunami warning would be of little value.

7. Using the accompanying map of the San Andreas Fault, complete the following:
   a. Which of the four segments of the San Andreas Fault do you think has the best chance of experiencing a major earthquake in the foreseeable future?
   b. Which segment do you think is experiencing *fault creep*?
   c. If major earthquakes occur along active segments of the San Andreas Fault about every 200 years, when should the section that ruptured during the Fort Tejon quake be expected to generate another major event?
   d. Do you think San Francisco or Los Angeles has the greater risk of experiencing major earthquake damage in the near future? Defend your selection.

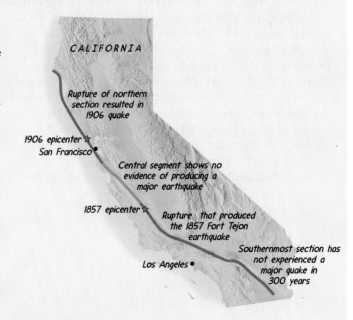

8. Describe the two different ways that Earth's layers are defined.

9. S waves temporarily change the shape of the material that transmits them. Can you identify a place in Earth's interior that would not transmit S waves? Why?

10. Based on the properties of Earth's layers, and the mode of travel of body waves, predict the location in Earth's interior where waves should (a) travel fastest, and (b) travel slowest. Is there an exception for these generalities? Explain your answers.

11. Using the Internet, compare and contrast the 2010 earthquakes in Chile and Haiti. Include magnitude, type of plate boundary, and extent of destruction. Explain why the Chile earthquake produced a tsunami, while the Haiti quake did not.

In Review   Chapter 8 Earthquakes and Earth's Interior

- *Earthquakes* are vibrations of Earth produced by the rapid release of energy from rocks that rupture because they have been subjected to stresses that exceed their strength. This energy, which takes the form of *seismic waves*, radiates in all directions from the earthquake's source, called the *focus*. The movements that produce most large earthquakes occur along large fractures, called *faults*, that are usually associated with plate boundaries.

- Along a fault, rocks store energy as they are bent. As slippage occurs at the weakest point (the focus), displacement will exert stress farther along a fault, where additional slippage will occur until most of the built-up strain is released. An earthquake occurs as the rock elastically returns to its original shape. The "springing back" of the rock is termed *elastic rebound*. Small earthquakes, called *foreshocks*, often precede a major earthquake. The adjustments that follow a major earthquake often generate smaller earthquakes called *aftershocks*.

- Two main types of *seismic waves* are generated during an earthquake: (1) *surface waves*, which travel along the outer layer of Earth, and (2) *body waves*, which travel through Earth's interior. Body waves are further divided into *primary (P) waves*, which push (squeeze) and pull (stretch) rocks in the direction the wave is traveling, and *secondary (S) waves*, which "shake" the particles in rock at right angles to their direction of travel. P waves can travel through solids, liquids, and gases. Fluids (gases and liquids) will not transmit S waves. In any solid material, P waves travel about 1.7 times faster than S waves.

- The location on Earth's surface directly above the focus of an earthquake is the *epicenter*. Using the difference in arrival times between P and S waves, the distance separating a recording station from the earthquake epicenter can be determined. When the distances are known from three or more seismic stations, the epicenter can be located using a method called *triangulation*.

- Seismologists use two fundamentally different measures to describe the size of an earthquake: intensity and magnitude. *Intensity* is a measure of the degree of ground shaking at a given locale based on the observed effects. The *Modified Mercalli Intensity Scale* is divided into 12 levels of severity based on observed effects such as people awakened from sleep, furniture moving, plaster cracking and falling, and finally—total destruction. *Magnitude* is calculated from seismic records and estimates the amount of energy released at the source of an earthquake. Using the *Richter scale*, the magnitude of an earthquake is estimated by measuring the *amplitude* (maximum displacement) of the largest seismic wave recorded. A logarithmic scale is used to express magnitude, in which a tenfold increase in ground shaking corresponds to an increase of 1 on the magnitude scale. *Moment magnitude* is currently used to estimate the size of moderate and large earthquakes. It can be calculated using the amount of slip on the fault surface, the area of the fault surface, and the strength of the faulted rock.

- *A close correlation exists between earthquake epicenters and plate boundaries.* The greatest energy is released by earthquakes along the margin of the Pacific Ocean, known as the *circum-Pacific Belt*, and the mountainous regions that flank the Mediterranean Sea and continue past the Himalayan complex. Another zone of comparatively weak seismicity runs through the world's oceans along the *oceanic ridge system*.

- The primary factors that determine the amount of destruction accompanying an earthquake are the magnitude of the earthquake and the proximity of the quake to a populated area. Structural damage attributable to ground shaking depends on several factors, including (1) the intensity and (2) the duration of ground shaking, (3) the nature of the material upon which the structure rests, and (4) the design of the structure. Secondary effects of earthquakes include *landslides, ground subsidence, fire,* and *tsunami damage*.

- Substantial research to predict earthquakes is under way in Japan, the United States, China, and Russia—countries where earthquake risk is high. No reliable method of short-range prediction has yet been devised. Long-range forecasts are based on the premise that earthquakes are repetitive or cyclical. Seismologists study the history of earthquakes for patterns so their occurrences might be predicted. Long-range forecasts are important because they provide information used to develop the Uniform Building Code and to assist in land-use planning.

- As indicated by the behavior of P and S waves as they travel through Earth, the four major zones of Earth's interior are: (1) *crust* (the very thin outer layer, 5–40 kilometers); (2) *mantle* (a rocky layer located below the crust with a thickness of 2,900 kilometers); (3) *outer core* (a layer about 2,270 kilometers thick, which exhibits the characteristics of a mobile liquid); and (4) *inner core* (a solid metallic sphere with a radius of about 1,216 kilometers).

Key Terms

aftershock (p. 230)
asthenosphere (p. 251)
body wave (p. 234)
core (p. 252)
crust (p. 250)
earthquake (p. 228)
elastic rebound (p. 230)
epicenter (p. 229)
fault (p. 228)
focus (p. 229)
foreshock (p. 231)

inner core (p. 252)
intensity (p. 236)
liquefaction (p. 241)
lithosphere (p. 251)
magnitude (p. 236)
mantle (p. 251)
Modified Mercalli Intensity Scale (p. 237)
moment magnitude (p. 239)
outer core (p. 252)
primary (P) waves (p. 234)
Richter scale (p. 237)

secondary (S) waves (p. 234)
seismic gaps (p. 247)
seismic sea wave (p. 243)
seismic waves (p. 229)
seismogram (p. 234)
seismograph (p. 233)
seismology (p. 233)
surface wave (p. 234)
tsunami (p. 243)

Examining the Earth System

What potentially disastrous phenomenon often occurs when the energy of an earthquake is transferred from the solid earth to the hydrosphere (ocean) at their interface on the floor of the ocean?

When the energy from this event is expended along a coast, how might coastal lands and the biosphere be altered?

Mastering Geology

Looking for additional review and test prep materials? Visit the Self Study area in **www.masteringgeology.com** to find practice quizzes, study tools, and multimedia that will aid in your understanding of this chapter's content. In **MasteringGeology™** you will find:

- **GEODe: Earth Science: An interactive visual walkthrough of key concepts**

- **Geoscience Animation Library: More than 100 animations illuminating many difficult-to-understand Earth science concepts**
- **In The News RSS Feeds: Current Earth science events and news articles are pulled into the site with assessment**
- **Pearson eText**
- **Optional Self Study Quizzes**
- **Web Links**
- **Glossary**
- **Flashcards**

Volcanoes and Other Igneous Activity

A recent eruption of Italy's Mount Etna.
(Photo by Marco Fulle)

The significance of igneous activity may not be obvious at first glance. However, because volcanoes extrude molten rock that formed at great depth, they provide the only windows we have for direct observation of processes that occur many kilometers below Earth's surface. Furthermore, the atmosphere and oceans are thought to have evolved from gases emitted during volcanic eruptions. Either of these facts is reason enough for igneous activity to warrant our attention.

FOCUS ON CONCEPTS

To assist you in learning **the important concepts in this chapter, focus on the following questions:**

- What primary factors determine the nature of volcanic eruptions? How do these factors affect a magma's viscosity?
- What materials are associated with a volcanic eruption?
- What are the eruptive patterns and basic characteristics of the three types of volcanoes generally recognized by volcanologists?
- What destructive forces are associated with composite volcanoes?
- How do calderas form?
- What is the source of magma for flood basalts?
- In what ways can magma be generated from solid rock?
- What is meant by *partial melting*?
- What is the relation between volcanic activity and plate tectonics?
- What changes in a volcanic landscape can be monitored to detect the movement of magma?

Mount St. Helens Versus Kilauea

On Sunday, May 18, 1980, the largest volcanic eruption to occur in North America in historic times transformed a picturesque volcano into a decapitated remnant (**Figure 9.1**). On this date in southwestern Washington State, Mount St. Helens erupted with tremendous force. The blast blew out the entire north flank of the volcano, leaving a gaping hole. In one brief moment, a prominent volcano whose summit had been more than 2,900 meters (9,500 feet) above sea level was lowered by more than 400 meters (1,350 feet).

The event devastated a wide swath of timber-rich land on the north side of the mountain (**Figure 9.2**). Trees within a 400-square-kilometer area lay intertwined and flattened, stripped of their branches and appearing from the air like toothpicks strewn about. The accompanying mudflows carried ash, trees, and water-saturated rock debris 29 kilometers (18 miles) down the Toutle River. The eruption claimed 59 lives, some dying from the intense heat and the suffocating cloud of ash and gases, others from being hurled by the blast, and still others from entrapment in the mudflows.

The eruption ejected nearly a cubic kilometer of ash and rock debris. Following the devastating explosion, Mount St. Helens continued to emit great quantities of hot gases and ash. The force of the blast was so strong that some ash was propelled more than 18,000 meters (over 11 miles) into the stratosphere. During the next few days, this very fine-grained material was carried around Earth by strong upper-air winds. Measurable deposits were reported in Oklahoma and Minnesota, with crop damage into central Montana. Meanwhile, ash fallout in the immediate vicinity exceeded 2 meters in depth. The air over

Students Sometimes Ask...

After all the destruction during the eruption of Mount St. Helens, what does the area look like today?

The area continues to make a slow recovery. Surprisingly, many organisms survived the blast, including animals that live underground and plants (particularly those protected by snow or near streams, where erosion quickly removed the ash). More than 30 years after the blast, plants have revegetated the area, first-growth forests are beginning to be established, and many animals have returned.

The volcano itself is rebuilding, too. A large lava dome is forming inside the summit crater, suggesting that the mountain will build up again. Many volcanoes similar to Mount St. Helens exhibit this behavior: rapid destruction followed by slow rebuilding. If you really want to see what it looks like, go to the Mount St. Helens National Volcanic home page at **http://www.fs.fed.us/gpnf/mshnvm/**, where they have a "volcanocam" with real-time images of the mountain.

Spirit Lake

FIGURE 9.1 Before-and-after photographs show the transformation of Mount St. Helens caused by the May 18, 1980, eruption. (Photo courtesy of U.S. Geological Survey)

Yakima, Washington (130 kilometers to the east), was so filled with ash that residents experienced midnightlike darkness at noon.

Not all volcanic eruptions are as violent as the 1980 Mount St. Helens event. Some volcanoes, such as Hawaii's Kilauea volcano, generate relatively quiet outpourings of fluid lavas. These "gentle" eruptions are not without some fiery displays; occasionally fountains of incandescent lava spray hundreds of meters into the air. Nevertheless, during Kilauea's most recent active phase, which began in 1983, more than 180 homes and a national park visitor center were destroyed.

Testimony to the "quiet nature" of Kilauea's eruptions is the fact that the Hawaiian Volcanoes Observatory has operated on its summit since 1912. This, despite the fact that Kilauea has had more than 50 eruptive phases since record keeping began in 1823.

CONCEPT CHECK 9.1

1. Briefly compare the May 18, 1980, eruption of Mount St. Helens to a typical eruption of Hawaii's Kilauea volcano.

The Nature of Volcanic Eruptions

Forces Within
▶ **Volcanoes and Other Igneous Activity**

Volcanic activity is commonly perceived as a process that produces a picturesque, cone-shaped structure that periodically erupts in a violent manner, like Mount St. Helens. Although some eruptions may be very explosive, many are not. What determines whether a volcano extrudes magma violently or "gently"? The primary factors include the magma's *composition*, its *temperature*, and the amount of *dissolved gases* it contains. To varying degrees, these factors affect the magma's mobility, or **viscosity** (*viscos* = sticky). The more viscous the material, the greater its resistance to flow. For example, compare syrup to water—syrup is more viscous and thus, more resistant to flow, than water. Magma associated with an explosive eruption may be five times more viscous than magma that is extruded in a quiescent manner.

A magma's viscosity is directly related to its silica content—*the more silica in magma, the greater its viscosity*. Silica impedes the flow of magma because silicate structures start to link together into long chains early in the crystallization process. Consequently, rhyolitic (felsic) lavas are very viscous and tend to form comparatively short, thick flows. By contrast, basaltic lavas which contain less silica are relatively fluid and have been known to travel 150 kilometers (90 miles) or more before congealing.

The amount of **volatiles** (the gaseous components of magma, mainly water) contained in magma also affects its mobility. Other factors being equal, water dissolved in the magma tends to increase fluidity because it reduces polymerization (formation of long silicate chains) by breaking silicon–oxygen bonds. It follows, therefore, that the loss of gases renders magma (lava) more viscous.

Why Do Volcanoes Erupt?

Most magma is generated by partial melting in the upper mantle to form molten material having a basaltic composition. Once formed, the buoyant molten rock will rise toward the surface (Figure 9.3). Because the density of crustal rocks decreases the closer they are to the surface, ascending basaltic magma may reach a level where the rocks above are less dense. Should this occur, the molten material begins to collect or pond, forming a magma chamber. As the magma body cools, minerals having high melting temperatures crystallize first, leaving the remaining melt enriched in silica and other less dense components. Some of this molten material may ascend to the surface to produce a volcanic eruption. In most tectonic settings, only a fraction of magma generated at depth ever reaches the surface.

Triggering Hawaiian-Type Eruptions Eruptions that involve very fluid basaltic magmas are often triggered by the arrival of a new batch of melt into a near-surface magma reservoir. This can be detected because the summit of the volcano begins to inflate months, or even years, before an eruption begins. The injection of a fresh supply of melt causes the magma chamber to swell and fracture the rock above. This, in turn, mobilizes the magma, which quickly moves upward along the newly

FIGURE 9.2 Douglas fir trees were snapped off or uprooted by the lateral blast of Mount St. Helens on May 18, 1980. (Large photo by Lyn Topinka/AP Photo/U.S. Geological Survey; inset photo by John M. Burnley/Photo Researchers, Inc.)

Factors Affecting Viscosity

The effect of temperature on viscosity is easily seen. Just as heating syrup makes it more fluid (less viscous), the mobility of lava is strongly influenced by temperature. As lava cools and begins to congeal, its mobility decreases and eventually the flow halts.

Another significant factor influencing volcanic behavior is the chemical composition of the magma. Recall that a major difference among various igneous rocks is their silica (SiO_2) content (Table 9.1). Magmas that produce mafic rocks such as basalt contain about 50 percent silica, whereas magmas that produce felsic rocks (granite and its extrusive equivalent, rhyolite) contain more than 70 percent silica. Intermediate rock types—andesite and diorite—contain about 60 percent silica.

TABLE 9.1 Magmas' Different Compositions Cause Properties to Vary

Composition	Silica Content	Viscosity	Gas Content	Tendency to Form Pyroclastics	Volcanic Landform
Basaltic (Mafic)	Least (~50%)	Least	Least (1–2%)	Least	Shield Volcanoes Basalt Plateaus Cinder Cones
Andesitic (Intermediate)	Intermediate (~60%)	Intermediate	Intermediate (3–4%)	Intermediate	Composite Cones
Rhyolitic (Felsic)	Most (~70%)	Greatest	Most (4–6%)	Greatest	Pyroclastic Flows Volcanic Domes

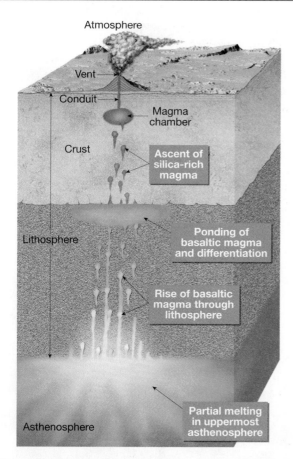

FIGURE 9.3 Schematic drawing showing the movement of magma from its source in the upper asthenosphere through the continental crust. During its ascent, mantle-derived basaltic magma evolves through the process of magmatic differentiation and by melting and incorporating continental crust. Magmas that feed volcanoes in a continental setting tend to be silica-rich (viscous) and have a high gas content.

formed openings, often generating outpourings of lava for weeks, months, or even years.

The Role of Volatiles in Explosive Eruptions   All magmas contain some water and other volatiles that are held in solution by the immense pressure of the overlying rock. Volatiles tend to be most abundant near the tops of magma reservoirs containing silica-rich melts. When magma rises (or the rocks confining the magma fail) a reduction in pressure occurs and the dissolved gases begin to separate from the melt, forming tiny bubbles. This is analogous to opening a warm soda and allowing the carbon dioxide bubbles to escape.

When fluid basaltic magmas erupt, the pressurized gases escape with relative ease. At temperatures of 1000° C and low near-surface pressures, these gases can quickly expand to occupy hundreds of times their original volumes. On some occasions, these expanding gases propel incandescent lava hundreds of meters into the air, producing lava fountains (**Figure 9.4**). Although spectacular, these fountains are mostly harmless and not generally associated with major explosive events that cause great loss of life and property.

At the other extreme, highly viscous, silica-rich magmas may produce explosive clouds of hot ash and gases that evolve into

buoyant plumes called **eruption columns** that extend thousands of meters into the atmosphere (**Figure 9.5**). Because of its high viscosity, a significant portion of the volatiles remain dissolved until the magma reaches a shallow depth, where tiny bubbles begin to form and grow. Bubbles grow by two processes: continued separation of gases from the melt and expansion of bubbles as the confining pressure drops. Should the pressure of the expanding magma body exceed the strength of the overlying rock, fracturing occurs. As magma moves up the fractures, a further drop in confining pressure causes more gas bubbles to form and grow. This chain reaction may generate an explosive event in which magma is literally blown into fragments (ash and pumice) that are carried to great heights by the hot expanding gases. (As exemplified by the

FIGURE 9.4 Fluid basaltic lava erupting from Kilauea Volcano, Hawaii. (Photo by Douglas Peebles/Photolibrary)

1980 eruption of Mount St. Helens, the collapse of a volcano's flank can also trigger an energetic explosive eruption.)

When magma in the uppermost portion of the magma chamber is forcefully ejected by the escaping gases, the confining pressure on the molten rock directly below drops suddenly. Thus, rather than a single "bang," volcanic eruptions are really a series of explosions. This process might logically continue until the entire magma chamber is emptied, much like a geyser empties itself of water (see Chapter 5). However, this is generally not the case. It is typically only the magma in the upper part of a magma chamber that has a sufficiently high gas content to trigger a steam-and-ash explosion.

To summarize, the viscosity of magma, plus the quantity of dissolved gases and the ease with which they can escape, largely determine the nature of a volcanic eruption. In general, hot basaltic magmas contain a smaller gaseous component and permit these gases to escape with relative ease as compared to more silica-rich magmas. This explains the contrast between "gentle" outflows of fluid basaltic lavas in Hawaii and the explosive and sometimes catastrophic eruptions of viscous lavas from volcanoes such as Mount St. Helens (1980), Mount Pinatubo in the Philippines (1991), and Soufriere Hills on the island of Montserrat (1995).

CONCEPT CHECK 9.2

1 List three factors that determine the nature of a volcanic eruption. What role does each play?

2 Generally, what triggers a Hawaiian-type eruption?

3 The eruption of what type of magma may produce an eruption column?

4 Why is a volcano fed by highly viscous magma likely to be a greater threat to life and property than a volcano supplied with very fluid magma?

FIGURE 9.5 Steam and ash eruption column from Mount Augustine, Cook Inlet, Alaska. (Photo by Steve Kaufman/Peter Arnold, Inc.)

Materials Extruded During an Eruption

Forces Within

▶ Volcanoes and Other Igneous Activity

Volcanoes extrude lava, large volumes of gas, and pyroclastic materials (broken rock, lava "bombs," fine ash, and dust). In this section we examine each of these materials.

Lava Flows

The vast majority of lava on Earth, more than 90 percent of the total volume, is estimated to be basaltic in composition. Andesites and other lavas of intermediate composition account for most of the rest, while rhyolitic (felsic) flows make up as little as 1 percent of the total.

Hot basaltic lavas, which are usually very fluid, generally flow in thin, broad sheets or streamlike ribbons. On the island of Hawaii, these lavas have been clocked at 30 kilometers (19 miles) per hour down steep slopes. However, flow rates of 10–300 meters (30–1,000 feet) per hour are more common. By contrast, the movement of silica-rich, rhyolitic lava may be too slow to perceive. Furthermore, most rhyolitic lavas seldom travel more than a few kilometers from their vents. As you might expect, andesitic lavas, which are intermediate in composition, exhibit characteristics that are between the extremes.

Aa and Pahoehoe Flows   Two types of lava flows are known by their Hawaiian names. The most common of these, **aa** (pronounced ah-ah) **flows**, have surfaces of rough jagged blocks with dangerously sharp edges and spiny projections (**Figure 9.6A**). Crossing an aa flow can be a trying and miserable experience. By contrast, **pahoehoe** (pronounced pah-hoy-hoy) **flows** exhibit smooth surfaces that often resemble the twisted braids of ropes (Figure 9.6B). Pahoehoe means "on which one can walk."

Aa and pahoehoe lavas can erupt from the same vent. However, pahoehoe lavas form at higher temperatures and are more fluid than aa flows. In addition, a pahoehoe lava flow can change

FIGURE 9.6 Lava flows **A.** A typical slow-moving, basaltic, aa flow. **B.** A typical fluid pahoehoe (ropy) lava. Both of these lava flows erupted from a rift on the flank of Hawaii's Kilauea Volcano. (Photo **A** by J. D. Griggs, U.S. Geological Survey; photo **B** by Philip Rosenberg/Photolibrary)

into an aa lava flow, although the reverse (aa to pahoehoe) does not occur.

Lava Tubes   Hardened basaltic flows commonly contain cave-like tunnels called **lava tubes** that were once conduits carrying lava from the volcanic vent to the flow's leading edge (Figure 9.7). These conduits develop in the interior of a flow where temperatures remain high long after the surface hardens. Lava tubes are important features because they serve as insulated pathways that facilitate the advance of lava great distances from its source.

Gases

Magmas contain varying amounts of dissolved gases (*volatiles*) held in the molten rock by confining pressure, just as carbon dioxide is held in cans and bottles of soft drinks. As with soft drinks, as soon as the pressure is reduced, the gases begin to escape. Obtaining gas samples from an erupting volcano is difficult and dangerous, so geologists usually must estimate the amount of gas originally contained within the magma.

The gaseous portion of most magmas makes up from 1 to 6 percent of the total weight, with most of this in the form of water vapor. Although the percentage may be small, the actual quantity of emitted gas can exceed thousands of tons per day. Occasionally, eruptions emit colossal amounts of volcanic gases that rise high into the atmosphere, where they may reside for several years. Some of these eruptions may have an impact on Earth's climate, a topic we consider later in this chapter.

The composition of volcanic gases is important because they contribute significantly to our planet's atmosphere. Analyses of samples taken during Hawaiian eruptions indicate that the gas component is about 70 percent water vapor, 15 percent carbon dioxide, 5 percent nitrogen, and 5 percent sulfur dioxide, with lesser amounts of chlorine, hydrogen, and argon. (The relative proportion of each gas varies significantly from one volcanic

FIGURE 9.7 Lava flows often develop a solid crust while the molten lava below continues to advance in conduits called *lava tubes*. View of an active lava tube as seen through the collapsed roof. (Photo by G. Brad Lewis/SPL/Photo Researchers, Inc.)

region to another.) Sulfur compounds are easily recognized by their pungent odor. Volcanoes are also natural sources of air pollution—some emit large quantities of sulfur dioxide, which readily combines with atmospheric gases to form sulfuric acid and other sulfate compounds.

Pyroclastic Materials

When volcanoes erupt energetically they eject pulverized rock, lava, and glass fragments from the vent. The particles produced are referred to as **pyroclastic materials** (*pyro* = fire, *clast* = fragment). These fragments range in size from very fine dust and sand-sized volcanic ash (less than 2 millimeters) to pieces that weigh several tons (Figure 9.8).

Ash and *dust* particles are produced when gas-rich viscous magma erupts explosively (Figure 9.8A). As magma moves up in the vent, the gases rapidly expand, generating a melt that resembles the froth that flows from a bottle of champagne. As the hot gases expand explosively, the froth is blown into very fine glassy fragments. When the hot ash falls, the glassy shards often fuse to form a rock called *welded tuff*. Sheets of this material, as well as ash deposits that later consolidate, cover vast portions of the western United States.

Somewhat larger pyroclasts that range in size from small beads to walnuts are known as *lapilli* ("little stones"). These ejecta are commonly called *cinders* (2–64 millimeters). Particles larger than 64 millimeters (2.5 inches) in diameter are called *blocks* when they are made of hardened lava and *bombs* when they are ejected as incandescent lava (Figure 9.8B and C). Because bombs are semimolten upon ejection, they often take on a streamlined shape as they hurtle through the air (Figure 9.9). Because of their size, bombs and blocks usually fall near the vent; however, they are occasionally propelled great distances. For instance, bombs 6 meters (20 feet) long and weighing about 200 tons were blown 600 meters (2,000 feet) from the vent during an eruption of the Japanese volcano Asama.

So far we have distinguished various pyroclastic materials based largely on the size of the fragments. Some materials are also identified by their texture and composition. In particular, **scoria** is the name applied to vesicular ejecta that is a product of basaltic magma (Figure 9.10A). These black to reddish-brown fragments are generally found in the size range of lapilli and resemble cinders and clinkers produced by furnaces used to smelt iron. When magmas with intermediate (andesitic) or felsic (rhyolitic) compositions erupt explosively, they emit ash and the vesicular rock **pumice** (Figure 9.10B). Pumice is usually lighter in color and less dense than scoria, and many pumice fragments have so many vesicles that they are light enough to float.

A.

B.

C.

FIGURE 9.8 Pyroclastic materials. **A.** *Volcanic ash* and small pumice fragments (*lapilli*) that erupted from Mount St. Helens in 1980. Inset photo is an image obtained using a scanning electron microscope (SEM). This vesicular ash particle exhibits a glassy texture and is roughly the diameter of a human hair. **B.** Volcanic block. Volcanic blocks are solid fragments that were ejected from a volcano during an explosive eruption. **C.** These basaltic bombs were erupted by Hawaii's Mauna Kea Volcano. Volcanic bombs are blobs of lava that are ejected while still molten and often acquire rounded, aerodynamic shapes as they travel through the air. (Photos courtesy of U.S. Geological Survey)

Volcanic Structures and Eruptive Styles

GEODe
Forces Within
EARTH SCIENCE   ▶ Volcanoes and Other Igneous Activity

The popular image of a volcano is that of a solitary, graceful, snowcapped cone, such as Mount Hood in Oregon or Japan's Fujiyama. These picturesque, conical mountains are produced by volcanic activity that occurred intermittently over thousands, or even hundreds of thousands, of years. However, many volcanoes do not fit this image. Cinder cones are quite small and form

---

CONCEPT CHECK 9.3

1 Describe pahoehoe and aa lava flows.
2 How do lava tubes form?
3 List the main gases released during a volcanic eruption. What role do gases play in eruptions?
4 How do volcanic bombs differ from blocks of pyroclastic debris?
5 What is scoria? How is scoria different from pumice?

A. Scoria

B. Pumice

FIGURE 9.10 Scoria and pumice are volcanic rocks that exhibit a vesicular texture. Vesicles are small holes left by escaping gas bubbles. **A.** Scoria is usually a product of mafic (basaltic) magma. **B.** Pumice forms during explosive eruptions of viscous magmas having an intermediate (andesitic) or felsic (rhyolitic) composition. (Photos by E. J. Tarbuck)

FIGURE 9.9 Volcanic bombs forming during an eruption of Hawaii's Kilauea Volcano. Ejected lava masses take on a streamlined shape as they sail through the air. The bomb in the insert is about 10 centimeters long. (Photo by Arthur Roy/National Audubon Society; inset photo by E. J. Tarbuck)

during a single eruptive phase that lasts a few days to a few years. Other volcanic landforms are not volcanoes at all. For example, Alaska's Valley of Ten Thousand Smokes is a flat-topped deposit consisting of 15 cubic kilometers of ash that erupted in less than 60 hours and blanketed a section of river valley to a depth of 200 meters (600 feet).

Volcanic landforms come in a wide variety of shapes and sizes, and each structure has a unique eruptive history. Nevertheless, volcanologists have been able to classify volcanic landforms and determine their eruptive patterns. In this section we consider the general anatomy of a volcano and look at three major volcanic types: shield volcanoes, cinder cones, and composite cones.

Anatomy of a Volcano

Volcanic activity frequently begins when a fissure (crack) develops in the crust as magma moves forcefully toward the surface. As the gas-rich magma moves up through a fissure, its path is usually localized into a circular conduit, or **pipe**, that terminates at a surface opening called a **vent** (**Figure 9.11**). Successive eruptions of lava, pyroclastic material, or frequently a combination of both, often separated by long periods of inactivity, eventually build the cone-shaped structure we call a **volcano**.

Located at the summit of most volcanoes is a somewhat funnel-shaped depression, called a **crater** (*crater* = a bowl). Volcanoes that are built primarily of pyroclastic materials typically have craters that form by gradual accumulation of volcanic debris on the surrounding rim. Other craters form during explosive eruptions as the rapidly ejected particles erode the crater walls. Craters also form when the summit area of a volcano collapses following an eruption (**Figure 9.12**). Some volcanoes have very large circular

FIGURE 9.11 Anatomy of a "typical" composite cone (see also Figures 9.13 and 9.16) for a comparison with a shield and cinder cone, respectively).

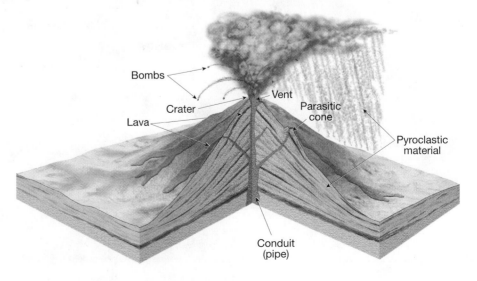

Bombs

Crater

Lava

Vent

Parasitic cone

Pyroclastic material

Conduit (pipe)

Shield Volcanoes

Shield volcanoes are produced by the accumulation of fluid basaltic lavas and exhibit the shape of a broad, slightly domed structure that resembles a warrior's shield (**Figure 9.13**). Most shield volcanoes begin on the ocean floor as seamounts, a few of which grow large enough to form volcanic islands. In fact, with the exception of the volcanic islands that form above subduction zones, most other oceanic islands are either a single shield volcano, or more often, the coalescence of two or more shields built upon massive amounts of pillow lavas. Examples include the Canary Islands, the Hawaiian Islands, the Galapagos, and Easter Island. In addition, some shield volcanoes form on continental crust. Included in this group are several volcanic structures located in East Africa.

Mauna Loa: A Classic Shield Volcano Extensive study of the Hawaiian Islands confirms that they are constructed of myriad thin basaltic lava flows averaging a few meters thick intermixed with relatively minor amounts of pyroclastic ejecta. Mauna Loa is one of five overlapping shield volcanoes that together comprise the Big Island of Hawaii (Figure 9.13). From its base on the floor of the Pacific Ocean to its summit, Mauna Loa is over 9 kilometers (6 miles) high, exceeding the height of Mount Everest. This massive pile of basaltic rock has a volume of 80,000 cubic kilometers that was extruded over a span of about one million years. The volume of material composing Mauna Loa is roughly 200 times greater than the amount composing a large composite cone such as Mount Rainier (**Figure 9.14**). Although the shield volcanoes that comprise islands are often quite large, some are more modest in size. In addition, an estimated one million

FIGURE 9.12 Crater versus caldera. **A.** The crater of Mount Vesuvius, Italy, is about 0.5 kilometers in diameter. The city of Naples is located northwest of Vesuvius, whereas Pompeii, the Roman town that was buried by an eruption in A.D. 79, is located southeast of the volcano. **B.** The huge caldera—6 kilometers in diameter—formed when Tambora's peak was removed during an explosive eruption in 1815. (Photos courtesy of NASA)

depressions called *calderas* that have diameters greater than one kilometer and in rare cases can exceed 50 kilometers. We consider the formation of various types of calderas later in this chapter.

FIGURE 9.13 Shield volcanoes. **A.** Shield volcanoes are built primarily of fluid basaltic lava flows and exhibit the shape of a broad, slightly dome-shaped structure that resembles a warrior's shield. **B.** Mauna Loa is one of five shield volcanoes that collectively make up the island of Hawaii. (Photo by Greg Vaughn/Alamy)

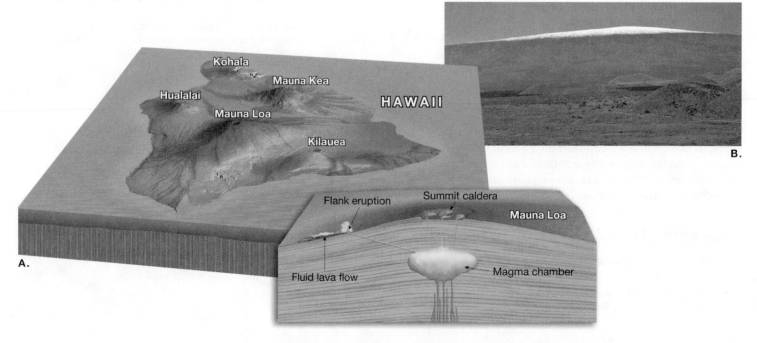

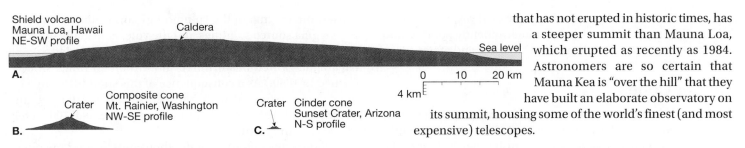

FIGURE 9.14 Profiles comparing scales of different volcanoes.
A. Profile of Mauna Loa, Hawaii, the largest shield volcano in the
Hawaiian chain. Note size comparison with Mount Rainier,
Washington, a large composite cone. **B.** Profile of Mount Rainier,
Washington. Note how it dwarfs a typical cinder cone. **C.** Profile of
Sunset Crater, Arizona, a typical steep-sided cinder cone.

basaltic submarine volcanoes (seamounts) of various sizes dot
the ocean floor.

The flanks of Mauna Loa have gentle slopes of only a few
degrees. The low angle results because very hot, fluid lava travels
"fast and far" from the vent. In addition, most of the lava (perhaps 80 percent) flows through a well-developed system of lava
tubes (see Figure 9.7). This greatly increases the distance lava can
travel before it solidifies. Thus, lava emitted near the summit often
reaches the sea, thereby adding to the width of the cone at the
expense of its height.

Another feature common in many active shield volcanoes is
a large, steep-walled caldera that occupies the summit. Calderas
on large shield volcanoes form when the roof above the magma
chamber collapses. This usually occurs as the magma reservoir
empties following a large eruption, or as magma migrates to the
flank of a volcano to feed a fissure eruption.

In the final stage of growth, shield volcanoes are more sporadic and pyroclastic ejections are more common. Furthermore,
lavas increase in viscosity, resulting in thicker, shorter flows.
These eruptions tend to steepen the slope of the summit area,
which often becomes capped with clusters of cinder cones. This
may explain why Mauna Kea, which is a more mature volcano
that has not erupted in historic times, has
a steeper summit than Mauna Loa,
which erupted as recently as 1984.
Astronomers are so certain that
Mauna Kea is "over the hill" that they
have built an elaborate observatory on
its summit, housing some of the world's finest (and most
expensive) telescopes.

Kilauea, Hawaii: Eruption of a Shield Volcano   Kilauea, the
most active and intensely studied shield volcano in the world, is
located on the island of Hawaii in the shadow of Mauna Loa. More
than 50 eruptions have been witnessed here since record keeping
began in 1823. Several months before each eruptive phase,
Kilauea inflates as magma gradually migrates upward and accumulates in a central reservoir located a few kilometers below the
summit. For up to 24 hours in advance of an eruption, swarms of
small earthquakes warn of the impending activity.

Most of the recent activity on Kilauea has occurred along the
flanks of the volcano in a region called the East Rift Zone. A rift
eruption here in 1960 engulfed the coastal village of Kapoho,
located nearly 30 kilometers (20 miles) from the source. The
longest and largest rift eruption ever recorded on Kilauea began
in 1983 and continues to this day, with no signs of abating.

The first discharge began along a 6-kilometer (4-mile) fissure
where a 100-meter (300-foot) high "curtain of fire" formed as red-
hot lava was ejected skyward (**Figure 9.15**). When the activity
became localized, a cinder and spatter cone, given the Hawaiian name *Puu Oo*, was built. Over the next 3 years the general
eruptive pattern consisted of short periods (hours to days) when
fountains of gas-rich lava sprayed skyward. Each event was followed by nearly a month of inactivity.

By the summer of 1986 a new vent opened 3 kilometers down-
rift. Here smooth-surfaced pahoehoe lava formed a lava lake.
Occasionally the lake overflowed, but more often lava escaped
through tunnels to feed flows that moved down the southeastern
flank of the volcano toward the sea. These flows destroyed nearly

FIGURE 9.15 Lava "curtain" extruded along the East Rift Zone, Kilauea, Hawaii. (Photo by Greg Vaughn/Alamy)

a hundred rural homes, covered a major roadway, and eventually reached the sea. Lava has been intermittently pouring into the ocean ever since, adding new land to the island of Hawaii.

Cinder Cones

As the name suggests, **cinder cones** (also called **scoria cones**) are built from ejected lava fragments that take on the appearance of cinders or clinkers as they begin to harden in flight (see Figure 9.9). These pyroclastic fragments range in size from fine ash to bombs that may exceed a meter in diameter. However, most of the volume of a cinder cone consists of pea- to walnut-sized lapilli that are markedly vesicular and have a black to reddish-brown color. Although cinder cones are composed mostly of loose pyroclastic material, they sometimes extrude lava. On such occasions the discharges most often come from vents located at or near the base rather than from the summit crater.

Cinder cones have very simple, distinctive shapes determined by the slope that loose pyroclastic material maintains as it comes to rest (**Figure 9.16**). Because cinders have a high angle of repose (the steepest angle at which material remains stable), cinder cones are steep-sided, having slopes between 30 and 40 degrees. In addition, cinder cones have large, deep craters in relation to the overall size of the structure. Although relatively symmetrical, many cinder cones are elongated, and higher on the side that was downwind during the eruptions.

Most cinder cones are produced by a single, short-lived eruptive event. One study found that half of all cinder cones examined were constructed in less than one month, and that 95 percent formed in less than one year. However, in some cases, they remain active for several years. Once the event ceases, the magma in the "plumbing" connecting the vent to the magma source solidifies and the volcano usually does not erupt again. (One exception is Cerro Negro, a cinder cone in Nicaragua, which has erupted more than 20 times since it formed in 1850.) As a consequence of this short life span, cinder cones are small, usually between 30 meters (100 feet) and 300 meters (1,000 feet). A few rare examples exceed 700 meters (2,100 feet) in height.

Cinder cones number in the thousands around the globe. Some occur in volcanic fields such as the one near Flagstaff, Arizona, which consists of about 600 cones. Others are parasitic cones that are found on the flanks of larger volcanoes.

Parícutin: Life of a Garden-Variety Cinder Cone One of the very few volcanoes studied by geologists from its very beginning is the cinder cone called Parícutin, located about 320 kilometers (200 miles) west of Mexico City. In 1943 its eruptive phase began in a cornfield owned by Dionisio Pulido, who witnessed the event as he prepared the field for planting.

For 2 weeks prior to the first eruption, numerous Earth tremors caused apprehension in the nearby village of Parícutin. Then, on February 20, sulfurous gases began billowing from a small depression that had been in the cornfield for as long as people could remember. During the night, hot, glowing rock fragments were ejected from the vent, producing a spectacular fireworks display. Explosive discharges continued, throwing hot fragments and ash occasionally as high as 6,000 meters (20,000 feet) into the air. Larger fragments fell near the crater, some remaining incandescent as they rolled down the slope. These built an aesthetically pleasing cone, while finer ash fell over a much larger area, burning and eventually covering the village of

FIGURE 9.16 SP Crater, a cinder cone located in the San Francisco Peaks volcanic field north of Flagstaff, Arizona. Cinder cones are built from ejected lava fragments (mostly cinders and bombs) and are usually less than 300 meters (1,000 feet) in height. The lava flow originated from the base of the cinder cone. (Photo by Michael Collier)

Lava flow

Crater

Pyroclastic material

Central vent filled with rock fragments

FIGURE 9.17 The village of San Juan Parangaricutiro engulfed by aa lava from Parícutin. Only the church towers remain. (Photo by Michael Collier)

Parícutin. In the first day the cone grew to 40 meters (130 feet), and by the fifth day it was more than 100 meters (330 feet) high.

The first lava flow came from a fissure that opened just north of the cone, but after a few months flows began to emerge from the base of the cone itself. In June 1944, a clinkery aa flow 10 meters (30 feet) thick moved over much of the village of San Juan Parangaricutiro, leaving only the church steeple exposed (**Figure 9.17**). After 9 years of intermittent pyroclastic explosions and nearly continuous discharge of lava from vents at its base, the activity ceased almost as quickly as it had begun. Today, Parícutin is just another one of the scores of cinder cones dotting the landscape in this region of Mexico. Like the others, it will not erupt again.

Composite Cones

Earth's most picturesque yet potentially dangerous volcanoes are **composite cones** or **stratovolcanoes**. Most are located in a relatively narrow zone that rims the Pacific Ocean, appropriately called the *Ring of Fire* (see Figure 9.35). This active zone consists of a chain of continental volcanoes that are distributed along the west coast of the Americas, including the large cones of the Andes in South America and the Cascade Range of the western United States and Canada. The latter group includes Mount St. Helens, Mount Shasta, and Mount Garibaldi. The most active regions in the Ring of Fire are located along curved belts of volcanic cones situated adjacent to the deep-ocean trenches of the northern and western Pacific. This nearly continuous chain of volcanoes stretches from the Aleutian Islands to Japan and the Philippines and to the North Island of New Zealand. These impressive volcanic structures are manifestations of processes that occur in the mantle in association with *subduction zones*.

The classic composite cone is a large, nearly symmetrical structure consisting of alternating layers of explosively erupted cinders and ash interbedded with lava flows. A few composite cones, notably Italy's Etna and Stromboli, display very persistent eruption activity, and molten lava has been observed in their summit craters for decades. Stromboli is so well known for eruptions that eject incandescent blobs of lava that it has been referred to as the "Lighthouse of the Mediterranean." Mount Etna, on the other hand, has erupted, on average, once every 2 years since 1979.

Just as shield volcanoes owe their shape to fluid basaltic lavas, composite cones reflect the viscous nature of the material from which they are made. In general, composite cones are the product of gas-rich magma having an andesitic composition. However, many composite cones also emit various amounts of fluid basaltic lava and occasionally, pyroclastic material having rhyolitic composition. Relative to shields, the silica-rich magmas typical of composite cones generate thick viscous lavas that travel less than a few kilometers. In addition, composite cones are noted for generating explosive eruptions that eject huge quantities of pyroclastic material.

A conical shape, with a steep summit area and more gradually sloping flanks, is typical of many large composite cones. This classic profile, which adorns calendars and postcards, is partially a consequence of the way viscous lavas and pyroclastic ejecta contribute to the growth of the cone. Coarse fragments ejected from the summit crater tend to accumulate near their source. Because of their high angle of repose, coarse materials contribute to the steep slopes of the summit area. Finer ejecta, on the other hand, are deposited as a thin layer over a large area. This acts to flatten the flank of the cone. In addition, during the early stages of growth, lavas tend to be more abundant and flow greater distances from the vent than lavas do later in the volcano's history. This contributes to the cone's broad base. As the volcano matures, the shorter flows that come from the central vent serve to armor and strengthen the summit area. Consequently, steep slopes exceeding 40 degrees are sometimes possible. Two of the most perfect cones—Mount Mayon in the Philippines and Fujiyama in Japan—exhibit the classic form we expect of a composite cone, with its steep summit and gently sloping flanks (**Figure 9.18**).

Despite the symmetrical forms of many composite cones, most have complex histories. Huge mounds of volcanic debris surrounding these structures provide evidence that large sections of these volcanoes slid downslope as massive landslides. Others develop horseshoe-shaped depressions at their summits as a result of explosive lateral eruptions—as occurred during the 1980 eruption of Mount St. Helens. Often, so much rebuilding has occurred since these eruptions that no trace of the amphitheater-shaped scars remain.

Many composite cones have numerous small, parasitic cones on their flanks, while others, such as Crater Lake, have been truncated by the collapse of their summit (see Figure 9.22). Still others have a lake in their crater, which may be hot and muddy. Such lakes are often highly acidic because of the influx of sulfur and

FIGURE 9.18 Japan's Fujiyama exhibits the classic form of a composite cone—steep summit and gently sloping flanks. (Photo by Koji Nakano/Getty Images/ Sebun)

chlorine gases that react with water to produce sulfuric (H_2SO_4) and hydrochloric acid (HCl).

CONCEPT CHECK 9.4

1. Compare a volcanic crater to a caldera.
2. Compare and contrast the three main types of volcanoes (consider size, composition, shape, and eruptive style).
3. Name a prominent volcano for each of the three types of volcanoes.
4. Briefly compare the eruptions of Kilauea and Parícutin.

Living in the Shadow of a Composite Cone

More than 50 volcanoes have erupted in the United States in the past 200 years. Fortunately, the most explosive of these eruptions occurred in sparsely inhabited regions of Alaska. On a global scale many destructive eruptions have occurred during the past few thousand years, a few of which may have influenced the course of human civilization.

Nuée Ardente: A Deadly Pyroclastic Flow

One of the most destructive forces of nature are **pyroclastic flows**, which consist of hot gases infused with incandescent ash and larger lava fragments. Also referred to as **nuée ardentes** (*glowing avalanches*), these fiery flows are capable of racing down steep volcanic slopes at speeds that can exceed 200 kilometers (125 miles) per hour (**Figure 9.19**). Nuée ardentes are composed of two parts: a low-density cloud of hot expanding gases containing fine ash particles and a ground-hugging portion that contains most of the material in the flow.

Driven by gravity, pyroclastic flows tend to move in a manner similar to snow avalanches. They are mobilized by volcanic gases released from the lava fragments and by the expansion of heated air that is overtaken and trapped in the moving front. These gases reduce friction between the fragments and the ground. Strong turbulent flow is another important mechanism that aids in the transport of ash and pumice fragments downslope in a nearly frictionless environment (Figure 9.19). This helps explain why some nuée ardente deposits are found more than 100 kilometers (60 miles) from their source.

Sometimes, powerful hot blasts that carry small amounts of ash separate from the main body of a pyroclastic flow. These low-density clouds, called *surges*, can be deadly, but seldom have sufficient force to destroy buildings in their paths. Nevertheless, on June 3, 1991, a hot ash cloud from Japan's Unzen Volcano engulfed and burned hundreds of homes and moved cars as much as 80 meters (250 feet).

Pyroclastic flows may originate in a variety of volcanic settings. Some occur when a powerful eruption blasts pyroclastic material out of the side of a volcano—the lateral eruption of Mount St. Helens in 1980, for example. More frequently, however, nuée ardentes are generated by the collapse of tall eruption columns during an explosive event. When gravity eventually overcomes the initial upward thrust provided by the escaping gases, the ejecta begin to fall, sending massive amounts of incandescent blocks, ash, and pumice cascading downslope.

In summary, pyroclastic flows are a mixture of hot gases and pyroclastic materials moving along the ground, driven primarily by gravity. In general, flows that are fast and highly turbulent can transport fine particles for distances of 100 kilometers or more.

The Destruction of St. Pierre   In 1902, an infamous nuée ardente and associated surge from Mount Pelée, a small volcano on the Caribbean island of Martinique, destroyed the port town of St. Pierre. Although the main pyroclastic flow was largely confined to the valley of Riviere Blanche, the fiery surge spread south of the river and quickly engulfed the entire city. The destruction happened in moments and was so devastating that almost all of St. Pierre's 28,000 inhabitants were killed. Only one person on the outskirts of town—a prisoner protected in a dungeon—and a few people on ships in the harbor were spared (**Figure 9.20**).

Within days of this calamitous eruption, scientists arrived on the scene. Although St. Pierre was mantled by only a thin layer of

FIGURE 9.19 Pyroclastic flows. **A.** Illustration of a fiery ash and pumice flow racing down the slope of a volcano. **B.** Pyroclastic flow moving rapidly down the forested slopes of Mt. Unzen toward a Japanese village. (Photo by Yomiuri/AP Photo)

FIGURE 9.20 The photo on the left shows St. Pierre as it appeared shortly after the eruption of Mount Pelée, 1902. (Reproduced from the collection of the Library of Congress) The photo on the right shows St. Pierre before the eruption. Many vessels are anchored offshore, as was the case on the day of the eruption. (Photo courtesy of The Granger Collection, New York)

volcanic debris, they discovered that masonry walls nearly a meter thick were knocked over like dominoes; large trees were uprooted and cannons were torn from their mounts. A further reminder of the destructive force of this nuée ardente is preserved in the ruins of the mental hospital. One of the immense steel chairs that had been used to confine alcoholic patients can be seen today, contorted, as though it were made of plastic.

Lahars: Mudflows on Active and Inactive Cones

In addition to violent eruptions, large composite cones may generate a type of very fluid mudflow referred to by its Indonesian name **lahar**. These destructive flows occur when volcanic debris becomes saturated with water and rapidly moves down steep volcanic slopes, generally following gullies and stream valleys. Some lahars may be triggered when magma is emplaced near the surface,

PEOPLE AND THE ENVIRONMENT

Eruption of Vesuvius A.D. 79

One well-documented volcanic eruption of historic proportions was the A.D. 79 eruption of the Italian volcano we now call Vesuvius. Prior to this eruption, Vesuvius had been dormant for centuries and had vineyards adorning its sunny slopes. On August 24, however, the tranquility ended, and in less than 24 hours the city of Pompeii (near Naples) and more than 2,000 of its 20,000 residents perished. Some were entombed beneath a layer of pumice nearly 3 meters (10 feet) thick, while others were encased within a layer of ash (**Figure 9.A** bottom). They remained this way for nearly 17 centuries, until the city was excavated, giving archaeologists a superbly detailed picture of ancient Roman life (Figure 9.A top).

By reconciling historical records with detailed scientific studies of the region, volcanologists have pieced together the chronology of the destruction of Pompeii. The eruption most likely began as steam discharges on the morning of August 24. By early afternoon fine ash and pumice fragments formed a tall eruptive cloud. Shortly thereafter, debris from this cloud began to shower Pompeii, which was located 9 kilometers (6 miles) downwind of the volcano. Many people fled during this early phase of the eruption. For the next several hours pumice fragments as large as 5 centimeters (2 inches) fell on Pompeii. One historical record of the eruption states that some people tied pillows to their heads in order to fend off the flying fragments.

The rain of pumice continued for several hours, accumulating at the rate of 12–15 centimeters (5–6 inches) per hour. Most of the roofs in Pompeii eventually gave way. Despite the accumulation of more than 2 meters of pumice, many of the people that had not evacuated Pompeii were probably still alive the next morning. Then, suddenly and unexpectedly, a surge of searing hot ash and gas swept rapidly down the flanks of Vesuvius. This deadly pyroclastic flow killed an estimated 2,000 people who had somehow managed to survive the pumice fall. Most died instantly as a result of inhaling the hot, ash-laden gases. Their remains were quickly buried by the falling ash. Rain then caused the ash to become rock hard before their bodies had time to decay. The subsequent decomposition of the bodies produced cavities in the hardened ash that replicated their forms and, in some cases, even preserved facial expressions. Nineteenth-century excavators found these cavities and created casts of the corpses by pouring plaster of Paris into the voids (Figure 9.A bottom).

Today, Vesuvius towers over the Naples skyline. Such an image should prompt us to consider how volcanic crises might be managed in the future.

FIGURE 9.A The Roman city of Pompeii was destroyed in A.D. 79 during an eruption of Mount Vesuvius. The top photo shows ruins of Pompeii. Excavation began in the 18th century and continues today. The lower photo shows plaster casts of several victims of the A.D. 79 eruption of Mount Vesuvius. (Photo **A** by Roger Ressmeyer, Photo **B** by Leonard von Matt/Photo Researchers, Inc.)

causing large volumes of ice and snow to melt. Others are generated when heavy rains saturate weathered volcanic deposits. Thus, lahars may occur even when a volcano is *not* erupting.

When Mount St. Helens erupted in 1980, several lahars were generated. These flows and accompanying flood waters raced down nearby river valleys at speeds exceeding 30 kilometers per hour. These raging rivers of mud destroyed or severely damaged nearly all the homes and bridges along their paths. Fortunately, the area was not densely populated (**Figure 9.21**).

In 1985, deadly lahars were produced during a small eruption of Nevado del Ruiz, a 5,300-meter (17,400-foot) volcano in the Andes Mountains of Colombia. Hot pyroclastic material melted ice and snow that capped the mountain (*nevado* means *snow* in Spanish) and sent torrents of ash and debris down three major river valleys that flank the volcano. Reaching speeds of 100 kilometers (60 miles) per hour, these mudflows tragically took 25,000 lives.

Mount Rainier, Washington, is considered by many to be America's most dangerous volcano because, like Nevado del Ruiz, it has a thick, year-round mantle of snow and glacial ice. Adding to the risk is the fact that more than 100,000 people live in the valleys around Rainier, and many homes are built on deposits left by lahars that flowed down the volcano hundreds or thousands of years ago. A future eruption, or perhaps just a period of extraordinary rainfall, may produce lahars that could take similar paths.

FIGURE 9.21 Lahars are mudflows that originate on volcanic slopes. This lahar raced down the Muddy River, located southeast of Mount St. Helens, following the May 18, 1980, eruption. Notice the former height of this fluid mudflow as recorded by the mudflow line on the tree trunks. Note person (circled) for scale. (Photo by Lyn Topinka/U.S. Geological Survey)

CONCEPT CHECK 9.5

1. Describe the nature of a pyroclastic flow.
2. Contrast the destruction of Pompeii (see Box 9.1) with the destruction of St. Pierre (discuss time frame, volcanic material, and nature of destruction).
3. Briefly describe a lahar.
4. Why do some people consider Mount Rainier America's most dangerous volcano?

Students Sometimes Ask...

If volcanoes are so dangerous, why do people live on or near them?

Realize that many who live near volcanoes did not choose the location; they were simply born there. Their ancestors may have lived in the region for generations. Historically, many have been drawn to volcanic regions because of their fertile soils. Not all volcanoes have explosive eruptions, but all active volcanoes are dangerous. Certainly, choosing to live close to an active composite cone like Mount St. Helens or Italy's Mount Vesuvius has a high inherent risk. However, the time interval between successive eruptions might be several decades or more—plenty of time for generations of people to forget the last eruption and consider the volcano to be dormant (*dormin* = to sleep) and therefore safe. Many people that choose to live near an active volcano have the belief that the *relative* risk is no higher than in other hazard-prone places. In essence, they are gambling that they will be able to live out their lives before the next major eruption.

Other Volcanic Landforms

The most obvious volcanic structure is a cone. But other distinctive and important landforms are also associated with volcanic activity: calderas, fissure eruptions, basalt plateaus, and volcanic pipes and necks.

Calderas

Calderas (*caldaria* = a cooking pot) are large depressions with diameters that exceed one kilometer and have a somewhat circular form. (Those less than a kilometer across are called *collapse pits* or *craters*.) Most calderas are formed by one of the following processes: (1) the collapse of the summit of a large composite volcano following an explosive eruption of silica-rich pumice and ash fragments (*Crater Lake–type calderas*); (2) the collapse of the top of a shield volcano caused by subterranean drainage from a central magma chamber (*Hawaiian-type calderas*); and (3) the collapse of a large area, caused by the discharge of colossal volumes of silica-rich pumice and ash along ring fractures (*Yellowstone-type calderas*).

Crater Lake–Type Calderas   Crater Lake, Oregon, is situated in a caldera that has a maximum diameter of 10 kilometers (6 miles) and is 1,175 meters (more than 3,800 feet) deep. This caldera formed about 7,000 years ago when a composite cone, later named Mount Mazama, violently extruded 50–70 cubic kilometers of pyroclastic material (**Figure 9.22**). With the loss of support, 1,500 meters (nearly a mile) of the summit of this once prominent cone collapsed. After the collapse, rainwater filled the caldera (Figure 9.22). Later volcanic activity built a small cinder cone in the caldera. Today, this cone, called Wizard Island, provides a mute reminder of past activity.

Hawaiian-Type Calderas   Although some calderas are produced by a *collapse following an explosive eruption*, many are not. For example, Hawaii's active shield volcanoes, Mauna Loa and Kilauea, both have large calderas at their summits. Kilauea's measures 3.3 by 4.4 kilometers (about 2 by 3 miles) and is 150 meters (500 feet) deep. The walls of this caldera are almost vertical, and as a result it looks like a vast, nearly flat-bottomed pit. Kilauea's caldera formed by gradual subsidence as magma slowly drained laterally from the underlying magma chamber to the East Rift Zone, leaving the summit unsupported.

Yellowstone-Type Calderas   Historic and destructive eruptions such as Mount St. Helens and Vesuvius pale in comparison to what happened 630,000 years ago in the region now occupied by Yellowstone National Park, when approximately 1,000 cubic kilometers of pyroclastic material erupted. This super-eruption sent showers of ash as far as the Gulf of Mexico and resulted in the eventual development of a caldera 70 kilometers (43 miles) across. It also gave rise to the Lava Creek Tuff, a hardened ash deposit that is 400 meters (more than 1,200 feet) thick in some places. Vestiges of this event are the many hot springs and geysers in the Yellowstone region.

Based on the extraordinary volume of erupted material, researchers have determined that the magma chambers associated with Yellowstone-type calderas must also be similarly monstrous. As more and more magma accumulates, the pressure within the magma chamber begins to exceed the pressure exerted by the weight of the overlying rocks. An eruption occurs when the gas-rich magma raises the overlying strata enough to create vertical fractures that extend to the surface. Magma surges upward along these cracks, forming a ring-shaped eruption. With a loss of support, the roof of the magma chamber collapses, forcing even more gas-rich magma toward the surface.

Caldera-forming eruptions are of colossal proportions, ejecting huge volumes of pyroclastic materials, mainly in the form of ash and pumice fragments. Typically, these materials form pyroclastic flows that sweep across the landscape, destroying most living things in their paths. Upon coming to rest, the hot fragments of ash and pumice fuse together, forming a welded tuff that closely resembles a solidified lava flow. Despite the immense size of these calderas, their eruptions are brief, lasting hours to perhaps a few days.

Unlike calderas associated with shield volcanoes or composite cones, these depressions are so large and poorly defined that many remained undetected until high-quality aerial and satellite images became available. Other examples of large calderas located in the United States are California's Long Valley Caldera; LaGarita Caldera, located in the San Juan Mountains of southern Colorado; and the Valles Caldera west of Los Alamos, New Mexico.

Fissure Eruptions and Basalt Plateaus

The greatest volume of volcanic material is extruded from fractures in the crust called **fissures** (*fissura* = to split). Rather than building a cone, these long, narrow cracks tend to emit low-viscosity basaltic lavas that blanket a wide area (**Figure 9.23**).

The Columbia Plateau in the northwestern United States is the product of this type of activity (**Figure 9.24**). Numerous **fissure eruptions** have buried the landscape, creating a lava plateau nearly a mile thick. Some of the lava remained molten long enough to flow 150 kilometers (90 miles) from its source. The term **flood basalts** appropriately describes these deposits.

Massive accumulations of basaltic lava, similar to those of the Columbia Plateau, occur elsewhere in the world. One of the largest examples is the Deccan Traps, a thick sequence of flat-lying basalt flows covering nearly 500,000 square kilometers (195,000 square miles) of west central India. When the Deccan Traps formed about 66 million years ago, nearly 2 million cubic

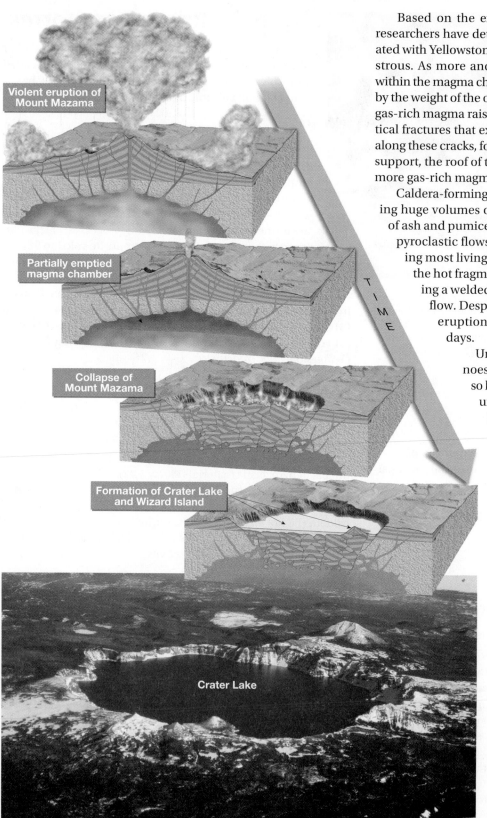

FIGURE 9.22 Sequence of events that formed Crater Lake, Oregon. About 7,000 years ago a violent eruption partly emptied the magma chamber, causing the summit of former Mount Mazama to collapse. Rainfall and groundwater contributed to form Crater Lake, the deepest lake in the United States. Subsequent eruptions produced the cinder cone called Wizard Island. (After H. Williams, *The Ancient Volcanoes of Oregon*. Photo courtesy of the U.S. Geological Survey)

Labels within figure:
- Violent eruption of Mount Mazama
- Partially emptied magma chamber
- Collapse of Mount Mazama
- Formation of Crater Lake and Wizard Island
- Crater Lake
- TIME

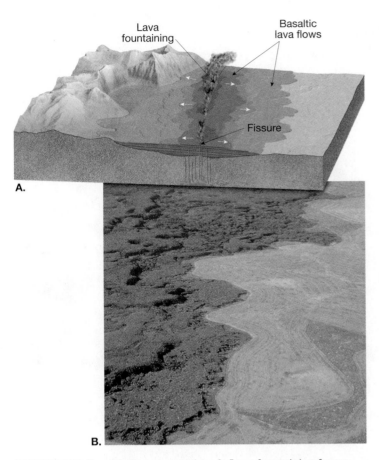

FIGURE 9.23 Basaltic fissure eruption. **A.** Lava fountaining from a fissure and formation of fluid lava flows called *flood basalts*. **B.** These basalt flows are near Idaho Falls. (Photo by John S. Shelton)

kilometers of lava were extruded in less than 1 million years. Several other huge deposits of flood basalts, including the Ontong Java Plateau, are found on the floor of the ocean.

Volcanic Pipes and Necks

Most volcanoes are fed magma through short conduits, called *pipes*, that connect a magma chamber to the surface. One rare type of pipe, called a *diatreme*, extends to depths that exceed 200 kilometers (125 miles). Magmas that migrate upward through diatremes travel rapidly enough that they undergo very little alteration during their ascent. Geologists consider these unusually deep pipes to be "windows" into Earth that allow us to view rock normally found only at great depths.

The best-known volcanic pipes are the diamond-bearing structures of South Africa. The rocks filling these pipes originated at depths of at least 150 kilometers (90 miles), where pressure is high enough to generate diamonds and other high-pressure minerals. The process of transporting essentially unaltered magma (along with diamond inclusions) through 150 kilometers of solid rock is exceptional. This fact accounts for the scarcity of natural diamonds.

Volcanoes on land are continually being lowered by weathering and erosion. Cinder cones are easily eroded because they are composed of unconsolidated materials. However, all volcanoes will eventually succumb to erosion. As erosion progresses, the rock occupying a volcanic pipe is often more resistant and may remain standing above the surrounding terrain long after most of the cone has vanished. Shiprock, New Mexico, is a classic example of this structure, which geologists

FIGURE 9.24 Volcanic areas that comprise the Columbia Plateau in the Pacific Northwest. **A.** The Columbia River basalts cover an area of nearly 200,000 square kilometers (80,000 square miles). Activity began about 17 million years ago as lava poured out of large fissures, eventually producing a basalt plateau with an average thickness of more than 1 kilometer. **B.** Basalt flows exposed along Dry Falls in eastern Washington State. (Photo by Wolfgang Kaehler/Alamy)

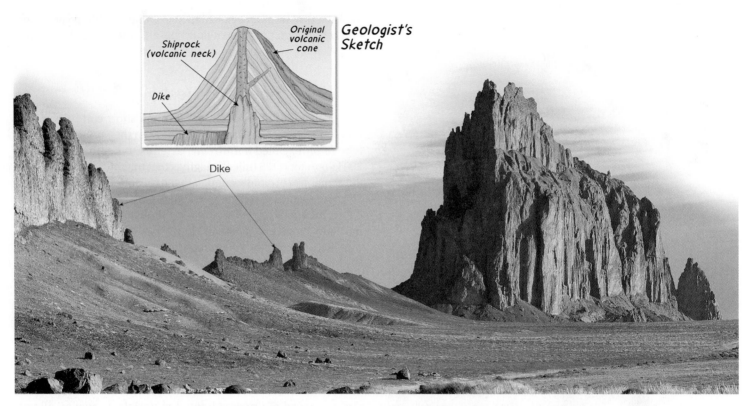

FIGURE 9.25 Shiprock, New Mexico, is a volcanic neck. This structure, which stands over 420 meters (1,380 feet) high, consists of igneous rock that crystallized in the vent of a volcano that has long since been eroded away. (Photo by Dennis Tasa)

call a **volcanic neck** (Figure 9.25). Higher than many skyscrapers, Shiprock is but one of many such landforms that protrude conspicuously from the red desert landscapes of the American Southwest.

CONCEPT CHECK 9.6

1 Describe the formation of Crater Lake. Compare it to the formation of a caldera found on shield volcanoes, such as Kilauea.

2 Extensive pyroclastic flow deposits are associated with which volcanic structure?

3 How do the eruptions that created the Columbia Plateau differ from eruptions that create large composite cones?

4 What is Shiprock, New Mexico, and how did it form?

Intrusive Igneous Activity

 Forces Within
▶ Volcanoes and Other Igneous Activity

Although volcanic eruptions can be violent and spectacular events, most magma is emplaced and crystallizes at depth, without fanfare. Therefore, understanding the igneous processes that occur deep underground is as important to geologists as the study of volcanic events.

When magma rises through the crust, it forcefully displaces preexisting crustal rocks referred to as *host* or *country rock*. Invariably, some of the magma will not reach the surface, but instead crystallize or "freeze" at depth where it becomes an intrusive igneous rock. Much of what is known about intrusive igneous activity has come from the study of old, now solid, magma bodies exhumed by erosion.

Nature of Intrusive Bodies

The structures that result from the emplacement of magma into preexisting rocks are called **intrusions** or **plutons**. Because all intrusions form out of view beneath Earth's surface, they are studied primarily after uplifting and erosion have exposed them. The challenge lies in reconstructing the events that generated these structures millions or even hundreds of millions of years ago.

Intrusions are known to occur in a great variety of sizes and shapes. Some of the most common types are illustrated in Figure 9.26. Notice that some plutons have a tabular (tabletop) shape, whereas others are best described as massive. Also, observe that some of these bodies cut across existing structures, such as sedimentary strata, whereas others form when magma is injected between sedimentary layers. Because of these differences, intrusive igneous bodies are generally classified according to their shape as either **tabular** (*tabula* = table) or **massive** and by their orientation with respect to the host rock. Igneous bodies are said to be **discordant** (*discordare* = to disagree) if they cut across existing structures and **concordant** (*concordare* = to agree) if they form parallel to features such as sedimentary strata.

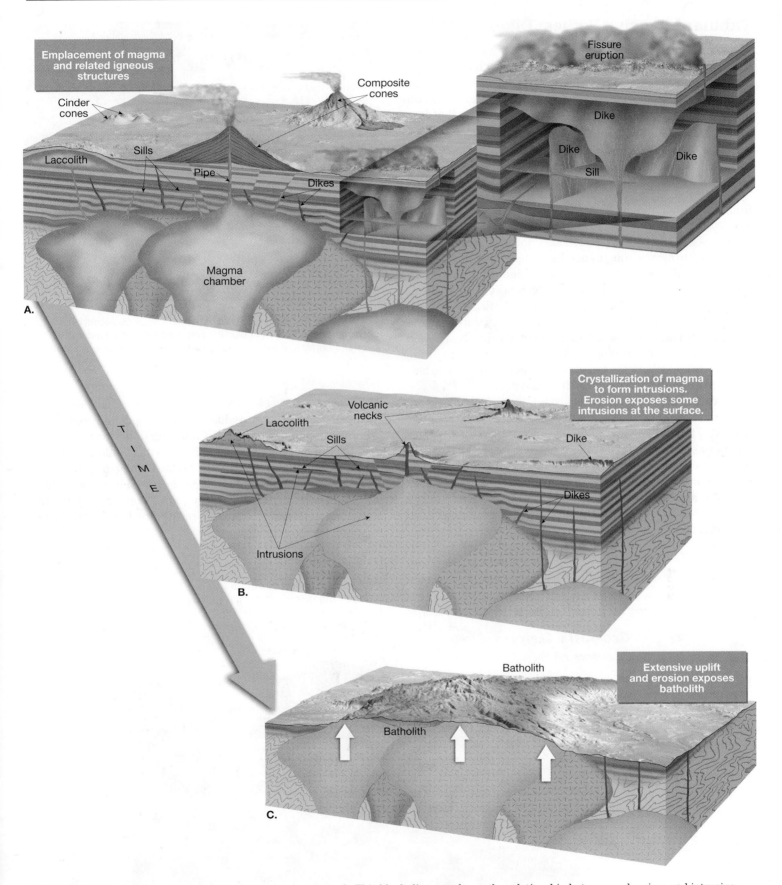

FIGURE 9.26 Illustrations showing basic igneous structures. **A.** This block diagram shows the relationship between volcanism and intrusive igneous activity. **B.** This view illustrates the basic intrusive igneous structures, some of which have been exposed by erosion long after their formation. **C.** After millions of years of uplifting and erosion, a batholith is exposed at the surface.

Tabular Intrusive Bodies: Dikes and Sills

Tabular intrusive bodies are produced when magma is forcibly injected into a fracture or zone of weakness, such as a bedding surface (Figure 9.26). **Dikes** are discordant bodies that cut across bedding surfaces or other structures in the host rock. By contrast, **sills** are nearly horizontal, concordant bodies that form when magma exploits weaknesses between sedimentary beds (Figure 9.27). In general, dikes serve as tabular conduits that transport magma, whereas sills store magma. Dikes and sills are typically shallow features, occurring where the host rocks are sufficiently brittle to fracture. Although they can range in thickness from less than a millimeter to over a kilometer, most are in the 1- to 20-meter range.

Dikes and sills can occur as solitary bodies, but dikes in particular tend to form in roughly parallel groups called *dike swarms*. These multiple structures reflect the tendency for fractures to form in sets when tensional forces stretch brittle country rock. Dikes can also occur radiating from an eroded volcanic neck, like spokes on a wheel. In these situations the active ascent of magma-generated fissures in the volcanic cone out of which lava flowed.

Dikes frequently weather more slowly than the surrounding rock. Consequently, when exposed by erosion, dikes tend to have a wall-like appearance, as shown in Figure 9.25.

Because dikes and sills are relatively uniform in thickness and can extend for many kilometers they are assumed to be the product of very fluid, and therefore, mobile magmas. One of the largest and most studied of all sills in the United States is the Palisades Sill. Exposed for 80 kilometers along the west bank of the Hudson River in southeastern New York and northeastern New Jersey, this sill is about 300 meters thick. Because it is resistant to erosion, the Palisades Sill forms an imposing cliff that can be easily seen from the opposite side of the Hudson.

FIGURE 9.27 Salt River Canyon, Arizona. The dark, essentially horizontal band is a sill of basaltic composition that intruded between horizontal layers of sedimentary rock. (Photo by E. J. Tarbuck)

In many respects, sills closely resemble buried lava flows. Both are tabular and can have a wide aerial extent and both may exhibit columnar jointing (Figure 9.28). **Columnar joints** form as igneous rocks cool and develop shrinkage fractures that produce elongated, pillar-like columns. Furthermore, because sills generally form in near-surface environments and may be only a few meters thick, the emplaced magma often cools quickly enough to generate a fine-grained texture. (Recall that most intrusive igneous bodies have a coarse-grained texture.)

FIGURE 9.28 Columnar jointing in basalt, Giants Causeway National Park, Northern Ireland. These five- to seven-sided columns are produced by contraction and fracturing that results as a lava flow or sill gradually cools. (Photo by John Lawrence/Getty Images)

Geologist's Sketch

Columnar jointing

Rapid cooling from the outside causes shrinkage cracks

Columnar jointing tends to produce 6-sided columns

Massive Intrusive Bodies: Batholiths, Stocks, and Laccoliths

Batholiths   By far the largest intrusive igneous bodies are **batholiths** (*bathos* = depth, *lithos* = stone). Batholiths occur as mammoth linear structures several hundreds of kilometers long and up to 100 kilometers wide (**Figure 9.29**). The Sierra Nevada batholith, for example, is a continuous granitic structure that forms much of the Sierra Nevada, in California. An even larger batholith extends for over 1,800 kilometers (1,100 miles) along the Coast Mountains of western Canada and into southern Alaska. Although batholiths can cover a large area, recent gravitational studies indicate that most are less than 10 kilometers (6 miles) thick. Some are even thinner. The Coastal batholith of Peru, for example, is essentially a flat slab with an average thickness of only 2–3 kilometers (1-2 miles).

Batholiths are almost always made up of granitic (felsic) and intermediate rock types and are often referred to as "granite batholiths." Large granite batholiths consist of hundreds of plutons that intimately crowd against or penetrate one another. These bulbous masses were emplaced over spans of millions of years. The intrusive activity that created the Sierra Nevada batholith, for example, occurred nearly continuously over a 130-million-year period that ended about 80 million years ago (**Figure 9.30**).

Stocks   By definition, a plutonic body must have a surface exposure greater than 100 square kilometers (40 square miles) to be considered a batholith. Smaller plutons of this type are termed **stocks**. However, many stocks appear to be portions of much larger intrusive bodies that would be called batholiths if they were fully exposed.

Laccoliths   A 19th century study by G. K. Gilbert of the U.S. Geological Survey in the Henry Mountains of Utah produced the first clear evidence that igneous intrusions can lift the sedimentary strata they penetrate. Gilbert named the igneous intrusions he observed **laccoliths**, which he envisioned as molten rock forcibly injected between sedimentary strata, so as to arch the beds above, while leaving those below relatively flat. It is now known that the five major peaks of the Henry Mountains are not laccoliths, but stocks. However, these central magma bodies are the source material for branching offshoots that are true laccoliths, as Gilbert defined them (**Figure 9.31**).

Numerous other granitic laccoliths have since been identified in Utah. The largest is a part of the Pine Valley Mountains located north of St. George, Utah. Others are found in the La Sal Mountains near Arches National Park and in the Abajo Mountains directly to the south.

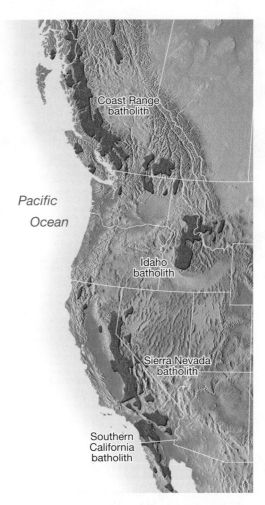

FIGURE 9.29 Granitic batholiths that occur along the western margin of North America. These gigantic, elongated bodies consist of numerous plutons that were emplaced during the last 150 million years of Earth's history.

FIGURE 9.30 Half Dome in Yosemite National Park, California, is part of the Sierra Nevada Batholith (Photo by Enrique R. Aguirre/agefotostock)

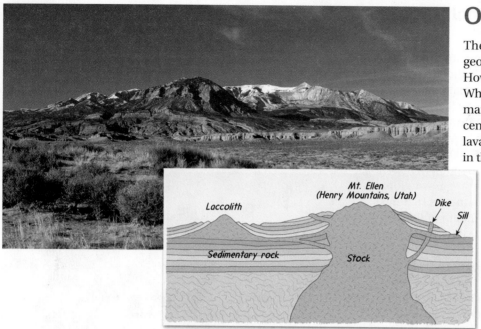

Geologist's Sketch

FIGURE 9.31 Mount Ellen, the northernmost of five peaks that make up Utah's Henry Mountains. Although the main intrusions in the Henry Mountains are stocks, numerous laccoliths formed as offshoots of these structures. (Photo by Michael DeFreitas North America/Alamy)

CONCEPT CHECK 9.7

① Describe each of the four basic intrusive features (dike, sill, batholith, and laccolith).

② What is the largest of all intrusive igneous bodies? Is it tabular or massive? Concordant or discordant?

Students Sometimes Ask...

Some of the larger volcanic eruptions, like the eruption of Krakatau, must have been impressive. What was it like?

On August 27, 1883, in what is now Indonesia, the volcanic island of Krakatau exploded and was nearly obliterated. The sound of the explosion was heard an incredible 4,800 kilometers (3,000 miles) away at Rodriguez Island in the western Indian Ocean. Dust from the explosion was propelled into the atmosphere and circled Earth on high-altitude winds. This dust produced unusual and beautiful sunsets for nearly a year.

Not many were killed directly by the explosion, because the island was uninhabited. However, the displacement of water from the explosion was enormous. The resulting *tsunami* exceeded 35 meters (116 feet) in height. It devastated the coastal region of the Sunda Strait between the nearby islands of Sumatra and Java, taking more than 36,000 lives. The energy carried by this wave reached every ocean basin and was detected by tide-recording stations as far away as London and San Francisco.

Origin of Magma

The origin of magma has been controversial in geology, almost from the beginning of the science. How do magmas of different compositions form? Why do volcanoes in the deep-ocean basins primarily extrude basaltic lava, whereas those adjacent to oceanic trenches extrude mainly andesitic lava? These are some of the questions we address in the following sections.

Generating Magma from Solid Rock

Based on evidence from the study of earthquake-generated waves, *Earth's crust and mantle are composed primarily of solid, not molten, rock.* Although the outer core is a fluid, its iron-rich material is very dense and remains deep within Earth. So what is the source of magma that produces igneous activity?

Increase in Temperature   Most magma originates when essentially solid rock, located in the crust and upper mantle, melts. The most obvious way to generate magma from solid rock is to raise the temperature above the rock's melting point.

Workers in underground mines know that temperatures get higher as they go deeper. Although the rate of temperature change varies considerably from place to place, it *averages* about 25° C per kilometer in the *upper* crust. This increase in temperature with depth, known as the **geothermal gradient**, is somewhat higher beneath the oceans than beneath the continents. As shown in Figure 9.32, when a typical geothermal gradient is compared to the melting point curve for the mantle rock peridotite, the temperature at which peridotite melts is everywhere higher than the geothermal gradient. Thus, under normal conditions, the mantle is solid. As you will see, tectonic processes exist that can increase the geothermal gradient sufficiently to trigger melting. In addition, other mechanisms exist that trigger melting by reducing the temperature at which peridotite begins to melt.

Decrease in Pressure: Decompression Melting   If temperature were the only factor that determined whether or not rock melts, our planet would be a molten ball covered with a thin, solid outer shell. This, of course, is not the case. The reason is that pressure also increases with depth.

Melting, which is accompanied by an increase in volume, *occurs at higher temperatures at depth* because of greater confining pressure. Consequently, an increase in confining pressure causes an increase in the rock's melting temperature. Conversely, reducing confining pressure lowers a rock's melting temperature. When confining pressure drops sufficiently, **decompression melting** is triggered.

Decompression melting occurs where hot, solid mantle rock ascends in zones of convective upwelling, thereby moving into regions of lower pressure. This process is responsible for generating magma along divergent plate boundaries (oceanic ridges)

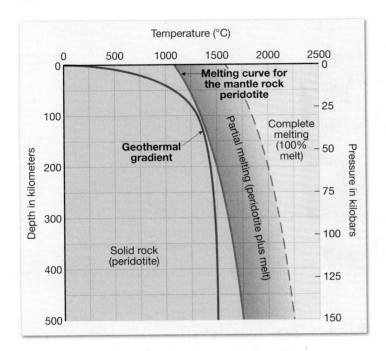

FIGURE 9.32 A schematic diagram illustrating a typical geothermal gradient (increase in temperature with depth) for the crust and upper mantle. Also illustrated is an idealized curve that depicts the melting point temperatures for the mantle rock peridotite. Notice that when the geothermal gradient is compared to the melting point curve for peridotite, the temperature at which peridotite melts is everywhere higher than the geothermal gradient. Thus, under normal conditions the mantle is solid. Special circumstances are required to generate magma.

where plates are rifting apart (**Figure 9.33**). Below the ridge crest, hot mantle rock rises and melts replacing the material that shifted horizontally away from the ridge axis. Decompression melting also occurs within ascending mantle plumes.

Addition of Volatiles   Another important factor affecting the melting temperature of rock is its water content. Water and other volatiles, such as carbon dioxide, act as salt does to melt ice. That is, volatiles cause rock to melt at lower temperatures. Furthermore, the effect of volatiles is magnified by increased pressure. Deeply buried "wet" rock has a much lower melting temperature than "dry" rock of the same composition. Therefore, in addition to a rock's composition, its temperature, depth (confining pressure), and water content determine whether it exists as a solid or liquid.

Volatiles play an important role in generating magma at convergent plate boundaries where cool slabs of oceanic lithosphere descend into the mantle (**Figure 9.34**). As an oceanic plate sinks, both heat and pressure drive water from the subducting crustal rocks. These fluids, which are very mobile, migrate into the wedge of hot mantle that lies directly above. The addition of water lowers the melting temperature of peridotite sufficiently to generate some melt. Laboratory studies have shown that the temperature at which peridotite begins to melt can be lowered by as much as 100° C by the addition of only 0.1 percent water.

Melting of peridotite generates basaltic magma having a temperature of 1200° C or higher. When enough mantle-derived basaltic magma forms, it will

buoyantly rise toward the surface. In a continental setting, basaltic magma may "pond" beneath crustal rocks, which have a lower density and are already near their melting temperature. This may result in some melting of the crust and the formation of a secondary, silica-rich magma.

In summary, magma can be generated three ways: (1) when an *increase in temperature* causes a rock to exceed its melting point; (2) in zones of upwelling *a decrease in pressure* (without the addition of heat) can result in *decompression melting*; and (3) the *introduction of volatiles* (principally water) can lower the melting temperature of hot mantle rock sufficiently to generate magma.

Partial Melting and Magma Compositions

An important difference exists between the melting of a substance that consists of a single compound, such as ice, and melting igneous rocks, which are mixtures of several different minerals. Ice melts at a specific temperature, whereas igneous rocks melt over a temperature range of about 200° C. As rock is heated, minerals with the lowest melting points tend to start melting first. Should melting continue, minerals with higher melting points begin to melt, and the composition of the magma steadily approaches the overall composition of the rock from which it was derived. Most often, melting is not complete. This process, known as **partial melting**, produces most, if not all, magma.

An important consequence of partial melting is the *production of a magma with a higher silica content than the original rock*. Recall from the discussion of Bowen's reaction series that basaltic (mafic) rocks contain mostly high-melting-temperature minerals that are comparatively low in silica, whereas granitic (felsic) rocks are composed primarily of low-melting-temperature silicates that are enriched in silica (see Chapter 3). Because silica-rich minerals melt first, magmas generated by partial melting are nearer to the granitic end of the compositional spectrum than are the rocks from which they formed.

FIGURE 9.33 As hot mantle rock ascends, it continually moves into zones of lower pressure. This drop in confining pressure can trigger melting, even without additional heat.

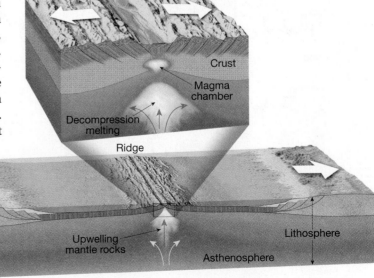

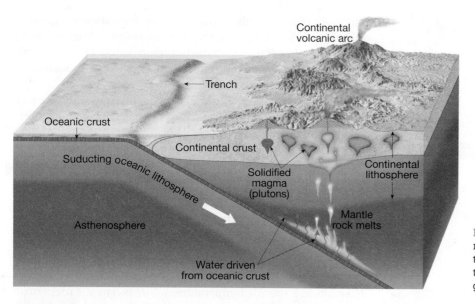

FIGURE 9.34 As an oceanic plate descends into the mantle, water and other volatiles are driven from the subducting crustal rocks. These volatiles lower the melting temperature of mantle rock sufficiently to generate magma.

CONCEPT CHECK 9.8

1 Define *geothermal gradient*.
2 Describe the three ways that solid rock in the upper mantle and crust may melt to become magma.
3 In which two settings does decompression melting occur?
4 What is partial melting?
5 How does the composition of a melt produced by partial melting compare with the composition of the parent rock?

Plate Tectonics and Volcanic Activity

Geologists have known for decades that the global distribution of volcanism is not random. Most active volcanoes are located along the margins of the ocean basins—notably within the circum-Pacific belt known as the *Ring of Fire* (**Figure 9.35**). These volcanoes consist mainly of composite cones that emit volatile-rich

FIGURE 9.35 The Ring of Fire contains the largest concentration of Earth's major volcanoes. Inset shows Ecuador's Cotopaxi volcano. (Photo by Patrick Esudero/Photolibrary)

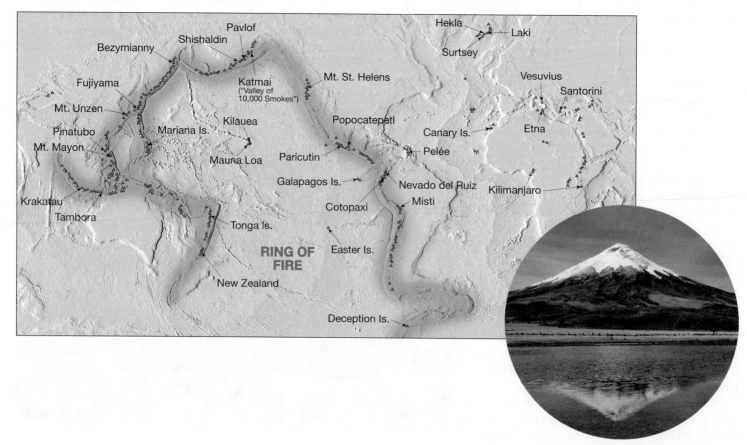

Chris Eisinger: Studying Active Volcanoes

Chris Eisinger describes volcanoes the same way he might discuss old, temperamental friends. "For the most part, volcanoes are very approachable," he says, "as long as you know their cycles." As a graduate student in the Department of Geological Sciences at Arizona State University, Eisinger had 10 years of firsthand experience with volcanoes under his belt. He began as an undergraduate and spent time at the Hawaiian Volcano Observatory as well as in Indonesia, a region rife with volcanic activity.

But the summit of an active volcano is anything but a breezy locale. On the contrary, volcanoes emit malodorous gases, as Eisinger can attest.

"The most distinct things you notice are the fumes," he says. "You get a strong sulfur smell."

If the volcano is near enough to the ocean for lava to flow into the sea, Eisinger also says the runoff creates a steam that is highly acidic, thanks to the high chlorine content of seawater. "Your eyes will water," he said. "It's typically difficult to breathe and you end up coughing if you're not wearing a gas mask."

Rainstorms can further cloud the air, creating steam when the falling water strikes the hot surface of the lava. At times, Eisinger says, visibility for volcanologists studying at the summit is reduced to a few feet.

The heavy influx of gases doesn't affect only the eyes and nose, either. The taste, if one isn't wearing a mask, can be "pretty nasty," Eisinger says, adding that the smell adheres to one's clothing as well.

For volcanoes that are in a constant state of eruption, Eisinger said there are distinct patterns in the emission cycles. Volcanoes he has visited in Indonesia, for example, would emit a stream of gaseous material every 20 to 60 minutes.

But while most volcanologists are able to remain safe by paying close attention to eruption cycles, fatalities have happened. Recently, in August 2000, two volcanologists died at the summit of Semeru on the island of Java in Indonesia when the volcano erupted with no warning.

Active volcanoes also emit low, rumbling sounds, Eisinger said, due largely to subterranean explosions that tend to be very muffled. Hawaiian volcanoes are unique in that they also emit an intense hissing sound, "like a jet engine," from the expulsion of gases. Eruptions also are often foreshadowed

"For the most part, volcanoes are very approachable as long as you know their cycles."

by seismic activity that contributes to the rumble of the explosions. Though Eisinger himself has never witnessed a volcanic eruption of historic magnitude, he noted that accounts of such colossal eruptions as that of Krakatau in 1883 produced reports of a boom that could be heard thousands of miles away.

—Chris Wilson

Volcanologist Chris Eisinger uses a crowbar to hit some molten sulfur on Kawah Ijen in Indonesia. The 200° C (390° F) molten sulfur is still red hot and will cool to a yellow or green color. (Courtesy of Chris Eisinger)

magma having an intermediate (andesitic) composition and that occasionally produce awe-inspiring eruptions.

A second group includes the basaltic shields that emit very fluid lavas. These volcanic structures comprise most of the islands of the deep ocean basins, including the Hawaiian Islands, the Galapagos Islands, and Easter Island. In addition, this group includes many active submarine volcanoes that dot the ocean floor; particularly notable are the innumerable small seamounts that occur along the axis of the mid-ocean ridge.

A third group includes volcanic structures that appear to be somewhat randomly distributed in the interiors of the continents. None are found in Australia nor in the eastern two-thirds of North and South America. Africa is notable because it has many potentially active volcanoes including Mount Kilimanjaro, the highest point on the continent (5,895 meters [19,454 feet]). When compared

to volcanism in the ocean basin, volcanism on continents is more diverse, ranging from eruptions of very fluid basaltic lavas, like those that generated the Columbia Plateau, to explosive eruptions of silica-rich rhyolitic magma as occurred in Yellowstone.

Until the late 1960s, geologists had no explanation for the apparently haphazard distribution of continental volcanoes, nor were they able to account for the almost continuous chain of volcanoes that circles the margin of the Pacific basin. With the development of the theory of plate tectonics, the picture was greatly clarified. Recall that most magma originates in the upper mantle and that the mantle is essentially solid, *not molten* rock. The basic connection between plate tectonics and volcanism is that *plate motions provide the mechanisms by which mantle rocks melt to generate magma.*

We now examine three zones of igneous activity and their relationship to plate boundaries (Figure 9.36). These active areas

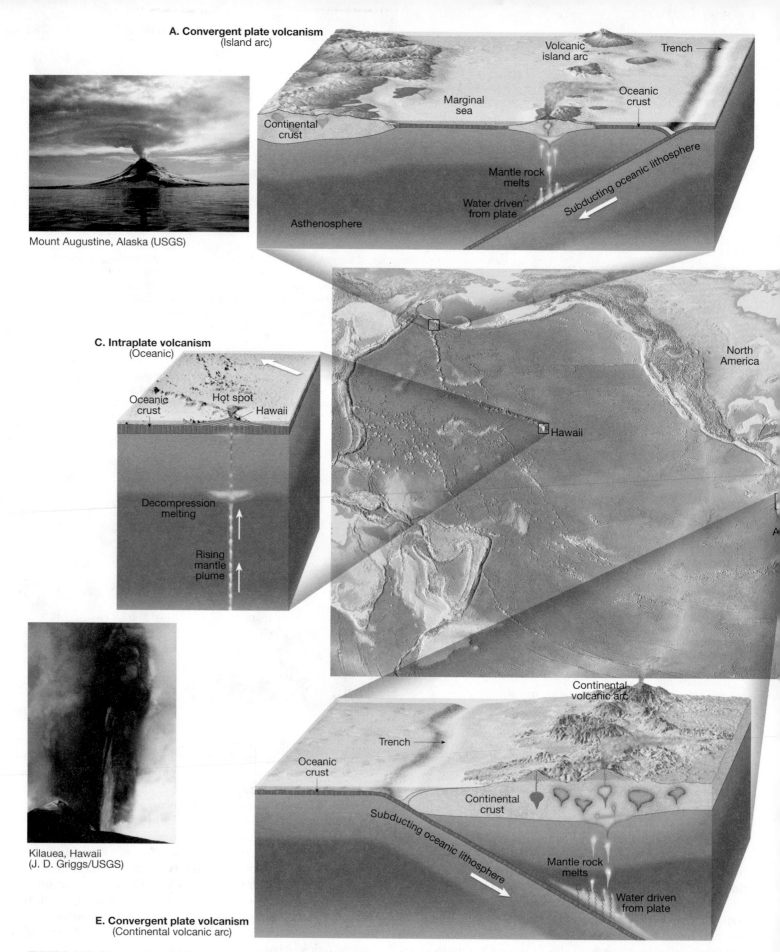

A. Convergent plate volcanism
(Island arc)

Volcanic island arc

Trench

Marginal sea

Oceanic crust

Continental crust

Mantle rock melts

Water driven from plate

Subducting oceanic lithosphere

Asthenosphere

Mount Augustine, Alaska (USGS)

C. Intraplate volcanism
(Oceanic)

Oceanic crust

Hot spot

Hawaii

Decompression melting

Rising mantle plume

Kilauea, Hawaii
(J. D. Griggs/USGS)

North America

Hawaii

Continental volcanic arc

Trench

Oceanic crust

Continental crust

Subducting oceanic lithosphere

Mantle rock melts

Water driven from plate

E. Convergent plate volcanism
(Continental volcanic arc)

FIGURE 9.36 Three zones of volcanism. Two of the zones are associated with plate boundaries. The third zone includes those volcanic structures that are irregularly distributed in the interiors of plates.

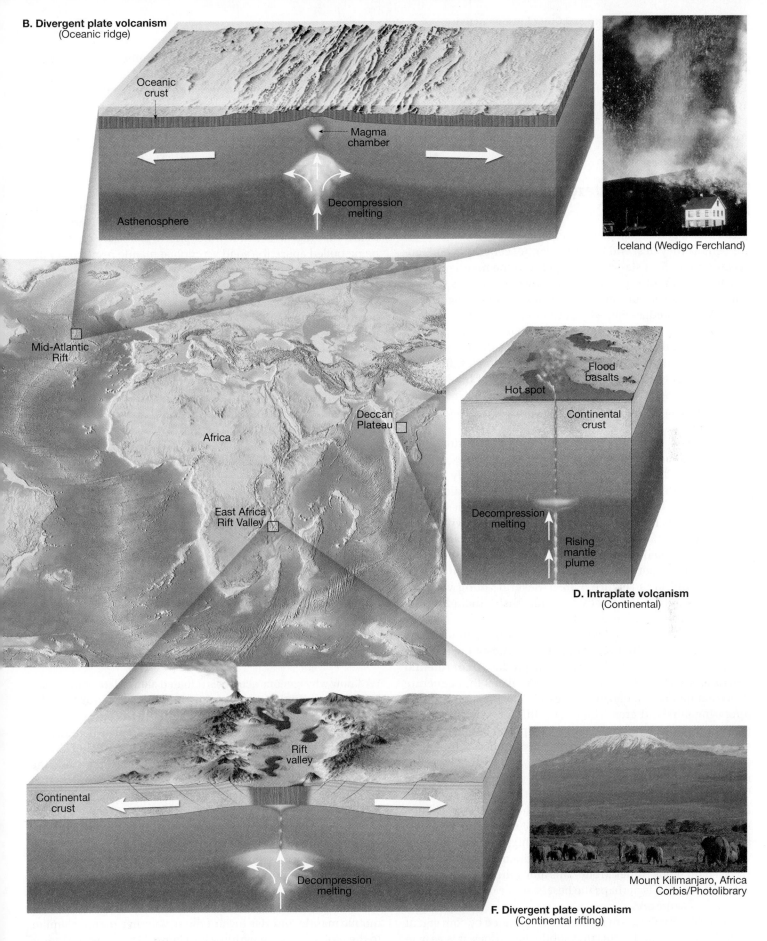

B. Divergent plate volcanism
(Oceanic ridge)

Oceanic crust

Magma chamber

Decompression melting

Asthenosphere

Iceland (Wedigo Ferchland)

Mid-Atlantic Rift

Africa

Deccan Plateau

East Africa Rift Valley

D. Intraplate volcanism
(Continental)

Hot spot

Flood basalts

Continental crust

Decompression melting

Rising mantle plume

Rift valley

Continental crust

Decompression melting

Mount Kilimanjaro, Africa
Corbis/Photolibrary

F. Divergent plate volcanism
(Continental rifting)

FIGURE 9.36 Continued

285

are located (1) along convergent plate boundaries where plates move toward each other and one sinks beneath the other; (2) along divergent plate boundaries, where plates move away from each other and new seafloor is created; and (3) areas within the plates proper that are not associated with any plate boundary.

Volcanism at Convergent Plate Boundaries

Recall that at convergent plate boundaries slabs of oceanic crust are bent as they descend into the mantle, generating a deep-ocean trench. As a slab sinks deeper into the mantle, the increase in temperature and pressure drives volatiles (mostly water) from the oceanic crust. These mobile fluids migrate upward into the wedge-shaped piece of mantle located between the subducting slab and the overriding plate. Once the sinking slab reaches a depth of about 100 kilometers, these water-rich fluids reduce the melting point of hot mantle rock sufficiently to trigger some melting. The partial melting of mantle rock (peridotite) generates magma with a basaltic composition. After a sufficient quantity of magma has accumulated, it slowly migrates upward.

Volcanism at a convergent plate margin results in the development of a slightly curved chain of volcanoes called a *volcanic arc*. These volcanic chains develop roughly parallel to the associated trench—at distances of 200–300 kilometers (100–200 miles). Volcanic arcs can be constructed on oceanic, or continental, lithosphere. Those that develop within the ocean and grow large enough for their tops to rise above the surface are labeled *island archipelagos* in most atlases. Geologists prefer the more descriptive term **volcanic island arcs**, or simply **island arcs** (Figure 9.36A). Several young volcanic island arcs border the western Pacific basin, including the Aleutians, the Tongas, and the Marianas.

Volcanism associated with convergent plate boundaries may also develop where slabs of oceanic lithosphere are subducted under continental lithosphere to produce a **continental volcanic arc** (Figure 9.36E). The mechanisms that generate these mantle-derived magmas are essentially the same as those operating at island arcs. The major difference is that continental crust is much thicker and is composed of rocks having a higher silica content than oceanic crust. Hence, through the assimilation of silica-rich crustal rocks, plus extensive magmatic differentiation, a mantle-derived magma may become highly evolved as it rises through continental crust. Stated another way, the magmas generated in the mantle may change from a comparatively dry, fluid basaltic magma to a viscous andesitic or rhyolitic magma having a high concentration of volatiles as it moves up through the continental crust. The volcanic chain of the Andes Mountains along the western margin of South America is perhaps the best example of a mature continental volcanic arc.

Since the Pacific basin is essentially bordered by convergent plate boundaries and associated subduction zones, it is easy to see why the irregular belt of explosive volcanoes we call the Ring of Fire formed in this region (see Figure 9.35). The volcanoes of the Cascade Range in the northwestern United States, including Mount Hood, Mount Rainier, and Mount Shasta, are included in this group.

Volcanism at Divergent Plate Boundaries

The greatest volume of magma is produced along the oceanic ridge system in association with seafloor spreading (Figure 9.36B). Below the ridge axis where lithospheric plates are continually being pulled apart, the solid yet mobile mantle responds to the decrease in overburden and rises to fill the rift. Recall that as rock rises, it experiences a decrease in confining pressure and undergoes melting without the addition of heat. This process, called *decompression melting*, is the most common process by which mantle rocks melt.

Partial melting of mantle rock at spreading centers produces basaltic magma. Because this newly formed magma is less dense than the mantle rock from which it was derived, it rises and collects in reservoirs located just beneath the ridge crest. About 10 percent of this melt eventually migrates upward along fissures to erupt on the ocean floor. This activity continuously adds new basaltic rock to plate margins, temporarily welding them together, only to break again as spreading continues. Along some ridges, outpourings of bulbous pillow lavas build numerous small seamounts.

Although most spreading centers are located along the axis of an oceanic ridge, some are not. In particular, the East African Rift is a site where continental lithosphere is being pulled apart (Figure 9.36F). In this setting, magma is generated by decompression melting in the same manner as along the oceanic ridge system. Vast outpourings of fluid lavas as well as basaltic shield volcanoes are common in this region.

Intraplate Volcanism

We know why igneous activity is initiated along plate boundaries, but why do eruptions occur in the interiors of plates? Hawaii's Kilauea is considered the world's most active volcano, yet it is situated thousands of kilometers from the nearest plate boundary in the middle of the vast Pacific plate (Figure 9.36C). Other sites of **intraplate volcanism** (meaning "within the plate") include the Canary Islands, Yellowstone, and several volcanic centers that you may be surprised to learn are located in the Sahara Desert of Africa.

Geologists now recognize that most intraplate volcanism occurs where a mass of hotter than normal mantle material called a **mantle plume** ascends toward the surface (Figure 9.36C). Although the depth at which (at least some) mantle plumes originate is still hotly debated, some appear to form deep within Earth at the core–mantle boundary. These plumes of solid yet mobile mantle rock rise toward the surface in a manner similar to the blobs that form within a lava lamp. (These are the lamps

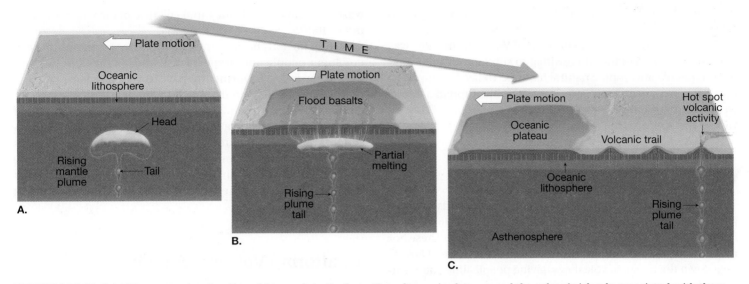

FIGURE 9.37 Model of hot-spot volcanism thought to explain the formation of oceanic plateaus and the volcanic islands associated with these features. **A.** A rising mantle plume with large bulbous head and narrow tail. **B.** Rapid decompression melting of the head of a mantle plume produces vast outpourings of basalt to generate the oceanic plateau. Large basaltic plateaus can also form on continental crust—examples include the Columbia Plateau in the northwestern United States and India's Deccan Plateau. **C.** Later, less voluminous activity caused by the rising plume tail produces a linear volcanic chain on the seafloor.

that contain two non-mixing liquids in a glass container. As the base of the lamp is heated, the denser liquid at the bottom becomes buoyant and forms blobs that rise to the top.) Like the blobs in a lava lamp, a mantle plume has a bulbous head that draws out a narrow stalk beneath it as it rises. Once the plume head nears the top of the mantle, decompression melting generates basaltic magma that may eventually trigger volcanism at the surface.

The result is a localized volcanic region a few hundred kilometers across called a **hot spot** (Figure 9.36C). More than 40 hot spots have been identified, and most have persisted for millions of years. The land surface surrounding a hot spot is often elevated because it is buoyed up by the rising plume of warm low-density material. Furthermore, by measuring the heat flow in these regions, geologists have determined that the mantle beneath hot spots must be 100–150° C hotter than normal mantle material.

Mantle plumes are responsible for the vast outpourings of basaltic lava that created the large basalt plateaus including the Siberian Traps in Russia, India's Deccan Plateau, and the Ontong Java Plateau in the western Pacific. The most widely accepted explanation for these eruptions, which emit extremely large volumes of basaltic lava over relatively short time intervals, involves a plume with a monstrous head and a long, narrow tail (Figure 9.37A). Upon reaching the base of the lithosphere, these unusually hot, massive heads begin to melt. Melting progresses rapidly, causing the burst of volcanism that emits voluminous outpourings of lava to form a huge basalt plateau in a matter of a million or so years (Figure 9.37B). The comparatively short initial eruptive phase is followed by tens of millions of years of less voluminous activity, as the plume tail slowly rises to the surface. Extending away from most large flood basalt provinces is a chain of volcanic structures, similar to the Hawaiian chain, that terminates over an active hot spot, marking the current position of the remaining tail of the plume (Figure 9.37C).

CONCEPT CHECK 9.9

❶ Are volcanoes in the Ring of Fire generally described as relatively quiet or violent? Name a volcano that would support your answer.

❷ How is magma generated along convergent plate boundaries?

❸ Volcanism at divergent plate boundaries is associated with which rock type? What causes rocks to melt in these regions?

❹ What is the source of magma for intraplate volcanism?

❺ At which type of plate boundary is the greatest quantity of magma generated?

Living with Volcanoes

About 10 percent of Earth's population lives in the vicinity of an active volcano. In fact, several major cities including Seattle, Washington; Mexico City, Mexico; Tokyo, Japan; Naples,

Italy; and Quito, Ecuador, are located on or near a volcano (**Figure 9.38**).

Until recently, the dominant view of Western societies was that humans possess the wherewithal to subdue volcanoes and other types of catastrophic natural hazards. Today, it is apparent that volcanoes are not only very destructive but unpredictable as well. With this awareness, a new attitude is developing—"How do we live with volcanoes?"

Volcanic Hazards

Volcanoes produce a wide variety of potential hazards that can kill people and wildlife, as well as destroy property (**Figure 9.39**). Perhaps the greatest threats to life are pyroclastic flows. These hot mixtures of gas, ash, and pumice that sometimes exceed 800° C race down the flanks of volcanoes, giving people little chance to escape.

Lahars, which can occur even when a volcano is quiet, are perhaps the next most dangerous volcanic hazard. These mixtures of volcanic debris and water can flow for tens of kilometers down steep volcanic slopes at speeds that may exceed 100 kilometers (60 miles) per hour. Lahars pose a potential threat to many communities downstream from glacier-clad volcanoes such as Mount Rainier. Other potentially destructive mass-wasting events include the rapid collapse of the volcano's summit or flank.

Other obvious hazards include explosive eruptions that can endanger people and property hundreds of miles from a volcano (**Figure 9.40**). During the past 15 years, at least 80 commercial jets have been damaged by inadvertently flying into clouds of volcanic ash. One of these was a near crash that occurred in 1989 when a Boeing 747, with more than 300 passengers aboard, encountered an ash cloud from Alaska's Redoubt Volcano. All four engines stalled after they became clogged with ash. Fortunately, the engines were restarted at the last minute and the aircraft managed to land safely in Anchorage.

Monitoring Volcanic Activity

Today, a number of volcano monitoring techniques are employed, with most of them aimed at detecting the movement of magma from a subterranean reservoir (typically several kilometers deep) toward the surface. The four most noticeable changes in a volcanic landscape caused by the migration of magma are (1) changes in the pattern of volcanic earthquakes; (2) expansion of a near-surface magma chamber, which leads to inflation of the volcano; (3) changes in the amount and/or com-

FIGURE 9.38 Seattle, Washington, with Mount Rainier in the background. (Photo by Ken Straiton/CORBIS)

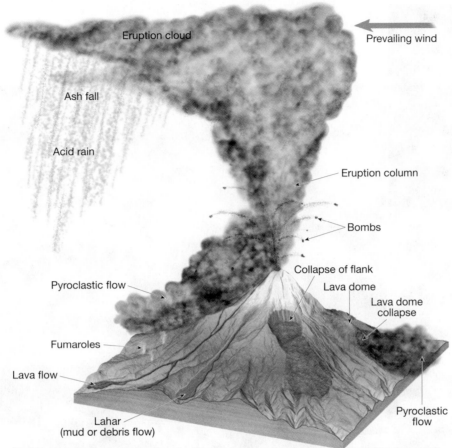

Eruption cloud

Prevailing wind

Ash fall

Acid rain

Eruption column

Bombs

Collapse of flank

Lava dome

Lava dome collapse

Pyroclastic flow

Fumaroles

Lava flow

Pyroclastic flow

Lahar (mud or debris flow)

FIGURE 9.39 Simplified drawing showing a wide variety of natural hazards associated with volcanoes. (After U.S. Geological Survey)

position of the gases that are released from a volcano; and (4) an increase in ground temperature caused by the implacement of new magma.

Almost a third of all volcanoes that have erupted in historic times are now monitored using seismographs, instruments that detect earthquake tremors. In general, a sharp increase in seismic unrest followed by a period of relative quiet has been shown to be a precursor for many volcanic eruptions. However, some large volcanic structures have exhibited lengthy periods of seismic unrest. For example, Rabaul Caldera in New Guinea recorded a strong increase in seismicity in 1981. This activity lasted 13 years and finally culminated with an eruption in 1994. Occasionally, a large earthquake triggers a volcanic eruption, or at least disturbs the volcano's plumbing. Kilauea, for example, began to erupt after the Kalapana earthquake of 1975.

The roof of a volcano may rise as new magma accumulates in its interior—a phenomenon that precedes many volcanic eruptions. Because the accessibility of many volcanoes is limited, remote sensing devices, including lasers, Doppler radar, and Earth-orbiting satellites, are often used to determine whether or not a volcano is swelling. The recent discovery of ground doming at Three Sisters Volcanoes in Oregon was first detected using radar images obtained from satellites.

Volcanologists also frequently monitor the gases that are released from volcanoes in an effort to detect even minor changes

in their amount and/or composition. Some volcanoes show an increase in sulfur dioxide (SO_2) emissions months or years prior to an eruption. On the other hand, a few days prior to the 1991 eruption of Mount Pinatubo, emissions of carbon dioxide (CO_2) dropped dramatically.

The development of remote sensing devices has greatly increased our ability to monitor volcanoes. These instruments and techniques are particularly useful for monitoring eruptions in progress. Photographic images and infrared (heat) sensors can detect lava flows and volcanic columns rising from a volcano. Furthermore, satellites can detect ground deformation as well as monitor SO_2 emissions.

The overriding goal of all monitoring is to discover precursors that may warn of an imminent eruption. This is accomplished by first diagnosing the current condition of a volcano and then using this baseline data to predict its future behavior. Stated another way, a volcano must be observed over an extended period to recognize significant changes from its "resting state."

CONCEPT CHECK 9.10

❶ Describe four natural hazards associated with volcanoes.
❷ What are the four changes in a volcanic area that are monitored in order to detect the migration of magma?

B.

A.

FIGURE 9.40 Volcanic eruptions can endanger people and property far from a volcano. **A.** The eruption of Iceland's Eyjafjallajökull volcano sent ash high into the atmosphere on April 16, 2010. The thick plume of ash drifted over Europe, causing airlines to cancel thousands of flights, leaving hundreds of thousands of travelers stranded. (AP Photo by Brynjar Gauti) **B.** Satellite image of the ash plume from Eyjafjallajökull volcano. (NASA)

Box 9.2

EARTH AS A SYSTEM

Can Volcanoes Change Earth's Climate?

The idea that explosive volcanic eruptions might alter Earth's climate was first proposed many years ago. It is still regarded as a plausible explanation for some aspects of climatic variability. Explosive eruptions emit huge quantities of gases and fine-grained debris high into the atmosphere, where it spreads around the globe and remains for many months or even years (**Figure 9.B**).

The Basic Premise

The basic premise is that this suspended volcanic material will filter out a portion of the incoming solar radiation, which in turn will drop temperatures in the lowest layer of the atmosphere. More than 200 years ago Benjamin Franklin used this idea to argue that material from the eruption of a large Icelandic volcano could have reflected sunlight back to space and therefore might have been responsible for the unusually cold winter of 1783–1784.

Mount Tambora

Perhaps the most notable cool period linked to a volcanic event is the "year without a summer" that followed the 1815 eruption of Mount Tambora in Indonesia. The eruption of Tambora is the largest of modern times. During April 7–12, 1815, this nearly 4,000-meter-high (13,000-foot) volcano violently expelled more than 100 cubic kilometers (24 cubic miles) of volcanic debris. The impact of the volcanic aerosols on climate is believed to have been widespread in the Northern Hemisphere. From May through September 1816 an unprecedented series of cold spells affected the northeastern United States and adjacent portions of Canada. There was heavy snow in June and frost in July and August. Abnormal cold was also experienced in much of western Europe.

Two Modern Examples

Two major volcanic events have provided considerable data and insight regarding the impact of volcanoes on global temperatures. The eruptions of Washington State's Mount St. Helens in 1980 and the Mexican volcano El Chichón in 1982 have given scientists an opportunity to study the atmospheric effects of volcanic eruptions with the aid of more sophisticated technology than had been available in the past. Satellite images and remote-sensing instruments allowed scientists to monitor closely the effects of the clouds of gases and ash that these volcanoes emitted.

Mount St. Helens

When Mount St. Helens erupted, there was immediate speculation about the possible effects on our climate. Could such an eruption cause our climate to change? There is no doubt that the large quantity of volcanic ash emitted by the explosive eruption had significant local and regional effects for a short period. Still, studies indicated that any longer-term lowering of hemispheric temperatures was negligible. The cooling was so slight, probably less than 0.1° C (0.2° F), that it could not be distinguished from other natural temperature fluctuations.

El Chichón

Two years of monitoring and studies following the 1982 El Chichón eruption indicated that its cooling effect on global mean temperature was greater than that of Mount St. Helens, on the order of 0.3–0.5° C (0.5–0.9° F). The eruption of El Chichón was *less explosive* than the Mount St. Helens blast, so why did it have a greater impact on global temperatures? The reason is that the material emitted by Mount St. Helens was largely fine ash that settled out in a relatively short time. El Chichón, on the other hand, emitted far greater quantities of sulfur dioxide gas (an estimated 40 times more) than Mount St. Helens. This gas combines with water vapor high in the atmosphere to produce a dense cloud of tiny sulfuric-acid particles. The particles, called *aerosols*, take several years to settle out completely. Like fine ash, these aerosols lower the atmosphere's mean temperature because they reflect solar radiation back to space.

We now understand that volcanic clouds that remain in the stratosphere for a year or more are composed largely of sulfuric-acid droplets and not of ash, as was once thought. Thus, the volume of fine debris emitted during an explosive event is not an accurate criterion for predicting the global atmosphere effects of an eruption.

It may be true that the impact on global temperature of eruptions like El Chichón and Mount St. Helens is relatively minor, but many scientists agree that the cooling produced could alter the general pattern of atmospheric circulation. Such a change, in turn, could influence the weather in some regions. Predicting or even identifying specific regional effects still presents a considerable challenge to atmospheric scientists.

The preceding examples illustrate that the impact on climate of a single volcanic eruption, no matter how great, is relatively small and short-lived. Therefore, if volcanism is to have a pronounced impact over an extended period, many great eruptions, closely spaced in time, need to occur. Although no such extended period of explosive volcanism is known to have occurred in historic times, such events may have altered climates in the geologic past. For example, massive eruptions of basaltic lava that began about 250 million years ago and lasted for a million years or more may have contributed to one of Earth's most profound mass extinctions. A discussion of a possible link between volcanic activity and the *Great Permian Extinction* is found in Chapter 12.

FIGURE 9.B Mount Etna, a volcano on the island of Sicily, erupting in late October 2002. Mount Etna is Europe's largest and most active volcano. **Upper.** This photo of Mount Etna looking southeast was taken by a crew member aboard the International Space Station. It shows a plume of volcanic ash streaming southeastward from the volcano. **Lower.** This image from the Atmospheric Infrared Sounder on NASA's *Aqua* satellite shows the sulfur dioxide (SO_2) plume in shades of purple and black. (Images courtesy of NASA)

GIVE IT SOME THOUGHT

1. Match each of these volcanic regions with one of the three zones of volcanism (convergent plate boundaries, divergent plate boundaries, or intraplate volcanism):
   a. Crater Lake
   b. Hawaii's Kilauea
   c. Mount St. Helens
   d. East African Rift
   e. Yellowstone
   f. Vesuvius
   g. Deccan Plateau
   h. Mount Etna

2. Examine the accompanying photo and complete the following:
   a. What type of volcano is it? What features helped you make a decision?
   b. What is the eruptive style of such volcanoes? Describe the likely composition and viscosity of the magma.
   c. Which one of the three zones of volcanism is the likely setting for this volcano?
   d. Name a city that is vulnerable to the effects of a volcano of this type.

Simon Phipps

3. Divergent boundaries, such as the Mid-Atlantic ridge, are characterized by outpourings of basaltic lava. Answer the following questions about divergent boundaries and their associated lavas:
   a. What is the source of these lavas?
   b. What causes the source rocks to melt?
   c. Describe a divergent boundary that would be associated with lava other than basalt. Why did you choose it and what type of lava would you expect to erupt there?

4. Explain why volcanic activity occurs in places other than plate boundaries.

5. For each of the accompanying four sketches, identify the geologic setting (zone of volcanism). Which of these settings will most likely generate explosive eruptions? Which will produce outpouring of fluid basaltic lavas?

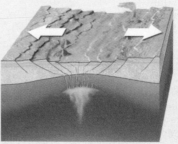

A.   B.   C.   D.

6. Assume you want to monitor a volcano that has erupted several times in the recent past, but appears to be quiet now. How might you determine if magma were actually moving through the crust beneath the volcano? Suggest at least two phenomena you would observe or measure.

7. Imagine you are a geologist charged with the task of choosing three sites where state-of-the-art volcano monitoring systems will be deployed. The sites can be anywhere in the world, but the budget and number of experts you can employ to oversee the operations are limited. What criteria would you use to select these sites? List some potential choices and your reasons for considering them.

8. Explain why an eruption of Mount Rainier, similar to the one that occurred at Mount St. Helens in 1980, would be considerably more destructive.

9. Each statement describes how an intrusive feature appears when exposed at Earth's surface by erosion. Name the feature.
   a. A dome-shaped mountainous structure flanked by upturned layers of sedimentary rocks.
   b. A vertical wall-like feature a few meters wide and hundreds of meters long.
   c. A huge expanse of granitic rock forming a mountainous terrain tens of kilometers wide.
   d. A relatively thin layer of basalt sandwiched between layers of sedimentary rocks exposed on the side of a canyon.

10. During a field trip with your geology class you visit an exposure of rock layers similar to the one sketched here. A fellow student suggests that the layer of basalt is a sill. You, however, disagree. Why do you think the other student is incorrect? What is a more likely explanation for the basalt layer?

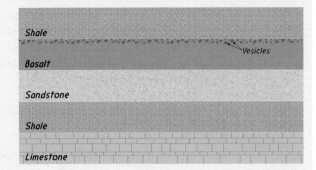

In Review   Chapter 9 Volcanoes and Other Igneous Activity

- The primary factors that determine the nature of volcanic eruptions include the magma's *composition*, its *temperature*, and the *amount of dissolved gases* it contains. As lava cools, it begins to congeal and, as *viscosity* increases, its mobility decreases. The *viscosity of magma is also directly related to its silica content*. Rhyolitic (felsic) lava, with its high silica content (over 70 percent), is very viscous and forms short, thick flows. Basaltic (mafic) lava, with a lower silica content (about 50 percent), is more fluid and may travel a long distance before congealing. Dissolved gases tend to make magma more fluid and, as they expand, provide the force that propels molten rock from the volcano.

- The materials associated with a volcanic eruption include (1) *lava flows* (*pahoehoe* flows, which resemble twisted braids; and *aa* flows, consisting of rough, jagged blocks; both form from basaltic lavas); (2) *gases* (primarily *water vapor*); and (3) *pyroclastic material* (pulverized rock and lava fragments blown from the volcano's vent, which include *ash, pumice, lapilli, cinders, blocks,* and *bombs*).

- Successive eruptions of lava from a central vent result in a mountainous accumulation of material known as a *volcano*. Located at the summit of many volcanoes is a steep-walled depression called a *crater*. *Shield cones* are broad, slightly domed volcanoes built primarily of fluid, basaltic lava. *Cinder cones* have steep slopes composed of pyroclastic material. *Composite cones*, or *stratovolcanoes*, are large, nearly symmetrical structures built of interbedded lavas and pyroclastic

deposits. Composite cones produce some of the most violent volcanic activity. Often associated with a violent eruption is a *nuée ardente*, a fiery cloud of hot gases infused with incandescent ash that races down steep volcanic slopes. Large composite cones may also generate a type of mudflow known as a *lahar*.

- Most volcanoes are fed by *conduits* or *pipes*. As erosion progresses, the rock occupying the pipe, which is often more resistant, may remain standing above the surrounding terrain as a *volcanic neck*. The summits of some volcanoes have large, nearly circular depressions called *calderas* that result from collapse. Calderas also form on shield volcanoes by subterranean drainage from a central magma chamber, and the largest calderas form by the discharge of colossal volumes of silica-rich pumice along ring fractures. Although volcanic eruptions from a central vent are the most familiar, by far the largest amounts of volcanic material are extruded from cracks in the crust called *fissures*. The term *flood basalts* describes the fluid basaltic lava flows that cover an extensive region in the northwestern United States known as the Columbia Plateau. When silica-rich magma is extruded, *pyroclastic flows*, consisting largely of ash and pumice fragments, usually result.

- Magma originates from essentially solid rock of the crust and mantle. In addition to a rock's composition, its temperature, depth (confining pressure), and water content determine whether it exists as a solid or liquid. Thus, magma can be generated by *increasing a rock's temperature*, as occurs when a hot mantle plume "ponds" beneath crustal rocks. A *decrease in pressure* can cause *decompression melting*. Furthermore, the *introduction of volatiles* (water) can lower a rock's melting point sufficiently to generate magma. A process called *partial melting* produces a melt made of the low-melting-temperature minerals, which are higher in silica than the original rock. Thus, magmas generated by partial melting are nearer to the granitic (felsic) end of the compositional spectrum than are the rocks from which they formed.

● Intrusive igneous bodies are classified according to their *shape* and by their *orientation with respect to the country or host rock*, generally sedimentary or metamorphic rock. The two general shapes are *tabular* (sheet-like) and *massive*. Intrusive igneous bodies that cut across existing sedimentary beds are said to be *discordant*; those that form parallel to existing sedimentary beds are *concordant*.

● *Dikes* are tabular, discordant igneous bodies produced when magma is injected into fractures that cut across rock layers. Nearly horizontal, tabular, concordant bodies, called *sills*, form when magma is injected along the bedding surfaces of sedimentary rocks. In many respects, sills closely resemble buried lava flows. *Batholiths*, the largest intrusive igneous bodies, sometimes make up large linear mountains, as exemplified by the Sierra Nevada. *Laccoliths* are similar to sills but form from less fluid magma that collects as a lens-shaped mass that arches overlying strata upward.

● *Most active volcanoes are associated with plate boundaries.* Active areas of volcanism are found along mid-ocean ridges where seafloor spreading is occurring (*divergent plate boundaries*), in the vicinity of ocean trenches where one plate is being subducted beneath another (*convergent plate boundaries*), and in the interiors of plates themselves (intraplate volcanism). Rising plumes of hot mantle rock are the source of most intraplate volcanism.

Key Terms

aa flow (p. 262)
batholith (p. 279)
caldera (p. 273)
cinder cone (p. 268)
columnar joint (p. 278)
composite cone (p. 269)
concordant (p. 276)
conduit (p. 265)
continental volcanic arc (p. 286)
crater (p. 265)
decompression melting (p. 280)
dike (p. 278)
discordant (p. 276)
eruption column (p. 261)
fissure (p. 274)
fissure eruption (p. 274)

flood basalt (p. 274)
geothermal gradient (p. 280)
hot spot (p. 287)
intraplate volcanism (p. 286)
intrusions (p. 276)
island arc (p. 286)
laccolith (p. 279)
lahar (p. 271)
lava tube (p. 263)
mantle plume (p. 286)
massive (p. 276)
nuée ardente (p. 270)
pahoehoe flow (p. 262)
partial melting (p. 281)
pipe (p. 265)
plutons (p. 276)

pumice (p. 264)
pyroclastic flow (p. 270)
pyroclastic material (p. 264)
scoria (p. 264)
scoria cone (p. 268)
shield volcano (p. 266)
sill (p. 278)
stock (p. 279)
stratovolcano (p. 269)
tabular (p. 276)
vent (p. 265)
viscosity (p. 259)
volatiles (p. 260)
volcanic island arc (p. 286)
volcanic neck (p. 276)
volcano (p. 265)

Examining the Earth System

1. Speculate about some of the possible consequences that a great and prolonged increase in explosive volcanic activity might have on each of Earth's four spheres.

2. Despite the potential for devastating destruction, humans live, work, and play on or near many active volcanoes. What are some of the benefits that a volcano or volcanic region might offer? (List some volcanoes and the assets they provide.)

Mastering Geology

Looking for additional review and test prep materials? Visit the Self Study area in **www.masteringgeology.com** to find practice quizzes, study tools, and multimedia that will aid in your understanding of this chapter's content. In **MasteringGeology™** you will find:

- **GEODe: Earth Science: An interactive visual walkthrough of key concepts**

- **Geoscience Animation Library: More than 100 animations illuminating many difficult-to-understand Earth science concepts**
- **In The News RSS Feeds: Current Earth science events and news articles are pulled into the site with assessment**
- **Pearson eText**
- **Optional Self Study Quizzes**
- **Web Links**
- **Glossary**
- **Flashcards**

Crustal Deformation and Mountain Building

10

Folded sedimentary layers exposed on the face of Mount Kidd, Alberta, Canada. (Photo by Tom Neversely/Photo Library)

Mountains provide some of the most spectacular scenery on our planet. This splendor has been captured by poets, painters, and songwriters alike. Geologists understand that at some time all continental regions were mountainous masses and that continents grow by the addition of mountains to their flanks. Consequently, as geologists unravel the secrets of mountain formation, they also gain a deeper understanding of the evolution of Earth's continents. If continents do indeed grow by adding mountains to their flanks, how do geologists explain the existence of mountains that are located in the interior of landmasses? This chapter attempts to piece together the sequence of events that generate these lofty structures in order to answer this and related questions.

FOCUS ON CONCEPTS

To assist you in learning **the important concepts in this chapter, focus on the following questions:**

- What is rock deformation? What factors influence how rock deforms?
- How are elastic, brittle, and ductile deformation different?
- What structures form as a result of brittle deformation?
- What structures form as a result of ductile deformation?
- What are the most common types of faults? How does each type form?
- How are Aleutian- and Andean-type mountain building similar? How are they different?
- How does continental accretion relate to mountain building?
- What are the stages in the formation of a major mountain belt?
- How is the concept of isostasy related to mountain building?
- What is meant by gravitational collapse?

Crustal Deformation

Forces Within
▶ **Mountain Building**

Earth is a dynamic planet. Shifting lithospheric plates continually change the face of our planet by moving continents across the globe. The results of this tectonic activity are perhaps most strikingly apparent in Earth's major mountain belts. Rocks containing fossils of marine organisms are found thousands of meters above sea level, and massive rock units are bent, contorted, overturned, and sometimes rife with fractures.

We begin our look at mountain building by examining the process of rock deformation and the structures that result (Figure 10.1). Every mass of rock, no matter how strong, has a point at which it will fracture or flow. **Deformation** is a general term that refers to all changes in the original shape, size (volume), or orientation of a rock body. Geologists use the term **stress** to describe the forces that deform rocks. Stress that squeezes and shortens a rock mass is called *compressional stress*, whereas stress that pulls apart or elongates a rock body is known as *tensional stress*. Stress can also cause a rock to *shear*, which is similar to the slippage that occurs between individual playing cards when the top of the deck is moved relative to the bottom.

When rocks are subjected to stresses greater than their strength, they begin to deform, usually by flowing or fracturing. It is easy to visualize how rocks break, because we normally think of them as being brittle. But how can masses of rock be *bent* into intricate folds without fracturing in the process? To determine this, geologists performed laboratory experiments in which rocks were subjected to differential stresses under conditions that simulate those existing at various depths within the crust.

Elastic, Brittle, and Ductile Deformation

Although each rock type deforms somewhat differently, the general characteristics of rock deformation were determined from such experiments. Geologists learned that when stress is gradually applied, rocks first respond by deforming elastically. Changes that result from **elastic deformation** are recoverable; that is, like a rubber band, the rock will return to nearly its original size and shape when the stress is removed. During elastic deformation the chemical bonds of the minerals within a rock are stretched but do not break. As you saw in Chapter 8, the energy for most earthquakes comes from stored elastic energy that is released as rock snaps back to its original shape.

Once the elastic limit (strength) of a rock is surpassed, it either flows or fractures. Rocks that break into smaller pieces exhibit

brittle deformation. From our everyday experience, we know that glass objects, wooden pencils, china plates, and even our bones exhibit brittle failure once their strength is surpassed. Brittle deformation occurs when stress causes the chemical bonds that hold a material together to break.

Ductile deformation, on the other hand, is a type of solid-state flow that produces a change in the shape of an object without fracturing. Ordinary objects that display ductile behavior include modeling clay, beeswax, taffy, and some metals. For example, a copper penny placed on a railroad track will be flattened and deformed (without breaking) by the force applied by a passing train. In rocks, ductile deformation is the result of some chemical bonds breaking while others are forming, allowing minerals to change shape.

Factors that Affect Rock Strength

The major factors that influence the strength of a rock and how it will deform include temperature, confining pressure, rock type, and time.

Temperature   The effect of temperature on the strength of a material can be easily demonstrated with a piece of glass tubing commonly found in a chemistry lab. If the tubing is dropped on a hard surface, it will shatter. However, if the tubing is heated over a Bunsen burner, it can be easily bent into a variety of shapes. Rocks respond similarly to heat. Where temperatures are high (deep in Earth's crust), rocks tend to deform ductilely and flow.

Likewise, where temperatures are low (at or near the surface), rocks tend to behave like brittle solids and fracture.

Confining Pressure   Recall from Chapter 7 that pressure, like temperature, increases with depth as the thickness of the overlying rock increases. Buried rocks are subjected to confining pressure, which is much like water pressure, where the forces are applied equally in all directions. The deeper you go in the ocean, the greater the confining pressure. The same is true for rock that is buried. Confining pressure "squeezes" the materials in Earth's crust. Therefore, rocks that are deeply buried are "held together" by the immense pressure and tend to flow, rather than fracture.

Rock Type   In addition to the physical environment, the mineral composition and texture of rock greatly influence how it will deform. For example, crystalline rocks composed of minerals that have strong internal molecular bonds tend to fail by brittle fracture. By contrast, sedimentary rocks that are weakly cemented, or metamorphic rocks that contain zones of weakness such as foliation, are more susceptible to ductile deformation.

Weak rocks that are most likely to behave in a ductile manner (flow or fold) when subjected to differential stress include rock salt, shale, limestone, and schist. Igneous and some metamorphic rocks tend to be strong and brittle. In a near-surface environment, strong, brittle rocks will fail by fracturing when subjected to stresses that exceed their strength. At increasing depths, however, the strength of all rock types decreases significantly.

FIGURE 10.1 A portion of the Sneffels Range near Ouray, Colorado. (Photo by James Hagar/Robert Harding)

Time One key factor that researchers are unable to duplicate in the laboratory is how rocks respond to small stresses applied gradually over long spans of *geologic time*. However, insights into the effects of time on deformation are provided in everyday settings. For example, marble benches have been known to sag under their own weight over a span of 100 years or so, and wooden bookshelves may bend after being loaded with books for a relatively short period.

In general, when tectonic forces are applied slowly over long time spans, rocks tend to display ductile behavior and deform by flowing and folding. An analogous situation occurs when you take a taffy bar and slowly move the two ends together. The taffy will deform by folding. However, if you swiftly hit the taffy against the edge of a table, it will break into two or more pieces, exhibiting brittle failure.

Likewise, rocks tend to deform in a ductile manner by folding when stress builds gradually. These same rocks may fracture if the stress increases suddenly. As a consequence, folding and faulting may occur simultaneously in the same rock body (**Figure 10.2**).

To review, the processes by which rocks deform occur along a continuum that ranges from brittle fracture at one end to ductile flow at the other. The processes of deformation generate geologic changes on many scales. At one extreme are Earth's major mountain systems. At the other extreme are minor fractures in bedrock created by highly localized stresses. All of these phenomena, from the largest folds in the Alps to the smallest fractures in a slab of rock, are considered to be *rock* or *tectonic structures*.

Geologist's Sketch

FIGURE 10.2 Deformed sedimentary strata exposed in a road cut near Palmdale, California. In addition to the obvious folding, light-colored beds are offset along a fault located on the right side of the photograph (Photo by E. J. Tarbuck)

CONCEPT CHECK 10.1

1. What is rock deformation? How might a rock body change during deformation?
2. Describe elastic deformation.
3. How is brittle deformation different from ductile deformation?
4. List and describe the four factors that affect rock strength.

Structures Formed by Ductile Deformation

GEODe
EARTH SCIENCE

Forces Within
▶ **Mountain Building**

Because folds are common features of deformed sedimentary rocks, we know rocks can bend without breaking. Ductile deformation is often accomplished by gradual slippage along planes

---

Students Sometimes Ask...

Who was the first person to climb to the top of Mount Everest?

Edmund Hillary, a New Zealand mountaineer and explorer, and Sherpa Tenzing Norgay of Nepal were the first to reach the summit of Mount Everest on May 29, 1953. Not one to rest on his laurels, Hillary later went on to lead the first crossing of Antarctica and also traveled to the North Pole. Hillary devoted much of his life to helping the Sherpa people of Nepal through the Himalayan Trusts, which he founded.

---

of weakness within the atomic structure of mineral grains. This microscopic form of gradual solid-state flow involves slippage facilitated by chemical bonds breaking in one location as new ones form at another site. Rocks that display evidence of ductile flow usually were deformed at great depth and may exhibit contorted folds that give the impression that the strength of the rock was akin to soft putty.

Folds

During mountain building, flat-lying sedimentary and volcanic rocks are often bent into a series of wavelike undulations called **folds**. Folds in sedimentary strata are much like those that would form if you were to hold the ends of a sheet of paper and then

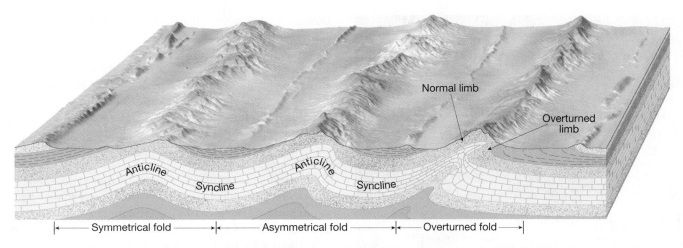

FIGURE 10.3 Block diagram of principal types of folded strata. The upfolded, or arched, structures are *anticlines*. The downfolds, or troughs, are *synclines*. Notice that the limb of an anticline is also the limb of the adjacent syncline.

push them together. In nature, folds come in a wide variety of sizes and configurations. Some folds are broad flexures in which strata hundreds of meters thick have been slightly warped. Others are very tight microscopic structures found in metamorphic rocks. Size differences notwithstanding, most folds are the result of *compressional stresses that result in a shortening and thickening of the crust.*

Anticlines and Synclines   The two most common types of folds are anticlines and synclines (**Figure 10.3**). **Anticlines** usually arise by upfolding, or arching, of sedimentary layers and are sometimes spectacularly displayed along highways that have been cut through deformed strata (**Figure 10.4A**).[23] Almost always found in association with anticlines are downfolds, or troughs, called **synclines** (Figure 10.4B). Notice in Figure 10.3 that the limb of an anticline is also a limb of the adjacent syncline.

Depending on their orientation, these basic folds are described as *symmetrical* when the limbs are mirror images of each other and *asymmetrical* when they are not. An asymmetrical fold is said to be *overturned* if one or both limbs are tilted beyond the vertical (**Figure 10.5**). An overturned fold can also "lie on its side" so a plane extending through the axis of the fold is horizontal. These *recumbent* folds are common in highly deformed mountainous regions such as the Alps.

Folds do not continue forever; rather, their ends die out, much like the wrinkles in cloth. Some folds *plunge* because the axis of the fold penetrates the ground. **Figure 10.6** shows an example of a plunging anticline and the pattern produced when erosion removes the upper layers of the structure and exposes its interior. Note that the outcrop pattern of an anticline points in the direction it is plunging. The opposite is true for a syncline. A good example of the kind of topography that results when erosional forces attack folded sedimentary strata is found in the Valley and Ridge Province of the Appalachians (see Figure 10.25).

It is important to realize that ridges are not necessarily associated with anticlines, nor are valleys related to synclines. Rather, ridges and valleys result because of differential weathering and erosion. For example, in the Valley and Ridge Province, resistant sandstone beds remain as imposing ridges separated by valleys cut into more easily eroded shale or limestone beds.

FIGURE 10.4 Anticline and syncline. **A.** An asymmetrical anticline in which one limb dips more steeply than the other. **B.** A nearly symmetrical syncline formed in limestone and siltstone strata. (Photos by E. J. Tarbuck)

A.

B.

[23] By strict definition, an anticline is a structure in which the oldest strata are found in the center. A syncline is a structure in which the youngest strata are found in the center.

FIGURE 10.5 Overturned fold, East Fork of Toklat River, Alaska. Overturned folds have one or both limbs tilted beyond vertical. (Photo by Michael Collier)

Domes and Basins   Broad upwarps in basement rock may deform the overlying cover of sedimentary strata and generate large folds. When this upwarping produces a circular or slightly elongated structure, the feature is called a **dome**. Downwarped structures having a similar shape are termed **basins**.

The Black Hills of western South Dakota is a large domed structure generated by upwarping. Here erosion has stripped away the highest portions of the overlying sedimentary beds, exposing older igneous and metamorphic rocks in the center (Figure 10.7).

Several large basins exist in the United States (Figure 10.8). The basins of Michigan and Illinois have gently sloping beds similar to saucers. These basins are thought to be the result of large accumulations of sediment, whose weight caused the crust to subside (see the section on isostasy later in this chapter). A few structural basins may have been the result of giant asteroid impacts.

Because large basins contain sedimentary beds sloping at low angles, they are usually identified by the age of the rocks composing them. The youngest rocks are found near the center, and the oldest rocks are at the flanks. This is just the opposite order of a domed structure, such as the Black Hills, where the oldest rocks form the core.

Monoclines   Although we have separated our discussion of folds and faults, in the real world folds can be intimately coupled

FIGURE 10.6 Plunging anticline, Sheep Mountain, Wyoming. In a plunging anticline the outcrop pattern "points" in the direction of plunge; the opposite is true of plunging synclines. (Photo by Michael Collier)

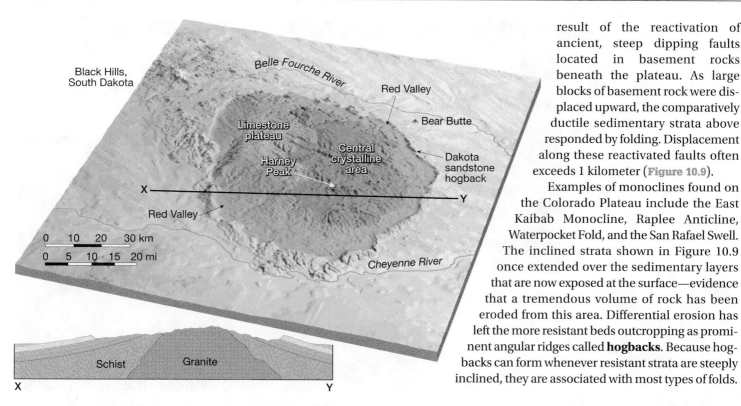

FIGURE 10.7 The Black Hills of South Dakota, a large domal structure with resistant igneous and metamorphic rocks exposed in the core.

with faults. Examples of this close association are broad, regional features called *monoclines*. Particularly prominent features of the Colorado Plateau, **monoclines** are large, steplike folds in otherwise horizontal sedimentary strata. These folds appear to be the

result of the reactivation of ancient, steep dipping faults located in basement rocks beneath the plateau. As large blocks of basement rock were displaced upward, the comparatively ductile sedimentary strata above responded by folding. Displacement along these reactivated faults often exceeds 1 kilometer (**Figure 10.9**).

Examples of monoclines found on the Colorado Plateau include the East Kaibab Monocline, Raplee Anticline, Waterpocket Fold, and the San Rafael Swell. The inclined strata shown in Figure 10.9 once extended over the sedimentary layers that are now exposed at the surface—evidence that a tremendous volume of rock has been eroded from this area. Differential erosion has left the more resistant beds outcropping as prominent angular ridges called **hogbacks**. Because hogbacks can form whenever resistant strata are steeply inclined, they are associated with most types of folds.

CONCEPT CHECK 10.2

❶ Distinguish between anticlines and synclines, domes and basins, anticlines and domes.
❷ The Black Hills of South Dakota is a good example of what type of structural feature?
❸ Describe the formation of a monocline.

FIGURE 10.8 The bedrock geology of the Michigan Basin. The youngest rocks are centrally located, whereas the oldest beds flank this structure.

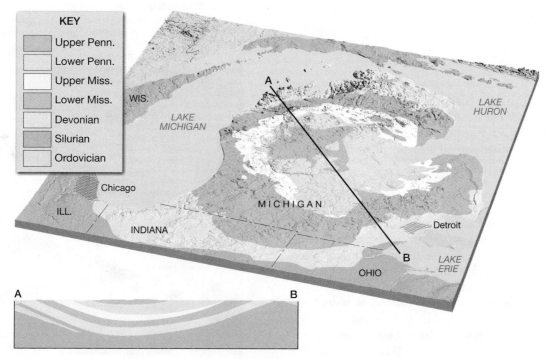

KEY

- Upper Penn.
- Lower Penn.
- Upper Miss.
- Lower Miss.
- Devonian
- Silurian
- Ordovician

FIGURE 10.9 The East Kaibab monocline in northern Arizona. This monocline consists of bent sedimentary beds that were deformed by faulting in the bedrock below. The thrust fault in this sketch is called a *blind thrust* because it does not reach the surface. (Photo by Michael Collier)

Structures Formed by Brittle Deformation

GEODe
EARTH
SCIENCE

Forces Within
▶ **Mountain Building**

You have probably seen a drinking glass drop on a hard surface, shattering into pieces. What you witnessed is analogous to brittle deformation that occurs in Earth's crust. In nature, brittle deformation occurs when stresses exceed the strength of a rock, causing it to break or fracture.

Faults

In the upper crust, to depths of about 10–15 kilometers (6–10 miles), rocks tend to exhibit brittle behavior by fracturing, or by faulting. **Faults** are fractures in the crust along which appreciable displacement has taken place. Occasionally, small faults can be recognized in road cuts where sedimentary beds have been offset a few meters, as shown in **Figure 10.10**. Faults of this scale usually occur as single discrete breaks. By contrast, large faults, like the San Andreas Fault in California, have displacements of hundreds of kilometers and consist of many interconnecting fault surfaces. These structures, best

FIGURE 10.10 Faulting caused the vertical displacement of these strata. The rock immediately above a fault surface is the *hanging wall block*, and that below is called the *footwall block*. (Photo by John S. Shelton)

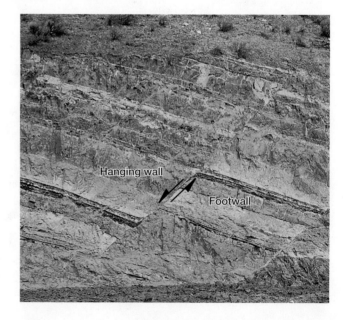

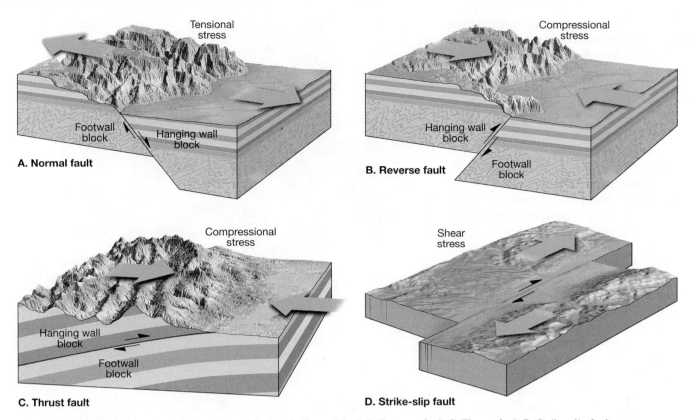

FIGURE 10.11 Block diagrams of four types of faults. **A.** Normal fault **B.** Reverse fault **C.** Thrust fault **D.** Strike-slip fault.

described as *fault zones,* can be several kilometers wide and are often easier to identify from high-altitude photographs than at ground level.

Dip-Slip Faults

Faults in which movement is primarily parallel to the *dip* (or inclination) of the fault surface are called **dip-slip faults**. It has become common practice to call the rock surface that is immediately above the fault the *hanging wall block* and to call the rock surface below, the *footwall block* (Figure 10.10). This nomenclature arose from prospectors and miners who excavated shafts and tunnels along fault zones that were sites of ore deposits. In these tunnels, the miners would walk on the rocks below the mineralized fault zone (the footwall block) and hang their lanterns on the rocks above (the hanging wall block).

Normal Faults   Dip-slip faults are classified as **normal faults** when the hanging wall block moves down relative to the footwall block (**Figure 10.11A**). Because of the downward motion of the hanging wall block, normal faults accommodate lengthening, or extension, of the crust.

Normal faults are found in a variety of sizes: some are small, having displacements of only a meter or so, like the one shown in the road cut in Figure 10.10. Others extend for tens of kilometers where they may sinuously trace the boundary of a mountain front. Most large, normal faults have relatively steep dips that tend to flatten out with depth. Vertical displacements along dip-slip faults may produce long, low cliffs called **fault scarps** (**Figure 10.12**).

In the western United States, large normal faults are associated with structures called **fault-block mountains**. Excellent examples of fault-block mountains are found in the Basin and Range Province, a region that encompasses Nevada and portions

FIGURE 10.12 This fault scarp formed north of Landers, California, during an earthquake in 1992. It is the largest of several scarps that formed during this event. Note the geologist for scale. The scarp was created when land on one side of the fault moved down relative to the land on the opposite side. (Photos by Roger Ressmeyer/CORBIS)

of the surrounding states. Here the crust has been elongated and broken to create more than 200 relatively small mountain ranges (Figure 10.13). Averaging about 80 kilometers in length, the ranges rise 900–1,500 meters above the adjacent down-faulted basins.

The topography of the Basin and Range Province evolved in association with a system of roughly north–south trending normal faults. Movements along these faults produced alternating uplifted fault blocks called **horsts** and down-dropped blocks called **grabens**. Horsts generate elevated topography, whereas grabens form basins. As Figure 10.13 illustrates, structures called **half-grabens**, which are titled fault blocks, also contribute to the alternating topographic highs and lows in the Basin and Range Province. The horsts and higher ends of the tilted fault blocks are the source of sediments that have accumulated in the basins created by the grabens and lower ends of the tilted blocks.

Notice in Figure 10.13 that the slopes of the normal faults decrease with depth and eventually join to form a nearly horizontal fault called a *detachment fault*. These faults form a major boundary between the rocks below, which exhibit ductile deformation, and the rocks above, which exhibit brittle deformation.

Fault motion provides geologists with a method of determining the nature of the tectonic forces at work within Earth. Normal faults are associated with tensional forces that pull the crust apart. This "pulling apart" can be accomplished either by uplifting that causes the surface to stretch and break or by opposing horizontal forces.

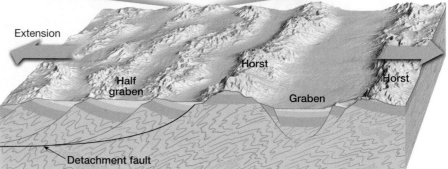

FIGURE 10.13 Normal faulting in the Basin and Range Province. Here, tensional stresses have elongated and fractured the crust into numerous blocks. Movement along these fractures has tilted the blocks, producing parallel mountain ranges called fault-block mountains. The down-faulted blocks (grabens) form basins, whereas the upfaulted blocks (horsts) erode to form rugged mountainous topography. In addition, numerous tilted blocks (half-grabens) form both basins and mountains. (Photo by Michael Collier)

Reverse and Thrust Faults   **Reverse faults** are dip-slip faults in which the hanging wall block moves up relative to the footwall block (Figure 10.11B). **Thrust faults** are reverse faults having dips less than 45 degrees, so the overlying block moves nearly horizontally over the underlying block (Figure 10.11C).

Whereas normal faults occur in tensional environments, reverse faults result from strong compressional stresses. Because the hanging wall block moves up and over the footwall block, reverse and thrust faults accommodate horizontal shortening of the crust.

Most high-angle reverse faults are small and accommodate local displacements in regions dominated by other types of faulting. Thrust faults, on the other hand, exist at all scales with some large thrust faults having displacements on the order of tens to hundreds of kilometers.

Thrust faulting is most pronounced along convergent plate boundaries. Compressional forces associated with colliding plates generally create folds as well as thrust faults that thicken and shorten the crust to produce mountainous topography (see Figure 10.22).

Strike-Slip Faults

A fault in which the dominant displacement is horizontal and parallel to the *trend* or *strike* of the fault surface is called a **strike-slip fault** (Figure 10.14A). The earliest scientific records of strike-slip

---

Students Sometimes Ask...

How do geologists determine which side of a fault has moved?

Surprisingly, for many faults it cannot be definitively established. For example, in the picture of a fault shown in Figure 10.10, did the left side move down, or did the right side move up? Since the surface (at the top of the photo) has been eroded flat, either side could have moved, or both could have moved, with one side moving more than the other (for instance, they both may have moved up, but the right side moved up more than the left side). That's why geologists talk about *relative* motion across faults. In this case, the left side moved down *relative to* the right side, and the right side moved up *relative to* the left side (note arrows on photo).

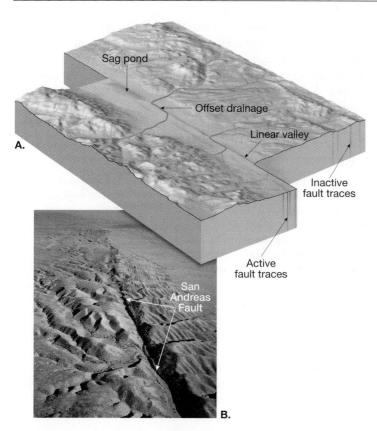

FIGURE 10.14 Strike-slip faulting. **A.** The block diagram illustrates the features associated with large strike-slip faults. Notice how the stream channels have been offset by fault movement. **B.** Aerial view of the San Andreas Fault. (Photo by D. Parker/Photo Researchers)

faulting were made following surface ruptures that produced large earthquakes. One of the most noteworthy of these was the great San Francisco earthquake of 1906. During this strong earthquake, structures such as fences that were built across the San Andreas Fault were displaced as much as 4.7 meters (15 feet). Because movement along the San Andreas causes the crustal block on the opposite side of the fault to move to the right as you face the fault, it is called a *right-lateral* strike-slip fault.

The Great Glen Fault in Scotland is a well-known example of a *left-lateral* strike-slip fault, which exhibits the opposite sense of displacement. The total displacement along the Great Glen Fault is estimated to exceed 100 kilometers (60 miles). Also associated with this fault trace are numerous lakes, including Loch Ness, home of the legendary monster.

Some strike-slip faults cut through the lithosphere and accommodate motion between two large tectonic plates. Recall this special kind of strike-slip fault is called a **transform fault**. Numerous transform faults cut the oceanic lithosphere and link spreading oceanic ridges. Others accommodate displacement between continental plates that move horizontally with respect to each other. One of the best-known transform faults is California's San Andreas Fault (Figure 10.14B). This plate-bounding fault can be traced for about 950 kilometers (600 miles) from the Gulf of California to a point along the Pacific Coast north of San Francisco, where it heads out to sea. Since its formation

about 30 million years ago, displacement along the San Andreas Fault has exceeded 560 kilometers. This movement has accommodated the northward displacement of southwestern California and the Baja Peninsula of Mexico in relation to the remainder of North America.

Joints

Among the most common rock structures are fractures called joints. Unlike faults, **joints** are fractures along which no appreciable displacement has occurred. Although some joints have a random orientation, most occur in roughly parallel groups (**Figure 10.15**).

We have already considered two types of joints. In Chapter 9 we learned that *columnar joints* form when igneous rocks cool and develop shrinkage fractures that produce elongated, pillar-like columns (see Figure 9.28). Also recall from Chapter 4 that sheeting produces a pattern of gently curved joints that develop more or less parallel to the surface of large exposed igneous bodies such as batholiths. Here the jointing results from the gradual expansion that occurs when erosion removes the overlying load (see Figure 4.5, p. 88).

Most joints are produced when rocks in the outermost crust are deformed as tensional stresses cause the rock to fail by brittle fracture. Extensive joint patterns often develop in response to relatively subtle and often barely perceptible regional upwarping and downwarping of the crust. In many cases, the cause for jointing at a particular locale is not readily apparent.

Many rocks are broken by two or even three sets of intersecting joints that slice the rock into numerous regularly shaped blocks. These joint sets often exert a strong influence on other geologic processes. For example, chemical weathering tends to be concentrated along joints, and in many areas groundwater movement and the resulting dissolution in soluble rocks is controlled by the joint pattern. Moreover, a system

FIGURE 10.15 The Fins in Utah's Arches National Park were created by weathering and erosion along parallel joints. (Photo by Michael Collier)

Students Sometimes Ask...

Has anyone ever seen a fault scarp forming?

Amazingly, yes. There have been several instances where people have been at the fortuitously appropriate place and time to observe the creation of a fault scarp—and have lived to tell about it. In Idaho a large earthquake in 1983 created a 3-meter (10-foot) fault scarp that was witnessed by several people, many of whom were knocked off their feet. More often, though, fault scarps are noticed *after* they form. For example, a 1999 earthquake in Taiwan created a fault scarp that formed a new waterfall and destroyed a nearby bridge.

of joints can influence the direction stream courses follow. The rectangular drainage pattern described in Chapter 5 is such a case.

Highly jointed rocks present a risk to the construction of engineering projects, including highways and dams. On June 5, 1976, 14 lives were lost and nearly $1 billion in property damage occurred when the Teton Dam in Idaho failed. This earthen dam was constructed of very erodible clays and silts and was situated on highly jointed volcanic rocks. Although attempts were made to fill the voids in the jointed rock, water gradually penetrated the subsurface fractures and undermined the dam's foundation. Eventually the moving water cut a tunnel into the easily erodible clays and silts. Within minutes the dam failed, sending a 20-meter-high wall of water down the Teton and Snake rivers.

CONCEPT CHECK 10.3

1. Contrast the movements that occur along normal and reverse faults. What type of stress is indicated by each fault?
2. What type of faults are associated with fault-block mountains?
3. How are reverse faults different from thrust faults? In what way are they the same?
4. Describe the relative movement along a strike-slip fault.
5. How are joints different from faults?

Mountain Building

Mountain building has occurred in the recent geologic past at several locations around the world (Figure 10.16). Young mountain belts include the American Cordillera, which runs along the

FIGURE 10.16 This peak is in Pakistan's Karakoram Range which is part of the Himalayan System. (Photo by Art Wolfe/Getty)

western margin of the Americas from Cape Horn at the tip of South America to Alaska and includes the Andes and Rocky Mountains; the Alpine–Himalaya chain, that extends along the margin of the Mediterranean, through Iran to northern India, and into Indochina; and the mountainous terrains of the western Pacific, which include volcanic island arcs that comprise Japan, the Philippines, and Sumatra. Most of these young mountain belts have come into existence within the last 100 million years. Some, including the Himalayas, began their growth as recently as 50 million years ago.

In addition to these young mountain belts, there are several chains of Paleozoic-age mountains found on Earth. Although these older structures are deeply eroded and topographically less prominent, they exhibit the same structural features found in younger mountains. The Appalachians in the eastern United States and the Urals in Russia are classic examples of this group of older and well-worn mountain belts.

The term for the processes that collectively produce a mountain belt is **orogenesis**. Most major mountain belts display striking visual evidence of great horizontal forces that have shortened and thickened the crust. These **compressional mountains** contain large quantities of preexisting sedimentary and crystalline rocks that have been faulted and contorted into a series of folds. Although folding and thrust faulting are often the most conspicuous signs of orogenesis, varying degrees of metamorphism and igneous activity are always present.

How do mountain belts form? As early as the ancient Greeks, this question has intrigued some of the greatest philosophers and scientists. One early proposal suggested that mountains are simply wrinkles in Earth's crust, produced as the planet cooled from its original semimolten state. According to this idea, Earth contracted and shrank as it lost heat, which caused the crust to deform in a manner similar to how an orange peel wrinkles as the fruit dries out. However, neither this nor any other early hypothesis withstood scientific scrutiny.

CONCEPT CHECK 10.4

❶ Define *orogenesis*.
❷ In the plate tectonics model, which type of plate boundary is most directly associated with Earth's major mountain belts?

Mountain Building at Subduction Zones

With the development of the theory of plate tectonics, a model for orogenesis with excellent explanatory power has emerged. According to this model, most mountain building occurs at convergent plate boundaries. Here, the subduction of oceanic lithosphere triggers partial melting of mantle rock, providing a source of magma that intrudes the crustal rocks that form the margin of the overlying plate. In addition, colliding plates provide the tectonic forces that fold, fault, and metamorphose the thick accumulations of sediments that have

been deposited along the flanks of landmasses. Together, these processes thicken and shorten the continental crust, thereby elevating rocks that may have formed near the ocean floor to lofty heights.

To unravel the events that produce mountains, researchers examine ancient mountain structures as well as sites where orogenesis is currently active. Of particular interest are active subduction zones, where lithospheric plates are converging. Here the subduction of oceanic lithosphere generates Earth's strongest earthquakes and most explosive volcanic eruptions, as well as playing a pivotal role in generating many of Earth's mountain belts.

The subduction of oceanic lithosphere gives rise to two different types of tectonic structures. Where oceanic lithosphere subducts beneath an oceanic plate, a **volcanic island arc** and related tectonic features develop. Subduction beneath a continental block, on the other hand, results in the formation of a volcanic arc along the margin of a continent. Plate boundaries that generate continental volcanic arcs are referred to as **Andean-type plate margins**.

Volcanic Island Arcs

Island arcs result from the steady subduction of oceanic lithosphere, which may last for 200 million years or more (Figure 10.17). Periodic volcanic activity, the emplacement of igneous plutons at depth, and the accumulation of sediment that is scraped from the subducting plate gradually increase the volume of crustal material capping the upper plate. Some large volcanic island arcs, such as Japan, owe their size to having been built upon a preexisting fragment of continental crust.

The continued growth of a volcanic island arc can result in the formation of mountainous topography consisting of belts of igneous and metamorphic rocks. This activity, however, is viewed as just one phase in the development of a major mountain belt. As you will see later, some volcanic arcs are carried by a subducting plate to the margin of a large continental block, where they become involved in a large-scale mountain-building episode.

FIGURE 10.17 Volcanic island arcs develop where one oceanic plate subducts beneath another. Continuous subduction results in the development of thick units of continental-type crust.

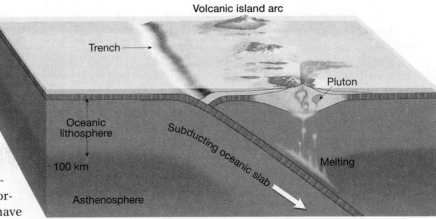

Mountain Building along Andean-Type Margins

The first stage in the development of an Andean-type mountain belt occurs along a **passive continental margin** prior to the formation of the subduction zone. The East Coast of the United States provides a modern example of a passive continental margin where sedimentation has produced a thick platform of shallow-water sandstones, limestones, and shales (Figure 10.18A). At some point, the forces that drive plate motions change and a subduction zone develops along the margin of the continent. It is along these **active continental margins** that the structural elements of a developing mountain belt gradually take form.

A good place to examine an active continental margin is the west coast of South America. Here the Nazca plate is being subducted beneath the South American plate along the Peru–Chile trench. This subduction zone probably formed prior to the breakup of the supercontinent of Pangaea.

In an idealized Andean-type subduction, convergence of the continental block and the subducting oceanic plate leads to deformation and metamorphism of the continental margin. Once the oceanic plate descends to about 100 kilometers (60 miles), partial melting of mantle rock above the subducting slab generates magma that migrates upward (Figure 10.18B).

Thick continental crust greatly impedes the ascent of magma. Consequently, a high percentage of the magma that intrudes the crust never reaches the surface. Instead, it crystallizes at depth to form plutons. Eventually, uplifting and erosion exhume these igneous bodies and associated metamorphic rocks. Once they are exposed at the surface, these massive structures are called *batholiths* (Figure 10.18C). Composed of numerous plutons, batholiths form the core of the Sierra Nevada in California and are prevalent in the Peruvian Andes.

During the development of this continental volcanic arc, sediment derived from the land and scraped from the subducting plate is plastered against the landward side of the trench like piles of dirt in front of a bulldozer. This chaotic accumulation of sedimentary and metamorphic rocks with occasional scraps of ocean crust is called an **accretionary wedge** (Figure 10.18B). Prolonged subduction can build an accretionary wedge that is large enough to stand above sea level (Figure 10.18C).

Andean-type mountain belts are composed of two roughly parallel zones. The volcanic arc develops on the continental block. It consists of volcanoes and large intrusive

bodies intermixed with high-temperature metamorphic rocks. The seaward segment is the accretionary wedge. It consists of folded and faulted sedimentary and metamorphic rocks (Figure 10.18C).

Sierra Nevada and Coast Ranges One of the best examples of an inactive Andean-type orogenic belt is found in the

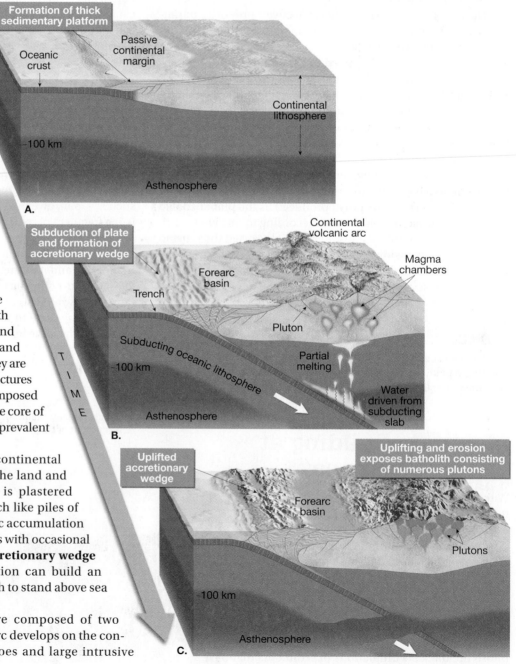

FIGURE 10.18 Mountain building along an Andean-type subduction zone. **A.** Passive continental margin with an extensive platform of sediments and sedimentary rocks. **B.** Plate convergence generates a subduction zone, and partial melting produces a volcanic arc. Continued convergence and igneous activity further deform and thicken the crust, elevating the mountain belt, while an accretionary wedge grows. **C.** Subduction ends and is followed by a period of erosion and uplift.

Students Sometimes Ask...

What is the highest elevation at which humans can survive for any extended period of time?

Individuals have lived for as long as 2 years at an altitude of nearly 6,000 meters (19,680 feet). The highest permanently inhabited village is believed to be La Rinconada, a mining town in southern Peru, at an altitude of about 5,100 kilometers (16,700 feet). The highest city in the world, La Paz, Bolivia, is located 3,570 meters (11,900 feet) above sea level, more than twice as high as Denver, Colorado—the highest major city in the United States.

western United States. It includes the Sierra Nevada and the Coast Ranges in California. These parallel mountain belts were produced by the subduction of a portion of the Pacific Basin under the western edge of the North American plate. The Sierra Nevada batholith is a remnant of a portion of the continental volcanic arc that was produced by several surges of magma over tens of millions of years. Subsequent uplifting and erosion have removed most of the evidence of past volcanic activity and exposed a core of crystalline, igneous, and associated metamorphic rocks.

In the trench region, sediments scraped from the subducting plate, plus those provided by the eroding continental volcanic arc, were intensely folded and faulted into an accretionary wedge. This chaotic mixture of rocks presently constitutes the Franciscan Formation of California's Coast Ranges. Uplifting of the Coast Ranges took place only recently, as evidenced by the young, unconsolidated sediments that still mantle portions of these highlands.

In summary, *the growth of mountain belts at subduction zones is a response to crustal thickening caused by the addition of mantle-derived igneous rocks and sediments scraped from the descending oceanic slab.*

Collisional Mountain Belts

Forces Within
▶ **Mountain Building**

Most major mountain belts are generated when one or more buoyant crustal fragments collide with a continental margin as a result of subduction. Oceanic lithosphere, which is relatively dense, readily subducts. Continental lithosphere, which contains significant amounts of low-density crustal rocks, is too buoyant to undergo subduction. Consequently, the arrival of a crustal fragment at a trench results in a collision with the margin of the adjacent continental block and an end to subduction.

Terranes and Mountain Building

The process of collision and accretion (joining together) of comparatively small crustal fragments to a continental margin has generated many of the mountainous regions that rim the Pacific. Geologists refer to these accreted crustal blocks as terranes. **Terrane** refers to any crustal fragment that has a geologic history distinct from that of the adjoining terranes.

The Nature of Terranes What is the nature of these crustal fragments, and where did they originate? Research suggests that prior to their accretion to a continental block, some of these fragments may have been **microcontinents** similar to the modern-day island of Madagascar, located east of Africa in the Indian Ocean. Many others were island arcs similar to Japan, the Philippines, and the Aleutian Islands. Still others may have been submerged oceanic plateaus created by massive outpourings of basaltic lavas associated with mantle plumes (**Figure 10.19**). More than 100 of these relatively small crustal fragments are presently known to exist.

Accretion and Orogenesis As oceanic plates move, they carry embedded oceanic plateaus, volcanic island arcs, and microcontinents to an Andean-type subduction zone. When an oceanic plate contains small seamounts, these structures are generally subducted along with the descending oceanic slab. However, large thick units

FIGURE 10.19 Distribution of present-day oceanic plateaus and other submerged crustal fragments. (Data from Ben-Avraham and others)

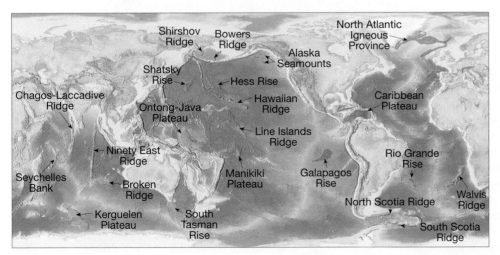

of oceanic crust, such as the Ontong Java Plateau, which is the size of Alaska, or an island arc composed of abundant "light" igneous rocks, render the oceanic lithosphere too buoyant to subduct. In these situations, a collision between the crustal fragment and the continental margin occurs.

The sequence of events that happen when an island arc reaches an Andean-type margin is shown in Figure 10.20. Rather than subduct, the upper crustal layers of these thickened zones are peeled from the descending plate and thrust in relatively thin sheets upon the adjacent continental block. Because subduction often continues for 100 million years or longer, several crustal fragments can be transported to the continental margin. Each collision displaces earlier accreted terranes further inland, adding to the zone of deformation as well as to the thickness and lateral extent of the continental margin.

The North American Cordillera   The relationship between mountain building and the accretion of crustal fragments arose primarily from studies conducted in the North American Cordillera (Figure 10.21). Researchers determined that some of the rocks in the orogenic belts of Alaska and British Columbia contained fossil and paleomagnetic evidence that indicated these strata previously lay much closer to the equator.

It is now known that many of the terranes that make up the North American Cordillera were scattered throughout the eastern Pacific, like the island arcs and oceanic plateaus currently distributed in the western Pacific. During the breakup of Pangaea, the eastern portion of the Pacific basin (Farallon plate) began to subduct under the western margin of North America. This activity resulted in the piecemeal addition of crustal fragments to the entire Pacific margin of the continent—from Mexico's Baja Peninsula to northern Alaska (Figure 10.21). Geologists expect that many modern microcontinents will likewise be accreted to active continental margins, producing new orogenic belts.

Continental Collisions

The Himalayas, Appalachians, Urals, and Alps represent mountain belts that were formed by the closure of major ocean basins. Continental collisions result in the development of mountains characterized by shortened and thickened crust achieved through folding and faulting. Some mountainous regions have crustal thicknesses that exceed 70 kilometers (40 miles).

Next, we take a closer look at two examples of collision mountains: the Himalayas and the Appalachians. The Himalayas are the youngest collision mountains on Earth and are still rising. The Appalachians are a much older mountain belt, in which active mountain building ceased about 250 million years ago.

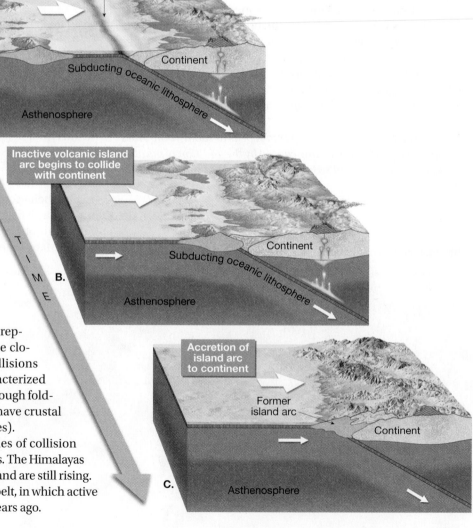

FIGURE 10.20 Sequence of events showing the collision and accretion of an island arc to a continental margin.

ARCTIC
OCEAN

ALASKA

Wrangellia

Yukon Tanana

Cache Creek

Stikinia

Wrangellia

PACIFIC
OCEAN

CANADA

Oceanic
terranes

Franciscan

UNITED STATES

MEXICO

FIGURE 10.21 Map showing terranes that have been added to western North America during the past 200 million years. Paleomagnetic studies and fossil evidence indicate that some of these terranes originated thousands of kilometers to the south of their present locations. (After D. R. Hutchinson and others)

The Himalayas

The mountain-building episode that created the Himalayas began roughly 50 million years ago when India began to collide with Asia. Prior to the breakup of Pangaea, India was located between Africa and Antarctica in the Southern Hemisphere (see Figure 7.A, p. 217). As Pangaea fragmented, India moved rapidly, geologically speaking, a few thousand kilometers in a northward direction.

The subduction zone that facilitated India's northward migration was near the southern margin of Asia (Figure 10.22A).

Continued subduction along Asia's margin created an Andean-type plate margin that contained a well-developed continental volcanic arc and an accretionary wedge. India's northern margin, on the other hand, was a passive continental margin consisting of a thick platform of shallow-water sediments and sedimentary rocks.

Geologists have determined that one or more small continental fragments were positioned on the subducting plate somewhere between India and Asia. During the closing of the intervening ocean basin, a small crustal fragment, which now forms southern Tibet, reached the trench. This event was followed by the docking of India itself. The tectonic forces involved in the collision of India with Asia were immense and caused the more deformable materials located on the seaward edges of these landmasses to become highly folded and faulted (see Figure 10.22B). The shortening and thickening of the crust elevated great quantities of crustal material, thereby generating the spectacular Himalayan mountains.

In addition to uplift, crustal shortening caused rocks at the "bottom of the pile" to become deeply buried—an environment where they experienced elevated temperatures and pressures (Figure 10.22B). Partial melting within the deepest and most deformed region of the developing mountain belt produced magmas that intruded the overlying rocks. It is in these environments that the metamorphic and igneous cores of collisional mountains are generated.

The formation of the Himalayas was followed by a period of uplift that raised the Tibetan Plateau. Seismic evidence suggests that a portion of the Indian subcontinent was thrust beneath Tibet, a distance of perhaps 400 kilometers. If so, the added crustal thickness would account for the lofty landscape of southern Tibet, which has an average elevation higher than Mount Whitney, the highest point in the contiguous United States.

The collision with Asia slowed but did not stop the northward migration of India, which has since penetrated at least 2,000 kilometers (1,200 miles) into the mainland of Asia. Crustal shortening accommodated some of this motion. Much of the remaining penetration into Asia caused lateral displacement of large blocks of the Asian crust by a mechanism described as *continental escape*. As shown in Figure 10.23, when India continued its northward trek, parts of Asia were "squeezed" eastward out of the collision zone. These displaced crustal blocks include much of present-day Indochina and sections of mainland China.

The Appalachians

The Appalachian Mountains provide great scenic beauty near the eastern margin of North America from Alabama to Newfoundland. In addition, mountains of similar origin that formed during the same period are found in the British Isles, Scandinavia, northwestern Africa, and Greenland (see Figure 7.6, p. 198). The orogeny that generated this extensive mountain system that presently lies on both sides of the North Atlantic lasted a few hundred million years and was one of the stages in assembling the supercontinent of Pangaea. Detailed studies of the Appalachians indicate that the formation of this mountain belt was complex and resulted from three distinct episodes of mountain building.

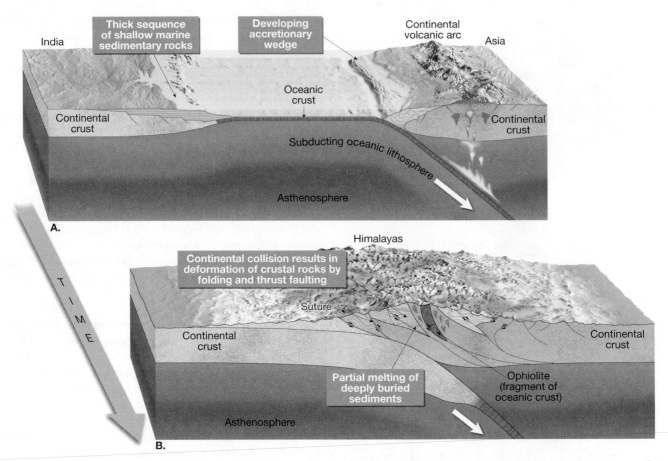

FIGURE 10.22 Diagrams illustrating the collision of India with the Eurasian plate, producing the spectacular Himalayas.

FIGURE 10.23 The collision between India and Asia that generated the Himalayas and Tibetan Plateau also severely deformed much of Southeast Asia. **A.** Map view of some of the major structural features of Southeast Asia thought to be related to this episode of mountain building. **B.** Re-creation of the deformation of Asia, with a rigid block representing India pushed into a mass of deformable modeling clay.

Our simplified overview begins roughly 750 million years ago with the breakup of a pre-Pangaea supercontinent called Rodinia, which rifted North America from Europe and Africa. This episode of continental rifting and seafloor spreading generated the ancestral North Atlantic. Located within this developing ocean basin was a fragment of continental crust that had been rifted from North America (**Figure 10.24A**).

About 600 million years ago, plate motion dramatically changed and the ancestral North Atlantic began to close. Two subduction zones likely formed: one located seaward of the coast of Africa that gave rise to a volcanic arc and another that developed along the margin of the continental fragment that lays off the coast of North America (Figure 10.24A).

Between 450 and 500 million years ago, the marginal sea between the crustal fragment and North America began to close. The collision that ensued, called the *Taconic Orogeny*, deformed the continental shelf and sutured the crustal fragment to the North American plate. The metamorphosed remnants of the continental fragment are recognized today as the crystalline rocks of the Blue Ridge and western Piedmont regions of the Appalachians (Figure 10.24B). In addition to the pervasive regional metamorphism, numerous magma bodies intruded the crustal rocks

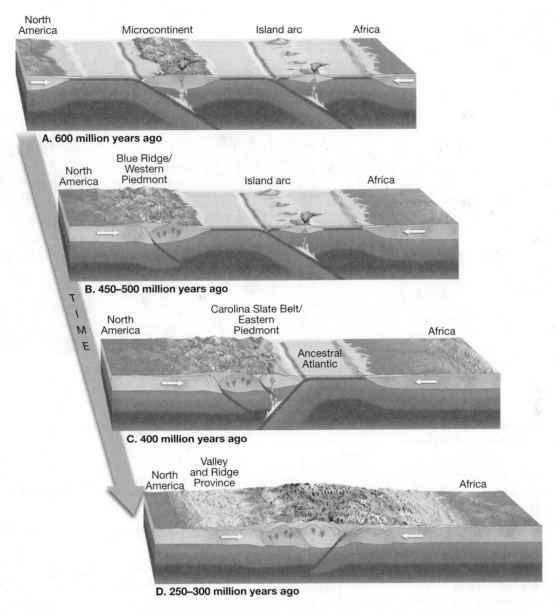

North America   Microcontinent   Island arc   Africa

A. 600 million years ago

North America   Blue Ridge/ Western Piedmont   Island arc   Africa

B. 450–500 million years ago

North America   Carolina Slate Belt/ Eastern Piedmont   Ancestral Atlantic   Africa

C. 400 million years ago

North America   Valley and Ridge Province   Africa

D. 250–300 million years ago

T I M E

FIGURE 10.24 These simplified diagrams depict the development of the southern Appalachians as the ancient North Atlantic was closed during the formation of Pangaea. Three separate stages of mountain-building activity spanned more than 300 million years. (After Zve Ben-Avraham, Jack Oliver, Larry Brown, and Frederick Cook)

along the entire continental margin, particularly in what we know today as New England.

A second episode of mountain building, called the *Acadian Orogeny*, began about 400 million years ago. The continued closing of the ancestral North Atlantic resulted in the collision of the developing island arc with North America (Figure 10.24C). Evidence for this event is visible in the Carolina Slate Belt of the eastern Piedmont, which contains metamorphosed sedimentary and volcanic rocks characteristic of an island arc.

The final orogeny occurred between 250 and 300 million years ago, when Africa collided with North America. The result was landward displacement of the Blue Ridge and Piedmont provinces by as much as 250 kilometers (155 miles). This event also displaced and further deformed the shelf sediments and sedimentary rocks that had once flanked the eastern margin of North America (Figure 10.24D). Today, these folded and thrust-faulted

sandstones, limestones, and shales make up the largely unmetamorphosed rocks of the Valley and Ridge Province. Outcrops of the folded and thrust-faulted structures that characterize collision mountains are found as far inland as central Pennsylvania and western Virginia (Figure 10.25).

Following the collision of Africa and North America, the Appalachians lay in the interior of Pangaea. About 180 million years ago, this newly formed supercontinent began to break into smaller fragments, a process that ultimately created the modern Atlantic Ocean. Because this new zone of rifting occurred east of the suture that formed when Africa and North America collided, remnants of Africa remain "welded" to the North American plate (Figure 10.24D).

Other mountain ranges that exhibit evidence of continental collisions include the Alps and the Urals. The Alps formed as Africa and Europe collided during the closing of the Tethys Sea.

Box 10.1

UNDERSTANDING EARTH

Fault-Block Mountains

Most mountain belts form in compressional environments, as evidenced by the predominance of large thrust faults and folded strata. However, other tectonic processes, such as continental rifting, can also produce topographic mountains.

The mountains that form in these settings, termed *fault-block mountains*, are bounded by high-angle normal faults that gradually flatten with depth. Most fault-block mountains form in response to broad uplifting, which causes elongation and faulting.

The Teton Range in western Wyoming is an excellent example of fault-block mountains. This lofty structure was faulted and uplifted along its eastern flank as the block tilted downward to the west. Looking west from Jackson Hole, Wyoming, the eastern front of these mountains rise more than 2 kilometers above the valley making it one of the most imposing mountain fronts in the United States (**Figure 10.A**).

Basin and Range Province

Located between the Sierra Nevada and Rocky Mountains is one of Earth's largest regions of fault-block mountains—the Basin and Range Province (see Figure 10.13). This region extends in a roughly north–south direction for nearly 3,000 kilometers (2,000 miles) and encompasses all of Nevada and portions of the surrounding states, as well as parts of southern Canada and western Mexico. In this region, the brittle upper crust has literally been broken into

FIGURE 10.A The mountains of Wyoming's Teton Range are fault-block mountains. (Photo by Stafano Amantini/Atlantide Phototravel/CORBIS)

hundreds of fault blocks. Tilting of these faulted structures, called *half-grabens*, gave rise to nearly parallel mountain ranges, averaging about 80 kilometers in length, which rise above adjacent sediment-filled basins.

Extension in the Basin and Range Province began about 20 million years ago and appears to have "stretched" the crust as much as twice its original width. High heat flow in the region, three times average, and several episodes of volcanism provide strong evidence that mantle upwelling caused doming of the crust, which in turn contributed to extension and faulting in the region.

Similarly, the Urals were uplifted during the assembly of Pangaea when northern Europe and northern Asia collided, forming a major portion of Eurasia.

CONCEPT CHECK 10.6

1. How is *terrane* different from *terrain*?
2. In addition to microcontinents, what other structures are carried by the oceanic lithosphere and eventually accreted to a continent?
3. How does the plate tectonics theory help explain the existence of fossil marine life in rocks atop compressional mountains?

Students Sometimes Ask...

What's the difference between a terrane and a terrain?

The term *terrane* is used to designate a distinct and recognizable series of rock formations that has been transported by plate tectonic processes. Since geologists who mapped these rocks were unsure where they came from, these rocks were sometimes called "exotic," "suspect," "accreted," or "foreign" terranes. Don't confuse this with the term *terrain*, which describes the shape of the surface topography or "lay of the land."

FIGURE 10.25 The Valley and Ridge Province. This portion of the Appalachian Mountains consists of folded and faulted sedimentary strata that were displaced landward with the closing of the proto-Atlantic.
(LANDSAT image, courtesy of Phillips Petroleum Company, Exploration Projects Section)

What Causes Earth's Varied Topography?

The causes of Earth's varied topography are complex and cumulative. Geologists know that colliding plates provide the tectonic forces that thicken and elevate crustal rocks during mountain building. Simultaneously, weathering and erosion sculpt Earth's surface into a vast array of landforms. In addition, up-and-down motions in the mantle can change the elevation of a region.

The Principle of Isostasy

During the 1840s, researchers discovered that Earth's low-density crust "floats" on top of the high-density, deformable rocks of the mantle. The concept of a floating crust in gravitational balance is called **isostasy**. The principle of isostasy helps us understand many large-scale variations on Earth' surface—from towering mountains to deep-ocean basins.

One way to explore the concept of isostasy is to envision a series of wooden blocks of different heights floating in water, as shown in **Figure 10.26A**. Note that the thicker wooden blocks float higher than the thinner blocks. Similarly, compressional mountains stand high above the surrounding terrain because crustal thickening creates buoyant crustal "roots" that extend deep into the supporting material below. Thus, lofty mountains such as the Himalayas are much like the thicker wooden blocks shown in Figure 10.26A.

How Is Isostasy Related to Changes in Elevation?   Visualize what would happen if another small block of wood were placed atop one of the blocks in Figure 10.26A. The combined block would sink until it reached a new isostatic (gravitational) balance. At this point, the top of the combined block would be higher than before, and the bottom would be lower. This process of establishing a new level of gravitational equilibrium by loading or unloading is called **isostatic adjustment**.

FIGURE 10.26 Principle of isostasy. **A.** Illustration showing how wooden blocks of different thicknesses float in water. In a similar manner, thick sections of crustal material float higher than thinner crustal slabs. **B.** A pine block floats higher than an oak block because it is less dense. **C.** Continental crust is less dense and thicker than oceanic crust, which explains the elevation differences between the continents and the ocean basins.

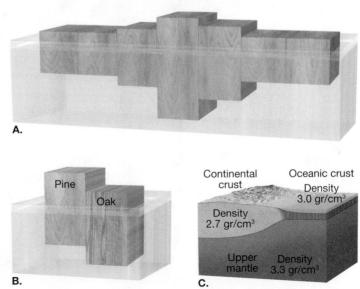

Applying the concept of isostatic adjustment, we should expect that when weight is added to the crust, it will respond by subsiding, and when weight is removed, it will rebound. (Visualize what happens when a ship's cargo is loaded or unloaded.) Evidence for crustal subsidence followed by crustal rebound is provided by Ice Age glaciers. When continental ice sheets occupied portions of North America during the Pleistocene epoch, the added weight of 3-kilometer-thick masses of ice caused downwarping of Earth's crust by hundreds of meters. In the 8,000 years since this last ice sheet melted, gradual uplift of as much as 330 meters (1,000 feet) has occurred in Canada's Hudson Bay region, where the thickest ice had accumulated.

One of the consequences of isostatic adjustment is that, as erosion lowers the summits of mountains, the crust rises in response to the reduced load (Figure 10.27). The processes of uplift and erosion continue until the mountain block reaches "normal" crustal thickness. When this occurs, these once-elevated structures will be near sea level, and the once-deeply buried interior of the mountain will be exposed at the surface. In addition, as mountains are worn down, the eroded sediment is deposited on adjacent landscapes, causing these areas to subside (Figure 10.27).

How High Is Too High?   Where compressional forces are great, such as those driving India into Asia, lofty mountains such as the Himalayas result. Is there a limit on how high a mountain can rise? As mountaintops are elevated, gravity-driven processes such as erosion and mass wasting are accelerated, carving the deformed strata into rugged landscapes. Equally important, however, is the

FIGURE 10.27 This sequence illustrates how the combined effects of erosion and isostatic adjustment result in a thinning of the crust in mountainous regions. **A.** When mountains are young, the continental crust is thickest. (Photo by Michael Collier) **B.** As erosion lowers the mountains, the crust rises in response to the reduced load. (Photo by Mark Karrass/CORBIS) **C.** Erosion and uplift continue until the mountains reach "normal" crustal thickness. (Photo by Pat & Chuck Blackley/Alamy)

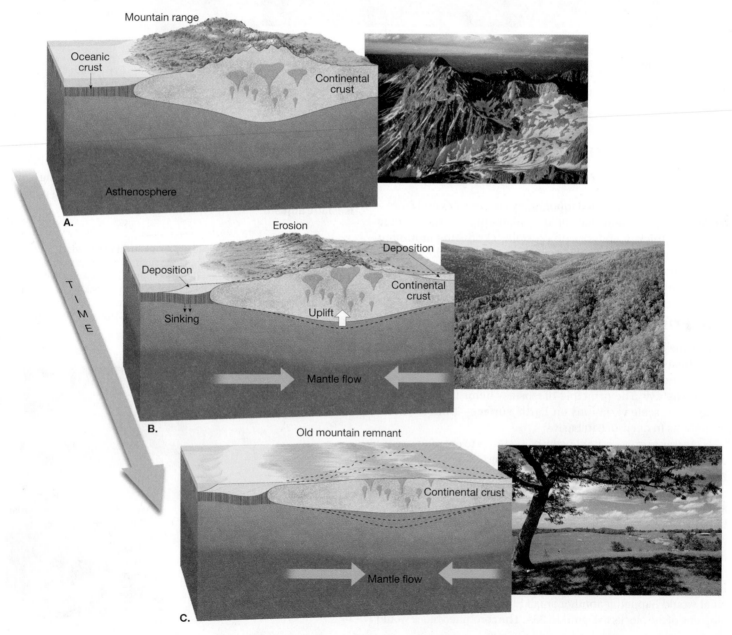

fact that gravity also acts on the rocks within these massive structures. The higher the mountain, the greater the downward force on rocks near the base. Eventually, the rocks deep within the developing mountain, which are relatively warm and weak, begin to flow laterally, as shown in Figure 10.28. This is analogous to what happens when a ladle of very thick pancake batter is poured onto a hot griddle. Similarly, mountains are altered by a process called **gravitational collapse**, which involves ductile spreading at depth and normal faulting and subsidence in the upper, brittle portion of the crust.

Considering these factors, a seemingly logical question follows—"What keeps the Himalayas standing?" Simply, the horizontal compressional forces that are driving India into Asia are greater than the vertical force of gravity. However, once India's northward trek ends, the downward pull of gravity, as well as weathering and erosion, will become the dominant forces acting on this mountainous region.

Continents Versus Ocean Basins

Why do the continents have an average elevation of nearly 1 kilometer *above* sea level, whereas the ocean basins have an average elevation of nearly 4 kilometers *below* sea level? Because continental crust is considerably thicker than oceanic crust—35 kilometers compared to 7 kilometers on average—we would expect the ocean floor to lie at a lower elevation. However, the differences in crustal thickness provides only part of the answer.

Figure 10.26C illustrates that a block of pine floats higher than the same-sized block of oak because the pine is less dense. Continental rocks have an average density of about 2.7 grams per cubic centimeter, whereas oceanic crust is somewhat more dense (about 3.0 grams per cubic centimeter). Thus, continental crust is less dense and thicker than oceanic crust, which largely explains the elevation differences between the continents and the ocean basins.

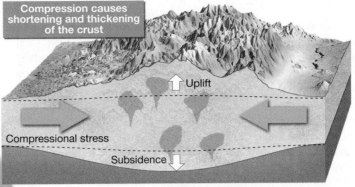

A. Horizontal compressional forces dominate

Compression causes shortening and thickening of the crust
Uplift
Compressional stress
Subsidence

T I M E

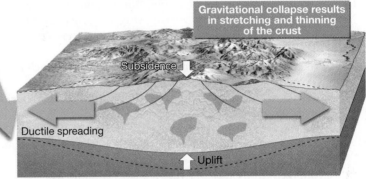

Gravitational collapse results in stretching and thinning of the crust
Subsidence
Ductile spreading
Uplift

B. Gravitational forces dominate

FIGURE 10.28 Block diagram of a mountain belt that is collapsing under its own "weight." Gravitational collapse involves normal faulting in the upper, brittle portion of the crust and ductile spreading in the warm, weak rocks at depth.

CONCEPT CHECK 10.7

1 Define *isostasy*.
2 Give one example of evidence that supports the concept of crustal uplift.
3 What happens to a floating object when weight is added? Subtracted?
4 Briefly describe how the principle of isostatic adjustment applies to changes in the elevations of mountains.

GIVE IT SOME THOUGHT

1. Describe an environment within Earth where you might expect rocks to experience ductile deformation. Suggest a scenario in which brittle rather than ductile deformation might occur.
2. Which of these rocks, *granite* or *mica schist*, is more likely to fold or flow rather than fracture when subjected to differential stress? Explain.
3. Refer to the accompanying diagrams to answer the following:
   a. What type of dip-slip fault is shown in Diagram 1? Were the dominant forces during faulting tensional, compressional, or shear?
   b. What type of dip-slip fault is shown in Diagram 2? Were the dominant forces during faulting tensional, compressional, or shear?
   c. Match the correct pair of arrows in Diagram 3 to the faults in Diagrams 1 and 2.
4. Refer to Figure 10.19 showing the distribution of present-day oceanic plateaus and other submerged crustal fragments. Which feature, the *Glapagos Rise* or the *Rio Grande Rise*, will

Diagram 1

Diagram 2

a. *b.*

Diagram 3

likely end up accreted to a continent? Explain your choice. In the future, how might a geologist determine that this accreted terrane is distinct from the continental crust to which it accreted?

5. The Ural Mountains exhibit a north–south orientation through Eurasia (see Figure 1.18). How does the theory of plate tectonics explain the existence of this mountain belt in the interior of an expansive landmass?

6. Explain why the Appalachian Mountains, which are similar in age to the Urals, are found along the margin of North America and not in the interior.

7. Briefly describe the major differences between the evolution of the Appalachian Mountains and the North American Cordillera.

8. Suppose a sliver of oceanic crust was discovered in the interior of a continent. Would this support or refute the theory of plate tectonics? Explain.

9. What processes (besides formation and melting of large ice sheets) could cause isostatic adjustments?

10. Mountains are long-lived features, but they do not retain their elevation forever. Describe or sketch a process other than weathering and erosion that eventually lowers a mountain range.

11. Ice has a density of roughly 0.9 g/cm^3 and water has a density of about 1.0 g/cm^3. Use that information to complete the following:

    a. Using the principle of isostasy, approximate how much of a 100-meter-thick iceberg would be located above sea level.

    b. Assuming that this iceberg melts evenly, how much of it will be above sea level after half of its mass has melted?

    c. Do you think icebergs make good examples for the principle of isostasy? Explain.

    d. Compare the melting of an iceberg to the process of isostatic adjustment that occurs when a mountain erodes.

12. In your own words, describe the geologic evolution of a "typical" collisional mountain using the Appalachian Mountains as a guide.

In Review   Chapter 10 Crustal Deformation and Mountain Building

- *Deformation* refers to changes in the shape, size (volume), or orientation of a rock body and is most pronounced along plate margins. Rocks deform differently depending on the environment (temperature and confining pressure), the composition and texture of the rock, and the length of time stress is maintained. Rocks first respond by deforming *elastically* and will return to their original shape when the stress is removed. Once their elastic limit (strength) is surpassed, rocks either deform by ductile flow or they fracture. *Ductile deformation* is a solid-state flow that changes the shape of an object without fracturing. Ductile flow may be accomplished by gradual slippage and recrystallization along planes of weakness within the crystal lattice of mineral grains. Ductile deformation tends to occur in a high-temperature/high-pressure environment. In a near-surface environment, most rocks deform by *brittle failure*.

- Among the most basic geologic structures associated with rock deformation are *folds* (flat-lying sedimentary and volcanic rocks bent into a series of wavelike undulations). The two most common types of folds are *anticlines*, formed by the upfolding, or arching, of rock layers, and *synclines*, which are downfolds. Most folds are the result of horizontal *compressional stresses*. Folds can be *symmetrical*, *asymmetrical*, or, if one limb has been tilted beyond the vertical, *overturned*. *Domes* (upwarped structures) and *basins* (downwarped structures) are circular or somewhat elongated folds formed by vertical displacements of strata.

- *Faults* are fractures in the crust along which appreciable displacement has occurred. Faults in which the movement is primarily vertical are called *dip-slip faults*. Dip-slip faults include both *normal* and *reverse faults*. Low-angle reverse faults are called *thrust faults*. Normal faults indicate *tensional stresses* that pull the crust apart. Reverse and thrust faulting indicate that *compressional forces* are at work. Large *thrust faults* are associated with subduction zones and other convergent boundaries where plates collide. *Strike-slip faults* exhibit mainly horizontal displacement parallel to the strike of the fault surface. Large strike-slip faults, called *transform faults*, accommodate displacement between plate boundaries.

- *Joints* are fractures along which no appreciable displacement has occurred. Joints generally occur in groups with roughly parallel orientations and are the result of brittle failure of rock units located in the outermost crust.

- The name for the processes that collectively produce a *compressional mountain belt* is orogenesis. Most compressional mountains consist of folded and faulted sedimentary and volcanic rocks, portions of which have been strongly metamorphosed and intruded by younger igneous bodies.

- Subduction of oceanic lithosphere beneath a continental block gives rise to an *Andean-type plate margin* that is characterized by a continental volcanic arc and associated igneous plutons. In addition, sediment derived from the land, as well as material scraped from the subducting plate, becomes plastered against the landward side of the trench, forming an *accretionary wedge*.

- Mountain belts can develop as a result of the collision and merger of one or more crustal fragments, including island arcs and oceanic plateaus to a continental block. Many of the mountain belts of the North American Cordillera were generated in this manner.

- Continued subduction of oceanic lithosphere beneath an Andean-type continental margin will eventually close an ocean basin. The result will be a *continental collision* and the development of compressional mountains that are characterized by shortened and thickened crust. The development of a major mountain belt is often complex, involving two or more distinct episodes of mountain building. Continental collisions have generated most of Earth's major mountain belts, including the Himalayas, Alps, Urals, and Appalachians.

- Earth's crust floats on top of the deformable rocks of the mantle, much like wooden blocks float in water. The concept of a floating crust in gravitational balance is called *isostasy*. Most mountains consist of crustal rocks that have been shortened and thickened, producing deep crustal roots that isostatically support them. As erosion lowers the peaks, *isostatic adjustment* gradually raises the mountains in response. The processes of uplift and erosion will continue until the mountain block reaches "normal" crustal thickness. Gravity can also cause these elevated structures to collapse under their own "weight."

Key Terms

accretionary wedge (p. 310)
active continental margin (p. 310)
Andean-type plate margin (p. 309)
anticline (p. 301)
basin (p. 302)
brittle deformation (p. 299)
compressional mountain (p. 309)
deformation (p. 298)
dip-slip fault (p. 305)
dome (p. 302)
ductile deformation (p. 299)
elastic deformation (p. 298)
fault (p. 304)

fault-block mountains (p. 305)
fault scarp (p. 305)
fold (p. 300)
graben (p. 306)
gravitational collapse (p. 319)
half-graben (p. 306)
hogback (p. 303)
horst (p. 306)
isostasy (p. 317)
isostatic adjustment (p. 317)
joint (p. 307)
microcontinent (p. 311)
monocline (p. 303)

normal fault (p. 305)
orogenesis (p. 309)
passive continental margin (p. 310)
reverse fault (p. 306)
stress (p. 298)
strike-slip fault (p. 306)
syncline (p. 301)
terrane (p. 311)
thrust fault (p. 306)
transform fault (p. 307)
volcanic island arc (p. 309)

Examining the Earth System

1. A good example of the interaction among Earth's spheres is the influence of mountains on climate. Examine the temperature graph for the cities of Seattle and Spokane, Washington in Figure 16.30, p. 482. Notice on the inset map that mountains (the Cascades) separate the two cities. The prevailing wind direction in the region is from west to east. (a) Contrast the summer and winter temperatures that occur at each city. Why are they different? *Hint*: Check out the sections on "Land and Water" and "Geographic Position" in Chapter 16.

(b) The annual rainfall at Spokane (16.6 inches) is less than half that for Seattle (37.1, inches). Can you explain why? *Hint*: A glance at Figure 17.11, p. 500 might be helpful.

2. The Cascades have had a profound effect on the amount and type of plant and animal life (biosphere) that inhabit the region around Spokane and Seattle. Provide several specific examples to verify this statement.

Mastering Geology

Looking for additional review and test prep materials? Visit the Self Study area in **www.masteringgeology.com** to find practice quizzes, study tools, and multimedia that will aid in your understanding of this chapter's content. In **MasteringGeology™** you will find:

- **GEODe: Earth Science: An interactive visual walkthrough of key concepts**

- **Geoscience Animation Library: More than 100 animations illuminating many difficult-to-understand Earth science concepts**
- **In The News RSS Feeds: Current Earth science events and news articles are pulled into the site with assessment**
- **Pearson eText**
- **Optional Self Study Quizzes**
- **Web Links**
- **Glossary**
- **Flashcards**

Deciphering Earth's History

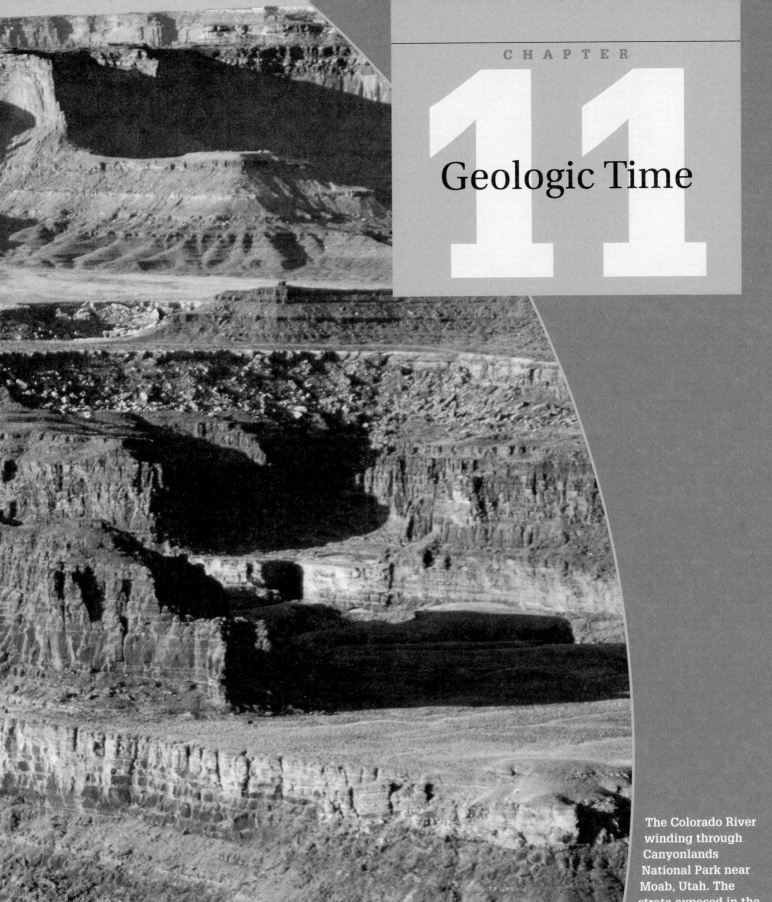

11

Geologic Time

The Colorado River winding through Canyonlands National Park near Moab, Utah. The strata exposed in the canyon walls contain clues to millions of years of Earth history.

(Photo by John T. Parkinson/Mira.com)

In the 18th century, James Hutton recognized the immensity of Earth history and the importance of time as a component in all geological processes. In the 19th century, others effectively demonstrated that Earth had experienced many episodes of mountain building and erosion, which must have required great spans of geologic time. Although these pioneering scientists understood that Earth was very old, they had no way of knowing its true age. Was it tens of millions, hundreds of millions, or even billions of years old? Consequently a geologic time scale was developed that showed the sequence of events based on relative dating principles. What are these principles? What part do fossils play? With the discovery of radioactivity and the development of radiometric dating techniques, geologists can now assign fairly accurate dates to many of the events in Earth history. What is radioactivity? Why is it a good "clock" for dating the geologic past? In this chapter we answer these questions.

FOCUS ON CONCEPTS

To assist you in learning **the important concepts in this chapter, focus on the following questions:**

- What is the basic difference between uniformitarianism and catastrophism?
- What are the two types of dates used by geologists in the study of Earth history?
- What principles, laws, and techniques are used to establish relative dates?
- What is a fossil? What are the conditions that favor the preservation of organisms as fossils?
- How are fossils used to correlate rocks of similar age that are in different places?
- What is radioactivity and how are radioactive isotopes used in obtaining numerical dates?
- What is the geologic time scale and what are its principal subdivisions?
- Why is it difficult to assign reliable numerical dates to samples of sedimentary rock?

Geology Needs a Time Scale

In 1869 John Wesley Powell, who was later to head the U.S. Geological Survey, led a pioneering expedition down the Colorado River and through the Grand Canyon (**Figure 11.1**). Writing about the rock layers that were exposed by the downcutting of the river, Powell said, "The canyons of this region would be a Book of Revelations in the rock-leaved Bible of geology." He was undoubtedly impressed with the millions of years of Earth history exposed along the walls of the Grand Canyon and the other canyons of the Colorado Plateau (see chapter-opening photo and **Figure 11.2**).

Powell realized that the evidence for an ancient Earth is concealed in its rocks. Like the pages in a long and complicated history book, rocks record the geological events and changing life forms of the past. The book, however, is not complete. Many pages, especially in the early chapters, are missing. Others are tattered, torn, or smudged. Yet enough of the book remains to allow much of the story to be deciphered.

Interpreting Earth history is a prime goal of the science of geology. Like a modern-day sleuth, the geologist must interpret clues found preserved in the rocks. By studying rocks, especially sedimentary rocks, and the features they contain, geologists can unravel the complexities of the past.

Geological events by themselves, however, have little meaning until they are put into a time perspective. Studying history, whether it be the Civil War or the Age of Dinosaurs, requires a calendar. Among geology's major contributions to human knowledge is the *geologic time scale* and the discovery that Earth history is exceedingly long.

The geologists who developed the geologic time scale revolutionized the way people think about time and how they perceive our planet. They learned that Earth is much older than anyone had previously imagined and that its surface and interior have been changed over and over again by the same geological processes that operate today.

CONCEPT CHECK 11.1

❶ Why is a geologic time scale a useful tool?

A Brief History of Geology

In the mid-1600s, James Ussher, Anglican Archbishop of Armagh, Primate of all Ireland, published a work that had immediate and profound influence on people's view of Earth's age. A respected scholar of the Bible, Ussher constructed a chronology of human and Earth history in which he determined that Earth was only a few thousand years old, having been created in 4004 B.C. Ussher's treatise earned widespread acceptance among Europe's scientific and religious leaders, and his chronology was soon printed in the margins of the Bible itself.

FIGURE 11.1 The start of the Powell expedition from Green River station, Wyoming, is depicted in this drawing from Powell's 1875 book. The inset photo is of Major John Wesley Powell, pioneering geologist and the second director of the U.S. Geological Survey. (Courtesy of the U.S. Geological Survey, Denver)

presently shaping our planet have been at work for a very long time. Thus, to understand ancient rocks, we must first understand present-day processes and their results. This idea is commonly expressed by saying, "The present is the key to the past."

Prior to Hutton's *Theory of the Earth*, no one had effectively demonstrated that geological processes occur over extremely long periods of time. However, Hutton persuasively argued that weak, slow-acting processes could, over long spans of time, produce effects just as great as those resulting from sudden catastrophic events. Unlike his predecessors, Hutton carefully cited verifiable observations to support his ideas.

For example, when he argued that mountains are sculpted and ultimately destroyed by weathering and the work of running water, and that their wastes are carried to the oceans by processes that can be observed, Hutton said, "We have a chain of facts which clearly demonstrates that the materials of the wasted mountains have traveled through the rivers"; and

FIGURE 11.2 The Grand Canyon of the Colorado River in northern Arizona. The rocks exposed here represent hundreds of millions of years of Earth history. It also took millions of years of weathering, mass wasting and erosion to create the canyon. Geologic processes often act so slowly that changes may not be visible during an entire human lifetime. Geologists must routinely deal with ancient materials and events that occurred in the distant geologic past. The youngest rocks are on top, and the oldest are at the bottom. (Photo by Jens Hilberger/age fotostock)

During the 17th and 18th centuries the doctrine of **catastrophism** strongly influenced people's thinking about Earth. Briefly stated, catastrophists believed that Earth's landscapes had been developed primarily by great catastrophes. Features such as mountains and canyons, which today we know take great periods of time to form, were explained as having been produced by sudden and often worldwide disasters triggered by unknowable causes that no longer operate. This philosophy was an attempt to fit the rate of Earth processes to the prevailing ideas on the age of Earth.

Birth of Modern Geology

Modern geology began in the late 1700s when James Hutton, a Scottish physician and gentleman farmer, published his *Theory of the Earth*. In this work, Hutton put forth a fundamental principle that is a pillar of geology today: **uniformitarianism**. It simply states that *the physical, chemical, and biological laws that operate today have also operated in the geologic past*. This means that the forces and processes that we observe

further, "There is not one step in all this progress that is not to be actually perceived." He then went on to summarize this thought by asking a question and immediately providing the answer: "What more can we require? Nothing but time."

Geology Today

Today, the basic tenets of uniformitarianism are just as viable as in Hutton's day. Indeed, we realize more strongly than ever that the present gives us insight into the past and that the physical, chemical, and biological laws that govern geological processes remain unchanging through time. However, we also understand that the doctrine should not be taken too literally. To say that geological processes in the past were the same as those occurring today is not to suggest that they always had the same relative importance or that they operated at precisely the same rate. Moreover, some important geologic processes are not currently observable, but evidence that they occur is well established. For example, we know that Earth has experienced impacts from large meteorites even though we have no human witnesses. Such events altered Earth's crust, modified its climate, and strongly influenced life on the planet.

The acceptance of uniformitarianism meant the acceptance of a very long history for Earth. Although Earth's processes vary in intensity, they still take a very long time to create or destroy major landscape features.

For example, geologists have established that mountains once existed in portions of present-day Minnesota, Wisconsin, Michigan, and Manitoba, Canada. Today, the region consists of low hills and plains because erosion gradually destroyed these peaks. The rock record contains evidence that shows Earth has experienced *many* cycles of mountain building and erosion. Concerning the ever-changing nature of Earth through great expanses of geologic time, Hutton made a statement that was to become his most famous. In concluding his classic 1788 paper published in the *Transactions of the Royal Society of Edinburgh*, he stated, "The results, therefore, of our present enquiry is, that we find no vestige of a beginning—no prospect of an end."

It is important to remember that although many features of our physical landscape may seem to be unchanging over our lifetimes, they are nevertheless changing, but on time scales of hundreds, thousands, or even many millions of years.

CONCEPT CHECK 11.2

❶ Contrast catastrophism and uniformitarianism. How did each philosophy view the age of Earth?

Relative Dating: Key Principles

Deciphering Earth's History
▶ A Relative Dating: Key Principles

The geologists who developed the geologic time scale revolutionized the way people think about time and perceive our planet. They learned that Earth is much older than anyone had previously

imagined and that its surface and interior have been changed over and over again by the same geological processes that operate today.

During the late 1800s and early 1900s, various attempts were made to determine the age of Earth. Although some of the methods appeared promising at the time, none proved reliable. What these scientists were seeking was a **numerical date**. Such dates specify the actual number of years that have passed since an event occurred—for example, the extinction of the dinosaurs about 65 million years ago. Today, our understanding of radioactivity allows us to accurately determine numerical dates for rocks that represent important events in Earth's distant past. We will study radioactivity later in this chapter. Prior to the discovery of radioactivity, geologists had no accurate and dependable method of numerical dating and had to rely solely on relative dating.

Relative dating means placing rocks in their proper *sequence of formation*, first, second, third, and so on. Relative dating cannot tell us how long ago something took place, only that it followed one event and preceded another. The relative dating techniques that were developed are valuable and still widely used. Numerical dating methods did not replace these techniques; they simply supplemented them. To establish a relative time scale, a few basic principles or rules had to be discovered and applied. Although they may seem obvious to us today, they were major breakthroughs in thinking at the time, and their discovery and acceptance was an important scientific achievement.

Law of Superposition

Nicolaus Steno, a Danish anatomist, geologist, and priest (1636–1686), is credited with being the first to recognize a sequence of historical events in an outcrop of sedimentary rock layers. Working in the mountains of western Italy, Steno applied a very simple rule that has come to be the most basic

Students Sometimes Ask...

You mentioned early attempts at determining Earth's age that proved unreliable. How did 19th-century scientists go about making such calculations?

One method that was attempted several times involved the rate at which sediment is deposited. Some reasoned that if they could determine the rate that sediment accumulates and could further ascertain the total thickness of sedimentary rock that had been deposited during Earth history, they could estimate the length of geologic time. All that was necessary was to divide the rate of sediment accumulation into the total thickness of sedimentary rock.

Estimates of Earth's age varied each time this method was attempted. The age of Earth as calculated by this method ranged from 3 million to 1.5 billion years! Obviously this method was riddled with difficulties. Can you suggest what some might have been?

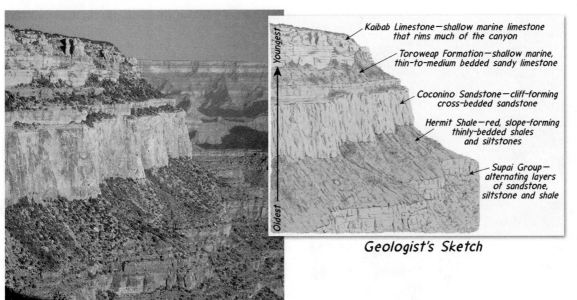

Kaibab Limestone—shallow marine limestone that rims much of the canyon

Toroweap Formation—shallow marine, thin-to-medium bedded sandy limestone

Coconino Sandstone—cliff-forming cross-bedded sandstone

Hermit Shale—red, slope-forming thinly-bedded shales and siltstones

Supai Group— alternating layers of sandstone, siltstone and shale

Geologist's Sketch

FIGURE 11.3 Applying the law of superposition to these layers exposed in the upper portion of the Grand Canyon, the Supai Group is oldest and the Kaibab Limestone is youngest. (Photo by E. J. Tarbuck)

principle of relative dating—the **law of superposition**. The law simply states that in an undeformed sequence of sedimentary rocks, each bed is older than the one above it and younger than the one below. Although it may seem obvious that a rock layer could not be deposited unless it had something older beneath it for support, it was not until 1669 that Steno clearly stated the principle.

This rule also applies to other surface-deposited materials, such as lava flows and beds of ash from volcanic eruptions. Applying the law of superposition to the beds exposed in the upper portion of the Grand Canyon, you can easily place the layers in their proper order (**Figure 11.3**). Among those that are shown, the sedimentary rocks in the Supai Group must be the oldest, followed in order by the Hermit Shale, Coconino Sandstone, Toroweap Formation, and Kaibab Limestone.

Principle of Original Horizontality

Steno is also credited with recognizing the importance of another basic principle, called the **principle of original horizontality**. Simply stated, it means that layers of sediment are generally deposited in a horizontal position. Thus, if we observe rock layers that are flat, it means they have not been disturbed and thus still have their *original* horizontality. The layers in Canyonlands National Park (see chapter-opening photo) and the Grand Canyon (see Figure 11.2) illustrate this. However, if the layers are folded or inclined at a steep angle, they must have been moved into that position by crustal disturbances sometime *after* their deposition (**Figure 11.4**).

Principle of Cross-Cutting Relationships

When a fault cuts through other rocks, or when magma intrudes and crystallizes, we can assume that the fault or intrusion is younger than the rocks affected. For example, in **Figure 11.5**, the faults and dikes clearly must have occurred *after* the sedimentary layers were deposited.

FIGURE 11.4 Most layers of sediment are deposited in a nearly horizontal position. Thus, when we see rock layers that are folded or tilted, we can assume that they must have been moved into that position by crustal disturbances *after* their deposition. These folded layers are exposed at Agio Pavlos on the Mediterranean Island of Crete. (Photo by Marco Simoni/Robert Harding)

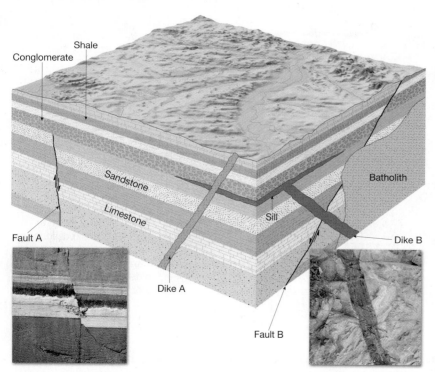

Shale
Conglomerate
Sandstone
Limestone
Fault A
Dike A
Batholith
Sill
Dike B
Fault B

FIGURE 11.5 Cross-cutting relationships are an important principle used in relative dating. An intrusive rock body is younger than the rocks it intrudes. A fault is younger than the rock layers it cuts.

This is the **principle of cross-cutting relationships**. By applying the cross-cutting principle, you can see that fault *A* occurred *after* the sandstone layer was deposited because it "broke" the layer. However, fault *A* occurred *before* the conglomerate was laid down because that layer is unbroken.

We can also state that dike *B* and its associated sill are older than dike *A* because dike *A* cuts the sill. In the same manner, we know that the batholith was emplaced after movement occurred along fault *B*, but before dike *B* was formed. This is true because the batholith cuts across fault *B*, and dike *B* cuts across the batholith.

Inclusions

Sometimes inclusions can aid the relative dating process. **Inclusions** are pieces of one rock unit that are contained within another. The basic principle is logical and straightforward.

FIGURE 11.6 These diagrams illustrate two ways that inclusions can form, as well as a type of unconformity termed a *nonconformity*. In part **A**, the inclusions in the igneous mass represent unmelted remnants of the surrounding host rock that were broken off and incorporated at the time the magma was intruded. In part **C**, the igneous rock must be older than the overlying sedimentary beds because the sedimentary beds contain inclusions of the igneous rock. When older intrusive igneous rocks are overlain by younger sedimentary layers, a noncomformity is said to exist. The inset in part **C** shows an inclusion of dark igneous rock in a lighter-colored and younger host rock.

The rock mass adjacent to the one containing the inclusions must have been there first in order to provide the rock fragments. Therefore, the rock mass containing inclusions is the younger of the two. Figure 11.6 provides an example. Here the inclusions of intrusive igneous rock in the adjacent sedimentary layer indicate that the sedimentary layer was deposited on top of a weathered igneous mass rather than being intruded from below by magma that later crystallized.

Unconformities

When we observe layers of rock that have been deposited essentially without interruption, we call them **conformable**. Particular sites exhibit conformable beds representing certain spans of geologic time. However, no place on Earth has a complete set of conformable strata.

Throughout Earth history, the deposition of sediment has been interrupted again and again. All such breaks in the rock record are termed unconformities. An **unconformity** represents a long period during which deposition ceased, erosion removed previously

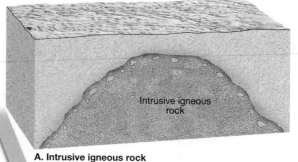

A. Intrusive igneous rock

Intrusive igneous rock

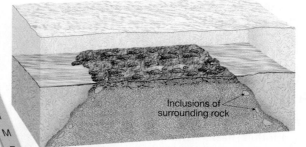

B. Exposure and weathering of intrusive igneous rock

Inclusions of surrounding rock

TIME

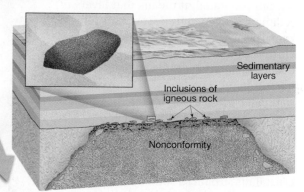

C. Deposition of sedimentary layers

Sedimentary layers
Inclusions of igneous rock
Nonconformity

formed rocks, and then deposition resumed. In each case uplift and erosion are followed by subsidence and renewed sedimentation. Unconformities are important features because they represent significant geologic events in Earth history. Moreover, their recognition helps us identify what intervals of time are not represented by strata and thus are missing from the geologic record.

The rocks exposed in the Grand Canyon of the Colorado River represent a tremendous span of geologic history. It is a wonderful place to take a trip through time. The canyon's colorful strata record a long history of sedimentation in a variety of environments—advancing seas, rivers and deltas, tidal flats, and sand dunes. But the record is not continuous. Unconformities represent vast amounts of time that have not been recorded in the canyon's layers. **Figure 11.7** is a geologic cross section of the Grand Canyon. Refer to it as you read about the three basic types of unconformities: angular unconformities, disconformities, and nonconformities.

Angular Unconformity   Perhaps the most easily recognized unconformity is an **angular unconformity**. It consists of tilted or folded sedimentary rocks that are overlain by younger, more flat-lying strata. An angular unconformity indicates that during the pause in deposition, a period of deformation (folding or tilting) and erosion occurred (**Figure 11.8**).

When James Hutton studied an angular unconformity in Scotland more than 200 years ago, it was clear to him that it represented a major episode of geologic activity (**Figure 11.9**). He also

appreciated the immense time span implied by such relationships. When a companion later wrote of their visit to the site, he stated that "the mind seemed to grow giddy by looking so far into the abyss of time."

Disconformity   When contrasted with angular unconformities, **disconformities** are more common, but usually far less conspicuous because the strata on either side are essentially parallel. For example, look at the disconformities in the cross section of the Grand Canyon in Figure 11.7. Many disconformities are difficult to identify because the rocks above and below are similar and there is little evidence of erosion. Such a break often resembles an ordinary bedding plane. Other disconformities are easier to identify because the ancient erosion surface is cut deeply into the older rocks below.

Nonconformity   The third basic type of unconformity is a **nonconformity**. Here the break separates older metamorphic or intrusive igneous rocks from younger sedimentary strata (Figures 11.6 and 11.7). Just as angular unconformities and disconformities imply crustal movements, so too do nonconformities. Intrusive igneous masses and metamorphic rocks originate far below the surface. Thus, for a nonconformity to develop, there must be a period of uplift and the erosion of overlying rocks. Once exposed at the surface, the igneous or metamorphic rocks are subjected to weathering and erosion prior to subsidence and the renewal of sedimentation.

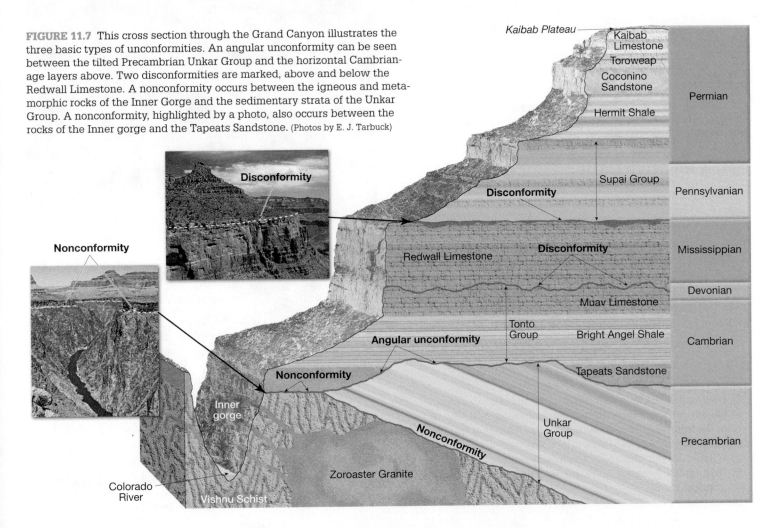

FIGURE 11.7 This cross section through the Grand Canyon illustrates the three basic types of unconformities. An angular unconformity can be seen between the tilted Precambrian Unkar Group and the horizontal Cambrian-age layers above. Two disconformities are marked, above and below the Redwall Limestone. A nonconformity occurs between the igneous and metamorphic rocks of the Inner Gorge and the sedimentary strata of the Unkar Group. A nonconformity, highlighted by a photo, also occurs between the rocks of the Inner gorge and the Tapeats Sandstone. (Photos by E. J. Tarbuck)

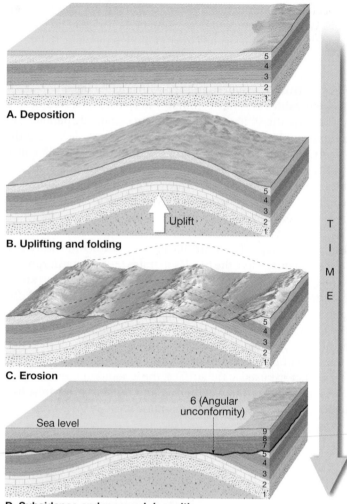

A. Deposition

B. Uplifting and folding

Uplift

C. Erosion

D. Subsidence and renewed deposition

Sea level

6 (Angular unconformity)

TIME

FIGURE 11.8 Formation of an angular unconformity. An angular unconformity represents an extended period during which deformation and erosion occurred.

Using Relative Dating Principles

If you apply the principles of relative dating to the hypothetical geologic cross section in **Figure 11.10**, you can place in proper sequence the rocks and the events they represent. The statements within the figure summarize the logic used to interpret the cross section.

In this example, we establish a relative time scale for the rocks and events in the area of the cross section. Remember that this method gives us no indication as to how many years of Earth history are represented, for we have no numerical dates. Nor do we know how this area compares to any other.

CONCEPT CHECK 11.3

1. Distinguish between numerical dates and relative dates.
2. Sketch and label three simple diagrams that illustrate each of the following: superposition, original horizontality, and cross-cutting relationships.
3. What is the significance of an unconformity?
4. Distinguish among angular unconformity, disconformity, and nonconformity.

Correlation of Rock Layers

To develop a geologic time scale that is applicable to the entire Earth, rocks of similar age in different regions must be matched up. Such a task is referred to as **correlation**.

Within a limited area, correlating the rocks of one locality with those of another may be done simply by walking along the outcropping edges. However, this might not be possible when the

FIGURE 11.9 This angular unconformity at Siccar Point, Scotland, was studied by James Hutton more than 200 years ago. (Photo by Marli Miller)

Above the unconformity lie gently dipping beds of reddish sandstone and conglomerate

Angular unconformity

Rock hammer

Below the unconformity lie nearly vertical sandstones and shales

Geologist's Sketch

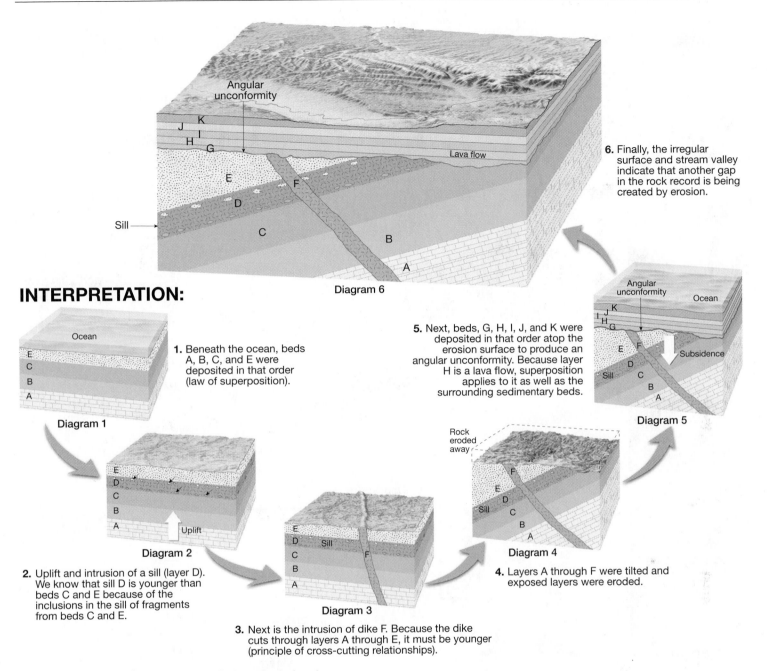

INTERPRETATION:

1. Beneath the ocean, beds A, B, C, and E were deposited in that order (law of superposition).

Diagram 1

2. Uplift and intrusion of a sill (layer D). We know that sill D is younger than beds C and E because of the inclusions in the sill of fragments from beds C and E.

Diagram 2

3. Next is the intrusion of dike F. Because the dike cuts through layers A through E, it must be younger (principle of cross-cutting relationships).

Diagram 3

4. Layers A through F were tilted and exposed layers were eroded.

Diagram 4

5. Next, beds, G, H, I, J, and K were deposited in that order atop the erosion surface to produce an angular unconformity. Because layer H is a lava flow, superposition applies to it as well as the surrounding sedimentary beds.

Diagram 5

6. Finally, the irregular surface and stream valley indicate that another gap in the rock record is being created by erosion.

Diagram 6

FIGURE 11.10 Geologic cross section of a hypothetical region.

rocks are mostly concealed by soil and vegetation. Correlation over short distances is often achieved by noting the position of a distinctive rock layer in a sequence of strata. Or, a layer may be identified in another location if it is composed of very distinctive or uncommon minerals.

By correlating the rocks from one place to another, a more comprehensive view of the geologic history of a region is possible. **Figure 11.11**, for example, shows the correlation of strata at three sites on the Colorado Plateau in southern Utah and northern Arizona. No single locale exhibits the entire sequence, but correlation reveals a more complete picture of the sedimentary rock record.

Many geologic studies involve relatively small areas. Such studies are important in their own right, but their full value is realized only when the rocks are correlated with those of other regions. Although the methods just described are sufficient to trace a rock formation over relatively short distances, they are not adequate for matching rocks that are separated by great distances. When correlation between widely separated areas or between continents is the objective, geologists must rely on fossils.

CONCEPT CHECK 11.4

❶ What is the goal of correlation?

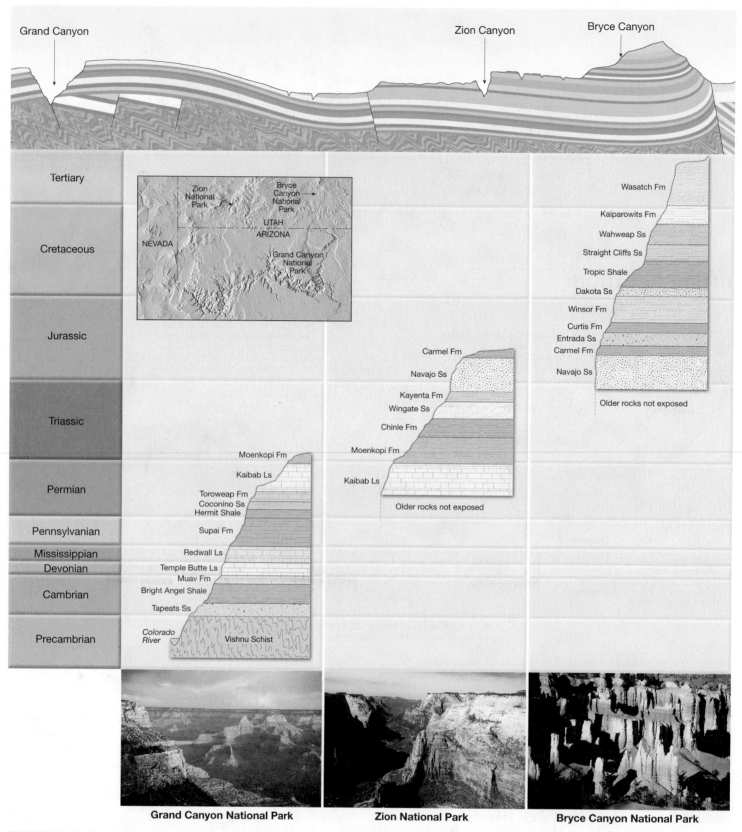

FIGURE 11.11 Correlation of strata at three locations on the Colorado Plateau provides a more complete picture of sedimentary rocks in the region. The diagram at the top is a geologic cross section of the region. (After U.S. Geological Survey; photos by E. J. Tarbuck)

Fossils: Evidence of Past Life

Fossils, the remains or traces of prehistoric life, are important inclusions in sediment and sedimentary rocks. They are important basic tools for interpreting the geologic past. The scientific study of fossils is called **paleontology**. It is an interdisciplinary science that blends geology and biology in an attempt to understand all aspects of the succession of life over the vast expanse of geologic time. Knowing the nature of the life-forms that existed at a particular time helps researchers understand past environmental conditions. Furthermore, fossils are important time indicators and play a key role in correlating rocks of similar ages that are from different places.

Types of Fossils

Fossils are of many types. The remains of relatively recent organisms may not have been altered at all (Figure 11.12). Such objects as teeth, bones, and shells are common examples. Far less common are entire animals, flesh included, that have been preserved because of rather unusual circumstances. Remains of prehistoric elephants called mammoths that were frozen in the Arctic tundra of Siberia and Alaska are examples, as are the mummified remains of sloths preserved in a dry cave in Nevada.

Given enough time, the remains of an organism are likely to be modified. Often, fossils become *petrified* (literally, "turned into stone"), meaning that the small internal cavities and pores of the original structure are filled with precipitated mineral matter (Figure 11.13A). In other instances *replacement* may occur. Here the cell walls and other solid material are removed and replaced with mineral matter. Sometimes the microscopic details of the replaced structure are faithfully retained.

Molds and casts constitute another common class of fossils. When a shell or other structure is buried in sediment and then dissolved by underground water, a *mold* is created. The mold faithfully reflects only the shape and surface marking of the organism; it does not reveal any information concerning its internal structure. If these hollow spaces are subsequently filled with mineral matter, *casts* are created (Figure 11.13B).

A type of fossilization called *carbonization* is particularly effective in preserving leaves and delicate animal forms. It occurs when fine sediment encases the remains of an organism. As time passes, pressure squeezes out the liquid and gaseous components and leaves behind a thin residue of carbon (Figure 11.13C). Black shales deposited as organic-rich mud in oxygen-poor environments often contain abundant carbonized remains. If the film of carbon is lost from a fossil preserved in fine-grained sediment, a

A.

B.

FIGURE 11.12 A. Excavating bones from Pit 91 at the La Brea tar pits in Los Angeles. It is a site rich in the unaltered remains of Ice Age organisms. Scientists have been digging here since 1915. (AP/Wide World Photo) **B.** Fossils of many relatively recent organisms are unaltered remains. The skeleton of the mammoth from the La Brea tar pits is a spectacular example. (Martin Shields/Alamy)

FIGURE 11.13 There are many types of fossilization. Six examples are shown here. **A.** Petrified wood in Petrified Forest National Park, Arizona. **B.** This trilobite photo illustrates mold and cast. **C.** A fossil bee preserved as a thin carbon film. **D.** Impressions are common fossils and often show considerable detail. **E.** Insect in amber. **F.** Coprolite is fossil dung. (Photo A by Bernhard Edmaier/Photo Researchers, Inc.; Photos B, D, and F by E .J. Tarbuck; Photo C courtesy of Florissant Fossil Beds National Monument; Photo E by Colin Keates/Dorling Kindersly Media Library)

replica of the surface, called an *impression*, may still show considerable detail (Figure 11.13D).

Delicate organisms, such as insects, are difficult to preserve, and consequently they are relatively rare in the fossil record. Not only must they be protected from decay but they must also not be subjected to any pressure that would crush them. One way in which some insects have been preserved is in *amber*, the hardened resin of ancient trees. The fly in Figure 11.13E was preserved after being trapped in a drop of sticky resin. Resin sealed off the insect from the atmosphere and protected the remains from damage by water and air. As the resin hardened, a protective pressure-resistant case was formed.

In addition to the fossils already mentioned, there are numerous other types, many of them only traces of prehistoric life. Examples of such indirect evidence include:

1. Tracks—animal footprints made in soft sediment that was later lithified.
2. Burrows—tubes in sediment, wood, or rock made by an animal. These holes may later become filled with mineral matter and preserved. Some of the oldest-known fossils are believed to be worm burrows.
3. Coprolites—fossil dung and stomach contents that can provide useful information pertaining to food habits of organisms (Figure 11.13F).

4. Gastroliths—highly polished stomach stones that were used in the grinding of food by some extinct reptiles.

Conditions Favoring Preservation

Only a tiny fraction of the organisms that have lived during the geologic past have been preserved as fossils. Normally the remains of an animal or plant are destroyed. Under what circumstances are they preserved? Two special conditions appear to be necessary: rapid burial and the possession of hard parts.

When an organism perishes, its soft parts usually are quickly eaten by scavengers or decomposed by bacteria. Occasionally, however, the remains are buried by sediment. When this occurs, the remains are protected from the environment, where destructive processes operate. Therefore, rapid burial is an important condition favoring preservation.

In addition, animals and plants have a much better chance of being preserved as part of the fossil record if they have hard parts. Although traces and imprints of soft-bodied animals such as jellyfish, worms, and insects exist, they are not common. Flesh usually decays so rapidly that preservation is exceedingly unlikely. Hard parts such as shells, bones, and teeth predominate in the record of past life.

Students Sometimes Ask...

How is paleontology different from archaeology?

People frequently confuse these two areas of study because a common perception of both paleontologists and archaeologists is of scientists carefully extracting important clues about the past from layers of rock or sediment. While it is true that scientists in both disciplines "dig" a lot, the focus of each is different. Paleontologists study fossils and are concerned with *all* life-forms in the geologic past. By contrast, archaeologists focus on the material remains of past human life. These remains include both the objects used by people long ago, called *artifacts*, and the buildings and other structures associated with where people lived, called *sites*. Archaeologists help us learn about how our human ancestors met the challenges of life in the past.

Because preservation is contingent on special conditions, the record of life in the geologic past is biased. The fossil record of those organisms with hard parts that lived in areas of sedimentation is quite abundant. However, we get only an occasional glimpse of the vast array of other life-forms that did not meet the special conditions favoring preservation.

Fossils and Correlation

The existence of fossils had been known for centuries, yet it was not until the late 1700s and early 1800s that their significance as geologic tools was made evident. During this period an English engineer and canal builder, William Smith, discovered that each rock formation in the canals on which he worked contained fossils unlike those in the beds either above or below. Furthermore, he noted that sedimentary strata in widely separated areas could be identified and correlated by their distinctive fossil content.

Based on Smith's classic observations and the findings of many geologists that followed, one of the most important and basic principles in historical geology was formulated: *Fossil organisms succeed one another in a definite and determinable order, and therefore any time period can be recognized by its fossil content*. This has come to be known as the **principle of fossil succession**. In other words, when fossils are arranged according to their age by applying the law of superposition to the rocks in which they are found, they do not present a random or haphazard picture. To the contrary, fossils show changes that document the evolution of life through time.

For example, an Age of Trilobites is recognized quite early in the fossil record. Then, in succession, paleontologists recognize an Age of Fishes, an Age of Coal Swamps, an Age of Reptiles, and an Age of Mammals. These "ages" pertain to groups that were especially plentiful and characteristic during particular time periods. Within each of the ages, there are many subdivisions based, for example, on certain species of trilobites and certain types of fish, reptiles, and so on. This same succession of dominant organisms, never out of order, is found on every continent.

Once fossils were recognized as time indicators, they became the most useful means of correlating rocks of similar age in different regions. Geologists pay particular attention to certain fossils called **index fossils**. These fossils are widespread geographically and are limited to a short span of geologic time, so their presence provides an important method of matching rocks of the same age. Rock formations, however, do not always contain a specific index fossil. In such situations, groups of fossils are used to establish the age of the bed. **Figure 11.14** illustrates how an assemblage of fossils can be used to date rocks more precisely than could be accomplished by the use of only one of the fossils.

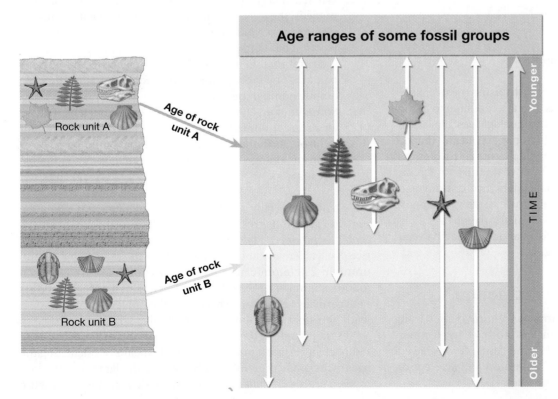

FIGURE 11.14 Overlapping ranges of fossils help date rocks more exactly than using a single fossil.

In addition to being important and often essential tools for correlation, fossils are important environmental indicators. Although much can be deduced about past environments by studying the nature and characteristics of sedimentary rocks, a close examination of any fossils present can usually provide a great deal more information.

For example, when the remains of certain clam shells are found in limestone, the geologist can assume that the region was once covered by a shallow sea, because that is where clams live today. Also, by using what we know of living organisms, we can conclude that fossil animals with thick shells capable of withstanding pounding and surging waves must have inhabited shorelines. Conversely, the remains of animals with thin, delicate shells probably indicate deep, calm offshore waters. Hence, by noting carefully the types of fossils, the approximate position of an ancient shoreline may be identified.

Furthermore, fossils can indicate the former temperature of the water. Certain present-day corals require warm and shallow tropical seas such as those around Florida and the Bahamas. When similar corals are found in ancient limestones, they indicate that a Florida-like marine environment must have existed when the corals were alive. These examples illustrate how fossils can help unravel the complex story of Earth history.

CONCEPT CHECK 11.5

1. Describe several ways that an animal or plant can be preserved as a fossil.
2. List two examples of trace fossils.
3. What conditions favor the preservation of an organism as a fossil?

Dating with Radioactivity

GEODe

Deciphering Earth's History

▶ A Dating with Radioactivity

EARTH SCIENCE

In addition to establishing relative dates by using the principles described in the preceding sections, it is also possible to obtain reliable numerical dates for events in the geologic past. For example, we know that Earth is about 4.6 billion years old and that the dinosaurs became extinct about 65 million years ago. Dates that are expressed in millions and billions of years truly stretch our imagination because our personal calendars involve time measured in hours, weeks, and years. Nevertheless, the vast expanse of geologic time is a reality, and it is radiometric dating that allows us to measure it. In this section you learn about radioactivity and its application in radiometric dating.

Reviewing Basic Atomic Structure

Recall from Chapter 2 that each atom has a *nucleus* containing protons and neutrons and that the nucleus is orbited by electrons. *Electrons* have a negative electrical charge, and *protons* have a positive charge. A *neutron* is actually a proton and an electron combined, so it has no charge (it is neutral).

The *atomic number* (each element's identifying number) is the number of protons in the nucleus. Every element has a different number of protons and thus a different atomic number. Atoms of the same element always have the same number of protons, so the atomic number stays constant.

Practically all of an atom's mass (99.9%) is in the nucleus, indicating that electrons have virtually no mass at all. So, by adding the protons and neutrons in an atom's nucleus, we derive the atom's *mass number*. The number of neutrons can vary, and these variants, or *isotopes*, have different mass numbers.

To summarize with an example, uranium's nucleus always has 92 protons, so its atomic number is always 92. But its neutron population varies, so uranium has three isotopes: uranium-234, uranium-235, and uranium-238. All three isotopes are mixed in nature. They look the same and behave the same in chemical reactions.

Radioactivity

The forces that bind protons and neutrons together in the nucleus usually are strong. However, in some isotopes, the nuclei are unstable because the forces binding protons and neutrons together are not strong enough. As a result, the nuclei spontaneously break apart (decay), a process called **radioactivity**.

What happens when unstable nuclei break apart? Three common types of radioactive decay are illustrated in Figure 11.15 and are summarized as follows:

1. Alpha (α) particles may be emitted from the nucleus. An alpha particle consists of two protons and two neutrons. Consequently, the emission of an alpha particle means that the mass number of the isotope is reduced by four and the atomic number is decreased by two.
2. When a beta (β) particle, or electron, is given off from a nucleus, the mass number remains unchanged because electrons have practically no mass. However, because the electron has come from a neutron (remember, a neutron is a combination of a proton and an electron), the nucleus contains one more proton than before. Therefore, the atomic number increases by one.
3. Sometimes an electron is captured by the nucleus. The electron combines with a proton and forms an additional neutron. As in the last example, the mass number remains unchanged. However, as the nucleus now contains one less proton, the atomic number decreases by one.

An unstable (radioactive) isotope of an element is called the *parent*. The isotopes resulting from the decay of the parent are the *daughter products*. Figure 11.16 provides an example of radioactive decay. Here it can be seen that when the radioactive parent, uranium-238 (atomic number 92, mass number 238), decays, it follows a number of steps, emitting eight alpha particles and six beta particles before finally becoming the stable daughter product lead-206 (atomic number 82, mass number 206).

Certainly among the most important results of the discovery of radioactivity is that it provided a reliable means of calculating the ages of rocks and minerals that contain particular radioactive isotopes, a procedure called **radiometric dating**. Why is radiometric

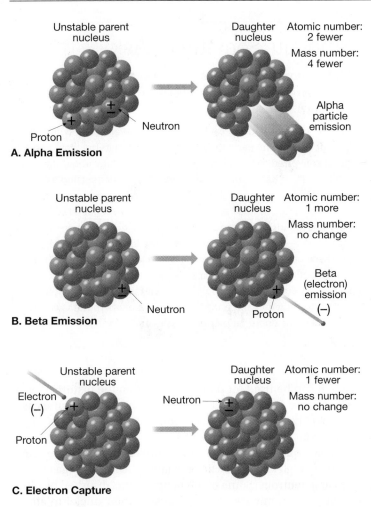

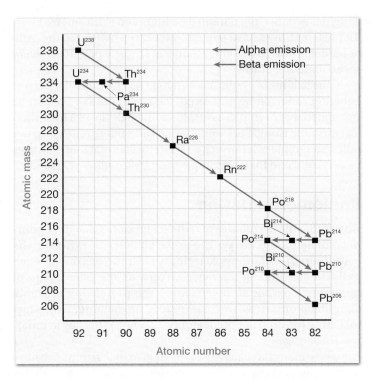

FIGURE 11.16 The most common isotope of uranium (U-238) is an example of a radioactive decay series. Before the stable end product (Pb-206) is reached, many different isotopes are produced as intermediate steps.

FIGURE 11.15 Common types of radioactive decay. Notice that in each case the number of protons (atomic number) in the nucleus changes, thus producing a different element.

FIGURE 11.17 The radioactive decay curve shows change that is exponential. Half of the radioactive parent remains after one half-life. After a second half-life one-quarter of the parent remains, and so forth.

dating reliable? Because the rates of decay for many isotopes have been precisely measured and do not vary under the physical conditions that exist in Earth's outer layers. Therefore, each radioactive isotope used for dating has been decaying at a fixed rate since the formation of the rocks in which it occurs, and the products of decay have been accumulating at a corresponding rate. For example, when uranium is incorporated into a mineral that crystallizes from magma, there is no lead (the stable daughter product) from previous decay. The radiometric "clock" starts at this point. As the uranium in this newly formed mineral disintegrates, atoms of the daughter product are trapped, and measurable amounts of lead eventually accumulate.

Half-Life

The time required for one half of the nuclei in a sample to decay is called the **half-life** of the isotope. Half-life is a common way of expressing the rate of radioactive disintegration. **Figure 11.17** illustrates what occurs when a radioactive parent decays directly into its stable daughter product. When the quantities of parent and daughter are equal (ratio 1:1), we know that one half-life has transpired. When one-quarter of the original parent atoms remain and three-quarters have decayed to the

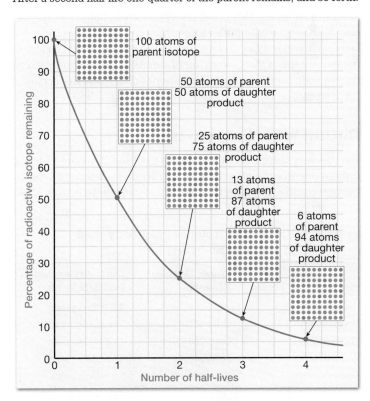

daughter product, the parent/daughter ratio is 1:3 and we know that two half-lives have passed. After three half-lives, the ratio of parent atoms to daughter atoms is 1:7 (one parent for every seven daughter atoms).

If the half-life of a radioactive isotope is known and the parent/daughter ratio can be measured, the age of the sample can be calculated. For example, assume that the half-life of a hypothetical unstable isotope is 1 million years and the parent/daughter ratio in a sample is 1:15. Such a ratio indicates that four half-lives have passed and that the sample must be 4 million years old.

Radiometric Dating

Notice that the *percentage* of radioactive atoms that decay during one half-life is always the same: 50 percent. However, the *actual number* of atoms that decay with the passing of each half-life continually decreases. Thus, as the percentage of radioactive parent atoms declines, the proportion of stable daughter atoms rises, with the increase in daughter atoms just matching the drop in parent atoms. This fact is the key to radiometric dating.

Of the many radioactive isotopes that exist in nature, five have proved particularly useful in providing radiometric ages for ancient rocks (**Table 11.1**). Rubidium-87, thorium-232, and the two isotopes of uranium are used only for dating rocks that are millions of years old, but potassium-40 is more versatile. Although the half-life of potassium-40 is 1.3 billion years, analytical techniques make possible the detection of tiny amounts of its stable daughter product, argon-40, in some rocks that are younger than 100,000 years.

It is important to realize that an accurate radiometric date can be obtained only if the mineral remained a closed system during the entire period since its formation. A correct date is not possible unless there was neither the addition nor loss of parent or daughter isotopes. This is not always the case. In fact, an important limitation of the potassium–argon method arises from the fact that argon is a gas, and it may leak from minerals, throwing off measurements. Cross-checking of samples, using two different radiometric methods, is done where possible to ensure accurate age determinations.

Dating with Carbon-14

To date very recent events, carbon-14 is used. Carbon-14 is the radioactive isotope of carbon. The process is often called **radiocarbon dating**. Because the half-life of carbon-14 is only 5,730

Students Sometimes Ask...

With radioactive decay, is there ever a time that all of the parent material is converted to the daughter product?

Theoretically, no. During each half-life, half of the parent material is converted to the daughter product. Then half again is converted after another half-life, and so on. (Figure 11.17 shows how this logarithmic relationship works—notice that the red line becomes nearly parallel to the horizontal axis after several half-lives.) By converting only half of the remaining parent material to the daughter product, there is never a time when all the parent material would be converted. Think about it this way. If you kept cutting a cake in half and eating only half, would you ever eat all of it? (The answer is no, assuming you had a sharp enough knife to slice the cake at an atomic scale!) However, after many half-lives, the parent material can exist in such small amounts that it is essentially undetectable.

years, it can be used for dating events from the historic past as well as those from very recent geologic history. In some cases carbon-14 can be used to date events as far back as 75,000 years.

Carbon-14 is continuously produced in the upper atmosphere as a consequence of cosmic-ray bombardment. Cosmic rays, which are high-energy particles, shatter the nuclei of gas atoms, releasing neutrons. Some of the neutrons are absorbed by nitrogen atoms (atomic number 7), causing their nuclei to emit a proton. As a result, the atomic number decreases by 1 (to 6), and a different element, carbon-14, is created (**Figure 11.18A**). This isotope of carbon quickly becomes incorporated into carbon dioxide, which circulates in the atmosphere and is absorbed by living matter. As a result, all organisms contain a small amount of carbon-14, including you.

FIGURE 11.18 A. Production and **B.** decay of carbon-14. These sketches represent the nuclei of the respective atoms.

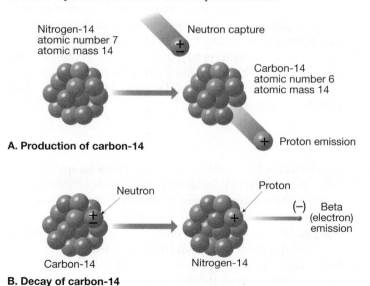

A. Production of carbon-14

Nitrogen-14
atomic number 7
atomic mass 14

Neutron capture

Carbon-14
atomic number 6
atomic mass 14

Proton emission

Neutron

Proton

Carbon-14

Nitrogen-14

(–) Beta (electron) emission

B. Decay of carbon-14

TABLE 11.1 Radioactive Isotopes Frequently Used in Radiometric Dating

Radioactive Parent	Stable Daughter Product	Currently Accepted Half-Life Values
Uranium-238	Lead-206	4.5 billion years
Uranium-235	Lead-207	713 million years
Thorium-232	Lead-208	14.1 billion years
Rubidium-87	Strontium-87	47.0 billion years
Potassium-40	Argon-40	1.3 billion years

While an organism is alive, the decaying radiocarbon is continually replaced, and the proportions of carbon-14 and carbon-12 remain constant. Carbon-12 is the stable and most common isotope of carbon. However, when any plant or animal dies, the amount of carbon-14 gradually decreases as it decays to nitrogen-14 by beta emission (Figure 11.18B). By comparing the proportions of carbon-14 and carbon-12 in a sample, radiocarbon dates can be determined.

Although carbon-14 is useful in dating only the last small fraction of geologic time, it has become a very valuable tool for anthropologists, archaeologists, and historians, as well as for geologists who study very recent Earth history. (Box 11.1 explores another method of studying and dating recent events.) In fact, the development of radiocarbon dating was considered so important that the chemist who discovered this application, Willard F. Libby, received a Nobel prize in 1960.

Importance of Radiometric Dating

Although the basic principle of radiometric dating is simple, the actual procedure is quite complex. The chemical analysis that determines the quantities of parent and daughter must be painstakingly precise. In addition, some radioactive materials do not decay directly into the stable daughter product. As you saw in Figure 11.16, uranium-238 produces 13 intermediate unstable daughter products before the 14th and final daughter product, the stable isotope lead-206, is produced.

Radiometric dating methods have produced literally thousands of dates for events in Earth history. Rocks exceeding 3.5 billion years in age are found on all of the continents. Earth's oldest rocks (so far) are gneisses from northern Canada near Great Slave Lake that have been dated at 4.03 billion years (b.y.). Rocks from western Greenland have been dated at 3.7–3.8 b.y. and rocks nearly as old are found in the Minnesota River Valley and northern Michigan (3.5–3.7 b.y.), in southern Africa (3.4–3.5 b.y.), and in western Australia (3.4–3.6 b.y.). It is important to point out that these ancient rocks are not from any sort of "primordial crust" but originated as lava flows, igneous intrusions, and sediments deposited in shallow water—an indication that Earth history began *before* these rocks formed. Even older mineral grains have been dated. Tiny crystals of the mineral zircon having radiometric ages as old as 4.3 b.y. have been found in younger sedimentary rocks in western Australia. The source rocks for these tiny durable grains either no longer exist or have not yet been found.

Radiometric dating has vindicated the ideas of James Hutton, Charles Darwin, and others who inferred that geologic time must be immense. Indeed, modern dating methods have proved that there has been enough time for the processes we observe to have accomplished tremendous tasks.

CONCEPT CHECK 11.6

1. List three types of radioactive decay. For each type, describe how the atomic number and mass number change.
2. Sketch a simple diagram as a way of explaining the idea of half-life.
3. Why is radiometric dating a reliable method for determining numerical ages?

The Geologic Time Scale

Deciphering Earth's History
▶ A Geologic Time Scale

Geologists have divided the whole of geologic history into units of varying magnitude. Together, they comprise the **geologic time scale** of Earth history (Figure 11.19). The major units of the time scale were delineated during the 19th century, principally by scientists working in western Europe and Great Britain. Because radiometric dating was unavailable at that time, the entire time scale was created using methods of relative dating. It was only in the 20th century that radiometric dating permitted numerical dates to be added.

Structure of the Time Scale

The geologic time scale subdivides the 4.6-billion-year history of Earth into many different units and provides a meaningful time frame within which the events of the geologic past are arranged. As shown in Figure 11.19, **eons** represent the greatest expanses of time. The eon that began about 542 million years ago is the **Phanerozoic**, a term derived from Greek words meaning *visible life*. It is an appropriate description because the rocks and deposits of the Phanerozoic eon contain abundant fossils that document major evolutionary trends.

Another glance at the time scale reveals that eons are divided into **eras**. The three eras within the Phanerozoic are the **Paleozoic**, the **Mesozoic**, and the **Cenozoic**. As the names imply, these eras are bounded by profound worldwide changes in life-forms. Major changes in life-forms are discussed in Chapter 12 "Earth's Evolution through Geologic Time." Each era of the Phanerozoic eon is subdivided into **periods**. Each period is characterized by a somewhat less profound change in life-forms as compared with the eras.

Each period is divided into still smaller units called **epochs**. As you can see in Figure 11.19, seven epochs have been named for the periods of the Cenozoic. The epochs of other periods usually are simply termed *early*, *middle*, and *late*.

Precambrian Time

Notice that the detail of the geologic time scale does not begin until about 542 million years ago, the date for the beginning of the Cambrian period. The more than 4 billion years prior to the Cambrian is divided into three eons, the **Hadean**, the **Archean**, and the **Proterozoic**. These in turn, are divided into seven eras. It is also common for this vast expanse of time to simply be referred to as the **Precambrian**.[24] Although it represents about 88 percent of Earth history, the Precambrian is not divided into nearly as many smaller time units as is the Phanerozoic eon.

The quantity of information geologists have deciphered about Earth's past is somewhat analogous to the detail of human history. The further back we go, the less we know. Certainly more data and information exist about the past 10 years than for the

---

[24] The terms *Hadean* and *Precambrian* are informal names that are in common use. There is more about this in the section "Terminology and the Geologic Time Scale."

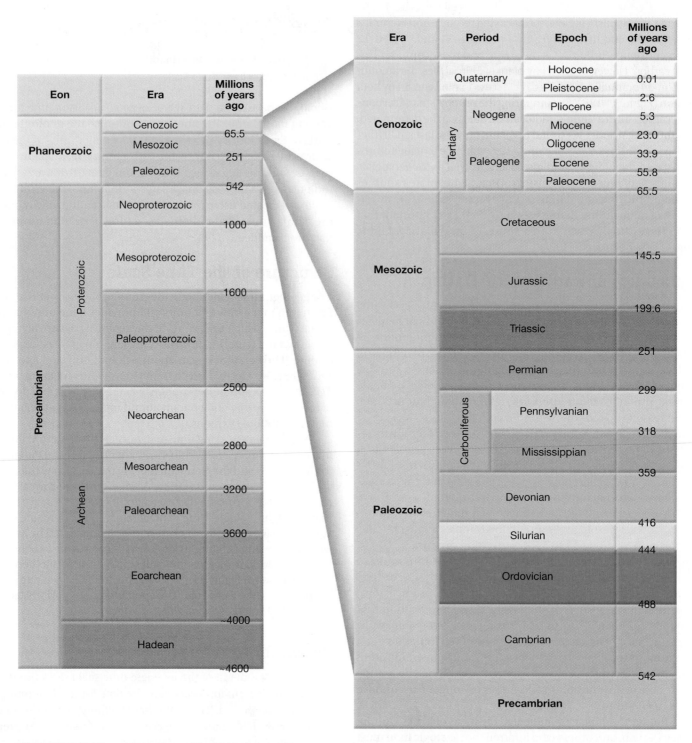

FIGURE 11.19 The geologic time scale. The numerical dates were added long after the time scale had been established using relative dating techniques. The time scale is a dynamic tool. Advances in the geosciences require that updates be made from time to time.

UNDERSTANDING EARTH

Using Tree Rings to Date and Study the Recent Past

If you look at the top of a tree stump or at the end of a log, you will see that it is composed of a series of concentric rings. Each of these *tree rings* becomes larger in diameter outward from the center (**Figure 11.A**). Every year in temperate regions trees add a layer of new wood under the bark. Characteristics of each tree ring, such as size and density, reflect the environmental conditions (especially climate) that prevailed during the year when the ring formed. Favorable growth conditions produce a wide ring; unfavorable ones produce a narrow ring. Trees growing at the same time in the same region show similar tree-ring patterns.

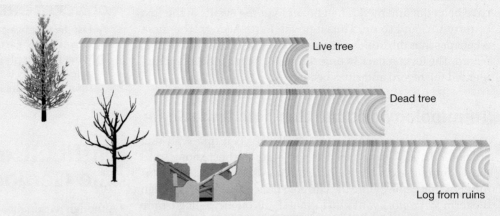

FIGURE 11.B Cross dating is a basic principle in dendrochronology. Here it was used to date an archaeological site by correlating tree-ring patterns for wood from trees of three different ages. First, a tree-ring chronology for the area is established using cores extracted from living trees. This chronology is extended further back in time by matching overlapping patterns from older, dead trees. Finally, cores taken from beams inside the ruin are dated using the chronology established from the other two sites.

Because a single growth ring is usually added each year, the age of the tree when it was cut can be determined by counting the rings. If the year of cutting is known, the age of the tree and the year in which each ring formed can be determined by counting back from the outside ring.* This procedure can be used to determine the dates of recent geologic events: for example, the minimum number of years since a new land surface was created by a landslide or a flood. The dating and study of annual rings in trees is called *dendrochronology*.

To make the most effective use of tree rings, extended patterns known as ring chronologies are established. They are produced by comparing the patterns of rings among trees in an area. If the same pattern can be identified in two samples, one of which has been dated, the second sample can be dated from the first by matching the ring pattern common to both. This technique, called *cross dating*, is illustrated in **Figure 11.B**. Tree-ring chronologies extending back for thousands of years have been established for some regions. To date a timber sample of unknown age, its ring pattern is matched against the reference chronology.

Tree-ring chronologies are unique archives of environmental history and have important applications in such disciplines as climate, geology, ecology, and archaeology. For example, tree rings are used to reconstruct climate variations within a region for spans of thousands of years prior to human historical records. Knowledge of such long-term variations is of great value in making judgments regarding the recent record of climate change.

In summary, dendrochronology provides useful numerical dates for events in the historic and recent prehistoric past. Moreover, because tree rings are a storehouse of data, they are a valuable tool in the reconstruction of past environments.

FIGURE 11.A Each year a growing tree produces a layer of new cells beneath the bark. If the tree is felled and the trunk examined (or if a core is taken, to avoid cutting the tree), each year's growth can be seen as a ring. Because the amount of growth (thickness of a ring) depends on precipitation and temperature, tree rings are useful records of past climates. (Photo by Stephen J. Krasemann/Photo Researchers, Inc)

*Scientists are not limited to working with trees that have been cut down. Small, nondestructive core samples can be taken from living trees.

first decade of the 20th century; the events of the 19th century have been documented much better than the events of the first century A.D., and so on. Thus it is with Earth history. The more recent past has the freshest, least disturbed, and most observable record. The further back in time the geologist goes, the more fragmented the record and clues become.

Terminology and the Geologic Time Scale

There are some terms associated with the geologic time scale that are not "officially" recognized as part of it. The best known, and most common, example is *Precambrian*—the informal name for the eons that came before the current Phanerozoic eon. Although the term *Precambrian* has no formal status on the geologic time scale, it has been traditionally used as though it did.

Hadean is another informal term that is found on some versions of the geologic time scale and is used by some geologists. It refers to the earliest interval (eon) of Earth history—before the oldest known rocks. When the term was coined in 1972, the age of Earth's oldest rocks was about 3.8 billion years. Today, that number stands at slightly greater than 4 billion and, of course, is subject to revision. The name *Hadean* derives from *Hades*, Greek for *underworld*—a reference to the "hellish" conditions that prevailed on Earth early in its history.

Effective communication in the geosciences requires that the geologic time scale consist of standardized divisions and dates. So, who determines which names and dates on the geologic time scale are "official"? The organization that is largely responsible for maintaining and updating this important information is the International Commission on Stratigraphy (ICS), which is associated with the International Union of Geological Sciences.[25] Advances in the geosciences require that the scale be periodically updated to include changes in unit names and boundary age estimates.

For example, the geologic time scale shown in Figure 11.19 was updated as recently as July 2009. After considerable dialogue among geologists who focus on very recent Earth history, the ICS changed the date for the start of the Quaternary period and the Pleistocene epoch from 1.8 million to 2.6 million years ago. Perhaps by the time you read this, other changes will have been made.

If you were to examine a geologic time scale from just a few years ago, it is quite possible that you would see the Cenozoic era divided into the Tertiary and Quaternary periods. However, on more recent versions the space formerly designated as Tertiary is divided into the Paleogene and Neogene periods. As our understanding of this time span changed, so too did its designation on the geologic time scale. Today, the Tertiary period is considered a "historic" name and is given no official status on the ICS version of the time scale. Many time scales still contain references to the Tertiary period, including Figure 11.19. One reason for this is that a great deal of past (and some current) geological literature uses this name.

For those who study historical geology, it is important to realize that the geologic time scale is a dynamic tool that continues to be refined as our knowledge and understanding of Earth history evolves.

[25] *Stratigraphy* is the branch of geology that studies rock layers (strata) and layering (stratification), thus its primary focus is sedimentary and layered volcanic rocks.

① List the four basic units that make up the geologic time scale.
② Why is "zoic" part of so many names on the geologic time scale?
③ What term applies to *all* of geologic time prior to the Phanerozoic eon? Why is this span *not* divided into as many smaller time units as the Phanerozoic eon?

Difficulties in Dating the Geologic Time Scale

Although reasonably accurate numerical dates have been worked out for the periods of the geologic time scale (Figure 11.19), the task is not without difficulty. The primary problem in assigning numerical dates to units of time is the fact that not all rocks can be dated by radiometric methods. Recall that for a radiometric date to be useful, all minerals in the rock must have formed at about the same time. For this reason, radioactive isotopes can be used to determine when minerals in an igneous rock crystallized and when pressure and heat created new minerals in a metamorphic rock.

However, samples of sedimentary rock can only rarely be dated directly by radiometric means. A sedimentary rock may include particles that contain radioactive isotopes, but the rock's age cannot be accurately determined because the grains making up the rock are not the same age as the rock in which they occur. Rather, the sediments have been weathered from rocks of diverse ages (**Figure 11.20**).

Radiometric dates obtained from metamorphic rocks may also be difficult to interpret because the age of a particular mineral in a metamorphic rock does not necessarily represent the time when the rock initially formed. Instead, the date may indicate any one of a number of subsequent metamorphic phases.

FIGURE 11.20 A useful numerical date for this conglomerate is not possible because the gravel that composes it was derived from rocks of diverse ages. (Photo by E. J. Tarbuck)

FIGURE 11.21 Numerical dates for sedimentary layers are often determined by examining their relationship to igneous rocks. (After U.S. Geological Survey)

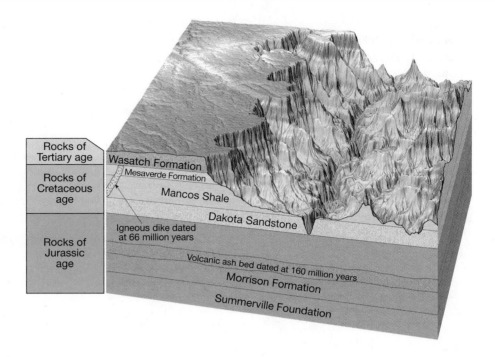

If samples of sedimentary rocks rarely yield reliable radiometric ages, how can numerical dates be assigned to sedimentary layers? Usually the geologist must relate them to datable igneous masses, as in Figure 11.21. In this example, radiometric dating has determined the ages of the volcanic ash bed within the Morrison Formation and the dike cutting the Mancos Shale and Mesaverde Formation. The sedimentary beds below the ash are obviously older than the ash, and all the layers above the ash are younger (principle of superposition). The dike is younger than the Mancos Shale and the Mesaverde Formation but older than the Wasatch Formation because the dike does not intrude the Tertiary rocks (cross-cutting relationships).

From this kind of evidence, geologists estimate that a part of the Morrison Formation was deposited about 160 million years ago, as indicated by the ash bed. Furthermore, they conclude that the Tertiary period began after the intrusion of the dike, 66 million years ago. This is one example of literally thousands that illustrates how datable materials are used to *bracket* the various episodes in Earth history within specific time periods. It shows the necessity of combining laboratory dating methods with field observations of rocks.

CONCEPT CHECK 11.8

1 Briefly explain why it is often difficult to assign a reliable numerical date to a sample of sedimentary rock.

GIVE IT SOME THOUGHT

1. Refer to Figure 11.5 and answer the following questions. Briefly explain each answer.
   a. Is fault A older or younger than the sandstone layer?
   b. Is dike A older or younger than the sandstone layer?
   c. Was the conglomerate deposited before or after fault A?
   d. Was the conglomerate deposited before or after fault B?
   e. Which fault is older, A or B?
   f. Is dike A older or younger than the batholith?
2. A mass of granite is in contact with a layer of sandstone. Using a principle described in this chapter, explain how you might determine whether the sandstone was deposited on top of the granite or whether the magma that formed the granite was intruded after the sandstone was deposited.
3. If a radioactive isotope of thorium (atomic number 90, mass number 232) emits six alpha particles and four beta particles during the course of radioactive decay, what is the atomic number and mass number of the stable daughter product?
4. A hypothetical radioactive isotope has a half-life of 10,000 years. If the ratio of radioactive parent to stable daughter product is 1:3, how old is the rock containing the radioactive material?

5. Solve the problems below. To make calculations easier, round the age of Earth to 5 billion years.

   a. What percentage of geologic time is represented by recorded history? (Assume 5,000 years for the length of recorded history.)

   b. Human-like ancestors (hominids) have been around for roughly 5 million years. What percentage of geologic time is represented by the existence of the ancestors?

   c. The first abundant fossil evidence does not appear until the beginning of the Cambrian period, about 540 million years ago. What percentage of geologic time is represented by abundant fossil evidence?

6. The accompanying block diagram depicts a hypothetical area in the American Southwest. Place the lettered features in the proper sequence, from oldest to youngest. Identify an angular unconformity and a nonconformity.

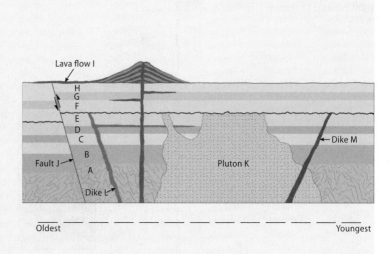

In Review   Chapter 11 Geologic Time

- During the 17th and 18th centuries, *catastrophism* influenced the formulation of explanations about Earth. Catastrophism states that Earth's landscapes have been developed primarily by great catastrophes. By contrast, *uniformitarianism*, one of the fundamental principles of modern geology advanced by *James Hutton* in the late 1700s, states that the physical, chemical, and biological laws that operate today have also operated in the geologic past. The idea is often summarized as "The present is the key to the past." Hutton argued that processes that appear to be slow-acting could, over long spans of time, produce effects that were just as great as those resulting from sudden catastrophic events.

- The two types of dates used by geologists to interpret Earth history are (1) *relative dates*, which put events in their *proper sequence of formation*, and (2) *numerical dates*, which pinpoint the *time in years* when an event took place.

- Relative dates can be established using the *law of superposition*, the *principle of original horizontality*, the *principle of cross-cutting relationships*, *inclusions*, and *unconformities*.

- *Correlation*, the matching-up of two or more geologic phenomena in different areas, is used to develop a geologic time scale that applies to the entire Earth.

- *Fossils* are the remains or traces of prehistoric life. The special conditions that favor preservation are *rapid burial* and the possession of *hard parts* such as shells, bones, or teeth.

- Fossils are used to *correlate* sedimentary rocks from different regions by using the rocks' distinctive fossil content and applying the *principle of fossil succession*. It states that fossil organisms succeed one another in a definite and determinable order, and therefore any time period can be recognized by its fossil content.

- Each atom has a nucleus containing *protons* (positively charged particles) and *neutrons* (neutral particles). Orbiting the nucleus are negatively charged *electrons*. The *atomic number* of an atom is the number of protons in the nucleus. The *mass number* is the number of protons plus the number of neutrons in an atom's nucleus. *Isotopes* are variants of the same atom, but with a different number of neutrons and hence a different mass number.

- *Radioactivity* is the spontaneous breaking apart (decay) of certain unstable atomic nuclei. Three common types of radioactive decay are (1) emission of alpha particles from the nucleus, (2) emission of beta particles (electrons) from the nucleus, and (3) capture of electrons by the nucleus.

- An unstable *radioactive isotope*, called the *parent*, will decay and form stable *daughter products*. The length of time for half of the nuclei of a radioactive isotope to decay is called the *half-life* of the isotope. If the half-life of the isotope is known and the parent/daugher ratio can be measured, the age of a sample can be calculated.

- The *geologic time scale* divides Earth's history into units of varying magnitude. It is commonly presented in chart form, with the oldest time and event at the bottom and the youngest at the top. The principal subdivisions of the geologic time scale, called *eons*, include the *Archean* and *Proterozoic* (together, these eons are commonly referred to as the *Precambrian*), and, beginning about 542 million years ago, the *Phanerozoic*. The Phanerozoic (meaning "visible life") eon is divided into the following *eras*: *Paleozoic* ("ancient life"), *Mesozoic* ("middle life"), and *Cenozoic* ("recent life").

- A significant problem in assigning numerical dates to units of time is that *not all rocks can be dated radiometrically*. A sedimentary rock may contain particles of many ages that have been weathered from different rocks that formed at various times. One way geologists assign numerical dates to sedimentary rocks is to relate them to datable igneous masses, such as dikes and volcanic ash beds.

Key Terms

angular unconformity (p. 329)
Archean eon (p. 339)
catastrophism (p. 325)
Cenozoic era (p. 339)
conformable (p. 328)
correlation (p. 330)
cross-cutting relationships, principle
   of (p. 328)
disconformity (p. 329)
eon (p. 339)
epoch (p. 339)
era (p. 339)

fossil (p. 333)
fossil succession, principle of (p. 335)
geologic time scale (p. 339)
Hadean eon (p. 339)
half-life (p. 337)
inclusions (p. 328)
index fossil (p. 335)
Mesozoic era (p. 339)
nonconformity (p. 329)
numerical date (p. 326)
original horizontality, principle of (p. 327)
paleontology (p. 333)

Paleozoic era (p. 339)
period (p. 339)
Phanerozoic eon (p. 339)
Precambrian (p. 339)
Proterozoic eon (p. 339)
radioactivity (p. 336)
radiocarbon dating (p. 338)
radiometric dating (p. 336)
relative dating (p. 326)
superposition, law of (p. 327)
unconformity (p. 328)
uniformitarianism (p. 325)

Examining the Earth System

1. Figure 11.13A is a large petrified log in Arizona's Petrified Forest National Park. Describe the transition of this tree from being part of the biosphere to being a component of the solid Earth. How might the hydrosphere and/or atmosphere have played a role in the transition?

2. The famous angular unconformity at Scotland's Siccar Point shown in Figure 11.9 was originally studied by James Hutton in the late 1700s.

   a. Describe in a general way what occurred to produce this feature.

   b. Suggest ways in which all four spheres of the Earth system could have been involved.

   c. The Earth system is powered by energy from two sources. How are both sources represented in the Siccar Point unconformity?

Mastering Geology

Looking for additional review and test prep materials? Visit the Self Study area in **www.masteringgeology.com** to find practice quizzes, study tools, and multimedia that will aid in your understanding of this chapter's content. In **MasteringGeology™** you will find:

- **GEODe: Earth Science: An interactive visual walkthrough of key concepts**

- **Geoscience Animation Library: More than 100 animations illuminating many difficult-to-understand Earth science concepts**
- **In The News RSS Feeds: Current Earth science events and news articles are pulled into the site with assessment**
- **Pearson eText**
- **Optional Self Study Quizzes**
- **Web Links**
- **Glossary**
- **Flashcards**

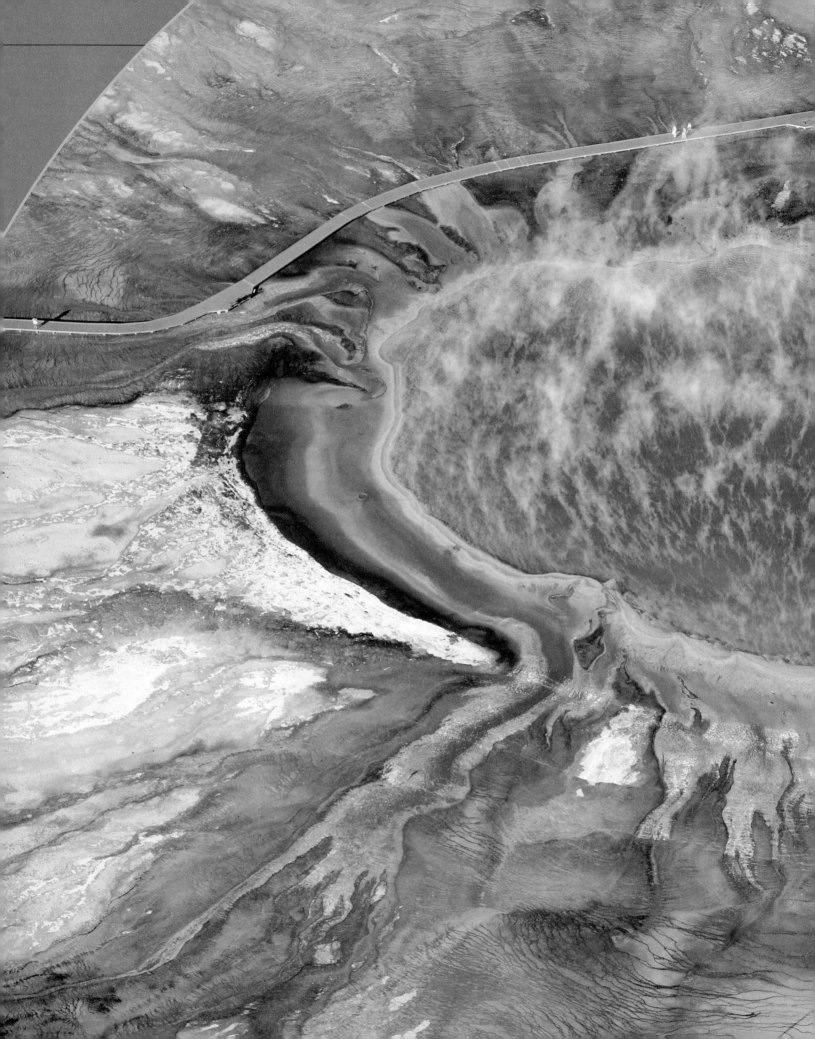

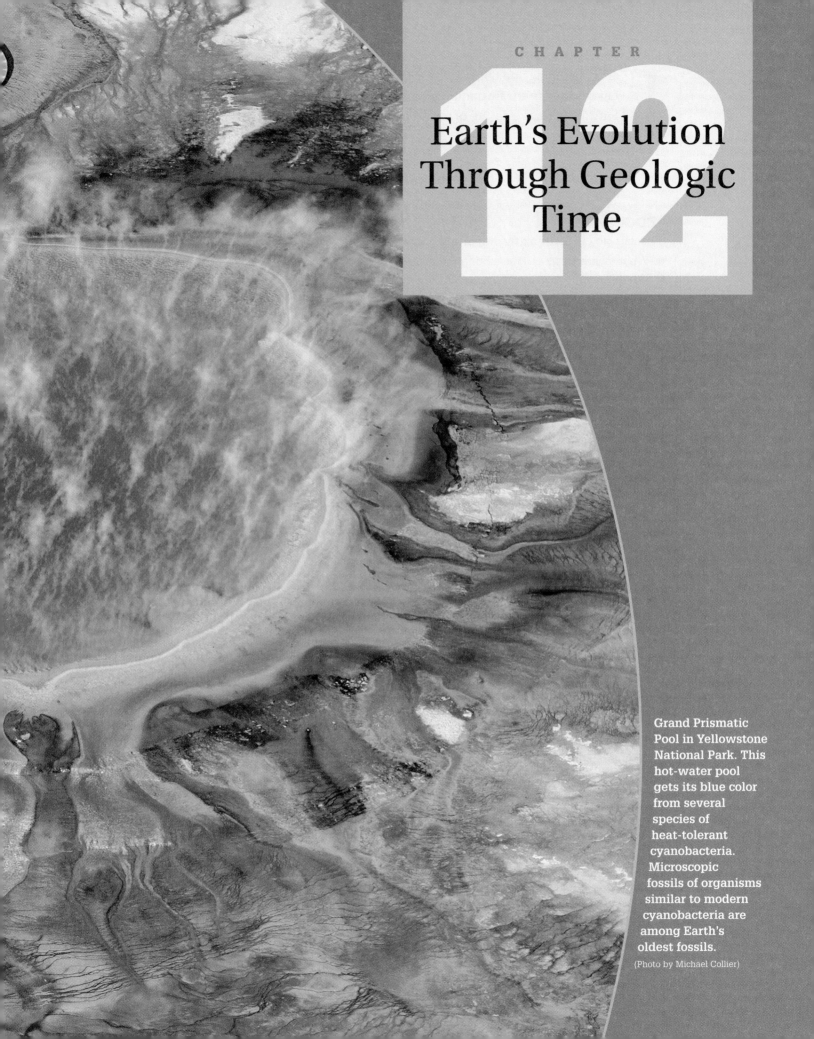

Earth's Evolution Through Geologic Time

12

Grand Prismatic Pool in Yellowstone National Park. This hot-water pool gets its blue color from several species of heat-tolerant cyanobacteria. Microscopic fossils of organisms similar to modern cyanobacteria are among Earth's oldest fossils.

(Photo by Michael Collier)

E arth has a long and complex history. Time and again, the splitting and colliding of continents has resulted in the formation of new ocean basins and the creation of great mountain ranges. Furthermore, the nature of life on our planet has experienced dramatic changes through time.

FOCUS ON CONCEPTS

To assist you in learning the important concepts in this chapter, focus on the following questions:

- What makes Earth unique among the planets?
- What is a supernova and how are ancient supernova events important to inhabitants of Earth?
- What is the relationship between planetesimals and protoplanets?
- How did Earth's atmosphere and oceans form and evolve through time?
- How did the first continents form?
- What is the supercontinent cycle?
- What was Pangaea and how did it form?
- What are some notable geological events in Earth history?
- What was the Cambrian explosion?
- What is the significance of lobe-finned fish?
- What important evolutionary advancement differentiates the reptiles from the amphibians?
- What event likely caused the demise of the dinosaurs?
- How did the KT extinction influence human evolution?
- What may have hastened the extinction of large mammals such as mastodons in the late Pleistocene?

Is Earth Unique?

There is only one place in the universe, as far as we know, that can support life—a modest-sized planet called Earth that orbits an average-sized star, the Sun. Life on Earth is ubiquitous; it is found in boiling mudpots and hot springs, in the deep abyss of the ocean, and even under the Antarctic ice sheet. Living space on our planet, however, is significantly limited when we consider the needs of individual organisms, particularly humans. The global ocean covers 71 percent of Earth's surface, but only a few hundred meters below the water's surface pressures are so intense that human lungs cannot function. In addition, many continental areas are too steep, too high, or too cold for us to inhabit (Figure 12.1). Nevertheless, based on what we know about other bodies in the solar system—and the hundreds of planets recently discovered orbiting around other stars—Earth is still, by far, the most accommodating.

What fortuitous events produced a planet so hospitable to life? Earth was not always as we find it today. During its formative years, our planet became hot enough to support a magma ocean. It also survived a several-hundred-million-year period of extreme bombardment, to which the heavily cratered surfaces of Mars, the Moon, and asteroids testify. The oxygen-rich atmosphere that makes higher life-forms possible developed relatively recently. Serendipitously, Earth seems to be the right planet, in the right location, at the right time.

The Right Planet

What are some of the characteristics that make Earth unique among the planets? Consider the following:

1. If Earth were considerably larger (more massive) its force of gravity would be proportionately greater. Like the giant planets, Earth would have retained a thick, hostile atmosphere consisting of ammonia and methane, and possibly hydrogen and helium.
2. If Earth were much smaller, oxygen, water vapor, and other volatiles would escape into space and be lost forever. Thus, like the Moon and Mercury, both of which lack an atmosphere, Earth would be void of life.
3. If Earth did not have a rigid lithosphere overlaying a weak asthenosphere, plate tectonics would not operate. The continental crust (Earth's "highlands") would not have formed without the recycling of plates. Consequently, the entire planet would likely be covered by an ocean a few kilometers deep. As author Bill Bryson so aptly stated, "There might be life in that lonesome ocean, but there certainly wouldn't be baseball."[26]
4. Most surprisingly, perhaps, is the fact that if our planet did not have a molten metallic core, most of the life-forms on Earth would not exist. Fundamentally, without the flow of iron in the core, Earth could not support a mag-

[26]*A Short History of Nearly Everything* (Broadway Books, 2003).

FIGURE 12.1 Climbers near the top of Mount Everest. At this altitude the level of oxygen is only one-third the amount available at sea level. (Photo courtesy of Woodfin Camp and Associates)

netic field. It is the magnetic field that prevents lethal cosmic rays (the solar wind) from showering Earth's surface and from stripping away our atmosphere.

The Right Location

One of the primary factors that determine whether a planet is suitable for higher life-forms is its location in the solar system. The following scenarios substantiate Earth's right position:

1. If Earth were about 10 percent closer to the Sun, like Venus, our atmosphere would consist mainly of the greenhouse gas carbon dioxide. As a result, Earth's surface temperature would be too hot to support higher life-forms.
2. If Earth were about 10 percent farther from the Sun, the problem would be reversed—it would be too cold. The oceans would freeze over and Earth's active water cycle would not exist. Without liquid water all life would perish.
3. Earth is near a star of modest size. Stars like the Sun have a life span of roughly 10 billion years. During most of this time, radiant energy is emitted at a fairly constant level. Giant stars, on the other hand, consume their nuclear fuel at very high rates and "burn out" in a few hundred million

years. Therefore, Earth's proximity to a modest-size star allowed enough time for the evolution of humans, who first appeared on this planet only a few million years ago.

The Right Time

The last, but certainly not the least, fortuitous factor is timing. The first organisms to inhabit Earth were extremely primitive and came into existence roughly 3.8 billion years ago. From this point in Earth's history innumerable changes occurred—life-forms came and went along with changes in the physical environment of our planet. Two of many timely, Earth-altering events include:

1. The development of our modern atmosphere. Earth's primitive atmosphere is thought to have been composed mostly of water vapor and carbon dioxide, with small amounts of other gases, but no free oxygen. Fortunately, microorganisms evolved that released oxygen into the atmosphere by the process of *photosynthesis*. About 2.2 billion years ago an atmosphere with free oxygen came into existence. The result was the evolution of the forbearers of the vast array of organisms that occupy Earth today.

FIGURE 12.2 This paleontologist is excavating a dinosaur fossil. (Photo by Michael Gunther/Photolibrary)

2. About 65 million years ago our planet was struck by an asteroid 10 kilometers in diameter. This impact likely caused a mass extinction during which nearly three-quarters of all plant and animal species were obliterated—including dinosaurs (Figure 12.2). Although this may not seem fortuitous, the extinction of dinosaurs opened new habitats for small mammals that survived the impact. These habitats, along with evolutionary forces, led to the development of many large mammals that occupy our modern world. Without this event, mammals may not have evolved beyond the small rodent-like creatures that live in burrows.

As various observers have noted, Earth developed under "just right" conditions to support higher life-forms. Astronomers refer to this as the *Goldilocks scenario*. Like the classic Goldilocks and the Three Bears fable, Venus is too hot (Papa Bear's porridge), Mars is too cold (Mama Bear's porridge), but Earth is just right (Baby Bear's porridge).

The remainder of this chapter focuses on the origin and evolution of planet Earth—the one place in the Universe we know fosters life. As you learned in Chapter 11, researchers utilize many tools to interpret the clues about Earth's past. Using these tools, and clues contained in the rock record, scientists continue to unravel many complex events of the geologic past. The goal of this chapter is to provide a brief overview of the history of our planet and its life-forms—a journey that takes us back about 4.6 billion years to the formation of Earth and its atmosphere. Later, we consider how our physical world assumed its present state and how Earth's inhabitants changed through time. We suggest that you reacquaint yourself with the *geologic time scale* presented in Figure 12.3 and refer to it throughout the chapter.

CONCEPT CHECK 12.1

1. Explain why Earth is just the right size.
2. Why is Earth's molten, metallic core important to humans living today?
3. Why is Earth's location in the solar system ideal for the development of higher life-forms?

Birth of a Planet

According to the Big Bang theory, the formation of our planet began about 13.7 billion years ago with a cataclysmic explosion that created all matter and space (Figure 12.4). Initially, sub-atomic particles (protons, neutrons, and electrons) formed. Later, as this debris cooled, atoms of hydrogen and helium, the two lightest elements, began to form. Within a few hundred million years, clouds of these gases condensed and coalesced into stars that compose the galactic systems we now observe.

As these gases contracted to become the first stars, heating triggered the process of *nuclear fusion*. Within stars' interiors, hydrogen nuclei convert to helium nuclei, releasing enormous amounts of radiant energy (heat, light, cosmic rays). Astronomers have determined that in stars more massive than our Sun, other thermonuclear reactions occur that generate all the elements on the periodic table up to number 26, iron. The heaviest elements (beyond number 26) are only created at extreme temperatures during the explosive death of a star 8 or more times as massive as the Sun. During these cataclysmic **supernova** events, exploding stars produce all of the elements heavier than iron and spew them into interstellar space. It is from such debris that our Sun and solar system formed. According to the Big Bang scenario, atoms in your body were produced billions of years ago in the hot interior of now defunct stars, and the formation of the gold in your jewelry was triggered by a supernova explosion that occurred trillions of miles away.

From Planetesimals to Protoplanets

Recall that the solar system, including Earth, formed about 4.6 billion years ago from the **solar nebula**, a large rotating cloud of interstellar dust and gas (see Figure 1.8, page 12). As the solar nebula contracted, most of the matter collected in the center to create the hot *protosun*, while the remainder became a flattened spinning disk. Within this spinning disk, matter gradually coalesced into clumps that collided and stuck together to become asteroid-size objects called **planetesimals**. The composition of each planetesimal was determined largely by its distance from the hot protosun.

Near the orbit of Mercury only metallic grains condensed from the solar nebula. Near Earth's orbit, metallic as well as rocky substances condensed, and beyond Mars ices of water, carbon dioxide, methane, and ammonia formed. It was from these clumps of matter that the planetesimals were made and, through repeated collisions and accretion (sticking together), grew into eight **protoplanets** and their moons (see Figure 12.4).

At some point in Earth's evolution a giant impact occurred between a Mars-sized object and a young, semimolten Earth. This collision ejected huge amounts of debris into space, some of which coalesced to form the Moon (Figure 12.4J, K, L).

Earth's Early Evolution

As material continued to accumulate, the high-velocity impact of interplanetary debris (planetesimals) and the decay of radioactive elements caused the temperature of our planet to steadily

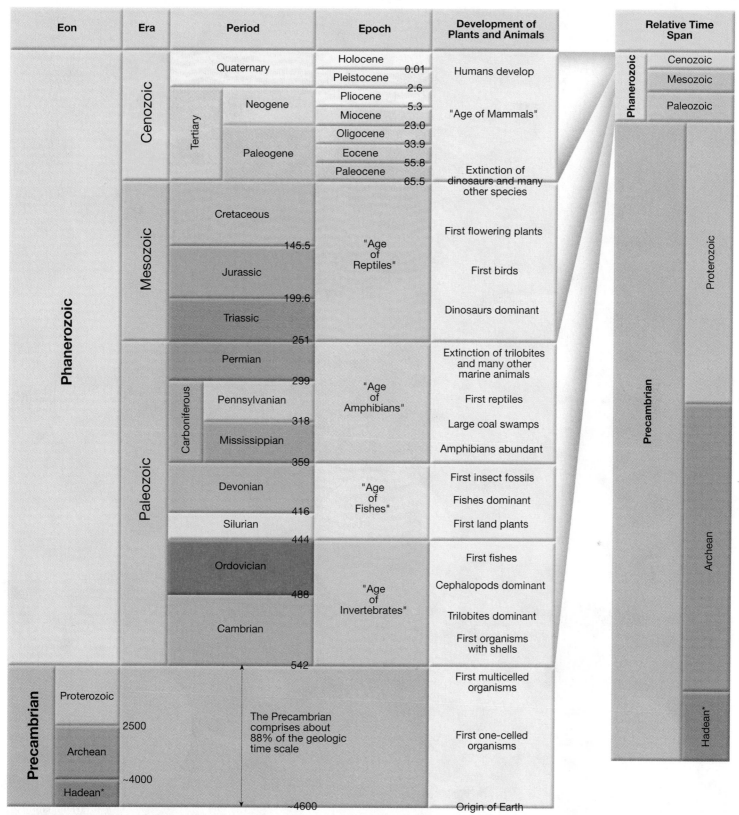

Eon	Era	Period	Epoch	Development of Plants and Animals
Phanerozoic	Cenozoic	Quaternary	Holocene / Pleistocene 0.01 / 2.6	Humans develop
		Tertiary — Neogene	Pliocene 5.3 / Miocene 23.0	"Age of Mammals"
		Tertiary — Paleogene	Oligocene 33.9 / Eocene 55.8 / Paleocene 65.5	Extinction of dinosaurs and many other species
	Mesozoic	Cretaceous 145.5	"Age of Reptiles"	First flowering plants
		Jurassic 199.6		First birds
		Triassic 251		Dinosaurs dominant
	Paleozoic	Permian 299	"Age of Amphibians"	Extinction of trilobites and many other marine animals
		Carboniferous — Pennsylvanian 318		First reptiles
		Carboniferous — Mississippian 359		Large coal swamps / Amphibians abundant
		Devonian 416	"Age of Fishes"	First insect fossils / Fishes dominant
		Silurian 444		First land plants
		Ordovician 488	"Age of Invertebrates"	First fishes / Cephalopods dominant / Trilobites dominant
		Cambrian 542		First organisms with shells
Precambrian	Proterozoic			First multicelled organisms
	2500 / Archean / ~4000 / Hadean*		The Precambrian comprises about 88% of the geologic time scale	First one-celled organisms
			~4600	Origin of Earth

Relative Time Span

Phanerozoic	Cenozoic
	Mesozoic
	Paleozoic
Precambrian	Proterozoic
	Archean
	Hadean*

* Hadean is the informal name for the span that begins at Earth's formation and ends with Earth's earliest-known rocks.

FIGURE 12.3 The geologic time scale. Numbers represent time in millions of years before the present. These dates were added long after the time scale had been established using relative dating techniques. The Precambrian accounts for about 88 percent of geologic time.

FIGURE 12.4 Major events that led to the formation of early Earth.

increase. During this period of intense heating, Earth became hot enough that iron and nickel began to melt. Melting produced liquid blobs of heavy metal that sank under their own weight. This process occurred rapidly on the scale of geologic time and produced Earth's dense iron-rich core. The formation of a molten iron core was the first of many stages of chemical differentiation in which Earth converted from a homogeneous body, with roughly the same matter at all depths, to a layered planet with material sorted by density (Figure 12.4).

This early period of heating also resulted in a magma ocean, perhaps several hundred kilometers deep. Within the magma ocean, buoyant masses of molten rock rose toward the surface, and eventually solidified to produce a thin, primitive crust.

This period of chemical differentiation established the three major divisions of Earth's interior—the iron-rich *core*, the thin *primitive crust*, and Earth's thickest layer, the *mantle*, located between the core and the crust. In addition, the lightest materials—including water vapor, carbon dioxide, and other gases—escaped to form a primitive atmosphere and shortly thereafter the oceans (Figure 12.5).

FIGURE 12.5 Artistic depiction of Earth more than 4 billion years ago. This was a time of intense volcanic activity that produced Earth's primitive atmosphere and oceans, while early life-forms produced mound-like structures called stromatolites.

CONCEPT CHECK 12.2

❶ What two elements made up most of the very early universe?
❷ What is the name of the cataclysmic event in which an exploding star produces all of the elements heavier than iron?
❸ Briefly describe the formation of the planets from the solar nebula.

Origin of the Atmosphere and Oceans

We can be thankful for our atmosphere; without it there would be no greenhouse effect and Earth would be nearly 60° F colder. Earth's water bodies would be frozen and the hydrologic cycle would be nonexistent.

The air we breathe is a stable mixture of 78 percent nitrogen, 21 percent oxygen, about 1 percent argon (an inert gas), and small amounts of gases such as carbon dioxide and water vapor. However, our planet's original atmosphere 4.6 billion years ago was substantially different.

Earth's Primitive Atmosphere

Early in Earth's formation, its atmosphere likely consisted of gases most common in the early solar system: hydrogen, helium, methane, ammonia, carbon dioxide, and water vapor. The lightest of these gases, hydrogen and helium, apparently escaped into space because Earth's gravity was too weak to hold them. Most of the remaining gases were probably scattered into space by strong *solar winds* (a vast stream of particles) from a young, active Sun. (All stars, including the Sun, apparently experience a highly active stage early in their evolution known as the *T-Tauri phase*, during which their solar winds are very intense.)

Earth's first enduring atmosphere was generated by a process called **outgassing**, through which gases trapped in the planet's interior are released. Outgassing from hundreds of active volcanoes still remains an important planetary function worldwide (Figure 12.6). However, early in Earth's history, when massive heating and fluidlike motion occurred in the mantle, the gas output must have been immense. Based on our understanding of modern volcanic eruptions, Earth's primitive atmosphere probably consisted of mostly water vapor, carbon dioxide, and sulfur dioxide with minor amounts of other gases, and minimal nitrogen. Most important, free oxygen was not present.

Oxygen in the Atmosphere

As Earth cooled, water vapor condensed to form clouds, and torrential rains began to fill low-lying areas, which became the oceans. In those oceans, nearly 3.5 billion years ago, photosynthesizing bacteria began to release oxygen into the water. During *photosynthesis*, the Sun's energy is used by organisms to produce organic material (energetic molecules of sugar containing hydrogen and carbon) from carbon dioxide (CO_2) and water (H_2O). The first bacteria probably used hydrogen sulfide (H_2S) as the source of hydrogen rather than water. One of the earliest

FIGURE 12.6 Earth's first enduring atmosphere was formed by a process called *outgassing*, which continues today from hundreds of active volcanoes worldwide. (Photo by Game McGimsey/CORBIS)

bacteria, *cyanobacteria* (once called blue-green algae), began to produce oxygen as a by-product of photosynthesis.

Initially, the newly released oxygen was readily consumed by chemical reactions with other atoms and molecules (particularly iron) in the ocean. It seems that large quantities of iron were released into the early ocean through submarine volcanism and associated hydrothermal vents. Iron has tremendous affinity for oxygen. When these two elements join, they become iron oxide (rust). As it accumulated on the seafloor, these early iron oxide deposits created alternating layers of iron-rich rocks and chert, called **banded iron formations** (Figure 12.7). Most banded iron deposits accumulated in the Precambrian between 3.5 and 2 billion years ago and represent the world's most important reservoir of iron ore.

As the number of oxygen-generating organisms increased, oxygen began to build in the atmosphere. Chemical analysis of rocks suggest that a significant amount of oxygen appeared in the atmosphere as early as 2.2 billion years ago and increased steadily until it reached stable levels about 1.5 billion years ago. Obviously, the availability of free oxygen had a positive impact on the development of life.

Another significant benefit of the "oxygen explosion" is that oxygen (O_2) molecules readily absorb ultraviolet radiation and rearrange themselves to form *ozone* (O_3). Today, ozone is concentrated above the surface in a layer called the *stratosphere* where it absorbs much of the ultraviolet radiation that strikes the upper atmosphere. For the first time, Earth's surface was protected from this type of solar radiation, which is particularly harmful to DNA. Marine organisms had always been shielded from ultraviolet radiation by the oceans, but the development of the atmosphere's protective ozone layer made the continents more hospitable.

Evolution of the Oceans

About 4 billion years ago, as much as 90 percent of the current volume of seawater was contained in the ocean basins. Because the primitive atmosphere was rich in carbon dioxide, sulfur dioxide, and hydrogen sulfide, the earliest rainwater was highly acidic—to an even greater degree than the acid rain that damaged lakes and streams in eastern North America during the latter part of the 20th century. Consequently, Earth's rocky surface weathered at an accelerated rate. The products released by chemical weathering included atoms and molecules of various substances, including sodium, calcium, potassium, and silica, that were carried into the newly formed oceans. Some of

FIGURE 12.7 These layered, iron-rich rocks, called *banded iron formations*, were deposited during the Precambrian. Much of the oxygen generated as a by-product of photosynthesis was readily consumed by chemical reactions with iron to produce these rocks. (Courtesy of Spencer R. Titley)

these dissolved substances precipitated to become chemical sediment that mantled the ocean floor. Other substances formed soluble salts, which increased the salinity of seawater. Research suggests that the salinity of the oceans increased rapidly at first but has remained constant over the last 2 billion years.

Earth's oceans also serve as a depository for tremendous volumes of carbon dioxide, a major constituent in the primitive atmosphere. This is significant because carbon dioxide is a greenhouse gas that strongly influences the heating of the atmosphere. Venus, once thought to be very similar to Earth, has an atmosphere composed of 97 percent carbon dioxide that produced a "runaway" greenhouse effect. As a result, its surface temperature is 475° C (900° F).

Carbon dioxide is readily soluble in seawater where it often joins other atoms or molecules to produce various chemical precipitates. The most common compound generated by this process is calcium carbonate ($CaCO_3$), which makes up limestone, the most abundant chemical sedimentary rock. About 542 million years ago, marine organisms began to extract calcium carbonate from seawater to make their shells and other hard parts. Included were trillions of tiny marine organisms such as foraminifera, whose shells were deposited on the seafloor at the end of their life cycle. Today, some of these deposits can be observed in the chalk beds exposed along the White Cliffs of Dover, England, shown in Figure 12.8. By "locking up" carbon dioxide, these limestone deposits store this greenhouse gas so it cannot easily reenter the atmosphere.

CONCEPT CHECK 12.3

1. What is meant by outgassing, and what modern phenomenon serves that role today?
2. Outgassing produced Earth's early atmosphere, which was rich in what two gases?
3. Why is the evolution of a type of bacteria that employed photosynthesis to produce food important to most modern organisms?
4. What was the source of water for the first oceans?
5. How does the ocean remove carbon dioxide from the atmosphere? What role do tiny marine organisms, such as foraminifera, play?

Precambrian History: The Formation of Earth's Continents

Earth's first 4 billion years are encompassed in the time span called the *Precambrian*. Representing nearly 90 percent of Earth's history, the Precambrian is divided into the *Archean eon* ("ancient age") and the *Proterozoic eon* ("early life"). Our knowledge of this ancient time is limited because much of the early rock record has been obscured by the very Earth processes you have been studying, especially plate tectonics, erosion, and deposition. Most Precambrian rocks lack fossils, which hinders correlation of rock units. In addition, rocks this old are metamorphosed and deformed, extensively eroded, and frequently concealed by younger strata. Indeed, Precambrian history is written in scattered, speculative episodes, like a long book with many missing chapters.

FIGURE 12.8 Prominent chalk deposit, the White Chalk Cliffs, Kent, England. Similar deposits are also found in northern France. (Photo by David Hughes/Robert Harding)

FIGURE 12.9 Rift pattern on lava lake. The crust covering this lava lake is continually being replaced with fresh lava from below, similar to how Earth's crust was recycled early in its history. (Photo by Juerg Alean/www.stromboli.net)

Earth's First Continents

More than 95 percent of Earth's human population lives on the continents—the other 5 percent are people living on volcanic islands such as the Hawaiian Islands and Iceland. These islanders inhabit pieces of oceanic crust that were thick enough to rise above sea level.

What differentiates continental crust from oceanic crust? Recall that oceanic crust is a relatively dense (3.0 g/cm^3), homogeneous layer of basaltic rocks derived from partial melting of the rocky upper mantle. In addition, oceanic crust is thin, averaging only 7 kilometers in thickness. Continental crust, on the other hand, is composed of a variety of rock types, has an average thickness of nearly 40 kilometers, and contains a large percentage of low-density (2.7 g/cm^3), silica-rich rocks such as granite.

The significance of these differences cannot be overstated in a review of Earth's geologic evolution. Oceanic crust, because it is relatively thin and dense, is found several kilometers below sea level—unless of course it has been pushed onto a landmass by tectonic forces. Continental crust, because of its great thickness and lower density, extends well above sea level. Also, recall that oceanic crust of normal thickness readily subducts, whereas thick, buoyant blocks of continental crust resist being recycled into the mantle.

Making Continental Crust Earth's first crust was probably ultramafic in composition, but because physical evidence no longer exists we are not certain. The hot, turbulent mantle that most likely existed during the Archean eon recycled most of this material back into the mantle. In fact, it may have been continuously recycled, in much the same way that the "crust" that forms on a lava lake is repeatedly replaced with fresh lava from below (**Figure 12.9**).

The oldest preserved continental rocks occur as small, highly deformed terranes, which are incorporated within somewhat younger blocks of continental crust (**Figure 12.10**). The oldest of these is the 4 billion-year-old Acasta gneiss located in the Slave Province of Canada's Northwest Territories.

The formation of continental crust is simply a continuation of the gravitational segregation of Earth materials that began during the final accretionary stage of our planet. After the metallic core and rocky mantle evolved, low-density, silica-rich minerals were gradually extracted from the mantle to form continental crust. This is an ongoing, multistage process during which partial melting of ultramafic mantle rocks (peridotite) generates basaltic rocks. Next, the melting of basaltic rocks produces magmas that crystallize to form felsic, quartz-bearing rocks (see Chapter 3). However, little is known about the details of the mechanisms that generated these silica-rich rocks during the Archean.

Most geologists agree that some type of platelike motion operated early in Earth's history. In addition, hot-spot volcanism likely played a role as well. However, because the mantle was hotter in the Archean than it is today, both of these phenomena would have progressed at higher rates than their modern counterparts. Hot-spot volcanism is thought to have created immense shield volcanoes as well as oceanic plateaus. Simultaneously, subduction of oceanic crust generated volcanic island arcs. Collectively, these relatively small crustal fragments represent the first phase in creating stable, continent-size landmasses.

From Continental Crust to Continents According to one model, the growth of large continental masses was accomplished through collision and accretion of various types of terranes, as illustrated in **Figure 12.11**. This type of collision tectonics deformed and metamorphosed sediments caught between converging crustal fragments, thereby shortening and thickening the developing crust. Within the deepest regions of these collision

FIGURE 12.10 These rocks at Isua, Greenland, some of the world's oldest, have been dated at 3.8 billion years. (Photo courtesy of James L. Amos/CORBIS)

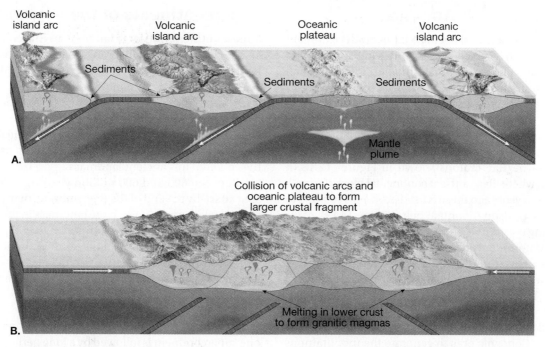

FIGURE 12.11 The growth of large continental masses was accomplished through the collision and accretion of various types of terranes.

zones, partial melting of the thickened crust generated silica-rich magmas that ascended and intruded the rocks above. The result was the formation of large crustal provinces that, in turn, accreted with others to form even larger crustal blocks called **cratons**. (The portion of a modern craton that is exposed at the surface is referred to as a *shield*.) The assembly of a large craton involves the accretion of several crustal blocks that cause major mountain-building episodes similar to India's collision with Asia. **Figure 12.12** shows the extent of crustal material that was produced during the Archean and Proterozoic eons. This was accomplished by the collision and accretion of many thin, highly mobile terranes into nearly recognizable continental masses.

Although the Precambrian was a time when much of Earth's continental crust was generated, a substantial amount of crustal material was destroyed as well. Crust can be lost by either weathering and erosion or by direct reincorporation into the mantle through subduction. Evidence suggests that during much of the Archean, thin slabs of continental crust were eliminated, mainly by subduction into the mantle. However, by about 3 billion years ago, cratons grew sufficiently large and thick to resist subduction. After that time, weathering and erosion became the primary processes of crustal destruction. By the close of the Precambrian, an estimated 85 percent of the modern continental crust had formed.

FIGURE 12.12 Illustration showing the extent of crustal material remaining from the Archean and Proterozoic eons.

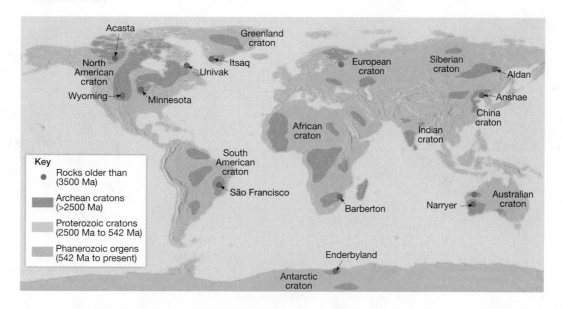

The Making of North America

North America provides an excellent example of the development of continental crust and its piecemeal assembly into a continent. Notice in **Figure 12.13** that very little continental crust older than 3.5 billion years remains. In the late Archean, between 3 and 2.5 billion years ago, there was a period of major continental growth. During this span, the accretion of numerous island arcs and other fragments generated several large crustal provinces. North America contains some of these crustal units, including the Superior and Hearne/Rae cratons shown in Figure 12.13. It remains unknown where these ancient continental blocks formed.

About 1.9 billion years ago these crustal provinces collided to produce the Trans-Hudson mountain belt (Figure 12.13). (This mountain-building episode was not restricted to North America, because ancient deformed strata of similar age are also found on other continents.) This event built the North American craton, around which several large and numerous small crustal fragments were later added. These late arrivals include Blue Ridge and Piedmont provinces of the Appalachians. Additionally, several terranes were added to the western margin of North America during the Mesozoic and Cenozoic eras to generate the mountainous North American Cordillera.

FIGURE 12.13 Map showing the major geological provinces of North America and their ages in billions of years (Ga). It appears that North America was assembled from crustal blocks that were joined by processes very similar to modern plate tectonics. These ancient collisions produced mountainous belts that include remnant volcanic island arcs trapped by the colliding continental fragments.

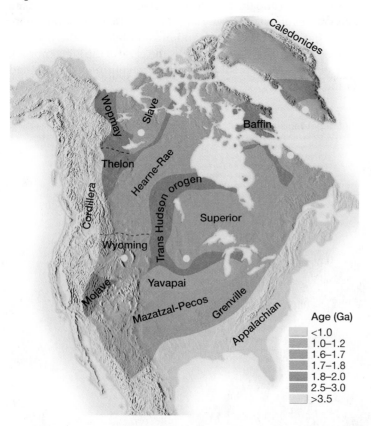

Age (Ga)
<1.0
1.0–1.2
1.6–1.7
1.7–1.8
1.8–2.0
2.5–3.0
>3.5

Supercontinents of the Precambrian

Supercontinents are large landmasses that contain all, or nearly all, the existing continents. Pangaea was the most recent, but certainly not the only, supercontinent to exist in the geologic past. The earliest well-documented supercontinent, *Rodinia*, formed during the Proterozoic eon about 1.1 billion years ago. Although its reconstruction is still being researched, it is clear that Rodinia's configuration was quite different from Pangaea's (**Figure 12.14**). One obvious distinction is that North America was located near the center of this ancient landmass.

Between 800 and 600 million years ago, Rodinia gradually split apart. By the end of the Precambrian, many of the fragments reassembled, producing a large landmass in the Southern Hemisphere called *Gondwana*, comprised mainly of present-day South America, Africa, India, Australia, and Antarctica (**Figure 12.15**). Other continental fragments also developed—North America, Siberia, and northern Europe. We consider the fate of these Precambrian landmasses later in the chapter.

Supercontinent Cycle   The idea that rifting and dispersal of one supercontinent is followed by a long period during which the fragments are gradually reassembled into a new supercontinent with a different configuration is called the **supercontinent cycle**. The assembly and dispersal of supercontinents had a profound impact on the evolution of Earth's continents. In addition, this phenomenon greatly influenced global climates and contributed to periodic episodes of rising and falling sea level.

Supercontinents and Climate   As continents move, the patterns of ocean currents and global winds change, which influences the global distribution of temperature and precipitation. One example of how a supercontinent's dispersal influenced climate is the formation of the Antarctic ice sheet. Although eastern Antarctica remained over the South Pole for more than 100 million

FIGURE 12.14 Simplified drawing showing one of several possible configurations of the supercontinent Rodinia. For clarity, the continents are drawn with somewhat modern shapes, not the shapes of 1 billion years ago. (After P. Hoffman, J. Rogers, and others)

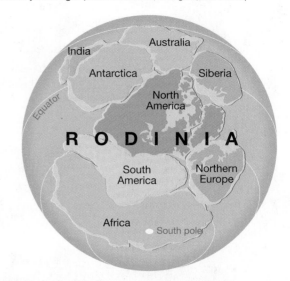

A. Continent of Gondwana

B. Continents not a part of Gondwana

FIGURE 12.15 Reconstruction of Earth as it may have appeared in late Precambrian time. The southern continents were joined into a single landmass called Gondwana. Other landmasses that were not part of Gondwana include North America, northwestern Europe, and northern Asia. (After P. Hoffman, J. Rogers, and others)

FIGURE 12.16 Comparison of the oceanic circulation pattern 50 million years ago with that of the present. When South America separated from Antarctica, the West Wind Drift developed, which effectively isolated the entire Antarctic coast from the warm, poleward-directed currents in the southern oceans. This led to the eventual covering of much of Antarctica with glacial ice.

years, it was not glaciated until about 25 million years ago. Prior to this period of glaciation, South America was connected to the Antarctic Peninsula. This arrangement of landmasses helped maintain a circulation pattern in which warm ocean currents reached the coast of Antarctica, as shown in **Figure 12.16A**. This is similar to the way in which the modern Gulf Stream keeps Iceland mostly ice free, despite its name. However, as South America separated from Antarctica, it moved northward, permitting ocean circulation to flow from west to east around the entire continent of Antarctica (Figure 12.16B). This cold current, called the West Wind Drift, effectively isolated the entire Antarctic coast from the warm, poleward-directed currents in the southern oceans. As a result, most of the Antarctic landmass became covered with glacial ice.

Local and regional climates have also been impacted by large mountain systems created by the collision of large cratons. Because of their high elevations, mountains exhibit markedly lower average temperatures than surrounding lowlands. In addition, when air rises over these lofty structures, lifting "squeezes" moisture from the air, leaving the region downwind relatively dry. A modern analogy is the wet, heavily forested western slopes of the Sierra Nevada compared to the dry climate of the Great Basin desert that lies directly downwind (see Figure 17.11, p. 500)

Supercontinents and Sea Level Changes   Significant and numerous sea level changes have been documented in geologic history, many of which appear related to the assembly and dispersal of supercontinents. If sea level rises, shallow seas advance onto the continents. Evidence for periods when the seas advanced onto the continents include thick sequences of ancient marine sedimentary rocks that blanket large areas of modern landmasses—including much of the eastern two-thirds of the United States.

The supercontinent cycle and sea level changes are directly related to rates of *seafloor spreading*. When the rate of spreading is rapid, as it is along the East Pacific Rise today, the production of warm oceanic crust is also high. Because warm oceanic crust is less dense (takes up more space) than cold crust, fast spreading ridges occupy more volume in the ocean basins than do slow spreading centers. (Think of getting into a tub filled with water.)

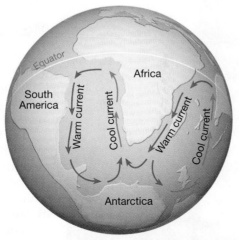

A. 50 million years ago

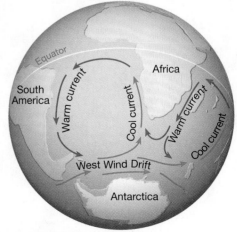

B. Present

As a result, when the rates of seafloor spreading increase, sea level rises. This, in turn, causes shallow seas to advance onto the low-lying portions of the continents.

CONCEPT CHECK 12.4

❶ Explain why Precambrian history is more difficult to decipher than more recent geological history.

❷ Briefly describe how cratons come into being.

❸ What is the supercontinent cycle?

❹ How can the movement of continents trigger climate change?

Geologic History of the Phanerozoic: The Formation of Earth's Modern Continents

The time span since the close of the Precambrian, called the *Phanerozoic eon*, encompasses 542 million years and is divided into three eras: *Paleozoic, Mesozoic,* and *Cenozoic*. The beginning of the Phanerozoic is marked by the appearance of the first life-forms with hard parts such as shells, scales, bones, or teeth—all of which greatly enhance the possibility of an organism being preserved in the fossil record (Figure 12.17).[27] Thus, the study of Phanerozoic crustal history was aided by the availability of fossils, which improved our ability to date and correlate geologic events. Moreover, because every organism is associated with its own particular niche, the greatly improved fossil record provided invaluable information for deciphering ancient environments.

[27]For more about this, see the section "Conditions Favoring Preservation" in Chapter 11, p. 334)

FIGURE 12.17 Trilobites were common Paleozoic life forms. (Photo by Ed Reschke/Photolibrary)

Paleozoic History

As the Paleozoic era opened, North America hosted no living things, neither plant nor animal. There were no Appalachian or Rocky Mountains; the continent was largely a barren lowland. Several times during the early Paleozoic, shallow seas moved inland, then receded from the continental interior and left behind the thick deposits of limestone, shale, and clean sandstone that mark the shorelines of these previous mid-continent shallow seas.

Formation of Pangaea One of the major events of the Paleozoic was the formation of the supercontinent of Pangaea, which began with a series of collisions that gradually joined North America, Europe, Siberia, and other smaller crustal fragments (Figure 12.18). These events eventually generated a large northern continent called *Laurasia*. This tropical landmass supported warm wet conditions that led to the formation of vast swamps that eventually converted to coal.

Simultaneously, the vast southern continent of *Gondwana* encompassed five continents—South America, Africa, Australia, Antarctica, India—and perhaps portions of China. Evidence of extensive continental glaciation places this landmass near the South Pole. By the end of the Paleozoic, Gondwana had migrated northward and collided with Laurasia, culminating in the formation of the supercontinent *Pangaea*.

The accretion of Pangaea spans more than 300 million years and resulted in the formation of several mountain belts. The collision of northern Europe (mainly Norway) with Greenland produced the Caledonian Mountains, whereas the joining of northern Asia (Siberia) and Europe created the Ural Mountains. Northern China is also thought to have accreted to Asia by the end of the Paleozoic, whereas southern China may not have become part of Asia until after Pangaea had begun to rift. (Recall that India did not begin to accrete to Asia until about 50 million years ago.)

Pangaea reached its maximum size about 250 million years ago as Africa collided with North America (Figure 12.18D). This event marked the final episode of growth in the long history of the Appalachian Mountains (see Chapter 10).

Mesozoic History

Spanning about 186 million years, the Mesozoic era is divided into three periods: the *Triassic, Jurassic,* and *Cretaceous*. Major geologic events of the Mesozoic include the breakup of Pangaea and the evolution of our modern ocean basins.

The Mesozoic era began with much of the world's continents above sea level. The exposed Triassic strata are primarily red sandstones and mudstones that lack marine fossils, features that indicate a terrestrial environment. (The red color in sandstone comes from the oxidation of iron.)

As the Jurrasic period opened, the sea invaded western North America. Adjacent to this shallow sea, extensive continental sediments were deposited on what is

Today, the Cretaceous coal deposits in the western United States and Canada are economically important. For example, on the Crow Native American reservation in Montana, there are nearly 20 billion tons of high-quality, Cretaceous-age coal.

Another major event of the Mesozoic era was the breakup of Pangaea. About 165 million years ago a rift developed between what is now North America and western Africa, which marked the birth of the Atlantic Ocean. As Pangaea gradually broke apart, the westward-moving North American plate began to override the Pacific basin. This tectonic event triggered a continuous wave of deformation that moved inland along the entire western margin of North America. By Jurassic times, subduction of the Farallon plate had begun to produce the chaotic mixture of rocks that exist today in the Coast Ranges of California. Further inland, igneous activity was widespread, and for more than 100 million years volcanism was rampant as huge masses of magma rose within a few miles of Earth's surface. The remnants of this activity include the granitic plutons of the Sierra Nevada, as well as the Idaho batholith and British Columbia's Coast Range batholith.

Tectonic activity initiated in the Jurassic continued throughout the Cretaceous. Compressional forces moved huge rock units in a shingle-like fashion toward the east. Across much of North America's western margin, older rocks were thrust eastward over younger strata, for distances exceeding 150 kilometers (90 miles). Ultimately, this activity was responsible for creating the vast Northern Rockies that extend from Wyoming to Alaska.

Toward the end of the Mesozoic, the southern portions of the Rocky Mountains developed. This mountain-building event, called the *Laramide Orogeny*, occurred when large blocks of deeply buried Precambrian rocks were lifted nearly vertically along steeply dipping faults, upwarping the overlying younger sedimentary strata. The mountain ranges produced by the Laramide Orogeny include Colorado's Front Range, the Sangre de Cristo of New Mexico and Colorado, and the Bighorns of Wyoming.

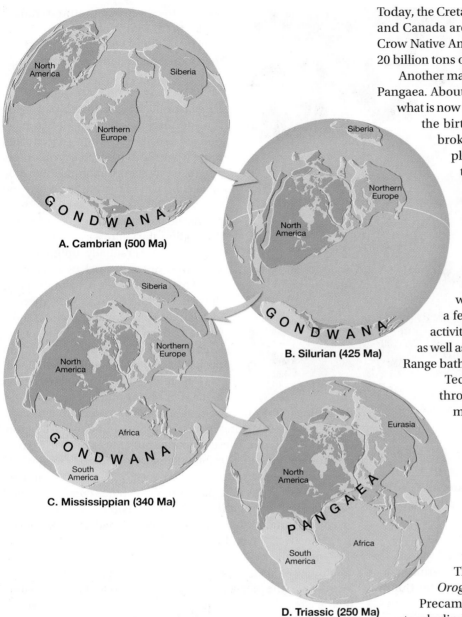

FIGURE 12.18 During the late Paleozoic, plate movements were joining the major landmasses to produce the supercontinent of Pangaea. (After P. Hoffman, J. Rogers, and others)

Cenozoic History

The Cenozoic era, or "era of recent life," encompasses the last 65.5 million years of Earth history. It was during this span that the physical landscapes and life-forms of our modern world came into existence. The Cenozoic era represents a considerably smaller fraction of geologic time than either the Paleozoic or the Mesozoic. Nevertheless, much more is known about this time span because the rock formations are more widespread and less disturbed than those of any preceding era.

Most of North America was above sea level during the Cenozoic era. However, the eastern and western margins of the continent experienced markedly contrasting events because of their different plate boundary relationships. The Atlantic and Gulf coastal regions, far removed from an active plate boundary,

now the Colorado Plateau. The most prominent is the Navajo Sandstone, a cross-bedded, quartz-rich layer, that in some places approaches 300 meters (1,000 feet) thick. These remnants of massive dunes indicate that an enormous desert occupied much of the American Southwest during early Jurassic times (**Figure 12.19**). Another well-known Jurassic deposit is the Morrison Formation—the world's richest storehouse of dinosaur fossils. Included are the fossilized bones of massive dinosaurs such as Apatosaurus (formerly Brontosaurus), Brachiosaurus, and Stegosaurus.

As the Jurassic period gave way to the Cretaceous, shallow seas again encroached upon much of western North America, as well as the Atlantic and Gulf coastal regions. This led to the formation of "coal swamps" similar to those of the Paleozoic era.

FIGURE 12.19 These massive, cross-bedded sandstone cliffs in Zion National Park are the remnants of ancient sand dunes. (Photo by Ruth Tomlinson/Robert Harding)

were tectonically stable. By contrast, western North America was the leading edge of the North American plate. As a result, plate interactions during the Cenozoic account for many events of mountain building, volcanism, and earthquakes.

Eastern North America   The stable continental margin of eastern North America was the site of abundant marine sedimentation. The most extensive deposition surrounded the Gulf of Mexico, from the Yucatan Peninsula to Florida, where a massive buildup of sediment caused the crust to downwarp. In many instances, faulting created structures in which oil and natural gas accumulated. Today, these and other petroleum traps are the Gulf Coast's most economically important resource, evidenced by numerous offshore drilling platforms.

By early Cenozoic time, most of the original Appalachians had eroded to create a low plain. Later, isostatic adjustments again raised the region and rejuvenated its rivers. Streams eroded with renewed vigor, gradually sculpting the surface into its present-day topography. Sediments from this erosion were deposited along the eastern continental margin, where they accumulated to a thickness of many kilometers. Today, portions of the strata deposited during the Cenozoic are exposed as the gently sloping Atlantic and Gulf coastal plains, where a large percentage of the eastern and southeastern United States population resides.

Western North America   In the West, the Laramide Orogeny responsible for building the southern Rocky Mountains was coming to an end. As erosional forces lowered the mountains,

the basins between uplifted ranges began to fill with sediment. East of the Rockies, a large wedge of sediment from the eroding mountains created the Great Plains.

Beginning in the Miocene epoch about 20 million years ago, a broad region from northern Nevada into Mexico experienced crustal extension that created more than 150 fault-block mountain ranges. Today, they rise abruptly above the adjacent basins, creating the Basin and Range Province (see Chapter 10).

As the Basin and Range Province was developing, the entire western interior of the continent gradually uplifted. This event reelevated the Rockies and rejuvenated many of the West's major rivers. As the rivers became incised, many spectacular gorges were created, including the Grand Canyon of the Colorado River, the Grand Canyon of the Snake River, and the Black Canyon of the Gunnison River.

Volcanic activity was also common in the West during much of the Cenozoic. Beginning in the Miocene epoch, great volumes of fluid basaltic lava flowed from fissures in portions of present-day Washington, Oregon, and Idaho. These eruptions built the extensive (1.3 million square miles) Columbia Plateau. Immediately west of the Columbia Plateau, volcanic activity was different in character. Here, more viscous magmas with higher silica content erupted explosively, creating the Cascades, a chain of stratovolcanoes extending from northern California into Canada, some of which are still active (**Figure 12.20**).

A final episode of deformation occurred in the late Cenozoic, creating the Coast Ranges that stretch along the Pacific coast. Meanwhile, the Sierra Nevada were faulted and uplifted along their eastern flank, forming the impressive mountains we see today.

As the Cenozoic was drawing to a close, the effects of mountain building, volcanic activity, isostatic adjustments, and extensive erosion and sedimentation created the physical landscape we know today. All that remained of Cenozoic time was the final 2.6-million-year episode called the Quaternary period. During this most recent, and ongoing, phase of Earth's history, in which humans evolved, the action of glacial ice, wind, and running water added the finishing touches to our planet's long, complex geologic history.

CONCEPT CHECK 12.5

❶ During which period of geologic history did the supercontinent of Pangaea come into existence?

❷ During which period of geologic history did Pangaea begin to break apart?

❸ Describe the climate during the early Jurassic period.

FIGURE 12.20 Mount Hood, Oregon. This volcano is one of several large composite cones that comprise the Cascade Range. (Photo by John M. Roberts/CORBIS/Stock Market)

Earth's First Life

The oldest fossils provide evidence that life on Earth was established at least 3.5 billion years ago. Microscopic fossils similar to modern cyanobacteria (formerly known as blue-green algae) have been found in silica-rich chert deposits worldwide. Notable examples include southern Africa, where rocks date to more than 3.1 billion years, and the Lake Superior region of western Ontario and northern Minnesota where the Gunflint Chert contains some fossils that are older than 2 billion years. Chemical traces of organic matter in rocks of greater age have led paleontologists to strongly suggest that life may have existed 3.8 billion years ago.

How did life begin? This question sparks considerable debate, and hypotheses abound. Requirements for life, assuming the presence of a hospitable environment, include the chemical raw materials that form the essential molecules of DNA, RNA, and proteins. These substances require organic compounds called *amino acids*. The first amino acids may have been synthesized from methane and ammonia, both of which were plentiful in Earth's primitive atmosphere. Some scientists suspect these gases could have been easily reorganized into useful organic molecules by ultraviolet light, while others consider lightning to have been the impetus, as the well-known experiments conducted by Stanley Miller and Harold Urey attempted to demonstrate.

Other researchers suggest that amino acids arrived "ready-made," delivered by asteroids or comets that collided with a young Earth. A group of meteorites (debris from asteroids and comets that strike Earth) called *carbonaceous chrondrites*, known to contain amino acid–like organic compounds, led some to hypothesize that early life may have had an extraterrestrial beginning.

Yet another hypothesis proposes that the organic material needed for life came from the methane and hydrogen sulfide that spews from deep-sea hydrothermal vents (black smokers). Is it also possible that life originated within a hot spring similar to those in Yellowstone National Park? Some origin-of-life researchers consider this scenario highly improbable as the scalding temperatures would have destroyed any early types of self-replicating molecules. Their supposition is that life would have first appeared along sheltered stretches of ancient beaches, where waves and tides brought together various organic materials formed in the Precambrian oceans.

Regardless of where or how life originated, it is clear that the journey from "then" to "now" involved change (**Figure 12.21**). The first known organisms were single-cell bacteria called **prokaryotes**, which means their genetic material (DNA) is *not separated* from the rest of the cell by a nucleus. Because oxygen was absent from Earth's early atmosphere and oceans, the first organisms employed anaerobic (without oxygen) metabolism to extract energy from "food." Their food source was likely organic molecules in their surroundings, but the supply was very limited. Later, bacteria evolved that used solar energy to synthesize organic compounds (sugars). This event was an important turning point in evolution—for the first time organisms had the capability of producing food for themselves as well as for other life-forms.

Recall that photosynthesis by ancient cyanobacteria, a type of prokaryote, contributed to the gradual rise in the level of oxygen, first in the ocean and later in the atmosphere. Thus, these early organisms dramatically transformed our planet. Fossil evidence for the existence of these microscopic bacteria includes distinctively layered mounds of calcium carbonate, called **stromatolites** (**Figure 12.22A**). Stromatolites are limestone mats built up by lime-accreting bacteria. What is known about these ancient fossils comes mainly from modern stromatolites like those found in Shark Bay, Australia (Figure 12.22B).

The oldest fossils of more advanced organisms, called **eukaryotes**, are about 2.1 billion years old. The first eukaryotes were microscopic, water-dwelling organisms. Unlike prokaryotes, the cellular structures of eukaryotes contain nuclei. This distinctive structure is what all multicellular organisms that now inhabit our planet—trees, birds, fishes, reptiles, and humans—have in common.

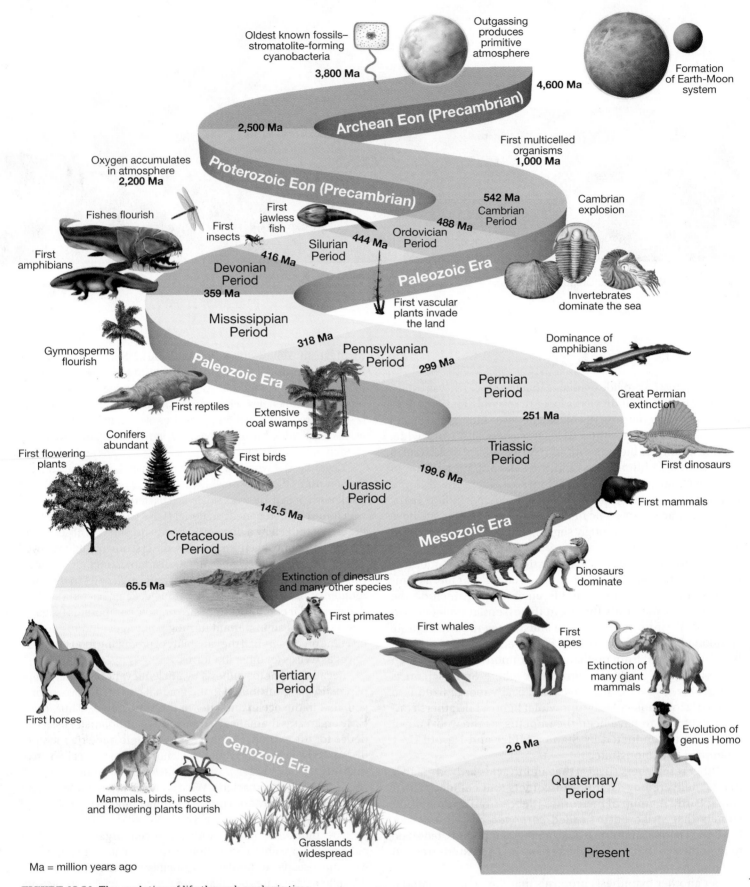

Oldest known fossils–
stromatolite-forming
cyanobacteria
3,800 Ma

Outgassing
produces
primitive
atmosphere

4,600 Ma

Formation
of Earth-Moon
system

Archean Eon (Precambrian)

2,500 Ma

Oxygen accumulates
in atmosphere
2,200 Ma

Proterozoic Eon (Precambrian)

First multicelled
organisms
1,000 Ma

542 Ma
Cambrian
Period

Cambrian
explosion

Fishes flourish

First
jawless
fish

First
insects

488 Ma
Ordovician
Period

444 Ma

First
amphibians

Silurian
Period

416 Ma

Devonian
Period

359 Ma

Paleozoic Era

First vascular
plants invade
the land

Invertebrates
dominate the sea

Mississippian
Period

318 Ma

Pennsylvanian
Period

299 Ma

Dominance of
amphibians

Gymnosperms
flourish

Paleozoic Era

Permian
Period

Great Permian
extinction

First reptiles

Extensive
coal swamps

251 Ma

Triassic
Period

First dinosaurs

First flowering
plants

Conifers
abundant

First birds

199.6 Ma

Jurassic
Period

Mesozoic Era

First mammals

Cretaceous
Period

145.5 Ma

Extinction of dinosaurs
and many other species

First primates

Dinosaurs
dominate

First whales

First
apes

Extinction of
many giant
mammals

65.5 Ma

First horses

Tertiary
Period

Evolution of
genus Homo

2.6 Ma

Quaternary
Period

Cenozoic Era

Mammals, birds, insects
and flowering plants flourish

Grasslands
widespread

Present

Ma = million years ago

FIGURE 12.21 The evolution of life through geologic time.

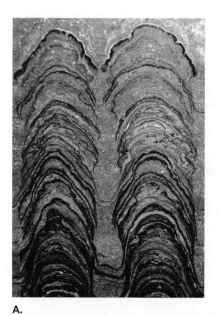

A.   **B.**

FIGURE 12.22 Stromatolites are among the most common Precambrian fossils. **A.** Precambrian fossil stromatolites composed of calcium carbonate deposited by algae. (Photo by Sinclair Stammers/Photo Researchers, Inc.) **B.** Modern stromatolites growing in shallow seas, western Australia. (Photo by Bill Bachman/Photo Researchers, Inc.)

During much of the Precambrian, life consisted exclusively of single-celled organisms. It wasn't until perhaps 1.5 billion years ago that multicelled eukaryotes evolved. Green algae, one of the first multicelled organisms, contained chloroplasts (used in photosynthesis) and were the ancestors of modern plants. The first primitive marine animals did not appear until somewhat later, perhaps 600 million years ago.

Fossil evidence suggests that organic evolution progressed at an excruciatingly slow pace until the end of the Precambrian. At this time, Earth's continents were barren, and the oceans were populated primarily by organisms too small to be seen with the naked eye. Nevertheless, the stage was set for the evolution of larger and more complex plants and animals.

CONCEPT CHECK 12.6

1. Why do some researchers think that asteroids, or *carbonaceous chrondrites*, played an important role in the development of life on Earth?
2. Compare *prokaryotes* with *eukaryotes*. Within which of these two groups do all multicelled organisms belong?

Paleozoic Era: Life Explodes

The Cambrian period marks the beginning of the Paleozoic era, a time span that saw the emergence of a spectacular variety of new life-forms. All major invertebrate (animals lacking backbones) groups made their appearance, including jellyfish, sponges, worms, mollusks (clams and snails), and arthropods (insects and crabs). This huge expansion in biodiversity is often referred to as the *Cambrian explosion*.

But did the Cambrian explosion really happen? Recent research suggests that these life-forms may have gradually diversified late in the Precambrian but were not preserved in the fossil record. Considering the Cambrian period marked the first time organisms developed hard parts, is it possible that the Cambrian event was simply an explosion of animal forms that grew in size and became "hard" enough to be fossilized?

Paleontologists may never answer that question definitively. They know, however, that hard parts clearly served many useful purposes and aided lifestyle adaptations. Sponges, for example, developed a network of fine, interwoven silica spicules that allowed them to grow larger and more erect, and thus capable of extending above the seafloor in search of food. Clams and snails secreted external shells of calcium carbonate that provided protection and allowed bodies to function in a more controlled environment. The successful trilobites developed a flexible exoskeleton of a protein called chitin (similar to a human fingernail), which permitted them to be mobile and search for food by burrowing through soft sediment (see Figure 12.17, page 360).

Early Paleozoic Life-Forms

The Cambrian period was the golden age of *trilobites*. More than 600 genera of these mud-burrowing scavengers flourished worldwide. The Ordovician marked the appearance of abundant cephalopods—mobile, highly developed mollusks that became the major predators of their time (**Figure 12.23**). Descendants of these cephalopods include the squid, octopus, and chambered nautilus that inhabit our modern oceans. Cephalopods were the first truly large organisms on Earth, including one species that reached a length of nearly 10 meters (30 feet).

Neil Shubin: Paleontologist

The process of *discovery* relies on people with relentlessly inquisitive minds. Neil Shubin is such a person. One focus of Shubin's work is directed toward trying to "discover one of the key stages in the shift from fish to land-living animals." Furthermore, he understands that such findings "offer clues about the fundamental structure of our bodies."

Recognizing that over "99 percent of all species that ever lived are now extinct, and that only a very small fraction are preserved

> **"One focus of Shubin's work is directed toward trying to discover one of the key stages in the shift from fish to land-living animals."**

as fossils, and that an even smaller fraction are ever found," Shubin has pursued a calculated yet daunting path. The path led Shubin and a research team to Ellesmer Island, in the Canadian Arctic, where their efforts resulted in a significant discovery, but only after several expensive and exhausting attempts. Shubin and his team knew their efforts had to focus on rocks of a specific age (375 million years old); rocks that would likely contain fossils; and rocks that were exposed at the surface. Challenges faced by the team included extreme Arctic weather, wildlife threats, space and time restrictions, and the knowledge that they needed to pinpoint a very small site within a 1,500-kilometer-wide expanse of Canadian tundra. Over the span of 6 years and four expeditions, Shubin and his colleagues were deeply engaged in a search for a fossil that would bridge a major evolutionary gap. Their hard work paid off.

The fossil they discovered, named *Tiktaalik* (meaning "large freshwater fish"),

exhibits characteristics of both a water-dwelling fish and a land-living animal. Its scales and fins led some to identify it as a fish, whereas its flat head with eyes on the top led others to see it as akin to a crocodile or lizard. *Tiktaalik*'s head moved freely from its shoulder, in contrast to fish in which bones attach the skull to the shoulder. (A simple nod or twist of the head demonstrates the concept of the head moving independently from the rest of the body.)

> **"99 percent of all species that ever lived are now extinct and . . . only a very small fraction are preserved as fossils."**

Shubin's research also demonstrated *Tiktaalik*'s links to amphibians, reptiles, birds, and mammals (including humans). Skeletal features such as wrists, ribs, and ears can be traced to the evolutionary transition observed in *Tiktaalik*.

As professionals in many fields will attest, expertise is often associated with the ability to "wear many hats." This is certainly true for Neil Shubin. At the University of Chicago he holds an endowed professorship and is Associate Dean in the Department of Organismal Biology and Anatomy. He is also Provost at Chicago's Field Museum. An unexpected assignment "directing the human anatomy course" at

the UC medical school, coupled with his research on *Tiktaalik*, led Shubin to "explore a profound connection" between fossils, genes, embryos, and human anatomy. Wearing the hats of research paleontologist, professor, and author, Shubin blended these seemingly unrelated topics into a monograph, *Your Inner Fish*. A national bestseller, this book provides readers with a remarkable and captivating explanation of the human body as it relates to fossils, genes, and embryos. A

good measure of humor enhances the book's appeal to a diverse audience.

In the afterword to *Your Inner Fish*, Shubin explains that the study of *Tiktaalik* is ongoing and that fieldwork continues in an effort to obtain additional fossils. As is true in many scientific endeavors, Shubin readily acknowledges that questions are posed more frequently than they are answered. Advancements in science rely on relentlessly inquisitive minds like those of Neil Shubin and his research team.

Additional information regarding Neil Shubin and *Tiktaalik* can be found at http://www.neilshubin.com and http://tiktaalik.uchicago.edu/

Neil Shubin stands with a cast of *Taktaalik* at The Field Museum in Chicago, Illinois. Neil and Ted Daeschler co-conceived this project in 1999 and have been working the Arctic with Farish A. Jenkins, Jr. since then. (Photo by John Weinstein, The Field Museum)

FIGURE 12.23 During the Ordovician period (488–444 million years ago), the shallow waters of an inland sea over central North America contained an abundance of marine invertebrates. Shown in this reconstruction are straight-shelled cephalopods, trilobites, brachiopods, snails, and corals. (© The Field Museum, Neg. # GEO80820c, Chicago)

The early diversification of animals was driven, in part, by the emergence of predatory lifestyles. The larger mobile cephalopods preyed on trilobites that were typically smaller than a child's hand. The evolution of efficient movement was often associated with the development of greater sensory capabilities and more complex nervous systems. These early animals developed sensory devices for detecting light, odor, and touch.

Approximately 400 million years ago, green algae that had adapted to survive at the water's edge gave rise to the first multicellular land plants. The primary difficulty in sustaining plant life on land was obtaining water and staying upright despite gravity and winds. These earliest land plants were leafless, vertical spikes about the size of a human index finger (**Figure 12.24**). The fossil record indicates that by the beginning of the Mississippian period there were forests with trees tens of meters tall—clear evidence that evolutionary processes were in full swing (**Figure 12.25**).

In the ocean, fishes perfected an internal skeleton as a new form of support, and were the first creatures to have jaws. Armorplated fishes that evolved during the Ordovician continued to adapt. Their armor plates thinned to lightweight scales that increased their speed and mobility. Other fishes evolved during the Devonian, including primitive sharks with cartilage skeletons, and bony fishes—the groups in which many modern fishes are classified. Fishes, the first large vertebrates, proved to be faster swimmers than invertebrates and possessed more acute senses and larger brains. They became the dominant predators of the sea, which is why the Devonian period is often referred to as the "Age of the Fishes."

Vertebrates Move to Land

During the Devonian, a group of fishes called the *lobe-finned fish* began to adapt to terrestrial environments (**Figure 12.26**). Lobe-finned fishes had sacks that could be filled with air to supplement their "breathing" through gills. The first lobe-finned fish probably occupied freshwater tidal flats or small ponds near the ocean. Some began to use their fins to move from one pond to another in search of food, or to evacuate a deteriorating pond. This favored the evolution of a group of animals able to stay out of water longer and move on land more efficiently. By the late Devonian, lobe-finned fish had evolved into air-breathing amphibians with strong legs, yet retained a fish-like head and tail (**Figure 12.27**).

Modern amphibians, such as frogs, toads, and salamanders, are small and occupy limited biological niches. However, conditions during the late Paleozoic were ideal for these newcomers to land. Large tropical swamps that were teeming with large insects and millipedes extended across North America, Europe, and Siberia (see Figure 12.25). With virtually no predatory risks, amphibians diversified rapidly. Some even took on lifestyles and forms similar to modern reptiles such as crocodiles.

Despite their success, amphibians were not fully adapted to life out of water. In fact, amphibian means "double life," because these animals need both the water from where they came and the land to which they moved. Amphibians are born in water, as exemplified by tadpoles, complete with gills and tails. These features disappear during the maturation process, resulting in air-breathing adults with legs.

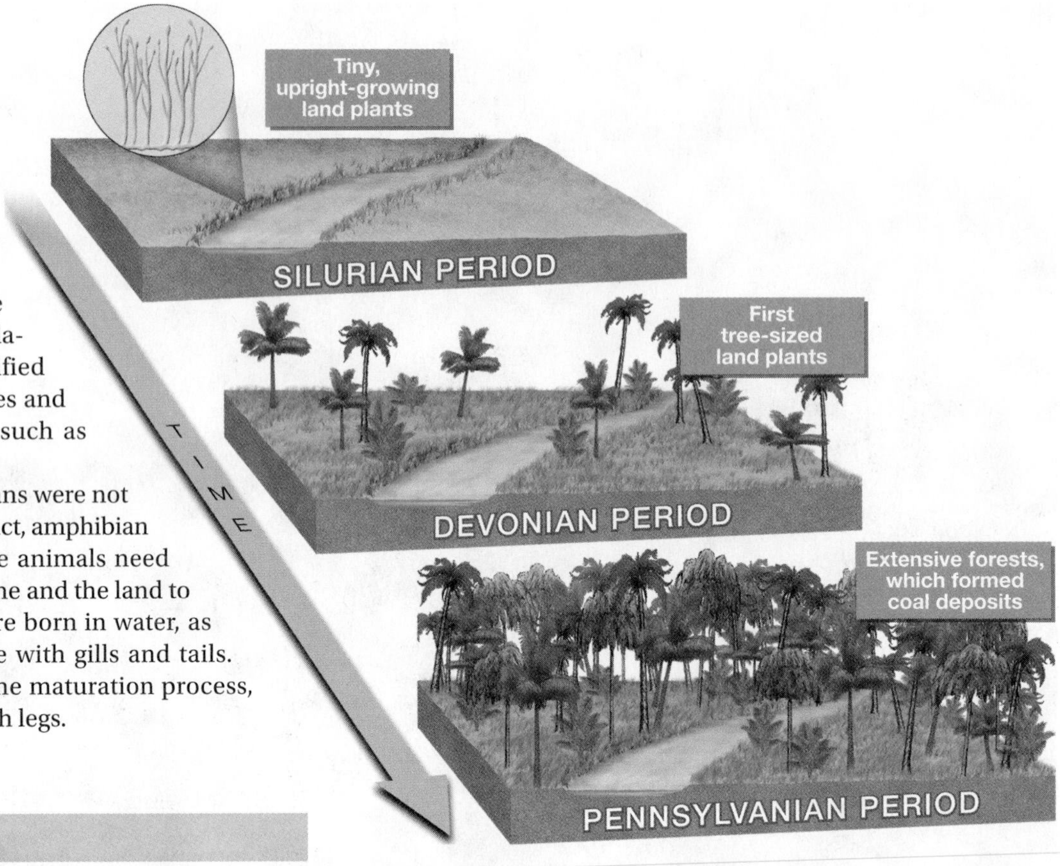

CONCEPT CHECK 12.7

1. What is the *Cambrian explosion*?
2. What animal group was dominant in Cambrian seas?
3. What did plants have to overcome to move onto land?
4. What group of animals is thought to have left the ocean to become the first amphibians?
5. Why are amphibians not considered "true" land animals?

FIGURE 12.24 Land plants of the Paleozoic. The Silurian saw the first upright-growing (vascular) plants. Plant fossils became increasingly common from the Devonian onward.

FIGURE 12.25 Restoration of a coal swamp (318 million to 299 million years ago). Shown are scale trees (left), seed ferns (lower left), and scouring rushes (right). Also note the large dragonfly. (© The Field Museum, Neg. # GEO85637c, Chicago. Photographer John Weinstein)

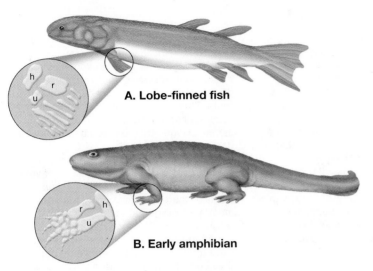

A. Lobe-finned fish

B. Early amphibian

FIGURE 12.26 Comparison of the anatomical features of the lobe-finned fish and early amphibians. **A.** The fins on the lobe-finned fish contained the same basic elements (*h*, humerus, or upper arm; *r*, radius; and *u*, ulna, or lower arm) as those of the amphibians. **B.** This amphibian is shown with the standard five toes, but early amphibians had as many as eight toes. Eventually the amphibians evolved to have a standard toe count of five.

Mesozoic Era: Age of the Dinosaurs

As the Mesozoic era dawned, its life-forms were the survivors of the great Permian extinction. These organisms diversified in many ways to fill the biological voids created at the close of the Paleozoic. On land, conditions favored those that could adapt to drier climates. Among plants, gymnosperms were one such group. Unlike the first plants to invade the land, seed-bearing gymnosperms did not depend on free-standing water for fertilization. Consequently, these plants were not restricted to a life near the water's edge.

The gymnosperms quickly became the dominant trees of the Mesozoic. They included cycads that resembled a large pineapple plant; ginkgoes that had fan-shaped leaves, much like their modern relatives; and the largest plants, the conifers, whose modern descendants include the pines, firs, and junipers. The best-known fossil occurrence of these ancient trees is in northern Arizona's Petrified Forest National Park. Here, huge petrified logs lie exposed at the surface, having been weathered from rocks of the Triassic Chinle Formation (**Figure 12.28**).

Reptiles: The First True Terrestrial Vertebrates

Among the animals, reptiles readily adapted to the drier Mesozoic environment, thereby relegating amphibians to the wetlands where most remain today. Reptiles were the first true terrestrial

FIGURE 12.27 Relationships of various vertebrates and their evolution from a fish-like ancestor.

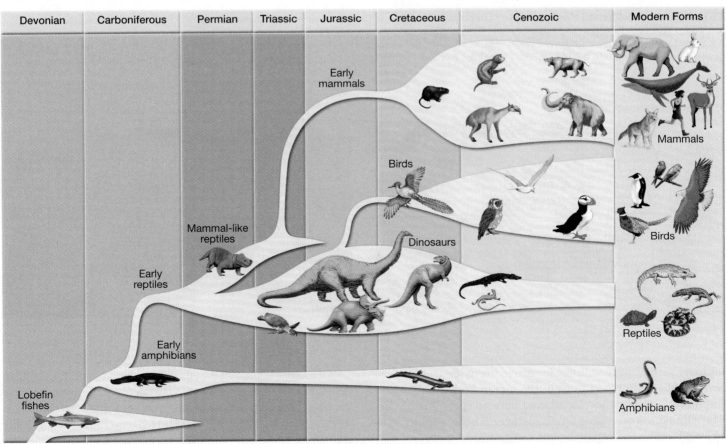

Box 12.1

UNDERSTANDING
EARTH

The Great Permian Extinction

By the close of the Permian period, a mass extinction destroyed 70 percent of all land-dwelling vertebrate species, and perhaps 90 percent of all marine organisms. The late Permian extinction was the most significant of at least five mass extinctions that occurred over the past 500 million years. Each extinction wreaked havoc with the existing biosphere, wiping out large numbers of species. In each case, however, survivors entered new biological communities that were ultimately more diverse. Therefore, mass extinctions actually invigorated life on Earth, as the few hardy survivors eventually filled more environmental niches than those left behind by the victims.

Several mechanisms have been proposed to explain these ancient mass extinctions. Initially, paleontologists believed they were gradual events caused by a combination of climate change and biological forces, such as predation and competition. In the 1980s, a research team proposed the mass extinction of 65 million years ago was caused by the explosive impact of a large asteroid. This event, which caused the swift extinction of dinosaurs, is described later in the chapter.

Was the Permian extinction caused by a giant impact? For many years, researchers thought this was a possible explanation. However, scant evidence has been found to substantiate this hypothesis.

Another possible mechanism to explain the Permian extinction was voluminous eruptions of basaltic lavas, which began about 250 million years ago and covered thousands of square kilometers of the continents. (This period of volcanism produced the Siberian Traps located in northern Russia.) The release of carbon dioxide would certainly have enhanced greenhouse warming, and the emissions of sulfur dioxide probably resulted in copious amounts of acid rain.

A group of researchers have taken the global warming hypothesis still further. Although they agree that rapid greenhouse warming occurred, it alone would not have destroyed most plants because they are heat tolerant and consume CO_2 in photosynthesis. They contend, instead, that the trouble began in the ocean rather than on land.

Most organisms, including humans, use oxygen to metabolize food. However, some forms of bacteria employ *anaerobic* (without oxygen) metabolism. Under normal conditions, oxygen from the atmosphere is readily dissolved in seawater and is then evenly distributed to all depths by deep-water currents. This "oxygen-rich" water relegates "oxygen-hating" anaerobic bacteria to anoxic (oxygen-free) environments found in deep-water sediments.

An intense period of greenhouse warming caused by an extreme episode of volcanism would have heated the ocean surface, thereby significantly reducing the amount of oxygen absorbed by seawater (**Figure 12.A**). This condition favors deep-sea anaerobic bacteria, which generate toxic hydrogen sulfide as a waste gas. As these organisms proliferated, the amount of hydrogen sulfide dissolved in seawater would have steadily increased. Eventually, the concentration of hydrogen sulfide would reach a critical threshold and release large toxic bubbles into the atmosphere (Figure 12.A). On land, hydrogen sulfide is lethal to both plants and animals, but it is most destructive to oxygen-breathing marine life.

How plausible is this scenario? Remember that these ideas represent a *hypothesis*, a tentative explanation regarding a particular set of observations. Additional research about this and other hypotheses that relate to the Permian extinction continues.

FIGURE 12.A Model for the "Great Permian Extinction." Extensive volcanism released greenhouse gases, which resulted in extreme global warming. This condition reduced the amount of oxygen dissolved by seawater. This, in turn, favored "oxygen-hating" anaerobic bacteria, which generated toxic hydrogen sulfide as a waste gas. Eventually, the concentration of hydrogen sulfide reached a critical threshold and great bubbles of this toxin exploded into the atmosphere, wreaking havoc on organisms on land, but oxygen-breathing marine life was hit the hardest.

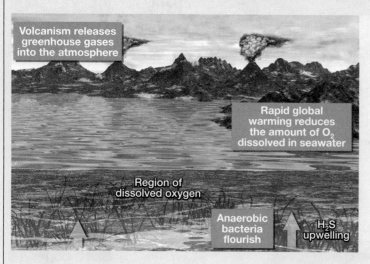
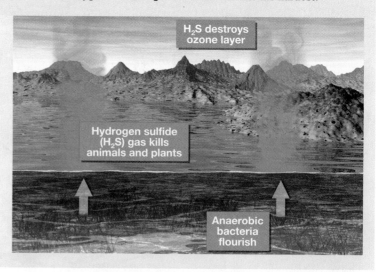

animals with improved lungs for active lifestyles and "waterproof" skin that helped prevent the loss of body fluids. Most importantly, reptiles developed shell-covered eggs laid on land. The elimination of a water-dwelling stage (like the tadpole stage in frogs) was an important evolutionary advancement.

Of interest is the fact that the watery fluid within the reptilian egg closely resembles seawater in chemical composition. Because the reptile embryo develops in this watery environment, the shelled egg has been characterized as a "private aquarium" in which the embryos of these land vertebrates spend their water-dwelling stage

FIGURE 12.28 Petrified log of Triassic age in Arizona's Petrified Forest National Park. (Photo by Michael Collier)

of life. With this "sturdy egg," the remaining ties to the oceans were broken, and reptiles moved inland.

The first reptiles were small, but larger forms evolved rapidly, particularly the dinosaurs. One of the largest was *Apatosaurus*, which weighed more than 30 tons and measured over 25 meters (80 feet) from head to tail. Some of the largest dinosaurs were carnivorous (*Tyrannosaurus*), whereas others were herbivorous (like ponderous *Apatosaurus*).

Reptiles made perhaps the most spectacular adaptive radiations in all of Earth history. One group, the pterosaurs, became airborne. These "dragons of the sky" possessed huge membranous wings that allowed them rudimentary flight. How the largest pterosaurs (some had wing spans of 26 feet and weighed 200 pounds) took flight is still unknown. Another group, exemplified by the fossil *Archaeopteryx*, led to more successful flyers—birds (**Figure 12.29**). Other reptiles returned to the sea, including fish-eating plesiosaurs and ichthyosaurs (**Figure 12.30**). These reptiles became proficient swimmers but retained their reptilian teeth and breathed by means of lungs.

For nearly 160 million years, dinosaurs reigned supreme. However, by the close of the Mesozoic, like many reptiles they became extinct. Select reptile groups survived to recent times, including turtles, snakes, crocodiles, and lizards. The huge, land-dwelling dinosaurs, the marine plesiosaurs, and the flying pterosaurs are known only through the fossil record. What caused this great extinction?

CONCEPT CHECK 12.8

❶ What major development allowed reptiles to move inland?
❷ What event is thought to have ended the reign of the dinosaurs?

Cenozoic Era: Age of Mammals

During the Cenozoic, mammals replaced reptiles as the dominant land animals. At nearly the same time, angiosperms (flowering plants with covered seeds) replaced gymnosperms as the

FIGURE 12.29 Paleontologists think that flying reptiles similar to *Archaeopteryx* were the ancestors of modern birds. Fossil evidence indicates that *Archaeopteryx* was capable of powerful flight but retained many characteristics of nonflight reptiles. (Photo by Michael Collier)

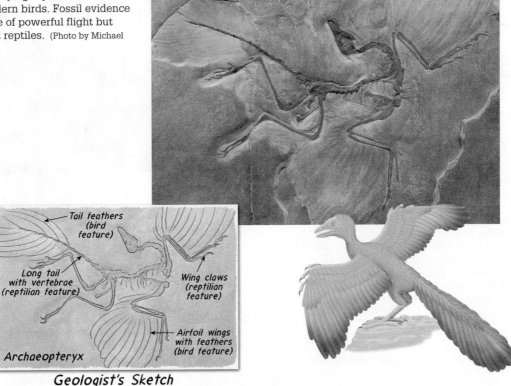

Tail feathers (bird feature)

Long tail with vertebrae (reptilian feature)

Wing claws (reptilian feature)

Airfoil wings with feathers (bird feature)

Archaeopteryx

Geologist's Sketch

FIGURE 12.30 Marine reptiles such as this *Ichthyosaur* were the most spectacular of sea animals. (Photo by Chip Clark)

Box 12.2
UNDERSTANDING
EARTH

Demise of the Dinosaurs

The boundaries between divisions on the geologic time scale represent times of significant geological and/or biological change. Of special interest is the boundary between the Mesozoic era ("middle life") and the Cenozoic era ("recent life"), about 65 million years ago. During this transition about three-quarters of all plant and animal species died out in a *mass extinction*. This boundary marks the end of the era in which dinosaurs and other reptiles dominated the landscape and the beginning of the era when mammals assumed that role (**Figure 12.B**). Because the last period of the Mesozoic is the Cretaceous (abbreviated K to avoid confusion with other "C" periods), and the first period of the Cenozoic is the Tertiary (abbreviated T), the time of this mass extinction is called the *Cretaceous–Tertiary*, or *KT, boundary.**

*The term *Paleogene* is now the preferred name for the first period of the Cenozoic era (see Figure 12.3). However, when this discovery was made, *Tertiary* was still widely used.

FIGURE 12.B Dinosaurs dominated the Mesozoic landscape until their extinction at the close of the Cretaceous period. **A.** Dinosaurs such as Allosaurus were fearsome predators. (Image by Joe Tucciarone/Photo Researchers, Inc.) **B.** These dinosaur footprints near Cameron, Arizona, were originally made in mud that eventually became sedimentary rock. (Photo by Tom Bean/CORBIS)

dominant plants. The Cenozoic is often called the "Age of Mammals" but can also be considered the "Age of Flowering Plants" because, in the plant world, angiosperms enjoy a status similar to that of mammals in the animal world.

The development of flowering plants strongly influenced the evolution of both birds and mammals that feed on seeds and fruits. During the middle of the Cenozoic, another type of angiosperm, grasses, developed rapidly and spread over the plains (**Figure 12.31**). This fostered the emergence of herbivorous (plant-eating) mammals, which, in turn, provided the evolutionary foundation for large, predatory mammals.

During the Cenozoic, the ocean was teeming with modern fish such as tuna, swordfish, and barracuda. In addition, some mammals, including seals, whales, and walruses, returned to the sea.

From Reptiles to Mammals

The earliest mammals coexisted with dinosaurs for nearly 100 million years but were small rodent-like creatures that gathered food at night when dinosaurs were less active. Then, about 65 million years ago, fate intervened when a large asteroid collided with Earth and dealt a crashing blow to the reign of the dinosaurs. This transition, during which one dominant group is replaced by another, is clearly visible in the fossil record.

Mammals are distinct from reptiles in that they give birth to live young that suckle on milk and are warm-blooded. This latter adaptation allowed mammals to lead more active lives and to occupy more diverse habitats than reptiles because they could survive in cold regions. (Most modern reptiles are dormant during cold weather.) Other mammalian adaptations included the development of insulating body hair and more efficient organs such as hearts and lungs.

With the demise of the large Mesozoic reptiles, Cenozoic mammals diversified rapidly. The many forms that exist today evolved from small primitive mammals that were characterized by short legs; flat, five-toed feet; and small brains. Their development and specialization took four principal directions: (1) increase in size, (2) increase in brain capacity, (3) specialization of teeth to better accommodate their diet, and (4) specialization of limbs to be better equipped for a particular lifestyle or environment.

Marsupial and Placental Mammals   Two groups of mammals, the marsupials and the placentals, evolved and diversified during the Cenozoic. The groups differ principally in their modes of reproduction. Young marsupials are born live at a very early

---

The extinction of the dinosaurs is generally attributed to their collective inability to adapt to radical change in environmental conditions. What event could have triggered the rapid extinction of one of the most successful groups of land animals? The most strongly supported hypothesis proposes that about 65 million years ago, our planet was struck by a large carbonaceous meteorite, a relic from the formation of the solar system. The errant mass of rock was approximately 10 kilometers in diameter and was traveling at about 90,000 kilometers per hour at the time of impact. It collided with the southern portion of North America in a shallow tropical sea—now Mexico's Yucatan Peninsula (**Figure 12.C**). The energy released by the impact is estimated to have been equivalent to 100 million megatons (*mega* = million) of explosives.

Following the impact, suspended dust greatly reduced the amount of sunlight reaching Earth's surface, which resulted in global cooling ("impact winter"), and inhibited photosynthesis, disrupting food production. Long after the dust settled, carbon dioxide, water vapor, and sulfur oxides, added to the atmosphere by the blast, remained. Sulfate aerosols, because of their high reflectivity, perpetuated the cooler surface temperatures for a few more years. Eventually, sulfate aerosols dissipated from the atmosphere as acid precipitation. With the aerosols gone, but carbon dioxide still present in large quantities, an enhanced

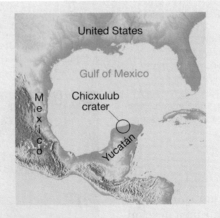

FIGURE 12.C **Chicxulub crater is a giant impact crater that formed about 65 million years ago and has since been filled with sediments. About 180 kilometers (110 miles) in diameter, Chicxulub crater is regarded by some researchers to be the impact site that resulted in the demise of the dinosaurs.**

greenhouse effect would have led to a long-term rise in average global temperatures. The likely result was that some of the plant and animal life that survived the initial impact eventually fell victim to stresses associated with global cooling, followed by acid precipitation and global warming.

The extinction of the dinosaurs opened habitats for the small mammals that survived. These new habitats, along with evolutionary forces, led to the development of the large mammals that occupy our modern world.

What evidence points to such a catastrophic collision 65 million years ago? First, a thin layer of sediment nearly 1 centimeter thick has been discovered at the KT boundary worldwide. This sediment contains a high level of the element *iridium*, rare in Earth's crust but found in high proportions in stony meteorites. Could this layer be the scattered remains of the meteorite responsible for the environmental changes that led to the demise of many reptile groups?

Despite substantial evidence and significant scientific support, some researchers disagree with the impact hypothesis. They suggest instead that huge volcanic eruptions may have led to a breakdown in the food chain. To support this hypothesis, they cite enormous outpourings of lavas in the Deccan Plateau of northern India about 65 million years ago. Regardless of what caused the KT extinction, its outcomes provide valuable lessons in understanding the role that catastrophic events play in shaping our planet's physical landscape and life-forms.

camels, and wolves migrated southward. Many animals that had been unique to South America disappeared completely after this event, including hoofed mammals, rhino-sized rodents, and a number of carnivorous marsupials. Because this period of extinction coincided with the formation of the Panamanian land-bridge, it was thought that advanced carnivores from North America were responsible. However, recent research suggests that other factors, including climatic changes, may have played a significant role.

Large Mammals and Extinction

During the rapid mammal diversification of the Cenozoic era, some groups became very large. For example, by the Oligocene epoch, a hornless rhinoceros evolved that stood nearly 5 meters (16 feet) high. It is the largest land mammal known to have existed. As time passed, many other mammals evolved to larger forms—more, in fact, than now exist. Many of these large forms were common as recently as 11,000 years ago. However, a wave of late Pleistocene extinctions rapidly eliminated these animals from the landscape.

North America experienced the extinction of mastodons and mammoths, both huge relatives of the modern elephant (**Figure 12.33**). In addition, saber-toothed cats, giant beavers, large ground sloths, horses,

FIGURE 12.31 Angiosperms, commonly known as flowering plants, are seed-plants that have reproductive structures called flowers and fruits. **A.** The most diverse and widespread of modern plants, many angiosperms display easily recognizable flowers. (Photo by WDG Photo/Shutterstock) **B.** Some angiosperms, including grasses, have very tiny flowers. The expansion of the grasslands during the Cenozoic era greatly increased the diversity of grazing mammals and the predators that feed on them. (Photo by Torleif/CORBIS)

FIGURE 12.32 After the breakup of Pangaea, the Australian marsupials evolved differently from their relatives in the Americas. (Photo by Martin Harvey/Peter Arnold Inc.)

stage of development. At birth, the tiny and immature young enter the mother's pouch to suckle and complete their development. Today, marsupials are found primarily in Australia, where they underwent a separate evolutionary expansion largely isolated from placental mammals. Modern marsupials include kangaroos, opossums, and koalas (**Figure 12.32**). Placental mammals, conversely, develop within the mother's body for a much longer period, so birth occurs when the young are comparatively mature. Most modern mammals, including humans, are placental.

In South America, primitive marsupials and placentals coexisted in isolation for about 40 million years after the breakup of Pangaea. Evolution and specialization of both groups continued undisturbed until about 3 million years ago when the Panamanian land-bridge connected the two American continents. This event permitted the exchange of fauna between the two continents—monkeys, armadillos, sloths, and opossums arrived in North America, while various types of horses, bears, rhinos, giant bison, and others died out. In Europe, late Pleistocene extinctions included woolly rhinos, large cave bears, and Irish elk. Scientists remain puzzled about the reasons for this recent wave of extinctions that targeted large animals. Having survived several major glacial advances and interglacial periods, it is difficult to ascribe extinctions of these animals to climate change. Some scientists hypothesize that early humans hastened the decline of these mammals by selectively hunting large forms (**Figure 12.34**).

CONCEPT CHECK 12.9

① What animal group became the dominant land animals of the Cenozoic era?

② How did the demise of the large Mesozoic reptiles impact the development of mammals?

③ Describe one hypothesis that explains the extinction of large mammals in the late Pleistocene.

FIGURE 12.33 Mammoths, related to modern elephants, were among the large mammals that became extinct at the close of the Ice Age. (Image courtesy of INTERFOTO/Alamy)

FIGURE 12.34 Cave painting of animals that early humans encountered about 17,000 years ago. (Photo courtesy of Sisse Brimberg/National Geographic Society)

GIVE IT SOME THOUGHT

1. Refer to the geologic time scale in Figure 12.3. The Precambrian accounts for nearly 90 percent of geologic time. Why do you think it has fewer divisions than the rest of the time scale?

2. Referring to Figure 12.4, list the major events that led to the formation of Earth.

3. What does the existence of iron-rich rocks during the Precambrian indicate about the evolution of Earth's atmosphere?

4. What are two ways that the appearance of a significant amount of oxygen in the atmosphere relates to life?

5. Presently, about 71 percent of Earth's surface is ocean covered. This percentage was much higher early in Earth history. Explain.

6. Refer to Figure 12.12. Explain why there are so few rocks older than 3,500 million years old on Earth.

7. Contrast the eastern and western margins of North America during the Cenozoic era in terms of their relationships to plate boundaries.

8. Why do you think plants moved onto land before animals?

9. Refer to Figure 12.21. How has biodiversity changed with time? How did mass extinctions contribute to this trend?

10. Some scientists have proposed that the environments around black smokers may be similar to the extreme conditions that existed early in Earth history. Therefore, they look to the unusual life that exists around black smokers for clues about how earliest life may have survived. Compare and contrast the environment of a black smoker to the environment on Earth approximately 3–4 billion years ago. Do you think there are parallels between the two, and if so, do you think black smokers are good examples of the environment that earliest life may have experienced? Explain.

In Review   Chapter 12 Earth's Evolution Through Geologic Time

- The history of Earth began about 13.7 billion years ago when the first elements were created during the *Big Bang*. It was from this material, plus other elements ejected into interstellar space by now-defunct stars, that Earth, along with the rest of the solar system, formed. As material collected, high-velocity impacts of chunks of matter called *planetesimals* and the decay of radioactive elements caused the temperature of our planet to steadily increase. Iron and nickel melted and sank to form the metallic core, while rocky material rose to form the mantle and Earth's initial crust.

- Earth's primitive atmosphere, which consisted mostly of water vapor and carbon dioxide, formed by a process called *outgassing*, which resembles the steam eruptions of modern volcanoes. About 3.5 billion years ago, photosynthesizing bacteria began to release oxygen, first into the oceans and then into the atmosphere. This began the evolution of our modern atmosphere. The oceans formed early in Earth's history as water vapor condensed to form clouds, and torrential rains filled low-lying areas. The salinity in seawater came from volcanic outgassing and from elements weathered and eroded from Earth's primitive crust.

- The *Precambrian*, which is divided into the *Archean* and *Proterozoic eons*, spans nearly 90 percent of Earth's history, beginning with the formation of Earth about 4.6 billion years ago and ending approximately 542 million years ago. During this time, much of Earth's stable continental crust was created through a multistage process. First, partial melting of the mantle generated magma that rose to form volcanic island arcs and oceanic plateaus. These thin crustal fragments collided and accreted to form larger crustal provinces, which in turn assembled into larger blocks called *cratons*. Cratons, which form the core of modern continents, were created mainly during the Precambrian.

- Supercontinents are large landmasses that consist of all, or nearly all, existing continents. *Pangaea* was the most recent supercontinent, but other massive continents including an even larger one, *Rodinia*, preceded it. The splitting and reassembling of supercontinents have generated most of Earth's major mountain belts. In addition, the movements of these crustal blocks have profoundly affected Earth's climate and caused sea level to rise and fall.

- The time span following the close of the Precambrian, called the *Phanerozoic eon*, encompasses 542 million years and is divided into three eras: *Paleozoic, Mesozoic*, and *Cenozoic*. The Paleozoic era was dominated by continental collisions as the supercontinent of Pangaea assembled—forming the Caledonian, Appalachian, and Ural Mountains. Early in the Mesozoic, much of the land was above sea level. However, by the middle Mesozoic, seas invaded western North America. As Pangaea began to break up, the westward-moving North American plate began to override the Pacific plate, causing crustal deformation along the entire western margin of North America. Owing to their different relations with plate boundaries, the eastern and western margins of the continent experienced contrasting events. The stable eastern margin was the site of abundant sedimentation as isostatic adjustment raised the modern Appalachians, causing streams to erode with renewed vigor and deposit their sediment along the continental margin. In the West, the *Laramide Orogeny* (responsible for building the Rocky Mountains) was coming to an end, the Basin and Range Province was forming, and volcanic activity was extensive.

- The first known organisms were single-celled bacteria, *prokaryotes*, which lack a nucleus. One group of these organisms, called cyanobacteria, used solar energy to synthesize organic compounds (sugars). For the first time, organisms had the ability to produce their own food. Fossil evidence for the existence of these bacteria includes layered mounds of calcium carbonate called *stromatolites*.

- The beginning of the Paleozoic is marked by the *appearance of the first life-forms with hard parts* such as shells. Therefore, abundant fossils occur, and a far more detailed record of Paleozoic events can be constructed. Life in the early Paleozoic was restricted to the seas and consisted of several invertebrate groups, including trilobites, cephalopods, sponges, and corals. During the Paleozoic, organisms diversified dramatically. Insects and plants moved onto land, and lobe-finned fishes that adapted to land became the first amphibians. By the Pennsylvanian period, large tropical swamps, which became the major coal deposits of today, extended across North America, Europe, and Siberia. At the close of the Paleozoic, a mass extinction destroyed 70 percent of all vertebrate species on land and 90 percent of all marine organisms.

- The Mesozoic era, literally the era of middle life, is often called the "*Age of Reptiles*." Organisms that survived the extinction at the end of the Paleozoic began to diversify in spectacular ways. *Gymnosperms* (cycads, conifers, and ginkgoes) became the dominant trees of the Mesozoic because they could adapt to the drier climates. Reptiles became the dominant land animals. The most awesome of the Mesozoic reptiles were the *dinosaurs*. At the close of the Mesozoic, many large reptiles, including the dinosaurs, became extinct.

- The Cenozoic is often called the "*Age of Mammals*" because these animals replaced the reptiles as the dominant vertebrate life-forms on land. Two groups of mammals, the marsupials and the placentals, evolved and expanded during this era. One tendency was for some mammal groups to become very large. However, a wave of late *Pleistocene* extinctions rapidly eliminated these animals from the landscape. Some scientists suggest that early humans hastened their decline by selectively hunting the larger animals. The Cenozoic could also be called the "*Age of Flowering Plants*." As a source of food, flowering plants (angiosperms) strongly influenced the evolution of both birds and herbivorous (plant-eating) mammals throughout the Cenozoic era.

Key Terms

banded iron formations (p. 354)
cratons (p. 357)
eukaryotes (p. 363)
outgassing (p. 353)

planetesimals (p. 350)
prokaryotes (p. 363)
protoplanets (p. 350)
solar nebula (p. 350)

stromatolites (p. 363)
supercontinent (p. 358)
supercontinent cycle (p. 358)
supernova (p. 350)

Examining the Earth System

1. The Earth system has been responsible for both the conditions that favored the evolution of life on this planet and for the mass extinctions that have occurred throughout geologic time. Describe the role of the biosphere, hydrosphere, and solid Earth in forming the current level of atmospheric oxygen. How did Earth's nearspace environment interact with the atmosphere and biosphere to contribute to the great mass extinction that marked the end of the dinosaurs?

2. Most of the vast North American coal resources located from Pennsylvania to Illinois began forming during the Pennsylvanian and Mississippian periods of Earth history. (This time period is also referred to as the Carboniferous period.) Using Figure 12.25, a restoration of a Pennsylvania period coal swamp, describe the climatic and biological conditions associated with this unique environment. Next, examine Figure 12.18B and C. The maps show the formation of the supercontinent of Pangaea and illustrate the geographic position of North America during the period of coal formation. Where, relative to the equator, was North America located during the time of coal formation? What role did plate tectonics play in determining the conditions that eventually produced North America's eastern coal reserves? Why is it unlikely that the coal-forming environment will repeat itself in North America in the near future? (You may find it helpful to visit the University of California Time Machine Exhibit at **http://www.ucmp.berkeley.edu/ carboniferous/carboniferous.html**.)

Mastering Geology

Looking for additional review and test prep materials? Visit the Self Study area in **www.masteringgeology.com** to find practice quizzes, study tools, and multimedia that will aid in your understanding of this chapter's content. In **MasteringGeology™** you will find:

- **GEODe: Earth Science: An interactive visual walkthrough of key concepts**

- **Geoscience Animation Library:** More than 100 animations illuminating many difficult-to-understand Earth science concepts
- **In The News RSS Feeds:** Current Earth science events and news articles are pulled into the site with assessment
- **Pearson eText**
- **Optional Self Study Quizzes**
- **Web Links**
- **Glossary**
- **Flashcards**

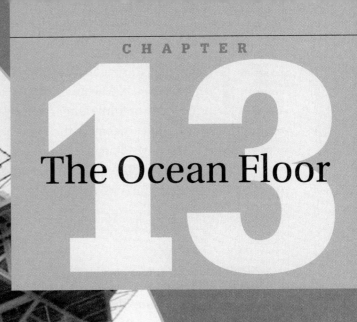

The Ocean Floor

The *Chikyu* (meaning "Earth" in Japanese) is the world's most advanced scientific drilling vessel. It can drill as deep as 7,000 meters (nearly 23,000 feet) below the seabed in water as deep as 2,500 meters (8,200 feet). It is part of the Integrated Ocean Drilling Program (IODP).

(AP Photo/Itsuo Inouye)

How deep is the ocean? How much of Earth is covered by the global sea? What does the seafloor look like? Humans have long been interested in finding answers to these questions, but it was not until rather recently that these seemingly simple questions could be answered. Suppose, for example, that all of the water were drained from the ocean. What would we see? Plains? Mountains? Canyons? Plateaus? Indeed, the ocean conceals all of these features, and more. And what about the carpet of sediment that covers much of the seafloor? Where did it come from, and what can be learned by examining it? This chapter provides answers to these questions.

FOCUS ON CONCEPTS

To assist you in learning the important concepts in this chapter, focus on the following questions:

- What is oceanography?
- What is the extent and distribution of the world's oceans?
- What techniques are used to map the ocean floor?
- How does a passive continental margin differ from an active continental margin?
- What are the major features of the deep-ocean basin?
- How are mid-ocean ridges and deep-ocean trenches related to plate tectonics boundaries?
- What are the various types of seafloor sediments? How can these sediments be used to study worldwide climate change?

The Vast World Ocean

Calling Earth the "water planet" is certainly appropriate, because nearly 71 percent of its surface is covered by the global ocean (Figure 13.1). Although the ocean comprises a much greater percentage of Earth's surface than the continents, it has only been in the relatively recent past that the ocean became an important focus of study. **Oceanography** is an interdisciplinary science that draws on the methods and knowledge of geology, chemistry, physics, and biology to study all aspects of the world ocean.

Geography of the Oceans

The area of Earth is about 510 million square kilometers (197 million square miles). Of this total, approximately 360 million square kilometers (140 million square miles), or 71 percent, is represented by oceans and marginal seas (meaning seas around the ocean's margin, like the Mediterranean Sea and Caribbean Sea). Continents and islands comprise the remaining 29 percent, or 150 million square kilometers (58 million square miles).

Clearly, oceans dominate Earth's surface. But is the distribution of land and water similar in the Northern and Southern hemispheres? By studying a world map or globe (Figure 13.1), it is readily apparent that the continents and oceans are not evenly divided between the two hemispheres. In the Northern Hemisphere, for instance, nearly 61 percent of the surface is water, whereas about 39 percent is land. In the Southern Hemisphere, on the other hand, almost 81 percent of the surface is water, and only

19 percent is land. It is no wonder then that the Northern Hemisphere is called the *land hemisphere* and the Southern Hemisphere the *water hemisphere*.

Figure 13.2A shows the distribution of land and water in the Northern and Southern hemispheres. Between latitudes 45 degrees north and 70 degrees north, there is actually more land than water, whereas between 40 degrees south and 65 degrees south there is almost no land to interrupt the oceanic and atmospheric circulation.

The world ocean can be divided into four main ocean basins (Figure 13.2B):

1. The *Pacific Ocean*, which is the largest ocean and the largest single geographic feature on the planet, accounts for over half of the ocean surface area on Earth. In fact, the Pacific Ocean is so large that all of the continents could fit into the space occupied by it—with room left over! It is also the world's deepest ocean, with an average depth of 3,940 meters (12,927 feet or about 2.5 miles).
2. The *Atlantic Ocean*, which is about half the size of the Pacific Ocean and not quite as deep. It is a relatively narrow ocean as compared to the Pacific and is bounded by almost parallel continental margins.
3. The *Indian Ocean*, which is slightly smaller than the Atlantic Ocean but has about the same average depth. Unlike the Pacific and Atlantic oceans, it is largely a Southern Hemisphere water body.
4. The *Arctic Ocean*, which is about 7 percent the size of the Pacific Ocean and is only a little more than one-quarter as deep as the rest of the oceans.

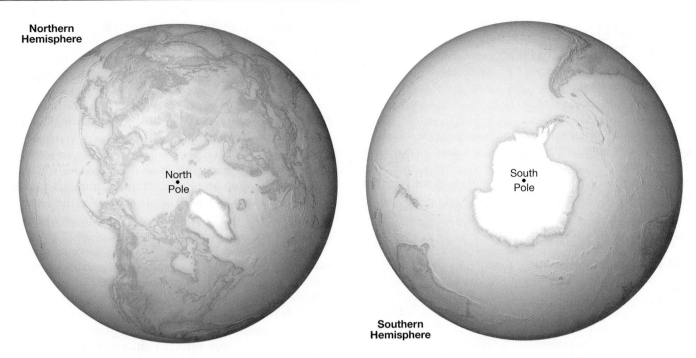

FIGURE 13.1 These views of Earth show the uneven distribution of land and water between the Northern and Southern Hemispheres. Almost 81 percent of the Southern Hemisphere is covered by the oceans—20 percent more than the Northern Hemisphere.

Oceanographers also recognize an additional ocean near the continent of Antarctica in the Southern Hemisphere (Figure 13.2B). Defined by the meeting of currents near Antarctica called the Antarctic Convergence, the *Southern Ocean* or *Antarctic Ocean* is actually those portions of the Pacific, Atlantic, and Indian Oceans south of about 50 degrees south latitude.

Comparing the Oceans to the Continents

A major difference between continents and the ocean basins is their relative levels. The average elevation of the continents above sea level is about 840 meters (2,756 feet), whereas the average depth of the oceans is nearly four and a half times this amount—3,729 meters (12,234 feet). The volume of ocean water is so large

that if Earth's solid mass were perfectly smooth (level) and spherical, the oceans would cover Earth's entire surface to a uniform depth of more than 2,000 meters (1.2 miles)!

CONCEPT CHECK 13.1

1. How does the area of Earth's surface covered by the oceans compare with that of the continents? Contrast the distribution of land and water in the Northern Hemisphere and the Southern Hemisphere.
2. Excluding the Southern Ocean, name the four main ocean basins. Contrast them in terms of area and depth.
3. How does the average depth of the oceans compare to the average elevation of the continents?

FIGURE 13.2 Distribution of land and water. **A.** The graph shows the amount of land and water in each 5-degree latitude belt. **B.** The world map provides a more familiar view.

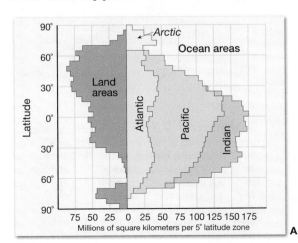

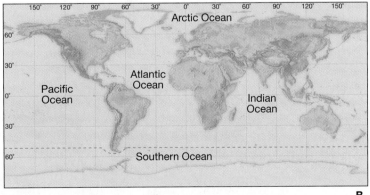

An Emerging Picture of the Ocean Floor

The Global Ocean
▶ Floor of the Ocean

If all water were drained from the ocean basins, a great variety of features would be seen, including broad volcanic peaks, deep trenches, extensive plains, linear mountain chains, and large plateaus. In fact, the scenery would be nearly as diverse as that on the continents.

Mapping the Seafloor

The complex nature of ocean-floor topography did not unfold until the historic 3-1/2 year voyage of the HMS *Challenger* (Figure 13.3). From December 1872 to May 1876, the *Challenger* expedition made the first comprehensive study of the global ocean ever attempted. During the 127,500-kilometer (79,200-mile) voyage, the ship and its crew of scientists traveled to every ocean except the Arctic. Throughout the voyage, they sampled a multitude of ocean properties, including water depth, which was accomplished by laboriously lowering long weighted lines overboard. The knowledge gained by the *Challenger* of the ocean's great depth and varied topography expanded with the laying of transatlantic telegraph cables. A far better understanding of the seafloor emerged with the development of modern instruments that measure ocean depths. **Bathymetry** (*bathos* = depth, *metry* = measurement) is the measurement of ocean depths and the charting of the shape or topography of the ocean floor.

Modern Bathymetric Techniques   Today, sound energy is used to measure water depths. The basic approach employs **sonar**, an acronym for *so*und *na*vigation and *r*anging. The first

devices that used sound to measure water depth, called **echo sounders**, were developed early in the 20th century. Echo sounders work by transmitting a sound wave (called a *ping*) into the water in order to produce an echo when it bounces off any object, such as a large marine organism or the ocean floor (Figure 13.4A). A sensitive receiver intercepts the echo reflected from the bottom, and a clock precisely measures the travel time to fractions of a second. By knowing the speed of sound waves in water—about 1,500 meters (4,900 feet) per second—and the time required for the energy pulse to reach the ocean floor and return, depth can be calculated. Depths determined from continuous monitoring of these echoes are plotted to create a profile of the ocean floor. By laboriously combining profiles, a chart of the seafloor can be produced.

Following World War II, the U.S. Navy developed *sidescan sonar* to look for explosive devices that had been deployed in shipping lanes. These torpedo-shaped instruments, can be towed behind a ship where they send out a fan of sound extending to either side of the ship's track. By combining swaths of sidescan sonar data, researchers produced the first photograph-like images of the seafloor. Although sidescan sonar provides valuable views of the seafloor, it does not provide bathymetric (water depth) data.

This drawback was resolved in the 1990s with the development of *high-resolution multibeam sonar* instruments. These systems use hull-mounted sound sources that send out a fan of sound, then record reflections from the seafloor through a set of narrowly focused receivers aimed at different angles (Figure 13.4B). Rather than obtaining the depth of a single point every few seconds, this technique makes it possible for a survey ship to map the features of the ocean floor along a strip tens of kilometers wide. These systems can collect bathymetric data of such high resolution that they can distinguish depths that differ by less than a meter (Figure 13.5). When multibeam sonar is used to make a map of a section of seafloor, the ship travels through the area in a regularly spaced back-and-forth pattern known as "mowing the lawn."

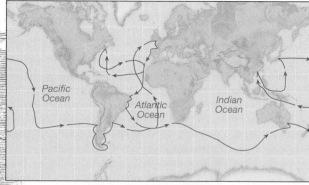

FIGURE 13.3 The first systematic bathymetric measurements of the ocean were made aboard the HMS *Challenger* during its historic 3-1/2 year voyage. Inset shows the route of the HMS *Challenger*, which departed England in December 1872 and returned in May 1876. (From C. W. Thompson and Sir John Murray, *Report on the Scientific Results of the Voyage of the* HMS *Challenger*, Vol. 1, Great Britain: Challenger Office, 1895, Plate 1. Library of Congress.)

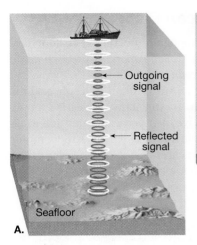 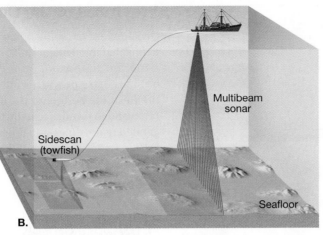

FIGURE 13.4 Types of sonar. **A.** An echo sounder determines the water depth by measuring the time interval required for an acoustic wave to travel from a ship to the seafloor and back. The speed of sound in water is 1,500 m/sec. Therefore, depth = $\frac{1}{2}$ (1,500 m/sec × echo travel time). **B.** Modern multibeam sonar and sidescan sonar obtain an "image" of a narrow swath of seafloor every few seconds.

Despite their greater efficiency and enhanced detail, research vessels equipped with multibeam sonar travel at a mere 10–20 kilometers (6–12 miles) per hour. It would take at least 100 vessels outfitted with this equipment hundreds of years to map the entire seafloor. This explains why only about 5 percent of the seafloor has been mapped in detail—and why large portions of the seafloor have not yet been mapped with sonar at all.

Viewing the Ocean Floor from Space   Another technological breakthrough that has led to an enhanced understanding of the seafloor involves measuring the shape of the ocean surface from space. After compensating for waves, tides, currents, and atmospheric effects, it was discovered that the ocean surface is not perfectly flat because gravity attracts water toward regions where massive seafloor features occur. Therefore, mountains and ridges produce elevated areas on the ocean surface, and,

FIGURE 13.5 Color-enhanced perspective map of the seafloor and coastal landforms in the Los Angeles area of California. The ocean floor portion of this map was constructed from data collected using a high-resolution mapping system. (U.S. Geological Survey)

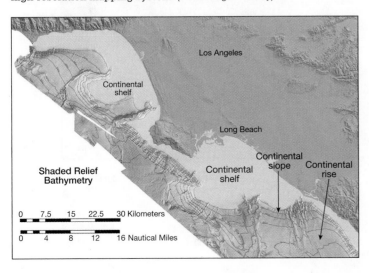

conversely, canyons and trenches cause slight depressions. In essence, subtle ocean surface irregularities mimic the shape of the underlying seafloor. Satellites equipped with *radar altimeters* are able to measure these subtle differences by bouncing microwaves off the sea surface (Figure 13.6). These devices can measure variations as small as a few centimeters. Such data have added greatly to the knowledge of ocean-floor topography. Combined with traditional sonar depth measurements, the data are used to produce detailed ocean-floor maps, such as the one shown in Figure 1.18 (pp. 20–21).

Seismic Reflection Profiles

Marine geologists are also interested in viewing the rock structure beneath the sediments that blanket much of the seafloor. This can be accomplished by making a **seismic reflection profile**. To construct such a profile, strong low-frequency sounds are produced by explosions (depth charges) or air guns. These sound waves penetrate beneath the seafloor and reflect off the contacts

FIGURE 13.6 A satellite altimeter measures the variation in sea surface elevation, which is caused by gravitational attraction and mimics the shape of the seafloor. The sea surface anomaly is the difference between the measured and theoretical ocean surface.

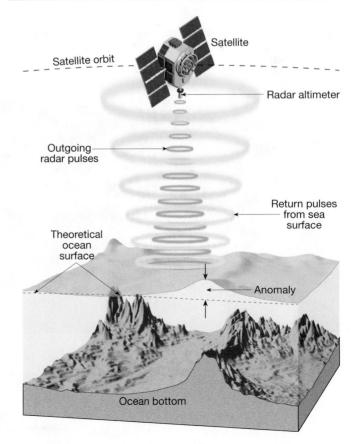

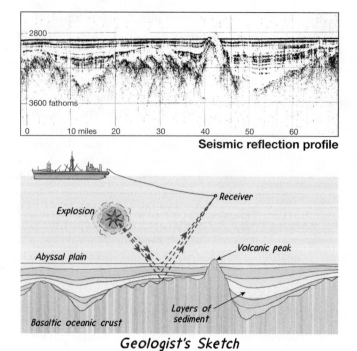

Seismic reflection profile

Geologist's Sketch

FIGURE 13.7 Seismic cross section and matching sketch across a portion of the Madeira abyssal plain in the eastern Atlantic Ocean, showing the irregular oceanic crust buried by sediments. (Image courtesy of Charles Hollister, Woods Hole Oceanographic Institution)

between rock layers and fault zones. **Figure 13.7** shows a seismic profile of a portion of the Madeira abyssal plain in the eastern Atlantic. Although the seafloor is flat, notice the irregular ocean crust buried by a thick accumulation of sediments.

Provinces of the Ocean Floor

Oceanographers studying the topography of the ocean floor have delineated three major units: *continental margins*, the *deep-ocean basin*, and the *oceanic (mid-ocean) ridge*. The map in **Figure 13.8** outlines these provinces for the North Atlantic Ocean, and the profile at the bottom of the illustration shows the varied topography. Such profiles usually have their vertical dimension exaggerated many times—40 times in this case—to make topographic features more conspicuous. Vertical exaggeration, however, makes slopes shown in seafloor profiles appear to be *much* steeper than they actually are.

CONCEPT CHECK 13.2

1 What is *bathymetry*?

2 Describe how satellites can be used to map the seafloor without being able to directly observe it.

3 What are the three major provinces of the ocean floor?

Continental Margins

The Global Ocean
▶ Floor of the Ocean

Two types of **continental margins** have been identified: *passive* and *active*. Passive margins are found along most of the coastal areas that surround the Atlantic Ocean, including the east coasts of North and South America, as well as the coastal areas of western Europe and Africa. Passive margins are *not* associated with

FIGURE 13.8 Map view (*above*) and corresponding profile view (*below*) showing the major topographic divisions of the North Atlantic Ocean. On the profile, the vertical scale has been expanded (exaggerated) by 40 times to make topographic features more conspicuous.

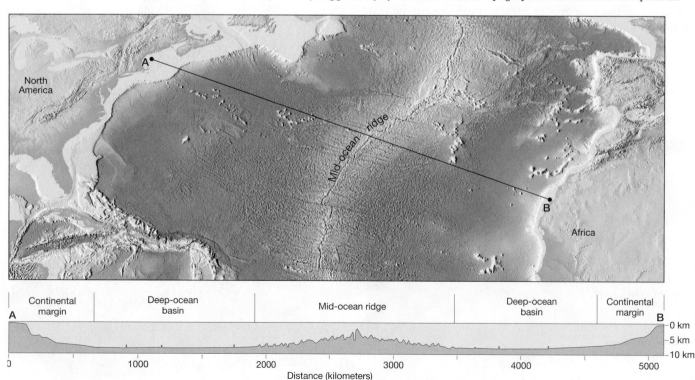

plate boundaries and therefore experience very little volcanism and few earthquakes. Here, weathered materials eroded from the adjacent landmass accumulate to form a thick, broad wedge of relatively undisturbed sediments.

By contrast, active continental margins occur where oceanic lithosphere is being subducted beneath the edge of a continent. The result is a relatively narrow margin, consisting of highly deformed sediments that were scraped from the descending lithospheric slab and plastered against the margin of the over-riding continent. Active continental margins are common around the Pacific Rim, where they parallel deep oceanic trenches.

Passive Continental Margins

The features comprising a **passive continental margin** include the continental shelf, the continental slope, and the continental rise (**Figure 13.9**).

Continental Shelf   The **continental shelf** is a gently sloping, submerged surface extending from the shoreline toward the deep-ocean basin. Because it is underlain by continental crust, it is clearly a flooded extension of the continents.

The continental shelf varies greatly in width. Although almost nonexistent along some continents, the shelf may extend seaward as far as 1,500 kilometers (930 miles) along others. On average, the continental shelf is about 80 kilometers (50 miles) wide and 130 meters (425 feet) deep at its seaward edge. The average incli-nation of the continental shelf is only about one-tenth of 1 degree, a drop of only about 2 meters per kilometer (10 feet per mile). In fact, the slope is so gentle that it would appear to an observer to be a horizontal surface.

Although continental shelves represent only 7.5 percent of the total ocean area, they have economic and political signifi-cance because they contain important mineral deposits, includ-ing large reservoirs of oil and natural gas, as well as huge sand and gravel deposits. The waters of the continental shelf also contain many important fishing grounds, which are significant sources of food.

The continental shelf tends to be relatively featureless; how-ever, some areas are mantled by extensive glacial deposits and thus are quite rugged. In addition, other continental shelves are dissected by large valleys extending from the coastline into deeper waters. Many of these *shelf valleys* are the seaward extensions of river valleys on the adjacent landmass. They were excavated during the Pleistocene epoch (Ice Age) when enormous quan-tities of water were stored in vast ice sheets on the continents. This caused sea level to drop by 100 meters (330 feet) or more, exposing large areas of the continental shelves (see Figure 6.21, p. 172). Because of this drop in sea level, rivers extended their courses, and land-dwelling plants and animals inhabited the newly exposed portions of the continents.

Most continental shelves associated with passive margins, such as along the East Coast of the United States, consist of shallow-water deposits that can reach several kilometers in thickness. Such deposits have led researchers to conclude that these thick accu-mulations of sediment are produced along a gradually subsiding continental margin.

Continental Slope   Marking the seaward edge of the conti-nental shelf is the **continental slope**, a relatively steep zone (as compared with the shelf) that marks the boundary between con-tinental crust and oceanic crust (Figure 13.9). Although the incli-nation of the continental slope varies greatly from place to place, it averages about 5 degrees and in places may exceed 25 degrees. Furthermore, the continental slope is a relatively narrow feature, averaging only about 20 kilometers (12 miles) in width.

Continental Rise   In regions where trenches do not exist, the steep continental slope merges into a more gradual incline known as the **continental rise** where the slope drops to about one-third

FIGURE 13.9 Features of a passive continental margin, including the continental shelf, continental slope, and continental rise. The Atlantic coast of North America is a good example of a passive margin. Note that the steepness of the slopes shown for the continental shelf and continental slope are greatly exaggerated. The continental shelf has an average slope of one-tenth of 1 degree, whereas the continental slope has an average slope of about 5 degrees.

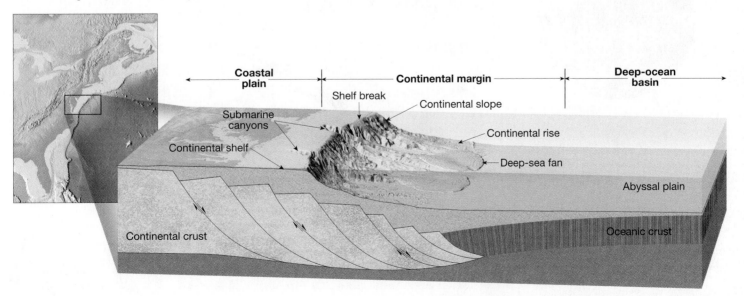

degree, or about 6 meters per kilometer (32 feet per mile). Whereas the width of the continental slope averages about 20 kilometers (12 miles), the continental rise may extend for hundreds of kilometers into the deep-ocean basin.

The continental rise consists of a thick accumulation of sediment that moved downslope from the continental shelf to the deep-ocean floor. The sediments are delivered to the base of the continental slope by turbidity currents that periodically flow down submarine canyons. When these dense muddy currents emerge from the mouth of a canyon onto the relatively flat ocean floor, they deposit sediment that forms a **deep-sea fan** (Figure 13.9). As fans from adjacent submarine canyons grow, they merge laterally with one another to produce a continuous covering of sediment at the base of the continental slope that we call the continental rise.

Submarine Canyons and Turbidity Currents   Deep, steep-sided valleys known as **submarine canyons** are cut into the continental slope and may extend across the entire continental rise to the deep-ocean basin (Figure 13.10). Although some of these canyons appear to be the seaward extensions of river valleys many others do not line up in this manner. Furthermore, submarine canyons extend far below the maximum lowering of sea level during the Ice Age, so we cannot attribute their formation to stream erosion.

These submarine canyons have probably been excavated by turbidity currents (Figure 13.10). **Turbidity currents** are downslope movements of dense, sediment-laden water. They are created when sand and mud on the continental shelf and slope are dislodged and thrown into suspension. Because such mud-choked water is denser than normal seawater, it flows downslope, eroding

and accumulating more sediment as it goes. The erosional work accomplished by these muddy torrents is thought to be the major force in the excavation of most submarine canyons.

Narrow continental margins, such as the one located along the California coast, are dissected by numerous submarine canyons. Here erosion has extended many of these canyons landward into shallow water. Sediments carried to the coasts by rivers are transported along the shore by wave activity until they reach a submarine canyon. This steady supply of sediment collects until it becomes unstable and moves as a massive underwater landslide (turbidity current) to the floor of the deep-ocean basin.

Turbidity currents eventually lose momentum and come to rest along the floor of the ocean basin. As these currents slow, the suspended sediments begin to settle out. First, the coarser, heavier sand is dropped, followed by successively finer deposits of silt and then clay. These deposits, called *turbidites*, are characterized by a decrease in sediment grain size from bottom to top, a phenomenon known as *graded bedding* (Figure 13.10).

Turbidity currents are an important mechanism of sediment transport in the ocean. By the action of turbidity currents, submarine canyons are eroded and sediments are deposited on the deep-ocean floor.

Active Continental Margins

Along active continental margins the continental shelf is very narrow, if it exists at all, and the continental slope descends abruptly into a deep-ocean trench. In these settings, the landward wall of a trench and the continental slope are essentially the same feature.

FIGURE 13.10 Turbidity currents are downslope movements of dense, sediment-laden water. They are created when sand and mud on the continental shelf and slope are dislodged and thrown into suspension. Because such mud-choked water is denser than normal seawater, it hugs the seafloor as it flows downslope, eroding and accumulating more sediment. Beds deposited by these currents are called *turbidites*. Each event produces a single bed characterized by a decrease in sediment size from bottom to top, a feature known as a *graded bed*. (Photo by Marli Miller)

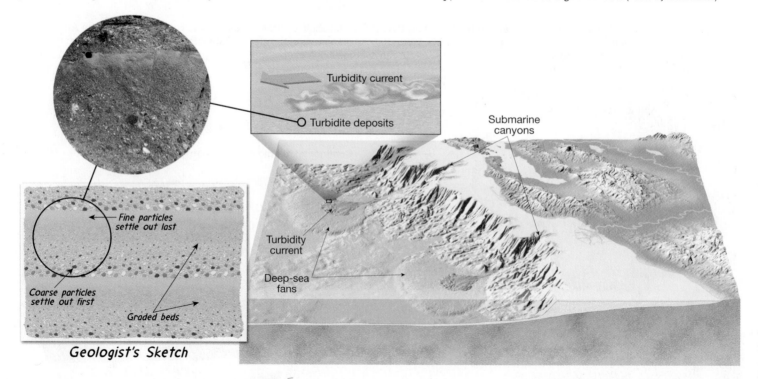

Turbidity current

Turbidite deposits

Submarine canyons

Turbidity current

Deep-sea fans

Fine particles settle out last

Coarse particles settle out first

Graded beds

Geologist's Sketch

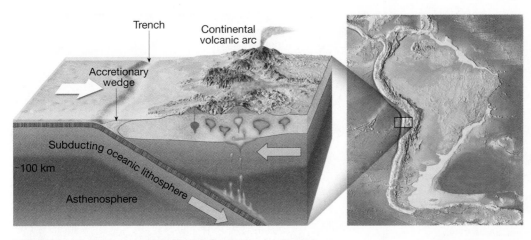

FIGURE 13.11 Active continental margin. Sediments from the ocean floor are scraped from the descending plate and added to the continental crust as an *accretionary wedge*. The Pacific margin of South America is an excellent example.

Active continental margins are located primarily around the Pacific Ocean in areas where oceanic lithosphere is being subducted beneath the leading edge of a continent (**Figure 13.11**). In such settings, sediments from the ocean floor and pieces of oceanic crust are scraped from the descending oceanic plate and plastered against the edge of the overriding continent. This chaotic accumulation of deformed sediment and scraps of oceanic crust is called an *accretionary wedge*. Prolonged plate subduction can produce massive accumulations of sediment along active continental margins.

Some active margins have little or no sediment accumulation, indicating that material is being carried into the mantle with the subducting plate. In these locations the continental margin is very narrow, as the trench may lie a mere 50 kilometers (31 miles) offshore.

CONCEPT CHECK 13.3

1. List the three major features that comprise a passive continental margin. Which one is considered a flooded extension of the continent? Which has the steepest slope?
2. Contrast active and passive continental margins. Provide a specific geographic example of each.
3. What is a *turbidity current*? What is a *turbidite*? What is meant by the term *graded bedding*?

The Deep-Ocean Basin

The Global Ocean
▶ Floor of the Ocean

Between the continental margin and the oceanic ridge lies the **deep-ocean basin** (see Figure 13.8). The size of this region—almost 30 percent of Earth's surface—is roughly comparable to the percentage of land that presently projects above sea level. This region includes remarkably flat areas known as *abyssal*

plains; tall volcanic peaks called *seamounts* and *guyots*; *deep-ocean trenches*, which are extremely deep linear depressions in the ocean floor; and extensive areas of lava flows piled one atop the other called *oceanic plateaus*.

Deep-Ocean Trenches

Deep-ocean trenches are long, relatively narrow troughs that are the deepest parts of the ocean. Most trenches are located along the margins of the Pacific Ocean (**Figure 13.12**) where many exceed 10,000 meters (33,000 feet) in depth. A portion of one—the Challenger Deep in the Mariana Trench—has been measured at a record 11,022 meters (36,163 feet) below sea level, making it the deepest known part of the world ocean. Only two trenches are located in the Atlantic: the Puerto Rico Trench and the South Sandwich Trench.

Although deep-ocean trenches represent only a very small portion of the area of the ocean floor, they are nevertheless significant geologic features. Trenches are sites of plate convergence where slabs of oceanic lithosphere subduct and plunge back into the mantle. In addition to earthquakes being created as one plate "scrapes" beneath another, volcanic activity is also associated with these regions. Thus, trenches are often paralleled by an arc-shaped row of active volcanoes called a **volcanic island arc**. Furthermore, **continental volcanic arcs**, such as those making up portions of the Andes and Cascades, are located parallel to trenches that lie adjacent to continental margins (see Figure 13.11). The volcanic activity associated with the trenches that surround the Pacific Ocean explains why the region is called the *Ring of Fire*.

Abyssal Plains

Abyssal (*a* = without, *byssus* = bottom) **plains** are deep, incredibly flat features; in fact, these regions are likely the most level places on Earth. The abyssal plain found off the coast of Argentina, for example, has less than 3 meters (10 feet) of relief over a distance exceeding 1,300 kilometers (800 miles). The monotonous topography of abyssal plains is occasionally interrupted by the protruding summit of a partially buried volcanic peak.

Using *seismic profilers*, instruments that generate signals designed to penetrate far below the ocean floor, researchers have determined that abyssal plains owe their relatively featureless topography to thick accumulations of sediment that have buried an otherwise rugged ocean floor (see Figure 13.7). The nature of the sediment indicates that these plains consist primarily of sediments transported far out to sea by turbidity currents.

Abyssal plains occur in all of the oceans. However, the Atlantic Ocean has the most extensive abyssal plains because it has few trenches to act as traps for sediment carried down the continental slope.

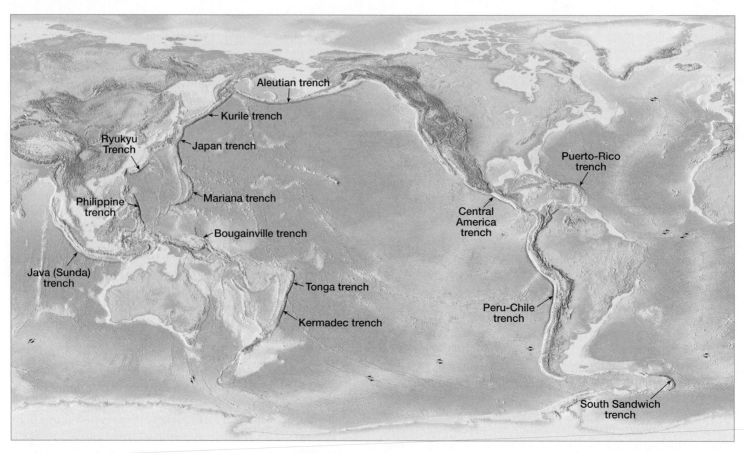

FIGURE 13.12 Distribution of the world's deep-ocean trenches and oceanic ridges.

Seamounts, Guyots, and Oceanic Plateaus

Dotting the ocean floor are submarine volcanoes called **seamounts**, which may rise hundreds of meters above the surrounding topography. It is estimated that more than a million exist. Some grow large enough to become oceanic islands, but most do not have a sufficiently long eruptive history to build a structure above sea level. Although seamounts are found on the floors of all the oceans, they are most common in the Pacific.

Some, like the Hawaiian Island-Emperor Seamount chain that stretches from the Hawaiian Islands to the Aleutian trench, form over volcanic hot spots in association with mantle plumes (see Figure 7.24, p. 213). Others are born near oceanic ridges. If the volcano is large enough before it is carried from the magma source by plate movement, the structure may emerge as an island. Examples in the Atlantic include Azores, Ascension, Tristan da Cunha, and St. Helena.

During the time they exist as islands, some of these volcanic structures are lowered to near sea level by the forces of weathering and erosion. In addition, islands gradually sink and disappear below the water surface as the moving plate slowly carries them away from the elevated oceanic ridge or hot-spot where they originated (Box 13.1). Submerged, flat topped seamounts that formed in this manner are called **guyots**.[28]

The ocean floor also contains several massive **oceanic plateaus**, which resemble flood basalt provinces on the continents. Oceanic plateaus, which in some cases are more than 30 kilometers thick, were generated from vast outpourings of fluid basaltic lavas. Some oceanic plateaus appear to have formed

Students Sometimes Ask...

Have humans ever explored the deepest ocean trenches? Could anything live there?

Humans have indeed visited the deepest part of the oceans—where there is crushing high pressure, complete darkness, and near-freezing water temperatures—more than 50 years ago! In January 1960, U.S. Navy Lt. Don Walsh and explorer Jacques Piccard descended to the bottom of the Challenger Deep region of the Mariana Trench in the deep-diving bathyscaphe *Trieste*. At 9,906 meters (32,500 feet), the men heard a loud cracking sound that shook the cabin. They were unable to see that a 7.6-centimeter (3-inch) Plexi-glas viewing port had cracked (miraculously, it held for the rest of the dive). More than 5 hours after leaving the surface, they reached the bottom at 10,912 meters (35,800 feet)—a record depth of human descent that has not been broken since. They did see some life-forms that are adapted to life in the deep: a small flatfish, a shrimp, and some jellyfish.

[28]"The term *guyot* is named after Princeton University's first geology professor. It is pronounced "GEE-oh" with a hard g as in "give.""

Box 13.1

UNDERSTANDING EARTH

Explaining Coral Atolls—Darwin's Hypothesis

Coral *atolls* are ring-shaped structures that often extend from slightly above sea level to depths of several thousand meters (**Figure 13.A**). What causes atolls to form, and how do they attain such thicknesses?

Corals are tiny animals that generally appear in large numbers, that link together to form colonies. Most corals create a hard external skeleton made of calcium carbonate. Some build large calcium carbonate structures, called *reefs*, where new colonies grow atop the strong skeletons of previous colonies. Sponges and algae also attach to the reef, enlarging it further.

Reef-building corals grow best in waters with an average annual temperature of about 24° C (75° F). They cannot survive prolonged exposure to temperatures below 18° C (64° F) or above 30° C (86° F). In addition, reef-builders require clear, sunlit water. Consequently, the depth of most active reef growth is limited to no more than about 45 meters (150 feet).

The environmental conditions required for coral growth create an interesting paradox: How can corals—which require warm, shallow, sunlit water no deeper than a few dozen meters—create thick structures such as coral atolls that extend to great depths?

Naturalist Charles Darwin was one of the first to formulate a hypothesis on the origin of ringed-shaped atolls. From 1831 to 1836 he sailed aboard the British ship HMS *Beagle* during its famous global circumnavigation. In various places Darwin noticed a progression of stages in coral reef development from (1) a *fringing reef* along the margins of a volcano to (2) a *barrier reef* with a volcano in the middle to (3) an *atoll*, consisting of a continuous or broken ring of coral reef surrounding a central lagoon (**Figure 13.B**). The essence of Darwin's hypothesis, illustrated in Figure 13.B, was that as a volcanic island slowly sinks, corals continue to build the reef complex upward.

FIGURE 13.A An aerial view of Tetiaroa Atoll in the Pacific. The light blue waters of the relatively shallow lagoon contrast with the dark blue color of the ocean surrounding the atoll. (Photo by Douglas Peebles Photography/Alamy)

During Darwin's time, however, there was no plausible mechanism to account for how an island might sink.

Currently, plate tectonics helps explain how volcanic islands become extinct and sink to great depths over long periods of time. Some volcanic islands form over a relatively stationary mantle plume, which causes the lithosphere to be buoyantly uplifted. Over a span of millions of years, these volcanic islands become inactive and gradually sink as the moving plate carries them away from the region of hot-spot volcanism (Figure 13.B).

FIGURE 13.B Formation of a coral atoll due to the gradual sinking of oceanic crust and upward growth of the coral reef. A fringing coral reef forms around an active volcanic island. As the volcanic island moves away from the region of hot-spot activity it sinks, and the fringing reef gradually becomes a barrier reef farther from shore. Eventually, the volcano is completely submerged and an atoll remains.

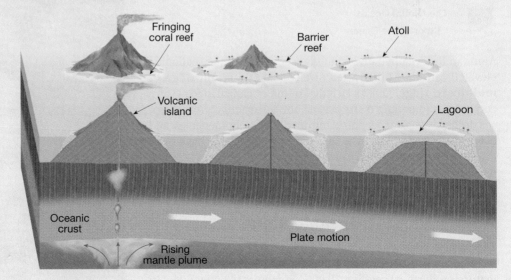

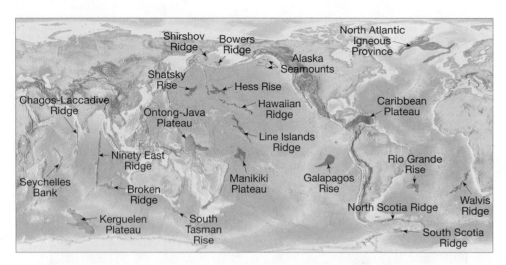

FIGURE 13.13 Distribution of oceanic plateaus and other submerged crustal fragments.

quickly in geologic terms. Examples include the Ontong Java, which formed in less than 3 million years, and the Kerguelen Plateau in 4.5 million years (Figure 13.13).

CONCEPT CHECK 13.4

1. Briefly describe the size and extent of the oceanic ridge system. With which type of plate boundary is it associated?
2. How do oceanic ridges differ from most mountain ranges on the continents?
3. Why are oceanic ridges elevated?

The Oceanic Ridge

GEODe The Global Ocean
EARTH SCIENCE ▶ Floor of the Ocean

Along well-developed divergent plate boundaries, the seafloor is elevated, forming a broad linear swell called the **oceanic ridge**, or **mid-ocean ridge**. Our knowledge of the oceanic ridge system comes from soundings of the ocean floor, core samples from deep-sea drilling, visual inspection using deep-diving submersibles, and even firsthand inspection of slices of ocean floor that have been thrust onto dry land during continental collisions (Figure 13.14). At oceanic ridges we find extensive normal and strike-slip faulting, earthquakes, high heat flow, and volcanism.

Anatomy of the Oceanic Ridge

The oceanic ridge system winds through all major oceans in a manner similar to the seam on a baseball, and is the longest topographic feature on Earth, exceeding 70,000 kilometers (43,000 miles) in length (Figure 13.15). The crest of the ridge typically stands 2–3 kilometers above the adjacent deep-ocean basins and marks the plate boundary where new oceanic crust is created.

Notice in Figure 13.15 that large sections of the oceanic ridge system have been named based on their locations within the

various ocean basins. Some ridges run through the middle of ocean basins, where they are appropriately called *mid-ocean* ridges. The Mid-Atlantic Ridge and the Mid-Indian Ridge are examples. By contrast, the East Pacific Rise is *not* a "mid-ocean" feature. Rather, as its name implies, it is located in the eastern Pacific, far from the center of the ocean.

The term *ridge* is somewhat misleading, because these features are not narrow and steep as the term implies, but have widths of from 1,000 to 4,000 kilometers and the appearance of broad, elongated swells that exhibit varying degrees of ruggedness. Furthermore, the ridge system is broken into segments that range from a few tens to hundreds of kilometers in length. Each segment is offset from the adjacent segment by a transform fault.

FIGURE 13.14 The deep-diving submersible *Alvin* is 7.6 meters long, weighs 16 tons, has a cruising speed of 1 knot, and can reach depths as great as 4,000 meters (2.5 miles). A pilot and two scientific observers are along during a normal 6- to 10-hour dive. (Courtesy of Rod Catanach/Woods Hole Oceanographic Institution)

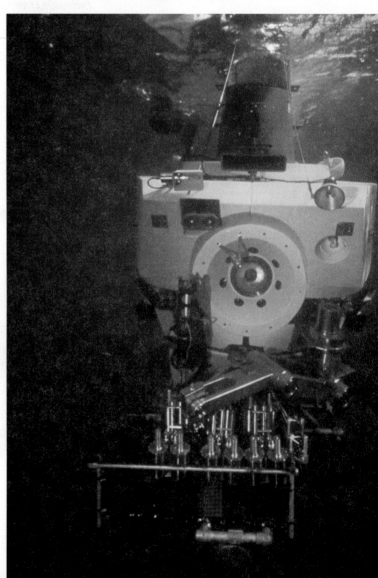

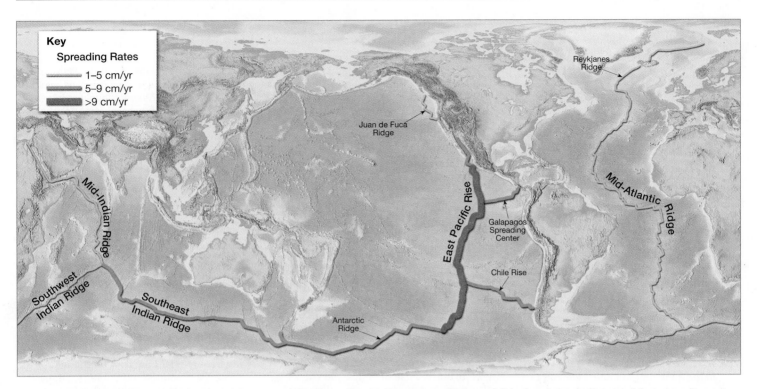

FIGURE 13.15 Distribution of the oceanic ridge system. The map shows ridge segments that exhibit slow, intermediate, and fast spreading rates.

Oceanic ridges are as high as some mountains on the continents; but the similarities end there. Whereas most mountain ranges on land form when the compressional forces associated with continental collisions fold and metamorphose thick sequences of sedimentary rocks, oceanic ridges form where upwelling from the mantle generates new oceanic crust. Oceanic ridges consist of layers and piles of newly formed basaltic rocks that are buoyantly uplifted by the hot mantle rocks from which they formed.

Along the axis of some segments of the oceanic ridge system are deep, down-faulted structures called **rift valleys** because of their striking similarity to the continental rift valleys found in East Africa (**Figure 13.16**). Some rift valleys, including those along the rugged Mid-Atlantic Ridge, are typically 30–50 kilometers wide and have walls that tower 500–2,500 meters above the valley floor. This makes them comparable to the deepest and widest part of Arizona's Grand Canyon.

Why Is the Oceanic Ridge Elevated?

The primary reason for the elevated position of the ridge system is that newly created oceanic lithosphere is hot and therefore less dense than cooler rocks of the deep-ocean basin. As the newly formed basaltic crust travels away from the ridge crest, it is cooled from above as seawater circulates through the pore spaces and fractures in the rock. In addition, it cools because it gets farther and farther from the zone of hot mantle upwelling. As a result, the lithosphere gradually cools, contracts, and becomes denser. This thermal contraction accounts for the greater ocean depths that occur away from the ridge.

It takes almost 80 million years of cooling and contraction for rock that was once part of an elevated ocean-ridge system to relocate to the deep-ocean basin.

FIGURE 13.16 The axis of some segments of the oceanic ridge system contain deep downfaulted structures called *rift valleys* that may exceed 30–50 kilometers in width and from 500 to 2,500 meters in depth.

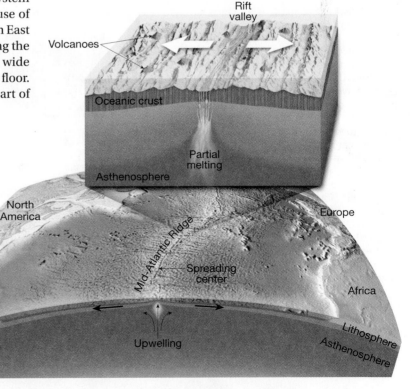

PROFESSIONAL PROFILE

Susan DeBari: A Career in Geology

I discovered geology the summer I worked doing trail maintenance in the North Cascade mountains of Washington State. I had just finished my freshman year in college and had never before studied Earth science. But a coworker (now my best friend) began to describe the geological features of the mountains that we were hiking in—the classic cone shape of Mount Baker volcano, the U-shaped glacial valleys, the advance of active glaciers, and other wonders. I was hooked and went back to college that fall with a passion for geology that hasn't abated. As an undergraduate, I worked as a field assistant to a graduate student and did a senior thesis project on rocks from the Aleutian island arc.

> **"I was hooked and went back to college that fall with a passion for geology that hasn't abated."**

From that initial spark, island arcs have remained my top research interest, on through Ph.D. research at Stanford University, postdoctoral work at the University of Hawaii, and as a faculty member at San Jose State University and Western Washington University. Of most interest was the deep crust of arcs, the material that lies close to the Mohorovičić discontinuity (fondly known as the Moho). What kinds of processes are occurring down there at the base of the crust in island arcs? What is the source of magmas that make their way to the surface—the mantle, or the deep crust itself? How do these magmas interact with the crust as they make their way upward? What do these magmas look like chemically? Are they very different from what erupts at the surface? Obviously, geologists cannot go down to the base of the crust (typically 20 to 40 kilometers beneath Earth's surface). So what they do

is play a bit of a detective game. They must use rocks that are *now exposed at the surface* that were originally formed in the deep crust of an island arc. The rocks must have been brought to the surface rapidly along fault zones to preserve their original features. Thus, I can walk on rocks of the deep crust without really leaving Earth's surface! There are a few places around the world where these rare rocks are exposed.

> **"Each dive lasted nine hours, and was spent in a space no bigger than the front seat of a Honda, shared with two Japanese pilots . . . "**

Some of the places that I have worked are the Chugach Mountains of Alaska, the Sierras Pampeanas of Argentina, the Karakorum Range in Pakistan, Vancouver Island's west coast, and the North Cascades of Washington. Fieldwork has most commonly involved hiking on foot, along with extensive use of mules and trucks. I also went looking for exposed pieces of the deep crust of island arcs in a less obvious place—the Izu Bonin trench, one of the deepest oceanic trenches of the world. Here I dove into the ocean in a submersible called the *Shinkai 6500* (pictured to my right in the background). The *Shinkai 6500* is a Japanese submersible that has the capability to dive to 6500 meters below the surface of the ocean (approximately 4 miles). My plan was to take rock samples from the wall of the trench at its deepest levels using the submersible's mechanical arm. Because preliminary data suggested that vast amounts of rock were exposed for several kilometers in a vertical sense, this could be a great way to sample the deep arc basement. I dove in the submersible three times, reaching a maximum depth of 6497 meters. Each dive lasted nine hours, and was spent in a space no bigger than the front seat of a Honda, shared with two Japanese pilots who controlled the submersible's movements. It was an exhilarating experience! I am now on the faculty at Western Washington University, where I continue to do research on the deep roots of volcanic arcs, and get students involved as well. I am also involved in science education training for K–12 teachers, hoping to get young people motivated to ask questions about the fascinating world that surrounds them!

—Susan DeBari

Susan DeBari photographed with the Japanese submersible, *Shinkai 6500*, which she used to collect rock samples from the Izu Bonin trench. (Photo courtesy of Susan DeBari)

Seafloor Sediments

Most of the ocean floor is covered with a blanket of sediment. Part of this material has been deposited by turbidity currents, and the rest has slowly settled onto the seafloor from above. The thickness of this carpet of debris varies greatly. In some trenches, which act as traps for sediment originating on the continental margin, accumulations may approach 10 kilometers (6 miles) in thickness. In general, however, sediment accumulations are considerably less. In the Pacific Ocean, for example, sediment thickness is about 600 meters (2,000 feet) or less, whereas on the floor of the Atlantic, the thickness varies from about 500 to 1,000 meters (1,500–3,000 feet).

Types of Seafloor Sediments

Seafloor sediments can be classified according to their origin into three broad categories: (1) **terrigenous** (*terra* = land, *generare* = to produce); (2) **biogenous** (*bio* = life, *generare* = to produce); and (3) **hydrogenous** (*hydro* = water, *generare* = to produce). Although each category is discussed separately, remember that all seafloor sediments are mixtures. No mass of sediment comes from a single source.

Terrigenous Sediment Terrigenous sediment consists primarily of mineral grains that were weathered from continental rocks and transported to the ocean. Larger particles (gravel and sand) usually settle rapidly near shore, whereas finer particles (microscopic clay-size particles) can take years to settle to the ocean floor and may be carried thousands of kilometers by ocean currents or transported far out to sea by the wind. As a consequence, virtually every part of the ocean receives some terrigenous sediment. The rate at which this sediment accumulates on the deep-ocean floor, however, is very slow. To form a 1-centimeter (0.4-inch) *abyssal clay* layer, for example, requires as much as 50,000 years. Conversely, on the continental margins near the mouths of large rivers, terrigenous sediment accumulates rapidly and forms thick deposits.

Biogenous Sediment Biogenous sediment consists of shells and skeletons of marine animals and algae (**Figure 13.17**). This debris is produced mostly by microscopic organisms living in sunlit waters near the ocean surface. Once these organisms die, their hard *tests* (*testa* = shell) constantly "rain" down and accumulate on the seafloor.

The most common biogenous sediment is *calcareous* ($CaCO_3$) *ooze*, which, as its name implies, has the consistency of thick mud. This sediment is produced from the tests of organisms such as *coccolithophores* (single-celled algae) and *foraminifera* (small organisms also called *forams*, for short) that inhabit warm surface waters. When calcareous tests slowly sink into deeper parts of the ocean, they begin to dissolve. This occurs because the deeper, cold seawater is richer in carbon dioxide and is thus more acidic than warm water. In seawater deeper than about 4,500 meters (15,000 feet), calcareous tests will completely dissolve before they reach bottom. Consequently, calcareous ooze does not accumulate at these greater depths.

Other biogenous sediments include *siliceous* (SiO_2) *ooze* and phosphate-rich material. The former is composed primarily of tests

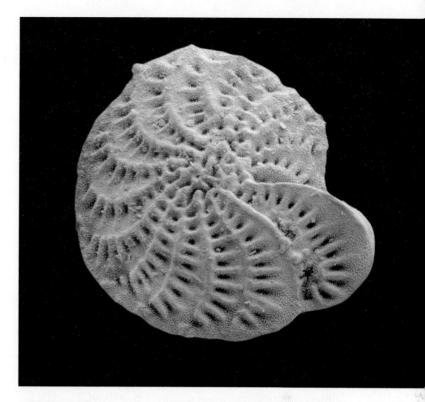

FIGURE 13.17 Image from a scanning electron microscope of a shell (test) of foraminifera, an example of biogenous sediment. These tiny, single-celled organisms are sensitive to even small fluctuations in temperature. Seafloor sediments containing fossils such as this are useful recorders of climate change. (Photo by Andrew Syred/Photo Researchers, Inc.)

of *diatoms* (single-celled algae) and *radiolarians* (single-celled animals) that prefer cooler surface waters, whereas the latter is derived from the bones, teeth, and scales of fish and other marine organisms.

Hydrogenous Sediment Hydrogenous sediment consists of minerals that crystallize directly from seawater through various chemical reactions. Hydrogenous sediments represent a relatively small portion of the overall sediment in the ocean. They do, however, have many different compositions and are distributed in diverse environments of deposition.

Students Sometimes Ask...

Are diatoms an ingredient in diatomaceous earth, which is used in swimming pool filters?

Not only are diatoms used as swimming pool filters, they are also used in a variety of everyday products, including toothpaste (yes, you're brushing your teeth with the remains of dead microscopic organisms!). Diatoms secrete walls of silica in a great variety of forms that accumulate as sediments in enormous quantities. Because it is lightweight, chemically stable, has high surface area, and is highly absorbent, diatomaceous earth has many practical uses. The main uses of diatoms include filters (for refining sugar, straining yeast from beer, and filtering swimming pool water); mild abrasives (in household cleaning and polishing compounds and facial scrubs); and absorbents (for chemical spills).

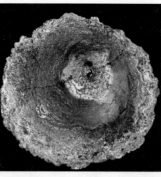

FIGURE 13.18 Manganese nodules, including some that are cut in half, revealing their central nucleation object and layered internal structure. (Left photo by Charles A. Winters/Photo Researchers, Inc. Right photo by J. and L. Weber/Peter Arnold, Inc.)

Some of the most common types of hydrogenous sediment include:

- *Manganese nodules*, which are rounded, hard lumps of manganese, iron, and other metals that precipitate in concentric layers around a central object (such as a volcanic pebble or a grain of sand) (Figure 13.18). The nodules can be up to 20 centimeters (8 inches) in diameter and are often littered across large areas of the deep seafloor.
- *Calcium carbonates*, which form by precipitation directly from seawater in warm climates. If this material is buried and hardened, it forms limestone. Most limestone, however, is composed of biogenous sediment.
- *Metal sulfides*, which are usually precipitated as coatings on rocks near black smokers associated with the crest of the mid-ocean ridge. (see Box 14.2, p. 416). These deposits contain iron, nickel, copper, zinc, silver, and other metals in varying proportions.
- *Evaporites*, which form where evaporation rates are high and there is restricted open-ocean circulation. As water evaporates from such areas, the remaining seawater becomes saturated with dissolved minerals, which then begin to precipitate. Heavier than seawater, they sink to the bottom or form a characteristic white crust of evaporite minerals around the edges of these areas. Collectively termed "salts," some evaporite minerals taste salty, such as *halite* (common table salt, NaCl), and some do not, such as the calcium sulfate minerals *anhydrite* ($CaSO_2$) and *gypsum* ($CaSO_4 \cdot 2\,H_2O$).

Seafloor Sediment—A Storehouse of Climate Data

We know that the parts of the Earth system are linked so that a change in one part can produce changes in any or all of the other parts. In this example you will see how changes in atmospheric and oceanic temperatures are reflected in the nature of life in the sea.

Most seafloor sediments contain the remains of organisms that once lived near the sea surface (the ocean–atmosphere interface). When such near-surface organisms die, their shells slowly settle to the floor of the ocean, where they become part of the sedimentary record. These seafloor sediments are useful recorders of worldwide climate change because the numbers and types of organisms living near the sea surface change with the climate:

> We would expect that in any area of the ocean/atmosphere interface the average annual temperature of the surface water of the ocean would approximate that of the contiguous atmosphere. The temperature equilibrium established between surface seawater and the air above it should mean that . . . changes in climate should be reflected in changes in organisms living near the surface of the deep sea . . . . When we recall that the seafloor sediments in vast areas of the ocean consist mainly of shells of pelagic foraminifers, and that these animals are sensitive to variations in water temperature, the connection between such sediments and climate change becomes obvious.[29]

Thus, in seeking to understand climate change as well as other environmental transformations, scientists are tapping the huge reservoir of data in seafloor sediments (Figure 13.19). The sediment cores gathered by drilling ships and other research vessels have provided invaluable data that have significantly expanded our knowledge and understanding of past climates (see Chapter-opening photo).

One notable example of the importance of seafloor sediments to our understanding of climate change relates to unraveling the fluctuating atmospheric conditions of the Ice Age. The records of temperature changes contained in cores of sediment from the ocean floor have proven critical to our present understanding of this recent span of Earth history.[30]

CONCEPT CHECK 13.5

1 Distinguish among the three basic types of seafloor sediments. Give an example of each.

2 Why are seafloor sediments useful in studying past climates?

[29]Richard F. Flint, *Glacial and Quatenary Geology* (New York: John Wiley & Sons, 1971), p. 718.

[30]For more information on this topic, see "Causes of Glaciation" in Chapter 6, page 174.

FIGURE 13.19 Scientists working with a core of seafloor sediment. Analysis of such cores provides useful data about the geologic past and Earth's climate history. (Photo by Science Source/Photo Researchers, Inc.)

Students Sometimes Ask...

Are there any areas of the ocean floor where no sediment is being deposited?

Actually, there are a few places in the ocean where very little sediment accumulates. One such place is along the continental slope, where there is active erosion by turbidity and other deep-ocean currents. Another place where very little sediment can be found is along the mid-ocean ridge. The seafloor along the crest of the mid-ocean ridge is so young (because of seafloor spreading) and the rates of sediment accumulation far from land are so slow that there hasn't been enough time for sediments to accumulate.

Resources from the Seafloor

The seafloor is rich in mineral and organic resources. Most, however, are not easily accessible, and recovery involves technological challenges and high cost. Nevertheless, certain resources have high value and thus make appealing exploration targets.

Energy Resources

Among the nonliving resources extracted from the oceans, more than 95 percent of the economic value comes from energy products. The main energy products are oil and natural gas, which are currently being extracted, and gas hydrates, which are not yet utilized but have vast potential.

Oil and Natural Gas   The ancient remains of microscopic organisms, buried within marine sediments before they could completely decompose, are the source of today's deposits of oil and natural gas. The percentage of world oil produced from offshore regions has increased from trace amounts in the 1930s to more than 30 percent today. Most of this increase results from continuing technological advancements employed by offshore drilling platforms (Figure 13.20).

Major offshore reserves exist in the Persian Gulf, in the Gulf of Mexico, off the coast of southern California, in the North Sea, and in the East Indies. Additional reserves are probably located off the north coast of Alaska and in the Canadian Arctic, Asian seas, Africa, and Brazil. Because the likelihood of finding major new reserves on land is small, future offshore exploration will continue to be important, especially in deeper waters of the continental margins. A major environmental concern about offshore petroleum exploration is the possibility of oil spills caused by inadvertent leaks or blowouts during the drilling process.

Gas Hydrates   **Gas hydrates** are unusually compact chemical structures made of water and natural gas. The most common type of natural gas is methane, which produces *methane hydrate*. Gas hydrates occur beneath permafrost areas on land and under the ocean floor at depths below 525 meters (1,720 feet).

Most oceanic gas hydrates are created when bacteria break down organic matter trapped in seafloor sediments, producing

FIGURE 13.20 Offshore drilling rigs are used to tap the oil and natural gas reserves of the continental shelf. This platform is in the North Sea. (Photo by Peter Bowater/Photo Researchers, Inc.)

methane gas with minor amounts of ethane and propane. These gases combine with water in deep-ocean sediments (where pressures are high and temperatures are low) in such a way that the gas is trapped inside a latticelike cage of water molecules.

Vessels that have drilled into gas hydrates have retrieved cores of mud mixed with chunks or layers of gas hydrates (Figure 13.21A) that fizzle and evaporate quickly when they are exposed to the relatively warm, low-pressure conditions at the ocean surface. Gas hydrates resemble chunks of ice but ignite when lit by a flame because methane and other flammable gases are released as gas hydrates vaporize (Figure 13.21B).

Some estimates indicate that as much as 20 quadrillion cubic meters (700 quadrillion cubic feet) of methane are locked up in sediments containing gas hydrates. This is equivalent to about *twice* as much carbon as Earth's coal, oil, and conventional gas reserves combined, so gas hydrates would seem to have great potential. However, research indicates that the potential is far more modest than these quantities would suggest. An article that focused on this potential resource states that " . . . all but a few percent of the great vastness of gas hydrates will likely remain beyond reach indefinitely. Most deposits are simply spread too thinly for economical recovery."[31] The article goes on to say that commercial production of gas from hydrates may begin within the next 10 to 15 years but will probably not make a significant contribution for at least 30 years.

[31]Kerr, Richard A., "Gas Hydrate Resource: Smaller But Sooner," *Science* 303 (13 February 2004): 946–947.

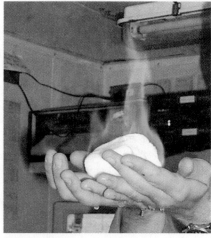

A.

B.

FIGURE 13.21 Gas hydrates. **A.** A sample retrieved from the ocean floor shows layers of white icelike gas hydrate mixed with mud. **B.** Gas hydrates evaporate when exposed to surface conditions and release natural gas, which can be ignited. (Photos courtesy of GEOMAR Research Center, Kiel, Germany)

Other Resources

Other major resources from the seafloor include sand and gravel, evaporative salts, and manganese nodules.

Sand and Gravel   The offshore sand-and-gravel industry is second in economic value only to the petroleum industry. Sand and gravel, which includes rock fragments that are washed out to sea and shells of marine organisms, are mined by offshore barges using suction dredges. Sand and gravel are primarily used as an aggregate in concrete, as a fill material in grading projects, and on recreational beaches.

In some cases, materials of high economic value are associated with offshore sand and gravel deposits. Gem-quality diamonds, for example, are recovered from gravels on the continental shelf offshore of South Africa and Australia. Sediments rich in tin have been mined from some offshore areas of Southeast Asia. Platinum and gold have been found in deposits in gold-mining areas throughout the world, and some Florida beach sands are rich in titanium.

Evaporative Salts   When seawater evaporates, the salts increase in concentration until they can no longer remain dissolved, so they precipitate out of solution and form salt deposits, which can then be harvested. The most economically important salt is *halite* (common table salt). Halite is widely used for seasoning, curing, and preserving foods. It is also used in water conditioners, in agriculture, in the clothing industry for dying fabric, and to de-ice roads. Ever since ancient times, the ocean has been an important source of salt for human consumption, and the sea remains a significant supplier (Figure 13.22).

Manganese Nodules   Manganese nodules contain significant concentrations of manganese, iron, and smaller concentrations of copper, nickel, and cobalt, all of which have a variety of economic uses. Cobalt, for example, is deemed "strategic" (essential to U.S. national security) because it is required to produce dense, strong alloys with other metals and is used in high-speed cutting tools, powerful permanent magnets, and jet engine parts. Technologically, mining the deep-ocean floor for manganese nodules is possible but economically not profitable.

Nodules are widely distributed, but not all regions have the same potential for mining. Good locations have abundant nodules that contain the economically optimum mix of copper, nickel, and cobalt. Sites meeting these criteria, however, are relatively limited. Additionally, there are political problems of establishing mining rights far from land and environmental concerns about disturbing large portions of the deep-ocean floor.

CONCEPT CHECK 13.6

1. Which seafloor resource is presently most valuable?
2. What are gas hydrates? Will they likely be a significant energy source in the next 10 years?
3. Which non energy seafloor resource is most valuable?

FIGURE 13.22 Each year about 30 percent of the world's supply of salt is extracted from seawater. In this process, salt water is held in shallow ponds, while solar energy evaporates the water. The nearly pure salt deposits that eventually form are essentially artificial evaporate deposits. At the southern end of San Francisco Bay, it takes nearly 38,000 liters (10,000 gallons) of water to produce 900 kilograms (1 ton) of salt. (Photo by William E. Townsend Jr./Photo Researchers, Inc.)

GIVE IT SOME THOUGHT

1. Refer to Figure 13.2 to answer the following questions:
   a. Water dominates Earth's surface, but not everywhere. In what Northern Hemisphere latitude belt is there more land than water?
   b. In what latitude belt is there no land at all?
2. Assuming the average speed of sound waves in water is 1,500 meters per second, determine the water depth if a signal sent out by an echo sounder on a research vessel requires 6 seconds to strike bottom and return to the recorder aboard the ship.
3. Refer to the accompanying map showing the Eastern Seaboard of the United States to complete the following:
   a. Which letter is associated with each of the following? Continental shelf; continental slope; and shelf-break.
   b. How does the size of the continental shelf surrounding Florida compare to the size of the Florida peninsula?
   c. Why are there no deep-ocean trenches on this map?

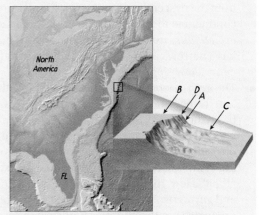

4. Are the continental margins surrounding the Atlantic Ocean primarily active or passive? How about the margins surrounding the Pacific Ocean? Based on your response to the foregoing questions, indicate whether each ocean basin is getting larger, shrinking, or staying the same size. Explain your answer.
5. Examine the accompanying sketch showing three sediment layers on the ocean floor. What term is applied to such layers? What process was responsible for creating these layers? Are these layers more likely part of a deep-sea fan or an accretionary wedge?
6. Imagine that while you and a passenger are in a deep-diving submersible in the North Pacific near Alaska's Aleutian Islands you encounter a long, narrow depression on the ocean floor. Your passenger asks whether you think it is a submarine canyon, a rift valley, or a deep-ocean trench. How would you respond? Explain your choice.
7. Reef-building corals are responsible for creating atolls—ring-shaped structures that extend from the surface of the ocean to depths of thousands of meters. These corals, however, can only live in warm, sunlit water no more than about 45 meters deep. This presents a paradox: How can corals, which require warm, sunlit water, create structures that extend to great depths? Explain the apparent contradiction.

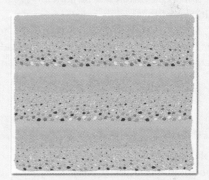

In Review   Chapter 13 The Ocean Floor

- *Oceanography is an interdisciplinary science* that draws on the methods and knowledge of geology, chemistry, physics, and biology to study all aspects of the world's ocean.
- *Earth is a planet dominated by oceans.* Nearly 71 percent of Earth's surface area consists of oceans and marginal seas. In the Southern Hemisphere, often called the *water hemisphere*, about 81 percent of the surface is water. The world ocean can be divided into four main ocean basins: the *Pacific Ocean* (largest and deepest ocean), the *Atlantic Ocean* (about half the size of the Pacific), the *Indian Ocean* (slightly smaller than the Atlantic and mostly in the Southern Hemisphere), and the *Arctic Ocean* (smallest and shallowest ocean). Oceanographers also recognize one additional ocean, the *Southern* or *Antarctic Ocean*, which occurs south of about 50 degrees south latitude. The average depth of the oceans is 3,729 meters (12,234 feet).

- *Ocean bathymetry is determined using echo sounders and multibeam sonars*, which bounce sonic signals off the ocean floor. Ship-based receivers record the reflected echoes and accurately measure the time interval of the signals. With this information, ocean depths are calculated and plotted to produce maps of ocean-floor topography. Recently, *satellite measurements* of the shape of the ocean surface have added data for mapping ocean-floor features.
- The zones that collectively make up a *passive continental margin* include the *continental shelf* (a gently sloping, submerged surface extending from the shoreline toward the deep-ocean basin); the *continental slope* (the true edge of the continent, which has a steep slope that leads from the continental shelf into deep water); and in regions where trenches do not exist, the steep continental slope merges into a more gradual incline known as the *continental rise* (which consists

of sediments that have moved downslope from the continental shelf to the deep-ocean floor).

- *Submarine canyons* are deep, steep-sided valleys that originate on the continental slope and may extend to the deep-ocean basin. Many submarine canyons have been excavated by *turbidity currents*, which are downslope movements of dense, sediment-laden water.

- *Active continental margins* are located primarily around the Pacific Rim in areas where the leading edge of a continent is overrunning oceanic lithosphere. Here sediment scraped from the descending oceanic plate is plastered against the continent to form a collection of sediments called an *accretionary wedge*. An active continental margin generally has a narrow continental shelf, which grades into a steep continental slope and deep-ocean trench.

- The *deep-ocean basin* lies between the continental margin and the oceanic ridge system. The features of the deep-ocean basin include *deep-ocean trenches* (the deepest parts of the ocean, where moving crustal plates descend into the mantle), *abyssal plains* (among the most level places on Earth, consisting of thick accumulations of sediments that were deposited atop the low, rough portions of the ocean floor), *seamounts* and *guyots* (isolated volcanic peaks on the ocean floor that originate near the mid-ocean ridge or in association with volcanic hot spots), and *oceanic plateaus* (vast accumulations of basaltic lava flows).

- *Atolls* form as corals and other organisms build a reef on the flanks of sinking volcanic islands. They gradually build the reef complex upward as the island slowly sinks.

- The *oceanic (mid-ocean) ridge* winds through the middle of most ocean basins. Seafloor spreading occurs along this broad feature, which is characterized by an elevated position, extensive faulting, and volcanic structures that have developed on newly formed oceanic crust. Much of the geologic activity associated with ridges occurs along a narrow region on the ridge crest, called the *rift valley*, where magma moves upward to create new slivers of oceanic crust.

- *There are three broad categories of seafloor sediments.* *Terrigenous sediment* consists primarily of mineral grains that were weathered from continental rocks and transported to the ocean; *biogenous sediment* consists of shells and skeletons of marine animals and plants; and *hydrogenous sediment* includes minerals that crystallize directly from seawater through various chemical reactions.

- *Seafloor sediments are helpful when studying worldwide climate changes* because they often contain the remains of organisms that once lived near the sea surface. The numbers and types of these organisms change as the climate changes, and their remains in seafloor sediments record these changes.

- *Energy resources* from the seafloor include *oil and natural gas* and large untapped deposits of *gas hydrates*. Other seafloor resources include *sand and gravel*, *evaporative salts*, and metals within *manganese nodules*.

Key Terms

abyssal plain (p. 389)
active continental margin (p. 389)
bathymetry (p. 384)
biogenous sediment (p. 395)
continental margin (p. 386)
continental rise (p. 387)
continental shelf (p. 387)
continental slope (p. 387)
continental volcanic arc (p. 389)
deep-ocean basin (p. 389)

deep-ocean trench (p. 389)
deep-sea fan (p. 388)
echo sounder (p. 384)
gas hydrate (p. 397)
guyot (p. 390)
hydrogenous sediment (p. 395)
mid-ocean ridge (p. 392)
oceanic plateau (p. 390)
oceanic ridge (p. 392)
oceanography (p. 382)

passive continental margin (p. 387)
rift valley (p. 393)
seamount (p. 390)
seismic reflection profile (p. 385)
sonar (p. 384)
submarine canyon (p. 388)
terrigenous sediment (p. 395)
turbidity current (p. 388)
volcanic island arc (p. 389)

Examining the Earth System

1. Describe some of the material and energy exchanges that take place at the interface between the (a) ocean surface and atmosphere, (b) ocean water and ocean floor, and (c) ocean biosphere and ocean water.

2. Sediment on the seafloor often leaves clues about various conditions that existed during deposition. What do the following layers in a seafloor core indicate about the environment in which each layer was deposited?

   - Layer #5 (top): A layer of fine clays
   - Layer #4: Siliceous ooze

   - Layer #3: Calcareous ooze
   - Layer #2: Fragments of coral reef
   - Layer #1 (bottom): Rocks of basaltic composition with some metal sulfide coatings

   Explain how one area of the seafloor could experience such varied conditions of deposition.

Mastering Geology

Looking for additional review and test prep materials? Visit the Self Study area in **www.masteringgeology.com** to find practice quizzes, study tools, and multimedia that will aid in your understanding of this chapter's content. In **MasteringGeology™** you will find:

- **GEODe: Earth Science: An interactive visual walkthrough of key concepts**

- **Geoscience Animation Library: More than 100 animations illuminating many difficult-to-understand Earth science concepts**
- **In The News RSS Feeds: Current Earth science events and news articles are pulled into the site with assessment**
- **Pearson eText**
- **Optional Self Study Quizzes**
- **Web Links**
- **Glossary**
- **Flashcards**

Ocean Water and Ocean Life*

14

Sea ice in the Arctic Ocean. Sea ice is frozen seawater. The formation and melting of sea ice causes the salinity of seawater in the Arctic to change seasonally. For more about Arctic sea ice, see Figure **14.4** (Photo by Wolfgang/Bechtold/Photolibrary))

*This chapter was prepared with the assistance of Professor Alan P. Trujillo, Palomar College.

What is the difference between pure water and seawater? One of the most obvious differences is that seawater contains dissolved substances that give it a distinctly salty taste. These dissolved substances are not simply sodium chloride (table salt)—they include various other salts, metals, and even dissolved gases. In fact, every known naturally occurring element is found dissolved in at least trace amounts in seawater. Unfortunately, the salt content of seawater makes it unsuitable for drinking or for irrigating most crops and causes it to be highly corrosive to many materials. Yet, many parts of the ocean are teeming with life that is superbly adapted to the marine environment.

There is an amazing variety of life in the ocean, from microscopic bacteria and algae to the largest organism alive today (the blue whale). Water is the major component of nearly every life-form on Earth, and our own body fluid chemistry is remarkably similar to the chemistry of seawater.

FOCUS ON CONCEPTS

To assist you in learning the important concepts in this chapter, focus on the following questions:

- What are the principal elements that contribute to the ocean's salinity? What are the sources of these elements?
- How do temperature, salinity, and density change with depth in the open ocean?
- On what basis are marine organisms classified?
- What factors influence the distributions of marine organisms?
- How does ocean productivity differ in polar, tropical, and temperate oceans?
- What are trophic levels? How efficient is energy transfer between various trophic levels in the ocean?

Composition of Seawater

Seawater consists of about 3.5 percent (by weight) dissolved mineral substances that are collectively termed "salts." Although the percentage of dissolved components may seem small, the actual quantity is huge because the ocean is so vast.

Salinity

Salinity (*salinus* = salt) is the total amount of solid material dissolved in water. More specifically, it is the ratio of the mass of dissolved substances to the mass of the water sample. Many common quantities are expressed in percent (%), which is really *parts per hundred*. Because the proportion of dissolved substances in seawater is such a small number, oceanographers typically express salinity in *parts per thousand* (‰). Thus, the average salinity of seawater is 3.5%, or 35‰.

Figure 14.1 shows the principal elements that contribute to the ocean's salinity. If one wanted to make artificial seawater, it could be approximated by following the recipe shown in **Table 14.1**. From this table it is evident that most of the salt in seawater is sodium chloride—common table salt. Sodium chloride together with the next four most abundant salts comprise over 99 percent of all dissolved substances in the sea. Although only eight elements make up these five most abundant salts, seawater contains all of Earth's

FIGURE 14.1 Relative proportions of water and dissolved components in seawater. Components shown by chemical symbol are chlorine (Cl^-), sodium (Na^+), sulfate (SO_4^{2-}), magnesium (Mg^{2+}), calcium (Ca^{2+}), potassium (K^+), strontium (Sr^{2+}), bromine (Br^-), and carbon (C).

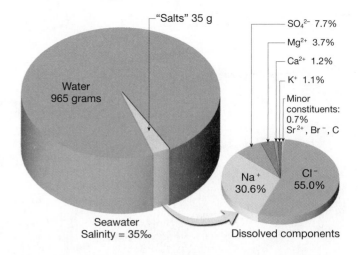

other naturally occurring elements. Despite their presence in minute quantities, many of these elements are very important in maintaining the necessary chemical environment for life in the sea.

TABLE 14.1 Recipe for Artificial Seawater

To make Seawater, Combine:	Amount (grams)
Sodium chloride (NaCl)	23.48
Magnesium chloride ($MgCl_2$)	4.98
Sodium sulfate (Na_2SO_4)	3.92
Calcium chloride ($CaCl_2$)	1.10
Potassium chloride (KCl)	0.66
Sodium bicarbonate ($NaHCO_3$)	0.192
Potassium bromide (KBr)	0.096
Hydrogen borate (H_3BO_3)	0.026
Strontium chloride ($SrCl_2$)	0.024
Sodium fluoride (NaF)	0.003

Then Add:

Pure water (H_2O) to form 1,000 grams of solution

Sources of Sea Salts

What are the primary sources for the vast quantities of dissolved substances in the ocean? Chemical weathering of rocks on the continents is one source. These dissolved materials are delivered to the oceans by streams at an estimated rate of more than 2.3 billion metric tons (2.5 billion short tons) annually.[32] The second major source of elements found in seawater is from Earth's interior. Through volcanic eruptions, large quantities of water vapor

[32]A metric ton equals 1,000 kilograms, or 2,205 pounds. Thus, it is larger than the "ton" that most Americans are familiar with (called a *short ton*). There are 2,000 pounds in a short ton.

and other gases have been emitted during much of geologic time. This process, called *outgassing*, is the principal source of water in the oceans. Certain elements—notably chlorine, bromine, sulfur, and boron—were outgassed along with water and exist in the ocean in much greater abundance than could be explained by weathering of rocks alone.

Although rivers and volcanic activity continually contribute dissolved substances to the oceans, the salinity of seawater is not increasing. In fact, evidence suggests that the composition of seawater has been relatively stable for millions of years. Why doesn't the sea get saltier? The answer is because material is being removed just as rapidly as it is added. For example, some dissolved components are withdrawn from seawater by organisms as they build hard parts. Other components are removed when they chemically precipitate out of the water as sediment. Still others are exchanged at the oceanic ridge by *hydrothermal* (*hydro* = water, *thermos* = hot) *activity*. The net effect is that the overall makeup of seawater has remained relatively constant through time (Box 14.1).

Processes Affecting Seawater Salinity

Because the ocean is well mixed, the relative abundances of the major components in seawater are essentially constant, no matter where the ocean is sampled. Variations in salinity, therefore, are primarily a consequence of changes in the water content of the solution.

Various surface processes alter the amount of water in seawater, thereby affecting salinity (**Figure 14.2**). Processes that add large amounts of freshwater to seawater—and thereby decrease salinity—include precipitation, runoff from land, icebergs melting, and sea ice melting. Processes that remove large amounts of freshwater from

FIGURE 14.2 Processes affecting seawater salinity. Processes that *decrease* seawater salinity include precipitation, runoff, icebergs melting, and sea ice melting. Processes that *increase* seawater salinity include formation of sea ice and evaporation. (Upper left, Bernhard Edmaier/Photo Researchers, Inc.; upper right, Photoshot Holdings Ltd/Alamy; lower left, NASA; lower right, Paul Steele/CORBIS/Stock Market)

Icebergs

Sea ice

Runoff

Evaporation

Box 14.1

PEOPLE AND THE ENVIRONMENT

Desalination of Seawater—Freshwater from the Sea

Earth's growing population uses freshwater in greater volumes each year. As freshwater becomes increasingly scarce, several countries have begun using the ocean as a source of water. The removal of salts and other chemicals to extract low-salinity ("fresh") water from seawater is termed *desalination*.

Worldwide, there are more than 12,500 desalination plants (**Figure 14.A**), with the majority located in arid regions of the Middle East, Caribbean, and Mediterranean. The United States produces only about 10 percent of the world's desalted water, primarily in Florida. To date, only a limited number of desalination plants have been built along the California coast, primarily because the cost of desalination is generally higher than the costs of other water supply alternatives available in California (such as water transfers and groundwater pumping). However, as drought conditions occur and concern over water availability increases, desalination projects are being proposed at numerous locations in the state.

Because desalinated water is expensive to produce, most desalination plants are small-scale operations. In fact, desalination plants provide only about 1 percent of the world's drinking water. More than half of the world's desalination plants use *distillation* to purify water, while most of the remaining plants use *membrane processes*.

In distillation, saltwater is evaporated, and the resulting water vapor is collected and condensed to produce freshwater. This simple procedure is very efficient at purifying seawater. For instance, distillation of 35‰ seawater produces freshwater with a salinity of only 0.03‰, which is about 10 times fresher than bottled water.

Membrane processes such as *electrolysis* or *reverse osmosis* use specialized semipermeable membranes to separate dissolved components from water molecules, thereby purifying water. Worldwide, at least 30 countries are operating reverse osmosis units. The world's largest plant, which is in Saudi Arabia, produces 485 million liters (128 million gallons) of desalted water daily. The largest plant in the United States is in Florida's Tampa Bay area and produces about 95 million liters (25 million gallons) of freshwater per day. A new facility planned for Carlsbad, California, is designed to produce twice as much freshwater as the Tampa Bay plant. Reverse osmosis is also used in many household water purification units and aquariums.

Other methods of desalination include freeze separation, crystallization of dissolved components directly from seawater, solvent demineralization using chemical catalysts, and even making use of salt-eating bacteria!

Although freshwater produced by various desalination methods is becoming more important as a source of water for human and even industrial use, it is unlikely to be an important supply for agricultural purposes because of the enormous quantities of water necessary to support agriculture. Consequently, making the deserts "bloom" by irrigating them with freshwater produced by desalination is only a dream and, for economic reasons, is likely to remain so for the foreseeable future.

FIGURE 14.A This desalination plant is located on the island of Curacao, in the Netherlands Antilles off the northwest coast of Venezuela. (Photo by Peter Guttman/CORBIS/Bettmann)

seawater—and thereby increase seawater salinity—include evaporation and the formation of sea ice. High salinities, for example, are found where evaporation rates are high, as is the case in the dry subtropical regions (roughly between 25 and 35 degrees north or south latitude). Conversely, where large amounts of precipitation dilute ocean waters, as in the midlatitudes (between 35 and 60 degrees north or south latitude) and near the equator, lower salinities prevail (Figure 14.3).

Surface salinity in polar regions varies seasonally due to the formation and melting of sea ice. When seawater freezes in winter, sea salts do not become part of the ice. Therefore, the salinity of the remaining seawater increases. In summer when sea ice melts, the addition of the relatively fresh water dilutes the solution and salinity decreases (Figure 14.4).

Surface salinity variation in the open ocean normally ranges from 33‰ to 38‰. Some marginal seas, however, demonstrate extraordinary extremes. For example, in the restricted waters of the Middle East's Persian Gulf and Red Sea—where evaporation far exceeds precipitation—salinity may exceed 42‰. Conversely, very low salinities occur where large quantities of freshwater are supplied by rivers and precipitation. Such is the case for northern Europe's Baltic Sea, where salinity is often below 10‰.

CONCEPT CHECK 14.1

❶ What is salinity and how is it usually expressed? What is the average salinity of the ocean?

❷ What are the six most abundant elements dissolved in seawater? What is produced when the two most abundant elements combine?

❸ What are the two primary sources for the elements that comprise the dissolved components in seawater?

❹ List several factors that cause salinity to vary from place to place and from time to time.

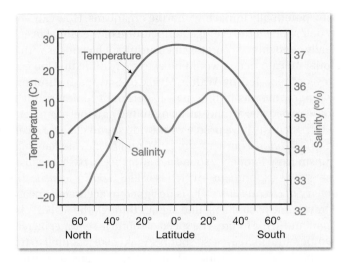

FIGURE 14.3 This graph shows variations in the ocean's surface temperature (top curve) and surface salinity (lower curve) with latitude. As you might expect, average temperatures are highest near the equator and get colder moving poleward. An important factor influencing differences in surface salinity is variations in rainfall and evaporation rates. For example, in the dry subtropics in the vicinity of the Tropics of Cancer and Capricorn, high evaporation rates remove more water than is replaced by the meager rainfall, resulting in high surface salinities. In the wet equatorial region, abundant rainfall reduces surface salinities.

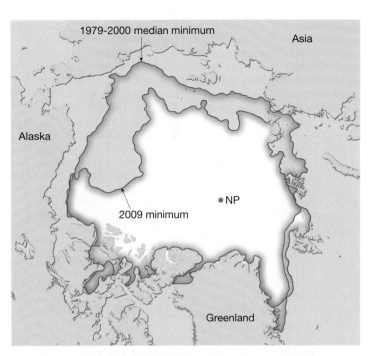

FIGURE 14.4 Sea ice is frozen seawater. In winter the Arctic Ocean is completely ice covered. In summer some of the sea ice melts. When sea ice forms, the salinity of the water beneath it increases; when the ice melts, the salinity of the water decreases. This map shows the extent of sea ice at the end of the summer melting period in 2009 compared to the average for the period 1979–2000. In 2009 the extent of sea ice was 24 percent below the long-term average for 1979–2000. This trend is likely linked to global climate change. For more about this, see the section "Some Possible Consequences of Global Warming" in Chapter 20.

Students Sometimes Ask...

What would happen to a person if he or she drank seawater?

It depends on the quantity. The salinity of seawater is about four times greater than that of your body fluids. In your body, seawater causes your internal membranes to lose water through *osmosis*, which transports water molecules from higher concentrations (the normal body chemistry of your internal fluids) to areas of lower concentrations (your digestive tract containing seawater). Thus, your natural body fluids would move into your digestive tract and eventually be expelled, causing dehydration if seawater is consumed in large amounts. If you inadvertently swallow a little bit of seawater, don't worry too much.

Ocean Temperature Variation

The ocean's surface water temperature varies with the amount of solar radiation received, which is primarily a function of latitude (Figure 14.3). The intensity of solar radiation in high latitudes is significantly less than that received in tropical latitudes.[33] Therefore, much lower sea surface temperatures are found in high-latitude regions, and much higher sea surface temperatures are found in low-latitude regions.

[33]For more on this subject, see the section "Earth–Sun Relationships" in Chapter 16.

Temperature Variation with Depth

If a thermometer were lowered from the surface of the ocean into deeper water, what temperature pattern would be found? Surface waters are warmed by the Sun, so they generally have higher temperatures than deeper waters. However, the observed temperature pattern depends on the latitude.

Figure 14.5 shows two graphs of temperature versus depth: one for high-latitude regions and one for low-latitude regions. The low-latitude curve begins at the surface with high temperature, but the temperature decreases rapidly with depth because of the inability of the Sun's rays to penetrate very far into the ocean. At a depth of about 1,000 meters (3,300 feet), the temperature remains just a few degrees above freezing and is relatively constant from this level down to the ocean floor. The layer of ocean water between about 300 meters (980 feet) and 1,000 meters (3,300 feet), where there is a rapid change of temperature with depth, is called the **thermocline** (*thermo* = heat, *cline* = slope). The thermocline is a very important zone in the ocean because it creates a vertical barrier to many types of marine life.

The high-latitude curve in Figure 14.5 displays a pattern quite different from the low-latitude curve. Surface water temperatures in high latitudes are much cooler than in low latitudes, so the curve begins at the surface with low temperature. Deeper in the ocean, the temperature of the water is similar to that at the surface (just a few degrees above freezing), so the curve remains vertical and there is no rapid change of temperature with depth. A thermocline

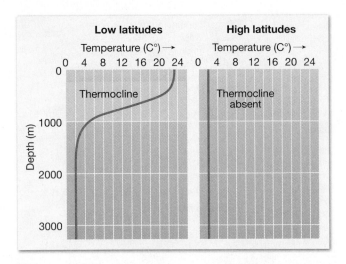

FIGURE 14.5 Variations in ocean water temperature with depth for low- and high-latitude regions. The layer of rapidly changing temperature, called the *thermocline*, is not present in the high latitudes.

is not present in high latitudes; instead, the water column is *isothermal* (*iso* = same, *thermo* = heat).

Some high-latitude waters can experience minor warming during the summer months. Thus, certain high-latitude regions experience an extremely weak seasonal thermocline. Midlatitude waters, on the other hand, experience a more dramatic seasonal thermocline and exhibit characteristics intermediate between high- and low-latitude regions.

Ocean Temperature Change over Time

Seawater has many unique thermal properties that make it resistant to changes in temperature.[34] Marine species are well adapted to life in the ocean, and many cannot withstand rapid changes in temperature. In fact, researchers have concluded that the ocean's stability as a habitat has been instrumental in the development of life on Earth.

Studies on the effect of global warming on the ocean indicate that atmospheric warming will eventually be transferred to the

Students Sometimes Ask...

Where is the saltiest water in the world?

Some of the most highly saline water in the world is found in arid regions that have inland lakes, which are often called "seas" because they are so salty. The Great Salt Lake in Utah, for example, has a salinity of 280‰, and the Dead Sea on the border of Israel and Jordan has a salinity of 330‰. The water in the Dead Sea, therefore, contains 33 percent dissolved solids and is almost *10 times saltier than seawater*. As a result, these waters have such high density and are so buoyant that while lying down in the water, you can easily float—with arms and legs sticking up above water level! Take a look at Figure 14.6.

---

[34]This is discussed more fully in the section "Controls of Temperature" in Chapter 16.

ocean, potentially impacting marine organisms. How can scientists determine if the average temperature of the global ocean is actually changing? Accurately measuring such changes with traditional thermometers would be extremely difficult. One group of researchers, led by Walter Munk of Scripps Institution of Oceanography, has initiated an experiment to determine the amount of ocean warming by using sound as a thermometer.

The scientists have used a worldwide sound channel—called the *SOFAR channel* (an acronym for *SOund Fixing And Ranging*)—to transmit low-frequency sound across an ocean basin to a distant receiver. The experiment—which is called *ATOC (Acoustic Thermometry of Ocean Climate)*—is designed to accurately measure the travel time of similar sound signals through the SOFAR channel now and in the future. The speed of sound in seawater increases as temperature increases, so sound should take less time to travel the same distance in the future if, in fact, the oceans are warming.

The group successfully tested the experiment by transmitting signals across the Pacific Ocean in 1991 and 1995. The transmissions were halted, however, because of concern about the sound's effect on whales, some of which may also use the SOFAR channel for communication. Nonetheless, the transmissions that have been conducted have established an important baseline for comparison with future measurements.

CONCEPT CHECK 14.2

❶ What one factor is primarily responsible for influencing seawater temperature?

❷ Describe temperature variation with depth in both high and low latitudes. Why do high-latitude waters generally lack a thermocline?

Ocean Density Variation

Density is defined as mass per unit volume but can be thought of as a measure of *how heavy something is for its size*. For instance, an object that has low density is lightweight for its size, such as a dry sponge, foam packing, or a surfboard. Conversely, an object that has high density is heavy for its size, such as cement and many metals.

Density is an important property of ocean water because it determines the water's vertical position in the ocean. Furthermore, density differences cause large areas of ocean water to sink or float. For example, when high-density seawater is added to low-density freshwater, the denser seawater sinks below the freshwater.

Factors Affecting Seawater Density

Seawater density is influenced by two main factors: *salinity* and *temperature*. An increase in salinity adds dissolved substances and results in an increase in seawater density (**Figure 14.6**). An increase in temperature, on the other hand, causes thermal expansion and results in a decrease in seawater density. Such a relationship where one variable decreases as a result of another variable's increase is known as an *inverse relationship*, where one variable is *inversely proportional* with the other.

Temperature has the greatest influence on surface seawater density because variations in surface seawater temperature are greater than salinity variations. In fact, only in the extreme polar areas of

versus depth: one for high-latitude regions and one for low-latitude regions. Not surprisingly, the curves in Figure 14.7 are a mirror image of the temperature curves in Figure 14.5. This similarity demonstrates that temperature is the most important factor affecting seawater density and that temperature is inversely proportional with density.

The low-latitude curve in Figure 14.7 begins at the surface with low density (related to high surface water temperatures). However, density increases rapidly with depth because the water temperature is getting colder. At a depth of about 1,000 meters (3,300 feet), seawater density reaches a maximum value related to the water's low temperature. From this depth to the ocean floor, density remains constant and high. The layer of ocean water between about 300 meters (980 feet) and 1,000 meters (3,300 feet), where there is a rapid change of density with depth, is called the **pycnocline** (*pycno* = density, *cline* = slope). A pycnocline has a high gravitational stability and presents a significant barrier to mixing between low-density water above and high-density water below.

The high-latitude curve in Figure 14.7 is also related to the temperature curve for high latitudes shown in Figure 14.5. Figure 14.7 shows that in high latitudes, there is high-density (cold) water at the surface and high-density (cold) water below. Thus, the high-latitude density curve remains vertical, and there is no rapid change of density with depth. A pycnocline is not present in high latitudes; instead, the water column is *isopycnal* (*iso* = same, *pycno* = density).

Ocean Layering

The ocean, like Earth's interior, is layered according to density. Low-density water exists near the surface, and higher-density water occurs below. Except for some shallow inland seas with a high rate of evaporation, the highest-density water is found at the greatest ocean depths. Oceanographers generally recognize a three-layered structure in most parts of the open ocean: a shallow surface mixed zone, a transition zone, and a deep zone (**Figure 14.8**).

Because solar energy is received at the ocean surface, it is here that water temperatures are warmest. The mixing of these waters by waves as well as the turbulence from currents and tides creates a rapid vertical heat transfer. Hence, this *surface mixed zone* has nearly uniform temperatures. The thickness and temperature of this layer vary, depending on latitude and season. The zone usually extends to about 300 meters (980 feet) but may attain a thickness of 450 meters (1,500 feet). The surface mixed zone accounts for only about 2 percent of ocean water.

Below the sun-warmed zone of mixing, the temperature falls abruptly with depth (see Figure 14.5). Here, a distinct layer called the *transition zone* exists between the warm surface layer above and the deep zone of cold water below. The transition zone includes a prominent thermocline and associated pycnocline and accounts for about 18 percent of ocean water.

Below the transition zone is the *deep zone*, where sunlight never reaches and water temperatures are just a few degrees above freezing. As a result, water density remains constant and high. Remarkably, the deep zone includes about 80 percent of ocean water, indicating the immense depth of the ocean (the average depth of the ocean is 3,729 meters, or 12,234 feet).

FIGURE 14.6 The Dead Sea, which has a salinity of 330‰ (almost 10 times the average salinity of seawater), has high density. As a result, it also has high buoyancy that allows swimmers to float easily. (Photo by Peter Guttman/CORBIS/Bettmann)

the ocean, where temperatures are low and remain relatively constant, does salinity significantly affect density. Cold water that also has high salinity is some of the highest-density water in the world.

Density Variation with Depth

By extensively sampling ocean waters, oceanographers have learned that temperature and salinity—and the water's resulting density—vary with depth. **Figure 14.7** shows two graphs of density

FIGURE 14.7 Variations in ocean water density with depth for low- and high-latitude regions. The layer of rapidly changing density, called the *pycnocline*, is present in the low latitudes but absent in the high latitudes.

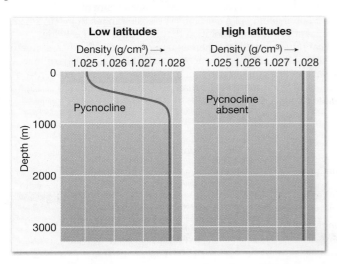

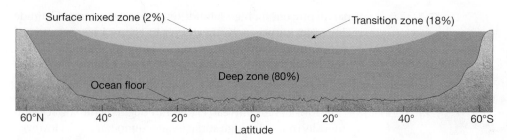

FIGURE 14.8 Oceanographers recognize three main layers in the ocean based on water density, which varies with temperature and salinity. The warm surface mixed layer accounts for only 2 percent of ocean water; the transition zone includes the thermocline and pycnocline and accounts for 18 percent of ocean water; the deep zone contains cold, high-density water that accounts for 80 percent of ocean water.

In high latitudes, the three-layer structure of ocean layering does not exist because the water column is isothermal and isopycnal, which means that there is no rapid change in temperature or density with depth. Consequently, good vertical mixing between surface and deep waters can occur in high-latitude regions. Here, cold high-density water forms at the surface, sinks, and initiates deep-ocean currents, which are discussed in Chapter 15.

CONCEPT CHECK 14.3

❶ Which two factors influence seawater density? Which one has the greater influence on surface seawater density?
❷ Describe density variation with depth in both high and low latitudes. Why do high-latitude waters generally lack a pycnocline?
❸ Describe the ocean's layered structure. Why does the three-layer structure not exist in high latitudes?

Recent Increase in Ocean Acidity

Human activities, especially burning fossil fuels and deforestation, have been adding ever-increasing quantities of carbon dioxide to the atmosphere for many decades. This change in atmospheric composition is linked in a cause-and-effect way to global climate change, a topic addressed at some length in Chapter 20. Rising levels of atmospheric carbon dioxide also have some serious implications for ocean chemistry and for marine life. About one-third of the human-generated carbon dioxide ends up dissolved in the oceans. This causes the ocean's pH to drop, making seawater more acidic. The pH scale is shown and briefly described in **Figure 14.9**.

When carbon dioxide (CO_2) from the atmosphere dissolves in seawater (H_2O) it forms carbonic acid (H_2CO_3). This lowers the ocean's pH and changes the balance of certain chemicals found naturally in seawater. In fact, the oceans have already absorbed enough carbon dioxide for surface waters to have experienced a pH decrease

of 0.1 since preindustrial times, with an additional pH decrease likely in the future. Moreover, if the current trend of human-induced carbon dioxide emissions continues, by the year 2100 the ocean will experience a pH decrease of at least 0.3, which represents a change in ocean chemistry that has not occurred for millions of years. This shift toward acidity and the changes in ocean chemistry that result make it more difficult for certain marine creatures to build hard parts out of calcium carbonate. The decline in pH thus threatens a variety of calcite-secreting organisms as diverse as microbes and corals, which concerns marine scientists because of the potential consequences for other sea life that depend on the health and availability of these organisms.

CONCEPT CHECK 14.4

❶ Is the pH of the ocean increasing or decreasing?
❷ What is responsible for the changing pH of the ocean?

The Diversity of Ocean Life

A wide variety of organisms inhabit the marine environment. These organisms range in size from microscopic bacteria and algae to blue whales, which are as long as three buses lined up end to end. Marine biologists have identified over 250,000 marine species, a number that is constantly increasing as new organisms are discovered.

Most marine organisms live within the sunlit surface waters of the ocean. Strong sunlight supports **photosynthesis** (*photo* = light, *syn* = with, *thesis* = an arranging) by marine algae, which either directly or indirectly provide food for the vast majority of marine organisms. All marine algae live near the surface because they need sunlight; most marine animals also live near the surface because this is where food can be obtained. In shallow water areas close to land, sunlight reaches all the way to the bottom, resulting in an abundance of marine life on the ocean floor.

FIGURE 14.9 The *pH scale* is a common measure of the degree of acidity or alkalinity of a solution. The scale ranges from 0 to 14, with a value of 7 denoting a solution that is neutral. Values below 7 indicate greater acidity, whereas numbers above 7 indicate greater alkalinity. It is important to note that the pH scale is logarithmic; that is, each whole number increment indicates a tenfold difference. Thus, pH 4 is 10 times more acidic than pH 5 and 100 times (10×10) more acidic than pH 6.

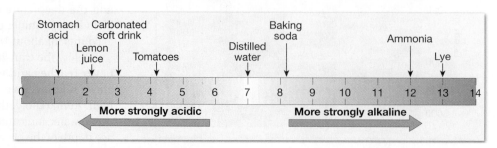

There are advantages and disadvantages to living in the marine environment. One advantage is that there is an abundance of water available, which is necessary for supporting all types of life. One disadvantage is that maneuvering in water, which has high density and impedes movement, can be difficult. The success of marine organisms depends on their ability to avoid predators, find food, and cope with the physical challenges of their environment.

Classification of Marine Organisms

Marine organisms can be classified according to where they live (their habitat) and how they move (their mobility). Organisms that inhabit the water column can be classified as either *plankton* (floaters) or *nekton* (swimmers). All other organisms are *benthos* (bottom dwellers).

Plankton (Floaters)   **Plankton** (*planktos* = wandering) include all organisms—algae, animals, and bacteria—that drift with ocean currents. Just because plankton drift does not mean they are unable to swim. Many plankton can swim but either move very weakly or move only vertically within the water column.

Among plankton, the algae (photosynthetic cells, most of which are microscopic) are called **phytoplankton** and the animals are called **zooplankton** (*zoo* = animal, *planktos* = wandering). Representative members of each group are shown in Figure 14.10.

Plankton are extremely abundant and very important within the marine environment. In fact, most of Earth's **biomass**—the mass of all living organisms—consists of plankton adrift in the oceans. Even though 98 percent of marine species are bottom-dwelling, the vast majority of the ocean's biomass is planktonic.

Nekton (Swimmers)   **Nekton** (*nektos* = swimming) include all animals capable of moving independently of the ocean currents, by swimming or other means of propulsion. They are capable not only of determining their position within the ocean but also, in many cases, of long migrations. Nekton include most adult fish and squid, marine mammals, and marine reptiles (Figure 14.11).

Although nekton move freely, they are unable to move throughout the breadth of the ocean. Gradual changes in temperature, salinity, density, and availability of nutrients effectively limit their lateral range. The deaths of large numbers of fish, for example, can be caused by temporary shifts of water masses in the ocean. High water pressure at depth normally limits the vertical range of nekton.

Fish may appear to exist everywhere in the oceans, but they are more abundant near continents and islands and in colder

FIGURE 14.10 A. A variety of live *phytoplankton* from the Atlantic Ocean. They include various diatoms and dinoflagellates. (Copyright Norman T. Nicoll/Natural Visions) **B.** This image of *zooplankton* includes the most common forms known as copepods and their larvae. The picture also includes larval forms of other common marine organisms. (Copyright Norman T. Nicoll/Natural Visions)

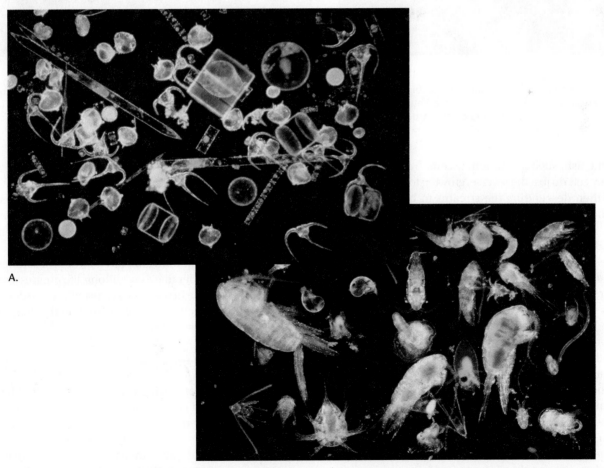

A.

B.

A.

B.

C.

D.

FIGURE 14.11 Nekton includes all animals capable of moving independently of ocean currents. **A.** Gray reef shark, Bikini Atoll. (Photo by Yann Hubert/BIOSphoto/Specialist Stock); **B.** California market squid. (Photo by Tom McHugh/Photo Researchers, Inc.); **C.** School of grunts, Florida Keys. (Photo by Georgie Holland/age footstock); **D.** Yellow-head moray eel. (© by David B. Fleetham/Seapics.com)

waters. Some fish, such as salmon, ascend freshwater rivers to spawn. Many eels do just the reverse, growing to maturity in freshwater and then descending the streams to breed in the great depths of the ocean.

Benthos (Bottom Dwellers)   The term **benthos** (*benthos* = bottom) describes organisms living on or in the ocean bottom. *Epifauna* (*epi* = upon, *fauna* = animal) live on the surface of the seafloor, either attached to rocks or moving along the bottom. *Infauna* (*in* = inside, *fauna* = animal) live buried in the sand or mud. Some benthos, called *nektobenthos*, live on the bottom but also swim or crawl through the water above the ocean floor. Examples of benthos are shown in Figure 14.12.

The shallow coastal ocean floor contains a wide variety of physical conditions and nutrient levels, both of which have allowed a great number of species to evolve. Moving across the bottom from the shore into deeper water, the number of benthos species may remain relatively constant, but the biomass of benthos organisms decreases. In addition, shallow coastal areas are the only locations where large marine algae (often called "seaweeds") are found

attached to the bottom. This is the case because these are the only areas of the seafloor that receive sufficient sunlight.

Throughout most of the deeper parts of the seafloor, animals live in perpetual darkness, where photosynthesis cannot occur. They must feed on each other, or on whatever nutrients fall from the productive surface waters above.

The deep-sea bottom is an environment of coldness, stillness, and darkness. Under these conditions, life progresses slowly, and organisms that live in the deep sea usually are widely distributed because physical conditions vary little on the deep-ocean floor, even over great distances.

Marine Life Zones

The distribution of marine organisms is affected by the chemistry, physics, and geology of the oceans. Marine organisms are influenced by a variety of physical oceanographic factors. Some of these factors—such as availability of sunlight, distance from shore, and water depth—are used to divide the ocean into distinct marine life zones (**Table 14.2** and Figure 14.13).

FIGURE 14.12 Benthos describes organisms living on or in the ocean bottom. **A.** Sea star. (Photo by David Hall/Photo Researchers, Inc.); **B.** Yellow tube sponge. (Photo by Andrew Martinez/Photo Researchers, Inc.); **C.** Green sea urchin. (Photo by Andrew Martinez/Photo Researchers, Inc.); **D.** Coral crab. (Photo by Images & Stories/Alamy)

TABLE 14.2 Marine Life Zones

Basis	Marine Life Zone	Subdivision	Characteristics
Available sunlight	Photic		Sunlit surface waters
		Euphotic	Has enough sunlight to support photosynthesis
	Aphotic		No sunlight; many organisms have bioluminescent capabilities
Distance from shore	Intertidal		Narrow strip of land between high and low tides; dynamic area
	Neritic		Above continental shelf; high biomass and diversity of species
	Oceanic		Open ocean beyond the continental shelf; low nutrient concentrations
Depth	Pelagic		All water above the ocean floor; organisms swim or float
	Benthic		Bottom of ocean; organisms attach to, burrow into, or crawl on seafloor
		Abyssal	Deep-sea bottom; dark, cold, high pressure; sparse life

Availability of Sunlight   The upper part of the ocean into which sunlight penetrates is called the **photic** (*photos* = light) **zone**. The clarity of seawater is affected by many factors, such as the amount of plankton, suspended sediment, and decaying organic particles in the water. In addition, the amount of sunlight varies with atmospheric conditions, time of day, season of the year, and latitude.

The **euphotic** (*eu* = good, *photos* = light) **zone** is the portion of the photic zone near the surface where light is strong enough for photosynthesis to occur. In the open ocean, this zone can reach a depth of 100 meters (330 feet), but the zone will be much shallower close to shore where water clarity is typically reduced. In the euphotic zone, phytoplankton use sunlight to produce food molecules and become the basis of most oceanic food webs.

Different wavelengths of sunlight are absorbed as they pass through seawater. The longer wavelength red and orange colors are absorbed first, the greens and yellows next, and the shorter wavelength blues and violets penetrate the farthest. In fact, faint traces of blue and violet light can still be measured in extremely clear water at depths of 1,000 meters (3,300 feet).

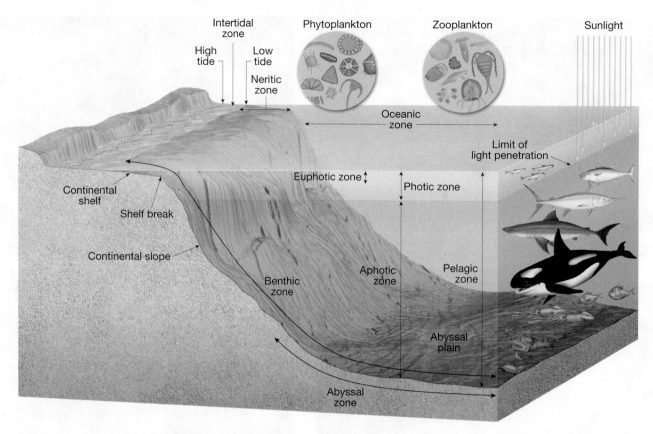

FIGURE 14.13 The ocean is divided into marine life zones, based on availability of light, distance from shore, and water depth.

Although photosynthesis cannot occur much below 100 meters (330 feet), there is enough light in the lower photic zone for marine animals to avoid predators, find food, recognize their species, and locate mates. Even deeper is the **aphotic** (*a* = without, *photos* = light) **zone**, where there is no sunlight.

Distance from Shore   Marine life zones can also be subdivided based on distance from shore. The area where the land and ocean meet and overlap is called the **intertidal zone**. This narrow strip of land between high and low tides is alternately covered and uncovered by seawater with each tidal change. Even though it appears to be a harsh place to live with crashing waves, periodic drying out, and rapid changes in temperature, salinity, and oxygen concentrations, many species live here that are superbly adapted to the dramatic environmental changes.

Seaward from the low-tide line is the **neritic** (*neritos* = of the coast) **zone**. This covers the gently sloping continental shelf out to the *shelf break* (see Figure 13.9, p. 387). This zone can be very narrow or may extend hundreds of kilometers from shore. The neritic zone is often shallow enough for sunlight to reach all the way to the ocean floor, putting it entirely within the photic zone.

Although the neritic zone covers only about 5 percent of the world's oceans, it is rich in both biomass and number of species. Many organisms find the conditions here ideal because photosynthesis occurs readily, nutrients wash in from the land, and the bottom provides shelter and habitat. This zone is so rich, in fact, that it supports 90 percent of the world's commercial fisheries.

Beyond the continental shelf is the **oceanic zone**. The open ocean reaches great depths, and as a result, surface waters typically have lower nutrient concentrations because nutrients tend to sink out of the photic zone to the deep-ocean floor. This low nutrient concentration usually results in much smaller populations than the more productive neritic zone.

Water Depth   A third method of classifying marine habitats is based on water depth. Open ocean of *any* depth is called the **pelagic** (*pelagios* = of the sea) **zone**. Animals in this zone swim or float freely. The photic part of the pelagic zone is home to phytoplankton, zooplankton, and nekton, such as tuna, sea turtles, and dolphins. The aphotic part has strange species like viperfish and giant squid that are adapted to life in deep water.

Students Sometimes Ask...

Do any deep-sea organisms produce light themselves?

Yes. Well over half of deep-sea organisms—including fish, jellies, crustaceans, and deep-sea squid—can *bioluminesce*, which means they can produce light organically. These organisms create light through a chemical reaction in specially designed structures or cells called *photophores*, some of which contain luminescent bacteria that live symbiotically within the organism. In a world of darkness, the ability to produce light can be used to attract prey, define territory, communicate with others, or avoid predators.

Benthos organisms such as giant kelp, sponges, crabs, sea anemones, sea stars, and marine worms that attach to, crawl upon, or burrow into the seafloor occupy parts of the **benthic** (*benthos* = bottom) **zone**. The benthic zone includes any sea-bottom surface regardless of its distance from shore and is mostly inhabited by benthos organisms.

The **abyssal** (*a* = without, *byssus* = bottom) **zone** is a subdivision of the benthic zone and includes the deep-ocean floor, such as *abyssal plains*. This zone is characterized by extremely high water pressure, consistently low temperature, no sunlight, and sparse life. Three food sources exist at abyssal depths: (1) tiny decaying particles steadily "raining" down from above, which provide food for filter-feeders, brittle stars, and burrowing worms; (2) large fragments or entire dead bodies falling at scattered sites, which supply meals for actively searching fish, such as the grenadier, tripodfish, and hagfish, which locate food by chemical sensing; and (3) hot springs on the seafloor, called *hydrothermal vents*. Box 14.2 describes these unique biocommunities.

CONCEPT CHECK 14.5

❶ Describe the lifestyles of plankton, nekton, and benthos, and give examples of each. Which group comprises the largest biomass?

❷ List three physical factors that are used to divide the ocean into marine life zones. How does each factor influence the abundance and distribution of marine life?

❸ Why are there greater numbers and types of organisms in the neritic zone than in the oceanic zone?

Oceanic Productivity

Why are some regions of the ocean teeming with life, while other areas seem barren? The answer is related to the amount of primary productivity in various parts of the ocean. **Primary productivity** is the amount of carbon fixed by organisms through the synthesis of organic matter using energy derived from solar radiation (*photosynthesis*) or chemical reactions (*chemosynthesis*). Although chemosynthesis supports hydrothermal vent biocommunities along the oceanic ridge (Box 14.2), it is much less significant than photosynthesis in worldwide oceanic productivity.

Two factors influence a region's photosynthetic productivity: *availability of nutrients* (such as nitrates, phosphorus, iron, and silica) and the *amount of solar radiation* (sunlight). Thus, the most abundant marine life exists where there are ample nutrients and good sunlight. Oceanic productivity, however, varies dramatically because of the uneven distribution of nutrients throughout the photosynthetic zone and seasonal changes in the availability of solar energy.

A permanent *thermocline* (and resulting *pycnocline*) develops nearly everywhere in the oceans (see Figure 14.8). This layer forms a barrier to vertical mixing and prevents the resupply of nutrients to sunlit surface waters. In the midlatitudes, a thermocline develops only during the summer season, and in polar regions a thermocline does not usually develop at all. The degree to which waters develop a thermocline profoundly affects the amount of productivity observed at different latitudes.

Productivity in Polar Oceans

Polar regions such as (the Arctic Ocean's Barents Sea, which is off the northern coast of Europe, experience continuous darkness for about three months of winter and continuous illumination for about three months during summer. Productivity of phytoplankton—mostly single-celled algae called *diatoms*—peaks there during May (**Figure 14.14**), when the Sun rises high enough in the sky so that there is deep penetration of sunlight into the water. As soon as the diatoms develop, zooplankton—mostly small crustaceans called *copepods* (Figure 14.14) and larger *krill*—begin feeding on them. The zooplankton biomass peaks in June and continues at a relatively high level until winter darkness begins in October.

Recall that temperature and density change very little with depth in polar regions (see Figures 14.5 and 14.7), so these waters are *isothermal*, and there is no barrier to mixing between surface waters and deeper, nutrient-rich waters. In the summer, however, melting ice creates a thin, low-salinity layer that does not readily mix with the deeper waters. This stratification is crucial to summer production, because it helps prevent phytoplankton from being carried into deeper, darker waters. Instead, they are concentrated in the sunlit surface waters where they reproduce continuously.

Because of the constant supply of nutrients rising from deeper waters below, high-latitude surface waters typically have high nutrient concentrations. The availability of solar energy, however, is what limits photosynthetic productivity in these areas.

Productivity in Tropical Oceans

You may be surprised to learn that productivity is low in tropical regions of the open ocean. Because the Sun is more directly overhead, light penetrates much deeper into tropical oceans than in temperate and polar waters, and solar energy is available year-round. However, productivity is low in tropical regions of the open ocean because a permanent thermocline produces a stratification of water masses that prevents mixing between surface waters and nutrient-rich deeper waters (**Figure 14.15**). In essence, the thermocline is a barrier that eliminates the supply of nutrients from deeper waters below. So, productivity in tropical regions is limited by the lack of nutrients (unlike polar regions, where

FIGURE 14.14 One example of productivity in polar oceans is illustrated by the Barents Sea. A springtime increase in diatom mass is followed closely by an increase in zooplankton abundance. (Photo by N.T. Nicoll/Natural Visions)

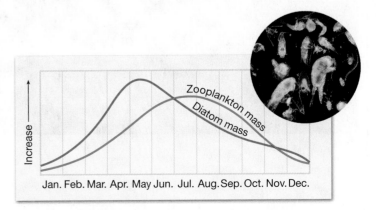

Box 14.2

EARTH AS A SYSTEM

Deep-Sea Hydrothermal Vent Biocommunities—Earth's First Life?

Deep-sea hydrothermal vents form along many active rift zones of the oceanic ridge. Here seawater percolates down into the hot and newly formed oceanic crust. During its journey, the water can become saturated with minerals before it spews back into the ocean as a *black smoker* (**Figure 14.B**). Black smokers usually emit from tall chimneys composed of metal sulfide minerals that have precipitated as the hot ventwater is exposed to cold seawater.

Water temperatures at some vents reach as high as 350° C (660° F), yet the dark water has recently been shown to be rich in microbes. Nearby, water temperatures of 100° C (212° F) or lower nourish dense, exotic *hydrothermal vent biocommunities* of organisms found nowhere else in the world. In fact, hundreds of new species—and even new genera and families—have been discovered surrounding these deep-sea habitats since their discovery by scientists along the Galápagos Rift in 1977. Additional hydrothermal vent biocommunities exist in discrete patches along the oceanic ridge and have been visited by scientists in deep-sea submersibles along the East Pacific Rise, Mid-Atlantic Ridge, Mid–Indian Ocean Ridge, and the Juan de Fuca Ridge.

How do these organisms survive in this dark, hot, sulfur-rich environment where photosynthesis cannot occur? Study of hydrothermal vent organisms shows that microscopic bacteria-like organisms called *archaea* (*archaeo* = ancient) living in and near the vents perform *chemosynthesis* (*chemo* = chemistry, *syn* = with, *thesis* = an arranging) and constitute the base of the food web. Hydrothermal vents provide heat energy for archaea to oxidize hydrogen sulfide (H_2S), which is formed by the reaction of hot water with dissolved sulfate (SO_4^{-2}). Through chemosynthesis, archaea produce sugars and other foods that enable them and many other organisms to live in this very unusual and extreme environment.

Some archaea live symbiotically inside giant, gutless tube-dwelling worms (**Figure 14.C**). These archaea provide food for the tube worms to grow as rapidly as 1 meter (3.3 feet) each year and up to 3 meters (10 feet) in length. Other archaea are consumed by specialized yellow mussels, giant white clams, and pink sea urchins. These in turn are eaten by unique crabs and fishes. Thus, archaea are the foundation of a living ecosystem that does not require sunlight.

It is very likely that environments similar to those of hydrothermal vents were present

FIGURE 14.C Tube worms up to 3 meters (10 feet) in length are among the organisms found in the extreme environment of hydrothermal vents along the crest of the oceanic ridge where sunlight is nonexistent. These organisms obtain their food from internal microscopic bacteria-like archaea, which acquire their nourishment and energy through the processes of chemosynthesis. (Photo by Al Giddings Images, Inc.)

during the early history of the planet. Some scientists have suggested that the uniformity of conditions and abundant energy of the vents would have provided an ideal habitat for the origin of life. In fact, hydrothermal vents may represent one of the oldest life-sustaining environments, because hydrothermal activity occurs wherever there are both volcanoes and water. An additional line of evidence in support of hydrothermal vents harboring some of Earth's first life exists in the fact that archaea contain ancient genetic makeup.

FIGURE 14.B View from the submersible *Alvin* of a black smoker spewing hot, mineral-rich water along the East Pacific Rise. When heated solutions meet cold seawater, metal sulfides precipitate and form mounds of minerals around these hydrothermal vents. (Photo by Dudley Foster, © Woods Hole Oceanographic Institution)

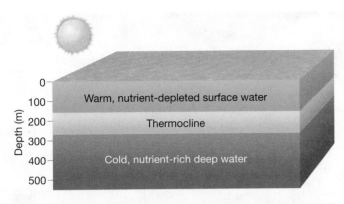

FIGURE 14.15 Productivity in tropical oceans. Although tropical regions receive adequate sunlight year-round, a permanent thermocline prevents the mixing of surface and deep water. As phytoplankton consume nutrients in the surface layer, productivity is limited because the thermocline prevents replenishment of nutrients from deeper water. Thus, productivity remains at a steady, low level.

productivity is limited by the lack of sunlight). In fact, these areas have so few organisms that they are considered biological deserts.

Productivity in Temperate Oceans

Productivity is limited by available sunlight in polar regions and by nutrient supply in the tropics. In temperate (midlatitude) regions, a combination of these two limiting factors controls productivity, as shown in **Figure 14.16** (which shows the pattern for the Northern Hemisphere; in the Southern Hemisphere, the seasons are reversed).

Winter Productivity in temperate oceans is very low during winter, even though nutrient concentration is highest at this time. The reason is that solar energy is limited because days are short and the Sun angle is low. As a result, the depth at which photosynthesis can occur is so shallow that phytoplankton do not grow much.

Spring The Sun rises higher in the sky during spring, creating a greater depth at which photosynthesis can occur. A *spring bloom* of phytoplankton occurs because solar energy and nutrients are available, and a seasonal thermocline develops (due to increased solar heating) that traps algae in the euphotic zone (**Figure** 14.16 and **14.17**). This creates a tremendous demand for nutrients in the euphotic zone, so the supply is quickly depleted, causing productivity to decrease sharply. Even though the days are lengthening and sunlight is increasing, productivity during the spring bloom is limited by the lack of nutrients.

Summer The Sun rises even higher in the summer, so surface waters in temperate parts of the ocean continue to warm. A strong seasonal thermocline is created that in turn prevents vertical mixing, so nutrients depleted from surface waters cannot be replaced by

those from deeper waters. Throughout summer, the phytoplankton population remains relatively low (Figure 14.16).

Fall Solar radiation diminishes in the fall as the Sun moves lower in the sky, so surface temperatures drop and the summer thermocline breaks down. Nutrients return to the surface layer as increased wind strength mixes surface waters with deeper waters. These conditions create a *fall bloom* of phytoplankton, which is much less dramatic than the spring bloom (Figure 14.16). The fall bloom is very short-lived because sunlight (not nutrient supply, as in the spring bloom) becomes the limiting factor as winter approaches to repeat the seasonal cycle.

Figure 14.18 compares the seasonal variation in phytoplankton biomass of tropical, north polar, and north temperate regions, where the total area under each curve represents photosynthetic productivity. The figure shows the dramatic peak in productivity in polar oceans during the summer; the steady, low rate of productivity year-round in the tropical oceans; and the seasonal productivity that occurs in temperate oceans. It also shows that the highest overall productivity occurs in temperate regions.

CONCEPT CHECK 14.6

1. List two methods by which primary productivity is accomplished in the ocean. Which one is most significant? What two factors influence it?
2. Compare the biological productivity of polar, temperate, and tropical regions of the ocean.

Oceanic Feeding Relationships

Marine algae, plants, bacteria, and bacteria-like archaea are the main oceanic producers. As these producers make food (organic matter) available to the consuming animals of the

FIGURE 14.16 Productivity in temperate oceans (Northern Hemisphere). The graph shows the relationship among phytoplankton, zooplankton, amount of sunshine, and nutrient levels for surface waters.

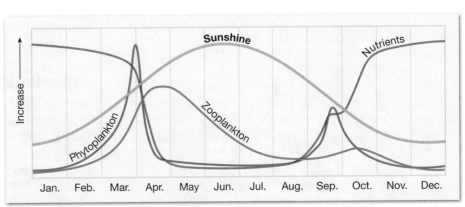

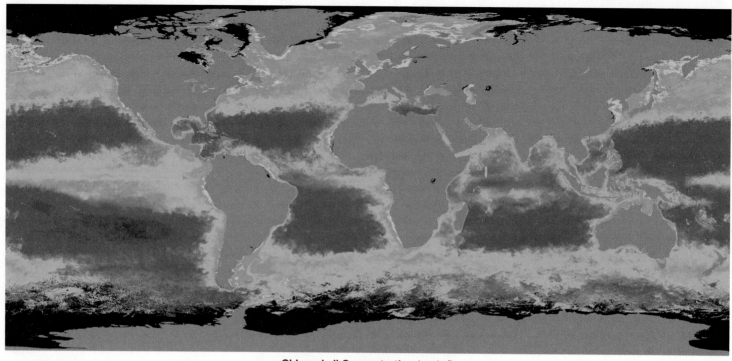

Chlorophyll Concentration (mg/m³)

0.01   0.1   1.0   10 20

FIGURE 14.17 Bright greens, yellows, and reds indicate that the northern oceans were alive with plant life in the spring of 2006. This image shows chlorophyll concentrations as measured by satellite instruments. High chlorophyll concentrations indicate high amounts of photosynthesis. Observations of global chlorophyll patterns tell scientists where ocean surface plants (phytoplankton) are growing, which is an indicator of where marine ecosystems are thriving. Such global maps also give scientists an idea of how much carbon the plants are soaking up, which is important in understanding the global carbon budget. (NASA)

ocean, it passes from one feeding population to the next. Only a small percentage of the energy taken in at any level is passed on to the next because energy is consumed and lost at each level. As a result, the producers' biomass in the ocean is many times greater than the mass of the top consumers, such as sharks or whales.

FIGURE 14.18 Comparison of productivity in tropical, temperate, and polar oceans (Northern Hemisphere), showing seasonal variation in phytoplankton biomass. The total area under each curve represents annual photosynthetic productivity.

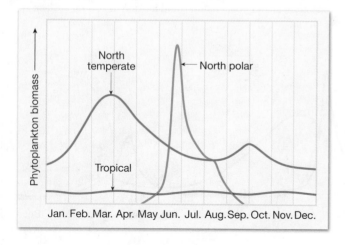

Trophic Levels

Chemical energy stored in the mass of the ocean's algae (the "grass of the sea") is transferred to the animal community mostly through feeding. Zooplankton are *herbivores* (*herba* = grass, *vora* = eat), so they eat diatoms and other microscopic marine algae. Larger herbivores feed on the larger algae and marine plants that grow attached to the ocean bottom near shore.

The herbivores (grazers) are then eaten by larger animals, the *carnivores* (*carni* = meat, *vora* = eat). They in turn are eaten by another population of larger carnivores, and so on. Each of these feeding stages is called a **trophic** (*tropho* = nourishment) **level**.

In the ocean, individual members of a feeding population are generally larger—but not too much larger—than the organisms they eat. There are conspicuous exceptions, however, such as the blue whale. Up to 30 meters (100 feet) long, it is possibly the largest animal that has ever existed on Earth, yet it feeds mostly on krill, which have a maximum length of only 6 centimeters (2.4 inches).

Transfer Efficiency

The transfer of energy between trophic levels is very inefficient. The efficiencies of different algal species vary, but the average is only about *2 percent*, which means that 2 percent of the light energy absorbed by algae is ultimately synthesized into food and made available to herbivores.

Figure 14.19 shows the passage of energy between trophic levels through an entire ecosystem, from the solar energy assimi-

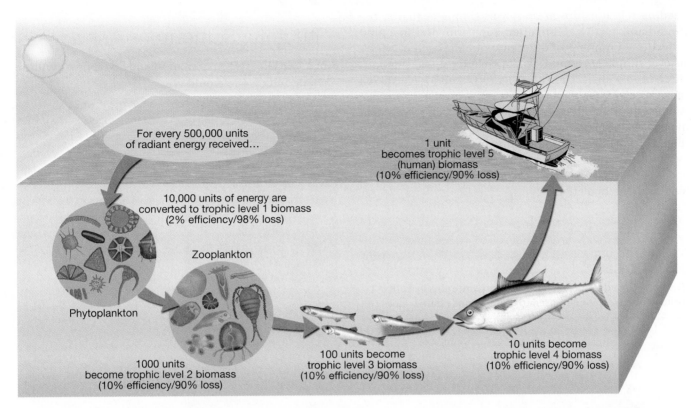

FIGURE 14.19 Ecosystem energy flow and efficiency. For every 500,000 units of radiant energy input available to the producers (phytoplankton), only one unit of mass is added to the fifth trophic level (humans). Average phytoplankton transfer efficiency is 2 percent (98 percent loss) and all other trophic levels average 10 percent efficiency (90 percent loss). The ultimate effect of energy transfer between trophic levels is that the number of individuals and the total biomass decrease at successive trophic levels because the amount of available energy decreases.

lated by phytoplankton through all trophic levels to the ultimate carnivore—humans. Because energy is lost at each trophic level, it takes thousands of smaller marine organisms to produce a single fish that is so easily consumed during a meal!

Food Chains and Food Webs

A **food chain** is a sequence of organisms through which energy is transferred, starting with an organism that is the primary producer, then an herbivore, then one or more carnivores, finally culminating with the "top carnivore," which is not usually preyed upon by any other organism.

Because energy transfer between trophic levels is inefficient, it is advantageous for fishers to choose a population that feeds as close to the primary producing population as possible. This increases the biomass available for food and the number of individuals available to be taken by the fishery. Newfoundland herring, for example, are an important fishery that usually represents the third trophic level in a food chain. They feed primarily on small copepods that feed, in turn, upon diatoms (Figure 14.20A).

FIGURE 14.20 Comparison between a food chain and a food web. **A.** A food chain is the passage of energy along a single path, such as from diatoms to copepods to Newfoundland herring. Feeding relationships are rarely this simple. **B.** A food web, showing multiple paths for food sources of the North Sea herring.

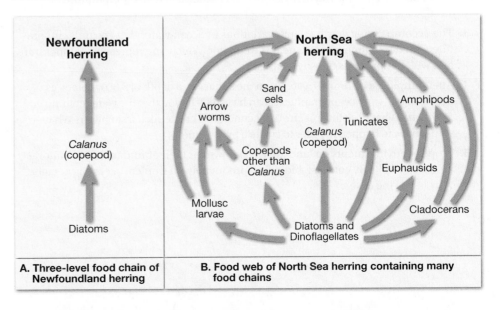

A. Three-level food chain of Newfoundland herring

B. Food web of North Sea herring containing many food chains

Feeding relationships are rarely as simple as that of the Newfoundland herring. More often, top carnivores in a food chain feed on a number of different animals, each of which has its own simple or complex feeding relationships. This constitutes a **food web**, as shown in Figure 14.20B for North Sea herring.

Animals that feed through a food web rather than a food chain are more likely to survive because they have alternative foods to eat should one of their food sources diminish in quantity or even disappear. Newfoundland herring, on the other hand, eat only copepods, so the disappearance of copepods would catastrophically affect their population.

CONCEPT CHECK 14.7

❶ Describe the advantage that a top carnivore gains by eating from a food web as compared to a single food chain.

Students Sometimes Ask...

In a battle between a killer whale and a great white shark, which one would win?

Although many people who are fascinated with large and powerful wild animals have often wondered which of the two would win such a fight, there was little evidence to settle the question until a remarkable video was taken in waters off the northern California coast in 1997. It documented a battle between a 6-meter (20-foot) juvenile killer whale (*Orcinus orca*) and a 3.6-meter (12-foot) adult great white shark (*Carcharodon carcharias*). The video clearly shows the killer whale completely severing the shark's head! If this is representative of the way these two animals interact in the wild, then the killer whale is the top carnivore in the ocean. It is thought that the killer whale's superior maneuverability and use of echolocation helped it defeat the shark.

GIVE IT SOME THOUGHT

1. Assume someone brings several water samples to your laboratory. His problem is that the labels are incomplete. He knows samples A and B are from the Atlantic Ocean and that one came from near the equator and the other from near the Tropic of Cancer. But, he does not know which one is which. He has a similar problem with samples C and D. One is from the Red Sea and the other is from the Baltic Sea. Applying your knowledge of ocean salinity, how would you identify the location of each sample? How were you able to figure this out?

2. You are swimming in the open ocean near the equator. The thermocline in this location is about 1° C per 50 meters of depth. If the sea surface temperature is 24° C, how deep must you dive before you encounter a water temperature of 19° C?

3. The accompanying graph depicts variations in ocean water density and temperature with depth for a location near the equator. Which line represents temperature and which represents density? Explain.

4. After sampling a column of water from the surface to a depth of 3,000 meters, a colleague aboard an oceanographic research vessel tells you that the water column is *isopycnal*. What does that mean? What conditions create such a situation? What would have to happen in order to create a pycnocline?

5. Tropical environments on land are well known for their abundant life—rainforests are an example. By contrast, biological productivity in tropical oceans is meager. Why is this the case?

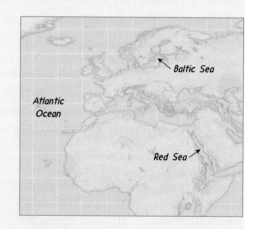

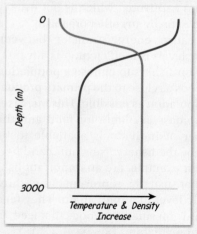

6. The accompanying graph relates to the abundance of ocean life (productivity) in a polar region of the Northern Hemisphere. Which line represents phytoplankton and which represents zooplankton? How did you figure this out? Why are the curves so low from November through February?

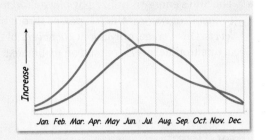

7. A storm near the coast produces sediment-rich runoff that causes water in the euphotic zone to become cloudy. Describe how this would affect the (a) plankton, (b) nekton, and (c) benthos.

8. Refer to Figure 14.19. What is the average efficiency of energy transfer between trophic levels? Use this efficiency to determine how much phytoplankton mass is required to add 1 gram of mass to a killer whale, which is a third-level carnivore.

9. How might the removal of the top carnivore affect a food web? How would the removal of the primary producer affect the food web? Which change would be more significant than the other?

In Review   Chapter 14 Ocean Water and Ocean Life

- *Salinity* is the amount of dissolved substances in water, usually expressed in parts per thousand (‰). Seawater salinity in the open ocean *averages* 35‰. The principal elements that contribute to the ocean's salinity are *chlorine* (55 percent) and *sodium* (31 percent), which combine to produce table salt. The primary *sources of the elements in sea salt* in the ocean are *chemical weathering* of rocks on the continents and *volcanic outgassing*.

- *Variations in seawater salinity* are primarily caused by changing the *water content*. Natural processes that add large amounts of freshwater to seawater and *decrease salinity* include *precipitation, runoff from land, icebergs melting,* and *sea ice melting*. Processes that remove large amounts of freshwater from seawater and *increase salinity* include the *formation of sea ice* and *evaporation*. Seawater salinity in the open ocean *ranges from 33‰ to 38‰*, with some marginal seas experiencing considerably more variation.

- The ocean's *surface temperature* is related to the amount of solar energy received and *varies as a function of latitude*. Low-latitude regions have distinctly colder water at depth, creating a *thermocline*, which is a layer of rapidly changing temperature. No thermocline exists in high-latitude regions, because the water column is *isothermal*.

- Water's unique thermal properties have caused the *ocean's temperature to remain stable* for long periods of time, facilitating the development of life on Earth. Experiments have been conducted that *send sound through the ocean* to determine if the ocean's temperature is increasing as a result of global warming.

- Seawater *density is mostly affected by water temperature* but also by salinity. Low-latitude regions have distinctly denser (colder) water at depth, creating a *pycnocline*, which is a layer of rapidly changing density. No pycnocline exists in high-latitude regions because the water column is *isopycnal*.

- Most open-ocean regions exhibit a *three-layered structure based on water density*. The shallow *surface mixed zone* has warm and nearly uniform temperatures. The *transition zone* includes a prominent thermocline and associated pycno-

cline. The *deep zone* is continually dark and cold and accounts for 80 percent of the water in the ocean. In high latitudes, the three-layered structure does not exist.

- The *increasing amount of atmospheric carbon dioxide released by the burning of fossil fuels is causing the ocean to become more acidic*, with serious implications for marine chemistry and marine life.

- *Marine life is superbly adapted to the oceans*. Marine organisms can be *classified into* one of three groups based on *habitat* and *mobility*. *Plankton* are free-floating forms with little power of locomotion, *nekton* are swimmers, and *benthos* are bottom dwellers. *Most of the ocean's biomass is planktonic*.

- Three criteria are frequently used to establish *marine life zones*. Based on *availability of sunlight*, the ocean can be divided into the *photic zone* (which includes the *euphotic zone*) and the *aphotic zone*. Based on *distance from shore*, the ocean can be divided into the *intertidal zone*, the *neritic zone*, and the *oceanic zone*. Based on *water depth*, the ocean can be divided into the *pelagic zone* and the *benthic zone* (which includes the *abyssal zone*).

- *Primary productivity* is the amount of carbon fixed by organisms through the synthesis of organic matter using energy derived from solar radiation (*photosynthesis*) or chemical reactions (*chemosynthesis*). Chemosynthesis is much less significant than photosynthesis in worldwide oceanic productivity. Photosynthetic productivity in the ocean varies due to the *availability of nutrients* and *amount of solar radiation*.

- *Oceanic photosynthetic productivity* varies at different latitudes because of *seasonal changes* and the *development of a thermocline*. In *polar oceans*, the availability of *solar radiation limits productivity* even though nutrient levels are high. In *tropical oceans*, a strong *thermocline* exists year-round, so the *lack of nutrients* generally limits productivity. In *temperate oceans*, productivity peaks in the spring and fall and is *limited by the lack of solar radiation in winter* and by the *lack of nutrients in summer*.

- The Sun's energy is utilized by *phytoplankton* and converted to *chemical energy*, which is passed through different *trophic levels*. On average, only about *10 percent* of the mass taken in at one trophic level is passed on to the next. As a result, the *size of individuals increases* but the *number of individuals decreases* with each trophic level of the *food chain* or *food web*. Overall, the total biomass of populations decreases at successive trophic levels.

Key Terms

abyssal zone (p. 415)
aphotic zone (p. 414)
benthic zone (p. 415)
benthos (p. 412)
biomass (p. 411)
density (p. 408)
euphotic zone (p. 413)
food chain (p. 418)

food web (p. 420)
intertidal zone (p. 414)
nekton (p. 411)
neritic zone (p. 414)
oceanic zone (p. 414)
pelagic zone (p. 414)
photic zone (p. 413)
photosynthesis (p. 410)

phytoplankton (p. 411)
plankton (p. 411)
primary productivity (p. 415)
pycnocline (p. 409)
salinity (p. 404)
thermocline (p. 407)
trophic level (p. 418)
zooplankton (p. 411)

Examining the Earth System

Discuss the ocean's importance in the hydrologic cycle [see Figures 1.11 (p. 14) and 5.2 (p. 119)]. If Earth did not have oceans, how might this affect (a) the global biosphere and (b) the rock cycle?

Mastering Geology

Looking for additional review and test prep materials? Visit the Self Study area in **www.masteringgeology.com** to find practice quizzes, study tools, and multimedia that will aid in your understanding of this chapter's content. In **MasteringGeology™** you will find:

- **GEODe: Earth Science: An interactive visual walkthrough of key concepts**

- **Geoscience Animation Library: More than 100 animations illuminating many difficult-to-understand Earth science concepts**
- **In The News RSS Feeds: Current Earth science events and news articles are pulled into the site with assessment**
- **Pearson eText**
- **Optional Self Study Quizzes**
- **Web Links**
- **Glossary**
- **Flashcards**

15

The Dynamic Ocean

Hatteras Island, North Carolina, is a barrier island. When storm waves strike this beach, these homes are obviously vulnerable.

(Photo by Michael Collier)

The restless waters of the ocean are constantly in motion, powered by many different forces. Winds, for example, generate surface currents, which influence coastal climate and provide nutrients that affect the abundance of algae and other marine life in surface waters. Winds also produce waves that carry energy from storms to distant shores, where their impact erodes the land. In some areas, density differences create deep-ocean circulation, which is important for ocean mixing and recycling nutrients. In addition, the Moon and the Sun produce tides, which periodically raise and lower average sea level. This chapter examines these movements of ocean waters and their effect upon coastal regions (Figure 15.1).

FOCUS ON CONCEPTS

To assist you in learning the important concepts in this chapter, focus on the following questions:

- What is the force that drives the ocean's surface currents? What is the basic pattern of surface currents?
- How do surface ocean currents influence climate?
- What is *thermohaline circulation*?
- Why is the shoreline considered a dynamic interface?
- What factors influence the height, length, and period of a wave? Can you describe the motion of water *within* a wave?
- How do waves erode?
- What are some typical features produced by wave erosion and from sediment deposited by beach drift and longshore currents?
- What are the local factors that influence shoreline erosion, and what are some basic responses to shoreline erosion problems?
- How do emergent and submergent coasts differ in their formation and characteristic features?
- How are tides produced?

Surface Circulation of the Ocean

Ocean currents are masses of ocean water that flow from one place to another. The amount of water can be large or small, currents can be at the surface or deep below, and the phenomena that create them can be simple or complex. In all cases, however, the currents that are generated involve water masses in motion.

Surface currents develop from friction between the ocean and the wind that blows across its surface. Some of these currents are short-lived and affect only small areas. Such water movements are responses to local or seasonal influences. Other surface currents are relatively permanent phenomena that extend over large portions of the oceans. These major horizontal movements of surface waters are closely related to the global pattern of prevailing winds.[35] As an example, Figure 15.2 shows how the wind belts known as the *trade winds* and the *westerlies* create large, circular-moving loops of water in the Atlantic Ocean basin. The same wind belts influence the other ocean basins as well, so that a similar pattern of currents can also be seen in the Pacific and Indian oceans. Essentially, the pattern of surface ocean circulation closely matches the

pattern of global winds, but is also strongly influenced by the distribution of major landmasses. Other factors that affect surface current patterns include gravity, friction, and the Coriolis effect.

The Pattern of Ocean Currents

Huge, circular-moving current systems dominate the surfaces of the oceans. These large whirls of water within an ocean basin are called **gyres** (*gyros* = a circle). Figure 15.2 shows the world's five main gyres: the *North Pacific Gyre*, the *South Pacific Gyre*, the *North Atlantic Gyre*, the *South Atlantic Gyre*, and the *Indian Ocean Gyre* (which exists mostly within the Southern Hemisphere). The center of each gyre coincides with the subtropics at about 30 degrees north or south latitude, so they are often called *subtropical gyres*.

As shown in Figure 15.2, subtropical gyres rotate clockwise in the Northern Hemisphere and counterclockwise in the Southern Hemisphere. Why do the gyres flow in different directions in the two hemispheres? Although wind is the force that generates surface currents, other factors also influence the movement of ocean waters. The most significant of these is the **Coriolis effect**. Because of Earth's rotation, currents are deflected to the *right* in the Northern Hemisphere and to the *left* in the Southern Hemisphere.[36] As

---

[35]Details about the global pattern of atmospheric winds appear in Chapter 18.

[36]The Coriolis effect is more fully explained in Chapter 18.

a consequence, gyres flow in opposite directions in the two different hemispheres.

Four main currents generally exist within each gyre (Figure 15.2). The North Pacific Gyre, for example, consists of the North Equatorial Current, the Kuroshio Current, the North Pacific Current, and the California Current. The tracking of floating objects that are released into the ocean intentionally or accidentally reveal that it takes about 6 years for the objects to go all the way around the loop (see Box 15.1).

The North Atlantic Ocean has four main currents, too (Figure 15.2). Beginning near the equator, the North Equatorial Current is deflected northward through the Caribbean, where it becomes the Gulf Stream. As the Gulf Stream moves along the East Coast of the United States, it is strengthened by the prevailing westerly winds and is deflected to the east (to the right) offshore of North Carolina into the North Atlantic. As it continues northeastward, it gradually widens and slows until it becomes a vast, slowly moving current known as the North Atlantic Current, which, because of its sluggish nature, is also known as the North Atlantic Drift.

As the North Atlantic Current approaches Western Europe, it splits, part of it moving northward past Great Britain, Norway, and Iceland, carrying heat to these otherwise chilly areas. The other part is deflected southward as the cool Canary Current. As the Canary Current moves southward, it eventually merges into the North Equatorial Current, completing the gyre. Because the North Atlantic Ocean basin is about half the size of the North Pacific, it takes floating objects about 3 years to go completely around this gyre.

The circular motion of gyres leaves a large central area that has no well-defined currents. In the North Atlantic, this zone of calmer waters is known as the Sargasso Sea, named for the large quantities of *Sargassum*, a type of floating seaweed encountered there.

The ocean basins in the Southern Hemisphere exhibit a similar pattern of flow as the Northern Hemisphere basins, with surface currents that are influenced by wind belts, the position of continents, and the Coriolis effect. In the South Atlantic and South Pacific, for example, surface ocean circulation is very much the same as in their Northern Hemisphere counterparts except that the direction of flow is counterclockwise (Figure 15.2).

The Indian Ocean exists mostly in the Southern Hemisphere, so it follows a surface circulation pattern similar to other Southern Hemisphere ocean basins (Figure 15.2). The small portion of the Indian Ocean in the Northern Hemisphere, however,

FIGURE 15.1 Wind is not only responsible for creating ocean waves but it also provides the force that drives the ocean's surface circulation. (Photo by Novastock/F1 ONLINE/age fotostock)

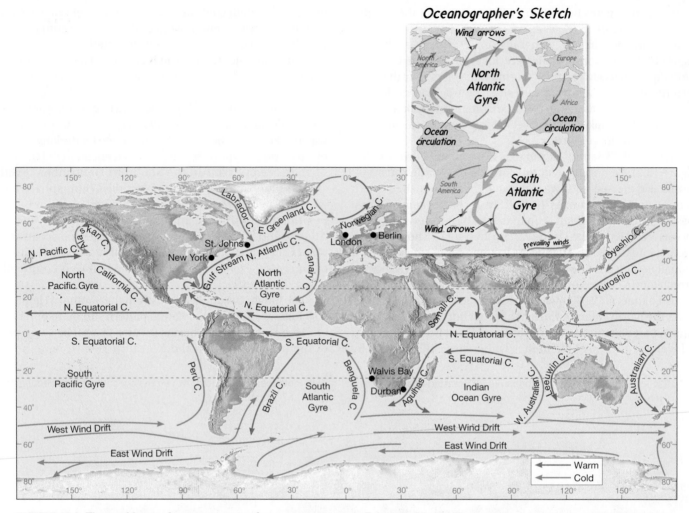

FIGURE 15.2 The world map shows average surface ocean currents in February–March. The ocean's circulation is organized into five major gyres, which occur in the North Pacific, South Pacific, North Atlantic, South Atlantic, and Indian oceans. Poleward-moving currents are warm and equatorward-moving currents are cold. Ocean currents play an important role in redistributing heat around the globe. Note that cities mentioned in the text discussion are shown on this map. In the oceanographer's sketch, broad arrows show the idealized surface circulation for the Atlantic Ocean. Prevailing winds are responsible for the circular-moving loops called gyres.

is influenced by the seasonal wind shifts known as the summer and winter *monsoons* (*mausim* = season). When the winds change direction, the surface currents also reverse direction.

The West Wind Drift is the only current that completely encircles Earth (Figure 15.2). It flows around the ice-covered continent of Antarctica, where no large landmasses are in the way, so its cold surface waters circulate in a continuous loop. It moves in response to the Southern Hemisphere prevailing westerly winds, and portions of it split off into the adjoining southern ocean basins. Its strong flow also helps define the Southern Ocean or Antarctic Ocean, which is really the portions of the Pacific, Atlantic, and Indian oceans south of about 50 degrees south latitude.

The Gulf Stream

Anyone who navigates the oceans needs to be aware of currents. By understanding the direction and strength of ocean currents, sailors soon realize that their voyage time can be reduced if they travel with a current, or increased if they travel against a current.

Such is the case for the strong Gulf Stream current, which is the best known and most studied of all ocean currents.

The Gulf Stream flows northward along the East Coast of the United States. It was given the name Gulf Stream because it carries warm water from the Gulf of Mexico and because it was narrow and well defined—similar to a stream, but in the ocean. As deputy postmaster general for the new colonies, Benjamin Franklin was the first to produce a map of the Gulf Stream (**Figure 15.3A**). His interest in the North Atlantic began when it was brought to his attention that ships carrying the mail took 2 weeks longer going from England to America than in the other direction. P. L. Richardson[37] describes how Franklin discovered that an ocean current was causing the delay:

> While he was in London as Deputy Postmaster General for the American colonies, Franklin was consulted on the

[37]P. L. Richardson, "Benjamin Franklin and Timothy Folger's First Printed Chart of the Gulf Stream," *Science* 207(1980): 643.

Box 15.1

UNDERSTANDING EARTH

*Running Shoes as Drift Meters—Just Do It**

Any floating object can serve as a makeshift drift meter, as long as it is known where the object entered the ocean and where it was retrieved. The path of the object can then be inferred, providing information about the movement of surface currents. Oceanographers have long used *drift bottles* (a floating "message in a bottle" or a radio-transmitting device set adrift in the ocean) to track the movement of currents and to refine computer models of ocean circulation.

Sometimes objects have inadvertently become drift meters when ships have lost some (or all) of their cargo at sea. In May 1990 when a ship en route from Korea to Seattle, Washington, encountered a severe North Pacific storm, it lost 21 deck containers overboard, including five that held Nike athletic shoes. The shoes that were released from their containers floated and were carried eastward by the North Pacific Current. Within 6 months thousands of the shoes began to wash up along the beaches of Alaska, Canada, Washington, and Oregon (**Figure 15.A**), more than 2,400 kilometers (1,500 miles) from the site of the spill. A few shoes were found on beaches in northern California, and over two years later shoes from the spill were even recovered from the north end of the Big Island of Hawaii! With help from the beachcombing public and

*This box was originally prepared by Professor Alan P. Trujillo, Palomar College.

remotely based lighthouse operators, information on the location and number of shoes collected was compiled during the months following the spill.

In January 1992 another cargo ship had 12 containers washed overboard during a storm to the north of where the shoes had previously spilled. One of these containers held 29,000 packages of small, floatable plastic bathtub toys. The floating toys began to come ashore in southeast Alaska 10 months later, helping to verify computer models of North Pacific circulation.

Oceanographers continue to study ocean currents by tracking floating items spilled from cargo ships. Examples include 34,000 hockey gloves, 5 million plastic Lego pieces, 262,500 plastic soap dispensers, and even 17,000 cans of noodles.

FIGURE 15.A Large ships crossing the ocean have lost entire containers overboard. If the containers release floating items, inadvertent float meters are launched that help oceanographers track ocean surface currents. The map shows the path of drifting shoes and recovery locations from a spill in 1990; inset shows recovered shoes from the 1990 spill and plastic bathtub toys from a spill in 1992. (Copyright by the American Geophysical Union)

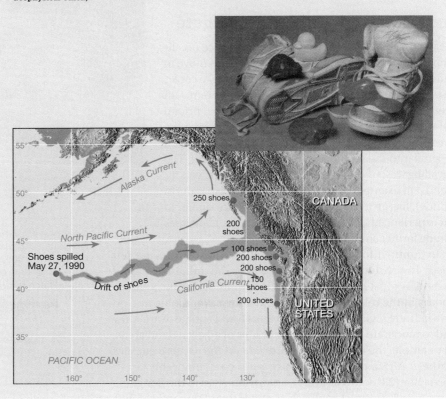

question of why the mail packets took a fortnight longer to sail to America than the merchant ships. In October 1768 Franklin discussed this problem with his cousin Timothy Folger, a Nantucket ship captain then visiting London. Folger told him the packet captains were ignorant of the Gulf Stream and frequently sailed in this current. . . . Folger sketched the Gulf Stream on a chart and added written notes on how to avoid the Gulf Stream and Franklin had the chart printed in 1769 or 1770.

Even though the Franklin–Folger chart of the Gulf Stream was widely distributed, it was initially ignored by the captains of the mail ships because they thought that simple American fishers could not possess superior knowledge of the sea.

Nearly 100 years later Matthew Fontaine Maury, the founder of physical oceanography, described the Gulf Stream in his 1855 book, *The Physical Geography of the Sea*:

> There is a river in the ocean. In the severest droughts it never fails, and in the mightiest floods . . . it never overflows. Its banks and its bottoms are of cold water, while its current is of warm. The Gulf of Mexico is its fountain, and its mouth is in the Arctic Sea. It is the Gulf Stream. There is in the world no other such majestic flow of waters.

Today the warm waters of the Gulf Stream can be mapped by Earth-orbiting satellites (Figure 15.3B). Satellite images reveal that the Gulf Stream has many complexities, including prominent bends called meanders, which produce large circular eddies that spin off from the main current and can last for up to 2 years before dissipating. Although the Franklin–Folger chart was created long before the advent of modern technology, which allows navigators to map ocean currents from space, it still remains a good summary of the average path of the Gulf Stream.

A.

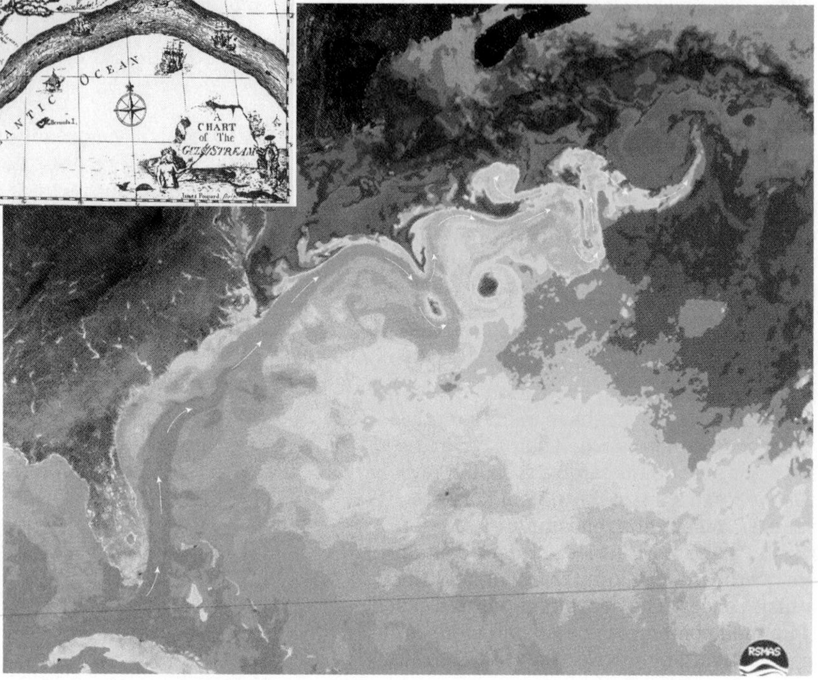

B.

Ocean Currents Influence Climate

Ocean currents can have a significant influence on climate. When currents from low-latitude regions move to higher latitudes, they transfer heat from warmer to cooler areas on Earth. In fact, the North Atlantic Current—an extension of the warm Gulf Stream—keeps Great Britain and much of northwestern Europe warmer during the winter than one would expect for their latitudes, which are similar to the latitudes of Alaska and Labrador. The prevailing westerly winds carry the moderating effects far inland. For example, Berlin, Germany (52 degrees north latitude), has an average January temperature similar to that experienced at New York City, which lies 12 degrees latitude farther south.

In contrast to warm ocean currents whose effects are felt mostly in the middle latitudes in winter, the influence of cold currents is most pronounced in the tropics or during summer months in the middle latitudes. Cold currents originate in cold high-latitude regions. As these currents travel equatorward, they tend to moderate the warm temperatures of adjacent land areas. For example, the cool Benguela Current off the western coast of southern Africa moderates the tropical heat along this coast. Walvis Bay (23° south latitude), a town adjacent to the Benguela Current, is 5° C (9° F) cooler in summer than Durban, which is 6° latitude farther poleward but on the eastern side of South Africa, away from the influence of the current (Figure 15.2). The east and west coasts of South America provide another example. Figure 15.4 shows monthly mean temperatures for Rio de Janeiro, Brazil, which is influenced by the warm Brazil Current and Arica, Chile, which is adjacent to the cold Peru Current. Closer to home, because of the cold California current, summer temperatures in subtropical coastal southern California are lower by 6° C (10.8° F) or more compared to East Coast stations.

In addition to influencing temperatures of adjacent land areas, cold currents have other climatic influences. For example, where tropical deserts exist along the west coasts of continents, cold ocean currents have a dramatic impact. The principal west coast deserts are the Atacama in Peru and Chile, and the Namib in southwestern Africa. The aridity along these

FIGURE 15.3 A. To aid ships crossing the Atlantic, Benjamin Franklin, with the assistance of his cousin, Timothy Folger, produced the first detailed chart of the Gulf Stream in 1769 or 1770. (Courtesy NOAA) **B.** Today, satellite images such as this false-color image of sea–surface temperature provide views of the Gulf Stream's complexities. The warm waters of the Gulf Stream are shown in red and orange; colder waters are shown in green, blue, and purple. As the Gulf Stream meanders northward, some of its meanders pinch off to form large circular eddies. (Courtesy of O. Brown, R. Evans, and M. Carle/ University of Miami Rosenstiel School of Marine and Atmospheric Science, Miami, Florida)

coasts is intensified because the lower atmosphere is chilled by cold offshore waters. When this occurs, the air becomes very stable and resists the upward movement necessary to create precipitation-producing clouds. In addition, the presence of cold currents causes temperatures to approach and often reach the dew point, the temperature at which water vapor condenses. As a result, these areas are characterized by high relative humidities and much fog. Thus, not all subtropical deserts are hot with low humidities and clear skies. Rather, the presence of cold currents transforms some coastal deserts into extremely dry places that are relatively cool and damp and frequently shrouded in fog (Figure 15.5).

Ocean currents also play a major role in maintaining Earth's heat balance. They accomplish this task by transferring heat from the tropics, where there is an excess of heat, to the polar regions,

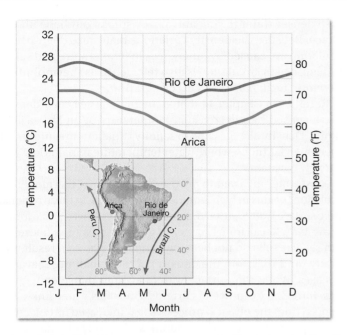

FIGURE 15.4 Monthly mean temperatures for Rio de Janeiro, Brazil, and Arica, Chile. Both are coastal cities near sea level. Even though Arica is closer to the equator than Rio de Janeiro, its temperatures are cooler. Arica is influenced by the cold Peruvian Current, whereas Rio de Janeiro is adjacent to the warm Brazilian Current.

where a heat deficit exists. Ocean water movement accounts for about a quarter of this heat transport, and winds transport the remaining three-quarters.

Upwelling

In addition to producing surface currents, winds can also cause *vertical* water movements. **Upwelling**, the rising of cold water from deeper layers to replace warmer surface water, is a common wind-induced vertical movement. One type of upwelling, called *coastal upwelling*, is most characteristic along the west coasts of continents, most notably along California, western South America, and West Africa.

Coastal upwelling occurs in these areas when winds blow toward the equator and parallel to the coast (**Figure 15.6**). Coastal winds combined with the Coriolis effect cause surface water to move away from shore. As the surface layer moves away from the coast, it is replaced by water that "upwells" from below the surface. This slow upward movement of water from depths of 50–300 meters (165–1,000 feet) brings water that is cooler than the original surface water and results in lower surface water temperatures near the shore.

For swimmers who are accustomed to the warm waters along the mid-Atlantic shore of the United States, a swim in the Pacific off the coast of central California can be a chilling surprise. In August, when temperatures in the Atlantic are 21° C (70° F) or higher, central California's surf is only about 15° C (60° F).

Upwelling brings greater concentrations of dissolved nutrients, such as nitrates and phosphates, to the ocean surface. These nutrient-enriched waters from below promote the growth of microscopic plankton, which in turn support extensive populations of fish and other marine organisms. Figure 15.6 is a satellite image that shows high productivity due to coastal upwelling off the southwest coast of Africa.

CONCEPT CHECK 15.1

1. What is the primary driving force of surface ocean currents? List two additional factors that influence surface currents.
2. Name the five subtropical gyres and identify the main surface currents in each.
3. How do ocean currents influence climate? Provide at least three examples.
4. Describe the process of coastal upwelling. Why is an abundance of marine life associated with these areas?

FIGURE 15.5 Chile's Atacama Desert is the driest desert on Earth. Average rainfall at the *wettest* locations is not more than 3 millimeters (0.12 inch) per year. Stretching nearly 1,000 kilometers (600 miles), the Atacama is situated between the Pacific Ocean and the towering Andes Mountains. The cold Peru Current makes this slender zone cooler and drier than it would otherwise be. (Photo by Leonid Plotkin/Alamy)

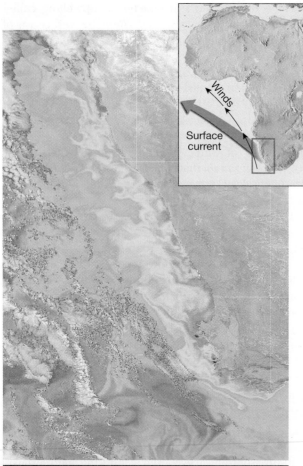

Chlorophyll a Concentration
mg/m³

FIGURE 15.6 Coastal upwelling occurs along the west coasts of continents where winds blow toward the equator and parallel to the coast. Owing to the Coriolis effect (deflection to the left in the Southern Hemisphere), surface water moves away from the shore, which brings cold, nutrient-rich water to the surface. This image from the SeaStar satellite shows chlorophyll concentration along the southwest coast of Africa (February 21, 2001). An instrument aboard the satellite detects changes in seawater color caused by changing concentrations of chlorophyll. High chlorophyll concentrations indicate high amounts of photosynthesis, which is linked to the upwelling nutrients. Red indicates high concentrations and blue, low concentrations. (Provided by the SeaWiFS Project, NASA/Goddard Space Flight Center and ORBIMAGE)

Deep-Ocean Circulation

In contrast to the largely horizontal movements of surface currents, deep-ocean circulation has a significant vertical component and accounts for the thorough mixing of deep-water masses. This component of ocean circulation is a response to density differences among water masses that cause denser water to sink and slowly spread out beneath the surface. Because the density variations that cause deep-ocean circulation are caused by differences in temperature and salinity, deep-ocean circulation is also referred to as **thermohaline** (*thermo* = heat, *haline* = salt) **circulation**.

Recall from Chapter 14 that an increase in seawater density can be caused by a decrease in temperature or an increase in salinity. Density changes due to salinity variations are important in very high latitudes, where water temperature remains low and relatively constant.

Most water involved in deep-ocean currents (thermohaline circulation) begins in high latitudes at the surface. In these regions, where surface waters are cold, salinity increases when sea ice forms (**Figure 15.7**). Recall from Chapter 14 that when seawater freezes to form sea ice, salts do not become part of the ice. As a result, the salinity (and therefore the density) of the remaining seawater increases. When this surface water becomes dense enough, it sinks, initiating deep-ocean currents. Once this water sinks, it is removed from the physical processes that increased its density in the first place, and so its temperature and salinity remain largely unchanged for the duration it spends in the deep ocean.

Near Antarctica, surface conditions create the highest-density water in the world. This cold saline brine slowly sinks to the seafloor, where it moves throughout the ocean basins in sluggish currents. After sinking from the surface of the ocean, deep waters will not reappear at the surface for an average of 500–2,000 years.

A simplified model of ocean circulation is similar to a conveyor belt that travels from the Atlantic Ocean through the Indian and Pacific oceans and back again (**Figure 15.8**). In this model, warm water in the ocean's upper layers flows poleward, converts to dense water, and returns equatorward as cold deep water that

FIGURE 15.7 Sea ice in the water surrounding Antartica. When seawater freezes, sea salts do not become part of the ice. Consequently, the salt content of the remaining seawater becomes more concentrated, which makes it denser and prone to sink. (Photo by John Higdon/agefotostock)

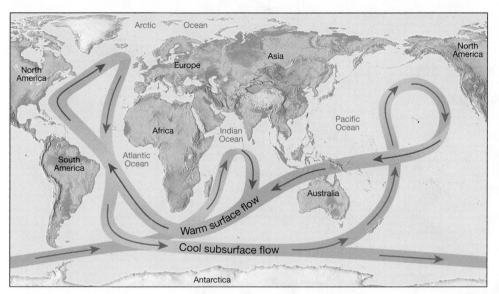

FIGURE 15.8 Schematic map of conveyor-belt circulation of the world ocean. Source areas exist in high-latitude regions. Here surface water cools, becomes denser, and sinks, providing the high-density waters that flow into all of the oceans. This water slowly ascends throughout the oceans and completes the conveyor by returning to the source areas as warm surface currents.

eventually upwells to complete the circuit. As this "conveyor belt" moves around the globe, it influences global climate by converting warm water to cold and liberating heat to the atmosphere.

CONCEPT CHECK 15.2

❶ Why is deep-ocean circulation referred to as *thermohaline circulation*?

❷ Describe or make a simple sketch of the ocean's conveyor-belt circulation.

The Shoreline: A Dynamic Interface

Nowhere is the restless nature of the ocean's water more noticeable than along the shore—the dynamic interface among air, land, and sea. An *interface* is a common boundary where different parts of a system interact. This is certainly an appropriate designation for the coastal zone. Here we can see the rhythmic rise and fall of tides and observe waves rolling in and breaking. Sometimes the waves are low and gentle. At other times they pound the shore with awesome fury.

Shorelines are dynamic environments. Their topography, geologic makeup, and climate vary greatly from place to place. Continental and oceanic processes converge along coasts to create landscapes that frequently undergo rapid change. When it comes to the deposition of sediment, they are transition zones between marine and continental environments.

Although it may not be obvious, the shoreline is constantly being modified by waves. Crashing surf can erode the adjacent land. Wave activity also moves sediment toward and away from the shore

as well as along it. Such activity sometimes produces narrow sandbars that frequently change size and shape as storm waves come and go.

The nature of present-day shorelines is not just the result of the relentless attack of the land by the sea. The shore has a complex character that results from multiple geologic processes. For example, practically all coastal areas were affected by the worldwide rise in sea level that accompanied the melting of glaciers at the close of the Pleistocene epoch. As the sea encroached landward, the shoreline retreated, becoming superimposed upon existing landscapes that had resulted from such diverse processes as stream erosion, glaciation, volcanic activity, and the forces of mountain building.

Today, the coastal zone is experiencing intensive human activity. Unfortunately, people often treat the shoreline as if it were a stable platform on which structures can safely be built. This attitude inevitably leads to conflicts between people and nature (see the chapter-opening photo and Figure 15.9). As you will see, many coastal landforms, especially beaches and barrier islands, are relatively fragile, short-lived features that are inappropriate sites for development.

Coastal Zone Features and Terminology

In general conversation several terms are used when referring to the boundary between land and sea. In the preceding section, the terms *shore*, *shoreline*, *coastal zone*, and *coast* were all used. Moreover, when many think of the land–sea interface, the word *beach* comes to mind. Let's take a moment to clarify these terms and introduce some other terminology used by those who study the land–sea boundary zone. You will find it helpful to refer to Figure 15.10, which is an idealized profile of the coastal zone.

Basic Features

The **shoreline** is the line that marks the contact between land and sea. Each day, as tides rise and fall, the position of the shoreline migrates. Over longer time spans, the average position of the shoreline gradually shifts as sea level rises or falls.

The **shore** is the area that extends between the lowest tide level and the highest elevation on land that is affected by storm waves. By contrast, the **coast** extends inland from the shore as far as ocean-related features can be found. The **coastline** marks the coast's seaward edge, whereas the inland boundary is not always obvious or easy to determine.

As Figure 15.10 illustrates, the shore is divided into the *foreshore* and the *backshore*. The **foreshore** is the area exposed when the tide is out (low tide) and submerged when the tide is in (high tide). The **backshore** is landward of the high-tide shoreline. It is usually dry, being affected by waves only during storms. Two other zones are commonly identified. The **nearshore zone** lies between the low-tide shoreline and the line where waves break at low tide. Seaward of the nearshore zone is the **offshore zone**.

FIGURE 15.9 Many shoreline areas are intensively developed. Often the shifting shoreline sands and the desires of people to occupy these areas are in conflict. This aerial view shows Crystal Beach, Texas, on September 16, 2008, 3 days after Hurricane Ike came ashore. The extraordinary storm surge caused much of the damage pictured here. There is more about hurricanes and hurricane damage in Chapter 19. (Photo by Earl Nottingham/Associated Press)

FIGURE 15.10 The coastal zone consists of several parts. The beach is an accumulation of sediment on the landward margin of the ocean or a lake. It can be thought of as material in transit along the shore.

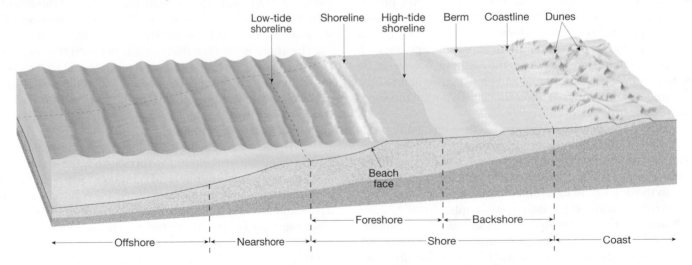

Beaches

For many a beach is the sandy area where people lie in the sun and walk along the water's edge. Technically, a **beach** is an accumulation of sediment found along the landward margin of the ocean or a lake. Along straight coasts, beaches may extend for tens or hundreds of kilometers. Where coasts are irregular, beach formation may be confined to the relatively quiet waters of bays.

Beaches consist of one or more **berms**, which are relatively flat platforms often composed of sand that are adjacent to coastal dunes or cliffs and marked by a change in slope at the seaward edge. Another part of the beach is the **beach face**, which is the wet sloping surface that extends from the berm to the shoreline. Where beaches are sandy, sunbathers usually prefer the berm, whereas joggers prefer the wet, hard-packed sand of the beach face.

Beaches are composed of whatever material is locally abundant. The sediment for some beaches is derived from the erosion of adjacent cliffs or nearby coastal mountains. Other beaches are built from sediment delivered to the coast by rivers.

Although the mineral makeup of many beaches is dominated by durable quartz grains, other minerals may be dominant. For example, in areas such as southern Florida, where there are no mountains or other sources of rock-forming minerals nearby, most beaches are composed of shell fragments and the remains of organisms that live in coastal waters (Figure 15.11A). Some beaches on volcanic islands in the open ocean are composed of weathered grains of the basaltic lava that comprise the islands, or of coarse debris eroded from coral reefs that develop around islands in low latitudes (Figure 15.11B).

Regardless of the composition, the material that comprises the beach does not stay in one place. Instead, crashing waves are constantly moving it. Thus, beaches can be thought of as material in transit along the shore.

FIGURE 15.11 Beaches are composed of whatever material is locally abundant. **A.** This beach on Florida's Sanibel Island consists of shells and shell fragments. (Photo by David R. Frazier Photo library, Inc./Alamy) **B.** The black sands at this beach on Hawaii were derived from dark volcanic rock. (Photo by E. J. Tarbuck)

A.

B.

Waves

The Global Ocean
▶ Coastal Processes

Ocean waves are energy traveling along the interface between ocean and atmosphere, often transferring energy from a storm far out at sea over distances of several thousand kilometers. That's why even on calm days the ocean still has waves that travel across its surface. When observing waves, always remember that you are watching *energy* travel through a medium (water). If you make waves by tossing a pebble into a pond, splashing in a pool, or blowing across the surface of a cup of coffee, you are imparting *energy* to the water, and the waves you see are just the visible evidence of the energy passing through.

Wind-generated waves provide most of the energy that shapes and modifies shorelines. Where the land and sea meet, waves that may have traveled unimpeded for hundreds or thousands of kilometers suddenly encounter a barrier that will not allow them to advance farther and must absorb their energy. Stated another way, the shore is the location where a practically irresistible force confronts an almost immovable object. The conflict that results is never-ending and sometimes dramatic.

Wave Characteristics

Most ocean waves derive their energy and motion from the wind. When a breeze is less than 3 kilometers (2 miles) per hour, only wavelets appear. At greater wind speeds, more stable waves gradually form and advance with the wind.

Characteristics of ocean waves are illustrated in Figure 15.12, which shows a simple, nonbreaking waveform. The tops of the waves are the *crests*, which are separated by *troughs*. Halfway between the crests and troughs is the *still water level*, which is the level that the water would occupy if there were no waves. The vertical distance between trough and crest is called the **wave height**, and the horizontal distance between successive crests (or troughs) is the **wavelength**. The time it takes one full wave—one wavelength—to pass a fixed position is the **wave period**.

The height, length, and period that are eventually achieved by a wave depend on three factors: (1) wind speed; (2) length of time the wind has blown; and (3) *fetch*, the distance that wind has traveled across open water. As the quantity of energy transferred from the wind to the water increases, both the height and steepness of the waves also increase. Eventually, a critical point is reached where waves grow so tall that they topple over, forming ocean breakers called *whitecaps*.

For a particular wind speed, there is a maximum fetch and duration of wind beyond which waves will no longer increase in size. When the maximum fetch and duration are reached for a given wind velocity, the waves are said to be "fully developed." The reason that waves can grow no further is that they are losing as much energy through the breaking of whitecaps as they are receiving energy from the wind.

When the wind stops or changes direction, or the waves leave the storm area where they were created, they continue on without relation to local winds. The waves also undergo a gradual change to *swells*, a term that describes any wave that has traveled out of its area of origination. Swells are lower in height and longer in length and may carry a storm's energy to distant shores. Because many independent wave systems exist at the same time, the sea surface acquires a complex and irregular pattern, sometimes producing very large waves. The sea waves that are seen from shore are usually a mixture of swells from faraway storms and waves created by local winds.

Circular Orbital Motion

Waves can travel great distances across ocean basins. In one study, waves generated near Antarctica were tracked as they traveled through the Pacific Ocean basin. After more than 10,000 kilometers (over 6,000 miles), the waves finally expended their energy a week later along the shoreline of the Aleutian Islands of Alaska. The water itself doesn't travel the entire distance, but the wave form does. As the wave travels, the water passes the energy along by moving in a circle. This movement is called *circular orbital motion*.

Observation of an object floating in waves shows that it moves not only up and down but also slightly forward and backward with each successive wave. Figure 15.13 shows that a floating object moves up and backward as the crest approaches, up and forward as the crest passes, down and forward after the crest, down and backward as the trough approaches, and rises and

FIGURE 15.12 Diagrammatic view of an idealized non-breaking ocean wave showing the basic parts of a wave as well as the movement of water particles at depth. Negligible water movement occurs below a depth equal to one-half the wavelength (*lower dashed line*).

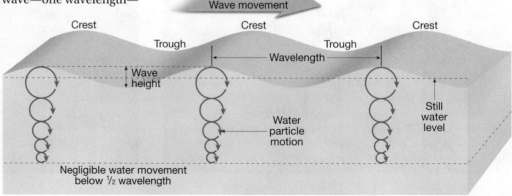

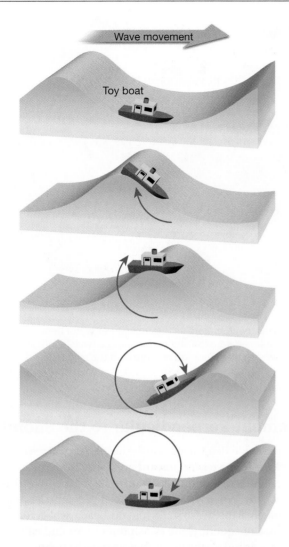

Wave movement

Toy boat

FIGURE 15.13 The movements of the toy boat show that the waveform advances, but the water does not advance appreciably from its original position. In this sequence, the wave moves from left to right as the boat (and the water in which it is floating) rotates in an imaginary circle.

moves backward again as the next crest advances. When the movement of the floating toy boat shown in Figure 15.13 is traced as a wave passes, it can be seen that the boat moves in a circle and it returns to essentially the same place. Circular orbital motion allows a waveform (the wave's shape) to move forward *through the water* while the individual water particles that transmit the wave move around in a circle. Wind moving across a field of wheat causes a similar phenomenon: The wheat itself doesn't travel across the field, but the waves do.

The energy contributed by the wind to the water is transmitted not only along the surface of the sea but also downward. However, beneath the surface the circular motion rapidly diminishes until, at a depth equal to one-half the wavelength measured from still water level, the movement of water particles becomes negligible. This depth is known as the *wave base*. The dramatic decrease of wave energy with depth is shown by the rapidly diminishing diameters of water-particle orbits in Figure 15.12.

Waves in the Surf Zone

As long as a wave is in deep water, it is unaffected by water depth (Figure 15.14, left). However, when a wave approaches the shore, the water becomes shallower and influences wave behavior. The wave begins to "feel bottom" at a water depth equal to its wave base. Such depths interfere with water movement at the base of the wave and slow its advance (Figure 15.14, center).

As a wave advances toward the shore, the slightly faster waves farther out to sea catch up, decreasing the wavelength. As the speed and length of the wave diminish, the wave steadily grows higher. Finally, a critical point is reached when the wave is too steep to support itself and the wave front collapses, or *breaks* (Figure 15.14, right), causing water to advance up the shore.

The turbulent water created by breaking waves is called **surf**. On the landward margin of the surf zone, the turbulent sheet of water from collapsing breakers, called *swash*, moves up the slope of the beach. When the energy of the swash has been expended, the water flows back down the beach toward the surf zone as *backwash*.

CONCEPT CHECK 15.4

❶ List three factors that determine the height, length, and period of a wave.
❷ Describe the motion of a floating object as a wave passes.
❸ How does a wave's speed, length, and height change as it moves into shallow water and breaks?

Wave Erosion

GEODe
EARTH
SCIENCE

The Global Ocean
▸ **Coastal Processes**

During calm weather, wave action is minimal. During storms, however, waves are capable of causing much erosion. The impact of large, high-energy waves against the shore can be awesome in its violence. Each breaking wave may hurl thousands of tons of water against the land, sometimes causing the ground literally to tremble. The pressures exerted by Atlantic waves in wintertime, for example, average nearly 10,000 kilograms per square meter (more than 2,000 pounds per square foot). The force during storms is even greater.

It is no wonder that cracks and crevices are quickly opened in cliffs, coastal structures, and anything else that is subjected to these enormous shocks (Figure 15.15). Water is forced into every opening, causing air in the cracks to become highly compressed by the thrust of crashing waves. When the wave subsides, the air expands rapidly, dislodging rock fragments and enlarging and extending fractures.

In addition to the erosion caused by wave impact and pressure, **abrasion**—the sawing and grinding action of the water armed with rock fragments—is also important. In fact, abrasion is probably more intense in the surf zone than in any other environment. Smooth, rounded stones and pebbles along the shore are obvious reminders of the relentless grinding action of rock against rock in the

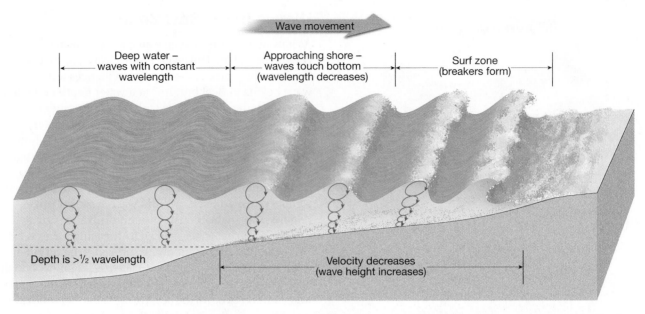

Wave movement

Deep water –
waves with constant
wavelength

Approaching shore –
waves touch bottom
(wavelength decreases)

Surf zone
(breakers form)

Depth is >½ wavelength

Velocity decreases
(wave height increases)

FIGURE 15.14 Changes that occur when a wave moves onto shore. The waves touch bottom as they encounter water depths less than half a wavelength. The wave speed decreases, and the waves stack up against the shore, causing the wavelength to decrease. This results in an increase in wave height to the point where the waves pitch forward and break in the surf zone.

surf zone (Figure 15.16A). Rock fragments are also used as "tools" by the waves as they cut horizontally into the land (Figure 15.16B).

CONCEPT CHECK 15.5

① Describe two ways in which waves cause erosion.

Sand Movement on the Beach

Beaches are sometimes called "rivers of sand." The reason is that the energy from breaking waves often causes large quantities of sand to move along the beach face and in the surf zone roughly parallel to the shoreline. Wave energy also causes sand to move perpendicular to (toward and away from) the shoreline.

Movement Perpendicular to the Shoreline

If you stand ankle-deep in water at the beach, you will see that swash and backwash move sand toward and away from the shoreline. Whether there is a net loss or addition of sand depends on the level of wave activity. When wave activity is relatively light (less energetic waves), much of the swash soaks into the beach, which reduces the backwash. Consquently, the swash dominates and causes a net movement of sand up the beach face toward the berm.

When high-energy waves prevail, the beach is saturated from previous waves, so much less of the swash soaks in. As a result, the berm erodes because backwash is strong and causes a net movement of sand down the beach face.

Along many beaches, light wave activity is the rule during the summer. Therefore, a wide sand berm gradually develops. During winter, when storms are frequent and more powerful, strong

Seawall

FIGURE 15.15 The force of breaking waves can be powerful. Here waves batter a seawall at Sea Bright, New Jersey. A seawall 5–6 meters (16–20 feet) high and 8 kilometers (5 miles) long was built to protect the town and the railroad that brought tourists to the beach. As you can see, after the wall was built, the beach narrowed dramatically. (Photo by Rafael Macia/Photo Researchers, Inc.)

A.

B.

FIGURE 15.16 **A.** Abrasion can be intense in the surf zone. Smooth rounded stones along the shore are an obvious reminder of this fact. (Photo by Michael Collier) **B.** Sandstone cliff undercut by wave erosion at Gabriola Island, British Columbia, Canada. (Photo by Fletcher and Baylis/Photo Researchers, Inc.)

wave activity erodes and narrows the berm. A wide berm that may have taken months to build can be dramatically narrowed in just a few hours by the high-energy waves created by a strong winter storm.

Wave Refraction

The bending of waves, called **wave refraction** (*refringere* = to break up) plays an important part in shoreline processes (Figure 15.17). It affects the distribution of energy along the shore and thus strongly influences where and to what degree erosion, sediment transport, and deposition will take place.

Waves seldom approach the shore truly straight on. Rather, most waves move toward the shore at a slight angle. However, when they reach the shallow water of a smoothly sloping bottom, the wave crests are refracted (bent) and tend to become parallel to the

Students Sometimes Ask...

During heavy wave activity, where does the sand from the berm go?

The orbital motion of waves is too shallow to move sand very far offshore. Consequently, the sand accumulates just beyond where the surf zone ends and forms one or more offshore sandbars called *longshore bars*.

shore. Bending occurs because the part of the wave nearest the shore touches bottom and slows first, whereas the part of the wave that is still in deep water continues forward at its full speed. The net result is a wave that approaches nearly parallel to the shore regardless of its original orientation.

Because of refraction, wave energy is concentrated against the sides and ends of headlands that project into the water, whereas wave attack is weakened in bays. This differential wave attack along irregular coastlines is illustrated in Figure 15.17. Because the waves reach the shallow water in front of the headland before they reach adjacent bays, they are bent more nearly parallel to the protruding land and strike it from all three sides. By contrast, refraction in the bays causes waves to diverge and expend less energy. In these zones of weakened wave activity, sediments can accumulate and form sandy beaches. Over a long period, erosion of the headlands and deposition in the bays will straighten an irregular shoreline.

Longshore Transport

Although waves are refracted, most still reach the shore at a slight angle. Consequently, the uprush of water from each breaking wave (the swash) is at a slight angle to the shoreline. However, the backwash is straight down the slope of the beach. The effect of this pattern of water movement is to transport sediment in a zigzag pattern along the beach face (Figure 15.18). This movement is called **beach drift**, and it can transport sand and pebbles hundreds or even thousands of meters daily. However, a more typical rate is 5–10 meters per day.

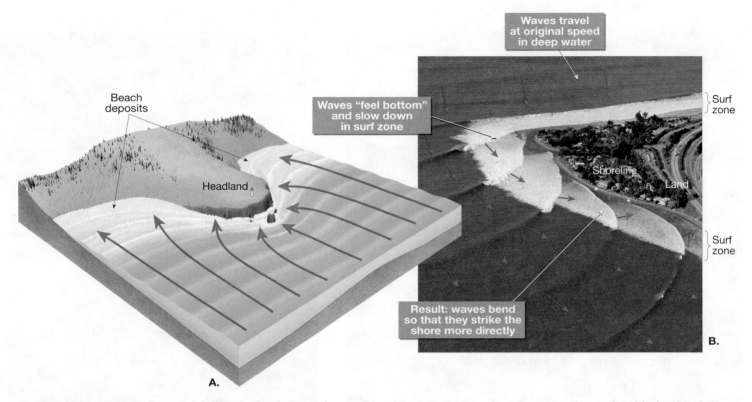

FIGURE 15.17 As waves first touch bottom in the shallows along an irregular coast, they are slowed, causing them to bend (refract) and align nearly parallel to the shoreline. **A.** In this diagram, waves approach nearly straight on. Refraction causes wave energy to be concentrated at headlands (resulting in erosion) and dispersed in bays (resulting in deposition). **B.** Wave refraction at Rincon Point, California. (Photo by Woody Woodworth/Creation Captured)

Waves that approach the shore at an angle also produce currents within the surf zone that flow parallel to the shore and move substantially more sediment than beach drift (Figure 15.18). Because the water here is turbulent, these **longshore currents** easily move the fine suspended sand and roll larger sand and gravel along the bottom. When the sediment transported by longshore currents is added to the quantity moved by beach drift, the total amount can be impressive. At Sandy Hook, New Jersey, for example, the quantity of sand transported along the shore over a 48-year period averaged almost 680,000 metric tons (750,000 short tons) annually. For a 10-year period at Oxnard, California, more than 1.4 million metric tons (1.5 million short tons) of sediment moved along the shore each year.

Both rivers and coastal zones move water and sediment from one area (*upstream*) to another (*downstream*). As a result, the beach has often been characterized as a "river of sand." Beach drift and longshore currents, however, move in a zigzag pattern, whereas rivers flow mostly in a turbulent, swirling fashion. Additionally, the direction of flow of longshore currents along a shoreline can change, whereas rivers flow in the same direction (downhill). Longshore currents change direction because the direction that waves approach

FIGURE 15.18 Beach drift and longshore currents are created by obliquely breaking waves. Beach drift occurs as incoming waves carry sand obliquely up the beach, while the water from spent waves carries it directly down the slope of the beach. Similar movements occur offshore in the surf zone to create the longshore current. These processes transport large quantities of material along the beach and in the surf zone. In the photo, waves approaching the beach at a slight angle near Oceanside, California, produce a longshore current moving from left to right. (Photo by John S. Shelton)

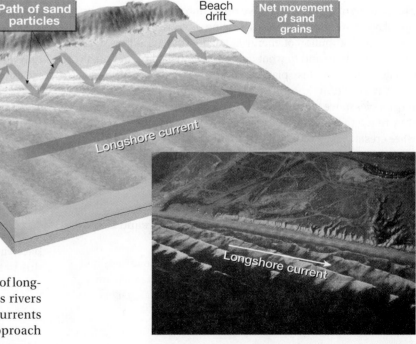

the beach changes seasonally. Nevertheless, longshore currents generally flow southward along both the Atlantic and Pacific shores of the United States.

Rip Currents

Rip currents are concentrated movements of water that flow in the *opposite* direction from breaking waves.[38] Most of the backwash from spent waves finds its way back to the open ocean as an unconfined flow across the ocean bottom called *sheet flow*. However, sometimes a portion of the returning water moves seaward in the form of surface rip currents. These currents do not travel far beyond the surf zone before breaking up and can be recognized by the way they interfere with incoming waves or by the sediment that is often suspended within the rip current (**Figure 15.19**). They can be a hazard to swimmers, who, if caught in them, can be carried out away from shore. The best strategy for exiting a rip current is to swim *parallel* to the shore for a few tens of meters.

FIGURE 15.19 A rip current (red arrow) extends outward from shore and interferes with incoming waves. As the warning sign indicates, rip currents can be dangerous. (Photo by A. P. Trujillo/APTPhotos)

CONCEPT CHECK 15.6

1. Why do waves that are approaching the shoreline often bend?
2. What is the effect of wave refraction along an irregular coastline?
3. Describe the two processes that contribute to longshore transport.

Shoreline Features

The Global Ocean
▸ Coastal Processes

A fascinating assortment of shoreline features can be observed along the world's coastal regions. These shoreline features vary depending on the type of rocks exposed along the shore, the intensity of waves, the nature of coastal currents, and whether the coast is stable, sinking, or rising. Features that owe their origin primarily to the work of erosion are called *erosional features*, whereas deposits of sediment produce *depositional features*.

Erosional Features

Many coastal landforms owe their origin to erosional processes. Such erosional features are common along the rugged and irregular New England coast and along the steep shorelines of the West Coast of the United States.

Wave-Cut Cliffs, Wave-Cut Platforms, and Marine Terraces
Wave-cut cliffs, as the name implies, originate due to the cutting action of the surf against the base of coastal land. As erosion progresses, rocks overhanging the notch at the base of the cliff crumble into the surf and the cliff retreats. A relatively flat, benchlike surface, called a **wave-cut platform**, is left behind by the receding cliff (**Figure 15.20**, left). The platform broadens as wave attack continues. Some debris produced by the breaking waves may remain along the water's edge as sediment on the beach, while the remainder is transported farther seaward. If a wave-cut platform is uplifted above sea level by tectonic forces, it becomes a

FIGURE 15.20 Wave-cut platform and marine terrace. A wave-cut platform is exposed at low tide along the California coast at Bolinas Point near San Francisco. On the right is an elevated wave-cut platform, called a marine terrace. (Photo by John S. Shelton)

[38]Sometimes rip currents are incorrectly called *rip tides*, although they are unrelated to tidal phenomena.

PROFESSIONAL PROFILE

Rob Thieler: Marine Geologist

Growing up on the New Jersey shore, Rob Thieler learned that beaches are literally as changeable as the weather. As a teenage lifeguard, he watched entire beaches run short on sand, leaving waves to break perilously close to buildings and streets.

> "I love going to sea. There's nothing like exploring a part of the ocean no one has seen before."

Today, Thieler's coastal observations have expanded considerably. As a research geologist with the U.S. Geological Survey, he studies how sandy coastlines shift and change over time.

"We don't have good models to predict where the shoreline's going to be in a hundred years. A beach can lose 30 meters of sand in a single storm, and we certainly can't predict when and where storms will hit for the next century. We have a really incomplete understanding of how sand moves around on beaches and the inner continental shelf," Thieler states. The problems aren't simply academic; the answers have major implications for the estimated 160 million Americans who live near the coast.

In past centuries, Thieler claims, residents threatened by disappearing beaches merely moved their houses further up the dunes. But for modern homeowners, that's no longer an option—higher ground generally means impinging upon someone else's property.

Most years, Thieler spends about a month at sea gathering information about coastal geology. The ship tows sidescan sonar, seismic reflection profilers, and bathymetric gear over the shallow waters of the continental shelf in a back-and-forth pattern that resembles the path of a lawnmower. The echoes reveal the texture, topography, and structure of bottom sediments in exquisite detail.

"I love going to sea. There's nothing like exploring a part of the ocean no one has seen before," Thieler says. Much of the technology Thieler uses to study ocean topography was only invented in the last few decades. As a result, the majority of the seafloor remains uncharted territory.

Thieler combines his marine data with land observations such as the historic locations of shorelines and the elevation of bluffs and beaches. His results can help explain why waves break in certain locations, and how currents shape nearshore marine topography.

Thieler has also put this data to use analyzing how vulnerable the U.S. coastlines are to sea-level rise. Already some planners are using his findings to adapt to rising seas. "Before, we've often chosen to hold the line to protect property. As we improve our understanding of how beaches and coasts evolve in a time of rising sea levels and changing climate, the choice is no longer clear," he says.

—Kathleen Wong

Dr. Rob Thieler, marine geologist, standing along the Massachusetts shore near his U.S. Geological Survey office in Woods Hole. Dr. Thieler is in one of his favorite places— the coast. Dr. Thieler studies coastal processes as part of a team of scientists assessing the vulnerability of the United States coastline to rising sea level. (Courtesy of Rob Thieler and the U.S. Geological Survey)

---

marine terrace (Figure 15.20, right). Marine terraces are easily recognized by their gentle seaward-sloping shape and are often desirable sites for coastal roads, buildings, or agriculture.

Sea Arches and Sea Stacks Headlands that extend into the sea are vigorously attacked by waves because of refraction. The surf erodes the rock selectively, wearing away the softer or more highly fractured rock at the fastest rate. At first, sea caves may form. When caves on opposite sides of a headland unite, a **sea arch** results (Figure 15.21). Eventually, the arch falls in, leaving an isolated remnant, or **sea stack**, on the wave-cut platform (Figure 15.21). In time, it too will be consumed by the action of the waves.

northern Florida to avoid the rough waters of the North Atlantic.

Barrier islands probably formed in several ways. Some originated as spits that were subsequently severed from the mainland by wave erosion or by the general rise in sea level following the last episode of glaciation. Others were created when turbulent waters in the line of breakers heaped up sand that had been scoured from the bottom. Finally, some barrier islands may be former sand-dune ridges that originated along the shore during the last glacial period, when sea level was lower. As the ice sheets melted, sea level rose and flooded the area behind the beach-dune complex.

FIGURE 15.21 Sea arch and sea stack at the tip of Mexico's Baja Peninsula. (Photo by Lew Robertson/Getty Images)

FIGURE 15.22 **A.** High-altitude image of a well-developed spit and baymouth bar along the coast of Martha's Vineyard, Massachusetts. Also notice the tidal delta in the lagoon adjacent to the inlet through the baymouth bar. (Photo courtesy of USDA-ASCS) **B.** This photograph, taken from the International Space Station, shows Provincetown Spit located at the tip of Cape Cod. (NASA)

Depositional Features

Sediment eroded from the beach is transported along the shore and deposited in areas where wave energy is low. Such processes produce a variety of depositional features.

Spits, Bars, and Tombolos   Where beach drift and longshore currents are active, several features related to the movement of sediment along the shore may develop. A **spit** (*spit* = spine) is an elongated ridge of sand that projects from the land into the mouth of an adjacent bay. Often the end in the water hooks landward in response to the dominant direction of the longshore current. Both images in **Figure 15.22** show spits. The term **baymouth bar** is applied to a sandbar that completely crosses a bay, sealing it off from the open ocean (Figure 15.22A). Such a feature tends to form across bays where currents are weak, allowing a spit to extend to the other side. A **tombolo** (*tombolo* = mound), a ridge of sand that connects an island to the mainland or to another island, forms in much the same manner as a spit.

Barrier Islands   The Atlantic and Gulf Coastal Plains are relatively flat and slope gently seaward. The shore zone is characterized by **barrier islands**. These low ridges of sand parallel the coast at distances from 3 to 30 kilometers (2 to 19 miles) offshore. From Cape Cod, Massachusetts, to Padre Island, Texas, nearly 300 barrier islands rim the coast (**Figure 15.23**).

Most barrier islands are 1–5 kilometers (0.6–3 miles) wide and between 15 and 30 kilometers (9–18 miles) long. The tallest features are sand dunes, which usually reach heights of 5–10 meters (16–33 feet). The lagoons that separate these narrow islands from the shore are zones of relatively quiet water that allow small craft traveling between New York and

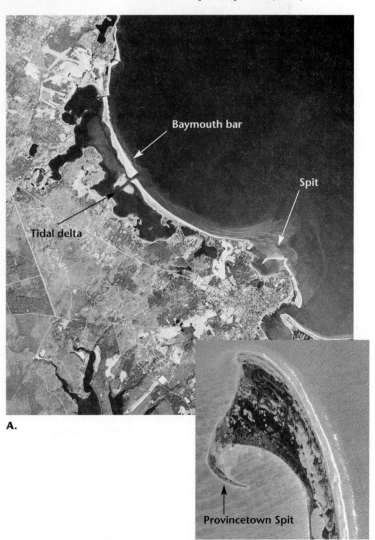

A.

B.

FIGURE 15.23 Nearly 300 barrier islands rim the Gulf and Atlantic coasts. The islands along the coast of North Carolina are excellent examples. (Photo by Michael Collier)

The Evolving Shore

A shoreline continually undergoes modification regardless of its initial configuration. At first most coastlines are irregular, although the degree of and reason for the irregularity may differ considerably from place to place. Along a coastline that is characterized by varied geology, the pounding surf may initially increase its irregularity because the waves will erode the weaker rocks more easily than the stronger ones. However, if a shoreline remains stable, marine erosion and deposition will eventually produce a straighter, more regular coast.

Figure 15.24 illustrates the evolution of an initially irregular coast that remains relatively stable and shows many of the coastal

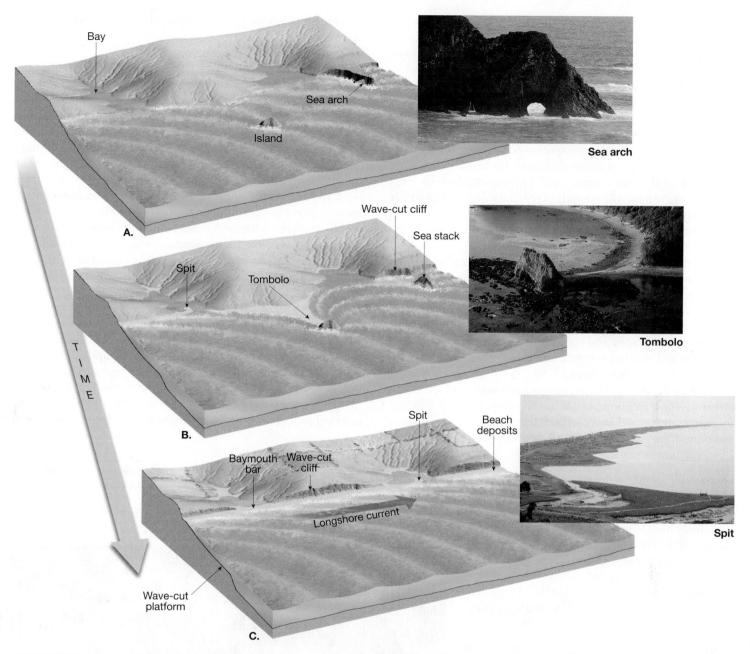

FIGURE 15.24 These diagrams illustrate the changes that can take place through time along an initially irregular coastline that remains relatively stable. The coastline shown in part **A** gradually evolves to **B**, and then **C**. The diagrams also serve to illustrate many of the features described in the section on shoreline features. (Photos A and C by E. J. Tarbuck; Photo B by Michael Collier)

features discussed in the previous section. As headlands are eroded and erosional features such as wave-cut cliffs and wave-cut platforms are created, sediment is produced that is carried along the shore by beach drift and longshore currents. Some material is deposited in the bays, while other debris is formed into depositional features such as spits and baymouth bars. At the same time, rivers fill the bays with sediment. Ultimately, a smooth coast results.

CONCEPT CHECK 15.7

1. How is a marine terrace related to a wave-cut platform?
2. Describe the formation of the features labeled in Figure 15.21 and Figure 15.22.
3. List three ways that barrier islands may form.

Stabilizing the Shore

 The Global Ocean

EARTH SCIENCE ▶ **Coastal Processes**

Today, the coastal zone teems with human activity. Unfortunately, people often treat the shoreline as if it were a stable platform on which structures can be built safely. This approach jeopardizes both people and the shoreline because many coastal landforms are relatively fragile, short-lived features that are easily damaged by development. As anyone who has endured a tsunami or a strong coastal storm knows, the shoreline is not always a safe place to live. The chapter-opening photo and Figure 15.9 illustrate this point.

Compared with other natural hazards, such as earthquakes, volcanic eruptions, and landslides, shoreline erosion appears to be a more continuous and predictable process that causes relatively modest damage to limited areas. In reality, the shoreline is one of Earth's most dynamic places that changes rapidly in response to natural forces. Storms, for example, are capable of eroding beaches and cliffs at rates that far exceed the long-term average. Such bursts of accelerated erosion not only have a significant impact on the natural evolution of a coast but can also have a profound impact on people who reside in the coastal zone. Erosion along the coast causes significant property damage. Huge sums are spent annually not only to repair damage but also in an attempt to prevent or control erosion. Already a problem at many sites, shoreline erosion is certain to become increasingly serious as extensive coastal development continues.

Although the same processes cause change along every coast, not all coasts respond in the same way. Interactions among different processes and the relative importance of each process depend on local factors. The factors include (1) the proximity of a coast to sediment-laden rivers, (2) the degree of tectonic activity, (3) the topography and composition of the land, (4) prevailing winds and weather patterns, and (5) the configuration of the coastline and nearshore areas.

During the past 100 years, growing affluence and increasing demands for recreation have brought unprecedented development to many coastal areas. As both the number and the value of buildings have increased, so too have efforts to protect property from storm waves by stabilizing the shore.

Hard Stabilization

Structures built to protect a coast from erosion or to prevent the movement of sand along a beach are known as **hard stabilization**. Hard stabilization can take many forms and often results in predictable yet unwanted outcomes. Hard stabilization includes groins, breakwaters, and seawalls.

Groins   To maintain or widen beaches that are losing sand, groins are sometimes constructed. A **groin** (*groin* = ground) is a barrier built at a right angle to the beach to trap sand that is moving parallel to the shore. Groins are usually constructed of large rocks but may also be composed of wood. These structures often do their job so effectively that the longshore current beyond the groin becomes sand-starved. As a result, the current erodes sand from the beach on the downstream side of the groin.

To offset this effect, property owners downstream from the structure may erect a groin on their property. In this manner, the number of groins multiplies, resulting in a *groin field* (Figure 15.25). An example of such proliferation is the shoreline of New Jersey, where hundreds of these structures have been built. Because it has been shown that groins often do not provide a satisfactory solution,

they are no longer the preferred method of keeping beach erosion in check.

Breakwaters and Seawalls   Hard stabilization can also be built parallel to the shoreline. One such structure is a **breakwater**, the purpose of which is to protect boats from the force of large breaking waves by creating a quiet water zone near the shore. However, when this is done, the reduced wave activity along the shore behind the structure may allow sand to accumulate. If this happens, the boat anchorage will eventually fill with sand, while the downstream beach erodes and retreats. This situation occurred at Santa Monica, California, where a breakwater was built to provide a boat anchorage safe from incoming waves. A bulge in the beach soon formed behind (inshore of) the breakwater and severe erosion occurred downstream (Figure 15.26). To compensate for this problem, the city dredged sand from the protected quiet water zone and deposited it farther downstream where longshore currents continued to move the sand down the coast.

Another type of hard stabilization built parallel to the shore is a **seawall**, which is designed to armor the coast and defend property from the force of breaking waves (see Figure 15.15, p. 438). Waves expend much of their energy as they move across an open beach. Seawalls cut this process short by reflecting the force of unspent waves seaward. As a consequence, the beach to the seaward side of the seawall experiences significant erosion and

FIGURE 15.25 A series of groins along the shoreline near Chichester, Sussex, England. Because groins trap sand on the upcurrent side, the movement of sand along this coast caused by the longshore current must be from lower right toward upper left. (Photo by Sandy Stockwell/London Aerial Photo Library/CORBIS)

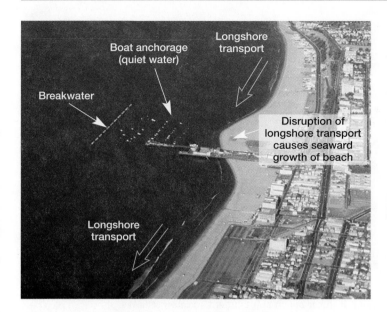

FIGURE 15.26 Aerial view of a breakwater at Santa Monica, California. The breakwater appears as a faint line in the water behind which many boats are anchored. The construction of the breakwater disrupted longshore transport and caused the seaward growth of the beach. The pier had no effect on the formation of the bulge. (Photo by John S. Shelton)

may, in some instances, be eliminated entirely. Once the width of the beach is reduced, the seawall is subjected to even greater pounding by the waves. Eventually this battering takes its toll on the seawall such that the seawall will fail and a larger, more expensive structure must be built to take its place.

The wisdom of building temporary protective structures along shorelines is questionable. The opinion of many coastal scientists and engineers is expressed in the following excerpt from a position paper that was developed as a result of a conference on America's Eroding Shoreline:

It is now clear that halting the receding shoreline with protective structures benefits only a few and seriously degrades or destroys the natural beach and the value it holds for the majority. Protective structures divert the ocean's energy temporarily from private properties, but usually refocus the energy on the adjacent natural beaches. Many interrupt the natural sand

Students Sometimes Ask...

Is it true that sea level is rising due to global warming?

Yes. One probable impact of a human-induced global warming is a rise in sea level. Some of the rise comes from water added by shrinking glaciers. Another significant factor is that a warmer atmosphere causes an increase in ocean volume due to thermal expansion. Higher air temperatures warm the upper layers of the ocean, which causes the water to expand and sea level to rise. For more on this, see the discussion of "Sea Level Rise" in Chapter 20 (pages 602–604).

flow in coastal currents, robbing many beaches of vital sand replacement.[39]

Alternatives to Hard Stabilization

Armoring the coast with hard stabilization has several potential drawbacks, including the cost of the structure and the loss of sand on the beach. Alternatives to hard stabilization include beach nourishment and relocation.

Beach Nourishment **Beach nourishment** represents an approach to stabilizing shoreline sands without hard stabilization. As the term implies, this practice simply involves the addition of large quantities of sand to the beach system (**Figure 15.27**). By building the beaches seaward, storm protection is improved. Beach nourishment, however, is not a permanent solution to the problem of shrinking beaches because the same processes that removed the sand in the first place will eventually remove the replacement sand as well. In addition, it can be very expensive because huge volumes of sand must be transported to the beach from offshore areas, nearby rivers, or other source areas. Orrin Pilkey, a respected coastal scientist, described the situation this way:

Nourishment has been carried out on beaches on both sides of the continent, but by far the greatest effort in dollars and sand volume has been expended on the East Coast barrier islands between the southern shore of Long Island, N.Y., and southern Florida. Over this shoreline reach, communities have added a total of 500 million cubic yards of sand on 195 beaches in 680 separate instances. Some beaches . . . have been renourished more than 20 times each since 1965. Virginia Beach has been renourished more than 50 times. Nourished beaches typically cost between $2 million and $10 million per mile.[40]

In some instances, beach nourishment can lead to unwanted environmental effects. For example, beach replenishment at Waikiki Beach, Hawaii, involved replacing the natural coarse calcareous beach sand with softer, muddier sand. Destruction of the softer sand by breaking waves increased the water's turbidity and killed offshore coral reefs. Similar damage to local coral communities occurred after beach replenishment at Miami Beach.

Beach nourishment appears to be an economically viable long-range solution to the beach-preservation problem only in areas where there exists dense development, large supplies of sand, relatively low wave energy, and reconcilable environmental issues. Unfortunately, few areas possess all these attributes.

Relocation Instead of building structures such as groins and seawalls to hold the shoreline in place or adding sand to replenish eroding beaches, another option is also available. Many coastal scientists and planners are calling for a policy shift from defending and rebuilding beaches and coastal property in high hazard areas to *relocating* storm-damaged or at-risk buildings and letting nature reclaim the beach. This approach is similar to that

[39]"Strategy for Beach Preservation Proposed," *Geotimes* 30 (12, December 1985): 15.

[40]"Beaches Awash with Politics," *Geotimes* (July 2005): 38–39.

A.   **B.**

FIGURE 15.27 Miami Beach. **A.** Before beach nourishment, **B.** After beach nourishment. (Courtesy of the U.S. Army Corps of Engineers, Vicksburg District)

adopted by the federal government for river floodplains following the devastating 1993 Mississippi River floods in which vulnerable structures are abandoned and relocated on higher, safer ground.

Such proposals, of course, are controversial. People with significant shoreline investments shudder at the thought of not rebuilding and defending coastal developments from the erosional wrath of the sea. Others, however, argue that with sea level rising, the impact of coastal storms will only get worse in the decades to come. This group advocates that oft-damaged structures be relocated or abandoned to improve personal safety and to reduce costs. Such ideas will no doubt be the focus of much study and debate as states and communities evaluate and revise coastal land-use policies.

CONCEPT CHECK 15.8

❶ List at least two examples of *hard stabilization* and describe what each is intended to do. How does each affect sand distribution on a beach?

❷ What are two alternatives to hard stabilization and potential problems associated with each?

Erosion Problems Along U.S. Coasts

The shoreline along the Pacific Coast of the United States is strikingly different from that characterizing the Atlantic and Gulf Coast regions. Some of the differences are related to plate tectonics. The West Coast represents a boundary of the North American plate, and because of this, it has experienced and continues to experience active uplift and deformation. By contrast, the East Coast is a tectonically quiet region that is far from any active plate margin. Because of this basic geological difference, the nature of shoreline erosion problems along America's opposite coasts is different.

Atlantic and Gulf Coasts

Much of the coastal development along the Atlantic and Gulf coasts has occurred on barrier islands. Typically, barrier islands consist of a wide beach that is backed by dunes and separated from the mainland by marshy lagoons. The broad expanses of sand and exposure to the ocean have made barrier islands exceedingly attractive sites for development. Unfortunately, development has taken place more rapidly than our understanding of barrier island dynamics has grown.

Because barrier islands face the open ocean, they receive the full force of major storms that strike the coast. When a storm occurs, the barriers absorb the energy of the waves primarily through the movement of sand (**Figure 15.28**). This process and the dilemma that results have been described as follows:

Waves may move sand from the beach to offshore areas or, conversely, into the dunes; they may erode the dunes, depositing sand onto the beach or carrying it out to sea; or they may carry sand from the beach and the dunes into the marshes behind the barrier, a process known as overwash. The common factor is movement. Just as a flexible reed may survive a wind that destroys an oak tree, so the barriers survive hurricanes and nor'easters not through unyielding strength but by giving before the storm.

This picture changes when a barrier is developed for homes or a resort. Storm waves that previously rushed harmlessly through gaps between the dunes now encounter buildings and roadways. Moreover, since the dynamic nature of the barriers is readily perceived only during storms, homeowners tend to attribute damage to a particular storm, rather than to the basic mobility of coastal barriers. With their homes or investments at stake, local residents are more likely to seek to hold the sand in place and the waves at bay than to admit that development was improperly placed to begin with.[41]

[41]Frank Lowenstein, "Beaches or Bedrooms—The Choice as Sea Level Rises," *Oceanus* 28 (3, Fall 1985): 22.

FIGURE 15.28 When the lighthouse at Cape Hatteras, North Carolina, was built in 1870, it was situated 457 meters (1,500 feet) from the shoreline. By 1970, waves began to lap just 37 meters (120 feet) from its base. In order to save the historic structure, it was moved 488 meters (1,600 feet) back from the shore. At the current rate of shoreline retreat, the lighthouse should be safe for another 50 years. (Photo by Reuters/Stringer/Getty Images, Inc.—Hulton Archive Photos)

quantities of water to the slope. This water percolates downward toward the base of the cliff, where it may emerge in small seeps. This action reduces the slope's stability and facilitates mass wasting.

Shoreline erosion along the Pacific Coast varies considerably from one year to the next, largely because of the sporadic occurrence of storms. As a consequence, when the infrequent but serious episodes of erosion occur, the damage is often blamed on the unusual storms and not on coastal development or the sediment-trapping dams that may be great distances away. If, as predicted, sea level rises at an increasing rate in the years to come, increased shoreline erosion and sea-cliff retreat should be expected along many parts of the Pacific Coast. This issue is addressed in the section "Sea-Level Rise" in Chapter 20.

> **CONCEPT CHECK** 15.9
> ❶ Briefly describe what happens when storm waves strike an undeveloped barrier island.
> ❷ How might building a dam on a river that flows to the sea affect a beach?

Coastal Classification

The great variety of shorelines demonstrates their complexity. Indeed, to understand any particular coastal area, many factors must be considered, including the composition and durability of the bedrock, size and direction of waves, frequency of storms, tidal range, and offshore topography. In addition, practically all coastal areas were affected by the worldwide rise in sea level that accompanied the melting of Ice Age glaciers at the close of the Pleistocene epoch. Finally, tectonic events that cause uplift or subsidence of the land or change the volume of ocean basins must be taken into account. The large number of factors that influence coastal areas make shoreline classification difficult.

Many geologists classify coasts based on changes that have occurred with respect to sea level. This commonly used classification system divides coasts into two very general categories: emergent and submergent. **Emergent coasts** develop either because an area experiences uplift or as a result of a drop in sea level. Conversely, **submergent coasts** are created when sea level rises or the land adjacent to the sea subsides.

Pacific Coast

In contrast to the broad, gently sloping coastal plains of the Atlantic and Gulf coasts, much of the Pacific Coast is characterized by relatively narrow beaches that are backed by steep cliffs and mountain ranges. Recall that America's western margin is a more rugged and tectonically active region than the eastern margin. Because uplift continues, the rise in sea level in the West is not so readily apparent. Nevertheless, like the shoreline erosion problems facing the East's barrier islands, West Coast difficulties also stem largely from the alteration of natural systems by people.

A major problem facing the Pacific shoreline—particularly along southern California—is a significant narrowing of many beaches. The bulk of the sand on many of these beaches is supplied by rivers that transport it from the mountainous regions to the coast. Over the years this natural flow of material to the coast has been interrupted by dams built for irrigation and flood control. The reservoirs effectively trap the sand that would otherwise nourish the beach environment. When the beaches were wider, they served to protect the cliffs behind them from the force of storm waves. Now, however, the waves move across the narrowed beaches without losing much energy and cause more rapid erosion of the sea cliffs.

Although the retreat of the cliffs provides material to replace some of the sand impounded behind dams, it also endangers homes and roads built on the bluffs. In addition, development atop the cliffs aggravates the problem. Urbanization increases runoff, which, if not carefully controlled, can result in serious bluff erosion. Watering lawns and gardens adds significant

Emergent Coasts

In some areas, the coast is clearly emergent because rising land or falling water levels expose wave-cut cliffs and marine terraces above sea level. Excellent examples include portions of coastal California where uplift has occurred in the recent geological past. The elevated wave-cut platform in Figure 15.20 (p. 441) also illustrates this situation. In the case of the Palos Verdes Hills, south of Los Angeles, California, seven different terrace levels exist, indicating at least seven episodes of uplift. The ever persistent sea is now cutting a new platform at the base of the cliff. If uplift follows, it too will become an elevated marine terrace.

Other examples of emergent shores include regions that were once buried beneath great ice sheets. When glaciers were present, their weight depressed the crust, and when the ice melted, the crust began gradually to spring back. Consequently, prehistoric shoreline features may now be found high above sea level. The Hudson Bay region of Canada is such an area, portions of which are still rising at a rate of more than a centimeter (0.4 inch) annually.

Submergent Coasts

In contrast to the preceding examples, other coastal areas show definite signs of submergence. Shorelines that have been submerged in the relatively recent past are often highly irregular because the sea typically floods the lower reaches of river valleys flowing into the ocean. The ridges separating the valleys, however, remain above sea level and project into the sea as headlands. These drowned river mouths, which are called **estuaries** (*aestus* = tide), characterize many coasts today. Along the Atlantic coastline, Chesapeake and Delaware bays are examples of estuaries created by submergence (Figure 15.29). The picturesque coast of Maine, particularly in the vicinity of Acadia National Park, is another excellent example of an area that was flooded by the postglacial rise in sea level and transformed into a highly irregular submerged coastline.

Keep in mind, however, that most coasts have a complicated geologic history. With respect to sea level, many coasts have at various times emerged and then submerged again. Each time they retain some of the features created during the previous situation.

CONCEPT CHECK 15.10

1. What observable features would lead you to classify a coastal area as emergent?
2. Are estuaries associated with submergent or emergent coasts? Explain.

Tides

Tides are daily changes in the elevation of the ocean surface. Their rhythmic rise and fall along coastlines have been known since antiquity. Other than waves, they are the easiest ocean movements to observe (Figure 15.30).

Although known for centuries, tides were not explained satisfactorily until Sir Isaac Newton applied the law of gravitation to them in 1686. Newton showed that there is a mutual attractive force between two bodies, as between Earth and the Moon. Because both the atmosphere and the ocean are fluids and are free to move, both are deformed by this force. Hence, ocean tides result from the gravitational attraction exerted upon Earth by the Moon and, to a lesser extent, by the Sun.

Causes of Tides

To illustrate how tides are produced, consider an idealized case in which Earth is a rotating sphere covered to a uniform depth with water (Figure 15.31). Furthermore, ignore the effect of the

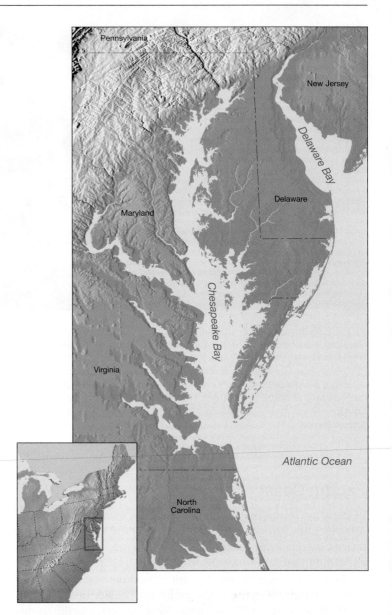

FIGURE 15.29 Major estuaries along the East Coast of the United States. The lower portions of many river valleys were flooded by the rise in sea level that followed the end of the last Ice Age, creating large estuaries such as Chesapeake and Delaware bays.

Sun for now. It is easy to see how the Moon's gravitational force can cause the water to bulge on the side of Earth nearer the Moon. In addition, however, an equally large tidal bulge is produced on the side of Earth directly *opposite* the Moon.

Both tidal bulges are caused, as Newton discovered, by the pull of gravity. Gravity is inversely proportional to the square of the distance between two objects, meaning simply that it quickly weakens with distance. In this case, the two objects are the Moon and Earth. Because the force of gravity decreases with distance, the Moon's gravitational pull on Earth is slightly greater on the near side of Earth than on the far side. The result of this differential pulling is to stretch (elongate) the "solid" Earth very slightly. In contrast, the world ocean, which is mobile, is deformed quite dramatically by this effect, producing the two opposing tidal bulges.

Because the position of the Moon changes only moderately in a single day, the tidal bulges remain in place while Earth rotates

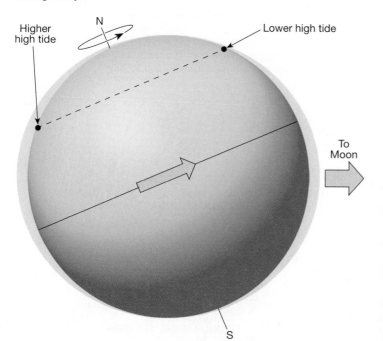

"through" them. For this reason, if you stand on the seashore for 24 hours, Earth will rotate you through alternating areas of higher and lower water. As you are carried into each tidal bulge, the tide rises, and as you are carried into the intervening troughs between the tidal bulges, the tide falls. Therefore, most places on Earth experience two high tides and two low tides each day.

In addition, the tidal bulges migrate as the Moon revolves around Earth about every 29 days. As a result, the tides—like the time of moonrise—shift about 50 minutes later each day. In essence, the tidal bulges exist in fixed positions relative to the Moon, which slowly moves progressively eastward as it orbits Earth.

In many locations, there may be an inequality between the high tides during a given day. Depending on the Moon's position, the tidal bulges may be inclined to the equator as in Figure 15.31. This figure illustrates that one high tide experienced by an observer in the Northern Hemisphere is considerably higher than the high tide half a day later. In contrast, a Southern Hemisphere observer would experience the opposite effect.

Monthly Tidal Cycle

The primary body that influences the tides is the Moon, which makes one complete revolution around Earth every 27.3 days. The Sun, however, also influences the tides. It is far larger than the Moon, but because it is much farther away, its effect is considerably less. In fact, the Sun's tide-generating effect is only about 46 percent (roughly one-half) that of the Moon's.

Near the times of new and full moons, the Sun and Moon are aligned and their forces are added together (Figure 15.32A). Accordingly, the combined gravity of these two tide-producing bodies causes larger tidal bulges (higher high tides) and larger tidal troughs (lower low tides), producing a large tidal range.

FIGURE 15.30 High tide and low tide on Nova Scotia's Minas Basin in the Bay of Fundy. The areas exposed during low tide and flooded during high tide are called *tidal flats*. Tidal flats here are extensive. (Courtesy of Nova Scotia Department of Tourism and Culture)

FIGURE 15.31 Idealized tidal bulges on Earth caused by the Moon. If Earth were covered to a uniform depth with water, there would be two tidal bulges: one on the side of Earth facing the Moon (right) and the other on the opposite side of Earth (left). Depending on the Moon's position, tidal bulges may be inclined relative to Earth's equator. In this situation, Earth's rotation causes an observer to experience two unequal high tides during a day.

Students Sometimes Ask...

Where are the world's largest tides?

The world's largest *tidal range* (the difference between successive high and low tides) is found in the northern end of Nova Scotia's 258-kilometer- (160-mile-) long Bay of Fundy. During maximum spring tide conditions, the tidal range at the mouth of the bay (where it opens to the ocean) is only about 2 meters (6.6 feet). However, the tidal range progressively increases from the mouth of the bay northward because the natural geometry of the bay concentrates tidal energy. In the northern end of Minas Basin, the maximum spring tidal range is about 17 meters (56 feet). This extreme tidal range leaves boats high and dry during low tide (see Figure 15.30).

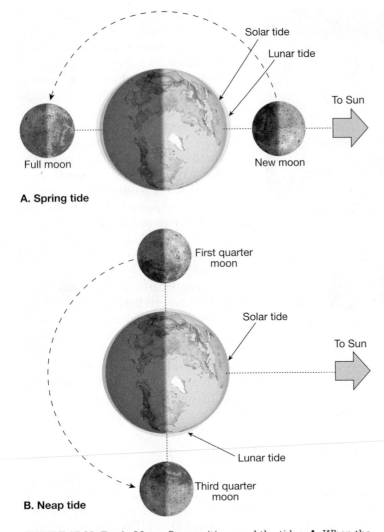

A. Spring tide

B. Neap tide

These are called the **spring** (*springen* = to rise up) **tides**, which have no connection with the spring season but occur twice a month during the time when the Earth–Moon–Sun system is aligned. Conversely, at about the time of the first and third quarters of the Moon, the gravitational forces of the Moon and Sun act on Earth at right angles, and each partially offsets the influence of the other (Figure 15.32B). As a result, the daily tidal range is less. These are called **neap** (*nep* = scarcely or barely touching) **tides**, which also occur twice each month. Each month, then, there are two spring tides and two neap tides, each about one week apart.

Tidal Patterns

So far, the basic causes and types of tides have been explained. Keep in mind, however, that these theoretical considerations cannot be used to predict either the height or the time of actual tides at a particular place. This is because many factors—including the shape of the coastline, the configuration of ocean basins, the Coriolis effect, and water depth—greatly influence the tides. Consequently, tides at various locations respond differently to the tide-producing forces. This being the case, the nature of the tide at any coastal location can be determined most accurately by actual observation. The predictions in tidal tables and tidal data on nautical charts are based on such observations.

Three main tidal patterns exist worldwide. A **diurnal** (*diurnal* = daily) **tidal pattern** is characterized by a single high tide and a single low tide each tidal day (**Figure 15.33**). Tides of this type occur along the northern shore of the Gulf of Mexico, among other locations. A **semidiurnal** (*semi* = twice, *diurnal* = daily) **tidal pattern** exhibits two high tides and two low tides each tidal day, with the two highs about the same height and the two lows about the same height (Figure 15.33). This type of tidal pattern is common along the Atlantic Coast of the United States. A **mixed tidal pattern** is similar to a semidiurnal pattern except that it is characterized by a large inequality in high water heights, low water heights, or both (Figure 15.33). In this case, there are usually two high and two low tides each day, with high tides of different heights and low

FIGURE 15.32 Earth–Moon–Sun positions and the tides. **A.** When the Moon is in the full or new position, the tidal bulges created by the Sun and Moon are aligned, there is a large tidal range on Earth, and spring tides are experienced. **B.** When the Moon is in the first- or third-quarter position, the tidal bulges produced by the Moon are at right angles to the bulges created by the Sun. Tidal ranges are smaller, and neap tides are experienced.

FIGURE 15.33 Tidal patterns and their occurrence along North and Central American coasts. A diurnal tidal pattern (lower right) shows one high and low tide each tidal day. A semidiurnal pattern (upper right) shows two highs and lows of approximately equal heights during each tidal day. A mixed tidal pattern (left) shows two highs and lows of unequal heights during each tidal day.

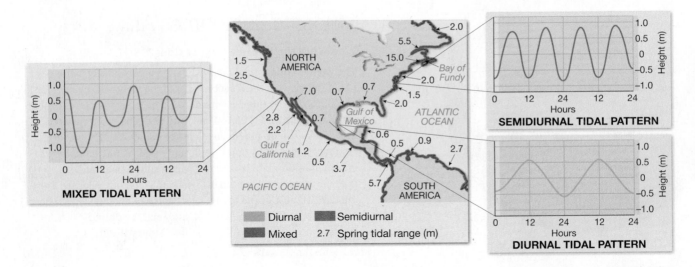

tides of different heights. Such tides are prevalent along the Pacific Coast of the United States and in many other parts of the world.

Tidal Currents

Tidal current is the term used to describe the *horizontal* flow of water accompanying the rise and fall of the tides. These water movements induced by tidal forces can be important in some coastal areas. Tidal currents that advance into the coastal zone as the tide rises are called *flood currents*. As the tide falls, seaward-moving water generates *ebb currents*. Periods of little or no current, called *slack water*, separate flood and ebb. The areas affected by these alternating tidal currents are called **tidal flats** (see Figure 15.30). Depending on the nature of the coastal zone, tidal flats vary from narrow strips seaward of the beach to zones that may extend for several kilometers.

Although tidal currents are not important in the open sea, they can be rapid in bays, river estuaries, straits, and other narrow places. Off the coast of Brittany in France, for example, tidal currents that accompany a high tide of 12 meters (40 feet) may attain a speed of 20 kilometers (12 miles) per hour. Tidal currents are not generally considered to be major agents of erosion and sediment transport, but notable exceptions occur where tides move through narrow inlets. Here they scour the narrow entrances to many harbors that would otherwise be blocked.

Sometimes deposits called **tidal deltas** are created by tidal currents (Figure 15.34). They may develop either as *flood deltas* landward of an inlet or as *ebb deltas* on the seaward side of an inlet. Because wave activity and longshore currents are reduced on the sheltered landward side, flood deltas are more common and are actually more prominent (see Figure 15.22A, p. 443). They form after the tidal current moves rapidly through an inlet. As the current emerges into more open waters from the narrow passage, it slows and deposits its load of sediment.

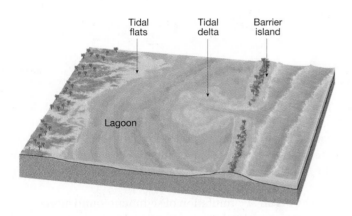

FIGURE 15.34 As a rapidly moving tidal current moves through a barrier island's inlet into the quiet waters of the lagoon, the current slows and deposits sediment, creating a tidal delta. Because this tidal delta has developed on the landward side of the inlet, it is called a *flood delta*. Such a tidal delta is also shown in Figure 15.22A, p. 443.

Students Sometimes Ask...

What are tidal waves?

Tidal waves, more accurately known as *tsunami* (*tsu* = harbor, *nami* = wave), have *nothing* to do with the tides. They are long-wavelength, fast-moving, often large, and sometimes destructive waves that originate from sudden changes in the topography of the seafloor. Tsunami are initiated by underwater faulting, landslides, or volcanic eruptions. Since the mechanisms that trigger these dangerous waves are frequently seismic events, tsunami are appropriately termed *seismic sea waves*. For more information about tsunami and their destructive effects, see "What Is a Tsunami?" in Chapter 8.

GIVE IT SOME THOUGHT

1. In this chapter you learned that global winds are the force that drive surface ocean currents. A glance at the accompanying map, however, shows a surface current that does not exactly coincide with the prevailing wind. Provide an explanation.

2. Palm trees associated with the subtropics can be found at the northern tip of the British Isles in Scotland. Suggest a possible explanation for how they can survive at such a high latitude.

3. During a visit to the beach, you get in a small rubber raft and paddle out *beyond* the surf zone. Tiring, you stop and take a rest. Describe the movement of your raft during your rest. How does this movement differ, if at all, from what you would have experienced if you had stopped paddling while *in* the surf zone?

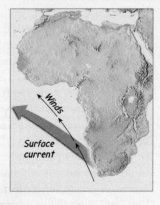

4. You and a friend set up an umbrella and chairs at a beach. Your friend then goes into the surf zone to play Frisbee with another person. Several minutes later your friend looks back toward the beach and is surprised to see that she is no longer near where the umbrella and chairs were set up. Although she is still in the surf zone, she is 30 or 40 yards away from where she started. How would you explain to your friend why she moved along the shore?

5. Examine the accompanying aerial photo that shows a portion of the New Jersey shoreline. What term is applied to the wall-like structures that extend into the water? What is their purpose? In what direction are beach drift and longshore currents moving sand? Is it moving toward the top or toward the bottom of the photo?

6. A friend wants to purchase a vacation home on a barrier island. If consulted, what advice would you give your friend?

7. The force of gravity plays a critical role in creating ocean tides. The more massive an object, the stronger the pull of gravity. Explain why the Sun's influence is only half that of the Moon's, even though the Sun is much more massive than the Moon.

In Review   Chapter 15 The Dynamic Ocean

- *The ocean's surface currents follow the general pattern of the world's major wind belts.* Surface currents are parts of huge, slowly moving loops of water called *gyres* that are centered in the subtropics of each ocean basin. The *positions of the continents* and the *Coriolis effect* also influence the movement of ocean water within gyres. Because of the Coriolis effect, subtropical gyres move *clockwise in the Northern Hemisphere* and *counterclockwise in the Southern Hemisphere.* Generally, *four main currents* comprise each subtropical gyre.

- *Ocean currents are important in navigation and for the effect they have on climates.* Poleward-moving *warm ocean currents moderate winter temperatures* in the middle latitudes. Cold currents exert their greatest influence during summer in middle latitudes and year-round in the tropics. In addition to cooler temperatures, *cold currents are associated with greater fog frequency and drought.*

- *Upwelling,* the rising of colder water from deeper layers, is a wind-induced movement that brings cold, nutrient-rich water to the surface. *Coastal upwelling* is most characteristic along the west coasts of continents.

- In contrast to surface currents, *deep-ocean circulation* is governed by *gravity* and driven by *density differences.* The two factors that are most significant in creating a dense mass of water are *temperature* and *salinity,* so the movement of deep-ocean water is often termed *thermohaline circulation.* Most water involved in thermohaline circulation begins in high latitudes at the surface when the salinity of the cold water increases due to sea ice formation. This dense water sinks, initiating deep-ocean currents.

- The *shore* is the area extending between the lowest tide level and the highest elevation on land that is affected by storm waves. The *coast* extends inland from the shore as far as ocean-related features can be found. The shore is divided into the *foreshore* and *backshore.* Seaward of the foreshore are the *nearshore* and *offshore* zones.

- A *beach* is an accumulation of sediment found along the landward margin of the ocean or a lake. Among its parts are one or more *berms* and the *beach face.* Beaches are composed of whatever material is locally abundant and should be thought of as material in transit along the shore.

- *Waves are moving energy* and *most ocean waves are initiated by the wind.* The three factors that influence the *height, wavelength,* and *period* of a wave are (1) *wind speed,* (2) *length of time the wind has blown,* and (3) *fetch,* the distance that the wind has traveled across open water. Once waves leave a storm area, they are termed *swells,* which are symmetrical, longer-wavelength waves.

- As waves travel, *water particles transmit energy by circular orbital motion,* which extends to a depth equal to one-half the wavelength. When a wave travels into shallow water, it experiences physical changes that can cause the wave to collapse, or *break,* and form *surf.*

- Wave erosion is caused by *wave impact pressure* and *abrasion* (the sawing and grinding action of water armed with rock fragments). The bending of waves is called *wave refraction.* Owing to refraction, wave impact is concentrated against the sides and ends of headlands.

- Most waves reach the shore at a slight angle. The uprush (swash) and backwash of water from each breaking wave moves the sediment in a zigzag pattern along the beach. This movement is called *beach drift*. Oblique waves also produce *longshore currents* within the surf zone that flow parallel to the shore and transport more sediment than beach drift. *Rip currents* occur perpendicular to the coast and move in the opposite direction of breaking waves.

- *Erosional features* include *wave-cut cliffs* (which originate from the cutting action of the surf against the base of coastal land), *wave-cut platforms* (relatively flat, benchlike surfaces left behind by receding cliffs), and *marine terraces* (uplifted wave-cut platforms). Erosional features also include *sea arches* (formed when a headland is eroded and sea caves from opposite sides unite) and *sea stacks* (formed when the roof of a sea arch collapses).

- Some of the *depositional features* that form when sediment is moved by beach drift and longshore currents are *spits* (elongated ridges of sand that project from the land into the mouth of an adjacent bay), *baymouth bars* (sandbars that completely cross a bay), and *tombolos* (ridges of sand that connect an island to the mainland or to another island). Along the Atlantic and Gulf Coastal Plains, the coastal region is characterized by offshore *barrier islands*, which are low ridges of sand that parallel the coast.

- Local *factors that influence shoreline erosion* are (1) the proximity of a coast to sediment-laden rivers, (2) the degree of tectonic activity, (3) the topography and composition of the land, (4) prevailing winds and weather patterns, and (5) the configuration of the coastline and nearshore areas.

- *Hard stabilization* involves building hard, massive structures in an attempt to protect a coast from erosion or prevent the movement of sand along the beach. Hard stabilization includes *groins* (short walls constructed at a right angle to the shore to trap moving sand), *breakwaters* (structures built parallel to the shore to protect it from the force of large breaking waves), and *seawalls* (armoring the coast to prevent waves from reaching the area behind the wall). Alternatives to hard stabilization include *beach nourishment*, which involves the addition of sand to replenish eroding beaches, and *relocation* of damaged or threatened buildings.

- Because of basic geological differences, the *nature of shoreline erosion problems along America's Pacific and Atlantic/ Gulf Coasts is very different*. Much of the development along the Atlantic and Gulf Coasts has occurred on barrier islands, which receive the full force of major storms. Much of the Pacific Coast is characterized by narrow beaches backed by steep cliffs and mountain ranges. A major problem facing the Pacific shoreline is the narrowing of beaches caused by irrigation and flood control dams that interrupt the natural flow of sand to the coast.

- One commonly used classification of coasts is based on changes that have occurred with respect to sea level. *Emergent coasts* often exhibit wave-cut cliffs and marine terraces and develop either because an area experiences uplift or as a result of a drop in sea level. Conversely, *submergent coasts* commonly display drowned river mouths called *estuaries* and are created when sea level rises or the land adjacent to the sea subsides.

- *Tides*, the daily rise and fall in the elevation of the ocean surface at a specific location, are caused by the *gravitational attraction* of the Moon and, to a lesser extent, the Sun. The Moon and the Sun each produce a pair of *tidal bulges* on Earth. These tidal bulges remain in fixed positions relative to the generating bodies as Earth rotates through them, resulting in alternating high and low tides. *Spring tides* occur near the times of new and full moons when the Sun and Moon are aligned and their bulges are added together to produce especially high and low tides (a *large daily tidal range*). Conversely, *neap tides* occur at about the times of the first and third quarters of the Moon when the bulges of the Moon and Sun are at right angles, producing a *smaller daily tidal range*.

- *Three main tidal patterns exist worldwide*. A *diurnal tidal pattern* exhibits one high and low tide daily; a *semidiurnal tidal pattern* exhibits two high and low tides daily of about the same height; and a *mixed tidal pattern* usually has two high and low tides daily of different heights.

- *Tidal currents* are horizontal movements of water that accompany the rise and fall of the tides. *Tidal flats* are the areas that are affected by the advancing and retreating tidal currents. When tidal currents slow after emerging from narrow inlets, they deposit sediment that may eventually create *tidal deltas*.

Key Terms

abrasion (p. 437)

backshore (p. 433)

barrier island (p. 443)

baymouth bar (p. 443)

beach (p. 435)

beach drift (p. 439)

beach face (p. 435)

beach nourishment (p. 447)

berm (p. 435)

breakwater (p. 446)

coast (p. 433)

coastline (p. 433)

Coriolis effect (p. 426)

diurnal tidal pattern (p. 452)

emergent coast (p. 449)

estuary (p. 450)

foreshore (p. 433)

groin (p. 446)

gyre (p. 426)

hard stabilization (p. 446)

longshore current (p. 440)

marine terrace (p. 442)

mixed tidal pattern (p. 452)

neap tide (p. 452)

nearshore zone (p. 433)

offshore zone (p. 433)

rip current (p. 441)

sea arch (p. 442)

sea stack (p. 442)

seawall (p. 446)

semidiurnal tidal pattern (p. 452)

shore (p. 433)

shoreline (p. 433)

spit (p. 443)

spring tide (p. 452)

submergent coast (p. 449)

surf (p. 437)

thermohaline circulation (p. 432)

tidal current (p. 453)

tidal delta (p. 453)

tidal flat (p. 453)

tide (p. 450)

tombolo (p. 443)

upwelling (p. 431)

wave-cut cliff (p. 441)

wave-cut platform (p. 441)

wave height (p. 436)

wavelength (p. 436)

wave period (p. 436)

wave refraction (p. 439)

Examining the Earth System

1. Describe the exchanges of energy between the ocean and the atmosphere that are responsible for (1) surface ocean currents and (2) the ocean's deep circulation.

2. If the warm North Atlantic Drift were to cease, how might the climate of Western Europe change? Speculate about how such a climate change might influence the biosphere.

3. In this chapter, the shoreline was described as a "dynamic interface." What is an interface? List and briefly describe some other interfaces in the Earth system. You need not confine yourself to examples from this chapter.

4. It is very likely that global temperatures will be increasing in the decades to come. How can a warmer atmosphere lead to a rise in sea level? How could geologic events associated with the solid Earth contribute to a rise in sea level? Suggest some possible changes in coastal areas if sea level were to rise.

Mastering Geology

Looking for additional review and test prep materials? Visit the Self Study area in **www.masteringgeology.com** to find practice quizzes, study tools, and multimedia that will aid in your understanding of this chapter's content. In **MasteringGeology™** you will find:

- **GEODe: Earth Science: An interactive visual walkthrough of key concepts**

- **Geoscience Animation Library: More than 100 animations illuminating many difficult-to-understand Earth science concepts**
- **In The News RSS Feeds: Current Earth science events and news articles are pulled into the site with assessment**
- **Pearson eText**
- **Optional Self Study Quizzes**
- **Web Links**
- **Glossary**
- **Flashcards**

16

The Atmosphere: Composition, Structure, and Temperature

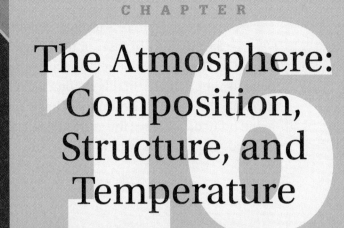

Solar radiation provides practically all of the energy that heats Earth's surface and atmosphere. This scene is from LaDigue Island, near the equator in the Indian Ocean.

(Photo by Chad Ehlers/ Photolibrary)

Earth's atmosphere is unique. No other planet in our solar system has an atmosphere with the exact mixture of gases or the heat and moisture conditions necessary to sustain life as we know it. The gases that make up Earth's atmosphere and the controls to which they are subject are vital to our existence. In this chapter we begin our examination of the ocean of air in which we all must live. We try to answer a number of basic questions: What is the composition of the atmosphere? At what point do we leave the atmosphere and enter outer space? What causes the seasons? How is air heated? What factors control temperature variations over the globe?

FOCUS ON CONCEPTS

To assist you in learning **the important concepts in this chapter, focus on the following questions:**

- What is weather? How is it different from climate?
- What are the basic elements of weather and climate?
- What are the major components of clean, dry air?
- What is ozone and why is it important to life on Earth?
- How do pressure and temperature change from Earth's surface to the top of the atmosphere?
- What causes the seasons?
- How do the noon sun angle and the length of daylight change between an equinox and a solstice?
- What paths does solar radiation take once it is intercepted by Earth?
- How is the atmosphere heated?
- What causes temperatures to vary from place to place?

Focus on the Atmosphere

Weather influences our everyday activities, our jobs, and our health and comfort. Many of us pay little attention to the weather unless we are inconvenienced by it or when it adds to our enjoyment outdoors. Nevertheless, there are few other aspects of our physical environment that affect our lives more than the phenomena we collectively call the weather.

Weather in the United States

The United States occupies an area that stretches from the tropics of Hawaii to beyond the Arctic circle in Alaska. It has thousands of miles of coastline and extensive regions that are far from the influence of the ocean. Some landscapes are mountainous and others are dominated by plains. It is a place where Pacific storms strike the West Coast while the East is sometimes influenced by events in the Atlantic and the Gulf of Mexico. For those in the center of the country, it is common to experience weather events triggered when frigid southward-bound Canadian air masses clash with northward-moving ones from the Gulf of Mexico.

Stories about weather are a routine part of the daily news. Articles and items about the effects of heat, cold, floods, drought, fog, snow, ice, and strong winds are commonplace. Of course, storms of all kinds are frequently front-page news. **Figure 16.1** provides some examples.

The United States has the greatest variety of weather of any country in the world. Severe weather events such as tornadoes, flash floods, and intense thunderstorms, as well as hurricanes and blizzards, are collectively more frequent and more damaging in the United States than in any other nation. Beyond its direct impact on the lives of individuals, the weather has a strong effect on the world economy by influencing agriculture, energy use, water resources, transportation, and industry.

Weather clearly influences our lives a great deal. Yet it is also important to realize that people influence the atmosphere and its behavior as well. There are, and will continue to be, significant political and scientific decisions that must be made involving these impacts. Answers to questions regarding air pollution and its control and the effects of various emissions on global climate and the atmosphere's ozone layer are important examples. So there is a need for increased awareness and understanding of our atmosphere and its behavior.

Weather and Climate

Acted on by the combined effects of Earth's motions and energy from the Sun, our planet's formless and invisible envelope of air reacts by producing an infinite variety of weather, which in turn creates the basic pattern of global climates. Although not identical, weather and climate have much in common.

Weather is constantly changing, sometimes from hour to hour and at other times from day to day. It is a term that refers to the state of the atmosphere at a given time and place. Whereas changes in the weather are continuous and sometimes seemingly erratic, it is nevertheless possible to arrive at a generalization of these variations. Such a description of aggregate weather conditions is termed **climate**. It is based on observations that have been accumulated over many years. Climate is often defined simply as

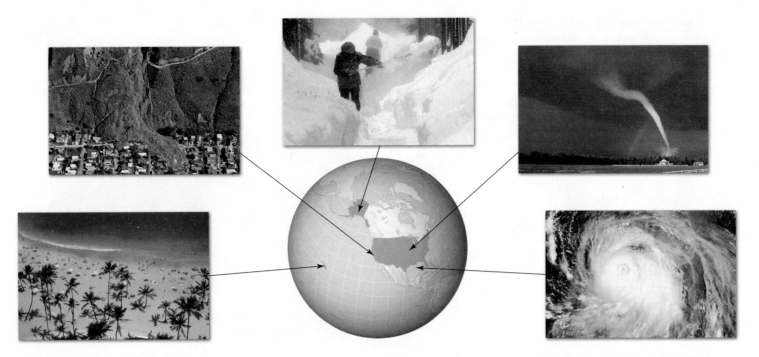

FIGURE 16.1 There are few aspects of our physical environment that influence our daily lives more than the weather. The United States experiences a remarkable variety of weather events. (Photo credits clockwise from left are Tomas del Amo/Photolibrary; G. Delaurentis; AP Photo/Michael Dwyer; Eric Nguyen/Photo Researchers; NASA)

"average weather," but this is an inadequate definition. In order to more accurately portray the character of an area, variations and extremes must also be included, as well as the probabilities that such departures will take place. For example, it is not only necessary for farmers to know the average rainfall during the growing season, but it is also important to know the frequency of extremely wet and extremely dry years. Thus, climate is the sum of all statistical weather information that helps describe a place or region.

Suppose you were planning a vacation trip to an unfamiliar place. You would probably want to know what kind of weather to expect. Such information would help as you selected clothes to pack and could influence decisions regarding activities you might engage in during your stay. Unfortunately, weather forecasts that go beyond a few days are not very dependable. Thus, it would not be possible to get a reliable weather report about the conditions you are likely to encounter during your vacation.

Instead, you might ask someone who is familiar with the area about what kind of weather to expect. "Are thunderstorms common?" "Does it get cold at night?" "Are the afternoons sunny?" What you are seeking is information about the climate, the conditions that are typical for that place. Another useful source of such information is the great variety of climate tables, maps, and graphs that are available. For example, the graph in **Figure 16.2** shows average daily high and low temperatures for each month, as well as extremes for New York City.

Such information could no doubt help as you planned your trip. But it is important to realize that *climate data cannot predict the weather*. Although the place may usually (climatically) be warm, sunny, and dry during the time of your planned vacation, you may actually experience cool, overcast, and rainy weather. There is a well-known saying that summarizes this idea: "Climate is what you expect, but weather is what you get."

FIGURE 16.2 Graph showing daily temperature data for New York City. In addition to the average daily maximum and minimum temperatures for each month, extremes are also shown. As this graph shows, there can be significant departures from average.

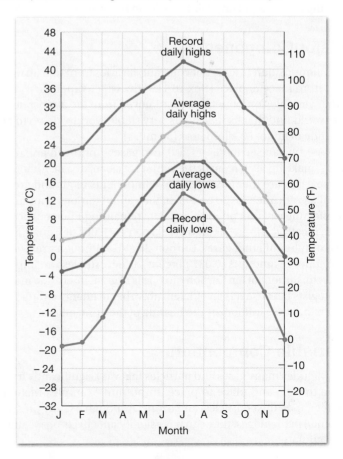

The nature of weather and climate is expressed in terms of the same basic **elements**, those quantities or properties that are measured regularly. The most important are (1) air temperature, (2) humidity, (3) type and amount of cloudiness, (4) type and amount of precipitation, (5) air pressure, and (6) the speed and direction of the wind. These elements are the major variables from which weather patterns and climate types are deciphered. Although you will study these elements separately at first, keep in mind that they are very much interrelated. A change in any one of the elements will often bring about changes in the others.

CONCEPT CHECK 16.1

1 Distinguish between weather and climate.

2 List the basic elements of weather and climate

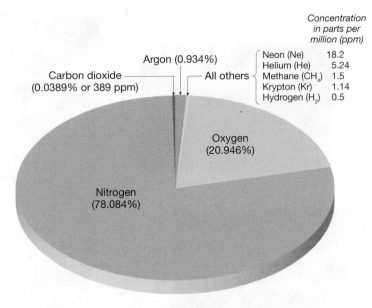

FIGURE 16.3 Proportional volume of gases composing dry air. Nitrogen and oxygen clearly dominate.

Composition of the Atmosphere

Earth's Dynamic Atmosphere
▶ Introduction to the Atmosphere

Air is *not* a unique element or compound. Rather, **air** is a *mixture* of many discrete gases, each with its own physical properties, in which varying quantities of tiny solid and liquid particles are suspended.

Major Components

The composition of air is not constant; it varies from time to time and from place to place (Box 16.1). If the water vapor, dust, and other variable components were removed from the atmosphere, we would find that its makeup is very stable worldwide up to an altitude of about 80 kilometers (50 miles).

As you can see in **Figure 16.3**, two gases—nitrogen and oxygen—make up 99 percent of the volume of clean, dry air. Although these gases are the most plentiful components of air and are of great significance to life on Earth, they are of minor importance in affecting weather phenomena. The remaining 1 percent of dry air is mostly the inert gas argon (0.93 percent) plus tiny quantities of a number of other gases. Carbon dioxide, although present in only minute amounts (0.037 percent), is nevertheless an important constituent of air. Carbon dioxide is of great interest to meteorologists because it is an efficient absorber of energy emitted by Earth and thus influences the heating of the atmosphere.

Variable Components

Air includes many gases and particles that vary significantly from time to time and place to place. Important examples include water vapor, dust particles, and ozone. Although usually present in small percentages, they can have significant effects on weather and climate.

Water Vapor   The amount of water vapor in the air varies considerably, from practically none at all up to about 4 percent by volume. Why is such a small fraction of the atmosphere so significant? Certainly the fact that water vapor is the source of all clouds and precipitation would be enough to explain its importance. However, water vapor has other roles. Like carbon dioxide, it has the ability to absorb heat given off by Earth as well as some solar energy. It is therefore important when we examine the heating of the atmosphere.

When water changes from one state to another (see Figure 17.2, p. 491), it absorbs or releases heat. This energy is termed *latent heat*, which means "hidden heat." As we shall see in later chapters, water vapor in the atmosphere transports this latent heat from one region to another, and it is the energy source that helps drive many storms.

Aerosols   The movements of the atmosphere are sufficient to keep a large quantity of solid and liquid particles suspended within it. Although visible dust sometimes clouds the sky, these relatively large particles are too heavy to stay in the air for very long. Still, many particles are microscopic and remain suspended for considerable periods of time. They may originate from many sources, both natural and human made, and include sea salts from breaking waves, fine soil blown into the air, smoke and soot from fires, pollen and microorganisms lifted by the wind, ash and dust from volcanic eruptions, and more (**Figure 16.4A**). Collectively, these tiny solid and liquid particles are called **aerosols**.

From a meteorological standpoint, these tiny, often invisible particles can be significant. First, many act as surfaces on which water vapor can condense, an important function in the formation of clouds and fog. Second, aerosols can absorb, reflect, and scatter incoming solar radiation. Thus, when an air-pollution episode is occurring or when ash fills the sky following a volcanic eruption, the amount of sunlight reaching Earth's surface can be measurably reduced. Finally, aerosols contribute to an optical phenomenon we have all observed—the varied hues of red and orange at sunrise and sunset (Figure 16.4B).

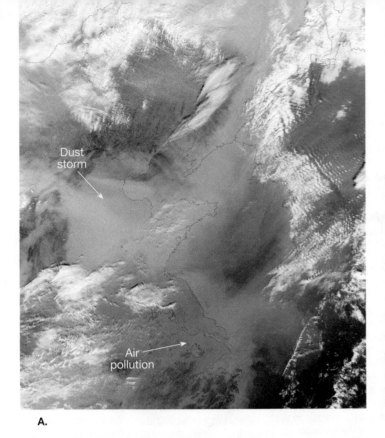

A.

B.

FIGURE 16.4 **A.** This satellite image from November 11, 2002, shows two examples of aerosols. First, a large dust storm is blowing across northeastern China toward the Korean Peninsula. Second, a dense haze toward the south (bottom center) is human-generated air pollution. (NASA Image) **B.** Dust in the air can cause sunsets to be especially colorful. (Photo by Steve Elmore/CORBIS/The Stock Market)

Ozone Another important component of the atmosphere is **ozone**. It is a form of oxygen that combines three oxygen atoms into each molecule (O_3). Ozone is not the same as oxygen we breathe, which has two atoms per molecule (O_2). There is very little of this gas in the atmosphere, and its distribution is not uniform. In the lowest portion of the atmosphere, ozone represents less than one part in 100 million. It is concentrated well above the surface in a layer called the *stratosphere*, between 10 and 50 kilometers (6 and 31 miles).

In this altitude range, oxygen molecules (O_2) are split into single atoms of oxygen (O) when they absorb ultraviolet radiation emitted by the Sun. Ozone is then created when a single atom of oxygen (O) and a molecule of oxygen (O_2) collide. This must happen in the presence of a third, neutral molecule that acts as a *catalyst* by allowing the reaction to take place without itself being consumed in the process. Ozone is concentrated in the 10- to 50-kilometer height range because a crucial balance exists there: The ultraviolet radiation from the Sun is sufficient to produce single atoms of oxygen, and there are enough gas molecules to bring about the required collisions.

The presence of the ozone layer in our atmosphere is crucial to those of us who dwell on Earth. The reason is that ozone absorbs the potentially harmful ultraviolet (UV) radiation from the Sun. If ozone did not filter a great deal of the ultraviolet radiation, and if the Sun's UV rays reached the surface of Earth undiminished, our planet would be uninhabitable for most life as we know it. Thus, anything that reduces the amount of ozone in the atmosphere could affect the well-being of life on Earth. Just such a problem exists and is described in the next section.

Ozone Depletion-A Global Issue

Although stratospheric ozone is 10–50 kilometers above Earth's surface, it is vulnerable to human activities. Chemicals we produce are breaking up ozone molecules in the stratosphere, weakening our shield against UV rays. This loss of ozone is a serious global-scale environmental problem. Measurements over the past three decades confirm that ozone depletion is occurring worldwide and is especially pronounced above Earth's poles. You can see this effect over the South Pole in **Figure 16.5**.

Over the past 60 years, people have unintentionally placed the ozone layer in jeopardy by polluting the atmosphere. The most significant of the offending chemicals are known as *chlorofluorocarbons* (CFCs for short). Over the decades, many uses were developed for CFCs: coolants for air-conditioning and refrigeration equipment, cleaning solvents for electronic components, propellants for aerosol sprays, and the production of certain plastic foams.

Because CFCs are practically inert (i.e. not chemically active) in the lower atmosphere, some of these gases gradually make their way to the ozone layer, where sunlight separates the chemicals into their constituent atoms. The chlorine atoms released this way break up some of the ozone molecules.

Because ozone filters out most of the UV radiation from the Sun, a decrease in its concentration permits more of these harmful wavelengths to reach Earth's surface. The most serious threat to human health is an increased risk of skin cancer. An increase in damaging UV radiation also can impair the human immune system as well as promote cataracts, a clouding of the eye lens that reduces vision and may cause blindness if not treated.

Box 16.1

PEOPLE AND THE ENVIRONMENT

Altering the Atmosphere's Composition—Sources and Types of Air Pollution

Air pollutants are airborne particles and gases that occur in concentrations that endanger the health and well-being of organisms or disrupt the orderly functioning of the environment. (**Figure 16.A**). One category of pollutants, the *primary pollutants*, are emitted directly from identifiable sources. They pollute the air immediately upon being emitted. The most significant

primary pollutants are carbon monoxide (CO), nitrogen oxides (NO_x), sulfur dioxide (SO_2), volatile organic compounds (VOC), and particulate matter (PM). **Figure 16.B** shows percentages calculated on the basis of weight. Important sources include industrial processes, electrical generation, solid waste disposal, and transportation (cars, trucks, trains, airplanes, etc.). In the United States the tens of millions of cars and trucks on the roads are the greatest contributors.

Sometimes the direct impact of primary pollutants on human health and the environment is less severe than the effects of the secondary pollutants they form. *Secondary pollutants* are not emitted directly into the air, but form in the atmosphere when reactions take place among primary pollutants. The chemicals that make up smog are important examples, as is the sulfuric acid that falls as acid precipitation. After the primary pollutant, sulfur dioxide, is emitted into the atmosphere, it combines with oxygen to produce sulfur trioxide, which then combines with water to create this irritating and corrosive acid.

Many reactions that produce secondary pollutants are triggered by strong sunlight and so are called *photochemical reactions*. One common example occurs when nitrogen oxides absorb solar radiation, initiating a chain of complex reactions. When certain volatile organic compounds are present, the result is the formation of a number of undesirable secondary products that are very reactive, irritating, and toxic. Collectively,

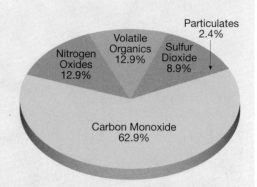

FIGURE 16.B Emissions estimates of primary pollutants for the United States in 2008. Percentages are calculated on the basis of weight. The total for 2008 was 124 million tons. (Data from the U.S. Environmental Protection Agency)

this noxious mixture of gases and particles is called *photochemical smog*.

Between 1980 and 2008, emissions of the five major primary pollutants dropped significantly. The aggregate decrease amounted to about 54 percent. During that same span, U.S. population increased by more than one-third, energy consumption was up by 29 percent, and vehicle miles traveled increased 91 percent (**Figure 16.C**). Despite this progress, about 124 million tons of pollutants were emitted into U.S. skies in 2008 and nearly 127 million people lived in counties where monitored air quality was unhealthy at times because of high levels of photochemical smog or one of the primary air pollutants.

FIGURE 16.A This crowded freeway reminds us that the transportation category is a major contributor to air pollution. Emissions from an individual vehicle are generally low, relative to the smokestack image many people associate with air pollution. However, in numerous cities across the country, the personal automobile is the single greatest polluter, when emissions from millions of vehicles are added together. (Photo by Visions of America/Purestock/SuperStock)

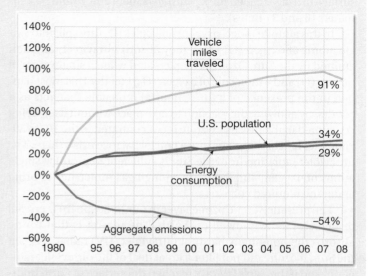

FIGURE 16.C Between 1980 and 2008 vehicle miles traveled increased 91 percent, energy consumption increased 29 percent, and U.S. population increased 34 percent. During that same time span, total emissions of the principal pollutants decreased 54 percent. (After U.S. Environmental Protection Agency)

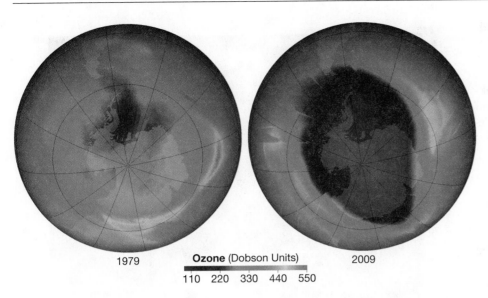

1979   **Ozone** (Dobson Units)   2009

110   220   330   440   550

FIGURE 16.5 These satellite images show ozone distribution in the Southern Hemisphere in 1979 and 2009. The dark blue colors over Antarctica correspond to the region with the sparsest ozone, called the *ozone hole*, which forms during the Southern Hemisphere spring. In 2009 the ozone hole covered an area about the size of North America. (NASA)

Vertical Structure of the Atmosphere

 Earth's Dynamic Atmosphere
▶ **Introduction to the Atmosphere**

To say that the atmosphere begins at Earth's surface and extends upward is obvious. However, where does the atmosphere end and outer space begin? There is no sharp boundary; the atmosphere rapidly thins as you travel away from Earth, until there are too few gas molecules to detect.

Pressure Changes

To understand the vertical extent of the atmosphere, let us examine the changes in the atmospheric pressure with height. Atmospheric pressure is simply the weight of the air above. At sea level, the average pressure is slightly more than 1,000 millibars. This corresponds to a weight of slightly more than 1 kilogram per square centimeter (14.7 pounds per square inch). Obviously the pressure at higher altitudes is less (**Figure 16.6**).

Students Sometimes Ask...

Isn't ozone some sort of pollutant?

Yes, you're right. Although the naturally occurring ozone in the stratosphere is critical to life on Earth, it is regarded as a pollutant when produced at ground level because it can damage vegetation and be harmful to human health.

Ozone is a major component in a noxious mixture of gases and particles called photochemical smog. It forms as a result of reactions triggered by sunlight that occur among pollutants emitted by motor vehicles and industries.

In response to this problem, an international agreement known as the *Montreal Protocol* was developed under the sponsorship of the United Nations to eliminate the production and use of CFCs. More than 180 nations eventually ratified the treaty.

Although relatively strong action has been taken, CFC levels in the atmosphere will not drop rapidly. Once in the atmosphere, CFC molecules can take many years to reach the ozone layer, and once there, they can remain active for decades. This does not promise a near-term reprieve for the ozone layer.

According to the U.S. Environmental Protection Agency, the ozone layer has not grown thinner since 1998 over most of the world. Between 2060 and 2075 the abundance of ozone-depleting gases is projected to fall to values that existed before the ozone hole began to form in the 1980s.

FIGURE 16.6 Atmospheric pressure variation with altitude. The rate of pressure decrease with an increase in altitude is not constant. Rather, pressure decreases rapidly near Earth's surface and more gradually at greater heights.

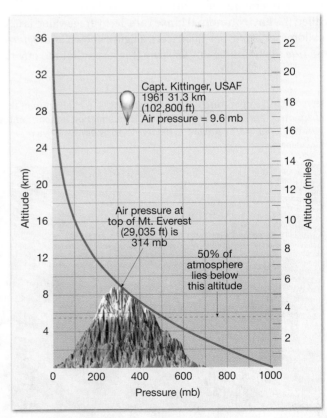

CONCEPT CHECK 16.2

1. What are the two major components of clean, dry air?
2. Why are water vapor and aerosols important constituents of Earth's atmosphere?
3. What is ozone? Why is ozone important to life on Earth?

One half of the atmosphere lies below an altitude of 5.6 kilometers (3.5 miles). At about 16 kilometers (10 miles), 90 percent of the atmosphere has been traversed, and above 100 kilometers (62 miles), only 0.00003 percent of all the gases making up the atmosphere remains. Even so, traces of our atmosphere extend far beyond this altitude, gradually merging with the emptiness of space.

Temperature Changes

By the early 20 century, much had been learned about the lower atmosphere. The upper atmosphere was partly known from indirect methods. Data from balloons and kites had revealed that the air temperature dropped with increasing height above Earth's surface. This phenomenon is felt by anyone who has climbed a high mountain and is obvious in pictures of snowcapped mountaintops rising above snow-free lowlands (**Figure 16.7**). We divide the atmosphere vertically into four layers on the basis of temperature (**Figure 16.8**).

Troposphere The bottom layer in which we live, where temperature decreases with an increase in altitude, is the **troposphere**. The term literally means the region where air "turns over," a reference to the appreciable vertical mixing of air in this lowermost zone. The troposphere is the chief focus of meteorologists, because it is in this layer that essentially all important weather phenomena occur.

The temperature decrease in the troposphere is called the **environmental lapse rate**. Its average value is 6.5° C per kilometer (3.5° F per 1,000 feet), a figure known as the *normal lapse rate*. It should be emphasized, however, that the environ-

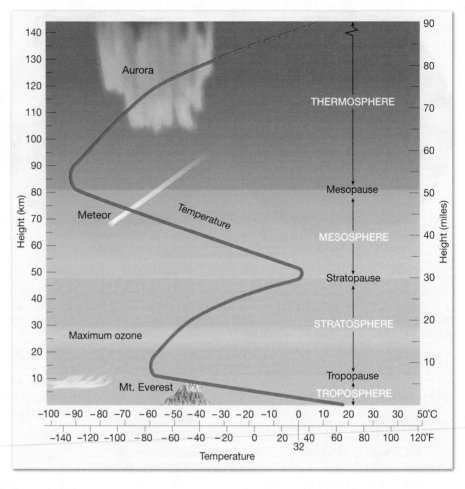

FIGURE 16.8 Thermal structure of the atmosphere.

FIGURE 16.7 Temperatures drop with an increase in altitude in the troposphere. Therefore, it is possible to have snow on a mountaintop and warmer, snow-free lowlands below. (Photo by David Wall/www.DanitaDelimont.com)

mental lapse rate is not a constant, but rather can be highly variable, and must be regularly measured. To determine the actual environmental lapse rate as well as gather information about vertical changes in pressure, wind, and humidity, radiosondes are used. The *radiosonde* is an instrument package that is attached to a balloon and transmits data by radio as it ascends through the atmosphere (**Figure 16.9**).

The thickness of the troposphere is not the same everywhere; it varies with latitude and the season. On the average, the temperature drop continues to a height of about 12 kilometers (7.4 miles). The outer boundary of the troposphere is the *tropopause*.

Stratosphere Beyond the tropopause is the **stratosphere**. In the stratosphere, the temperature remains constant to a height of about 20 kilometers (12 miles) and then begins a gradual increase that continues until the *stratopause*, at a height of nearly 50 kilometers (30 miles) above Earth's surface. Below the tropopause, atmospheric properties like temperature and humidity are readily transferred by large-scale turbulence and mixing. Above the tropopause, in the stratosphere, they are not. Temperatures increase in the stratosphere because it is in this layer that the atmosphere's ozone is concentrated. Recall that ozone

speeds, the temperature is very high. But the gases are so sparse that, collectively, they possess only an insignificant quantity of heat. For this reason, the temperature of a satellite orbiting Earth in the thermosphere is determined chiefly by the amount of solar radiation it absorbs and not by the high temperature of the almost nonexistent surrounding air. If an astronaut inside were to expose his or her hand, it would not feel hot.

CONCEPT CHECK 16.3

1. The atmosphere is divided vertically into four layers on the basis of temperature. List the names of these layers in order (from lowest to highest) and describe how temperature changes in each layer.
2. What is the environmental lapse rate and how is it determined?
3. Why do temperatures increase in the stratosphere?

Earth–Sun Relationships

Earth's Dynamic Atmosphere
▶ Heating Earth's Surface and Atmosphere

Always remember that nearly all of the energy that drives Earth's variable weather and climate comes from the Sun. Earth intercepts only a minute percentage of the energy given off by the Sun—less than one two-billionth. This may seem to be an insignificant amount until we realize that it is several hundred thousand times the electrical-generating capacity of the United States.

Solar energy is not distributed evenly over Earth's land–sea surface. The amount of energy received varies with latitude, time of day, and season of the year. Contrasting images of polar bears on ice rafts and palm trees along a remote tropical beach serve to illustrate the extremes. It is the unequal heating of Earth that creates winds and drives the ocean's currents. These movements, in turn, transport heat from the tropics toward the poles in an unending attempt to balance energy inequalities. The consequences of these processes are the phenomena we call weather. If the Sun were "turned off," global winds and ocean currents would quickly cease. Yet as long as the Sun shines, the winds *will* blow and weather *will* persist. So to understand how the atmosphere's dynamic weather machine works, we must first know why different latitudes receive varying quantities of solar energy and why the amount of solar energy changes to produce the seasons. As you will see, the variations in solar heating are caused by the motions of Earth relative to the Sun and by variations in Earth's land–sea surface.

Earth's Motions

Earth has two principal motions—rotation and revolution. **Rotation** is the spinning of Earth about its axis. The axis is an imaginary line running through the poles. Our planet rotates once every 24 hours, producing the daily cycle of daylight and darkness. At any moment, half of Earth is experiencing daylight, and the other half darkness. The line separating the dark half of Earth from the lighted half is called the **circle of illumination**.

absorbs ultraviolet radiation from the Sun. As a consequence, the stratosphere is heated.

Mesosphere   In the third layer, the **mesosphere**, temperatures again decrease with height until, at the *mesopause*, more than 80 kilometers (50 miles) above the surface, the temperature approaches −90° C (−130° F). The coldest temperatures anywhere in the atmosphere occur at the mesopause. Because accessibility is difficult, the mesosphere is one of the least explored regions of the atmosphere. The reason is that it cannot be reached by the highest research balloons nor is it accessible to the lowest orbiting satellites. Recent technological developments are just beginning to fill this knowledge gap.

Thermosphere   The fourth layer extends outward from the mesopause and has no well-defined upper limit. It is the **thermosphere**, a layer that contains only a tiny fraction of the atmosphere's mass. In the extremely rarefied air of this outermost layer, temperatures again increase, owing to the absorption of very short-wave, high-energy solar radiation by atoms of oxygen and nitrogen.

Temperatures rise to extremely high values of more than 1000° C in the thermosphere. But such temperatures are not comparable to those experienced near Earth's surface. Temperature is defined in terms of the average speed at which molecules move. Because the gases of the thermosphere are moving at very high

Revolution refers to the movement of Earth in its orbit around the Sun. Hundreds of years ago, most people believed that Earth was stationary in space and that the Sun and stars revolved around our planet. Today, we know that Earth is traveling at nearly 113,000 kilometers (70,000 miles) per hour in an eliptical orbit about the Sun.

Seasons

We know that it is colder in winter than in summer. But why? Length of daylight certainly accounts for some of the difference. Long summer days expose us to more solar radiation, whereas short winter days expose us to less.

Furthermore, a gradual change in the angle of the noon Sun above the horizon is quite noticeable (Figure 16.10). At midsummer, the noon Sun is seen high above the horizon. But as summer gives way to autumn, the noon Sun appears lower in the sky and sunset occurs earlier each evening. What we observe here is the annual shifting of the solar angle or *altitude* of the Sun.

The seasonal variation in the altitude of the Sun affects the amount of energy received at Earth's surface in two ways. First, when the Sun is high in the sky, the solar rays are most concentrated (you can see this in Figure 16.11A). The lower the angle, the more spread out and less intense is the solar radiation reaching the surface (Figure 16.11B, C). To illustrate this principle, hold a flashlight at a right angle to a surface and then change the angle.

Second, and of lesser importance, the angle of the Sun determines the amount of atmosphere the rays must penetrate (Figure 16.12). When the Sun is directly overhead, the rays pass through a thickness of only 1 atmosphere, whereas rays entering at a 30-degree angle travel through twice this amount, and 5-degree rays travel through a thickness roughly equal to 11 atmospheres. The longer the path, the greater the chances for absorption, reflection, and scattering by the atmosphere, all of which reduce the intensity at the surface. The same effects account for the fact that we cannot look directly at the midday Sun, but we can enjoy gazing at a sunset.

It is also important to remember that Earth has a spherical shape. Hence, on any given day only places located at a particular latitude receive vertical (90-degree) rays from the Sun. As we move either north or south of this location, the Sun's rays strike at an ever decreasing angle. Thus, the nearer a place is to the latitude receiving vertical rays of the Sun, the higher will be its noon Sun and the more intense will be the radiation it receives.

Earth's Orientation

What causes the fluctuations in the Sun angle and length of daylight that occur during the course of a year? They occur because *Earth's orientation to the Sun continually changes as it travels along its orbit.* Earth's axis is not perpendicular to the plane of its orbit around the Sun. Instead, it is tilted $23\frac{1}{2}$ degrees from the perpendicular, as shown in Figure 16.13. This is called the **inclination of the axis**. As you will see, if the axis were not inclined, we would have no seasonal changes. In addition, because the axis remains pointed in the same direction (toward the North Star) as Earth journeys around the Sun, the orientation of Earth's axis to the Sun's rays is constantly changing (Figure 16.13).

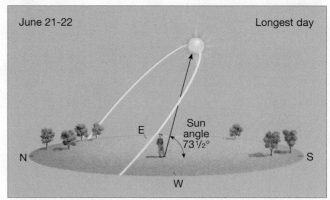

A. Summer solstice

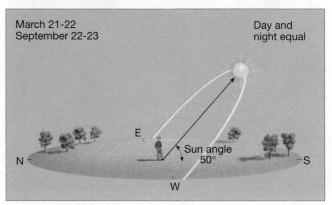

B. Spring or fall equinox

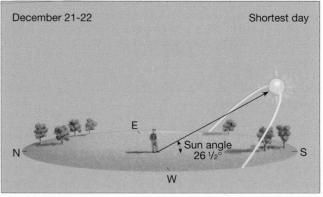

C. Winter solstice

FIGURE 16.10 Daily paths of the Sun for a place located at 40°N latitude for **A.** summer solstice, **B.** spring or fall equinox, and **C.** winter solstice. As we move from summer to winter, the angle of the noon Sun decreases from $73\frac{1}{2}$ to $26\frac{1}{2}$ degrees—a difference of 47 degrees. Notice also how the location of sunrise (east) and sunset (west) change during a year.

For example, on one day in June each year the axis is such that the Northern Hemisphere is "leaning" $23\frac{1}{2}$ degrees *toward* the Sun. Six months later, in December, when Earth has moved to the opposite side of its orbit, the Northern Hemisphere "leans" $23\frac{1}{2}$ degrees *away* from the Sun. On days between these extremes, Earth's axis is leaning at amounts less than $23\frac{1}{2}$ degrees to the rays of the Sun. This change in orientation causes the spot where the Sun's rays are vertical to make an annual migration from $23\frac{1}{2}$ degrees north of the equator to $23\frac{1}{2}$ degrees south of the equator.

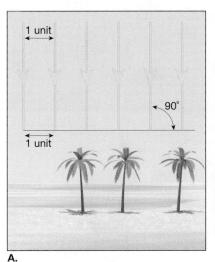

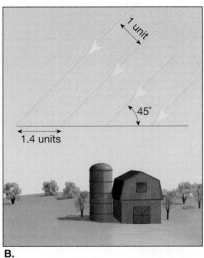

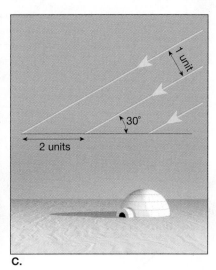

A.   **B.**   **C.**

FIGURE 16.11 Changes in the Sun's angle cause variations in the amount of solar energy reaching Earth's surface. The higher the angle, the more intense the solar radiation.

In turn, this migration causes the angle of the noon Sun to vary by up to 47 degrees $\left(23\frac{1}{2} + 23\frac{1}{2}\right)$ for many locations during the year. For example, a mid-latitude city like New York (about 40 degrees north latitude) has a maximum noon Sun angle of $73\frac{1}{2}$ degrees when the Sun's vertical rays reach their farthest northward location in June and a minimum noon Sun angle of $26\frac{1}{2}$ degrees 6 months later.

Solstices and Equinoxes

Historically, 4 days a year have been given special significance based on the annual migration of the direct rays of the Sun and its importance to the yearly weather cycle. On June 21 or 22, Earth is in a position where the axis in the Northern Hemisphere is tilted $23\frac{1}{2}$ degrees toward the Sun (**Figure 16.14A**). At this time the ver-

tical rays of the Sun strike $23\frac{1}{2}$ degrees north latitude ($23\frac{1}{2}$ degrees north of the equator), a latitude known as the **Tropic of Cancer**. For people in the Northern Hemisphere, June 21 or 22 is known as the **summer solstice**, the first "official" day of summer.

Six months later, on about December 21 or 22, Earth is in an opposite position, with the Sun's vertical rays striking at $23\frac{1}{2}$ degrees south latitude (Figure 16.14B). This parallel is known as the **Tropic of Capricorn**. For those in the Northern Hemisphere, December 21 and 22 is the **winter solstice**, the first day of winter. However, at the same time in the Southern Hemisphere, people are experiencing just the opposite: the summer solstice.

The equinoxes occur midway between the solstices. September 22 or 23 is the date of the **autumnal equinox** in the Northern Hemisphere, and March 21 or 22 is the date of the **spring equinox**. On these dates, the vertical rays of the Sun strike the equator (0 degrees latitude) because Earth is in such a position in its orbit that the axis is tilted neither toward nor away from the Sun (Figure 16.14C).

The length of daylight versus darkness is also determined by Earth's position in orbit. The length of daylight on June 21, the summer solstice in the Northern Hemisphere, is greater than the length of night. This fact can be established from Figure 16.14A by comparing the fraction of a given latitude that is on the "day" side of the circle of illumination with the fraction on the "night" side. The opposite is true for the winter solstice, when the nights are longer than the days. Again for comparison let us consider New York City, which has 15 hours of daylight on June 21 and only 9 hours on December 21 (you can see this in Figure 16.14 and **Table 16.1**). Also note from Table 16.1 that the farther you are north of the equator on June 21, the longer the period of daylight. When you reach the Arctic Circle $66\frac{1}{2}°$N, the length of daylight is 24 hours. This is the land of the "midnight Sun," which does not set for about 6 months at the North Pole (**Figure 16.15**).

During an equinox (meaning "equal night"), the length of daylight is 12 hours everywhere on Earth, because the circle of illumination passes directly through the poles, dividing the latitudes in half.

FIGURE 16.12 Notice that rays striking Earth at a low angle (toward the poles) must travel through more of the atmosphere than rays striking at a high angle (around the equator) and thus are subject to greater depletion by reflection and absorption.

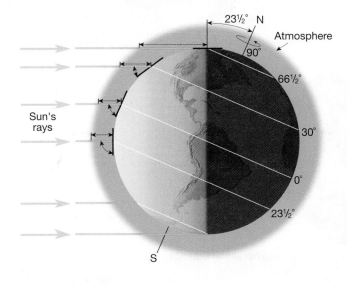

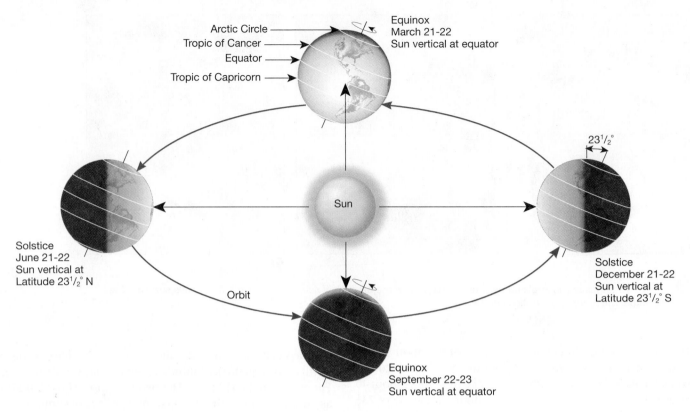

FIGURE 16.13   Earth–Sun relationships.

FIGURE 16.14   Characteristics of the solstices and equinoxes.

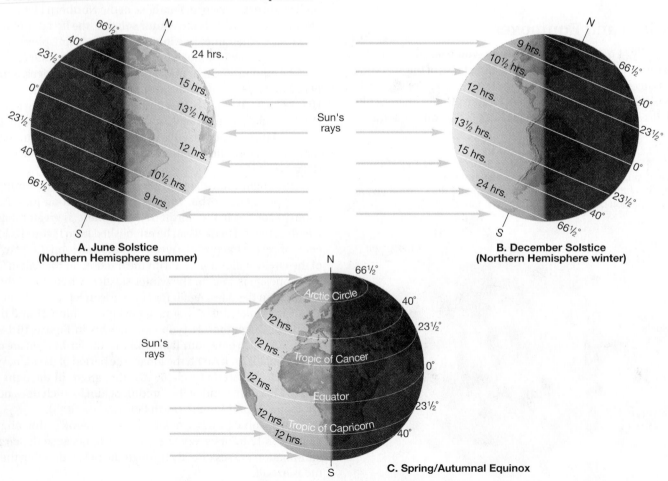

TABLE 16.1 Length of Daylight

Latitude (degrees)	Summer Solstice	Winter Solstice	Equinoxes
0	12 h	12 h	12 h
10	12 h 35 min	11 h 25 min	12
20	13 h 12 min	10 h 48 min	12
30	13 h 56 min	10 h 04 min	12
40	14 h 52 min	9 h 08 min	12
50	16 h 18 min	7 h 42 min	12
60	18 h 27 min	5 h 33 min	12
70	24 h (for 2 mo)	0 h 00 min	12
80	24 h (for 4 mo)	0 h 00 min	12
90	24 h (for 6 mo)	0 h 00 min	12

As a review of the characteristics of the summer solstice for the Northern Hemisphere, examine Figure 16.14A and Table 16.1 and consider the following facts:

1. The solstice occurs on June 21 or 22.
2. The vertical rays of the Sun are striking the Tropic of Cancer ($23\frac{1}{2}$ degrees north latitude).
3. Locations in the Northern Hemisphere are experiencing their greatest length of daylight (opposite for the Southern Hemisphere).
4. Locations north of the Tropic of Cancer are experiencing their highest noon Sun angles (opposite for places south of the Tropic of Capricorn).
5. The farther you are north of the equator, the longer the period of daylight, until the Arctic Circle is reached, where daylight lasts for 24 hours (opposite for the Southern Hemisphere).

The facts about the winter solstice are just the opposite. It should now be apparent why a midlatitude location is warmest in the summer. It is then that the days are longest and the Sun's altitude is highest.

In summary, seasonal variations in the amount of solar energy reaching places on Earth's surface are caused by the migrating vertical rays of the Sun and the resulting variations in Sun angle and length of daylight.

These changes in turn cause the month-to-month variations in temperature observed at most locations outside the tropics. Figure 16.16 shows mean monthly temperatures for selected cities at different latitudes. Notice that the cities located at more poleward latitudes experience larger temperature differences from summer to winter than do cities located nearer the equator. Also notice that temperature minimums for Southern Hemisphere locations occur in July, whereas they occur in January for most places in the Northern Hemisphere.

All places at the same latitude have identical Sun angles and lengths of daylight. If the Earth–Sun relationships just described were the only controls of temperature, we would expect these places to have identical temperatures as well. Obviously this is

FIGURE 16.15 Multiple exposures of the midnight Sun in late June or July in high northern latitudes—Alaska, Scandinavia, northern Canada, etc. (Photo by Brian Stablyk/Getty Images, Inc.—Stone Allstock)

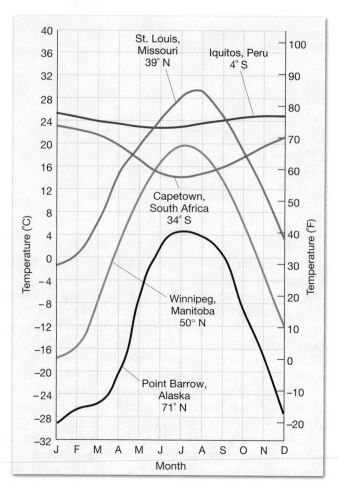

FIGURE 16.16 Mean monthly temperatures for six cities located at different latitudes. Note that Capetown, South Africa, experiences winter in June, July, and August.

not the case. Although the altitude of the Sun is an extremely important control of temperature, it is not the only control, as you will see.

1. Use a simple sketch to show why the intensity of solar radiation striking Earth's surface changes when the sun angle changes.
2. Briefly explain the primary cause of the seasons.
3. What is the significance of the Tropic of Cancer and the Tropic of Capricorn?
4. After examining Table 16.1, write a general statement that relates the season, the latitude, and the length of daylight.

Energy, Heat, and Temperature

Earth's Dynamic Atmosphere
▶ **Heating Earth's Surface and Atmosphere**

The universe is made up of a combination of matter and energy. The concept of matter is easy to grasp because it is the "stuff" we can see, smell, and touch. Energy, on the other hand, is abstract

and therefore more difficult to describe. For our purposes we define energy simply as *the capacity to do work*. We can think of work as being accomplished whenever matter is moved. You are likely familiar with some of the common forms of energy, such as thermal, chemical, nuclear, radiant (light), and gravitational energy. One type of energy is described as *kinetic energy*, which is energy of motion. Recall that matter is composed of atoms or molecules that are constantly in motion and therefore possesses kinetic energy.

Heat is a term that is commonly used synonymously with *thermal energy*. In this usage, heat is energy possessed by a material arising from the internal motions of its atoms or molecules. Whenever a substance is heated, its atoms move faster and faster, which leads to an increase in its heat content. **Temperature**, on the other hand, is related to the average kinetic energy of a material's atoms or molecules. Stated another way, the term *heat* generally refers to the quantity of energy present, whereas the word *temperature* refers to the intensity, that is, the degree of "hotness."

Heat and temperature are closely related concepts. Heat is the energy that flows because of temperature differences. In all situations, *heat is transferred from warmer to cooler objects*. Thus, if two objects of different temperature are in contact, the warmer object will become cooler and the cooler object will become warmer until they both reach the same temperature.

Three mechanisms of heat transfer are recognized: conduction, convection, and radiation. Although we present them separately, all three processes go on simultaneously in the atmosphere. In addition, these mechanisms operate to transfer heat between Earth's surface (both land and water) and the atmosphere.

Mechanism of Heat Transfer: Conduction

Conduction is familiar to all of us. Anyone who has touched a metal spoon that was left in a hot pan has discovered that heat was conducted through the spoon. **Conduction** is *the transfer of heat through matter by molecular activity*. The energy of mole-

Students Sometimes Ask...

The "official" first day of winter isn't until December 21 or 22. Realistically, doesn't winter start sooner?

The most publicized dates for the seasons are based on the astronomical definition, which uses the dates of the solstices and equinoxes as the "first day" of each of the seasons. Because the weather phenomena we normally associate with each season do not coincide well with this definition, meteorologists prefer to divide the year into four 3 month periods based primarily on temper-ature. Thus, winter is defined as December, January, and February, the three coldest months of the year in the Northern Hemisphere. Summer is defined as the three warmest months, June, July, and August. Spring and autumn are the transitional periods between these two seasons. This definition is more useful for meteorological purposes.

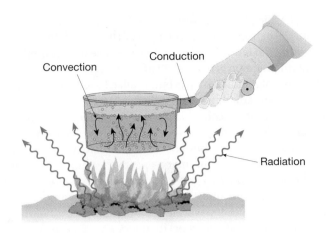

FIGURE 16.17 The three mechanisms of heat transfer: conduction, convection, and radiation.

Students Sometimes Ask...

In the morning when I get out of bed, why does the tile flooring in the bathroom feel much colder than the carpeted area, even though both materials are the same temperature?

The difference you feel is due mainly to the fact that floor tile is a much better conductor of heat than carpet. Hence, heat is more rapidly conducted from your bare feet when you are standing on the tile floor than when you are on the carpeted floor. Even at room temperature (20° C or 68° F) objects that are good conductors can feel chilly to the touch. (Remember, body temperature is about 98.6° F.)

cules is transferred through collisions from one molecule to another, with the heat flowing from the higher temperature to the lower temperature.

The ability of substances to conduct heat varies considerably. Metals are good conductors, as those of us who have touched hot metal have quickly learned (**Figure 16.17**). Air, conversely, is a very poor conductor of heat. Consequently, conduction is important only between Earth's surface and the air directly in contact with the surface. As a means of heat transfer for the atmosphere as a whole, conduction is the least significant.

Mechanism of Heat Transfer: Convection

Much of the heat transport that occurs in the atmosphere is carried on by convection. **Convection** is *the transfer of heat by mass movement or circulation within a substance.* It takes place in fluids (e.g., liquids like the ocean and gases like air) where the atoms and molecules are free to move about.

The pan of water in Figure 16.17 illustrates the nature of simple convective circulation. Radiation from the fire warms the bottom of the pan, which conducts heat to the water near the bottom of the container. As the water is heated, it expands and becomes less dense than the water above. Because of this new buoyancy, the warmer water rises. At the same time, cooler, denser water near the top of the pan sinks to the bottom, where it becomes heated. As long as the water is heated unequally—that is, from the bottom up—the water will continue to "turn over," producing a *convective circulation*. In a similar manner, most of the heat acquired in the lowest layer of the atmosphere by way of radiation and conduction is transferred by convective flow.

On a global scale, convection in the atmosphere creates a huge, worldwide air circulation. This is responsible for the redistribution of heat between hot equatorial regions and the frigid poles. This important process is discussed in detail in Chapter 18.

Mechanism of Heat Transfer: Radiation

The third mechanism of heat transfer is radiation. As shown in Figure 16.17, radiation travels out in all directions from its source. Unlike conduction and convection, which need a medium to travel through, radiant energy readily travels through the vacuum of space. Thus, radiation is the heat-transfer mechanism by which solar energy reaches our planet.

Because the Sun is the ultimate source of energy that creates our weather, we consider the nature of solar radiation in more detail. From our everyday experience we know that the Sun emits light and heat as well as the ultraviolet rays that cause suntan. Although these forms of energy comprise a major portion of the total energy that radiates from the Sun, they are only part of a large array of energy called **radiation** or **electromagnetic radiation**. This array or spectrum of electromagnetic energy is shown in **Figure 16.18**. All radiation, whether X-rays, radio waves, or heat waves, travel through the vacuum of space at 300,000 kilometers (186,000 miles) per second and only slightly slower through our atmosphere.

FIGURE 16.18 The electromagnetic spectrum, illustrating the wavelengths and names of various types of radiation.

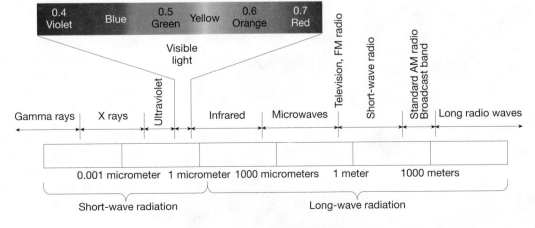

It helps to understand radiant energy by picturing ripples made in a pond by tossing in a pebble or by observing ocean waves. Like these waves, *electromagnetic waves*, as waves of radiant energy are called, come in various sizes. For our purpose, the most important difference among electromagnetic waves is their *wavelength*, or the distance from one crest to the next. Radio waves have the longest wavelengths, ranging to tens of kilometers. Gamma waves are the shortest, being less than a billionth of a centimeter long.

Visible light, as the name implies, is the only portion of the spectrum we can see. We often refer to visible light as "white" light because it appears "white" in color. However, it is easy to show that white light is really a mixture of colors, each corresponding to a specific wavelength (Figure 16.19). By using a prism, white light can be divided into the colors of the rainbow, from violet with the shortest wavelength—0.4 micrometer (1 micrometer is 0.0001 centimeter)—to red with the longest wavelength—0.7 micrometer.

Located adjacent to red, and having a longer wavelength, is **infrared** radiation, which we cannot see but which we can detect as heat. The closest invisible waves to violet are called **ultraviolet** rays. They are responsible for the sunburn that can occur after intense exposure to the Sun. Although we divide radiant energy into groups based on our ability to perceive them, all forms of radiation are basically the same. When any form of radiant energy is absorbed by an object, the result is an increase in molecular motion, which causes a corresponding increase in temperature.

To understand better how the atmosphere is heated, it is useful to have a general understanding of the basic laws governing radiation.

1. *All objects, at whatever temperature, emit radiant energy.* Thus, not only hot objects like the Sun but also Earth, including its polar ice caps, continually emit energy.
2. *Hotter objects radiate more total energy per unit area than do colder objects.*
3. *The hotter the radiating body, the shorter the wavelength of maximum radiation.* The Sun, with a surface temperature of nearly 6000° C, radiates maximum energy at 0.5 microm-

eter, which is in the visible range. The maximum radiation for Earth occurs at a wavelength of 10 micrometers, well within the infrared (heat) range. Because the maximum Earth radiation is roughly 20 times longer than the maximum solar radiation, it is often called long-wave radiation, and solar radiation is called short-wave radiation.

4. *Objects that are good absorbers of radiation are good emitters as well.* Earth's surface and the Sun are nearly perfect radiators because they absorb and radiate with nearly 100 percent efficiency for their respective temperatures. On the other hand, *gases are selective absorbers and radiators.* Thus, the atmosphere, which is nearly transparent to (does not absorb) certain wavelengths of radiation, is nearly opaque (a good absorber) to others. Our experience tells us that the atmosphere is transparent to visible light; hence, it readily reaches Earth's surface. This is not the case for the longer wavelength radiation emitted by Earth.

CONCEPT CHECK 16.5

1. Distinguish between heat and temperature.
2. Describe the three basic mechanisms of heat transfer. Which mechanism is *least* important as a means of heat transfer in the atmosphere?
3. Describe the relationship between the temperature of a radiating body and the wavelengths it emits.

The Fate of Incoming Solar Radiation

Earth's Dynamic Atmosphere
▶ Heating Earth's Surface and Atmosphere

When radiation strikes an object, there are usually three different results. First, some of the energy is *absorbed* by the object. Recall that when radiant energy is absorbed, it is converted to heat, which causes an increase in temperature. Second, substances such as water and air are transparent to certain wavelengths of radiation. Such materials simply *transmit* this energy. Radiation that is transmitted does not contribute energy to the object. Third, some radiation may "bounce off" the object without being absorbed or transmitted. *Reflection* and *scattering* are responsible for redirecting incoming solar radiation. In summary, *radiation may be absorbed, transmitted, or redirected (reflected or scattered).*

Figure 16.20 shows the fate of incoming solar radiation averaged for the entire globe. Notice that the atmosphere is quite transparent to incoming solar radiation. On average, about 50 percent of the solar energy reaching the top of the atmosphere is absorbed at Earth's surface. Another 30 percent is reflected back to space by the atmosphere, clouds, and reflective surfaces such as snow and water. The remaining 20 percent is absorbed by clouds and the atmosphere's gases.

What determines whether solar radiation will be transmitted to the surface, scattered, reflected outward, or absorbed by the atmosphere? As you will see, it depends greatly on the wavelength of the energy being transmitted, as well as on the nature of the intervening material.

FIGURE 16.19 Visible light consists of an array of colors we commonly call the "colors of the rainbow." Rainbows are relatively common optical phenomena produced by the bending and reflection of light by drops of water. (Photo by Michael Giannechini/Photo Researchers, Inc.)

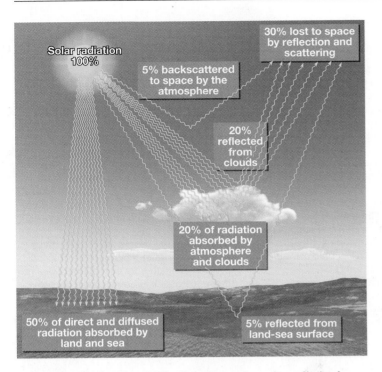

FIGURE 16.20 Average distribution of incoming solar radiation by percentage. More solar energy is absorbed by Earth's surface than by the atmosphere. Consequently, the air is not heated directly by the Sun, but is heated indirectly from Earth's surface.

Reflection and Scattering

Reflection is the process whereby light bounces back from an object at the same angle at which it encounters a surface and with the same intensity (Figure 16.21A). By contrast, **scattering** produces a larger number of weaker rays that travel in different directions. Although scattering disperses light both forward and backward (*backscattering*), more energy is dispersed in the forward direction (Figure 16.21B).

Reflection and Earth's Albedo   Energy is returned to space from Earth in two ways: reflection and emission of radiant energy. The portion of solar energy that is reflected back to space leaves in the same short wavelengths in which it came to Earth. About 30 percent of the solar energy reaching the outer atmosphere is reflected back to space. Included in this figure is the amount sent skyward by backscattering. This energy is lost to Earth and does not play a role in heating the atmosphere.

The fraction of the total radiation that is reflected by a surface is called its **albedo**. Thus, the albedo for Earth as a whole (the *planetary albedo*) is 30 percent. However, the albedo from place to place as well as from time to time in the same locale varies considerably, depending on the amount of cloud cover and particulate matter in the air, as well as on the angle of the Sun's rays and the nature of the surface. A lower Sun angle means that more atmosphere must be penetrated, thus making the "obstacle course" longer and therefore the loss of solar radiation greater (see Figure 16.12, p. 469). Figure 16.22 gives the albedo for various surfaces. Note that the angle at which the Sun's rays strike a water surface greatly affects its albedo.

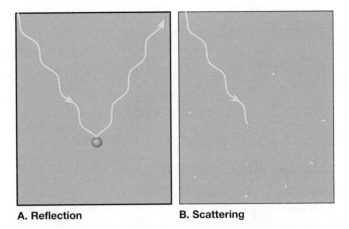

FIGURE 16.21 Reflection and scattering. **A**. Reflected light bounces back from a surface at the same angle at which it strikes that surface and with the same intensity. **B**. When a beam of light is scattered, it results in a larger number of weaker rays, traveling in all different directions. Usually more energy is scattered in the forward direction than is backscattered.

Scattering   Although incoming solar radiation travels in a straight line, small dust particles and gas molecules in the atmosphere scatter some of this energy in all directions. The result, called **diffused light**, explains how light reaches into the area beneath a shade tree, and how a room is lit in the absence of direct sunlight. Furthermore, scattering accounts for the brightness and even the blue color of the daytime sky (see Box 16.2). In contrast, bodies like the Moon and Mercury, which are without atmospheres, have dark skies and "pitch-black" shadows, even during daylight hours. Overall, about half of the solar radiation that is absorbed at Earth's surface arrives as diffused (scattered) light.

FIGURE 16.22 Albedo (reflectivity) of various surfaces. In general, light-colored surfaces tend to be more reflective than dark-colored surfaces and thus have higher albedos.

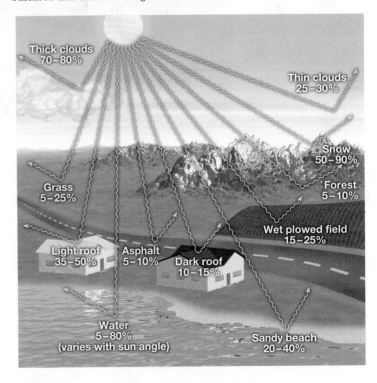

Box 16.2

UNDERSTANDING EARTH

Blue Skies and Red Sunsets

Gas molecules more effectively scatter the shorter wavelengths (blue and violet) of visible light than the longer wavelengths (red and orange). This fact explains the blue color of the sky and the orange and red colors seen at sunrise and sunset (see Figure 16.4B). Remember, sunlight is composed of all colors. When the Sun is overhead you can look in any direction away from the direct Sun and see predominantly blue light, which is the wavelength more readily scattered by the atmosphere.

Conversely, the Sun appears to have an orangish-to-reddish tint when viewed near the horizon (**Figure 16.D**). This is because solar radiation must travel through a greater thickness of atmosphere before it reaches your eyes. As a consequence, most of the blue and violet wavelengths will be scattered out, leaving light that consists mostly of reds and oranges. The reddish appearance of clouds during sunrise and sunset also results because the clouds are illuminated by light from which the blue color has been subtracted by scattering.

The most spectacular sunsets occur when large quantities of fine dust or smoke particles penetrate into the stratosphere. For 3 years after the great eruption of the Indonesian volcano Krakatau in 1883, brilliant sunsets occurred worldwide. The European summer that followed this colossal explosion was cooler than normal, a fact that has been attributed to the greater loss of radiation caused by backscattering.

Large particles associated with haze, fog, or smog scatter light more equally in all wavelengths. Because no color is predominant over any other, the sky appears white or gray on days when large particles are abundant.

In summary, the color of the sky gives an indication of the number of large or small particles present. Lots of small particles produce red sunsets, whereas large particles produce a white sky. Furthermore, the bluer the sky, the cleaner the air.

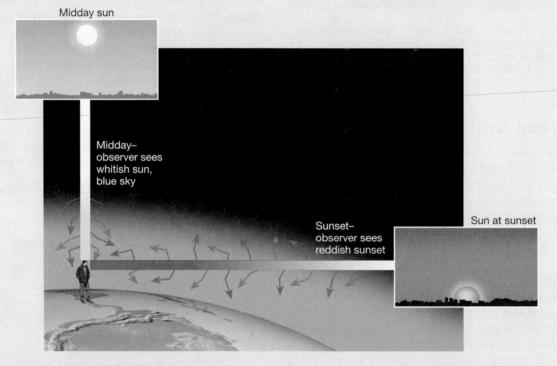

Midday sun

Midday– observer sees whitish sun, blue sky

Sunset– observer sees reddish sunset

Sun at sunset

FIGURE 16.D Short wavelengths (blue and violet) of visible light are scattered more effectively than are longer wavelengths (red, orange). Therefore, when the Sun is overhead an observer can look in any direction and see predominantly blue light that was selectively scattered by the gases in the atmosphere. By contrast, at sunset, the path that light must take through the atmosphere is much longer. Consequently, most of the blue light is scattered before it reaches an observer. Thus, the Sun appears reddish in color.

Absorption

As stated earlier, gases are selective absorbers, meaning that they absorb strongly in some wavelengths, moderately in others, and only slightly in still others. When a gas molecule absorbs light waves, this energy is transformed into internal molecular motion, which is detectable as a rise in temperature.

Nitrogen, the most abundant constituent in the atmosphere, is a poor absorber of all types of incoming radiation. Oxygen and ozone are efficient absorbers of ultraviolet radiation. Oxygen removes most of the shorter ultraviolet radiation high in the atmosphere, and ozone absorbs most of the remaining ultraviolet rays in the stratosphere. The absorption of UV radiation in the stratosphere accounts for the high temperatures experienced there. The only other significant absorber of incoming solar radiation is water vapor, which, along with oxygen and ozone, accounts for most of the solar radiation absorbed within the atmosphere.

For the atmosphere as a whole, none of the gases are effective absorbers of visible radiation. This explains why most visible radiation reaches Earth's surface and why we say that the atmosphere is *transparent* to incoming solar radiation. Thus, the atmosphere does not acquire the bulk of its energy directly from the Sun. Rather, it is heated chiefly by energy that is first absorbed by Earth's surface and then reradiated to the sky.

CONCEPT CHECK 16.6

❶ Prepare and label a simple sketch that shows what happens to incoming solar radiation.

❷ What factors cause albedo to vary from time to time and from place to place?

Heating the Atmosphere: The Greenhouse Effect

Earth's Dynamic Atmosphere
▸ Heating Earth's Surface and Atmosphere

Approximately 50 percent of the solar energy that strikes the top of the atmosphere reaches Earth's surface and is absorbed. Most of this energy is then reradiated skyward. Because Earth has a much lower surface temperature than the Sun, the radiation that it emits has longer wavelengths than solar radiation.

The atmosphere as a whole is an efficient absorber of the longer wavelengths emitted by Earth (*terrestrial radiation*). Water vapor and carbon dioxide are the principal absorbing gases. Water vapor absorbs roughly five times more terrestrial radiation than do all other gases combined and accounts for the warm temperatures found in the lower troposphere, where it is most highly concentrated. Because the atmosphere is quite transparent to shorter-wavelength solar radiation and more readily absorbs longer-wavelength terrestrial radiation, the atmosphere is heated from the ground up rather than vice versa. This explains the general drop in temperature with increasing altitude experienced in the troposphere. The farther from the "radiator," the colder it becomes.

When the gases in the atmosphere absorb terrestrial radiation, they warm, but they eventually radiate this energy away. Some travels skyward, where it may be reabsorbed by other gas molecules, a possibility less likely with increasing height because the concentration of water vapor decreases with altitude. The remainder travels Earthward and is again absorbed by Earth. For this reason, Earth's surface is continually being supplied with heat from the atmosphere as well as from the Sun. Without these absorptive gases in our atmosphere, Earth would not be a suitable habitat for humans and numerous other life-forms.

This very important phenomenon has been termed the **greenhouse effect** because it was once thought that greenhouses were heated in a similar manner (**Figure 16.23**). The gases of our atmosphere, especially water vapor and carbon dioxide, act very much like the glass in a greenhouse. They allow shorter-wavelength solar radiation to enter, where it is absorbed by the objects inside. These objects in turn reradiate the heat, but at longer wavelengths, to which glass is nearly opaque. The heat therefore is trapped in the greenhouse. However, a more important factor in

FIGURE 16.23 The heating of the atmosphere. Most of the short-wavelength radiation from the Sun passes through the atmosphere and is absorbed by Earth's land–sea surface. This energy is then emitted from the surface as longer-wavelength radiation, much of which is absorbed by certain gases in the atmosphere. Some of the energy absorbed by the atmosphere will be reradiated Earthward. This process, called the *greenhouse effect*, is responsible for keeping Earth's surface and lower atmosphere much warmer than it would be otherwise.

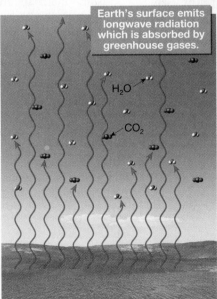
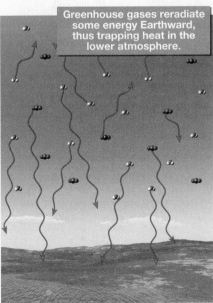

keeping a greenhouse warm is the fact that the greenhouse itself prevents mixing of air inside with cooler air outside. Nevertheless, the term *greenhouse effect* is still used.

For the Record: Air Temperature Data

Earth's Dynamic Atmosphere
▸ **Temperature Data and the Controls of Temperature**

Temperature is one of the basic elements of weather and climate. When someone asks what it is like outside, air temperature is often the first element we mention. At a weather station, the temperature is read on a regular basis from instruments mounted in an instrument shelter (**Figure 16.24**). The shelter protects the instruments from direct sunlight and allows a free flow of air. In addition to a standard mercury thermometer, the shelter is likely to contain a thermograph to continuously record temperature and a set of maximum–minimum thermometers. As their name

FIGURE 16.24 This modern shelter contains an electrical thermometer called a *thermistor*. A shelter protects instruments from direct sunlight and allows for the free flow of air. (Photo by Bobbé Christopherson)

implies, these thermometers record the highest and lowest temperatures during a measurement period, usually 24 hours.

The daily maximum and minimum temperatures are the bases for many of the temperature data compiled by meteorologists:

1. By adding the maximum and minimum temperatures and then dividing by two, the *daily mean temperature* is calculated.
2. The *daily range* of temperature is computed by finding the difference between the maximum and minimum temperatures for a given day.
3. The *monthly mean* is calculated by adding together the daily means for each day of the month and dividing by the number of days in the month.
4. The *annual mean* is an average of the 12 monthly means.
5. The *annual temperature range* is computed by finding the difference between the highest and lowest monthly means.

Mean temperatures are particularly useful for making comparisons, whether on a daily, monthly, or annual basis. It is common to hear a weather reporter state, "Last month was the hottest July on record," or "Today Chicago was 10 degrees warmer than Miami." Temperature ranges are also useful statistics because they give an indication of extremes.

To examine the distribution of air temperatures over large areas, isotherms are commonly used. An **isotherm** is a line that connects points on a map that have the same temperature (*iso* = equal, *therm* = temperature). Therefore, all points through which an isotherm passes have identical temperatures for the time period indicated. Generally, isotherms representing 5° or 10° temperature differences are used, but any interval may be chosen. **Figure 16.25** illustrates how isotherms are drawn on a map. Notice that most isotherms do not pass directly through the observing stations, because the station readings may not coincide with the values chosen for the isotherms. Only an occasional station temperature will be exactly the same as the value of the isotherm, so it is usually necessary to draw the lines by estimating the proper position between stations.

Isothermal maps are valuable tools because they clearly make temperature distribution visible at a glance. Areas of low and high temperatures are easy to pick out. In addition, the amount of temperature change per unit of distance, called the *temperature gradient*, is easy to visualize. Closely spaced isotherms indicate a rapid rate of temperature change, whereas more widely spaced lines indicate a more gradual rate of change. You can see this in Figure 16.25. The isotherms are closer in Colorado and Utah (steeper temperature gradient), whereas the isotherms are spread farther in Texas (gentler temperature gradient). Without isotherms, a map would be covered with numbers representing temperatures at dozens or hundreds of places, which would make patterns difficult to see.

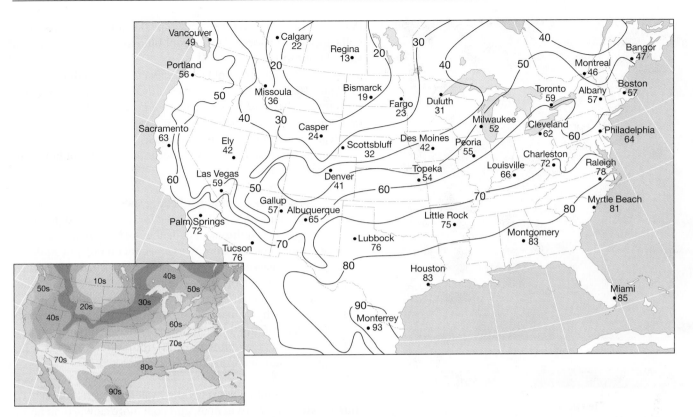

FIGURE 16.25 Temperature distribution using isotherms. Isotherms are lines that connect points of equal temperature. Showing temperature distribution in this way makes patterns easier to see. On television, and in many newspapers, temperature maps are in color. Rather than labeling isotherms, the area *between* isotherms is labeled. For example, the zone between the 60° and 70° isotherms is labeled "60s."

Why Temperatures Vary: The Controls of Temperature

Earth's Dynamic Atmosphere
▶ **Temperature Data and the Controls of Temperature**

A *temperature control* is any factor that causes temperature to vary from place to place and from time to time. Earlier in this chapter we examined the most important cause for temperature variations—differences in the receipt of solar radiation. Because variations in Sun angle and length of daylight depend on latitude, they are responsible for warm temperatures in the tropics and colder temperatures at more poleward locations. Of course, seasonal temperature changes at a given latitude occur as the Sun's vertical rays migrate toward and away from a place during the year.

However, latitude is not the only control of temperature. If it were, we would expect all places along the same parallel of latitude to have identical temperatures. This is clearly not the case. For example, Eureka, California, and New York City are both coastal cities at about the same latitude and both have an annual mean temperature of 11° C (52° F). However, New York City is 9° C (16° F) warmer than Eureka in July and 10° C (18° F) cooler in January. In another example, two cities in Ecuador—Quito and Guayaquil—are relatively close to each other, yet the annual mean temperatures of these two cities differ by 12° C (21° F). To explain these situations and countless others, we must realize that factors other than latitude also exert a strong influence on temper-

Students Sometimes Ask...

What's the hottest city in the United States?

It depends on how you want to define "hottest." If average annual temperature is used, then Key West, Florida, is the hottest, with an annual mean of 25.6° C (78° F) for the 30-year span 1971–2000. However, if we look at cities with the highest July maximums during the 1971–2000 period, then the desert community of Palm Springs, California, has the distinction of being hottest. Its average daily high in July is a blistering 42.2° C (108.3° F)! Yuma, Arizona (41.7° C/107° F), Phoenix, Arizona (41.4° C/106° F), and Las Vegas, Nevada (40° C/104.1° F) aren't far behind.

ature. In the next sections, we examine these other controls, which include differential heating of land and water, altitude, geographic position, cloud cover and albedo, and ocean currents.[42]

Land and Water

The heating of Earth's surface controls the heating of the air above it. Therefore, to understand variations in air temperature, we must examine the nature of the surface. Different land surfaces absorb

---

[42]For a discussion of the effects of ocean currents on temperature, see Chapter 15.

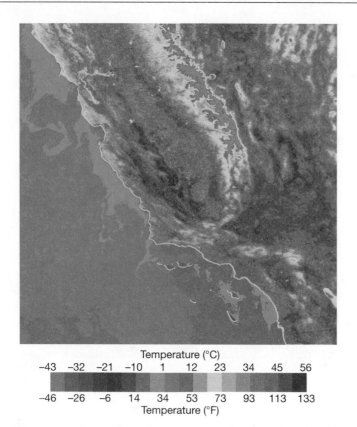

FIGURE 16.26 The differential heating of land and water is an important control of air temperature. In this satellite image from the afternoon of May 2, 2004, water-surface temperatures in the Pacific Ocean are much lower than land-surface temperatures in California and Nevada. (NASA)

varying amounts of incoming solar energy, which in turn cause variations in the temperature of the air above. The largest contrast, however, is not between different land surfaces, but between land and water. Figure 16.26 illustrates this idea nicely. This satellite image shows surface temperatures in portions of Nevada, California, and the adjacent Pacific Ocean on the afternoon of May 2, 2004, during a spring heat wave. Land-surface temperatures are clearly much higher than water-surface temperatures. The image shows the extreme high surface temperatures in southern California and Nevada in dark red.[43] Surface temperatures in the

Pacific Ocean are much lower. The peaks of the Sierra Nevada, still capped with snow, form a cool blue line down the eastern side of California.

In side-by-side areas of land and water, such as those shown in Figure 16.26, *land heats more rapidly and to higher temperatures than water, and it cools more rapidly and to lower temperatures than water.* Variations in air temperatures, therefore, are much greater over land than over water.

Among the reasons for the differential heating of land and water are the following:

1. The *specific heat* (amount of energy needed to raise 1 gram of a substance 1° C) is far greater for water than for land. Thus, water requires a great deal more heat to raise its temperature the same amount as an equal quantity of land.
2. Land surfaces are opaque, so heat is absorbed only at the surface. Water, being more transparent, allows some solar radiation to penetrate to a depth of many meters.
3. The water that is heated often mixes with water below, thus distributing the heat through an even larger mass.
4. Evaporation (a cooling process) from water bodies is greater than that from land surfaces.

All these factors collectively cause water to warm more slowly, store greater quantities of heat, and cool more slowly than land.

Monthly temperature data for two cities will demonstrate the moderating influence of a large water body and the extremes associated with land (Figure 16.27). Vancouver, British Columbia, is located along the windward Pacific coast, whereas Winnipeg, Manitoba, is in a continental position far from the influence of water. Both cities are at about the same latitude and thus experience similar Sun angles and lengths of daylight. Winnipeg, how-

FIGURE 16.27 Mean monthly temperatures for Vancouver, British Columbia, and Winnipeg, Manitoba. Vancouver has a much smaller annual temperature range owing to the strong marine influence of the Pacific Ocean. Winnipeg illustrates the greater extremes associated with an interior location.

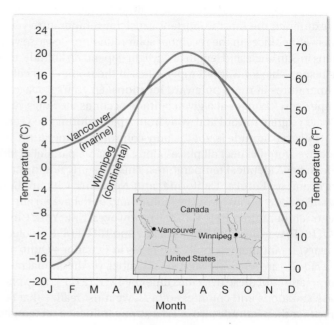

Students Sometimes Ask...

What are the highest and lowest temperatures ever recorded at Earth's surface?

The world's record-high temperature is nearly 59° C (136° F). It was recorded on September 13, 1922, at Azizia, Libya, in North Africa's Sahara Desert. The lowest recorded temperature is −89° C (−129° F). It should come as no surprise that this incredibly frigid temperature was recorded in Antarctica, at the Russian Vostok Station, on August 24, 1960.

[43]Realize that when a land surface is hot, the air above is cooler. For example, the surface of a sandy beach can be painfully hot, but the air temperature above the surface is more comfortable.

ever, has a mean January temperature that is 20° C lower than Vancouver's. Conversely, Winnipeg's July mean is 2.6° C higher than Vancouver's. Although their latitudes are nearly the same, Winnipeg, which has no water influence, experiences much greater temperature extremes than does Vancouver. The key to Vancouver's moderate year-round climate is the Pacific Ocean.

On a different scale, the moderating influence of water may also be demonstrated when temperature variations in the Northern and Southern hemispheres are compared. In the Northern Hemisphere, 61 percent is covered by water, and land accounts for the remaining 39 percent. However, in the Southern Hemisphere, 81 percent is covered by water and 19 percent by land. The Southern Hemisphere is correctly called the *water hemisphere* (see Figure 13.1, p. 383). **Table 16.2** portrays the considerably smaller annual temperature variations in the water-dominated Southern Hemisphere as compared with the Northern Hemisphere.

Altitude

The two cities in Ecuador mentioned earlier—Quito and Guayaquil—demonstrate the influence of altitude on mean temperature (**Figure 16.28**). Both cities are near the equator and relatively close to one another, but the annual mean temperature at Guayaquil is 25° C (77° F) compared to Quito's mean of 13° C (55° F). The difference may be understood when the cities' elevations are noted. Guayaquil is only 12 meters (40 feet) above sea level, whereas Quito is high in the Andes Mountains at 2,800 meters (9,200 feet).

Recall that temperatures drop an average of 6.5° C per kilometer in the troposphere; thus, cooler temperatures are to be expected at greater heights (see Figures 16.7 and 16.8, p. 466). Yet, the magnitude of the difference is not explained completely by the normal lapse rate. If the normal lapse rate is used, we would expect Quito to be about 18° C cooler than Guayaquil, but the difference is only 12° C. The fact that high-altitude places such as Quito are warmer than the value calculated using the normal lapse rate results from the absorption and reradiation of solar energy by the ground surface.

Geographic Position

The geographic setting can greatly influence the temperatures experienced at a specific location. A coastal location where prevailing winds blow from the ocean onto the shore (a *windward*

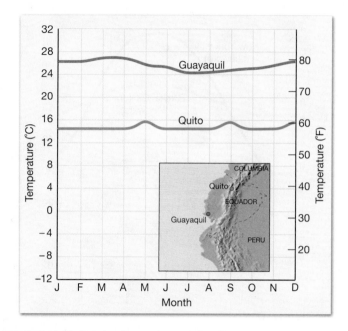

FIGURE 16.28 Graph comparing monthly mean temperatures at Quito and Guayaquil, Ecuador. Because Quito is high in the Andes, it experiences much cooler temperatures than Guayaquil, which is near sea level. Although Guayaquil is not far from Quito, it is near sea level and therefore significantly warmer.

coast) experiences considerably different temperatures than does a coastal location where the prevailing winds blow from the land toward the ocean (a *leeward* coast). In the first situation, the windward coast will experience the full moderating influence of the ocean—cool summers and mild winters—compared to an inland station at the same latitude.

A leeward coast, on the other hand, will have a more continental temperature pattern because the winds do not carry the ocean's influence onshore. Eureka, California, and New York City, the two cities mentioned earlier, illustrate this aspect of geographic position. The annual temperature range at New York City is 19° C (34° F) greater than Eureka's (**Figure 16.29**).

Seattle and Spokane, both in the state of Washington, illustrate a second aspect of geographic position—mountains that act as barriers. Although Spokane is only about 360 kilometers (220 miles) east of Seattle, the towering Cascade Range separates the cities. Consequently, Seattle's temperatures show a marked marine influence, but Spokane's are more typically continental (**Figure 16.30**). Spokane is 7° C (13° F) cooler than Seattle in January and 4° C (7° F) warmer than Seattle in July. The annual range at Spokane is 11° C (20° F) greater than at Seattle. The Cascade Range effectively cuts Spokane off from the moderating influence of the Pacific Ocean.

Cloud Cover and Albedo

The extent of cloud cover is a factor that influences temperatures in the lower atmosphere. Cloud cover is important because many clouds have a high albedo and therefore reflect a significant portion of the sunlight that strikes them back to space. By reducing the amount of incoming solar radiation, daytime temperatures

TABLE 16.2 Variation in Annual Mean Temperature Range (°C) with Latitude

Latitude	Northern Hemisphere	Southern Hemisphere
0	0	0
15	3	4
30	13	7
45	23	6
60	30	11
75	32	26
90	40	31

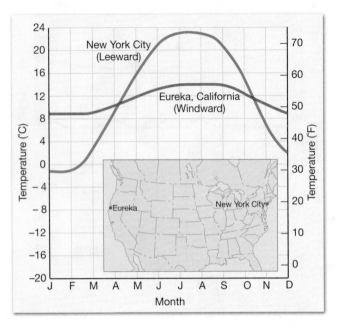

FIGURE 16.29 Monthly mean temperatures for Eureka, California, and New York City. Both cities are coastal and located at about the same latitude. Because Eureka is strongly influenced by prevailing winds from the ocean and New York City is not, the annual temperature range at Eureka is much smaller.

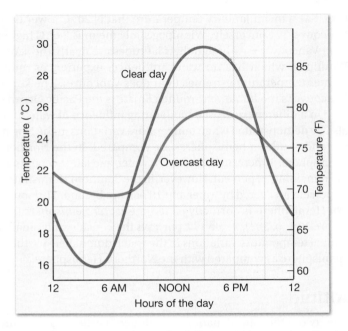

FIGURE 16.31 The graph shows the daily temperature cycle at Peoria, Illinois, for two July days—one clear and the other overcast. On a clear day the maximum temperature is higher, and the minimum temperature is lower than if the day had been overcast. On an overcast day, the daily temperature range is less than if it had been sunny.

will be lower than if the clouds were absent and the sky was clear (Figure 16.31).

At night, clouds have the opposite effect as during daylight. They act as a blanket by absorbing outgoing terrestrial radiation and reradiating a portion of it back to the surface. Consequently, some of the heat that otherwise would have been lost remains near the ground. Thus, nighttime air temperatures do not drop as low as they would on a clear night. The effect of cloud cover is

FIGURE 16.30 Monthly mean temperatures for Seattle and Spokane, Washington. Because the Cascade Mountains cut off Spokane from the moderating influence of the Pacific Ocean, its annual temperature range is greater than Seattle's.

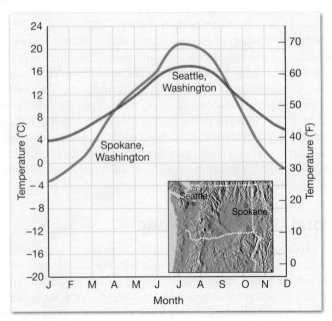

to reduce the daily temperature range by lowering the daytime maximum and raising the nighttime minimum.

Clouds are not the only phenomenon that increase albedo and thereby reduce air temperatures. We also recognize that snow- and ice-covered surfaces have high albedos. This is one reason why mountain glaciers do not melt away in the summer and why snow may still be present on a mild spring day. In addition, during the winter, when snow covers the ground, daytime maximums on a sunny day are less than they otherwise would be because energy that the land would have absorbed and used to heat the air is reflected and lost.

CONCEPT CHECK 16.9

❶ List the factors that cause land and water to heat and cool differently.

❷ Quito, Ecuador, is located on the equator and is *not* a coastal city. It has an average annual temperature of only 13° C (55° F). What is the likely cause for this low average temperature?

❸ In what ways can geographic position be considered a control of temperature?

❹ How does cloud cover influence the maximum temperature on an overcast day? How is the nighttime minimum influenced by clouds?

World Distribution of Temperature

Take a moment to study the two world isothermal maps (Figures 16.32 and 16.33). From hot colors near the equator to cool colors toward the poles, these maps portray sea-level tempera-

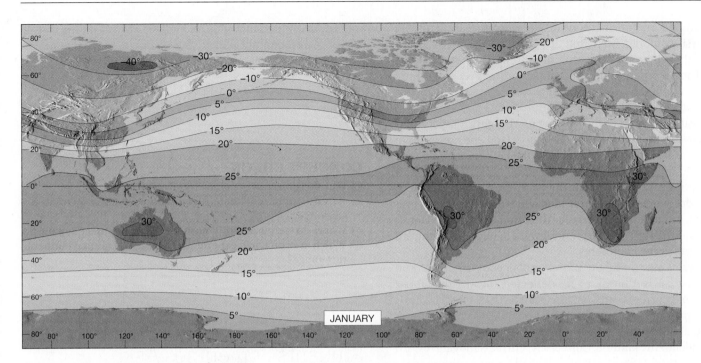

FIGURE 16.32 World mean sea-level temperatures in January in degrees Celsius.

tures in the seasonally extreme months of January and July. On these maps you can study global temperature patterns and the effects of the controlling factors of temperature, especially latitude, the distribution of land and water, and ocean currents. Like most isothermal maps of large regions, all temperatures on these world maps have been reduced to sea level to eliminate the complications caused by differences in altitude.

On both maps, the isotherms generally trend east and west and show a decrease in temperatures poleward from the tropics.

They illustrate one of the most fundamental aspects of world temperature distribution: that the effectiveness of incoming solar radiation in heating Earth's surface and the atmosphere above it is largely a function of latitude.

Moreover, there is a latitudinal shifting of temperatures caused by the seasonal migration of the Sun's vertical rays. To see this, compare the color bands by latitude on the two maps. For example, on the January map, the hot spots of 30° C are *south* of the equator, but in July they have shifted *north* of the equator.

FIGURE 16.33 World mean sea-level temperatures in July in degrees Celsius.

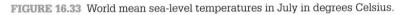

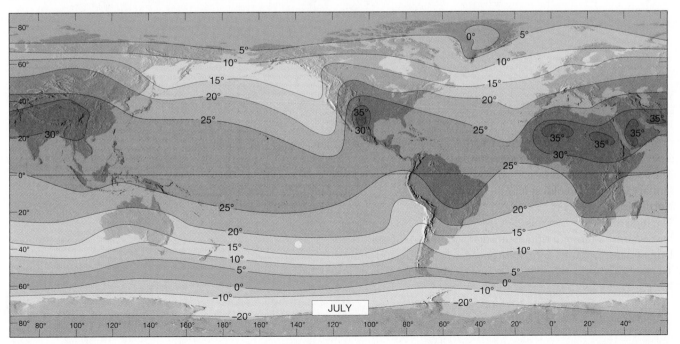

If latitude were the only control of temperature distribution, our analysis could end here, but this is not the case. The added effect of the differential heating of land and water is clearly reflected on the January and July temperature maps. The warmest and coldest temperatures are found over land—note the coldest area, a purple oval in Siberia, and the hottest areas, the deep orange ovals—all over land. Consequently, because temperatures do not fluctuate as much over water as over land, the north–south migration of isotherms is greater over the continents than over the oceans.

In addition, it is clear that the isotherms in the Southern Hemisphere, where there is little land and where the oceans predominate, are much more regular than in the Northern Hemisphere, where they bend sharply northward in July and southward in January over the continents.

Isotherms also reveal the presence of ocean currents. Warm currents cause isotherms to be deflected toward the poles, whereas cold currents cause an equatorward bending. The horizontal transport of water poleward warms the overlying air and results in air temperatures that are higher than would otherwise be expected for the latitude. Conversely, currents moving toward the equator produce cooler than expected air temperatures.

Figures 16.32 and 16.33 show the seasonal extremes of temperature, so comparing them enables us to see the annual range of temperature from place to place. Comparing the two maps shows that a station near the equator has a very small annual range because it experiences little variation in the length of daylight and it always has a relatively high Sun angle. A station in the middle latitudes, however, experiences wide variations in Sun angle and length of daylight and hence large variations in temperature. Therefore, we can state that the annual temperature range increases with an increase in latitude.

Moreover, land and water also affect seasonal temperature variations, especially outside the tropics. A continental location must endure hotter summers and colder winters than a coastal location. Consequently, outside the tropics the annual temperature range will increase with an increase in continentality.

CONCEPT CHECK 16.10

❶ Why do isotherms generally trend east–west?

❷ Why do isotherms shift north and south from season to season?

❸ Where do isotherms shift most, over land or water? Explain.

❹ Which area on Earth experiences the highest annual temperature range?

Students Sometimes Ask...

Where in the world would I experience the greatest contrast between summer and winter temperatures?

Among places for which records exist, it appears as though Yakutsk, in the heart of Siberia, is the best candidate. The latitude of Yakutsk is 62° north, just a few degrees south of the Arctic Circle. Moreover, it is far from the influence of water. The January mean at Yakutsk is a frigid −43° C (−45° F), whereas its July mean is a pleasant 20° C (68° F). The result is an average annual temperature range of 63° C (113° F), among the highest anywhere on the globe.

GIVE IT SOME THOUGHT

1. Determine which statements refer to weather and which refer to climate. (*Note:* One statement includes aspects of *both* weather and climate.)
   a. The baseball game was rained out today.
   b. January is Omaha's coldest month.
   c. North Africa is a desert.
   d. The high this afternoon was 25° C.
   e. Last evening a tornado ripped through central Oklahoma.
   f. I am moving to southern Arizona because it is warm and sunny.
   g. Thursday's low of −20° C is the coldest temperature ever recorded for that city.
   h. It is partly cloudy.

2. The accompanying photo shows the explosive 1991 eruption of Mt. Pinatubo in the Philippines. How would you expect global temperatures to respond to the ash and debris that this volcano spewed high into the atmosphere?

3. Refer to Figure 16.6 to answer the following questions.
   a. If you were to climb to the top of Mt. Everest, how many breaths of air would you have to take at that altitude to equal one breath at sea level?
   b. If you are flying in a commercial jet at an altitude of 12 kilometers, about what percentage of the atmosphere's mass is below you?

4. If you were ascending from the surface of Earth to the top of the atmosphere, which of the following would be most useful for determining the layer of the atmosphere you were in? Explain.
   a. Doppler radar
   b. Hygrometer (humidity)
   c. Weather satellite
   d. Barometer (air pressure)
   e. Thermometer (temperature)

(US Geological Survey)

5. The circumference of Earth at the equator is 24,900 miles. Calculate how fast someone at the equator is rotating in miles per hour. If the rotational speed of Earth were to slow down, how might this impact daytime highs and nighttime lows?

6. If you were to see the Sun directly overhead in New York City (40° N lat.) on the summer solstice, at what angle would Earth's axis be tilted? If Earth were inclined at this angle, what might the seasons be like?

7. The Sun shines continually at the North Pole for 6 months, from the spring equinox until the fall equinox, yet temperatures never get very warm. Explain why this is the case.

8. Rank the following according to the wavelength of radiant energy each emits from the shortest wavelength to the longest.
   a. A light bulb with a filament glowing at 4000° C
   b. A rock at room temperature
   c. A car engine at 140° C

9. Figure 16.20 shows that about 30 percent of the Sun's energy is reflected and scattered back to space. If Earth's albedo were to increase to 50 percent, how would you expect average surface temperatures to change?

10. The accompanying graph shows average monthly high temperatures for Urbana, Illinois, and San Francisco, California. Although both cities are located at about the same latitude, the temperatures they experience are quite different. Which line on the graph represents Urbana and which represents San Francisco? How did you figure this out?

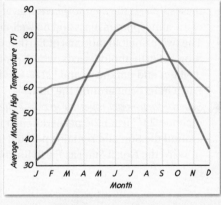

11. On which summer day would you expect the *greatest* temperature range? Which would have the *smallest* range in temperature? Explain your choices.
   a. Cloudy skies during the day and clear skies at night
   b. Clear skies during the day and cloudy skies at night
   c. Clear skies during the day and clear skies at night
   d. Cloudy skies during the day and cloudy skies at night

12. Refer to the chapter-opening photo. Describe how latitude, altitude, and the differential heating of land and water influence the climate of the place in the image.

In Review   Chapter 16 The Atmosphere: Composition, Structure, and Temperature

- *Weather* is the state of the atmosphere at a particular place for a short period of time. *Climate*, on the other hand, is a generalization of the weather conditions of a place over a long period of time.

- The most important *elements*, those quantities or properties that are measured regularly, of weather and climate are (1) *air temperature*, (2) *humidity*, (3) type and amount of *cloudiness*, (4) type and amount of *precipitation*, (5) *air pressure*, and (6) the speed and direction of the *wind*.

- If water vapor, dust, and other variable components of the atmosphere were removed, clean, dry air would be composed almost entirely of *nitrogen* (N), about 78% of the atmosphere by volume, and *oxygen* (O_2) about 21%. *Carbon dioxide* (CO_2), although present only in minute amounts (0.036%), is important because it has the ability to absorb heat radiated by Earth and thus helps keep the atmosphere warm. Among the variable components of air, *water vapor* is very important because it is the source of all clouds and precipitation and, like carbon dioxide, it is also a heat absorber.

- *Ozone* (O_3) the triatomic form of oxygen, is concentrated in the 10- to 50-kilometer altitude range of the atmosphere, and is important to life because of its ability to absorb potentially harmful ultraviolet radiation from the Sun.

- Because the atmosphere gradually thins with increasing altitude, it has no sharp upper boundary but simply blends into outer space. Based on temperature, the atmosphere is divided vertically into four layers. The *troposphere* is the lowermost layer. In the troposphere, temperature usually decreases with increasing altitude. This *environmental lapse rate* is variable, but averages about 6.5° C per kilometer (3.5° F per 1,000 feet). Essentially all important weather phenomena occur in the troposphere. Above the troposphere is the *stratosphere*, which exhibits warming because of absorption of ultraviolet radiation by ozone. In the *mesosphere*, temperatures again decrease. Upward from the mesosphere is the *thermosphere*, a layer with only a tiny fraction of the atmosphere's mass and no well-defined upper limit.

- The two principal motions of Earth are (1) *rotation*, the spinning of Earth about its axis, which produces the daily cycle of daylight and darkness; and (2) *revolution*, the movement of Earth in its orbit around the Sun.

- *Several factors act together to cause the seasons.* Earth's axis is inclined $23\frac{1}{2}$° from the perpendicular to the plane of its orbit around the Sun and remains pointed in the same direction (toward the North Star) as Earth journeys around the Sun. As a consequence, Earth's orientation to the Sun continually changes. The yearly fluctuations in the angle of the Sun and length of daylight brought about by Earth's changing orientation to the Sun cause the seasons.

- The three mechanisms of heat transfer are (1) *conduction*, the transfer of heat through matter by molecular activity; (2) *convection*, the transfer of heat by the movement of a mass or substance from one place to another; and (3) *radiation*, the transfer of heat by electromagnetic waves.

- *Electromagnetic radiation* is energy emitted in the form of rays, called electromagnetic waves. All radiation is capable of transmitting energy through the vacuum of space. An important difference among electromagnetic waves is their *wavelengths*, which range from very long *radio waves* to very short *gamma rays*. *Visible light* is the only portion of the electromagnetic spectrum we can see. Some of the basic laws that govern radiation as it heats the atmosphere are (1) all objects emit radiant energy; (2) hotter objects radiate more total energy than do colder objects; (3) the hotter the radiating body, the shorter the wavelengths of maximum radiation; and (4) objects that are good absorbers of radiation are good emitters as well.

- The general drop in temperature with increasing altitude in the troposphere supports the fact that the *atmosphere is heated from the ground up*. Approximately 50 percent of the solar energy that strikes the top of the atmosphere is ultimately absorbed at Earth's surface. Earth emits the absorbed radiation in the form of long-wave radiation. The atmospheric absorption of this long-wave *terrestrial radiation*, primarily by water vapor and carbon dioxide, is responsible for heating the atmosphere.

- The factors that cause temperature to vary from place to place, also called the *controls of temperature*, are (1) *differences in the receipt of solar radiation*—the greatest single cause; (2) the unequal heating and cooling of *land and water*, in which land heats more rapidly and to higher temperatures than water and cools more rapidly and to lower temperatures than water; (3) *altitude*; (4) *geographic position*; (5) *cloud cover and albedo*; and (6) *ocean currents*.

- Temperature distribution is shown on a map by using *isotherms*, which are lines that connect equal temperatures. Differences between January and July temperatures around the world can be explained in terms of the basic controls of temperature.

Key Terms

aerosols (p. 462)
air (p. 462)
albedo (p. 475)
autumnal equinox (p. 469)
circle of illumination (p. 467)
climate (p. 460)
conduction (p. 472)
convection (p. 473)
diffused light (p. 475)
electromagnetic radiation (p. 473)
element (of weather and climate) (p. 462)
environmental lapse rate (p. 466)

greenhouse effect (p. 477)
heat (p. 472)
inclination of the axis (p. 468)
infrared (p. 474)
isotherm (p. 478)
mesosphere (p. 467)
ozone (p. 463)
radiation (p. 473)
reflection (p. 475)
revolution (p. 468)
rotation (p. 467)
scattering (p. 475)

spring equinox (p. 469)
stratosphere (p. 466)
summer solstice (p. 469)
temperature (p. 472)
thermosphere (p. 467)
Tropic of Cancer (p. 469)
Tropic of Capricorn (p. 469)
troposphere (p. 466)
ultraviolet (p. 474)
visible light (p. 474)
weather (p. 460)
winter solstice (p. 469)

Examining the Earth System

1. Give an example or two of how the Earth system might be affected if Earth's axis were perpendicular to the plane of its orbit instead of being inclined $23\frac{1}{2}°$.

2. Speculate on the changes in global temperatures that might occur if Earth had substantially more land area and less ocean area than at present. How might such changes influence the biosphere?

Mastering Geology

Looking for additional review and test prep materials? Visit the Self Study area in **www.masteringgeology.com** to find practice quizzes, study tools, and multimedia that will aid in your understanding of this chapter's content. In **MasteringGeology™** you will find:

- **GEODe: Earth Science: An interactive visual walkthrough of key concepts**

- **Geoscience Animation Library: More than 100 animations illuminating many difficult-to-understand Earth science concepts**
- **In The News RSS Feeds: Current Earth science events and news articles are pulled into the site with assessment**
- **Pearson eText**
- **Optional Self Study Quizzes**
- **Web Links**
- **Glossary**
- **Flashcards**

Moisture, Clouds, and Precipitation

Desert thunderstorm at sunset. Cumulonimbus clouds are associated with lightning and thunder. (Photo by Karl Kost/Alamy)

Water vapor is an odorless, colorless gas that mixes freely with the other gases of the atmosphere. Unlike oxygen and nitrogen—the two most abundant components of the atmosphere—water can change from one state of matter to another (solid, liquid, or gas) at the temperatures and pressures experienced on Earth. Because of this unique property, water freely leaves the oceans as a gas and returns again as a liquid.

As you observe day-to-day weather changes, you might ask: Why is it generally more humid in the summer than in the winter? Why do clouds form on some occasions but not on others? Why do some clouds look thin and harmless whereas others form gray and ominous towers? Answers to these questions involve the role of water vapor in the atmosphere, the central theme of this chapter.

FOCUS ON CONCEPTS

To assist you in learning the important concepts in this chapter, focus on the following questions:

- What are the processes that cause water to change from one state of matter to another?
- What is latent heat?
- What is humidity? What is the most common method used to express humidity?
- What processes cause relative humidity to change?
- What four mechanisms cause air to rise?
- How is atmospheric stability determined?
- What are the necessary conditions for condensation?
- Which two criteria are used for cloud classification?
- What is fog? How do the various types of fog form?
- What two processes produce precipitation in a cloud?
- What are the different types of precipitation and how does each form?
- How is precipitation measured?

Water's Changes of State

Earth's Dynamic Atmosphere
▸ Moisture and Cloud Formation

Water is the only substance that exists in the atmosphere as a solid, liquid, and gas (**Figure 17.1**). It is made of hydrogen and oxygen atoms that are bonded together to form water molecules (H_2O). In all three states of matter (even ice) these molecules are in constant motion—the higher the temperature, the more vigorous the movement. The chief difference among liquid water, ice, and water vapor is the arrangement of the water molecules.

Ice, Liquid Water, and Water Vapor

Ice is composed of water molecules that are held together by mutual molecular attractions. Here the molecules form a tight, orderly network, as shown in **Figure 17.2**. As a consequence, the water molecules in ice are not free to move relative to each other but rather vibrate about fixed sites. When ice is heated, the molecules oscillate more rapidly. When the rate of molecular movement increases sufficiently, the bonds between some of the water molecules are broken, resulting in melting.

In the liquid state, water molecules are still tightly packed but are moving fast enough that they are able to slide past one another. As a result, liquid water is fluid and will take the shape of its container.

As liquid water gains heat from its environment, some of the molecules will acquire enough energy to break the remaining molecular attractions and escape from the surface, becoming water vapor. Water-vapor molecules are widely spaced compared to liquid water and exhibit very energetic random motion. What distinguishes a gas from a liquid is its compressibility (and expandability). For example, you can easily put more and more air into a tire and increase its volume only slightly. However, don't try to put 10 gallons of gasoline into a five-gallon can.

To summarize, when water changes state, it does not turn into a different substance; only the distances and interactions among the water molecules change.

Latent Heat

Whenever water changes state, heat is exchanged between water and its surroundings. When water evaporates, heat is absorbed (Figure 17.2). Meteorologists often measure heat energy in calories. One **calorie** is the amount of heat required to raise the temperature of 1 gram of water 1° C (1.8° F). Thus, when 10 calories

FIGURE 17.1 Caught in a downpour. (Photo by ImageState/Alamy)

energy go? In this case, the added energy went to break the molecular attractions between the water molecules in the ice cubes.

Because the heat used to melt ice does not produce a temperature change, it is referred to as **latent heat**. (Latent means "hidden," like the latent fingerprints hidden at a crime scene.) This energy can be thought of as being stored in liquid water, and it is not released to its surroundings as heat until the liquid returns to the solid state.

It requires 80 calories to melt one gram of ice, an amount referred to as *latent heat of melting. Freezing*, the reverse process, releases these 80 calories per gram to the environment as *latent heat of fusion*.

Evaporation and Condensation   We saw that heat is absorbed when ice is converted to liquid water. Heat is also absorbed during **evaporation**, the process of converting a liquid to a gas (vapor). The energy absorbed by water molecules during evaporation is used to give them the motion needed to escape the surface of the liquid and become a gas. This energy is referred to as the *latent heat of vaporization*. During the process of evaporation, it is the higher-temperature (faster-moving) molecules that escape the surface. As a result, the average molecular motion (temperature) of the remaining water is reduced—hence the common expression, "evaporation is a cooling process." You have undoubtedly experienced this cooling effect on stepping dripping wet from a swimming pool or bathtub. In this situation the energy used to evaporate water comes from your skin—hence, you feel cool.

of heat are absorbed by 1 gram of water, the molecules vibrate faster and a 10° C (18° F) temperature rise occurs.

Under certain conditions, heat may be added to a substance without an accompanying rise in temperature. For example, when a glass of ice water is warmed, the temperature of the ice-water mixture remains a constant 0° C (32° F) until all the ice has melted. If adding heat does not raise the temperature, where does this

FIGURE 17.2 Changes of state always involve an exchange of heat. The numbers shown here are the approximate number of calories either absorbed or released when 1 gram of water changes from one state of matter to another.

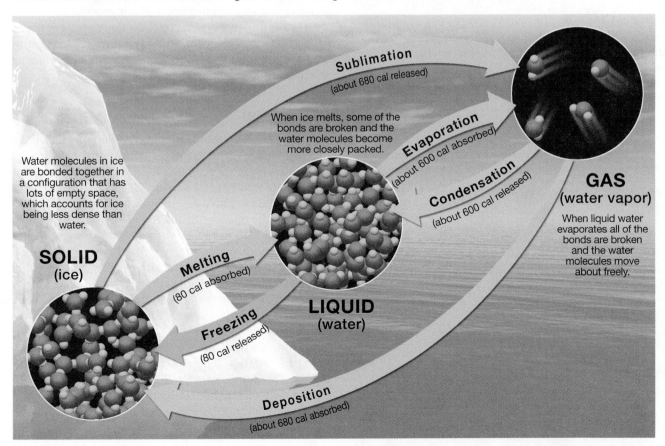

Sublimation (about 680 cal released)

When ice melts, some of the bonds are broken and the water molecules become more closely packed.

Evaporation (about 600 cal absorbed)

Condensation (about 600 cal released)

Water molecules in ice are bonded together in a configuration that has lots of empty space, which accounts for ice being less dense than water.

GAS (water vapor)

When liquid water evaporates all of the bonds are broken and the water molecules move about freely.

SOLID (ice)

Melting (80 cal absorbed)

Freezing (80 cal released)

LIQUID (water)

Deposition (about 680 cal absorbed)

Condensation, the reverse process, occurs when water vapor changes to the liquid state. During condensation, water-vapor molecules release energy (*latent heat of condensation*) in an amount equivalent to what was absorbed during evaporation. When condensation occurs in the atmosphere, it results in the formation of such phenomena as fog and clouds.

As you will see, latent heat plays an important role in many atmospheric processes. In particular, when water vapor condenses to form cloud droplets, latent heat of condensation is released, warming the surrounding air and giving it buoyancy. When the moisture content of air is high, this process can spur the growth of towering storm clouds.

Sublimation and Deposition   You are probably least familiar with the last two processes illustrated in Figure 17.2—sublimation and deposition. **Sublimation** is the conversion of a solid directly to a gas without passing through the liquid state. Examples you may have observed include the gradual shrinking of unused ice cubes in the freezer and the rapid conversion of dry ice (frozen carbon dioxide) to wispy clouds that quickly disappear.

Deposition refers to the reverse process, the conversion of a vapor directly to a solid. This change occurs, for example, when water vapor is deposited as ice on solid objects such as grass or windows (**Figure 17.3**). These deposits are called *white frost* or *hoar frost* and are frequently referred to simply as *frost*. A household example of the process of deposition is the "frost" that accumulates in a freezer. As shown in Figure 17.2, deposition releases an amount of energy equal to the total amount released by condensation and freezing.

CONCEPT CHECK 17.1

1. Summarize the processes by which water changes from one state of matter to another. Indicate whether energy is absorbed or released.
2. What is *latent heat*?
3. What is a common example of sublimation?
4. How does frost form?

FIGURE 17.3 Frost on a window pane is an example of deposition. (Photo by Stockxpert/Jupiter Images Unlimited)

Humidity: Water Vapor in the Air

 Earth's Dynamic Atmosphere
▶ **Moisture and Cloud Formation**

Water vapor constitutes only a small fraction of the atmosphere, varying from as little as one-tenth of 1 percent up to about 4 percent by volume. But the importance of water in the air is far greater than these small percentages would indicate. Indeed, scientists agree that *water vapor* is the most important gas in the atmosphere when it comes to understanding atmospheric processes.

Humidity is the general term for the amount of water vapor in air. Meteorologists employ several methods to express the water-vapor content of the air; we examine three: mixing ratio, relative humidity, and dew-point temperature.

Saturation

Before we consider these humidity measures further, it is important to understand the concept of **saturation**. Imagine a closed jar containing water overlain by dry air, both at the same temperature. As the water begins to evaporate from the water surface, a small increase in pressure can be detected in the air above. This increase is the result of the motion of the water-vapor molecules that were added to the air through evaporation. In the open atmosphere, this pressure is termed **vapor pressure** and is defined as that part of the total atmospheric pressure that can be attributed to the water-vapor content.

In the closed container, as more and more molecules escape from the water surface, the steadily increasing vapor pressure in the air above forces more and more of these molecules to return to the liquid. Eventually the number of vapor molecules returning to the surface will balance the number leaving. At that point, the air is said to be *saturated*. If we add heat to the container, thereby increasing the temperature of the water and air, more water will evaporate before a balance is reached. Consequently, at higher temperatures, more moisture is required for saturation. The amount of water vapor required for saturation at various temperatures is shown in **Table 17.1**.

Mixing Ratio

Not all air is saturated, of course. Thus, we need ways to express how humid a parcel of air is. One method specifies the amount of water vapor contained in a unit of air. The **mixing ratio** is the mass of water vapor in a unit of air compared to the remaining mass of dry air.

$$\text{mixing ratio} = \frac{\text{mass of water vapor (grams)}}{\text{mass of dry air (kilograms)}}$$

Table 17.1 shows the mixing ratios of saturated air at various temperatures. For example, at 25° C (77° F) a saturated parcel of air (one kilogram) would contain 20 grams of water vapor.

Because the mixing ratio is expressed in units of mass (usually in grams per kilogram), it is not affected by changes in pressure or temperature. However, the mixing ratio is time-consuming to measure by direct sampling. Thus, other methods are employed to express the moisture content of the air. These include relative humidity and dew-point temperature.

Relative Humidity

The most familiar and, unfortunately, the most misunderstood term used to describe the moisture content of air is relative humidity. **Relative humidity** *is a ratio of the air's actual water-vapor content compared with the amount of water vapor required for saturation at that temperature (and pressure).* Thus, relative humidity indicates how near the air is to saturation, rather than the actual quantity of water vapor in the air (see Box 17.1).

To illustrate, we see from Table 17.1 that at 25° C (77° F), air is saturated when it contains 20 grams of water vapor per kilogram

TABLE 17.1 Amount of Water Vapor Needed to Saturate a Kilogram of Air at Various Temperatures

Temperature °C (°F)	Water-Vapor Content at Saturation (grams)
−40 (−40)	0.1
−30 (−22)	0.3
−20 (−4)	0.75
−10 (14)	2
0 (32)	3.5
5 (41)	5
10 (50)	7
15 (59)	10
20 (68)	14
25 (77)	20
30 (86)	26.5
35 (95)	35
40 (104)	47

of air. Thus, if the air contains 10 grams per kilogram on a 25° C day, the relative humidity is expressed as 10/20, or 50 percent. If air with a temperature of 25° C had a water-vapor content of 20 grams per kilogram, the relative humidity would be expressed as 20/20, or 100 percent. When the relative humidity reaches 100 percent, the air is said to be saturated.

Because relative humidity is based on the air's water-vapor content, as well as the amount of moisture required for saturation, it can be changed in either of two ways. First, relative humidity can be changed by the addition or removal of water vapor. Second, because the amount of moisture required for saturation is a function of air temperature, relative humidity varies with temperature. (Recall that the amount of water vapor required for saturation is temperature dependent, such that at higher temperatures it takes more water vapor to saturate air than at lower temperatures.)

Adding or Subtracting Moisture Notice in **Figure 17.4** that when water vapor is added to a parcel of air, its relative humidity increases until saturation occurs (100 percent relative humidity). What if even more moisture is added to this parcel of saturated air? Does the relative humidity exceed 100 percent? Normally, this situation does not occur. Instead, the excess water vapor condenses to form liquid water.

In nature, moisture is added to the air mainly via evaporation from the oceans. However, plants, soil, and smaller bodies of water do make substantial contributions.

Changes with Temperature The second condition that affects relative humidity is air temperature. Examine **Figure 17.5** carefully. Note in Part A that when air at 20° C contains 7 grams of water vapor per kilogram, it has a relative humidity of 50 percent. This can be verified by referring to Table 17.1. Here we can see that at 20° C, air is saturated when it contains 14 grams of water vapor per kilogram of air. Because the air in Figure 17.5A contains 7 grams of water vapor, its relative humidity is 7/14, or 50 percent.

When the flask is cooled from 20° to 10° C, as shown in Figure 17.5B, the relative humidity increases from 50 to 100 percent. We can conclude from this that when the water-vapor content remains constant, *a decrease in temperature results in an increase in relative humidity.*

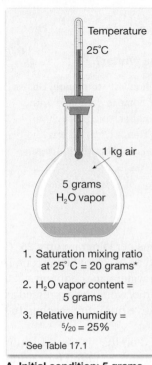

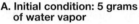

1. Saturation mixing ratio at 25° C = 20 grams*

2. H₂O vapor content = 5 grams

3. Relative humidity = ⁵/₂₀ = 25%

*See Table 17.1

A. Initial condition: 5 grams of water vapor

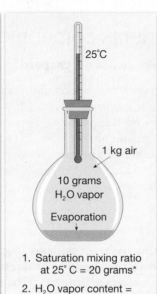

1. Saturation mixing ratio at 25° C = 20 grams*

2. H₂O vapor content = 10 grams

3. Relative humidity = ¹⁰/₂₀ = 50%

B. Addition of 5 grams of water vapor = 10 grams

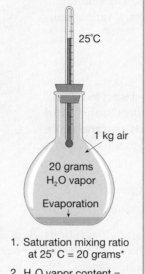

1. Saturation mixing ratio at 25° C = 20 grams*

2. H₂O vapor content = 20 grams

3. Relative humidity = ²⁰/₂₀ = 100%

C. Addition of 10 grams of water vapor = 20 grams

FIGURE 17.4 At a constant temperature, the relative humidity will increase as water vapor is added to the air. In this example, the capacity remains constant at 20 grams per kilogram and the relative humidity rises from 25 to 100 percent as the water-vapor content increases.

What happens when the air is cooled below the temperature at which saturation occurs? Part C in Figure 17.5 illustrates this situation. Notice from Table 17.1 that when the flask is cooled to 0° C, the air is saturated at 3.5 grams of water vapor per kilogram of air. Because this flask originally contained 7 grams of water vapor, 3.5 grams of water vapor will condense to form liquid droplets that collect on the walls of the container. The relative humidity of the air inside remains at 100 percent. This raises an important concept. When air aloft is cooled below its saturation level, some of the water vapor condenses to form clouds. As

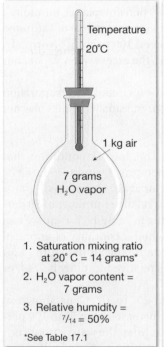

1. Saturation mixing ratio at 20° C = 14 grams*

2. H₂O vapor content = 7 grams

3. Relative humidity = ⁷/₁₄ = 50%

*See Table 17.1

A. Initial condition: 20°C

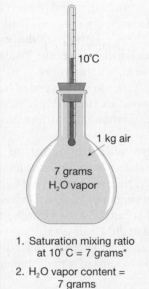

1. Saturation mixing ratio at 10° C = 7 grams*

2. H₂O vapor content = 7 grams

3. Relative humidity = ⁷/₇ = 100%

B. Cooled to 10°C

1. Saturation mixing ratio at 0° C = 3.5 grams*

2. H₂O vapor content = 3.5 grams

3. Relative humidity = ³·⁵/₃.₅ = 100%

C. Cooled to 0°C

FIGURE 17.5 Relative humidity varies when temperatures change. When the water-vapor content (mixing ratio) remains constant, the relative humidity can be changed by increasing or decreasing the air temperature. In this example, when the temperature of the air in the flask was lowered from 20° to 10° C, the relative humidity increased from 50 to 100 percent. Further cooling (from 10° to 0° C) causes one-half of the water vapor to condense. In nature, cooling of air below its saturated mixing ratio generally causes condensation in the form of clouds, dew, or fog.

clouds are made of liquid droplets (or ice crystals), they are no longer part of the *water-vapor* content of the air. (Clouds are not water vapor; they are composed of liquid water droplets or ice crystals too tiny to fall to Earth.)

We can summarize the effects of temperature on relative humidity as follows. When the water-vapor content of air remains at a constant level, a decrease in air temperature results in an increase in relative humidity and an increase in temperature causes a decrease in relative humidity. In **Figure 17.6** the variations in temperature and relative humidity during a typical day demonstrate the relationship just described.

Dew-Point Temperature

Another important measure of humidity is the dew-point temperature. The **dew-point temperature** or simply the **dew point** is the temperature to which a parcel of air would need to be cooled to reach saturation. For example, in Figure 17.5, the unsaturated air in the flask had to be cooled to 10° C before saturation occurred. Therefore, 10° C is the dew-point temperature for this air. In nature, cooling below the dew point causes water vapor to condense, typically as dew, fog, or clouds. The term *dew point* stems from the fact that during nighttime hours, objects near the

FIGURE 17.7 Condensation, or "dew," occurs when a cold drinking glass chills the surrounding layer of air below the dew-point temperature. (Photo by Elena Elisseeva/Shutterstock)

FIGURE 17.6 Typical daily variations in temperature and relative humidity during a spring day at Washington, D.C. When temperature increases, relative humidity drops (see mid-afternoon) and vice versa.

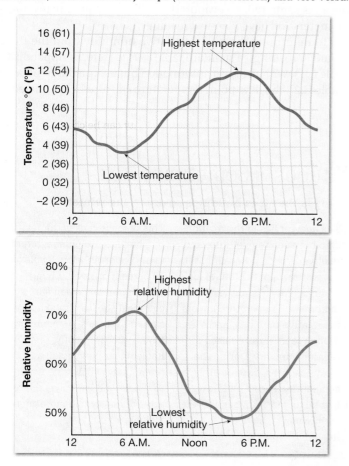

ground often cool below the dew-point temperature and become coated with dew (**Figure 17.7**).

Unlike relative humidity, which is a measure of how near the air is to being saturated, dew-point temperature is a measure of its *actual moisture* content. Because the dew-point temperature is directly related to the amount of water vapor in the air, and because it is easy to determine, it is one of the most useful measures of humidity.

The amount of water vapor needed for saturation is temperature dependent and that for every 10° C (18° F) increase in temperature, the amount of water vapor needed for saturation doubles (see Table 17.1). Therefore, relatively cold *saturated air* (0° C or 32° F) contains about half the water vapor of *saturated air* having a temperature of 10° C (50° F) and roughly one-fourth that of hot *saturated air* with a temperature of 20° C (68° F). Because the dew point is the temperature at which saturation occurs, we can conclude that high dew-point temperatures equate to moist air, and low dew-point temperatures indicate dry air. More precisely, based on what we have learned about vapor pressure and saturation, we can state that for every 10° C (18° F) increase in the dew-point temperature, the air contains about twice as much water vapor. Therefore, we know that air over Fort Myers, Florida, with a dew point of 25° C (77° F), contains about twice the water vapor of air situated over St. Louis, Missouri, with a dew point of

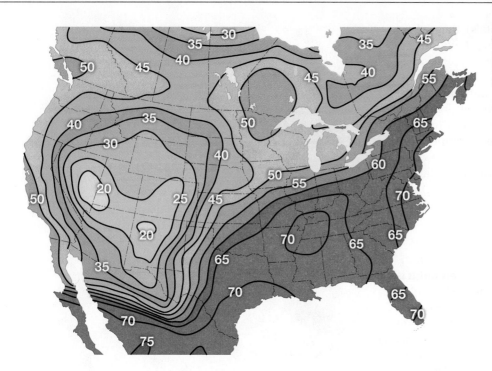

FIGURE 17.8 Surface map showing dew-point temperatures for September 15, 2005. Dew-point temperatures above 60° F dominate the southeastern United States, indicating that this region is blanketed with humid air.

15° C (59° F), and four times that of Tucson, Arizona, with a dew point of 5° C (41° F).

Because the dew-point temperature is a good measure of the amount of water vapor in the air, it is the measure of atmospheric moisture that appears on a variety of weather maps. Notice on the map in **Figure 17.8** that most of the places located near the warm Gulf of Mexico have dew-point temperatures that exceed 70° F (21° C). When the dew point exceeds 65° F (18° C), it is considered humid by most people, and air with a dew point of 75° F (24° C) or higher is oppressive. Also notice in Figure 17.8 that although the Southeast is dominated by humid conditions (dew points above 65° F), most of the remainder of the country is experiencing drier air.

Measuring Humidity

Relative humidity is commonly measured using a **hygrometer** (*hygro* = moisture, *metron* = measuring instrument). One type of hygrometer, called a **psychrometer**, consists of two identical thermometers mounted side by side (**Figure 17.9**). One thermometer, the *dry bulb*, gives the current air temperature. The other, called the *wet-bulb* thermometer, has a thin muslin wick tied around the end (see end of thermometers in the photo).

To use the psychrometer, the cloth sleeve is saturated with water and a continuous current of air is passed over the wick (Figure 17.9). This is done either by swinging the instrument freely in the air or by fanning air past it. As a consequence, water evaporates from the wick, and the heat absorbed by the evaporating water makes the temperature of the wet bulb drop. The loss of heat that was required to evaporate water from the wet bulb lowers the thermometer reading.

The amount of cooling that takes place is directly proportional to the dryness of the air. The drier the air, the more moisture evap-

orates. The more heat the evaporating water absorbs, the greater the cooling. Therefore, the larger the difference that is observed between the thermometer readings, the lower the relative humidity; the smaller the difference, the higher the relative humidity. If the air is saturated, no evaporation will occur, and the two thermometers will have identical readings.

To determine the precise relative humidity from the thermometer readings, a standard table is used (refer to Appendix C, Table C.1). With the same information but using a different table (Table C.2), the dew-point temperature can also be calculated.

Students Sometimes Ask...

Why is the air in buildings so dry in the winter?

The answer lies in the relationship between temperature and relative humidity. Recall that if the water-vapor content of air remains at a constant level, an increase in temperature lowers the relative humidity, and a drop in temperature raises the relative humidity. During the winter months, outside air is comparatively cool and has a low mixing ratio. When this air is drawn into the home, it is heated to room temperature. This process causes the relative humidity to plunge, often to uncomfortably low levels of 10 percent or lower. Living with dry air can mean static electrical shocks, dry skin, sinus headaches, or even nosebleeds. Consequently, the homeowner may install a humidifier, which adds water to the air and increases the relative humidity to a more comfortable level.

Box 17.1

Dry Air at 100 Percent Relative Humidity?

A common misconception relating to meteorology is the notion that air with a high relative humidity must have a greater water-vapor content than air with a lower relative humidity. Frequently, this is not the case (Figure 17.A). To illustrate, consider a typical January day at two different locations—International Falls, Minnesota, and Phoenix, Arizona. On this hypothetical day the temperature in International Falls is a cold −10° C (14° F) and the relative humidity is 100 percent. By referring to Table 17.1, we can see that saturated −10° C (14° F) air has a water-vapor content (mixing ratio) of 2 grams per kilogram (g/kg). By contrast, the desert air at Phoenix is a warm 25° C (77° F), and the relative humidity is 20 percent. A look at Table 17.1 reveals that the 25° C (77° F) air has a saturation mixing ratio of 20 g/kg. Therefore, with a relative humidity of 20 percent, the air at Phoenix has a water-vapor content of 4 g/kg (20 grams × 20 percent). Consequently, the "dry" air at Phoenix actually contains twice the water vapor as the "moist" air at International Falls.

This comparison illustrates why places that are very cold are also very dry. The low water-vapor content of frigid air (even when saturated) helps explain why many Arctic areas receive only meager amounts of precipitation and are referred to as "polar deserts." Furthermore, the dry skin and chapped lips people frequently experience during the winter months in the United States can be attributed to dry air. The water vapor content of the cold air is low, even when compared to some hot, arid regions.

FIGURE 17.A Moisture content of hot air versus frigid air. Hot desert air with a low relative humidity generally has a higher water-vapor content than frigid air with a high relative humidity. (Top photo by Matt Duvall, bottom photo by E. J. Tarbuck)

Another instrument used for measuring relative humidity, the *hair hygrometer*, can be read directly without the use of tables. The hair hygrometer operates on the principle that hair or certain synthetic fibers change their length in proportion to changes in the relative humidity, lengthening as relative humidity increases and shrinking as the relative humidity drops. The tension of a bundle of hairs is linked mechanically to an indicator that is calibrated between 0 and 100 percent. Thus, we need only glance at the dial to determine the relative humidity. Unfortunately, the hair hygrometer is less accurate than the psychrometer. Furthermore, it requires frequent calibration and is slow in responding to changes in humidity, especially at low temperatures.

A different type of hygrometer is used in remote-sensing instrument packages such as radiosondes that transmit upper-air observations back to ground stations. The *electric hygrometer* contains an electrical conductor coated with a moisture-absorbing chemical. It works on the principle that the passage of current varies as the relative humidity varies.

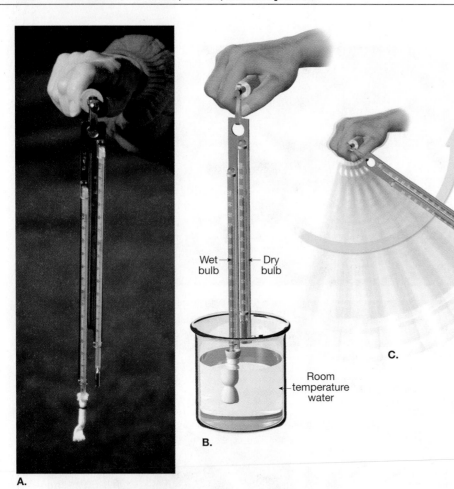

FIGURE 17.9 Sling psychrometer. **A.** Sling psychrometers are used to determine both relative humidity and dew point. **B.** The dry-bulb thermometer gives the current air temperature. The web-bulb thermometer is covered with a cloth wick that is dipped in water. **C.** The thermometers are spun until the temperature of the wet-bulb thermometer stops declining. Then the thermometers are read and the data used in conjunction with the tables in Appendix C. (Photo by E. J. Tarbuck)

Wet bulb — Dry bulb

Room temperature water

B.

C.

A.

CONCEPT CHECK 17.2

1. Describe how air temperature and saturation are related.
2. List three measures that are used to express humidity.
3. How do relative humidity and mixing ratio differ?
4. If the temperature remains unchanged and the amount of water vapor in the air decreases, how will relative humidity change?
5. Explain the principle of the sling psychrometer and the hair hygrometer.
6. On a warm summer day when the relative humidity is high, it may seem even warmer than the thermometer indicates. Why do we feel so uncomfortable on these muggy days?

The Basis of Cloud Formation: Adiabatic Cooling

Earth's Dynamic Atmosphere
▶ Moisture and Cloud Formation

We have considered basic properties of water vapor and how its variability is measured. This section examines some of the important roles that water vapor plays in weather, especially the formation of clouds.

Fog and Dew versus Cloud Formation

Recall that condensation occurs when water vapor changes to a liquid. Condensation may form dew, fog, or clouds. Although these three forms are different, all require that air reach saturation. As indicated earlier, saturation occurs either when sufficient water vapor is added to the air or, more commonly, when the air is cooled to its dew point.

Near Earth's surface, heat is readily exchanged between the ground and the air above. During evening hours, the surface radiates heat away, causing the surface and adjacent air to cool rapidly. This radiation cooling accounts for the formation of dew and some types of fog. Thus, surface cooling that occurs after sunset accounts for some condensation. However, cloud formation often takes place during the warmest part of the day. Clearly some other mechanism must operate aloft that cools air sufficiently to generate clouds.

Adiabatic Temperature Changes

The process that is responsible for most cloud formation is easily demonstrated if you have ever pumped up a bicycle tire and noticed that the pump barrel became quite warm. The heat you felt was the consequence of the work you did on the air to compress it. When energy is used to compress air, the motion of the gas molecules increases and therefore the temperature of the air rises. Conversely, air that is allowed to escape from a bicycle tire *expands and cools*. This results because the expanding air pushes (does work on) the surrounding air and must cool by an amount equivalent to the energy expended.

You have probably felt the cooling effect of the propellant gas expanding as you applied hairspray or spray deodorant. As the compressed gas in the aerosol can is released, it quickly expands and cools. This drop in temperature occurs *even though heat is neither added nor subtracted*. Such variations are known as **adiabatic temperature changes** and result when air is compressed or allowed to expand. In summary, *when air is allowed to expand, it cools, and when it is compressed, it warms*.

Adiabatic Cooling and Condensation

To simplify the following discussion, it helps to imagine a volume of air enclosed in a thin elastic cover. Meteorologists call this imaginary volume of air a **parcel**. Typically, we consider a parcel to be a few hundred cubic meters in volume, and we assume that it acts independently of the surrounding air. It is also assumed that no heat is transferred into, or out of, the parcel. Although highly idealized, over short time spans a parcel of air behaves in a manner much like an actual volume of air moving vertically in the atmosphere.

Dry Adiabatic Rate   As you travel from Earth's surface upward through the atmosphere, the atmospheric pressure rapidly diminishes, because there are fewer and fewer gas molecules. Thus, any time a parcel of air moves upward, it passes through regions of successively lower pressure, and the ascending air expands. As it expands, it cools adiabatically. Unsaturated air cools at the constant rate of 10° C for every 1,000 meters of ascent (5.5° F per 1,000 feet).

Conversely, descending air comes under increasingly higher pressures, compresses, and is heated 10° C for every 1,000 meters of descent. This rate of cooling or heating applies only to *unsaturated air* and is known as the **dry adiabatic rate**.

Wet Adiabatic Rate   If a parcel of air rises high enough, it will eventually cool to its dew point, where the process of condensation begins. From this point on along its ascent, *latent heat of condensation* stored in the water vapor will be liberated. Although the air will continue to cool after condensation begins, the released latent heat works against the adiabatic process, thereby reducing the rate at which the air cools. This slower rate of cooling caused by the addition of latent heat is called the **wet adiabatic rate** of cooling. Because the amount of latent heat released depends on the quantity of moisture present in the air, the wet adiabatic rate varies from 5° C per 1,000 meters for air with a high moisture content to 9° C per 1,000 meters for dry air.

Figure 17.10 illustrates the role of adiabatic cooling in the formation of clouds. Note that from the surface up to the condensation level the air cools at the dry adiabatic rate. The wet adiabatic rate begins at the condensation level.

CONCEPT CHECK 17.3

1. How is the formation of dew different than the formation of a cloud? How are they similar?
2. Why does air cool when it rises through the atmosphere?
3. What do meteorologists mean by the word *parcel*?
4. Why does the adiabatic rate of cooling change when condensation begins? Why is the wet adiabatic rate not a constant figure?
5. The contents of an aerosol can are under very high pressure. When you push the nozzle on such a can, the spray feels cold. Explain.

FIGURE 17.10 Rising air cools at the dry adiabatic rate of 10° per 1,000 meters, until the air reaches the dew point and condensation (cloud formation) begins. As air continues to rise, the latent heat released by condensation reduces the rate of cooling. The wet adiabatic rate is therefore always less than the dry adiabatic rate.

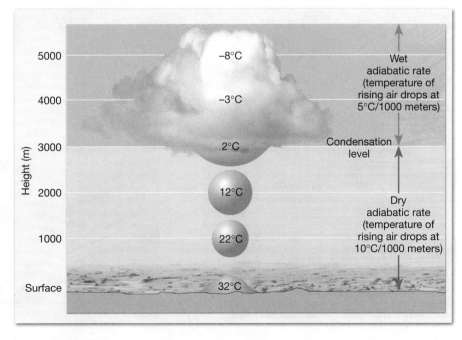

Processes that Lift Air

Earth's Dynamic Atmosphere
▶ Moisture and Cloud Formation

To review, when air rises, it expands and cools adiabatically. If air is lifted sufficiently, it will eventually cool to its dew-point temperature, saturation will occur, and clouds will develop. But why does air rise on some occasions and not on others?

It turns out that, in general, air tends to resist vertical movement. Therefore, air located near the surface tends to stay near the surface, and air aloft tends to remain aloft. Exceptions to this rule, as we shall see, include conditions in the atmosphere that give air sufficient buoyancy to rise without the aid of outside forces. In many situations, however, when you see clouds forming there is some mechanical phenomenon at work that forces the air to rise (at least initially).

There are four mechanisms that force air to rise:

1. *Orographic lifting*—air is forced to rise over a mountainous barrier.
2. *Frontal wedging*—warmer, less dense air is forced over cooler, denser air.
3. *Convergence*—a pileup of horizontal air flow results in upward movement.
4. *Localized convective lifting*—unequal surface heating causes localized pockets of air to rise because of their buoyancy.

In later chapters we consider other mechanisms that cause air to rise. In particular, horizontal airflow at upper levels significantly contributes to vertical lifting across the middle latitudes.

Orographic Lifting

Orographic lifting occurs when elevated terrains, such as mountains, act as barriers to the flow of air (Figure 17.11). As air ascends a mountain slope, adiabatic cooling often generates clouds and copious precipitation. In fact, many of the rainiest places in the world are located on windward mountain slopes.

By the time air reaches the leeward side of a mountain, much of its moisture has been lost. If the air descends, it warms adiabatically, making condensation and precipitation even less likely. As shown in Figure 17.11, the result can be a **rain shadow desert**. The Great Basin Desert of the western United States lies only a few hundred kilometers from the Pacific Ocean, but it is effectively cut off from the ocean's moisture by the imposing Sierra Nevada (Figure 17.11). The Gobi Desert of Mongolia, the Takla Makan of China, and the Patagonia Desert of Argentina are other examples of deserts that exist because they are on the leeward sides of mountains (for a map showing deserts, see Figure 20.8, p. 589).

Frontal Wedging

If orographic lifting were the only mechanism that forced air aloft, the relatively flat central portion of North America would be an expansive desert instead of the nation's breadbasket. Fortunately, this is not the case.

In central North America, masses of warm and cold air collide, producing a **front**. Here the cooler, denser air acts as a barrier over which the warmer, less dense air rises. This process, called **frontal wedging**, is illustrated in Figure 17.12.

It should be noted that weather-producing fronts are associated with storm systems called *middle-latitude cyclones*. Because these storms are responsible for producing a high percentage of the precipitation in the middle latitudes, we examine them closely in Chapter 19.

Convergence

We saw that the collision of contrasting air masses forces air to rise. In a more general sense, whenever air in the lower atmosphere flows together, lifting results. This phenomenon is called **convergence**. When air flows in from more than one direction, it must go somewhere. Because it cannot go down, it goes up (Figure 17.13). This, of course, leads to adiabatic cooling and possibly cloud formation.

Convergence can also occur whenever an obstacle slows or restricts horizontal air flow (wind).

FIGURE 17.11 Orographic lifting leads to precipitation on windward slopes. By the time air reaches the leeward side of the mountains, much of the moisture has been lost. The Great Basin desert is a rain shadow desert that covers nearly all of Nevada and portions of adjacent states. (Photo on left by Dean Pennala/Shutterstock, photo on right by Dennis Tasa)

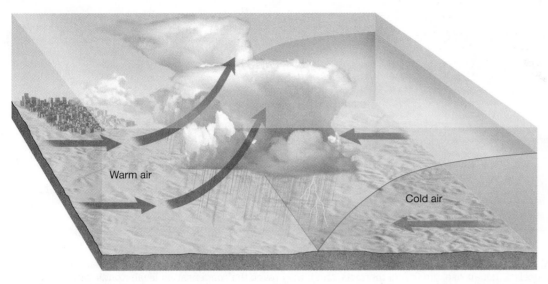

FIGURE 17.12 Frontal wedging. Colder, denser air acts as a barrier over which warmer, less dense air rises.

When air moves from a relatively smooth surface, such as the ocean, onto an irregular landscape, its speed is reduced. The result is a pileup of air (convergence). This is similar to what happens when people leave a well-attended sporting event and a pileup results at the exits. When air converges, the air molecules do not simply squeeze closer together (like people); rather, there is a net upward flow.

The Florida peninsula provides an excellent example of the role that convergence can play in initiating cloud development and precipitation. On warm days, the airflow is from the ocean to the land along both coasts of Florida. This leads to a pileup of air along the coasts and general convergence over the peninsula. This pattern of air movement and the uplift that results is aided by intense solar heating of the land. This is why the peninsula of Florida experiences the greatest frequency of mid-afternoon thunderstorms in the United States.

More importantly, convergence as a mechanism of forceful lifting is a major contributor to the weather associated with middle-latitude cyclones and hurricanes. The low-level horizontal air flow associated with these systems is inward and upward around their centers. These important weather producers are covered in more detail later, but for now remember that convergence near the surface results in a general upward flow.

FIGURE 17.13 Convergence. When surface air converges, the column of air increases in height to allow for the decreased area it occupies. This photo shows southern Florida viewed from the space shuttle. On warm days, airflow from the Atlantic Ocean and Gulf of Mexico onto the Florida peninsula generates many mid-afternoon thunderstorms. (Photo by NASA/Media Services)

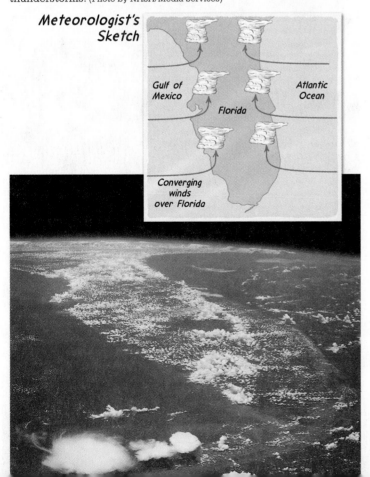

Localized Convective Lifting

On warm summer days, unequal heating of Earth's surface may cause pockets of air to be warmed more than the surrounding air. For instance, air above a paved parking lot will be warmed more than the air above an adjacent wooded park. Consequently, the parcel of air above the parking lot, which is warmer (less dense) than the surrounding air, will be buoyed upward (Figure 17.14). These rising parcels of warmer air are called *thermals*. Birds such as hawks and eagles use these thermals to carry them to great heights where they can gaze down on unsuspecting prey. People have learned to employ these rising parcels using hang gliders as a way to "fly."

The phenomenon that produces rising thermals is called **localized convective lifting**. When these warm parcels of air rise above the lifting condensation level, clouds form, which can produce mid-afternoon rain showers. The accompanying rains, although occasionally heavy, are of short duration and widely scattered.

Although localized convective lifting by itself is not a major producer of precipitation, the added buoyancy that results from surface heating contributes significantly to the lifting initiated by the other mechanisms. It should also be remembered that even though the other mechanisms force air to rise, convective lifting occurs because the air is warmer (less dense) than the surrounding air and rises for the same reasons as does a hot-air balloon.

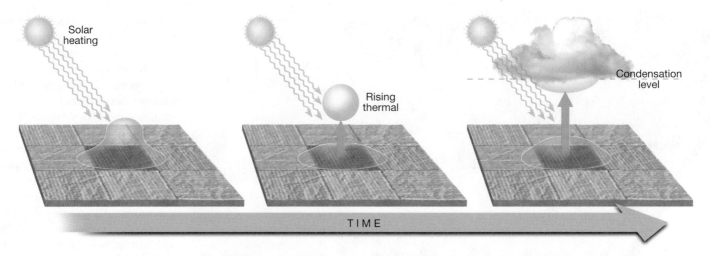

FIGURE 17.14 Localized convective lifting. Unequal heating of Earth's surface causes pockets of air to be warmed more than the surrounding air. These buoyant parcels of hot air rise, producing thermals, and if they reach the condensation level, clouds form.

CONCEPT CHECK 17.4

1. List the four mechanisms that cause air to rise.
2. How do orographic lifting and frontal wedging force air to rise?
3. Explain why the Great Basin area of the western United States is so dry. What term is applied to this situation?
4. What causes the Florida peninsula to experience the greatest frequency of mid-afternoon thunderstorms in the United States?

The Weathermaker: Atmospheric Stability

 Earth's Dynamic Atmosphere
▸ **Moisture and Cloud Formation**

When air rises, it cools and eventually produces clouds. Why do clouds vary so much in size, and why does the resulting precipitation vary so much? The answers are closely related to the *stability* of the air.

Recall that a parcel of air can be thought of as having a thin, flexible cover that allows it to expand but prevents it from mixing with the surrounding air (picture a hot-air balloon). If this parcel were forced to rise, *its temperature will decrease because of expansion*. By comparing the parcel's temperature to that of the surrounding air, we can determine the stability of the bubble. If the parcel's temperature is lower than that of the surrounding environment, it will be denser; and if allowed to move freely, it will sink to its original position. Air of this type, called **stable air**, resists vertical movement.

If, however, our imaginary rising parcel is *warmer* and hence less dense than the surrounding air, it will continue to rise until it reaches an altitude where its temperature equals that of its surroundings. This is exactly how a hot-air balloon works, rising as long as it is warmer and less dense than the surrounding air (**Figure 17.15**). This type of air is classified as **unstable air**. In summary, stability is a property of air that describes its tendency to remain in its original position (stable) or to rise (unstable).

Types of Stability

The stability of air is determined by measuring the temperature of the atmosphere at various heights. Recall from Chapter 16 that this measure is termed the *environmental lapse rate*. The environmental lapse rate is the temperature of the atmosphere as determined from observations made by radiosondes and aircraft.

FIGURE 17.15 As long as air is warmer and therefore less dense than its surroundings, it will rise. Hot-air balloons rise up through the atmosphere for this reason. (Photo by Steve Vidler/SuperStock)

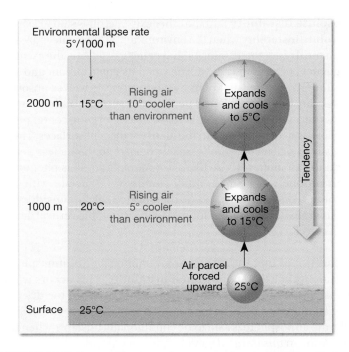

FIGURE 17.16 In a stable atmosphere, as an unsaturated parcel of air is lifted, it expands and cools at the dry adiabatic rate of 10° C per 1,000 meters. Because the temperature of the rising parcel of air is lower than the surrounding environment, it will be heavier and, if allowed to do so, will sink to its original position.

It is important not to confuse this with *adiabatic temperature changes*, which are changes in temperature caused by expansion or compression as a parcel of air rises or descends.

To illustrate, we examine a situation in which the environmental lapse rate is 5° C per 1,000 meters (**Figure 17.16**). Under this condition, when air at the surface has a temperature of 25° C, the air at 1,000 meters will be 5 degrees cooler, or 20° C, the air at 2,000 meters will have a temperature of 15° C, and so forth. At first glance it appears that the air at the surface is less dense than the air at 1,000 meters, because it is 5 degrees warmer. However, if the air near the surface is unsaturated and were to rise to 1,000 meters, it would expand and cool at the dry adiabatic rate of 10° C per 1,000 meters. Therefore, upon reaching 1,000 meters, its temperature would have dropped 10° C. Being 5 degrees cooler than its environment, it would be denser and tend to sink to its original position. Hence, we say that the air near the surface is potentially cooler than the air aloft and therefore will not rise on its own. The air just described is *stable* and resists vertical movement.

Absolute Stability   Stated quantitatively, **absolute stability** prevails when the environmental lapse rate is less than the wet adiabatic rate. **Figure 17.17** depicts this situation using an environmental lapse rate of 5° C per 1,000 meters and a wet adiabatic rate of 6° C per 1,000 meters. Note that at 1,000 meters the temperature of the surrounding air is 15° C, while the rising parcel of air has cooled to 10° C and is therefore the denser air. Even if this stable air were to be forced above the condensation level, it would remain cooler and denser than its environment, and thus it would tend to return to the surface.

FIGURE 17.17 *Absolute stability* prevails when the environmental lapse rate is less than the wet adiabatic rate. **A.** The rising parcel of air is always cooler and heavier than the surrounding air, producing stability. **B.** Graphic representation of the conditions shown in part (**A**).

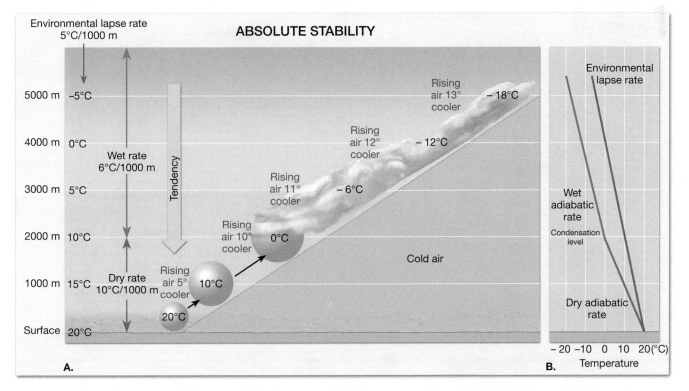

Students Sometimes Ask...

What are the wettest places on Earth?

Many of the rainiest places in the world are located on windward mountain slopes. A station at Mount Waialeale, Hawaii, records the highest average annual rainfall, some 1,234 centimeters (486 inches). The greatest recorded rainfall for a single 12-month period occurred at Cherrapunji, India, where an astounding 2,647 centimeters (1,042 inches, over 86 feet) fell. Much of this rainfall occurred in the month of July—a record 930 centimeters (366 inches, over 30 feet). This is 10 times more rain than Chicago receives in an average year.

The most stable conditions occur when the temperature in a layer of air actually increases with altitude. When such a reversal occurs, a *temperature inversion* is said to exist. Temperature inversions frequently occur on clear nights as a result of radiation cooling of Earth's surface. Under these conditions, an inversion is created because the ground and the air immediately above will cool more rapidly than the air aloft. When warm air overlies cooler air, it acts as a lid and prevents appreciable vertical mixing. Because of this, temperature inversions are responsible for trapping pollutants in a narrow zone near Earth's surface.

Absolute Instability At the other extreme, air is said to exhibit **absolute instability** when the environmental lapse rate is greater than the dry adiabatic rate. As shown in Figure 17.18, the ascending parcel of air is always warmer than its environment and will continue to rise because of its own buoyancy. However, absolute instability is generally found near Earth's surface. On hot, sunny days the air above some surfaces, such as shopping center parking lots, is heated more than the air over adjacent surfaces. These invisible pockets of more intensely heated air, being less dense than the air aloft, will rise like a hot-air balloon. This phenomenon produces the small, fluffy clouds we associate with fair weather. Occasionally, when the surface air is considerably warmer than the air aloft, clouds with considerable vertical development can form.

Conditional Instability A more common type of atmospheric instability is called **conditional instability**. This occurs when moist air has an environmental lapse rate between the dry and wet adiabatic rates (between 5 and 10° C per 1,000 meters). Simply, the atmosphere is said to be conditionally unstable when it is *stable* for an *unsaturated* parcel of air, but *unstable* for a *saturated* parcel of air. Notice in Figure 17.19 that the rising parcel of air is cooler than the surrounding air for nearly 3,000 meters. With the addition of latent heat above the lifting condensation level, the parcel becomes warmer than the surrounding air. From this point along its ascent the parcel will continue to rise because of its own buoyancy, without an outside force. Thus, conditional instability depends on whether or not the rising air is saturated. The word

FIGURE 17.18 Illustration of *absolute instability*. **A.** Absolute instability develops when solar heating causes the lowermost layer of the atmosphere to be warmed to a higher temperature than the air aloft. The result is a steep environmental lapse rate that renders the atmosphere unstable. **B.** Graphic representation of the conditions shown in part (**A**).

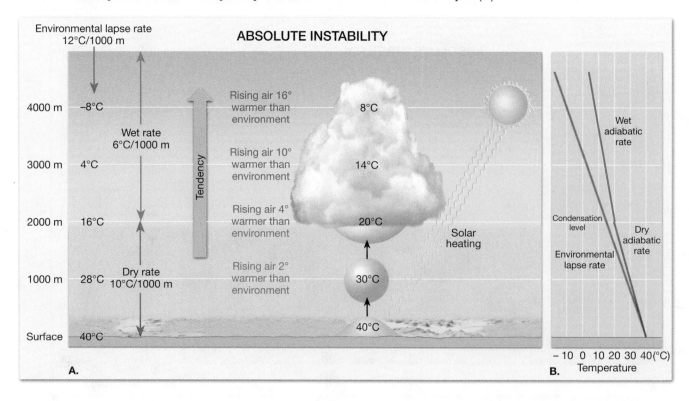

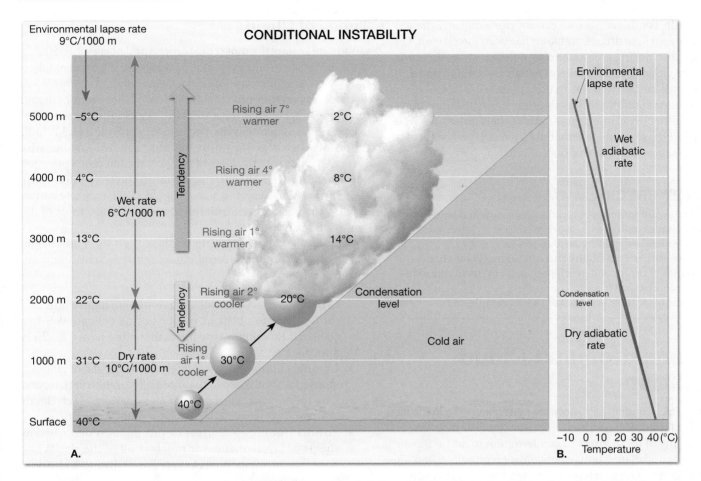

FIGURE 17.19 Illustration of *conditional instability*, where warm air is forced to rise along a frontal boundary. **A.** Note that the environmental lapse rate of 9° C per 1,000 meters lies between the dry and wet adiabatic rates. The parcel of air is cooler than the surrounding air up to nearly 3,000 meters, where its tendency is to sink toward the surface (stable). Above this level, however, the parcel is warmer than its environment and will rise because of its own buoyancy (unstable). Thus, when conditionally unstable air is forced to rise, the result can be towering cumulus clouds. **B.** Graphic representation of the conditions shown in part (**A**).

conditional is used because the air must be forced upward, such as over mountainous terrain, before it becomes unstable and rises because of its own buoyancy.

In summary, the stability of air is determined by measuring the temperature of the atmosphere at various heights. In simple terms, a column of air is deemed unstable when the air near the bottom of this layer is significantly warmer (less dense) than the air aloft, indicating a steep environmental lapse rate. Under these conditions the air actually turns over, as the warm air below rises and displaces the colder air aloft. Conversely, the air is considered to be stable when the temperature drops gradually with increasing altitude. The most stable conditions occur during a temperature inversion when the temperature actually increases with height. Under these conditions, there is very little vertical air movement.

Stability and Daily Weather

From the previous discussion, we can conclude that stable air resists vertical movement, whereas unstable air ascends freely because of its own buoyancy. But how do these facts manifest themselves in our daily weather?

Because stable air resists upward movement, we might conclude that clouds will not form when stable conditions prevail in the atmosphere. Although this seems reasonable, recall that processes exist that *force* air aloft. These include orographic lifting, frontal wedging, and convergence. When stable air is forced aloft, the clouds that form are widespread and have little vertical thickness when compared to their horizontal dimension, and precipitation, if any, is light to moderate.

By contrast, clouds associated with the lifting of unstable air are towering and often generate thunderstorms and occasionally

Students Sometimes Ask...

What is the driest place on Earth?

The record for the lowest average annual precipitation belongs to Arica, Chile, which receives 0.03 inch of rainfall per year. Over a span of 59 years this region in South America received a total of less than 2 inches of precipitation.

even a tornado. For this reason, we can conclude that on a dreary, overcast day with light drizzle, stable air has been forced aloft. On the other hand, during a day when cauliflower-shaped clouds appear to be growing as if bubbles of hot air are surging upward, we can be fairly certain that the ascending air is unstable.

In summary, stability plays an important role in determining our daily weather. To a large degree, stability determines the type of clouds that develop and whether precipitation will come as a gentle shower or a heavy downpour.

CONCEPT CHECK 17.5

❶ Explain the difference between the environmental lapse rate and adiabatic cooling.

❷ Describe absolute stability in your own words.

❸ Compare absolute instability and conditional instability.

❹ What types of clouds and precipitation, if any, form when stable air is forced aloft?

❺ What type of weather is associated with unstable air?

Condensation and Cloud Formation

To review briefly, condensation occurs when water vapor in the air changes to a liquid. The result of this process may be dew, fog, or clouds. For any of these forms of condensation to occur, the air must be saturated. Saturation occurs most commonly when air is cooled to its dew point, or less often when water vapor is added to the air.

Generally, there must be a surface on which the water vapor can condense. When dew occurs, objects at or near the ground such as grass and car windows serve this purpose. But when condensation occurs in the air above the ground, tiny bits of particulate matter, known as **condensation nuclei**, serve as surfaces for water-vapor condensation. These nuclei are very important, for in their absence a relative humidity well in excess of 100 percent is needed to produce clouds.

Condensation nuclei such as microscopic dust, smoke, and salt particles (from the ocean) are profuse in the lower atmosphere. Because of this abundance of particles, relative humidity rarely exceeds 101 percent. Some particles, such as ocean salt, are particularly good nuclei because they absorb water. These particles are termed **hygroscopic** (*hygro* = moisture, *scopic* = to seek) **nuclei**. When condensation takes place, the initial growth rate of cloud droplets is rapid. It diminishes quickly because the excess water vapor is quickly absorbed by the numerous competing particles. This results in the formation of a cloud consisting of millions upon millions of tiny water droplets, all so fine that they remain suspended in air. When cloud formation occurs at below-freezing temperatures, tiny ice crystals form. Thus, a cloud might consist of water droplets, ice crystals, or both.

The slow growth of cloud droplets by additional condensation and the immense size difference between cloud droplets and raindrops suggest that condensation alone is not responsible for the formation of drops large enough to fall as rain. We first examine clouds and then return to the question of how precipitation forms.

Types of Clouds

Clouds are among the most conspicuous and observable aspects of the atmosphere and its weather. **Clouds** are a form of condensation best described as *visible aggregates of minute droplets of water or tiny crystals of ice*. In addition to being prominent and sometimes spectacular features in the sky, clouds are of continual interest to meteorologists, because they provide a visible indication of what is going on in the atmosphere. Anyone who observes clouds with the hope of recognizing different types often finds that there is a bewildering variety of these familiar white and gray masses streaming across the sky. Still, once one comes to know the basic classification scheme for clouds, most of the confusion vanishes.

Clouds are classified on the basis of their *form* and *height* (**Figure 17.20**). Three basic forms are recognized: cirrus, cumulus, and stratus.

- **Cirrus** (*cirrus* = a curl of hair) clouds are high, white, and thin. They can occur as patches or as delicate veil-like sheets or extended wispy fibers that often have a feathery appearance.
- **Cumulus** (*cummulus* = a pile) clouds consist of globular individual cloud masses. They normally exhibit a flat base and have the appearance of rising domes or towers. Such clouds are frequently described as having a cauliflower structure.
- **Stratus** (*stratum* = a layer) clouds are best described as sheets or layers that cover much or all of the sky. While there may be minor breaks, there are no distinct individual cloud units.

All other clouds reflect one of these three basic forms or are combinations or modifications of them.

Three levels of cloud heights are recognized: high, middle, and low (Figure 17.20). **High clouds** normally have bases above 6,000 meters (20,000 feet), **middle clouds** generally occupy heights from 2,000 to 6,000 meters (6,500–20,000 feet), and **low clouds** form below 2,000 meters (6,500 feet). The altitudes listed for each height category are not hard and fast. There is some seasonal as well as latitudinal variation. For example, at high latitudes or during cold winter months in the mid-latitudes, high clouds are often found at lower altitudes.

High Clouds   Three cloud types make up the family of high clouds (above 6,000 meters): cirrus, cirrostratus, and cirrocumulus. *Cirrus* clouds are thin and delicate and sometimes appear as hooked filaments called "mares' tails" (**Figure 17.21A**). As the names suggest, *cirrocumulus* clouds consist of fluffy masses (Figure 17.21B), whereas *cirrostratus* clouds are flat layers (Figure 17.21C). Because of the low temperatures and small quantities of water vapor present at high altitudes, all high clouds are thin and white and are made up of ice crystals. Furthermore, these clouds are not considered precipitation makers. However, when cirrus clouds are followed by cirrocumulus clouds and increased sky coverage, they may warn of impending stormy weather.

Middle Clouds   Clouds that appear in the middle range (2,000–6,000 meters) have the prefix *alto* as part of their name. *Altocumulus* clouds are composed of globular masses that differ

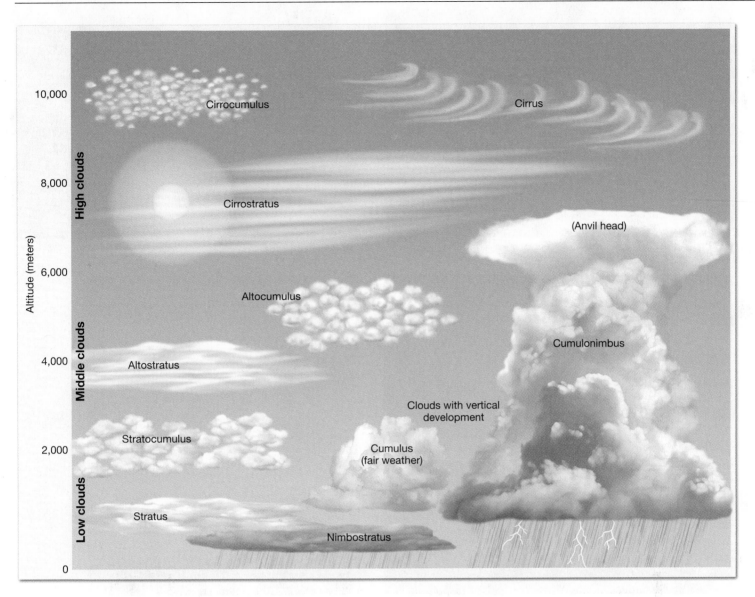

FIGURE 17.20 Classification of clouds according to height and form.

from cirrocumulus clouds in that they are larger and denser (Figure 17.21D). *Altostratus* clouds create a uniform white to grayish sheet covering the sky with the Sun or Moon visible as a bright spot (Figure 17.21E). Infrequent light snow or drizzle may accompany these clouds.

Students Sometimes Ask...

How much water is found in a cloud?

That depends a lot on the size of the cloud. Let's consider a modest-sized cumulonimbus cloud that is roughly 3,000 meters (about 2 miles) in width and depth and 10,000 meters high. If our hypothetical cloud contains an average of 0.5 cubic centimeters of liquid per cubic meter, it would contain 45 billion cubic centimeters of water (3,000 × 3,000 × 10,000 × 0.5). That equates to 13 million gallons, enough to fill a small pond.

Low Clouds There are three members in the family of low clouds: stratus, stratocumulus, and nimbostratus. *Stratus* are a uniform foglike layer of clouds that frequently covers much of the sky. On occasions these clouds may produce light precipitation. When stratus clouds develop a scalloped bottom that appears as long parallel rolls or broken globular patches, they are called *stratocumulus* clouds.

Nimbostratus clouds derive their name from the Latin *nimbus*, which means rainy cloud, and *stratus*, which means to cover with a layer (Figure 17.21F). As the name suggests, nimbostratus clouds are one of the chief precipitation producers. Nimbostratus clouds form in association with stable conditions. We might not expect clouds to grow or persist in stable air, yet cloud growth of this type is common when air is forced to rise, as occurs along a mountain range, a front, or near the center of a cyclone where converging winds cause air to ascend. Such forced ascent of stable air leads to the formation of a stratified cloud layer that is large horizontally compared to its depth.

A. Cirrus

B. Cirrocumulus

C. Cirrostratus

D. Altocumulus

FIGURE 17.21 These photos depict common forms of several different cloud types. (Photos **A**, **B**, **D**, **E**, **F**, and **G** by E. J. Tarbuck. Photo **C** Jung-Pang/Getty Images. Photo **H** by Doug Millar/Science Source/Photo Researchers, Inc.)

Clouds of Vertical Development Some clouds do not fit into any one of the three height categories mentioned. Such clouds have their bases in the low height range but often extend upward into the middle or high altitudes. Consequently, clouds in this category are called **clouds of vertical development**. They are all related to one another and are associated with unstable air. Although *cumulus* clouds are often connected with fair weather (Figure 17.21G), they may grow dramatically under the proper circumstances. Once upward movement is triggered, acceleration is powerful, and clouds with great vertical extent form. The end result is often a towering cloud, called a *cumulonimbus*, that may produce rain showers or a thunderstorm (Figure 17.21H).

Definite weather patterns can often be associated with particular clouds or certain combinations of cloud types, so it is important to become familiar with cloud descriptions and characteristics. **Table 17.2** lists the 10 basic cloud types that are recognized internationally and gives some characteristics of each.

CONCEPT CHECK 17.6

❶ As you drink an ice-cold beverage on a warm humid day, the outside of the glass or bottle becomes wet. Explain.

❷ What is the function of condensation nuclei in cloud formation?

❸ What is the basis for the classification of clouds?

❹ Why are high clouds always thin?

❺ Which cloud types are associated with the following characteristics: thunder, halos, precipitation, hail, mackerel sky, lightning, mares' tails?

E. Altostratus

F. Nimbostratus

FIGURE 17.21   (Continued)

G. Cumulus

H. Cumulonimbus

TABLE 17.2 Cloud Types and Characteristics

Cloud Family and Height	Cloud Type	Characteristics
High clouds—above 6,000 meters (20,000 feet)	Cirrus	Thin, delicate, fibrous, ice-crystal clouds. Sometimes appear as hooked filaments called "mares' tails." (Figure 17.21A)
	Cirrocumulus	Thin, white, ice-crystal clouds in the form of ripples, waves, or globular masses all in a row. May produce a "mackerel sky." Least common of the high clouds. (Figure 17.21B)
	Cirrostratus	Thin sheet of white, ice-crystal clouds that may give the sky a milky look. Sometimes produce halos around the Sun or Moon. (Figure 17.21C).
Middle clouds—2,000–6,000 meters (6,500–20,000 feet)	Altocumulus	White to gray clouds often composed of separate globules; "sheep-back" clouds. (Figure 17.21D)
	Altostratus	Stratified veil of clouds that are generally thin and may produce very light precipitation. When thin, the Sun or Moon may be visible as a bright spot, but no halos are produced. (Figure 17.21E)
Low clouds—below 2,000 meters (6,500 feet)	Stratocumulus	Soft, gray clouds in globular patches or rolls. Rolls may join together to make a continuous cloud.
	Stratus	Low uniform layer resembling fog but not resting on the ground. May produce drizzle.
	Nimbostratus	Amorphous layer of dark gray clouds. One of the chief precipitation-producing clouds. (Figure 17.21F)
Clouds of vertical development— 500–18,000 meters (1,600–60,000 feet)	Cumulus	Dense, billowy clouds often characterized by flat bases. May occur as isolated clouds or closely packed. (Figure 17.21G)
	Cumulonimbus	Towering cloud sometimes spreading out on top to form an "anvil head." Associated with heavy rainfall, thunder, lightning, hail, and tornadoes. (Figure 17.21H)

Fog

Fog is generally considered to be an atmospheric hazard. When it is light, visibility is reduced to 2 or 3 kilometers (1 or 2 miles). However, when it is dense, visibility may be cut to a few dozen meters or less, making travel by any mode not only difficult but often dangerous. Officially, visibility must be reduced to 1 kilometer or less before fog is reported. While this figure is arbitrary, it does provide an objective criterion for comparing fog frequencies at different locations.

Fog is defined as *a cloud with its base at or very near the ground*. There is no basic physical difference between a fog and a cloud; their appearance and structure are the same. The essential difference is the method and place of formation. Whereas clouds result when air rises and cools adiabatically, most fogs are the consequence of radiation cooling or the movement of air over a cold surface. (The exception is upslope fog.) In other circumstances, fogs form when enough water vapor is added to the air to bring about saturation (evaporation fogs). We examine both of these types.

Fogs Caused by Cooling

Three common fogs form when air at Earth's surface is chilled below its dew point. Often a fog is a combination of these types.

Advection Fog When warm, moist air moves over a cool surface, the result might be a blanket of fog called **advection fog** (Figure 17.22). Examples of such fogs are very common. The foggiest location in the United States, and perhaps in the world, is Cape Disappointment, Washington. The name is indeed appropriate because this station averages 2,552 hours of fog each year (there are about 8,760 hours in a year). The fog experienced at Cape Disappointment, and at other West Coast locations, is produced when warm, moist air from the Pacific Ocean moves over the cold California Current and is then carried onshore by the prevailing winds. Advection fogs are also relatively common in the winter season when warm air from the Gulf of Mexico moves across cold, often snow-covered surfaces of the Midwest and East.

Radiation Fog **Radiation fog** forms on cool, clear, calm nights, when Earth's surface cools rapidly by radiation. As the night progresses, a thin layer of air in contact with the ground is cooled below its dew point. As the air cools and becomes denser, it drains into low areas, resulting in "pockets" of fog. The largest pockets are often river valleys, where thick accumulations may occur (Figure 17.23).

Radiation fog may also form when the skies clear after a rainfall. In these situations the air near the surface is close to saturation and only a small amount of radiation cooling is needed to

FIGURE 17.22 Advection fog rolling into San Francisco Bay. (Photo by Ed Pritchard/Getty Images, Inc.–Stone Allstock)

Meteorologist's Sketch

San Francisco

Moist air is cooled by cold water below, forming fog

Advection fog

Cold water

FIGURE 17.23 A. Satellite image of dense fog in California's San Joaquin Valley on November 20, 2002. This early morning radiation fog was responsible for several car accidents in the region, including a 14-car pileup. The white areas to the east of the fog are the snowcapped Sierra Nevadas. (NASA image) **B.** Radiation fog can make the morning commute hazzardous. (Photo by Jeremy Walker/Getty Images Inc.–Stone Allstock)

promote condensation. Radiation fog of this type often occurs around sunset and can make driving hazardous.

Upslope Fog   As its name implies, **upslope fog** is created when relatively humid air moves up a gradually sloping plain or up the steep slopes of a mountain. As a result of this upward movement, air expands and cools adiabatically. If the dew point is reached, an extensive layer of fog may form. In the United States, the Great Plains offers an excellent example. When humid easterly or southeasterly winds blow westward from the Mississippi River upslope toward the Rocky Mountains, the air gradually rises, resulting in an adiabatic temperature decrease of about 13° C. When the difference between the air temperature and dew point of westward-moving air is less than 13° C, an extensive fog can result in the western plains.

Evaporation Fogs

When the saturation of air occurs primarily because of the addition of water vapor, the resulting fogs are called *evaporation fogs.* Two types of evaporation fogs are recognized: steam fog and frontal, or precipitation, fog.

Steam Fog   When cool air moves over warm water, enough moisture may evaporate from the water surface to produce saturation. As the rising water vapor meets the cold air, it immediately recondenses and rises with the air that is being warmed from below. Because the water has a steaming appearance, the phe-

nomenon is called **steam fog**. Steam fog is fairly common over lakes and rivers in the fall and early winter, when the water may still be relatively warm and the air is rather crisp (Figure 17.24). Steam fog is often shallow because as the steam rises, it reevaporates in the unsaturated air above.

Frontal Fog   When frontal wedging occurs, warm air is lifted over colder air. If the resulting clouds yield rain, and the cold air below is near the dew point, enough rain will evaporate to produce fog. A

FIGURE 17.24 Steam fog rising from Sparks Lake near Bend, Oregon. (Photo by Warren Morgan/CORBIS)

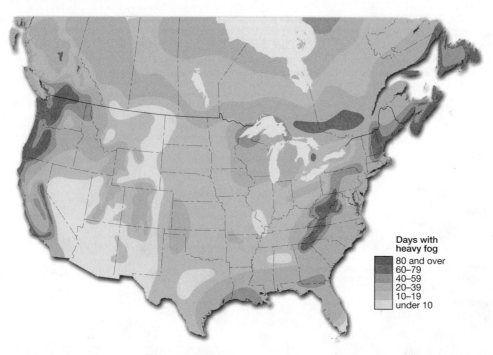

FIGURE 17.25 Map showing average number of days per year with heavy fog. Notice that the frequency of dense fog varies considerably from place to place. Coastal areas, particularly the Pacific Northwest and New England where cold currents prevail, have high occurrences of dense fog.

fog formed in this manner is called **frontal fog**, or **precipitation fog**. The result is a more or less continuous zone of condensed water droplets reaching from the ground up through the clouds.

Both steam fog and frontal fog result from the addition of moisture to a layer of air. The air is usually cool or cold and already near saturation. Because air's capacity for water vapor at low temperatures is small, only a relatively modest amount of evaporation is necessary to produce saturated conditions and fog.

The frequency of dense fog varies considerably from place to place (Figure 17.25). As might be expected, fog incidence is highest in coastal areas, especially where cold currents prevail, as along the Pacific and New England coasts. Relatively high frequencies are also found in the Great Lakes region and in the humid Appalachian Mountains of the East. By contrast, fogs are rare in the interior of the continent, especially in the arid and semiarid areas of the West.

Students Sometimes Ask...

Why does it often seem like the roads are slippery when it rains after a long dry period?

It appears that a buildup of debris on roads during dry weather causes more slippery conditions after a rainfall. One recent traffic study indicates that if it rains today, there will be no increase in risk of fatal crashes if it also rained yester- day. However, if it has been 2 days since the last rain, then the risk for a deadly accident increases by 3.7 percent. Fur- thermore, if it has been 21 days since the last rain, the risk increases to 9.2 percent.

How Precipitation Forms

Although all clouds contain water, why do some produce precipitation and others drift placidly overhead? This seemingly simple question perplexed meteorologists for many years. Before examining the processes that generate precipitation, we need to examine a couple of facts.

First, cloud droplets are very tiny, averaging under 20 micrometers (0.02 millimeter) in diameter (Figure 17.26). For comparison, a human hair is about 75 micrometers in diameter. The small size of cloud droplets results mainly because condensation nuclei are usually very abundant and the available water is distributed among numerous droplets rather than concentrated into fewer large droplets.

Second, because of their small size, the rate at which cloud droplets fall is incredibly slow. An average cloud droplet falling from a cloud base at 1,000 meters would require several hours to reach the ground. However, it would never complete its journey.

FIGURE 17.26 Comparative diameters of particles involved in condensation and precipitation processes.

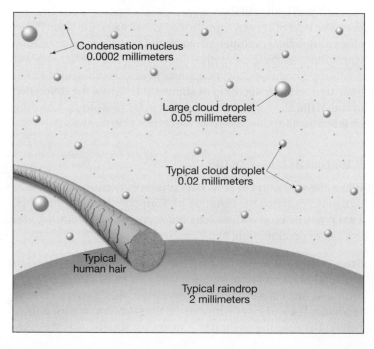

Condensation nucleus
0.0002 millimeters

Large cloud droplet
0.05 millimeters

Typical cloud droplet
0.02 millimeters

Typical human hair

Typical raindrop
2 millimeters

This cloud droplet would evaporate before it fell a few meters from the cloud base into the unsaturated air below.

How large must a droplet grow in order to fall as precipitation? A typical raindrop has a diameter of about 2,000 micrometers (2 millimeters) or 100 times that of the average cloud droplet having a diameter of 20 micrometers (0.02 millimeter). However, the *volume* of a typical raindrop is a million times that of a cloud droplet. Thus, for precipitation to form, cloud droplets must grow in volume by roughly one million times. Two mechanisms that give rise to these "massive" drops are the Bergeron process and the collision–coalescence process.

Precipitation from Cold Clouds: The Bergeron Process

You have probably watched a TV documentary in which mountain climbers brave intense cold and a ferocious snowstorm to scale an ice-covered peak. Although it is hard to imagine, very similar conditions exist in the upper portions of towering cumulonimbus clouds, even on sweltering summer days. It turns out that the frigid conditions high in the troposphere provide an ideal environment to initiate precipitation. In fact, in the middle latitudes much of the rain that falls begins with the birth of snowflakes high in the cloud tops where temperatures are considerably below freezing. Obviously, in the winter, even low clouds are cold enough to trigger precipitation.

The process that generates much of the precipitation in the middle latitudes is named the **Bergeron process** for its discoverer, the highly respected Swedish meteorologist, Tor Bergeron. The process relies on two interesting phenomena: supercooling and supersaturation.

Supercooling   Cloud droplets do not freeze at 0° C (32° F) as expected. In fact, pure water suspended in air does not freeze until it reaches a temperature of nearly −40° C (−40° F). Water in the liquid state below 0° C is referred to as **supercooled**. Supercooled water will readily freeze if it impacts an object. This explains why airplanes collect ice when they pass through a cloud composed of supercooled droplets.

In addition, supercooled water droplets will freeze upon contact with solid particles that have a crystal form closely resembling that of ice. Such materials are termed **freezing nuclei**. The need for freezing nuclei to initiate the freezing process is similar to the requirement for condensation nuclei in the process of condensation.

In contrast to condensation nuclei, freezing nuclei are very sparse in the atmosphere and do not generally become active until the temperature reaches −10° C (14° F) or less. Only at temperatures well below freezing will ice crystals begin to form in clouds, and even at that, they will be few and far between. Once ice crystals form, they are in direct competition with the supercooled droplets for the available water vapor.

Supersaturation   When air is saturated (100% relative humidity) with respect to water, it is **supersaturated** (relative humidity is greater than 100%) with respect to ice. **Table 17.3** shows that at −10° C (14° F), when the relative humidity is 100 percent with respect to water, the relative humidity with respect to ice is nearly 110 percent. Thus, ice crystals cannot coexist with water droplets because the air always "appears" supersaturated to the ice crystals. Hence, the ice crystals begin to consume the "excess" water vapor, which lowers the relative humidity near the surrounding droplets. In turn, the water droplets evaporate to replenish the diminishing water vapor, thereby providing a continual source of vapor for the growth of the ice crystals (**Figure 17.27**).

Because the level of supersaturation with respect to ice can be quite great, the growth of ice crystals is generally rapid enough to generate crystals large enough to fall. During their descent, these ice crystals enlarge as they intercept cloud drops, which freeze upon them. Air movement will sometimes break up these delicate crystals and the fragments will serve as freezing nuclei. A chain reaction develops, producing many ice crystals, which by accretion form into large crystals called snowflakes.

In summary, the Bergeron process can produce precipitation throughout the year in the middle latitudes, provided at least the upper portions of clouds are cold enough to generate ice crystals. The type of precipitation (snow, sleet, rain, or freezing rain) that reaches the ground depends on the temperature profile in the lower few kilometers of the atmosphere. When the surface

TABLE 17.3 Relative Humidity with Respect to Ice When Relative Humidity with Respect to Water is 100 Percent

Temperature (°C)	Relative Humidity with Respect to:	
	Water (%)	Ice (%)
0	100	100
−5	100	105
−10	100	110
−15	100	116
−20	100	121

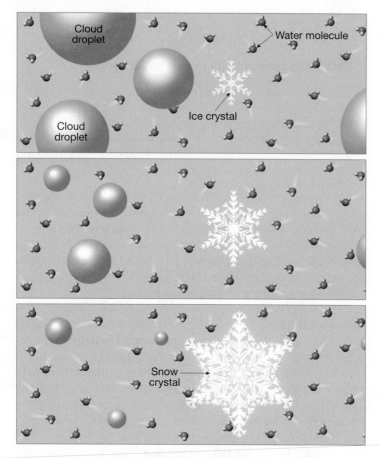

FIGURE 17.27 The Bergeron process. Ice crystals grow at the expense of cloud droplets until they are large enough to fall. The size of these particles has been greatly exaggerated.

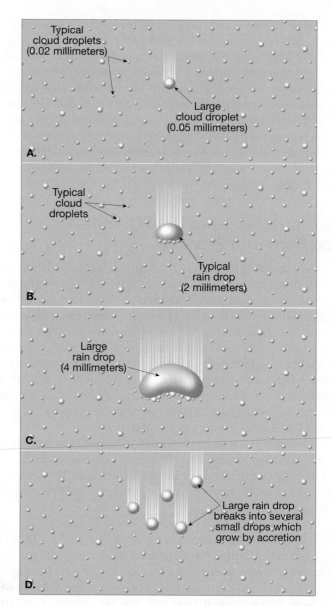

FIGURE 17.28 The collision–coalescence process. **A.** Most cloud droplets are so small that the motion of the air keeps them suspended. Because large cloud droplets fall more rapidly than smaller droplets, they are able to sweep up the smaller ones in their path and grow. **B.** As these drops increase in size, their fall velocity increases, resulting in increased air resistance, which causes the raindrop to flatten. **C.** As the raindrop approaches 4 millimeters in size, it develops a depression in the bottom. **D.** Finally, when the diameter exceeds about 5 millimeters, the depression grows upward almost explosively, forming a donutlike ring of water that immediately breaks into smaller drops. (Note that the drops are not drawn to scale—a typical raindrop has a volume equal to roughly 1 million cloud droplets.)

temperature is above 4° C (39° F), snowflakes usually melt before they reach the ground and continue their descent as rain. Even on a hot summer day, a heavy downpour may have begun as a snowstorm high in the clouds overhead.

Precipitation from Warm Clouds: The Collision–Coalescence Process

A few decades ago, meteorologists believed that the Bergeron process was responsible for the formation of most precipitation. However, it was discovered that copious rainfall can be associated with clouds located well below the freezing level (*warm clouds*), particularly in the tropics. This led to the proposal of a second mechanism thought to produce precipitation—the **collision–coalescence process**.

Research has shown that clouds composed entirely of liquid droplets must contain some droplets larger than 20 micrometers (0.02 millimeters) if precipitation is to form. These large droplets form when "giant" condensation nuclei are present, or when hygroscopic particles such as sea salt exist. Hygroscopic particles begin to remove water vapor from the air at relative humidities under 100 percent and can grow quite large. Because the rate at which drops fall is size-dependent, these "giant" droplets fall most rapidly. As they plummet, they collide with smaller, slower droplets and coalesce. Growing larger in the process, they fall

even more rapidly (or in an updraft, rise more slowly), increasing their chances of collision and rate of growth (**Figure 17.28**). After a great many such collisions they are large enough to fall to the surface without completely evaporating.

Because of the number of collisions required for growth to raindrop size, droplets in clouds that have great vertical thickness and abundant moisture have a better chance of reaching the required size. Updrafts also aid this process, for they allow the droplets to traverse the cloud repeatedly.

Raindrops can grow to a maximum size of 5 millimeters, at which point they fall at the rate of 33 kilometers (20 miles) per hour. At this size and speed, the water's surface tension, which holds the drop together, is overcome by the drag imposed by the air, which in turn pulls the drops apart. The resulting breakup of a large raindrop produces numerous smaller drops that begin anew the task of sweeping up cloud droplets. Drops that are less than 0.5 millimeter upon reaching the ground are termed *drizzle* and require about 10 minutes to fall from a cloud 1,000 meters (3,300 feet) overhead.

The collision–coalescence process is not quite as simple as described. First, as the large droplets descend, they produce an air stream around them similar to that produced by a fast-moving automobile. If an automobile is driven at night and we use the bugs that are often out to be analogous to the cloud droplets, it is easy to visualize how most cloud droplets are swept aside. The larger the cloud droplet (or bug), the better its chance of colliding with the giant droplet (or car). Second, collision does not guarantee coalescence. Experiments indicate that atmospheric electricity may hold these droplets together once they collide. If a droplet with a negative charge should collide with a positively charged droplet, electrical attraction of opposite charges may bind them together.

In summary, two mechanisms are known to generate precipitation: the *Bergeron process* and the *collision–coalescence process*. The Bergeron process is dominant in the middle latitudes where cold clouds (or cold cloud tops) are the rule. In the tropics, abundant water vapor and comparatively few condensation nuclei are the norm. This leads to the formation of fewer, larger drops with fast fall velocities that grow by collision and coalescence.

Forms of Precipitation

Much of the world's precipitation begins as snow crystals or other solid forms, such as hail or graupel (**Table 17.4**). Entering the warmer air below the cloud, these ice particles often melt and reach the ground as raindrops. In some parts of the world, particularly the subtropics, precipitation often forms in clouds that are warmer than 0° C (32° F). These rains frequently occur over the ocean where cloud condensation nuclei are not plentiful and those that exist vary in size. Under such conditions, cloud droplets can grow rapidly by the collision–coalescence process to produce copious amounts of rain.

Because atmospheric conditions vary greatly from place to place as well as seasonally, several different forms of precipitation are possible. Rain and snow are the most common and familiar, but the other forms of precipitation listed in Table 17.4 are important as well. The occurrence of sleet, glaze, or hail is often associated with important weather events. Although limited in occurrence and sporadic in both time and space, these forms, especially glaze and hail, can cause considerable damage.

TABLE 17.4 Forms of Precipitation

Type	Appropriate Size	State of Matter	Description
Mist	0.005–0.05 mm	Liquid	Droplets large enough to be felt on the face when air is moving 1 meter/second. Associated with stratus clouds.
Drizzle	Less than 0.5 mm	Liquid	Small uniform drops that fall from stratus clouds, generally for several hours.
Rain	0.5–5 mm	Liquid	Generally produced by nimbostratus or cumulonimbus clouds. When heavy, it can show high variability from one place to another.
Sleet	0.5–5 mm	Solid	Small, spherical to lumpy ice particles that form when raindrops freeze while falling through a layer of subfreezing air. Because the ice particles are small, damage, if any, is generally minor. Sleet can make travel hazardous.
Glaze	Layers 1 mm–2 cm thick	Solid	Produced when supercooled raindrops freeze on contact with solid objects. Glaze can form a thick coating of ice having sufficient weight to seriously damage trees and power lines.
Rime	Variable accumulations	Solid	Deposits usually consisting of ice feathers that point into the wind. These delicate, frostlike accumulations form as supercooled cloud or fog droplets encounter objects and freeze on contact.
Snow	1 mm–2 cm	Solid	The crystalline nature of snow allows it to assume many shapes including six-sided crystals, plates, and needles. Produced in supercooled clouds where water vapor is deposited as ice crystals that remain frozen during their descent.
Hail	5 mm–10 cm or larger	Solid	Precipitation in the form of hard, rounded pellets or irregular lumps of ice. Produced in large cumulonimbus clouds, where frozen ice particles and supercooled water coexist.
Graupel	2–5 mm	Solid	Sometimes called soft hail, graupel forms when rime collects on snow crystals to produce irregular masses of "soft" ice. Because these particles are softer than hailstones, they normally flatten out upon impact.

Rain

In meteorology, the term **rain** is restricted to drops of water that fall from a cloud and have a diameter of at least 0.5 millimeter (0.02 inch). Most rain originates either in nimbostratus clouds or in towering cumulonimbus clouds that are capable of producing unusually heavy rainfalls known as *cloudbursts*. Raindrops rarely exceed about 5 millimeters (0.2 inch) in diameter. Larger drops do not survive because surface tension, which holds the drops together, is exceeded by the frictional drag of the air. Consequently, large raindrops regularly break apart into smaller ones.

Fine, uniform drops of water having a diameter less than 0.5 millimeter (0.02 inch) are called *drizzle*. Drizzle can be so fine that the tiny drops appear to float and their impact is almost imperceptible. Drizzle and small raindrops generally are produced in stratus or nimbostratus clouds where precipitation can be continuous for several hours or, on rare occasions, for days.

Snow

Snow is precipitation in the form of ice crystals (snowflakes) or, more often, aggregates of crystals. The size, shape, and concentration of snowflakes depend to a great extent on the temperature at which they form.

Recall that at very low temperatures, the moisture content of air is small. The result is the formation of very light, fluffy snow made up of individual six-sided ice crystals. This is the "powder" that downhill skiers talk so much about. By contrast, at temperatures warmer than about −5° C (23° F), the ice crystals join together into larger clumps consisting of tangled aggregates of crystals. Snowfalls composed of these composite snowflakes are generally heavy and have a high moisture content, which makes them ideal for making snowballs.

Sleet and Glaze

Sleet is a wintertime phenomenon and refers to the fall of small particles of ice that are clear to translucent. For sleet to be produced, a layer of air with temperatures above freezing must overlie a subfreezing layer near the ground. When raindrops, which are often melted snow, leave the warmer air and encounter the colder air below, they freeze and reach the ground as small pellets of ice the size of the raindrops from which they formed.

Students Sometimes Ask...

What places in the United States receive the most snowfall?

Despite the impressive lake-effect snowstorms recorded in cities such as Buffalo and Rochester, New York, the greatest accumulations of snow generally occur in the mountainous regions of the western United States. The record for a single season goes to Mount Baker ski area north of Seattle, Washington, where 2,896 centimeters (1,140 inches) of snow fell during the winter of 1998–1999.

On some occasions, when the vertical distribution of temperatures is similar to that associated with the formation of sleet, freezing rain or **glaze** results instead. In such situations, the subfreezing air near the ground is not thick enough to allow the raindrops to freeze. The raindrops, however, do become supercooled as they fall through the cold air and turn to ice upon colliding with solid objects. The result can be a thick coating of ice having sufficient weight to break tree limbs, down power lines, and make walking or driving extremely hazardous (**Figure 17.29**).

Hail

Hail is precipitation in the form of hard, rounded pellets or irregular lumps of ice. Moreover, large hailstones often consist of a series of nearly concentric shells of differing densities and degrees of opaqueness (**Figure 17.30**). Most hailstones have diameters

FIGURE 17.29 Glaze forms when supercooled raindrops freeze on contact with objects. In January 1998 an ice storm of historic proportions caused enormous damage in New England and southeastern Canada. Nearly 5 days of freezing rain (glaze) left millions without electricity—some for as long as a month. (Photo by Syracuse Newspapers/The Image Works)

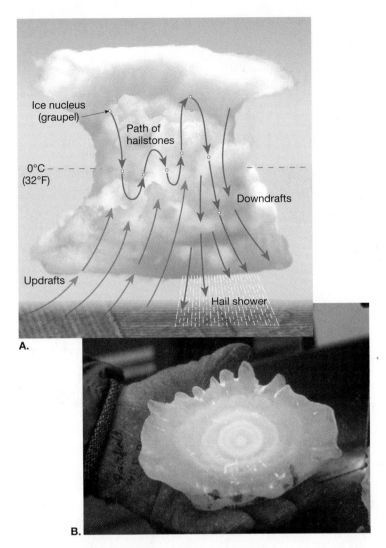

A.

B.

FIGURE 17.30 Hailstones. **A.** Hailstones begin as small ice pellets that grow by adding supercooled water droplets as they move through a cloud. Strong updrafts may carry stones upward in several cycles, increasing the size of the hail by adding a new layer with each cycle. Eventually, the hailstones encounter a downdraft or grow too large to be supported by the updraft. **B.** This cut hailstone fell at Coffeyville, Kansas, in 1970 and weighed 0.75 kilogram (1.67 pounds). Notice its layered structure. (Photo courtesy of University Corporation for Atmospheric Research/National Science Foundation/NCAR).

between 1 centimeter (0.4-inch, pea size) and 5 centimeters (2-inch golf ball–size), although some can be as big as an orange or larger. Occasionally, hailstones weighing a pound or more have been reported. Many of these were probably composites of several stones frozen together.

The record for the largest hailstone ever found in the United States was set on July 23, 2010, in Vivian, South Dakota. The stone was over 20 centimeters (8 inches) in diameter and weighed nearly 900 grams (2 pounds). The stone that held the previous record of 766 grams (1.69 pounds) fell in Coffeyville, Kansas, in 1970 (Figure 17.30B). The diameter of the stone found in South Dakota also surpassed the previous record of a 17.8-centimeter (7-inch) stone that fell in Aurora, Nebraska, in 2003. Even larger hailstones have reportedly been recorded in Bangladesh, where a 1987 hailstorm killed more than 90 people. It is estimated that

large hailstones hit the ground at speeds exceeding 160 kilometers (100 miles) per hour.

The destructive effects of large hailstones are well known, especially to farmers whose crops have been devastated in a few minutes and to people whose windows and roofs have been damaged (Figure 17.31). In the United States, hail damage each year is in the hundreds of millions of dollars.

Hail is produced only in large cumulonimbus clouds where updrafts can sometimes reach speeds approaching 160 kilometers (100 miles) per hour, and where there is an abundant supply of supercooled water. Hailstones begin as small embryonic ice pellets that grow by collecting supercooled water droplets as they fall through the cloud (Figure 17.30A). If they encounter a strong updraft, they may be carried upward again and begin the downward journey anew. Each trip through the supercooled portion of the cloud may be represented by an additional layer of ice. Hailstones can also form from a single descent through an updraft. Either way, the process continues until the hailstone encounters a downdraft or grows too heavy to remain suspended by the thunderstorm's updraft.

Rime

Rime is a deposit of ice crystals formed by the freezing of supercooled fog or cloud droplets on objects whose surface temperature is below freezing. When rime forms on trees, it adorns them with its characteristic ice feathers, which can be spectacular to behold (Figure 17.32). In these situations, objects such as pine needles act as freezing nuclei, causing the supercooled droplets to freeze on contact. When the wind is blowing, only the windward surfaces of objects will accumulate the layer of rime.

CONCEPT CHECK 17.9

❶ List the forms of precipitation and the circumstances of their formation.

❷ Explain why snow can sometimes reach the ground as rain, but the reverse does not occur.

❸ How is sleet different from glaze? Which is usually more hazardous?

FIGURE 17.31 Hail damage at an auto dealership in Amarillo, Texas, following a severe thunderstorm in June 2004. (Photo by Henry Bargas/*Amarillo Globe News*/AP Photo)

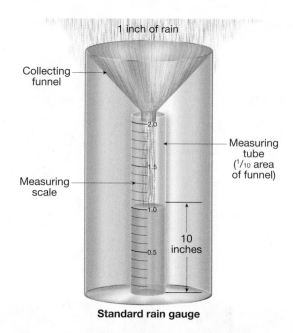

Standard rain gauge

FIGURE 17.33 Precipitation measurement. The standard rain gauge allows for accurate rainfall measurement to the nearest 0.025 centimeter (0.01 inch). Because the cross-sectional area of the measuring tube is only one-tenth as large as the collector, rainfall is magnified 10 times.

FIGURE 17.32 Rime consists of delicate ice crystals that form when supercooled fog or cloud droplets freeze on contact with objects. (Photo by Siepman/Photolibrary)

Measuring Precipitation

The most common form of precipitation—rain—is the easiest to measure. Any open container having a consistent cross section throughout can be a rain gauge. In general practice, however, more sophisticated devices are used so that small amounts of rainfall can be measured accurately and losses from evaporation can be reduced. The *standard rain gauge* (**Figure 17.33**) has a diameter of about 20 centimeters (8 inches) at the top. Once the water is caught, a funnel conducts the rain into a cylindrical measuring tube that has a cross-sectional area only one-tenth as large as the receiver. Consequently, rainfall depth is magnified 10 times, which allows for accurate measurements to the nearest 0.025 centimeter (0.01 inch). The narrow opening also minimizes evaporation. When the amount of rain is less than 0.025 centimeter, it is reported as a *trace* of precipitation.

Measurement Errors

In addition to the standard rain gauge, several types of recording gauges are routinely used. These instruments record not only the amount of rain but also its time of occurrence and intensity (amount per unit of time).

No matter which rain gauge is used, proper exposure is critical. Errors arise when the gauge is shielded from obliquely falling rain by buildings, trees, or other high objects. Hence, the instrument should be at least as far away from such obstructions as the objects are high. Another cause of error is the wind. It has been

shown that with increasing wind and turbulence, it becomes more difficult to collect a representative quantity of rain.

Measuring Snowfall

When snow records are kept, two measurements are normally taken: depth and water equivalent. Usually the depth of snow is measured with a calibrated stick. The actual measurement is simple, but choosing a *representative* spot often poses a dilemma. Even when winds are light or moderate, snow drifts freely. As a rule, it is best to take several measurements in an open place away from trees and obstructions and then average them. To obtain the water equivalent, samples can be melted and then weighed or measured as rain.

The quantity of water in a given volume of snow is not constant. A general ratio of 10 units of snow to 1 unit of water is often used when exact information is not available, but the actual water content of snow may deviate widely from this figure. It may take as much as 30 centimeters of light, fluffy dry snow or as little as 4 centimeters of wet snow to produce 1 centimeter of water.

Precipitation Measurement by Weather Radar

Today's TV weathercasts show helpful maps like the one in **Figure 17.34** to depict precipitation patterns. The instrument that produces these images is called *weather radar*.

The development of radar has given meteorologists an important tool to probe storm systems that may be up to a few hundred kilometers away. All radar units have a transmitter that sends out short pulses of radio waves. The specific wavelengths that are used depend on the objects the user wants to detect. When radar is used to monitor precipitation, wavelengths between 3 and 10 centimeters are employed.

These wavelengths can penetrate small cloud droplets, but are reflected by larger raindrops, ice crystals, or hailstones. The reflected signal, called an *echo*, is received and displayed on a TV monitor. Because the echo is "brighter" when the precipitation is more intense, modern radar is able to depict not only the regional extent of the precipitation but also the rate of rainfall. Figure 17.34 is a typical radar display in which colors show precipitation intensity. Weather radar is also an important tool for determining the rate and direction of storm movement.

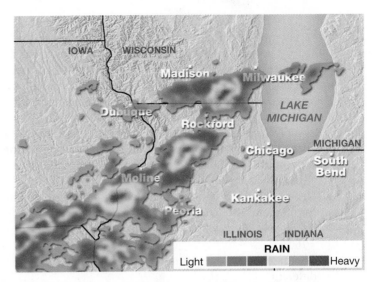

FIGURE 17.34 A color weather radar display, commonly seen on TV weathercasts. Colors indicate different intensities of precipitation.

CONCEPT CHECK 17.10

1 Sometimes, when rainfall is light, the amount is reported as a trace. When this occurs, how much (or little) rain has fallen?

2 Why is rainfall easier to measure than snowfall?

GIVE IT SOME THOUGHT

1. Refer to Figure 17.2 to complete the following:
   a. In which state of matter is water densest?
   b. In which state of matter are water molecules most energetic?
   c. In which state of matter is water compressible?
2. The accompanying photo shows a cup of hot coffee. What state of matter is the "steam" rising from the liquid? Explain your answer.
3. The primary mechanism by which the human body cools itself is perspiration.
   a. Explain how perspiring cools the skin.
   b. Referring to the data for Phoenix, Arizona, and Tampa, Florida (Table A), in which city would it be easier to stay cool by perspiring? Explain your choice.

Photo by Dmitry Kolmakov/Shutterstock

Table A

City	Temperature	Dew point temperature
Phoenix, AZ	101°F	47°F
Tampa, FL	101°F	77°F

4. During hot summer weather, many people put "koozies" around their beverages to keep the drinks cold. In addition to preventing a warm hand from heating the container through conduction, what other mechanism(s) slow the process of warming beverages?
5. Refer to Table 17.1 to answer this question. How much more water is contained in saturated air at a tropical location with a temperature of 40° C compared to a polar location with a temperature of $-10°$ C?
6. Refer to the data for Phoenix, Arizona, and Bismarck, North Dakota (Table B), to complete the following:
   a. Which city has a higher relative humidity?
   b. Which city has the greatest quantity of water vapor in the air?
   c. In which city is the air closest to its saturation point with respect to water vapor?
   d. In which city does the air have the greatest holding capacity for water vapor?

Table B

City	Temperature	Dew point temperature
Phoenix, AZ	101°F	47°F
Bismark, ND	39°F	38°F

7. The accompanying graph shows how air temperature and relative humidity change on a typical summer day in the Midwest. Assuming the dew-point temperature remained constant, what would be the best time of day to water a lawn to minimize evaporation of the water spread on the grass?

8. The accompanying diagram shows air flowing from the ocean over a coastal mountain range. Assume that the dew-point temperature remains constant in dry air (relative humidity less than 100%). If the air parcel becomes saturated, the dew-point temperature will cool at the wet adiabatic rate as it ascends, but will not change as the air parcel descends. Use this information to complete the following:

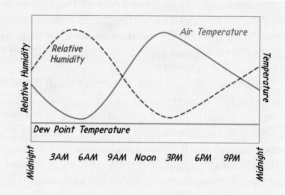

   a. Determine the air temperature and dew-point temperature for the air parcel at each location shown (B–G) on the diagram.
   b. At what elevation will clouds begin to form (relative humidity = 100%)?
   c. Compare the air temperatures at points A and G. Why are they different?
   d. How did the water vapor content of the air change as the parcel of air traversed the mountain? (*Hint:* Compare dew-point temperatures.)
   e. On which side of the mountain might you expect lush vegetation and on which side would you expect desert-like conditions?
   f. Where in the United States might you find such a situation?

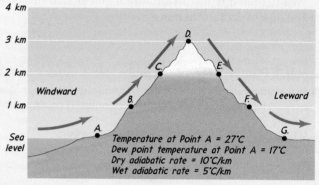

9. The dimensions of the cumulonimbus cloud pictured in Figure 17.21H are roughly 12 km tall, 8 km wide, and 8 km long. Assume that the droplets in every cubic meter of the cloud total 0.5 cubic centimeter. How much liquid does the cloud contain? How many gallons is this ($3785 \text{ cm}^3 = 1$ gallon)?

10. Cloud droplets form and grow as water vapor condenses onto a hygroscopic condensation nuclei. Research has shown that the maximum radius for cloud droplets is about 0.05 millimeter. However, typical raindrops have volumes thousands of times greater. How do cloud droplets become raindrops?

11. Why does radiation fog form mainly on clear nights as opposed to cloudy nights?

12. Which winter storm is likely to produce deeper snowfall: a low-pressure system that passes through the Midwestern states of Nebraska, Iowa, and Illinois (26° F average temperature at the time of the storm) or the *same exact system* passing through North Dakota, Minnesota, and Wisconsin (16° F average temperature at the time of the storm).

13. Weather radar provides information on the intensity of precipitation in addition to the total amount of precipitation that falls over a given time period. Table C shows the relationship between radar reflectivity values and rainfall rates. If radar measured a reflectivity value of 47 dBZ for 2½ hours over a location, how much rain will have fallen there?

14. What are the advantages and disadvantages of rain gauges compared to , weather radar in measuring rainfall?

Table C

Conversion of radar reflectivity to rainfall rate

Radar Reflectivity (dBZ)	Rainfall Rate (inches/hr)
65	16+
60	8.0
55	4.0
52	2.5
47	1.3
41	0.5
36	0.3
30	0.1
20	trace

In Review   Chapter 17 Moisture, Clouds and Precipitation

- *Water vapor*, an odorless, colorless gas, changes from one state of matter (solid, liquid, or gas) to another at the temperatures and pressures experienced near Earth's surface. The processes involved are *evaporation, condensation, melting, freezing, sublimation,* and *deposition.* During each change, *latent* (hidden) *heat* is either absorbed or released.

- *Humidity* is the general term to describe the amount of water vapor in the air. The methods used to express humidity quantitatively include (1) *mixing ratio,* the mass of water vapor in a unit of air compared to the remaining mass of dry air; (2) *vapor pressure,* that part of the total atmospheric pressure attributable to its water-vapor content; (3) *relative humidity,* the ratio of the air's actual water-vapor content compared to the amount of water vapor required for saturation at that temperature; and (4) *dew point,* the temperature to which a parcel of air would need to be cooled to reach saturation. When air is saturated, the pressure exerted by the water vapor, called the *saturation vapor pressure,* produces a balance between the number of water molecules leaving the surface of the water and the number returning. Because the saturation vapor pressure is temperature-dependent, at higher temperatures more water vapor is required for saturation to occur.

- *Relative humidity can be changed in two ways.* One is by *adding or subtracting water vapor.* The second is by *changing the air's temperature.* When air is cooled, its relative humidity increases.

- The cooling of air as it rises and expands owing to successively lower air pressure is the basic cloud-forming process. Temperature changes in air brought about by compressing or expanding the air are called *adiabatic temperature changes.* Unsaturated air warms by compression and cools by expansion at the rather constant rate of 10° C per 1,000 meters of altitude change, a figure called the *dry adiabatic rate.* If air rises high enough, it will cool sufficiently to cause condensation and form a cloud. From this point on, air that continues to rise will cool at the *wet adiabatic rate,* which varies from 5° C to 9° C per 1,000 meters of ascent. The difference in the wet and dry adiabatic rates is caused by the condensing water vapor releasing *latent heat,* thereby reducing the rate at which the air cools.

- Four mechanisms that can initiate the vertical movement of air are (1) *orographic lifting,* which occurs when elevated terrains, such as mountains, act as barriers to the flow of air; (2) *frontal wedging,* when cool air acts as a barrier over which warmer, less dense air rises; (3) *convergence,* which happens when air flows together and a general upward movement of air occurs; and (4) *localized convective lifting,* which occurs when unequal surface heating causes pockets of air to rise because of their buoyancy.

- The *stability of air* is determined by examining the temperature of the atmosphere at various altitudes. Air is said to be *unstable* when the *environmental lapse rate* (the rate of temperature decrease with increasing altitude in the troposphere) is greater than the *dry adiabatic rate.* Stated differently, a column of air is unstable when the air near the bottom is significantly warmer (less dense) than the air aloft.

- *For condensation to occur, air must be saturated.* Saturation takes place either when air is cooled to its dew point, which most commonly happens, or when water vapor is added to the air. There must also be a surface on which the water vapor can condense. In cloud and fog formation, tiny particles called *condensation nuclei* serve this purpose.

- *Clouds* are classified on the basis of their *appearance* and *height.* The three basic forms are *cirrus* (high, white, thin, wispy fibers), *cumulus* (globular, individual cloud masses), and *stratus* (sheets or layers that cover much or all of the sky). The four categories based on height are *high clouds* (bases normally above 6,000 meters), *middle clouds* (from 2,000 to 6,000 meters), *low clouds* (below 2,000 meters), and *clouds of vertical development.*

- *Fog* is defined as a cloud with its base at or very near the ground. Fogs form when air is cooled below its dew point or when enough water vapor is added to the air to bring about saturation. Various types of fog include *advection fog, radiation fog, upslope fog, steam fog,* and *frontal* (or *precipitation*) *fog.*

- For *precipitation* to form, millions of cloud droplets must somehow join together into large drops. Two mechanisms for the formation of precipitation have been proposed. (1) In clouds where the temperatures are below freezing, ice crystals form and fall as snowflakes. At lower altitudes the snowflakes melt and become raindrops before they reach the ground. (2) *Large* droplets form in warm clouds that contain large *hygroscopic* ("water-seeking") *nuclei,* such as salt particles. As these big droplets descend, they collide and join with smaller water droplets. After many collisions, the droplets are large enough to fall to the ground as rain.

- The forms of precipitation include *rain, snow, sleet, freezing rain (glaze), hail,* and *rime.*

Key Terms

absolute instability (p. 504)
absolute stability (p. 503)
adiabatic temperature change (p. 499)
advection fog (p. 510)
Bergeron process (p. 513)
calorie (p. 490)
cirrus (p. 506)
cloud (p. 506)
cloud of vertical development (p. 508)
collision–coalescence process (p. 514)
condensation (p. 492)
condensation nuclei (p. 506)
conditional instability (p. 504)
convergence (p. 500)
cumulus (p. 506)
deposition (p. 492)
dew-point temperature (p. 495)
dry adiabatic rate (p. 499)
evaporation (p. 491)

fog (p. 510)
freezing nuclei (p. 513)
front (p. 500)
frontal fog (p. 512)
frontal wedging (p. 500)
glaze (p. 516)
hail (p. 516)
high cloud (p. 506)
humidity (p. 492)
hygrometer (p. 496)
hygroscopic nuclei (p. 506)
latent heat (p. 491)
localized convective lifting (p. 501)
low cloud (p. 506)
middle cloud (p. 506)
mixing ratio (p. 493)
orographic lifting (p. 500)
parcel (p. 499)
precipitation fog (p. 512)

psychrometer (p. 496)
radiation fog (p. 510)
rain (p. 516)
rain shadow desert (p. 500)
relative humidity (p. 493)
rime (p. 517)
saturation (p. 492)
sleet (p. 516)
snow (p. 516)
stable air (p. 502)
steam fog (p. 511)
stratus (p. 506)
sublimation (p. 492)
supercooled (p. 513)
supersaturation (p. 513)
unstable air (p. 502)
upslope fog (p. 511)
vapor pressure (p. 492)
wet adiabatic rate (p. 499)

Examining the Earth System

1. The interrelationships among Earth's spheres have produced the Great Basin area of the western United States, which includes some of the driest areas in the world. Examine the map of the region in Appendix D. Although the area is only a few hundred miles from the Pacific Ocean, it is a desert. Why? Did any geologic factor(s) contribute to the formation of this desert environment? Do any major rivers have their source in the Great Basin? Explain. (For information about deserts in the United States, visit the United States Geological Survey [USGS] Deserts: Geology and Resources Website at http://pubs.usgs.gov/gip/deserts/.)

2. Phoenix and Flagstaff, Arizona, are located nearby each other in the southwestern United States. Using the climate diagrams for the two cities found in Figure 20.16, describe the impact that elevation has on the precipitation and tem- perature of each city. Compare the natural vegetation of these nearby locations. (To examine the current weather conditions and forecasts for Phoenix and Flagstaff, Arizona, contact *USA Today* at http://www.usatoday.com/weather/ and/or The Weather Channel at http://www.weather.com.)

3. The amount of precipitation that falls at any particular place and time is controlled by the quantity of moisture in the air and many other factors. How might each of the following alter the precipitation at a particular locale? (a) An increase in the elevation of the land. (b) A decrease in the area cov- ered by forests and other types of vegetation. (c) Lowering of average ocean-surface temperatures. (d) An increase in the percentage of time that the winds blow from an adjacent body of water. (e) A major episode of global volcanism last- ing for decades.

Mastering Geology

Looking for additional review and test prep materials? Visit the Self Study area in **www.masteringgeology.com** to find practice quizzes, study tools, and multimedia that will aid in your understanding of this chapter's content. In **MasteringGeology™** you will find:

- **GEODe: Earth Science: An interactive visual walkthrough of key concepts**

- **Geoscience Animation Library: More than 100 animations illuminating many difficult-to-understand Earth science concepts**
- **In The News RSS Feeds: Current Earth science events and news articles are pulled into the site with assessment**
- **Pearson eText**
- **Optional Self Study Quizzes**
- **Web Links**
- **Glossary**
- **Flashcards**

18

Air Pressure
and Wind

Since the year 2000, worldwide wind energy capacity has doubled every 3 years. These wind turbines are in California. There is more about wind energy in **Box 18.1, p. 536.** (Photo by Travel Pix/Robert Harding)

O f the various elements of weather and climate, changes in air pressure are the least noticeable. In listening to a weather report, generally we are interested in moisture conditions (humidity and precipitation), temperature, and perhaps wind. It is the rare person, however, who wonders about air pressure. Although the hour-to-hour and day-to-day variations in air pressure are not perceptible to human beings, they are very important in producing changes in our weather. For example, it is variations in air pressure from place to place that generate winds that in turn can bring changes in temperature and humidity (Figure 18.1). Air pressure is one of the basic weather elements and is a significant factor in weather forecasting. As you will see, air pressure is closely tied to the other elements of weather in a cause-and-effect relationship.

FOCUS ON CONCEPTS

To assist you in learning the important concepts in this chapter, focus on the following questions:

- What is air pressure and how is it measured?
- What force creates wind, and what other factors influence wind?
- What are the two types of pressure centers? What wind patterns and weather conditions are associated with each type?
- What is the idealized global circulation? How do continents complicate patterns of global circulation?
- What are the names and causes of some local winds?
- How is wind measured?
- What is El Niño and how is it different from La Niña?
- What factors control and influence the global distribution of precipitation?

Understanding Air Pressure

In Chapter 16 we noted that **air pressure** is simply the pressure exerted by the weight of air above. Average air pressure at sea level is about 1 kilogram per square centimeter, or 14.7 pounds per square inch. This is roughly the same pressure that is produced by a column of water 10 meters (33 feet) in height. With some simple arithmetic you can calculate that the air pressure exerted on the top of a small (50 centimeter by 100 centimeter) school desk exceeds 5,000 kilograms (11,000 pounds), or about the weight of a 50-passenger school bus. Why doesn't the desk collapse under the weight of the ocean of air above? Simply, air pressure is exerted in all directions—down, up, and sideways. Thus, the air pressure pushing down on the desk exactly balances the air pressure pushing up on the desk.

You might be able to visualize this phenomenon better if you imagine a tall aquarium that has the same dimensions as the desktop. When this aquarium is filled to a height of 10 meters (33 feet), the water pressure at the bottom equals 1 atmosphere (14.7 pounds per square inch). Now, imagine what will happen if this aquarium is placed on top of our student desk so that all the force is directed downward. Compare this to what results when the desk is placed inside the aquarium and allowed to sink to the bottom. In the latter situation the desk survives because the water pressure is exerted in all directions, not just downward as in our earlier example. The desk, like your body, is "built" to withstand the pressure of 1 atmosphere. It is important to note that although we do not generally notice the pressure exerted by the ocean of air around us, except when ascending or descending in an elevator or airplane, it is nonetheless substantial. The pressurized suits used by astronauts on space walks are designed to duplicate the atmospheric pressure experienced at Earth's surface. Without these protective suits to keep body fluids from boiling away, astronauts would perish in minutes.

The concept of air pressure can be better understood if we examine the behavior of gases. Gas molecules, unlike those of the liquid and solid phases, are not "bound" to one another but are freely moving about, filling all space available to them. When two gas molecules collide, which happens frequently under normal atmospheric conditions, they bounce off each other like very elastic balls. If a gas is confined to a container, this motion is restricted by its sides, much like the walls of a handball court redirect the motion of the handball. The continuous bombardment of gas molecules against the sides of the container exerts an outward push that we call air pressure. Although the atmosphere is without walls, it is confined from below by Earth's surface and effectively from above because the force of gravity prevents its escape. Here we define *air pressure* as the force exerted against a surface by the continuous collision of gas molecules.

FIGURE 18.1 Strong winds blowing snow during a blizzard. (Photo by AGRfoto/Alex Rowbotham/Alamy)

CONCEPT CHECK 18.1

❶ What is air pressure?
❷ Express air pressure in pounds per square inch and kilograms per square centimeter.

Measuring Air Pressure

 Earth's Dynamic Atmosphere
▶ Air Pressure and Wind

When meteorologists measure atmospheric pressure, they employ a unit called the *millibar*. Standard sea-level pressure is 1013.2 millibars. Although the millibar has been the unit of measure on all U.S. weather maps since January 1940, the media use "inches of mercury" to describe atmospheric pressure. In the United States, the National Weather Service converts millibar values to inches of mercury for public and aviation use.

Inches of mercury are easy to understand. The use of mercury for measuring air pressure dates from 1643, when Torricelli, a student of the famous Italian scientist Galileo, invented the **mercury barometer** (*bar* = pressure, *metron* = measuring instrument). Torricelli correctly described the atmosphere as a vast ocean of air that exerts pressure on us and all objects about us. To measure this force, he filled a glass tube, which was closed at one end, with mercury. He then inverted the tube into a dish of mercury (Figure 18.2).

FIGURE 18.2 Simple mercury barometer. The weight of the column of mercury is balanced by the pressure exerted on the dish of mercury by the air above. If the pressure decreases, the column of mercury falls; if the pressure increases, the column rises.

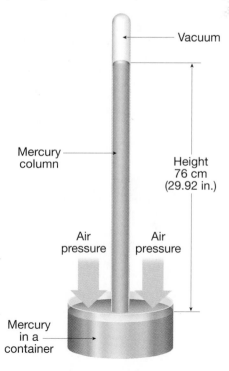

Torricelli found that the mercury flowed out of the tube until the weight of the column was balanced by the pressure that the atmo–sphere exerted on the surface of the mercury in the dish. In other words, the weight of mercury in the column equaled the weight of the same diameter column of air that extended from the ground to the top of the atmosphere.

When air pressure increases, the mercury in the tube rises. Conversely, when air pressure decreases, so does the height of the mercury column. With some refinements the mercurial barometer invented by Torricelli is still the standard pressure–measuring instrument used today. Standard atmospheric pressure at sea level equals 29.92 inches of mercury.

The need for a smaller and more portable instrument for measuring air pressure led to the development of the **aneroid** (*an* = without, *ner* = fluid) **barometer** (Figure 18.3). Instead of having a mercury column held up by air pressure, the aneroid barometer uses a partially evacuated metal chamber. The chamber, being very sensitive to variations in air pressure, changes shape, compressing as the pressure increases and expanding as the pressure decreases. A series of levers transmits the movements of the chamber to a pointer on a dial that is calibrated to read in inches of mercury and/or millibars.

As shown in Figure 18.3, the face of an aneroid barometer intended for home use is inscribed with words like *fair*, *change*, *rain*, and *stormy*. Notice that "fair weather" corresponds with high-pressure readings, whereas "rain" is associated with low pressures. Although barometric readings may indicate the present weather, this is not always the case. The dial may point to "fair" on a rainy day, or you may be experiencing "fair" weather when the dial indicates "rainy." If you want to "predict" the weather in a local area, the change in air pressure over the past few hours is more important than the current pressure reading. Falling pressure is often associated with increasing cloudiness and the possibility of precipitation, whereas rising air pressure generally indicates clearing conditions. It is useful to remember, however, that particular barometer readings or trends do not always correspond to specific types of weather.

One advantage of the aneroid barometer is that it can easily be connected to a recording mechanism. The resulting instrument is a **barograph**, which provides a continuous record of pressure changes with the passage of time (Figure 18.4). Another important adaptation of the aneroid barometer is its use to indicate altitude for aircraft, mountain climbers, and mapmakers.

CONCEPT CHECK 18.2

❶ What is standard sea level pressure in millibars? In inches of mercury?

❷ Describe the operating principles of a mercury barometer and an aneroid barometer.

Students Sometimes Ask...

What is the lowest barometric pressure ever recorded?

All of the lowest-recorded barometric pressures have been associated with strong hurricanes. The record for the United States is 882 millibars (26.12 inches) measured during Hurricane Wilma in October 2005. The world record, 870 millibars (25.70 inches), occurred during Typhoon Tip (a Pacific hurricane), in October 1979.

FIGURE 18.3 Aneroid Barometer. this instrument has a partially evacuated chamber that changes shape, compressing as atmospheric pressure increases, and expanding as pressure decreases.

FIGURE 18.4 An aneroid barograph makes a continuous record of pressure changes. (Photo courtesy of Qualimetrics, Inc., Sacramento, California)

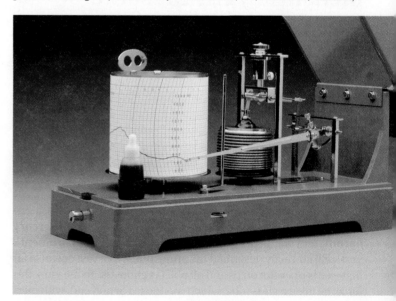

Factors Affecting Wind

Earth's Dynamic Atmosphere
▶ **Air Pressure and Wind**

In Chapter 17 we examined the upward movement of air and its role in cloud formation. As important as vertical motion is, far more air moves horizontally, the phenomenon we call **wind**. What causes wind?

Simply stated, wind is the result of horizontal differences in air pressure. *Air flows from areas of higher pressure to areas of lower pressure.* You may have experienced this when opening a vacuum-packed can of coffee. The noise you hear is caused by air rushing from the higher pressure outside the can to the lower pressure inside. Wind is nature's attempt to balance such inequalities in air pressure. Because unequal heating of Earth's surface generates these pressure differences, *solar radiation is the ultimate energy source for most wind.*

If Earth did not rotate, and if there were no friction between moving air and Earth's surface, air would flow in a straight line from areas of higher pressure to areas of lower pressure. But because both factors exist, wind is controlled by a combination of forces, including (1) the pressure-gradient force, (2) the Coriolis effect, and (3) friction. We now examine each of these factors.

Pressure-Gradient Force

Pressure differences create wind, and the greater these differences, the greater the wind speed. Over Earth's surface, variations in air pressure are determined from barometric readings taken at hundreds of weather stations. These pressure data are shown on a weather map using **isobars**, lines that connect places of equal air pressure (**Figure 18.5**). The spacing of isobars indicates the amount of pressure change occurring over a given distance and is expressed as the **pressure gradient** (*gradus* = slope).

You might find it easier to visualize a pressure gradient if you think of it as being similar to the slope of a hill. A steep pressure gradient, like a steep hill, causes greater acceleration of an air parcel than does a weak pressure gradient (a gentle hill). Thus, the relationship between wind speed and the pressure gradient is straightforward: *Closely spaced isobars indicate a steep pressure gradient and high winds, whereas widely spaced isobars indicate a weak pressure gradient and light winds.* Figure 18.5 illustrates the relationship between the spacing of isobars and wind speed. Notice that wind speeds are greater in Ohio, Kentucky, Michigan, and Illinois, where isobars are more closely spaced, than in the western states, where isobars are more widely spaced.

The pressure gradient is the driving force of wind, and it has both magnitude and direction. Its magnitude is determined from

FIGURE 18.5 Isobars are lines connecting places of equal sea-level pressure. They are used to show the distribution of pressure on daily weather maps. Isobars are seldom straight, but usually form broad curves. Concentric rings of isobars indicate cells of high and low pressure. The "wind flags" indicate the expected airflow surrounding pressure cells and are plotted as "flying" with the wind (i.e., the wind blows toward the station circle). Notice that where the isobars are more closely spaced, the wind speed is faster. Closely spaced isobars indicate a strong pressure gradient and high wind speeds, whereas widely spaced isobars indicate a weak pressure gradient and low wind speeds.

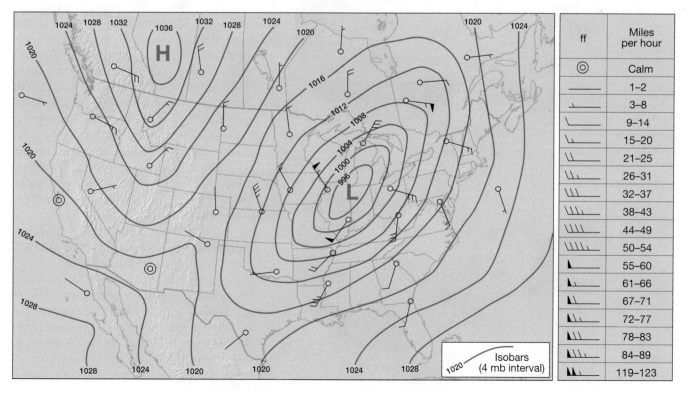

the spacing of isobars. The direction of force is always from areas of higher pressure to areas of lower pressure and at right angles to the isobars. Once the air starts to move, the Coriolis effect and friction come into play, but then only to modify the movement, not to produce it.

Coriolis Effect

The weather map in Figure 18.5 shows the typical air movements associated with high- and low-pressure systems. As expected, the air moves out of the regions of higher pressure and into the regions of lower pressure. However, the wind does not cross the isobars at right angles as the pressure-gradient force directs it. This deviation is the result of Earth's rotation and has been named the **Coriolis effect** after the French scientist who first thoroughly described it.

All free-moving objects or fluids, including the wind, are deflected to the *right* of their path of motion in the Northern Hemisphere and to the *left* in the Southern Hemisphere. The reason for this deflection can be illustrated by imagining the path of a rocket launched from the North Pole toward a target located on the equator (**Figure 18.6**). If the rocket took an hour to reach its

target, Earth would have rotated 15 degrees to the east during its flight. To someone standing on Earth it would look as if the rocket veered off its path and hit Earth 15 degrees west of its target. The true path of the rocket is straight and would appear so to someone out in space looking down at Earth. It was Earth turning under the rocket that gave it its *apparent* deflection.

Note that the rocket was deflected to the right of its path of motion because of the counterclockwise rotation of the Northern Hemisphere. In the Southern Hemisphere, the effect is reversed. Clockwise rotation produces a similar deflection, but to the left of the path of motion. The same deflection is experienced by wind regardless of the direction it is moving.

We attribute the apparent shift in wind direction to the Coriolis effect. This deflection (1) is always directed at right angles to the direction of airflow; (2) affects only wind direction, not wind speed; (3) is affected by wind speed (the stronger the wind, the greater the deflection); and (4) is strongest at the poles and weakens equatorward, becoming nonexistent at the equator.

It is of interest to point out that any "free-moving" object will experience a deflection caused by the Coriolis effect. This fact was dramatically discovered by the United States Navy in World War II. During target practice long-range guns on battleships continually missed their targets by as much as several hundred yards until ballistic corrections were made for the changing position of a seemingly stationary target. Over a short distance, however, the Coriolis effect is relatively small.

FIGURE 18.6 The Coriolis effect illustrated using a one-hour flight of a rocket traveling from the North Pole to a location on the equator. **A.** On a nonrotating Earth, the rocket would travel straight to its target. **B.** However, Earth rotates 15° each hour. Thus, although the rocket travels in a straight line, when we plot the path of the rocket on Earth's surface, it follows a curved path that veers to the right of the target.

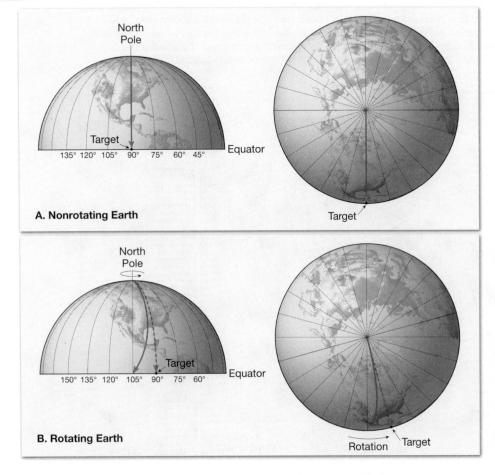

A. Nonrotating Earth

B. Rotating Earth

Friction with Earth's Surface

The effect of friction on wind is important only within a few kilometers of Earth's surface. We know that friction acts to slow the movement of air (**Figure 18.7**). As a consequence, wind direction is also affected. To illustrate friction's effect on wind direction, let us look at a situation in which it has no role. Above the friction layer, the pressure-gradient force and Coriolis effect work together to direct the flow of air. Under these conditions, the pressure-gradient force causes air to start moving across the isobars. As soon as the air starts to move, the Coriolis effect acts at right angles to this motion. The faster the wind speed, the greater the deflection.

Eventually, the Coriolis effect will balance the pressure-gradient force, and the wind will blow parallel to the isobars (**Figure 18.8**). Upper-air winds generally take this path and are called **geostrophic winds**. Because of the lack of friction with Earth's surface, geostrophic winds travel at higher speeds than do surface winds. This can be observed in **Figure 18.9** by noting the wind flags, many of which indicate winds of 50–100 miles per hour.

FIGURE 18.7 The effect of friction is to slow the wind. In this image, a snow fence reduces wind speed, thereby diminishing the ability of the moving air to carry snow. As a result, snow accumulates as a drift. (Photo by Garry Black/Superstock)

to wind speed. Friction lowers the wind speed, so it reduces the Coriolis effect. Because the pressure-gradient force is not affected by wind speed, it wins the tug of war shown in **Figure 18.10**. The result is a movement of air at an angle across the isobars toward the area of lower pressure.

The roughness of the terrain determines the angle of airflow across the isobars. Over the smooth ocean surface, friction is low and the angle is small. Over rugged terrain, where friction is higher, the angle that air makes as it flows across the isobars can be as great as 45 degrees.

In summary, upper airflow is nearly parallel to the isobars, whereas the effect of friction causes the surface winds to move more slowly and cross the isobars at an angle.

The most prominent features of upper-level flow are the **jet streams**. First encountered by high-flying bombers during World War II, these fast-moving rivers of air travel between 120 and 240 kilometers (75 and 150 miles) per hour in a west-to-east direction. One such stream is situated over the polar front, which is the zone separating cool polar air from warm subtropical air.

Below 600 meters (2,000 feet), friction complicates the airflow just described. Recall that the Coriolis effect is proportional

CONCEPT CHECK 18.3

1. What force is responsible for generating wind?
2. Write a generalization relating the spacing of isobars to the speed of wind.
3. How does the Coriolis effect modify air movement?
4. Contrast surface winds and upper-air winds in terms of speed and direction.

FIGURE 18.8 The geostrophic wind. The only force acting on a stationary parcel of air is the pressure-gradient force. Once air begins to accelerate, the Coriolis effect deflects it to the right in the Northern Hemisphere. Greater wind speeds result in a stronger Coriolis effect until the flow is parallel to the isobars. At this point the pressure-gradient force and Coriolis effect are in balance and the flow is called a *geostrophic wind*. In the "real" atmosphere, airflow is continually adjusting for variations in the pressure field. As a result, the adjustment to geostrophic equilibrium is much more irregular than shown.

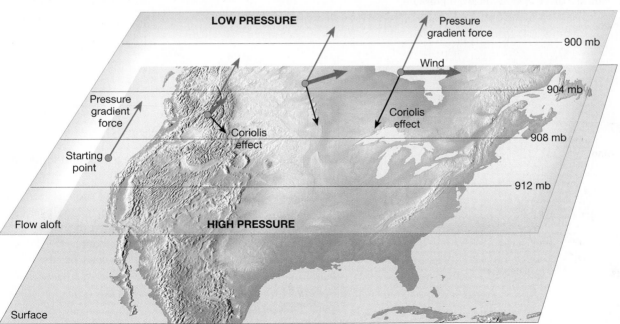

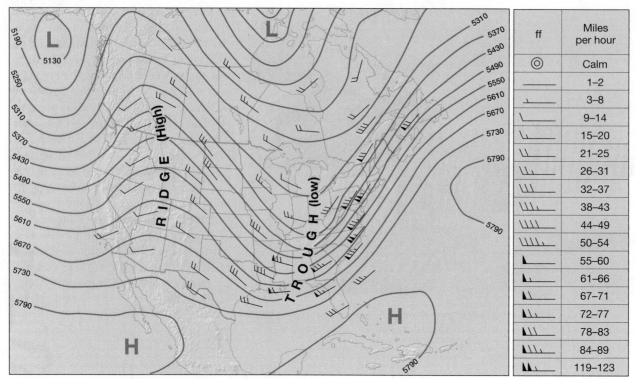

A. Upper-level weather chart

ff	Miles per hour
◎	Calm
—	1–2
	3–8
	9–14
	15–20
	21–25
	26–31
	32–37
	38–43
	44–49
	50–54
	55–60
	61–66
	67–71
	72–77
	78–83
	84–89
	119–123

B. Representation of upper-level chart

FIGURE 18.9 Upper-air winds. This map shows the direction and speed for the upper-air wind for a particular day. Note that the airflow is nearly parallel to the contours. These isolines are height contours for the 500-millibar level.

FIGURE 18.10 Comparison between upper-level winds and surface winds showing the effects of friction on airflow. Friction slows surface wind speed, which weakens the Coriolis effect, causing the winds to cross the isobars and move toward the lower pressure.

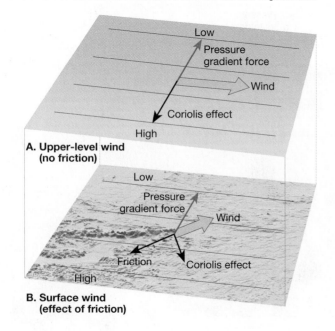

A. Upper-level wind (no friction)

B. Surface wind (effect of friction)

Students Sometimes Ask...

Why doesn't the Coriolis effect cause a baseball to be deflected when you are playing catch?

Over very short distances the Coriolis deflection is too small to be noticed. Nevertheless, in the middle latitudes the Coriolis effect is great enough to potentially affect the outcome of a baseball game. A ball hit a horizontal distance of 100 meters (330 feet) in 4 seconds down the right field line will be deflected 1.5 centimeters (more than 1/2 inch) to the right by the Coriolis effect. This could be just enough to turn a potential home run into a foul ball!

Highs and Lows

Earth's Dynamic Atmosphere
▶ **Air Pressure and Winds**

Among the most common features on any weather map are areas designated as pressure centers. **Lows**, or **cyclones** (*kyklon* = moving in a circle) are centers of low pressure, and **highs**, or **anticyclones**, are high-pressure centers. As **Figure 18.11** illustrates, the pressure decreases from the outer isobars toward the center in a low. In a high, just the opposite is the case—the values of the isobars increase from the outside toward the center. By knowing just a few basic facts about centers of high and low pressure, you can greatly increase your understanding of current and forthcoming weather.

Cyclonic and Anticyclonic Winds

From the preceding section, you learned that the two most significant factors that affect wind are the pressure-gradient force and the Coriolis effect. Winds move from higher pressure to lower pressure and are deflected to the right or left by Earth's rotation. When these controls of airflow are applied to pressure centers in the Northern Hemisphere, the result is that winds blow inward and counterclockwise around a low (**Figure 18.12A**). Around a high, they blow outward and clockwise (Figure 18.11).

In the Southern Hemisphere the Coriolis effect deflects the winds to the left, and therefore winds around a low blow clockwise (Figure 18.12B), and winds around a high move counterclockwise. In either hemisphere, friction causes a net inflow (**convergence**) around a cyclone and a net outflow (**divergence**) around an anticyclone.

Weather Generalizations about Highs and Lows

Rising air is associated with cloud formation and precipitation, whereas subsidence produces clear skies. In this section you will learn how the movement of air can itself create pressure change and hence generate winds. In addition, you will examine the relationship between horizontal and vertical flow and its effect on the weather.

Let us first consider the situation around a surface low-pressure system where the air is spiraling inward. Here the net inward transport of air causes a shrinking of the area occupied by the air mass, a process that is termed *horizontal convergence*. Whenever air converges horizontally, it must pile up, that is, increase in height to allow for the decreased area it now occupies. This generates a taller and therefore heavier air column. Yet a surface low can exist only as long as the column of air above exerts less pressure than that occurring in surrounding regions. We seem to have encountered a paradox—a low-pressure center causes a net accumulation of air, which increases its pressure. Consequently, a surface cyclone should quickly eradicate itself in a manner not unlike what happens when a vacuum-packed can is opened.

You can see that for a surface low to exist for very long, compensation must occur aloft. For example, surface convergence could be maintained if divergence (spreading out) aloft occurred at a rate equal to the inflow below. **Figure 18.13** shows the relationship between surface convergence (inflow) and divergence (outflow) aloft that is needed to maintain a low-pressure center.

Divergence aloft may even exceed surface convergence, thereby resulting in intensified surface inflow and accelerated vertical motion. Thus, divergence aloft can intensify storm centers as well as maintain them. On the other hand, inadequate divergence aloft permits surface flow to "fill" and weaken the accompanying cyclone.

FIGURE 18.11 Cyclonic and anticyclonic winds in the Northern Hemisphere. Arrows show that winds blow into and counterclockwise around a low. By contrast, around a high, winds blow outward and clockwise.

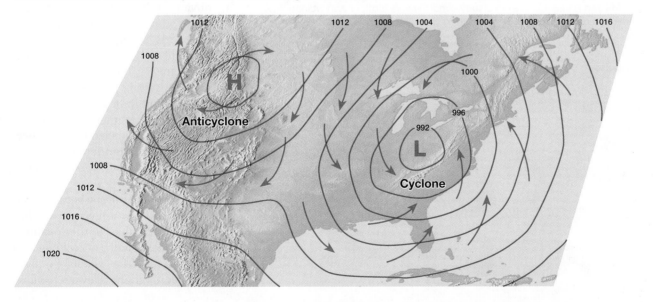

A.   B.

FIGURE 18.12 Cyclonic circulation in the Northern and Southern Hemispheres. The cloud patterns in these images allow us to "see" the circulation pattern in the lower atmosphere. **A.** This satellite image shows a large low-pressure center in the Gulf of Alaska on August 17, 2004. The cloud pattern clearly shows an inward and *counterclockwise* spiral. **B.** This image from March 26, 2004, shows a strong cyclonic storm in the South Atlantic near the coast of Brazil. The cloud pattern reveals an inward and *clockwise* circulation. (NASA images)

Note that surface convergence about a cyclone causes a net upward movement. The rate of this vertical movement is slow, generally less than 1 kilometer per day. Nevertheless, because rising air often results in cloud formation and precipitation, a low-pressure center is generally related to unstable conditions and stormy weather (**Figure 18.14A**).

As often as not, it is divergence aloft that creates a surface low. Spreading out aloft initiates upflow in the atmosphere directly below, eventually working its way to the surface, where inflow is encouraged.

Like their cyclonic counterparts, anticyclones must be maintained from above. Outflow near the surface is accompanied by convergence aloft and general subsidence of the air column (Figure 18.13). Because descending air is compressed and warmed, cloud formation and precipitation are unlikely in an anticyclone. Thus, "fair" weather can usually be expected with the approach of a high-pressure center (Figure 18.14B).

For reasons that should now be obvious, it has been common practice to print on household barometers the words "stormy" at the low-pressure end and "fair" on the high-pressure end. By noting whether the pressure is rising, falling, or steady, we have a good indication of what the forthcoming weather will be. Such a determination, called the **pressure**, or **barometric tendency**, is a very useful aid in short-range weather prediction.

You should now be better able to understand why television weather reporters emphasize the locations and projected paths of cyclones and anticyclones. The "villain" on these weather programs is always the low-

FIGURE 18.13 Airflow associated with surface cyclones and anticyclones. A low, or cyclone, has converging surface winds and rising air causing cloudy conditions. A high, or anticyclone, has diverging surface winds and descending air, which lead to clear skies and fair weather.

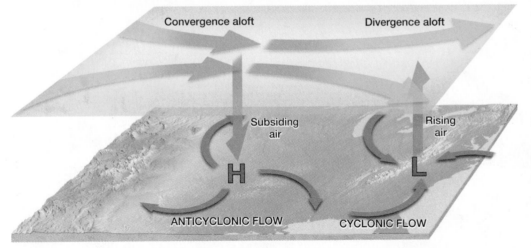

Convergence aloft   Divergence aloft

Subsiding air   Rising air

H   L

ANTICYCLONIC FLOW   CYCLONIC FLOW

A.

B.

FIGURE 18.14 These two photographs illustrate the basic weather generalizations associated with pressure centers. **A**. A rainy day in New York City. Centers of low pressure are frequently associated with cloudy conditions and precipitation. (Photo by David Grossma/Alamy) **B**. By contrast, clear skies and "fair" weather may be expected when an area is under the influence of high pressure. This scene is in New York City's Central Park. (Photo by Kevin C. Downs/Photolibrary)

pressure center, which produces "bad" weather in any season. Lows move in roughly a west-to-east direction across the United States and require a few days to more than a week for the journey. Because their paths can be somewhat erratic, accurate prediction of their migration is difficult, although essential, for short-range forecasting.

Meteorologists must also determine if the flow aloft will intensify an embryo storm or act to suppress its development. Because of the close tie between conditions at the surface and those aloft, a great deal of emphasis has been placed on the importance and understanding of the total atmospheric circulation, particularly in the mid-latitudes. We now examine the workings of Earth's general atmospheric circulation, and then again consider the structure of the cyclone in light of this knowledge.

CONCEPT CHECK 18.4

❶ Describe the weather that usually accompanies a drop in barometric pressure and a rise in barometric pressure.

❷ Sketch a simple diagram (including isobars and wind arrows) showing the winds associated with surface cyclones and anticyclones in both the Northern and Southern hemispheres.

General Circulation of the Atmosphere

As noted, the underlying cause of wind is unequal heating of Earth's surface (see Box 18.1). In tropical regions, more solar radiation is received than is radiated back to space. In polar regions the opposite is true: less solar energy is received than is lost. Attempting to balance these differences, the atmosphere acts as a giant heat-transfer system, moving warm air poleward and cool air equatorward. On a smaller scale, but for the same reason,

ocean currents also contribute to this global heat transfer. The general circulation is very complex. We can, however, develop a general understanding by first considering the circulation that would occur on a nonrotating Earth having a uniform surface. We then modify this system to fit observed patterns.

Circulation on a Nonrotating Earth

On a hypothetical nonrotating planet with a smooth surface of either all land or all water, two large thermally produced cells would form (Figure 18.15). The heated equatorial air would rise until it reached the tropopause, which, acting like a lid, would

FIGURE 18.15 Global circulation on a nonrotating Earth. A simple convection system is produced by unequal heating of the atmosphere.

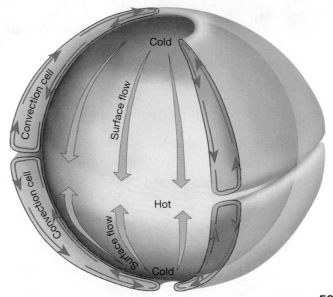

Box 18.1

PEOPLE AND THE ENVIRONMENT

Wind Energy—An Alternative with Potential

Air has mass, and when it moves (i.e. when the wind blows), it contains the energy of that motion—kinetic energy. A portion of that energy can be converted into other forms—mechanical force or electricity—that we can use to perform work (**Figure 18.A**).

Mechanical energy from wind is commonly used for pumping water in rural or remote places. The "farm windmill," still a familiar sight in many rural areas, is an example. Mechanical energy converted from wind can also be used for other purposes, such as sawing logs, grinding grain, and propelling sailboats. By contrast, wind-powered electric turbines generate electricity for homes, businesses, and for sale to utilities.

Today, modem wind turbines are being installed at break-neck speed. In fact, worldwide, in 2010 the installed wind power capacity was expected to exceed 203,000 megawatts, an increase of 28 percent over 2009.* This is equivalent to the total electrical demand of Italy—2 percent of global electricity production. Worldwide, wind energy installations have doubled every 3 years since 2000. The United States is the world's leading producer (22.3%), followed by China (16.3%), Germany (16.2%), and Spain (11.5%). Within the next decade, China is expected to produce the most wind-generated electricity.

Wind speed is a crucial element in determining whether a place is a suitable site for installing a wind-energy facility. Generally a

*One megawatt is enough electricity to supply 250–300 average American households.

FIGURE 18.A Farm windmills such as the one on the left are still familiar sights in some areas. Mechanical energy from wind is commonly used to pump water. (Photo by Mehmet Dilsiz/Shutterstock) The wind turbines on the right are operating near Palm Springs, California. Although California is the state where significant wind-power development got its start, by 2009 it had been surpassed by Texas and Iowa. (Photo by John Mead/Science Photo Library/Photo Researchers, Inc.)

Students Sometimes Ask...

What is the highest wind speed ever recorded?

The highest wind speed recorded at a surface station is 372 kilometers (231 miles) per hour, measured April 12, 1934, at Mount Washington, New Hampshire. Located at an elevation of 1,879 meters (6,262 feet), the observatory atop Mount Washington has an average wind speed of 56 kilometers (35 miles) per hour. Faster wind speeds have undoubtedly occurred on mountain peaks, but no instruments were in place to record them.

deflect the air poleward. Eventually, this upper-level airflow would reach the poles, sink, spread out in all directions at the surface, and move back toward the equator. Once there, it would be reheated and start its journey over again. This hypothetical circulation system has upper-level air flowing poleward and surface air flowing equatorward.

If we add the effect of rotation, this simple convection system will break down into smaller cells. **Figure 18.16** illustrates the three pairs of cells proposed to carry on the task of heat redistribution on a rotating planet. The polar and tropical cells retain the characteristics of the thermally generated convection described earlier. The nature of the mid-latitude circulation is more complex and is discussed in more detail in a later section.

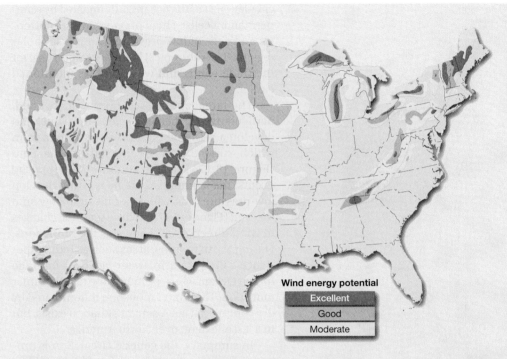

TABLE 18.A Leading States for Wind Energy Potential

Rank	State	Potential*
1	North Dakota	1,210
2	Texas	1,190
3	Kansas	1,070
4	South Dakota	1,030
5	Montana	1,020
6	Nebraska	868
7	Wyoming	747
8	Oklahoma	725
9	Minnesota	657
10	Iowa	551
11	Colorado	481
12	New Mexico	435
13	Idaho	73
14	Michigan	65
15	New York	62
16	Illinois	61
17	California	59
18	Wisconsin	58
19	Maine	56
20	Missouri	52

*The total amount of electricity that could potentially be generated each year, measured in *billions* of kilowatt hours. A typical American home would use several hundred kilowatt hours per month.

Source: U.S. Department of Energy

FIGURE 18.B Wind energy potential for the United States. Large wind energy systems require average wind speeds of about 6 meters per second (13 miles per hour). In the key, "moderate" refers to regions that experience wind speeds of 6.4–6.9 meters per second (m/s), "good" means 7–7.4 m/s, and "excellent" means 7.5 m/s or higher. (After U.S. Department of Energy)

minimum average wind speed of 21 kilometers (13 miles) per hour is necessary for a large-scale wind-power plant to be profitable. A small difference in wind speed results in a large difference in energy production, and therefore a large difference in the cost of the electricity generated. For example, a turbine operating on a site with an average wind speed of 12-mph would generate about 33 percent more electricity than one operating at 11-mph. Also, there is little energy to be harvested at low wind speeds—6-mph winds contain less than one eighth the energy of 12-mph winds.

The United States has tremendous wind energy resources (**Figure 18.B**). In 2010, 36 states had commercial facilities that produced electricity from wind power. The leading states in installed capacity were Texas, number one, followed by Iowa, California, and Minnesota. Despite the fact that California gave birth to the modem U.S. wind industry, 16 states have greater wind potential. As shown in **Table 18.A**, the top five states for wind energy potential include North Dakota, Texas, Kansas, South Dakota, and Montana. Although only a small fraction of U.S. electrical generation currently comes from wind energy, it has been estimated that wind energy potential equals more than twice the total electricity currently consumed.

The U.S. Department of Energy has announced a goal of obtaining 5 percent of U.S. electricity from wind by the year 2020—a goal that seems consistent with the current growth rate of wind energy nationwide. Thus wind-generated electricity appears to be shifting from being an "alternative" to a "mainstream" energy source.

Idealized Global Circulation

Near the equator, the rising air is associated with the pressure zone known as the **equatorial low**—a region marked by abundant precipitation. As the upper-level flow from the equatorial low reaches 20–30 degrees latitude, north or south, it sinks back toward the surface. This subsidence and associated adiabatic heating produce hot, arid conditions. The center of this zone of subsiding dry air is the **subtropical high**, which encircles the globe near 30 degrees latitude, north and south (Figure 18.16). The great deserts of Australia, Arabia, and North Africa exist because of the stable dry conditions associated with the subtropical highs.

At the surface, airflow is outward from the center of the subtropical high. Some of the air travels equatorward and is deflected by the Coriolis effect, producing the reliable **trade winds**. The remainder travels poleward and is also deflected, generating the prevailing **westerlies** of the mid-latitudes. As the westerlies move poleward, they encounter the cool **polar easterlies** in the region of the **subpolar low**. The interaction of these warm and cool winds produces the stormy belt known as the **polar front**. The source region for the variable polar easterlies is the **polar high**. Here, cold polar air is subsiding and spreading equatorward.

In summary, this simplified global circulation is dominated by four pressure zones. The subtropical and polar highs are areas of dry subsiding air that flows outward at the surface, producing the

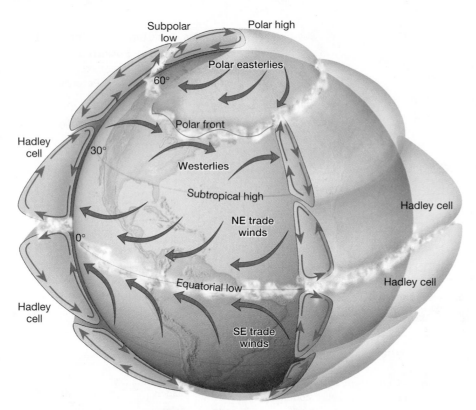

FIGURE 18.16 Idealized global circulation proposed for the three-cell circulation model of a rotating Earth.

prevailing winds. The low-pressure zones of the equatorial and subpolar regions are associated with inward and upward airflow accompanied by clouds and precipitation.

Influence of Continents

Up to this point, we have described the surface pressure and associated winds as continuous belts around Earth. However, the only truly continuous pressure belt is the subpolar low in the Southern Hemisphere. Here the ocean is uninterrupted by landmasses. At other latitudes, particularly in the Northern Hemisphere where landmasses break up the ocean surface, large seasonal temperature differences disrupt the pattern. **Figure 18.17** shows the resulting pressure and wind patterns for January and July. The

---

Students Sometimes Ask...

Does monsoon mean "rainy season"?

No. Regions that experience monsoons typically have both a wet and a dry season. Monsoon refers to a wind system that exhibits a pronounced seasonal reversal in direction. In general, winter is associated with winds that blow predominantly off the continents and produce a dry winter monsoon. By contrast, in summer, warm, moisture-laden air blows from the sea toward the land. Thus, the summer monsoon, which is usually associated with abundant precipitation, is the source of the misconception.

---

circulation over the oceans is dominated by semi-permanent cells of high pressure in the subtropics and cells of low pressure over the subpolar regions. The subtropical highs are responsible for the trade winds and westerlies, as mentioned earlier.

The large landmasses, on the other hand, particularly Asia, become cold in the winter and develop a seasonal high-pressure system from which surface flow is directed off the land (Figure 18.17A). In the summer, the opposite occurs; the landmasses are heated and develop low-pressure cells, which permit air to flow onto the land (Figure 18.17B). These seasonal changes in wind direction are known as the **monsoons**. During warm months, areas such as India experience a flow of warm, water-laden air from the Indian Ocean, which produces the rainy summer monsoon. The winter monsoon is dominated by dry continental air. A similar situation exists, but to a lesser extent, over North America.

In summary, the general circulation is produced by semipermanent cells of high and low pressure over the oceans and is complicated by seasonal pressure changes over land.

The Westerlies

Circulation in the mid-latitudes, the zone of the westerlies, is complex and does not fit the convection system proposed for the tropics. Between about 30 and 60 degrees latitude, the general west-to-east flow is interrupted by migrating cyclones and anticyclones. In the Northern Hemisphere these cells move from west to east around the globe, creating an anticyclonic (clockwise) flow or a cyclonic (counterclockwise) flow in their area of influence. A close correlation exists between the paths taken by these surface pressure systems and the position of the upper-level airflow, indicating that the upper air strongly influences the movement of cyclonic and anticyclonic systems.

Among the most obvious features of the flow aloft are the seasonal changes. The steep temperature gradient across the middle latitudes in the winter months corresponds to a stronger flow aloft. In addition, the polar jet stream fluctuates seasonally such that its average position migrates southward with the approach of winter and northward as summer nears. By midwinter, the jet core may penetrate as far south as central Florida.

Because the paths of low-pressure centers are guided by the flow aloft, we can expect the southern tier of states to experience more of their stormy weather in the winter season. During the hot summer months, the storm track is across the northern states, and some cyclones never leave Canada. The northerly storm track associated with summer applies also to Pacific storms, which move toward Alaska during the warm months, thus producing an extended dry season for much of the West Coast. The number of cyclones generated is seasonal as well, with the largest number occurring in the cooler months when the temperature gradients are greatest. This fact is in agreement with the role of cyclonic storms in the distribution of heat across the mid-latitudes.

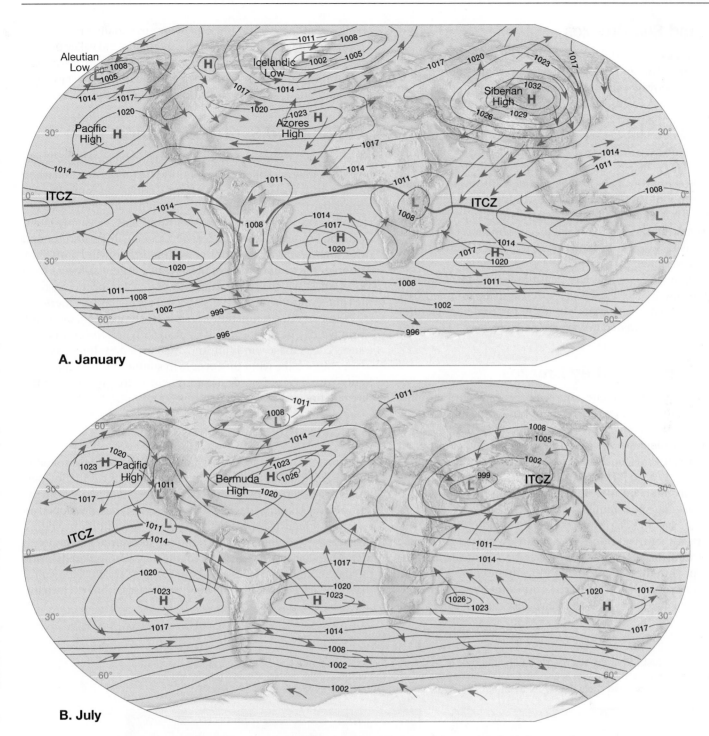

FIGURE 18.17 Average surface pressure in millibars for **A.** January and **B.** July, with associated winds.

CONCEPT CHECK 18.5

1. In which belt of prevailing winds is most of the United States situated?
2. The trade winds diverge from which pressure belt?
3. Which prevailing wind belts converge in the stormy region known as the polar front?
4. Which pressure belt is associated with the equator?
5. Describe the monsoon circulation of India.

Local Winds

Having examined Earth's large-scale circulation, let us turn briefly to winds that influence much smaller areas. Remember that all winds are produced for the same reason: pressure differences that arise because of temperature differences that are caused by unequal heating of Earth's surface. *Local winds* are simply small-scale winds produced by a locally generated pressure gradient. Those described here are caused either by topographic effects or variations in surface composition in the immediate area.

Land and Sea Breezes

In coastal areas during the warm summer months, the land surface is heated more intensely during the daylight hours than is the adjacent body of water (see the section "Land and Water" in Chapter 16). As a result, the air above the land surface heats, expands, and rises, creating an area of lower pressure. A **sea breeze** then develops, because cooler air over the water (higher pressure) moves toward the warmer land (lower pressure) (Figure 18.18A). The sea breeze begins to develop shortly before noon and generally reaches its greatest intensity during the mid- to late afternoon. These relatively cool winds can be a significant moderating influence on afternoon temperatures in coastal areas.

At night, the reverse may take place. The land cools more rapidly than the sea, and the **land breeze** develops (Figure 18.18B). Small-scale sea breezes can also develop along the shores of large lakes. People who live in a city near the Great Lakes, such as Chicago, recognize this lake effect, especially in the summer. They are reminded daily by weather reports of the cool temperatures near the lake as compared to warmer outlying areas.

Mountain and Valley Breezes

A daily wind similar to land and sea breezes occurs in many mountainous regions. During daylight hours, the air along the slopes of the mountains is heated more intensely than the air at the same elevation over the valley floor. Because this warmer air is less dense, it glides up along the slope and generates a **valley breeze** (Figure 18.19A). The occurrence of these daytime upslope breezes can often be identified by the cumulus clouds that develop on adjacent mountain peaks.

After sunset, the pattern may reverse. Rapid radiation cooling along the mountain slopes produces a layer of cooler air next to the ground. Because cool air is denser than warm air, it drains downslope into the valley. Such a movement of air is called a **mountain breeze** (Figure 18.19B). The same type of cool air drainage can occur in places that have very modest slopes. The

result is that the coldest pockets of air are usually found in the lowest spots. Like many other winds, mountain and valley breezes have seasonal preferences. Although valley breezes are most common during the warm season when solar heating is most intense, mountain breezes tend to be more dominant in the cold season.

Chinook and Santa Ana Winds

Warm, dry winds are common on the eastern slopes of the Rockies, where they are called **chinooks**. Such winds are created when air descends the leeward (sheltered) side of a mountain and warms by compression. Because condensation may have occurred as the air ascended the windward side, releasing latent heat, the air descending the leeward slope will be warmer and drier than it was at a similar elevation on the windward side. Although the temperature of these winds is generally less than 10° C (50° F), which is not particularly warm, they occur mostly in the winter and spring when the affected areas may be experiencing below-freezing temperatures. Thus, by comparison, these dry, warm winds often bring a drastic change. When the ground has a snow cover, these winds are known to melt it in short order.

A chinooklike wind that occurs in southern California is the **Santa Ana**. These hot, desiccating winds greatly increase the threat of fire in this already dry area (Figure 18.20).

Country Breeze

One type of local wind, called a **country breeze**, is associated with large urban areas. As the name implies, this circulation pattern is characterized by a light wind blowing into the city from the surrounding countryside. The country breeze is best developed on relatively clear, calm nights. Under these conditions, cities, because they contain massive buildings and surfaces composed of rocklike materials, tend to retain the heat accumulated during the day more than the less built-up outlying areas. The result is that the warm, less dense air over the city rises, which in turn initiates the country-to-city flow.

FIGURE 18.18 Illustration of a sea breeze and a land breeze. **A.** During the daylight hours the air above the land heats and expands, creating an area of lower pressure. Cooler and denser air over the water moves onto the land, generating a sea breeze. **B.** At night the land cools more rapidly than the sea, generating an offshore flow called a *land breeze*.

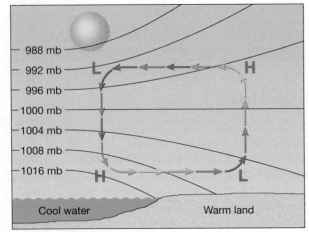

A. Sea breeze

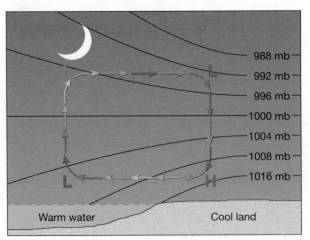

B. Land breeze

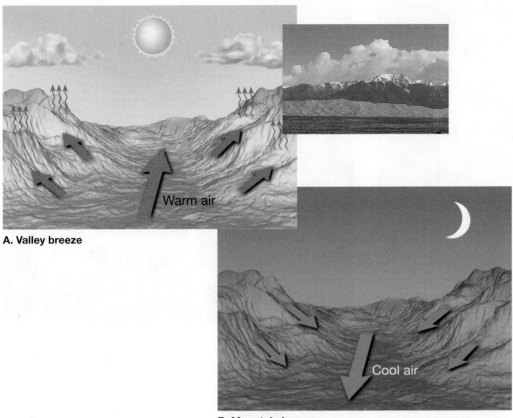

A. Valley breeze

B. Mountain breeze

FIGURE 18.19 Valley and mountain breezes. **A.** Heating during the daylight hours warms the air along the mountain slopes. This warm air rises, generating a valley breeze. In the photo, the occurrence of a daytime valley breeze is identified by cloud development on mountain peaks, sometimes leading to an afternoon rain shower. (Photo by James E. Patterson/James Patterson Collection) **B.** After sunset, cooling of the air near the mountain can result in cool air drainage into the valley, producing the mountain breeze.

One investigation in Toronto showed that heat accumulated within this city created a rural–city pressure difference that was sufficient to cause an inward and counterclockwise circulation centered on the downtown area. One of the unfortunate consequences of the country breeze is that pollutants emitted near the urban perimeter tend to drift in and concentrate near the city's center.

CONCEPT CHECK 18.6

1. What is a local wind?
2. Describe the formation of a sea breeze.
3. Does a land breeze blow toward or away from the shore?
4. During what time of day would you expect to experience a well-developed valley breeze—midnight, late morning, or late afternoon?

Measuring Wind

Two basic wind measurements—direction and speed—are important to the weather observer. One simple device for determining both measurements is the simple *wind sock* that is a common sight at small airports and landing strips (Figure 18.21A). The cone-shaped bag is open at both ends and is free to change position with shifts in wind direction. The degree to which the sock is inflated is an indication of wind speed.

Winds are always labeled by the direction from which they blow. A north wind blows *from* the north *toward* the south, an east wind *from* the east *toward* the west. The instrument most commonly used to determine wind direction is the **wind vane** (Figure 18.21B, upper right). This instrument, a common sight on many buildings, always points *into* the wind. Often the wind

FIGURE 18.20 This satellite image shows strong Santa Ana winds fanning the flames of several large wildfires in southern California on October 27, 2003. These fires scorched more than 740,000 acres and destroyed more than 3,000 homes. (NASA)

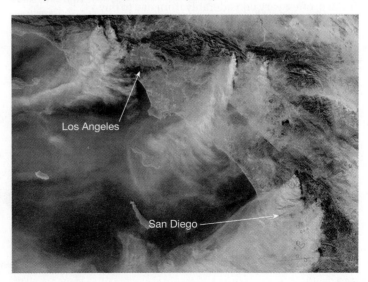

Los Angeles

San Diego

Students Sometimes Ask...

A friend who lives in Colorado talks about "snow eaters." What are they?

"Snow eaters" is a local term for chinooks, the warm, dry winds that descend the eastern slopes of the Rockies. These winds have been known to melt more than a foot of snow in a single day. A chinook that moved through Granville, North Dakota, on February 21, 1918, caused the temperature to rise from −33°F to 50°F, an increase of 83°F!

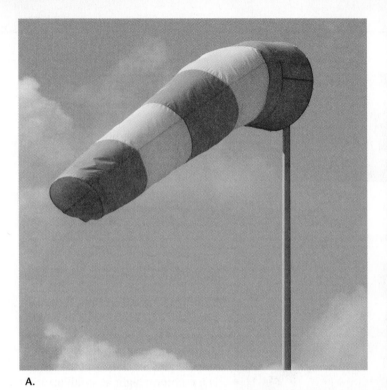

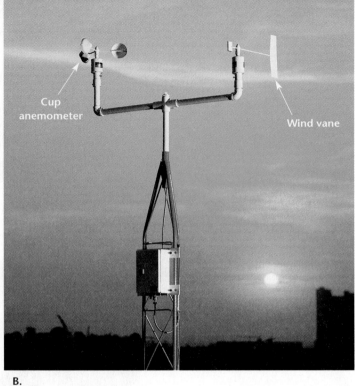

Cup anemometer

Wind vane

A.

B.

FIGURE 18.21 **A.** A *wind sock* is a common device used for determining wind direction and estimating wind speed. They are common sights at small airports and landing strips. (Photo by Lourens Smak/Alamy Images) **B.** Wind vane (right) and cup anemometer (left). The wind vane shows wind direction and the anemometer measures wind speed. (Photo by Belfort Instrument Company)

direction is shown on a dial that is connected to the wind vane. The dial indicates wind direction, either by points of the compass (N, NE, E, SE, etc.) or by a scale of 0° to 360°. On the latter scale, 0° or 360° are both north, 90° is east, 180° is south, and 270° is west.

When the wind consistently blows more often from one direction than from any other, it is called a **prevailing wind**. You may be familiar with the prevailing westerlies that dominate the circulation in the mid-latitudes. In the United States, for example, these winds consistently move the "weather" from west to east across the continent. Embedded within this general eastward flow are cells of high and low pressure with the characteristic clockwise and counterclockwise flow. As a result, the winds associated with the westerlies, as measured at the surface, often vary considerably from day to day and from place to place. By contrast, the direction of airflow associated with the belt of trade winds is much more consistent, as can be seen in Figure 18.22.

Wind speed is commonly measured using a **cup anemometer** (*anemo* = wind, *metron* = measuring instrument) (Figure 18.21B, upper left). The wind speed is read from a dial much like the speedometer of an automobile. Places where winds are steady and speeds are relatively high are potential sites for tapping wind energy.

By knowing the locations of cyclones and anticyclones in relation to where you are, you can predict the changes in wind direction that will be experienced as a pressure center moves past. Because changes in wind direction often bring

changes in temperature and moisture conditions, the ability to predict the winds can be very useful. In the Midwest, for example, a north wind may bring cool, dry air from Canada, whereas a south wind may bring warm, humid air from the Gulf of Mexico.

Recall that about 70 percent of Earth's surface is covered by the ocean, where conventional methods of gathering wind data are seldom possible. Ocean buoys and ships at sea provide very limited coverage. However, since the 1990s, weather forecasts have improved significantly due to the availability of satellite-derived

FIGURE 18.22 Wind roses showing the percentage of time airflow is coming from various directions. **A.** Wind frequency for the winter in the eastern United States. **B.** Wind frequency for the winter in northern Australia. Note the reliability of the southeast trades in Australia as compared to the westerlies in the eastern United States. (Data from G. T. Trewartha)

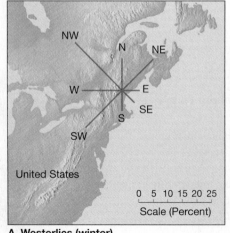

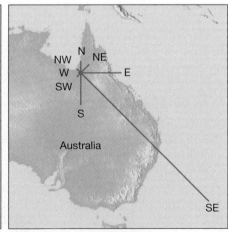

A. Westerlies (winter)

B. Southeast Trades (winter)

542

wind data. One way that wind speed and direction can be established is by using satellite images to track cloud movements.

CONCEPT CHECK 18.7

❶ What are the two basic wind measurements? What instruments are used to make these measurements?

❷ *From* what direction does a northeast wind blow? *Toward* what direction does a south wind blow?

El Niño and La Niña

As can be seen in **Figure 18.23A**, the cold Peruvian current flows equatorward along the coast of Ecuador and Peru. This flow encourages upwelling of cold nutrient-filled waters that serve as the primary food source for millions of fish, particularly anchovies. Near the end of each year, however, a warm current that flows southward along the coasts of Ecuador and Peru replaces the cold Peruvian current. During the 19th century the local residents named this warm countercurrent El Niño ("the child") after the Christ child because it usually appeared during the Christmas

FIGURE 18.23 The relationship between the Southern Oscillation and El Niño is illustrated on these simplified maps. **A.** Normally, the trade winds and strong equatorial currents flow toward the west. At the same time, the strong Peruvian current causes upwelling of cold water along the west coast of South America. **B.** When the Southern Oscillation occurs, the pressure over the eastern and western Pacific flip-flops. This causes the trade winds to diminish, leading to an eastward movement of warm water along the equator. As a result, the surface waters of the central and eastern Pacific warm, with far-reaching consequences to weather patterns.

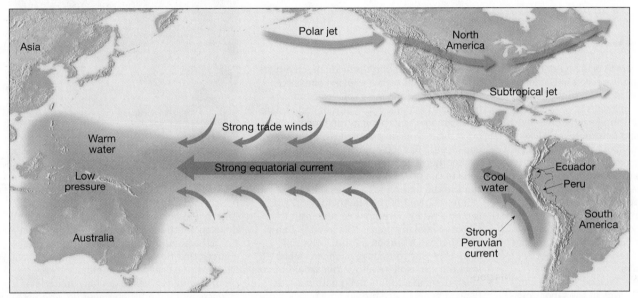

A. Normal conditions

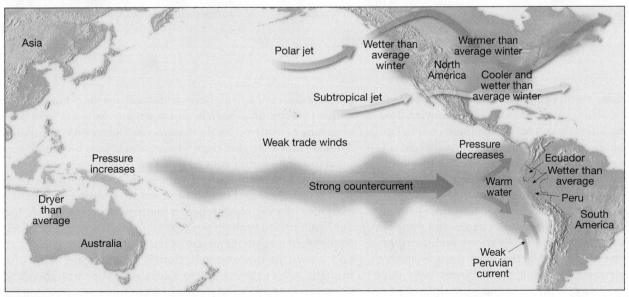

B. El Niño

PROFESSIONAL PROFILE

Sally Benson: Climate and Energy Scientist

Sally Benson has spent her career researching solutions to the most pressing environmental problems of our time. In the mid-1970s, as a young scientist with Lawrence Berkeley National Laboratory, she tackled the first oil shortage by investigating ways to harness the power of geothermal energy.

"I get to do important work . . . and have fun while I do it."

Ten years later, she was elbow-deep in the mud of Kesterson Reservoir in California's Central Valley. Irrigation runoff had caused selenium from local soil to accumulate in the water, causing local birds to hatch chicks with horrifying deformities. Benson's studies made her realize microbes could be environmentally friendly tools to clean up other sites with toxic metal contamination.

"The experience was so positive because I was doing this cutting-edge science at the same time regulators had to make a decision about how to clean up the site. I got an idea of the impact that science can make, and how research could get results," Benson says.

By the mid-1990's, Benson was directing all Earth science research at the laboratory. "I talked with many people, read a lot, and came to the conclusion that climate change

was the most significant issue facing the world." Benson organized research programs that developed regional models of climate change to help residents plan for droughts and temperature shifts, and studied the carbon cycle of the oceans, among other projects.

In 2007, Benson was appointed director of the Global Climate and Energy Project at Stanford University, which seeks to develop energy sources that release fewer greenhouse gases. "Energy efficiency in lighting, heating, and cooling systems, and autos, makes all the sense in the world. But at the end of the day, we need to do a lot more than that. If our current understanding is correct, we need to cut overall emissions by 80 percent of today's levels," Benson states.

Benson sees many promising ways to achieve that goal. One is renewable energy.

"Today's biofuels don't provide much advantage in terms of carbon dioxide emissions. But alternatives such as cellulosic ethanol (producing ethanol with plant fibers), where there are low emissions in the process of growing and making them, are incredibly important," Benson opines.

Another is to continue using some fossil fuels to run power plants and ships, but to capture the greenhouse gases they emit. "More than half of the electricity worldwide is produced by burning coal, a plentiful resource in many places. We need to find a way to make it carbon neutral," Benson says. One way to remove such emissions from the atmosphere altogether would be to inject them in aquifers deep within Earth. Benson herself has studied how to use technology developed by the oil and gas industry to select sites where the rock offers a

good seal, inject the gas, and monitor the area for leaks.

"I get to do important work solving critical problems, get to be outdoors, do experiments at scale, and have fun while I do it. Interacting with the huge number of people impacted by these issues makes the Earth sciences very rewarding."

—Kathleen Wong

Professor Sally Benson is Executive Director of Stanford University's Global Climate and Energy Project, a $225 million effort to develop energy resources that are not harmful to the environment, especially energy sources that release fewer greenhouse gases. Dr. Benson has a multidisciplinary background with degrees in geology, materials science, and minerals engineering, and with applied research in hydrology and reservoir engineering. (Courtesy Dr. Sally Benson)

season. Normally, these warm countercurrents last for at most a few weeks when they again give way to the cold Peruvian flow. However, at irregular intervals of 3 to 7 years, these countercurrents become unusually strong and replace normally cold offshore waters with warm equatorial waters (Figure 18.23B). Today, scientists use the term **El Niño** for these episodes of ocean warming that affect the eastern tropical Pacific.

The onset of El Niño is marked by abnormal weather patterns that drastically affect the economies of Ecuador and Peru. As shown in Figure 18.23B, these unusually strong undercurrents amass large quantities of warm water that block the upwelling of colder, nutrient-filled water. As a result, the anchovies starve, devastating the fishing industry. At the same time, some inland areas that are normally arid receive an abnormal amount of rain. Here,

pastures and cotton fields have yields far above the average. These climatic fluctuations have been known for years, but they were originally considered local phenomena. Today, we know that El Niño is part of the global circulation and affects the weather at great distances from Peru and Ecuador.

Two of the strongest El Niño events on record occurred between 1982–83 and 1997–98 and were responsible for weather extremes of a variety of types in many parts of the world. The 1997–98 El Niño brought ferocious storms that struck the California coast, causing unprecedented beach erosion, landslides, and floods. In the southern United States, heavy rains also brought floods to Texas and the Gulf states. The same energized jet stream that produced storms in the South, upon reaching the Atlantic, sheared off the northern portions of hurricanes, destroying the storms. It was one of the quietest Atlantic hurricane seasons in years.

Major El Niño events, such as the one in 1997–98, are intimately related to the large-scale atmospheric circulation. Each time an El Niño occurs, the barometric pressure drops over large portions of the southeastern Pacific, whereas in the western Pacific, near Indonesia and northern Australia, the pressure rises (Figure 18.24). Then, as a major El Niño event comes to an end, the pressure difference between these two regions swings back in the opposite direction. This seesaw pattern of atmospheric pressure between the eastern and western Pacific is called the **Southern Oscillation**. It is an inseparable part of the El Niño warmings that occur in the central and eastern Pacific every 3 to 7 years. Therefore, this phenomenon is often termed El Niño/Southern Oscillation, or ENSO for short.

Winds in the lower atmosphere are the link between the pressure change associated with the Southern Oscillation and the extensive ocean warming associated with El Niño. During a typical year, the trade winds converge near the equator and flow westward toward Indonesia (Figure 18.24A). This steady westward flow creates a warm surface current that moves from east to west along the equator. The result is a "piling up" of a thick layer of warm surface water that produces higher sea levels (by 30 centimeters) in the western Pacific. Meanwhile, the eastern Pacific is characterized by a strong Peruvian current, upwelling of cold water, and lower sea levels.

Then when the Southern Oscillation occurs, the normal situation just described changes dramatically. Barometric pressure rises in the Indonesian region, causing the pressure gradient along the equator to weaken or even to reverse. As a consequence, the once-steady trade winds diminish and may even change direction. This reversal creates a major change in the equatorial current system, with warm water flowing eastward (Figure 18.24B). With time, water temperatures in the central and eastern Pacific increase and sea level in the region rises. This eastward shift of the warmest surface water marks the onset of El Niño and sets up changes in atmospheric circulation that affect areas far outside the tropical Pacific.

When an El Niño began in the summer of 1997, forecasters predicted that the pool of warm water over the Pacific would displace the paths of both the subtropical and mid-latitude jet streams, which steer weather systems across North America (see Figure 18.23). As predicted, the subtropical jet brought rain to the Gulf Coast, where Tampa, Florida, received more than three times its normal winter precipitation. Furthermore, the mid-latitude jet pumped warm air far north into the continent. As a result, winter temperatures west of the Rockies were significantly above normal.

The effects of El Niño are somewhat variable depending in part on the temperatures and size of the warm pools. Nevertheless, some locales appear to be affected more consistently. In particular, during most El Niños, warmer-than-normal winters occur in the northern United States and Canada. In addition, normally arid portions of Peru and Ecuador, as well as the eastern United States, experience wet conditions. By contrast, drought conditions are generally observed in Indonesia, Australia, and the Philippines. One major benefit of the circulation associated with El Niño is a suppression of the number of Atlantic hurricanes.

The opposite of El Niño is an atmospheric phenomenon known as **La Niña**. Once thought to be the normal conditions that occur between two El Niño events, meteorologists now consider La Niña an important atmospheric phenomenon in its own right. Researchers have come to recognize that when surface temperatures in the eastern Pacific are *colder than average*, a La Niña event is triggered that has a distinctive set of weather patterns. A typical La Niña winter blows colder than normal air over the Pacific Northwest and the northern Great Plains while warming much of the rest of the United States. Furthermore, greater precipitation is expected in the Northwest. During the La Niña winter of 1998–99, a world-record snowfall for one season occurred in Washington State. Another La Niña impact is greater hurricane activity. A recent study concluded that the cost of hurricane damages in the United States is 20 times greater in La Niña years as compared to El Niño years.

FIGURE 18.24 Simplified illustration of the see-saw pattern of atmospheric pressure between the eastern and western pacific, called the *Southern Oscillation*. **A.** During average years, high pressure over the eastern pacific causes surface winds and warm equatorial waters to flow westward. The result is a pileup of warm water in the western Pacific, which promotes the lowering of pressure. **B.** An el Niño event begins as surface pressure increases in the western Pacific and decreases in the eastern Pacific. This air pressure reversal weakens, or may even reverse the trade winds, and results in an eastward movement of the warm waters that had accumulated in the western Pacific.

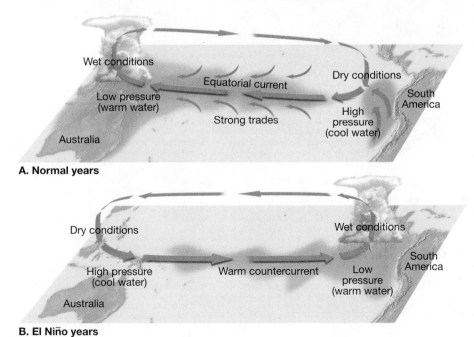

A. Normal years

B. El Niño years

CONCEPT CHECK 18.8

❶ How is La Niña different from El Niño?

❷ Describe the relationship between the Southern Oscillation and a major El Niño event.

Global Distribution of Precipitation

A casual glance at Figure 18.25 shows a relatively complex pattern for the distribution of precipitation. Although the map appears to be complicated, the general features of the map can be explained by applying our knowledge of global winds and pressure systems.

In general, regions influenced by high pressure, with its associated subsidence and diverging winds, experience relatively dry conditions. On the other hand, regions under the influence of low pressure and its converging winds and ascending air receive ample precipitation. This pattern is illustrated by noting that the tropical region dominated by the equatorial low is the rainiest region on Earth (Figure 18.26). It includes the rain forests of the Amazon basin in South America and the Congo basin in Africa. Here the warm, humid trade winds converge to yield abundant rainfall throughout the year. By way of contrast, areas dominated by the subtropical high-pressure cells clearly receive much smaller amounts of precipitation. These are regions of extensive deserts. In the Northern Hemisphere the largest is the Sahara. Examples in the Southern Hemisphere include the Kalahari in southern Africa and the dry lands of Australia.

If Earth's pressure and wind belts were the only factors controlling precipitation distribution, the pattern shown in Figure 18.25 would be simpler. The inherent nature of the air is also an important factor in determining precipitation potential. Because cold air has a low capacity for moisture compared with warm air, we would expect a latitudinal variation in precipitation, with low latitudes receiving the greatest amounts of precipitation and high latitudes receiving the smallest amounts. Figure 18.25 indeed reveals heavy rainfall in equatorial regions and meager precipitation in high-latitude areas. Recall that the dry region in the warm subtropics is explained by the presence of the subtropical high.

In addition to latitudinal variations in precipitation, the distribution of land and water complicates the precipitation pattern. Large landmasses in the middle latitudes commonly experience decreased precipitation toward their interiors. For example, central North America and central Eurasia receive considerably less precipitation than do coastal regions at the same latitude. Furthermore, the effects of mountain barriers alter the idealized precipitation patterns we would expect solely from global wind and pressure systems. Windward mountain slopes receive abundant rainfall resulting from orographic lifting, whereas leeward slopes and adjacent lowlands are usually deficient in moisture.

CONCEPT CHECK 18.9

❶ With which global pressure belt are the rain forests of Africa's Congo Basin associated? Which pressure system is linked to the Sahara Desert?

❷ Other than Earth's pressure and wind belts, list two other factors that exert a significant influence on the global distribution of precipitation.

FIGURE 18.25 Average annual precipitation. (**Note**: 400 mm is equal to 15.6 inches.)

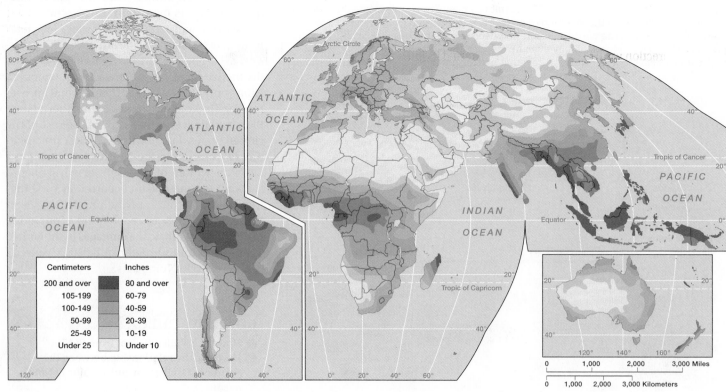

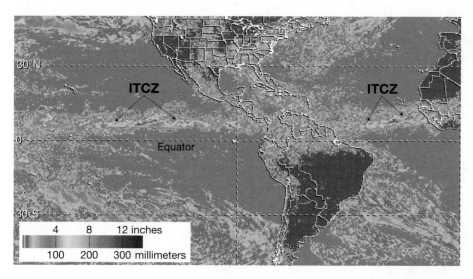

FIGURE 18.26 The Intertropical Convergence Zone (ITCZ) is associated with the pressure zone known as the *equatorial low*. In this satellite image, produced with data from the *Tropical Rainfall Measuring Mission (TRMM)*, the ITCZ is seen as a band of heavy rainfall shown in reds and yellows, which extends east–west just north of the equator. (Courtesy of NOAA)

GIVE IT SOME THOUGHT

1. Mercury is 13.5 times denser (heavier) than water. If you built a barometer using water rather than mercury, how tall (in inches) would it have to be to record standard sea-level pressure?

2. The accompanying map shows the distribution of air pressure at 4:00 P.M. CDT on May 24, 2010. Within which one of these states should wind speeds be greatest—New York, Texas, Illinois, Kansas, Idaho, or California? What guided your selection?

3. If you wanted to erect wind turbines to generate electricity, would you search for a location that typically experiences a strong pressure gradient or a weak pressure gradient? Explain.

4. If Earth did not rotate on its axis and was completely covered with water, in what direction would a sailboat move if it started its journey in the mid-latitudes of the Northern Hemisphere? How did you figure this out?

5. Refer to the accompanying satellite image of a hurricane. Is the storm located in the Northern Hemisphere or the Southern Hemisphere? How were you able to determine your answer?

6. If divergence in the jet stream above a surface low-pressure center exceeds convergence at the surface, will surface winds likely get stronger or weaker? Explain.

7. You and a friend are watching TV on a rainy day when the weather reporter states that, "The barometric pressure is 28.8 inches and rising." Hearing this, you say, "It looks like fair weather is on its way." Your friend responds with the following questions. "I thought air pressure had something to do with the weight of air. How does inches relate to weight? And, why do you think the weather is going to improve?" How would you respond to your friend's queries?

8. If you live in the Northern Hemisphere and are directly west of the center of a cyclone, what is the probable wind direction? What if you were west of an anti-cyclone?

9. It is late afternoon on a warm summer day and you are enjoying some time at the beach. Until the last hour or two, winds were calm. Then a breeze began to develop. Is it more likely a cool breeze from the water or a warm breeze from the adjacent land area? Explain.

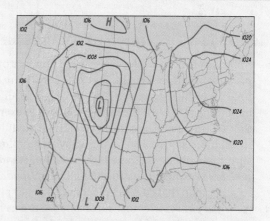

(Photo from NASA)

10. When designing an airport it is important to have the planes take off into the wind. Refer to the accompanying wind rose and discuss the orientation of the runway and the direction planes would travel when they took off. *Bonus: Where on Earth would you find a wind rose like this?*

11. The accompanying maps of Africa show the distribution of precipitation for July and January. Which map represents July and which represents January? How were you able to figure this out?

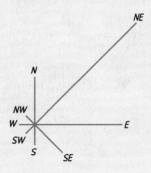

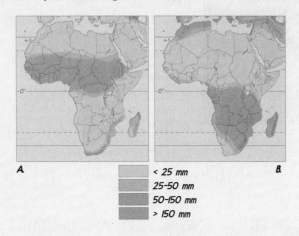

< 25 mm

25–50 mm

50–150 mm

> 150 mm

In Review   Chapter 18 Air Pressure and Wind

- *Air has weight.* At sea level it exerts a pressure of 1 kilogram per square centimeter (14.7 pounds per square inch). *Air pressure* is the force exerted by the weight of air above. With increasing altitude there is less air above to exert a force, and thus air pressure decreases with altitude, rapidly at first, then much more slowly. The unit used by meteorologists to measure atmospheric pressure is the *millibar. Standard sea-level pressure* is expressed as 1013.2 millibars. *Isobars* are lines on a weather map that connect places of equal air pressure.

- A *mercury barometer* measures air pressure using a column of mercury in a glass tube that is sealed at one end and inverted in a dish of mercury. As air pressure increases, the mercury in the tube rises; conversely, when air pressure decreases, so does the height of the column of mercury. A mercury barometer measures atmospheric pressure in *inches of mercury*, the height of the column of mercury in the barometer. Standard atmospheric pressure at sea level equals 29.92 inches of mercury. *Aneroid* (without liquid) *barometers* consist of partially evacuated metal chambers that compress as air pressure increases and expand as pressure decreases.

- *Wind* is the horizontal flow of air from areas of higher pressure toward areas of lower pressure. Winds are controlled by the following combination of forces: (1) the *pressure-gradient force* (amount of pressure change over a given distance); (2) the *Coriolis effect* (deflective effect of Earth's rotation to the right in the Northern Hemisphere and to the left in the Southern Hemisphere); and (3) *friction* with Earth's surface (slows the movement of air and alters wind direction).

- Upper-air winds, called *geostrophic winds*, blow parallel to the isobars and reflect a balance between the pressure-gradient force and the Coriolis effect. Upper-air winds are faster than surface winds because friction is greatly reduced aloft. Friction slows surface winds, which in turn reduces the Coriolis effect. The result is air movement at an angle across the isobars toward the area of lower pressure.

- The two types of pressure centers are (1) *cyclones*, or *lows* (centers of low pressure), and (2) *anticyclones*, or *highs* (high-pressure centers). In the Northern Hemisphere, winds around a low (cyclone) are counterclockwise and inward. Around a high (anticyclone), winds are clockwise and outward. In the Southern Hemisphere, the Coriolis effect causes winds to move clockwise around a low and counterclockwise around a high. Because air rises and cools adiabatically in low-pressure centers, cloudy conditions and precipitation are often associated with their passage. In high-pressure centers, descending air is compressed and warmed; therefore, cloud formation and precipitation are unlikely, and "fair" weather is usually expected.

- Earth's *global pressure zones* include the *equatorial low*, *subtropical high*, *subpolar low*, and *polar high*. The *global surface winds* associated with these pressure zones are the *trade winds*, *westerlies*, and *polar easterlies*.

- Particularly in the Northern Hemisphere, large seasonal temperature differences over continents disrupt the idealized, or zonal, global patterns of pressure and wind. In winter, large, cold landmasses develop a seasonal high-pressure system from which surface airflow is directed off the land. In summer, landmasses are heated and low pressure develops over them, which permits air to flow onto the land. The seasonal changes in wind direction are known as *monsoons*.

- In the middle latitudes, between 30 and 60 degrees latitude, the general west-to-east flow of the westerlies is interrupted by migrating cyclones and anticyclones. The paths taken by these pressure systems are closely related to upper-level airflow and the polar *jet stream*. The average position of the polar jet stream, and hence the paths followed by cyclones, migrates equatorward with the approach of winter and poleward as summer nears.

● *Local winds* are small-scale winds produced by a locally generated pressure gradient. Local winds include *sea* and *land breezes* (formed along a coast because of daily pressure differences caused by the differential heating of land and water); *valley* and *mountain breezes* (daily wind similar to sea and land breezes except in a mountainous area where the air along slopes heats differently from the air at the same elevation over the valley floor); and *chinook* and *Santa Ana winds* (warm, dry winds created when air descends the leeward side of a mountain and warms by compression).

● The two basic wind measurements are *direction* and *speed*. Winds are always labeled by the direction *from* which they blow. Wind direction is measured with a *wind vane*, and wind speed is measured using a *cup anemometer*.

● *El Niño* is the name given to the periodic warming of the ocean that occurs in the central and eastern Pacific. It is associated with periods when a weakened pressure gradient causes the trade winds to diminish. A major El Niño event triggers extreme weather in many parts of the world. When surface temperatures in the eastern Pacific are colder than average, a *La Niña* event is triggered. A typical La Niña winter blows colder-than-normal air over the Pacific Northwest and the northern Great Plains while warming much of the rest of the United States.

● The global distribution of precipitation is strongly influenced by the global pattern of air pressure and wind, latitude, and distribution of land and water.

Key Terms

air pressure (p. 526)
aneroid barometer (p. 528)
anticyclone (p. 533)
barograph (p. 528)
barometric tendency (p. 534)
chinook (p. 540)
convergence (p. 533)
Coriolis effect (p. 530)
country breeze (p. 540)
cup anemometer (p. 542)
cyclone (p. 533)
divergence (p. 533)
El Niño (p. 544)
equatorial low (p. 537)

geostrophic wind (p. 530)
high (p. 533)
isobar (p. 529)
jet stream (p. 531)
land breeze (p. 540)
La Niña (p. 545)
low (p. 533)
mercury barometer (p. 527)
monsoon (p. 538)
mountain breeze (p. 540)
polar easterlies (p. 537)
polar front (p. 537)
polar high (p. 537)

pressure gradient (p. 529)
pressure tendency (p. 534)
prevailing wind (p. 542)
Santa Ana (p. 540)
sea breeze (p. 540)
Southern Oscillation (p. 545)
subpolar low (p. 537)
subtropical high (p. 537)
trade winds (p. 537)
valley breeze (p. 540)
westerlies (p. 537)
wind (p. 529)
wind vane (p. 541)

Examining the Earth System

1. Examine the image of Africa in Figure 1.9B (p. 13) and pick out the region dominated by the equatorial low and the areas influenced by the subtropical highs in each hemisphere. What clue(s) did you use? Speculate on the differences in the biosphere between the regions dominated by high pressure and the zone influenced by low pressure.

2. How are global winds related to surface ocean currents? (Try comparing Figure 18.17, p. 539, with Figure 15.2,

p. 428.) What is the *ultimate* source of energy that drives both of these circulations?

3. Winds and ocean currents change in the tropical Pacific during an El Niño event. How might this impact the biosphere and geosphere in Peru and Ecuador? How about in Indonesia? (One useful Web site that deals with El Niño is NOAA's *El Niño Theme Page* at http://www.pmel.noaa.gov/tao/elnino/nino-home.html).

Mastering Geology

Looking for additional review and test prep materials? Visit the Self Study area in **www.masteringgeology.com** to find practice quizzes, study tools, and multimedia that will aid in your understanding of this chapter's content. In **MasteringGeology™** you will find:

● **GEODe: Earth Science: An interactive visual walkthrough of key concepts**

● **Geoscience Animation Library: More than 100 animations illuminating many difficult-to-understand Earth science concepts**

● **In The News RSS Feeds: Current Earth science events and news articles are pulled into the site with assessment**

● **Pearson eText**

● **Optional Self Study Quizzes**

● **Web Links**

● **Glossary**

● **Flashcards**

19

Weather Patterns and Severe Storms

This tornado near Mulvane, Kansas, was one of several to strike south-central Kansas on June 12, 2004.

(Photo by Eric Nguyen/CORBIS)

ornadoes and hurricanes rank among nature's most destructive forces. Each spring, newspapers report the death and destruction left in the wake of a band of tornadoes. During late summer and fall we hear occasional news reports about hurricanes. Storms with names such as Katrina, Rita, Charley, and Ike make front-page headlines (Figure 19.1). Thunderstorms, although less intense and far more common than tornadoes and hurricanes, are also part of our discussion on severe weather in this chapter. Before looking at violent weather, however, we will study those atmospheric phenomena that most often affect our day-to-day weather: air masses, fronts, and traveling middle-latitude cyclones. Here we will see the interplay of the elements of weather discussed in Chapters 16, 17, and 18.

FOCUS ON CONCEPTS

To assist you in learning the important concepts in this chapter, focus on the following questions:

- What is an air mass?
- How are air masses classified? What is the general weather associated with each air-mass type?
- What are fronts? How do warm fronts and cold fronts differ?
- What are the primary weather producers in the middle latitudes? What weather conditions are associated with these systems?
- What atmospheric conditions are associated with the formation of thunderstorms, tornadoes, and hurricanes?
- How is the profile of a tornado different from the profile of a hurricane?

Air Masses

Earth's Dynamic Atmosphere
▶ Basic Weather Patterns

For many people who live in the middle latitudes, which includes much of the United States, summer heat waves and winter cold spells are familiar experiences. In the first instance, several days of high temperatures and oppressive humidities may finally end when a series of thunderstorms pass through the area, followed by a few days of relatively cool relief. By contrast, the clear skies that often accompany a span of frigid subzero days may be replaced by thick gray clouds and a period of snow as temperatures rise to levels that seem mild when compared to those that existed just a day earlier. In both examples, what was experienced was a period of generally constant weather conditions followed by a relatively short period of change, and then the reestablishment of a new set of weather conditions that remained for perhaps several days before changing again.

What Is an Air Mass?

The weather patterns just described result from movements of large bodies of air, called air masses. An **air mass**, as the term implies, is an immense body of air, usually 1,600 kilometers (1,000 miles) or more across and perhaps several kilometers thick, that is characterized by a similarity of temperature and moisture at

any given altitude. When this air moves out of its region of origin, it will carry these temperatures and moisture conditions with it, eventually affecting a large portion of a continent.

An excellent example of the influence of an air mass is illustrated in Figure 19.2, which shows a cold, dry mass from northern Canada moving southward. With a beginning temperature of $-46°$ C ($-51°$ F) the air mass warms to $-33°$ C ($-27°$ F) by the time it reaches Winnipeg. It continues to warm as it moves southward through the Great Plains and into Mexico. Throughout its southward journey, the air mass becomes warmer. But it also brings some of the coldest weather of the winter to the places in its path. Thus, the air mass is modified, but it also modifies the weather in the areas over which it moves.

The horizontal uniformity of an air mass is not complete because it may extend through 20 degrees or more of latitude and cover hundreds of thousands to millions of square kilometers. Consequently, small differences in temperature and humidity are to be expected from one point to another at the same level. Still, the differences observed within an air mass are small in comparison to the rapid changes experienced across air-mass boundaries.

Since it may take several days for an air mass to move across an area, the region under its influence will probably experience fairly constant weather, a situation called **air-mass weather**. Certainly, some day-to-day variations may occur, but the events will be very unlike those in an adjacent air mass.

The air-mass concept is an important one because it is closely related to the study of atmospheric disturbances. Many of the dis-

FIGURE 19.1 Hurricane Rita became a category 5 hurricane late on September 21, 2005, with sustained winds of 275 kilometers (170 miles) per hour. It was the third most powerful Atlantic basin storm ever measured. When this image was taken about mid-day on September 22, the storm was slightly weaker.

turbances in the middle latitudes originate along *fronts*, the boundary zones that separate unlike air masses.

Source Regions

When a portion of the lower atmosphere moves slowly or stagnates over a relatively uniform surface, the air will assume the distinguishing features of that area, particularly with regard to temperature and moisture conditions.

The area where an air mass acquires its characteristic properties of temperature and moisture is called its **source region**. The source regions that produce air masses influencing North America are shown in **Figure 19.3**.

Air masses are classified according to their source region. **Polar (P)** and **arctic (A) air masses** originate in high latitudes toward Earth's poles, whereas those that form in low latitudes are called **tropical (T) air masses**. The designation *polar*, *arctic*, or *tropical* gives an indication of the temperature characteristics of an air mass. *Polar* and *arctic* indicate cold, and *tropical* indicates warm.

In addition, air masses are classified according to the nature of the surface in the source region. **Continental (c) air masses** form over land, and **maritime (m) air masses** originate over water. The designation *continental* or *maritime* thus suggests the moisture characteristics of the air mass. Continental air is likely to be dry, and maritime air, humid.

The basic types of air masses according to this scheme of classification are continental polar (cP), continental arctic (cA), continental tropical (cT), maritime polar (mP), and maritime tropical (mT).

Weather Associated with Air Masses

Continental polar and maritime tropical air masses influence the weather of North America most, especially east of the Rocky Mountains. Continental polar air masses originate in northern Canada, interior Alaska, and the Arctic areas that are uniformly cold and dry in winter and cool and dry in summer. In winter, an invasion of continental polar air brings the clear skies and cold temperatures we associate with a cold wave as it moves southward from Canada into the United States. In summer, this air mass may bring a few days of cooling relief.

Although cP air masses are not, as a rule, associated with heavy precipitation, those that cross the Great Lakes during late autumn and winter sometimes bring snow to the leeward shores. These localized storms often form when the surface weather map indicates no apparent cause for a snowstorm to occur. Known as

FIGURE 19.2 As this frigid Canadian air mass moved southward, it brought some of the coldest weather of the winter. (After Tom L. McKnight, *Physical Geography*, 5th ed. Upper Saddle River, NJ: Prentice Hall, 1996, p. 174)

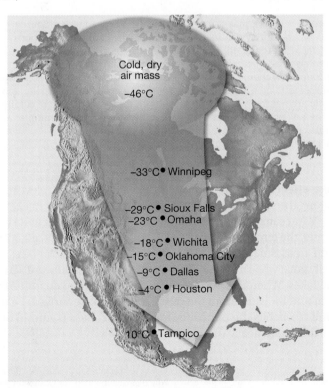

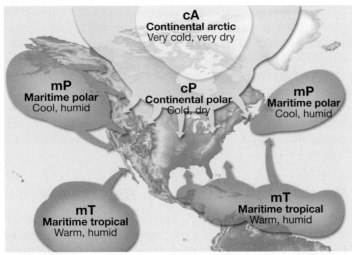

A. Winter pattern

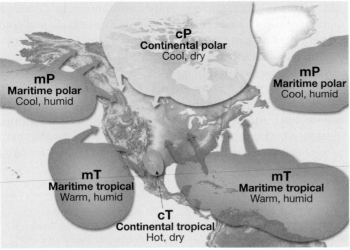

B. Summer pattern

FIGURE 19.3 Air-mass source regions for North America. Source regions are largely confined to subtropical and subpolar locations. The fact that the middle latitudes are the site where cold and warm air masses clash, often because the converging winds of a traveling cyclone draw them together, means that this zone lacks the conditions necessary to be a source region. The differences between polar and arctic are relatively small and serve to indicate the degree of coldness of the respective air masses in winter. By comparing the summer and winter maps, it is clear that the extent and temperature characteristics for various air masses fluctuate.

lake-effect snows, they make Buffalo and Rochester, New York, among the snowiest cities in the United States (**Figure 19.4**).

What causes lake-effect snow? During late autumn and early winter, the temperature contrast between the lakes and adjacent land areas can be large.[44] The temperature contrast can be especially great when a very cold cP air mass pushes southward across the lakes. When this occurs, the air acquires large quantities of heat and moisture from the relatively warm lake surface. By the time it reaches the opposite shore, the air mass is humid and unstable, and heavy snow showers are likely.

[44]Recall that land cools more rapidly and to lower temperatures than water. See the discussion of "Land and Water" in the section "Why Temperatures Vary: The Controls of Temperature" in Chapter 16.

Maritime tropical air masses affecting North America most often originate over the warm waters of the Gulf of Mexico, the Caribbean Sea, or the adjacent Atlantic Ocean. As you might expect, these air masses are warm, moisture-laden, and usually unstable. Maritime tropical air is the source of much, if not most, of the precipitation in the eastern two-thirds of the United States. In summer, when an mT air mass invades the central and eastern United States, and occasionally southern Canada, it brings the high temperatures and oppressive humidity typically associated with its source region.

Of the two remaining air masses, maritime polar and continental tropical, the latter has the least influence on the weather of North America. Hot, dry continental tropical air, originating in the Southwest and Mexico during the summer, only occasionally affects the weather outside its source region.

During the winter, maritime polar air masses coming from the North Pacific often originate as continental polar air masses in Siberia. The cold, dry cP air is transformed into relatively mild, humid, unstable mP air during its long journey across the North Pacific (**Figure 19.5**). As this mP air arrives at the western shore of North America, it is often accompanied by low clouds and shower activity. When this air advances inland against the western mountains, orographic uplift produces heavy rain or snow on the windward slopes of the mountains. Maritime polar air also originates in the North Atlantic off the coast of eastern Canada and occasionally influences the weather of the northeastern United States. In winter, when New England is on the northern or northwestern side of a passing low-pressure center, the counterclockwise cyclonic winds draw in maritime polar air. The result is a storm characterized by snow and cold temperatures, known locally as a *nor'easter*.

CONCEPT CHECK 19.1

1. Define *air mass*. What is air-mass weather?
2. On what basis are air masses classified?
3. Compare the temperature and moisture characteristics of the following air masses: cP, mP, mT, and cT.
4. Which air mass is associated with lake-effect snow? What causes lake-effect snow?

Students Sometimes Ask...

When a cold air mass moves south from Canada into the United States, how fast can temperatures change?

When a fast-moving frigid air mass advances into the northern Great Plains, temperatures have been known to plunge 20°–30° C (40°–50° F) in a matter of just a few hours. One notable example is a drop of 55.5° C (100° F), from 6.7° C to −48.8° C (44° F to −56° F), in 24 hours at Browning, Montana, on January 23–24, 1916. Another remarkable example occurred on Christmas Eve in 1924, when the temperature at Fairfield, Montana, dropped from 17° C (63° F) at noon to −29° C (−21° F) at midnight—an amazing 46° C (83° F) change in just 12 hours.

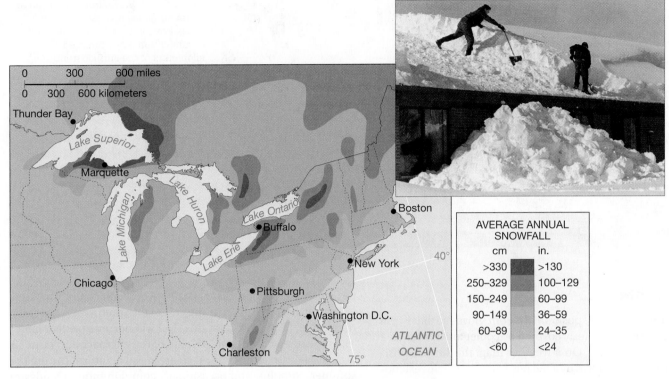

FIGURE 19.4 The snowbelts of the Great Lakes are easy to pick out on this snowfall map. (Data from NOAA) The photo was taken following a 6-day lake-effect snowstorm in November 1996 that dropped 175 centimeters (nearly 69 inches) of snow on Chardon, Ohio, setting a new state record. (Photo by Tony Dejak/AP/Wide World Photos)

Fronts

Earth's Dynamic Atmosphere
▶ **Basic Weather Patterns**

Fronts are boundaries that separate air masses of different densities. One air mass is usually warmer and contains more moisture than the other. Fronts can form between any two contrasting

FIGURE 19.5 During winter, maritime polar (mP) air masses in the North Pacific usually begin as continental polar (cP) air masses in Siberia. The cP air is modified to mP as it slowly crosses the ocean.

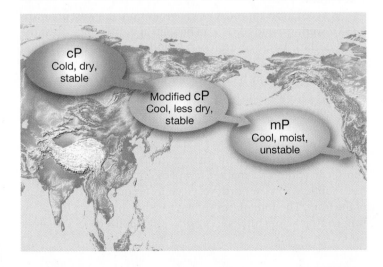

air masses. Considering the vast size of the air masses involved, fronts are relatively narrow, being 15- to 200-kilometer- (9- to 120-mile-) wide bands of discontinuity. On the scale of a weather map, they are generally narrow enough to be represented by a broad line.

Above Earth's surface, the frontal surface slopes at a low angle so that warmer air overlies cooler air, as shown in **Figure 19.6**. In the ideal case, the air masses on both sides of the front move in the same direction and at the same speed. Under this condition, the front acts simply as a barrier that travels along with the air masses and that neither mass can penetrate.

Generally, however, the distribution of pressure across a front is such that one air mass moves faster than the other. Thus, one air mass actively advances into another and "clashes" with it. In fact, the boundaries were tagged *fronts* during World War I by Norwegian meteorologists who visualized them as analogous to battle lines between two armies. It is along these "battlegrounds" that cyclonic circulation (centers of low pressure) develop and generate much of the precipitation and severe weather in the middle latitudes.

As one air mass moves into another, some mixing does occur along the frontal surface, but for the most part, the air masses retain their distinct identities as one is displaced upward over the other. No matter which air mass is advancing, *it is always the warmer, less dense air that is forced aloft, whereas the cooler, denser air acts as the wedge upon which lifting takes place.* The term **overrunning** is generally applied to warm air gliding up along a cold air mass. We now take a look at different types of fronts.

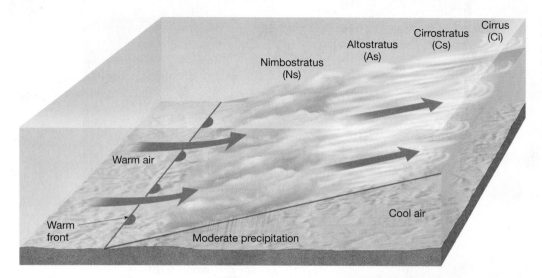

FIGURE 19.6 Warm front produced as warm air glides up over a cold air mass. Precipitation is moderate and occurs within a few hundred kilometers of the surface front.

Warm Fronts

When the surface (ground) position of a front moves so that warm air occupies territory formerly covered by cooler air, it is called a **warm front** (Figure 19.6). On a weather map, the surface position of a warm front is shown by a red line with semicircles extending into the cooler air.

East of the Rockies, warm tropical air often enters the United States from the Gulf of Mexico and overruns receding cool air. As the cold air retreats, friction with the ground greatly slows the advance of the surface position of the front compared to its position aloft. Stated another way, less dense, warm air has a hard time displacing denser cold air. For this reason, the boundary separating these air masses acquires a very gradual slope. The average slope of a warm front is about 1:200, which means that if you are 200 kilometers (120 miles) ahead of the surface location of a warm front, you will find the frontal surface at a height of 1 kilometer (0.6 mile).

As warm air ascends the retreating wedge of cold air, it cools adiabatically to produce clouds, and frequently, precipitation. The sequence of clouds shown in Figure 19.6 typically precedes a warm front. The first sign of the approaching warm front is the appearance of cirrus clouds. These high clouds form where the overrunning warm air has ascended high up the wedge of cold air, 1,000 kilometers (600 miles) or more ahead of the surface front.

As the front nears, cirrus clouds grade into cirrostratus, which blend into denser sheets of altostratus. About 300 kilometers (180 miles) ahead of the front, thicker stratus and nimbostratus clouds appear, and rain or snow begins.

Because of their slow rate of advance and very low slope, warm fronts usually produce light-to-moderate precipitation over a large area for an extended period. Warm fronts, however, are occasionally associated with cumulonimbus clouds and thunderstorms. This occurs when the overrunning air is unstable, and the temperatures on opposite sides of the front contrast sharply. At the other extreme, a warm front associated with a dry air mass could pass unnoticed at the surface.

A gradual increase in temperature occurs with the passage of a warm front. As you would expect, the increase is most apparent when a large contrast exists between adjacent air masses. Moreover, a wind shift from the east to the southwest is generally noticeable. (The reason for this shift will become evident later.) The moisture content and stability of the encroaching warm air mass largely determine when clear skies will return. During the summer, cumulus, and occasionally cumulonimbus, clouds are embedded in the warm unstable air mass that follows the front. These clouds can produce precipitation, but it is usually restricted in extent and of short duration.

Cold Fronts

When cold air actively advances into a region occupied by warmer air, the boundary is called a **cold front** (Figure 19.7). As with warm fronts, friction tends to slow the surface position of a cold front more so than its position aloft. However, because of the relative positions of the adjacent air masses, the cold front steepens as it moves. On the average, cold fronts are about twice as steep as warm fronts, having a slope of perhaps 1:100. In addition, cold fronts advance more rapidly than do warm fronts. These two differences—rate of movement and steepness of slope—largely account for the more violent nature of cold-front weather compared to the weather generally accompanying a warm front.

As a cold front approaches, generally from the west or northwest, towering clouds often can be seen in the distance. Near the front, a dark band of ominous clouds foretells the ensuing weather. The forceful lifting of air along a cold front is often so rapid that the latent heat released as water vapor condenses increases the air's buoyancy appreciably. The heavy downpours and vigorous wind gusts associated with mature cumulonimbus clouds frequently result. Because a cold front produces roughly the same amount of lifting as a warm front, but over a shorter distance, the precipitation intensity is greater, but the duration is shorter. In addition, a marked temperature drop and a wind shift from south to west or northwest usually accompany the passage of the front. The sometimes violent weather and sharp temperature contrast along the cold front are symbolized on a weather

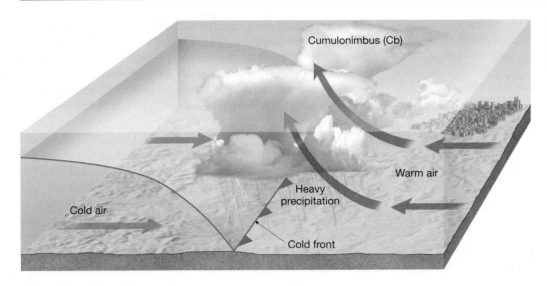

FIGURE 19.7 Fast-moving cold front and cumulonimbus clouds. Often thunderstorms occur if the warm air is unstable.

map by a blue line with triangle-shaped points that extend into the warmer air mass (Figure 19.7).

The weather behind a cold front is dominated by a subsiding and relatively cold air mass. Thus, clearing usually begins soon after the front passes. Although the compression of air due to subsidence causes some adiabatic heating, the effect on surface temperatures is minor. In winter, the long, cloudless nights that often follow the passage of a cold front allow for abundant radiation cooling that reduces surface temperatures. When a cold front moves over a relatively warm area, surface heating of the air can produce shallow convection. This, in turn, may generate low cumulus or stratocumulus clouds behind the front.

Stationary Fronts and Occluded Fronts

Occasionally the flow on both sides of a front is neither toward the cold air mass nor toward the warm air mass, but almost parallel to the line of the front. Thus, the surface position of the front does not move. This condition is called a **stationary front**. On a weather map, stationary fronts are shown with blue triangular points on one side of the front and red semicircles on the other. At times, some overrunning occurs along a stationary front, most likely causing gentle to moderate precipitation.

The fourth type of front is the **occluded front**. An active cold front overtakes a warm front, as shown in Figure 19.8. As the advancing cold air wedges the warm front upward, a new front emerges between the advancing cold air and the air over which the warm front is gliding. The weather of an occluded front is generally complex. Most precipitation is associated with the warm air being forced aloft. When conditions are suitable, however, the newly formed front is capable of initiating precipitation of its own.

A word of caution is in order concerning the weather associated with various fronts. Although the preceding discussion will help you recognize the weather patterns associated with fronts, remember that these descriptions are generalizations. The weather generated along any individual front may or may not conform fully to this idealized picture. Fronts, like all aspects of nature, never lend themselves to classification as easily as we would like.

FIGURE 19.8 Stages in the formation of an occluded front.

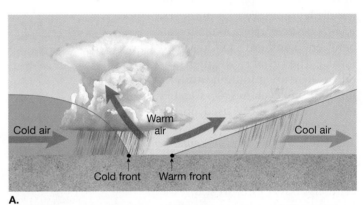

A.

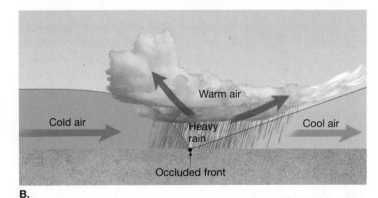

B.

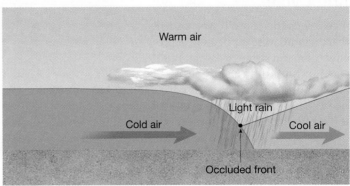

C.

The Middle-Latitude Cyclone

Earth's Dynamic Atmosphere
▶ Basic Weather Patterns

So far we have examined the basic elements of weather as well as the dynamics of atmospheric motions. We are now ready to apply our knowledge of these diverse phenomena to an understanding of day-to-day weather patterns in the middle latitudes. For our purposes, *middle latitudes* refers to the region between southern Florida and Alaska. The primary weather producers here are **middle-latitude** or **mid-latitude cyclones**. On weather maps, such as those used on the Weather Channel, they are shown by an *L*, meaning *low-pressure center*.

Middle-latitude cyclones are large centers of low pressure that generally travel from west to east. Lasting from a few days to more than a week, these weather systems have a counterclockwise circulation with an airflow inward toward their centers. Most middle-latitude cyclones also have a cold front extending from the central area of low pressure, and frequently a warm front as well. Convergence and forceful lifting initiate cloud development and often cause abundant precipitation.

Life Cycle

As early as the 1800s it was known that cyclones were the bearers of precipitation and severe weather. But it was not until the early part of the 20th century that a model was developed that described how cyclones form and evolve. It was formulated by a group of Norwegian scientists and published in 1918. The model was created primarily from near-surface observations.

Years later, as data from the middle and upper troposphere and from

satellite images became available, modifications were necessary. Yet this model is still an accepted working tool for interpreting the weather. If you keep this model in mind when you observe changes in the weather, the changes will no longer come as a surprise. You should begin to see some order in what had appeared to be disorder, and you might even occasionally "predict" the impending weather.

Formation: The Clash of Two Air Masses   Cyclones form along fronts and proceed through a generally predictable life cycle. Figure 19.9 shows six stages in the life of a typical mid-latitude cyclone. In part A, the stage is set for cyclone formation. Here two air masses of different densities (temperatures) are moving roughly parallel to the front, but in opposite directions. (In the classic model, this would be continental polar air associated with the polar easterlies on the north side of the front and

FIGURE 19.9 Stages in the life cycle of a middle-latitude cyclone.

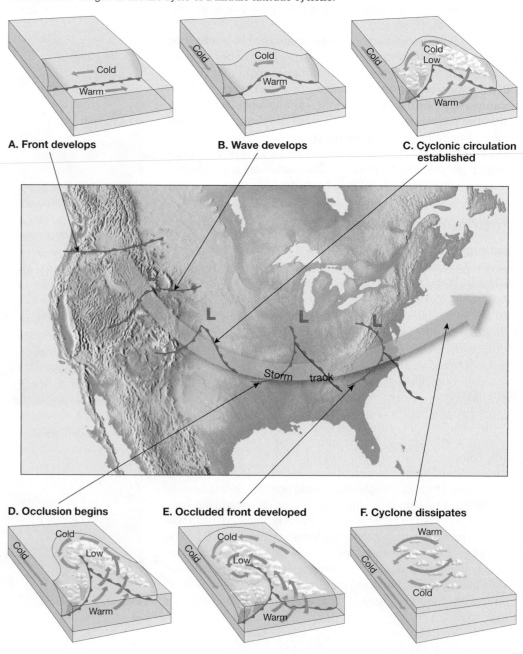

A. Front develops

B. Wave develops

C. Cyclonic circulation established

D. Occlusion begins

E. Occluded front developed

F. Cyclone dissipates

maritime tropical air driven by the westerlies on the south side of the front.)

Under suitable conditions, the frontal surface that separates these two contrasting air masses will take on a wave shape that is usually several hundred kilometers long. These waves are analogous to the waves produced on water by moving air, except that the scale is different. Some waves tend to dampen, or die out, whereas others grow in amplitude. Those storms that intensify develop waves that change in shape over time, much like a gentle ocean swell does as it moves into shallow water and becomes a tall, breaking wave.

Development of Cyclonic Flow   As the wave develops, warm air advances poleward invading the area formerly occupied by colder air, while cold air moves equatorward. This change in the direction of the surface flow is accompanied by a readjustment in the pressure pattern that results in nearly circular isobars, with the low pressure centered at the apex of the wave. The resulting flow is a counterclockwise cyclonic circulation that can be seen clearly on the weather map shown in Figure 19.10. Once the cyclonic circulation develops, we would expect general convergence to result in vertical lifting, especially where warm air is overrunning colder air. You can see in Figure 19.10 that the air in the warm sector (over the southern states) is flowing northeastward toward colder air that is moving toward the northwest. Because the warm air is moving in a direction perpendicular to this front, we can conclude that the warm air is invading a region formerly occupied by cold air. Therefore, this must be a warm front. Similar reasoning indicates that to the left (west) of the cyclonic disturbance, cold air from the northwest is displacing the air of the warm sector and generating a cold front.

Occlusion: The Beginning of the End   Usually, the position of the cold front advances faster than the warm front and begins to close (lift) the warm front, as shown in Figure 19.9D, E. This process, known as **occlusion**, forms an *occluded front*, which grows in length as it displaces the warm sector aloft. As occlusion begins, the storm often intensifies. Pressure at the storm's center falls, and wind speeds increase. In the winter, heavy snowfalls and blizzard like conditions are possible during this phase of the storm's evolution.

As more of the sloping discontinuity (front) is forced aloft, the pressure gradient weakens. In a day or two, the entire warm sector is displaced and cold air surrounds the cyclone at low levels (Figure 19.9F). Thus, the horizontal temperature (density) difference that existed between the two contracting air masses has been eliminated. At this point, the cyclone has exhausted its source of energy. Friction slows the surface flow, and the once highly organized counterclockwise flow ceases to exist.

Idealized Weather

The middle-latitude cyclone model provides a useful tool for examining the weather patterns of the middle latitudes. Figure 19.10 illustrates the distribution of clouds and thus the regions of possible precipitation associated with a mature system. Compare this drawing to the satellite image of a cyclone shown in Figure 19.11.

FIGURE 19.10 Cloud patterns typically associated with a mature middle-latitude cyclone. The middle section is a map view. Note the cross-section lines (F–G, A–E). Above the map is a vertical cross section along line F–G. Below the map is a section along A–E. For cloud abbreviations, refer to Figures 19.6 and 19.7.

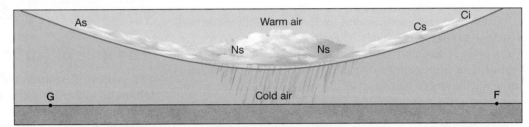

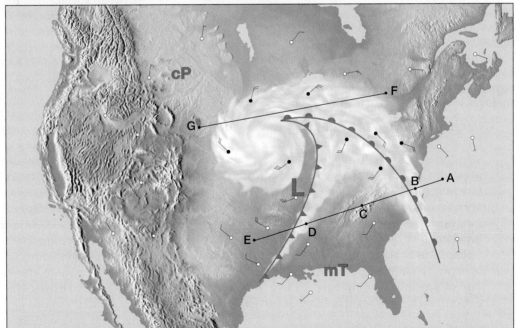

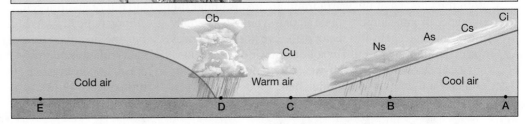

FIGURE 19.11 Satellite view of a mature cyclone over the eastern United States. It is easy to see why we often refer to the cloud pattern of a cyclone as having a "comma" shape. (Courtesy of John Jensenius/National Weather Service)

Guided by the westerlies aloft, cyclones generally move eastward across the United States. Therefore, we can expect the first signs of a cyclone's arrival to appear in the western sky. In the region of the Mississippi Valley, however, cyclones often begin a more northeasterly path and occasionally move directly northward. Typically, a mid-latitude cyclone requires 2–4 days to pass over a region. During that brief period, abrupt changes in atmospheric conditions may occur, particularly in the spring of the year when the greatest temperature contrasts occur across the mid-latitudes.

Using Figure 19.10 as a guide, let us examine these weather producers and the changes we can expect as they pass in the spring of the year. To facilitate our discussion, Figure 19.10 includes two profiles along lines *A–E* and *F–G*.

- Imagine the change in weather as you move from right to left along profile *A–E* (bottom of figure). At point *A*, the sighting of high cirrus clouds is the first sign of the approaching cyclone. These high clouds can precede the surface front by 1,000 kilometers (600 miles) or more, and they are normally accompanied by falling pressure. As the warm front advances, a lowering and thickening of the cloud deck is noticed.
- Within 12 to 24 hours after the first sighting of cirrus clouds, light precipitation begins (point *B*). As the front nears, the rate of precipitation increases, a rise in temperature is noticed, and winds begin to change from east or southeast to south or southwest.
- With the passage of the warm front, an area is under the influence of a maritime tropical air mass (point *C*). Usually the region affected by this sector of the cyclone experiences warm temperatures, south or southwest winds, and generally clear skies, although fair-weather cumulus or altocumulus clouds are not uncommon here.

- The relatively warm, humid weather of the warm sector passes quickly in the spring of the year and is replaced by gusty winds and precipitation generated along the cold front. The approach of a rapidly advancing cold front is marked by a wall of dark clouds (point *D*). Severe weather accompanied by heavy precipitation, hail, and an occasional tornado is a definite possibility at this time of year. The passage of the cold front is easily detected by a wind shift; the southwest winds are replaced by winds from the west to northwest and by a pronounced drop in temperature. Also, rising pressure hints of the subsiding cool, dry air behind the front.
- Once the front passes, skies clear as cooler air invades the region (point *E*). Often a day or two of almost cloudless deep-blue skies occurs unless another cyclone is edging into the region.

A very different set of weather conditions will prevail in those regions north of the storm's center along profile *F–G* in Figure 19.10. Often the storm reaches its greatest intensity in this zone, and the area along profile *F–G* receives the brunt of the storm's fury. Temperatures remain cold during the passage of the system, and heavy snow, sleet, and/or freezing rain may develop during the winter months.

The Role of Airflow Aloft

When the earliest studies of cyclones were made, little was known about the nature of the airflow in the middle and upper troposphere. Since then, a close relationship has been established between surface disturbances and the flow aloft. Airflow aloft plays an important role in maintaining cyclonic and anticyclonic circulation. In fact, more often than not, these rotating surface wind systems are actually generated by upper-level flow.

Recall that the airflow around a surface low is inward, a fact that leads to mass convergence, or coming together (**Figure 19.12**). The resulting accumulation of air must be accompanied by a corresponding increase in surface pressure. Consequently, we might expect a low-pressure system to "fill" rapidly and be eliminated, just as the vacuum in a coffee can is quickly dissipated when we open it. However, this does not occur.

On the contrary, cyclones often exist for a week or longer. For this to happen, surface convergence must be offset by a mass outflow at some level aloft (Figure 19.12). As long as divergence (spreading out) aloft is equal to, or greater than, the surface inflow, the low pressure and its accompanying convergence can be sustained.

Because cyclones are bearers of stormy weather, they have received far more attention than anticyclones. Nevertheless, a close relation exists, which makes it difficult to separate any discussion of these two types of pressure systems. The surface air that feeds a cyclone, for example, generally originates as air flowing out of an anticyclone. Consequently, cyclones and anticyclones typically are found adjacent to one another. Like the cyclone, an anticyclone depends on the flow far above to maintain its circulation. In this instance, divergence at the surface is balanced by convergence aloft and general subsidence of the air column (Figure 19.12).

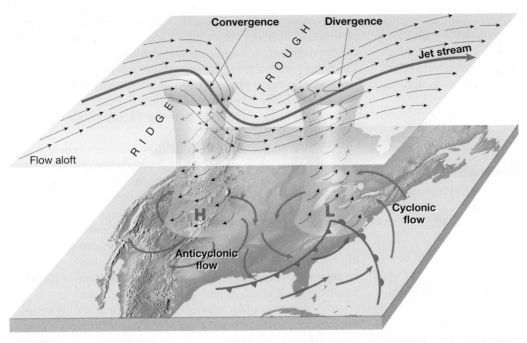

FIGURE 19.12 Idealized diagram depicting the support that divergence and convergence aloft provide to cyclonic and anticyclonic circulation at the surface. Divergence aloft initiates upward air movement, reduced surface pressure, and cyclonic flow. In contrast, convergence along the jet stream results in general subsidence of the air column, increased surface pressure, and anticyclonic surface winds.

CONCEPT CHECK 19.3

1. Describe the changes that an observer experiences when a middle-latitude cyclone passes with its center to the north of the observer. Include wind direction, pressure tendency, cloud type, precipitation, and temperature.
2. Describe the weather conditions an observer would experience if the center of a middle-latitude cyclone passed to the south.
3. Briefly explain how the flow aloft aids the formation of cyclones at the surface.

ricanes and tornadoes are, in fact, cyclones, the vast majority of cyclones are *not* hurricanes or tornadoes. The term *cyclone* simply refers to the circulation around any low-pressure center, no matter how large or intense it is.

Tornadoes and hurricanes are both smaller and more violent than middle-latitude cyclones. Middle-latitude cyclones can have a diameter of 1,600 kilometers (1,000 miles) or more. By contrast, hurricanes average only 600 kilometers (375 miles) across, and tornadoes, with a diameter of just 0.25 kilometer (0.16 mile), are much too small to show up on a weather map.

The thunderstorm, a much more familiar weather event, hardly needs to be distinguished from tornadoes, hurricanes, and mid-latitude cyclones. Unlike the flow of air about these latter storms, the circulation associated with thunderstorms is characterized by strong up-and-down movements. Winds in the vicinity of a thunderstorm do not follow the inward spiral of a cyclone, but they are typically variable and gusty.

Although thunderstorms form "on their own" away from cyclonic storms, they also form in conjunction with cyclones. For

FIGURE 19.13 In parts of the Great Plains, cyclone is a synonym for tornado. The nickname for the athletic teams at Iowa State University is the *Cyclones*. (Iowa State University is the only Division I school to use Cyclones as its team name. The pictured logo was created to better communicate the school's image by combining the mascot, a cardinal bird named Cy, and the Cyclone team name, a swirling twister in the bird's tail.) (Image courtesy of Iowa State University)

What's in a Name?

Up to now we have examined the middle-latitude cyclones, which play such an important role in causing day-to-day weather changes. Yet the use of the term *cyclone* is often confusing. To many people, the term implies only an intense storm, such as a tornado or a hurricane. When a hurricane unleashes its fury on India or Bangladesh, for example, it is usually reported in the media as a cyclone (the term denoting a hurricane in that part of the world).

Similarly, tornadoes are referred to as cyclones in some places. This custom is particularly common in portions of the Great Plains of the United States. Recall that in *The Wizard of Oz*, Dorothy's house was carried from her Kansas farm to the land of Oz by a cyclone. Indeed, the nickname for the athletic teams at Iowa State University is the *Cyclones* (Figure 19.13). Although hur-

instance, thunderstorms are frequently spawned along the cold front of a mid-latitude cyclone, where on rare occasions a tornado may descend from the thunderstorm's cumulonimbus tower. Hurricanes also generate widespread thunderstorm activity. Thus, thunderstorms are related in some manner to all three types of cyclones mentioned here.

Thunderstorms

Thunderstorms are the first of three severe weather types we examine in this chapter. Sections on tornadoes and hurricanes follow.

Severe weather has a fascination that everyday weather phenomena cannot provide. The lightning display and booming thunder generated by a severe thunderstorm can be a spectacular event that elicits both awe and fear (Figure 19.14). Of course, hurricanes and tornadoes also attract a great deal of much-deserved attention. A single tornado outbreak or hurricane can cause billions of dollars in property damage as well as many deaths.

Thunderstorm Occurrence

Almost everyone has observed a small-scale phenomenon that is the result of the vertical movements of relatively warm, unstable air. Perhaps you have seen a dust devil over an open field on a hot day whirl its dusty load to great heights, or seen a bird glide

effortlessly skyward on an invisible thermal of hot air. These examples illustrate the dynamic thermal instability that occurs during the development of a *thunderstorm*. A **thunderstorm** is simply a storm that generates lightning and thunder. It frequently produces gusty winds, heavy rain, and hail. A thunderstorm may be produced by a single cumulonimbus cloud and influence only a small area or it may be associated with clusters of cumulonimbus clouds covering a large area.

Thunderstorms form when warm, humid air rises in an unstable environment. Various mechanisms can trigger the upward air movement needed to create thunderstorm-producing cumulonimbus clouds. One mechanism, the unequal heating of Earth's surface, significantly contributes to the formation of *air-mass thunderstorms*. These storms are associated with the scattered puffy cumulonimbus clouds that commonly form *within* maritime tropical air masses and produce scattered thunderstorms on summer days. Such storms are usually short-lived and seldom produce strong winds or hail.

In contrast, another type of thunderstorm not only benefits from uneven surface heating but is associated with the lifting of warm air, as occurs along a front or a mountain slope. Moreover, diverging winds aloft frequently contribute to the formation of these storms because they tend to draw air from lower levels upward beneath them. Some of the thunderstorms of this type may produce high winds, damaging hail, flash floods, and tornadoes. Such storms are described as *severe*.

At any given time an estimated 2,000 thunderstorms are in progress on Earth (Figure 19.15). As we would expect, the greatest number occur in the tropics where warmth, plentiful moisture, and instability are always present. About 45,000 thunderstorms take place each day, and more than 16 million occur annually around the world. The lightning from these storms strikes Earth 100 times each second (Figure 19.15A)!

Annually, the United States experiences about 100,000 thunderstorms and millions of lightning strikes. A glance at Figure 19.15B shows that in the United States, thunderstorms are most frequent in Florida and the eastern Gulf Coast region, where such activity is recorded between 70 and 100 days each year. The region on the east side of the Rockies in Colorado and New Mexico is next, with thunderstorms occurring on 60–70 days each year. Most of the rest of the nation experiences thunderstorms on 30–50 days annually. Clearly, the western margin of the United States has little thunderstorm activity. The same is true for the northern tier of states and for Canada, where warm, moist, unstable mT air seldom penetrates.

Stages of Thunderstorm Development

All thunderstorms require warm, moist air, which, when lifted, will release suffi-

FIGURE 19.14 This lightning display occurred near Colorado Springs, Colorado. Lightning produces thunder. The electrical discharge of lightning superheats the air. In less than a second, the temperature rises by as much as 33,000° C. When air is heated this quickly, it expands explosively and produces the sound we hear as thunder. (Photo by Sean Cayton/The Image Works)

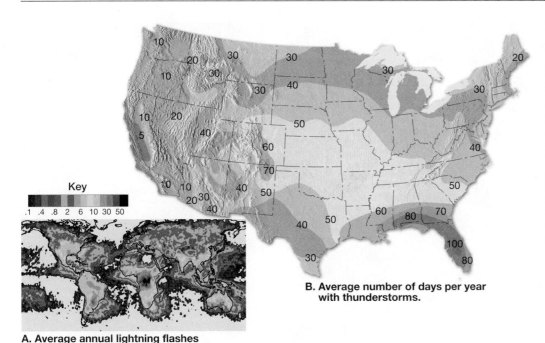

Key
.1 .4 .8 2 6 10 30 50

A. Average annual lightning flashes
per square kilometer.

B. Average number of days per year
with thunderstorms.

FIGURE 19.15 A. Worldwide distribution of lightning, with color variations indicating the average annual number of lightning flashes per square kilometer. Data comes from satellite-based sensors that use high-speed cameras capable of detecting brief lightning flashes even under daytime conditions. (NASA) **B.** Average number of days each year with thunderstorms. The humid subtropical climate that dominates the southeastern United States receives much of its precipitation in the form of thunderstorms. Most of the Southeast averages 50 or more days each year with thunderstorms. (Environmental Data Service, NOAA)

cient latent heat to provide the buoyancy necessary to maintain its upward flight. This instability and associated buoyancy are triggered by a number of different processes, yet all thunderstorms have a similar life history.

Because instability and buoyancy are enhanced by high surface temperatures, thunderstorms are most common in the afternoon and early evening (**Figure 19.16A**). Surface heating is generally not sufficient in itself to cause the growth of towering cumulonimbus clouds. A solitary cell of rising hot air produced by surface heating alone can, at best, produce a small cumulus cloud, which would evaporate within 10–15 minutes.

The development of 12-kilometer (40,000-foot) or on rare occasions 18-kilometer (60,000-foot) cumulonimbus towers requires a continuous supply of moist air (Figure 19.16B). Each

FIGURE 19.16 A. Buoyant thermals often produce fair-weather cumulus clouds that soon evaporate into the surrounding air, making it more humid. As this process of cumulus development and evaporation continues, the air eventually becomes sufficiently humid, so that newly forming clouds do not evaporate but continue to grow. (Photo by Henry Lansford/Photo Researchers, Inc.) **B.** This Developing cumulonimbus cloud became a towering August thunderstorm over central Illinois. (Photo by E. J. Tarbuck)

A.

B.

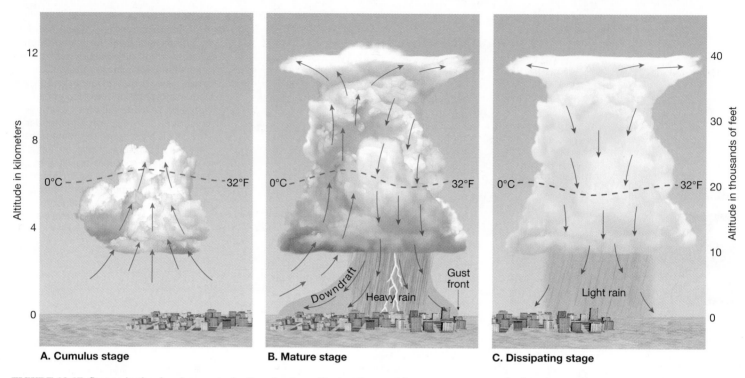

A. Cumulus stage **B. Mature stage** **C. Dissipating stage**

FIGURE 19.17 Stages in the development of a thunderstorm. During the cumulus stage, strong updrafts act to build the storm. The mature stage is marked by heavy precipitation and cool downdrafts in part of the storm. When the warm updrafts disappear completely, precipitation becomes light, and the cloud begins to evaporate.

Students Sometimes Ask...

How far away from a storm can lightning strike?

According to the National Severe Storms Laboratory (NSSL) in Norman, Oklahoma, it is not certain what the maximum possible distance might be. Lightning has been known to strike more than 16 kilometers (10 miles) from a storm in an area of clear skies.

new surge of warm air rises higher than the last, adding to the height of the cloud (**Figure 19.17**). These updrafts must occasionally reach speeds over 100 kilometers (60 miles) per hour to accommodate the size of hailstones they are capable of carrying upward.

Usually within an hour, the amount and size of precipitation that has accumulated is too much for the updrafts to support, and in one part of the cloud downdrafts develop, releasing heavy precipitation. This represents the most active stage of the thunderstorm. Gusty winds, lightning, heavy precipitation, and sometimes hail are experienced.

Eventually downdrafts dominate throughout the cloud. The cooling effect of falling precipitation, coupled with the influx of colder air aloft, marks the end of the thunderstorm activity. The life span of a single cumulonimbus cell within a thunderstorm complex is only an hour or two, but as the storm moves, fresh supplies of warm, water-laden air generate new cells to replace those that are dissipating.

CONCEPT CHECK 19.5

❶ Where are thunderstorms most common on Earth? In the United States?

❷ What is the primary requirement for the formation of thunderstorms?

Tornadoes

Tornadoes are local storms of short duration that must be ranked high among nature's most destructive forces (see chapter opening photo). Their sporadic occurrence and violent winds cause many deaths each year. The nearly total destruction in some stricken areas has led many to liken their passage to bombing raids during war.

Tornadoes are violent windstorms that take the form of a rotating column of air or *vortex* that extends downward from a cumulonimbus cloud. Pressures within some tornadoes have been estimated to be as much as 10 percent lower than immediately outside the storm. Drawn by the much lower pressure in the center of the vortex, air near the ground rushes into the tornado from all directions. As the air streams inward, it is spiraled upward around the core until it eventually merges with the airflow of the parent thunderstorm deep in the cumulonimbus tower. Because of the tremendous pressure gradient associated with a strong tornado, maximum winds can sometimes approach 480 kilometers (300 miles) per hour.

Some tornadoes consist of a single vortex, but within many stronger tornadoes are smaller whirls called *suction vortices* that rotate within the main vortex (**Figure 19.18**). Suction vortices have diameters of only about 10 meters (30 feet) and rotate very rapidly.

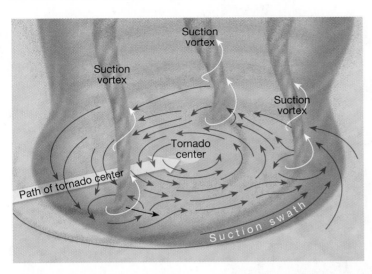

FIGURE 19.18 Some tornadoes have multiple suction vortices. These small and very intense vortices are roughly 10 meters (30 feet) across and move in a counterclockwise path around the tornado center. Because of this multiple vortex structure, one building might be heavily damaged and another one, just 10 meters away, might suffer little damage. (After Fujita)

This structure accounts for occasional observations of virtually total destruction of one building while another one, just 10 or 20 meters away, suffers little damage.

Tornado Occurrence and Development

Tornadoes form in association with severe thunderstorms that produce high winds, heavy (sometimes torrential) rainfall, and often damaging hail. Fortunately, less than 1 percent of all thunderstorms produce tornadoes. Nevertheless, a much higher number must be monitored as potential tornado producers. Although meteorologists are still not sure what triggers tornado formation, it is apparent that they are the product of the interaction between strong updrafts in the thunderstorm and the winds in the troposphere.

Tornadoes can form in any situation that produces severe weather, including cold fronts and tropical cyclones (hurricanes). The most intense tornadoes are usually those that form in association with huge thunderstorms called *supercells*. An important precondition linked to tornado formation in severe thunderstorms is the development of a *mesocyclone*—a vertical cylinder of rotating air, typically about 3–10 kilometers (2–6 miles) across, that develops in the updraft of a severe thunderstorm (**Figure 19.19**). The formation of this large vortex often precedes tornado formation by 30 minutes or so.

The formation of a mesocyclone does not necessarily mean that tornado formation will follow. Only about half of all mesocyclones produce tornadoes. Forecasters cannot determine in advance which mesocyclones will spawn tornadoes.

General Atmospheric Conditions   Severe thunderstorms and hence tornadoes are most often spawned along the cold front of a middle-latitude cyclone or in association with a supercell thunderstorm such as the one pictured in Figure 19.19D, E. Throughout spring, air masses associated with middle-latitude cyclones are most likely to have greatly contrasting conditions. Continen-

tal polar air from Canada may still be very cold and dry, whereas maritime tropical air from the Gulf of Mexico is warm, humid, and unstable. The greater the contrast when these air masses meet, the more intense the storm tends to be.

These two contrasting air masses are most likely to meet in the central United States, because there is no significant natural barrier separating the center of the country from the Arctic or the Gulf of Mexico. Consequently, this region generates more tornadoes than any other area of the country or, in fact, the world. **Figure 19.20**, which depicts tornado incidence in the United States for a 27-year period, readily substantiates this fact.

Tornado Climatology   An average of between 1,200 and 1,300 tornadoes were reported annually in the United States between 2000 and 2009. Still, the actual numbers that occur from one year to the next vary greatly. During the span just mentioned, for example, yearly totals ranged from a low of 938 in the year 2002 to a high of 1,820 in 2004. Tornadoes occur during every month of the year. April through June is the period of greatest tornado frequency in the United States, and December and January are the months of lowest activity (Figure 19.20, graph inset). Of the 40,522 confirmed tornadoes reported over the contiguous 48 states during the 50-year period from 1950 through 1999, an average of almost six per day occurred during May. At the other extreme, a tornado was reported only about every other day in December and January.

Profile of a Tornado   An average tornado has a diameter of between 150 and 600 meters (500 and 2,000 feet), travels across the landscape at approximately 45 kilometers (30 miles) per hour, and cuts a path about 10 kilometers (6 miles) long.[45] Because tornadoes usually occur slightly ahead of a cold front, in the zone of southwest winds, most move toward the northeast. The Illinois example demonstrates this movement (**Figure 19.21**). Figure 19.21 also shows that many tornadoes do not fit the description of the "average" tornado.

Of the hundreds of tornadoes reported in the United States each year, more than half are comparatively weak and short-lived. Most of these small tornadoes have lifetimes of 3 minutes or less and paths that seldom exceed 1 kilometer (0.6 mile) in length and 100 meters (330 feet) wide. Typical wind speeds are on the order of 150 kilometers (90 miles) per hour or less. On the other end of the tornado spectrum are the infrequent and often long-lived violent tornadoes. Although large tornadoes constitute only a small percentage of the total reported, their effects are often devastating. Such tornadoes may exist for periods in excess of 3 hours and produce an essentially continuous damage path more than 150 kilometers (90 miles) long and perhaps a kilometer or more wide. Maximum winds range beyond 500 kilometers (310 miles) per hour.

Tornado Destruction

The potential for tornado destruction depends largely on the strength of the winds generated by the storm. Because tornadoes generate the strongest winds in nature, they have accomplished

[45]The 10-kilometer figure applies to documented tornadoes. Because many small tornadoes go undocumented, the real average path of all tornadoes is unknown but shorter than 10 kilometers.

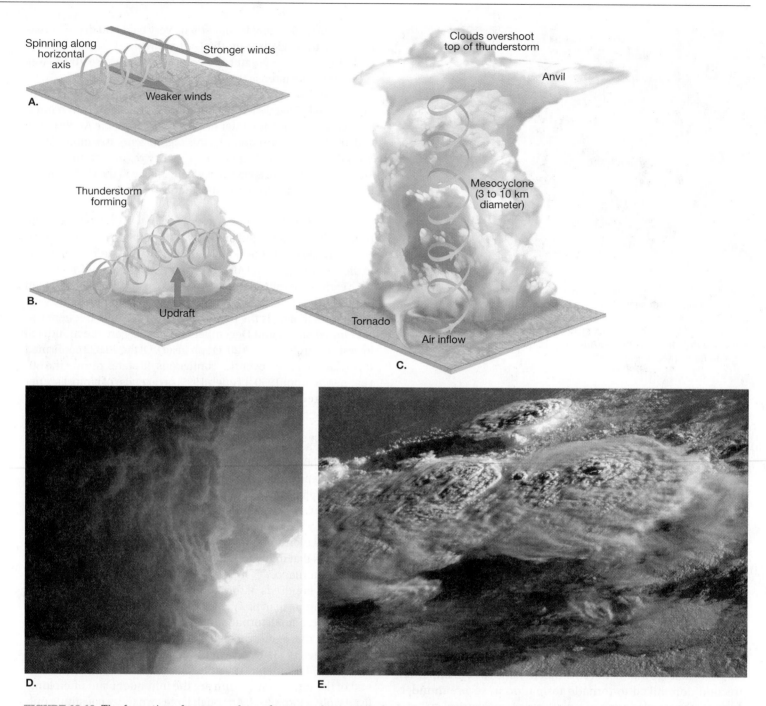

FIGURE 19.19 The formation of a mesocyclone often precedes tornado formation. **A.** Winds are stronger aloft than at the surface (called *speed wind shear*), producing a rolling motion about a horizontal axis. **B.** Strong thunderstorm updrafts tilt the horizontally rotating air to a nearly vertical alignment. **C.** The mesocyclone, a vertical cylinder of rotating air, is established. **D.** If a tornado develops it will descend from a slowly rotating wall cloud in the lower portion of the mesocyclone. (Photo © Howard B. Bluestein, Professor of Meteorology) **E.** This photo of a cluster of supercell thunderstorms along the Manitoba–Minnesota border in September 1994 was taken from space by an astronaut. (NASA)

many seemingly impossible tasks, such as driving a piece of straw through a thick wooden plank and uprooting huge trees. Although it may seem impossible for winds to cause some of the fantastic damage attributed to tornadoes, tests in engineering facilities have repeatedly demonstrated that winds in excess of 320 kilometers (200 miles) per hour are capable of incredible feats (Figure 19.22).

One commonly used guide to tornado intensity is the *Enhanced Fujita (EF) intensity scale* (Table 19.1). Because tornado winds cannot be measured directly, a rating on the EF-scale is determined by assessing the worst damage produced by a storm.

Although the greatest part of tornado damage is caused by violent winds, most tornado injuries and deaths result from flying debris. For the United States, the average annual death toll from tornadoes is about 65 people. However, the actual number of deaths each year can depart significantly from the average. On April 3–4, 1974, for example, an outbreak of 148 tornadoes brought death and destruction to a 13-state region east of the Mississippi River. More than 300 people died and nearly 5,500 people were injured in this worst disaster in half a century. Most tornadoes, however, do not result in a loss of life. In one statistical study that

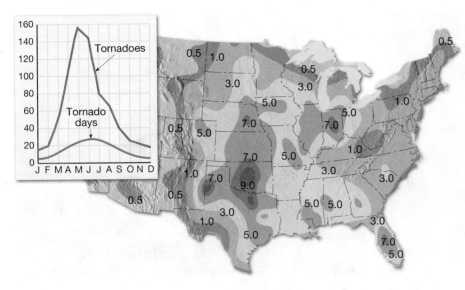

FIGURE 19.20 The map shows average annual tornado incidence per 10,000 square miles (26,000 square kilometers) for a 27-year period. The graph shows the average number of tornadoes and tornado days each month in the United States for the same period.

FIGURE 19.21 Paths of Illinois tornadoes (1916–1969). Because most tornadoes occur slightly ahead of a cold front, in the zone of southwest winds, they tend to move toward the northeast. Tornadoes in Illinois verify this. Over 80 percent exhibited directions of movement toward the northeast through east. (After John W. Wilson and Stanley A. Changnon, Jr., *Illinois Tornadoes*, Illinois State Water Survey Circular 103, 1971, pp. 10, 24)

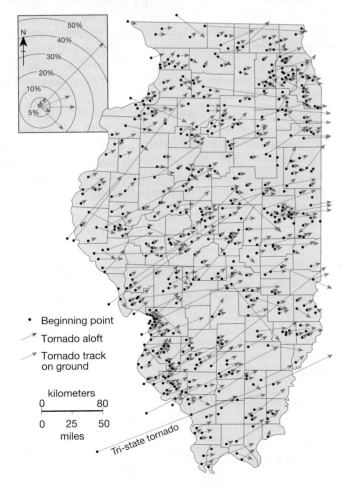

examined a 29-year period, there were 689 tornadoes that resulted in deaths. This figure represents slightly less than 4 percent of the total 19,312 reported storms.

Although the percentage of tornadoes that result in death is small, each tornado is potentially lethal. When tornado fatalities and storm intensities are compared, the results are quite interesting: The majority (63 percent) of tornadoes are weak (EF-0 and EF-1), and the number of storms decreases as tornado intensity increases. The distribution of tornado fatalities, however, is just the opposite. Although only 2 percent of tornadoes are classified as violent (EF-4 and EF-5), they account for nearly 70 percent of the deaths. If there is some question as to the causes of tornadoes, there certainly is no question about the destructive effects of these violent storms.

Tornado Forecasting

Because severe thunderstorms and tornadoes are small and relatively short-lived phenomena, they are among the most difficult weather features to forecast precisely. Nevertheless, the prediction, detection, and monitoring of such storms are among the most important services provided by professional meteorologists. The timely issuance and dissemination of watches and warnings are both critical to the protection of life and property (Box 19.1).

The Storm Prediction Center (SPC) located in Norman, Oklahoma, is part of the National Weather Service (NWS) and the National Centers for Environmental Prediction (NCEP). Its mission is to provide timely and accurate forecasts and watches for severe thunderstorms and tornadoes.

Severe thunderstorm outlooks are issued several times daily. *Day 1* outlooks identify those areas likely to be affected by severe thunderstorms during the next 6–30 hours, and *day 2* outlooks extend the forecast through the following day. Both outlooks describe the type, coverage, and intensity of the severe weather expected. Many local NWS field offices also issue severe weather

Students Sometimes Ask...

Someone told me that my house could explode if I don't open windows when a tornado is approaching. Is that true?

No. The drop in atmospheric pressure associated with the passage of a tornado plays a minor role in the damage process. Most structures have sufficient venting to allow for the sudden drop in pressure. Opening windows, once thought to be a way of minimizing damage by allowing inside and outside atmospheric pressure to equalize, is no longer recommended. In fact, if a tornado gets close enough to a structure for the pressure drop to be experienced, the strong winds will have already caused significant damage.

A.

B.

FIGURE 19.22 **A.** The force of the wind during a tornado near Wichita, Kansas, in April 1991 was enough to drive this piece of metal into a utility pole. (Photo by John Sokich/NOAA) **B.** The remains of a truck wrapped around a tree in Bridge Creek, Oklahoma, on May 4, 1999, following a major tornado outbreak. (AP Photo/L. M. Otero)

outlooks that provide a more local description of the severe weather potential for the next 12–24 hours.

Tornado Watches and Warnings **Tornado watches** alert the public to the possibility of tornadoes over a specified area for a particular time interval. Watches serve to fine-tune forecast areas already identified in severe weather outlooks. A typical watch covers an area of about 65,000 square kilometers (25,000 square miles) for a 4- to 6-hour period. A tornado watch is an important part of the tornado alert system because it sets in motion the procedures necessary to deal adequately with detection, tracking, warning, and response. Watches are generally reserved for organized severe weather events where the tornado threat will affect at least 26,000 square kilometers (10,000 square miles) and/or persist for at least 3 hours. Watches typically are not issued when the threat is thought to be isolated and/or short-lived.

Whereas a tornado watch is designed to alert people to the possibility of tornadoes, a **tornado warning** is issued by local offices of the National Weather Service when a tornado has actually been sighted in an area or is indicated by weather radar. It warns of a high probability of imminent danger. Warnings are issued for much smaller areas than watches, usually covering portions of a county or counties. In addition, they are in effect for much shorter periods, typically 30–60 minutes. Because a tornado warning may be based on an actual sighting, warnings are occasionally issued after a tornado has already developed. However, most warnings are issued prior to tornado formation, sometimes by several tens of minutes, based on Doppler radar data and/or spotter reports of funnel clouds.

If the direction and the approximate speed of the storm are known, an estimate of its most probable path can be made. Because tornadoes often move erratically, the warning area is fan-shaped downwind from the point where the tornado has been spotted. Improved forecasts and advances in technology have contributed to a significant decline in tornado deaths over the past 50 years.

TABLE 19.1 Enhanced Fujita Intensity Scale*

	Wind Speed		
Scale	Km/Hr	Mi/Hr	Damage
EF-0	105–137	65–85	*Light.* Some damage to siding and shingles.
EF-1	138–177	86–110	*Moderate.* Considerable roof damage. Winds can uproot trees and overturn single-wide mobile homes. Flagpoles bend.
EF-2	178–217	111–135	*Considerable.* Most single-wide homes destroyed. Permanent homes can shift off foundations. Flagpoles collapse. Softwood trees debarked.
EF-3	218–265	136–165	*Severe.* Hardwood trees debarked. All but small portions of houses destroyed.
EF-4	266–322	166–200	*Devastating.* Complete destruction of well-built residences, large sections of school buildings.
EF-5	>322	>200	*Incredible.* Significant structural deformation of mid- and high-rise buildings.

* The original Fujita scale was developed by T. Theodore Fujita in 1971 and put into use in 1973. The Enhanced Fujita Scale is a revision that was put into use in February 2007. Wind speeds are estimates (not measurements) based on damage.

Box 19.2

PEOPLE AND THE ENVIRONMENT

Surviving a Violent Tornado

About 11:00 A.M. on Tuesday, July 13, 2004, much of northern and central Illinois was put on a tornado watch. A large supercell had developed in the northwestern part of the state and was moving southeast into a very unstable environment (**Figure 19.A**). A few hours later, as the supercell entered Woodford County, rain began to fall and the storm showed signs of becoming severe. The National Weather Service (NWS) issued a *severe thunderstorm warning* at 2:29 P.M. CDT. Minutes afterward a tornado developed. Twenty-three minutes later, the quarter-mile-wide twister had carved a 9.6-mile-long path across the rural Illinois countryside.

What, if anything, made this storm special or unique? After all, it was just one of a record-high 1,819 tornadoes that were reported in the United States in 2004. For one, this tornado attained EF-4 status for a portion of its life. The NWS estimated that maximum winds reached 240 miles per hour.

FIGURE 19.A On July 13, 2004, an EF-4 tornado cut a 23-mile path through the rural Illinois countryside near the Woodford County town of Roanoke. The Parsons Manufacturing plant was just west of town.

Fewer than 1 percent of tornadoes attain this level of severity. However, what was most remarkable is that no one was killed or injured when the Parsons Manufacturing facility west of the small town of Roanoke took a direct hit while the storm was most intense. At the time, 150 people were in three buildings that comprised the plant. The 250,000-square-foot facility was flattened, cars were twisted into gnarled masses, and debris was strewn for miles (**Figure 19.B**).

How did 150 people escape death or injury? The answer is foresight and planning. More than 30 years earlier, company owner Bob Parsons was inside his first factory when a small tornado passed close enough to blow windows out. Later, when he built a new plant, he made sure that the restrooms were constructed to double as tornado shelters with steel-reinforced concrete walls and eight-inch-thick concrete ceilings. In addition, the company developed a severe weather plan. When the severe thunderstorm warning was issued at 2:29 P.M. on July 13, the emergency response team leader at the Parsons plant was immediately notified. A few moments later he went outside and observed a rotating wall cloud with a developing funnel cloud. He radioed back to the office to institute the company's severe weather plan. Employees were told to immediately go to their designated storm shelter. Everyone knew where to go and what to do because the plant conducted semi-annual tornado drills. All 150 people reached a shelter in less than 4 minutes. The emergency response team leader was the last person to reach shelter, less than 2 minutes before the tornado destroyed the plant at 2:41 P.M.

The total number of tornado deaths in 2004 for the entire United States was just 36. The toll could have been much higher. The building of tornado shelters and the development of an effective severe storm plan made the difference between life and death for 150 people at Parsons Manufacturing.

FIGURE 19.B The quarter-mile-wide tornado had wind speeds reaching 240 miles per hour. The destruction at Parsons Manufacturing was devastating. (Photos courtesy of NOAA)

Doppler Radar Many of the difficulties that once limited the accuracy of tornado warnings have been reduced or eliminated by an advancement in radar technology called **Doppler radar**. Doppler radar not only performs the same tasks as conventional radar but also has the ability to detect motion directly (**Figure 19.23**). Doppler radar can detect the initial formation and subsequent development of a *mesocyclone*. Almost all mesocyclones produce damaging hail, severe winds, or tornadoes. Those that produce tornadoes (about 50%) can sometimes be distinguished by their stronger wind speeds and their sharper gradients of wind speeds.

It should also be pointed out that not all tornado-bearing storms have clear-cut radar signatures and that other storms can give false signatures. Detection, therefore, is sometimes a subjective process and a given display could be interpreted in several ways. Consequently, trained observers will continue to form an important part of the warning system in the foreseeable future.

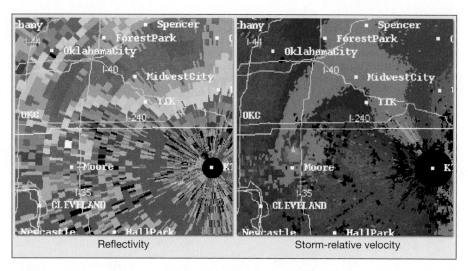

Reflectivity | Storm-relative velocity

FIGURE 19.23 This is a dual Doppler radar image of an EF-5 tornado near Moore, Oklahoma, on May 3, 1999. The left image (reflectivity) shows precipitation in the supercell thunderstorm. The right image shows motion of the precipitation along the radar beam, that is, how fast rain or hail is moving toward or away from the radar. In this example, the radar was unusually close to the tornado—close enough to make out the signature of the tornado itself (usually only the weaker and larger mesocyclone is detected). (After NOAA)

Students Sometimes Ask...

What is the most destructive tornado on record?

One tornado easily ranks above all others as the single most dangerous and destructive. Known as the Tri-state tornado, it occurred on March 18, 1925. Its path is labeled on Figure 19.21. Starting in southeastern Missouri, the tornado remained on the ground for 352 kilometers (219 miles), finally ending in Indiana. The losses included 695 dead and 2,027 injured. Property losses were also great, with several small towns almost totally destroyed.

Although some operational problems exist, the benefits of Doppler radar are many. As a research tool, it is not only providing data on the formation of tornadoes but is also helping meteorologists gain new insights into thunderstorm development, the structure and dynamics of hurricanes, and air-turbulence hazards that plague aircraft. As a practical tool for tornado detection, Doppler radar has significantly improved our ability to track thunderstorms and issue warnings.

CONCEPT CHECK 19.6

1. Why do tornadoes have such high wind speeds?
2. What atmospheric conditions are most conducive to the formation of tornadoes?
3. During what months is tornado activity most pronounced in the United States?
4. Distinguish between a tornado watch and a tornado warning.

Hurricanes

Most of us view the weather in the tropics with favor. Places such as Hawaii and the islands of the Caribbean are known for their lack of significant day-to-day variations. Warm breezes, steady temperatures, and rains that come as heavy but brief tropical showers are expected. It is ironic that these relatively tranquil regions sometimes produce the most violent storms on Earth.

The whirling tropical cyclones that on occasion have wind speeds attaining 300 kilometers (185 miles) per hour are known in the United States as **hurricanes** and are the greatest storms on Earth. Hurricane Rita, shown in Figure 19.1 (p. 553), and Hurricane Katrina (**Figure 19.24**) are two examples from the record-breaking 2005 Atlantic hurricane season. Should a hurricane smash onto land, strong winds coupled with extensive flooding can impose billions of dollars in damage and great loss of life (see Figure 1.2, p. 6).

The vast majority of hurricane-related deaths and damage are caused by relatively infrequent, yet powerful, storms. A storm that pounded an unsuspecting Galveston, Texas, in 1900 was responsible for at least 8,000 deaths. It was not just the deadliest U.S. hurricane ever, but the deadliest natural disaster of *any kind* to affect the United States. Of course, the deadliest and most costly storm in recent memory occurred in August 2005, when Hurricane Katrina devastated the Gulf Coast

FIGURE 19.24 Satellites are invaluable tools for tracking storms and gathering atmospheric data. This color-enhanced infrared image from the GOES-East satellite shows Hurricane Katrina several hours before making landfall on August 29, 2005. The most intense activity is associated with the colors red and orange. (NOAA)

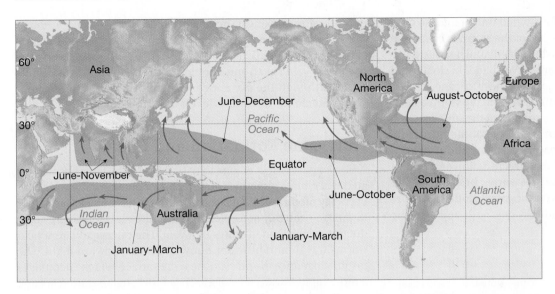

FIGURE 19.25 This world map shows the regions where most hurricanes form as well as their principal months of occurrence and the most common tracks they follow. Hurricanes do not develop within about 5° of the equator because the Coriolis effect is too weak. Because warm surface ocean temperatures are necessary for hurricane formation, they seldom form poleward of 20° latitude nor over cool waters of the South Atlantic and the eastern South Pacific.

of Louisiana, Mississippi, and Alabama and took an estimated 1,300 lives. Although hundreds of thousands fled before the storm made landfall, thousands of others were caught by the storm. In addition to the human suffering and tragic loss of life that were left in the wake of Hurricane Katrina, the financial losses caused by the storm are practically incalculable. Although imprecise, some suggest that the final accounting exceeded $100 billion.

Profile of a Hurricane

Most hurricanes form between the latitudes of 5 degrees and 20 degrees over all the tropical oceans except those of the South Atlantic and eastern South Pacific (**Figure 19.25**). The North Pacific has the greatest number of storms, averaging 20 per year. Fortunately for those living in the coastal regions of the southern and eastern United States, fewer than five hurricanes, on the average, develop annually in the warm sector of the North Atlantic.

These intense tropical storms are known in various parts of the world by different names. In the western Pacific, they are called *typhoons*, and in the Indian Ocean, including the Bay of Bengal and Arabian Sea, they are simply called *cyclones*. In the following discussion, these storms are referred to as hurricanes. The term *hurricane* is derived from Huracan, a Carib god of evil.

Although many tropical disturbances develop each year, only a few reach hurricane status. By international agreement, a hurricane has wind speeds in excess of 119 kilometers (74 miles) per hour and a rotary circulation. Mature hurricanes average 600 kilometers (375 miles) in diameter and often extend 12,000 meters (40,000 feet) above the ocean surface. From the outer edge to the center, the barometric pressure has on occasion dropped 60 millibars, from 1,010 millibars to 950 millibars. The lowest pressures ever recorded in the Western Hemisphere are associated with these storms.

A steep pressure gradient generates the rapid, inward-spiraling winds of a hurricane (**Figure 19.26**). As the air rushes toward the center of the storm, its velocity increases. This occurs for the same reason that skaters with their arms extended spin faster as they pull their arms in close to their bodies.

As the inward rush of warm, moist surface air approaches the core of the storm, it turns upward and ascends in a ring of cumulonimbus

FIGURE 19.26 Weather maps showing Hurricane Fran at 7:00 A.M. EST on two successive days, September 5 and 6, 1996. On September 5, winds exceeded 190 kilometers (118 miles) per hour. As the storm moved inland, heavy rains caused flash floods, killed 30 people, and caused more than $3 billion in damages. The station information plotted off the Gulf and Atlantic coasts is from data buoys, which are remote floating instrument packages. The small boxes extending southeast from the storm's center show the position of the eye at 6-hour intervals.

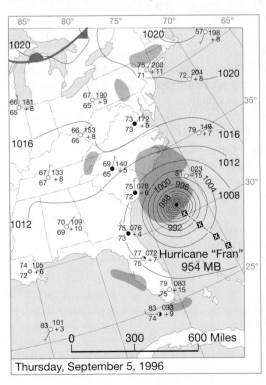

Thursday, September 5, 1996

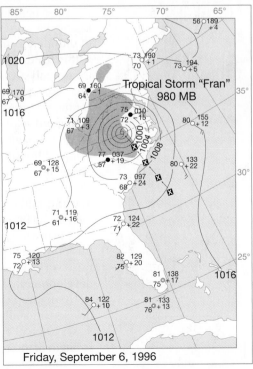

Friday, September 6, 1996

towers (Figure 19.27). This doughnut-shaped wall of intense convective activity surrounding the center of the storm is called the **eye wall**. It is here that the greatest wind speeds and heaviest rainfall occur. Surrounding the eye wall are curved bands of clouds that trail away in a spiral fashion. Near the top of the hurricane, the airflow is outward, carrying the rising air away from the storm center, thereby providing room for more inward flow at the surface.

At the very center of the storm is the **eye** of the hurricane (Figure 19.27). This well-known feature is a zone about 20 kilometers (12.5 miles) in diameter where precipitation ceases and winds subside. It offers a brief but deceptive break from the extreme weather in the enormous curving wall clouds that surround it. The air within the eye gradually descends and heats by

FIGURE 19.27 A. Cross section of a hurricane. Note that the vertical dimension is greatly exaggerated. The eye, the zone of relative calm at the center of the storm, is a distinctive hurricane feature. Sinking air in the eye warms by compression. Surrounding the eye is the eye wall, the zone where winds and rain are most intense. Tropical moisture spiraling inward creates rain bands that pinwheel around the storm center. Outflow of air at the top of the hurricane is important because it prevents the convergent flow at lower levels from "filling in" the storm. (After NOAA) **B.** Measurements of surface pressure and wind speed during the passage of Cyclone Monty at Mardie Station in Western Australia between February 29 and March 2, 2004. The strongest winds are associated with the eye wall, and the weakest winds and lowest pressure are found in the eye. (Data from World Meteorological Organization)

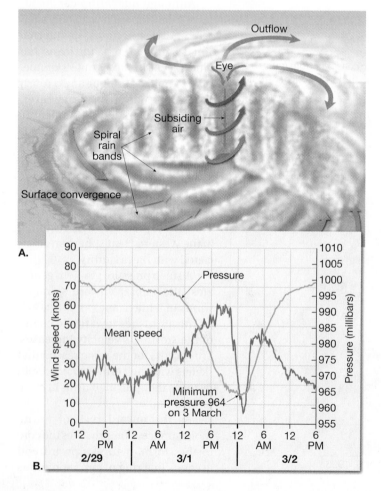

compression, making it the warmest part of the storm. Although many people believe that the eye is characterized by clear blue skies, such is usually not the case because the subsidence in the eye is seldom strong enough to produce cloudless conditions. Although the sky appears much brighter in this region, scattered clouds at various levels are common.

Hurricane Formation and Decay

A hurricane is a heat engine that is fueled by the latent heat liberated when huge quantities of water vapor condense. The amount of energy produced by a typical hurricane in just a single day is truly immense—roughly equivalent to the entire electrical energy production of the United States in a year. The release of latent heat warms the air and provides buoyancy for its upward flight. The result is to reduce the pressure near the surface, which encourages a more rapid inward flow of air. To get this engine started, a large quantity of warm, moisture-laden air is required, and a continual supply is needed to keep it going.

Hurricanes develop most often in the late summer when water temperatures have reached 27° C (80° F) or higher and thus are able to provide the necessary heat and moisture to the air. This ocean-water temperature requirement accounts for the fact that hurricanes do not form over the relatively cool waters of the South Atlantic and the eastern South Pacific. For the same reason, few hurricanes form poleward of 20 degrees of latitude. Although water temperatures are sufficiently high, hurricanes do not form within 5 degrees of the equator, because the Coriolis effect is too weak to initiate the necessary rotary motion.

Many tropical storms begin as disorganized arrays of clouds and thunderstorms that develop weak pressure gradients but exhibit little or no rotation. Such areas of low-level convergence and lifting are called *tropical disturbances*. Most of the time these zones of convective activity die out. However, tropical disturbances occasionally grow larger and develop a strong cyclonic rotation.

What happens on those occasions when conditions favor hurricane development? As latent heat is released from the clusters of thunderstorms that make up the tropical disturbance, areas within the disturbance get warmer. As a consequence, air density lowers and surface pressure drops, creating a region of weak low pressure and cyclonic circulation. As pressure drops at the storm center, the pressure gradient steepens. If you were watching an animated weather map of the storm, you would see the isobars get closer together. In response, surface wind speeds increase and bring additional supplies of moisture to nurture storm growth. The water vapor condenses, releasing latent heat, and the heated air rises. Adiabatic cooling of rising air triggers more condensation and the release of more latent heat, which causes a further increase in buoyancy. And so it goes.

Meanwhile, at the top of the storm, air is diverging. Without this outward flow up top, the inflow at lower levels would soon raise surface pressures (i.e., fill in the low) and thwart storm development.

Many tropical disturbances occur each year, but only a few develop into full-fledged hurricanes. By international agreement, lesser tropical cyclones are given different names based on the strength of their winds. When a cyclone's strongest winds do not exceed 61 kilometers (38 miles) per hour, it is called a **tropical**

depression. When winds are between 61 and 119 kilometers (38 and 74 miles) per hour, the cyclone is termed a **tropical storm**. It is during this phase that a name is given (Andrew, Mitch, Opal, etc.). Should the tropical storm become a hurricane, the name remains the same. Each year between 80 and 100 tropical storms develop around the world. Of these, usually half or more eventually reach hurricane status.

Hurricanes diminish in intensity whenever they (1) move over ocean waters that cannot supply warm, moist tropical air; (2) move onto land; or (3) reach a location where the large-scale flow aloft is unfavorable. Whenever a hurricane moves onto land, it loses its punch rapidly. For example, in Figure 19.26 notice how the isobars show a much weaker pressure gradient on September 6 after Hurricane Fran moved ashore than on September 5 when it was over the ocean. The most important reason for this rapid demise is the fact that the storm's source of warm, moist air is cut off. When an adequate supply of water vapor does not exist, condensation and the release of latent heat must diminish. In addition, friction from the increased roughness of the land surface rapidly slows surface wind speeds. This factor causes the winds to move more directly into the center of the low, thus helping to eliminate the large pressure differences.

Hurricane Destruction

A location only a few hundred kilometers from a hurricane—just a day's striking distance away—may experience clear skies and virtually no wind. Before the age of weather satellites, such a situation made it very difficult to warn people of impending storms. The deadly hurricane that devastated Galveston, Texas, in 1900 is a tragic example (**Figure 19.28**). It was a complete surprise to the unsuspecting residents of this vulnerable barrier island.

Once a storm develops cyclonic flow and the spiraling bands of clouds characteristic of a hurricane, it receives continuous monitoring. Satellites are able to identify and track the storms long before they make landfall. Moreover, reconnaissance aircraft are sent to evaluate storms. When a hurricane is within range, specially instrumented aircraft can fly directly into a threatening storm and accurately measure details of its position and current state of development. When a hurricane nears the coast, it is monitored by land-based Doppler weather radar. Data buoys represent another source of information. These remote floating instrument packages are positioned in fixed locations all along the Gulf Coast and Atlantic Coast of the United States. If you examine the weather maps in Figure 19.26, you can see data buoy information plotted at several offshore stations.

In the United States, early warning systems have greatly reduced the number of deaths caused by hurricanes. At the same time, however, there has been an astronomical rise in the amount of property damage. The primary reason for this trend has been the rapid population growth in coastal areas.

Although the amount of damage caused by a hurricane depends on several factors, including the size and population density of the area affected and the shape of the ocean bottom near the shore, the most significant factor is the strength of the storm itself.

Based on the study of past storms, the *Saffir–Simpson scale* was established to rank the relative intensities of hurricanes (**Table 19.2**). Predictions of hurricane severity and damage are usually expressed in terms of this scale. When a tropical storm becomes a hurricane, the National Weather Service assigns it a scale (category) number. As conditions change, the category of the storm is reevaluated, so that public-safety officials can be kept informed. By using the Saffir–Simpson scale, the disaster potential of a hurricane can be monitored and appropriate precautions can be planned and implemented. A rating of 5 on the scale represents the worst storm possible, and a 1 is least severe. Storms that fall into category 5 are rare. Only three storms this powerful are known to have hit the United States. Andrew struck Florida in 1992, Camille pounded Mississippi in 1969, and a Labor Day hurricane struck the Florida Keys in 1935.

Students Sometimes Ask...

Why are hurricanes given names, and who picks the names?

Actually, the names are given once the storms reach tropical-storm status (winds between 61 and 119 kilometers per hour). Tropical storms are named to provide ease of communication between forecasters and the general public regarding forecasts, watches, and warnings. Tropical storms and hurricanes can last a week or longer, and two or more storms can be occurring in the same region at the same time. Thus, names can reduce the confusion about what storm is being described.

The World Meteorological Organization (affiliated with the United Nations) creates the lists of names. The names for Atlantic storms are used again at the end of a 6-year cycle unless a hurricane was particularly noteworthy. Such names are retired to prevent confusion when storms are discussed in future years.

FIGURE 19.28 Aftermath of the Galveston hurricane that struck an unsuspecting and unprepared city on September 8, 1900. It was the worst natural disaster in U.S. history. Entire blocks were swept clean, while mountains of debris accumulated around the few remaining buildings. (AP Photo)

TABLE 19.2 Saffir–Simpson Hurricane Scale

Scale Number (category)	Central Pressure (millibars)	Winds (km/hr)	Storm Surge (meters)	Damage
1	≥980	119–153	1.2–1.5	Minimal
2	965–979	154–177	1.6–2.4	Moderate
3	945–964	178–209	2.5–3.6	Extensive
4	920–944	210–250	3.7–5.4	Extreme
5	<920	>250	>5.4	Catastrophic

Damage caused by hurricanes can be divided into three categories: (1) storm surge, (2) wind damage, and (3) inland flooding.

Storm Surge The most devastating damage in the coastal zone is caused by the storm surge (see Figure 19.28). It not only accounts for a large share of coastal property losses but is also responsible for 90 percent of all hurricane-caused deaths. A **storm surge** is a dome of water 65–80 kilometers (40–50 miles) wide that sweeps across the coast near the point where the eye makes landfall. If all wave activity were smoothed out, the storm surge is the height of the water above normal tide level. Thus, a storm surge commonly adds 2–3 meters (6–10 feet) to normal tide heights—to say nothing of tremendous wave activity superimposed atop the surge.

The most important factor responsible for the development of a storm surge is the piling up of ocean water by strong onshore winds. The hurricane's winds gradually push water toward the shore, causing sea level to elevate while also churning up violent wave activity.

As a hurricane advances toward the coast in the Northern Hemisphere, storm surge is always most intense on the right side of the eye where winds are blowing *toward* the shore. In addition, on this side of the storm, the forward movement of the hurricane also contributes to the storm surge. In Figure 19.29, assume a hurricane with peak winds of 175 kilometers (109 miles) per hour is moving toward the shore at 50 kilometers (31 miles) per hour. The net wind speed on the right side of the advancing storm is 225 kilometers (140 miles) per hour. On the left side, the hurricane's winds are blowing opposite the direction of storm movement, so the net winds are blowing *away* from the coast at 125 kilometers (78 miles) per hour. Along

the shore facing the left side of the oncoming hurricane, the water level may actually decrease as the storm makes landfall.

Wind Damage Destruction caused by wind is perhaps the most obvious of the classes of hurricane damage. For some structures, the force of the wind is sufficient to cause total ruin. Mobile homes are particularly vulnerable. In addition, the strong winds can create a dangerous barrage of flying debris. In regions with good building codes, wind damage is usually not as catastrophic as storm-surge damage. However, hurricane-force winds affect a much larger area than does storm surge and can cause huge economic losses.

Hurricanes sometimes produce tornadoes that contribute to the storm's destructive power. Studies have shown that more than half of the hurricanes that make landfall produce at least one tor-

FIGURE 19.29 Winds associated with a Northern Hemisphere hurricane that is advancing toward the coast. This hypothetical storm, with peak winds of 175 kilometers (109 miles) per hour, is moving toward the coast at 50 kilometers (31 miles) per hour. On the right side of the advancing storm, the 175-kilometer-per-hour winds are in the same direction as the movement of the storm (50 kilometers per hour). Therefore, the *net* wind speed on the right side of the storm is 225 kilometers (140 miles) per hour. On the left side, the hurricane's winds are blowing opposite the direction of storm movement, so the net winds of 125 kilometers (78 miles) per hour are away from the coast. Storm surge will be greatest along that part of the coast hit by the right side of the advancing hurricane.

Students Sometimes Ask...

When is hurricane season?

Hurricane season is different in different parts of the world. People in the United States are usually most interested in Atlantic storms. The Atlantic hurricane season officially extends from June through November. More than 97 percent of tropical activity in that region occurs during this 6 month span. The "heart" of the season is August through October. During these 3 months, 87 percent of the minor hurricane (category 1 and 2) days and 96 percent of the major hurricane (category 3, 4, and 5) days occur. Peak activity is in early- to mid-September.

nado. In 2004, the number of tornadoes associated with tropical storms and hurricanes was extraordinary. Tropical Storm Bonnie and five landfalling hurricanes—Charley, Frances, Gaston, Ivan, and Jeanne— produced nearly 300 tornadoes that affected the southeast and mid-Atlantic states. Hurricane Frances produced 117 tornadoes—the most ever reported from one hurricane. The large number of hurricane-generated tornadoes in 2004 helped make it a record-breaking year—surpassing the previous record by more than 300.

Inland Flooding   The torrential rains that accompany most hurricanes represent a third significant threat—flooding. While the effects of storm surge and strong winds are concentrated in coastal areas, heavy rains may affect places hundreds of kilometers from the coast for several days after the storm has lost its hurricane-force winds.

Hurricanes weaken rapidly as they move inland, yet the remnants of the storm can still yield 15–30 centimeters (6–12 inches) or more of rain as they move inland. A good example of such destruction was Hurricane Floyd. In September 1999, this storm dumped more than 48 centimeters (19 inches) of rain on Wilmington, North Carolina, 33.98 centimeters (13.38 inches) of it in a single 24-hour span. In August and September 2004, when the remnants of hurricanes Charley, Frances, Ivan, and Jeanne moved northward from Florida and the Gulf Coast, they brought huge rains and floods from Alabama and Georgia northward through the Carolinas and beyond.

To summarize, extensive damage and loss of life in the coastal zone can result from storm surge, torrential rains, and strong winds. When loss of life occurs, it is commonly caused by the storm surge, which can devastate entire barrier islands and low-lying land along the coast. Although wind damage is usually not as catastrophic as the storm surge, it affects a much larger area. Where building codes are inadequate, economic losses can be especially severe. Because hurricanes weaken as they move inland, most wind damage occurs within 200 kilometers of the coast. Far from the coast a weakening storm can produce extensive flooding long after the winds have diminished below hurricane levels. Sometimes the damage from inland flooding exceeds storm-surge destruction.

Tracking Hurricanes

Today, we have the benefit of numerous observational tools for tracking tropical storms and hurricanes. Using input from satellites, aircraft reconnaissance, coastal radar, and remote data buoys in conjunction with sophisticated computer models, meteorologists monitor and forecast storm movements and intensity. The goal is to issue timely watches and warnings.

The track forecast is probably the most basic information because accurate prediction of other storm characteristics is of little value if there is significant uncertainty as to where the storm is going. Accurate track forecasts are important because they can lead to timely evacuations from the surge zone where the greatest number of deaths usually occur. Fortunately, track forecasts have been steadily improving. During the span 2001–2005, forecast errors were roughly half of what they were in 1990. During the very active 2004 and 2005 Atlantic hurricane seasons, 12- to 72-hour track forecast accuracy was at or near record levels. Consequently, the length of official track forecasts issued by the National Hurricane Center was extended from 3 days to 5 days (**Figure 19.30**). Current 5-day track forecasts are now as accurate as 3-day forecasts were 15 years ago. Much of the progress is due to improved computer models and to a dramatic increase in the quantity of satellite data from over the oceans.

Despite improvements in accuracy, forecast uncertainty still requires that hurricane warnings be issued for relatively large coastal areas. During the span 2000–2005, the average length of

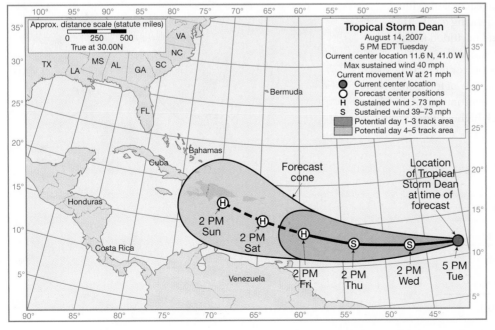

FIGURE 19.30 Five-day track forecast for Tropical Storm Dean issued at 5 P.M. EDT, Tuesday, August 14, 2007. When a hurricane track forecast is issued by the National Hurricane Center, it is termed a *forecast cone*. The cone represents the probable track of the center of the storm and is formed by enclosing the area swept out by a set of circles along the forecast track (at 12, 24, 36 hours, etc.). The size of each circle gets larger with time. Based on statistics from 2003 to 2007, the entire track of Atlantic tropical cyclones can be expected to remain entirely within the cone roughly 60–70 percent of the time. (National Weather Service/National Hurricane Center)

coastline under a hurricane warning in the United States was 510 kilometers (316 miles). This represents a significant improvement over the preceding decade when the average was 730 kilometers (452 miles). Nevertheless, only about one quarter of an average warning area experiences hurricane conditions.

CONCEPT CHECK 19.6

1. Which has stronger winds, a tropical storm or a tropical depression?

2. During what time of year do most of the hurricanes that affect North America form? Why is hurricane formation favored at this time?

3. Why does the intensity of a hurricane diminish rapidly when it moves onto land?

4. Hurricane damage can be divided into three broad categories. List them. Which category is responsible for the highest percentage of hurricane-related deaths?

GIVE IT SOME THOUGHT

1. Refer to Figure 19.4 to answer these questions.
   a. Thunder Bay and Marquette are both on the shore of Lake Superior yet Marquette gets much more snow than Thunder Bay. Why is this the case?
   b. Notice the narrow, north–south oriented zone of relatively heavy snow east of Pittsburgh and Charleston. This region is too far from the Great Lakes to receive lake-effect snowfall. Speculate on a likely reason for the higher snowfalls here. Does your answer explain the shape of this snowy zone?

2. During winter, polar air masses are cold. Which should be colder, a wintertime mP air mass or a wintertime cP air mass? Explain.

3. Apply your knowledge of fronts to explain the following weather proverb:
   Rain long foretold, long last;
   Short notice, soon past.

4. If you hear that a cyclone is approaching, should you immediately seek shelter? Why or why not?

5. Refer to the accompanying weather map and answer the following questions.
   a. What is the probable wind direction at each of the cities?
   b. Identify the likely air mass that is influencing each city.
   c. Identify the cold front, warm front, and occluded front.
   d. What is the barometric tendency at city A and city C?
   e. Which of the three cities is coldest? Which one is warmest?

6. The accompanying table lists the total number of tornadoes reported in the United States by decade. Propose a reason to explain why the totals for the 1990s and 2000s are so much higher than for the 1950s and 1960s.

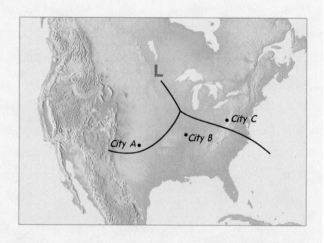

1950–59	4,796
1960–69	6,613
1970–79	8,579
1980–89	8,196
1990–99	12,138
2000–09	12,914

7. Assume it is late September 2016 and Hurricane Gaston, a category 5 storm, is projected to follow the path shown on the accompanying map. Answer the following questions.

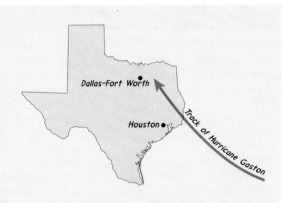

a. If this were the first category 5 hurricane to make landfall since Andrew (1992), what would be the *total* category 5 hurricanes to have hit the United States?

b. Name the stages of development that Gaston must have gone through to become a hurricane? When did it receive its name?

c. Should the city of Houston prepare for Gaston's fastest winds and greatest storm surge? Explain why or why not.

d. What is the greatest threat to life and property if this storm approaches the Dallas–Fort Worth area?

8. Refer to Figure 19.26 to answer the following questions.

a. On which of the two days were Fran's wind speeds probably highest? How were you able to determine this?

b. How far (in miles) did the eye of the hurricane move in the 24-hour period represented by these maps? At what rate (in miles per hour) did the storm move during this 24-hour span?

In Review   Chapter 19 Weather Patterns and Severe Storms

- An *air mass* is a large body of air, usually 1,600 kilometers (1,000 miles) or more across, which is characterized by a *sameness of temperature and moisture* at any given altitude. When this air moves out of its region of origin, called the *source region*, it will carry these temperatures and moisture conditions elsewhere, perhaps eventually affecting a large portion of a continent.

- Air masses are classified according to (1) the nature of the surface in the source region and (2) the latitude of the source region. *Continental (c)* designates an air mass of land origin, with the air likely to be dry; whereas a *maritime (m)* air mass originates over water, and therefore will be humid. *Polar (P) and arctic (A)* air masses originate in high latitudes and are cold. *Tropical (T)* air masses form in low latitudes and are warm. According to this classification scheme, the *basic types of air masses are continental polar (cP), continental arctic (cA), continental tropical (cT), maritime polar (mP), and maritime tropical (mT)*. Continental polar (cP) and maritime tropical (mT) air masses influence the weather of North America most, especially east of the Rocky Mountains. Maritime tropical air is the source of much, if not most, of the precipitation received in the eastern two-thirds of the United States.

- *Fronts* are boundaries that separate air masses of different densities, one warmer and often higher in moisture content than the other. A *warm front* occurs when the surface position of the front moves so that warm air occupies territory formerly covered by cooler air. Along a warm front, a warm air mass overrides a retreating mass of cooler air. As the warm air ascends, it cools adiabatically to produce clouds and, frequently, light-to-moderate precipitation over a large area. A *cold front* forms where cold air is actively advancing into a region occupied by warmer air. Cold fronts are about twice as steep and move more rapidly than do warm fronts. Because of these two differences, precipitation along a cold front is usually more intense and of shorter duration than is precipitation associated with a warm front.

- The primary weather producers in the middle latitudes are *large centers of low pressure* that generally travel from *west to east*, called *middle-latitude cyclones*. These *bearers of stormy weather*, which last from a few days to a week, have a *counterclockwise circulation* pattern in the Northern Hemisphere, with an *inward flow of air* toward their centers. Most middle-latitude cyclones have a *cold front and frequently a warm front* extending from the central area of low pressure. *Convergence and forceful lifting along the fronts* initiate cloud development and frequently cause precipitation. As a middle-latitude cyclone with its associated fronts passes over a region, it often brings with it abrupt changes in the weather. The particular weather experienced by an area depends on the path of the cyclone.

- *Thunderstorms* are caused by the upward movement of warm, moist, unstable air, triggered by a number of different processes. They are associated with cumulonimbus clouds that generate heavy rainfall, thunder, lightning, and occasionally hail and tornadoes.

- *Tornadoes*—destructive, local storms of short duration—are violent windstorms associated with severe thunderstorms that take the form of a rotating column of air that extends downward from a cumulonimbus cloud. Tornadoes are most often spawned along the cold front of a middle-latitude cyclone, most frequently during the spring months.

- *Hurricanes*, the greatest storms on Earth, are tropical cyclones with wind speeds in excess of 119 kilometers (74 miles) per hour. These complex tropical disturbances develop over tropical ocean waters and are fueled by the latent heat liberated when huge quantities of water vapor condense. Hurricanes form most often in late summer when ocean-surface temperatures reach 27° C (80° F) or higher and thus are able to provide the necessary heat and moisture to the air. Hurricanes diminish in intensity whenever they (1) move over cool ocean water that cannot supply adequate heat and moisture, (2) move onto land, or (3) reach a location where large-scale flow aloft is unfavorable.

Key Terms

air-mass (p. 552)
air-mass weather (p. 552)
arctic (A) air mass (p. 553)
cold front (p. 556)
continental (c) air mass (p. 553)
Doppler radar (p. 569)
eye (p. 572)
eye wall (p. 572)
front (p. 555)
hurricane (p. 570)

lake-effect snow (p. 554)
maritime (m) air mass (p. 553)
middle-latitude or mid-latitude cyclone
  (p. 558)
occluded front (p. 557)
occlusion (p. 559)
overrunning (p. 555)
polar (P) air mass (p. 553)
source region (p. 553)
stationary front (p. 557)

storm surge (p. 574)
thunderstorm (p. 562)
tornado (p. 564)
tornado warning (p. 568)
tornado watch (p. 568)
tropical (T) air mass (p. 553)
tropical depression (p. 572)
tropical storm (p. 573)
warm front (p. 556)

Examining the Earth System

1. Which spheres of the Earth system interact in the Great Lakes region of North America to produce the high snowfalls on the downwind shores of the lakes? What term is applied to these heavy snows? Describe what creates this effect.

2. Hurricanes are among the most severe storms experienced on Earth. When a hurricane makes landfall in the southeastern United States, what impact might it have on coastal lands (geosphere), drainage networks (hydrosphere), and natural vegetation (biosphere)?

Mastering Geology

Looking for additional review and test prep materials? Visit the Self Study area in **www.masteringgeology.com** to find practice quizzes, study tools, and multimedia that will aid in your understanding of this chapter's content. In **MasteringGeology™** you will find:

- **GEODe: Earth Science: An interactive visual walkthrough of key concepts**

- **Geoscience Animation Library:** More than 100 animations illuminating many difficult-to-understand Earth science concepts
- **In The News RSS Feeds:** Current Earth science events and news articles are pulled into the site with assessment
- **Pearson eText**
- **Optional Self Study Quizzes**
- **Web Links**
- **Glossary**
- **Flashcards**

20

World Climates and Global Climate Change

Mauna Loa Observatory is an important atmospheric research facility on Hawaii's Big Island. It has been collecting data and monitoring atmospheric change since the 1950s. Its remote location atop a 3,397-meter (11,140-foot) volcano make it ideal for monitoring atmospheric constituents that can cause climate change.

(Photo by Forrest M. Mims III)

The focus of this chapter is *climate*, the long-term aggregate of weather. Climate is more than just an expression of average atmospheric conditions. In order to accurately portray the character of a place or an area, variations and extremes must also be included. Climate strongly influences the nature of plant and animal life, the soil, and many external geological processes. Climate influences people as well.

Although climate has a significant impact on people, we are learning that people also have a strong influence on climate. In fact, today global climate change caused by humans is making headlines. Unlike changes in the geologic past, which represented natural variations, modem climate change is dominated by human influences that are sufficiently large that they exceed the bounds of natural variability (Figure 20.1). Moreover, these changes are likely to continue for many centuries. The effects of this venture into the unknown with climate could be very disruptive not only to humans but to many other life-forms as well.

FOCUS ON CONCEPTS

To assist you in learning the important concepts in this chapter, focus on the following questions:

- What is the climate system and what are its five major parts?
- What climate data are needed to classify climates using the Köppen system?
- What are the five principal climate groups and their characteristics?
- In what ways are humans changing the composition of the atmosphere? How is the atmosphere responding to these changes?
- What are climate feedback mechanisms? What part might they play in global climate change?
- What are some possible consequences of global warming?

The Climate System

Throughout this book you have been reminded that Earth is a multidimensional system that consists of many interacting parts. A change in any one part can produce changes in any or all of the other parts—often in ways that are neither obvious nor immediately apparent. This fact is certainly true when it comes to the study of climate and climate change.

To understand and appreciate climate, it is important to realize that climate involves more than just the atmosphere.

The atmosphere is the central component of the complex, connected, and interactive global environmental system upon which all life depends. Climate may be broadly defined as the long-term behavior of this environmental system. To understand fully and to predict changes in the atmospheric component of the climate system, one must understand the Sun, oceans, ice sheets, solid Earth, and all forms of life.[46]

Indeed, we must recognize that there is a **climate system** that includes the atmosphere, hydrosphere, geosphere, biosphere, and cryosphere. (The *cryosphere* is the ice and snow that exist at Earth's surface.) Powered by energy from the Sun, the climate system involves the exchanges of energy and moisture that occur among the five spheres. These exchanges link the atmosphere to the other parts of the system to produce an integrated and

FIGURE 20.1 Ancient bristlecone pines in California's White Mountains. The study of tree-growth rings is one way that scientists reconstruct past climates. Some of these trees are more than 4,000 years old. (Photo by Bill Stevenson/Photolibrary)

[46]The American Meteorological Society and the University Corporation for Atmospheric Research, "Weather and the Nation's Well-Being," *Bulletin of the American Meteorological Society* 73(12) (December 1991): 2038.

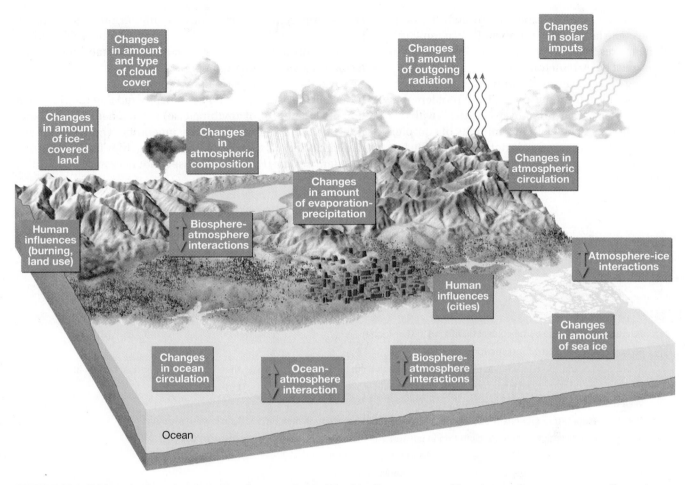

FIGURE 20.2 Schematic view showing several components of Earth's climate system. Many interactions occur among the various components on a wide range of space and time scales, making the system extremely complex.

extremely complex interactive unit. The major components of the climate system are shown in **Figure 20.2**.

The climate system provides a framework for the study of climate. The interactions and exchanges among the parts of the climate system create a complex network that links the five spheres. Changes to the system do not occur in isolation. Rather, when one part of this interactive unit changes, the other components also react. This well-established relationship is demonstrated often when we study the world's climates and global climate change.

CONCEPT CHECK 20.1

❶ **List the five parts of the climate system.**

World Climates

Previous chapters have already presented the spatial and seasonal variations of the major elements of weather and climate. Chapter 16 examined the controls of temperature and the world distribution of temperature. In Chapter 18, you studied the general circulation of the atmosphere and the global distribution of precipitation. You are now ready to investigate the *combined* effects of these variations in different parts of the world. The varied nature of Earth's surface and the many interactions that occur among atmospheric processes give every location on our planet a distinctive, even unique, climate. However, we are not going to describe the unique climatic character of countless different locales. Instead, the purpose is to introduce the major climate regions of the world. The discussion examines large areas and uses particular places only to illustrate the characteristics of these major climate regions.

Temperature and precipitation are the most important elements in a climate description because they have the greatest influence on people and their activities and also have an important impact on the distribution of such phenomena as vegetation and soils. Nevertheless, other factors are also important for a complete climatic description. When possible, some of these factors are introduced into our discussion of world climates.

Climate Classification

The distribution of the major atmospheric elements is, to say the least, complex. Because of the many differences from place to place as well as from time to time at a particular locale, it is unlikely that any two sites on Earth experience exactly the same weather conditions. The fact that the number of places on Earth is virtually infinite makes it readily apparent that the number of different climates must be extremely large.

Of course, having a great diversity of information to investigate is not unique to the study of the atmosphere; it is a problem that is basic to all science. Consider astronomy, which deals with billions of stars, and biology, which studies millions of complex organisms. To cope with such variety, it is not only desirable but essential to devise some means of classifying the vast array of data to be studied. By establishing groups consisting of items that have certain important characteristics in common, order and simplicity are introduced. Bringing order to large quantities of information not only aids comprehension and understanding but also facilitates analysis and explanation.

Over the past 100 years, many climate-classification systems have been devised. It should be remembered that the classification of climates (or of anything else) is not a natural phenomenon but the product of human ingenuity. The value of any particular classification is determined largely by its intended use. A system designed for one purpose is not necessarily applicable to another.

TABLE 20.1 Köppen System of Climate Classification[*]

Letter Symbol			
1st	**2nd**	**3rd**	
A			Average temperature of the coldest month is 18° C or higher.
	f		Every month has 6 cm of precipitation or more.
	m		Short, dry season; precipitation in driest month less than 6 cm but equal to or greater than 10–R/25 (R is annual rainfall in cm).
	w		Well-defined winter dry season; precipitation in driest month less than 10–R/25.
	s		Well-defined summer dry season (rare).
B			Potential evaporation exceeds precipitation. The dry–humid boundary is defined by the following formulas:
			(Note: R is the average annual precipitation in cm and T is the average annual temperature in °C.)
			$R < 2T + 28$ when 70% or more of rain falls in warmer 6 months.
			$R < 2T$ when 70% or more of rain falls in cooler 6 months.
			$R < 2T + 14$ when neither half year has 70% or more of rain.
	S	Steppe	The BS–BW boundary is 1/2 the dry–humid boundary.
	W	Desert	
		h	Average annual temperature is 18° C or greater.
		k	Average annual temperature is less than 18° C.
C			Average temperature of the coldest month is under 18° C and above –3° C.
	w		At least 10 times as much precipitation in a summer month as in the driest winter month.
	s		At least three times as much precipitation in a winter month as in the driest summer month; precipitation in driest summer month less than 4 cm.
	f		Criteria for w and s cannot be met.
		a	Warmest month is over 22° C; at least 4 months over 10° C.
		b	No month above 22° C; at least 4 months over 10° C.
		c	One to 3 months above 10° C.
D			Average temperature of coldest month is –3° C or below; average temperature of warmest month is greater than 10° C.
	s		Same as under C.
	w		Same as under C.
	f		Same as under C.
		a	Same as under C.
		b	Same as under C.
		c	Same as under C.
		d	Average temperature of the coldest month is –38° C or below.
E			Average temperature of the warmest month is below 10° C.
	T		Average temperature of the warmest month is greater than 0° C and less than 10° C.
	F		Average temperature of the warmest month is 0° C or below.

[*] When classifying climatic data using Table 20.1, you should first determine whether the data meet the criteria for the E climates. If the station is not an E climate, proceed to the criteria for B climates. If your data do not fit into either the E or B groups, check the data against the criteria for A, C, and D climates, in that order.

The Köppen Classification

In this chapter, we use a classification devised by Russian-born German climatologist Wladimir Köppen (1846–1940). As a tool for presenting the general world pattern of climates, the **Köppen classification** has been the best-known and most used system for decades. It is widely accepted for many reasons. For one, it uses only easily obtained data: mean monthly and annual values of temperature and precipitation. Furthermore, the criteria are unambiguous, relatively simple to apply, and divide the world into climate regions in a realistic way.

Köppen believed that the distribution of natural vegetation was an excellent expression of the totality of climate. Consequently, the boundaries he chose were largely based on the limits of certain plant associations. Five principal groups were recognized; each group was designated by a capital letter as follows:

A. *Humid tropical.*   Winterless climates; all months having a mean temperature above 18° C (64° F).
B. *Dry.*   Climates where evaporation exceeds precipitation; there is a constant water deficiency.
C. *Humid middle-latitude.*   Mild winters; the average temperature of the coldest month is below 18° C (64° F) but above –3° C (27° F).
D. *Humid middle-latitude.*   Severe winters; the average temperature of the coldest month is below –3° C (27° F), and the warmest monthly mean exceeds 10° C (50° F).
E. *Polar.*   Summerless climates; the average temperature of the warmest month is below 10° C (50° F).

Notice that four of the major groups (A, C, D, and E) are defined on the basis of temperature characteristics, and the fifth, the B group, has precipitation as its primary criterion. Each of the five groups is further subdivided by using the criteria and symbols presented in **Table 20.1**. A strength of the Köppen system is the relative ease with which boundaries are determined. However, these boundaries cannot be viewed as fixed. On the contrary, all climate boundaries shift from year to year (**Figure 20.3**). The boundaries shown on climate maps are simply average locations based on data collected over many years. Thus, a climate boundary should be regarded as a broad transition zone and not a sharp line.

The world distribution of climates according to the Köppen classification is shown in **Figure 20.4**. You will refer to this map several times as Earth's climates are discussed in the following pages.

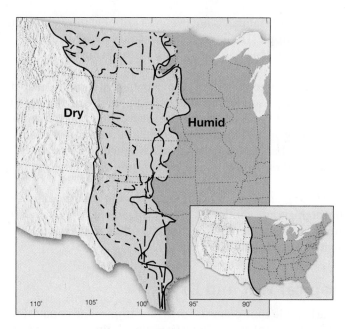

FIGURE 20.3 Yearly fluctuations in the dry–humid boundary during a 5-year period. The small inset shows the average position of the dry–humid boundary.

Humid Tropical (A) Climates

Within the A group of climates, two main types are recognized: wet tropical climates (Af and Am) and tropical wet and dry (Aw).

The Wet Tropics

The constantly high temperatures and year-round rainfall in the wet tropics combine to produce the most luxuriant vegetation found in any climatic realm: the **tropical rain forest** (**Figure 20.5**).

The environment of the wet tropics characterizes almost 10 percent of Earth's land area. An examination of Figure 20.4 shows that Af and Am climates form a discontinuous belt astride the equator that typically extends 5–10° into each hemisphere. The poleward margins are most often marked by diminishing rainfall, but occasionally decreasing temperatures mark the boundary. Because of the general decrease in temperature with height in the troposphere, this climate region is restricted to elevations below 1,000 meters. Consequently, the major interruptions near the equator are principally cooler highland areas.

Data for some representative stations in the wet tropics are shown in **Figure 20.6A**, B. A brief examination reveals the most obvious features that characterize the climate in these areas.

1. Temperatures usually average 25° C (77° F) or more each month. Consequently, not only is the annual mean temperature high, but the annual temperature range is also very small.
2. The total precipitation for the year is high, often exceeding 200 centimeters (80 inches).

CONCEPT CHECK 20.2
1 Why is classification often a helpful or even necessary task in science?
2 What climate data are needed in order to classify a climate using the Köppen system?

3. Although rainfall is not evenly distributed throughout the year, tropical rain forest stations are generally wet in all months. If a dry season exists, it is very short.

Because places with an Af or Am designation lie near the equator, the reason for the uniform temperature rhythm experienced in such locales is clear: The intensity of solar radiation is consistently high. The vertical rays of the Sun are always relatively close, and changes in the length of daylight throughout the year are slight; therefore, seasonal temperature variations are minimal.

The region is strongly influenced by the equatorial low. Its converging trade winds and the accompanying ascent of warm, humid, unstable air produce conditions that are ideal for the formation of precipitation.

Tropical Wet and Dry

In the latitude zone poleward of the wet tropics and equatorward of the subtropical deserts lies a transitional climatic region called **tropical wet and dry**. Here the rain forest gives way to the *savanna*, a tropical grassland with scattered drought-tolerant trees (**Figure 20.7**). Because temperature characteristics among all A climates are quite similar, the primary factor that distinguishes the Aw climate from Af and Am is precipitation. Although the overall amount of precipitation in the tropical wet and dry realm is often considerably less than in the wet tropics, the most distinctive feature of this climate is not the annual rainfall total but the markedly seasonal character of the rainfall. The climatic diagram for Normanton, Australia (Figure 20.6C), clearly illustrates this trait. As the equatorial low advances poleward in summer, the rainy season commences and features weather patterns typical of the wet tropics. Later, with the retreat of the equatorial low, the subtropical high advances into the region and brings with it pronounced dryness. In some Aw regions such as India, Southeast Asia, and portions of Australia, the alternating periods of rainfall and dryness are associated with a well-established monsoon circulation (see Chapter 18).

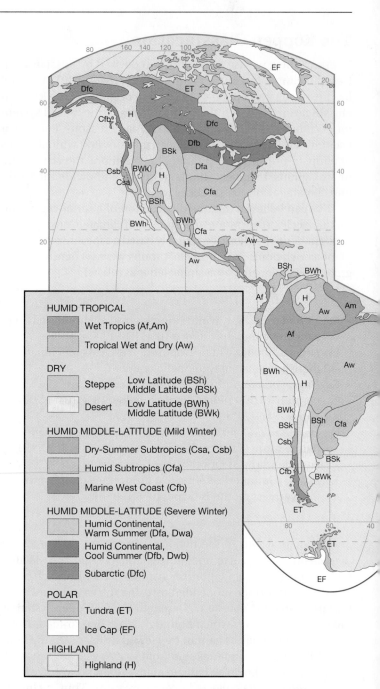

FIGURE 20.4 Climates of the world based on the Köppen classification.

CONCEPT CHECK 20.3

❶ What is the main factor that distinguishes Aw climates from Af and Am? How is this difference reflected in the vegetation of these climate regions?

❷ How do the equatorial low and the subtropical high influence the seasonal distribution of rainfall in the Aw climate?

Dry (B) Climates

It is important to realize that the concept of dryness is a relative one and refers to any situation in which a water deficiency exists. Climatologists define a dry climate as one in which the yearly precipitation is not as great as the potential loss of water by evapo-

Students Sometimes Ask...

Where is the driest desert on Earth?

The Atacama Desert of Chile has the distinction of being the world's driest desert. This relatively narrow belt of arid land extends for about 1,200 kilometers (750 miles) along South America's Pacific Coast (see **Figure 20.8**). It is said that some portions of the Atacama have not received rain for more than 400 years! One must view such pronouncements skeptically. Nevertheless, for places where records have been kept, Arica, Chile, in the northern part of the Atacama, has experienced a span of 14 years without measurable rainfall.

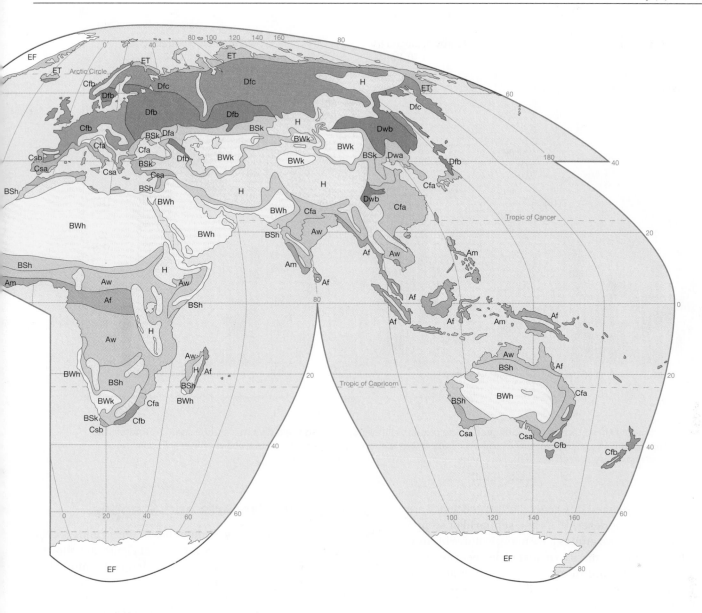

FIGURE 20.5 Unexcelled in luxuriance and characterized by hundreds of different species per square kilometer, the tropical rain forest is a broadleaf evergreen forest that dominates the wet tropics. (Photo by I. K. Lee/Alamy)

ration. Thus, dryness is not only related to annual rainfall totals but is also a function of evaporation, which in turn is closely dependent upon temperature.

To establish the boundary between dry and humid climates, the Köppen classification uses formulas that involve three variables: average annual precipitation, average annual temperature, and seasonal distribution of precipitation. The use of average annual temperature reflects its importance as an index of evaporation. The amount of rainfall defining the humid–dry boundary increases as the annual mean temperature increases. The use of seasonal precipitation as a variable is also related to this idea. If rain is concentrated in the warmest months, loss to evaporation is greater than if the precipitation were concentrated in the cooler months.

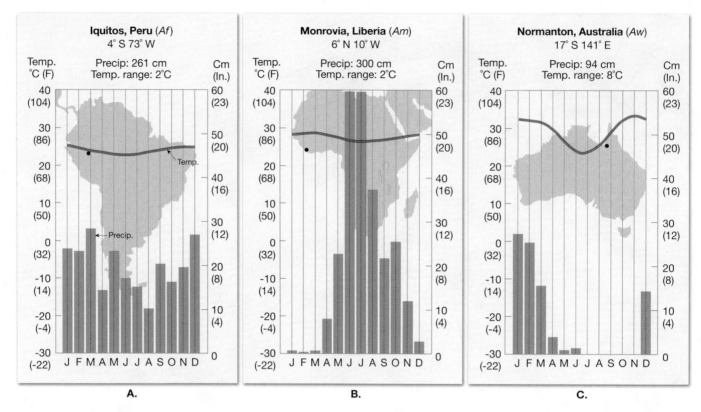

FIGURE 20.6 By comparing these three climatic diagrams, the primary differences among the a climates can be seen. **A.** Iquitos, the Af station, is wet throughout the year. **B.** Monrovia, the Am station, has a short, dry season. **C.** As is true for all Aw stations, Normanton has an extended dry season and a higher annual temperature range than the others.

Within the regions defined by a general water deficiency are two climatic types: **arid** or **desert** (BW) and **semiarid** or **steppe** (BS) (Figure 20.8). These two groups have many features in common; their differences are primarily a matter of degree. The semiarid is a marginal and more humid variant of the arid and represents a transition zone that surrounds the desert and separates it from the bordering humid climates.

FIGURE 20.7 Tropical savanna grassland in Tanzania's Serengeti National Park. This tropical savanna, with its stunted, drought-resistant trees, was probably strongly influenced by seasonal burnings carried out by native human populations. (Photo by Kondrachov Vladimir/Shutterstock)

Low-Latitude Deserts and Steppes

The heart of low-latitude dry climates lies in the vicinities of the Tropics of Cancer and Capricorn. A glance at Figure 20.8 shows a virtually unbroken desert environment stretching for more than 9,300 kilometers (nearly 6,000 miles) from the Atlantic coast of North Africa to the dry lands of northwestern India. In addition to this single great expanse, the Northern Hemisphere contains another, much smaller area of subtropical desert and steppe in northern Mexico and the southwestern United States. In the Southern Hemisphere, dry climates dominate Australia. Almost 40 percent of the continent is desert, and much of the remainder is

Students Sometimes Ask...

Are deserts always hot?

Actually, this is a misconception. Deserts can certainly be hot places. In fact, the record high temperature for the United States, 57° C (134° F), was set at Death Valley, California. Despite this and other remarkably high figures, deserts also experience cold temperatures. For example, the average daily minimum in January at Phoenix, Arizona, is 1.7° C (35° F), just barely above freezing. At Ulan Bator in Mongolia's Gobi Desert, the average *high* temperature in January is only −19° C (−2° F)! Although subtropical deserts lack a cold season, mid-latitude deserts do experience seasonal temperature changes.

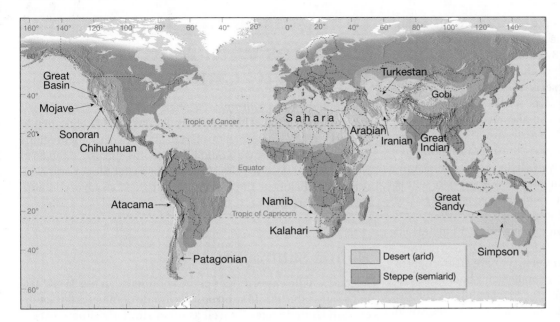

FIGURE 20.8 Arid and semiarid climates cover about 30 percent of Earth's land surface. No other climatic group covers so large an area.

subtropical high-pressure belts (see Figures 6.27, p. 177 and 18.16, p. 538). Here, air is subsiding. When air sinks, it is compressed and warmed. Such conditions are just opposite of what is needed for cloud formation and precipitation. Therefore, clear skies, a maximum of sunshine, and drought are to be expected. The climate diagrams for Cairo, Egypt, and Monterrey, Mexico (Figure 20.9A, B), illustrate the characteristics of low-latitude dry climates.

Middle-Latitude Deserts and Steppes

Unlike their low-latitude counterparts, middle-latitude deserts and steppes are not controlled by the subsiding air masses associated with high pressure. Instead, these

steppe. In addition, arid and semiarid areas are found in southern Africa and make a limited appearance in coastal Chile and Peru.

The existence of this dry subtropical realm is primarily the result of the prevailing global distribution of air pressure and winds. Earth's low-latitude deserts and steppes coincide with the

dry lands exist principally because of their positions in the deep interiors of large landmasses far removed from the oceans, which are the ultimate source of moisture for cloud formation and precipitation. In addition, the presence of high mountains across the paths of prevailing winds further acts to separate these areas from waterbearing, maritime air masses.

FIGURE 20.9 Climatic diagrams for representative arid and semiarid stations. Stations **A.** and **B.** are in the subtropics, whereas **C.** is in the middle latitudes. Cairo and Lovelock are classified as deserts; Monterrey is a steppe. Lovelock, Nevada, may also be called a rain shadow desert.

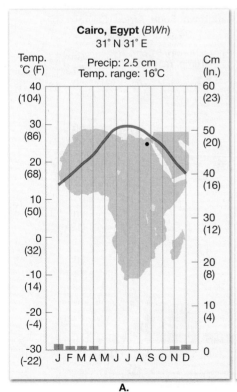

A.

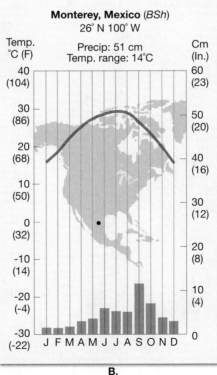

B.

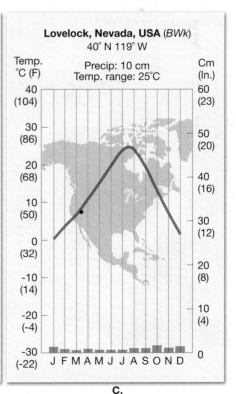

C.

Windward sides of mountains are often wet. As prevailing winds meet mountain barriers, the air is forced to ascend, producing clouds and precipitation. By contrast, the leeward sides of mountains are usually much drier and are often arid enough to be referred to as **rain shadow deserts** (see Figure 17.11, p. 500). Because many middle-latitude deserts occupy sites on the leeward sides of the mountains, they can also be classified as rain shadow deserts (Figure 20.9C). In North America, the Coast Ranges, Sierra Nevada, and Cascades are the foremost mountain barriers. In Asia, the great Himalayan chain prevents the summertime monsoon flow of moist Indian Ocean air from reaching the interior. Because the Southern Hemisphere lacks extensive land areas in the middle latitudes, only a small area of desert and steppe is found in this latitude range, existing primarily in the rain shadow of the towering Andes.

In the case of middle-latitude deserts, we have an example of the impact of tectonic processes on climate. Rain shadow deserts exist by virtue of the mountains produced when plates collide. Without such mountain-building episodes, wetter climates would prevail where many dry regions exist today.

CONCEPT CHECK 20.4

1. Why is the amount of precipitation that defines the boundary between humid and dry climates variable?
2. What is the primary reason (control) for the existence of the dry subtropical realm (BWh and BSh)?
3. What is the primary reason for the existence of middle-latitude deserts and steppes?

Humid Middle-Latitude Climates with Mild Winters (C Climates)

Although the term *subtropical* is often used for the C climates, it can be misleading. Although many areas with C climates do indeed possess some near-tropical characteristics, other regions do not. For example, we would be stretching the use of the term *subtropical* to describe the climates of coastal Alaska and Norway, which belong to the C group. Within the C group of climates, several subgroups are recognized.

Humid Subtropics

Located on the eastern sides of the continents, in the 25–40° latitude range, the **humid subtropical climate** dominates the southeastern United States, as well as other similarly situated areas around the world (Figures 20.4 and 20.10A). In the summer, the humid subtropics experience hot, sultry weather of the type one expects to find in the rainy tropics. Daytime temperatures are generally high, and because both mixing ratio and relative humidity are high, the night brings little relief. An afternoon or evening thunderstorm is also possible, for these areas experience such storms on an average of 40–100 days each year, the majority during the summer months.

As summer turns to autumn, the humid subtropics lose their similarity to the rainy tropics. Although winters are mild, frosts

FIGURE 20.10 Each of these climatic diagrams represents one of the three main types of C climates: **A.** humid subtropical, **B.** marine west coast, and **C.** dry-summer subtropical.

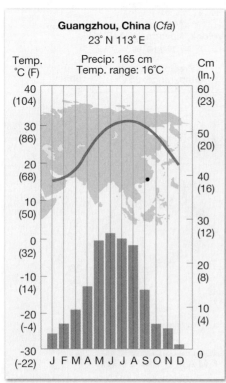

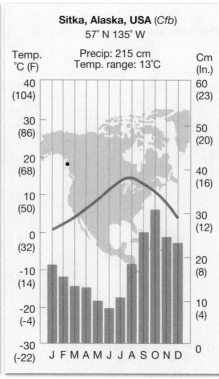

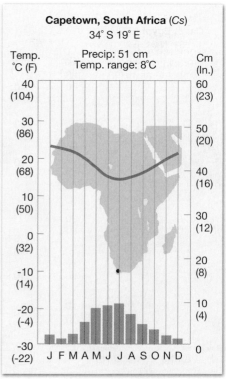

are common in the higher-latitude Cfa areas and occasionally plague the tropical margins as well. The winter precipitation is also different in character from the summer. Some is in the form of snow, and most is generated along fronts of the frequent middle-latitude cyclones that sweep over these regions.

Marine West Coast

Situated on the western (windward) side of continents, from about 40–65° north and south latitude, is a climatic region dominated by the onshore flow of oceanic air (Figure 20.11). In North America, the **marine west coast climate** extends from near the U.S.–Canadian border northward as a narrow belt into southern Alaska (Figure 20.10B). The largest area of Cfb climate is found in Europe, for here there is no mountain barrier blocking the movement of cool maritime air from the North Atlantic.

The prevalence of maritime air masses means that mild winters and cool summers are the rule, as is an ample amount of rainfall throughout the year. Although there is no pronounced dry period, there is a drop in monthly precipitation totals during the summer. The reason for the reduced summer rainfall is the poleward migration of the oceanic subtropical highs. Although the areas of marine west coast climate are situated too far poleward to be dominated by these dry anticyclones, their influence is sufficient to cause a decrease in warm season rainfall.

Dry-Summer Subtropics

The **dry-summer subtropical climate** is typically located along the west sides of continents between latitudes 30 and 45°. Situated between the marine west coast climate on the poleward side and the subtropical steppes on the equatorward side, this climatic region is best described as transitional in character. It is unique because it is the only humid climate that has a strong winter rainfall maximum, a feature that reflects its intermediate position (Figure 20.10C). In summer, the region is dominated by stable conditions associated with the oceanic subtropical highs. In win-

ter, as the wind and pressure systems follow the Sun equatorward, it is within range of the cyclonic storms of the polar front. Thus, during the course of a year, these areas alternate between becoming a part of the dry tropics and an extension of the humid middle latitudes. Although middle-latitude changeability characterizes the winter, tropical constancy describes the summer.

As was the case for the marine west coast climate, mountain ranges limit the dry-summer subtropics to a relatively narrow coastal zone in both North and South America. Because Australia and southern Africa barely reach to the latitudes where dry-summer climates exist, the development of this climatic type is limited on these continents as well. Consequently, because of the arrangement of the continents, and of their mountain ranges, inland development occurs only in the Mediterranean basin. Here the zone of subsidence extends far to the east in summer; in winter, the sea is a major route of cyclonic disturbances. Because the dry-summer climate is particularly extensive in this region, the name *Mediterranean climate* is often used as a synonym.

CONCEPT CHECK 20.5

❶ Describe and explain the differences between summertime and wintertime precipitation in the humid subtropical climate (Cfa).

❷ Why is the marine west coast climate (Cfb) represented by only slender strips of land in North and South America, and why is it very extensive in western Europe?

❸ The dry-summer subtropics were described as transitional. Explain why this is true.

FIGURE 20.11 Fog is common along the rocky Pacific coastline. This area is classified as a *marine west coast* climate. As the name implies, the ocean exerts a strong influence. (Photo by Ed Reschke/Photolibrary)

Humid Middle-Latitude Climates with Severe Winters (D Climates)

The C climates that were just described characteristically have mild winters. By contrast, D climates experience severe winters. Two types of D climates are recognized, the humid continental and the subarctic. Climatic diagrams of representative locations are shown in Figure 20.12. The D climates are land-controlled climates, the result of broad continents in the middle latitudes. Because continentality is a basic feature, D climates are absent in the Southern Hemisphere where the middle-latitude zone is dominated by the oceans.

Humid Continental

The **humid continental climate** is confined to the central and eastern portions of North America and Eurasia in the latitude range between approximately 40 and 50° north latitude. It may at first seem unusual that a continental climate should extend eastward to the margins of the ocean. However, because the prevailing

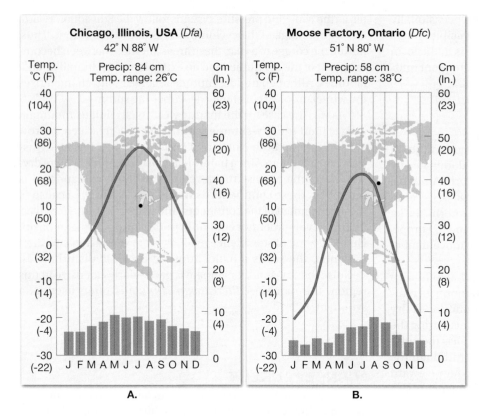

A.

Chicago, Illinois, USA *(Dfa)*
42° N 88° W

Temp. °C (F)   Precip: 84 cm   Cm (In.)
Temp. range: 26°C

B.

Moose Factory, Ontario *(Dfc)*
51° N 80° W

Temp. °C (F)   Precip: 58 cm   Cm (In.)
Temp. range: 38°C

FIGURE 20.12 D climates are associated with the interiors of large landmasses in the mid-to-high latitudes of the Northern Hemisphere. Although winters can be harsh in Chicago's humid continental (Dfa) climate, the subarctic environment (Dfc) of Moose Factory is more extreme.

atmospheric circulation is from the west, deep and persistent incursions of maritime air from the east are not likely to occur.

Both winter and summer temperatures in the humid continental climate can be characterized as relatively severe. Consequently, annual temperature ranges are high throughout the climate.

Precipitation is generally greater in summer than in winter. Precipitation totals generally decrease toward the interior of the continents as well as from south to north, primarily because of increasing distance from the sources of mT air. Furthermore, the more northerly stations are also influenced for a greater part of the year by drier polar air masses.

Wintertime precipitation in humid continental climates is chiefly associated with the passage of fronts connected with traveling middle-latitude cyclones. Part of this precipitation is in the form of snow, the proportion increasing with latitude. Although precipitation is often considerably less during the cold season, it is usually more conspicuous than the greater amounts that fall during summer. An obvious reason is that snow remains on the ground, often for extended periods, and rain, of course, does not.

Subarctic

Situated north of the humid continental climate and south of the polar tundra is an extensive **subarctic climate** region covering broad, uninterrupted expanses from western Alaska to Newfoundland in North America and from Norway to the Pacific coast of Russia in Eurasia. It is often referred to as the *taiga* climate, for its extent closely corresponds to the northern coniferous forest

region of the same name (**Figure 20.13**). Although scrawny, the spruce, fir, larch, and birch trees in the taiga represent the largest stretch of continuous forest on the surface of Earth.

Here in the source regions of continental polar air masses, the outstanding feature is certainly the dominance of winter. Not only is it long but temperatures are also bitterly cold. Winter minimum temperatures are among the lowest ever recorded outside the ice sheets of Greenland and Antarctica. In fact, for many years, the world's coldest temperature was attributed to Verkhoyansk in east central Siberia, where the temperature dropped to −68° C (−90° F) on February 5 and 7, 1892. Over a 23-year period, this same station had an average monthly minimum of −62° C (−80° F) during January. Although exceptional temperatures, they illustrate the extreme cold that envelops the taiga in winter.

By contrast, summers in the subarctic are remarkably warm, despite their short duration. However, when compared with regions farther south, this short season must be characterized as cool. The extremely cold winters and relatively warm summers combine to produce the highest annual temperature ranges on Earth. Because these far northerly continental interiors are the source regions for cP air masses, there is very limited moisture available throughout the year. Precipitation totals are therefore small, with a maximum occurring during the warmer summer months.

CONCEPT CHECK 20.6

❶ Why is the humid continental climate confined to the Northern Hemisphere?

❷ Describe and explain the annual temperature range you should expect in the realm of the *taiga*.

FIGURE 20.13 The northern coniferous forest is also called the *taiga*. (Photo by Martin Shields/Alamy)

A.

Polar (E) Climates

Polar climates are those in which the mean temperature of the warmest month is below 10° C (50° F). Thus, just as the tropics are defined by their year-round warmth, the polar realm is known for its enduring cold. As winters are periods of perpetual night, or nearly so, temperatures at most polar locations are understandably bitter. During the summer months temperatures remain cool despite the long days, because the Sun is so low in the sky that its oblique rays are not effective in bringing about a genuine warming. Although polar climates are classified as humid, precipitation is generally meager. Evaporation, of course, is also limited. The scanty precipitation totals are easily understood in view of the temperature characteristics of the region. The amount of water vapor in the air is always small because low mixing ratios must accompany low temperatures. Usually precipitation is most abundant during the warmer summer months when the moisture content of the air is highest.

Two types of polar climates are recognized. The **tundra climate** (ET) is a treeless climate found almost exclusively in the Northern Hemisphere (Figure 20.14A). Because of the combination of high latitude and continentality, winters are severe, summers are cool, and annual temperature ranges are high (Figure 20.15A). Furthermore, yearly precipitation is small, with a modest summer maximum.

The **ice cap climate** (EF) does not have a single monthly mean above 0° C (32° F) (Figure 20.15B). Consequently, because the average temperature for all months is below freezing, the growth of vegetation is prohibited and the landscape is one of permanent ice and snow (Figure 20.14B). This climate of perpetual frost covers a surprisingly large area of more than 15.5 million square kilometers (6 million square miles), or about 9 percent of Earth's land area. Aside from scattered occurrences in high mountain areas, it is confined to the ice sheets of Greenland and Antarctica (see Figure 6.2, p. 160).

B.

FIGURE 20.14 **A.** The tundra is a region almost completely devoid of trees. Bogs and marshes are common, and plant life frequently consists of mosses, low shrubs, and flowering herbs. (Photo by Arcticphoto/Alamy) **B.** The ice sheets on Greenland and Antarctica are the primary examples of places with an *ice cap climate*. (Photo by Graham Neden/ Ecoscene/CORBIS)

CONCEPT CHECK 20.7

❶ Why are annual precipitation totals low in polar climates? Which season has the most precipitation? Why?

❷ Where are EF climates most extensively developed?

Highland Climates

It is a well-known fact that mountains have climate conditions that are distinctively different from those found in adjacent lowlands. Compared to nearby places at lower elevations, sites with **highland climates** are cooler and usually wetter. Unlike the world climate types already discussed, which consist of large, relatively homogeneous regions, the outstanding characteristic of highland climates is the great diversity of climatic conditions that occur.

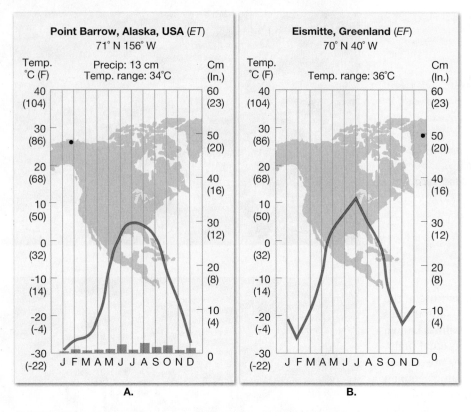

FIGURE 20.15 These climatic diagrams represent the two basic types of polar climates. **A.** Barrow, Alaska, exhibits a tundra (ET) climate. **B.** Eismitte, Greenland, a station located on a massive ice sheet, is classified as an ice cap (EF) climate.

able. Mountains create various obstacles to winds. Locally, winds may be funneled through valleys or forced over ridges and around mountain peaks. When weather conditions are fair, mountain and valley breezes are created by the topography itself.[47]

We know that climate strongly influences vegetation, which is the basis for the Köppen system. Thus, where there are vertical differences in climate, we should expect a vertical zonation of vegetation as well. Ascending a mountain can let us view dramatic vegetation changes that otherwise might require a poleward journey of thousands of kilometers (Figure 20.16). This occurs because altitude duplicates, in some respects, the influence of latitude on the distribution of vegetation.

Perhaps the terms *variety* and *changeability* best describe mountain climates. Because atmospheric conditions fluctuate rapidly with changes in altitude and exposure, a nearly limitless variety of local climates occur in mountainous regions. The climate in a protected valley is very different from that of an exposed peak. Conditions on windward slopes contrast sharply with those on the leeward sides, whereas slopes facing the Sun are unlike those that lie mainly in the shadows.

The best-known climatic effect of increased altitude is lower temperatures. In addition, an increase in precipitation due to orographic lifting usually occurs at higher elevations. Despite the fact that mountain stations are colder and often wetter than locations at lower elevations, highland climates are often very similar to those in adjacent lowlands in terms of seasonal temperature cycles and precipitation distribution. **Figure 20.16** illustrates this relationship.

Phoenix, at an elevation of 338 meters (1,109 feet), lies in the desert lowlands of southern Arizona. By contrast, Flagstaff is located at an altitude of 2,100 meters (7,000 feet) on the Colorado Plateau in northern Arizona. When summer averages climb to 34° C (93° F) in Phoenix, Flagstaff is experiencing a pleasant 19° C (66° F), which is a full 15° C (27° F) cooler. Although the temperatures at each city are quite different, the pattern of monthly temperature changes for each place is similar. Both experience their minimum and maximum monthly means in the same months. When precipitation data are examined, both places have a similar seasonal pattern, but the amounts at Flagstaff are higher in every month. In addition, owing to its higher altitude, much of Flagstaff's winter precipitation is in the form of snow. By contrast, all of the precipitation at Phoenix is rain.

Because topographic variations are pronounced in mountains, every change in slope with respect to the Sun's rays produces a different microclimate. In the Northern Hemisphere, south-facing slopes are warmer and dryer because they receive more direct sunlight than do north-facing slopes and deep valleys. Wind direction and speed in mountains can be highly vari-

CONCEPT CHECK 20.8

① The Arizona cities of Flagstaff and Phoenix are relatively close to one another yet have contrasting climates. Briefly describe the differences and why they occur.

Human Impact on Global Climate

The proposals to explain global climate change are many and varied. In Chapter 6, we examined some possible causes for ice-age climates. These hypotheses, which included the movement of lithospheric plates and variations in Earth's orbit, involved natural forcing mechanisms. Another natural forcing mechanism, discussed in Chapter 9, is the possible role of explosive volcanic eruptions in modifying the atmosphere. It is important to remember that these mechanisms, as well as others, not only have contributed to climate change in the geologic past but will also be responsible for future shifts in climate. When relatively recent and future changes in our climate are considered, we must also examine the impact of human beings. In this section, we examine the major way in which humans are contributing to global climatic change. This impact results from the addition of carbon dioxide and other gases to the atmosphere.

[47]Mountain and valley breezes are described in the section "Local Winds" in Chapter 18. Also see Figure 18.19, p. 541.

Phoenix, AZ

Flagstaff, AZ

FIGURE 20.16 Climate diagrams for two stations in Arizona illustrate the general influence of elevation on climate. Flagstaff is cooler and wetter because of its position on the Colorado Plateau, nearly 1,800 meters (6,000 feet) higher than Phoenix. Only scanty drought-tolerant natural vegetation can survive in the hot, dry climate of southern Arizona, near Phoenix. (Photo by Design Pics/Superstock) The natural vegetation associated with the cooler, wetter highlands near Flagstaff, Arizona, is much different from the desert lowlands. (Photo by Jim Cole/Alamy)

Human influence of regional and global climate did not just begin with the onset of the modern industrial period. There is solid evidence that humans have been modifying the environment over extensive areas for thousands of years. The use of fire as well as the overgrazing of marginal lands by domesticated animals have negatively affected the abundance and distribution of vegetation. By altering ground cover, such important climatological factors as surface albedo, evaporation rates, and surface winds have been, and continue to be, modified. Commenting on this aspect of human-induced climate modification, the authors of one study observed:

> In contrast to the prevailing view that only modern humans are able to alter climate, we believe it is more likely that the human species has made a substantial and continuing impact on climate since the invention of fire.[48]

Subsequently, these ideas were reinforced and expanded upon by a study that used data collected from Antarctic ice cores. This research suggested that humans may have started to have a significant impact on atmospheric composition and global temperatures thousands of years ago.

> Humans started slowly ratcheting up the thermostat as early as 8000 years ago, when they began clearing forests for agriculture, and 5000 years ago with the arrival of wet-rice cultivation. The greenhouse gases carbon dioxide and methane given off by these changes would have warmed the world. . . . "[49]

Students Sometimes Ask...

Do cities influence climate?

Yes, in lots of ways, including more fog and rain, as well as less sunshine and slower winds. The most studied and well-documented effect is the *urban heat island*, which refers to the fact that city temperatures are generally higher than surrounding rural areas. Several factors contribute to the heat island. The city's concrete and asphalt surfaces absorb and store more solar energy than the natural rural landscape. At night these stonelike materials help keep the city warmer. Urban temperatures are also higher because cities generate a great deal of waste heat (home heating, factories, cars, etc.). The "blanket" of pollutants also contributes by "trapping" heat that would otherwise escape.

CONCEPT CHECK 20.9

① How might people have altered climate thousands of years ago?

[48]Sagan, Carl, et al., Anthropogenic Albedo Changes and the Earth's Climate, *Science* 206 (4425) (1980): 1367.

[49]An Early Start for Greenhouse Warming?", *Science* 303 (January 16, 2004). This article is a report on a paper given by paleoclimatologist William Ruddiman at a meeting of the American Geophysical Union in December 2003.

Michael Mann
Climate Change
Scientist

In the last few decades of the 20th century, evidence of global warming was growing too obvious to ignore. Vast ice shelves in the Antarctic were breaking into floating isles, glaciers worldwide were in full retreat, and each turn of the calendar seemed to bring a new record for the hottest year yet measured. But whether this warming was due to humans or nature, no one was sure. Researchers needed to study a longer history of Earth's climate than just the 20th century.

Michael Mann, of Pennsylvania State University, knew where to find the answers. Ice cores, sediments, coral skeletons, and tree rings all contain natural records of past temperature fluctuations. However, Mann, a meteorology professor, points out that "it's a fuzzy record, not as precise as thermometer readings." For example, while individual annual layers of ice in ice cores provide a

Dr. Michael Mann is a member of the Penn State University faculty, holding joint appointments in the Departments of Meteorology and Geosciences, as well as the Earth and Environmental Systems Institute. He is also Director of the Penn State Earth System Science Center. (Photo courtesy of Jon Golden)

starting around 1900—the same period when humans began burning large quantities of fossil fuels.

Known as the "hockey stick graph" for its distinctive shape, Mann's 1998 paper caused a sensation. Such vivid evidence of human contributions to global warming helped catapult both Mann and the issue of climate change onto the public stage. Meanwhile, some commentators encouraged their peers to "break the hockey stick" and discredit the idea of a warming Earth. The attacks convinced Mann to spread the word about climate change research. "I realized I had a responsibility to use that spotlight to educate the public and policy-makers, to try to do some good with that opportunity," he says.

In 2001, Mann's research became a centerpiece of the Third Report of the Intergovernmental Panel on Climate Change (IPCC). These reports, written by scientists, are summaries of the most current and important climate research. Mann served as a lead author of the Third Report—and received a singular honor for his labors. For their work to raise awareness of the threat of human-caused climate change, the IPCC and former Vice President Al Gore were awarded the 2007 Nobel Peace Prize.

Mann continues to spread the word about global warming today. In addition to his scientific research, he has helped found the website RealClimate.org, featuring commentary on climate news by climate researchers themselves. Mann also coauthored the book *Dire Predictions*, which explains the complexities of climate-change science to the public.

Mann says that humans can still steer a course that could evade the worst consequences of a warming globe. "I like to think that together, those of us who are engaging the public on this issue are making some positive steps forward," he says. His main message is that "we hold the future in our own hands."

> **"I realized I had a responsibility to use that spotlight to educate the public and policy makers . . . "**

direct reading of ancient atmospheric conditions, deciphering which layers belonged to which years can be difficult. And snow might have fallen only in winter in some areas, versus year round in others, making it difficult to compare cores from different areas.

These variables can make interpreting such proxy data a complex affair. Even so, with the help of modern statistical methods and climate models, Mann was able to piece together a reconstruction of the average temperature of the Northern Hemisphere

going back 1,000 years. Mann was most interested in the surface temperature patterns he uncovered. But colleagues suggested he also plot out the average temperature of the Northern Hemisphere.

"It was that single result—the item we thought was least interesting in our work—that got the most attention. It spoke to the question of how anomalous recent warming was," Mann says. The graph revealed an alarming trend. Temperatures stayed relatively warm during medieval times, cooled after that, but zoomed sharply higher

> **"It was that single result— the item we thought was least interesting in our work—that got the most attention."**

Carbon Dioxide, Trace Gases, and Global Climate Change

In Chapter 16, you learned that although carbon dioxide (CO_2) represents only about 0.039 percent of the gases that make up clean, dry air, it is nevertheless a meteorologically significant com-

ponent. The importance of carbon dioxide lies in the fact that it is transparent to incoming short-wavelength solar radiation, but it is not transparent to some of the longer-wavelength outgoing radiation emitted by Earth. A portion of the energy leaving the ground is absorbed by carbon dioxide and subsequently reemitted, part of it toward the surface, thereby keeping the air near the ground warmer than it would be without carbon dioxide.

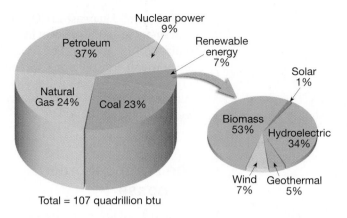

Total = 107 quadrillion btu

FIGURE 20.17 Paralleling the rapid growth of industrialization, which began in the 19th century, has been the combustion of fossil fuels, which has added great quantities of carbon dioxide to the atmosphere. The graph shows energy consumption in the United States in 2008. The total was about 107 quadrillion Btu. A quadrillion is 10 raised to the 12th power, or a million million. A quadrillion Btu is a convenient unit for referring to U.S. energy use as a whole. Fossil fuels (petroleum, coal, and natural gas) represent about 85 percent of the total. (Source: U.S. Department of Energy, Energy Information Administration)

ural gas, and petroleum (Figure 20.17). Combustion of these fuels has added great quantities of carbon dioxide to the atmosphere.

The use of coal and other fuels is the most prominent means by which humans add CO_2 to the atmosphere, but it is not the only way. The clearing of forests also contributes substantially because CO_2 is released as vegetation, is burned, or decays. Deforestation is particularly pronounced in the tropics, where vast tracts are cleared for ranching and agriculture or subjected to inefficient commercial logging operations (Figure 20.18). According to United Nations estimates, nearly 10.2 million hectares (25.1 million acres) of tropical forest were permanently destroyed each year during the decade of the 1990s. Between the years 2000 and 2005, the average figure increased to 10.4 million hectares (25.7 million acres) per year.

Some of the excess CO_2 is taken up by plants or is dissolved in the ocean. It is estimated that 45–50 percent remains in the atmosphere. Figure 20.19 is a graphic record of changes in atmospheric CO_2 extending back 400,000 years. Over this long span, natural fluctuations varied from about 180 to 300 parts per million (ppm). As a result of human activities, the present CO_2 level is about 30 percent higher than its highest level over at least the last 650,000

Thus, along with water vapor, carbon dioxide is largely responsible for the *greenhouse effect* of the atmosphere.[50] Carbon dioxide is an important heat absorber, and it follows logically that any change in the air's carbon dioxide content could alter temperatures in the lower atmosphere.

CO_2 Levels Are Rising

Earth's tremendous industrialization of the past two centuries has been fueled—and still is fueled—by burning fossil fuels: coal, nat-

[50]To review the greenhouse effect, see Figure 16.23 on p. 477.

Students Sometimes Ask...

Figure 20.17 shows biomass as a form of renewable energy. What exactly is biomass?

Biomass refers to organic matter that can be burned directly as fuel or converted into a different form and then burned. Biomass is a relatively new name for the oldest human fuels. Examples include firewood, charcoal, crop residues, and animal waste. Biomass burning is especially important in emerging economies.

FIGURE 20.18 Clearing the tropical rain forest is a serious environmental problem. In addition to the loss of biodiversity, tropical deforestation is a significant source of carbon dioxide. **A.** This August 2007 satellite image shows deforestation in the Amazon rain forest in western Brazil. Intact forest is dark green, whereas cleared areas are tan (bare ground) or light green (crops and pasture). (NASA) **B.** Fires are frequently used to clear the land. This scene is also in Brazil's Amazon rain forest. (Photo by Pete Oxford/nature Picture Library)

A.

B.

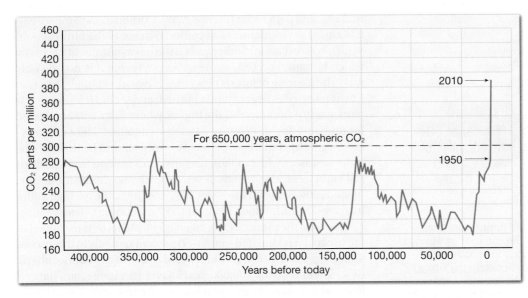

FIGURE 20.19 Carbon dioxide concentrations over the past 400,000 years. Most of the data comes from analysis of air bubbles trapped in cores of glacial ice. The record since 1958 comes from direct measurements of atmospheric CO_2 taken at Mauna Loa Observatory, Hawaii. The rapid increase in CO_2 concentrations since the onset of the industrial revolution is obvious.

FIGURE 20.20 Americans are responsible for about 25 percent of the world's greenhouse-gas emissions. That amounts to 24,300 kilograms (54,000 pounds) of carbon dioxide each year, which is about five times the emissions of the *average* global citizen. This diagram represents some of the ways we contribute. (Data from various U.S. government agencies; Photo by Jerry Schad/Photo Researchers, Inc.)

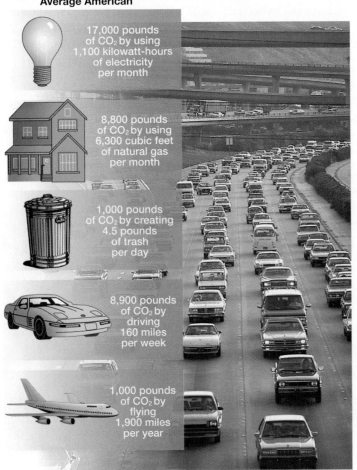

Annual CO₂ Contribution of an Average American

17,000 pounds of CO₂ by using 1,100 kilowatt-hours of electricity per month

8,800 pounds of CO₂ by using 6,300 cubic feet of natural gas per month

1,000 pounds of CO₂ by creating 4.5 pounds of trash per day

8,900 pounds of CO₂ by driving 160 miles per week

1,000 pounds of CO₂ by flying 1,900 miles per year

years. The rapid increase in CO_2 concentrations since the onset of industrialization is obvious. The annual rate at which atmospheric CO_2 concentrations are growing has been increasing over the last several decades (Figure 20.20).

The Atmosphere's Response

Given the increase in the atmosphere's carbon dioxide content, have global temperatures actually increased? The answer is yes. According to a 2007 report by the Intergovernmental Panel on Climate Change (IPCC), "Warming of the climate system is unequivocal, as is now evident from observations of increases in global average air and ocean temperatures, widespread melting of snow and ice, and rising global sea level."[51] It is very likely that most of the observed increase in global average temperatures since the mid-20th century is due to the observed increase in human-generated greenhouse gas concentrations. (As used by the IPCC, *very likely* indicates a probability of 90–99%.) Global warming since the mid-1970s is now about 0.6° C (1° F), and total warming in the past century is about 0.8° C (1.4° F). The upward trend in surface temperatures is shown in Figure 20.21A. The world map in Figure 20.21B compares surface temperatures for 2009 to the base period (1951–1980). You can see that the greatest warming has been in the Arctic and neighboring high-latitude regions. Here are some related facts:

- When we consider the time span for which there are instrumental records (since 1850), 14 of the last 15 years (1995–2009) rank among the 14 warmest (Figure 20.22).
- Global mean temperature is now higher than at any time in at least the past 500–1,000 years.
- The average temperature of the global ocean has increased to depths of at least 3,000 meters (10,000 feet).

Are these temperature trends caused by human activities, or would they have occurred anyway? The scientific consensus of the IPCC is that human activities were very *likely* responsible for most of the temperature increase since 1950.

What about the future? Projections for the years ahead depend In part on the quantities of greenhouse gases that are emitted. Figure 20.23 shows the best estimates of global warming for several different scenarios. The 2007 IPCC report also states that if there is a doubling of the preindustrial level of carbon dioxide (280 ppm) to 560 ppm, the "likely" temperature

---

[51]IPCC, Summary for Policy Makers. In *Climate Change 2007: The Physical Science Basis* (Cambridge University Press, Cambridge, United Kingdom and New York, NY), 4.

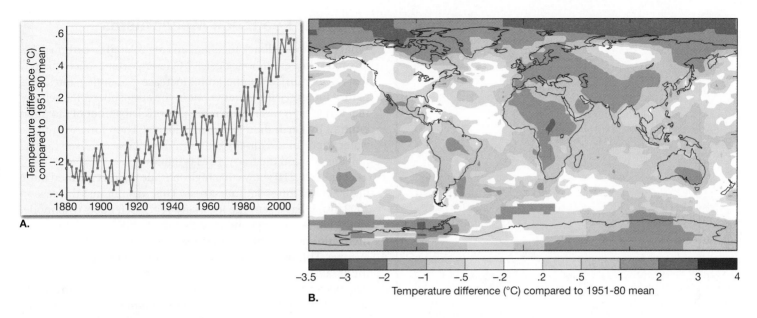

FIGURE 20.21 **A.** The graph depicts global temperature change in °C since the year 1880. **B.** The world map shows how temperatures in 2009 deviated from the mean for the 1951–1980 base period. The high latitudes in the Northern Hemisphere clearly stand out. (After NASA/Goddard Institute for Space Studies)

increase will be in the range of 2–4.5° C (3.5–8.1° F). The increase is "very unlikely" (1–10% probability) to be less than 1.5° C (2.7° F) and values higher than 4.5° C (8.1° F) cannot be excluded.

The Role of Trace Gases

Carbon dioxide is not the only gas contributing to a possible global increase in temperature. In recent years, atmospheric scientists have come to realize that the industrial and agricultural activities of people are causing a buildup of several trace gases that may also play a significant role. The substances are called *trace gases* because their concentrations are so much smaller than that of carbon dioxide. The trace gases that appear to be most important are methane (CH_4), nitrous oxide (N_2O), and certain types of chlorofluorocarbons (CFCs). These gases absorb wavelengths of outgoing Earth radiation that would otherwise escape into space. Although individually their impact is modest, taken together the effects of these trace gases play a significant role in warming the troposphere.

Sophisticated computer models of the atmosphere show that the warming of the lower atmosphere caused by CO_2 and trace gases will not be the same everywhere. Rather, the temperature response in polar regions could be two to three times greater than the global average. One reason for such a response is an expected reduction in sea ice. This topic is explored more fully in the following section.

FIGURE 20.22 Annual average global temperature variations for the period 1860–2009. The basis for comparison is the average for the 1961–1990 period (the 0.0 line on the graph). Each narrow bar on the graph represents the departure of the global mean temperature from the 1961–1990 average for 1 year. For example, the global mean temperature for 1862 was more than 0.5° C (1° F) *below* the 1961–1990 average, whereas the global mean for 1998 was more than 0.5° C above. The bar graph clearly indicates that there can be *significant variations from year to year*. The graph also shows a trend. Estimated global mean temperatures have been above the 1961–1990 average nearly every year since 1978. (Data from Climate Research Unit/University of East Anglia)

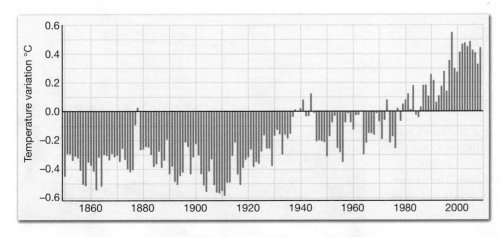

CONCEPT CHECK 20.10

❶ Why has the CO_2 level of the atmosphere been increasing for more than 150 years?

❷ How are temperatures in the lower atmosphere likely to change as CO_2 levels continue to increase?

❸ Aside from CO_2 what trace gases are contributing to global warming?

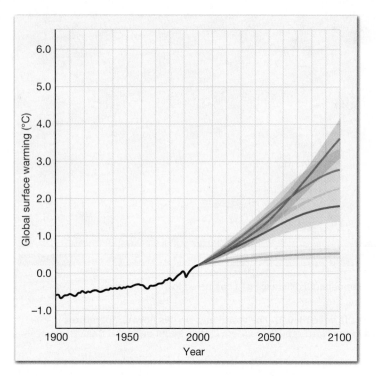

FIGURE 20.23 The left half of the graph shows global temperature changes for the 20th century. The right half shows projected global warming in different emissions scenarios. The shaded zone adjacent to each colored line shows the uncertainty range for each scenario. The basis for comparison (0.0 on the vertical axis) is the global average for the period 1980–1999. The orange line represents the scenario in which carbon dioxide concentrations were held constant at the values for the year 2000. (After IPCC, 2007)

Students Sometimes Ask...

What is the Intergovernmental Panel on Climate Change?

Recognizing the problem of potential global climate change, the World Meteorological Organization and the United Nations Environment Program established the *Intergovernmental Panel on Climate Change* (*IPCC*, for short). The IPCC assesses the scientific, technical, and socioeconomic information that is relevant to an understanding of human-induced climate change. This authoritative group provides advice to the world community through periodic reports that assess the state of knowledge of causes of climate change.

More than 1250 authors and 2500 scientific reviewers from more than 130 countries contributed to the IPCC's most recent report.

Climate-Feedback Mechanisms

Climate is a very complex interactive physical system. Thus, when any component of the climate system is altered, scientists must consider many possible outcomes. These possible outcomes are called **climate-feedback mechanisms**. They complicate climate-modeling efforts and add greater uncertainty to climate predictions.

Types of Feedback Mechanisms

What climate-feedback mechanisms are related to carbon dioxide and other greenhouse gases? One important mechanism is that warmer surface temperatures increase evaporation rates. This in turn increases the water vapor in the atmosphere. Remember that water vapor is an even more powerful absorber of radiation emitted by Earth than is carbon dioxide. Therefore, with more water vapor in the air, the temperature increase caused by carbon dioxide and the trace gases is reinforced.

Figure 20.21B shows that high-latitude regions warmed more in 2009 than lower-latitude areas. This was not just the case in 2009, but it has been true for many years. Scientists who model global climate change indicate that the temperature increase at high latitudes may be two to three times greater than the global average. This assumption is based in part on the likelihood that the area covered by sea ice will decrease as surface temperatures rise. Because ice reflects a much larger percentage of incoming solar radiation than does open water, the melting of the sea ice replaces a highly reflecting surface with a relatively dark surface (**Figure 20.24**). The result is a substantial increase in the solar energy absorbed at the surface. This in turn feeds back to the atmosphere and magnifies the initial temperature increase created by higher levels of greenhouse gases.

So far the climate-feedback mechanisms discussed have magnified the temperature rise caused by the buildup of carbon dioxide. Because these effects reinforce the initial change, they are called **positive-feedback mechanisms**. However, other effects must be classified as **negative-feedback mechanisms** because they produce results that are just the opposite of the initial change and tend to offset it.

One probable result of a global temperature rise would be an accompanying increase in cloud cover due to the higher moisture content of the atmosphere. Most clouds are good reflectors

FIGURE 20.24 This satellite image shows the springtime breakup of sea ice near Antarctica. The inset shows a likely feedback loop. A reduction in sea ice acts as a positive-feedback mechanism because surface reflectivity (albedo) decreases and the amount of solar energy absorbed at the surface increases. (Photo by George Holton/Photo Researchers, Inc.)

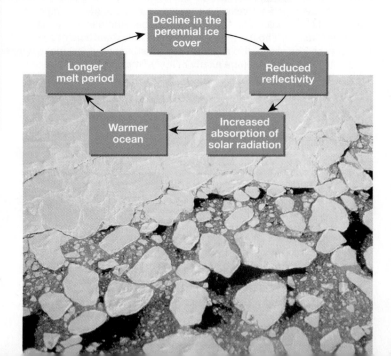

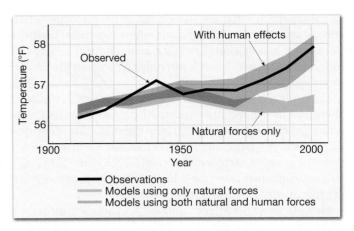

FIGURE 20.25 The blue band shows how global average temperatures would have changed due to natural forces only, as simulated by climate models. The red band shows model projections of the effects of human and natural forces combined. The black line shows actual observed global average temperatures. As the blue band indicates, without human influences, temperatures over the past century would actually have first warmed and then cooled slightly over recent decades. Bands of color are used to express the range of uncertainty. (After U.S. Global Change Research Program)

of solar radiation. At the same time, however, they are also good absorbers and emitters of radiation emitted by Earth. Consequently, clouds produce two opposite effects. They are a negative-feedback mechanism because they increase albedo and thus diminish the amount of solar energy available to heat the atmosphere. On the other hand, clouds act as a positive-feedback mechanism by absorbing and emitting radiation that would otherwise be lost from the troposphere.

Which effect, if either, is stronger? Atmospheric modeling shows that the negative effect of a higher albedo is stronger. Therefore, the net result of an increase in cloudiness should be a decrease in air temperature. The magnitude of this negative feedback, however, is not believed to be as great as the positive feedback caused by added moisture and decreased sea ice. Thus, although increases in cloud cover may partly offset a global temperature increase, climate models show that the ultimate effect of the projected increase in CO_2 and trace gases will still be a temperature increase.

The problem of global warming caused by human-induced changes in atmospheric composition continues to be one of the most studied aspects of climate change. Although no models yet incorporate the full range of potential factors and feedbacks, the scientific consensus is that the increasing levels of atmospheric carbon dioxide and trace gases will lead to a warmer planet with a different distribution of climate regimes.

Computer Models of Climate: Important Yet Imperfect Tools

Earth's climate system is amazingly complex. Comprehensive state-of-the-science climate simulation models are among the basic tools used to develop possible climate change scenarios. They are based on fundamental laws of physics and chemistry and incorporate human and biological interactions. The models simulate many variables, including temperature, rainfall, snow cover, soil moisture, winds, clouds, sea ice, and ocean circulation over the entire globe through the seasons and over spans of decades.

In many other fields of study, hypotheses can be tested by direct experimentation in the laboratory or by observations and measurements in the field. However, this is often not possible in the study of climate. Rather, scientists must construct computer models of how our planet's climate system works. If we understand the climate system correctly and construct the model appropriately, then the behavior of the model climate system should mimic the behavior of Earth's climate system (Figure 20.25).

What factors influence the accuracy of climate models? Clearly, mathematical models are *simplified* versions of the real Earth and cannot capture its full complexity, especially at smaller geographic scales. Moreover, when computer models are used to simulate future climate change, many assumptions have to be made that significantly influence the outcome. They must consider a wide range of possibilities for future changes in population, economic growth, consumption of fossil fuels, technological development, improvements in energy efficiency, and more.

Despite many obstacles, our ability to use supercomputers to simulate climate continues to improve. Although today's models are far from infallible, they are powerful tools for understanding what Earth's future climate might be like.

CONCEPT CHECK 20.11

1. Distinguish between positive and negative feedback mechanisms.
2. Describe at least one example of each type of feedback mechanism.

How Aerosols Influence Climate

Increasing the levels of carbon dioxide and other greenhouse gases in the atmosphere is the most direct human influence on global climate, but it is not the only impact. Global climate is also affected by human activities that contribute to the atmosphere's aerosol content. *Aerosols* are the tiny, often microscopic, liquid and solid particles that are suspended in the air. Atmospheric aerosols are composed of many different materials, including soil, smoke, sea salt, and sulfuric acid. Natural sources are numerous and include such phenomena as dust storms and volcanoes.

Presently the human contribution of aerosols to the atmosphere *equals* the quantity emitted by natural sources. Most human-generated aerosols come from the sulfur dioxide emitted during the combustion of fossil fuels and as a consequence of burning vegetation to clear agricultural land. Chemical reactions in the atmosphere convert the sulfur dioxide into sulfate aerosols, the same material that produces acid precipitation.

How do aerosols affect climate? Aerosols act directly by reflecting sunlight back to space and indirectly by making clouds "brighter" reflectors. The second effect relates to the fact that sulfuric acid aerosols attract water and thus are especially effective as cloud condensation nuclei (tiny particles upon which water vapor condenses). The large quantity of aerosols produced by human activities (especially industrial emissions) trigger an increase in the number of cloud droplets that form within a cloud. A greater number of small droplets increases the cloud's brightness—that is, more sunlight is reflected back to space.

By reducing the amount of solar energy available to the climate system, aerosols have a net cooling effect. Studies indicate that the cooling effect of human-generated aerosols could offset a portion of the global warming caused by the growing quantities of greenhouse gases in the atmosphere. The magnitude and extent of the cooling effect of aerosols is uncertain and represents a hurdle in advancing our understanding of how humans alter Earth's climate.

It is important to point out some significant differences between global warming by greenhouse gases and aerosol cooling. After being emitted, greenhouse gases such as carbon dioxide remain in the atmosphere for many decades. By contrast, aerosols released into the lower atmosphere remain there for only a few days or, at most, a few weeks before they are "washed out" by precipitation. Because of their short lifetime in the atmosphere, human-generated aerosols are distributed unevenly over the globe. As expected, they are concentrated near the areas that produce them, namely industrialized regions that burn fossil fuels and land areas where vegetation is burned (**Figure 20.26**).

Because the lifetime of human-generated aerosols in the atmosphere is short, the effect on today's climate is determined by the quantity of material emitted during the preceding couple of weeks. By contrast, the carbon dioxide released into the atmosphere remains for much longer spans and thus influences climate for many decades.

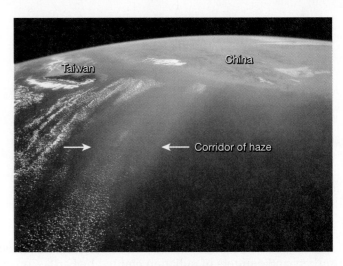

FIGURE 20.26 Human-generated aerosols are concentrated near the areas that produce them. Because aerosols reduce the amount of solar energy available to the climate system, they have a net cooling effect. This satellite image shows a dense blanket of pollution moving away from the coast of China. The plume is about 200 kilometers wide and more than 600 kilometers long. (NASA Image)

Some Possible Consequences of Global Warming

What consequences can be expected if the carbon dioxide content of the atmosphere reaches a level that is twice what it was early in the 20th century? Because the climate system is so complex, predicting the distribution of particular regional changes is speculative. Nevertheless, plausible scenarios can be given for larger scales of space and time.

As noted, the magnitude of the temperature increase will not be the same everywhere. The temperature rise will probably be smallest in the tropics and increase toward the poles. As for precipitation, the models indicate that some regions will experience significantly more precipitation and runoff. However, other regions will experience a decrease in runoff due to reduced precipitation or because of greater evaporation caused by higher temperatures.

Table 20.2 summaries some of the more likely effects and their possible consequences. The table also provides the IPCC's estimate of the probability of each effect. Levels of confidence for these projections vary from *"likely"* (67–90% probability) to *"very likely"* (90–99% probability) to *"virtually certain"* (greater than 99% probability).

Sea-Level Rise

A significant impact of human-induced global warming is a rise in sea level (**Figure 20.27A**). As this occurs, coastal cities, wetlands, and low-lying islands could be threatened with more frequent flooding, increased shoreline erosion, and saltwater encroachment into coastal rivers and aquifers.

How is a warmer atmosphere related to a global rise in sea level? The most obvious connection, the melting of glaciers, is

TABLE 20.2 Projected Changes and Effects of Global Warming in the 21st Century

Projected changes and estimated probability*	Examples of projected impacts
Higher maximum temperatures; more hot days and heat waves over nearly all land areas (*virtually certain*)	Increased incidence of death and serious illness in older age groups and urban poor
	Increased heat stress in livestock and wildlife
	Shift in tourist destinations
	Increased risk of damage to a number of crops
	Increased electric cooling demand and reduced energy-supply reliability
Higher minimum temperatures; fewer cold days, frost days, and cold waves over nearly all land areas (*virtually certain*)	Decreased cold-related human morbidity and mortality
	Decreased risk of damage to a number of crops, and increased risk to others
	Extended range and activity of some pest and disease vectors
	Reduced heating energy demand
More intense precipitation events (*very likely*)	Increased flood, landslide, avalanche, and debris flow damage
	Increased soil erosion
	Increased flood runoff could increase recharge of some floodplain aquifers
	Increased pressure on government and private flood insurance systems and disaster relief
Area affected by drought increases (*likely*)	Decreased crop yields
	Increased damage to building foundations caused by ground shrinkage
	Decreased water-resource quantity and quality
	Increased risk of forest fire
Intense tropical cyclone activity increases (*likely*)	Increased risk to human life, risk of infectious-disease epidemics, and many other risks
	Increased coastal erosion and damage to coastal buildings and infrastructure
	Increased damage to coastal ecosystems, such as coral reefs and mangroves

* *Virtually certain* indicates a probability greater than 99 percent, *very likely* indicates a probability of 90–99 percent, and *likely* indicates a probability of 67–90 percent.
Source: IPCC (2001, 2007).

FIGURE 20.27 Research indicates that sea level has risen between 10 and 23 centimeters (4 and 8 inches) over the past century and that the trend will continue at an accelerated rate. **A.** Using data from satellites and floats, it was determined that sea level rose an average of 3 millimeters (0.1 inch) per year between 1993 and 2005. Researchers attribute about half the rise to melting glacial ice and the other half to thermal expansion. Rising sea level can adversely affect some of the most densely populated regions on Earth. (NASA/Jet Propulsion Laboratory) **B.** The slope of a shoreline is critical to determining the degree to which sea-level changes will affect it. When the slope is gentle, small changes in sea level cause a substantial shift. C. The same sea-level rise along a steep coast results in only a small shoreline shift.

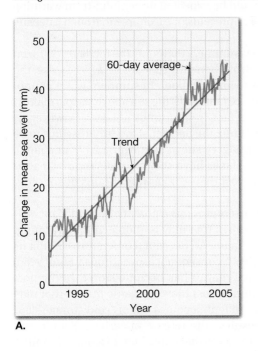

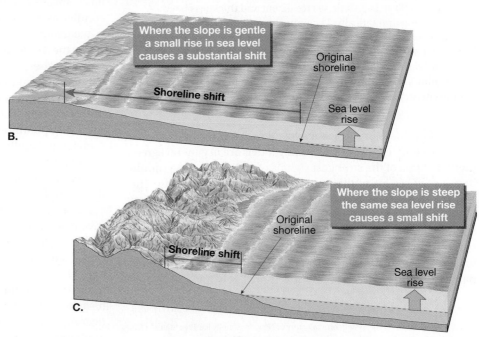

important, but not the only factor.[52] An equally important factor is that a warmer atmosphere causes an increase in ocean volume due to thermal expansion. Higher air temperatures warm the adjacent upper layers of the ocean, which in turn causes the water to expand and sea level to rise.

Research indicates that sea level has risen between 10 and 25 centimeters (4 and 8 inches) over the past century and that this trend will continue at an accelerated rate. Some models indicate that the rise may approach or even exceed 50 centimeters (20 inches) by the end of the 21st century. Such a change may seem modest, but scientists realize that any rise in sea level along a *gently* sloping shoreline, such as the Atlantic and Gulf coasts of the United States, will lead to significant erosion and severe, permanent inland flooding (Figure 20.27B, C). If this happens, many beaches and wetlands will be eliminated and coastal civilization would be severely disrupted.

Because rising sea level is a gradual phenomenon, it may be overlooked by coastal residents as an important contributor to shoreline erosion problems. Rather, the blame may be assigned to other forces, especially storm activity. Although a given storm may be the immediate cause, the magnitude of its destruction may result from the relatively small sea-level rise that allowed the storm's power to cross a much greater land area.

As mentioned, a warmer climate will cause glaciers to melt. In fact, a portion of the 10- to 25-centimeter (4- to 8-inch) rise in sea level over the past century is attributed to the melting of glaciers, especially mountain glaciers. This contribution is projected to continue through the 21st century. Of course, if the Greenland and Antarctic ice sheets were to experience a significant increase in melting, it would lead to a much greater rise in sea level and a major encroachment by the sea in coastal zones.

The Changing Arctic

A recent study of climate change in the Arctic began with the following statement:

> For nearly 30 years, Arctic sea ice extent and thickness have been falling dramatically. Permafrost temperatures are rising and coverage is decreasing. Mountain glaciers and the Greenland ice sheet are shrinking. Evidence suggests we are witnessing the early stage of an anthropogenically induced global warming superimposed on natural cycles, reinforced by reductions in Arctic ice.[53]

Arctic Sea Ice   Arctic sea ice has long been recognized as a sensitive climate indicator. Climate models are in general agreement that one of the strongest signals of global warming should be a loss of sea ice in the Arctic. This is indeed occurring. The graph in **Figure 20.28** shows the decline in the extent of sea ice at the end of the summer melting period for the span 1979–2010. September represents the end of the melt period when the area covered by sea ice is at a minimum. The 2010 minimum was third lowest, after 2007 and 2008. Is it possible that this trend may be part of a natural cycle? Yes, but it is more likely that the sea-ice

[52]For more about the behavior of glaciers, see the section "Budget of a Glacier" in Chapter 6.

[53]J. T. Overpeck, et al. "Arctic System on Trajectory to New, Seasonally Ice-Free States," *EOS, Transactions, American Geophysical Union* 86(34), (August 2005): 309.

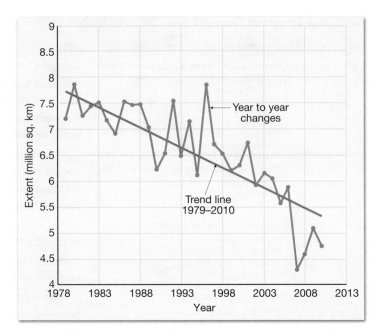

FIGURE 20.28   This graph shows the decline in the area covered by Arctic sea ice at the end of the summer melting for the span 1979–2010. In 2010 the minimum occurred on September 10th. In other years this point might be reached on other dates in September. After this date, the extent of sea ice begins its cycle of growth in response to autumn cooling. The 2010 minimum ice extent was the third lowest on record. (National Snow and Ice Data Center)

decline represents a combination of natural variability and human-induced global warming, with the latter being increasingly evident in coming decades. As was noted in the section "Climate Feedback Mechanisms," a reduction in sea ice represents a positive feedback mechanism that reinforces global warming.

Permafrost   During the past decade, evidence has mounted to indicate that the extent of permafrost in the Northern Hemisphere has decreased, as would be expected under long-term warming conditions. **Figure 20.29** presents one example that such a decline is occurring.

In the Arctic, short summers thaw only the top layer of frozen ground. The permafrost beneath this *active layer* is like the cement bottom of a swimming pool. In summer, water cannot percolate downward, so it saturates the soil above the permafrost and collects on the surface in thousands of lakes. However, as Arctic temperatures climb, the bottom of the "pool" seems to be "cracking." Satellite imagery shows that over a 20-year span, a significant number of lakes have shrunk or disappeared altogether. As the permafrost thaws, lake water drains deeper into the ground.

Thawing permafrost represents a potentially significant positive feedback mechanism that may reinforce global warming. When vegetation dies in the Arctic, cold temperatures inhibit its total decomposition. As a consequence, over thousands of years a great deal of organic matter has become stored in the permafrost. When the permafrost thaws, organic matter that may have been frozen for millennia comes out of "cold storage" and decomposes. The result is the release of carbon dioxide and methane—greenhouse gases that contribute to global warming.

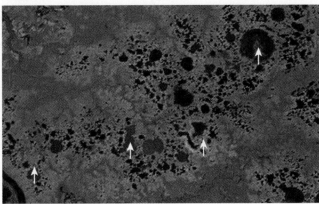

A. June 27, 1973

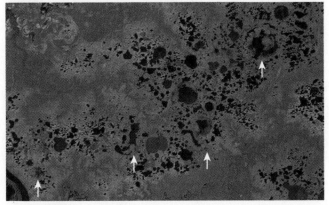

B. July 2, 2002

FIGURE 20.29 This image pair shows lakes dotting the tundra in northern Siberia in 1973 and 2002. The tundra vegetation is colored a faded red, whereas lakes appear blue or blue-green. Many lakes have clearly disappeared or shrunk considerably between 1973 and 2002. Compare the areas highlighted by the white arrows in each image. After studying satellite imagery of about 10,000 large lakes in a 500,000-square-kilometer area in northern Siberia, scientists documented an 11 percent decline in the number of lakes, with at least 125 disappearing completely. (NASA)

The Potential for "Surprises"

In summary, you have seen that climate in the 21st century, unlike the preceding thousand years, is not expected to be stable. Rather, a constant state of change is very likely. Many of the changes will probably be gradual environmental shifts, imperceptible from year to year. Nevertheless, the effects, accumulated over decades, will have powerful economic, social, and political consequences.

Despite our best efforts to understand future climate shifts, there is also the potential for "surprises." This simply means that due to the complexity of Earth's climate system, we might experience relatively sudden, unexpected changes or see some aspects of climate shift in an unexpected manner. The report *Climate Change Impacts on the United States* describes the situation like this:

> Surprises challenge humans' ability to adapt, because of how quickly and unexpectedly they occur. For example, what if the Pacific Ocean warms in such a way that El Niño events become much more extreme? This could reduce the frequency, but perhaps not the strength, of hurricanes along the East Coast, while on the West Coast, more severe winter storms, extreme precipitation events, and damaging winds could become common. What if large quantities of methane, a potent greenhouse gas currently frozen in icy Arctic tundra and sediments, began to be released to the atmosphere by warming, potentially creating an amplifying "feedback loop" that would cause even more warming? We simply do not know how far the climate system or other systems it affects can be pushed before they respond in unexpected ways.
>
> There are many examples of potential surprises, each of which would have large consequences. Most of these potential outcomes are rarely reported, in this study or elsewhere. Even if the chance of any particular surprise happening is small, the chance that at least one such surprise will occur is much greater. In other words, while we can't know which of these events will occur, it is likely that one or more will eventually occur.[54]

The impact on climate of an increase in atmospheric carbon dioxide and trace gases is obscured by some uncertainties. Yet, climate scientists continue to improve our understanding of the climate system and the potential impacts and effects of global climate change. Policy makers are confronted with responding to the risks posed by emissions of greenhouse gases knowing that our understanding is imperfect. However, they are also faced with the fact that climate-induced environmental changes cannot be reversed quickly, if at all, owing to the lengthy time scales associated with the climate system.

CONCEPT CHECK 20.13

❶ Describe at least four potential consequences of global warming.

[54]National Assessment Synthesis Team. *Climate Change Impacts on the United States: The Potential Consequences of Climate Variability and Change.* Washington, DC: U.S. Global Research Program, 2000, p. 19.

GIVE IT SOME THOUGHT

1. Refer to Figure 20.2, which illustrates various components of Earth's climate system. Boxes represent interactions or changes that occur in the climate system. Select three boxes and provide an example of an interaction or change associated with each. Explain how these interactions or changes may influence temperature.

2. Describe one way in which changes in the biosphere can cause changes in the climate system. Next, suggest one way in which the biosphere is affected by changes in some other part of the climate system. Finally, indicate one way in which the biosphere records changes in the climate system.

3. Refer to the monthly rainfall data (in millimeters) for three cities in Africa. Their locations are shown on the accompanying map. Match the data for each city to the correct location (1, 2, or 3) on the map. How were you able to figure this out? *Bonus*: Which figure in Chapter 18 would be especially useful in explaining or illustrating why these places have rainfall maximums and minimums when they do?

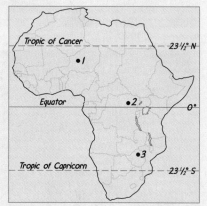

	J	F	M	A	M	J	J	A	S	O	N	D
CITY A	81	102	155	140	133	119	99	109	206	213	196	122
CITY B	0	2	0	0	1	15	88	249	163	49	5	6
CITY C	236	168	86	46	13	8	0	3	8	38	94	201

4. Refer to Figure 20.4, which shows climates of the world. Humid continental (Dfb and Dwb) and subarctic (Dfc) climates are usually described as being "land controlled," that is, they lack marine influence. Nevertheless, these climates are found along the margins of the North Atlantic and the North Pacific oceans. Explain why this occurs.

5. Explain these two apparent paradoxes associated with polar and subarctic climates.
   a. Although generally associated with small precipitation totals, these climates are classified as humid and not dry.
   b. Although polar regions experience extended periods of almost perpetual sunlight in the summer, temperatures remain cool.

6. If a fellow student who, unlike you, had not studied climate were to ask, "Isn't the greenhouse effect a bad thing because it's responsible for global warming?", how would you respond?

7. During a conversation, an acquaintance indicates that he is skeptical about global warming. When you ask why he feels that way, he says, "The past couple of years in this area have been among the coolest I can remember." While you assure this person that it is useful to question scientific findings, you suggest to him that his reasoning in this case may be flawed. Use your understanding of the definition of climate along with one or more graphs in the chapter to persuade this person to reevaluate his reasoning.

In Review   Chapter 20 World Climates and Global Climate Change

- *Climate* is the aggregate of weather conditions for a place or region over a long period of time. Earth's *climate system* involves the exchanges of energy and moisture that occur among the atmosphere, hydrosphere, solid Earth, biosphere, and *cryosphere* (the ice and snow that exist at Earth's surface).

- Climate classification brings order to large quantities of information, which aids comprehension and understanding, and facilitates analysis and explanation. *Temperature and precipitation are the most important elements in a climatic description.* Many climate classifications have been devised, with the value of each determined by its intended use. The *Köppen classification*, which uses mean monthly and annual values of temperature and precipitation, is a widely used system. The boundaries Köppen chose were largely based on the limits of certain plant associations. Five principal climate groups, each with subdivisions, were recognized. Each group is designated by a capital letter. Four of the climate groups (A, C, D, and E) are defined on the basis of temperature characteristics, and the fifth, the B group, has precipitation as its primary criterion.

- *Humid tropical (A) climates* are winterless, with all months having a mean temperature above 18° C. *Wet tropical climates* (Af and Am), which lie near the equator, have constantly high temperatures and enough rainfall to support the most luxuriant vegetation (tropical rain forest) found in any climatic realm. *Tropical wet and dry climates* (Aw) are found poleward of the wet tropics and equatorward of the subtropical deserts, where the rain forest gives way to the tropical grasslands and scattered drought-tolerant trees of the savanna. The most distinctive feature of this climate is the seasonal character of the rainfall.

- *Dry (B) climates*, in which the yearly precipitation is less than the potential loss of water by evaporation, are subdivided into two types: *arid* or *desert* (BW) and *semiarid* or *steppe* (BS). Their differences are primarily a matter of degree, with semiarid being a marginal and more humid variant of arid. Low-latitude deserts and steppes coincide with the clear skies caused by subsiding air beneath the subtropical high-pressure belts. Middle-latitude deserts and steppes exist principally because of their position in the deep interiors of large landmasses far removed from the ocean. Because many middle-latitude deserts occupy sites on the leeward sides of mountains, they can also be classified as *rain shadow deserts*.

- *Middle-latitude climates with mild winters* (C climates) occur where the average temperature of the coldest month is below 18° C but above −3° C. Several C climate subgroups exist. *Humid subtropical climates* (Cfa) are located on the eastern sides of the continents, in the 25–40° latitude range. Summer weather is hot and sultry, and winters are mild. In North America, the *marine west coast climate* (Cfb, Cfc) extends from near the U.S.–Canada border northward as a narrow belt into southern Alaska. The prevalence of maritime air masses means that mild winters and cool summers are the rule. *Dry-summer subtropical climates* (Csa, Csb) are typically located along the west sides of continents between latitudes 30–45°.

In summer, the regions are dominated by stable, dry conditions associated with the oceanic subtropical highs. In winter they are within range of the cyclonic storms of the polar front.

- *Humid middle-latitude climates with severe winters* (D climates) are land-controlled climates that are absent in the Southern Hemisphere. The D climates have severe winters. The average temperature of the coldest month is −3° C or below, and the warmest monthly mean exceeds 10° C. *Humid continental climates* (Dfa, Dfb, Dwa, Dwb) are confined to the eastern portions of North America and Eurasia in the latitude range between approximately 40 and 50° north latitude. Both winter and summer temperatures can be characterized as relatively severe. Precipitation is generally greater in summer than in winter. *Subarctic climates* (Dfc, Dfd, Dwc, Dwd) are situated north of the humid continental climates and south of the polar tundras. The outstanding feature of subarctic climates is the dominance of winter. By contrast, summers in the subarctic are remarkably warm, despite their short duration. The highest annual temperature ranges on Earth occur here.

- *Polar (E) climates* are summerless, with the average temperature of the warmest month below 10° C. Two types of polar climates are recognized. The *tundra climate* (ET) is a treeless climate found almost exclusively in the Northern Hemisphere. The *ice cap climate* (EF) does not have a single monthly mean above 0° C. As a consequence, the growth of vegetation is prohibited, and the landscape is one of permanent ice and snow.

- Compared to nearby places of lower elevation, *highland climates* are cooler and usually wetter. Because atmospheric conditions fluctuate rapidly with changes in altitude and exposure, these climates are best described by their variety and changeability.

- Humans have been modifying the environment for thousands of years. By altering ground cover with the use of fire and the overgrazing of land, people have modified such important climatological factors as surface albedo, evaporation rates, and surface winds.

- By adding carbon dioxide and other trace gases (methane, nitrous oxide, and chlorofluorocarbons) to the atmosphere, humans are likely contributing significantly to global warming.

- When any component of the climate system is altered, scientists must consider the many possible outcomes, called *climate-feedback mechanisms*. Changes that reinforce the initial change are called *positive-feedback mechanisms*. On the other hand, *negative-feedback mechanisms* produce results that are the opposite of the initial change and tend to offset it.

- Global climate is also affected by human activities that contribute to the atmosphere's *aerosol* (tiny, often microscopic, liquid and solid particles that are suspended in air) content. By reflecting sunlight back to space, aerosols have a net cooling effect. The effect of aerosols on today's climate is determined by the amount emitted during the preceding couple of weeks, while carbon dioxide remains for much longer spans and influences climate for many decades.

- Because the climate system is very complex, predicting specific regional changes that may occur as the result of increased levels of carbon dioxide in the atmosphere is highly speculative. However, some potential consequences of global warming include: (1) altering the distribution of the world's water resources; (2) a probable rise in sea level; (3) a change in weather patterns, such as a greater intensity of tropical cyclones; and (4) changes in the extent of Arctic sea ice and permafrost.

Key Terms

arid climate (p. 588)
climate-feedback mechanisms (p. 600)
climate system (p. 582)
desert climate (p. 588)
dry-summer subtropical climate (p. 591)
highland climate (p. 593)
humid continental climate (p. 591)

humid subtropical climate (p. 590)
ice cap climate (p. 593)
Köppen classification (p. 585)
marine west coast climate (p. 591)
negative-feedback mechanism (p. 600)
polar climate (p. 593)
positive-feedback mechanism (p. 600)

rain shadow desert (p. 590)
semiarid climate (p. 588)
steppe climate (p. 588)
subarctic climate (p. 592)
tropical rain forest (p. 585)
tropical wet and dry climate (p. 586)
tundra climate (p. 593)

Examining the Earth System

1. The Köppen climate classification is based on the fact that there is an excellent association between natural vegetation (biosphere) and climate (atmosphere). Briefly describe the climate conditions (temperature and precipitation) and natural vegetation associated with each of the following Köppen climates: Af, BWh, Dfc, and ET.

2. How might the burning of fossil fuels, such as the gasoline to run your car, influence global temperature? If such a temperature change occurs, how might sea level be affected? How might the intensity of hurricanes change? How might these changes impact people who live on a beach or barrier island along the Atlantic or Gulf coasts?

Mastering Geology

Looking for additional review and test prep materials? Visit the Self Study area in **www.masteringgeology.com** to find practice quizzes, study tools, and multimedia that will aid in your understanding of this chapter's content. In **MasteringGeology™** you will find:

- **GEODe: Earth Science: An interactive visual walkthrough of key concepts**

- **Geoscience Animation Library: More than 100 animations illuminating many difficult-to-understand Earth science concepts**
- **In The News RSS Feeds: Current Earth science events and news articles are pulled into the site with assessment**
- **Pearson eText**
- **Optional Self Study Quizzes**
- **Web Links**
- **Glossary**
- **Flashcards**

21

Origins of Modern Astronomy*

England's Stonehenge is one of the most famous prehistoric sites in the world. One of its functions was as an astronomical observatory.

(Photo by Ian Waldie/Reuters/CORBIS)

*This chapter was revised with the assistance of Professors Mark Watry and Teresa Tarbuck, Spring Hill College.

The science of astronomy is a rational way of knowing and understanding the origins of Earth, the solar system, and the universe. Earth was once thought to be unique, different in every way from anything else in the universe. However, through the science of astronomy, we have discovered that Earth and the Sun are similar to other objects in the universe and that the physical laws that apply on Earth seem to apply everywhere in the universe.

How did our understanding of the universe change so drastically? In this chapter we examine the transformation from the ancient view of the universe, which focused on the positions and movements of celestial objects, to the modern perspective, which focuses on *understanding how* these objects came to be and *why* they move the way they do.

FOCUS ON CONCEPTS

To assist you in learning the important concepts in this chapter, focus on the following questions:

- What is the geocentric view of the universe and how does it differ from the heliocentric view?
- What occurred during the "Golden Age" of early astronomy and where was it located?
- How does Ptolemy's model account for the observed motions of the celestial bodies including retrograde motion?
- Who was the first modern astronomer to advocate a heliocentric model for the solar system?
- What were the contributions to modern astronomy of Nicolaus Copernicus, Tycho Brahe, Johannes Kepler, Galileo Galilei, and Issac Newton?
- What are perturbations?
- How does modern astronomy use constellations?
- What is the equatorial system?
- What are some of the primary motions of Earth?
- What is the difference between a *synodic month* and a *sidereal month*?
- What causes the phases of the moon?
- What causes a solar eclipse? What causes a lunar eclipse?

Ancient Astronomy

Long before recorded history, people were aware of the close relationship between events on Earth and the positions of heavenly bodies. They realized that changes in the seasons and floods of great rivers such as the Nile in Egypt occurred when certain celestial bodies, including the Sun, Moon, planets, and stars, reached particular places in the heavens. Early agrarian cultures, whose survival depended on seasonal change, believed that if these heavenly objects could control the seasons, they could also strongly influence all Earthly events. These beliefs undoubtedly encouraged early civilizations to begin keeping records of the positions of celestial objects.

The origin of astronomy began more than 5,000 years ago when humans began to track the motion of celestial objects so they knew when to plant their crops or prepare to hunt migrating herds (Figure 21.1). The ancient Chinese, Egyptians, and Babylonians are well known for their record keeping. These cultures recorded the locations of the Sun, Moon, and the five visible planets as these objects moved slowly against the background of "fixed" stars. Eventually, it was not enough to track the motions of celes-

FIGURE 21.1 Chomsung Dae Observatory in Kyongju, Korea. This simple structure, with a central opening in the roof, resembles a number of ancient observatories found around the world. (Photo by Steven Vidler/Eurasia Press/CORBIS)

FIGURE 21.2 The Bayeux Tapestry that hangs in Bayeux, France, shows the apprehension caused by Halley's comet in A.D. 1066. This event preceded the defeat of King Harold by William the Conqueror. ("Sighting of a comet." Detail from Bayeux Tapestry. Musee de la Tapisserie, Bayeux. "With special authorization of the City of Bayeux." Bridgeman-Giraudon/Art Resource, NY)

tial objects; predicting their future positions (to avoid getting married at an unfavorable time, for example) became important.

A study of Chinese archives shows that the Chinese recorded every appearance of the famous Halley's Comet for at least 10 centuries. However, because this comet appears only once every 76 years, they were unable to link these appearances to establish that what they saw was the same object multiple times. Like most ancients, the Chinese considered comets to be mystical. Generally, comets were seen as bad omens and were blamed for a variety of disasters, from wars to plagues (**Figure 21.2**). In addition, the Chinese kept quite accurate records of "guest stars." Today we know that a "guest star" is a normal star, usually too faint to be visible, which increases its brightness as it explosively ejects gases from its surface, a phenomenon we call a *nova* (*novus* = new) or *supernova* (**Figure 21.3**).

The Golden Age of Astronomy

The "Golden Age" of early astronomy (600 B.C.–A.D. 150) was centered in Greece. Although the early Greeks have been criticized, and rightly so, for using purely philosophical arguments to explain natural phenomena, they employed observational data as well. The basics of geometry and trigonometry, which they developed, were used to measure the sizes of and distances to the largest-appearing bodies in the heavens—the Sun and the Moon.

The early Greeks held the incorrect **geocentric** (*geo* = Earth, *centric* = centered) view of the universe—which professed that Earth was a sphere that remained motionless at the center of the universe. Orbiting Earth were the Moon, Sun, and known planets—Mercury, Venus, Mars, Jupiter, and Saturn. The Sun and

Moon were thought to be perfect crystal spheres. Beyond the planets was a transparent, hollow **celestial sphere** on which the stars were attached and traveled daily around Earth. (Although it appears that the stars and planets move across the sky, this effect is actually caused by Earth's rotation on its axis.) Some early Greeks realized that the motion of the stars could be explained just as easily by a rotating Earth, but they rejected that idea because Earth exhibits no sense of motion and seemed too large to be movable. In fact, proof of Earth's rotation was not demonstrated until 1851.

To the Greeks, all of the heavenly bodies, except seven, appeared to remain in the same relative position to one another. These seven wanderers (*planetai* in Greek) included the Sun, the Moon, Mercury, Venus, Mars, Jupiter, and Saturn. Each was thought to have a circular orbit around Earth. Although this system was incorrect, the Greeks refined it to the point that it explained the apparent movements of all celestial bodies.

The famous Greek philosopher Aristotle (384–322 B.C.) concluded that Earth is spherical because it always casts a curved shadow when it eclipses the moon. Although most of Aristotle's teachings were considered infallible by many for centuries after his death, his belief in a spherical Earth was lost during the Middle Ages.

Measuring the Earth's Circumference   The first successful attempt to establish the size of Earth is credited to Eratosthenes (276–194 B.C.). Eratosthenes observed the angles of the noonday Sun in two Egyptian cities that were roughly north and south of each other—Syene (presently Aswan) and Alexandria (**Figure 21.4**). Finding that the angles of the noonday sun differed by 7 degrees, or 1/50 of a complete circle, he concluded that the

FIGURE 21.3   The Chinese recorded the sudden appearance of a "guest star" in 1054 A.D. The scattered remains of that supernova is the Crab Nebula in the constellation Taurus. This image comes from the Hubble Space Telescope. (NASA)

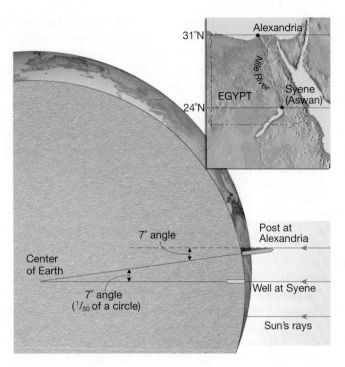

FIGURE 21.4 Orientation of the Sun's rays at Syene (Aswan) and Alexandria in Egypt on June 21 when Eratosthenes calculated Earth's circumference.

A Sun-Centered Universe?   The first Greek to profess a *Sun-centered*, or **heliocentric** (*helios* = Sun, *centric* = centered), universe was Aristarchus (312–230 B.C.). Aristarchus also used simple geometric relations to calculate the relative distances from Earth to the Sun and the Moon. He later used these data to calculate their sizes. As a result of an observational error beyond his control, he came up with measurements that were much too small. However, he did discover that the Sun was many times more distant than the Moon and many times larger than Earth. The latter fact may have prompted him to suggest a Sun-centered universe. Nevertheless, because of the strong influence of Aristotle's writings, the Earth-centered view dominated Western thought for nearly 2,000 years.

Mapping the Stars   Probably the greatest of the early Greek astronomers was Hipparchus (2nd century B.C.), best known for his star catalogue. Hipparchus determined the location of almost 850 stars, which he divided into six groups according to their brightness. (This system is still used today.) He measured the length of the year to within minutes of the modern value and developed a method for predicting the times of lunar eclipses to within a few hours.

Although many of the Greek discoveries were lost during the Middle Ages, the Earth-centered view that the Greeks proposed became entrenched in Europe. Presented in its finest form by Claudius Ptolemy, this geocentric outlook became known as the **Ptolemaic System**.

Ptolemy's Model

circumference of Earth must be 50 times the distance between these two cities. The cities were 5,000 *stadia* apart, giving him a measurement of 250,000 *stadia*. Many historians believe the *stadia* was 157.6 meters (517 feet), which would make Eratosthenes's calculation of Earth's circumference—39,400 kilometers (24,428 miles)— very close to the modern value of 40,075 kilometers (24,902 miles).

Much of our knowledge of Greek astronomy comes from a 13-volume treatise, *Almagest* (the great work), which was compiled by Ptolemy in A.D. 141. In addition to presenting a summary of Greek astronomical knowledge, Ptolemy is credited with developing a model of the universe that accounted for the observable motions of the celestial bodies (**Figure 21.5**).

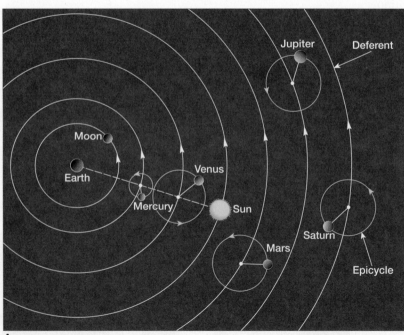

A.

B.

FIGURE 21.5 The universe according to Ptolemy, second century A.D. **A.** Ptolemy believed that the star-studded celestial sphere made a daily trip around a motionless Earth. In addition, he proposed that the Sun, Moon, and planets made trips of various lengths along individual orbits. **B.** A three-dimensional model of an Earth-centered system. Ptolemy likely utilized something similar to this to calculate the motions of the heavens. (Photo by The Bridgeman Art Library)

FIGURE 21.6 Retrograde (backward) motion of Mars as seen against the background of distant stars. When viewed from Earth, Mars moves eastward among the stars each day, then periodically appears to stop and reverse direction. This apparent westward drift is a result of the fact that Earth has a faster orbital speed than Mars and overtakes it. As this occurs, Mars appears to be moving backward, that is, it exhibits retrograde motion.

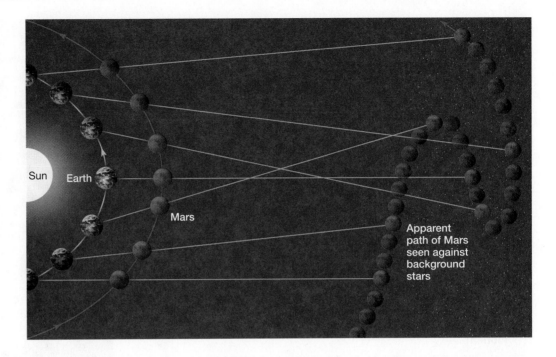

In the Greek tradition, the Ptolemaic model had the planets moving in perfect circular orbits around a motionless Earth. (The Greeks considered the circle to be the pure and perfect shape.) However, the motion of the planets, as seen against the background of stars, is not so simple. Each planet, if watched night after night, moves slightly eastward among the stars. Periodically, each planet appears to stop, reverse direction for a period of time, and then resume an eastward motion. The apparent westward drift is called **retrograde** (*retro* = to go back, *gradus* = walking) **motion**. This rather odd apparent motion results from the combination of the motion of Earth and the planet's own motion around the Sun.

The retrograde motion of Mars is shown in **Figure 21.6**. Because Earth has a faster orbital speed than Mars, it overtakes its neighbor. While doing so, Mars *appears* to be moving backward, in retrograde motion. This is analogous to what a driver sees out the side window when passing a slower car. The slower planet, like the slower car, appears to be going backward, although its actual motion is in the same direction as the faster-moving body.

It is difficult to accurately represent retrograde motion using the incorrect Earth-centered model, but that is what Ptolemy was able to accomplish (**Figure 21.7**). Rather than using a single circle for each planet's orbit, he proposed that the planets orbited on small circles (*epicycles*), revolving along large circles (*deferents*). By trial and error, he found the right combination of circles to produce the amount of retrograde motion observed for each planet. (An interesting note is that almost any closed curve can be produced by the combination of two circular motions, a fact that can be verified by anyone who has used the Spirograph™ design-drawing toy.)

It is a tribute to Ptolemy's genius that he was able to account for the planets' motions as well as he did, considering that he used an incorrect model. The precision with which his model was able to predict planetary motion is attested to by the fact that it went virtually unchallenged, in principle if not in detail, until the 17th century. When Ptolemy's predicted positions for the planets became out of step with the observed positions (which took 100 years or more), his model was simply recalibrated using the new observed positions as a starting point.

With the decline of the Roman Empire around the 4th century, much of the accumulated knowledge disappeared as libraries were destroyed. After the decline of Greek and Roman civilizations, the center of astronomical study moved east to Baghdad where, fortunately, Ptolemy's work was translated into Arabic. Later, Arabic astronomers expanded Hipparchus's star catalog

FIGURE 21.7 Ptolemy's explanation of retrograde motion—the backward motion of planets against the background of fixed stars. In Ptolemy's model, the planets move on small circles (epicycles) while they orbit Earth on larger circles (deferents). Through trial and error, Ptolemy discovered the right combination of circles to produce the retrograde motion observed for each planet.

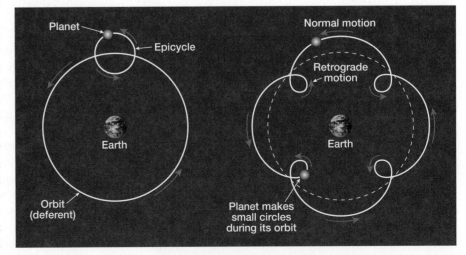

and divided the sky into 48 constellations—the foundation of our present-day constellation system. It wasn't until some time after the 10th century that the ancient Greeks' contributions to astronomy were reintroduced to Europe through the Arabic community. The Ptolemaic model soon dominated European thought as the correct representation of the heavens, which created problems for anyone who found errors in it.

CONCEPT CHECK 21.1

① Why did the ancients believe that celestial objects had some influence over their lives?

② What is the modern explanation of the "guest stars" that suddenly appeared in the night sky?

③ Explain the *geocentric* view of the universe.

④ In the Greek model of the universe what were the seven wanderers or planets? How were they different from stars?

⑤ Describe what produces the retrograde motion of Mars. What geometric arrangements did Ptolemy use to explain this motion?

The Birth of Modern Astronomy

Ptolemy's Earth-centered universe was not discarded overnight. Modern astronomy's development was more than a scientific endeavor, it required a break from deeply entrenched philosophical and religious views that had been a basic part of Western society for thousands of years. Its development was brought about by the discovery of a new and much larger universe governed by discernible laws. We examine the work of five noted scientists involved in this transition from an astronomy that merely describes what is observed, to an astronomy that tries to explain what is observed and more importantly why the universe behaves the way it does. They include Nicolaus Copernicus, Tycho Brahe, Johannes Kepler, Galileo Galilei, and Sir Isaac Newton.

Nicolaus Copernicus

For almost 13 centuries after the time of Ptolemy, very few astronomical advances were made in Europe—some were even lost, including the notion of a spherical Earth. The first great astronomer to emerge after the Middle Ages was Nicolaus Copernicus (1473–1543) from Poland (**Figure 21.8**). After discovering Aristarchus's writings, Copernicus became convinced that Earth is a planet, just like the other five then-known planets. The daily motions of the heavens, he reasoned, could be more simply explained by a rotating Earth.

Having concluded that Earth is a planet, Copernicus constructed a *heliocentric* model for the solar system, with the Sun at the center and the planets Mercury, Venus, Earth, Mars, Jupiter, and Saturn orbiting it. This was a major break from the ancient and prevailing idea that a motionless Earth lies at the center of all movement in the universe. However, Copernicus retained a

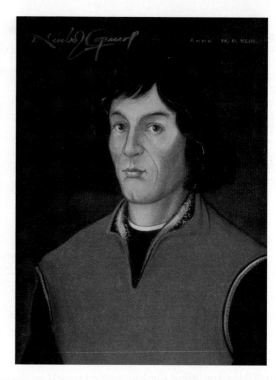

FIGURE 21.8 Polish astronomer Nicolaus Copernicus (1473–1543) believed that Earth was just another planet. (Detlev van Ravenswaay/Photo Researchers, Inc.)

link to the past and used circles to represent the orbits of the planets. Because of this Copernicus was unable to accurately predict the future locations of the planets. Copernicus found it necessary to add smaller circles (epicycles) like those used by Ptolemy. The discovery that the planets actually have *elliptical* orbits occurred a century later and is credited to Johannes Kepler.

Like his predecessors, Copernicus also used philosophical justifications to support his point of view: " . . . In the midst of all stands the Sun. For who could in this most beautiful temple place this lamp in another or better place than that from which it can at the same time illuminate the whole?"

Students Sometimes Ask...

If Ptolemy's theory was so successful, why was it rejected?

The discovery of Jupiter's moons showed that there was a fundamental flaw in the Ptolemaic theory, which described motion in the universe. According to Ptolemy's Earth-centered model, *all* heavenly bodies revolved around Earth. When Galileo, using a crude telescope, saw four moons revolving around Jupiter, he demon-strated that Earth was not the center of all motion. Consequently, at least one of the basic tenets of the Ptolemaic model had to be incorrect. Astronomers soon demon-strated that the other basic assumptions of the Earth-centered model were also inconsistent with observations.

Copernicus's monumental work, *De Revolutionibus, Orbium Coelestium* (*On the Revolution of the Heavenly Spheres*), which set forth his controversial Sun-centered solar system, was published as he lay on his deathbed. Hence, he never suffered the criticisms that fell on many of his followers. Although Copernicus's model was a vast improvement over Ptolemy's, it did not attempt to explain how planetary motions occurred or why.

The greatest contribution of the Copernican system to modern science is its challenge of the primacy of Earth in the universe. At the time this was considered heretical by many Europeans. Professing the Sun-centered model cost at least one person his life. Giordano Bruno was seized by the Inquisition, a Church tribunal, in 1600, and, refusing to denounce the Copernican theory, was burned at the stake.

Tycho Brahe

Tycho Brahe (1546–1601) was born of Danish nobility 3 years after the death of Copernicus. Reportedly, Tycho became interested in astronomy while viewing a solar eclipse that had been predicted by astronomers. He persuaded King Frederick II to establish an observatory near Copenhagen, which Tycho headed. There he designed and built pointers (the telescope would not be invented for a few more decades), which he used for 20 years to systematically measure the locations of the heavenly bodies in an effort to disprove the Copernican theory (**Figure 21.9**). His observations, particularly of Mars, were far more precise than any made previously and are his legacy to astronomy.

Tycho did not believe in the Copernican model because he was unable to observe an apparent shift in the position of stars that should result if Earth traveled around the Sun. His argument went like this: If Earth orbits the Sun, the position of a nearby star, when observed from two locations in Earth's orbit 6 months apart, should shift with respect to the more distant stars. Tycho was correct, but his measurements did not have great enough precision to show any displacement. The apparent shift of the stars is called *stellar parallax* and today it is used to measure distances to the nearest stars. (Stellar parallax is discussed in Apendix D, page 720.)

The principle of parallax is easy to visualize: Close one eye, and with your index finger vertical, use your eye to line up your finger with some distant object. Now, without moving your finger, view the object with your other eye and notice that the object's position appears to change. The farther away you hold your finger, the less the object's position seems to shift. Herein lay the flaw in Tycho's argument. He was right about parallax, but the distance to even the nearest stars is enormous compared to the width of Earth's orbit. Consequently, the shift that Tycho was looking for is too small to be detected without the aid of a telescope—an instrument that had not yet been invented.

With the death of his patron, the King of Denmark, Tycho was forced to leave his observatory. Known for his arrogance and extravagant nature, Tycho was unable to continue his work under Denmark's new ruler. As a result, Tycho moved to Prague in the present-day Czech Republic, where, in the last year of his life, he acquired an able assistant, Johannes Kepler. Kepler retained most of the observations made by Tycho and put them to exceptional use. Ironically, the data Tycho collected to refute the Copernican view of the solar system would later be used by Kepler to support it.

FIGURE 21.9 Tycho Brahe (1546–1601) in his observatory, in Uraniborg, on the Danish island of Hveen. Tycho (central figure) and the background are painted on the wall of the observatory within the arc of the sighting instrument called a quadrant. In the far right, Tycho can be seen "sighting" a celestial object through the "hole" in the wall. Tycho's accurate measurements of Mars enabled Johannes Kepler to formulate his three laws of planetary motion. (Courtesy of The Bridgeman Library International)

Johannes Kepler

If Copernicus ushered out the old astronomy, Johannes Kepler (1571–1630) ushered in the new (**Figure 21.10**). Armed with Tycho's data, a good mathematical mind, and, of greater importance, a strong belief in the accuracy of Tycho's work, Kepler derived three basic laws of planetary motion. The first two laws resulted from his inability to fit Tycho's observations of Mars to a circular orbit. Unwilling to concede that the discrepancies were a result of observational error, he searched for another solution. This endeavor led him to discover that the orbit of Mars is not a perfect circle but is slightly elliptical (**Figure 21.11**). About the same time, he realized that the orbital speed of Mars varies in a predictable way. As it approaches the Sun, it speeds up, and as it moves away, it slows down.

FIGURE 21.10 German astronomer Johannes Kepler (1571–1630) helped establish the era of modern astronomy by deriving three laws of planetary motion. (Photo by Imagno/Getty Images)

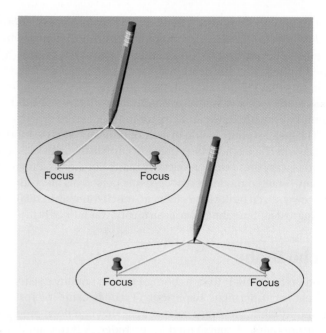

FIGURE 21.11 Drawing ellipses with various eccentricities. Using two straight pins for foci and a loop of string, trace out a curve while keeping the string taut, and you will have drawn an ellipse. The farther the pins (the foci) are moved apart, the more flattened (more eccentric) is the resulting ellipse.

In 1609, after nearly a decade of work, Kepler proposed his first two laws of planetary motion:

1. The path of each planet around the Sun, while almost circular, is actually an ellipse, with the Sun at one focus (Figure 21.11).

2. Each planet revolves so that an imaginary line connecting it to the Sun sweeps over equal areas in equal intervals of time (**Figure 21.12**). This *law of equal areas* geometrically expresses the variations in orbital speeds of the planets.

Figure 21.12 illustrates the second law. Note that in order for a planet to sweep equal areas in the same amount of time, it must travel more rapidly when it is nearer the Sun and more slowly when it is farther from the Sun.

Kepler was devout and believed that the Creator made an orderly universe and that this order would be reflected in the positions and motions of the planets. The uniformity he tried to find eluded him for nearly a decade. Then in 1619, Kepler published his third law in *The Harmony of the Worlds*:

3. The orbital periods of the planets and their distances to the Sun are proportional.

In its simplest form, the orbital period is measured in Earth years, and the planet's distance to the Sun is expressed in terms of Earth's mean distance to the Sun. The latter "yardstick" is called the **astronomical unit (AU)** and is equal to about 150 million kilometers (93 million miles). Using these units, Kepler's third law states that the planet's orbital period squared is equal to its mean solar distance cubed. Consequently, the solar distances of the planets can be calculated when their periods of revolution are known. For example, Mars has an orbital period of 1.88 years, which squared equals 3.54. The cube root of 3.54 is 1.52, and that

FIGURE 21.12 Kepler's law of equal areas. A line connecting a planet (Earth) to the Sun sweeps out an area in such a manner that equal areas are swept out in equal times. Thus, Earth revolves slower when it is farther from the Sun (aphelion) and faster when it is closest (perihelion). The eccentricity of Earth's orbit is greatly exaggerated in this diagram.

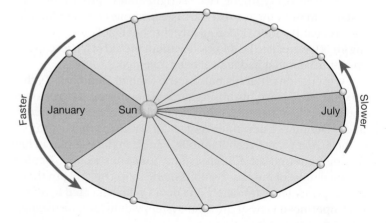

is the average distance from Mars to the Sun, in astronomical units (**Table 21.1**).

Kepler's laws assert that the planets revolve around the Sun, and therefore support the Copernican theory. Kepler, however, did not determine the *forces* that act to produce the planetary motion he had so ably described. That task would remain for Galileo Galilei and Sir Isaac Newton.

Galileo Galilei

Galileo Galilei (1564–1642) was the greatest Italian scientist of the Renaissance (**Figure 21.13**). He was a contemporary of Kepler and, like Kepler, strongly supported the Copernican theory of a Sun-centered solar system. Galileo's greatest contributions to science were his descriptions of the behavior of moving objects, which he derived from experimentation. The method of using experiments to determine natural laws had essentially been lost since the time of the early Greeks.

All astronomical discoveries before Galileo's time were made without the aid of a telescope. In 1609, Galileo heard that a Dutch lens maker had devised a system of lenses that magnified objects. Apparently without ever seeing a telescope, Galileo constructed his own, which magnified distant objects three times the size seen by the unaided eye. He immediately made others, the best having a magnification of about 30 (**Figure 21.14**).

With the telescope, Galileo was able to view the universe in a new way. He made many important discoveries that supported the Copernican view of the universe, including the following:

1. The discovery of Jupiter's four largest satellites, or moons (**Figure 21.15**). This find dispelled the old idea that Earth was the sole center of motion in the universe; for here, plainly visible, was another center of motion—Jupiter. It also countered the frequently used argument that the Moon would be left behind if Earth revolved around the Sun.

2. The discovery that the planets are circular disks rather than just points of light, as was previously thought. This indicated that the planets must be Earth-like as opposed to star-like.

FIGURE 21.13 Italian scientist Galileo Galilei (1564–1642) used a new invention, the telescope, to observe the Sun, Moon, and planets in more detail than ever before. (Nimatallah/Art Resource, N.Y.)

FIGURE 21.14 One of Galileo's telescopes. Although Galileo did not invent the telescope, he built several—the largest of which had a magnification of 30. (Photo by Gianni Tortoli/Photo Researchers, Inc.)

TABLE 21.1 Period of Revolution and Solar Distances of Planets

Planet	Solar Distance (AU)*	Period (years)	Ellipticity 0 = circle
Mercury	0.39	0.24	0.205
Venus	0.72	0.62	0.007
Earth	1.00	1.00	0.017
Mars	1.52	1.88	0.094
Jupiter	5.20	11.86	0.049
Saturn	9.54	29.46	0.057
Uranus	19.18	84.01	0.046
Neptune	30.06	164.80	0.011

*AU = astronomical unit

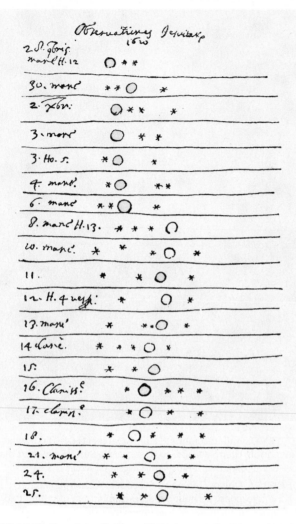

FIGURE 21.15 Sketch by Galileo of how he saw Jupiter and its four largest satellites through his telescope. The positions of Jupiter's four largest Moons (drawn as stars) change nightly. You can observe these same changes with binoculars. (Yerkes Observatory Photograph/University of Chicago)

3. The discovery that Venus exhibits phases just as the Moon does and that Venus appears smallest when it is in full phase and thus is farthest from Earth (Figure 21.16B, C). This observation demonstrates that Venus orbits its source of light—the Sun. In the Ptolemaic system, shown in Figure 21.16A, the orbit of Venus lies between Earth and the Sun, which means that only the crescent phases of Venus should ever be seen from Earth.

4. The discovery that the Moon's surface is not a smooth glass sphere, as the ancients had proclaimed. Rather, Galileo saw mountains, craters, and plains, indicating that the Moon was Earth-like. He thought the plains might be bodies of water, and this idea was strongly promoted by others, as we can tell from the names given to these features (Sea of Tranquility, Sea of Storms, etc.).

5. The discovery that the Sun (the viewing of which may have caused the eye damage that later blinded him) had sunspots—dark regions caused by slightly lower temperatures. He tracked the movement of these spots and esti-

mated the rotational period of the Sun as just under a month. Hence, another heavenly body was found to have both "blemishes" and rotational motion.

Each of these observations eroded a bedrock principle held by the prevailing view on the nature of the universe.

In 1616, the Church condemned the Copernican theory as contrary to Scripture because it did not put humans at their rightful place at the center of Creation, and Galileo was told to abandon this theory. Undeterred, Galileo began writing his most famous work, *Dialogue of the Great World Systems*. Despite poor health, he completed the project and in 1630 went to Rome, seeking permission from Pope Urban VIII to publish. Since the book was a dialogue that expounded both the Ptolemaic and Copernican systems, publication was allowed. However, Galileo's detractors were quick to realize that he was promoting the Copernican view at the expense of the Ptolemaic system. Sale of the book was quickly halted, and Galileo was called before the Inquisition. Tried and convicted of proclaiming doctrines contrary to religious teachings, he was sentenced to permanent house arrest, under which he remained for the last 10 years of his life.

Despite this restriction, and his grief following the death of his eldest daughter, Galileo continued to work. In 1637 he became totally blind, yet during the next few years he completed his finest scientific work, a book on the study of motion in which he stated that the natural tendency of an object in motion is to remain in motion. Later, as more scientific evidence in support of the Copernican system was discovered, the Church allowed Galileo's works to be published.

Sir Isaac Newton

Sir Isaac Newton (1642–1727) was born in the year of Galileo's death (Figure 21.17). His many accomplishments in mathematics and physics led a successor to say, "Newton was the greatest genius that ever existed."

Students Sometimes Ask...

Did Galileo drop balls of iron and wood from the Leaning Tower of Pisa?

Through experimentation, Galileo discovered that the acceleration of falling objects does not depend on their weight. According to some accounts, Galileo made this discovery by dropping balls of iron and wood from the Leaning Tower of Pisa to show that they would fall together and hit the ground at the same time. Despite the popularity of this legend, Galileo probably did not attempt this experiment. In fact, it would have been inconclusive because of the effect of air resistance. However, nearly four centuries later, this experiment was dramatically performed on the airless Moon when David Scott, an *Apollo 15* astronaut, demonstrated that a feather and a hammer do, indeed, fall at the same rate.

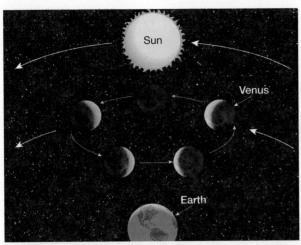

A. Phases of Venus as seen from Earth in the Earth-centered model.

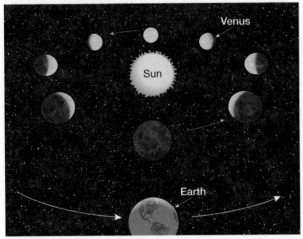

B. Phases of Venus as seen from Earth in the sun-centered model.

C.

FIGURE 21.16 Using a telescope, Galileo discovered that Venus has phases just like the Moon. **A.** In the Ptolemaic (Earth-centered) system, the orbit of Venus lies between the Sun and Earth, as shown in Figure 21.5A. Thus, in an Earth-centered solar system, only the crescent phase of Venus would be visible from Earth. **B.** In the Copernican (Sun-centered) system, Venus orbits the Sun and hence all of the phases of Venus should be visible from Earth. **C.** As Galileo observed, Venus goes through a series of Moonlike phases. Venus appears smallest during the full phase when it is farthest from Earth and largest in the crescent phase when it is closest to Earth. This verified Galileo's belief that the Sun was the center of the solar system. (Photo courtesy of Lowell Observatory)

Although Kepler and those who followed attempted to explain the forces involved in planetary motion, their explanations were less than satisfactory. Kepler believed that some force pushed the planets along in their orbits. Galileo, however, correctly reasoned that no force is required to keep an object in motion. Instead, Galileo proposed that the natural tendency for a moving object that is unaffected by an outside force is to continue moving at a uniform speed and in a straight line. This concept, *inertia*, was later formalized by Newton as his first law of motion.

The problem, then, was not to explain the force that keeps the planets moving but rather to determine the force that *keeps them from going in a straight line out into space*. It was to this end that Newton conceptualized the force of gravity. At the early age of 23, he envisioned a force that extends from Earth into space and holds the Moon in orbit around Earth. Although others had theorized the existence of such a force, he was the first to formulate and test the *law of universal gravitation*. It states:

Every body in the universe attracts every other body with a force that is directly proportional to their masses and inversely proportional to the square of the distance between them.

Thus, gravitational force decreases with distance, so that two objects 3 kilometers apart have 3^2 or 9, times less gravitational attraction than if the same objects were 1 kilometer apart.

The law of gravitation also states that the greater the mass of an object, the greater its gravitational force. For example, the large mass of the Moon has a gravitational force strong enough to cause ocean tides on Earth, whereas the tiny mass of a communications satellite has very little effect on Earth.

With his laws of motion, Newton proved that the force of gravity—combined with the tendency of a planet to remain in straight-line motion—would result in a planet having an elliptical orbit as established by Kepler. Earth, for example, moves forward in its orbit about 30 kilometers (18.5 miles) each second, and during the same second, the force of gravity pulls it toward the Sun about 0.5 centimeter (1/8 inch). Therefore, as Newton concluded, it is the combination of Earth's forward motion and its "falling" motion that defines its orbit (**Figure 21.18**). If gravity were somehow eliminated, Earth would move in a straight line out into space. Conversely, if Earth's forward motion suddenly stopped, gravity would pull it, crashing into the Sun.

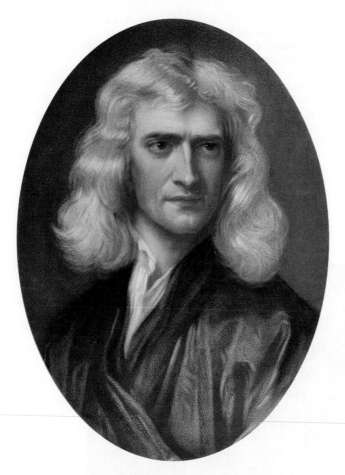

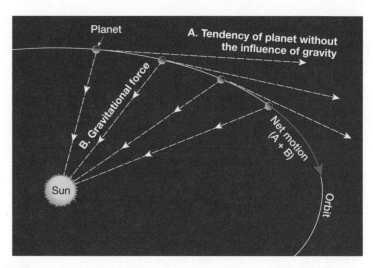

FIGURE 21.18 Orbital motion of Earth and other planets.

FIGURE 21.17 English scientist Sir Isaac Newton (1642–1727) explained gravity as the force that holds planets in orbit around the Sun. (The Granger Collection)

only if the bodies noticeably affect the orbit of a neighboring body, or of a nearby artificial satellite.

CONCEPT CHECK 21.2

1. What major change did Copernicus make in the Ptolemaic system? Why was this change philosophically different?
2. What data did Tycho Brahe collect that was useful to Johannes Kepler in his quest to describe planetary motion?
3. Who discovered that planetary orbits were ellipses rather than circles?
4. Does Earth move faster in its orbit near perihelion (January) or near aphelion (July)?
5. Explain why Galileo's discovery of a rotating Sun supports the Copernican view of a Sun-centered universe.
6. Newton discovered that the orbits of the planets are the result of opposing forces. Briefly explain these forces.

Thus far, we have discussed Earth as if the only forces involved in its motion were caused by its gravitational relationship with the Sun. However, all bodies in the solar system have gravitational effects on Earth and on each other. For this reason, the orbit of Earth is not the perfect ellipse determined by Kepler. There are slight variances in the orbits of the planets from their predicted paths. These are called **perturbations** (*perturb* = disturb). For example, Jupiter's gravitational pull on Saturn shortens Saturn's orbital period by nearly one week from what it would be if Jupiter did not exist. The application of this concept led to the discovery of the planet Neptune. When astronomers applied Newton's laws to the orbit of Uranus, it became clear that a large, unknown body (Neptune) was gravitationally affecting the motion of Uranus.

Newton used the law of universal gravitation to express Kepler's third law, which defines the relationship between the orbital periods of the planets and their solar distances. In its new form, Kepler's third law takes into account the masses of the bodies involved and thereby provides a method for determining the mass of a body when the orbit of one of its satellites is known. For example, the mass of the Sun is known from Earth's orbit, and Earth's mass has been determined from the orbit of the Moon. In fact, the mass of any body with a satellite can be determined. The masses of bodies that do not have satellites can be determined

Positions in the Sky

If you gaze at the stars away from city lights, you will get the distinct impression that the stars produce a spherical shell surrounding Earth. This impression seems so real that it is easy to understand why the early Greeks regarded the stars as being fixed to a crystalline celestial sphere. Although we realize that no such sphere exists, it is convenient to use this concept to map the stars and other celestial objects. We describe two mapping systems that use the concept of celestial sphere: (1) the division of the sky into areas called *constellations* and (2) the extension of Earth's lines of longitude and latitude into space (the *equatorial system*).

Constellations

The natural fascination people have with the star-studded skies led them to name the patterns they saw (see Box 21.1). These configurations, called **constellations** (*con* = with, *stella* = star), were named in honor of mythological characters or great heroes,

Box 21.1

UNDERSTANDING EARTH

Astrology—the Forerunner of Astronomy

Many people confuse astrology and astronomy to the point of believing these terms to be synonymous. *Astronomy* is a scientific probing of the universe aiming to determine the properties of celestial objects and the laws under which the universe operates. *Astrology*, on the other hand, is based on ancient superstitions that hold that an individual's actions and personality are based on the positions of the planets and stars now, and at the person's birth. Scientists do not accept astrology, regarding it as a pseudoscience (false science). Today, many people read horoscopes as a pastime and do not let them influence daily living.

Apparently astrology had its origin more than 5,000 years ago when the positions of the planets were plotted as they regularly migrated against the background of the "fixed" stars. Because the solar system is "flat," like a whirling Frisbee, the planets orbit the Sun along nearly the same plane.

FIGURE 21.B Stonehenge, an ancient observatory in England. On June 21–22 (summer solstice), the Sun can be observed rising above the heelstone. (Robin Scagell/Science Photo Library/Photo Researchers, Inc.)

Therefore, the planets, Sun, and Moon all appear to move along a band around the sky known as the *zodiac*. Because Earth's Moon cycles through its phases about 12 times each year, the Babylonians divided the zodiac into 12 constellations (**Figure 21.A**). Thus, each successive full Moon can be seen against the backdrop of the next constellation of the zodiac.

The dozen constellations of the zodiac ("Zone of Animals," so named because some constellations represent animals) are Aries, Taurus, Gemini, Cancer, Leo, Virgo, Libra, Scorpio, Sagittarius, Capricorn, Aquarius, and Pisces. These names may be familiar to you as the astrological signs of the zodiac. When first established, the first

day of spring (vernal equinox) occurred when the Sun was viewed against the constellation Aries. However, during each succeeding vernal equinox, the position of the Sun shifts very slightly against the background stars. Now, over 2,000 years later, the first day of spring occurs when the Sun is in Pisces. In about 600 years, it will occur when the Sun appears in the constellation Aquarius. (Hence, the "Age of Aquarius" is coming.)

Although astrology is not a science and has no basis in fact, it did contribute to the science of astronomy. The positions of the Moon, Sun, and planets at the time of a person's birth (sign of the zodiac) were considered to have great influence on that person's life. Even the great astronomer Kepler was required to make horoscopes part of his duties. To make forward-looking horoscopes, astrologers attempted to predict the future positions of the celestial bodies. Consequently, astronomical observatories were built in order to obtain more accurate predictions of events such as eclipses, which were considered highly significant in a person's life.*

Even prehistoric people built observatories. The structure known as Stonehenge, in England, was undoubtedly an attempt at better solar predictions (**Figure 21.B**). At the time of midsummer in the Northern Hemisphere (June 21–22, the summer solstice), the rising Sun emerges directly above the heel stone of Stonehenge. Besides keeping the calendar, Stonehenge may also have provided a method of determining eclipses. The remnants of other early observatories exist elsewhere in the Americas, Europe, Asia, and Africa.

FIGURE 21.A The 12 constellations of the zodiac. Earth is shown in its autumn (September) position in orbit, from which the Sun is seen against the background of the constellation Virgo.

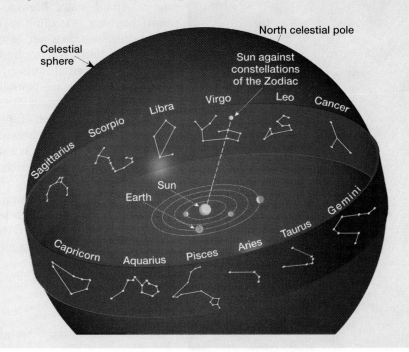

*It is interesting to note that 2,000 years ago a person born on July 28 was considered a Leo because the Sun was in that constellation. During modern times the Sun appears in the constellation Cancer on this date, but individuals born during this time are still dubbed Leos.

such as Orion the hunter (Figure 21.19). Sometimes it takes a bit of imagination to make out the intended subjects, as most constellations were probably not thought of as likenesses in the first place. Although we inherited many of the constellations from the Greeks and their names from Greek mythology, it is believed that Greeks acquired most of their constellations from the Babylonians, Egyptians, and Mesopotamians.

Although the stars that make up constellations all appear to be the same distance from Earth, this is not the case. Some are many times farther away than others. Thus, the stars in a particular constellation are not associated with each other in any important physical way. In addition, various cultural groups, including Native Americans and the Chinese, attached their own names, pictures, and stories to the constellations. For example, the constellation Orion the hunter was known as the White Tiger to ancient Chinese observers.

Today, 88 constellations are recognized, and they are used to divide the sky into units, just as state boundaries divide the United States. Every star in the sky is within the boundaries of one of these constellations. Astronomers use constellations when they want to roughly identify the area of the heavens they are observing. For the student, constellations provide a good way to become familiar with the night sky.

Some of the brightest stars in the heavens were given proper names, such as Sirius, Arcturus, and Betelgeuse. In addition, the brightest stars in a constellation are generally named in order of their brightness by the letters of the Greek alphabet—alpha (α), beta (β), and so on—followed by the name of the parent constellation. For example, Sirius, the brightest star in the constellation Canis Major (Larger Dog), is also called Alpha (α) Canis Majoris.

The Equatorial System

The **equatorial system** divides the celestial sphere into coordinates that are similar to the latitude and longitude system we use for establishing locations on Earth's surface (Figure 21.20).[55] Because the celestial sphere appears to rotate around an imaginary line extending from Earth's axis, the north and south celestial poles are aligned with the terrestrial North Pole and South Pole. The north celestial pole happens to be very near the bright star whose various names reflect its location: "pole star," Polaris, and North Star. To an observer in the Northern Hemisphere, the stars appear to circle Polaris, because it, like the North Pole, is in the center of motion (Figure 21.21). (Figure 21.22 shows how to locate the North Star using two stars in the easily located constellation the Big Dipper.)

Now, imagine a plane through Earth's equator, a plane that extends outward from Earth and intersects the celestial sphere. The intersection of this plane with the celestial sphere is called the *celestial equator* (Figure 21.20). In the equatorial system, the term *declination* is analogous to latitude, and the term *right ascension* is analogous to longitude (Figure 21.20). **Declination** (*declinare* = to turn away), like latitude, is the angular distance north or south of the celestial equator. **Right ascension** (*ascendere* = to climb up) is the angular distance measured eastward along the celestial equator from the position of the vernal equinox. (The *vernal equinox* is at the point in the sky where the Sun crosses the celestial equator, at the onset of spring.) While declination is expressed in degrees, right ascension is usually expressed in hours, where each hour is equivalent to 15 degrees. (Earth rotates 15 degrees each hour.) To visualize distances on the celestial sphere, it helps to remember that the Moon and Sun have an apparent width of about 0.5 degree.

FIGURE 21.19 Constellation Orion the hunter. **A.** Artist's depiction of Orion based on descriptions from Greek mythology. **B.** Photo showing the brightest stars in Orion. The bright star in the upper left is named Betegeuse—a red supergiant. (Photo by John Chumack/Photo Researchers, Inc.)

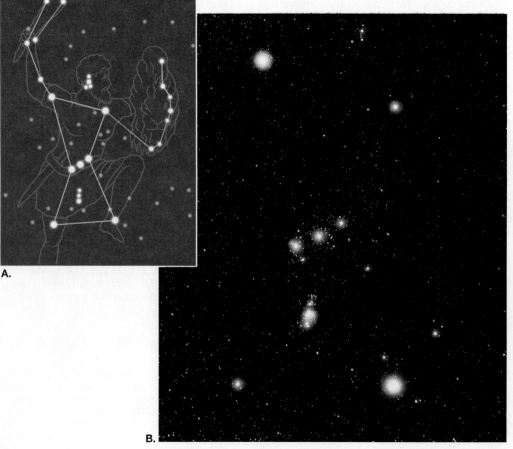

[55]Latitude and longitude are described in Appendix B, "Earth's Grid System."

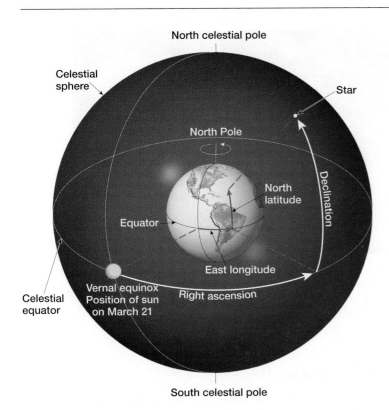

FIGURE 21.20 Astronomical coordinate system on the celestial sphere.

North celestial pole

Celestial sphere

Star

North Pole

North latitude

Declination

Equator

East longitude

Celestial equator

Vernal equinox
Position of sun
on March 21

Right ascension

South celestial pole

FIGURE 21.22 Locating the North Star (Polaris) from the pointer stars in the Big Dipper, which is part of the constellation Ursa Major. The Big Dipper is shown soon after sunset in December (lower figure), April (upper figure), and August (left).

North Star

Pointer stars

The Motions of Earth

The two primary motions of Earth are *rotation* and *revolution*. A lesser motion is *axial precession*. **Rotation** is the turning, or spinning, of a body on its axis. **Revolution** is the motion of a body, such as a planet or moon, along a path around some point in space. For example, Earth *revolves* around the Sun, and the Moon *revolves* around Earth. Earth also has another very slow motion known as **axial precession**, which is the gradual change in the orientation of Earth's axis over a period of 26,000 years.

Rotation

The main consequences of Earth's rotation are day and night. Earth's rotation has become a standard method of measuring time because it is so dependable and easy to use. You may be surprised to learn that Earth's rotation is measured in two ways, making two kinds of days. Most familiar is the **mean solar day**, the time interval from one noon to the next, which averages about 24 hours. Noon is when the Sun has reached its highest point in the sky.

The **sidereal** (*sider* = star, *at* = pertaining to) **day**, on the other hand, is the time it takes for Earth to make one complete rotation (360 degrees) with respect to a star other than our Sun. The sidereal day is measured by the time required for a star to reappear at the identical position in the sky. The sidereal day has a period of 23 hours, 56 minutes, and four seconds (measured in solar time), which is almost 4 minutes shorter than the mean solar day. This difference results because the direction to distant stars changes only infinitesimally, whereas the direction to the Sun changes by almost 1 degree each day. This difference is shown in **Figure 21.23**.

CONCEPT CHECK 21.3

❶ How do modern astronomers use constellations?
❷ How many constellations are currently recognized?
❸ How are the brightest stars in a constellation denoted?
❹ Briefly describe the *equatorial system*.

FIGURE 21.21 Star trails in the region of Polaris (north celestial pole) on a time exposure. (Photo by Douglas Kirkland/CORBIS)

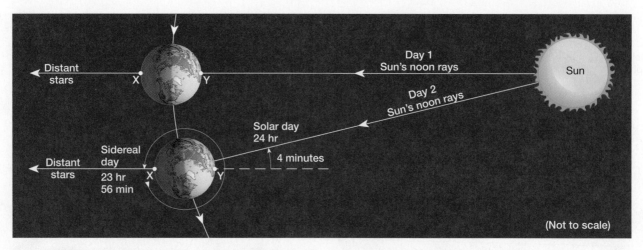

FIGURE 21.23 The difference between a solar day and a sidereal day. Locations X and Y are directly opposite each other. It takes Earth 23 hours and 56 minutes to make one rotation with respect to the stars (sidereal day). However, notice that after Earth has rotated once with respect to the stars, point Y is not yet returned to the "noon position" with respect to the Sun. Earth has to rotate another 4 minutes to complete the solar day.

Why do we use the mean solar day to measure time rather than the sidereal day? Consider the fact that in sidereal time, "noon" occurs 4 minutes earlier each day. Therefore, after a span of 6 months, "noon" would occur at "midnight." However, observatories use clocks that keep sidereal time because the stars appear to move through the sky in sidereal time. Simply, if a star is sighted directly south of an observatory at 9:00 P.M. (sidereal time), it will appear in the same direction at that time every (sidereal) day.

Revolution

Earth revolves around the Sun in an elliptical orbit at an average speed of 107,000 kilometers (66,000 miles) per hour. Its average distance from the Sun is 150 million kilometers (93 million miles), but because its orbit is an ellipse, Earth's distance from the Sun varies. At **perihelion** (peri = near, *helios* = sun) it is 147 million kilometers (91.5 million miles) distant, which occurs about January 3 each year. At **aphelion** (*apo* = away, *helios* = sun) Earth is 152 million kilometers (94.5 million miles) distant, which occurs about July 4.

Because of Earth's orbital movement the Sun appears to be displaced among the constellations at a distance equal to about twice its width, or 1 degree each day. The apparent annual path of the Sun against the backdrop of the celestial sphere is called the **ecliptic** (Figure 21.24). The planets and the Moon travel in nearly the same plane as Earth. Hence, their paths on the celestial sphere also lie near the ecliptic.

The imaginary plane that connects points along the ecliptic is called the **plane of the ecliptic**. As measured from this imaginary plane, Earth's axis is tilted about $23\frac{1}{2}$ degrees (Figure 21.24). This angle is very important to Earth's inhabitants because the inclination of Earth's axis causes the yearly cycle of seasons, a topic discussed in detail in Chapter 16.

Precession

A third and very slow movement of Earth is called *axial precession*. Although Earth's axis maintains approximately the same angle of tilt, the direction in which the axis points continually changes (Figure 21.25A). As a result, the axis traces a circle on the

FIGURE 21.24 Earth's orbital motion causes the apparent position of the Sun to shift about 1 degree each day on the celestial sphere.

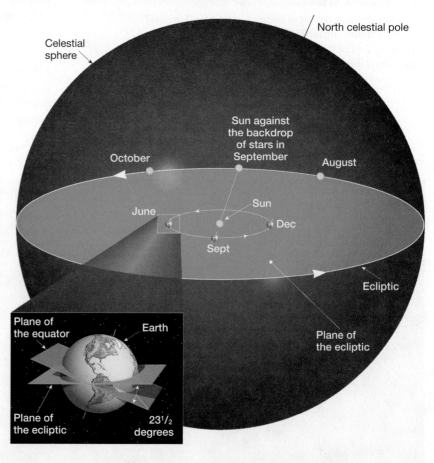

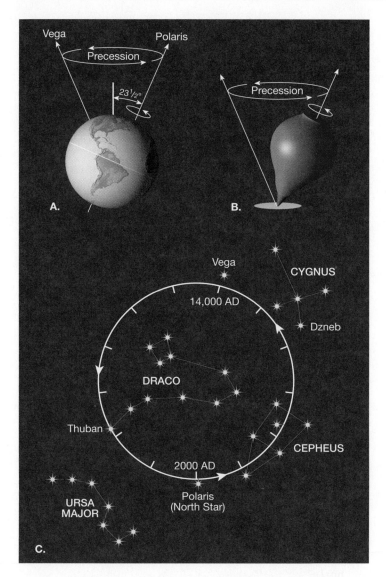

FIGURE 21.25 Precession of Earth's axis. **A.** The precession of Earth's axis causes the North Pole to "trace" a circle through the sky during a 26,000-year cycle. Currently, the North Pole points toward Polaris (North Star). In about 12,000 years, Vega will be the North Star. Around 3000 BC, the North Star was Thuban, a bright star in the constellation Draco. **B.** Precession illustrated by a spinning toy top. **C.** The circle shows the path of the North Pole among some prominent stars and constellations in the northern sky.

Students Sometimes Ask...

Our book states that Earth is farthest from the Sun in July and closest to the Sun in January. What would the seasons be like if this situation were reversed?

The situation you proposed will, in fact, occur in about 12,000 years because of axial precession. (Recall that variations in Earth–Sun distance are not the primary cause of the seasons. Nevertheless, they do affect average seasonal temperatures.) Gradually, the position of Earth's axis will change so that in 12,000 years the Northern Hemisphere will experience winter when Earth is farthest from the Sun (aphelion), and summer will occur when our planet is closest to the Sun (perihelion). This is just the opposite of the current situation. Thus, 12,000 years from now, average summer temperatures in the Northern Hemisphere will be warmer than they presently are. A summer in Montreal, Canada, might be more akin to a typical summer in Washington, D.C., today. However, northern latitudes will experience winter temperatures that are, on average, colder than they presently are.

motion. We are presently approaching one of our nearest galactic neighbors, the Great Galaxy in Andromeda.

In summary, the motions of Earth are many and complex. Fortunately, one rarely has to consider all the motions at once. For example, since the solar system moves as a unit in the galaxy, and the galaxy moves as a unit through the universe, we do not have to consider these motions when discussing the motions of the Earth and Moon around the Sun.

CONCEPT CHECK 21.4

1. Describe the three primary motions of Earth.
2. Explain the difference between the mean solar day and the sidereal day.
3. Define the *ecliptic*.
4. Why does the axial precession have little effect on the seasons?

sky. This movement is similar to the "wobble" of a spinning top (Figure 21.25B). At the present time, the axis points toward the bright star Polaris. In AD 14,000, it will point toward the bright star Vega, which will then be the North Star for about a thousand years or so (Figure 21.25C). The period of precession is 26,000 years. By the year 28,000, Polaris will once again be the North Star. Precession has only a minor effect on the seasons because Earth's angle of tilt changes only slightly.

In addition to its own movements, Earth accompanies the Sun as it speeds in the direction of the bright star Vega at 20 kilometers (12 miles) per second. Also, the Sun, like other nearby stars, revolves around the galaxy, a trip that requires 230 million years to complete at speeds approaching 250 kilometers (150 miles) per second. In addition, the galaxies themselves are in

Motions of the Earth–Moon System

Earth has one natural satellite, the Moon. In addition to accompanying Earth in its annual trek around the Sun, our Moon orbits Earth about once each month. When viewed from a Northern Hemisphere perspective, the Moon moves counterclockwise (eastward) around Earth. The Moon's orbit is elliptical, causing the Earth–Moon distance to vary by about 6 percent, averaging 384,401 kilometers (238,329 miles).

The motions of the Earth–Moon system constantly change the relative positions of the Sun, Earth, and Moon. The results are some of the most noticeable astronomical phenomena, namely the *phases of the Moon* and the occasional *eclipses of the Sun and Moon*.

Lunar Motions

The cycle of the Moon through its phases requires $29\frac{1}{2}$ days—a time span called the **synodic month**. This cycle was the basis for the first Roman calendar. However, this is the *apparent period* of the Moon's revolution around Earth and not the true period, which takes only $27\frac{1}{3}$ days and is known as the **sidereal month**. The reason for the difference of nearly 2 days each cycle is shown in **Figure 21.26**. Notice that as the Moon orbits Earth, the Earth–Moon system also moves in an orbit around the Sun. Consequently, even after the Moon has made a complete revolution around Earth, it has not yet reached its starting position with respect to the Sun, which is directly between the Sun and Earth (new-Moon phase). This motion takes an additional 2 days.

An interesting fact concerning the motions of the Moon is that its period of rotation around its axis and its revolution around Earth are the same—$27\frac{1}{3}$ days. Because of this, the same lunar hemisphere always faces Earth. All of the landings of the manned *Apollo* missions were confined to the Earth-facing side. Only orbiting satellites and astronauts have seen the "back" side of the Moon.

Because the Moon rotates on its axis only once every $27\frac{1}{3}$ days, any location on its surface experiences periods of daylight and darkness lasting about 2 weeks. This, along with the absence of an atmosphere, accounts for the high surface temperature of 127° C (261° F) on the day side of the Moon and the low surface temperature of –173° C (–280° F) on its night side.

Phases of the Moon

The first astronomical phenomenon to be understood was the regular cycle of the **phases of the Moon**. On a monthly basis, we observe the phases as a systematic change in the amount of the Moon that appears illuminated (**Figure 21.27**). We will choose the "new-Moon" position in the cycle as a starting point. About 2 days after the new Moon, a thin sliver (*crescent phase*) can be seen with the naked eye low in the western sky just after sunset. During the following week, the illuminated portion of the Moon that is visible from Earth increases (*waxing*) to a half-circle (*first-quarter phase*) that can be seen from about noon to midnight. In another week, the complete disk (*full-Moon phase*) can be seen rising in the east as the Sun sinks in the west. During the next 2 weeks, the percentage of the Moon that can be seen steadily declines (*waning*), until the Moon disappears altogether (*new-Moon phase*). The cycle soon begins anew with the reappearance of the crescent Moon.

The lunar phases are a consequence of the motion of the Moon and the sunlight that is reflected from its surface (Figure 21.27B). Half of the Moon is illuminated at all times (note the inner group of Moon sketches in Figure 21.27A). But to an earthbound observer, the percentage of the bright side that is visible depends on the location of the Moon with respect to the Sun and Earth. When the Moon lies *between* the Sun and Earth, none of its bright side faces Earth, so we see the new-Moon ("no-Moon") phase. Conversely, when the Moon lies on the side of Earth opposite the Sun, all of its lighted side faces Earth, so we see the full Moon. At all positions between these extremes, an intermediate amount of the Moon's illuminated side is visible from Earth.

CONCEPT CHECK 21.5

❶ Compare the *synodic month* with the *sidereal month*.

❷ What is the approximate length of the cycle of the phases of the Moon?

❸ What phenomenon results from the fact that the Moon's period of rotation and revolution are the same?

❹ The Moon rotates very slowly (once in $27\frac{1}{3}$ days) on its axis. How does this affect the lunar surface temperature?

❺ What is different about the crescent phase that precedes the new-Moon phase and that which follows the new-Moon phase?

❻ What phase of the Moon occurs approximately one week after the new Moon? Two weeks?

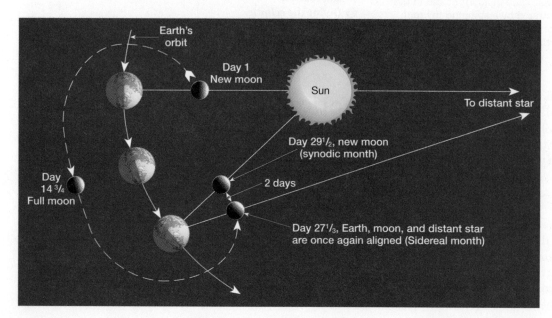

FIGURE 21.26 The difference between the sidereal month ($27\frac{1}{3}$ days) and the synodic month ($29\frac{1}{2}$ days). Distances and angles are not shown to scale.

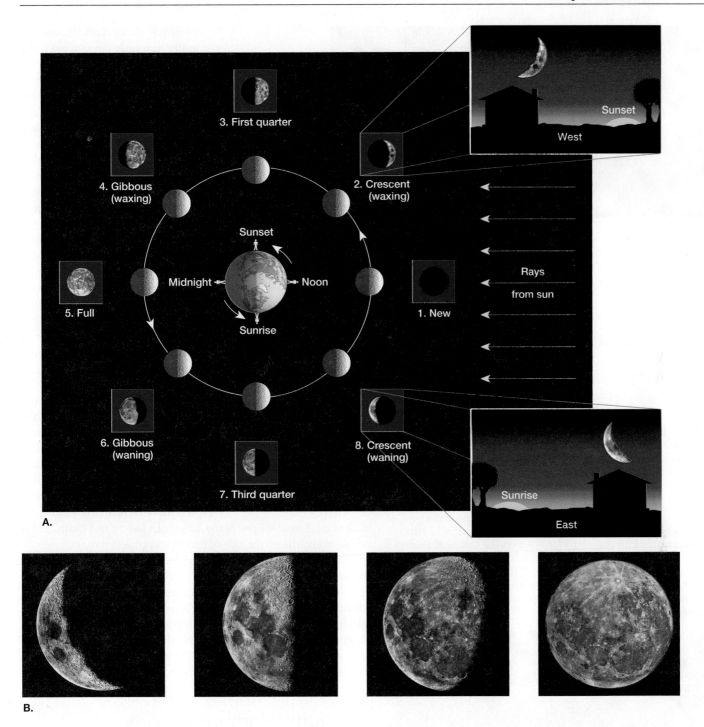

FIGURE 21.27 Phases of the Moon. **A.** The outer figures show the phases as seen from Earth. **B.** Compare these photographs with the diagram. (Photos © UC Regents/Lick Observatory)

Eclipses of the Sun and Moon

Along with understanding the Moon's phases, the early Greeks also realized that eclipses are simply shadow effects. When the Moon moves in a line directly between Earth and the Sun, which can occur only during the new-Moon phase, it casts a dark shadow on Earth, producing a **solar eclipse** (*eclipsis* = failure to appear) (**Figure 21.28**). Conversely, the Moon is eclipsed (**lunar eclipse**) when it moves within Earth's shadow, a situation that is possible only during the full-Moon phase (**Figure 21.29**).

Why does a solar eclipse not occur with every new-Moon phase and a lunar eclipse with every full Moon? They would, if the orbit of the Moon lay exactly along the plane of Earth's orbit. However, the Moon's orbit is inclined about 5 degrees to the plane of the ecliptic. Thus, during most new-Moon phases, the shadow of the Moon passes either above or below Earth; and during most full-Moon phases, the shadow of Earth misses the Moon. An eclipse can only take place when a new- or full-Moon phase occurs while the Moon's orbit crosses the plane of the ecliptic.

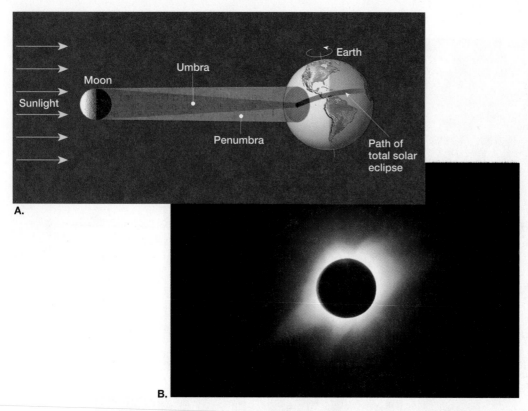

A.

B.

FIGURE 21.28 Solar eclipse. **A.** Observers in the zone of the umbral shadow see a total solar eclipse. Those located in the penumbra only see a partial eclipse. The path of the solar eclipse moves eastward across the Earth. **B.** During a total solar eclipse, the blotted-out solar disk is surrounded by an irregularly shaped halo called the *corona*. (Photo by Roger Ressmeyer/CORBIS)

FIGURE 21.29 Lunar eclipse. **A.** During a total lunar eclipse the Moon's orbit carries it into the dark shadow of Earth (umbra). During a partial eclipse only a portion of the Moon enters the umbra. **B.** During a total lunar eclipse a dark, copper-colored Moon is observed. The color is a result of a small amount of sunlight that is reddened by Earth's atmosphere—for the same reason sunsets appear red. This light is refracted (bent) toward the Moon's surface. (Photo by Eckhard Slawik/Photo Researchers, Inc.)

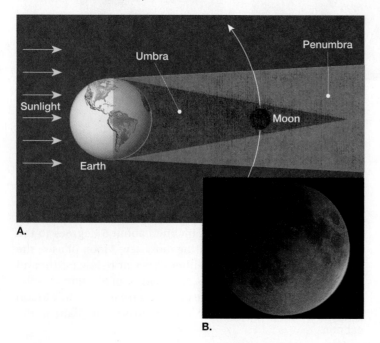

A.

B.

Because these conditions are normally met only twice a year, the usual number of eclipses is four. These occur as a set of one solar and one lunar eclipse, followed 6 months later with another set. Occasionally the alignment is such that three eclipses can occur in a one-month period—at the beginning, middle, and end. These occur as a solar eclipse flanked by two lunar eclipses, or vice versa. Furthermore, it occasionally happens that the first set of eclipses for the year occurs at the very beginning of a year, the second set in the middle, and a third set occurs before the calendar year ends, resulting in six eclipses in that year. More rarely, if one of these sets consists of three eclipses, the total number of eclipses in a year can reach seven, which is the maximum.

During a total lunar eclipse, Earth's circular shadow moves slowly across the disk of the full Moon. When totally eclipsed, the Moon is completely within Earth's shadow but is still visible as a coppery disk, because Earth's atmosphere bends some long-wavelength light (red) into its shadow. Some of this light reflects off the Moon and back to us. A total eclipse of the Moon can last up to 4 hours and is visible to anyone on the side of Earth facing the Moon.

During a total solar eclipse, the Moon casts a circular shadow that is never wider than 275 kilometers (170 miles), about the size of South Carolina. This shadow traces a stripe on Earth's surface. Anyone observing in this region will see the Moon slowly block the Sun from view and the sky darken (**Figure 21.30**). Near totality, a sharp drop in temperature of a few degrees is experienced. The solar disk is completely blocked for a maximum of only 7 minutes, because the Moon's shadow is so small. At totality, the dark

Students Sometimes Ask...

Why do we sometimes see the Moon in the daytime?

During the full-Moon phase, the Moon and the Sun are on opposite sides of Earth, which causes the Moon to rise around sunset and set at sunrise. Thus, the full moon tends to be visible only at night. However, during the other phases of the lunar cycle, the Moon and the Sun are not directly opposite each other, and the lit portion of the Moon is visible in the daytime sky. For example, the waning-gibbous phase can be seen in the early morning hours and the waxing-gibbous Moon in the afternoon (see Figure 21.27). Although the crescent Moon is "out" shortly before sunset and after sunrise, you probably won't see it in the daytime. Why not?

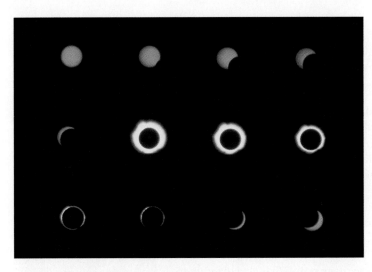

FIGURE 21.30 This sequence of photos starting from the upper left to the lower right shows the stages of a total solar eclipse. (From *Foundations of Astronomy*, Third Edition, p. 54, by Michael Seeds. © 1992. Reprinted with permission of Brooks/Cole Publishing, a division of Thomson Learning.)

Moon is seen covering the complete solar disk, and only the Sun's brilliant white outer atmosphere is visible (Figure 21.28). Total solar eclipses are visible only to people in the dark part of the Moon's shadow (*umbra*), while a partial eclipse is seen by those in the light portion (*penumbra*) (Figure 21.28).

Partial solar eclipses are most common in the Polar regions, because it is these areas that the penumbra blankets when the dark umbra of the Moon's shadow just misses Earth. A total solar eclipse is a rare event at any given location. The next one that will be visible from the contiguous United States will occur on August 21, 2017.

CONCEPT CHECK 21.6

❶ Sketch the locations of the Sun, Moon, and Earth during a solar eclipse and during a lunar eclipse.

❷ How many eclipses normally occur each year?

❸ Solar eclipses are slightly more common than lunar eclipses. Why, then, is it more likely that your region of the country will experience a lunar eclipse?

❹ How long can a total eclipse of the Moon last? How about a total eclipse of the Sun?

GIVE IT SOME THOUGHT

1. Refer to Figure 21.4 and imagine that Eratosthenes had measured the difference in the angles of the noonday Sun between Syene and Alexandria to be 10 degrees instead of 7 degrees. Consider how this new measurement would have affected his calculation of Earth's circumference to answer the following questions.
   a. Would this new measurement lead to a more accurate calculation?
   b. Would this new measurement lead to an estimate for the circumference of Earth that is larger or smaller than Eratosthenes's original estimate?

2. Use Kepler's third law to answer the following questions:
   a. Determine the period of a planet with a solar distance of 10AU.
   b. Determine the distance between the Sun and a planet with a period of 5 years.
   c. Imagine two bodies, one twice as large as the other, orbiting the Sun at the same distance. Which of the bodies, if either, would move faster than the other?

3. Galileo used his telescope to observe the planets and moons in our solar system. These observations allowed him to determine the positions and relative motions of the Sun, Earth, and other objects in the solar system. Refer to Figure 21.16A, which shows an Earth-centered solar system, and Figure 21.16B, which shows a Sun-centered solar system, to complete the following:
   a. Describe the phases of Venus an observer on Earth would see for the Earth-centered model of the solar system.
   b. Describe the phases of Venus an observer on Earth would see for the Sun-centered model of the solar system.
   c. Explain how Galileo used observations of the phases of Venus to determine the correct positions of the Sun, Earth, and Venus.

4. Refer to the accompanying diagram, which shows three asteroids (A, B, and C). They are being pulled by the gravitational force exerted on them by their partner asteroid shown on the left. How will the strength of the gravitational force felt by each asteroid (A, B, and C) compare? (Assume all of these asteroids are composed of the same material.)

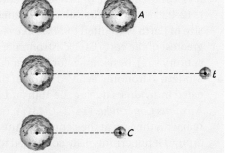

5. Refer to the accompanying diagram, which shows two pairs of asteroids, Pair A and Pair B. Is it possible for the asteroids in Pair A to be experiencing the same degree of gravitational force as the asteroids in Pair B? Explain your answer.

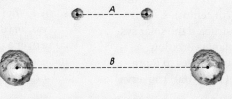

6. Imagine that Earth rotates on its axis at half its current rate. How much time would be required to capture the photo shown in Figure 21.21?

7. If we were able to reverse the direction of Earth's rotation, would the solar day be longer, shorter, or stay the same?

8. Refer to Figure 21.A to complete the following:
   a. What constellation is located in the sky near the Sun on the day illustrated?
   b. After Earth has been moving around the Sun for 5 months, which constellation will be near the Sun at noon?
   c. Which constellation will be high in the sky at midnight? Explain your reasoning.

9. Imagine that today is your birthday. Would you be able to see the stars of your astrological birth sign? Explain your answer.

10. Refer to the accompanying photo to complete the following.
    a. When you observe the phase of the moon shown, is the moon waxing or waning?
    b. What time of day can this phase of the moon be observed?

11. Imagine you are looking up at a full moon. At the same time, an astronaut on the Moon is viewing Earth. In what phase will Earth appear to be from the astronaut's vantage point? Sketch a diagram to illustrate your answer.

12. If the moon's orbit were precisely aligned with the plane of Earth's orbit, how many eclipses (solar and lunar) would occur in a 6-month period of time? If the moon's orbit were tilted 90 degrees with respect to the plane of Earth's orbit, how many eclipses (solar and lunar) would occur in a 6-month period?

In Review   Chapter 21 Origins of Modern Astronomy

- Early Greeks held the *geocentric* (Earth-centered) view of the universe, believing that Earth was a sphere that stayed motionless at the center of the universe. Orbiting Earth were the seven wanderers (*planetai* in Greek), which included the Moon, Sun, and the known planets Mercury, Venus, Mars, Jupiter, and Saturn. To the early Greeks, the stars traveled daily around Earth on a transparent, hollow sphere called the *celestial sphere*. In A.D. 141, *Claudius Ptolemy* presented the geocentric outlook of the Greeks in its most sophisticated form in a model that became known as the *Ptolemaic system*. The Ptolemaic model had the planets moving in circular orbits around a motionless Earth. To explain the *retrograde motion* of planets (the apparent westward or opposite motion that planets exhibit for a period of time as Earth overtakes and passes them), Ptolemy proposed that the planets orbited in small circles (*epicycles*), revolving along large circles (*deferents*).

- In the 5th century B.C., the Greek *Anaxagoras* reasoned that the Moon shines by reflected sunlight, and because it is a sphere, only half is illuminated at one time. *Aristotle* (384–322 B.C.) concluded that Earth is spherical. The first Greek to profess a Sun-centered, or *heliocentric*, universe was *Aristarchus* (312–230 B.C.). The first successful attempt to establish the size of Earth is credited to *Eratosthenes* (276–194 B.C.). The greatest of the early Greek astronomers was *Hipparchus* (2nd century B.C.), best known for his star catalogue.

- Modern astronomy evolved through the work of many dedicated individuals during the 16th and 17th centuries. *Nicolaus Copernicus* (1473–1543) reconstructed the solar system with the Sun at the center and the planets orbiting around it but erroneously continued to use circles to represent the orbits of planets. *Tycho Brahe*'s (1546–1601) observa-

tions were far more precise than any made previously and are his legacy to astronomy. *Johannes Kepler* (1571–1630) ushered in the new astronomy with his three laws of planetary motion. After constructing his own telescope, *Galileo Galilei* (1564–1642) made many important discoveries that supported the Copernican view of a Sun-centered solar system. *Sir Isaac Newton* (1642–1727) was the first to formulate and test the law of universal gravitation, develop the laws of motion, and prove that the force of *gravity*, combined with the tendency of an object to move in a straight line (*inertia*), results in the elliptical orbits discovered by Kepler.

- As early as 5,000 years ago people began naming the configurations of stars, called *constellations*, in honor of mythological characters or great heroes. Today, 88 constellations are recognized that divide the sky into units, just as state boundaries divide the United States.

- One method for locating stars, called the *equatorial system*, divides the celestial sphere into a coordinate system similar to the latitude–longitude system used for locations on Earth's surface. *Declination*, like latitude, is the angular distance north or south of the *celestial equator*. *Right ascension* is the angular distance measured eastward from the position of the *vernal equinox* (the point in the sky where the Sun crosses the celestial equator at the onset of spring).

- The two primary motions of Earth are *rotation* (the turning, or spinning, of a body on its axis) and *revolution* (the motion of a body, such as a planet or moon, along a path around some point in space). Another very slow motion of Earth is *precession* (the slow motion of Earth's axis that traces out a cone over a period of 26,000 years). Earth's rotation can be measured in two ways, making two kinds of days. The *mean solar day* is the time interval from one noon to the next, which averages about 24 hours. In contrast, the *sidereal day* is the time it takes for Earth to make one complete rotation with respect to a star other than the Sun, a period of 23 hours, 56 minutes, and 4 seconds. Earth revolves around the Sun in an elliptical orbit at an average distance from the Sun of 150 million kilometers (93 million miles). At *perihelion* (closest to the Sun), which occurs in January, Earth is 147 million kilometers from the Sun. At *aphelion* (farthest from the Sun), which occurs in July, Earth is 152 million kilometers distant. The imaginary plane that connects Earth's orbit with the celestial sphere is called the *plane of the ecliptic.*

- One of the first astronomical phenomena to be understood was the regular cycle of the phases of the Moon. The cycle of the Moon through its phases requires $29\frac{1}{2}$ days, a time span called the *synodic month*. However, the true period of the Moon's revolution around Earth takes $27\frac{1}{3}$ days and is known

as the *sidereal month*. The difference of nearly 2 days is due to the fact that as the Moon orbits Earth, the Earth–Moon system also moves in an orbit around the Sun.

- In addition to understanding the Moon's phases, the early Greeks also realized that eclipses are simply shadow effects. When the Moon moves in a line directly between Earth and the Sun, which can occur only during the new-Moon phase, it casts a dark shadow on Earth, producing a *solar eclipse*. A *lunar eclipse* takes place when the Moon moves within the shadow of Earth during the full-Moon phase. Because the Moon's orbit is inclined about 5 degrees to the plane that contains the Earth and Sun (the plane of the ecliptic), during most new- and full-Moon phases no eclipse occurs. Only if a new- or full-Moon phase occurs as the Moon crosses the plane of the ecliptic can an eclipse take place. The usual number of eclipses is four per year.

Key Terms

aphelion (p. 626)
astronomical unit (AU) (p. 618)
axial precession (p. 625)
celestial sphere (p. 613)
constellations (p. 622)
declination (p. 624)
ecliptic (p. 626)
equatorial system (p. 624)
geocentric (p. 613)

heliocentric (p. 614)
lunar eclipse (p. 629)
mean solar day (p. 625)
perihelion (p. 626)
perturbation (p. 622)
phases of the Moon (p. 628)
plane of the ecliptic (p. 626)
Ptolemaic system (p. 614)

retrograde motion (p. 615)
revolution (p. 625)
right ascension (p. 624)
rotation (p. 625)
sidereal day (p. 625)
sidereal month (p. 628)
solar eclipse (p. 629)
synodic month (p. 628)

Examining the Earth System

1. Currently, Earth is closest to the Sun (perihelion) in January (147 million kilometers/91.5 million miles) and farthest from the Sun in July (152 million kilometers/94.5 million miles). As the result of the precession of Earth's axis, 12,000 years from now perihelion (closest) will occur in July and aphelion (farthest) will take place in January. Assuming no other changes, how might this change *average* summer temperatures for your location? What about *average* winter temperatures? What might the impact be on the biosphere and

hydrosphere? (To aid your understanding of the effect of Earth's orbital parameters on the seasons, you may want to review the section "Variations in Earth's Orbit" in Chapter 6, pp. 170–171.)

2. In what ways do the interactions between Earth and its Moon influence the Earth system? If Earth did not have a Moon, how might the atmosphere, hydrosphere, geosphere, and biosphere be different?

Mastering Geology

Looking for additional review and test prep materials? Visit the Self Study area in **www.masteringgeology.com** to find practice quizzes, study tools, and multimedia that will aid in your understanding of this chapter's content. In **MasteringGeology™** you will find:

- **GEODe: Earth Science: An interactive visual walkthrough of key concepts**

- **Geoscience Animation Library: More than 100 animations illuminating many difficult-to-understand Earth science concepts**
- **In The News RSS Feeds: Current Earth science events and news articles are pulled into the site with assessment**
- **Pearson eText**
- **Optional Self Study Quizzes**
- **Web Links**
- **Glossary**
- **Flashcards**

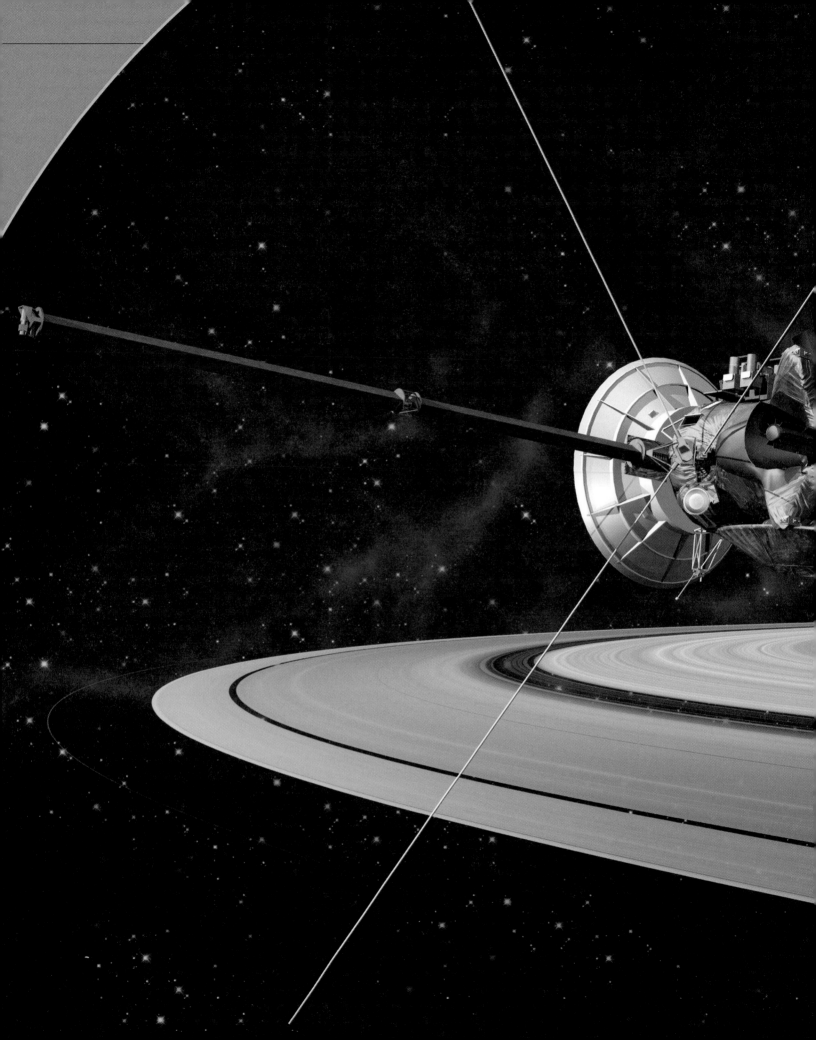

22

Touring Our Solar System*

Artists' concept of
Cassini Orbiter after
crossing Saturn's
rings.
(Courtesy of NASA)

*This chapter was revised with the
assistance of Professors Teresa Tarbuck and Mark Watry, Spring Hill College.

636

Planetary geology is the study of the formation and evolution of the bodies in our solar system—including the eight planets and myriad smaller objects: moons, asteroids, comets, and meteoroids. Studying these objects provides valuable insights into the dynamic processes that operate on Earth. Understanding how other atmospheres evolve helps scientists build better models for predicting climate change. Studying tectonic processes on other planets helps us appreciate how these complex interactions alter Earth. In addition, seeing how erosional forces work on other bodies allows us to observe the many ways landscapes are created. Finally, the uniqueness of Earth, a body that harbors life, is revealed through investigations of other planetary bodies.

FOCUS ON CONCEPTS

To assist you in learning the important concepts in this chapter, focus on the following questions:

- What is the nebular theory?
- What are the general characteristics that differentiate the terrestrial planets from the Jovian planets?
- What factors determine the variations observed among the atmospheres of planetary bodies?
- How did Earth's Moon form?
- What are the major features of the lunar surface?
- How is crater density used to date surface features on the Moon?
- Why is the Venusian surface so hot?
- What surface features do Mars and Earth have in common?
- Why is the discovery of subsurface ice on Mars important?
- What are the "spots" in the atmosphere of the Jovian planets?
- Which planets have rings?
- Where are most asteroids found?
- How are asteroids different from comets?
- What are dwarf planets?

Our Solar System: An Overview

The Sun is at the center of a revolving system, trillions of miles wide, consisting of eight planets, their satellites, and numerous smaller asteroids, comets, and meteoroids (Figure 22.1). An estimated 99.85 percent of the mass of our solar system is contained within the Sun. Collectively, the planets account for most of the remaining 0.15 percent. Starting from the Sun, the planets are Mercury, Venus, Earth, Mars, Jupiter, Saturn, Uranus, and Neptune (Figure 22.1). Pluto was recently reclassified as a member of a new class of solar system bodies called *dwarf planets*.

Tethered to the Sun by gravity, all of the planets travel in the same direction on slightly elliptical orbits (Table 22.1). Gravity causes objects nearest the Sun to travel fastest. Therefore, Mercury has the highest orbital velocity, 48 kilometers per second, and the shortest period of revolution around the Sun, 88 Earth-days. By contrast, the distant dwarf planet Pluto has an orbital speed of just 5 kilometers per second and requires 248 Earth-years to complete one revolution. Most large bodies orbit the Sun approximately in the same plane. The planets' inclination with

respect to the Earth–Sun orbital plane, known as the *ecliptic*, is shown in Table 22.1.

Nebular Theory: Formation of the Solar System

The **Nebular theory**, which explains the formation of the solar system, states that the Sun and planets formed from a rotating cloud of interstellar gases (mainly hydrogen and helium) and dust called the **solar nebula**. As the solar nebula contracted due to gravity, most of the material collected in the center to form the hot *protosun*. The remaining materials formed a thick, flattened, rotating disk, within which matter gradually cooled and condensed into grains and clumps of icy, rocky material. Repeated collisions resulted in most of the material eventually collecting into asteroid-sized objects called **planetesimals**.

The composition of planetesimals was largely determined by their proximity to the protosun. As you might expect, temperatures were highest in the inner solar system and decreased toward the outer edge of the disk. Therefore, between the present orbits of Mercury and Mars, the planetesimals were composed of mate-

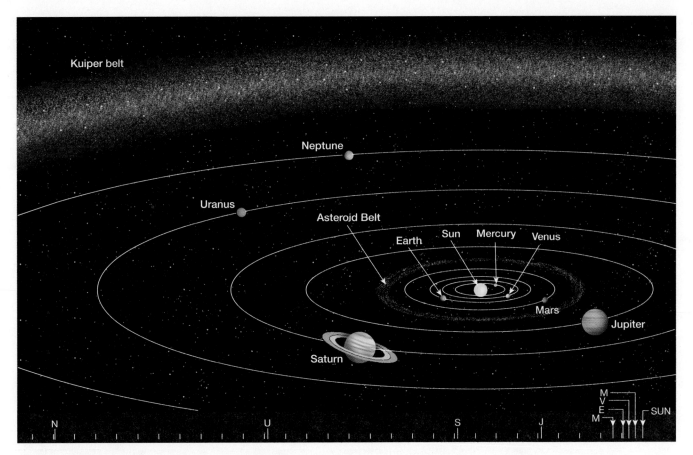

FIGURE 22.1 Orbits of the planets. Positions of the planets are shown to scale along bottom of diagram.

rials with high melting temperatures—metals and rocky substances. Then, through repeated collisions and accretion (sticking together) these asteroid-sized rocky bodies combined to form the four **protoplanets** that eventually became Mercury, Venus, Earth, and Mars.

The planetesimals that formed beyond the orbit of Mars, where temperatures are low, contained high percentages of ices—water, carbon dioxide, ammonia, and methane—as well as small amounts of rocky and metallic debris. It was mainly from these planetesimals that the four outer planets eventually formed. The accumulation of ices accounts, in part, for the large sizes and low densities of the outer planets. The two most massive planets, Jupiter and Saturn, had surface gravities sufficient to attract and retain large quantities of hydrogen and helium, the lightest elements.

It took roughly a billion years after the protoplanets formed for the planets to gravitationally accumulate most of the interplanetary debris. This was a period of intense bombardment as the planets cleared their orbits of much of the leftover material. The "scars" of this period are still evident on the Moon's surface. Small bodies were flung into planet-crossing orbits, or into interstellar space. The small fraction of interplanetary matter that escaped this violent period became asteroids, comets, and meteoroids. By comparison, the present-day solar system is a much quieter place, although many of these processes continue today at a reduced pace.

The Planets: Internal Structures and Atmospheres

The planets fall into two groups based on location, size, and density; the **terrestrial** (Earth-like) **planets** (Mercury, Venus, Earth, and Mars), and the **Jovian** (Jupiter-like) **planets** (Jupiter, Saturn, Uranus, and Neptune). Because of their relative locations, the four terrestrial planets are also known as "*inner planets*" and the four Jovian planets are known as "*outer planets*." A correlation exists between planetary locations and sizes—the inner planets are substantially smaller than the outer planets, also known as *gas giants*. For example, the diameter of Neptune (the smallest Jovian planet) is nearly four times larger than the diameter of Earth or Venus. Furthermore, Neptune's mass is 17 times greater than that of Earth or Venus (**Figure 22.2**).

Other properties that differ include densities, chemical compositions, orbital periods, and numbers of satellites. Variations in the chemical composition of planets are largely responsible for their density differences. Specifically, the average density of the terrestrial planets is about five times the density of water, whereas the average density of the Jovian planets is only 1.5 times that of water. Saturn has a density only 0.7 times that of water, which means that it would float if placed in a large enough tank of water. The outer planets are also characterized by long orbital periods and numerous satellites.

Internal Structures   Shortly after Earth formed, the segregation of material resulted in the formation of three major layers

TABLE 22.1 Planetary Data

Planet	Symbol	AU*	Mean Distance from Sun — Millions of Miles	Mean Distance from Sun — Millions of Kilometers	Period of Revolution	Inclination of Orbit	Orbital Velocity — mi/s	Orbital Velocity — km/s
Mercury	☿	0.39	36	58	88^d	7°00′	29.5	47.5
Venus	♀	0.72	67	108	225^d	3°24′	21.8	35.0
Earth	⊕	1.00	93	150	365.25^d	0°00′	18.5	29.8
Mars	♂	1.52	142	228	687^d	1°51′	14.9	24.1
Jupiter	♃	5.20	483	778	12^{yr}	1°18′	8.1	13.1
Saturn	♄	9.54	886	1427	30^{yr}	2°29′	6.0	9.6
Uranus	♅	19.18	1783	2870	84^{yr}	0°46′	4.2	6.8
Neptune	♆	30.06	2794	4497	165^{yr}	1°46′	3.3	5.3

Planet	Period of Rotation	Diameter — Miles	Diameter — Kilometers	Relative Mass (Earth = 1)	Average Density (g/cm^3)	Polar Flattening (%)	Eccentricity†	Number of Known Satellites††
Mercury	59^d	3015	4878	0.06	5.4	0.0	0.206	0
Venus	243^d	7526	12,104	0.82	5.2	0.0	0.007	0
Earth	23^{h}56^{m}04^s	7920	12,756	1.00	5.5	0.3	0.017	1
Mars	24^{h}37^{m}23^s	4216	6794	0.11	3.9	0.5	0.093	2
Jupiter	9^{h}56^m	88,700	143,884	317.87	1.3	6.7	0.048	63
Saturn	10^{h}30^m	75,000	120,536	95.14	0.7	10.4	0.056	61
Uranus	17^{h}14^m	29,000	51,118	14.56	1.2	2.3	0.047	27
Neptune	16^{h}07^m	28,900	50,530	17.21	1.7	1.8	0.009	13

*AU = astronomical unit, Earth's mean distance from the Sun.

†Eccentricity is a measure of the amount an orbit deviates from a circular shape. The larger the number, the less circular the orbit.

††Includes all satellites discovered as of December 2010.

FIGURE 22.2 The planets drawn to scale.

defined by their chemical composition—the crust, mantle, and core. This type of chemical separation occurred in the other planets as well. However, because the terrestrial planets are compositionally different than the Jovian planets, the nature of these layers differs between these two groups (**Figure 22.3**).

The terrestrial planets are dense, having relatively large cores of iron, iron compounds, and nickel. From their centers outward, the amount of metallic iron decreases while the amount of rocky silicate minerals increase. The outer cores of Earth and Mercury are liquid, whereas the cores of Venus and Mars are thought to be partially molten. This difference is attributable to Venus and Mars having lower internal temperatures than those of Earth and Mercury. Silicate minerals and other lighter compounds make up the mantles of the terrestrial planets. Finally, the silicate crusts of terrestrial planets are relatively thin compared to their mantles.

The two largest Jovian planets, Jupiter and Saturn, have small metallic inner cores consisting of iron compounds at extremely high temperatures and pressures. The outer cores of these two giants are thought to be liquid metallic hydrogen, whereas the mantles are comprised of liquid hydrogen and helium. The outermost layers are gases and ices of hydrogen, helium, water, ammonia, and methane—which account for the low densities of these planets. Uranus and Neptune also have small metallic cores but their mantles are likely hot dense water and ammonia. Above their mantles, the amount of hydrogen and helium increases, but exists in much lower concentrations than those of Jupiter and Saturn.

All planets, except Venus and Mars, have significant magnetic fields generated by flow in their liquid outer cores, or liquid mantles. Venus has a weak field due to the interaction between the solar wind and its uppermost atmosphere (ionosphere), while Mars' weak magnetic field is thought to be a remnant from when its interior was hotter. Magnetic fields play an important role in determining the nature of a planet's atmosphere. In addition, a planet's magnetic field can protect its surface from bombardment by charged particles of the solar wind—a necessary condition for the survival of life-forms.

The Atmospheres of the Planets   The Jovian planets have very thick atmospheres composed mainly of hydrogen and helium, with lesser amounts of water, methane, ammonia, and other hydrocarbons. The Jovian atmospheres are so thick that there is not a clear boundary between "atmosphere" and "planet." By contrast, the terrestrial planets, including Earth, have relatively meager atmospheres composed of carbon dioxide, nitrogen, and oxygen.

Two factors explain these significant differences—solar heating (temperature) and gravity (**Figure 22.4**). These variables determine what planetary gases, if any, were captured by planets during the formation of the solar system and which were ultimately retained.

During planetary formation, the inner regions of the developing solar system were too hot for ices and gases to condense. By

FIGURE 22.3 Comparing the internal structures of the planets.

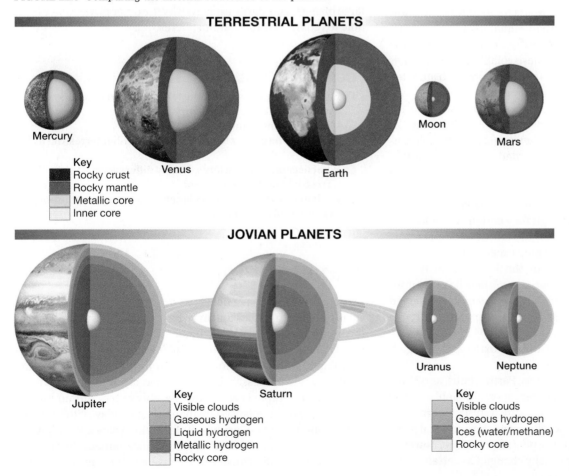

TERRESTRIAL PLANETS

Mercury

Venus

Earth

Moon

Mars

Key
Rocky crust
Rocky mantle
Metallic core
Inner core

JOVIAN PLANETS

Jupiter

Saturn

Uranus

Neptune

Key
Visible clouds
Gaseous hydrogen
Liquid hydrogen
Metallic hydrogen
Rocky core

Key
Visible clouds
Gaseous hydrogen
Ices (water/methane)
Rocky core

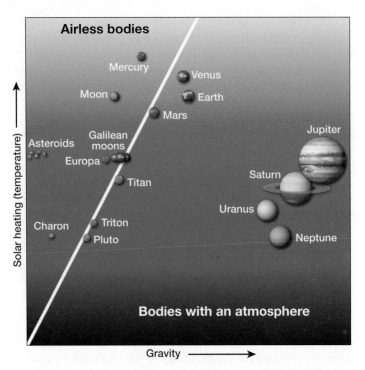

FIGURE 22.4 The factors that explain why some bodies have thick atmospheres, whereas others are airless include solar heating (temperature) and gravity. Airless worlds are relatively warm and have weak gravity. Bodies with significant atmospheres have weak heating and strong gravity.

contrast, the Jovian planets formed where temperatures were low and solar heating of planetesimals was minimal. This allowed water vapor, ammonia, and methane to condense into ices. Hence, the gas giants contain large amounts of these volatiles. As the planets grew, the largest Jovian planets, Jupiter and Saturn, also attracted large quantities of the lightest gases, hydrogen and helium.

How did Earth acquire water and other volatile gases? It seems that early in the history of the solar system, gravitational tugs by the developing protoplanets sent planetesimals into very eccentric orbits. As a result, Earth was bombarded with icy objects that originated beyond the orbit of Mars. This was a fortuitous event for organisms that currently inhabit our planet.

Mercury, our Moon, and numerous other small bodies lack significant atmospheres even though they certainly would have been bombarded by icy bodies early in their development. Airless bodies develop where solar heating exceeds a certain level, which depends on the strength of the body's gravity (see Figure 22.4). Simply stated, *less massive planets* have a better chance of losing their atmosphere because gas molecules need less speed to escape their weak gravities. Comparatively warm bodies with small surface gravity, such as our Moon, are unable to hold even heavy gases such as carbon dioxide and nitrogen. Mercury holds trace amounts of gas.

The slightly larger terrestrial planets, Earth, Venus, and Mars, retain some heavy gases including water vapor, nitrogen, and carbon dioxide. However, their atmospheres are miniscule compared to their total mass. Early in their development, the terrestrial planets probably had much thicker atmospheres. Over time, however, these primitive atmospheres gradually changed as certain gases

trickled away into space. For example, Earth's atmosphere continues to leak hydrogen and helium (the two lightest gases) into space. This phenomenon occurs near the top of Earth's atmosphere where air is so tenuous that nothing stops the fastest moving ions from flying off into space. The speed required to escape a planet's gravity is called **escape velocity**. Because hydrogen is the lightest gas, it most easily reaches the space needed to overcome Earth's gravity.

Sometime in the distant future, the loss of hydrogen (one of the components of water) will eventually "dry out" Earth's oceans, ending its hydrologic cycle. Life, however, may still hang on in Earth's polar regions.

Because Mars and Venus lack significant magnetic fields, their upper atmospheres are exposed to the brute force of the solar wind, which consists of fast-moving charged particles. Without a magnetic field to shield their atmospheres, the solar wind picks up ionic gases and carries them out to space. Mars' atmosphere is enriched in heavy isotopes of both nitrogen and carbon (carbon dioxide), suggesting that it has lost as much as 90 percent of its primitive atmosphere.

Because they have strong gravitational and magnetic fields, the massive Jovian planets have a better chance of retaining their atmospheres. Furthermore, because of their great distances from the Sun, the temperatures that occur in their upper atmospheres are incredibly cold. For example, at its cloud tops, Neptune's atmosphere has a temperature of about −218° C (−360° F)—one of the coldest places in the solar system. Because the molecular motion of a gas is temperature dependent, even hydrogen and helium move too slowly to escape the gravitational pull of these large planets. This partly explains why the outer planets have been able to retain their thick atmospheres.

CONCEPT CHECK 22.1

❶ Briefly outline the steps in the formation of our solar system according to the nebular theory.

❷ What are planetesimals?

❸ By what criteria are planets considered either terrestrial or Jovian?

❹ What accounts for the large density differences between the terrestrial and Jovian planets?

❺ Explain why the terrestrial planets have meager atmospheres, as compared to the Jovian planets.

Planetary Impacts

Planetary impacts have occurred throughout the history of the solar system. On bodies that have little or no atmosphere, such as the Moon and Mercury, even the smallest pieces of interplanetary debris (meteorites) can produce microscopic cavities on individual mineral grains. By contrast, large **impact craters** are the result of collisions with massive bodies, such as asteroids and comets.

Planetary impacts were considerably more common in the early history of the solar system than they are today, with the heaviest bombardment occurring 3.8–4.1 billion years ago. Following that period, the rate of cratering diminished dramatically and now remains essentially constant. Because weathering and erosion are almost nonexistent on the Moon and Mercury, evidence of their cratered past is clearly evident.

On larger bodies, thick atmospheres may cause the impacting objects to break up and/or decelerate. For example, Earth's atmosphere causes meteoroids with masses of less than 10 kilograms (22 pounds) to lose up to 90 percent of their speed as they penetrate the atmosphere. Therefore, impacts of low-mass bodies produce only small craters on Earth. Earth's atmosphere is much less effective in slowing large bodies—fortunately, they make very rare appearances.

The formation of a large impact crater is illustrated in Figure 22.5. The meteoroid's high-speed impact compresses the material it strikes, causing an almost instantaneous rebound, which ejects material from the surface. Craters excavated by objects that are several kilometers across often exhibit a central peak, such as the one in the large crater in Figure 22.6. Much of the material expelled, called *ejecta*, lands in or near the crater, where it accumulates to form a rim. Large meteoroids may generate sufficient heat to melt some of the impacted rock. Samples of glass beads produced in this manner, as well as rocks consisting of broken fragments welded by the heat of impacts, have been collected from the Moon, allowing planetary geologists to learn about such events.

CONCEPT CHECK 22.2

1. Why are impact craters more common on the Moon than on Earth, even though the Moon is a much smaller target and has a weaker gravitational field?
2. When did the solar system experience the period of heaviest planetary impacts?

Earth's Moon: A Chip Off the Old Block

The Earth–Moon system is unique because the Moon is the largest satellite relative to its planet. Mars is the only other terrestrial planet with moons, but its tiny satellites are likely captured asteroids. Most of the 150 or so satellites of the Jovian planets are composed of low-density rock–ice mixtures, none of which resemble the Moon. As we will see later, our unique planet-satellite system is closely related to the mechanism that created it.

The diameter of the Moon is 3,475 kilometers (2,160 miles), about one-fourth of Earth's 12,756 kilometers (7,926 miles). The Moon's surface temperature averages about 107° C (225° F) for daylight hours and −153° C (−243° F) for night. Because its period of rotation on its axis equals its period of revolution around Earth, the same lunar hemisphere always faces Earth. All of the landings of manned *Apollo* missions were confined to the side of the Moon facing Earth.

The Moon's density is 3.3 times that of water, comparable to that of *mantle* rocks on Earth, but considerably less than Earth's average density (5.5 times that of water). The Moon's relatively small iron core is thought to account for much of this difference.

The Moon's low mass relative to Earth results in a lunar gravitational attraction that is one-sixth that of Earth. A person

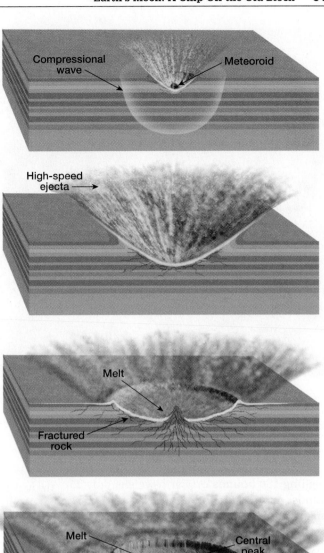

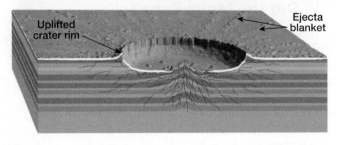

FIGURE 22.5 Formation of an impact crater. The energy of the rapidly moving meteoroid is transformed into heat and compressional waves. The rebound of the compressed rock causes debris to be ejected from the crater. Heat melts some material, producing glass beads. Small secondary craters are formed by the material "splashed" from the impact crater. (After E. M. Shoemaker)

FIGURE 22.6 The 20-kilometer-wide lunar crater Euler in the southwestern part of Mare Imbrium. Clearly visible are the bright rays, central peak, secondary craters, and the large accumulation of ejecta near the crater rim. (Courtesy of NASA)

from the rocky mantle of the impactor while its core became part of the growing Earth. This *impact model* is consistent with the Moon's low density and internal structure, which consists of a large mantle and a small iron-rich core.

The Lunar Surface When Galileo first pointed his telescope toward the Moon, he observed two different types of terrain: dark lowlands and brighter, highly cratered highlands (**Figure 22.7**). Because the dark regions appeared smooth, resembling seas on Earth, they were called **maria** (*mar* = sea, singular *mare*). The *Apollo 11* mission showed conclusively that the maria are exceedingly smooth plains composed of basaltic lavas. These vast plains are strongly concentrated on the side of the Moon facing Earth and cover about 16 percent of the lunar surface. The lack of large volcanic cones on these surfaces is evidence of high eruption rates of very fluid basaltic lavas similar to the Columbia Plateau flood basalts on Earth.

By contrast, the Moon's light-colored areas resemble Earth's continents, so the first observers dubbed them **terrae** (Latin for "land"). These areas are now generally referred to as the **lunar highlands** because they are elevated several kilometers above the maria. Rocks retrieved from the highlands are mainly breccias, pulverized by massive bombardment early in the Moon's history. The arrangement of terrae and maria result in the legendary "face" of the "man in the moon."

weighing 150 pounds on Earth weighs only 25 pounds on the Moon, although their mass remains the same. This difference allows an astronaut to carry a heavy life-support system with relative ease. If not burdened with such a load, an astronaut could jump six times higher on the Moon than on Earth. The Moon's small mass (and low gravity) is the primary reason it was not able to retain an atmosphere.

How Did the Moon Form?

Until recently, the origin of the Moon—our nearest planetary neighbor—was a topic of considerable debate among scientists. Current models show that Earth is too small to have formed with a moon, particularly one so large. Furthermore, a captured moon would likely have an eccentric orbit similar to the captured moons that orbit the Jovian planets.

The current consensus is that the Moon formed as the result of a collision between a Mars-sized body and a youthful, semi-molten Earth about 4.5 billion years ago. (Collisions of this type were probably frequent at that time.) During this explosive event, some of the ejected debris was thrown into Earth's orbit and gradually coalesced to form the Moon. Computer simulations show that most of the ejected material would have come

FIGURE 22.7 Telescopic view of the lunar surface from Earth. The major features are the dark maria and the light, highly cratered highlands. (UCO/Lick Observatory Image)

Some of the most obvious lunar features are impact craters. A meteoroid 3 meters (10 feet) in diameter can blast out a crater 50 times larger, or about 150 meters (500 feet) in diameter. The larger craters shown in Figure 22.7, such as Kepler and Copernicus (32 and 93 kilometers in diameter, respectively) were created from bombardment by bodies 1 kilometer or more in diameter. These two craters are thought to be relatively young because of the bright *rays* (lightly colored ejected material) that radiate from them for hundreds of kilometers.

History of the Lunar Surface The evidence used to unravel the history of the lunar surface comes primarily from radiometric dating of rocks returned from *Apollo* missions and studies of crater densities—counting the number of craters per unit area. The greater the crater density, the older the feature. Such evidence suggests that, after the Moon coalesced it passed through the following four phases: (1) formation of the original crust and lunar highlands, (2) excavation of the large impact basins, (3) filling of mare basins, and (4) formation of rayed craters.

During the late stages of its accretion, the Moon's outer shell was most likely completely melted—literally a magma ocean. Then, about 4.4 billion years ago, the magma ocean began to cool and underwent magmatic differentiation (see Chapter 3). Most of the dense minerals, olivine and pyroxene, sank, while less dense silicate minerals floated to form the Moon's primitive crust. The highlands are made of these igneous rocks that rose buoyantly like "scum" from the crystallizing magma. The most common highland rock type is *anorthosite*, which is composed mainly of calcium-rich plagioglase feldspar.

Once formed, the lunar crust was continually impacted as the Moon swept up debris from the solar nebula. During this time, several large impact basins were created. Then, about 3.8 billion years ago, the Moon, as well as the rest of the solar system, experienced a sudden drop in the rate of meteoritic bombardment.

The Moon's next major event was the filling of the large impact basins created at least 300 million years earlier (**Figure 22.8**). Radiometric dating of the maria basalts puts their age between 3.0 billion and 3.5 billion years, considerably younger than the initial lunar crust.

The Mare basalts are thought to have originated at depths between 200 and 400 kilometers. They were likely generated by a slow rise in temperature attributed to the decay of radioactive elements. Partial melting probably occurred in several isolated pockets as indicated by the diverse chemical makeup of the rocks retrieved during the *Apollo* missions. Recent evidence suggests

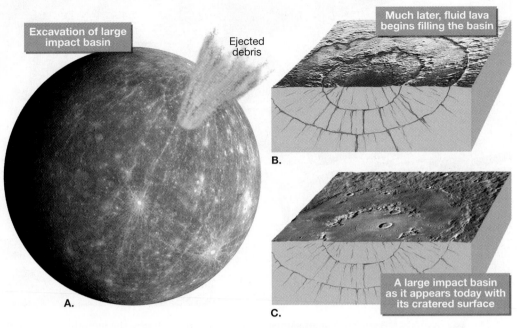

FIGURE 22.8 Formation and filling of large impact basins. **A.** Impact of an asteroid-size mass produced a huge crater hundreds of kilometers in diameter and disturbed the lunar crust far beyond the crater. **B.** Filling of the impact area with fluid basalts, perhaps derived from partial melting deep within the lunar mantle. **C.** Today, such basins make up the lunar maria and a few similar large structures on Mercury.

that some mare-forming eruptions may have occurred as recently as 1 billion years ago.

Other lunar surface features related to this period of volcanism include small shield volcanoes (8–12 kilometers in diameter), evidence of pyroclastic eruptions, rilles formed by localized lava channels, and grabens.

The last prominent features to form were rayed craters, as exemplified by the 90-kilometer-wide Copernicus crater shown in Figure 22.7. Material ejected from these younger depressions blankets the maria surfaces and many older, rayless craters. The relatively young Copernicus crater is thought to be about 1 billion years old. Had it formed on Earth, weathering and erosion would have long since obliterated it.

Today's Lunar Surface: Weathering and Erosion The Moon's small mass and low gravity account for its lack of an atmosphere or flowing water. Therefore, the processes of weathering and erosion that continually modify Earth's surface are absent on the Moon. In addition, tectonic forces are no longer active on the Moon, so quakes and volcanic eruptions have ceased. Because the Moon is unprotected by an atmosphere, erosion is dominated by the impact of tiny particles from space (*micrometeorites*) that continually bombard its surface and gradually smooth the landscape. This activity has crushed and repeatedly mixed the upper portions of the lunar crust.

Both the maria and terrae are mantled with a layer of gray, unconsolidated debris derived from a few billion years of meteoric bombardment (**Figure 22.9**). This soil-like layer, properly called **lunar regolith** (*rhegos* = blanket, *lithos* = stone), is composed of igneous rocks, breccia, glass beads, and fine *lunar*

FIGURE 22.9 Astronaut Harrison Schmitt sampling the lunar surface. Notice the footprint (inset) in the lunar "soil." (Courtesy of NASA)

dust. The lunar regolith is anywhere from 2 to 20 meters thick depending on the age of the surface.

CONCEPT CHECK 22.3

1. Briefly describe the origin of the Moon.
2. Compare and contrast Moon's maria and highlands.
3. How are maria on the Moon similar to the Columbia Plateau in the Pacific Northwest?
4. How is crater density used in the relative dating of surface features on the Moon?
5. List the major stages in the development of the modern lunar surface.
6. Compare and contrast the processes of weathering and erosion on Earth with the same processes on the Moon.

Terrestrial Planets

GEODe
EARTH SCIENCE
Earth's Place in the Universe
▸ A Brief Tour of the Planets

Mercury: The Innermost Planet

Mercury, the innermost and smallest planet, revolves around the Sun quickly (88 days) but rotates slowly on its axis. Mercury's day–night cycle, which lasts 176 Earth-days, is very long compared to Earth's 24-hour cycle. One "night" on Mercury is roughly equivalent to 3 months on Earth, and is followed by the same duration of daylight. Mercury has the greatest temperature extremes, which drop as low as −173° C (−280° F) whereas noontime temperatures exceed 427° C (800° F), hot enough to melt tin and lead. These extreme temperatures make life "as we know it" impossible on Mercury.

Mercury absorbs most of the sunlight that strikes it, reflecting only 6 percent into space, a characteristic of terrestrial bodies that have little or no atmosphere. The minuscule amount of gas present on Mercury may have originated from several sources: ionized gas emitted from the Sun; ices that vaporized during a recent comet impact; and/or outgassing of the planet's interior.

Although Mercury is small and scientists expected the planet's interior to have already cooled, the *Messenger* spacecraft detected a magnetic field. This finding suggests that Mercury has a large core that remains hot and fluid enough to generate a magnetic field.

Mercury resembles Earth's Moon in that it has very low reflectivity, no sustained atmosphere, numerous volcanic features, and a heavily cratered terrain (Figure 22.10). The largest (1,300 kilometers in diameter) known impact crater on Mercury is Caloris Basin. Images and other data gathered by *Mariner 10* show evidence of volcanism in and around Caloris Basin and a few other smaller basins. Also like our Moon, Mercury has smooth plains that cover nearly 40 percent of the area imaged by *Mariner 10*. Most of these smooth areas are associated with large impact basins, including Caloris Basin, where lava partially filled the basins and the surrounding lowlands. Consequently, these smooth plains appear to be similar in origin to lunar maria. Hopefully, data gathered by *Messenger* during its orbit around Mercury in 2011 will shed additional light on the relationship between cratering and volcanism.

Venus: The Veiled Planet

Venus, second only to the Moon in brilliance in the night sky, is named for the Roman "Goddess of Love and Beauty." It orbits the Sun in a nearly perfect circle once every 225 Earth-days. However, Venus rotates in the opposite direction of the other planets (*retrograde motion*) at an agonizingly slow pace—one Venus day

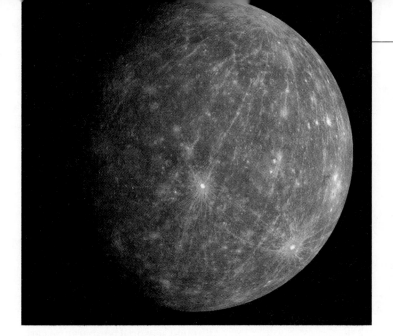

FIGURE 22.10 Mercury. This view of Mercury looks similar to Earth's Moon. (Courtesy of NASA)

is equivalent to about 244 Earth days. Venus has the densest atmosphere of the terrestrial planets, consisting mostly of carbon dioxide (97%)—the prototype for an extreme *greenhouse effect.* As a consequence, the surface temperature of Venus averages about 450° C (900° F). Temperature variations at the surface are generally minimal because of the intense mixing within the planet's dense atmosphere. Investigations of the extreme and uniform surface temperatures led scientists to more fully understand how the greenhouse effect operates on Earth.

The composition of the Venusian interior is probably similar to Earth's. However, Venus' weak magnetic field means its internal dynamics must be very different. Mantle convection is thought to operate on Venus, but the processes of plate tectonics, which recycle rigid lithosphere, do not appear to have contributed to the present Venusian topography.

The surface of Venus is completely hidden from view by a thick cloud layer composed mainly of tiny sulfuric acid droplets. In the 1970s, despite extreme temperatures and pressures, four Russian spacecraft landed successfully and obtained surface images. (As expected, however, all of the probes were crushed by the planet's immense atmospheric pressure within an hour of landing.) Using radar imaging, the unmanned spacecraft *Magellan* mapped Venus' surface in amazing detail (Figure 22.11).

Approximately 1,000 impact craters have been identified on Venus—far fewer than Mercury and Mars, but more than Earth. Researchers, who expected that Venus would show evidence of extensive cratering from the heavy

bombardment period, found instead that a period of extensive volcanism was responsible for resurfacing Venus. Its thick atmosphere also limits the number of impacts by breaking up large incoming meteoroids, and incinerating most of the small debris.

About 80 percent of the Venusian surface consists of lowlying plains covered by lava flows, some of which traveled along lava channels that extend for hundreds of kilometers (Figure 22.12). Venus' Baltis Vallis, the longest known lava channel in the solar system, meanders 6,800 kilometers (4,255 miles) across the planet. More than 100 large volcanoes have been identified on Venus. However, high surface temperatures and pressures inhibit explosive volcanism. In addition, Venus' extreme conditions result in volcanoes that tend to be shorter and wider than those on Earth or Mars (Figure 22.13). Maat Mons, the largest volcano on Venus, is about 8.5 kilometers high and 400 kilometers wide. By comparison, Mauna Loa, the largest volcano on Earth, is about 9 kilometers high and only 120 kilometers wide.

Venus also has major highlands that consist of plateaus, ridges, and topographic rises that stand above the plains. The rises are thought to have formed where hot mantle plumes encountered the base of the planet's crust, causing uplift. Much like mantle plumes on Earth, abundant volcanism is associated with mantle

FIGURE 22.11 This global view of the surface of Venus is computer-generated from years of investigations culminating with the *Magellan* mission. The twisting bright features that cross the globe are highly fractured ridges and canyons of the eastern Aphrodite highland. (Courtesy of NASA/JPL)

Planetary Radius (km)
6048 6050 6052 6054 6056 6058 6060 6062

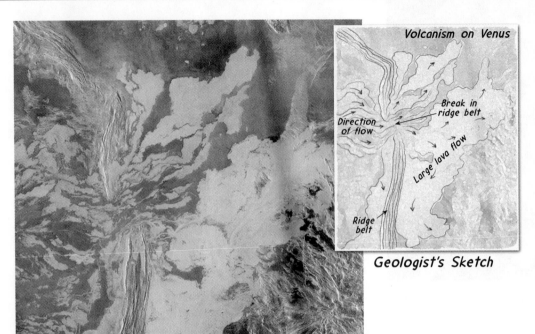

Geologist's Sketch

upwelling on Venus. Recent data collected by the European Space Agency's *Venus Express* suggest that Venus' highlands contain silica-rich granitic rock. As such, these elevated landmasses resemble Earth's continents, albeit on a much smaller scale.

Mars: The Red Planet

Mars, approximately one-half the diameter of Earth, revolves around the Sun in 687 Earth-days. Mean surface temperatures range from lows of $-140°$ C $(-220°$ F) in the winter to highs of $20°$ C ($68°$ F) in the summer. Although seasonal temperature variations are similar to Earth's, daily temperature variations are greater due to its very thin atmosphere (only 1% as dense as Earth's). The tenuous Martian atmosphere consists primarily of carbon dioxide (95%), with small amounts of nitrogen, oxygen, and water vapor.

FIGURE 22.12 Extensive lava flows on Venus. This *Magellan* radar image shows a system of lava flows that orignated from a volcano named Ammavaru, which lies approximately 300 kilometers (186 miles) west of the scene. The lava, which appears bright in this radar image, has rough surfaces, whereas the darker flows are smooth. Upon breaking through the ridge belt (left of center), the lava collected in a 100,000-square-kilometer pool. (Courtesy of NASA)

Topography   Mars, like the Moon, is pitted with impact craters. The smaller craters are usually filled with wind-blown dust—confirming that Mars is a dry, desert world. The reddish color of the Martian landscape is iron oxide (rust). Large impact craters provide information about the nature of the Martian surface. For example, if the surface is composed of dry dust and rocky debris, ejecta similar to that surrounding lunar craters is to be expected. But the ejecta surrounding some Martian craters has a different appearance—that of a muddy slurry that was splashed from the crater. Planetary geologists believe that a layer of permafrost (frozen, icy soil) lies below portions of the Martian surface and that impacts heated and melted the ice to produce the fluid-like appearance of these ejecta.

About two-thirds of the surface of Mars consists of heavily cratered highlands, concentrated mostly in its southern hemisphere (**Figure 22.14**). The period of extreme cratering occurred early in the planet's history and ended about 3.8 billion years ago, as it did in the rest of the solar system. Thus, Martian highlands are similar in age to the lunar highlands.

Based on relatively low crater counts, the northern plains, which account for the remaining one-third of the planet, are younger than the highlands (Figure 22.14). The plains' relatively flat topography is consistent with vast outpourings of fluid basaltic lavas. Visible on these plains are volcanic cones, some with summit pits and lava flows with wrinkled edges.

Located along the Martian equator is an enormous elevated region, about the size of North America, called the *Tharsis bulge* (Figure 22.14). This feature, about 10 kilometers high, appears to

FIGURE 22.13 Venus' Sapas Mons (center) is a broad volcano 400 kilometers (250 miles) wide. The bright areas in the foreground are lava flows. Another large volcano, Maat Mons, is in the background. This computer-generated view is constructed from data acquired by the *Magellan* spacecraft. (Courtesy of NASA/JPL)

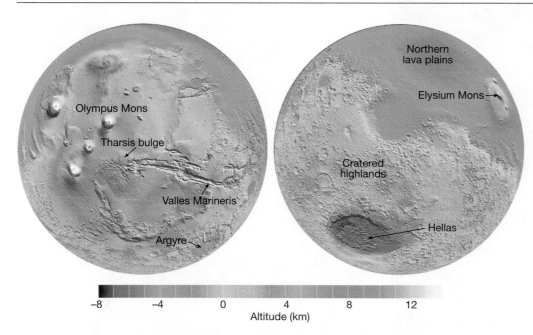

FIGURE 22.14 Two computer-generated globes of Mars with some major surface features labeled. Color represents height above (or below) the mean planetary radius—white is about 12 kilometers above average and dark blue 8 kilometers below average. (Courtesy of NASA/JPL)

constant motion. Consequently, mantle plumes tend to produce a chain of volcanic structures, like the Hawaiian Islands. By contrast, on Mars, plate tectonics is absent so successive eruptions build up in the same location. As a result, enormous volcanoes such as Olympus Mons form rather than a string of smaller ones.

Currently, the dominant force shaping the Martian surface is wind erosion. Extensive dust storms, with winds up to 270 kilometers (170 miles) per hour, can persist for weeks. Dust devils have also been photographed. Most of the Martian landscape resembles Earth's rocky deserts, with abundant dunes and low areas partially filled with dust.

Yes, Water Ice on Mars! Liquid water does not appear to exist anywhere on the Martian surface. However, poleward of about 30 degrees latitude ice can be found within a meter of the surface. In the polar regions it forms small permanent ice caps along with carbon dioxide ice. In addition, considerable evidence indicates that in the first billion years of the planet's history, liquid water flowed on the surface, creating stream valleys and related features.

One location where running water was involved in carving valleys can be seen in the Mars *Reconnaissance Orbiter* image in **Figure 22.16**. Researchers have proposed that melting of subsurface

have been uplifted and capped with a massive accumulation of volcanic rock that includes the planet's five largest volcanoes. A much smaller volcanic center (bulge) also exists.

The tectonic forces that created the Tharsis region also produced fractures that radiate from its center, like spokes on a bicycle wheel. Along the eastern flanks of the bulge, a series of vast canyons called *Valles Marineris* (Mariner Valleys) developed. Valles Marineris is so vast it can be seen in the image of Mars in Figure 22.14. This canyon network was largely created by downfaulting, not by stream erosion as is the case for Arizona's Grand Canyon. Thus, it consists of graben-like valleys similar to the East African Rift Valleys. Once formed, Valles Marineris grew by water erosion and collapse of the rift walls. The main canyon is more than 5000 kilometers long, 7 kilometers deep, and 100 kilometers wide (Figure 22.14).

Other prominent features on the Martian landscape are large impact basins. Hellas, the largest identifiable impact structure on the planet, is about 2,300 kilometers (1,400 miles) in diameter and is the planet's lowest elevation (Figure 22.14). Debris ejected from this basin contributed to the elevation of the adjacent highlands. Other buried crater basins that are even larger than Hellas probably exist.

Volcanism was prevalent on Mars during most of its history. The scarcity of impact craters on some volcanic surfaces suggests that the planet is still active. Mars has several of the largest known volcanoes in the solar system, including the largest, Olympus Mons, which is about the size of Arizona and stands nearly three times higher than Mount Everest. This gigantic volcano was last active about 100 million years ago and resembles Earth's Hawaiian shield volcanoes (**Figure 22.15**).

How did the volcanoes on Mars grow so much larger than similar structures on Earth? The largest volcanoes on the terrestrial planets tend to form where plumes of hot rock rise from deep within their interiors. On Earth, moving plates keep the crust in

FIGURE 22.15 Image of Olympus Mons, an inactive shield volcano on Mars that covers an area about the size of the state of Arizona. (Courtesy of the U.S. Geological Survey)

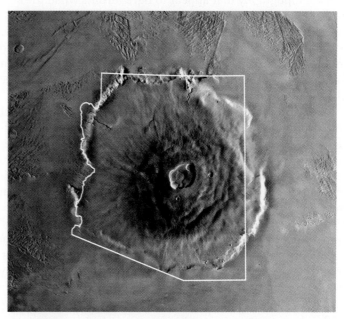

FIGURE 22.16 This image was obtained by the *Mars Reconnaissance Orbiter* and shows gullies emanating from rocky cliffs. The meandering and braided patterns are typical of water-carved channels. (Courtesy of NASA/JPL)

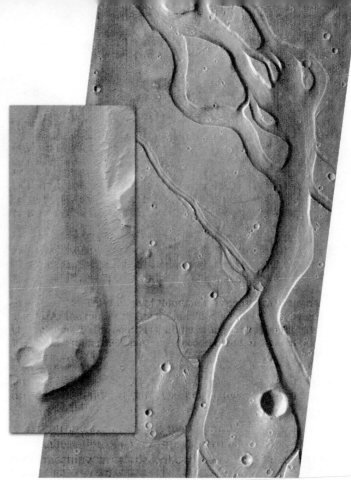

FIGURE 22.17 Stream-like channels are strong evidence that Mars once had flowing water. Inset shows close-up of streamlined island where running water encountered resistant material along its channel. (Courtesy of NASA/JPL)

ice caused spring-like seeps to emerge along the valley wall, slowly creating the gullies—a process that may still be active today.

Other channels have stream-like banks and contain numerous teardrop-shaped islands (Figure 22.17). These valleys appear to have been cut by catastrophic floods with discharge rates that were more than 1,000 times greater than those of the Mississippi River. Most of these large flood channels emerge from areas of chaotic topography that appear to have formed when the surface collapsed. The most likely source of water for these flood-created valleys was the melting of subsurface ice. If the melt water was trapped beneath a thick layer of permafrost, pressure could mount until a catastrophic release occurred. As the water escaped, the overlying surface would collapse, creating the chaotic terrain.

Not all Martian valleys appear to be the result of water released in this manner. Some exhibit branching, tree-like patterns that resemble dendritic drainage networks on Earth. In addition, the *Opportunity* rover investigated structures similar to features created by water on Earth. These included layered sedimentary rocks, playas (salt flats), and lake beds. Minerals that form only in the presence of water such as hydrated sulfates were also detected. Small spheres of hematite, dubbed "blueberries," were found that probably precipitated from water to form lake sediments. Nevertheless, with the exception of the polar regions, water does not appear to have significantly altered the topography of Mars for more than a billion years.

CONCEPT CHECK 22.4

1. What body in our solar system is most like Mercury?
2. Why are the surface temperatures so much higher on Venus than on Earth?
3. Venus was once referred to as "Earth's twin." How are these two planets similar? How do they differ?
4. What surface features do Mars and Earth have in common?
5. Why are the largest volcanoes on Earth so much smaller than the largest ones on Mars?
6. What evidence suggests that Mars had an active hydrologic cycle in the past?

Jovian Planets

Earth's Place in the Universe
▶ A Brief Tour of the Planets

Jupiter: Lord of the Heavens

The giant among planets, Jupiter has a mass two and a half times greater than the combined mass of all other planets, satellites, and asteroids in the solar system. However, it pales in comparison to the Sun, with only 1/800 of the Sun's mass.

Jupiter orbits the Sun once every 12 Earth-years, and rotates more rapidly than any other planet, completing one rotation in slightly less than 10 hours. When viewed telescopically, the effect of this fast spin is noticeable. The bulge of the equatorial region and the contraction of the polar dimension are evident (see the Polar Flattening column in Table 22.1).

Jupiter's appearance is mainly attributable to the colors of light reflected from its three main cloud layers (Figure 22.18). The warmest, and lowest, layer is composed mainly of water ice and appears blue-gray—generally not seen in visible-light images. A little higher, where temperatures are cooler, is a layer of brown to orange-brown clouds of ammonium hydrosulfide droplets. These colors are thought to be by-products of chemical reactions occur-

ring in Jupiter's atmosphere. Near the top of its atmosphere lie white wispy clouds of ammonia ice.

Because of its immense gravity, Jupiter is shrinking a few centimeters each year. This contraction generates most of the heat that drives Jupiter's atmospheric circulation. Thus, unlike winds on Earth, which are driven by solar energy, the heat emanating from Jupiter's interior produces the huge convection currents observed in its atmosphere.

Jupiter's convective flow produces alternating dark-colored *belts* and light-colored *zones*, as shown in Figure 22.18. The light clouds (*zones*) are regions where warm material is ascending and cooling, whereas the dark belts represent cool material that is sinking and warming. This convective circulation, along with Jupiter's rapid rotation, generates the high-speed, east–west flow observed between the belts and zones.

FIGURE 22.18 The structure of Jupiter's atmosphere. The areas of light clouds (*zones*) are regions where gases are ascending and cooling. Sinking dominates the flow in the darker cloud layers (*belts*). This convective circulation, along with the rapid rotation of the planet, generates the high-speed winds observed between the belts and zones.

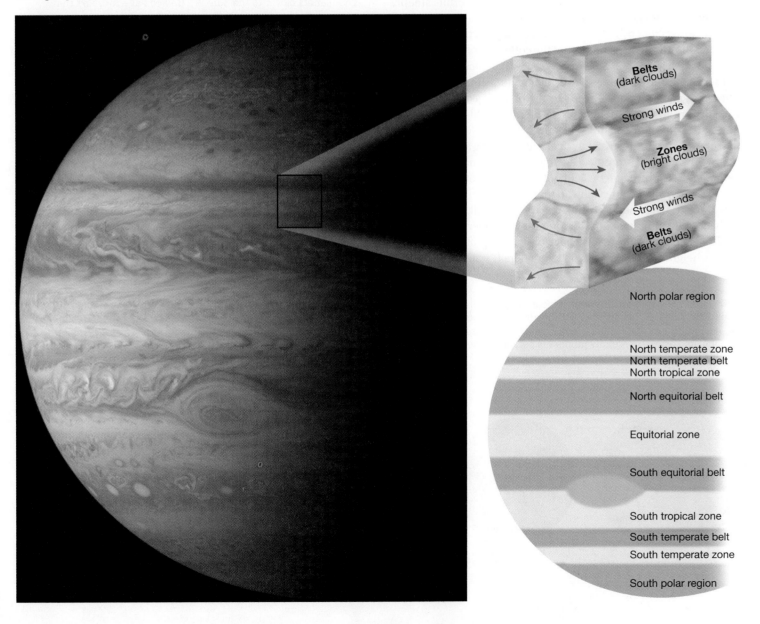

The largest storm on the planet is the Great Red Spot. This enormous, anticyclonic storm that is twice the size of Earth has been known for 300 years. In addition to the Great Red Spot, there are various white and brown oval-shaped storms (Figure 22.18). The white ovals are the cold cloud tops of huge storms many times larger than hurricanes on Earth. The brown storm clouds reside at lower levels in the atmosphere. Lightning in various white oval storms has been photographed by the *Cassini* spacecraft, but the strikes appear to be less frequent than on Earth.

Jupiter's Moons   Jupiter's satellite system, consisting of 63 moons discovered thus far, resembles a miniature solar system. Galileo discovered the four largest satellites, referred to as Galilean satellites, in 1610 (**Figure 22.19**). The two largest, Ganymede and Callisto, are roughly the size of Mercury, whereas the two smaller ones, Europa and Io, are about the size of Earth's Moon. The eight largest moons appear to have formed around Jupiter as the solar system condensed.

Jupiter also has many very small satellites (about 20 kilometers in diameter) that revolve in the opposite direction (*retrograde motion*) of the largest moons, and have eccentric orbits steeply inclined to the Jovian equator. These satellites appear to be asteroids or comets that passed near enough to be gravitationally captured by Jupiter, or are remnants of the collisions of larger bodies.

The Galilean moons can be observed with binoculars or a small telescope and are interesting in their own right. Images from *Voyagers 1* and *2* revealed, to the surprise of most geoscientists, that each of the four Galilean satellites is a unique world (Figure 22.19). The *Galileo* mission also unexpectedly revealed that the composition of each satellite is strikingly different, implying a different evolution for each. For example, Ganymede has a dynamic core that generates a strong magnetic field not observed in other satellites.

The innermost of the Galilean moons, Io, is perhaps the most volcanically active body in our solar system. In all, more than 80 active, sulfurous volcanic centers have been discovered. Umbrella-shaped plumes have been observed rising from Io's surface to heights approaching 200 kilometers (**Figure 22.20A**). The heat source for volcanic activity is tidal energy generated by a relentless "tug of war" between Jupiter and the other Galilean satellites—with Io as the rope. The gravitational field of Jupiter and the other nearby satellites pull and push on Io's tidal bulge as its slightly eccentric orbit takes it alternately closer to, then farther from, Jupiter. This gravitational flexing of Io is transformed into heat (similar to the back-and-forth bending of a piece of sheet metal) and results in Io's spectacular sulfurous volcanic eruptions. Moreover, lava, thought to be mainly composed of silicate minerals, regularly erupts on its surface (Figure 22.20B).

Jupiter's Rings   One of the surprising aspects of the *Voyager 1* mission was the discovery of Jupiter's ring system. More recently,

Students Sometimes Ask...

Besides Earth, do any other bodies in the solar system have liquid water?

The planets closer to the Sun than Earth are considered too warm to contain liquid water, and those farther from the Sun are generally too cold (although some features on Mars indicate that it probably had abundant liquid water at some point in its history). The best prospects of finding liquid water within our solar system lie beneath the icy surfaces of some of Jupiter's moons. For instance, an ocean of liquid water is possibly hidden under Europa's outer covering of ice. Detailed images from *Galileo* have revealed that Europa's icy surface is quite young and exhibits cracks apparently filled with dark fluid from below. This suggests that under its icy shell, Europa must have a warm, mobile interior—perhaps an ocean. Because liquid water is a necessity for life as we know it, there has been considerable interest in sending an orbiter to Europa—and eventually a lander capable of launching a robotic submarine—to determine if it harbors life.

FIGURE 22.19 Jupiter's four largest moons (from left to right) are called the Galilean moons because they were discovered by Galileo. **A.** The innermost moon, Io, is one of only three volcanically active bodies known to exist in the solar system. **B.** Europa, smallest of the Galilean moons, has an icy surface that is criss-crossed by many linear features. **C.** Ganymede, the largest Jovian satellite, exhibits cratered areas, smooth regions, and areas covered by numerous parallel grooves. **D.** Callisto, the outermost of the Galilean satellites, is densely cratered, much like Earth's moon. (Courtesy of NASA/NGS Image Collection)

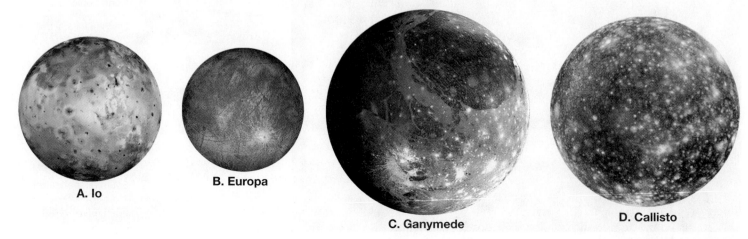

A. Io

B. Europa

C. Ganymede

D. Callisto

A.

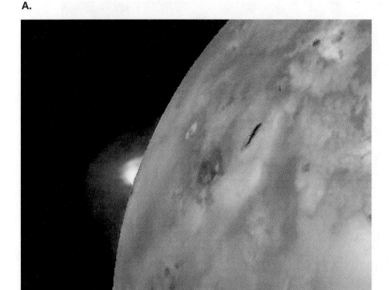

B.

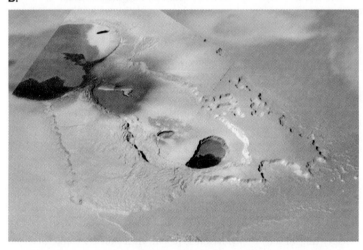

FIGURE 22.20 A volcanic eruption on Jupiter's Moon Io. **A.** This plume of volcanic gases and debris is rising more than 100 kilometers (60 miles) above Io's surface. **B.** The bright red area on the left side of the image is newly erupted lava. (Courtesy of NASA)

the ring system was thoroughly investigated by the *Galileo* mission. By analyzing how these rings scatter light, researchers determined that the rings are composed of fine, dark particles, similar in size to smoke particles. Furthermore, the faint nature of the rings indicates that these minute particles are widely dispersed. The main ring is composed of particles believed to be fragments blasted from the surfaces of Metis and Adrastea, two small moons of Jupiter. Impacts on Jupiter's moons Amalthea and Thebe are believed to be the source of the debris from which the outer Gossamer ring formed.

Saturn: The Elegant Planet

Requiring more than 29 Earth-years to make one revolution, Saturn is almost twice as far from the Sun as Jupiter, yet their atmo-

spheres, compositions, and internal structures are remarkably similar. The most striking feature of Saturn is its system of rings, first observed by Galileo in 1610 (Figure 22.21). With his primitive telescope, the rings appeared as two small bodies adjacent to the planet. Their ring nature was determined 50 years later by Dutch astronomer Christian Huygens.

Saturn's atmosphere, like Jupiter's, is dynamic (Figure 22.21). Although the bands of clouds are fainter and wider near the equator, rotating "storms" similar to Jupiter's Great Red Spot occur in Saturn's atmosphere, as does intense lightning. Although the atmosphere is nearly 75 percent hydrogen and 25 percent helium, the clouds are composed of ammonia, ammonia hydrosulfide, and water, each segregated by temperature. Like Jupiter, the atmosphere's dynamics are driven by the heat released by gravitational compression.

Saturn's Moons The Saturnian satellite system consists of 61 known moons of which 53 have been named. The moons vary significantly in size, shape, surface age, and origin. Twenty-three of the moons are "original" satellites that formed in tandem with their parent planet. At least two (Dione and Tethys) show evidence of tectonic activity, where internal forces have ripped apart their icy surfaces. Others, like Hyperion, are so porous that impacts punch into their surfaces, and Rhea may have its own rings (Figure 22.22). Many of Saturn's smallest moons have irregular shapes and are only a few tens of kilometers in diameter.

Saturn's largest moon, Titan, is larger than Mercury and is the second-largest satellite in the solar system. Titan, and Neptune's Triton, are the only satellites in the solar system known to have substantial atmospheres. Titan was visited and photographed by the *Huygens* probe in 2005. The atmospheric pressure at Titan's surface is about 1.5 times that at Earth's surface, and the atmospheric composition is about 98 percent nitrogen and 2 percent methane with trace organic compounds. Titan has Earth-like geological landforms and geological processes, such as dune formation and stream-like erosion caused by methane "rain." In addition, the northern latitudes appear to have lakes of liquid methane.

Enceladus is another unique satellite of Saturn—one of the few where active eruptions have been observed (Figure 22.23). The outgassing, comprised mostly of water, is thought to be the source replenishing the material in Saturn's E ring. The volcanic-like activity occurs in an area called "tiger stripes" that consists of four large fractures with ridges on either side.

Saturn's Ring System In the early 1980s, the nuclear-powered *Voyagers 1* and *2* explored Saturn within 100,000 miles of its surface. More information was collected about Saturn in that short time than had been acquired since Galileo first viewed this "elegant planet" in the early 1600s. More recently, observations from ground-based telescopes, the Hubble Space Telescope, and the *Cassini-Huygens* spacecraft have added to our knowledge of Saturn's ring system. In 1995 and 1996, when the positions of Earth and Saturn allowed the rings to be viewed edge-on, Saturn's faintest rings and satellites became visible. (The rings were visible edge-on again in 2009.)

Saturn's ring system is more like a large rotating disk of varying density and brightness than a series of independent ringlets.

FIGURE 22.21 Image taken by the Earth-orbiting Hubble Space Telescope shows Saturn's dynamic ring system. The two bright rings, called A ring (outer) and B ring (inner), are separated by the Cassini division. A second small gap (Encke gap) is also visible as a thin line in the outer portion of the A ring. (Courtesy of NASA)

Each ring is composed of individual particles—mainly water ice with lesser amounts of rocky debris—that circle the planet while regularly impacting one another. There are only a few gaps—most of the areas that look like empty space either contain fine dust particles, or coated ice particles that are inefficient reflectors of light.

FIGURE 22.22 Saturn's impact-pummeled satellite, Hyperion, imaged by the *Cassini Orbiter*. Planetary geologists think Hyperion's surface is so weak and porous that impacts punch into its surface. (Courtesy of NASA/JPL)

Most of Saturn's rings fall into one of two categories based on density. Saturn's main (bright) rings, designated A and B, are tightly packed and contain particles that range in size from a few centimeters (pebble-size) to tens of meters (house-size), with most of the particles being roughly the size of a large snowball (see Figure 22.21). In the dense rings, particles collide frequently as they orbit the planet. Although Saturn's main rings (A and B) are 40,000 kilometers wide, they are very thin, only 10–30 meters from top to bottom.

At the other extreme are Saturn's faint rings. Saturn's outermost ring (E ring), not visible in Figure 22.21, is composed of widely dispersed, tiny particles. Recall that volcanic-like activity on Saturn's satellite Enceladus is thought to be the source of material for the E ring.

Studies have shown that the gravitational tugs of nearby moons tend to shepherd the ring particles by gravitationally altering their orbits (**Figure 22.24**). For example, the F ring, which is very narrow, appears to be the work of satellites located on either side that confine the ring by pulling back particles that try to escape. Whereas the Cassini Division, a clearly visible gap in Figure 22.21, arises from the gravitational pull of Jupiter's moon, Mimas.

Some of the ring particles are believed to be debris ejected from the moons embedded in them. It is also possible that material is continually recycled between the rings and the ring moons. The ring moons gradually sweep up particles, which are subsequently ejected by collisions with large chunks of ring material, or perhaps by energetic collisions with other moons. It seems, then, that planetary rings are not the timeless features that we once thought—rather, they are continually recycled.

The origin of planetary ring systems is still being debated. Perhaps the rings formed simultaneously and from the same material as the planets and moons—condensing from a flattened cloud of dust and gases that encircled the parent planet. Or, perhaps the rings

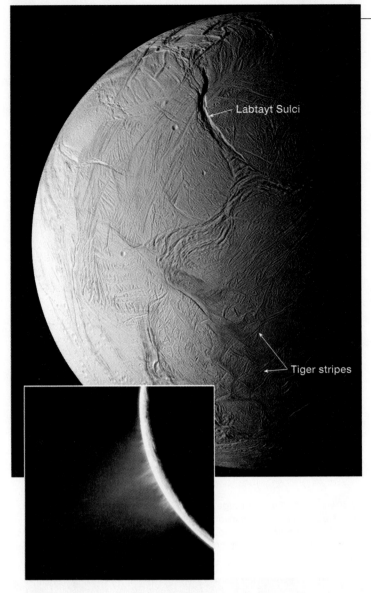

FIGURE 22.23 NASA's *Cassini Orbiter* captured this mosiac of Saturn's tectonically active, icy satellite Enceladus. The northern hemisphere contains a 1-kilometer deep chasm, while linear features, called tiger stripes, are visible in the lower right. Inset image shows jets spurting ice particles, water vapor, and organic compounds from the area of the tiger stripes. (Courtesy of NASA/JPL)

formed later, when a moon or large asteroid was gravitationally pulled apart after straying too close to a planet. Yet another hypothesis suggests that a foreign body collided catastrophically with one of the planet's moons, the fragments of which would tend to jostle one another and form a flat, thin ring. Researchers expect more light to be shed on the origin of planetary rings as the *Cassini* spacecraft continues its tour of Saturn.

Uranus and Neptune: Twins

Although Earth and Venus have many similar traits, Uranus and Neptune are perhaps more deserving of being called "twins." They are nearly equal in diameter (both about four times the size of Earth), and are both bluish in appearance—a result of methane in their atmospheres. Their days are nearly the same length and their cores are made of rocky silicates and iron—similar to the other gas giants. Their mantles, comprised mainly of water, ammonia, and methane, are thought to be very different from Jupiter and Saturn (see Figure 22.3). One of the most pronounced differences between Uranus and Neptune is the time they take to complete one revolution around the Sun—84 and 165 Earth-years, respectively.

Uranus: The Sideways Planet   Unique to Uranus is its axis of rotation, which lies nearly parallel to the ecliptic ("lying on its side"). Its rotational motion, therefore, resembles a rolling ball instead of a spinning toy top (**Figure 22.25**). This unusual characteristic of Uranus is likely due to a huge impact that essentially knocked the planet sideways from its original orbit early in its evolution.

Uranus, once thought to be weatherless, shows evidence of huge storm systems the size of the United States. Recent photographs from the Hubble Space Telescope also reveal banded clouds composed mainly of ammonia and methane ice—similar to the cloud systems of the other gas giants.

Uranus' Moons   Spectacular views from *Voyager 2* showed that Uranus' five largest moons have varied terrains. Some have long, deep canyons and linear scars, whereas others possess large, smooth areas on otherwise crater-riddled surfaces. Studies conducted at California's Jet Propulsion Laboratory suggest that Miranda, the innermost of the five largest moons, was recently geologically active—most likely driven by gravitational heating, as occurs on Io.

Uranus' Rings   A surprise discovery in 1977 showed that Uranus has a ring system. The find occurred as Uranus passed in front of a distant star and blocked its view, a process called

FIGURE 22.24 Two of Saturn's ring moons. **A.** Pan is a small moon about 30 kilometers in diameter that orbits in the Encke gap, located in the A ring. It is responsible for keeping the Encke gap open. **B.** Prometheus, a potato-shaped moon, acts as a ring shepard. Its gravity helps confine the moonlets in Saturn's thin F ring. (Courtesy of NASA/JPL)

A.

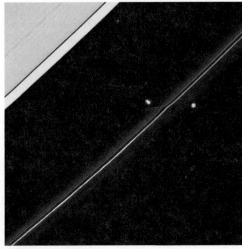

B.

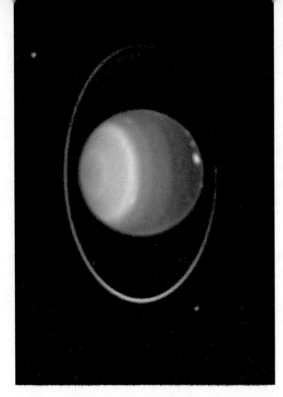

FIGURE 22.25 Uranus surrounded by its major rings and 10 of its 27 known moons. Also visible are cloud patterns and several oval storm systems. This false-color image was generated from data obtained by Hubble's Near Infrared Camera. (Image by Hubble Space Telescope courtesy of NASA)

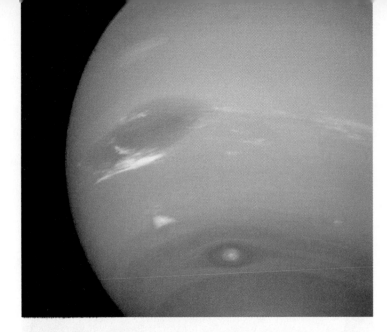

FIGURE 22.26 This image of Neptune was constructed from two images. At the top is the Great Dark Spot, accompanied by bright white clouds. To the south is a bright white area thought to be high cloud tops. Still farther south is a second dark spot with a bright core. (Courtesy of NASA/JPL)

occultation (*occult* = hidden). Observers saw the star "wink" briefly five times (meaning five rings) before the primary occultation and again five times afterward (see Figure 22.25). More recent ground- and space-based observations indicate that Uranus has at least 10 sharp-edged, distinct rings orbiting its equatorial region. Interspersed among these distinct structures are broad sheets of dust.

Neptune: The Windy Planet

Because of its great distance from Earth, astronomers knew very little about Neptune until 1989. Twelve years and nearly 3 billion miles of *Voyager 2* travel provided investigators an amazing opportunity to view the outermost planet in the solar system.

Neptune has a dynamic atmosphere, much like that of the other Jovian planets (Figure 22.26). Record wind speeds that exceed 2,400 kilometers (1,500 miles) per hour encircle the planet, making Neptune one of the windiest places in the solar system. Recall that Neptune exhibits large dark spots thought to be rotating storms similar to Jupiter's Great Red Spot. However, Neptune's storms appear to have comparatively short life spans—usually only a few years. Another feature that Neptune has in common with the other Jovian planets are layers of white, cirrus-like clouds (probably frozen methane) about 50 kilometers above the main cloud deck.

Neptune's Moons

Neptune has 13 known satellites, the largest of which is the moon Triton; the remaining 12 are small, irregularly shaped bodies. Triton is the only large moon in the solar system that exhibits retrograde motion, indicating that it most likely

formed independently, and was later gravitationally captured by Neptune (Figure 22.27).

Triton and a few other icy moons erupt "fluid" ices—an amazing manifestation of volcanism. **Cryovolcanism** (from the Greek *Kryos*, meaning "frost") describes the eruption of magmas derived

FIGURE 22.27 This montage shows Triton, Neptune's largest moon, with Neptune in the background. The bottom of the image shows Triton's wind and sublimation-eroded south polar cap. Sublimation is the process whereby a solid (ice) changes directly to a gas. (Courtesy of NASA/JPL)

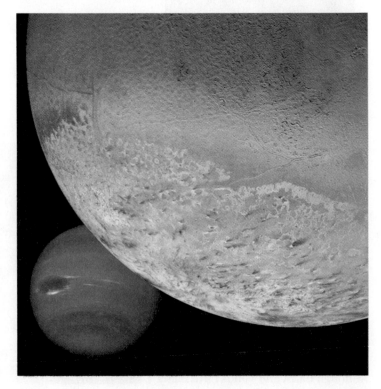

from the partial melting of ice instead of silicate rocks. Triton's icy magma is a mixture of water-ice, methane, and probably ammonia. When partially melted, this mixture behaves as molten rock does on Earth. In fact, upon reaching the surface these magmas can generate quiet outpourings of ice lavas, or occasionally, produce explosive eruptions. An explosive eruptive column can generate the ice equivalent of volcanic ash. In 1989, *Voyager 2* detected active plumes on Triton that rose 8 kilometers above the surface and were blown downwind for more than 100 kilometers. In other environments, ice lavas develop that can flow great distances from their source—similar to the fluid basaltic flows on Hawaii.

Neptune's Rings Neptune has five named rings, two of which are broad, and three that are narrow, perhaps no more than 100 kilometers wide. The outermost ring appears to be partially confined by the satellite Galatea. Neptune's rings are most similar to Jupiter's in that they appear faint, which suggests they are composed mostly of dust-size particles. Neptune's rings also display red colors that indicate the dust is composed of organic compounds.

CONCEPT CHECK 22.5

1. What is the nature of Jupiter's Great Red Spot?
2. Why are the Galilean satellites of Jupiter so named?
3. What is distinctive about Jupiter's satellite, Io?
4. Why are many of Jupiter's small satellites thought to have been captured?
5. How are Jupiter and Saturn similar?
6. What two roles do ring moons play in the nature of planetary ring systems?
7. How are Saturn's satellite Titan and Neptune's satellite Triton similar?
8. Name three bodies in the solar system that exhibit active volcanism.

Small Solar System Bodies

There are countless chunks of debris in the vast spaces separating the eight planets and in the outer reaches of the solar system. In 2006, the International Astronomical Union organized solar system objects not classified as planets or moons into two broad categories: (1) **small solar system bodies** that include *asteroids*, *comets*, and *meteoroids* and (2) **dwarf planets**. The newest grouping, dwarf planets, includes Ceres, the largest known object in the asteroid belt, and Pluto, a former planet.

Asteroids and meteoroids are composed of rocky and/or metallic material with compositions somewhat like the terrestrial planets. They are distinguished according to size: Asteroids are larger than 100 meters in diameter, whereas meteoroids have diameters less than 100 meters. Comets, on the other hand, are loose collections of ices, dust, and small rocky particles that originate in the outer reaches of the solar system.

Asteroids: Leftover Planetesimals

Asteroids are small bodies (planetesimals) remaining from the formation of the solar system, making them about 4.6 billion years old. Most asteroids orbit the Sun between Mars and Jupiter in the region known as the **asteroid belt** (Figure 22.28). There are only five that are more than 400 kilometers in diameter, but the solar system hosts an estimated 1–2 million asteroids larger than 1 kilometer, and many millions that are smaller. Some travel along eccentric orbits that take them very near the Sun, and others regularly pass close to Earth and the Moon (Earth-crossing asteroids). Many of the recent large-impact craters on the Moon and Earth were probably the result of collisions with asteroids. About 2,000 Earth-crossing asteroids are known, one-third of which are more than 1 kilometer in diameter. Inevitably, Earth–asteroid collisions will occur again (Box 22.1).

Because most asteroids have irregular shapes, planetary geologists initially speculated that they might be fragments of a broken planet that once orbited between Mars and Jupiter. However, the combined mass of all asteroids is now estimated to be only 1/1,000 of the modest-sized Earth. Today, most researchers agree that asteroids are leftover debris from the solar nebula. Asteroids have lower densities than scientists originally thought, suggesting they are porous bodies, like "piles of rubble," loosely bound together.

In February 2001 an American spacecraft became the first visitor to an asteroid. Although it was not designed for landing, *NEAR—Shoemaker* landed successfully on Eros and collected information that has planetary geologists both intrigued and perplexed. Images obtained as the spacecraft drifted toward the surface of Eros revealed a barren, rocky surface composed of particles ranging in size from fine dust to boulders up to 10 meters (30 feet) across (Figure 22.29). Researchers unexpectedly discovered that fine debris tends to concentrate in the low areas where it forms flat deposits resembling ponds. Surrounding the low areas, the landscape is marked by an abundance of large boulders.

One of several hypotheses to explain the boulder-strewn topography is seismic shaking, which would cause the boulders to move upward as the finer materials sink. This is analogous to what happens when a jar of sand and various sized pebbles is shaken—the larger pebbles rise to the top while the smaller sand grains settle to the bottom.

FIGURE 22.28 The orbits of most asteroids lie between Mars and Jupiter. Also shown are the orbits of a few known near-Earth asteroids.

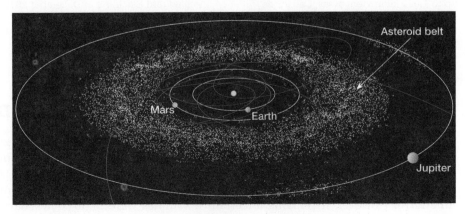

Box 22.1

EARTH AS A SYSTEM

Is Earth on a Collision Course?

The solar system is cluttered with asteroids, active comets, and extinct comets. These fragments travel at great speeds and can strike Earth with an explosive force many times greater than a powerful nuclear weapon.

In recent decades, it has become increasingly clear that comets and asteroids collide with Earth far more frequently than previously thought, as evidenced by the many large impact structures that have been identified (**Figure 22.A**). (Many impact craters were once mistaken for volcanic structures.) Most impact structures are so old and highly eroded that they were not discovered until satellite images became available (**Figure 22.B**). One notable exception is a very fresh-looking crater near Winslow, Arizona, known as Meteor Crater (see Figure 22.33, page 661). This crater was produced by a relatively small body, about the size of an Olympic swimming pool (50 meters in diameter).

About 65 million years ago, a large asteroid about 10 kilometers (6 miles) in diameter collided with Earth off the Yucatan Peninsula in Mexico. This impact is thought to have caused the demise of the dinosaurs, as well as the extinction of nearly 50 percent of all plant and animal species (see Chapter 12).

More recently, a spectacular explosion has been attributed to the collision of an asteroid or comet with our planet. In 1908, in a remote region of Siberia, a "fireball" that appeared more brilliant than the Sun exploded violently. The shock waves rattled windows and triggered reverberations heard up to 1,000 kilometers away. Called the *Tunguska event*, it scorched, delimbed, and flattened trees up to 30 kilometers from the point of impact. Surprisingly, expeditions to the area found no evidence of an impact

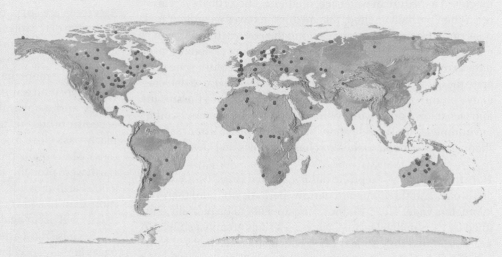

FIGURE 22.A World map of major impact structures. Additional structures are being identified every year. (Data from Griffith Observatory)

crater or any metallic fragments. Evidently, the explosion, which equaled at least a 10-megaton nuclear bomb, occurred several kilometers above the surface. Why it exploded prior to impact is uncertain.

The dangers of living with these small, but deadly objects from space came to public attention again in 1989 when an asteroid nearly 1 kilometer across shot past Earth in a "near miss." Traveling at 70,000 kilometers (44,000 miles) per hour, it could have produced a crater 10 kilometers (6 miles) wide, and perhaps 2 kilometers (1.2 miles) deep. As an observer noted, "Sooner or later it will be back." Statistics show that collisions with bodies larger than 1 kilometer should be expected every few hundred thousand years. Collisions with bodies larger than 6 kilometers, resulting in mass extinctions, are anticipated every 100 million years.

NASA scientists continually track *near-Earth objects*. When asteroids or comets pass closely to any large body in the solar system, their orbits may be altered by the gravitational interaction, which may send them toward Earth. As of December 2010, more than 7000 near-Earth objects have been discovered, of which slightly more than 1,000 have been classified as potentially hazardous asteroids.

FIGURE 22.B Manicouagan, Quebec, is a 200-million-year-old eroded impact structure. The lake outlines the crater remnant, which is 70 kilometers (42 miles) across. Fractures related to this event extend outward for an additional 30 kilometers. (Courtesy of U.S. Geological Survey)

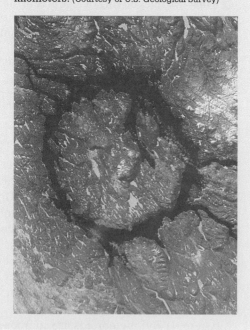

Indirect evidence from meteorites suggests that some asteroids might have been heated by a large impact event. A few large asteroids may have completely melted, causing them to differentiate into a dense iron core and a rocky mantle. In November 2005, the Japanese probe *Hayabusa* landed on a small near-Earth asteroid named 25143 Itokawa and returned to Earth in June 2010. However, it remains uncertain if samples were collected.

Comets: Dirty Snowballs

Comets, like asteroids, are leftover material from the formation of the solar system. They are loose collections of rocky material, dust, water ice, and frozen gases (ammonia, methane, and carbon dioxide), thus the nickname "dirty snowballs." Recent space missions to comets have shown their surfaces to be dry and dusty, which indicates their ices are hidden beneath a rocky layer.

Close-up of surface

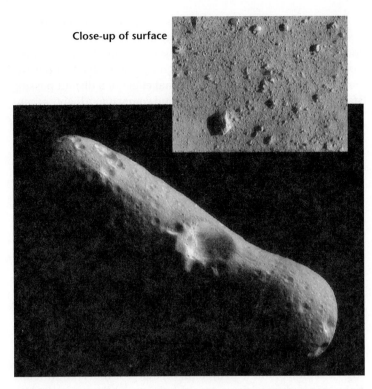

FIGURE 22.29 Image of asteroid Eros obtained by the *NEAR-Shoemaker* probe. Inset shows a close-up of Eros's barren rocky surface. (Courtesy of NASA)

Most comets reside in the outer reaches of the solar system and take hundreds of thousands of years to complete a single orbit around the Sun. However, a smaller number of *short-period comets* (those having orbital periods of less than 200 years), such as the famous Halley's Comet, make regular encounters with the inner solar system (**Figure 22.30**). The shortest period comet (Encke's Comet) orbits around the Sun once every three years.

FIGURE 22.30 Orientation of a comet's tail as it orbits the Sun.

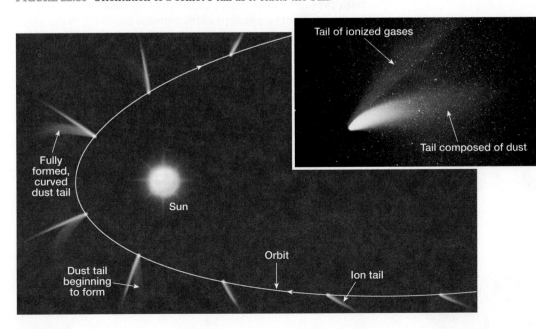

All of the phenomena associated with comets come from a small central body called the **nucleus**. These structures are typically 1–10 kilometers in diameter, but nuclei 40 kilometers across have been observed. When comets reach the inner solar system, solar energy begins to vaporize their ices. The escaping gases carry dust from the comet's surface, producing a highly reflective halo called a **coma** (**Figure 22.31**). Within the coma, the small glowing nucleus with a diameter of only a few kilometers can sometimes be detected.

As comets approach the Sun, most develop tails that can extend for millions of kilometers. The tail of a comet points away from the Sun in a slightly curved manner (see Figure 22.30), which led early astronomers to believe that the Sun has a repulsive force that pushes away particles of the coma to form the tail. Scientists have identified two solar forces known to contribute to tail formation. One is *radiation pressure* caused by radiant energy (light) emitted by the Sun, and the second is the *solar wind*, a stream of charged particles ejected from the Sun. Sometimes a single tail composed of both dust and ionized gases is produced, but two tails are often observed (Figure 22.30). The heavier dust particles produce a slightly curved tail that follows the comet's orbit, whereas the extremely light ionized gases are "pushed" directly away from the Sun, forming the second tail.

As a comet's orbit carries it away from the Sun, the gases forming the coma recondense, the tail disappears, and the comet returns to cold storage. Material that was blown from the coma to form the tail is lost forever. When all the gases are expelled, the inactive comet, which closely resembles an asteroid, continues its orbit without a coma or tail. It is believed that few comets remain active for more than a few hundred close orbits of the Sun.

The very first samples from a comet's coma (Comet Wild 2) were returned to Earth in January 2006 by NASA's *Stardust* spacecraft (**Figure 22.32**). Images from *Stardust* show that the comet's surface was riddled with flat-bottomed depressions and appeared dry, although at least 10 gas jets were active. Laboratory studies revealed that the coma contained a wide range of organic compounds and substantial amounts of silicate crystals.

Most comets originate in one of two regions: the *Kuiper belt* or the *Oort cloud*. Named in honor of astronomer Gerald Kuiper, who predicted its existence, the **Kuiper belt** hosts comets that orbit in the outer solar system, beyond Neptune (see Figure 22.1, page 637). This disc-shaped structure is thought to contain about a billion objects over 1 kilometer in size. However, most comets are too small and too distant to be observed from Earth, even using the Hubble Space Telescope. Like the asteroids in the inner solar system, most Kuiper belt comets move in slightly elliptical orbits that lie roughly in the same plane as the planets. A chance collision between two Kuiper belt comets, or the gravitational influence of

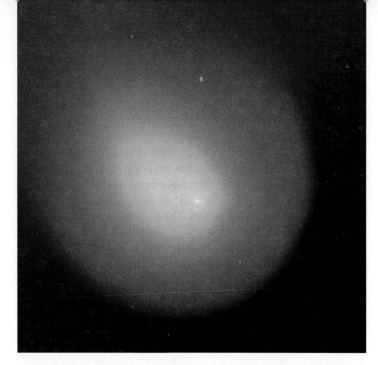

FIGURE 22.31 Coma of comet Holmes as it orbits the Sun. The nucleus of the comet is within the bright spot in the center. Comet Holmes, which orbits the Sun every 6 years, was uncharacteristically active during its most recent entry into the inner solar system. (Image by Spitzer Space Telescope, courtesy of NASA)

one of the Jovian planets, occasionally alters their orbits sufficiently to send them into our view.

Halley's Comet originated in the Kuiper belt. Its orbital period averages 76 years, and every one of its 29 appearances since 240 B.C. has been recorded by Chinese astronomers—testimony to their dedication as astronomical observers and the endurance of Chinese culture. In 1910, Halley's Comet made a very close approach to Earth, making for a spectacular display.

FIGURE 22.32 Comet Wild 2, as seen by NASA's *Stardust* spacecraft, during its flyby of the comet. Inset, artist's concept depicting jets of gas and dust erupting from Comet Wild 2. (Courtesy of NASA)

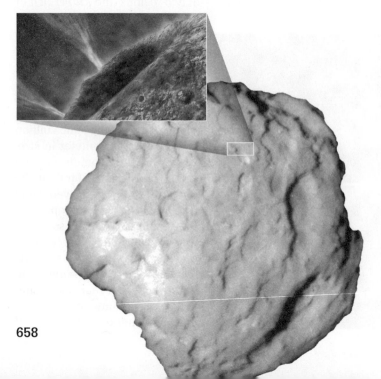

Named for Dutch astronomer Jan Oort, the **Oort cloud** consists of comets that are distributed in all directions from the Sun, forming a spherical shell around the solar system. Most Oort cloud comets orbit the Sun at distances greater than 10,000 times the Earth–Sun distance. The gravitational effect of a distant passing star may send an occasional Oort cloud comet into a highly eccentric orbit that carries it toward the Sun. However, only a tiny fraction of Oort cloud comets have orbits that bring them into the inner solar system.

Meteoroids: Visitors to Earth

Nearly everyone has seen a **meteor**, commonly (but inaccurately) called "shooting stars." These streaks of light can be observed in as little as the blink of an eye, or can last as "long" as a few seconds. They occur when a small solid particle, a **meteoroid**, enters Earth's atmosphere from interplanetary space. Heat, created by friction between the meteoroid and the air, produces the light we see trailing across the sky. Most meteoroids originate from one of the following three sources: (1) interplanetary debris missed by the gravitational sweep of the planets during formation of the solar system; (2) material that is continually being ejected from the asteroid belt; or (3) the rocky and/or metallic remains of comets that once passed through Earth's orbit. A few meteoroids are probably fragments of the Moon, Mars, or possibly Mercury, ejected by a violent asteroid impact. Before *Apollo* astronauts brought Moon rocks back to Earth, meteorites were the only extraterrestrial materials that could be studied in the laboratory.

Meteoroids less than about a meter in diameter generally vaporize before reaching Earth's surface. Some, called *micrometeorites*, are so tiny and their rate of fall so slow that they drift to Earth continually as space dust. Researchers estimate that thousands of meteoroids enter Earth's atmosphere every day. After sunset on a clear, dark night, many are bright enough to be seen with the naked eye anywhere on Earth.

Occasionally, meteor sightings increase dramatically to 60 or more per hour. These displays, called **meteor showers**, result when Earth encounters a swarm of meteoroids traveling in the same direction at nearly the same speed as Earth. The close association of these swarms to the orbits of some short-term comets strongly suggests that they represent material lost by these comets (**Table 22.2**). Some swarms, not associated with the orbits of known comets, are probably the scattered remains of the nucleus of a long-defunct comet. The notable *Perseid meteor shower* that occurs each year around August 12 is likely material ejected from the comet *Swift-Tuttle* on previous approaches to the Sun.

Most meteoroids large enough to survive probably originate among the asteroids, where chance collisions or gravitational interactions with Jupiter modify their orbits and project them toward Earth. Earth's gravity does the rest.

A few very large meteoroids have blasted craters on Earth's surface that strongly resemble those on our Moon. At least 40 terrestrial craters exhibit features that could only be produced by an explosive impact of a large asteroid, or perhaps even a comet nucleus. More than 250 others may be of impact origin. Most notable is Arizona's Meteor Crater, a huge cavity more than 1 kilo-

TABLE 22.2 Major Meteor Showers

Shower	Approximate Dates	Associated Comet
Quadrantids	January 4–6	
Lyrids	April 20–23	Comet 1861 I
Eta Aquarids	May 3–5	Halley's Comet
DeltaAquarids	July 30	
Perseids	August 12	Comet 1862 III
Draconids	October 7–10	Comet Giacobini-Zinner
Orionids	October 20	Halley's Comet
Taurids	November 3–13	Comet Encke
Andromedids	November 14	Comet Biela
Leonids	November 18	Comet 1866 I
Geminids	December 4–16	

meter wide, 170 meters (560 feet) deep, with an upturned rim that rises above the surrounding countryside (Figure 22.33). More than 30 tons of iron fragments have been found in the immediate area, but attempts to locate the main body have been unsuccessful. Based on the amount of erosion observed on the crater rim, the impact likely occurred within the last 50,000 years.

The remains of meteoroids, when found on Earth, are referred to as **meteorites** (Figure 22.34). Classified by their composition, meteorites are either (1) *irons*, mostly iron with 5–20 percent nickel; (2) *stony* (also called *chondrites*), silicate minerals with inclusions of other minerals; or (3) *stony–irons*, mixtures of the two. Although stony meteorites are the most common, irons are found in large numbers because metallic meteorites withstand impacts better, weather more slowly, and are easily distinguished from terrestrial rocks. Iron meteorites are probably fragments of once molten cores of large asteroids or small planets.

One type of stony meteorite, called a *carbonaceous chondrite*, contains organic compounds and occasionally simple amino acids, which are the basic building blocks of life. This discovery confirms similar findings in observational astronomy, which indicate that numerous organic compounds exist in interstellar space.

Data from meteorites have been used to ascertain the internal structure of Earth and the age of the solar system. If meteorites represent the composition of the terrestrial planets, as some planetary geologists suggest, our planet must contain a much larger percentage of iron than is indicated by surface rocks. This is one reason that geologists think Earth's core is mostly iron and nickel. In addition, radiometric dating of meteorites indicates the age of our solar system is about 4.6 billion years. This "old age" has been confirmed by data obtained from lunar samples.

FIGURE 22.33 Meteor Crater, near Winslow, Arizona. This cavity is about 1.2 kilometers (0.75 mile) across and 170 meters (560 feet) deep. The solar system is cluttered with meteoroids and other objects that can strike Earth with explosive force. (Photo by Michael Collier)

FIGURE 22.34 Iron meteorite found near Meteor Crater, Arizona. (Courtesy of Meteor Crater Enterprises, Inc.)

FIGURE 22.35 Pluto with its three known moons. Image by the Hubble Space Telescope. (Courtesy of NASA)

CONCEPT CHECK 22.6

1. Where are most asteroids found?
2. Compare and contrast asteroids and comets.
3. What do you think would happen if Earth passed through the tail of a comet?
4. Where are most comets thought to reside? What eventually becomes of comets that orbit close to the Sun?
5. Differentiate between the following solar bodies: meteoroid, meteor, and meteorite.
6. What are the three main sources of meteoroids?

Dwarf Planets

Since its discovery in 1930, Pluto has been a mystery to astronomers who were searching for another planet in order to explain irregularities in Neptune's orbit. At the time of its discovery, Pluto was thought to be the size of Earth—too small to significantly alter Neptune's orbit. Later, estimates of Pluto's diameter, adjusted because of improved satellite images, indicated that it was less than half Earth's diameter. Then, in 1978, astronomers realized that Pluto appeared much larger than it really is because of the brightness of its newly discovered satellite, Charon (Figure 22.35). Most recently, calculations based on images obtained by the Hubble Space Telescope show that Pluto's diameter is 2,300 kilometers (1,430 miles), about one-fifth the diameter of Earth and less than half that of Mercury (long considered the solar system's "runt"). In fact, seven moons in the solar system, including Earth's, are larger than Pluto.

Even more attention was given to Pluto's status as a planet when astronomers discovered another large icy body in orbit beyond Neptune. Soon, over a thousand of these *Kuiper belt objects* were discovered forming a band of objects—a second asteroid belt, but located at the outskirts of the solar system. The Kuiper belt objects are rich in ices and have physical properties similar to those of comets. Many other planetary objects, some perhaps larger than Pluto, are thought to exist in this belt of icy worlds beyond Neptune's orbit.

The International Astronomical Union, the group responsible for naming and classifying celestial objects, voted in 2006 to create a new class of solar system objects called dwarf planets. These are celestial bodies that orbit the Sun, are essentially spherical due to their own gravity, but are not large enough to sweep their orbits clear of other debris. By this definition, Pluto is recognized as a dwarf planet and the prototype of this new category of planetary objects. Other dwarf planets include Eris, a Kuiper belt object, and Ceres, the largest known asteroid.

Pluto's reclassification was not the first such "demotion." In the mid-1800s, astronomy textbooks listed as many as 11 planets in our solar system, including the asteroids Vesta, Juno, Ceres, and Pallas. Astronomers continued to discover dozens of other "planets," a clear signal that these small bodies represent a class of objects separate from the planets.

Researchers now recognize that Pluto was unique among the classical planets—completely different from the four rocky, innermost planets, as well as the four gaseous giants. The new classification will give a home to the hundreds of additional dwarf planets astronomers assume exist in the solar system. *New Horizons*, the first spacecraft designed to explore the outer solar system, was launched in January 2006. Scheduled to fly by Pluto in July 2015, and later explore the Kuiper belt, *New Horizons* carries tremendous potential for aiding researchers in further understanding the solar system.

CONCEPT CHECK 22.7

1. Define *dwarf planets*.
2. Why was Pluto demoted from the ranks of the classical planets?

GIVE IT SOME THOUGHT

1. Imagine that a solar system has been discovered in a nearby region of the Milky Way Galaxy. The accompanying table shows data that have been gathered about three of the planets orbiting the central star of this newly discovered solar system. Using Table 22.1 as a guide, classify each planet as either Jovian, terrestrial, or neither. Explain your reasoning.

	Planet 1	Planet 2	Planet 3
Relative Mass (Earth = 1)	1.2	15	0.1
Diameter (km)	15,000	52,000	5000
Mean Distance from Star (AU)	1.4	17	35
Density (g/cm3)	4.8	1.22	5.3
Orbital Eccentricity	0.01	0.05	0.23

2. In order to conceptualize the size and scale of Earth and Moon as they relate to the solar system, complete the following.
   a. Approximately how many Moons (diameter = 3,474 km) would fit side-by-side across the diameter of Earth (diameter = 12,756 km)?
   b. Given that the Moon's orbital diameter is 768,798 km, approximately how many Earths would fit side-by-side between Earth and the Moon?
   c. Approximately how many Earths would fit side-by-side across the Sun whose diameter is about 1,390,000 km?
   d. Approximately how many Suns would fit side-by-side between Earth and the Sun, a distance of about 150,000,000 km?

3. The accompanying graph shows the temperature at various distances from the Sun during the formation of our solar system. Use it to complete the following.
   a. Which planet(s) formed at locations where the temperature in the solar system was hotter than the boiling point of water?
   b. Which planet(s) formed at locations where the temperature in the solar system was cooler than the freezing point of water?

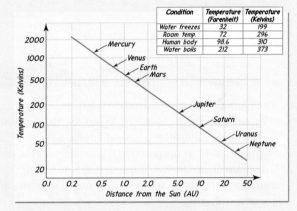

Condition	Temperature (Farenheit)	Temperature (Kelvins)
Water freezes	32	199
Room temp.	72	296
Human body	98.6	310
Water boils	212	373

4. The accompanying sketch shows four primary craters (A, B, C, and D). The impact that produced crater A produced two secondary craters (labeled "a") and three rays. Crater D has one secondary crater (labeled "d"). Rank the four primary craters from oldest to youngest and explain your ranking.

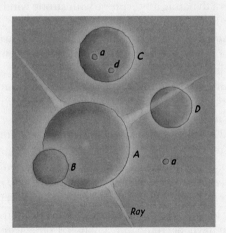

5. The accompanying diagram shows two of Uranus' moons, Ophelia and Cordelia, which act as shepherd moons for the Epsilon ring. Explain what would happen to the Epsilon ring if a large asteroid struck Ophelia, knocking it out of the Uranian system.

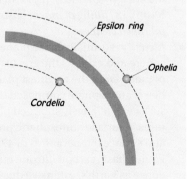

6. The accompanying diagram shows a comet traveling toward the Sun at the first position where it has both an ion tail and a dust tail. Refer to this diagram to complete the following.
   a. For each of the three numbered sites, indicate whether the comet will have no tails, one tail, or two tails. If one tail or two tails are present, in what direction will they point?
   b. Would your answers to the preceding question change if the Sun's energy output were to increase significantly? If so, how would they change?
   c. If the solar wind suddenly ceased, how would it affect this comet and its tails?

7. It has been estimated that Halley's Comet has a mass of 100 billion tons. Furthermore, it is estimated to lose about 100 million tons of material when its orbit brings it close to the Sun. With an orbital period of 76 years, calculate the maximum remaining life span of Halley's Comet.

8. Assume three irregularly shaped planet-like objects, each smaller than our Moon, have just been discovered orbiting the Sun at a distance of 35 AU. One of your friends argues the objects should be classified as planets because they are large and orbit the Sun. Another friend argues that the objects should be classified as dwarf planets, such as Pluto. State whether you agree or disagree with either or both of your friends. Explain your reasoning.

In Review   Chapter 22 Planetary Geology

- The planets are arranged into two groups: the *terrestrial* (Earth-like) *planets* (Mercury, Venus, Earth, and Mars) and the *Jovian* (Jupiter-like) *planets* (Jupiter, Saturn, Uranus, and Neptune). When compared to the Jovian planets, the terrestrial planets are smaller, more dense, contain proportionally more rocky material, have slower rates of rotation, and meager atmospheres.

- The lunar surface exhibits several types of features. *Impact craters* were produced by the collision of rapidly moving interplanetary debris (*meteoroids*). Bright, densely cratered *highlands* make up most of the lunar surface. The dark, fairly smooth lowlands are called *maria*. Maria basins are enormous impact craters that were later flooded with layer upon layer of very fluid basaltic lava. All lunar terrains are mantled with a soil-like layer of gray, unconsolidated debris, called *lunar regolith*, which has been derived from a few billion years of meteoric bombardment. One hypothesis for the Moon's origin suggests that a Mars-sized object collided with Earth to produce the Moon. Scientists conclude that the *lunar surface evolved in four stages:* (1) *formation of the original crust (highlands),* (2) *excavations of the large impact basins,* (3) *filling of maria basins,* and (4) *formation of youthful rayed craters.*

- *Mercury* is a small, dense planet that has virtually no atmosphere and exhibits the greatest temperature extremes of any planet. *Venus*, the brightest planet in the sky, has a thick, heavy atmosphere composed of 97 percent carbon dioxide, a surface of relatively subdued plains and inactive volcanic features, a surface atmospheric pressure 90 times that of Earth's, and surface temperatures of 475° C (900° F). *Mars*, the Red Planet, has a carbon dioxide atmosphere only 1 percent as dense as Earth's, extensive dust storms, numerous inactive volcanoes, many large canyons, and several valleys exhibiting drainage patterns similar to stream valleys on Earth.

- *Jupiter*, the largest planet, rotates rapidly, has a banded appearance caused by huge convection currents driven by the planet's interior heat, a *Great Red Spot* that varies in size, a thin ring system, and at least 63 moons (one of the moons, *Io*, is perhaps the most volcanically active body in the solar system). *Saturn* is best known for its system of rings. It also has a dynamic atmosphere with strong winds up to 930 miles per hour and storms similar to Jupiter's Great Red Spot. *Uranus* and *Neptune* are often called "the twins" because of their similar structure and composition. A unique feature of Uranus is the fact that it rotates on its side. Neptune has white, cirrus-like clouds above its main cloud deck and large dark spots, assumed to be large rotating storms similar to Jupiter's Great Red Spot.

- In 2006, the International Asronomical Union organized solar system objects not classified as planets or moons into two broad catergories: (1) *small solar system bodies* that include *asteroids, comets, meteoroids,* and (2) *dwarf planets*. Most asteroids lie between the orbits of Mars and Jupiter. Asteroids are leftover rocky and metallic debris from the solar nebula that never accreted into a planet. Comets are made of ices (water, ammonia, methane, carbon dioxide, and carbon monoxide) with small pieces of rocky and metallic material. Comets are thought to reside in the outer solar system in one of two locations: the *Kuiper belt* or the *Oort cloud*. Meteoroids, small solid particles that travel through interplanetary space, become *meteors* when they enter Earth's atmosphere and vaporize with a flash of light. *Meteor showers* occur when Earth encounters a swarm of meteoroids, probably material lost by a comet. *Meteorites* are the remains of meteoroids found on Earth. Recently, Pluto was placed into a new class of solar system objects called *dwarf planets*.

Key Terms

asteroids (p. 655)
asteroid belt (p. 655)
coma (p. 657)
comet (p. 656)
cryovolcanism (p. 654)
dwarf planet (p. 655)
escape velocity (p. 640)
impact craters (p. 640)
Jovian planet (p. 637)

Kuiper belt (p. 657)
lunar highlands (p. 642)
lunar regolith (p. 643)
maria (p. 642)
meteor (p. 658)
meteor shower (p. 658)
meteorite (p. 659)
meteoroid (p. 658)
Nebular theory (p. 636)

nucleus (p. 657)
Oort cloud (p. 658)
planetesimals (p. 636)
protoplanets (p. 637)
small solar system bodies (p. 655)
solar nebula (p. 636)
terrae (p. 642)
terrestrial planet (p. 637)

Examining the Earth System

1. On Earth the four major spheres (atmosphere, hydrosphere, geosphere, and biosphere) interact as a system with occasional influences from our near-space neighbors. Which of these spheres are absent, or nearly absent, on the Moon? Because the Moon lacks these spheres, list at least five processes that operate on Earth but are absent on the Moon.

2. Among the planets in our solar system, Earth is unique because water exists in all three states (solid, liquid, and gas) on and near its surface. In what state(s) of matter is water found on Mercury, Venus, and Mars? How would Earth's hydrologic cycle be different if (a) its orbit was inside the orbit of Venus? (b) its orbit was outside the orbit of Mars?

3. If a large meteorite were to strike Earth in the near future, what effect might this event have on the atmosphere (in particular, average temperatures and climate)? Assuming the conditions persisted, speculate about how the changes might influence the biosphere.

Mastering Geology

Looking for additional review and test prep materials? Visit the Self Study area in **www.masteringgeology.com** to find practice quizzes, study tools, and multimedia that will aid in your understanding of this chapter's content. In **MasteringGeology™** you will find:

- **GEODe: Earth Science: An interactive visual walkthrough of key concepts**

- **Geoscience Animation Library: More than 100 animations illuminating many difficult-to-understand Earth science concepts**
- **In The News RSS Feeds: Current Earth science events and news articles are pulled into the site with assessment**
- **Pearson eText**
- **Optional Self Study Quizzes**
- **Web Links**
- **Glossary**
- **Flashcards**

Light, Astronomical Observations, and the Sun*

23

Comet Hale-Bopp above the observatories on the summit of Mauna Kea, Hawaii. (Photo by David Nunuk/Photo Researchers, Inc.)

*This chapter was revised with the assistance of Mark Watry and Teresa Tarbuck, Spring Hill College.

Since astronomers cannot study the universe by bringing it into the lab, and because the vast majority of celestial objects are too far away to visit, astronomers collect and study those things that come to Earth from space. Overwhelmingly, this means collecting and studying light emitted or reflected by objects found in the universe. In fact, everything that is known about the universe beyond the solar system comes from the analysis of the light from distant sources. This chapter examines the properties and utility of light, some of the tools astronomers use to collect and study light, and what is known about the nearest source of light, the Sun.

FOCUS ON CONCEPTS

To assist you in learning the important concepts in this chapter, focus on the following questions:

- What is *electromagnetic radiation*?
- What can a continuous spectrum tell astronomers about stars?
- What can be learned about a star from its dark-line spectrum?
- What is the *Doppler effect*?
- What is the difference between a refracting telescope and a reflecting telescope?
- Why do all large telescopes use mirrors rather than lenses to collect light?
- What are some of the advantages of radio telescopes over optical telescopes?
- Why is the Sun important to the study of astronomy?
- What are the four major layers of the Sun?
- What phenomenon occurs on Earth as a result of solar flares on the Sun?
- What happens to the matter that is consumed in the proton–proton chain reaction?

Signals from Space

Although visible light is most familiar to us, it constitutes only a tiny sliver of an array of energy referred to as **electromagnetic radiation** (Figure 23.1). Included in this array are *gamma rays, X-rays, ultraviolet light, visible light, infrared radiation* (heat), *microwaves,* and *radio waves* (Figure 23.2). All forms of radiant energy travel through the vacuum of space in a straight line at the rate of 300,000 kilometers (186,000 miles) per second.[56] Over 24 hours, this is a staggering 26 billion kilometers. The light that we collect tells us about the processes that created it and about the matter lying between us and the source of the light.

Nature of Light

Experiments have demonstrated that light can be described in two ways. In some instances light behaves like waves, and in others like discrete particles. In the wave sense, light is analogous to swells in the ocean. This motion is characterized by *wavelength*—the distance from one wave crest to the next. Wavelengths vary from several kilometers for some radio waves to less than a billionth of a centimeter for gamma rays (Figure 23.2). Most of these waves are either too long or too short for our eyes to detect; how-

ever, the primary characteristics of all electromagnetic radiation can be described using visible light as an example.

The extremely narrow band of electromagnetic radiation we can see (which is labeled as *visible light* in Figure 23.2) is sometimes referred to as *white light*. White light consists of an array of waves having various wavelengths, a fact easily demonstrated with a prism (Figure 23.3A). As white light passes through a prism, the color with the shortest wavelength, violet, is bent more than blue, which is bent more than green, and so forth (Table 23.1). Thus, white light can be separated into its component colors, producing the familiar "rainbow of colors" (Figure 23.3A).

Wave theory, however, cannot explain some of the observed characteristics of light. In these cases, light acts like a stream of particles, analogous to infinitesimally small bullets fired from a machine gun. These particles, called **photons**, can exert a pressure (push) on matter, which is called **radiation pressure**. Recall that photons from the Sun are responsible for pushing material away from a comet to produce its dust tail. Each photon has a specific amount of energy, which is related to its wavelength in a simple way: *Shorter wavelengths* correspond to *more energetic photons*. Thus, blue light has more energetic photons than red light.

Which theory of light—the wave theory or the particle theory—is correct? The answer is that both are correct because each will predict the behavior of light for certain phenomena. As George Abell, a prominent astronomer, stated about all scientific laws, "The mistake is only to apply them to situations that are outside their range of validity."

[56]Light rays are "bent" slightly when they pass nearby a very massive object such as the Sun.

FIGURE 23.1 A face-on view of the galaxy NGC 1232, located in the southern constellation Eridanus. Despite being 100 million light-years away, modern telescopes allow astronomers to study its intricate details. Older, reddish stars are located mainly in the galaxy's central region, while young, hot blue stars make up the spiral arms. (Photo by European Southern Observatories)

FIGURE 23.2 Electromagnetic radiation. The electromagnetic spectrum ranges from long-wavelength radio waves to short-wavelength gamma radiation.

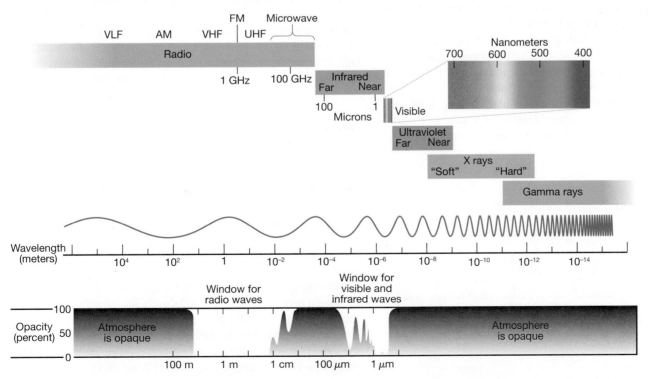

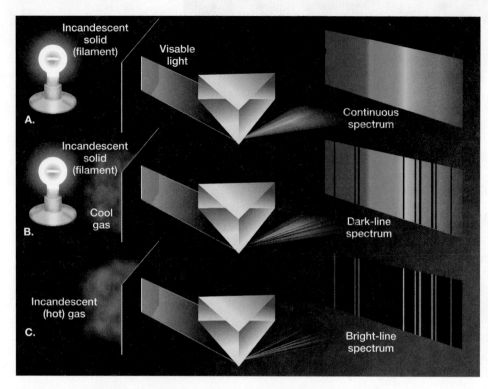

FIGURE 23.3 Formation of the three types of spectra. **A.** Continuous spectrum. **B.** Dark-line spectrum. **C.** Bright-line spectrum.

CONCEPT CHECK 23.1

❶ What term is used to describe the collection that includes gamma rays, X-rays, ultraviolet light, visible light, infrared radiation, microwaves, and radio waves?

❷ Which color has the longest wavelength? The shortest?

❸ How does the amount of energy contained in a photon relate to its wavelength?

Light and Processes

When violent events occur in the universe, large amounts of high-energy radiation are emitted. For example, when matter is engulfed by a black hole, the result is the emission of high-energy X-rays. By contrast, when less violent processes occur, small amounts of low-energy radiation are released. For example, when a shock wave moves through a gas cloud it heats the cloud, causing infrared (heat) energy to be emitted. The intensity of the light emitted and its wavelength distribution tells us a lot about the type of process that is occurring. This information can be used to support or refute scientific hypotheses. For example, theoretical studies predicted the existence of black holes long before observational evidence existed. The concept of black holes gained considerable support when X-rays matching the wavelengths predicted by the theory were detected around objects suspected of being black holes.

TABLE 23.1 Colors and Corresponding Wavelengths

Color	Wavelength (Nanometers[*])
Violet	380–440
Blue	440–500
Green	500–560
Yellow	560–590
Orange	590–640
Red	640–750

[*] One nanometer is 10^{-9} meter.

Spectroscopy

When Sir Isaac Newton used a prism to disperse white light into its component colors, he unknowingly initiated the field of **spectroscopy**—the study of those properties of light that are wavelength dependent. The rainbow of colors Newton produced is called a *continuous spectrum*, because all wavelengths of visible light are included. Later it was learned that two other types of spectra (*dark line* and *bright line*) exist and that each one is generated under somewhat different conditions (**Figure 23.3**). Just as visible light can produce a spectrum, other regions of the electromagnetic spectrum can be dispersed to produce spectra.

Continuous Spectrum

A **continuous spectrum** is produced by an incandescent (glowing) solid, liquid, or gas under high pressure. (Incandescent means "to emit light when hot.") It consists of a continuous band of wavelengths like that generated by a common 100-watt light bulb (Figure 23.3A). A continuous spectrum contains two important pieces of information about radiating bodies.

First, a continuous spectrum provides information about the total energy output of the radiating body. If the temperature of a radiating surface increases, the total amount of energy emitted increases. The rate of increase is stated in the Stefan-Boltzmann law: *The energy radiated by a body is directly proportional to the fourth power of its absolute temperature.* For example, if the temperature of a star is twice that of another star, the total radiation emitted by the hotter star is $2^4 = 2 \times 2 \times 2 \times 2$, or 16 times greater than that of the cooler star.

Second, a continuous spectrum contains information about the surface temperature of the radiating body. As the surface temperature of an object increases, a larger proportion of its energy is radiated at shorter wavelengths (higher energy). To illustrate, imagine a metal rod that is heated slowly. Initially, the rod appears dull red (longer wavelengths), then yellow, and later bluish-white (shorter wavelengths). All incandescent bodies show this behavior, so it follows that blue stars are hotter than yellow stars (like the Sun), which are hotter than red stars (Table 23.1).

Dark-Line Spectrum

If one collects the "continuous" spectrum from a star and passes it through an instrument called a **spectroscope** (which spreads out the wavelengths in a manner similar to a prism), a series of dark lines appear. A **dark-line** or **absorption spectrum** is produced when "white light" passes through a comparatively cool gas at low pressure (Figure 23.3B). The spectrum looks like a continuous spectrum with a series of dark lines (or "missing wavelengths"). This spectrum contains all the information present in the continuous spectrum in addition to information about the composition of matter present and the relative amounts of each kind of matter.

When visible light is passed through a glass jar containing hydrogen gas, the hydrogen atoms absorb specific wavelengths of light, resulting in a unique set of dark lines. Each set of spectral lines, like a set of fingerprints, identifies the matter present. Elements, such as iron, which exist in the gaseous state on the Sun, have been identified by studying their spectra. Even organic molecules have been discovered in distant interstellar clouds of dust and gases using this technique.

The spectra of most stars are of the dark-line type. Imagine the light produced in the Sun's interior passing outward through its atmosphere. The gas in the solar atmosphere is cooler than that inside, and when it absorbs some of the sunlight (and re-emits it in a random direction) we do not see it, resulting in a dark spot (line) in the spectrum. Although the lines appear black, they just look that way next to the bright parts of the spectrum. The relative intensities of the light in the dark lines contain information about the relative amounts of each kind of matter present.

Bright-Line Spectrum

A **bright-line** or **emission spectrum** is produced by hot (incandescent) matter at low pressure (Figure 23.3C). It is a series of bright lines (a fingerprint for the matter producing them) that appear in the same locations as the dark lines for the same gas. These spectra contain information about the temperature of the gas and the matter in it. Figure 23.4 shows the emission spectra of hydrogen and helium, the two most abundant elements in the universe.

Bright-line or emission spectra are produced by large interstellar clouds (nebula) consisting largely of hydrogen gas excited by extremely hot stars. Because the brightest emission line produced by hydrogen is red, these clouds tend to have a red glow that is characteristic of excited hydrogen gas. The Orion Nebula is a well-known emission nebula that is bright enough to be seen by the naked eye (Figure 23.5). It is located in the constellation Orion in the sword of the hunter.

CONCEPT CHECK 23.2

❶ What is *spectroscopy*?
❷ Describe a continuous spectrum. Give an example of a natural phenomenon that exhibits a continuous spectrum.
❸ What can a continuous spectrum tell astronomers about a star?
❹ What can be learned about a star (or other celestial objects) from a dark-line (absorption) spectrum?
❺ What produces emission lines (bright lines) in a spectrum?

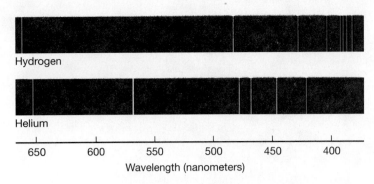

FIGURE 23.4 Bright-line spectra of the two most abundant elements in the universe.

The Doppler Effect

The positions of the bright and dark lines in the spectra described earlier shift when the source of energy moves relative to the observer. This effect is observed for all types of waves. You may have heard the change in pitch of a car horn or ambulance siren as it passes by. When it is approaching, the sound seems to have a higher-than-normal pitch, and when it is moving away, the pitch sounds lower than normal. This effect was first explained by Christian Doppler in 1842 and is called the **Doppler effect**. The reason for the difference in pitch is that it takes time for the wave to be emitted. If the source is moving away, the beginning of the wave is emitted nearer to you than the end, which stretches the wave—that is, gives it a longer wavelength (Figure 23.6). The opposite is true for an approaching source.

In the case of light, when a source is moving away, its light appears redder than it actually is because its waves are lengthened. Objects approaching have their light waves shifted toward the blue (shorter wavelength) end of the spectrum. Thus, if a source of red light approached you at a very high speed (near the speed of light), it would actually appear blue. The same effect is produced if you are moving and the light remains stationary.

The Doppler effect is important because it reveals whether Earth is approaching or receding from a star or another celestial body. In addition, the amount of shift allows us to calculate the rate at which the relative movement is occurring. Large Doppler shifts indicate high velocities; small Doppler shifts indicate low velocities. Doppler shifts are generally measured from the dark lines in the spectra of stars, by comparing them with a standard spectrum produced in the laboratory (Figure 23.7).

There are two types of Doppler shifts important in astronomy: those caused by local motions and those caused by the expansion of the universe. Doppler shifts due to local motions indicate how fast one star orbits another in a binary (two-star) system and how fast a pulsing star expands and contracts. Those shifts caused by the expansion of the universe (where space is continually being created between the galaxies) can tell us how far away distant objects are. These measurements coupled with the speed of light tell us how long ago the light left these distant objects, and as we look farther out, we can get a sense of the age of the universe.

FIGURE 23.5 The Orion Nebula is a well-known emission nebula. Bright enough to be seen by the naked eye, the Orion Nebula is located in the sword of the hunter in the constellation of the same name. (Courtesy of National Optical Astronomy Observatories)

CONCEPT CHECK 23.3

1 Briefly describe the Doppler effect.
2 Describe how astronomers determine whether a star is moving toward or away from Earth.

Light Collection

The light emitted from distant sources is collected and analyzed to determine the temperature, composition, relative motion, and distance to celestial objects. For nearby objects (bright objects in the solar system or Milky Way Galaxy), the tools required are relatively simple. But for faint or distant sources as much light as possible must be collected, and for the longest amount of time that is reasonable. This requires very large instruments with very sensitive detectors and little to no interference from other sources of electromagnetic energy.

The earliest tool used to collect light from the heavens was the human eye. Although early astronomers like Tycho Brahe were extremely successful using just their eyes, the human eye is a poor instrument for astronomical observation (see Chapter 21). The eye cannot collect much light, is not very sensitive to faint colors, collects only visible light, and refreshes itself many times each second. Early telescopes and photographic film were vast improvements, allowing for the collection of large amounts of light over extended periods of time. However, Earth's atmosphere is very turbulent, which turns faint points of light into very faint smudges, and photographic film collects only a small amount of all the light that strikes it.

The largest telescopes were built on mountaintops away from large cities to get above as much of the turbulent atmosphere as possible, and to reduce the effects of light pollution. This solved only part of the problem (Figure 23.8). Recent developments in electronics have helped modern earthbound astronomical instruments (computer-controlled telescopes and electronic

FIGURE 23.6 The Doppler effect, illustrating the apparent lengthening and shortening of wavelengths caused by the relative motion between a source and an observer.

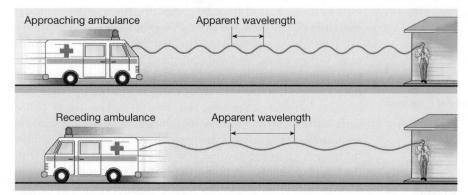

Approaching ambulance   Apparent wavelength

Receding ambulance   Apparent wavelength

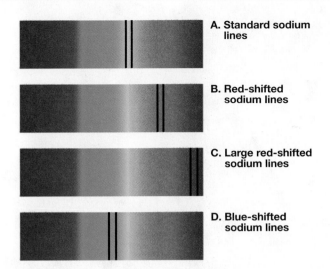

A. Standard sodium lines

B. Red-shifted sodium lines

C. Large red-shifted sodium lines

D. Blue-shifted sodium lines

FIGURE 23.7 The Doppler effect allows astronomers to determine whether Earth is approaching or receding from a star or other celestial body. **A.** Standard dark-line spectrum for sodium produced in the laboratory. **B.** and **C.** Sodium lines as they would appear when a light source is receding (red shift). **D.** Sodium lines produced by an approaching star (blue shift).

detectors) to get around these limitations. However, even these instruments are limited to collecting light in the visible or radio wavelength regions because other wavelengths do not penetrate Earth's atmosphere (see Figure 23.2).

Finally, with the dawn of the space age, even the wavelength limitations have been overcome. It has become practical to put astronomical observatories in space, avoiding the turbulent atmosphere and allowing for the collection of light at all wavelengths. Since the basic principles for detecting radiation were originally developed for visual observations, we consider optical telescopes first, followed by radio telescopes, and finally, orbiting observatories.

CONCEPT CHECK 23.4

1 Why is the human eye an ineffective tool for astronomical observation?
2 Provide two reasons why the largest telescopes are built on mountaintops away from large cities.

FIGURE 23.8 Kitt Peak Observatory on a starlit night. (Photo by Bryan Allen/CORBIS)

Optical Telescopes

Optical telescopes collect light with visible (or nearly visible) wavelengths and come in two basic types—*refracting* and *reflecting* telescopes.

Refracting Telescopes

Much like the one used by Galileo, **refracting telescopes** employ lenses to collect and focus light (Figure 23.9). The light coming from a distant object can be thought of as a ray or beam by the time it reaches Earth. Our eye, or a telescope lens, intercepts some portion of the incoming light. To collect more light, one simply uses a larger lens.

There are two major problems that prevent the manufacture of large refracting telescopes. First, the lens acts like a prism spreading out the colors in the light, an effect called **chromatic aberration**. This is a problem that grows quickly as lenses get larger. Second, large lenses weigh so much that they sag under their own weight, changing their shape, and hence, changing their focusing properties.

The world's largest refracting telescope is the 1-meter (40-inch) telescope at Yerkes Observatory in Williams Bay, Wisconsin (Figure 23.10). This telescope was successfully used for spectroscopic work (and other work), but it suffers from the problems described above.

Reflecting Telescopes

Although small refracting telescopes work very well for observations of objects in the solar system and for observing any other bright source, they have mostly been replaced by **reflecting telescopes**, which use a curved mirror to collect and focus the light. All large telescopes built today are of the reflecting type. Reflecting telescopes do not suffer from chromatic aberration because the light does not travel through glass, but is reflected from a coated surface instead (Figure 23.11). The mirror is generally made of glass and finely ground to a nearly perfect paraboloid (Figure 23.12). A parabola is the geometric shape that takes parallel lines—or parallel light rays—and focuses them to a point. The Hale telescope, with a 5-meter mirror, is ground to within a millionth of a centimeter of being a perfect paraboloid. (If you have the time and patience, you could grind your own 8-, 10-, or even 12-inch mirror.) Once ground, the surface of the mirror is coated with a highly reflective material.

Reflecting telescopes collect more light as the diameter of the mirror increases, just like a refracting telescope with larger lenses.

FIGURE 23.10 The largest refracting telescope, a 1-meter (40-inch) refractor, located at Yerkes Observatory, Williams Bay, Wisconsin. (Photo by Roger Ressmeyer/CORBIS)

However, there are difficulties in increasing the size of the mirror beyond several meters. These include supporting such a large mass, moving that mass to realign the telescope, warping of the mirror surface under its own weight, and the time required to grind a near perfect surface over such a large area.

These difficulties have recently been overcome in two ways. First, by using an array of smaller deformable mirrors under computer control to give the effect of one large mirror (see Box 23.1). Second, by using a single, very thin mirror mounted on actuators, and controlling the mirror shape by computer with inputs from an active optics system. Active optics are a recent development that corrects for distortions caused by turbulence in the atmosphere

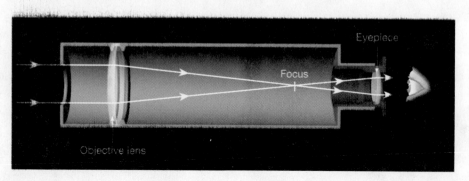

FIGURE 23.9 Simple refracting telescope.

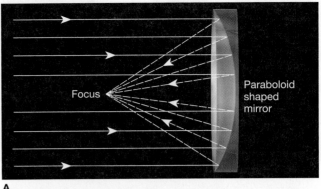

A.

FIGURE 23.11 Newton's reflector telescope. Through experimentation, Newton discovered that a large lens would cause white light to separate into its constituent parts, causing a halo of colored light to form around the object being viewed. By designing a telescope that used a mirror rather than a lens, he avoided this problem altogether. (Photo by Dave King/Dorling Kindersley Media Library)

and only became practical recently due to the availability of fast, relatively inexpensive computers.

Larger telescopes not only allow us to collect more light from faint nearby objects, they also allow us to collect more light from very distant objects. Since the speed of light is finite (about 300,000 kilometers per second), it takes time for the light to get to us. Even light from the Sun takes about 8½ minutes to reach Earth, and light from the nearest large galaxy takes 2 million years to reach us. Lit-

B.

FIGURE 23.12 Reflecting telescope. **A.** Diagram illustrating how paraboloidal mirrors, like those used in reflector telescopes, gather light. **B.** Preparation of the 2.4-meter mirror for the Hubble Space Telescope. (Courtesy of Space Telescope Science Institute)

Students Sometimes Ask...

Why do astronomers build observatories on mountaintops?

Observatories are most often located on mountaintops because sites above the densest part of the atmosphere provide better conditions for "seeing," and high mountaintops tend to be far from urban centers and their light pollution. At high elevations, there is less air to scatter and dim the incoming light, and less water vapor to absorb infrared radiation. Furthermore, the thin air on mountaintops causes less distortion of the images being observed due to density variations in the air. (Think of when you have seen the blurring effect caused by heat rising off the pavement on a hot summer day.) For all of these reasons, the Hubble Space Telescope is a very valuable instrument because it is completely outside the atmosphere.

erally, larger telescopes allow us to look back in time. Our desire to understand the nature and evolution of the universe has motivated us to develop telescopes that look farther and farther back in time. Larger telescopes also generally provide better resolution, or clarity (Figure 23.13).

Light Detection

Telescopes simply collect light. They become useful only when the collected light is detected and analyzed. The earliest detectors were the astronomer's eyes. Astronomers would look through telescopes and draw what they saw (Figure 23.14). Each person's eyes perceive light intensity and faint color differently (and each person's drawing talent is not the same), so that under the same seeing conditions, different images of the same object were produced. In addition, personal biases can easily creep in. For example, in the early 20th century, noted astronomer Percival Lowell (1855–1916) was convinced that there were canals on the surface of Mars. Thus,

A.

B.

FIGURE 23.13 Appearance of the galaxy in the constellation Andromeda using telescopes of different resolution. (Courtesy of AURA)

FIGURE 23.14 Percival Lowell believed that life existed on Mars and drew these canals, influenced perhaps by his personal biases. (Photo by Photo Researchers, Inc.)

CONCEPT CHECK 23.5

1. What is the major difference between reflecting and refracting telescopes?
2. Why do astronomers seek to design telescopes with larger and larger mirrors?
3. Why do all large optical telescopes use mirrors to collect light rather than lenses?
4. Explain the following statement: "Photography has extended the limits of our vision."
5. What are the advantages of *charge coupled devices* (CCD) over photographic film?

he "saw" them in his telescope, and drew them in his images. Subsequent studies did not support Lowell's observations.

Photographic film was a revolutionary improvement. It is not impeded by personal biases, it records reasonably accurate relative light intensities, and records faint colors more accurately than the human eye. However, only about 2 percent of the light that strikes film is recorded. This means that long exposure times are required when recording faint images. Furthermore, photographic film, like the human eye, is not equally sensitive to all wavelengths. There are also differences between individual batches or even between pieces of film that need to be accounted for when making quantitative comparisons.

Advances in semiconductor technology have produced the *charge coupled device* (CCD), which takes an electronic photograph and effectively uses the same piece of "film" over and over again. (Charged coupled devices are used in digital cameras as the light-sensing component.) CCD cameras offer a tremendous improvement over photographic film for detection of visible and near visible light. They typically detect 70 percent, or more, of all incoming light and are easily calibrated for variations in wavelength sensitivity. Using CCD cameras, astronomers can collect light from distant objects for hours, as long as the telescope is accurately steered. Light can also be collected over several nights and added together to make a single image.

Radio- and Space-Based Astronomy

Sunlight consists of more than the visible portion of the electromagnetic spectrum. Gamma rays, X-rays, ultraviolet radiation, infrared radiation, and radio waves are also produced by stars and other celestial objects. CCD cameras that are sensitive to ultraviolet and infrared radiation have been developed, thereby extending the limits of our vision. However, much of the radiation produced by celestial objects cannot penetrate our atmosphere or is not detectable by optical telescopes. As a result, astronomers have developed other observational techniques covering the remaining portions of the electromagnetic spectrum.

Radio Telescopes

Of great importance is a narrow band of radio waves that does penetrate the atmosphere (see Figure 23.2). One particular wavelength is the 21-centimeter line produced by neutral hydrogen (hydrogen atoms that still hold their electron). Measurement of

Box 23.1

UNDERSTANDING EARTH

The World's Largest Optical Telescopes

The main purpose of a telescope is to collect as much light as possible. The larger the telescope's lens or mirror, the more light it collects, allowing for viewing of fainter objects. Because an important objective of astronomy involves observing extremely distant and hence very faint cosmic sources, large telescopes are essential.

Until recently, the largest telescopes were limited to mirrors about 5 meters (200 inches) in diameter because the task of casting, cooling, and polishing large glass mirrors to very fine tolerances (less than the thickness of a human hair) was enormously time-consuming and expensive. For example, the construction of the 5-meter mirror for the Hale Telescope on Mount Palomar, California, began in 1934 and was not completed until 1948. However, during the last decade, with the aid of high-tech manufacturing techniques, several large-diameter telescopes have been built, and several more are being planned.

The world's largest telescope is the Gran Telescopio Canarias located in La Palma, Canary Islands, Spain, which boasts a 10.4-meter mirror composed of 36 1.8-meter segments that are positioned by computer. Nearly equal in size and of the same design are the Keck I and Keck II, which are located on Hawaii's Mauna Kea at an altitude of nearly 4,200 meters (13,800 feet). These 10-meter telescopes are capable of working independently or in tandem (**Figure 23.A**).

One of the largest optical telescopes, in terms of total light-gathering capability, is the European Southern Observatory's Very Large Telescope (VLT), located at Cerro Paranal, Chile. It consists of four separate 8.2-meter instruments that work independently, or in conjunction with one another. When working in tandem, these telescopes have 10 times the light-gathering capacity of the 5-meter Hale Telescope and therefore can "see" cosmic objects that are 10 times dimmer.

Several other large optical telescopes are under construction or in the planning stages. A United States–Korea–Australia consortium is developing a 24.5-meter (80-foot) instrument called the Giant Magellan Telescope. In addition, the consortium participating in the European Southern Observatory has approved funding for a telescope that boasts a 42-meter (138-foot) composite mirror. If these plans come to fruition, this new instrument will be over 100 times more powerful than the most powerful telescopes currently operational.

FIGURE 23.A Mirror of the 10-meter Keck Telescope. The mirror was constructed from 36 hexagonal segments. (Photo by Roger Ressmeyer/CORBIS)

this radiation has permitted us to map the galactic distribution of hydrogen, the material from which stars are made.

The detection of radio waves is accomplished by "big dishes" called **radio telescopes** (Figure 23.15A). In principle, the dish of a radio telescope operates in the same manner as the mirror of an optical telescope. It is parabolic in shape and focuses the incoming radio waves on an antenna, which collects and transmits these waves to an amplifier.

Because radio waves are about 100,000 times longer than visible radiation, the surface of a dish need not be as smooth as a mirror. In fact, except for the shortest radio waves, wire mesh is an adequate reflector. On the other hand, because radio signals from celestial sources are very weak, large dishes are necessary in order to intercept a signal that is strong enough to be detected. The largest radio telescope is a bowl-shaped antenna hung in a natural depression in Puerto Rico (Figure 23.16). It is 300 meters (1,000 feet) in diameter and has some directional flexibility because of its movable antenna. The largest steerable types have about 100-meter (330-foot) dishes. The National Radio Astronomy Observatory in Green Bank, West Virginia, provides an example (Figure 23.15A).

Radio telescopes have relatively poor resolution, making it difficult to pinpoint the radio source. Pairs or groups of telescopes are used to reduce this problem. When several radio telescopes are wired together, the resulting network is called a **radio interferometer** (Figure 23.15B).

Orbiting Observatories

Orbiting observatories circumvent all of the problems caused by Earth's atmosphere and have led to many significant discoveries in astronomy. NASA's series of "*Four Great Observatories*" provide a good illustration.

The Hubble Space Telescope Launched in 1990, the *Hubble Space Telescope* (*HST*) is an optical reflecting telescope in orbit around Earth (Figure 23.17). Its images are not distorted by the

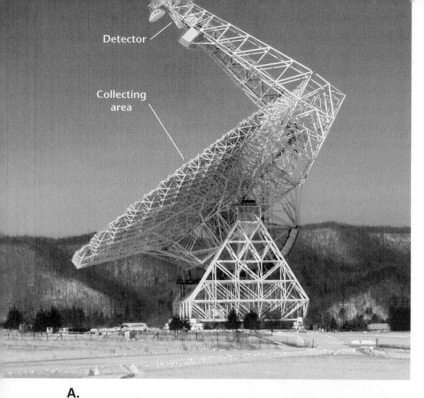

A.

B.

atmosphere and there is no atmospherically scattered light to drown out faint sources of light. In addition, it can collect ultraviolet light that is absorbed by Earth's ozone layer and thus, unavailable to ground-based telescopes. Hubble must be considered to be one of the most important instruments in the history of astronomy because of the large number of discoveries that have been made from its images. The 2.4-meter mirror has produced images with a sensitivity and resolution that are only now being matched by much larger (10-meter) ground-based telescopes.

Here are just a few of the many discoveries made with Hubble. HST provided visual proof that pancake-shaped disks of dust are common around young stars, providing support for the Nebular Hypothesis of solar system formation. Hubble provided decisive evidence that super massive black holes reside in the center of many galaxies by imaging the movements of dust and gas in the interiors of galaxies. The HST has also allowed us to look farther out into the universe (and farther back in time) than ever before, while producing the most "elusive" astronomical image ever taken, the *Ultra Deep Field* (Figure 23.18). This image was acquired by looking at a patch

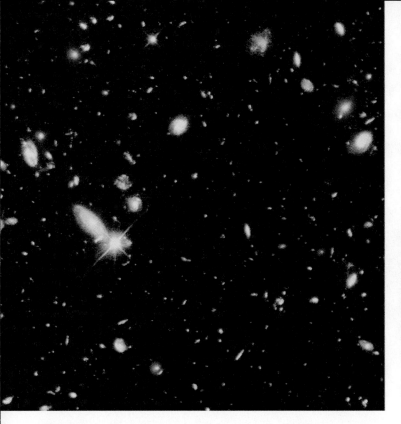

FIGURE 23.18 As we probe deep into space, we are really looking back in time. Due to NASA's orbiting observatories, we have the deepest, most detailed views of extragalactic space yet obtained. This image, called the *Hubble Deep Field*, was taken of what appears to be an "empty" part of the sky located near the Big Dipper. The colors are approximately what the human eye would see. What the image shows are numerous young galaxies and protogalaxies (faint smudges of blue light) that gave rise to galaxies that exist today. (Photo courtesy of NASA)

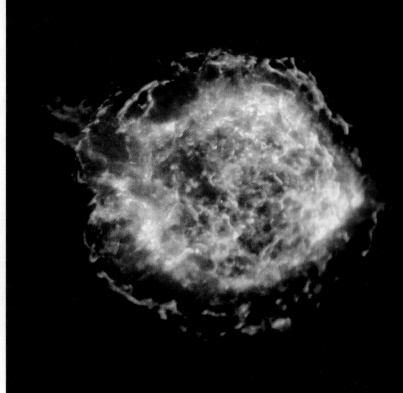

FIGURE 23.19 Chandra X-Ray Observatory captured this image of a supernova remnant—scattered glowing debris ejected from a massive star. This feature, barely visible in the optical part of the spectrum, is awash in brilliantly glowing gases emitting X-rays. (NASA)

of "empty" sky for a total of 1 million seconds with the faintest objects putting only one photon per minute into the exposure.

The Compton Gamma Ray Observatory Designed to collect data on some of the most violent physical processes in the universe, the *Compton Gamma Ray Observatory* (*CGRO*) was launched in 1991. It had a sensitivity 10 times greater than any previous gamma ray instrument and collected an incredible range of high-energy radiation. One of the main scientific discoveries made by CGRO was the uniform distribution of *gamma ray bursts*, which suggest that they are common events associated with ordinary objects.

Gamma ray bursts are flashes of gamma rays that come from seemingly random places deep in the universe at random times. They are probably the most luminous, and therefore, the most energetic events occurring in the universe since the Big Bang. It is quite likely that many of them are caused by rapidly rotating massive stars as they collapse to form black holes (see Chapter 24).

The Chandra X-Ray Observatory The *Chandra X-Ray Observatory* (*CXO*), launched in 1999, was designed to observe objects such as black holes, quasars, and high-temperature gases at X-ray wavelengths to better understand the structure and evolution of the universe. With a resolution 25 times greater than any other X-ray observatory, it only uses as much power as an ordinary hair dryer (Figure 23.19). The CXO has observed a black hole pulling in matter, and two black holes merging into one. In addition, it has provided an independent measurement of the age of

the universe, reinforcing the estimated age of 12–14 billion years. The CXO has shown what galaxies were like when the universe was only a few billion years old.

The Spitzer Space Telescope Launched in 2003, the *Spitzer Space Telescope* (*SST*) was designed to collect infrared (heat) energy that is mostly blocked by Earth's atmosphere. Its instruments must be cooled to near zero Kelvin so that heat from nearby objects (and the satellite itself) do not interfere with the measurements. The telescope is actually in an orbit around the Sun to keep it away from the thermal energy radiated by Earth, and it is outfitted with a shield to deflect solar radiation.

Spitzer's highly sensitive instruments give us unique views of the universe and allow us to peer into regions of space that are hidden from optical telescopes by vast, dense clouds of gas and dust (nebula). Fortunately, infrared light can pass through these clouds, allowing us to peer into regions of star formation, the centers of galaxies, and into newly forming planetary systems. Infrared light also brings us information about cool celestial objects, such as small stars that are too dim to be detected at visible wavelengths, planets that lay outside our solar system, and molecular clouds.

CONCEPT CHECK 23.6

❶ Why are radio telescopes much larger than optical telescopes?
❷ What are some of the advantages of radio telescopes over optical telescopes?
❸ Explain why the Moon would make a good site for an optical observatory.
❹ What can astronomers learn about the universe by studying it at multiple wavelengths?

FIGURE 23.20 The Sun is the source of more than 99 percent of all energy on Earth. (Photo by Jerry and Marcy Monkman/Danita Delimont)

The Sun

The Sun is one of the 200 billion stars that make up the Milky Way Galaxy. Although the Sun is of little significance to the universe as a whole, to those of us who inhabit Earth it is the primary source of energy. Everything from the food we eat to the fossil fuels we burn in our automobiles and power plants is ultimately derived from solar energy (Figure 23.20). The Sun is also important in astronomy, since it is the only star close enough to permit easy study of its surface. Even with the largest telescopes, most other stars appear only as points of light.

Because the Sun is so bright and emits eye-damaging radiation, it is not safe to observe it directly. However, it can be studied safely when a telescope is used to project the Sun's image on a piece of cardboard placed behind the telescope's eyepiece. This basic method is used by several telescopes around the world, which keep a constant vigil of the Sun. One of the finest is at the Kitt Peak National Observatory in southern Arizona (Figure 23.21). It consists of a 150-meter sloped enclosure that directs sunlight to a mirror situated below ground. From the mirror, an 85-centimeter (33-inch) image of the Sun is projected to an observing room, where it is studied.

FIGURE 23.21 The unique Robert J. McMath Solar Telescope at Kitt Peak, near Tucson, Arizona. Movable mirrors at the top follow the Sun, reflecting its light down the sloping tunnel. (Photo by Kent Wood/Photo Researchers, Inc.) Inset photo shows a view of the solar disk obtained by a solar telescope. (Photo by European Space Agency)

2002/06/09 13:19

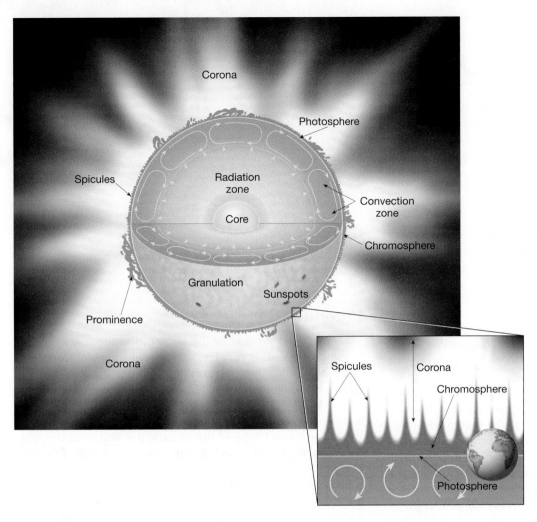

FIGURE 23.22 Diagram of solar structure in cutaway view. Earth is shown for scale.

(**Figure 23.22**). Because the Sun is gaseous throughout, no sharp boundaries exist between these layers. The Sun's interior makes up all but a tiny fraction of the solar mass, and unlike the outer three layers, it is not accessible to direct observation. We discuss the visible layers first.

Photosphere

The **photosphere** (*photos* = light, *sphere* = ball) is aptly named because it radiates most of the sunlight we see and therefore appears as the bright disk of the Sun. Although it is considered to be the Sun's "surface," it is unlike most surfaces to which we are accustomed. The photosphere consists of a layer of incandescent gas less than 500 kilometers (300 miles) thick, having a pressure less than 1/100 of our atmosphere. Furthermore, it is neither smooth nor uniformly bright, as the ancients had imagined. It has numerous blemishes.

When viewed telescopically, the photosphere's grainy texture is apparent. This is the result of numerous comparatively small, bright markings called **granules** (*granuum* = small grain that are surrounded by narrow, dark regions (**Figure 23.23**). Granules are typically the size of Texas, and owe their brightness to hotter gases that are rising from below. As this gas spreads laterally, cooling causes it to darken and sink back into the interior. Each granule survives for only 10–20 minutes, while the combined motion of old granules being replaced by new ones gives the photosphere the appearance of boiling. This

Compared to other stars of the universe, many of which are larger, smaller, hotter, cooler, more red, or more blue, the Sun is an "average star." However, on the scale of our solar system, it is truly gigantic, having a diameter equal to 109 Earth diameters (1.35 million kilometers) and a volume of 1.25 million times as great as that of Earth. Yet, because of its gaseous nature, the density is only one-quarter that of Earth, a little greater than the density of water.

FIGURE 23.23 Granules of the solar photosphere. Granules appear as yellowish-orange patches. Each granule is about the size of Texas and lasts for only 10–20 minutes before being replaced by a new granule. (Courtesy of National Optical Astronomy Observatories)

CONCEPT CHECK 23.7

❶ Why is the Sun so important to inhabitants of Earth?
❷ Why is the Sun significant to the study of astronomy?
❸ Describe the Sun in relationship to other stars in the universe.

Structure of the Sun

For convenience of discussion, we divide the Sun into four parts: the *solar interior*; the *visible surface*, or *photosphere*; and the two layers of its atmosphere, the *chromosphere* and the *corona*

Students Sometimes Ask...

If the Sun is an enormous ball of hot gas, how can it have a surface?

That's a good observation. Actually, the visible surface of the Sun, the photosphere, is a layer of gas about 500 kilometers (300 miles) thick. It is from this layer that we receive the most sunlight. Although more light is emitted from the layer below the photosphere, that light is absorbed in the overlying layers of gas. Above the photosphere, the gas is much less dense and does not radiate much light. The photosphere is the layer that is dense enough to emit ample light yet has a density low enough to allow light to escape. Since the photosphere emits most of the light we see, it appears as the outermost surface of the Sun.

FIGURE 23.24 Spicules of the chromosphere on the edge of the solar disk. (National Solar Observatory/Sacramento Peak)

up-and-down movement of gas, called *convection*, produces the grainy appearance of the photosphere and is responsible for the transfer of energy in the uppermost part of the Sun's interior (Figure 23.22).

The composition of the photosphere has been determined from the dark lines of its absorption spectrum (see Figure 23.3). When these fingerprints are compared to the spectra of known elements, they indicate that most of the elements found on Earth also exist on the Sun. When the strengths of the absorption lines are analyzed, the relative abundance of the elements can be determined. These studies show that 90 percent of the Sun's surface atoms are hydrogen and almost 10 percent are helium. That leaves only minor amounts of the other detectable elements. Other stars also show similar disproportionate percentages of these two lightest elements, a fact we consider later.

Chromosphere

Just above the photosphere lies the **chromosphere** (color sphere), a relatively thin layer of hot, incandescent gases a few thousand kilometers thick. The chromosphere is observable for a few moments during a total solar eclipse, or by using a special instrument that blocks out the light from the photosphere. Under such conditions, it appears as a thin red rim around the Sun. Because the chromosphere consists of hot, incandescent gases under low pressure, it produces a bright-line spectrum that is nearly the reverse of the dark-line spectrum of the photosphere. One of the bright lines of hydrogen contributes a good portion of its total output and accounts for this sphere's red color.

In 1868, a study of the chromospheric spectrum revealed the existence of an element unknown on Earth. It was named helium, from *helios*, the Greek word for Sun. Originally, helium was thought to be an element unique to the stars, but 27 years later it was discovered in a natural-gas well on Earth.

The top of the chromosphere contains numerous **spicules** (*spica* = point), flamelike structures that extend upward about 10,000 kilometers into the lower corona, almost like trees that reach into our atmosphere (Figure 23.24). Spicules are produced by the turbulent motion of the granules below.

Corona

The outermost portion of the solar atmosphere, the **corona** (*corona* = crown) is very tenuous and, like the chromosphere, is visible only when the brilliant photosphere is blocked (Figure 23.25). This envelope of ionized gases normally extends a million kilometers or so from the Sun and produces a glow about half as bright as the full Moon.

At the outer fringe of the corona, the ionized gases have speeds great enough to escape the gravitational pull of the Sun. The streams of protons and electrons that boil from the corona constitute the **solar wind**. The solar wind travels outward through the solar system at high speeds (250–800 kilometers a second) and much of it is lost to interstellar space. During its journey, the solar wind interacts with the bodies of the solar system, continually bombarding lunar rocks and altering their appearance. Although Earth's magnetic field prevents the solar winds from reaching the surface, these streams of charged particles interact with gases in our atmosphere—a topic we will discuss later.

Studies of the energy emitted from the photosphere indicate that its temperature averages about 6,000 K (10,000° F). Upward from the photosphere, the temperature unexpectedly increases,

FIGURE 23.25 Solar corona photographed during a total eclipse. (Photo by Jerry Lodriguss/Photo Researchers, Inc.)

exceeding 1 million K at the top of the corona. It should be noted that although the coronal temperature exceeds that of the photosphere by many times, it radiates much less energy overall because of its very low density.

Surprisingly, the high temperature of the corona is probably caused by sound waves generated by the convective motion of the photosphere. Just as boiling water makes noise, energetic sound waves generated in the photosphere are believed to be absorbed by the gases of the corona, thereby increasing its temperature.

CONCEPT CHECK 23.8

1. Describe the photosphere, chromosphere, and corona.
2. Why are there no distinct boundaries between the layers of the Sun?
3. Why is the photosphere considered the Sun's "surface"?
4. Briefly describe the *solar wind*.

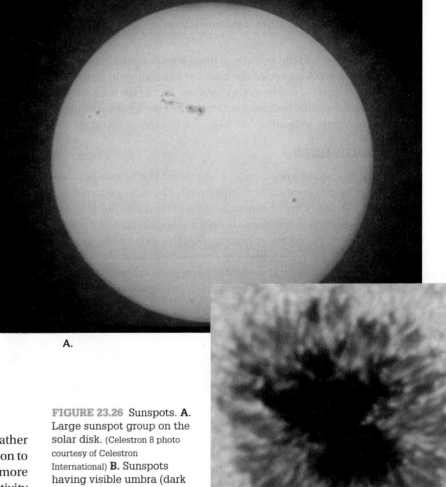

A.

The Active Sun

Most of the Sun's energy that reaches Earth is the result of a rather steady, continuous emission from the photosphere. In addition to this predictable aspect of our Sun's energy output is a much more irregular component, characterized by explosive surface activity that includes *sun spots*, *prominences*, and *solar flares*.

Sunspots

The most conspicuous features on the surface of the Sun are the dark blemishes called **sunspots** (Figure 23.26A). Although large sunspots were occasionally observed before the advent of the telescope, they were generally regarded as opaque objects located somewhere between the Sun and Earth. In 1610 Galileo concluded that they were residents of the solar surface, and from their motion he deduced that the Sun rotates on its axis about once a month.

Later observations indicated that the time required for one rotation varied by latitude. The Sun's equator rotates once in 25 days, whereas a place located 70 degrees from the solar equator, either north or south, requires 33 days for one rotation. If Earth rotated in a similar disjointed manner, imagine the consequences! The Sun's nonuniform rotation is a testimonial to its gaseous nature.

Sunspots begin as small, dark pores about 1,600 kilometers (1,000 miles) in diameter. Although most sunspots last for only a few hours, some grow into blemishes many times larger than Earth and last for a month or more. The largest sunspots often occur in pairs surrounded by several smaller sunspots. An individual spot contains a black center, the *umbra* (*umbra* = shadow), which is rimmed by a lighter region, the *penumbra* (*paene* = almost, *umbra* = shadow) (Figure 23.26B). Sunspots appear dark only by contrast with the brilliant photosphere, a fact accounted for by their temperature, which is about 1,500 K less than that of the solar surface. If these dark spots could be observed away from the Sun, they would appear many times brighter than the full moon.

FIGURE 23.26 Sunspots. **A.** Large sunspot group on the solar disk. (Celestron 8 photo courtesy of Celestron International) **B.** Sunspots having visible umbra (dark central area) and penumbra (lighter area surrounding umbra). (Courtesy of National Optical Astronomy Observatories)

B.

During the early 19th century, it was believed that a tiny planet named Vulcan orbited between Mercury and the Sun. In the search for Vulcan an accurate record of sunspot occurrences was kept. Although the planet was never found, the sunspot data showed that the number of sunspots on the solar disk varies in an 11-year cycle.

First, the number of sunspots increases to a maximum, with perhaps a hundred or more visible at a given time. Then, over a period of 5–7 years, their numbers decline to a minimum, when only a few or even none are visible. At the beginning of each cycle, the first sunspots form about 30 degrees from the solar equator, but as the cycle progresses and their numbers increase, they form nearer the equator. During the period when sunspots are most abundant, the majority form about 15 degrees from the equator. They rarely occur more than 40 degrees away from the Sun's equator, or within 5 degrees of it.

Another interesting characteristic of sunspots was discovered by astronomer George Hale, for whom the Hale Telescope is named. Hale deduced that the large spots are strongly magnetized, and when they occur in pairs, they have opposite magnetic poles. For instance, if one member of the pair is a north magnetic pole, then the other member is a south magnetic pole. Also, every pair located in the same hemisphere is magnetized in the same manner. However, all pairs in the *other* hemisphere are magnetized in the opposite manner. At the beginning of each sunspot cycle, the situation reverses, and the polarity of these sunspot pairs is opposite those of the previous cycle. The cause of this change in polarity—in fact, the cause of sunspots themselves—is not fully understood. However, other modes of solar activity vary in the same cyclic manner as sunspots, indicating the likelihood of a common origin.

Prominences

Among the more spectacular features of the active Sun are **prominences** (*prominere* = to jut out). These huge cloudlike structures, consisting of concentrations of chromospheric gases, are best observed when they are on the edge, or limb, of the Sun, where they often appear as bright arches that extend well into the corona (**Figure 23.27**). *Quiescent prominences* have the appearance of a fine tapestry and seem to hang motionless for days at a time, but motion pictures reveal that the material within them is continually falling like luminescent rain.

By contrast, *eruptive prominences* rise almost explosively away from the Sun. These active prominences reach velocities up to 1,000 kilometers (620 miles) per second and may leave the Sun entirely. Whether eruptive or quiescent, prominences are ionized chromospheric gases trapped by magnetic fields that extend from regions of intense solar activity.

Solar Flares

Solar flares are brief outbursts that normally last an hour or so and appear as a sudden brightening of the region above a sunspot cluster. During their existence, enormous quantities of energy are released across the entire electromagnetic spectrum, much of it in the form of ultraviolet, radio, and X-ray radiation. Simultaneously, fast-moving atomic particles are ejected, causing the *solar wind* to intensify noticeably. Although a major flare could conceivably endanger a manned space flight, these are relatively rare events. Within hours after a large outburst, the ejected particles reach Earth and disturb the ionosphere,[57] affecting long-distance radio communications.

The most spectacular effects of solar flares are the **auroras**, also called the Northern and Southern lights (**Figure 23.28**). Following a strong solar flare, Earth's upper atmosphere above the magnetic poles is set aglow for several nights. The auroras appear in a wide variety of forms. Sometimes the display consists of vertical streamers with considerable movement. At other times, the auroras appear as a series of luminous, expanding arcs or as a quiet, almost foglike, glow. Auroral displays, like other solar activities, vary in intensity with the 11-year sunspot cycle.

CONCEPT CHECK 23.9

1. What did Galileo learn about the Sun from his observations of sunspots?
2. Briefly describe the 11-year sunspot cycle.
3. What are prominences?
4. How do solar flares affect the solar wind?

The Solar Interior

The interior of the Sun cannot be observed directly. For that reason, what we know about it is based on information acquired from the energy radiated from the photosphere and solar atmosphere and from theoretical studies.

The source of the Sun's energy, **nuclear fusion** (*fusus* = to melt), was not discovered until the late 1930s. Deep in its interior, a nuclear reaction called the **proton–proton chain reaction** converts four hydrogen nuclei (protons) into the nucleus of a helium atom. The energy released from the proton–proton reaction results because some of the matter involved is actually converted to radiant energy. This can be illustrated by noting that four hydrogen atoms have a combined atomic mass of 4.032 (4×1.008), whereas the atomic mass of helium is 4.003, or 0.029 less than the combined mass of the hydrogen. The tiny missing mass is emitted as energy according to Einstein's formula $E = mc^2$ where E equals energy, m equals mass, and c equals the speed of light. Because the speed of light is very great, the amount of energy released from even a small amount of mass is enormous.

The conversion of just one pinhead's worth of hydrogen to helium generates more energy than burning thousands of tons of coal. (This process is often referred to as "hydrogen burning," but it is nothing like the type of burning to which we are accustomed.) Most of this energy is in the form of very high-energy photons that work their way toward the solar surface, being absorbed and re-emitted many times until they reach an opaque layer just below the photosphere. Here, convection currents serve to transport this energy to the solar surface, where it radiates through the nearly

FIGURE 23.27 A huge solar prominence. (SOHO/ESA/NASA/Photo Researchers, Inc.)

---

[57]The ionosphere is a complex atmospheric zone of ionized gases extending between about 80 and 400 kilometers (50 and 250 miles) above Earth's surface.

FIGURE 23.28 Aurora borealis (Northern lights) as seen from Alaska. The same phenomenon occurs toward the South Pole, where it is called the Aurora australis (Southern lights). (Photo by Daniel Cox/ Photolibrary)

transparent chromosphere and corona as mostly visible light (see Figure 23.22).

Only a small percentage (0.7%) of the hydrogen in the proton–proton reaction is actually converted to energy. Nevertheless, the Sun is consuming an estimated 600 million tons of hydrogen each second, with about 4 million tons of it being converted to energy. The by-product of hydrogen burning is helium, which forms the solar core. Consequently, the core continually grows in size.

How long can the Sun produce energy at its present rate before all of its fuel (hydrogen) is consumed? Even at the enormous rate of consumption, the Sun has enough fuel to easily last another 100 billion years. However, evidence from other stars indicates that the Sun will grow dramatically and engulf Earth long before all of its hydrogen is gone. It is likely that a star the size of the Sun can remain in a stable state for about 10 billion years. Since the Sun is already 5 billion years old, it is "middle-aged."

To initiate the proton–proton reaction, the Sun's internal temperature must have reached several million degrees. What was the source of this heat? As previously noted, the solar system formed from an enormous cloud of dust and gases (mostly hydrogen) that gravitationally collapsed. When a gas is squeezed (compressed) its temperature increases. Although all of the bodies in the solar system were heated in this manner, the Sun was the only one, because of its mass, that became hot enough to trigger the proton–proton reaction. Astronomers currently estimate its internal temperature at 15 million K.

The planet Jupiter is basically a hydrogen-rich gas ball. Why didn't it become a star? Although it is a huge planet, the lowest mass stars are between 75 and 80 times the size of Jupiter.

CONCEPT CHECK 23.10

1. What is the source of the Sun's energy?
2. What "fuel" does the Sun consume?
3. What happens to the matter that is consumed in the proton–proton chain reaction?

GIVE IT SOME THOUGHT

1. Refer to Figure 23.2 to answer the following questions.
    a. Is the atmosphere mostly transparent or opaque to visible light?
    b. Is the atmosphere mostly transparent or opaque to radio waves with a wavelength of 1 meter?
    c. Is the atmosphere mostly transparent or opaque to gamma rays?
2. Imagine that the composition of Earth's atmosphere was altered so that its ability to absorb visible and far infrared light was reversed.
    a. If you were outdoors when the Sun was at its highest point in the sky, how would the sky appear?
    b. Would there be an increase or decrease in Earth's average surface temperature?
3. Suppose a well-known scientist claimed that stars consist primarily of helium rather than hydrogen.
    a. What type of object in the galaxy could you study to investigate whether stars consist primarily of helium or hydrogen?
    b. How could spectroscopy help you verify or disprove the scientist's claim? Explain your reasoning.
4. Imagine that you are responsible for funding the construction of observatories. After considering the four proposals listed below, state whether you would or would not recommend funding for each proposal and explain your reasoning.
    Proposal A: A ground-based X-ray telescope on the top of a mountain in Arizona designed to observe supernovae in distant galaxies.
    Proposal B: A space-based 3-meter reflecting infrared telescope designed to observe very distant galaxies.
    Proposal C: A ground-based 8-meter refracting optical telescope located on the top of Mauna Kea in Hawaii designed to measure the spectra of binary stars in our galaxy.
    Proposal D: A ground-based 250-meter radio telescope array in New Mexico designed to measure the distribution of hydrogen gas clouds in the disk of our galaxy.

5. An important absorption line in the spectrum of stars occurs at a wavelength of 656 nm for stars not moving toward or away from Earth. Imagine that you observe four stars in our galaxy and discover that this absorption line is at the wavelength shown in the accompanying diagram. Using this data, complete the following questions. Explain the reasoning behind your answers. If you are unable to determine the answer to any of these questions from the given information, explain.

STAR	Wavelength of Absorption Line
A	649 nm
B	656 nm
C	658 nm
D	647 nm

   a. Which of these stars is moving the fastest toward Earth?
   b. Which of these stars is closest to Earth?
   c. Which of these stars is moving away from Earth?

6. Consider the following discussion among three of your classmates regarding why telescopes are put in space. Support or refute each statement.

   Student # 1:   "I think it is because the atmosphere distorts and magnifies light, which causes objects to look larger than they actually are."

   Student #2:   "I thought it was because some of the wavelengths of light being sent out from the telescopes can be blocked by Earth's atmosphere so the telescopes need to be above the atmosphere."

   Student #3:   "Wait, I thought it was because by moving the telescope above the atmosphere the telescope is closer to the objects, which makes them appear brighter."

7. Refer to the accompanying spectra which represent four identical stars in our galaxy. One star is not moving, another is moving away from you, and two stars are moving toward you. Determine which star is which and explain how you reached your conclusion.

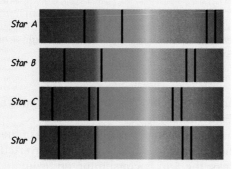

Star A

Star B

Star C

Star D

In Review   Chapter 23 Light, Astronomical Observations, and the Sun

- Visible light constitutes only a small part of an array of energy, generally referred to as *electromagnetic radiation*. Light, a type of electromagnetic radiation, can be described in two ways: (1) as waves and (2) as a stream of particles, called *photons*. The wavelengths of electromagnetic radiation vary from several kilometers for *radio waves* to less than a billionth of a centimeter for *gamma rays*. The shorter wavelengths correspond to more energetic photons.

- *Spectroscopy* is the study of the properties of light that depend on wavelength. When a prism is used to disperse visible light into its component parts (wavelengths), one of three possible types of *spectra* is produced (a *spectrum*, the singular form of *spectra*, is the light pattern produced by passing light through a prism). The three types of spectra are (1) *continuous spectrum*, (2) *dark-line (absorption) spectrum*, and (3) *bright-line (emission) spectrum*. The spectra of most stars are of the dark-line type. Spectroscopy can be used to determine (1) the composition of stars and other gaseous objects, (2) the temperature of a radiating body, and (3) the motion of an object. Motion (direction toward or away and velocity) is determined using the *Doppler effect*—the apparent change in the wavelength of radiation emitted by an object caused by the relative motions of the source and the observer.

- There are two types of optical telescopes: (1) the *refracting telescope*, which uses a *lens* to bend or refract light, and (2) the *reflecting telescope*, which uses a *concave mirror* to focus (gather) the light.

- Telescopes simply collect light. They become useful when the collected light is detected and analyzed. Historically, astronomers relied on their eyes as detectors. Then, photographic film was developed, which was a revolutionary advancement. Presently, light is collected using a *charged coupled device* (CCD). A CCD camera produces a "digital image" akin to that of a digital camera.

- The detection of *radio waves* is accomplished by "big dishes" known as *radio telescopes*. A parabolic-shaped dish, often consisting of wire mesh, operates in a manner similar to the mirror of a reflecting telescope. Of great importance is a narrow band of radio waves that is able to penetrate Earth's atmosphere. Because this radiation is produced by neutral hydrogen, it has permitted us to map the galactic distribution of the material from which stars are made.

- The *Sun* is one of the 200 billion stars that make up the Milky Way Galaxy. The Sun can be divided into four parts: (1) the *solar interior*, (2) the *photosphere* (visible surface) and the two layers of its atmosphere, (3) the *chromosphere*, and (4) *corona*. The photosphere radiates most of the light we see. Unlike most surfaces, it consists of a layer of incandescent gas less than 500 kilometers (300 miles) thick, with a grainy texture consisting of numerous relatively small, bright markings called *granules*. Just above the photosphere lies the chromosphere, a relatively thin layer of hot, incandescent gases a few thousand kilometers thick. At the edge of the uppermost portion of the solar atmosphere, called the *corona*, ionized gases escape the gravitational pull of the Sun and stream toward Earth at high speeds, producing the *solar wind*.

- Numerous features have been identified on the active Sun. *Sunspots* are dark blemishes with a black center, the *umbra*, which is rimmed by a lighter region, the *penumbra*. The number of sunspots observable on the solar disk varies in an 11-year cycle. *Prominences*, huge cloudlike structures best observed when they are on the edge, or limb, of the Sun, are produced by ionized chromospheric gases trapped by magnetic fields that extend from regions of intense solar activity. The most explosive events associated with sunspots are *solar flares*. Flares are brief outbursts that release enormous quantities of energy that appear as a sudden brightening of the region above sunspot clusters. During the event, radiation and fast-moving atomic particles are ejected, causing the solar wind to intensify. When the ejected particles reach Earth and disturb the ionosphere, radio communication is disrupted and the *auroras*, also called the Northern and Southern lights, occur.

- The source of the Sun's energy is *nuclear fusion*. Deep in the solar interior, at a temperature of 15 million K, a nuclear reaction called the *proton–proton chain* converts four hydrogen nuclei (protons) into the nucleus of a helium atom. During the reaction some of the matter is converted to the energy of the Sun. A star the size of the Sun can exist in its present stable state for 10 billion years. As the Sun is already 5 billion years old, it is a "middle-aged" star.

Key Terms

absorption spectrum (p. 669)
aurora (p. 682)
bright-line (emission) spectrum (p. 669)
chromatic aberration (p. 672)
chromosphere (p. 680)
continuous spectrum (p. 668)
corona (p. 680)
dark-line (absorption) spectrum (p. 669)
Doppler effect (p. 669)
electromagnetic radiation (p. 666)

emission spectrum (p. 669)
granules (p. 679)
nuclear fusion (p. 682)
photon (p. 666)
photosphere (p. 679)
prominence (p. 682)
proton–proton chain reaction (p. 682)
radiation pressure (p. 666)
radio interferometer (p. 675)

radio telescope (p. 675)
reflecting telescope (p. 672)
refracting telescope (p. 672)
solar flare (p. 682)
solar wind (p. 680)
spectroscope (p. 669)
spectroscopy (p. 668)
spicule (p. 680)
sunspot (p. 681)

Examining the Earth System

1. Of the two sources of energy that power the Earth system, the Sun is the main driver of Earth's external processes. If the Sun increased its energy output by 10 percent, what would happen to global temperatures? What effect would this temperature change have on the percentage of water that exists as ice? What would be the impact on the position of the ocean shoreline? Speculate as to whether the change in temperature might produce an increase or decrease in the amount of surface vegetation. In turn, what impact might this change in vegetation have on the level of atmospheric carbon dioxide? How would such a change in the amount of carbon dioxide in the atmosphere affect global temperatures?

Mastering Geology

Looking for additional review and test prep materials? Visit the Self Study area in **www.masteringgeology.com** to find practice quizzes, study tools, and multimedia that will aid in your understanding of this chapter's content. In **MasteringGeology™** you will find:

- **GEODe: Earth Science: An interactive visual walkthrough of key concepts**

- **Geoscience Animation Library: More than 100 animations illuminating many difficult-to-understand Earth science concepts**
- **In The News RSS Feeds: Current Earth science events and news articles are pulled into the site with assessment**
- **Pearson eText**
- **Optional Self Study Quizzes**
- **Web Links**
- **Glossary**
- **Flashcards**

24

Beyond Our Solar System*

The Orion Nebula, a huge mass of glowing interstellar gas and dust. (© Anglo-Australian Observatory/Royal Observatory, Photo by David Malin)

*Revised with the assistance of Teresa Tarbuck and Mark Watry, Spring Hill College.

stronomers and cosmologists study the nature of our vast universe, attempting to answer questions such as: Is our Sun a typical star? Do other stars have solar systems with planets similar to Earth? Are galaxies distributed randomly, or are they organized into groups? How do stars form? What happens when stars expend their fuel? If the early universe consisted mostly of hydrogen and helium, how did other elements come into existence? How large is the universe? Did it have a beginning? Will it have an end? This chapter explores these as well as other questions.

FOCUS ON CONCEPTS

To assist you in learning the important concepts in this chapter, focus on the following questions:

- What is cosmology?
- What role do Cepheid variable stars play in the study of astronomy?
- What was Edwin Hubble's most significant discovery about the universe?
- Why is interstellar matter often referred to as a stellar nursery?
- How are bright and dark nebulae different? In what way are they similar?
- What is a main-sequence star?
- How do astronomers determine that some stars are "giants"?
- What are the stages in the evolution of a typical Sun-like star?
- What is the final state of the most massive stars?
- What are the three major types of galaxies?
- How do large elliptical galaxies form?
- What is the big bang theory and what does it tell us about the universe?
- How might the universe end?

The Universe

The universe is more than a collection of dust clouds, stars, stellar remnants, and galaxies (Figure 24.1). It is an entity with its own properties. **Cosmology** is the study of the universe, including its properties, structure, and evolution. Over the years cosmologists have developed a comprehensive theory that describes the structure and evolution of the universe. Some of the questions cosmologists seek to answer with this theory include: Did the universe have a beginning? How did the universe evolve to its present state? How long has it existed and how will it end? Modern cosmology addresses these important issues and helps us understand the universe we inhabit.

How Large Is It?

For most of human existence, our universe was believed to be Earth-centered, containing only the Sun, Moon, five wandering stars, and about 6,000 fixed stars. Even after the Copernican view of a Sun-centered universe became widely accepted, the entire universe was believed to consist of a single galaxy, the Milky Way, composed of innumerable stars along with many faint "fuzzy patches," thought to be clouds of dust and gases.

In the mid-1700s, German philosopher Immanuel Kant proposed that many of the telescopically visible fuzzy patches of light scattered among the stars were actually distant galaxies similar to the Milky Way. Kant described them as "island universes." Each galaxy, he believed, contained billions of stars and was a universe in itself. In Kant's time, however, the weight of opinion favored the hypothesis that the faint patches of light occurred within our galaxy. Admitting otherwise would have implied a vastly larger universe, thereby diminishing the status of Earth, and likewise, humankind.

In 1919 Edwin Hubble arrived at the observatory at Mount Wilson, California, to conduct research using a 100-inch (2.5-meter) telescope, then the world's largest and most advanced astronomical instrument. Armed with this modern tool, Hubble set out to solve the mystery of the "fuzzy patches." (At that time, the debate was still raging as to whether the fuzzy patches were "island universes," as Kant had proposed more than 150 years earlier, or clouds of dust and gases [nebulae].) To accomplish this task, Hubble studied a group of pulsating stars known as *Cepheid variables*—extremely bright variable stars that increase and decrease in brightness in a repetitive cycle. This group of stars is significant because their "true" brightness, called **absolute magnitude**, can be determined by knowing the rate at which they pulsate (see Appendix D). When the absolute magnitude of a star is compared to its observed brightness, a reliable approximation of distance can be established. (This is similar to how we judge the distance of an oncoming vehicle when driving at night.) Thus,

FIGURE 24.1 The Trifid Nebula, in the constellation Sagittarius. This colorful nebula is a cloud consisting mostly of hydrogen and helium gases. These gases are excited by light emitted by hot, young stars within and produce a reddish glow. (Courtesy of National Optical Astronomy Observatories)

Cepheid variables are important because they can be used to measure large astronomical distances.

Using the telescope at Mount Wilson, Hubble found several Cepheid variables imbedded in one of the fuzzy patches. However, because these intrinsically bright stars appeared faint, Hubble concluded that they must lie outside the Milky Way. Indeed, one of the objects Hubble observed lies more than 2 million light-years away, and is now known as the *Andromeda Galaxy* (Figure 24.2).

Based on his observations, Hubble determined that the universe extended far beyond the limits of our imagination. Today, we know there are hundreds of billions of galaxies, each containing hundreds of billions of stars. For example, researchers estimate that a million galaxies exist in the portion of the sky bounded by the cup of the Big Dipper. Literally, there are more stars in the heavens than grains of sand in all the beaches on Earth.

A Brief History of the Universe

Large telescopes can literally "look back in time," which accounts for much of the knowledge astronomers have acquired regarding the history of the universe. Light from celestial objects that are great distances from Earth require millions or even billions of years to reach Earth. The distance light travels in one year is called a **light-year** (slightly less than 10 trillion kilometers, or 6 trillion miles). Therefore, the farther out telescopes can "see," the farther back in time astronomers are able to reach. Even the closest large

galaxy, the Andromeda Galaxy, is a staggering 2.5 million light-years away. Light that left Andromeda Galaxy 2.5 million years ago is just now reaching Earth, allowing scientists to observe the galaxy as it was when that light left the galaxy. Light from the furthest known objects, about 13 billion light-years away, came from stars that have long since burned out.

The timeline for the history of the universe, shown in Figure 24.3, highlights some of the major events in the evolution of matter and energy. The model that most accurately describes the birth and current state of the universe is the **big bang theory**. According to this theory, all of the energy and matter of the universe was compressed into an incomprehensibly hot and dense state. About 13.7 billion years ago our universe began as a cataclysmic explosion, which continued to expand, cool, and evolve to its current state. In the earliest moments of this expansion, only energy and quarks (subatomic particles that are the building blocks of protons and neutrons) existed. Not until 380,000 years after the initial expansion did the universe cool sufficiently for electrons and protons to combine to form hydrogen and helium atoms—the lightest elements in the universe. For the first time, light traveled through space. Eventually, temperatures decreased sufficiently to allow clumps of matter to collect. This material formed the first nebulae, which quickly evolved into the first stars and galaxies. Our Sun and planetary system, having formed about 5 billion years ago (nearly 9 billion years after the big bang), is a latecomer to the universe.

FIGURE 24.2 Andromeda is a nearby galaxy that is larger than our Milky Way. **A.** Photo of Andromeda Galaxy taken by an optical telescope aboard GALEX Orbiter (NASA). **B.** Andromeda Galaxy as it appears under low magnification. When viewed with the naked eye, Andromeda appears as a fuzzy patch surrounded by stars. (European Southern Observatory/ESO)

A.

B.

FIGURE 24.3 Timeline, according to the big bang theory, for the expansion of the universe over the past 13.7 billion years.

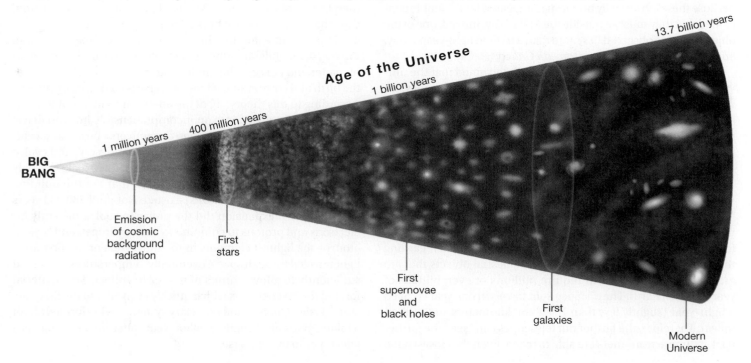

BIG BANG

1 million years

Emission of cosmic background radiation

400 million years

First stars

Age of the Universe

1 billion years

First supernovae and black holes

First galaxies

13.7 billion years

Modern Universe

Interstellar Matter: Nursery of the Stars

As the universe expanded, gravity caused matter to accumulate into large "clumps" and "strands" of **interstellar matter** known as **nebulae** (meaning clouds). Additionally, considerable amounts of interstellar matter once resided in the interiors of stars and was subsequently returned to space. Some stars ejected matter as part of their normal life cycle, some exploded when they died, and some formed black holes that ejected matter through jets.

Interstellar matter resides between stars within the galaxies and consists of roughly 90 percent hydrogen and 9 percent helium. The remainder, *interstellar dust*, is comprised of atoms, molecules, and larger dust grains of the heavier elements. These huge concentrations of interstellar dust and gases are extremely *diffuse*, similar to fog, with no distinct edges or boundaries. However, because nebulae are enormous, their total mass is many times that of the Sun. If a nebula is dense enough, it will contract due to gravity, leading to processes that form stars and planets.

When nebulae are in close proximity to very hot (blue) stars, they glow and are called **bright nebulae**. By contrast, when clouds of interstellar material are too far from bright stars to be illuminated, they are referred to as **dark nebulae**.

Bright Nebulae

There are three main types of bright nebulae: *emission, reflection,* and *planetary nebulae.* Emission and reflection nebulae consist of the material from which new stars are born—stellar nurseries. By contrast, planetary nebulae form when stars die and expel material into space. Thus, all stars form from nebulae and many ultimately return to the material of nebulae when their life cycle ends.

Emission Nebulae   Glowing clouds of hydrogen gas, called **emission nebulae**, are produced in active star-forming regions of galaxies. Energetic ultraviolet light emitted from hot, young stars ionizes the hydrogen atoms in the nebulae. Because these gases exist under extremely low pressure, they radiate, or emit, their energy as less energetic *visible light*. The conversion of ultraviolet light to visible light is known as fluorescence—the same phenomenon that causes neon lights to glow. Hydrogen emits much of its energy in the red portion of the spectrum, which accounts for the red glow from emission nebulae (**Figure 24.4**). When elements other

FIGURE 24.4 Emission nebulae. Lagoon Nebulae is a large emission nebula composed mainly of hydrogen. Its red color is attributed to ionized gases, which are excited by the energetic light emitted from young, hot stars imbedded in the nebula. (National Optical Astronomy Observatories)

FIGURE 24.5 A blue reflection nebula, in the Pleiades star cluster, is caused by the scattering of starlight by relatively large molecules and interstellar dust. The Pleiades star cluster, barely visible to the naked eye in the constellation Taurus, is spectacular when viewed through binoculars or a small telescope. (Palomar Observatories/California Institute of Technology)

than hydrogen are ionized, the glowing cloud may exhibit a broader range of colors. An emission nebula that is easily seen with binoculars is located in the constellation Orion, in the sword of the hunter (see Figure 23.5, page 670).

Reflection Nebulae   As the name implies, **reflection nebulae** merely reflect the light of nearby stars (Figure 24.5). Reflection nebulae are likely composed of significant amounts of comparatively large debris including grains of carbon compounds. This view is supported by the fact that atomic gases with low densities could not reflect light sufficiently to produce the glow observed. Reflection nebulae are usually blue because blue light (shorter wavelength) is scattered more efficiently than red light (longer wavelength)— a process that also produces the blue color of the sky. The blue wisps in Pleiades, shown in Figure 24.5, are reflection nebulae.

Planetary Nebulae **Planetary nebulae** are generally not as diffuse as other nebulae and originate from the remnants of dying sun-like stars (Figure 24.6). Planetary nebulae con-

sist of glowing clouds of dust and hot gases that have been expelled near the end of a star's life. When viewed through a small optical telescope, planetary nebulae are similar in appearance to giant planets such as Jupiter. It is because of this resemblance that planetary nebulae got their name. Although planetary nebulae may resemble giant planets, the two phenomena are not related. A good example of a planetary nebula is the Helix Nebula in the constellation Aquarius (Figure 24.6).

Dark Nebulae

Recall that when clouds of interstellar material are too distant from bright stars to be illuminated, they are referred to as a *dark nebula*. Exemplified by the Horsehead Nebula in Orion, dark nebulae appear as opaque objects silhouetted against bright backgrounds (Figure 24.7). In addition, dark nebulae can also easily be seen as starless regions—"holes in the heavens" when viewing the Milky Way (see Figure 24.14). Although dark nebulae often appear dense, they are made of the same matter as bright nebulae and consist of thinly scattered matter.

FIGURE 24.6 The Helix Nebula, the nearest planetary nebula to our solar system. A planetary nebula is the ejected outer envelope of a Sun-like star that formed during the star's collapse from a red giant to a white dwarf. (© Anglo-Australian Observatory, photograph by David Malin)

FIGURE 24.7 The Horsehead Nebula is a dark nebula in the constellation Orion. (Courtesy of the European Southern Observatory)

CONCEPT CHECK 24.2

❶ What role does interstellar matter (nebulae) play in stellar evolution?
❷ Compare and contrast bright and dark nebulae.
❸ Why are reflection nebulae generally blue?
❹ How are planetary nebulae different from other types of bright nebulae?

Classifying Stars: Hertzsprung–Russell Diagrams (H-R Diagrams)

Early in the 20th century, Einar Hertzsprung and Henry Russell independently studied the relationship between the true brightness (absolute magnitude) of stars and their respective temperatures. Their work resulted in the development of a graph, called a **Hertzsprung–Russell diagram (H-R diagram)**, that employs these intrinsic stellar properties. By studying H-R diagrams, we can learn a great deal of information about the relationships among the sizes, colors, and temperatures of stars (see Appendix D).

To produce an H-R diagram, astronomers survey a portion of the sky and plot each star according to its luminosity (true brightness) and temperature (Figure 24.8). Notice that the stars in Figure 24.8 are not uniformly distributed. Rather, about 90 percent of all stars fall along a band that runs from the upper-left corner to the lower-right corner of the H-R diagram. These ordinary stars are called **main-sequence stars**. As you can see in Figure 24.8, the hottest main-sequence stars are intrinsically the brightest, and conversely, the coolest are the dimmest.

The luminosity of main-sequence stars is also related to their mass. The hottest (blue) stars are about 50 times more massive than the Sun, whereas the coolest (red) stars are only 1/10 as massive. Therefore, on the H-R diagram, the main-sequence stars appear in decreasing order, from hotter, more massive blue stars to cooler, less massive red stars.

Note the location of the Sun in Figure 24.8. The Sun is a yellow main-sequence star with an absolute magnitude of about 5 (see Appendix D). Because the vast majority of main-sequence stars have magnitudes between –5 and 15, the Sun's midpoint position in this range results in its classification as an "average star."

Just as all humans do not fall into the normal size range, some stars differ significantly from main-sequence stars. Above and to the right of the main sequence stars lies a group of very luminous stars called *giants*, or, on the basis of their color, **red giants** (Figure 24.8). The size of these giants can be estimated by comparing them with stars of known size that have the same surface temperature. Scientists have discovered that objects having equal surface temperatures radiate the same amount of energy per unit area. Any difference in the brightness of two stars having the same surface temperature can be attributed to their relative sizes. Therefore, if one red star is 100 times more luminous than another red star, it must have a surface area that is 100 times larger. Stars with large radiating surfaces appear in the upper-right position of an H-R diagram and are appropriately called giants.

Some stars are so immense that they are called **supergiants**. Betelgeuse, a bright red supergiant in the constellation Orion, has a radius about 800 times that of the Sun. If this star were at the center of our solar system, it would extend beyond the orbit of Mars, and Earth would find itself buried inside this supergiant!

In the lower portion of the H-R diagram, opposite conditions are observed. These stars are much fainter than main-sequence stars of the same temperature, and likewise are much smaller. Some likely approximate Earth in size. These stars are called *white dwarfs*.

The development of H-R diagrams proved to be an important tool for interpreting stellar evolution. As with living things, stars are born, age, and die. Considering that almost 90 percent of stars lie on the main sequence, we can be relatively certain that they spend most of their active years as main-sequence stars. Only a few percent are giants, and perhaps 10 percent are white dwarfs.

CONCEPT CHECK 24.3

❶ On an H-R diagram, where do stars spend most of their life span?
❷ How does the Sun compare in size and brightness to other main-sequence stars?
❸ Describe how the H-R diagram is used to determine which stars are "giants."

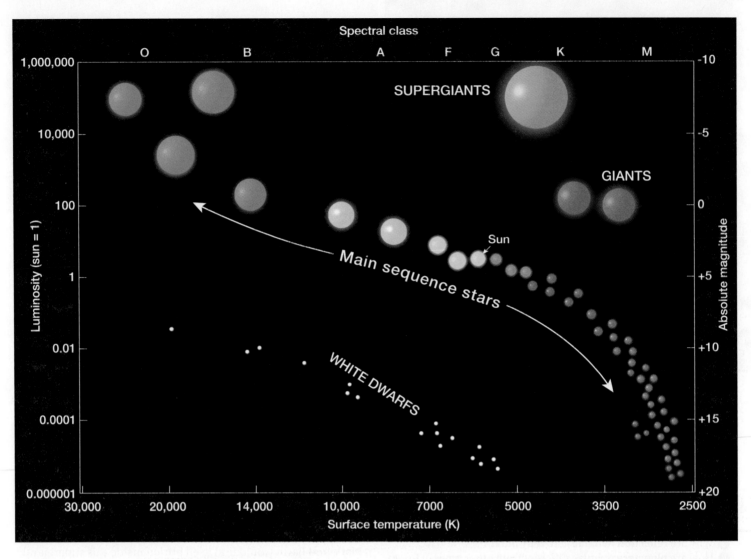

FIGURE 24.8 Idealized Hertzsprung–Russell diagram on which stars are plotted according to temperature and luminosity (absolute magnitude).

Stellar Evolution

The idea of describing how a star is born, ages, and dies may seem a bit presumptuous, for most stars have life spans that exceed billions of years. However, by studying stars of different ages, at different points in their life cycles, astronomers have been able to assemble a model for stellar evolution.

The method used to create this model is analogous to how an alien, upon reaching Earth, might determine the developmental stages of human life. By observing large numbers of humans, this stranger would witness the onset of life, the progression of life in children and adults, and the death of the elderly. From this information, the alien could put the stages of human development into their natural sequence. Based on the relative abundance of humans in each stage of development, it would even be possible to conclude that humans spend more of their lives as adults than as toddlers. Similarly, astronomers have pieced together the life story of stars.

The first stars probably formed about 300 million years after the big bang. The most massive nebulae were the birthplace of the first stars because their immense gravity caused them to collapse quickly. As a result, the stars that formed early in the history of the

universe, often referred to as *first-generation stars*, were very massive. First-generation stars consisted mostly of hydrogen with lesser amounts of helium, the primary elements formed during the big bang. Massive stars have relatively short lifetimes followed by violent explosive deaths. These explosions create the heavier elements and expel them into space. Some of this ejected matter was incorporated into subsequent generations of stars such as our Sun.

Every stage of a star's life is ruled by gravity. The mutual gravitational attraction of particles in a thin, gaseous nebula causes the cloud to collapse on itself. As the cloud is squeezed to unimaginable pressures, its temperature rises, igniting its nuclear furnace, and a star is born. A star is a ball of very hot gases, caught between the opposing forces of gravity trying to contract it and thermal nuclear energy trying to expand it. Eventually, all of a star's nuclear fuel will be exhausted and gravity will prevail, collapsing the stellar remnant into a small, dense body.

Stellar Birth

The birthplaces of stars are interstellar clouds, rich in dust and gases (see Figure 24.4). In the Milky Way, these gaseous clouds are about 92 percent hydrogen, 7 percent helium, and less than 1 percent heavier elements.

If these thin gaseous clouds become sufficiently concentrated, they begin to gravitationally contract (Figure 24.9). One of the mechanisms that triggers star formation is a shock wave from a catastrophic explosion (supernova) of a nearby star. Slow dissipation of thermal energy is also thought to cause nebulae to collapse. Regardless of how the process is initiated, once it begins, mutual gravitational attraction of the particles causes the cloud to contract, pulling every particle toward the center.

As the cloud collapses, gravitational energy (potential energy) is converted into energy of motion, or thermal energy, causing the contracting gases to gradually increase in temperature. When the temperature of these gaseous bodies increases sufficiently they begin to radiate energy in the form of long-wavelength red light. Because these large red objects are not hot enough to engage in nuclear fusion, they are not yet stars. The name **protostar** is applied to these bodies (Figure 24.9).

Protostar Stage

During the protostar stage, gravitational contraction continues, slowly at first, then much more rapidly. This collapse causes the core of the developing star to heat much faster than its outer envelope. When the core reaches a temperature of 10 million K, the pressure within is so intense that groups of four hydrogen nuclei (through a several-step process) fuse into single helium nuclei (see "The Solar Interior" in Chapter 23, p. 682). Astronomers refer to this nuclear reaction, in which hydrogen nuclei are fused into helium, as **hydrogen fusion**.

The immense heat released by hydrogen fusion causes the gases inside stars to move with increased vigor, raising the internal gas pressure. At a certain point, the increased atomic motion produces an outward force (pressure) that balances the inward-directed force of gravity. Upon reaching this balance, the stars become *stable main-sequence stars* (Figure 24.9). In other words, a main-sequence star is one in which the force of gravity, in an effort to squeeze the star into the smallest possible ball, is precisely balanced by gas pressure created by hydrogen fusion in the star's interior.

Main-Sequence Stage

During the main-sequence stage, stars experience minimal changes in size or energy output. Hydrogen is continually being converted into helium, and the energy released maintains the gas pressure sufficiently high to prevent gravitational collapse.

How long can stars maintain this balance? Hot, massive blue stars radiate energy at such an enormous rate that they substantially deplete their hydrogen fuel in only a few million years, approaching the end of their main-sequence stage rapidly. By contrast, the smallest (red) main-sequence stars may take hundreds of billions of years to burn their hydrogen, living practically forever. A yellow star, such as the Sun, typically remains a main-sequence star for about 10 billion years. Because the solar system is about 5 billion years old, the Sun is expected to remain a stable main-sequence star for another 5 billion years.

The average star spends 90 percent of its life as a hydrogen-burning main-sequence star. Once the hydrogen fuel in the star's core is depleted, it evolves rapidly and dies. However, with the

FIGURE 24.9 H-R diagram showing stellar evolution for a star about as massive as the Sun.

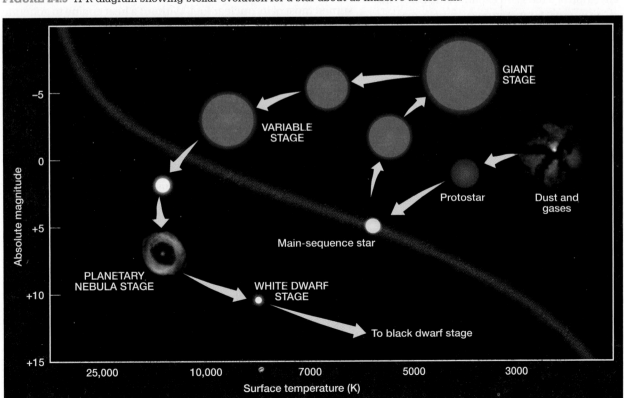

exception of the least-massive stars, death is delayed when another type of nuclear reaction is triggered and the star becomes a *red giant* (Figure 24.9).

Red Giant Stage

Evolution to the red giant stage begins when the usable hydrogen in the star's interior is consumed, leaving a helium-rich core. Although hydrogen fusion is still progressing in the star's outer shell, no fusion is taking place in the core. Without a source of energy, the core no longer has the gas pressure necessary to support itself against the inward force of gravity. As a result, the core begins to contract.

The collapse of the star's interior causes its temperature to rise rapidly as gravitational energy is converted into thermal energy. Some of this energy is radiated outward, initiating a more vigorous level of hydrogen fusion in the shell surrounding the star's core. The additional heat from the accelerated rate of hydrogen fusion expands the star's outer gaseous shell enormously. Sun-like stars become bloated **red giants**, while the most massive stars become *supergiants*, which are thousands of times larger than their main-sequence size.

As the star expands, its surface cools, which explains the star's color—relatively cool objects radiate more energy as long-wavelength radiation (nearer the red end of the visible spectrum). Eventually the star's gravitational force stops this outward expansion and the two opposing forces, gravity and gas pressure, again achieve balance. The star enters a stable state, but is much larger in size. Some red giants overshoot the equilibrium point and instead rebound like an overextended spring. These stars, which alternately expand and contract, never reach equilibrium and are known as **variable stars**.

While the outer envelope of a red giant expands, the core continues to collapse and the internal temperature eventually reaches 100 million K. This astonishing temperature triggers another nuclear reaction in which helium is converted to carbon. At this point, a red giant consumes both hydrogen and helium to produce energy. In stars more massive than the Sun, other thermonuclear reactions occur that generate the elements on the periodic table up to and including number 26, iron.

Burnout and Death

What happens to stars after the red giant phase? We know that stars, regardless of their size, must eventually exhaust their usable nuclear fuel and collapse in response to their immense gravity. Because the gravitational field of a star is dependent on its mass, low-mass stars have fates different than high-mass stars.

Death of Low-Mass Stars Stars less than one-half the mass of the Sun (0.5 solar mass) consume their fuel at relatively low rates (Figure 24.10A). Consequently, there are many small, *cool red stars* that may remain stable for as long as 100 billion years. Because the interiors of low-mass stars never attain sufficiently high temperatures and pressures to fuse helium, their only energy source is hydrogen fusion. Thus, low-mass stars never become bloated red giants. Rather, they remain a stable main-sequence star until they consume their usable hydrogen fuel and collapse into hot, dense *white dwarfs*.

Death of Medium-Mass (Sun-Like) Stars Stars with masses ranging between one-half and eight times that of the Sun have the same evolutionary history (Figure 24.10B). During their red giant phase, Sun-like stars fuse hydrogen and helium fuel at accelerated rates. Once this fuel is exhausted, these stars (like low-mass stars) collapse into Earth-size bodies of great density—white dwarfs. Without a source of nuclear energy, white dwarfs become cooler and dimmer as they continually radiate thermal energy into space.

During their collapse from red giants to white dwarfs, medium mass stars cast off their bloated outer atmosphere, creating an expanding spherical cloud of gas. The remaining hot, central white dwarf heats the gas cloud, causing it to glow. Recall that these spectacular, spherical clouds are called *planetary nebulae* (see Figure 24.6).

Death of Massive Stars In contrast to Sun-like stars, which expire nonviolently, stars exceeding eight solar masses have relatively short life spans and terminate in brilliant explosions called **supernovas** (Figure 24.10C). During supernova events, these stars become millions of times brighter than they were in prenova stages. If a star located near Earth produced such an outburst, its brilliance would surpass that of the Sun. Fortunately for us, supernovae are relatively rare events; none have been observed in our galaxy since the advent of the telescope, although Tycho Brahe and Johannes Kepler each recorded one, about 30 years apart, late in the 16th century. Chinese astronomers recorded an even brighter supernova in A.D. 1054. Today, the remnant of this great outburst is the Crab Nebula, shown in **Figure 24.11**.

A supernova event is triggered when a massive star has consumed most of its nuclear fuel. Without a heat engine to generate the gas pressure required to balance its immense gravitational field, it collapses. This implosion is enormous, resulting in a shock wave that rebounds out from the star's interior. This energetic shock wave blasts the star's outer shell into space, generating the supernova event.

Theoretical work predicts that during a supernova, the star's interior condenses into a incredibly hot object, possibly no larger than 20 kilometers in diameter. These incomprehensibly dense bodies have been named *neutron stars*. Some supernova events are thought to produce even smaller and more intriguing objects called *black holes*. We consider the nature of neutron stars and black holes in the following section on stellar remnants.

CONCEPT CHECK 24.4

❶ What element is the fuel for main-sequence stars?

❷ Describe how main-sequence stars become giants.

❸ Why are less massive stars thought to age more slowly than more massive stars, despite the fact they have much less "fuel"?

❹ List the steps thought to be involved in the evolution of Sun-like stars.

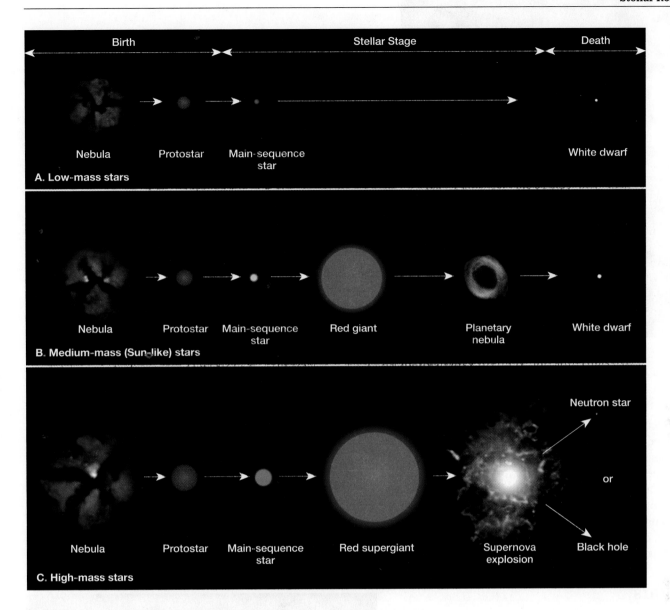

Birth | Stellar Stage | Death

Nebula   Protostar   Main-sequence star   White dwarf

A. Low-mass stars

Nebula   Protostar   Main-sequence star   Red giant   Planetary nebula   White dwarf

B. Medium-mass (Sun-like) stars

Nebula   Protostar   Main-sequence star   Red supergiant   Supernova explosion   Neutron star   or   Black hole

C. High-mass stars

FIGURE 24.10 The evolutionary stages of stars having various masses.

Students Sometimes Ask...

How will the Sun die?

In about 5 billion years, the Sun will exhaust the remaining hydrogen fuel in its core, an event that will trigger hydrogen fusion in the surrounding shell. As a result, the Sun's outer envelope will expand, producing a red giant that is hundreds of times larger and more luminous. The intense solar radiation will cause Earth's oceans to boil, and the solar winds will drive away Earth's atmosphere long before the Sun reaches its largest size and swallows Earth. After another billion years, the Sun will expel its outermost layer, producing a spectacular planetary nebula, while its interior will collapse to form a dense, small (planet-size) white dwarf. The Sun's energy output will be less than 1 percent of its current level because it will have consumed its nuclear fuel. Gradually, the Sun will emit its remaining thermal energy, eventually becoming a cold, nonluminous body.

Stellar Remnants

Eventually, all stars consume their nuclear fuel and collapse into one of three celestial objects— *white dwarfs, neutron stars,* or *black holes.* How a star's life ends, and what final form it takes, depends largely on the star's mass during its main-sequence stage.

White Dwarfs

After low- and medium-mass stars consume their remaining fuel, gravity causes them to collapse into **white dwarfs**. The density of these Earth-sized objects, having masses roughly equal to the Sun, may be a million times greater than water. A spoonful of such matter would weigh several tons on Earth. Densities of this magnitude are possible only when electrons are displaced inward from their regular orbits around an atom's nucleus. Material in this state is called **degenerate matter**.

The atoms in degenerate matter have been squeezed together so tightly that the electrons are pushed very close to the nucleus.

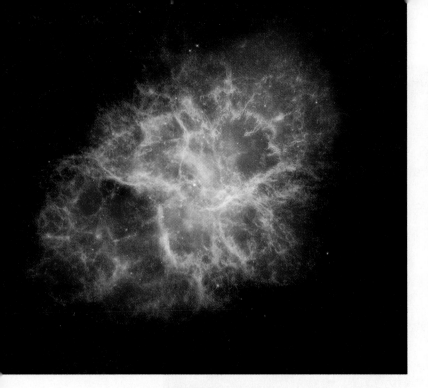

FIGURE 24.11 Crab Nebula in the constellation Taurus, the remains of the supernova of A.D. 1054. (Courtesy of NASA)

However, the electrical repulsion that occurs between the negatively charged electrons supports the star against complete gravitational collapse.

As main-sequence stars contract into white dwarfs, their surfaces become extremely hot, sometimes exceeding 25,000 K. Without sources of energy, main-sequence stars slowly cool and eventually become small, cold, burned-out embers called *black dwarfs*. However, our galaxy is not yet old enough for any black dwarfs to have formed.

Neutron Stars

A study of white dwarfs produced a surprising conclusion. The *smallest white dwarfs* are the *most massive*, and the *largest* are the *least massive*. Researchers have discovered that more massive stars, because of their greater gravitational fields, are squeezed into smaller, more densely packed objects than less massive stars. Thus, the smallest white dwarfs were produced from the collapse of larger, more massive main-sequence stars than are the largest white dwarfs.

This conclusion led to the prediction that stars even *smaller* and *more massive* than white dwarfs must exist. Named **neutron stars**, these objects are the remnants of explosive supernova events. In white dwarfs, the electrons are pushed close to the nucleus, whereas in neutron stars the electrons are forced to combine with protons in the nucleus to produce neutrons (hence the name *neutron star*). A pea-size sample of this matter would weigh 100 million tons. This is approximately the density of an atomic nucleus; thus, a neutron star can be thought of as a large atomic nucleus, composed entirely of neutrons.

During supernova implosions, the outer envelope of the star is ejected, while the core collapses into a very hot star that is only about 20–30 kilometers (12–18 miles) in diameter. Although

neutron stars have high surface temperatures, their small size greatly limits their luminosity, making them nearly impossible to locate visually.

Theoretical models predict that neutron stars have a very strong magnetic field and a high rate of rotation. As stars collapse, they rotate faster, for the same reason ice skaters rotate faster as they pull their arms in as they spin. Radio waves generated by the rotating magnetic fields of neutron stars are concentrated into two narrow zones that align with the star's magnetic poles. Consequently, these stars resemble rapidly rotating beacons emitting strong radio waves. If Earth happened to be in the path of these beacons, the star would appear to blink on and off, or pulsate, as the waves swept past.

In the early 1970s, a source that radiates short pulses of radio energy named a **pulsar** (*pulsating radio source*) was discovered in the Crab Nebula (Figure 24.12). Visual inspection of this radio source indicated that it was coming from a small star centered in the nebula. The pulsar found in the Crab Nebula is most likely the remains of the supernova of A.D. 1054 (see Figure 24.11).

Black Holes

Although neutron stars are extremely dense, they are not the densest objects in the universe. Stellar evolutionary theory predicts neutron stars cannot exceed three times the mass of the Sun. Above this mass, not even tightly packed neutrons can withstand the star's gravitational pull. Following supernova explosions, if the core of a remaining star exceeds the three solar mass limit, gravity prevails over pressure, and the stellar remnant collapses.

FIGURE 24.12 The Crab Pulsar is a young neutron star centered in the crab Nebula. This is the first pulsar to be associated with a supernova. The energy emitted from this star illuminates the Crab Nebula. (Courtesy of NASA)

Box 24.1

EARTH AS A SYSTEM

From Stardust to You

During a supernova implosion, the internal temperature of a star may reach 1 billion K, a condition that likely produces the heaviest elements such as gold and uranium. These elements, plus debris the stars emit during their lifetime, are continually returned to interstellar space where they are available for the formation of the next generation of stars (**Figure 24.A**).

The earliest stars were mostly hydrogen with lesser amounts of helium. Nuclear fusion during the life and death of stars in turn produced heavier elements including carbon, nitrogen, and oxygen, some of which were returned to space. Because the Sun contains some heavy elements but has not yet reached the stage in its evolution where it could have produced them, it must be at least a second-generation star. Thus, our Sun, as well as the rest of the solar system, is believed to have formed from debris scat-

tered from preexisting stars. If this is the case, the atoms in your body were produced billions of years ago inside a star, and the gold in your jewelry was formed during a

supernova event that occurred trillions of kilometers away. Without such events, the development of life on Earth would not have been possible.

FIGURE 24.A Eagle Nebula in the constellation Serpens. This gaseous nebula is the site of ongoing star formation. (Courtesy of National Optical Astronomy Observatories)

(The pre-supernova mass of such a star was likely 25 times that of our Sun.) The incredible objects, or celestial phenomena, created by such a collapse are called **black holes**.

Einstein's theory of general relativity predicts that even though black holes are extremely hot, their surface gravity is so immense that even light cannot escape. Consequently, they literally disappear from sight. Anything that moves too close to a black hole can be swept in and devoured by its immense gravitational field.

How do astronomers find objects whose gravitational field prevents the escape of all matter and energy? Theory predicts that as matter is pulled into a black hole, it should become extremely hot and emit a flood of X-rays before it is engulfed. Because *isolated* black holes do not have a source of matter to engulf, astronomers decided to look at binary-star systems for evidence

of matter emitting X-rays while being rapidly swept into a region of apparent nothingness.

X-rays cannot penetrate our atmosphere; therefore the existence of black holes was not confirmed until the advent of orbiting observatories. The first black hole to be identified, Cygnus X-1, orbits a massive supergiant companion once every 5.6 days. The gases that are pulled from the companion form an *accretion disk* that spirals around a "void" while emitting a steady stream of X-rays (**Figure 24.13**). Recent observations have determined that pairs of jets extend outward from these accretion disks and are thought to return some of this material back to space (Figure 24.13).

Cygnus X-1, which is about eight or nine times as massive as our Sun, probably formed from a star of approximately 40 solar

TABLE 24.1 Summary of Evolution for Stars of Various Masses

Initial Mass of Main-Sequence Star (Sun = 1)*	Main-Sequence Stage	Giant Phase	Evolution After Giant Phase	Terminal State (Final Mass)
0.001	None (Planet)	No	Not applicable	Planet (0.001)
0.1	Red	No	Not applicable	White dwarf (0.1)
1–3	Yellow	Yes	Planetary Nebula	White dwarf (<1.4)
8	White	Yes	Supernova	Neutron star (1.4–3)
25	Blue	Yes (Supergiant)	Supernova	Black hole (>3)

*These mass numbers are estimates.

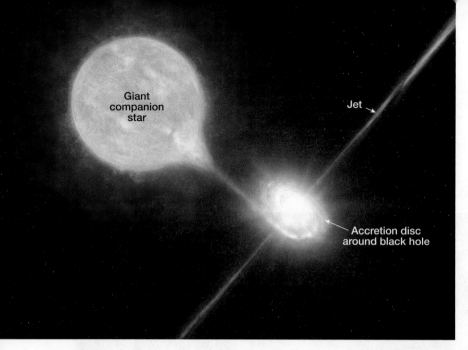

FIGURE 24.13 This illustration shows how a black hole and a giant companion star might appear. Note the accretion disk surrounding the black hole. (Courtesy of European Southern Observatory/L. Calcada/M. Kornmesser)

masses. Following the discovery of Cygnus X-1, many other X-ray sources have been discovered, resulting in the identification of numerous black holes.

Astronomers have established that black holes are common objects in the universe and vary considerably in size. Small black holes have masses approximately 10 times that of our Sun, but are only about 20 miles across, less than the distance of a marathon course. Intermediate black holes have masses 1,000 times our Sun, and the largest black holes (*supermassive black holes*) found in the center of galaxies are estimated to be millions of solar masses. Because the earliest stars were thought to be massive, their deaths could have provided the seeds that eventually formed the supermassive black holes at the centers of galaxies.

CONCEPT CHECK 24.5

1. Describe degenerate matter.
2. What is the final state of a medium-mass (Sun-like) star?
3. How does the "lives" of the most massive stars end? What are the two possible products of this event?
4. Explain how it is possible for the *smallest* white dwarfs to be the *most* massive.
5. Black holes are thought to be very abundant, yet they are hard to find. Explain why.

Galaxies and Galactic Clusters

On a clear and moonless night away from city lights, you can see a truly marvelous sight—a band of light stretching from horizon to horizon. With his tele-

scope, Galileo discovered that this band of light was composed of countless individual stars. Today, we realize that the Sun is actually a part of this vast system of stars, the Milky Way Galaxy (Figure 24.14).

Galaxies (*galaxias* = milky) are collections of interstellar matter, stars, and stellar remnants that are gravitationally bound (Figure 24.14). Recent observational data indicates that supermassive black holes may exist at the center of most galaxies. In addition, spherical halos of very tenuous gas and numerous star clusters (*globular clusters*) appear to surround many of the largest galaxies.

The first galaxies were small and composed mainly of massive stars and abundant interstellar matter. These galaxies grew quickly by accreting nearby interstellar matter and by colliding and merging with other galaxies. In fact, our galaxy is currently absorbing at least two tiny satellite galaxies.

Types of Galaxies

Among the hundreds of billions of galaxies, three basic types have been identified: *spiral, elliptical,* and *irregular.* Within each of these categories there are many variations, the causes of which are still a mystery.

Spiral Galaxies Our Milky Way Galaxy is an example of a large **spiral galaxy** (Figure 24.15). Spiral galaxies are flat, disk-shaped objects that range from 20,000 to about 125,000 light-years in diameter. Typically, spiral galaxies have a greater concentration of stars near their centers, but there are numerous variations. As shown in Figure 24.16, spiral galaxies have arms (usually two) extending from the central nucleus. Spiral galaxies rotate rapidly in the center, while the outermost stars rotate more slowly, which gives these galaxies the appearance of a fireworks pinwheel. Gen-

FIGURE 24.14 View of our Milky Way Galaxy at sunset. The dark patches in the "milky" band of light are caused by the presence of dark nebulae. (Courtesy of European Southern Observatory)

FIGURE 24.15 This dramatic image is of a spiral galaxy named Messier 83. Although smaller, Messier 83 is thought to be very similar to the Milky Way Galaxy. (European Space Observatory)

erally, the central bulge contains older stars that give it a yellowish color while younger hot stars are located in the arms. The young, hot stars in the arms are found in large groups that appear as bright patches of blue and violet light shown in Figure 24.16.

Many spiral galaxies have a band of stars extending outward from the central bulge that merges with the spiral arms. These are known as **barred spiral galaxies** (Figure 24.17). Recent investigations have found evidence that our galaxy probably has a bar structure. What produces these bar-shaped structures is a matter of ongoing research.

Elliptical Galaxies   As the name implies, **elliptical galaxies** have an ellipsoidal shape that can be nearly spherical, and they lack spiral arms (Figure 24.18). Generally, some of the largest and the smallest galaxies are elliptical. Some are so small that they are known as **dwarf galaxies**. The two small companions of Andromeda shown in Figure 24.2 are dwarf elliptical galaxies.

Although many elliptical galaxies are small, the very largest known galaxies (1 million light-years in diameter) are also elliptical. For comparison, the Milky Way is a large spiral galaxy that is one-half that

FIGURE 24.16 Spiral galaxy. **A.** Sprial galaxies typically have a greater concentration of older stars near their center, which gives the central bulge its yellowish color. By contrast, the arms of spiral galaxies contain numerous hot, young stars that give these structures a bluish or violet tint. **B.** Edge on view showing the central bulge. **C.** Surrounding most large galaxies are spherical halos of very tenuous gases and groups of stars called globular clusters. This large globular cluster contains an estimated 10 million stars. (Image **A.** Courtesy of NASA, **B.** and **C.** Courtesy of European Southern Observatory)

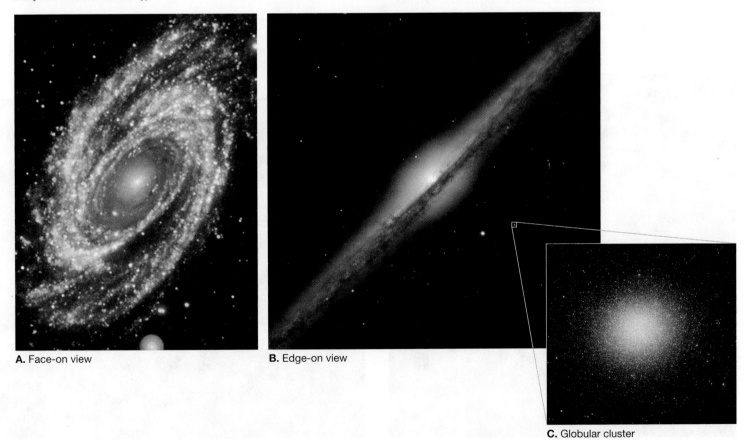

A. Face-on view

B. Edge-on view

C. Globular cluster

FIGURE 24.17 Barred spiral galaxy. (Courtesy of NASA)

diameter. Many large elliptical galaxies are believed to result from the merger of two or more smaller galaxies.

Large elliptical galaxies are generally composed of older, low-mass stars (red) and have minimal amounts of interstellar matter. Thus, unlike the arms of spiral galaxies, they have low rates of star formation. As a result, elliptical galaxies appear yellow to red in color, as compared to the bluish tint emanating from the young, hot stars in the arms of spiral galaxies.

Irregular Galaxies   Approximately 25 percent of known galaxies show no symmetry and are classified as **irregular galaxies**. Some were once spiral or elliptical galaxies that were subsequently distorted by the gravity of a larger neighbor. The best known irregular galaxies, the *Large* and *Small Magellanic Clouds*,

are named for explorer Ferdinand Magellan who observed them when he circumnavigated the globe in 1520. They are among our nearest galactic neighbors.

Recent images of the Large Magellanic Cloud revealed a central bar-shaped structure. Thus, the Large Magellanic Cloud was once a barred spiral galaxy that was subsequently distorted by gravity.

Galactic Clusters

Once astronomers discovered that stars occur in groups (galaxies), they set out to determine whether galaxies also occurred in groups or are just randomly distributed. They found that galaxies are grouped into gravitationally bound clusters (**Figure 24.19**). Some large **galactic clusters** contain thousands of galaxies. Our own galactic cluster, called the **Local Group**, consists of more than 40 galaxies and may contain many undiscovered dwarf galaxies. Of these, three are large spiral galaxies including the Milky Way and Andromeda Galaxies.

Galactic clusters also reside in huge groups called *super-clusters*. There are possibly 10 million superclusters; our Local Group is found in the Virgo Supercluster. From visual observations, it appears that superclusters may be the largest entities in the universe.

Galactic Collisions

Within galactic clusters, interactions between galaxies, often driven by one galaxy's gravity disturbing another, are common.

FIGURE 24.18 A very large elliptical galaxy belonging to the Fornax Cluster. Dark clouds of interstellar matter are visible within the central nucleus of this galaxy. Some of the star-like objects in this image are large groups of stars (globular clusters) that belong to the galaxy. (Courtesy of European Southern Observatory)

FIGURE 24.19 The Fornax Galaxy Cluster is one of the nearest groupings of galaxies to our Local Group. Although many of the galaxies shown are elliptical, an elegant barred spiral galaxy is visible in the lower right. (Courtesy of European Southern Observatory/ J. Emerson/VISTA)

For example, a large galaxy may engulf a dwarf satellite galaxy. In this case the larger galaxy will retain its form, while the smaller galaxy will at least be torn apart and assimilated into the larger galaxy. Recall that two dwarf satellite galaxies are currently merging with the Milky Way.

Galactic interactions may also involve two galaxies of similar size that pass through one another without merging. It is unlikely that the individual stars within these galaxies will collide because they are so widely dispersed. However, the interstellar matter will likely interact, triggering an intense period of star formation.

In an extreme case, two large galaxies may collide and merge into a single large galaxy (Figure 24.20). Many of the largest elliptical galaxies are thought to be the product of the merger of two large spiral galaxies. Some studies have predicted that in 2–4 billion years there is a 50 percent probability that the Milky Way and Andromeda galaxies may collide and merge.

CONCEPT CHECK 24.6

1. Compare the three main types of galaxies.
2. What type of galaxy is our Milky Way?
3. Describe a possible scenario for the formation of a large elliptical galaxy.

Students Sometimes Ask...

Is Earth home to the only life in the universe?

There are *hundreds of billions* of stars in our galaxy *alone*. Because the number of stars and, therefore potential planetary systems, is so large, there is a reasonable possibility that life exists elsewhere. The first search for intelligent life outside of our solar system was conducted in the 1960s by the SETI Institute (Search for Extraterrestrial Intelligence), which used large radio telescopes in Mountain View, California. Currently, NASA's Kepler mission is searching for habitable regions around stars where Earth-like planets may exist. As of 2010, more than 700 potential planets have been found but only a handful have been confirmed by ground-based observations.

The Big Bang Theory

The *big bang theory* describes the birth, evolution, and fate of the universe. According to the big bang theory, the universe was originally in an extremely hot, supermassive state that expanded rapidly in all directions. Based on astronomers' best calculations this expansion began about 13.7 million years ago. What scientific evidence exists to support this theory?

FIGURE 24.20 The collision of the Antennae Galaxies. When two galaxies collide, the stars generally do not. However, the clouds of dust and gas common to both do collide. During these galactic encounters there is a rapid birth of millions of stars, shown as the bright regions in this image. (Courtesy of NASA)

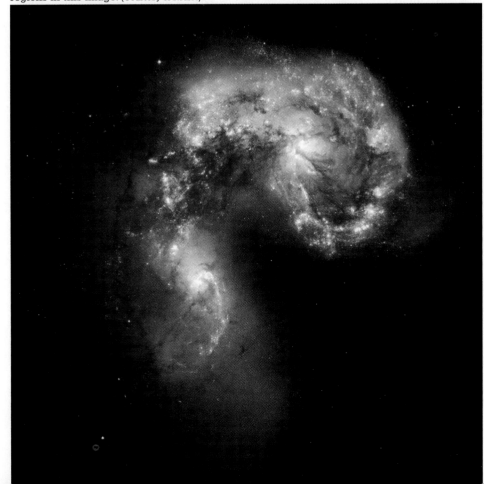

Evidence for an Expanding Universe

In 1912 Vesto Slipher, while working at the Lowell Observatory in Flagstaff, Arizona, was the first to discover that galaxies exhibit motion. The motions he detected were twofold: galaxies rotate and galaxies move relative to each other. Slipher's efforts focused on the shifts in the spectra of the "fuzzy patches," which astronomers soon concluded were galaxies similar to the Milky Way. Recall from Chapter 23 that when a source of light is moving away from an observer the spectral lines shift toward the red end of the spectrum (longer wavelengths). Conversely, when celestial objects approach the observer, the spectral lines shift to the blue end of the spectrum (shorter wavelengths).

In 1929, a study of galaxies conducted by Edwin Hubble expanded the groundwork established by Slipher. Hubble noticed that most galaxies have spectral shifts toward the red end of the spectrum—which occurs when an object emitting light is receding from an observer (Figure 24.21). Therefore, all galaxies (except those in the Local Group) appear to be moving away from the Milky Way. These patterns were later called

Standard spectral lines (unshifted)

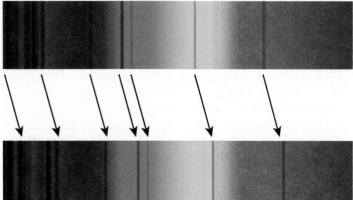

Redshift moves spectral lines to longer wavelengths

FIGURE 24.21 Illustration of the shift in spectral lines toward the red end of the spectrum, which occurs when an object emitting light is receding from an observer.

cosmological red shifts because they were deemed to be the result of the expansion of space.

Using his measurements of galactic distances along with Vesto Slipher's measurements of red shifts, Hubble determined that a relationship existed between the distance to a galaxy and its red shift. Simply stated, Hubble discovered that the red shifts of galaxies increase with distance and that the most distant galaxies are receding the fastest. This concept, now called **Hubble's law**, states that *galaxies recede at speeds proportional to their distances from the observer*.

Hubble was surprised by this discovery, because conventional wisdom was that the universe was unchanging and would likely remain unchanged. What cosmological theory could explain Hubble's findings? Researchers concluded that an expanding universe accounts for the observed red shifts.

To help visualize why Hubble's law implies an expanding universe, imagine a batch of raisin bread dough that has been set out to rise for a few hours (**Figure 24.22**). In this analogy, the raisins represent galaxies and the dough represents space. As the dough doubles in size, so does the distance between all of the raisins.

The distance between raisins that were originally 2 centimeters apart will become 4 centimeters, while the distance between raisins originally 6 centimeters apart will increase to 12 centimeters. The raisins that were originally farthest apart traveled greater distances than those located closer together. Therefore, in an expanding universe, as in our analogy, more space is created between two objects located farther apart than between two objects that are closer together.

Another feature of the expanding universe can be demonstrated using the same bread analogy. Regardless of which raisin you look at, it will move away from all the other raisins. Likewise, at any point in the universe, all other galaxies (except those in the same cluster) are receding from that location. Thus, Hubble's law implies a centerless universe that is expanding uniformly. The Hubble Space Telescope is named in honor of Edwin Hubble's invaluable contributions to the scientific understanding of the universe.

Students Sometimes Ask...

Who was the first person to suggest the idea of a big bang?

In 1931, Georges Lemaître, a Belgian physicist and Roman Catholic priest, was the first to propose what was later dubbed the big bang theory. Lemaître reasoned that since the universe is expanding and galaxies are moving apart, then, at a much earlier time, the universe must have been smaller. Thus, if a hypothetical universal clock could be turned back to the beginning of time—approximately 13.7 billion years ago— the *entire* universe would have been confined to a dense, hot, supermassive *point*. It is from this "primeval atom" that Lemaître suggested the fabric of time and space came into existence. Many scientists contributed to the development of the current big bang theory and others will refine it based on new scientific observations gleaned from increasingly powerful telescopes.

Predictions of the Big Bang Theory

Recall from Chapter 1 in order for a hypothesis to become an accepted component of scientific knowledge (a theory), it must incorporate predictions that can be tested. One prediction of the big bang model is that if the universe was initially unimaginatively hot, then researchers should be able to detect the remnant of that heat. The electromagnetic radiation (light) emitted by a white-hot universe would have extremely high energy and short wavelengths. However, according to the big bang theory, the continued expansion of the universe would have stretched the waves so

FIGURE 24.22 Illustration of the raisin bread analogy for an expanding universe. As the dough rises, raisins originally farthest apart travel greater distances than those located closer together. Thus, in an expanding universe (as with the raisins) more space is created between two objects that are farther apart than between two objects that are closer together.

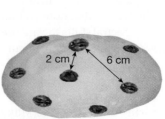

A. Raisin bread dough before it rises.

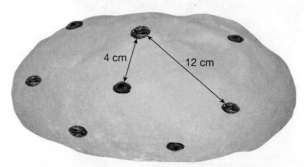

B. Raisin bread dough a few hours later.

that by now they should be detectable as long wavelength radio waves called microwave radiation (see Figure 23.2, p. 667). Scientists began to search for this "missing" radiation, which they named *cosmic microwave background radiation*. As predicted, in 1965 this microwave radiation was detected and found to fill the entire visible universe.

Detailed observations of the cosmic microwave background radiation since its original discovery, have confirmed many theoretical details of the big bang theory, including the order and timing of important events in the early history of the universe.

CONCEPT CHECK 24.7

1 In your own words, explain how astronomers determined that the universe is expanding.

2 What did the big bang theory predict that was finally confirmed years after it was formulated?

The Fate of the Universe

Will the universe come to an end, or will it expand forever? In one scenario, the stars will slowly burn out, being replaced by invisible degenerate matter and black holes that travel outward through an endless, dark, cold universe. Another possibility is that the outward flight of the galaxies will slow and eventually stop. Gravitational contraction would follow, causing all matter to eventually collide and coalesce into the high-energy, high-density state from which the universe began. This fiery death of the universe, the big bang operating in reverse, has been called the "*big crunch*."

Whether the universe will expand forever or eventually collapse upon itself is contingent on its density. If the average density of the universe is more than an amount known as its *critical density* (about one atom for every cubic meter), gravitational attraction would be sufficient to stop the outward expansion and cause the universe to collapse. On the other hand, if the density of the universe is less than the critical value, it will continue to expand forever. Current estimates place the known matter in the universe below the critical density, which predicts an ever expanding or *open universe*. Additional support for an open universe is based on research that indicates the universe is expanding faster now than in the past. Hence, the view currently favored by most cosmologists is an endless, expanding universe.

It should be noted, however, that the methods used to determine the density of the universe have substantial uncertainties. It is possible that previously undetected matter called **dark matter** exists in great quantities in the universe. Although the concept of *dark matter* may sound foreboding, it simply allows for the possibility that matter exists that does not interact with electromagnetic radiation. Recall that most knowledge about the universe comes to us via light. If there is a form of matter that does not interact with light, we cannot "see" it, hence, the term dark matter. However, that is not the end of the story. It is possible that dark matter will show itself by gravitational interactions with types of matter with which we are familiar. Many astronomers are presently looking for these interactions. If there is enough dark matter, the universe could, in fact, collapse in the "big crunch." Consider the following phrase as astronomers search for dark matter—"Absence of evidence is not evidence of absence."

CONCEPT CHECK 24.8

1 Which view of the fate of the universe is currently favored: a "big crunch" or an endless, ever expanding universe?

2 What property does the universe possess that will determine its final state?

Students Sometimes Ask...

I have a hard time buying into the universe starting as a "big bang." Did it really happen?

You're not the first to express such doubt. In fact, cosmologist Fred Hoyle originally coined the term big bang in 1949 as a sarcastic comment about the believability of the theory. After decades of experimentation and observation, scientists have gathered substantial evidence supporting this theory. Despite this overwhelming support, the big bang theory, like all other scientific theories, can never be proven. It is always possible that a future observation will require modification or abandonment of an accepted theory. Nevertheless, the big bang has replaced all alternative theories and remains the only widely accepted scientific model for the origin and evolution of the universe.

GIVE IT SOME THOUGHT

1. Assume NASA is sending a space probe to each of the following locations:
   a. Polaris (the North Star)
   b. A comet near the outer edge of our solar system
   c. Jupiter
   d. The far edge of the Milky Way Galaxy
   e. The near side of the Andromeda Galaxy
   f. The Sun
   List the locations in order, *from nearest to farthest*.

2. Use the information provided below about three main-sequence stars (A, B, and C) to complete the following and explain your reasoning.
   - Star A has a main-sequence life span of 5 billion years.
   - Star B has the same luminosity (absolute magnitude) as the Sun.
   - Star C has a surface temperature of 5,000 K.
   a. Rank the mass of these stars from *greatest to least*.
   b. Rank the energy output of these stars from *greatest to least*.
   c. Rank the main-sequence life span of these stars from *longest to shortest*.

3. The mass of three clouds of gas and dust (nebulae) are provided below. Imagine that each cloud will collapse to form a single star. Use this information to complete the following and explain your reasoning.
   - Cloud A is 60 times the mass of the Sun.
   - Cloud B is 7 times the mass of the Sun.
   - Cloud C is 2 times the mass of the Sun.
   a. Which cloud or clouds, if any, will evolve into a red main-sequence star?
   b. Which of the stars that will form from these clouds, if any, will reach the giant stage?
   c. Which of the stars that will form from these clouds, if any, will go through the supernova stage?

4. Refer to the accompanying images (A, B, C, and D) to complete the following:
   a. Which of these nebulae, if any, is an emission nebula?
   b. Which of these nebulae, if any, formed near the end of a star's lifetime?
   c. Which of these nebulae, if any, is a reflection nebula?

5. The accompanying photo shows the Trifid Nebula, which can be easily observed with a small telescope. What unique properties does this nebula exhibit?

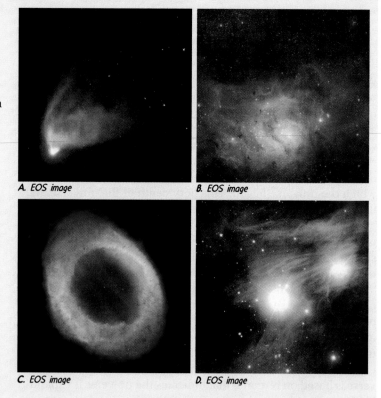

A. EOS image   B. EOS image

C. EOS image   D. EOS image

EOS image

6. How a star evolves is closely related to its mass as a main-sequence star. Complete the accompanying diagram by labeling the evolutionary stages for the three groups of main-sequence stars shown.

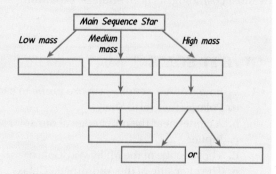

7. Refer to the accompanying photos of an elliptical galaxy and a spiral galaxy to complete the following.
   a. Which image (A or B) is an elliptical galaxy?
   b. Which of these galaxies appears to contain more young, hot massive stars? How did you determine your answer?
   c. When stars are born from a cloud of dust and gases, both large and small stars form at about the same time. Which group of stars, the large or the small, will die out first? Over time, how will this affect the color of the light we observe coming from this group of stars? Based on your response, which of these galaxies appears to be older? Explain.
8. Consider these three characteristics of the universe.
   - It does not have a center.
   - It does not have edges.
   - Its galaxies are all moving away from each other.
   a. Which of the three characteristics of the universe is/are depicted in the raisin bread dough analogy (Figure 24.22)?
   b. Which characteristics is/are not accurately depicted?

A. EOS image   B. NASA

In Review   Chapter 24 Beyond our Solar System

- *Cosmology* is the study of the universe, including its properties, structure, and evolution. The universe consists of hundreds of billions of galaxies, each containing billions of stars. The model that most accurately describes the birth and current state of the universe is the *big bang theory*. According to this model, the universe began about 13.7 billion years ago in a cataclysmic explosion, then continued to expand, cool, and evolve to its current state.

- *Hertzsprung–Russell diagrams* are constructed by plotting the absolute magnitudes and temperatures of stars on graphs. Significant information about stars and stellar evolution has been discovered through the use of H-R diagrams. Stars are positioned within H-R diagrams as follows: (1) 90 percent of all stars, *main-sequence stars*, are in the band that runs from the upper-left corner (massive, hot blue stars) to the lower-right corner (low mass, red stars); (2) *red giants* and *supergiants*, very luminous stars with large radii, are located in the upper-right position; and (3) *white dwarfs*, small, dense stars, are located in the lower portion.

- New stars are born out of enormous accumulations of dust and gases, called *nebulae*, which are scattered between existing stars. *Bright nebulae* glow because they are in close proximity to very hot (blue) stars. *Emission nebulae* derive their visible light from nearby hot stars or stars that are imbedded in them. *Reflection nebulae* contain comparatively large debris including grains of carbon compounds that reflect the light of nearby stars. *Planetary nebulae* consist of glowing clouds of dust and hot gases that have been expelled near the end of a star's life. Nebulae that are too distant from bright stars to be illuminated are referred to as *dark nebulae*.

- Stars are born when their nuclear furnaces are ignited by the unimaginable pressures and temperatures generated during the collapse of nebulae. Red star-like objects not yet hot enough for nuclear fusion are called *protostars*. When the core of a protostar reaches a temperature of about10 million K, a process called *hydrogen fusion* begins in which hydrogen nuclei fuse into a single helium nuclei. Two opposing forces acting on stars are *gravity*—trying to contract it—and *gas pressure* (*thermal nuclear energy*)—trying to expand it. When the forces are balanced, the star becomes a stable *main-sequence star*. Medium- and high-mass stars experience another type of nuclear fusion that causes their outer envelopes to expand enormously (hundreds to thousands of times larger), making them *red giants* or *supergiants*. When all the usable nuclear fuel in these giants is exhausted, gravity takes over and the stellar remnant collapses into a small, dense body.

- The *final fate of a star is determined by its mass*. Stars with less than one-half the mass of the Sun collapse into hot, dense *white dwarfs*. Medium-mass stars, like the Sun, become red giants, collapse, and end up as white dwarfs, often surrounded by expanding spherical clouds of glowing gas called *planetary nebulae*. Massive stars terminate in a brilliant explosion called a *supernova*. Supernovae events can produce small, extremely dense *neutron stars*, composed entirely of neutrons; or smaller, more dense *black holes*— objects that have such immense gravity that light cannot escape their surface.

- The various types of galaxies include (1) *irregular galaxies*, which lack symmetry and account for about 25 percent of the known galaxies; (2) *spiral galaxies*, which are disk-shaped with a greater concentration of stars near their centers, and arms extending from their central nucleus; and (3) *elliptical galaxies*, which have an ellipsoidal shape that may be nearly spherical.

- Galaxies are grouped in *galactic clusters*, some containing thousands of galaxies. Our own, called the *Local Group*, contains at least 40 galaxies.
- Evidence for an expanding universe came from the study of red shifts in the spectra of galaxies. Edwin Hubble concluded

that the observed red shifts, later called *cosmological red shifts*, were the result of the expansion of space. This evidence strongly supports the big bang model of an expanded universe. One question that remains is whether the universe will expand forever or gravitationally contract in the "big crunch."

Key Terms

absolute magnitude (p. 688)
barred spiral galaxy (p. 701)
big bang theory (p. 689)
black hole (p. 699)
bright nebulae (p. 691)
cosmological red shift (p. 704)
cosmology (p. 688)
dark matter (p. 705)
dark nebulae (p. 691)
degenerate matter (p. 697)
dwarf galaxy (p. 701)
elliptical galaxy (p. 701)

emission nebulae (p. 691)
galactic cluster (p. 702)
Hertzsprung–Russell (H-R) diagram (p. 693)
Hubble's law (p. 704)
hydrogen fusion (p. 695)
interstellar matter (p. 691)
irregular galaxy (p. 702)
light-year (p. 689)
Local Group (p. 702)
main-sequence stars (p. 693)
nebulae (p. 691)
neutron star (p. 698)

planetary nebulae (p. 692)
protostar (p. 695)
pulsar (p. 698)
red giant (p. 693)
reflection nebulae (p. 692)
spiral galaxy (p. 700)
supergiants (p. 693)
supernova (p. 696)
variable stars (p. 696)
white dwarfs (p. 697)

Examining the Earth System

1. Briefly describe how the atmosphere, hydrosphere, geosphere, and biosphere are each related to the death of stars that occurred billions of years ago.

2. If a supernova explosion were to occur within the immediate vicinity of our solar system, what might be some possible consequences of the intense X-ray and gamma radiation that would reach Earth?

3. Scientists are continuously searching the Milky Way Galaxy for other stars that may have planets. What types of stars would most likely have a planet or planets suitable for life as we know

it? If you would like to investigate extra-solar planets online, you might find these two Websites helpful: *NOVA Online* at **http://www.pbs.org/wgbh/nova/worlds/** and the *Electronic Universe Project* at **http://zebu.uoregon.edu/galaxy.html**

4. Based on your knowledge of the Earth system, the planets in our solar system, and the cosmos in general, speculate about the likelihood that extra-solar planets exist with atmospheres, hydrospheres, geospheres, and biospheres similar to Earth's. Explain your speculation.

Mastering Geology

Looking for additional review and test prep materials? Visit the Self Study area in **www.masteringgeology.com** to find practice quizzes, study tools, and multimedia that will aid in your understanding of this chapter's content. In **MasteringGeology™** you will find:

- **GEODe: Earth Science: An interactive visual walkthrough of key concepts**

- **Geoscience Animation Library: More than 100 animations illuminating many difficult-to-understand Earth science concepts**
- **In The News RSS Feeds: Current Earth science events and news articles are pulled into the site with assessment**
- **Pearson eText**
- **Optional Self Study Quizzes**
- **Web Links**
- **Glossary**
- **Flashcards**

Metric and English Units Compared

Units

1 kilometer (km)	= 1,000 meters (m)
1 meter (m)	= 100 centimeters
1 centimeter (cm)	= 0.39 inch (in.)
1 mile (mi)	= 5,280 feet (ft)
1 foot (ft)	= 12 inches (in.)
1 inch (in.)	= 2.54 centimeters (cm)
1 square mile (mi^2)	= 640 acres (a)
1 kilogram (kg)	= 1,000 grams (g)
1 pound (lb)	= 16 ounces (oz)
1 fathom	= 6 feet (ft)

Conversions

Length

When you want to convert:	multiply by:	to find:
inches	2.54	centimeters
centimeters	0.39	inches
feet	0.30	meters
meters	3.28	feet
yards	0.91	meters
meters	1.09	yards
miles	1.61	kilometers
kilometers	0.62	miles

Area

When you want to convert:	multiply by:	to find:
square inches	6.45	square centimeters
square centimeters	0.15	square inches
square feet	0.09	square meters
square meters	10.76	square feet
square miles	2.59	square kilometers
square kilometers	0.39	square miles

Volume

When you want to convert:	multiply by:	to find:
cubic inches	16.38	cubic centimeters
cubic centimeters	0.06	cubic inches
cubic feet	0.028	cubic meters
cubic meters	35.3	cubic feet
cubic miles	4.17	cubic kilometers
cubic kilometers	0.24	cubic miles
liters	1.06	quarts
liters	0.26	gallons
gallons	3.78	liters

Masses and Weights

When you want to convert:	multiply by:	to find:
ounces	28.35	grams
grams	0.035	ounces
pounds	0.45	kilograms
kilograms	2.205	pounds

Temperature

When you want to convert degrees Fahrenheit (°F) to degrees Celsius (°C), subtract 32 degrees and divide by 1.8.

When you want to convert degrees Celsius (°C) to degrees Fahrenheit (°F), multiply by 1.8 and add 32 degrees.

When you want to convert degrees Celsius (°C) to kelvins (K), delete the degree symbol and add 273. When you want to convert kelvins (K) to degrees Celsius (°C), add the degree symbol and subtract 273.

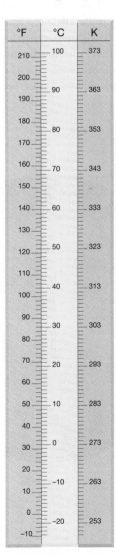

FIGURE A.1
Temperature scales.

APPENDIX B
Earth's Grid System

A glance at any globe reveals a series of north–south and east–west lines that together make up Earth's grid system, a universally used scheme for locating points on Earth's surface. The north–south lines of the grid are called **meridians** and extend from pole to pole (**Figure B.1**). All are halves of great circles. A **great circle** is the largest possible circle that may be drawn on a globe; if a globe were sliced along one of these circles, it would be divided into two equal parts called **hemispheres**. By viewing a globe or Figure B.1, you can see that meridians are spaced farthest apart at the equator and converge toward the poles. The east–west lines (circles) of the grid are called **parallels**. As their name implies, these circles are parallel to one another (Figure B.1). Whereas all meridians are parts of great circles, all parallels are not. In fact, only one parallel, the equator, is a great circle.

Latitude and Longitude

Latitude may be defined as distance, measured in degrees, *north* and *south* of the equator. Parallels are used to show latitude. Because all points that lie along the same parallel are an identical distance from the equator, they all have the same latitude designation. The latitude of the equator is 0 degrees, whereas the north and south poles lie 90 degrees N and 90 degrees S, respectively.

FIGURE B.1 Earth's grid system.

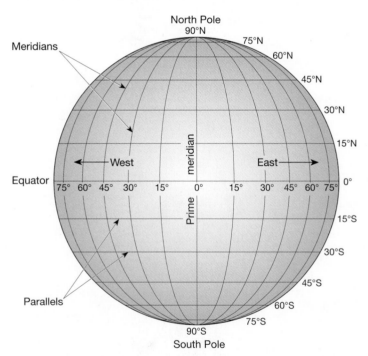

Longitude is defined as distance, measured in degrees, *east* and *west* of the zero or *prime meridian*. Because all meridians are identical, the choice of a zero line is obviously arbitrary. However, the meridian that passes through the Royal Observatory at Greenwich, England, is universally accepted as the reference meridian. Thus, the longitude for any place on the globe is measured east or west from this line. Longitude can vary from 0 degrees along the prime meridian to 180 degrees, halfway around the globe.

It is important to remember that when a location is specified, directions must be given—that is, north or south latitude and east or west longitude (**Figure B.2**). If this is not done, more than one point on the globe is being designated. The only exceptions, of course, are places that lie along the equator, the prime meridian, or the 180-degree meridian. It should also be noted that while it is not incorrect to use fractions, a degree of latitude or longitude is usually divided into minutes and seconds. A minute (′) is 1/60th of a degree, and a second (″) is 1/60th of a minute. When locating a place on a map, the degree of exactness will depend on the scale of the map. When using a small-scale world map or globe, it may be difficult to estimate latitude and longitude to the nearest whole degree or two. On the other hand, when a large-scale map of an area is used, it is often possible to estimate latitude and longitude to the nearest minute or second.

Distance Measurement

The length of a degree of longitude depends on where the measurement is taken. Along the equator, which is a great circle, a degree of east–west distance is equal to approximately 111 kilometers (69 miles). This figure is found by dividing Earth's circumference—40,075 kilometers (24,900 miles)—by 360. However, with an increase in latitude, the parallels become smaller, and the length of a degree of longitude diminishes (see **Table B.1**). Thus, at about latitude 60 degrees N and S, a degree of longitude has a value equal to about half of what it was at the equator.

As all meridians are halves of great circles, a degree of latitude is equal to about 111 kilometers (69 miles), just as a degree of longitude along the equator is. However, Earth is not a perfect sphere but is slightly flattened at the poles and bulges slightly at the equator. Because of this, there are small differences in the length of a degree of latitude.

Determining the shortest distance between two points on a globe can be done easily and fairly accurately using the globe-and-string method. It should be noted here that the arc of a great circle is the shortest distance between two points on a sphere. To determine the great circle distance (as well as observe the great circle route) between two places, stretch a piece of string between the locations in question. Then, measure the length of the string along the equator (since it is a great circle with degrees marked on it) to determine the number of degrees between the two points. To calculate the distance in kilometers or miles, simply multiply the number of degrees by 111 or 69, respectively.

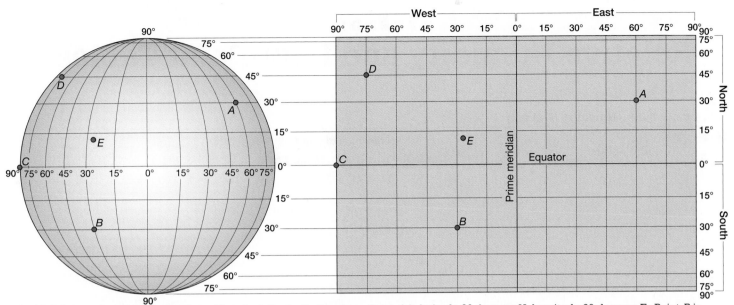

FIGURE B.2 Locating places using the grid system. For both diagrams: Point *A* is latitude 30 degrees N, longitude 60 degrees E; Point *B* is latitude 30 degrees S, longitude 30 degrees W; point *C* is latitude 0 degrees, longitude 90 degrees W; Point *D* is latitude 45 degrees N, longitude 75 degrees W; Point *E* is approximately latitude 10 degrees N, longitude 25 degrees W.

TABLE B.1 Longitude as Distance

Length of 1° Longitude			Length of 1° Longitude			Length of 1° Longitude		
° Latitude	km	miles	° Latitude	km	miles	° Latitude	km	miles
0	111.367	69.172	30	96.528	59.955	60	55.825	34.674
1	111.349	69.161	31	95.545	59.345	61	54.131	33.622
2	111.298	69.129	32	94.533	58.716	62	52.422	62.560
3	111.214	69.077	33	93.493	58.070	63	50.696	31.488
4	111.096	69.004	34	92.425	57.407	64	48.954	30.406
5	110.945	68.910	35	91.327	56.725	65	47.196	29.314
6	110.760	68.795	36	90.203	56.027	66	45.426	28.215
7	110.543	68.660	37	89.051	55.311	67	43.639	27.105
8	110.290	68.503	38	87.871	54.578	68	41.841	25.988
9	110.003	68.325	39	86.665	53.829	69	40.028	24.862
10	109.686	68.128	40	85.431	53.063	70	38.204	23.729
11	109.333	67.909	41	84.171	52.280	71	36.368	22.589
12	108.949	67.670	42	82.886	51.482	72	34.520	21.441
13	108.530	67.410	43	81.575	50.668	73	32.662	20.287
14	108.079	67.130	44	80.241	49.839	74	30.793	19.126
15	107.596	66.830	45	78.880	48.994	75	28.914	17.959
16	107.079	66.509	46	77.497	48.135	76	27.029	16.788
17	106.530	66.168	47	76.089	47.260	77	25.134	15.611
18	105.949	65.807	48	74.659	46.372	78	23.229	14.428
19	105.337	65.427	49	73.203	45.468	79	21.320	13.242
20	104.692	65.026	50	71.727	44.551	80	19.402	12.051
21	104.014	64.605	51	70.228	43.620	81	17.480	10.857
22	103.306	64.165	52	68.708	42.676	82	15.551	9.659
23	102.565	63.705	53	67.168	41.719	83	13.617	8.458
24	101.795	63.227	54	65.604	40.748	84	11.681	7.255
25	100.994	62.729	55	64.022	39.765	85	9.739	6.049
26	100.160	62.211	56	62.420	38.770	86	7.796	4.842
27	99.297	61.675	57	60.798	37.763	87	5.849	3.633
28	98.405	61.121	58	59.159	36.745	88	3.899	2.422
29	97.481	60.547	59	57.501	35.715	89	1.950	1.211
30	96.528	59.955	60	55.825	34.674	90	0.000	0.000

APPENDIX C
Relative Humidity and Dew-Point Tables*

TABLE C.1 Relative Humidity (Percent)

Dry bulb (°C)	Depression of Wet-Bulb Temperature (Dry-Bulb Temperature Minus Wet Bulb of the Wet Bulb)																					
	1	2	3	4	5	6	7	8	9	10	11	12	13	14	15	16	17	18	19	20	21	22
−20	28																					
−18	40																					
−16	48	0																				
−14	55	11																				
−12	61	23																				
−10	66	33	0																			
−8	71	41	13																			
−6	73	48	20	0																		
−4	77	54	32	11																		
−2	79	58	37	20	1																	
0	81	63	45	28	11																	
2	83	67	51	36	20	6																
4	85	70	56	42	27	14																
6	86	72	59	46	35	22	10	0														
8	87	74	62	51	39	28	17	6														
10	88	76	65	54	43	38	24	13	4													
12	88	78	67	57	48	38	28	19	10	2												
14	89	79	69	60	50	41	33	25	16	8	1											
16	90	80	77	62	54	45	37	29	21	14	7	1										
18	91	81	72	64	56	48	40	33	26	19	12	6	0									
20	91	82	74	66	58	51	44	36	30	23	17	11	5									
22	92	83	75	68	60	53	46	40	33	27	21	15	10	4	0							
24	92	84	76	69	62	55	49	42	36	30	25	20	14	9	4	0						
26	92	85	77	70	64	57	51	45	39	34	28	23	18	13	9	5						
28	93	86	78	71	65	59	53	45	42	36	31	26	21	17	12	8	4					
30	93	86	79	72	66	61	55	49	44	39	34	29	25	20	16	12	8	4				
32	93	86	80	73	68	62	56	51	46	41	36	32	27	22	19	14	11	8	4			
34	93	86	81	74	69	63	58	52	48	43	38	34	30	26	22	18	14	11	8	5		
36	94	87	81	75	69	64	59	54	50	44	40	36	32	28	24	21	17	13	10	7	4	
38	94	87	82	76	70	66	60	55	51	46	42	38	34	30	26	23	20	16	13	10	7	5
40	94	89	82	76	71	67	61	57	52	48	44	40	36	33	29	25	22	19	16	13	10	7

To determine the relative humidity, find the air (dry-bulb) temperature on the vertical axis (far left) and the depression of the wet bulb on the horizontal axis (top). Where the two meet, the relative humidity is found. For example, when the dry-bulb temperature is 20° C and a wet-bulb temperature is 14° C, then the depression of the wet bulb is 6° C (20° C −14° C). From Table C.1, the relative humidity is 51 percent and from Table C.2, the dew point is 10° C.

TABLE C.2 Dew-point Temperature (°C)

Dry bulb (°C)	(Dry-Bulb Temperature Minus Wet-Bulb Temperature = Depression of the Wet Bulb)																					
	1	2	3	4	5	6	7	8	9	10	11	12	13	14	15	16	17	18	19	20	21	22
−20	−33																					
−18	−28																					
−16	−24																					
−14	−21	−36																				
−12	−18	−28																				
−10	−14	−22																				
−8	−12	−18	−29																			
−6	−10	−14	−22																			
−4	−7	−12	−17	−29																		
−2	−5	−8	−13	−20																		
0	−3	−6	−9	−15	−24																	
2	−1	−3	−6	−11	−17																	
4	1	−1	−4	−7	−11	−19																
6	4	1	−1	−4	−7	−13	−21															
8	6	3	1	2	−5	−9	−14															
10	8	6	4	1	−2	−5	−9	−14	−18													
12	10	8	6	4	1	−2	−5	−9	−16													
14	12	11	9	6	4	1	−2	−5	−10	−17												
16	14	13	11	9	7	4	1	−1	−6	−10	−17											
18	16	15	13	11	9	7	4	2	−2	5	10	−19										
20	19	17	15	14	12	10	7	4	2	−2	−5	−10	−19									
22	21	19	17	16	14	12	10	8	5	3	−1	−5	−10	−19								
24	23	21	20	18	16	14	12	10	8	6	2	−1	−5	−10	−18							
26	25	23	22	20	18	17	15	13	11	9	6	3	0	−4	−9	−18						
28	27	25	24	22	27	19	17	16	14	11	9	7	4	1	−3	−9	16					
30	29	27	26	24	23	21	19	18	16	14	12	10	8	5	1	−2	−8	−15				
32	31	29	28	27	25	24	22	21	19	17	15	13	11	8	5	2	−2	−7	−14			
34	33	31	30	29	27	26	24	23	21	20	18	16	14	12	9	6	3	−1	−5	−12	−29	
36	35	33	32	31	29	28	27	25	24	22	20	19	17	15	13	10	7	4	0	−4	−10	
38	37	35	34	33	32	30	29	28	26	25	23	21	19	17	15	13	11	8	5	1	−3	9
40	39	37	36	35	34	32	31	30	28	27	25	24	22	20	18	16	14	12	9	6	2	−2

Dry-Bulb (Air) Temperature

Dew-Point Values

Stellar Properties

Measuring Distances to the Closest Stars

Measuring the distance to stars is difficult. Nevertheless, astronomers have developed some direct as well as indirect methods to measure stellar distances. One simple measurement, called *stellar parallax*, is effective in determining the distances to only the closest stars.

Stellar parallax is the slight back-and-forth shift of the apparent position of a nearby star due to the orbital motion of Earth around the Sun. The principle of parallax is easy to visualize. Close one eye, and with your index finger in a vertical position, use your open eye to align your finger with some distant object. Without moving your finger, view the object with your other eye and notice that its position appears to have changed. Now repeat the exercise holding your finger farther away, and notice that the farther away you hold your finger, the less its position seems to shift. In principle, this method of measuring stellar distances is elementary and was practiced by ancient Greek astronomers.

Modern cosmologists determine parallax by photographing a nearby star against the background of distant stars. Then, when Earth has moved halfway around its orbit six months later, the same star is photographed again. When these two photographs are compared, the position of the nearby star appears to have shifted with respect to the background stars. **Figure D.1** illustrates this shift and the parallax angle is determined from it. The nearest stars have the largest parallax angles, whereas those of distant stars are much too small to measure.

In practice, conducting parallax measurements are quite complex because of the miniscule angles being measured. The process is further complicated because both the Sun and the star being measured are moving relative to each other. The first accurate stellar parallax was not determined until 1838. Even today, parallax angles for only a few thousand of the nearest stars are known with certainty—nearly all others have such small parallax shifts that accurate measurements are not possible. Fortunately, other methods have been developed to estimate distances to more distant stars. In addition, the Hubble Space Telescope, which is not hindered by Earth's light-distorting atmosphere, has obtained accurate parallax distances for many more stars.

Stellar Brightness

The oldest means of classifying stars is based on their *brightness*, also called *luminosity* or *magnitude*. Three factors control the brightness of a star as seen from Earth: *how large* it is, *how hot* it is, and its *distance* from Earth. The stars in the night sky come in a grand assortment of sizes, temperatures, and distances, so their apparent brightness varies widely.

Apparent Magnitude

Stars have been classified according to their apparent brightness since at least the second century B.C., when Hipparchus placed about 850 of them into six categories based on his ability to see differences in brightness. Because he could only reliably see six different brightness levels, he created six categories. These categories were later called *magnitudes*, with first magnitude being the brightest and sixth magnitude the dimmest. Because some stars may appear dimmer than others only because they are farther away, a star's brightness, *as it appears when viewed from Earth*, is called its *apparent magnitude*. With the invention of the telescope, many stars fainter than the sixth magnitude were discovered.

In the mid-1800s, a method was developed to standardize the magnitude scale. An absolute comparison was made between the light coming from stars of the first magnitude and those of the sixth magnitude.

FIGURE D.1 Geometry of stellar parallax. The parallax angle shown here is enormously exaggerated to illustrate the principle. Because distances to even the nearest stars are thousands of times greater than the Earth–Sun distance, the triangles that astronomers work with are extremely long and narrow, making the angles that are measured very small.

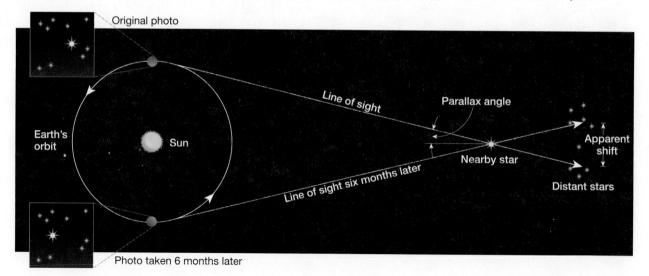

TABLE D.1 Ratios of Star Brightness

Difference in Magnitude	Brightness Ratio
0.5	1.6:1
1	2.5:1
2	6.3:1
3	16:1
4	40:1
5	100:1
10	10,000:1
20	100,000,000:1

*Calculations: 2.512 × 2.512 × 2.512 × 2.512 × 2.512 or 2.512 raised to the fifth power, equals 100.

It was determined that a first-magnitude star was about 100 times brighter than a sixth-magnitude star. On the scale that was devised, any two stars that differ by five magnitudes have a ratio in brightness of 100 to 1. Hence, a third-magnitude star is 100 times brighter than an eighth-magnitude star. It follows, then, that the brightness ratio of two stars differing by only one magnitude is about 2.5.* A star of the first magnitude is about 2.5 times brighter than a star of the second magnitude. **Table D.1** shows how differences in magnitude correspond to brightness ratios.

Because some celestial bodies are brighter than first-magnitude stars, zero and negative magnitudes were introduced. On this scale, the Sun has an apparent magnitude of –26.7. At its brightest, Venus has a magnitude of –4.3. At the other end of the scale, the Hubble Space Telescope can view stars with an apparent magnitude of 30, more than one billion times dimmer than stars that are visible to the unaided eye.

Absolute Magnitude

Apparent magnitudes were good approximations of the true brightness of stars when astronomers thought that the universe was very small—containing no more than a few thousand stars that were all at very similar distances from Earth. However, we now know that the universe is unimaginably large and contains innumerable stars at wildly varying distances. Since astronomers are interested in the "true" brightness of stars, they devised a measure called *absolute magnitude.*

Stars of the same apparent magnitude usually do not have the same brightness because their distances from us are not equal. Astronomers correct for distance by determining what brightness (magnitude) the stars would have if they were at a standard distance—about 32.6 light-

*The more negative, the brighter; the more positive, the dimmer.

years. For example, if the Sun, which has an apparent magnitude of –26.7, was located 32.6 light-years from Earth, it would have an absolute magnitude of about +5. Thus, stars with absolute magnitudes greater than 5 (smaller numerical value) are intrinsically brighter than the Sun, but appear much dimmer because of their distance from Earth. **Table D.2** lists the absolute and apparent magnitudes of some stars as well as their distances from Earth. Most stars have an absolute magnitude between –5 (very bright) and 15 (very dim). The Sun is near the midpoint of this range.

Stellar Color and Temperature

The next time you are outside on a clear night, look carefully at the stars and note their colors (**Figure D.2**). Because human eyes do not respond well to color in low-intensity light (when it is very dark, we only see in black and white), we tend to look at the brightest stars. Some that are quite colorful can be found in the constellation Orion. Of the two brightest stars in Orion, Rigel (β Orionis) appears blue, whereas Betelgeuse (α Orionis) is definitely red.

Very hot stars with surface temperatures above 30,000 K emit most of their energy in the form of short-wavelength light and therefore appear blue. On the other hand, cooler red stars, with surface temperatures generally less than 3,000 K, emit most of their energy as longer-wavelength red light. Stars, such as the Sun, with surface temperatures between 5,000 and 6,000 K, appear yellow. Because color is primarily a manifestation of a star's surface temperature, this characteristic provides astronomers with useful information. As you will see, combining temperature data with stellar magnitude tells us a great deal about the size and mass of stars.

Binary Stars and Stellar Mass

One of the night sky's best-known constellations, the Big Dipper, appears to consist of seven stars. But those with good eyesight can recognize that the second star in the handle is actually two stars. In the early 19th century, careful examination of numerous star pairs by William Herschel showed that many stars found in pairs actually orbit one another. In such cases, the two stars are in fact united by their mutual gravitation. These pairs, in which the members are far enough apart to be telescopically identified as two stars, are called *visual binaries* (*binaries* = double). The idea of one star orbiting another may seem unusual, but many stars in the universe exist in pairs or multiples.

Binary stars can be used to determine the star property most difficult to calculate—its *mass*. The mass of a body can be established if it is gravitationally attached to a partner. Binary stars orbit each other around a common point called the *center of mass* (**Figure D.3**). For stars of equal mass, the center of mass lies exactly halfway between them. When one star is more massive than its partner, their common center

TABLE D.2 Distance, Apparent Magnitude, and Absolute Magnitude of Some Stars

Name	Distance (light-years)	Apparent Magnitude*	Absolute Magnitude*
Sun	NA	−26.7	5.0
Alpha Centauri	4.27	0.0	4.4
Sirius	8.70	−1.4	1.5
Arcturus	36	−0.1	−0.3
Betelgeuse	520	0.8	−5.5
Deneb	1600	1.3	−6.9

FIGURE D.2 Time-lapse photograph of stars in the constellation Orion. These star trails show some of the various star colors. It is important to note that the human eye sees color somewhat differently than photographic film. (Courtesy of National Optical Astronomy Observatories)

will be located closer to the more massive one. Thus, if the sizes of their orbits can be observed, their individual masses can be determined. You can experience this relationship on a seesaw by trying to balance a person who has a much greater (or smaller) mass.

For illustration, when one star has an orbit half the size (radius) of its companion, it is twice as massive as its companion. If their combined masses are equal to three solar masses, then the larger will be twice as massive as the Sun, and the smaller will have a mass equal to that of the Sun. Most stars have a mass that falls in a range between 1/10 and 50 times the mass of the Sun.

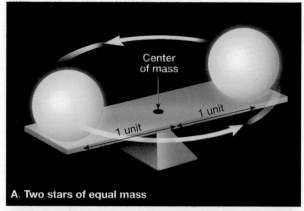

A. Two stars of equal mass

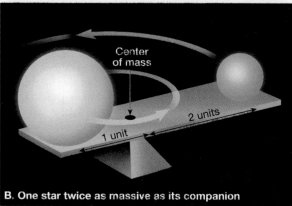

B. One star twice as massive as its companion

FIGURE D.3 Binary stars orbit each other around their common center of mass. **A.** For stars of equal mass, the center of mass lies exactly halfway between them. **B.** If one star is twice as massive as its companion, it is twice as close to their common center of mass. Therefore, more massive stars have proportionately smaller orbits than their less massive companions.

Glossary

Aa flow A type of lava flow that has a jagged, blocky surface.

Abrasion The grinding and scraping of a rock surface by the friction and impact of rock particles carried by water, wind, or ice.

Absolute humidity The weight of water vapor in a given volume of air (usually expressed in GRAMS/M³).

Absolute instability Air that has a lapse rate greater than the dry adiabatic rate.

Absolute magnitude The apparent brightness of a star if it were viewed from a distance of 10 parsecs (32.6 light-years). Used to compare the true brightness of stars.

Absolute stability Air with a lapse rate less than the wet adiabatic rate.

Absorption spectrum A continuous spectrum with dark lines superimposed.

Abyssal plain Very level area of the deep-ocean floor, usually lying at the foot of the continental rise.

Abyssal zone A subdivision of the benthic zone characterized by extremely high pressures, low temperatures, low oxygen, few nutrients, and no sunlight.

Accretionary wedge A large wedge-shaped mass of sediment that accumulates in subduction zones. Here, sediment is scraped from the subducting oceanic plate and accreted to the overriding crustal block.

Acid precipitation Rain or snow with a pH value that is less than the pH of unpolluted precipitation.

Active continental margin Usually narrow and consisting of highly deformed sediments. They occur where oceanic lithosphere is being subducted beneath the margin of a continent.

Adiabatic temperature change Cooling or warming of air caused when air is allowed to expand or is compressed, not because heat is added or subtracted.

Advection Horizontal convective motion, such as wind.

Advection fog A fog formed when warm, moist air is blown over a cool surface.

Aerosols Tiny solid and liquid particles suspended in the atmosphere.

Aftershocks Smaller earthquakes that follow the main earthquake.

Air A mixture of many discrete gases, of which nitrogen and oxygen are most abundant, in which varying quantities of tiny solid and liquid particles are suspended.

Air mass A large body of air that is characterized by a sameness of temperature and humidity.

Air-mass weather The conditions experienced in an area as an air mass passes over it. Because air masses are large and fairly homogenous, air-mass weather will be fairly constant and may last for several days.

Air pollutants Airborne particles and gases that occur in concentrations that endanger the health and well-being of organisms or disrupt the orderly functioning of the environment.

Air pressure The force exerted by the weight of a column of air above a given point.

Albedo The reflectivity of a substance, usually expressed as a percentage of the incident radiation reflected.

Alluvial fan A fan-shaped deposit of sediment formed when a stream's slope is abruptly reduced.

Alluvium Unconsolidated sediment deposited by a stream.

Alpine glacier A glacier confined to a mountain valley, which in most instances had previously been a stream valley.

Altitude (of the Sun) The angle of the Sun above the horizon.

Andean-type plate margin Plate boundaries that generate continental volcanic arcs.

Andesitic composition *See* Intermediate composition.

Anemometer An instrument used to determine wind speed.

Aneroid barometer An instrument for measuring air pressure that consists of evacuated metal chambers very sensitive to variations in air pressure.

Angle of repose The steepest angle at which loose material remains stationary without sliding downslope.

Angular unconformity An unconformity in which the strata below dip at an angle different from that of the beds above.

Annual mean temperature An average of the 12 monthly temperature means.

Annual temperature range The difference between the highest and lowest monthly temperature means.

Anthracite A hard, metamorphic form of coal that burns clean and hot.

Anticline A fold in sedimentary strata resembling an arch.

Anticyclone A high-pressure center characterized by a clockwise flow of air in the Northern Hemisphere.

Aphelion The place in the orbit of a planet where the planet is farthest from the Sun.

Aphotic zone That portion of the ocean where there is no sunlight.

Apparent magnitude The brightness of a star when viewed from Earth.

Aquifer Rock or soil through which groundwater moves easily.

Aquitard Impermeable beds that hinder or prevent groundwater movement.

Archean eon The second eon of Precambrian time, following the Hadean and preceding the Proterozoic. It extends between 3.8 billion and 2.5 billion years before the present.

Arctic (A) air mass A bitterly cold air mass that forms over the frozen Arctic Ocean.

Arête A narrow knifelike ridge separating two adjacent glaciated valleys.

Arid *See* Desert.

Arid climate *See* Dry climate.

Arkose A feldspar-rich sandstone.

Artesian well A well in which the water rises above the level where it was initially encountered.

Asteroid belt The region in which most asteroids orbit the Sun between Mars and Jupiter.

Asteroids Thousands of small planetlike bodies, ranging in size from a few hundred kilometers to less than a kilometer, whose orbits lie mainly between those of Mars and Jupiter.

Asthenosphere A subdivision of the mantle situated below the lithosphere. This zone of weak material exists below a depth of about 100 kilometers and in some regions extends as deep as 700 kilometers. The rock within this zone is easily deformed.

Astronomical theory A theory of climatic change first developed by Yugoslavian astronomer Milutin Milankovitch. It is based on changes in the shape of Earth's orbit, variations in the obliquity of Earth's axis, and the wobbling of Earth's axis.

Astronomical unit (AU) Average distance from Earth to the Sun; 1.5×10^8 km, or 93×10^6 miles.

Astronomy The scientific study of the universe; it includes the observation and interpretation of celestial bodies and phenomena.

Atmosphere The gaseous portion of a planet; the planet's envelope of air. One of the traditional subdivisions of Earth's physical environment.

Atoll A continuous or broken ring of coral reef surrounding a central lagoon.

Atom The smallest particle that exists as an element.

Atomic number The number of protons in the nucleus of an atom.

Atomic weight The average of the atomic masses of isotopes for a given element.

Aurora A bright display of ever-changing light caused by solar radiation interacting with the upper atmosphere in the region of the poles.

Autumnal equinox The equinox that occurs on September 21–23 in the Northern Hemisphere and on March 21–22 in the Southern Hemisphere.

Axial precession A slow motion of Earth's axis that traces out a cone over a period of 26,000 years.

Backshore The inner portion of the shore, lying landward of the high-tide shoreline. It is usually dry, being affected by waves only during storms.

Backswamp A poorly drained area on a floodplain that results when natural levees are present.

Banded iron formations A finely layered iron and silica-rich (chert) layer deposited mainly during the Precambrian.

Bar Common term for sand and gravel deposits in a stream channel.

Barchan dune A solitary sand dune shaped like a crescent with its tips pointing downward.

Barchanoid dune Dunes forming scalloped rows of sand oriented at right angles to the wind. This form is intermediate between isolated barchans and extensive waves of transverse dunes.

Barograph A recording barometer.

Barometer An instrument that measures atmospheric pressure.

Barometric tendency *See* Pressure tendency.

Barred spiral galaxy A galaxy having straight arms extending from its nucleus.

Barrier island A low, elongate ridge of sand that parallels the coast.

Basalt A fine-grained igneous rock of mafic composition.

Basaltic composition A compositional group of igneous rocks indicating that the rock contains substantial dark silicate minerals and calcium-rich plagioclase feldspar.

Base level The level below which a stream cannot erode.

Basin A circular downfolded structure.

Batholith A large mass of igneous rock that formed when magma was emplaced at depth, crystallized, and subsequently exposed by erosion.

Bathymetry The measurement of ocean depths and the charting of the shape or topography of the ocean floor.

Baymouth bar A sandbar that completely crosses a bay, sealing it off from the open ocean.

Beach An accumulation of sediment found along the landward margin of the ocean or a lake.

Beach drift The transport of sediment in a zigzag pattern along a beach caused by the uprush of water from obliquely breaking waves.

Beach face The wet, sloping surface that extends from the berm to the shoreline.

Beach nourishment The process by which large quantities of sand are added to the beach system to offset losses caused by wave erosion.

Bed load Sediment that is carried by a stream along the bottom of its channel.

Benioff zone Zone of inclined seismic activity that extends from a trench downward into the asthenosphere.

Benthic zone The marine life zone that includes *any* seabottom surface regardless of its distance from shore.

Benthos The forms of marine life that live on or in the ocean bottom.

Bergeron process A theory that relates the formation of precipitation to supercooled clouds, freezing nuclei, and the different saturation levels of ice and liquid water.

Berm The dry, gently sloping zone on the backshore of a beach at the foot of the coastal cliffs or dunes.

Big bang theory The theory that proposes that the universe originated as a single mass, which subsequently exploded.

Binary stars Two stars revolving around a common center of mass under their mutual gravitational attraction.

Biogenous sediment Seafloor sediments consisting of material of marine-organic origin.

Biomass The total mass of a defined organism or group of organisms in a particular area or ecosystem.

Biosphere The totality of life on Earth; the parts of the solid Earth, hydrosphere, and atmosphere in which living organisms can be found.

Bituminous The most common form of coal, often called soft, black coal.

Black dwarf A final state of evolution for a star, in which all of its energy sources are exhausted and it no longer emits radiation.

Black hole A massive star that has collapsed to such a small volume that its gravity prevents the escape of all radiation.

Blowout (deflation hollow) A depression excavated by the wind in easily eroded deposits.

Bode's law A sequence of numbers that approximates the mean distances of the planets from the Sun.

Body waves Seismic waves that travel through Earth's interior.

Bowen's reaction series A concept proposed by N. L. Bowen that illustrates the relationships between magma and the minerals crystallizing from it during the formation of igneous rocks.

Braided stream A stream consisting of numerous intertwining channels.

Breakwater A structure protecting a nearshore area from breaking waves.

Breccia A sedimentary rock composed of angular fragments that were lithified.

Bright-line spectrum The bright lines produced by an incandescent gas under low pressure.

Bright nebula A cloud of glowing gas excited by ultraviolet radiation from hot stars.

Brittle deformation Deformation that involves the fracturing of rock. Associated with rocks near the surface.

Cactolith A quasi-horizontal chonolith composed of anastomosing ductoliths, whose distal ends curl like a harpolith, thin like a sphenolith, or bulge discordantly like an akmolith or ethmolith.

Caldera A large depression typically caused by collapse or ejection of the summit area of a volcano.

Calorie The amount of heat required to raise the temperature of one gram of water 1° C.

Calving Wastage of a glacier that occurs when large pieces of ice break off into water.

Capacity The total amount of sediment a stream is able to transport.

Carbonate group Mineral group whose members contain the carbonate ion (CO_2^{2-}) and one or more kinds of positive ions. Calcite is a common example.

Cassini division A wide gap in the ring system of Saturn between the A ring and the B ring.

Catastrophism The concept that Earth was shaped by catastrophic events of a short-term nature.

Cavern A naturally formed underground chamber or series of chambers most commonly produced by solution activity in limestone.

Celestial sphere An imaginary hollow sphere upon which the ancients believed the stars were hung and carried around Earth.

Cenozoic era A span on the geologic time scale beginning about 65 million years ago following the Mesozoic era.

Cepheid variable A star whose brightness varies periodically because it expands and contracts. A type of pulsating star.

Chemical bond A strong attractive force that exists between atoms in a substance. It involves the transfer or sharing of electrons that allows each atom to attain a full valence shell.

Chemical compound A substance formed by the chemical combination of two or more elements in definite proportions and usually having properties different from those of its constituent elements.

Chemical sedimentary rock Sedimentary rock consisting of material that was precipitated from water by either inorganic or organic means.

Chemical weathering The processes by which the internal structure of a mineral is altered by the removal and/or addition of elements.

Chinook A wind blowing down the lee-ward side of a mountain and warming by compression.

Chromatic aberration The property of a lens whereby light of different colors is focused at different places.

Chromosphere The first layer of the solar atmosphere found directly above the photosphere.

Cinder cone A rather small volcano built primarily of pyroclastics ejected from a single vent.

Circle of illumination The great circle that separates daylight from darkness.

Cirque An amphitheater-shaped basin at the head of a glaciated valley produced by frost wedging and plucking.

Cirrus One of three basic cloud forms; also one of the three high cloud types. They are thin, delicate ice-crystal clouds often appearing as veil-like patches or thin, wispy fibers.

Clastic rock A sedimentary rock made of broken fragments of preexisting rock.

Cleavage The tendency of a mineral to break along planes of weak bonding.

Climate A description of aggregate weather conditions; the sum of all statistical weather information that helps describe a place or region.

Climate-feedback mechanism Because the atmosphere is a complex interactive physical system, several different possible outcomes may result when one of the system's elements is altered. These various possibilities are called *climate-feedback mechanisms*.

Climate system The exchanges of energy and moisture that occur among the atmosphere, hydrosphere, solid Earth, biosphere, and cryosphere.

Climatology The scientific study of climate.

Closed system A system that is self-contained with regards to matter—that is, no matter enters or leaves.

Cloud A form of condensation best described as a dense concentration of suspended water droplets or tiny ice crystals.

Clouds of vertical development A cloud that has its base in the low-height range but extends upward into the middle or high altitudes.

Cluster (star) A large group of stars.

Coarse-grained texture An igneous rock texture in which the crystals are roughly equal in size and large enough so that individual minerals can be identified with the unaided eye.

Coast A strip of land that extends inland from the coastline as far as ocean-related features can be found.

Coastline The coast's seaward edge. The landward limit of the effect of the highest storm waves on the shore.

Col A pass between mountain valleys where the headwalls of two cirques intersect.

Cold front A front along which a cold air mass thrusts beneath a warmer air mass.

Collision–coalescence process A theory of raindrop formation in warm clouds (above 0° C) in which large cloud droplets (giants) collide and join together with smaller droplets to form a raindrop. Opposite electrical charges may bind the cloud droplets together.

Color A phenomenon of light by which otherwise identical objects may be differentiated.

Column A feature found in caves that is formed when a stalactite and stalagmite join.

Columnar joints A pattern of cracks that form during cooling of molten rock to generate columns that are generally six-sided.

Coma The fuzzy, gaseous component of a comet's head.

Comet A small body that generally revolves about the Sun in an elongated orbit.

Competence A measure of the largest particle a stream can transport; a factor dependent on velocity.

Composite cone A volcano composed of both lava flows and pyroclastic material.

Compound A substance formed by the chemical combination of two or more elements in definite proportions and usually having properties different from those of its constituent elements.

Compressional mountains Mountains in which great horizontal forces have shortened and thickened the crust. Most major mountain belts are of this type.

Concordant A term used to describe intrusive igneous masses that form parallel to the bedding of the surrounding rock.

Condensation The change of state from a gas to a liquid.

Condensation nuclei Tiny bits of particulate matter that serve as surfaces on which water vapor condenses.

Conditional instability Moist air with a lapse rate between the dry and wet adiabatic rates.

Conduction The transfer of heat through matter by molecular activity. Energy is transferred through collisions from one molecule to another.

Conduit A pipelike opening through which magma moves toward Earth's surface. It terminates at a surface opening called a vent.

Cone of depression A cone-shaped depression in the water table immediately surrounding a well.

Conformable Layers of rock that were deposited without interruption.

Conglomerate A sedimentary rock composed of rounded, gravel-size particles.

Constellation An apparent group of stars originally named for mythical characters. The sky is presently divided into 88 constellations.

Contact metamorphism Changes in rock caused by the heat from a nearby magma body.

Continental (c) air mass An air mass that forms over land; it is normally relatively dry.

Continental drift A theory that originally proposed that the continents are rafted about. It has essentially been replaced by the plate tectonics theory.

Continental margin That portion of the seafloor adjacent to the continents. It may include the continental shelf, continental slope, and continental rise.

Continental rift A linear zone along which continental lithosphere stretches and pulls apart. Its creation may mark the beginning of a new ocean basin.

Continental rise The gently sloping surface at the base of the continental slope.

Continental shelf The gently sloping submerged portion of the continental margin, extending from the shoreline to the continental slope.

Continental slope The steep gradient that leads to the deep-ocean floor and marks the seaward edge of the continental shelf.

Continental volcanic arc Mountains formed in part by igneous activity associated with the subduction of oceanic lithosphere beneath a continent.

Continuous spectrum An uninterrupted band of light emitted by an incandescent solid, liquid, or gas under pressure.

Convection The transfer of heat by the movement of a mass or substance. It can take place only in fluids.

Convergence The condition that exists when the distribution of winds within a given area results in a net horizontal inflow of air into the area. Because convergence at lower levels is associated with an upward movement of air, areas of convergent winds are regions favorable to cloud formation and precipitation.

Convergent boundary A boundary in which two plates move together, causing one of the slabs of lithosphere to be consumed into the mantle as it descends beneath on an overriding plate.

Coral reef Structure formed in a warm, shallow, sunlit ocean environment that consists primarily of the calcite-rich remains of corals as well as the limy secretions of algae and the hard parts of many other small organisms.

Core Located beneath the mantle, it is the innermost layer of Earth. The core is divided into an outer core and an inner core.

Coriolis force (effect) The deflective force of Earth's rotation on all free-moving objects, including the atmosphere and oceans. Deflection is to the right in the Northern Hemisphere and to the left in the Southern Hemisphere.

Corona The outer, tenuous layer of the solar atmosphere.

Correlation Establishing the equivalence of rocks of similar age in different areas.

Cosmological red shift Changes in the spectra of galaxies that indicate they are moving away from the Milky Way as the result of the expansion of space.

Cosmology The study of the universe.

Country breeze A circulation pattern characterized by a light wind blowing into a city from the surrounding countryside. It is best developed on clear and otherwise calm nights when the urban heat island is most pronounced.

Covalent bond A chemical bond produced by the sharing of electrons.

Crater The depression at the summit of a volcano, or that which is produced by a meteorite impact.

Craton That part of the continental crust that has attained stability; that is, it has not been affected by significant tectonic activity during the Phanerozoic eon. It consists of the shield and stable platform.

Creep The slow downhill movement of soil and regolith.

Crevasse A deep crack in the brittle surface of a glacier.

Cross-bedding Structure in which relatively thin layers are inclined at an angle to the main bedding. Formed by currents of wind or water.

Cross-cutting A principle of relative dating. A rock or fault is younger than any rock (or fault) through which it cuts.

Crust The very thin outermost layer of Earth.

Cryovolcanism A type of volcanism that results from the eruption of magmas derived from the partial melting of ice.

Crystal An orderly arrangement of atoms.

Crystal form *See* Habit.

Crystal settling During the crystallization of magma, the earlier-formed minerals are denser than the liquid portion and settle to the bottom of the magma chamber.

Crystal shape *See* Habit.

Crystallization The formation and growth of a crystalline solid from a liquid or gas.

Cumulus One of three basic cloud forms; also the name given one of the clouds of vertical development. Cumulus are billowy individual cloud masses that often have flat bases.

Cup anemometer *See* Anemometer.

Curie point The temperature above which a material loses its magnetization.

Cut bank The area of active erosion on the outside of a meander.

Cutoff A short channel segment created when a river erodes through the narrow neck of land between meanders.

Cyclone A low-pressure center characterized by a counterclockwise flow of air in the Northern Hemisphere.

Daily mean The mean temperature for a day that is determined by averaging the 24 hourly readings or, more commonly, by averaging the maximum and minimum temperatures for a day.

Daily temperature range The difference between the maximum and minimum temperatures for a day.

Dark-line spectrum *See* Absorption spectrum.

Dark matter Undetected matter that is thought to exist in great quantities in the universe.

Dark nebula A cloud of interstellar dust that obscures the light of more distant stars and appears as an opaque curtain.

Daughter product An isotope resulting from radioactive decay.

Debris flow A relatively rapid type of mass wasting that involves a flow of soil and regolith containing a large amount of water. Also called *mudflows.*

Declination (stellar) The angular distance north or south of the celestial equator denoting the position of a celestial body.

Decompression melting Melting that occurs as rock ascends due to a drop in confining pressure.

Deep-ocean basin The portion of seafloor that lies between the continental margin and the oceanic ridge system. This region comprises almost 30 percent of Earth's surface.

Deep-ocean trench *See* Trench.

Deep-sea fan A cone-shaped deposit at the base of the continental slope. The sediment is transported to the fan by turbidity currents that follow submarine canyons.

Deflation The lifting and removal of loose material by wind.

Deformation General term for the processes of folding, faulting, shearing, compression, or extension of rocks as the result of various natural forces.

Degenerate matter Extremely dense solar material caused by electrons being displaced inward toward an atom's nucleus.

Delta An accumulation of sediment formed where a stream enters a lake or ocean.

Dendritic pattern A stream system that resembles the pattern of a branching tree.

Density Mass per unit volume of a substance, usually expressed as grams per cubic centimeter (g/cm^3)

Deposition The process by which water vapor is changed directly to a solid without passing through the liquid state.

Desalination The removal of salts and other chemicals from seawater.

Desert One of the two types of dry climate; the driest of the dry climates.

Desert pavement A layer of coarse pebbles and gravel created when wind removed the finer material.

Detrital sedimentary rock Rock formed from the accumulation of material that originated and was transported in the form of solid particles derived from both mechanical and chemical weathering.

Dew-point temperature The temperature to which air has to be cooled in order to reach saturation.

Differential weathering The variation in the rate and degree of weathering caused by such factors as mineral makeup, degree of jointing, and climate.

Diffused light Solar energy scattered and reflected in the atmosphere that reaches Earth's surface in the form of diffuse blue light from the sky.

Dike A tabular-shaped intrusive igneous feature that cuts through the surrounding rock.

Dip-slip fault A fault in which the movement is parallel to the dip of the fault.

Discharge The quantity of water in a stream that passes a given point in a period of time.

Disconformity A type of unconformity in which the beds above and below are parallel.

Discordant A term used to describe plutons that cut across existing rock structures, such as bedding planes.

Disseminated deposit Any economic mineral deposit in which the desired mineral occurs as scattered particles in the rock but in sufficient quantity to make the deposit an ore.

Dissolved load That portion of a stream's load carried in solution.

Distributary A section of a stream that leaves the main flow.

Diurnal tidal pattern A tidal pattern exhibiting one high tide and one low tide during a tidal day; a daily tide.

Divergence The condition that exists when the distribution of winds within a given area results in a net horizontal outflow of air from the region. In divergence at lower levels the resulting deficit is compensated for by a downward movement of air from aloft; hence, areas of divergent winds are unfavorable to cloud formation and precipitation.

Divergent boundary A region where the rigid plates are moving apart, typified by the mid-oceanic ridges.

Divide An imaginary line that separates the drainage of two streams; often found along a ridge.

Dome A roughly circular upfolded structure similar to an anticline.

Doppler effect The apparent change in wavelength of radiation caused by the relative motions of the source and the observer.

Doppler radar In addition to the tasks performed by conventional radar, this new generation of weather radar can detect motion directly and hence greatly improve tornado and severe storm warnings.

Drainage basin The land area that contributes water to a stream.

Drawdown The difference in height between the bottom of a cone of depression and the original height of the water table.

Drift *See* Glacial drift.

Drumlin A streamlined asymmetrical hill composed of glacial till. The steep side of the hill faces the direction from which the ice advanced.

Dry adiabatic rate The rate of adiabatic cooling or warming in unsaturated air. The rate of temperature change is 1° C per 100 meters.

Dry climate A climate in which yearly precipitation is not as great as the potential loss of water by evaporation.

Dry-summer subtropical climate A climate located on the west sides of continents between latitudes 30° and 45°. It is the only humid climate with a strong winter precipitation maximum.

Ductile deformation A type of solid state flow that produces a change in the size and shape of a rock body without fracturing. Occurs at depths where temperatures and confining pressures are high.

Dune A hill or ridge of wind-deposited sand.

Dwarf galaxy Very small galaxies, usually elliptical and lacking spiral arms.

Dwarf planets Celestial bodies that orbit stars, massive enough to be spherical but have not cleared their neighboring regions of planetesimals.

Earthflow The downslope movement of water-saturated, clay-rich sediment. Most characteristic of humid regions.

Earthquake The vibration of Earth produced by the rapid release of energy.

Earth science The name for all the sciences that collectively seek to understand Earth. It includes geology, oceanography, meteorology, and astronomy.

Earth system science An interdisciplinary study that seeks to examine Earth as a system composed of numerous interacting parts or subsystems.

Ebb current The movement of a tidal current away from the shore.

Eccentricity The variation of an ellipse from a circle.

Echo sounder An instrument used to determine the depth of water by measuring the time interval between emission of a sound signal and the return of its echo from the bottom.

Eclipse The cutting-off of the light of one celestial body by another passing in front of it.

Ecliptic The yearly path of the Sun plotted against the background of stars.

Elastic deformation Rock deformation in which the rock will return to nearly its original size and shape when the stress is removed.

Elastic rebound The sudden release of stored strain in rocks that results in movement along a fault.

Electromagnetic radiation *See* Radiation.

Electromagnetic spectrum The distribution of electromagnetic radiation by wavelength.

Electron A negatively charged subatomic particle that has a negligible mass and is found outside an atom's nucleus.

Element A substance that cannot be decomposed into simpler substances by ordinary chemical or physical means.

Elements of weather and climate Those quantities or properties of the atmosphere that are measured regularly and that are used to express the nature of weather and climate.

Elliptical galaxy A galaxy that is round or elliptical in outline. It contains little gas and dust, no disk or spiral arms, and few hot, bright stars.

El Niño The name given to the periodic warming of the ocean that occurs in the central and eastern Pacific. A major El Niño episode can cause extreme weather in many parts of the world.

Eluviation The washing-out of fine soil components from the horizon by downward-percolating water.

Emergent coast A coast where land that was formerly below sea level has been exposed either because of crustal uplift or a drop in sea level or both.

Emission nebula A gaseous nebula that derives its visible light from the fluorescence of ultraviolet light from a star in or near the nebula.

Emission spectrum *See* Bright-line spectrum

End moraine A ridge of till marking a former position of the front of a glacier.

Energy The capacity to do work.

Energy levels Spherically shaped, negatively charged zones that surround the nucleus of an atom.

Environment Everything that surrounds and influences an organism.

Environmental lapse rate The rate of temperature decrease with increasing height in the troposphere.

Eon The largest time unit on the geologic time scale, next in order of magnitude above era.

Ephemeral stream A stream that is usually dry because it carries water only in response to specific episodes of rainfall. Most desert streams are of this type.

Epicenter The location on Earth's surface that lies directly above the focus of an earthquake.

Epoch A unit of the geologic calendar that is a subdivision of a period.

Equatorial low A belt of low pressure lying near the equator and between the subtropical highs.

Equatorial system A method of locating stellar objects much like the coordinate system used on Earth's surface.

Equinox The time when the vertical rays of the Sun are striking the equator. The length of daylight and darkness is equal at all latitudes at equinox.

Era A major division on the geologic calendar; eras are divided into shorter units called periods.

Erosion The incorporation and transportation of material by a mobile agent, such as water, wind, or ice.

Eruption column Buoyant plumes of hot, ash-laden gases that can extend thousands of meters into the atmosphere.

Eruptive variable A star that varies in brightness.

Escape velocity The initial velocity an object needs to escape from the surface of a celestial body.

Esker Sinuous ridge composed largely of sand and gravel deposited by a stream flowing in a tunnel beneath a glacier near its terminus.

Estuary A partially enclosed coastal water body that is connected to the ocean. Salinity here is measurably reduced by the freshwater flow of rivers.

Eukaryotes An organism whose genetic material is enclosed in a nucleus; plants, animals, and fungi are eukaryotes.

Euphotic zone The portion of the photic zone near the surface where light is bright enough for photosynthesis to occur.

Evaporation The process of converting a liquid to a gas.

Evaporite deposits A sedimentary rock formed of material deposited from solution by evaporation of water.

Evapotranspiration The combined effect of evaporation and transpiration.

Evolution (Theory of) A fundamental theory in biology and paleontology that sets forth the process by which members of a population of organisms come to differ from their ancestors. Organisms evolve by means of mutations, natural selection, and genetic factors. Modern species are descended from related but different species that lived in earlier times.

Exfoliation dome Large, dome-shaped structure, usually composed of granite, formed by sheeting.

Exotic stream A permanent stream that traverses a desert and has its source in well-watered areas outside the desert.

External process Process such as weathering, mass wasting, or erosion that is powered by the Sun and transforms solid rock into sediment.

Extrusive Igneous activity that occurs outside the crust.

Eye A zone of scattered clouds and calm averaging about 20 kilometers in diameter at the center of a hurricane.

Eyepiece A short-focal-length lens used to enlarge the image in a telescope. The lens nearest the eye.

Eye wall The doughnut-shaped area of intense cumulonimbus development and very strong winds that surrounds the eye of a hurricane.

Fall A type of movement common to mass-wasting processes that refers to the free falling of detached individual pieces of any size.

Fault A break in a rock mass along which movement has occurred.

Fault-block mountain A mountain formed by the displacement of rock along a fault.

Fault creep Displacement along a fault that is so slow and gradual that little seismic activity occurs.

Fault scarp A cliff created by movement along a fault. It represents the exposed surface of the fault prior to modification by weathering and erosion.

Felsic The group of igneous rocks composed primarily of feldspar and quartz.

Filaments Dark, thin streaks that appear across the bright solar disk.

Fine-grained texture A texture of igneous rocks in which the crystals are too small for individual minerals to be distinguished with the unaided eye.

Fiord A steep-sided inlet of the sea formed when a glacial trough was partially submerged.

Fissure A crack in rock along which there is a distinct separation.

Fissure eruption An eruption in which lava is extruded from narrow fractures or cracks in the crust.

Flare A sudden brightening of an area on the Sun.

Flood basalts Flows of basaltic lava that issue from numerous cracks or fissures and commonly cover extensive areas to thicknesses of hundreds of meters.

Flood current The tidal current associated with the increase in the height of the tide.

Floodplain The flat, low-lying portion of a stream valley subject to periodic inundation.

Flow A type of movement common to mass-wasting processes in which water-saturated material moves downslope as a viscous fluid.

Fluorescence The absorption of ultraviolet light, which is reemitted as visible light.

Focal length The distance from the lens to the point where it focuses parallel rays of light.

Focus (earthquake) The zone within Earth where rock displacement produces an earthquake.

Focus (light) The point where a lens or mirror causes light rays to converge.

Fog A cloud with its base at or very near Earth's surface.

Fold A bent rock layer or series of layers that were originally horizontal and subsequently deformed.

Foliation A texture of metamorphic rocks that gives the rock a layered appearance.

Food chain A succession of organisms in an ecological community through which food energy is transferred from producers through herbivores and on to one or more carnivores.

Food web A group of interrelated food chains.

Foreshocks Small earthquakes that often precede a major earthquake.

Foreshore That portion of the shore lying between the normal high and low water marks; the intertidal zone.

Fossil The remains or traces of organisms preserved from the geologic past.

Fossil fuel General term for any hydrocarbon that may be used as a fuel, including coal, oil, and natural gas.

Fossil magnetism See *Paleomagnetism*.

Fossil succession Fossil organisms that succeed one another in a definite and determinable order, and any time period can be recognized by its fossil content.

Fracture zone Any break or rupture in rock along which no appreciable movement has taken place.

Freezing The change of state from a liquid to a solid.

Freezing nuclei Solid particles that serve as cores for the formation of ice crystals.

Front The boundary between two adjoining air masses having contrasting characteristics.

Frontal fog Fog formed when rain evaporates as it falls through a layer of cool air.

Frontal wedging Lifting of air resulting when cool air acts as a barrier over which warmer, lighter air will rise.

Frost wedging The mechanical breakup of rock caused by the expansion of freezing water in cracks and crevices.

Fumarole A vent in a volcanic area from which fumes or gases escape.

Galactic cluster Groups of gravitationally bound galaxies that sometimes contain thousands of galaxies.

Geocentric The concept of an Earth-centered universe.

Geologic time scale The division of Earth history into blocks of time—eons, eras, periods, and epochs. The time scale was created using relative dating principles.

Geology The science that examines Earth, its form and composition, and the changes it has undergone and is undergoing.

Geosphere The solid Earth, the largest of Earth's four major spheres.

Geostrophic wind A wind, usually above a height of 600 meters (2,000 feet), that blows parallel to the isobars.

Geothermal energy Natural steam used for power generation.

Geothermal gradient The gradual increase in temperature with depth in the crust. The average is 30° C per kilometer in the upper crust.

Geyser A fountain of hot water ejected periodically.

Giant (star) A luminous star of large radius.

Glacial drift An all-embracing term for sediments of glacial origin, no matter how, where, or in what shape they were deposited.

Glacial erratic An ice-transported boulder that was not derived from bedrock near its present site.

Glacial striations Scratches and grooves on bedrock caused by glacial abrasion.

Glacial trough A mountain valley that has been widened, deepened, and straightened by a glacier.

Glacier A thick mass of ice originating on land from the compaction and recrystallization of snow that shows evidence of past or present flow.

Glassy texture A term used to describe the texture of certain igneous rocks, such as obsidian, that contain no crystals.

Glaze A coating of ice on objects formed when supercooled rain freezes on contact.

Globular cluster A nearly spherically shaped group of densely packed stars.

Globule A dense, dark nebula thought to be the birthplace of stars.

Gondwanaland The southern portion of Pangaea consisting of South America, Africa, Australia, India, and Antarctica.

Graben A valley formed by the downward displacement of a fault-bounded block.

Graded bed A sediment layer that is characterized by a decrease in sediment size from bottom to top.

Gradient The slope of a stream; generally measured in feet per mile.

Granitic composition A compositional group of igneous rocks that indicates a rock is composed almost entirely of light-colored silicates.

Granules The fine structure visible on the solar surface caused by convective cells below.

Gravitational collapse The gradual subsidence of mountains caused by lateral spreading of weak material located deep within these structures.

Greenhouse effect The transmission of short-wave solar radiation by the atmosphere, coupled with the selective absorption of longer-wavelength terrestrial radiation, especially by water vapor and carbon dioxide.

Groin A short wall built at a right angle to the shore to trap moving sand.

Ground moraine An undulating layer of till deposited as the ice front retreats.

Groundwater Water in the zone of saturation.

Guyot A submerged flat-topped seamount.

Gyre The large circular surface current pattern found in each ocean.

Habit Refers to the common or characteristic shape of a crystal, or aggregate of crystals.

Hadean eon A term found on some versions of the geologic time scale. It refers to the earliest interval (eon) of Earth history, and ended 4 billion years ago.

Hail Nearly spherical ice pellets having concentric layers and formed by the successive freezing of layers of water.

Half graben A tilted fault block in which the higher side is associated with mountainous topography and the lower side is a basin that fills with sediment.

Half-life The time required for one-half of the atoms of a radioactive substance to decay.

Halocline A layer of water in which there is a high rate of change in salinity in the vertical dimension.

Hanging valley A tributary valley that enters a glacial trough at a considerable height above its floor.

Hard stabilization Any form of artificial structure built to protect a coast or to prevent the movement of sand along a beach. Examples include groins, jetties, breakwaters, and seawalls.

Hardness The resistance a mineral offers to scratching.

Heat The kinetic energy of random molecular motion.

Heliocentric The view that the Sun is at the center of the solar system.

Hertzsprung-Russell diagram *See* H-R diagram.

High A center of high pressure characterized by anticyclonic winds.

High cloud A cloud that normally has its base above 6,000 meters; the base may be lower in winter and at high-latitude locations.

Highland climate Complex pattern of climate conditions associated with mountains. Highland climates are characterized by large differences that occur over short distances.

Hogback A narrow, sharp-crested ridge formed by the upturned edge of a steeply dipping bed of resistant rock.

Horizon A layer in a soil profile.

Horn A pyramid-like peak formed by glacial action in three or more cirques surrounding a mountain summit.

Horst An elongated, uplifted block of crust bounded by faults.

Hot spot A concentration of heat in the mantle capable of producing magma, which in turn extrudes onto Earth's surface. The intraplate volcanism that produced the Hawaiian Islands is one example.

Hot spot track Chain of volcanic structures produced as a lithospheric plate moves over a mantle plume.

Hot spring A spring in which the water is 6–9° C (10–15° F) warmer than the mean annual air temperature of its locality.

H-R diagram A plot of stars according to their absolute magnitudes and spectral types.

Hubble's law Relates the distance to a galaxy and its velocity.

Humid continental climate A relatively severe climate characteristic of broad continents in the middle latitudes between approximately 40 and 50 degrees north latitude. This climate is not found in the Southern Hemisphere, where the middle latitudes are dominated by the oceans.

Humid subtropical climate A climate generally located on the eastern side of a continent and characterized by hot, sultry summers and cool winters.

Humidity A general term referring to water vapor in the air but not to liquid droplets of fog, cloud, or rain.

Humus Organic matter in soil produced by the decomposition of plants and animals.

Hurricane A tropical cyclonic storm having winds in excess of 119 kilometers (74 miles) per hour.

Hydrogen burning The conversion of hydrogen through fusion to form helium.

Hydrogen fusion The nuclear reaction in which hydrogen nuclei are fused into helium nuclei.

Hydrogenous sediment Seafloor sediments consisting of minerals that crystallize from seawater. An important example is manganese nodules.

Hydrosphere The water portion of our planet; one of the traditional subdivisions of Earth's physical environment.

Hydrothermal solution The hot, watery solution that escapes from a mass of magma during the later stages of crystallization. Such solutions may alter the surrounding country rock and are frequently the source of significant ore deposits.

Hygrometer An instrument designed to measure relative humidity.

Hygroscopic nuclei Condensation nuclei having a high affinity for water, such as salt particles.

Hypothesis A tentative explanation that is tested to determine if it is valid.

Ice cap A mass of glacial ice covering a high upland or plateau and spreading out radially.

Ice cap climate A climate that has no monthly means above freezing and supports no vegetative cover except in a few scattered high mountain areas. This climate, with its perpetual ice and snow, is confined largely to the ice sheets of Greenland and Antarctica.

Ice sheet A very large, thick mass of glacial ice flowing outward in all directions from one or more accumulation centers.

Ice shelf Forming where glacial ice flows into bays, it is a large, relatively flat mass of floating ice that extends seaward from the coast but remains attached to the land along one or more sides.

Igneous rock A rock formed by the crystallization of molten magma.

Immature soil A soil lacking horizons.

Impact craters Depressions that are the result of collisions with bodies such as asteroids and comets.

Incised meander Meandering channel that flows in a steep, narrow valley. They form either when an area is uplifted or when base level drops.

Inclination of the axis The tilt of Earth's axis from the perpendicular to the plane of Earth's orbit.

Inclusion A piece of one rock unit contained within another. Inclusions are used in relative dating. The rock mass adjacent to the one containing the inclusion must have been there first in order to provide the fragment.

Index fossil A fossil that is associated with a particular span of geologic time.

Inertia A property of matter that resists a change in its motion.

Infiltration The movement of surface water into rock or soil through cracks and pore spaces.

Infrared Radiation with a wavelength from 0.7 to 200 micrometers.

Inner core The solid innermost layer of Earth, about 1,300 kilometers (800 miles) in radius.

Inner planets See Terrestrial planets.

Inselberg An isolated mountain remnant characteristic of the late stage of erosion in an arid region.

Intensity (earthquake) A measure of the degree of earthquake shaking at a given locale based on the amount of damage.

Interface A common boundary where different parts of a system interact.

Interior drainage A discontinuous pattern of intermittent streams that do not flow to the ocean.

Intermediate composition The composition of igneous rocks lying between felsic and mafic.

Interstellar matter Dust and gases found between stars.

Intertidal zone The area where land and sea meet and overlap; the zone between high and low tides.

Intraplate volcanism Igneous activity that occurs within a tectonic plate away from plate boundaries.

Intrusion See Pluton.

Intrusive Igneous rock that formed below Earth's surface.

Ion An atom or molecule that possesses an electrical charge.

Ionic bond A chemical bond between two oppositely charged ions formed by the transfer of valence electrons from one atom to the other.

Ionosphere A complex zone of ionized gases that coincides with the lower portion of the thermosphere.

Iron meteorite One of the three main categories of meteorites. This group is composed largely of iron with varying amounts of nickel (5–20 percent). Most meteorite finds are irons.

Irregular galaxy A galaxy that lacks symmetry.

Island arc See Volcanic island arc.

Isobar A line drawn on a map connecting points of equal atmospheric pressure, usually corrected to sea level.

Isostasy The concept that Earth's crust is floating in gravitational balance upon the material of the mantle.

Isostatic adjustment Compensation of the lithosphere when weight is added or removed. When weight is added, the lithosphere will respond by subsiding, and when weight is removed, there will be uplift.

Isotherms Lines connecting points of equal temperature.

Isotope Varieties of the same element that have different mass numbers; their nuclei contain the same number of protons but different numbers of neutrons.

Jet stream Swift (120–240 kilometers per hour), high-altitude winds.

Jetties A pair of structures extending into the ocean at the entrance to a harbor or river that are built for the purpose of protecting against storm waves and sediment deposition.

Joint A fracture in rock along which there has been no movement.

Jovian planet The Jupiter-like planets: Jupiter, Saturn, Uranus, and Neptune. These planets have relatively low densities.

Kame A steep-sided hill composed of sand and gravel originating when sediment is collected in openings in stagnant glacial ice.

Karst A topography consisting of numerous depressions called *sinkholes*.

Kettle holes Depressions created when blocks of ice became lodged in glacial deposits and subsequently melted.

Köppen classification A system for classifying climates devised by Wladimir Köppen that is based on mean monthly and annual values of temperature and precipitation.

Kuiper belt A region outside the orbit of Neptune where most short-period comets are thought to originate.

La Niña An episode of strong trade winds and unusually low sea-surface temperatures in the central and eastern Pacific. The opposite of *El Niño*.

Laccolith A massive igneous body intruded between preexisting strata.

Lahar Mudflows on the slopes of volcanoes that result when unstable layers of ash and debris become saturated and flow downslope, usually following stream channels.

Lake-effect snow Snow showers associated with a cP air mass to which moisture and heat are added from below as the air mass traverses a large and relatively warm lake (such as one of the Great Lakes), rendering the air mass humid and unstable.

Laminar flow The movement of water particles in straight-line paths that are parallel to the channel. The water particles move downstream without mixing.

Land breeze A local wind blowing from land toward the water during the night in coastal areas.

Lapse rate (normal) The average drop in temperature (6.5° C per kilometer; 3.5° F per 1,000 feet) with increased altitude in the troposphere.

Latent heat The energy absorbed or released during a change in state.

Lateral moraine A ridge of till along the sides of an alpine glacier composed primarily of debris that fell to the glacier from the valley walls.

Laurasia The northern portion of Pangaea consisting of North America and Eurasia.

Lava Magma that reaches Earth's surface.

Lava tube Tunnel in hardened lava that acts as a horizontal conduit for lava flowing from a volcanic vent. Lava tubes allow fluid lavas to advance great distances.

Law of conservation of angular momentum The product of the velocity of an object around a center of rotation (axis), and the distance squared of the object from the axis is constant.

Leaching The depletion of soluble materials from the upper soil by downward-percolating water.

Lightning A sudden flash of light generated by the flow of electrons between oppositely charged parts of a cumulonimbus cloud or between the cloud and the ground.

Light-year The distance light travels in a year; about 6 trillion miles.

Liquefaction A phenomenon, sometimes associated with earthquakes, in which soils and other unconsolidated materials containing abundant water are turned into a fluid-like mass that is not capable of supporting buildings.

Lithification The process, generally cementation and/or compaction, of converting sediments to solid rock.

Lithosphere The rigid outer layer of Earth, including the crust and upper mantle.

Lithospheric plate A coherent unit of Earth's rigid outer layer that includes the crust and upper unit.

Local group The cluster of 20 or so galaxies to which our galaxy belongs.

Localized convective lifting Unequal surface heating that causes localized pockets of air (thermals) to rise because of their buoyancy.

Loess Deposits of windblown silt, lacking visible layers, generally buff-colored, and capable of maintaining a nearly vertical cliff.

Longitudinal (seif dunes) Long ridges of sand oriented parallel to the prevailing wind; these dunes form where sand supplies are limited.

Longshore current A nearshore current that flows parallel to the shore.

Low A center of low pressure characterized by cyclonic winds.

Low cloud A cloud that forms below a height of 2,000 meters.

Low-velocity zone *See* Asthenosphere.

Lower mantle The part of the mantle that extends from the core–mantle boundary to a depth of 660 kilometers.

Luminosity The brightness of a star. The amount of energy radiated by a star.

Lunar breccia A lunar rock formed when angular fragments and dust are welded together by the heat generated by the impact of a meteoroid.

Lunar eclipse An eclipse of the Moon.

Lunar highlands *See* Terrae.

Lunar regolith A thin, gray layer on the surface of the Moon, consisting of loosely compacted, fragmented material believed to have been formed by repeated meteoritic impacts.

Luster The appearance or quality of light reflected from the surface of a mineral.

Mafic Igneous rocks with a low silica content and a high iron–magnesium content.

Magma A body of molten rock found at depth, including any dissolved gases and crystals.

Magmatic differentiation The process of generating more than one rock type from a single magma.

Magnetic reversal A change in Earth's magnetic field from normal to reverse or vice versa.

Magnetic time scale A scale that shows the ages of magnetic reversals and is based on the polarity of lava flows of various ages.

Magnetometer A sensitive instrument used to measure the intensity of Earth's magnetic field at various points.

Magnitude (earthquake) The total amount of energy released during an earthquake.

Magnitude (stellar) A number given to a celestial object to express its relative brightness.

Main-sequence stars A sequence of stars on the Hertzsprung-Russell diagram, containing the majority of stars, that runs diagonally from the upper left to the lower right.

Manganese nodules Rounded lumps of hydrogenous sediment scattered on the ocean floor, consisting mainly of manganese and iron and usually containing small amounts of copper, nickel, and cobalt.

Mantle The 2,900-kilometer- (1,800-mile-) thick layer of Earth located below the crust.

Mantle plume A mass of hotter-than-normal mantle material that ascends toward the surface, where it may lead to igneous activity. These plumes of solid yet mobile material may originate as deep as the core–mantle boundary.

Maria The Latin name for the smooth areas of the Moon formerly thought to be seas.

Marine terrace A wave-cut platform that has been exposed above sea level.

Marine west coast climate A climate found on windward coasts from latitudes 40–65 degrees and

dominated by maritime air masses. Winters are mild and summers are cool.

Maritime (m) air mass An air mass that originates over the ocean. These air masses are relatively humid.

Mass number The number of neutrons and protons in the nucleus of an atom.

Mass wasting The downslope movement of rock, regolith, and soil under the direct influence of gravity.

Massive An igneous pluton that is not tabular in shape.

Mean solar day The average time between two passages of the Sun across the local celestial meridian.

Meander A looplike bend in the course of a stream.

Mechanical weathering The physical disintegration of rock, resulting in smaller fragments.

Medial moraine A ridge of till formed when lateral moraines from two coalescing alpine glaciers join.

Melt The liquid portion of magma, excluding the solid crystals.

Melting The change of state from a solid to a liquid.

Mercalli intensity scale *See* Modified Mercalli intensity scale.

Mercury barometer A mercury-filled glass tube in which the height of the mercury column is a measure of air pressure.

Mesocyclone An intense, rotating wind system in the lower part of a thunderstorm that precedes tornado development.

Mesopause The boundary between the mesosphere and the thermosphere.

Mesosphere The layer of the atmosphere immediately above the stratosphere and characterized by decreasing temperatures with height.

Mesozoic era A span on the geologic time scale between the Paleozoic and Cenozoic eras from about 248 million to 65 million years ago.

Metallic bond A chemical bond present in all metals that may be characterized as an extreme type of electron sharing in which the electrons move freely from atom to atom.

Metamorphic rock Rocks formed by the alteration of preexisting rock deep within Earth (but still in the solid state) by heat, pressure, and/or chemically active fluids.

Metamorphism The changes in mineral composition and texture of a rock subjected to high temperature and pressure within Earth.

Meteor The luminous phenomenon observed when a meteoroid enters Earth's atmosphere and burns up; popularly called a "shooting star."

Meteor shower Many meteors appearing in the sky caused when Earth intercepts a swarm of meteoritic particles.

Meteorite Any portion of a meteoroid that survives its traverse through Earth's atmosphere and strikes Earth's surface.

Meteoroid Small solid particles that have orbits in the solar system.

Meteorology The scientific study of the atmosphere and atmospheric phenomena; the study of weather and climate.

Microcontinents Relatively small fragments of continental crust that may lie above sea level, such as the island of Madagascar, or be submerged, as exemplified by the Campbell Plateau located near New Zealand.

Middle cloud A cloud occupying the height range from 2,000 to 6,000 meters.

Middle-latitude cyclone Large center of low pressure with an associated cold front and often a warm front. Frequently accompanied by abundant precipitation.

Mid-ocean ridge *See* Oceanic ridge system.

Mineral A naturally occurring, inorganic crystalline material with a unique chemical composition.

Mineral resource All discovered and undiscovered deposits of a useful mineral that can be extracted now or at some time in the future.

Mineralogy The study of minerals.

Mixed tidal pattern A tidal pattern exhibiting two high tides and two low tides per tidal day with a large inequality in high water heights, low water heights, or both. Coastal locations that experience such a tidal pattern may also show alternating periods of diurnal and semidiurnal tidal patterns. Also called mixed semidiurnal.

Mixing depth The height to which convectional movements extend above Earth's surface. The greater the mixing depth, the better the air quality.

Mixing ratio The mass of water vapor in a unit mass of dry air; commonly expressed as grams of water vapor per kilogram of dry air.

Model A term often used synonymously with hypothesis but is less precise because it is sometimes used to describe a theory as well.

Modified Mercalli intensity scale A 12-point scale developed to evaluate earthquake intensity based on the amount of damage to various structures.

Mohorovičić; discontinuity (Moho) The boundary separating the crust from the mantle, discernible by an increase in seismic velocity.

Mohs scale A series of 10 minerals used as a standard in determining hardness.

Moment magnitude A more precise measure of earthquake magnitude than the Richter scale that is derived from the amount of displacement that occurs along a fault zone.

Monocline A one-limbed flexure in strata. The strata are unusually flat-lying or very gently dipping on both sides of the monocline.

Monsoon Seasonal reversal of wind direction associated with large continents, especially Asia. In winter, the wind blows from land to sea; in summer, from sea to land.

Monthly mean temperature The mean temperature for a month that is calculated by averaging the daily means.

Mountain breeze The nightly downslope winds commonly encountered in mountain valleys.

Natural leeves The elevated landforms that parallel some streams and act to confine their waters, except during floodstage.

Neap tide Lowest tidal range, occurring near the times of the first- and third-quarter phases of the Moon.

Nearshore zone The zone of beach that extends from the low-tide shoreline seaward to where waves break at low tide.

Nebula A cloud of interstellar gas and/or dust.

Nebular theory The basic idea that the Sun and planets formed from the same cloud of gas and dust in interstellar space.

Negative feedback mechanism A feedback mechanism that tends to maintain a system as it is—that is, maintain the status quo.

Nekton Pelagic organisms that can move independently of ocean currents by swimming or other means of propulsion.

Neritic zone The marine-life zone that extends from the low tideline out to the shelf break.

Neutron A subatomic particle found in the nucleus of an atom. The neutron is electrically neutral and has a mass approximately that of a proton.

Neutron star A star of extremely high density composed entirely of neutrons.

Nonconformity An unconformity in which older metamorphic or intrusive igneous rocks are overlain by younger sedimentary strata.

Nonfoliated texture Metamorphic rocks that do not exhibit foliation.

Nonmetallic mineral resource Mineral resource that is not a fuel or processed for the metals it contains.

Nonrenewable resource Resource that forms or accumulates over such long time spans that it must be considered as fixed in total quantity.

Nonsilicates Mineral groups that lack silicas in their structures and account for less than 10 percent of Earth's crust.

Normal fault A fault in which the rock above the fault plane has moved down relative to the rock below.

Normal polarity A magnetic field that is the same as that which exists at present.

Nova A star that explosively increases in brightness.

Nuclear fusion The source of the Sun's energy.

Nucleus The small heavy core of an atom that contains all of its positive charge and most of its mass.

Nuée ardente Incandescent volcanic debris buoyed up by hot gases that moves downslope in an avalanche fashion.

Numerical date Date that specifies the actual number of years that have passed since an event occurred.

Obliquity The angle between the planes of Earth's equator and orbit.

Obsidian A volcanic glass of felsic composition.

Occluded front A front formed when a cold front overtakes a warm front. It marks the beginning of the end of a middle-latitude cyclone.

Occlusion The overtaking of one front by another.

Occultation An eclipse of a star or planet by the Moon or a planet.

Oceanic plateau An extensive region on the ocean floor composed of thick accumulations of pillow basalts and other mafic rocks that in some cases exceed 30 kilometers in thickness.

Oceanic ridge system A continuous elevated zone on the floor of all the major ocean basins and varying in width from 500 to 5,000 kilometers (300–3,000 miles). The rifts at the crests of ridges represent divergent plate boundaries.

Oceanic zone The marine-life zone beyond the continental shelf.

Oceanography The scientific study of the oceans and oceanic phenomena.

Octet rule Atoms combine in order that each may have the electron arrangement of a noble gas; that is, the outer energy level contains eight neutrons.

Offshore zone The relatively flat submerged zone that extends from the breaker line to the edge of the continental shelf.

Oort cloud A spherical shell composed of comets that orbit the Sun at distances generally greater than 10,000 times the Earth–Sun distance.

Open cluster A loosely formed group of stars of similar origin.

Open system One in which both matter and energy flow into and out of the system. Most natural systems are of this type.

Orbit The path of a body in revolution around a center of mass.

Ore Usually a useful metallic mineral that can be mined at a profit. The term is also applied to certain nonmetallic minerals such as fluorite and sulfur.

Ore deposit A naturally occurring concentration of one or more metallic minerals that can be extracted economically.

Original horizontality Layers of sediments are generally deposited in a horizontal or nearly horizontal position.

Orogenesis The processes that collectively result in the formation of mountains.

Orographic lifting Mountains acting as barriers to the flow of air, forcing the air to ascend. The air cools adiabatically, and clouds and precipitation may result.

Outer core A layer beneath the mantle about 2,200 kilometers (1,364 miles) thick that has the properties of a liquid.

Outer planet *See* Jovian planet.

Outgassing The escape of gases that had been dissolved in magma.

Outwash plain A relatively flat, gently sloping plain consisting of materials deposited by meltwater streams in front of the margin of an ice sheet.

Overrunning Warm air gliding up a retreating cold air mass.

Oxbow lake A curved lake produced when a stream cuts off a meander.

Ozone A molecule of oxygen containing three oxygen atoms.

Pahoehoe flow A lava flow with a smooth-to-ropey surface.

Paleomagnetism The natural remnant magnetism in rock bodies. The permanent magnetization acquired by rock that can be used to determine the location of the magnetic poles and the latitude of the rock at the time it became magnetized.

Paleontology The systematic study of fossils and the history of life on Earth.

Paleozoic era A span on the geologic time scale between the eons of the Precambrian and Mesozoic era from about 540 million to 248 million years ago.

Pangaea The proposed supercontinent that 200 million years ago began to break apart and form the present landmasses.

Parabolic dunes The shape of these dunes resembles barchans, except their tips point into the wind; they often form along coasts that have strong onshore winds, abundant sand, and vegetation that partly covers the sand.

Paradigm A theory that is held with a very high degree of confidence and is comprehensive in scope.

Parallax The apparent shift of an object when viewed from two different locations.

Parasitic cone A volcanic cone that forms on the flank of a larger volcano.

Parcel An imaginary volume of air enclosed in a thin elastic cover. Typically it is considered to be a few hundred cubic meters in volume and is assumed to act independently of the surrounding air.

Parent material The material upon which a soil develops.

Parsec The distance at which an object would have a parallax angle of 1 second of arc (3.26 light-years).

Partial melting The process by which most igneous rocks melt. Since individual minerals have different melting points, most igneous rocks melt over a temperature range of a few hundred degrees. If the liquid is squeezed out after some melting has occurred, a melt with a higher silica content results.

Passive continental margin Margins that consist of a continental shelf, continental slope, and continental rise. They are *not* associated with plate boundaries and therefore experience little volcanism and few earthquakes.

Pegmatite A very coarse-grained igneous rock (typically granite) commonly found as a dike associated with a large mass of plutonic rock that has smaller crystals. Crystallization in a water-rich environment is believed to be responsible for the very large crystals.

Pelagic zone Open ocean of *any* depth. Animals in this zone swim or float freely.

Penumbra The portion of a shadow from which only part of the light source is blocked by an opaque body.

Perched water table A localized zone of saturation above the main water table created by an impermeable layer (aquiclude).

Peridotite An igneous rock of ultramafic composition thought to be abundant in the upper mantle.

Perihelion The point in the orbit of a planet where it is closest to the Sun.

Period A basic unit of the geologic calendar that is a subdivision of an era. Periods may be divided into smaller units called epochs.

Periodic table The tabular arrangement of the elements according to atomic number.

Permeability A measure of a material's ability to transmit water.

Perturbation The gravitational disturbance of the orbit of one celestial body by another.

pH scale A common measure of the degree of acidity or alkalinity of a solution, it is a logarithmic scale ranging from 0 to 14. A value of 7 denotes a neutral solution, values below 7 indicate greater acidity, and numbers above 7 indicate greater alkalinity.

Phanerozoic eon That part of geologic time represented by rocks containing abundant fossil evidence. The eon extending from the end of the Proterozoic eon (about 540 million years ago) to the present.

Phases of the Moon The progression of changes in the Moon's appearance during the month.

Pheoncryst Conspicuously large crystals embedded in a matrix of finer-grained crystals.

Photic zone The upper part of the ocean into which any sunlight penetrates.

Photochemical reaction A chemical reaction in the atmosphere that is triggered by sunlight, often yielding a secondary pollutant.

Photon A discrete amount (quantum) of electromagnetic energy.

Photosphere The region of the Sun that radiates energy to space. The visible surface of the Sun.

Photosynthesis The process by which plants and algae produce carbohydrates from carbon dioxide and water in the presence of chlorophyll, using light energy and releasing oxygen.

Physical environment The part of the environment that encompasses water, air, soil, and rock, as well as conditions such as temperature, humidity, and sunlight.

Phytoplankton Algal plankton, which are the most important community of primary producers in the ocean.

Piedmont glacier A glacier that forms when one or more valley glaciers emerge from the confining walls of mountain valleys and spread out to create a broad sheet in the lowlands at the base of the mountains.

Pipe A vertical conduit through which magmatic materials have passed.

Placer Deposit formed when heavy minerals are mechanically concentrated by currents, most commonly streams and waves. Placers are sources of gold, tin, platinum, diamonds, and other valuable minerals.

Plane of the ecliptic The imaginary plane that connects Earth's orbit with the celestial sphere.

Planetary nebula A shell of incandescent gas expanding from a star.

Planetesimal A solid celestial body that accumulated during the first stages of planetary formation. Planetesimals aggregated into increasingly larger bodies, ultimately forming the planets.

Plankton Passively drifting or weakly swimming organisms that cannot move independently of ocean currents. Includes microscopic algae, protozoa, jellyfish, and larval forms of many animals.

Plate *See* Lithospheric plate.

Plate tectonics The theory that proposes that Earth's outer shell consists of individual plates that interact in various ways and thereby produce earthquakes, volcanoes, mountains, and the crust itself.

Playa A flat area on the floor of an undrained desert basin. Following heavy rain, the playa becomes a lake.

Playa lake A temporary lake in a playa.

Pleistocene epoch An epoch of the Quaternary period beginning about 1.8 million years ago and ending about 10,000 years ago. Best known as a time of extensive continental glaciation.

Plucking (quarrying) The process by which pieces of bedrock are lifted out of place by a glacier.

Pluton A structure that results from the emplacement and crystallization of magma beneath the surface of Earth.

Pluvial lake A lake formed during a period of increased rainfall. During the Pleistocene epoch this occurred in some nonglaciated regions during periods of ice advance elsewhere.

Point bar A crescent-shaped accumulation of sand and gravel deposited on the inside of a meander.

Polar (P) air mass A cold air mass that forms in a high-latitude source region.

Polar easterlies In the global pattern of prevailing winds, winds that blow from the polar high toward the subpolar low. These winds, however, should not be thought of as persistent winds, such as the trade winds.

Polar front The stormy frontal zone separating air masses of polar origin from air masses of tropical origin.

Polar high Anticyclones that are assumed to occupy the inner polar regions and are believed to be thermally induced, at least in part.

Polar wandering As the result of paleomagnetic studies in the 1950s, researchers proposed that either the magnetic poles migrated greatly through time or the continents had gradually shifted their positions.

Population I Stars rich in atoms heavier than helium. Nearly always relatively young stars found in the disk of the galaxy.

Population II Stars poor in atoms heavier than helium. Nearly always relatively old stars found in the halo, globular clusters, or nuclear bulge.

Porosity The volume of open spaces in rock or soil.

Porphyritic texture An igneous texture consisting of large crystals embedded in a matrix of much smaller crystals.

Positive feedback mechanism A feedback mechanism that enhances or drives change.

Precambrian All geologic time prior to the Paleozoic era.

Precession *See* Axial precession.

Precipitation fog Fog formed when rain evaporates as it falls through a layer of cool air.

Pressure gradient The amount of pressure change occurring over a given distance.

Pressure tendency The nature of the change in atmospheric pressure over the past several hours. It can be a useful aid in short-range weather prediction.

Prevailing wind A wind that consistently blows from one direction more than from another.

Primary pollutants Those pollutants emitted directly from identifiable sources.

Primary productivity The amount of organic matter synthesized by organisms from inorganic substances through photosynthesis or chemosynthesis within a given volume of water or habitat in a unit of time.

Primary (P) wave A type of seismic wave that involves alternating compression and expansion of the material through which it passes.

Principal shells *See* Energy levels.

Prokaryotes Refers to the cells or organisms such as bacteria whose genetic material is not enclosed in a nucleus.

Prominence A concentration of material above the solar surface that appears as a bright archlike structure.

Proterozoic eon The eon following the Archean and preceding the Phanerozoic. It extends between about 2,500 million (2.5 billion) and 540 million years ago.

Proton A positively charged subatomic particle found in the nucleus of an atom.

Proton–proton chain A chain of thermonuclear reactions by which nuclei of hydrogen are built up into nuclei of helium.

Protoplanets A developing planetary body that grows by the accumulation of planetesimals.

Protostar A collapsing cloud of gas and dust destined to become a star.

Psychrometer A device consisting of two thermometers (wet bulb and dry bulb) that is rapidly whirled and, with the use of tables, yields the relative humidity and dew point.

Ptolemaic system An Earth-centered system of the universe.

Pulsar A variable radio source of small size that emits radio pulses in very regular periods.

Pulsating variable A variable star that pulsates in size and luminosity.

Pycnocline A layer of water in which there is a rapid change of density with depth.

Pyroclastic An igneous rock texture resulting from the consolidation of individual rock fragments that are ejected during a violent eruption.

Pyroclastic flow A highly heated mixture, largely of ash and pumice fragments, traveling down the flanks of a volcano or along the surface of the ground.

Pyroclastic material The volcanic rock ejected during an eruption, including ash, bombs, and blocks.

Radial pattern A system of streams running in all directions away from a central elevated structure, such as a volcano.

Radiation The transfer of energy (heat) through space by electromagnetic waves.

Radiation fog Fog resulting from radiation heat loss by Earth.

Radiation pressure The force exerted by electromagnetic radiation from an object such as the Sun.

Radio interferometer Two or more radio telescopes that combine their signals to achieve the resolving power of a larger telescope.

Radio telescope A telescope designed to make observations in radio wavelengths.

Radioactive decay The spontaneous decay of certain unstable atomic nuclei.

Radioactivity The spontaneous emission of certain unstable atomic nuclei.

Radiocarbon (carbon-14) The radioactive isotope of carbon, which is produced continuously in the atmosphere and is used in dating events from the very recent geologic past (the last few tens of thousands of years).

Radiometric dating The procedure of calculating the absolute ages of rocks and minerals that contain radioactive isotopes.

Rain Drops of water that fall from clouds that have a diameter of at least 0.5 millimeter (0.02 inch).

Rainshadow desert A dry area on the lee side of a mountain range. Many middle-latitude deserts are of this type.

Rapids A part of a stream channel in which the water suddenly begins flowing more swiftly and turbulently because of an abrupt steepening of the gradient.

Ray (lunar) Any of a system of bright elongated streaks, sometimes associated with a crater on the Moon.

Recessional moraine An end moraine formed as the ice front stagnated during glacial retreat.

Rectangular pattern A drainage pattern characterized by numerous right-angle bends that develops on jointed or fractured bedrock.

Red giant A large, cool star of high luminosity; a star occupying the upper-right portion of the Hertzsprung-Russell diagram.

Reflecting telescope A telescope that concentrates light from distant objects by using a concave mirror.

Reflection The process whereby light bounces back from an object at the same angle at which it encounters a surface and with the same intensity.

Reflection nebula A relatively dense dust cloud in interstellar space that is illuminated by starlight.

Refracting telescope A telescope that employs a lens to bend and concentrate the light from distant objects.

Refraction The process by which the portion of a wave in shallow water slows, causing the wave to bend and tend to align itself with the underwater contours.

Regional metamorphism Metamorphism associated with large-scale mountain-building processes.

Regolith The layer of rock and mineral fragments that nearly everywhere covers Earth's surface.

Relative dating Rocks are placed in their proper sequence or order. Only the chronological order of events is determined.

Relative humidity The ratio of the air's water-vapor content to its water-vapor capacity.

Renewable resource A resource that is virtually inexhaustible or that can be replenished over relatively short time spans.

Reserve Already identified deposits from which minerals can be extracted profitably.

Residual soil Soil developed directly from the weathering of the bedrock below.

Resolving power The ability of a telescope to separate objects that would otherwise appear as one.

Retrograde motion The apparent westward motion of the planets with respect to the stars.

Reverse fault A fault in which the material above the fault plane moves up in relation to the material below.

Reverse polarity A magnetic field opposite to that which exists at present.

Revolution The motion of one body about another, as Earth about the Sun.

Richter scale A scale of earthquake magnitude based on the motion of a seismograph.

Ridge push A mechanism that may contribute to plate motion. It involves the oceanic lithosphere sliding down the oceanic ridge under the pull of gravity.

Rift valley A long, narrow trough bounded by normal faults. It represents a region where divergence is taking place.

Rift zone A region of Earth's crust along which divergence is taking place.

Right ascension An angular distance measured eastward along the celestial equator from the vernal equinox. Used with declination in a coordinate system to describe the position of celestial bodies.

Rime A thin coating of ice on objects produced when supercooled fog droplets freeze on contact.

Rip current A strong narrow surface or near-surface current of short duration and high speed flowing seaward through the breaker zone at nearly right angles to the shore. It represents the return to the ocean of water that has been piled up on the shore by incoming waves.

Rock A consolidated mixture of minerals.

Rock cycle A model that illustrates the origin of the three basic rock types and the interrelatedness of Earth materials and processes.

Rock flour Ground-up rock produced by the grinding effect of a glacier.

Rock-forming minerals The minerals that make up most of the rocks of Earth's crust.

Rockslide The rapid slide of a mass of rock downslope along planes of weakness.

Rotation The spinning of a body, such as Earth, about its axis.

Runoff Water that flows over the land rather than infiltrating into the ground.

Salinity The proportion of dissolved salts to pure water, usually expressed in parts per thousand (‰).

Saltation Transportation of sediment through a series of leaps or bounces.

Santa Ana The local name given a chinook wind in southern California.

Saturation The maximum quantity of water vapor that the air can hold at any given temperature and pressure.

Scattering The redirecting (in all directions) of light by small particles and gas molecules in the atmosphere. The result is diffused light.

Scoria Hardened lava that has retained the vesicles produced by escaping gases.

Scoria cone *See* Cinder cone.

Sea arch An arch formed by wave erosion when caves on opposite sides of a headland unite.

Sea breeze A local wind blowing from the sea during the afternoon in coastal areas.

Sea stack An isolated mass of rock standing just offshore, produced by wave erosion of a headland.

Seafloor spreading The process of producing new seafloor between two diverging plates.

Seamount An isolated volcanic peak that rises at least 1,000 meters (3,000 feet) above the deep-ocean floor.

Seawall A barrier constructed to prevent waves from reaching the area behind the wall. Its purpose is to defend property from the force of breaking waves.

Secondary enrichment The concentration of minor amounts of metals that are scattered through unweathered rock into economically valuable concentrations by weathering processes.

Secondary pollutants Pollutants that are produced in the atmosphere by chemical reactions that occur among primary pollutants.

Secondary (S) wave A seismic wave that involves oscillation perpendicular to the direction of propagation.

Sediment Unconsolidated particles created by the weathering and erosion of rock, by chemical precipitation from solution in water, or from the secretions of organisms and transported by water, wind, or glaciers.

Sedimentary rock Rock formed from the weathered products of preexisting rocks that have been transported, deposited, and lithified.

Seismic gap A segment of an active fault zone that has not experienced a major earthquake over a span when most other segments have. Such segments are probable sites for future major earthquakes.

Seismic waves A rapidly moving ocean wave generated by earthquake activity capable of inflicting heavy damage in coastal regions.

Seismogram The record made by a seismograph.

Seismograph An instrument that records earthquake waves.

Seismology The study of earthquakes and seismic waves.

Semiarid *See* Steppe.

Semidiurnal tidal pattern A tidal pattern exhibiting two high tides and two low tides per tidal day with small inequalities between successive highs and successive lows; a semi-daily tide.

Settling velocity The speed at which a particle falls through a still fluid. The size, shape, and specific gravity of particles influence settling velocity.

Shadow zone The zone between 104 and 143 degrees distance from an earthquake epicenter in which direct waves do not arrive because of refraction by Earth's core.

Sheeting A mechanical weathering process characterized by the splitting-off of slablike sheets of rock.

Shelf break The point where a rapid steepening of the gradient occurs, marking the outer edge of the continental shelf and the beginning of the continental slope.

Shield A large, relatively flat expanse of ancient metamorphic rock within the stable continental interior.

Shield volcano A broad, gently sloping volcano built from fluid basaltic lavas.

Shore Seaward of the coast, this zone extends from the highest level of wave action during storms to the lowest tide level.

Shoreline The line that marks the contact between land and sea. It migrates up and down as the tide rises and falls.

Sidereal day The period of Earth's rotation with respect to the stars.

Sidereal month A time period based on the revolution of the Moon around Earth with respect to the stars.

Silicate Any one of numerous minerals that have the oxygen and silicon tetrahedron as their basic structure.

Silicon-oxygen tetrahedron A structure composed of four oxygen atoms surrounding a silicon atom that constitutes the basic building block of silicate minerals.

Sill A tabular igneous body that was intruded parallel to the layering of preexisting rock.

Sinkhole A depression produced in a region where soluble rock has been removed by groundwater.

Slab pull A mechanism that contributes to plate motion in which cool, dense oceanic crust sinks into the mantle and "pulls" the trailing lithosphere along.

Sleet Frozen or semifrozen rain formed when raindrops freeze as they pass through a layer of cold air.

Slide A movement common to mass-wasting processes in which the material moving downslope remains fairly coherent and moves along a well-defined surface.

Slip face The steep, leeward slope of a sand dune; it maintains an angle of about 34 degrees.

Slump The downward slipping of a mass of rock or unconsolidated material moving as a unit along a curved surface.

Small solar system bodies Solar system objects not classified as planets or moons that include dwarf planets, asteroids, comets, and meteoroids.

Snow A solid form of precipitation produced by sublimation of water vapor.

Snowfield An area where snow persists year-round.

Snowline Lower limit of perennial snow.

Soil A combination of mineral and organic matter, water, and air; that portion of the regolith that supports plant growth.

Soil horizon A layer of soil that has identifiable characteristics produced by chemical weathering and other soil-forming processes.

Soil profile A vertical section through a soil showing its succession of horizons and the underlying parent material.

Soil taxonomy A soil classification system consisting of six hierarchical categories based on observable soil characteristics. The system recognizes 12 soil orders.

Soil texture The relative proportions of clay, silt, and sand in a soil. Texture strongly influences the soil's ability to retain and transmit water and air.

Solar constant The rate at which solar radiation is received outside Earth's atmosphere on a surface perpendicular to the Sun's rays when Earth is at an average distance from the Sun.

Solar eclipse An eclipse of the Sun.

Solar flare A sudden and tremendous eruption in the solar chromosphere.

Solar nebula The cloud of interstellar gas and/or dust from which the bodies of our solar system formed.

Solar winds Subatomic particles ejected at high speed from the solar corona.

Solifluction Slow, downslope flow of water-saturated materials common to permafrost areas.

Solstice The time when the vertical rays of the Sun are striking either the Tropic of Cancer or the Tropic of Capricorn. Solstice represents the longest or shortest day (length of daylight) of the year.

Solum The O, A, and B horizons in a soil profile. Living roots and other plant and animal life are largely confined to this zone.

Sorting The process by which solid particles of various sizes are separated by moving water or wind. Also, the degree of similarity in particle size in sediment or sedimentary rock.

Source region The area where an air mass acquires its characteristic properties of temperature and moisture.

Specific gravity The ratio of a substance's weight to the weight of an equal volume of water.

Spectral class A classification of a star according to the characteristics of its spectrum.

Spectroscope An instrument for directly viewing the spectrum of a light source.

Spectroscopy The study of spectra.

Spheroidal weathering Any weathering process that tends to produce a spherical shape from an initially blocky shape.

Spicule A narrow jet of rising material in the solar chromosphere.

Spiral galaxy A flattened, rotating galaxy with pinwheel-like arms of interstellar material and young stars winding out from its nucleus.

Spit An elongated ridge of sand that projects from the land into the mouth of an adjacent bay.

Spreading center *See* Divergent boundary.

Spring A flow of groundwater that emerges naturally at the ground surface.

Spring equinox The equinox that occurs on March 21–22 in the Northern Hemisphere and on September 21–23 in the Southern Hemisphere.

Spring tide Highest tidal range that occurs near the times of the new and full moons.

Stable air Air that resists vertical displacement. If it is lifted, adiabatic cooling will cause its temperature to be lower than the surrounding environment; if it is allowed, it will sink to its original position.

Stable platform That part of the craton that is mantled by relatively undeformed sedimentary rocks and underlain by a basement complex of igneous and metamorphic rocks.

Stalactite The icicle-like structure that hangs from the ceiling of a cavern.

Stalagmite The columnlike form that grows upward from the floor of a cavern.

Star dune Isolated hill of sand that exhibits a complex form and develops where wind directions are variable.

Stationary front A situation in which the surface position of a front does not move; the flow on either side of such a boundary is nearly parallel to the position of the front.

Steam fog Fog having the appearance of steam, produced by evaporation from a warm water surface into the cool air above.

Stellar parallax A measure of stellar distance.

Steppe One of the two types of dry climate. A marginal and more humid variant of the desert that separates it from bordering humid climates.

Stock A pluton similar to but smaller than a batholith.

Stony meteorite One of the three main categories of meteorites. Such meteorites are composed largely of silicate minerals with inclusions of other minerals.

Stony-iron meteorite One of the three main categories of meteorites. This group, as the name implies, is a mixture of iron and silicate minerals.

Storm surge The abnormal rise of the sea along a shore as a result of strong winds.

Strata Parallel layers of sedimentary rock.

Stratified drift Sediments deposited by glacial meltwater.

Stratopause The boundary between the stratosphere and the mesosphere.

Stratosphere The layer of the atmosphere immediately above the troposphere, characterized by increasing temperatures with height, owing to the concentration of ozone.

Stratovolcano *See* Composite cone.

Stratus One of three basic cloud forms; also, the name given one of the flow clouds. They are sheets or layers that cover much or all of the sky.

Streak The color of a mineral in powdered form.

Stream valley The channel, valley floor, and sloping valley walls of a stream.

Stress The force per unit area acting on any surface within a solid.

Striations (glacial) Scratches or grooves in a bedrock surface caused by the grinding action of a glacier and its load of sediment.

Strike-slip fault A fault along which the movement is horizontal.

Stromatolite Structures that are deposited by algae and consist of layered mounds of calcium carbonate.

Subarctic climate A climate found north of the humid continental climate and south of the polar climate and characterized by bitterly cold winters and short, cool summers. Places within this climatic realm experience the highest annual temperature ranges on Earth.

Subduction The process of thrusting oceanic lithosphere into the mantle along a convergent boundary.

Subduction zone A long, narrow zone where one lithospheric plate descends beneath another.

Sublimation The conversion of a solid directly to a gas without passing through the liquid state.

Submarine canyon A seaward extension of a valley that was cut on the continental shelf during a time when sea level was lower, or a canyon carved into the outer continental shelf, slope, and rise by turbidity currents.

Submergent coast A coast with a form that is largely the result of the partial drowning of a former land surface either because of a rise of sea level or subsidence of the crust or both.

Subpolar low Low pressure located at about the latitudes of the Arctic and Antarctic circles. In the Northern Hemisphere the low takes the form of individual oceanic cells; in the Southern Hemisphere there is a deep and continuous trough of low pressure.

Subsoil A term applied to the B horizon of a soil profile.

Subtropical high Not a continuous belt of high pressure but rather several semipermanent, anticyclonic centers characterized by subsidence and divergence located roughly between latitudes 25 and 35 degrees.

Summer solstice The solstice that occurs on June 21–22 in the Northern Hemisphere and on December 21–22 in the Southern Hemisphere.

Sunspot A dark spot on the Sun, which is cool by contrast to the surrounding photosphere.

Supercontinent A large landmass that contains all, or nearly all, of the existing continents.

Supercontinent cycle The idea that the rifting and dispersal of one supercontinent is followed by a long period during which the fragments gradually reassemble into a new supercontinent.

Supercooled The condition of water droplets that remain in the liquid state at temperatures well below 0° C.

Supergiant A very large star of high luminosity.

Supernova An exploding star that increases in brightness many thousands of times.

Superposition In any undeformed sequence of sedimentary rocks, each bed is older than the layers above and younger than the layers below.

Supersaturation The condition of being more highly concentrated than is normally possible under given temperature and pressure conditions. When describing humidity, it refers to a relative humidity that is greater than 100 percent.

Surf A collective term for breakers; also, the wave activity in the area between the shoreline and the outer limit of breakers.

Surface soil The uppermost layer in a soil profile: the A horizon.

Surface waves Seismic waves that travel along the outer layer of Earth.

Suspended load The fine sediment carried within the body of flowing water.

Swells Wind-generated waves that have moved into an area of weaker winds or calm.

Syncline A linear downfold in sedimentary strata; the opposite of anticline.

Synodic month The period of revolution of the Moon with respect to the Sun, or its cycle of phases.

System Any size group of interacting parts that form a complex whole.

Talus An accumulation of rock debris at the base of a cliff.

Tarn A small lake in a cirque.

Tectonic plate A coherent unit of Earth's rigid outer layer that includes the crust and upper unit.

Tectonics The study of the large-scale processes that collectively deform Earth's crust.

Temperature A measure of the degree of hotness or coldness of a substance; a measure of the *average* kinetic energy of individual atoms or molecules in a substance.

Temperature inversion A layer in the atmosphere of limited depth where the temperature increases rather than decreases with height.

Temporary (local) base level The level of a lake, resistant rock layer, or any other base level that stands above sea level.

Tenacity Describes a mineral's toughness or its resistance to breaking or deforming.

Terminal moraine The end moraine marking the farthest advance of a glacier.

Terrace A flat, benchlike structure produced by a stream, which was left elevated as the stream cut downward.

Terrae The extensively cratered highland areas of the Moon.

Terrane A crustal block bounded by faults, whose geologic history is distinct from the histories of adjoining crustal blocks.

Terrestrial planets Any of the Earth-like planets, including Mercury, Venus, Mars, and Earth.

Terrigenous sediment Seafloor sediments derived from terrestrial weathering and erosion.

Texture The size, shape, and distribution of the particles that collectively constitute a rock.

Theory A well-tested and widely accepted view that explains certain observable facts.

Thermal gradient The increase in temperature with depth. It averages 1° C per 30 meters (1–2° F per 100 feet) in the crust.

Thermal metamorphism *See* Contact metamorphism.

Thermocline A layer of water in which there is a rapid change in temperature in the vertical dimension.

Thermohaline circulation Movements of ocean water caused by density differences brought about by variations in temperature and salinity.

Thermosphere The region of the atmosphere immediately above the mesosphere and characterized by increasing temperatures due to absorption of very shortwave solar energy by oxygen.

Thrust fault A low-angle reverse fault.

Thunder The sound emitted by rapidly expanding gases along the channel of lightning discharge.

Thunderstorm A storm produced by a cumulonimbus cloud and always accompanied by lightning and thunder. It is of relatively short duration and usually accompanied by strong wind gusts, heavy rain, and sometimes hail.

Tidal current The alternating horizontal movement of water associated with the rise and fall of the tide.

Tidal delta A deltalike feature created when a rapidly moving tidal current emerges from a narrow inlet and slows, depositing its load of sediment.

Tidal flat A marshy or muddy area that is covered and uncovered by the rise and fall of the tide.

Tide Periodic change in the elevation of the ocean surface.

Till Unsorted sediment deposited directly by a glacier.

Tombolo A ridge of sand that connects an island to the mainland or to another island.

Tornado A small, very intense cyclonic storm with exceedingly high winds, most often produced along cold fronts in conjunction with severe thunderstorms.

Tornado warning A warning issued when a tornado has actually been sighted in an area or is indicated by radar.

Tornado watch A warning issued for areas of about 65,000 square kilometers (25,000 square miles), indicating that conditions are such that tornadoes may develop; it is intended to alert people to the possibility of tornadoes.

Trade winds Two belts of winds that blow almost constantly from easterly directions and are located on the equatorward sides of the subtropical highs.

Transform fault A major strike-slip fault that cuts through the lithosphere and accommodates motion between two plates.

Transform fault boundary A boundary in which two plates slide past one another without creating or destroying lithosphere.

Transpiration The release of water vapor to the atmosphere by plants.

Transported soil Soils that form on unconsolidated deposits.

Transverse dunes A series of long ridges oriented at right angles to the prevailing wind; these dunes form where vegetation is sparse and sand is very plentiful.

Travertine A form of limestone ($CaCO_3$) that is deposited by hot springs or as a cave deposit.

Trellis pattern A system of streams in which nearly parallel tributaries occupy valleys cut in folded strata.

Trench An elongated depression in the seafloor produced by bending of oceanic crust during subduction.

Trophic level A nourishment level in a food chain. Plant and algae producers constitute the lowest level, followed by herbivores and a series of carnivores at progressively higher levels.

Tropic of Cancer The parallel of latitude, 23½ degrees north latitude, marking the northern limit of the Sun's vertical rays.

Tropic of Capricorn The parallel of latitude, 23½ degrees south latitude, marking the southern limit of the Sun's verticalrays.

Tropical depression By international agreement, a tropical cyclone with maximum winds that do not exceed 61 kilometers (38 miles) per hour.

Tropical rain forest A luxuriant broadleaf evergreen forest; also, the name given the climate associated with this vegetation.

Tropical storm By international agreement, a tropical cyclone with maximum winds between 61 and 119 kilometers (38 and 74 miles) per hour.

Tropical wet and dry A climate that is transitional between the wet tropics and the subtropical steppes.

Tropopause The boundary between the troposphere and the stratosphere.

Troposphere The lowermost layer of the atmosphere. It is generally characterized by a decrease in temperature with height.

Tsunami The Japanese word for a seismic sea wave.

Tundra climate Found almost exclusively in the Northern Hemisphere or at high altitudes in many mountainous regions. A treeless climatic realm of sedges, grasses, mosses, and lichens that is dominated by a long, bitterly cold winter.

Turbidite Turbidity current deposit characterized by graded bedding.

Turbidity current A downslope movement of dense, sediment-laden water created when sand and mud on the continental shelf and slope are dislodged and thrown into suspension.

Turbulent flow The movement of water in an erratic fashion, often characterized by swirling, whirlpool-like eddies. Most streamflow is of this type.

Ultimate base level Sea level; the lowest level to which stream erosion could lower the land.

Ultramafic composition Igneous rocks composed mainly of iron and magnesium-rich minerals.

Ultraviolet Radiation with a wavelength from 0.2 to 0.4 micrometer.

Umbra The central, completely dark part of a shadow produced during an eclipse.

Unconformity A surface that represents a break in the rock record, caused by erosion or nondeposition.

Uniformitarianism The concept that the processes that have shaped Earth in the geologic past are essentially the same as those operating today.

Unsaturated zone The area above the water table where openings in soil, sediment, and rock are not saturated but filled mainly with air.

Unstable air Air that does not resist vertical displacement. If it is lifted, its temperature will not cool as rapidly as the surrounding environment, so it will continue to rise on its own.

Upslope fog Fog created when air moves up a slope and cools adiabatically.

Upwelling The rising of cold water from deeper layers to replace warmer surface water that has been moved away.

Urban heat island The fact that temperatures within a city are generally higher than in surrounding rural areas.

Valence electron The electrons involved in the bonding process; the electrons occupying the highest-principal energy level of an atom.

Valley breeze The daily upslope winds commonly encountered in a mountain valley.

Valley glacier *See* Alpine glacier.

Valley train A relatively narrow body of stratified drift deposited on a valley floor by meltwater streams that issue from a valley glacier.

Vapor pressure That part of the total atmospheric pressure attributable to water-vapor content.

Variable stars Red giants that overshoot equilibrium, then alternately expand and contract.

Vein deposit A mineral filling a fracture or fault in a host rock. Such deposits have a sheetlike, or tabular, form.

Vent The surface opening of a conduit or pipe.

Ventifact A cobble or pebble polished and shaped by the sandblasting effect of wind.

Vesicular texture A term applied to igneous rocks that contain small cavities called vesicles, which are formed when gases escape from lava.

Viscosity A measure of a fluid's resistance to flow.

Visible light Radiation with a wavelength from 0.4 to 0.7 micrometer.

Volatiles Gaseous components of magma dissolved in the melt. Volatiles will readily vaporize (form a gas) at surface pressures.

Volcanic bomb A streamlined pyroclastic fragment ejected from a volcano while molten.

Volcanic island arc A chain of volcanic islands generally located a few hundred kilometers from a trench where active subduction of one oceanic slab beneath another is occurring.

Volcanic neck An isolated, steep-sided, erosional remnant consisting of lava that once occupied the vent of a volcano.

Volcano A mountain formed of lava and/or pyroclastics.

Warm front A front along which a warm air mass overrides a retreating mass of cooler air.

Wash A common term for a desert stream course that is typically dry except for brief periods immediately following a rain.

Water table The upper level of the saturated zone of groundwater.

Wave-cut cliff A seaward-facing cliff along a steep shoreline formed by wave erosion at its base and mass wasting.

Wave-cut platform A bench or shelf in the bedrock at sea level, cut by wave erosion.

Wave height The vertical distance between the trough and crest of a wave.

Wave of oscillation A water wave in which the wave form advances as the water particles move in circular orbits.

Wave of translation The turbulent advance of water created by breaking waves.

Wave period The time interval between the passage of successive crests at a stationary point.

Wave refraction *See* Refraction.

Wavelength The horizontal distance separating successive crests or troughs.

Weather The state of the atmosphere at any given time.

Weathering The disintegration and decomposition of rock at or near Earth's surface.

Welded tuff A pyroclastic rock composed of particles that have been fused together by the combination of heat still contained in the deposit after it has come to rest and by the weight of overlying material.

Well An opening bored into the zone of saturation.

Westerlies The dominant west-to-east motion of the atmosphere that characterizes the regions on the poleward side of the subtropical highs.

Wet adiabatic rate The rate of adiabatic temperature change in saturated air. The rate of temperature change is variable, but it is always less than the dry adiabatic rate.

White dwarf A star that has exhausted most or all of its nuclear fuel and has collapsed to a very small size; believed to be near its final stage of evolution.

White frost Ice crystals instead of dew that form on surfaces when the dew point is below freezing.

Wind Air flowing horizontally with respect to Earth's surface.

Wind vane An instrument used to determine wind direction.

Winter solstice The solstice that occurs on December 21–22 in the Northern Hemisphere and on June 21–22 in the Southern Hemisphere.

Yazoo tributary A tributary that flows parallel to the main stream because a natural levee is present.

Zodiac A band along the ecliptic containing the 12 constellations of the zodiac.

Zone of accumulation The part of a glacier characterized by snow accumulation and ice formation. Its outer limit is the snowline.

Zone of fracture The upper portion of a glacier consisting of brittle ice.

Zone of saturation Zone where all open spaces in sediment and rock are completely filled with water.

Zone of wastage The part of a glacier beyond the zone of accumulation where all of the snow from the previous winter melts, as does some of the glacial ice.

Zooplankton Animal plankton.

Index